# A Handbook of Environmental Toxicology

## Human Disorders and Ecotoxicology

# A Handbook of Environmental Toxicology

## Human Disorders and Ecotoxicology

*Edited by*

**J.P.F. D'Mello**

*Formerly of SAC, University of Edinburgh King's Buildings Campus, Edinburgh, UK*

CABI

**CABI is a trading name of CAB International**

CABI
Nosworthy Way
Wallingford
Oxfordshire OX10 8DE
UK

Tel: +44 (0)1491 832111
Fax: +44 (0)1491 833508
E-mail: info@cabi.org
Website: www.cabi.org

CABI
745 Atlantic Avenue
8th Floor
Boston, MA 02111
USA

Tel: +1 (617)682-9015
E-mail: cabi-nao@cabi.org



A catalogue record for this book is available from the British Library, London, UK.

Library of Congress Cataloging-in-Publication Data

Names: D'Mello, J. P. Felix, editor.
Title: A handbook of environmental toxicology : human disorders and ecotoxicology / edited by J.P.F. D'Mello.
Description: Wallingford, Oxfordshire, UK ; Boston, MA : CAB International, [2020] | Includes bibliographical references and index.
Identifiers: LCCN 2019016396| ISBN 9781786394675 (hbk : alk. paper) | ISBN 9781786394682 (ePDF) | ISBN 9781786394699 (epub)
Subjects: | MESH: Environmental Pollutants | Environmental Pollution | Environmental Health | Ecotoxicology
Classification: LCC RA566 | NLM WA 670 | DDC 362.1969/8--dc23 LC record available at https://lccn.loc.gov/2019016396

ISBN-13: 9781786394675 (hardback)
9781786394682 (ePDF)
9781786394699 (ePub)

Commissioning Editor: Alex Lainsbury
Editorial Assistant: Tabitha Jay
Production Editor: Ali Thompson

Typeset by SPi, Pondicherry, India
Printed and bound in the UK by CPI Group (UK) Ltd, Croydon, CR0 4YY

# Contents

# Contributors

**Aburto-Medina, A.**, Centre for Environmental Sustainability and Remediation, School of Sciences, RMIT University, Bundoora, Victoria 3083, Australia. E-mail: arturoaburto.medina@rmit.edu.au

**Agullo-Chazarra, L.**, Instituto de Biologia Molecular y Celular (IBMC), Universidad Miguel Hernandez, Avenida de la Universidad s/n, 03202 Elche, Alicante, Spain. E-mail: lagullo@umh.es

**Ahmad, S.**, Department of Anesthesiology and Perioperative Medicine, School of Medicine, University of Alabama at Birmingham, USA. E-mail: shamaahmad@uabmc.edu

**Ahmad, A.**, Department of Material Science Engineering, Purdue University, West Lafayette, Indiana, USA. E-mail: ahmad24@purdue.edu

**Ahmad, A.**, Department of Anesthesiology and Perioperative Medicine, School of Medicine, University of Alabama at Birmingham, USA. E-mail: aftabahmad@uabmc.edu

**Akesson, A.**, Institute of Environmental Medicine, Karolinska Institutet, Stockholm, Sweden. E-mail: Agneta.Akesson@ki.se

**Arismendez, S.**, Texas Commission on Environmental Quality, 12100 Park Thirty Five Circle, Austin, TX78753 USA. E-mail: Sandra.Arismendez@gmail.com

**Ball, A.S.**, Centre for Environmental Sustainability and Remediation, School of Sciences, RMIT University, Bundoora, Victoria 3083, Australia. E-mail: andy.ball@rmit.edu.au

**Barrajon-Catalán, E.**, Instituto de Biologia Molecular y Celular (IBMC), Universidad Miguel Hernandez, Avenida de la Universidad s/n, 03202 Elche, Alicante, Spain. E-mail: e.barrajon@umh.es

**Caudle, W.M.**, Department of Environmental Health, Emory University, Atlanta, Georgia, USA. E-mail: william.m.caudle@emory.edu

**Crosby, L.M.**, US Geological Survey, Reston, Virginia, USA; present address: 849 Kenly Avenue, Hagerstown, Maryland 21740, USA. E-mail: Lynn.crosby@fda.hhs.gov

**Darbre, P.D.**, School of Biological Sciences, University of Reading, Reading, UK. E-mail: p.d.darbre@reading.ac.uk

**Del Risco Kollerud, R.**, Department of Community Medicine and Global Health, University of Oslo, Norway. E-mail: r.d.r.kollerud@medisin.uio.no

**D'Mello, J.P.F.**, Formerly of SAC, University of Edinburgh Campus, West Mains Road, Edinburgh EH9 3JG, UK. E-mail: jpfdmello@hotmail.co.uk

**Dopico, M.**, Universidad de Oviedo, Gijon, Asturias, Spain. E-mail: migueldopico89@gmail.com

**Everett, C.J.**, Medical University of South Carolina, Charleston, USA. E-mail: everettc@musc.edu

**Fuentes, N.**, The Pennsylvania State University College of Medicine, Hershey, PA, USA. E-mail: nfuentes1@pennstatehealth.psu.edu

**Gomez, A.**, Universidad de Oviedo, Gijon, Asturias, Spain. E-mail: albertogomez@uniovi.es

**Hachiya, N.**, National Institute for Minamata Disease, 4058-18 Hama, Minamata-shi, Kumamoto-ken 867-0008, Japan. E-mail: hachiya@nimd.go.jp

**Haleyur, N.**, Centre for Environmental Sustainability and Remediation, School of Sciences, RMIT University, Bundoora, Victoria 3083, Australia. E-mail: nagalakshmi.haleyur@rmit.edu.au

**Harrison, V.**, The Open University, Milton Keynes, UK. E-mail: gini.harrison@open.ac.uk

**Haycox, S.**, University of Alaska, Anchorage, Alaska, USA. E-mail: swhaycox@alaska.edu

**He, G.Z.**, State Key Laboratory of Urban and Regional Ecology, Research Centre for Eco-Environmental Sciences, Chinese Academy of Sciences, Beijing 100085, China and University of Chinese Academy of Sciences, Beijing 101407, China. E-mail: gzhe@rcees.ac.cn

**Herranz-Lopez, M.**, Instituto de Biologia Molecular y Celular (IBMC), Universidad Miguel Hernandez, Avenida de la Universidad s/n, 03202 Elche, Alicante, Spain. E-mail: mherranz@umh.es

**Huang, Y-C. T.**, Division of Pulmonary, Allergy and Critical Care Medicine, Duke University Medical Center, USA. E-mail: huang002@mc.duke.edu

**Jiang, H.**, Department of Chemistry, University of Science and Technology of China. No. 96, Jinzhai Road, Hefei 230026, Anhui, China. E-mail: jhong@ustc.edu.cn

**Jones, N.**, Centre for Environmental Sustainability and Remediation, School of Sciences, RMIT University, Bundoora, Victoria 3083, Australia. E-mail: nikki_9618@hotmail.com

**Katner, A.L.**, Environmental and Occupational Health Science Program, School of Public Health, Louisiana State University Health Sciences Center, New Orleans, Louisiana, USA. E-mail: akatn1@lsuhsc.edu

**Kippler, M.**, Institute of Environmental Medicine, Karolinska Institutet, Stockholm, Sweden. E-mail: Maria.Kippler@ki.se

**Kováts, N.**, Institute of Environmental Sciences, University of Pannonia, Veszprem, Hungary. E-mail: kovats@almos.uni-pannon.hu

**Lein, P.J.**, Department of Molecular Biosciences, University of California Davis, Davis, CA, USA. E-mail: pjlein@ucdavis.edu

**Liang, C.**, Jiangsu Key Laboratory of Anaerobic Biotechnology, School of Environmental and Civil Engineering, Jiangnan University, Wuxi, 214122, China. E-mail: cjliang78@yahoo.com or liangchanjuan@jiangnan.edu.cn

**Lundqvist, J.**, Swedish University of Agricultural Sciences, Uppsala, Sweden. E-mail: johan.lundqvist@slu.se

**Mackenzie Ross, S.J.**, University College London, London, UK. E-mail: s.mackenzie-ross@ucl.ac.uk

**Mansur, A.A.**, Centre for Environmental Sustainability and Remediation, School of Sciences, RMIT University, Bundoora, Victoria 3083, Australia. E-mail: mansour2001uk@yahoo.com

**Melloni, B.**, Pulmonary Diseases Department, Limoges University Hospital, Limoges, France. E-mail: boris.melloni@unilim.fr

**Metcalf, J.S.**, Brain Chemistry Labs., Institute of Ethnomedicine, Jackson, WY 83001, USA. E-mail: james@ethnomedicine.org

**Micol, V.**, Instituto de Biologia Molecular y Celular (IBMC), Universidad Miguel Hernandez, Avenida de la Universidad s/n, 03202 Elche, Alicante, Spain and CIBER, Fisiopatologia de la Obesidad y la Nutricion, CIBERobn, Instituto de Salud Carlos III (CB12/03/30038), Spain. E-mail: vmicol@umh.es

**Mielke, H.W.**, Department of Pharmacology, School of Medicine, Tulane University, New Orleans, Louisiana, USA. E-mail: hmielke@tulane.edu

**Montagna, P.A.**, Texas A&M University-Corpus Christi, Harte Research Institute for Gulf of Mexico Studies, 6300 Ocean Drive, Unit 5869, Corpus Christi, Texas, 78412 USA. E-mail: Paul.Montagna@tamucc.edu

**Noubactep, C.**, Angewandie Geologie, Universitat Gottingen, Goldschmidstrasse 3, D-37077 Gottingen, Germany. E-mail: cnoubac@gwdg.de

**Orem, W.H.**, US Geological Survey, Reston, Virginia, USA. E-mail: borem@usgs.gov

**Otaki, J.M.**, Department of Chemistry, Biology and Marine Science, Faculty of Science, University of Ryukyus, Okinawa, Japan. E-mail: otaki@sci.u-ryukyu.ac.jp

**Pampanin, D.M.**, University of Stavanger and International Research Institute of Stavanger, Stavanger, Norway. E-mail: Daniela.m.pampanin@uis.no

**Perez-Sanchez, A.**, Instituto de Biologia Molecular y Celular (IBMC), Universidad Miguel Hernandez, Avenida de la Universidad s/n, 03202 Elche, Alicante, Spain. E-mail: almudena.perez@umh.es

**Purser, D.A.**, Hartford Environmental Research, Hatfield, UK.

**Rivera, L.**, The Pennsylvania State University College of Medicine, Hershey, PA, USA. E-mail: lrivera@pennstatehealth.psu.edu

**Santana, M.F.M.**, Department of Science and Engineering, James Cook University and Australian Institute of Marine Science, Australia. E-mail: marina.santana@my.jcu.edu.au

**Schlenk, D.**, University of California Riverside, Riverside, California, USA. E-mail: dschlenk@ucr.edu

**Seneff, S.**, Computer Science and Artificial Intelligence Laboratory, MIT, Cambridge, MA, USA. E-mail: Seneff@csail.mit.edu

**Sethi, S.**, Department of Molecular Biosciences, University of California Davis, Davis, CA, USA. E-mail: sosethi@ucdavis.edu

**Shahsavari, E.**, Centre for Environmental Sustainability and Remediation, School of Sciences, RMIT University, Bundoora, Victoria 3083, Australia. E-mail: esmaeil.shahsavari@rmit.edu.au

**Silveyra, P.**, The University of North Carolina at Chapel Hill, School of Nursing, Chapel Hill, NC 27599, USA. E-mail: patry@email.unc.edu

**Sima, X-F.**, Department of Chemistry, University of Science and Technology of China. No. 96, Jinzhai Road, Hefei 230026, Anhui, China. E-mail: smxf2013@mail.ustc.edu.cn

**Souza, N.R.**, Brain Chemistry Labs, Institute of Ethnomedicine, Jackson,WY 83001, USA. E-mail: souza.nara.r@icloud.com

**Staessen, J.A.**, Studies Coordinating Centre, Research Unit Hypertension and Cardiovascular Epidemiology, KU Leuven Department of Cardiovascular Sciences, Leuven, Belgium. E-mail: jan.staessen@kuleuven.be

**Thiermann, H.**, Bundeswehr Institute of Pharmacology and Toxicology, Munich, Germany. E-mail: HorstThiermann@bundeswehr.org

**Thompson, K.C.**, Department of Biological Sciences and Institute of Structural and Molecular Biology, Birkbeck College, University of London, Malet Street, London, WC1E 7HX, UK. E-mail: k.thompson@mail.cryst.bbk.ac.uk

**Tucker, J.L.**, Division of Pulmonary, Allergy and Critical Care Medicine, Duke University Medical Center, USA. E-mail: jesse.tucker@duke.edu

**Tuomisto, J.**, National Institute for Health and Welfare, Kuopio, Finland. E-mail: j.tuomisto@dnainternet.net

**Turra, A.**, Oceanographic Institute, University of Sao Paulo, Sao Paulo, Brazil. E-mail: turra@usp.br

**Viluksela, M.**, National Institute for Health and Welfare, Kuopio, and the University of Eastern Finland, Kuopio, Finland. E-mail: matti.viluksela@uef.fi

**Wille, T.**, Bundeswehr Institute of Pharmacology and Toxicology, Munich, Germany. E-mail: TimoWille@Bundeswehr.org

**Wilson, M.J.**, Department of Global Environmental Health Sciences, Tulane University School of Public Health and Tropical Medicine, New Orleans, Louisiana, USA. E-mail: mwilson9@tulane.edu

**Worek, F.**, Bundeswehr Institute of Pharmacology and Toxicology, Munich, Germany. E-mail: FranzWorek@bundeswehr.org

**Yang, W.**, Studies Coordinating Centre, Research Unit Hypertension and Cardiovascular Epidemiology, KU Leuven Department of Cardiovascular Sciences, Leuven, Belgium. Present address: Department of Cardiology, Shanghai Jiao Tong University, School of Medicine, Shanghai, China. E-mail: wenyi.yang@shgh.cn

# Preface

## History

It will surprise many to learn that global environmental pollution can be traced back to Roman times, with respect to lead emissions following gold extraction processes (Hillman *et al.*, 2017). However, the emergence of environmental toxicology as a formal discipline in its own right occurred much later as a direct result of major pollution and contamination episodes around the world, including, for example, deployment of two atomic weapons, accidents at nuclear power stations, oil spills and vehicular emissions (see Table 1). Consistent with this development, the traditional approach based on lethality and chronic tests has been replaced in this volume, in favour of a more practical, relevant and contemporary presentation of toxicology.

## Major Contamination Incidents

A significant number of contamination events has occurred since 1945 and these are recorded for the benefit of readers who are new to the field of environmental toxicology. Details of some of these cases are included in the text of this volume to illustrate specific issues. Table 1 is not designed to be comprehensive but, rather, illustrative of the diverse nature of pollutants deliberately or accidentally released into the environment and the regular incidence of crude oil and radiological contamination. This list is arranged in chronological order to highlight the regular frequency of such events.

Two sets of events deserve special mention, due to profound long-term human health and ecological implications. The devastation caused by the detonation of two atomic devices at Hiroshima and Nagasaki (Fig. 1) will always represent an ignominious phase in human history. However, whereas the infrastructure has now been restored in both cities, the toxic legacy for survivors remains to this day, arguably the most cogent expression of 'Man's inhumanity to man'. For example, the incidence and analysis of myelodysplastic syndromes afflicting these subjects have been investigated in recent toxicological research (Horai *et al.*, 2018). In addition, accidents and regular radiation leaks from nuclear power stations present continuing worldwide risks (Table 1).

**Table 1.** Major environmental contamination incidents in recent history.

| Year/frequency | Incident | Principal contaminants |
|---|---|---|
| 1945 | Detonation of two nuclear weapons over Hiroshima and Nagasaki (Japan) | Radionuclides |
| 1952 | The Great Smog (London) | Particulates and gaseous mixture (Chapters 5–8, 40) |
| 1956 | Minamata disease outbreak (Japan) | Methylmercury (Chapter 24) |
| 1961–1971 | Vietnam War | Agent Orange containing a mixture of two herbicides and traces of dioxin (Chapter 13) |
| 1965 | San Jacinto river contamination (USA) | PCDD and PCDFs |
| 1967 | *Torrey Canyon* oil spill (UK) | Crude oil (Chapter 40) |
| 1976 | Seveso chemical plant explosion (Italy) | TCDD release (Chapters 12 and 13) |
| 1978 | *Amoco Cadiz* oil spill (France) | Crude oil |
| 1979 | Three Mile Island accident (USA) | Radionuclides |
| 1979 | Ixtoc oil well explosion (Mexico) | Crude oil |
| 1979–1985 | Gold-mining pollution (Brazil) | Mercury |
| 1983 | *Castillo de Bellver* oil spill (South Africa) | Crude oil |
| 1984 | Bhopal disaster (India) | Methyl isocyanate |
| 1986 | Chernobyl explosion (Ukraine) | Radionuclides |
| 1988 | *Piper Alpha* offshore oil and gas explosion (UK) | Crude oil |
| 1988 | *Odyssey* oil spill (Canada) | Crude oil |
| 1989 | *Exxon Valdez* oil spill (Alaska) | Crude oil (Chapter 22) |
| 1997–2004 | Endemic neurolathyrism (Ethiopia) | Dietary exposure to lathyrogenic amino acids (Chapter 1) |
| 2002 | *Prestige* oil spill (Spain) | Crude oil |
| 2004 | Acute aflatoxicosis (Kenya) | Dietary intake of aflatoxins (Chapter 2) |
| 2010 | *Deepwater Horizon* accident (Gulf of Mexico) | Crude oil (Chapters 20 and 21) |
| 2011 | Fukushima nuclear accident (Japan) | Radionuclides (Chapter 34) |
| 2014 | Flint water contamination (USA) | Lead (Chapter 25) |
| 2015 | 'Defeat software' fitted to certain vehicles to falsify results in emissions tests (USA) | Carbon dioxide and nitrogen dioxide (Chapters 7 and 40) |
| 2017 | Wastewater release from shale oil and gas exploration (fracking) (USA) | Diverse 'toxic chemical substances' (Chapter 23) |
| Regular | Vehicular emissions | Nitrogen dioxide and particulates (Chapters 7, 28 and 29) |
| Regular | Lake Erie hypoxia (North America) | Fertilizer pollution |
| Regular | Arable farming, worldwide | Pesticides (Chapters 16–19) |
| Regular | Radiation leaks from nuclear plants and storage: based on local reports on both sides of the Atlantic | Radionuclides (Chapter 34) |

In the case of crude oil pollution, the *Deepwater Horizon* debacle (Fig. 2) is destined to symbolize an iconic image of ecological catastrophe for future generations of environmentalists. The risks remain, driven by the unremitting demand for fuel on an industrial scale. Thus, emissions of noxious gases and particulates from vehicles are regularly linked with a diverse range of human health disorders. The recent headline that London breached annual air pollution limits for 2017 in just 5 days gives immense cause for concern, but does anybody care?

It is within the context of these events that I present a new volume on environmental toxicology, based on fundamentals and relevance to specific case-studies, both historical and contemporary.

**Fig. 1.** Toxic legacy I. Radionuclide contamination following the disproportionate and reckless detonation of two atomic weapons at Hiroshima and Nagasaki (an artist's composite interpretation; courtesy of Mr T.F. D'Mello). Although the infrastructure has now been restored, the toxic effects for survivors remains to this day (Horai *et al.*, 2018). Furthermore, accidents and regular radiation leaks from nuclear power stations and storage facilities present continuing risks worldwide (Table 1).

**Fig. 2.** Toxic legacy II. The *Deepwater Horizon* oil spill in the Gulf of Mexico (an artist's portrayal designed to reflect also the preceding *Piper Alpha* accident in North Sea; courtesy of Mr T.F. D'Mello). These events are destined to serve as an iconic beacon of ecotoxicity, resonating long into the future. Crude oil pollution is likely to continue, driven by regular accidents and consumer demand for fuel in transport, power generation and other activities. Nevertheless, it should be recognized that offshore oil and gas platforms in waters of between 25 and 30 m may provide niches for colonization by corals (Kolian *et al.*, 2017).

## A Handbook of Environmental Toxicology: Human Disorders and Ecotoxicology

### Background

The concept for a new edition of this volume emerged during my tenure as Academic Director of a first-degree course in 'Environmental Protection and Management' (EP&M) validated by the University of Edinburgh and delivered to undergraduates at the Scottish Agricultural College. In particular, my role as coordinator of a third-year EP&M module in Environmental Chemistry, including a series of lectures on 'Environmental Toxicology', crystallized in my mind the need for an up-to-date advanced text on this subject. Further impetus was provided through my research interests in the toxicology of natural compounds associated with the secondary metabolism of plants and fungi. In addition, my experience in the provision of consultancy services in mycotoxin contamination of primary foods and animal feeding-stuffs allowed me to develop practical credentials in a relevant branch of environmental toxicology. Subsequent to my retirement, the EP&M course was replaced by a postgraduate degree within the Masters programme.

### Context

Following recent declarations in Europe and North America, it has become patently clear that our elected leaders are unreliable custodians of the environment. The assertion in the USA that global climate change is an irrelevance compared with the regeneration of the 'rust belt' is evidence of a short-sighted attitude. Somewhere in the arguments currently being promulgated is the notion that the global warming issue is a myth invented by third-world agencies to present manufacturing in advanced economies as less competitive. Furthermore, the edict of protectionism in economic matters is gathering momentum on both sides of the Atlantic, with the likely effect that international cooperation on pollution research and adherence to the Paris Climate Agreement will be severely compromised. Insular policies will not only serve to exacerbate environmental disparities for communities in vulnerable regions but also jeopardize human health and biodiversity in affluent countries.

Against this backdrop of confused thinking, there is now an opportunity for environmental toxicologists to demonstrate leadership and it is envisaged that publication of this volume will contribute to an active debate on all aspects of human health and ecological conservation. The initial signs are encouraging. For example, the work of Landrigan (2017) linking air pollution with human morbidity has attracted wide attention in the media, due to the bold expression of current and projected data. *The Lancet* has also been at the forefront of this activity by highlighting key public health disorders associated with environmental pollution (see, for example, Samet *et al.*, 2013; Shah *et al.*, 2013; Beelen *et al.*, 2014; Guarnieri and Balmes, 2014; and Landrigan, 2017).

### Faltering progress

The history of environmental protection and management is marked by significant developments in risk assessment across a wide spectrum of pollutants, but the translation of results into policies of statutory regulation, interventions and compliance has been ponderous at best. Many observers would argue that intransigence and inaction have been the hallmarks of recent efforts to address global issues in the pollution–human-welfare–ecology axis. Thus, despite the occurrence of the Great Smog of 1952, limited steps have only recently been implemented to control traffic pollution in London. The legacy of persistent organic pollutants (POPs) is still with us notwithstanding all the advances in toxicological research and definitive evidence of human morbidity associated with dioxins, polychlorinated biphenyls (PCBs) and certain pesticides. For example, there are concerns over the pesticides Fipronil and pyrethroids in eggs while other surveillance has revealed the occurrence of neonicotinoids in global honey supplies. It is also disturbing that 60 years after identification of Minamata disease in Japan, steps are only now being undertaken to curb mercury contamination on a worldwide basis. For example, a report by Taylor and Williamson (2017) demonstrated an ongoing issue of mercury contamination in US coastal fisheries.

## Approach

In designing this *Handbook*, I have commissioned a team of experts from around the world to submit critical reviews highlighting the effects of diverse pollutants on human morbidity and on ecotoxicology. However, in addition to chapters presented in conventional format, I invited a few authors to submit short 'research communication'-style papers to demonstrate to advanced students the specialist aspects of current work. Environmental toxicology is increasingly perceived as a scientific evidence-based discipline. It is all too easy to dismiss evidence linking environmental contaminants to ill health in humans and habitat degradation as merely statistical inference lacking biochemical basis. However, several authors contributing to this *Handbook* are at pains to point to biologically plausible mechanisms underlying the said adverse effects. Furthermore, these authors have published detailed evidence in refereed papers in respected journals such as *The Lancet*, *European Respiratory Journal*, *Environmental Health Perspectives*, *Toxicological Research*, *Marine and Freshwater Research*, *Marine Pollution Bulletin* and many others.

## Overview and design

*A Handbook of Environmental Toxicology: Human Disorders and Ecotoxicology* is published specifically to promote debate and research in academic and corporate institutions in Europe, the USA, Canada, Japan, Australia and New Zealand but generally in all countries where the English language is a primary medium of communication. This *Handbook* should appeal to a wide readership, including advanced undergraduate and graduate students in addition to teaching and research staff in colleges, universities and state-funded institutes. *A Handbook of Environmental Toxicology: Human Disorders and Ecotoxicology* is recommended for courses in environmental protection and management, ecology, toxicology and the biological and medical sciences. A major objective is to encourage the incorporation of environmental toxicology into existing Masters programmes provided at UK and US universities to enhance the scientific base of advanced degrees. This volume should also be of interest to scientists and managers in the commercial sector, for example oil and energy companies. Furthermore, it is important, through this *Handbook*, to engage with local authorities, particularly with respect to rural and urban pollution, recycling measures and remediation technologies.

The chapters in this volume are arranged in nine sections, each representing a particular theme and representing diversity of biogenic compounds and pollutants impinging on human health and ecotoxicity. The nature of the subjects under review and the need for continuity necessarily involves a certain degree of overlap. This is not envisaged as a detraction, as individual chapters are self-contained as a consequence, thereby reducing the need for cross-referencing to other parts of the book. More importantly, this approach has also allowed authors increased flexibility in terms of emphasis and interpretation.

### *Part I Biogenic compounds*

Three chapters herein highlight important secondary metabolites, including phytotoxins, mycotoxins and cyanobacterial toxins synthesized by plants and microbes. Consideration of these compounds is justified, because their occurrence may be affected by environmental factors, for example global climate change, or the potential for the production of bio-pesticides to replace existing harmful synthetic insecticides, fungicides and herbicides. The toxicology of biogenic compounds is characterized by diverse and profound effects in humans and other vertebrates following intake via contaminated food and water or exposure in damp dwellings characterized by the 'sick building syndrome'. Both acute and chronic effects are exemplified in epidemiological reports of poisonings caused by biogenic metabolites. Plants and microbes may also exert physical effects affecting, for example, habitat selection by vectors (Binckley, 2017) or efficacy of oceans to extract carbon from the atmosphere (Kondrik *et al.*, 2018). Such issues are outside the scope of this volume but should contribute to the general model of biological–environmental interactions.

### *Part II Ambient gases affecting human health and adaptation in higher plants*

Recent research has confirmed the detrimental health hazards associated with ambient air pollution. The effects of ozone, nitrogen dioxide, sulfur dioxide, polyaromatic hydrocarbons and particulates are under regular scrutiny not only for individual contributions to human morbidity but also with respect to interactions within this group. Specific conditions currently under investigation include exacerbation of idiopathic pulmonary fibrosis, chronic obstructive pulmonary disease (COPD), asthma, cardiovascular disease (CVD), metabolic syndromes (particularly childhood obesity), type 2 diabetes and diverse forms of cancer. The cognitive and neurological effects of air pollutants are an additional source of concern and *in utero* exposure may be a contributory feature in certain cases, such as the incidence of autism. Interactions between certain gaseous pollutants and other ambient air contaminants, for example particulates, inevitably add to the difficulties in interpretation of emerging data for these conditions. Although much of the evidence is based on epidemiological approaches, several research scientists consistently point to biologically plausible mechanisms underlying morbidity caused by gaseous pollutants. The diverse effects of acid rain on adaptation mechanisms in higher plants are relevant here, since the main contributors to this type of precipitation include sulfur dioxide and nitrogen oxides. The toxic effects in plants include adaptations in morphology, photosynthesis, nutrient uptake and oxidative status.

### *Part III Persistent Organic Pollutants (POPs)*

This diverse group is of immense significance due to low environmental degradability of POPs, presenting long-term risks for human health and biodiversity in the major ecosystems. The anthropogenic derivation of these compounds, including PCBs, dioxins and endocrine disruptors, only adds to current disquiet over their distribution in different matrices such as food, water and sediments. Issues of particular concern with POPs in relation to human morbidity include the consequences of maternal exposure, transgenerational effects, endocrine disruption and relationship to carcinogenesis. Also under consideration is the possible link to asthma and diabetes. Furthermore, the biochemical mechanisms underlying the adverse effects of POPs remain complex issues despite advances in activation of receptors, signalling pathways and gene expression. The influence of trophic ecology on bioaccumulation of POPs in different wildlife species is under active investigation. Attempts are also under way to develop novel technologies for remediation employing microbial degradation.

Pollution in the rural environment remains an intractable problem due to multiple sources of contaminants, causing adverse human health and ecological effects. Of particular concern are pesticides and nutrient pollution with global implications for food safety and water quality. There are currently concerns over the pesticides Fipronil and pyrethroids in eggs and neonicotinoids in global honey supplies, but the issues of pesticide contamination of foods in general continue to be a matter of concern for a wide range of foods. In addition, the toxicology of organochlorine and organophosphate compounds continues to attract attention, justifying the inclusion of three chapters in this volume. Emerging issues relate to the effects of prenatal and early life exposures in the development and exacerbation of morbidity. Current studies are designed to elucidate effects on the aetiology of asthma, neurodegenerative disorders and specific forms of cancer. Methodological considerations are also presented in this section of the *Handbook*. Furthermore, there is, at present, an active debate on the use of a major herbicide, glyphosate.

### *Part IV Petroleum pollution*

Crude oil pollution has, unfortunately, become a regular occurrence in recent times, arising from accidental discharges into the sea. The *Torrey Canyon* oil spill of 1967 arguably remains the worst marine contamination incident in UK history. Images of the *Exxon Valdez* oil spill in 1989 in Prince William Sound, Alaska, will readily be recalled by many readers of this volume. However, implications of the 2010 *Deepwater Horizon* oil spill in the Gulf of Mexico will reverberate in regulatory, legal and ecological circles for many years to come, and rightly so. The effects of crude oil contamination on microbial and animal ecology are now emerging, although it is already apparent, from first principles, that detoxification pathways are limited in vertebrate species. Nevertheless, observations 24 years after the *Exxon Valdez* oil spill indicate that recovery is possible in the long term but remains somewhat patchy. Using biomarker evidence, efforts to establish timelines for different animal species exposed to crude oil contaminants are being undertaken.

With the emergence of novel exploration technologies in oil extraction such as hydraulic fracturing (fracking) come additional risks for human welfare, wildlife and associated habitats. Top predators may be exposed to or bioaccumulate via their macroinvertebrate food the chemical contaminants released into the ecosphere by fracking. Key areas of research and monitoring now need to be addressed to assist in the formulation of effective guidelines and policies to protect local communities and vulnerable animal species at fracking sites.

### *Part V Toxicology of heavy metals*

Of all heavy metals, mercury pollution and toxicity continue to attract attention, due to widespread contamination associated with mining, burning of fossil fuels, deforestation and accidental discharges. Instances of mercury pollution have been reported in Brazil, China and parts of Africa. However, it is now generally accepted that mercury contamination is a global issue, even extending to US coastal fisheries. It is of concern that, 60 years after identification of the Minamata poisoning incident in Japan, steps are only now being undertaken to curb mercury contamination on a worldwide basis. The long-term neurological and behavioural effects of mercury toxicity as a result of prenatal and postnatal exposure are worth reviewing. Lead is also associated with neuropsychological and functional decline in humans, as indicated by a variety of manifestations including difficulties in intelligence, memory, attention and mood. The relationship between lead and autistic behaviours in children has been investigated, leading to recommendations to further reduce such exposure. The effects of lead during the early stages of brain development remain a primary area of research with neural stem cells in gene expression studies. Cadmium has been linked with toxic effects in plants and animals. Oxidative stress may be an underlying feature in chronic cadmium-induced hepatotoxicity and nephrotoxicity, while other research highlights the role of glutathione as a first line of defence against cadmium toxicity. Exposure and human health effects of cadmium are reviewed in this volume, but there is scope also to explore cardiovascular effects in a separate chapter.

### *Part VI Particulates and plastics*

Particulates from fire incidents and combustion of diesel in vehicles are emerging as major pollutants worldwide. Although particulates are associated with long-term morbidity in humans, interactions with gaseous pollutants and other ambient air contaminants add to the difficulties in interpretation of published data for these conditions. An attempt will be made in a subsequent chapter to summarize these interactions. In addition, assessment of the ecotoxicity of airborne particulate matter remains largely unexplored.

Marine litter comprising discarded plastic packaging constitutes a significant problem, as highlighted in the media. However, there are scientific issues in addition to wildlife ingestion and entanglement. For example, there is increasing evidence that plastics may act as ligands for a variety of chemical compounds, thereby contributing to persistence of pollutants in the marine environment. It is too early to evaluate the long-term ecotoxicity of these adsorption phenomena.

### *Part VII Radiation risks*

The incidence of non-melanoma skin cancer in white populations is increasing in many countries. Exposure to UV radiation is believed to be the underlying cause, though the pattern of exposure that promotes the different types of malignancy varies. Controversially, exponentially increasing incidence of cutaneous malignant melanoma in Europe correlates with low personal annual UV doses. It is suggested that intermittent UV exposures result in low cutaneous levels of vitamin $D_3$ and viral infections may possibly predispose to this incidence. Regarding radon, residential exposure is definitively linked with lung cancer incidence. For example, there are indications that high concentrations of radon progeny induce lung cancer in both underground miners and experimentally exposed laboratory animals, suggesting that ambient radon represents an important health risk. The direction of current research is now focusing on possible links to leukaemia and interactions with ambient particulates. Recent assessments of the 30-year legacy of the Chernobyl nuclear accident indicate increased long-term risks of leukaemia and CVD among clean-up personnel as well as thyroid cancer in subjects exposed to radiation as children and adolescents. In addition, mental health effects appear to be the most significant public health issues in the heavily contaminated regions of Ukraine, Belarus and the Russian Federation. As might be anticipated, radioactive emissions following the Chernobyl nuclear accident continue to induce wildlife abnormalities. However, 30 years after the accident, there is still a lack of data relating to the genetic effects of radionuclide contamination on plant ecology. The human health and ecological impacts of the Fukushima nuclear accident are considered in this *Handbook*, supplementing evidence from the Chernobyl emissions.

### *Part VIII Remediation*

Diverse technologies are potentially available for environmental remediation of contaminated land and groundwater. Particular emphasis is currently being placed on the development of nanomaterials for removal of pollutants and biological contaminants. Nanomaterials are considered to be superior to other systems because of their higher surface-to-volume ratios. There is also interest in exploiting polymer-supported titanium dioxide photocatalysts for environmental remediation. Other methodologies include: the application of natural zeolites, by virtue of their ion-exchange properties in the separation, binding and chemical stabilization of hazardous substances; biochar; and iron-based applications in contaminated land and groundwater remediation. Of particular relevance also is the exploitation of plants for the removal of organics and other contaminants. Three chapters are included in this *Handbook* to exemplify the wide scope and likely constraints of new technologies currently available.

### *Part IX Outlook and conclusions*

The development and implementation of environmental regulations are under constant scrutiny, despite success in a number of high-profile cases. A potential concern relates to the impact of these regulations on industrial competitiveness, affecting international trade, employment, plant location and productivity, particularly in pollution-inducing and energy-intensive sectors. However, it is conceivable that environmental regulations may stimulate investment and innovation in clean technologies. It is also difficult to assess the relative stringency of international regulations, with the promotion, in some states, of lower standards in order to attract commercial investment. A case in point relates to a member of the World Trade Organization (WTO) who argued that the maximum residue limits (MRLs) for certain pesticides in food commodities were more stringent than the standards recommended by the *Codex Alimentarius* and were more trade-restrictive than necessary.

In addition, there are ongoing questions regarding auditing and compliance. It is maintained by some commentators that auditors' responsibilities in the context of detection and reporting of contraventions in environmental regulations are limited. Nevertheless, vigilance on the part of the regulatory agencies has reaped limited dividends for environmental protection. For example, the discharge of untreated sewage into the river Thames over several months by a major UK utility company has been labelled as an 'environmental disaster', deservedly attracting severe penalties. In 2016, a cruise-line operator incurred a substantial fine after illegally discharging oil and associated waste via a 'magic pipe' off the UK coast. The success of vigilance and surveillance is, arguably, best exemplified by the US Environmental Agency (EPA) findings that a number of German-manufactured cars were fitted with 'defeat software' designed to falsify emissions in performance tests.

The efficacy of regulatory agencies will depend upon the development of new methodologies to keep ahead of those who would seek to contravene environmental directives. A forward-looking chapter is, therefore, included to review future procedures for environmental toxicology and monitoring.

The aims in the concluding chapter are to collate and integrate the considerable advances linking pollution with human morbidity and ecological deterioration. This chapter is deliberately presented to appeal to a general audience in order to ensure wide dissemination of the central issues emerging from recent research in environmental toxicology. Accordingly, the use of jargon has been minimized. There is a risk that a number of key findings may be 'lost in translation' due to lack of clarity in presentation. It is imperative that significant environmental health and ecological implications are perceived as action points for individuals as much as for corporate organizations and regulatory institutions at international and local levels. The evidence reviewed in this chapter relies on data obtained in epidemiological investigations, case studies linked to specific contamination incidents and fundamental experimentation designed to discern the underlying mechanisms of action of pollutants. It is worth considering whether, in the absence of political leadership, research scientists should exert a more forthright role in all matters relating to environmental protection.

## Disclosure

It is self-evident that disclosure should be a central element in environmental protection policy. However, internet sources claim that communities on both sides of the Atlantic are unaware of radioactive contamination at nuclear power stations and naval bases. Phrases such as 'kept in the dark' and 'cover-up' are regularly associated with reports of such incidents. One press article referred to 'dangerous radiation leaks from three-quarters of US nuclear power plants'. Furthermore, according to Tkavc *et al.* (2018), highly contaminated radionuclide waste stored in a strongly acidic medium mixed with heavy metals at US Department of Energy sites has been leaking since the 1950s. The response of the authorities is that such occurrences represent a public relations problem rather than a genuine health hazard. Others with intuitive scepticism might well disagree and insist on an independent environmental audit. This is essential as nuclear power is increasingly portrayed as a clean-energy source.

## Expectations

With sustained funding and with participation of dedicated scientists, the discipline of environmental toxicology should evolve into a more exact science, underpinned by advances in medical research and quantitative methodologies, in place of anecdotal and qualitative risk assessments. Authors contributing to this edition are at the forefront of current developments to remove ambiguity, improve current models of analysis and elucidate biochemical mechanisms underlying epidemiological observations. Consequently, by reducing speculative theories and including more substantive evidence, future statements should become less tentative, leading to clarity in communication, improved regulatory measures and effective interventions. Going forward, however, I expect steady progress on environmental toxicology, akin to that in establishing the relationships between cigarette smoking and lung cancer/CVD or alcohol abuse and liver cirrhosis. Whatever the difficulties, we should not abandon our resolve but adopt an evangelical fervour in communication of research findings. In pursuing our objectives, it is salutary to recall the exhortation given by St Paul: 'The revelation of a mystery kept secret for endless ages, but now so clear that it must be broadcast' (*Romans* 16:25–27).

## Acknowledgements

The publication of this edition of *A Handbook of Environmental Toxicology: Human Disorders and Ecotoxicology* is only possible by virtue of the expertise and cooperation of my international team of authors. They have an established reputation acquired by publishing original papers in peer-reviewed research journals. I am indebted to them for devoting valuable time and effort in the presentation of authoritative and comprehensive chapters in a readable format. I am convinced that their work will inspire students and staff alike for many years to come. In deference to their expertise, I decided that authors should be given freedom of expression in the preparation of individual chapters, while recognizing that the scientific terminology becomes increasingly more complex with successive advances in research in the various disciplines embodied in environmental toxicology.

## Disclaimer

*A Handbook of Environmental Toxicology: Human Disorders and Ecotoxicology* necessarily contains references to and descriptors of experimental protocols and commercial products. Authors were asked to refrain from excessive use of trade names unless there were compelling reasons for doing so. It is emphasized that no endorsement or criticism of these procedures and products is implied or should be attributed to the Editor or to CAB International, the publisher of this volume. We confirm our absolute impartiality in the choice of chapter titles and in the appointment of authors. Nevertheless, I should confirm my personal ownership of shares in oil and utility companies traded in the London Stock Exchange. These shares were purchased more than 15 years before my appointment as Editor of this *Handbook*.

The information set out in this volume is presented in good faith and in accordance with our understanding of the principles of 'best practice' and 'due diligence'. Although every effort has been made to verify the facts and figures, neither the Editor nor CAB International can accept responsibility for the data and conclusions presented in individual chapters or for any consequences of their use. The submission of signed contracts confers on each contributing author absolute responsibility to check all figures, facts and conclusions. All participating authors have undertaken to abide by stringent rules concerning submission of their chapters so as to avoid any 'material that might be deemed to be libellous, obscene, defamatory or improper' or incorrect. A number of products and methodologies are described by my team of authors. However, publication of this volume should not be interpreted as a recommendation for our readers to use these compounds or techniques for whatever purpose. It is particularly emphasized that data in this volume should not be used to extol or discredit the efficacy or competitiveness of any proprietary product cited or reviewed in individual chapters. The opinions expressed in this volume are exclusively those of the contributing authors, based on their data published in refereed journals, and should not be attributed to the Editor or CAB International. In selecting chapter titles, we have been guided entirely by the need to pursue diverse issues in environmental toxicology wherever it takes us, however unpalatable, but always based on robust evidence available in reputable research journals. This *Handbook*, therefore, is intended exclusively for use as a text in education and research in our collective efforts to protect and enhance human welfare and natural habitats.

## References


Beelen, R., Raaschou-Nielsen, O., Stafoggia, M., Andersen, Z.J., Weinmayr, G., Frosberg, B., Modig, L., Havulinna, A.S. and Hoek, G. (2014) Effects of long-term exposure to air pollution on natural-cause mortality: an analysis of 22 European cohorts within the ESCAPE project. *The Lancet* 383, 785–795.

Binckley, C.A. (2017) Forest canopy, water level and biopesticide interact to determine oviposition habitat selection in *Aedes albopictus*. *Journal of Vector Ecology* 42, 319–324.

Guarnieri, M. and Balmes, J.R. (2014) Outdoor air pollution and asthma. *The Lancet* 383, 1581–1592.

Hillman, A.L., Abbott, M.B., Valero-Garces, B.L., Morellon, M., Barreiro-Lostres, F. and Bain, D.J. (2017) Lead pollution from Roman gold extraction in north-western Spain. *The Holocene* 27, 1465–1474.

Horai, M., Satoh, S., Matsuo, M., Takasaki, Y., Kanaguchi, Y. and Tsushima, H. (2018) Chromosomal analysis of myelodysplastic syndromes among atomic bomb survivors in Nagasaki. *The British Journal of Haematology* 180, 381–390.

Kolian, S.R., Sammarco, P.W. and Porter, S.A. (2017) Abundance of corals on offshore oil and gas platforms in the Gulf of Mexico. *Environmental Management* 60, 357–366.

Kondrik, D.V., Pozdnyakov, D.V. and Johannessen, O.M. (2018) Satellite evidence that *Emiliania huxleyi* phytoplankton blooms weaken marine carbon sinks. *Geophysical Research Letters* 45, 846–854.

Landrigan, P.J. (2017) Air pollution and health. *The Lancet Public Health* 2, e4–e5, January 2017.

Samet, J., Saldiva, P.H.N., Brauer, M., Chen, G., White, P., Huang, W., Knudsen, L.E., Heinrich, U., Yorifuji, T., Henderson, R., Laden, F. and Schauer, J.J. (2013) The carcinogenicity of outdoor air pollution. *The Lancet Oncology* 14, 1262.

Shah, A.S.V., Langrish, J.P., Nairn, H., McAllister, D.A., Donaldson, K., Newby, D.E. and Mills, N.L. (2013) Global association of air pollution and heart failure: a systematic review and meta-analysis. *The Lancet* 382, 1039–1048.

Taylor, D.L. and Williamson, P.R. (2017) Mercury contamination in Southern New England coastal fisheries and dietary habits of recreational anglers and their families: implications to human health and issuance of consumption advisories. *Marine Pollution Bulletin* 114, 144–156.

Tkavc, R., Matrosova, V.Y., Grichenko, O.E., Gostincar, C., Volpe, R.P., Dugan, L. and Daly, M.J. (2018) Prospects for fungal bioremediation of acidic radioactive waste sites: characterization and genome sequence of *Rhodotorula taiwanensis* MD 1149. *Frontiers in Microbiology*. doi: 10.3389/fmicb.2017.02528.

# Terms and Acronyms

## Introduction

Specific nomenclature and technical descriptors are now firmly embedded in the literature associated with environmental toxicology. Although many of the terms and acronyms appearing in this *Handbook* are already in the public domain, it is important to provide a comprehensive glossary to assist those readers who are new to this field. Additional definitions are available in a wide range of scientific dictionaries including, for example, the compilations of Parish *et al.* (2006), Singleton and Sainsbury (2006), Allaby (2010), Lackie (2013), Hodgson and Roe (2014) and Martin (2015). Readers will be aware that several free dictionaries are available and readily accessible online. Glossaries and other relevant information have also been provided in specialist monographs and handbooks such as those edited by D'Mello *et al.* (1991), D'Mello (1997), D'Mello (2012) and D'Mello (2015). These sources are recommended on the basis of a common vocabulary in the different disciplines of the life sciences. In addition, standard titles including works by Alberts *et al.* (2014), Lodish *et al.* (2013), Nelson and Cox (2013), Klug *et al.* (2014), Madigan *et al.* (2015) and Strelkauskas *et al.* (2016) are recommended as sources of in-depth information on different aspects of the biological sciences. Food toxicology is a recurring theme in this *Handbook* and the introduction by Shibamoto and Bjeldanes (2009) should be consulted for basic understanding of this expanding subject. Regarding epidemiology and clinical pathology, two volumes by Ward *et al.* (2016) and Carton (2017), respectively, may be relevant for the human health aspects of my *Handbook*. Fundamental principles of environmental toxicology and ecotoxicology are outlined for beginners in the works of Dong (2014) and Walker *et al.* (2012), respectively. It is assumed that readers will be conversant with general physiology, biochemistry, microbiology, pathology and molecular genetics to levels covered by these textbooks. However, the language associated with environmental toxicology is continually evolving and unremittingly complex, even to experienced researchers, but a highly effective way of updating information and relevant terminology is via the various research articles cited at the end of each chapter of this volume.

## Definition of Terms and Acronyms

The important terms and acronyms appearing in *A Handbook of Environmental Toxicology: Human Disorders and Ecotoxicology* are listed and explained in Table 2. This compilation includes definitions of standard as well as unique chapter-specific abbreviations or terms. Where appropriate, cross-referencing to individual chapters in this volume is included in order to facilitate a greater appreciation of the context of usage of particular terms. Alternative abbreviations for the same entry are listed here due to widespread use in research publications and in other literature.

**Table 2.** Explanation of acronyms and relevant terms used in *A Handbook of Environmental Toxicology: Human Disorders and Ecotoxicology*.

| Abbreviation or term | Definition |
|---|---|
| AA | Amino acid |
| AAP | American Academy of Pediatrics (Ch 25) |
| Aβ | Amyloid beta (Ch 19) |
| ABA | Abscisic acid (Ch 4) |
| ABC | ATP-binding cassette (transporter) (Ch 4) |
| ACC | 1-Amino-cyclopropane-1-carboxylic acid (Ch 4) |
| Ach | Acetylcholine (Ch 17) |
| AchE, AChE | Acetylcholine esterase (Ch 3, 17, 18) |
| ACGIH | American Conference of Governmental Industrial Hygienists(Ch 20) |
| AD | Alzheimer's disease (Ch 25 and 40) |
| *ADC* | Arginine decarboxylase (gene) (Ch 4) |
| ADHD | Attention deficit/hypersensitivity disorder (Ch 11, 26) |
| ADL | Activity of daily living (Ch 24) |
| ADME | Absorption, distribution, metabolism and excretion (Ch 17) |
| $AFB_1$ | Aflatoxin $B_1$ (Ch 2) |
| $AFB_2$ | Aflatoxin $B_2$ (Ch 2) |
| $AFG_1$ | Aflatoxin $G_1$ (Ch 2) |
| $AFG_2$ | Aflatoxin $G_2$ (Ch 2) |
| $AFM_1$ | Aflatoxin $M_1$ (Ch 2) |
| Ah, AH | Aryl hydrocarbon (Ch 13, 14) |
| AhR/AHR | Aryl hydrocarbon receptor (Ch 11, 13, 39) |
| AKT | Protein kinase B (Ch 27) |
| Al | Aluminium (Ch 4 and 9) |
| AL | Acute leukaemia (Ch 33) |
| ALA | American Lung Association (Ch 5) |
| ALA | δ-Aminolevulinic acid (Ch 19) |
| AlaAT/ ALT | Alanine aminotransferase (Ch 4, 34) |
| ALK | Anaplastic lymphoma kinase (Ch 32) |
| ALP | Alkaline phosphatase (Ch 34) |
| ALRI | Acute lower respiratory infections (Ch 40) |
| ALS/PDC | Amyotrophic lateral sclerosis/Parkinsonism dementia complex (Ch 3, 19, 25) |
| AM | Arithmetic mean (Ch 24) |
| AMY | Amylase (Ch 34) |
| AOP | Adverse outcome pathway (Ch 10) |
| AP-1 | Activator protein-1 (Ch 31) |
| APP | Amyloid precursor protein (Ch 19) |
| APX | Ascorbate peroxidase (Ch 9) |
| AQI | Air Quality Index (Ch 5) |
| AQSIQ | Administration of Quality Supervision, Inspection and Quarantine (China) (Ch 38) |
| AR | Acid rain (Ch 9) |
| AR | Androgen receptor (Ch 39) |
| ARDS | Acute respiratory distress syndrome (Ch 7) |
| ARE | Antioxidant-response element (Ch 31) |
| ARFY | Atherosclerosis risk factors in female youngsters (Ch 27) |
| ARGAH | Arginine amidohydrolase (Ch 4) |
| *ArgE* | *N*-Acetylornithine deacetylase gene in *E. coli* (Ch 4) |

*Continued*

**Table 2.** Continued.

| Abbreviation or term | Definition |
|---|---|
| ARIC | Atherosclerosis Risk in Communities (Ch 14) |
| ARISA | Automated ribosomal intergenic spacer analysis (Ch 35) |
| ARLIS | Alaska Resource Library and Information Services (Ch 22) |
| ARNT | AHR nuclear translocator (Ch 13) |
| As | Arsenic |
| ASD | Autism spectrum disorders (Ch 11) |
| AST | Aspartate aminotransferase (Ch 34) |
| ATP | Adenosine triphosphate (Ch 4, 8, 9) |
| ATSDR | Agency for Toxic Substances and Disease Registry (Ch 8, 25) |
| AWD | Atmospheric wet deposition (Ch 29) |
| Aze | Azetidine-2-carboxylic acid (Ch 19) |
| Ba | Barium |
| BA | Bioaugmentation (Ch 35) |
| BABA | β-Aminobutyrate (Ch 4) |
| BAL | Broncheoalveolara lavage (Ch 7) |
| BAM | Brewster Angle Mirror copy (Ch 6) |
| BaP | Benzo[a]pyrene (Ch10) |
| BAT | Best available technologies (Ch 12) |
| BBB | Blood brain barrier (Ch 25) |
| BChE | Butyrylcholinesterase (Ch 17) |
| BEIR | Biological Effects of Ionizing Radiation (conference) (Ch 32) |
| BEN | Balkan endemic nephropathy (Ch 2) |
| bHLH | Basic helix–loop–helix (Ch 13) |
| BHMSM | Bushnell–Haas mineral salt medium (Ch 35) |
| BLLs | Blood lead levels (Ch 25) |
| B2M | β-2-Microglobulin (Ch 26) |
| BMAA | β-*N*-Methylamino-l-alanine (Ch 3, 19) |
| BMI | Body mass index (Ch 34) |
| BOOP | *Bronchiolitis obliterans* organizing pneumonia (Ch 7) |
| BP | British Petroleum (Ch 20, 21) |
| Bq | Bequerel (Ch 32) |
| BR | Brassinosteroid (Ch 4) |
| BS | Biostimulation (Ch 35) |
| BTEX | Benzene, toluene, ethylbenzene and xylene (Ch 20, 23, 35) |
| BuCHE | Butyrylcholinesterase (Ch 18) |
| C | Carbon |
| Ca | Calcium |
| CAS | Chemical Abstract Service (Ch 23) |
| CAT | Catalase (Ch 9, 31) |
| CC | Case-control (study) (Ch 33) |
| Cd | Cadmium |
| CDC | Centers for Disease Control (and Prevention) (Ch 11, 14, 25) |
| CEC | Cation exchange capacity (Ch 37) |
| CFA | Coal fly ash (Ch 29) |
| ChB | Chafuroside B (Ch 31) |
| ChE | Choline esterase (Ch 34) |
| CI | Confidence interval (Ch 14, 26, 27, 32) |
| cJNKs | c-Jun N-terminal kinases (Ch 31) |
| CNS | Central nervous system (Ch 33, 40) |
| CO | Carbon monoxide |
| $CO_2$ | Carbon dioxide |
| CoA | Coenzyme A |

*Continued*

**Table 2.** Continued.

| Abbreviation or term | Definition |
|---|---|
| COD | Chemical oxygen demand (Ch 23) |
| COEH | Council on Environmental Health (Ch 25) |
| Cog | Cognitive (tests) (Chapter 18) |
| COHb | Carboxyhaemoglobin (Ch 28) |
| –COOH | Carboxyl group (Ch 37) |
| COPD | Chronic obstructive pulmonary disease (Ch 5, 7, 8, 19, 28, 40) |
| COT | Committee on Toxicity (of Chemicals in Food, Consumer Products and the Environment) (Ch 18) |
| COX | Cytochrome c oxidase (Ch 8) |
| COX-2 | Cyclooxygenase-2 (Ch 31) |
| CPC | Communist Party of China (Ch 38) |
| CPDs | Cyclobutane pyrimidine dimers (Ch 31) |
| CPK | Creatine phosphokinase (Ch 34) |
| Cr | Chromium |
| CRA | Chemical risk assessment (Ch 18) |
| CRD | Centre for Reviews and Dissemination (Ch 18) |
| CREB | Cyclic AMP response element binding protein (Ch 11) |
| CRP | C-reactive proteins (Ch 34) |
| Cs | Caesium |
| CSF | Cancer slope factor (Ch 23) |
| Cu | Copper |
| CVD | Cardiovascular disease (Ch 26, 40) |
| CvE | *Calluna vulgaris* extract (Ch 31) |
| CYPs | Cytochrome P450 enzymes (Ch 8, 13, 19) |
| DAS | Diacetoxyscirpenol (Ch 2) |
| DAT | Dopamine transporter (Ch 11, 16) |
| DCF | 2′,7′-dichlorofluorescein (Ch 31) |
| DCRLs | Derived consideration reference levels (Ch 34) |
| DCM | Dichloromethane (Ch 29) |
| DDD | Dichlorodiphenyldichloroethane (Ch 40) |
| DDE | Dichlorodiphenyldichloroethylene (Ch 16, 40) |
| DDT | Dichlorodiphenyltrichloroethane/1,1,1-trichloro-2,2-*bis*(4-chlorophenyl) ethane (Ch 15, 16, 35, 40) |
| DEHP | Di (2-ethylhexyl) phthalate (Ch 30) |
| DES | Diethylstilbestrol (Ch 15) |
| DGGE | Denaturant gradient gel electrophoresis (Ch 35) |
| DGOMB | Deep Gulf of Mexico Benthos (Ch 21) |
| DIVER | Data Integration Visualization Exploration and Reporting (Ch 21) |
| DL | Dioxin-like (Ch 11) |
| dl-PCBs | Dioxin-like PCBs (Ch 13) |
| DLN | Draining lymph nodes (Ch 31) |
| DMF | Dimethylformamide (Ch 23) |
| DMSO | Dimethylsulfoxide (Ch 29) |
| DMT-1 | Divalent metal transporter-1 (Ch 26) |
| DNA | Deoxyribonucleic acid (Ch 2, 3, 8, 10, 13, 25, 26, 30, 33, 34, 40) |
| DNT | Developmental neurotoxicity (Ch 11) |
| DOC | Dissolved organic carbon (Ch 23, 37) |
| DOM | Dissolved organic matter (Ch 37) |
| DON | Deoxynivalenol (Ch 2) |
| DOSS | dioctyl sulfosuccinate (Ch 21) |
| DPM | diesel exhaust particulate matter (Ch 29) |
| DPPC | Dipalmitoyl-*sn*-glycero-3-phosphtidylcholine (Ch 6) |

*Continued*

**Table 2.** Continued.

| Abbreviation or term | Definition |
|---|---|
| DPPH | 2,2-Diphenyl-1-picrylhydrazyl |
| DREs | Dioxin response elements (Ch 13) |
| DSM-IV | Diagnostic and Statistical Manual of Psychiatric Disorders (Ch 18) |
| DTNB | 5,5′-Dithio-bis-2-nitrobenzoic acid (Ch 17) |
| DWH | *Deepwater Horizon* (Ch 21) |
| EA | Ellagic acid (Ch 31) |
| EaE | *Epilobium angustifolium* extract (Ch 31) |
| EBL | Erythroblast (Ch 34) |
| $EC_{50}$ | Effect concentration required to elicit a negative 50% effect relative to control (Ch 21) |
| ECM | Extracellular matrix (Ch 31) |
| EDCs | Endocrine-disrupting chemicals (Ch 15) |
| EEG | Electroencephalogram (Ch 17) |
| EFSA | European Food and Safety Authority (Ch 10, 26) |
| EGCG | (-)-Epigallocatechin-3-gallate (Ch 31) |
| EGF | Epidermal growth factor (Ch 13) |
| EGFR | Epidermal growth factor receptor (Ch 31, 32) |
| eGFR | Estimated glomerular filtration rate (Ch 34) |
| EHQs | Exposure history questionnaires (Ch 18) |
| EIA | Environmental impact assessment (Ch 38) |
| EIARTR | Environmental impact assessment restriction targeting regions (Ch 38) |
| EIN2 | Ethylene insensitive2 (Ch 4) |
| ELISA | Enzyme-linked immunosorbent assay (Ch 3 and 10) |
| EPA | Environmental Protection Agency (USA) (Ch 5, 40, 18) |
| EPBs | Environmental Protection Bureaus (China) (Ch 38) |
| EPL | Environmental Protection Law (China) (Ch 38) |
| EP&M | Environmental Protection & Management (Preface) |
| EPSP | 5-Enolpyruvylshikimate-3-phosphate (Ch 19) |
| EREBP | Ethylene responsive-element binding protein (Ch 4) |
| ERF | Ethylene response factor (Ch 4) |
| ERKs | Extracellular signal-regulated kinases (Ch 31) |
| EROD | 7-ethoxyresorufin-O-deethylase (Ch 22) |
| ET | Ethylene (Ch 4) |
| EU | European Union (Ch 2) |
| FAO | Food and Agriculture Organization (United Nations) (Ch 10, 30) |
| $FB_1$ | Fumonisin $B_1$ (Ch 2) |
| $FB_2$ | Fumonisin $B_2$ (Ch 2) |
| $FB_3$ | Fumonisin $B_3$ (Ch 2) |
| FDA | Food and Drug Administration (USA) (Ch 17, 20) |
| FDNPP | Fukushima Dai-ichi Nuclear Power Plant (Ch 34) |
| Fe | Iron |
| $Fe^0$ | Metallic/elemental iron (Ch 36) |
| FET | Fish embryo test (Ch 39) |
| FEV1 | Forced expiratory volume in 1 second (Ch 5, 7) |
| FHB | Fusarium head blight |
| FPIA | Fluorescence polarization immunoassay (Ch 10) |
| $FT_4$ | Free thyroxine (Ch 34) |
| FTIR | Fourier transform infrared spectrometry (Ch 35) |
| FVC | Forced vital capacity (Ch 5) |
| FYPs | Five-year plans (China) (Ch 38) |
| GABA | $\gamma$-Aminobutyrate/ $\gamma$-aminobutyric acid (Ch 4, 16, 25) |
| GAD | Glutamic acid decarboxylase (Ch 4) |

*Continued*

**Table 2.** Continued.

| Abbreviation or term | Definition |
|---|---|
| Gas | Glycoalkaloids (Ch 1) |
| GB | Glycine betaine (Ch 4) |
| GC | Gas chromatography (Ch 35) |
| GC-MS | Gas chromatography-mass spectrometry (Ch 10, 35) |
| GDH | Glutamate dehydrogenase (Ch 4) |
| GESAMP | Group of Experts on the Scientific Aspects of Marine (Environmental) Protection (Ch 30) |
| GFP | Green fluorescent protein (Ch 39) |
| GGT | Gamma-glutamyl transferase (Ch 14) |
| GLP | Glucagon-like-peptide (Ch 5) |
| GM | Geometric mean (Ch 24) |
| GOGAT | Glutamate synthase (Ch 4) |
| G6PD | Glucose-6-phosphate dehydrogenase (Ch 1, 4) |
| GPX | Guaiacol peroxidase (Ch 9) |
| GPx | Glutathione peroxidase (Ch 31) |
| GR | Glucocorticoid receptor (Ch 39) |
| GR | Glutathione reductase (Ch 9) |
| GS | Glutamine synthetase (Ch 4) |
| GSA | Glutamate-semialdehyde (Ch 4) |
| GSD | Geometric standard deviation (Ch 28) |
| GSH | Glutathione (Ch 31) |
| GSLs | Glucosinolates (Ch 1) |
| GST | Glutathione *S*-transferase (Ch 31, 32) |
| GTP | Green tea polyphenols (Ch 31) |
| HAQTS | Hebei Provincial Administration of Quality and Technical Supervision (China) (Ch 38) |
| HAWS | Hebei Administration of Work Safety (China) (Ch 38) |
| H2AX | Histone H2A (Ch 31) |
| HBH | Hebei Bureau of Health (China) (Ch 38) |
| HBS | Hebei Bureau of Statistics (China) (Ch 38) |
| HCC | Hepatocellular carcinoma (Ch 2) |
| β-HCH | beta-Hexachlorocyclohexane (Ch 16) |
| HCl | Hydrogen chloride (Ch 28) |
| HCN | Hydrogen cyanide (Ch 1, 28) |
| HDA | Hebei Provincial Department of Agriculture (China) (Ch 38) |
| HDL | High-density lipoprotein (Ch 14, 19) |
| HDM | House dust mite (Ch 5) |
| HEIQB | Hebei Entry-Exit Inspection and Quarantine Bureau (China) (Ch 38) |
| HEPB | Hebei Environmental Protection Bureau (China) (Ch 38) |
| HF | hydrogen fluoride (Ch 28) |
| Hg | Mercury |
| HHE | Health hazard evaluation (Ch 20) |
| HIITD | Hebei Industry and Information Technology Department (China) (Ch 38) |
| $HNO_2$ | Nitrous oxide (Ch 7) |
| $HNO_3$ | Nitric acid (Ch 7) |
| HO-1 | Heme oxygenase-1 (Ch 31) |
| $H_2O_2$ | Hydrogen peroxide (Ch 4, 31) |
| HPEE | High pesticide exposure event (Ch 18) |
| HPLC | High-performance liquid chromatography (Ch 3, 6) |
| HR | Hazard ratio (Ch 26, 33) |
| $H_2SO_4$ | Sulfuric acid (Ch 8) |
| HSP | Heat shock protein (Ch 13) |

*Continued*

**Table 2.** Continued.

| Abbreviation or term | Definition |
|---|---|
| HSPGs | Heparin sulfate proteoglycans (Ch 19) |
| HULIS | Humic-like substances (Ch 29) |
| HWI | Hazardous waste incinerator (Ch 12) |
| HxCDF | Hexachlorodibenzofuran |
| HyT | Hydroxytyrosol (Ch 31) |
| I | Iodine |
| IAA | Indole acetic acid (Ch 4) |
| IAA-Trp | IAA-tryptophan conjugate (Ch 4) |
| IARC | International Agency for Research on Cancer (Ch 11, 13, 26, 32, 33) |
| IBI | International Biochar Initiative (Ch 37) |
| IC50 | Inhibitory concentration of 50% of enzyme activity (Ch 31) |
| ICP-MS | Inductively coupled plasma-mass spectrometry (Ch 35) |
| ICRP | International Commission on Radiological Protection (Ch 34) |
| IDEA | Individuals with Disabilities Education Act (USA) (Ch 19) |
| IFN | Interferon (Ch 31) |
| IFN$\gamma$ | Interferon gamma (Ch 31) |
| IKK$\alpha$ | I$\kappa$B kinase $\alpha$ (Ch 31) |
| IL | Interleukin (Ch 5 and 31) |
| IL-12-KO | Interleukin-12p40 knockout (Ch 31) |
| Ile | Isoleucine (Ch 4) |
| iNOS | Inducible nitric oxide synthase (Ch 31) |
| IPCS | International Programme on Chemical Safety (Ch 24) |
| IQ | Intelligence quotient (Ch 25) |
| IQR | Interquartile range (Ch 14) |
| IRIS | Integrated Risk Information System (Ch 23) |
| IRS | Indoor residual spraying (Ch 16) |
| IS | Immune system (Ch 31) |
| IWI | Industrial waste incinerator (Ch 12) |
| JA | Jasmonic acid (jasmonate) (Ch 4) |
| JAQTS | Jasmine Administration of Quality and Technical Supervision (China) (Ch 38) |
| JAWS | Jasmine Administration of Work Safety (China) (Ch 38) |
| JBA | Jasmine Bureau of Agriculture (China) (Ch 38) |
| JBDR | Jasmine Bureau of Development and Reform (China) (Ch 38) |
| JBH | Jasmine Bureau of Health (China) (Ch 38) |
| JBIIT | Jasmine Bureau of Industry and Information Technology (China) (Ch 38) |
| JBPS | Jasmine Bureau of Public Security (China) (Ch 38) |
| JBS | Jasmine Bureau of Statistics (China) (Ch 38) |
| JBTT | Jasmine Bureau of Traffic and Transport (China) (Ch 38) |
| JCG | Jasmine County Government (China) (Ch 38) |
| JECFA | Joint (FAO/WHO) Expert Committee on Food Additives (Ch 24) |
| JEPB | Jasmine Environmental Protection Bureau (China) (Ch 38) |
| JNKS | c-Jun N terminal kinase (Ch 31) |
| K | Potassium |
| Keap 1 | Kelch-like ECH-associated protein 1 (Ch 39) |
| KER | Key event relationship (Ch 10) |
| La | Lanthanum (Ch 9) |
| Laur-Hyt | Hydroxytyrosyl laurate (Ch 31) |
| LbE | Lemon balm extract (Ch 31) |
| LC | liquid chromotography (Ch 10) |
| $LC_{50}$ | Lethal concentration required to kill 50% of a population (Ch 10) |
| $LD_{50}$ | Lethal dose required to kill 50% of a population of organisms (Ch 2, 17) |
| LDH | Lactose dehydrogenase (Ch 34) |

*Continued*

**Table 2.** Continued.

| Abbreviation or term | Definition |
|---|---|
| LDL | Low-density lipoprotein (Ch 34) |
| L-E | Long-Evans rat strain (Ch 13) |
| *LeARG1* and *LeARG2* | Genes encoding arginase in tomato (Ch 4) |
| LMW | Low molecular weight (Ch 26) |
| LNT | Linear non-threshold (Ch 34) |
| LOAEL | Lowest observed adverse effect level (Ch 23, 24) |
| LOC | Level of concern (Ch 20) |
| LMW | Low molecular weight (Ch 26) |
| LPS | Lipopolysaccharide (Ch 3) |
| LRRK | Leucine-rich repeat receptor kinase (Ch 4) |
| LTLL | Long-term low-level (Ch 18) |
| LV | Left ventricle (Ch 8) |
| MAPK | Mitogen-activated protein kinases (Ch 26, 31) |
| MBP | Myelin basic protein (Ch 19) |
| MCHC | Mean corpuscular haemoglobin concentration (Ch 34) |
| MCL | Maximum contaminant limits (Ch 23) |
| MC-LR | Microcystin-LR (Ch 3) |
| MCT 1 | Monocarboxylase transporter 1 (Ch 19) |
| MCV | Mean corpuscular volume (Ch 34) |
| MD | Minimata disease (Ch 24) |
| MDA | Malondialdehyde (Ch 9) |
| MEE | Ministry of Ecology and Environment (China) (Ch 38) |
| MeHg | Methylmercury (Ch 11) |
| MEP | Ministry of Environmental Protection (China) (Ch 38) |
| MeV | Megaelectronvolt (Ch. 32) |
| Mg | Magnesium |
| mGy | Milligray (Ch 33) |
| MHW | Ministry of Health and Welfare (Japan) (Ch 24) |
| MIE | Molecular initiating event (Ch 10) |
| miRNA | MicroRNA |
| MITI | Minister of International Trade and Industry (Japan) (Ch 24) |
| MLOD | Maximum limit of detection (Ch 14) |
| MMAD | Mass median aerodynamic diameter (Ch 28) |
| MMP | Mitochondrial membrane potential (Ch 8) |
| MMPs | Matrix metalloproteinases (Ch 31) |
| MMPI | Minnesota multiphasic personality inventory (Ch 18) |
| MMSE | Mini mental state examination (Ch 18) |
| MoA | Ministry of Agriculture (China) (Ch 38) |
| MoA | Mode of action (Ch 10) |
| MOSSFA | Marine oil snow sedimentation and flocculent accumulation (Ch 21) |
| MPs | Microplastics (Ch 10) |
| MPTP | 1-Methyl-4-phenyl-1,2,3,6-tetrahydropyridine (Ch 16) |
| MRI | Magnetic resonance imaging (Ch 18) |
| MRLs | Maximum residue limits (Ch 40) |
| mRNA | Messenger ribonucleic acid (Ch 13, 25) |
| mTOR | Mammalian target of rapamycin (Ch 27) |
| MS | Mass spectrometer/spectrometry (Ch 3 and 10) |
| MS | Multiple sclerosis (Ch 40) |
| MSH | Melanocyte stimulating hormone (Ch 31) |
| MSL | multi-soil layering (Ch 36) |
| mSv | millisievert (Ch. 33) |

*Continued*

**Table 2.** Continued.

| Abbreviation or term | Definition |
|---|---|
| MSWI | municipal solid waste incinerator (Ch 12) |
| MT | Metallothionein (Ch 26) |
| Myr-Hyt | hydroxytyrosyl myristate (Ch 31) |
| N | Nitrogen |
| NAAQS | National Ambient Air Quality Standard (Ch 7) |
| NAcc | Nucleus accumbens (Ch 11) |
| NaCl | Sodium chloride (Chapter 4) |
| NADPH | Nicotinamide adenine dinucleotide phosphate (reduced) (Ch 4, 17, 40) |
| NALC | *N*-Acetyl-L-cysteine (Ch 8) |
| ncRNA | Non-coding RNA (Ch 25) |
| NDL | Non-dioxin-like (Ch 11) |
| NDRC | National Development and Reform Commission (China) (Ch 38) |
| NEC | No effect concentration (Ch 10) |
| NER | Nucleotide excision repair (Ch 31) |
| Neut | Neutrophil (Ch 34) |
| NF-κb | Nuclear factor kappa B (Ch 31) |
| NGO(s) | Non-governmental organisation(s) (Ch 15, 17, 38) |
| NGS | Next-generation sequence (Ch 32, 35) |
| $NH_3$ | Ammonia |
| NHANES | National Health and Nutrition Examination Survey (Ch 11, 14, 15, 20, 25, 27) |
| NHEK | Normal human epidermal keratinocytes (Ch 31) |
| Ni | Nickel |
| NIEHS | National Institute of Environmental Health Sciences (Ch 20) |
| NIH | National Institutes of Health (Ch 20, 24) |
| NIMD | National Institute for Minamata Disease (Japan) (Ch 24) |
| NIOSH | National Institute for Occupational Safety and Health (Ch 7, 20) |
| NIV | Nivalenol (Ch 2) |
| NK | Natural killer (Ch 31) |
| NMDA | *N*-Methyl-D-aspartate (Ch 19) |
| NMDAR | *N*-Methyl-D-aspartic acid receptor (Ch 11) |
| NO | Nitric oxide (Ch 4, 5, 7, 31) |
| $NO_x$ | Nitrogen oxides (Ch 5, 9) |
| $NO_2$ | Nitrogen dioxide (Ch 7, 40) |
| NOAA | National Oceanographic and Atmospheric Administration (Ch 20, 21) |
| NOAEL | No observable adverse effect level (Ch 20) |
| NOEC | No observed effect concentration (Ch 10) |
| NOEL | No observable effect level (Ch 20) |
| NORM | Naturally occurring radioactive materials (Ch 23) |
| NOS | Nitric oxide synthase (Ch 4) |
| NPC | National Peoples Congress (China) (Ch 38) |
| NQO1 | Quinone oxidoreductase 1 (Ch 31) |
| NR | Narrative review (Ch 18) |
| NRC | National Research Council (USA) (Ch 24 and 32) |
| NRDA | National Resource Damage Assessment (Ch 21) |
| NRF | Nuclear respiratory factor (Ch 8) |
| Nrf2 | Nuclear factor erythroid 2-related factor 2 (Ch 31, 39) |
| NSCLC | Non-small cell lung cancer (Ch 32) |
| NSL | Nuclear translocation site (Ch 13) |
| 3-NT | 3-nitrotyrosine (Ch 7) |
| NTP | National Toxicology Program (USA) (Ch 25) |
| $O_3$ | Ozone (Ch 6, 7, 9) |
| OAT | Ornithine-δ-aminotransferase (Ch 4) |

*Continued*

**Table 2.** Continued.

| Abbreviation or term | Definition |
|---|---|
| OC | Organic carbon (Ch 29) |
| OC | Organochlorine (Ch 40) |
| OCPs | Organochlorine pesticides (Ch 37) |
| ODC | Ornithine decarboxylase (Ch 4) |
| OECD | Organization of Economic Cooperation and Development (Ch 10, 39) |
| OELs | Occupational exposure limits (Ch 20) |
| ·OH | Hydroxyl radical (Ch 6, 37) |
| OP(s) | Organophosphorous / organophosphates (Ch 17, 18, 40) |
| OPAH | Oxygen-substituted PAH (Ch 40) |
| OR | Odds ratio (Ch 14, 33) |
| -OR | Ketone group (Ch 37) |
| OSAT | Operational Science Advisory Team (Ch 21) |
| OSHA | Occupational Safety and Health Administration (USA) (Ch 25) |
| OSIL | Ocean Scientific International Ltd (Ch 21) |
| OTA | Ochratoxin A (Ch 2) |
| OTB | Ochratoxin B (Ch 2) |
| 5-OZT | 5-Vinyl oxazolidinethione (Ch 1) |
| P | Phosphorus |
| PAC | Public Advisory Committee (Alaska) (Ch 22) |
| PAHs | Polycyclic aromatic hydrocarbons (Ch 10, 20, 21, 28, 29, 35, 40) |
| PAI-1 | Plasminogen activator inhibitor-1 (Ch 5) |
| PAL | Phenylalanine ammonia-lyase (Ch 4) |
| PAzPC | 1-Palmitoyl-2-azelaoyl-*sn*-glycero-3-phosphocholine (Ch 6) |
| Pb | Lead |
| PBDEs | Polybrominated diphenyl ethers (Ch 15, 37) |
| PBT | Persistent, bioaccumulative and toxic (Ch 25) |
| PCBs | Polychlorinated biphenyls (Ch 11, 13, 14, 15, 28, 35, 37, 40) |
| PCDDs | Polychlorinated dibenzo-*p*-dioxins (Ch 12, 13, 29, 37) |
| PCDFs | Polychlorinated dibenzofurans (Ch 12, 13, 15, 29, 37) |
| pCi | Picocurie (Ch 32) |
| PCP | Pentachlorophenol (Ch 37) |
| PCR | Polymerase chain reaction (Ch 3) |
| P5CR | Pyrroline-5-carboxylate reductase (Ch 4) |
| P5CS | $\Delta^1$-Pyrroline-5-carboxylate synthetase (Ch 4) |
| PD | Parkinson's disease (Ch 40) |
| PDA | Photodiode array (Ch 3) |
| PE | Polyethylene (Ch 30) |
| PEC | Predicted environmental concentration (Ch 10) |
| PEG | Polyethylene glycol (Ch 4) |
| PELs | Permissible exposure limits (Ch 20) |
| PEP | Phosphoenolpyruvate (Ch 19) |
| PET | Positon emission tomography (Ch 18) |
| PG | Prostaglandin (Ch 31) |
| PGC | Peroxisome proliferator-activated receptor gamma coactivator (Ch 8) |
| PgE | *Punica granatum* extract (Ch 31) |
| PGF2 alpha | Prostaglandin F2 alpha (Ch 8) |
| PH | Phenylalanine hydroxylase (Ch 31) |
| $PH_3$ | Phosgene (Ch 28) |
| PKC | Protein kinase C (Ch 11) |
| PLPC | 1-Palmitoyl-2-linoleoyl-*sn*-glycero-3-phosphatidylcholine (Ch 6) |
| PM | Particulate matter (Ch 29, 40) |
| PNEC | Predicted no-effect concentration (Ch 10) |

*Continued*

**Table 2.** Continued.

| Abbreviation or term | Definition |
|---|---|
| POD | Peroxidases (Ch 9) |
| POG | 1-Palmitoyl-2-oleoyl-*sn*-glycerol (Ch 6) |
| PON1 | Paraoxonase 1 (Ch 17, 18) |
| PonPC | 1-Palmitoyl-2-(9′-oxo-nonanoyl)-*sn*-glycero-3-phosphocholine (Ch 6) |
| POPC | 1-Palmitoyl-2-oleoyl-*sn*-glycero-3-phosphtidylcholine (Ch 6) |
| POPG | 1-Palmitoyl-2-oleoyl-*sn*-glycero-3-phosphtidylglycerol (Ch 6) |
| POPs | Persistent organic pollutants (Ch 12, 15, 40) |
| PP | Polypropylene (Ch 30) |
| PPAR | Peroxisome proliferator-activated receptor (Ch 14, 39) |
| PpE | Pomegranate polyphenolic extract (Ch 31) |
| PPE | Personal protection equipment (Ch 20) |
| Pref | Prefecture (in Japan) (Ch 24) |
| PRPs | Proline-rich proteins (Ch 1) |
| PSP | Progressive supranuclear palsy (Ch 3) |
| PST | Paralytic shellfish toxins (Ch 3) |
| PtE | *Passiflora tarminiana* extract (Ch 31) |
| PTFE | Polytetrafluoroethylene (Ch 28) |
| PTWI | Provisional tolerable weekly intake (Ch 24) |
| PVC | Polyvinyl chloride (Ch 30) |
| PXR | Pregnane-X-receptor (Ch 39) |
| QMS | Quadrupole spectrometer |
| qPCR | Quantitative polymerase chain reaction (Ch 3) |
| QSAR | Quantitative structure–activity relationship (Ch 23) |
| Ra | Radium |
| RADS | Reactive airway dysfunction syndrome (Ch 8, 28) |
| RANKL | Receptor activator of nuclear factor κb ligand (Ch 31) |
| RBP | Retinol-binding protein (Ch 26) |
| RBC | Red blood cell (Ch 8) |
| RDX | Hexahydro-1,3,5-trinitro-1,3,5-triazine (Ch 37) |
| REE | Rare-earth element (Ch 9) |
| REL | Recommended exposure limit (Ch 7) |
| REP | Relative effect potency (Ch 14) |
| RfC | Reference concentration (Ch 20, 23) |
| RfD | Reference dose (Chapter 20) |
| Rn | Radon |
| RNA | Ribonucleic acid (Ch 2) |
| RNS | Reactive nitrogen species (Ch 11) |
| ROS | Reactive oxygen species (Ch 4, 6, 9, 10, 11, 25, 29, 31, 39) |
| RR | Relative Risk (Ch 33) |
| RSDL | Reactive skin decontamination lotion (Ch 17) |
| RT | Response times (Ch 18) |
| RyR | Ryanodine receptors (Ch 11) |
| S | Sulfur (Ch 4) |
| SA | Salicylic acid (Ch 4) |
| SA-Asp | SA-aspartate (conjugate) (Ch 4) |
| SAM | *S*-Adenosylmethionine (Ch 4) |
| SAMDC | SAM decarboxylase (Ch 4) |
| SAR | Systemic acquired resistance (Ch 1) |
| SAWS | State Administration of Work Safety (China) (Ch 38) |
| SBS | Sick buildings syndrome (Ch 2) |
| SCD | Shijiazhuang Customs District (Chapter 38) |
| SCLC | Small cell lung cancer (Ch 32) |

*Continued*

**Table 2.** Continued.

| Abbreviation or term | Definition |
|---|---|
| SDG | Sustainable Development Goals, UN (Ch 38) |
| SDS-PAGE | Sodium dodecyl sulfate-polyacrylamide gel electrophoresis (Ch 6) |
| Se | Selenium |
| SETAC | Society of Environmental Toxicology and Chemistry (Ch 10) |
| SIDS | Sudden infant death syndrome (Ch 40) |
| siRNA | Small interfering ribonucleic acid (Ch 11) |
| SMCO | *S*-Methylcysteine sulfoxide (Ch 1) |
| SMEs | Small and medium enterprises (Ch 38) |
| SMR | Standardized mortality ratio (Ch 33) |
| SNpc | Substantia nigra pars compacta (Ch 16) |
| $SO_2$ | Sulfur dioxide (Ch 8, 9) |
| SOD | Superoxide dismutase (Ch 9, 31) |
| SP | Surfactant protein (Ch 6) |
| SPDS | Spermidine synthase (Ch 4) |
| SPMS | Spermine synthase (Ch 4) |
| SPR | Surface plasmon resonance (Ch 10) |
| SQT | Sediment quality triad (Ch 21) |
| Sr | Strontium |
| SR | Systematic review (Ch 18) |
| Src | Sarcoma |
| SSD | Species sensitivity distribution (Ch 10) |
| SSGM | Second Study Group on Minamata (Disease) (Ch 24) |
| STAT3 | Signal transducer and activator of transcription-3 (Ch 31) |
| STEL | Short-term exposure limit (Ch 28) |
| T3 | Triiodothyronine (Ch 15) |
| T4 | Thyroxine (Ch 15) |
| TBT | Tributyltin (Ch 15) |
| TCA | Tricarboxylic acid (as in TCA cycle) (Ch 4) |
| TCA | Trichloroacetic acid (Ch 19) |
| TCDD | 2,3,7,8-tetrachlorodibenzo-*p*-dioxin (Ch 12, 13, 14, 39, 40) |
| TCE | Trichloroethylene |
| TD | Threonine dehydratase (Ch 1) |
| TD | Toxico dynamics (Ch 10) |
| *TD2* | Threonine dehydratase 2 gene (Ch 1) |
| TDI | Tolerable daily intake (Ch 13) |
| TDS | Total dissolved solids (Ch 23) |
| TDP | Transactive-response DNA-binding protein (Ch 19) |
| Te | Tellurium (Ch 34) |
| TE | Trace element (Ch 29) |
| TEFs | Toxicity equivalent factors (Ch 10, 12, 13, 14) |
| TEL | Tetraethyl lead (Ch 25) |
| TENORM | Technologically advised NORM (Ch 23) |
| TEQ | Toxicity equivalency (Cha 12, 13, 14, 28) |
| TFAM | Mitochondrial transcription factor A (Ch 8) |
| TG | Triglyceride (Ch 34) |
| TgAb | Anti-thyroglobulin antibody (Ch 34) |
| TGF-$\alpha$ | Transforming growth factor alpha (Ch 31) |
| TGF$\beta$1 | Transforming growth factor-beta (Ch 8) |
| TGGE | Temperature gradient gel electrophoresis (Ch 35) |
| TGIR | Thermogravimetric infrared spectrometry (Ch 35) |
| TH | Tyrosine hydroxylase (Ch 16) |
| $T_h$ | T helper (cell) (Ch 8) |

*Continued*

**Table 2.** Continued.

| Abbreviation or term | Definition |
|---|---|
| TK | Toxico kinetics (Ch 10) |
| TLR4 | Toll-like receptor 4 (Ch 8) |
| TLVs | Threshold limit values (Ch 20) |
| TMB-4 | Trimedoxime (Ch 17) |
| TMPP | Trimethylolpropane phosphate (Ch 28) |
| $TNB^-$ | Thionitrobenzoate |
| TNF | Tumour necrosis factor (Ch 5, 31) |
| TOPKAT | Toxicity Prediction by K(c)omputer Assisted Technology (Ch 23) |
| tPA | Tissue-type plasminogen activator (Ch 5) |
| TPH | Total petroleum hydrocarbons (Ch 21 and 35) |
| TPOAb | Anti-thyroid-peroxidase antibody (Ch 34) |
| TRFLP | Terminal restriction fragment length polymorphism (Ch 35) |
| TRPV | Transient receptor potential vanilloid (Ch 8) |
| TSH | Thyroid-stimulating hormone (Ch 15, 34) |
| TTT | Thymol turbidity test (Ch 34) |
| TUNEL | Terminal deoxynucleotidyl transferase (TDT) dUTP Nick-End Labelling (Ch 11) |
| TWI | Tolerable weekly intake (Ch 26) |
| UCA | Urocanic acid (Ch 31) |
| UFPs | Ultrafine particles (Ch 29) |
| UIBC | Unsaturated iron-binding capacity (Ch 34) |
| UK | United Kingdom (Ch 15) |
| UNEP | United Nations Environment Programme (Ch 24, 30) |
| UNSCEAR | UN Scientific Committee on the Effects of Atomic Radiation (Ch 34) |
| UOG | Unconventional oil and gas (Ch 23) |
| UPLC | Ultra-performance liquid chromatography (Ch 3) |
| USEPA/US EPA | United States Environmental Protection Agency (Ch 7, 10, 11, 20, 30, 39) |
| UV | Ultraviolet (Ch 2, 3, 10, 15, 31, 40) |
| VaD | Vascular dementia (Ch 40) |
| VDCC | Voltage-dependent calcium channel (Ch 11) |
| $V_E$ | Volume (of air) exhaled (Ch 28) |
| VMAT/VMAT2 | Vesicular monoamine transporter 2 (Ch 11 and 16) |
| VmE | *Vaccinium myrtillus* fruit extract (Ch 31) |
| VOCs | Volatile organic compounds (Ch 5, 6, 20) |
| WHO | World Health Organization (of UN) (Ch 10, 13, 14, 17, 18, 25, 32) |
| WAIS | Wechsler adult intelligence scale (Ch 18) |
| WL | Working level (Ch 32) |
| WLM | Working level month (Ch 32) |
| WMS | Wechsler memory scale (Ch 18) |
| WOE | Weight of evidence (Ch 16) |
| WOR | Whole organism response (Ch 10) |
| WTC | World Trade Center |
| WTO | World Trade Organization |
| ZAWS | Zhangjiakou Administration of Work Safety (Chapter 38) |
| ZBIIT | Zhangjiakou Bureau of Industry and Information Technology (Chapter 38) |
| ZEN | Zearalenone (Chapter 2) |
| ZEPB | Zhangjiakou Environmental Protection Bureau (Chapter 38) |
| ZJK City | Zhangjiakou City (Chapter 38) |
| Zn | Zinc |
| ZVI | Zero-valent iron (Ch 36) |

## References

Alberts, B., Bray, D., Hopkin, K., Johnson, A., Lewis, J., Raff, M., Roberts, K. and Walter, P. (2014) *Essential Cell Biology,* 4th edn. Garland Science (Taylor & Francis Group) New York and Abingdon, UK.

Allaby, M. (2010) *A Dictionary of Ecology*, 4th edn. Oxford University Press, Oxford.

Carton, J. (ed.) (2017) *Oxford Handbook of Clinical Pathology*, 2nd edn. Oxford University Press, Oxford.

D'Mello, J.P.F. (ed.) (1997) *Handbook of Plant and Fungal Toxicants,* 1st edn. CRC Press, Boca Raton, Florida.

D'Mello, J.P.F. (ed.) (2012) *Amino Acids in Human Nutrition and Health*, 1st edn. CAB International, Wallingford, UK.

D'Mello, J.P.F. (ed.) (2015) *Amino Acids in Higher Plants*, 1st ed. CAB International, Wallingford, UK.

D'Mello, J.P.F., Duffus, C.M. and Duffus, J.H.(eds) (1991) *Toxic Substances in Crop Plants*, 1st edn. The Royal Society of Chemistry, Cambridge, UK.

Dong, M.H. (2014) *An Introduction to Environmental Toxicology*, 3rd edn. CreateSpace Publishing, North Charleston, USA.

Hodgson, E. and Roe, M. (2014) *Dictionary of Toxicology*, 3rd edn. Academic Press, New York and London.

Klug, W.S., Cummings, M.R., Spencer, C. and Palladino, M. (2014) *Concepts of Genetics*, 10th edn. Pearson Education Limited, Harlow, UK.

Lackie, J. (ed.) (2013) *The Dictionary of Cell and Molecular Biology*, 5th edn. Academic Press, New York and London.

Lodish, H., Berk, A., Kaiser, C.A., Krieger, M., Bretscher, A., Ploegh, H., Amon, A. and Scott, M.P. (2013) *Molecular Cell Biology*, 7th edn. W.H. Freeman and Company, New York.

Madigan, M., Martinko, J., Bender, K., Buckley, D. and Stahl, D. (2015) *Brock Biology of Microorganisms*, 14th edn. Pearson Education Limited, Harlow, UK.

Martin, E. (ed.) (2015) *Concise Medical Dictionary*, 9th edn. Oxford University Press, Oxford, UK.

Nelson, D.L. and Cox, M.M. (2013) *Lehinger Principles of Biochemistry*, 6th edn. Macmillan Higher Education, Basingstoke, UK.

Parish, H., Smith, T., Stirling, J. and Vella, F. (eds) (2006) *Oxford Dictionary of Biochemistry and Molecular Biology*, 2nd edn. Oxford University Press, Oxford, UK and New York.

Shibamoto, T. and Bjeldanes, L.F. (2009) *Introduction to Food Toxicology*, 2nd edn. Elsevier Science Publishing, Amsterdam, The Netherlands.

Singleton, P. and Sainsbury, D. (2006) *Dictionary of Microbiology and Molecular Biology*, 3rd edn. Wiley-Blackwell, Hoboken, New Jersey.

Strelkauskas, A., Fahnert, B., Pryor, G. and Strelkauskas, J. (2016) *Microbiology: a Clinical Approach*, 2nd edn. Garland Science (Taylor & Francis Group) New York and Abingdon, UK.

Walker, C.H., Sibly, R.M., Hopkin, S.P. and Peakall, D.B. (2012) *Principles of Ecotoxicology*, 4th edn. CRC Press, Taylor & Francis Group, Boca Raton, Florida, USA.

Ward, H., Toledano, M.B., Shaddick, G., Davies, B., Peacock, J. and Peacock, P. (2016) *Oxford Handbook of Epidemiology for Clinicians*, 2nd edn. Oxford University Press, Oxford, UK.

# Part I

# Biogenic Compounds

# 1 Phytotoxins

**J.P.F. D'Mello***
*Formerly of SAC, University of Edinburgh King's Buildings Campus, West Mains Road, Edinburgh, UK*

## 1.1 Abstract

Evidence of the existence of a wide range of secondary compounds and proteins in plants with pharmacological activity or with the potential to cause adverse effects in other living organisms is now firmly embedded in the scientific literature. The term 'phytotoxins' is conventionally used to describe these substances, but there are issues associated with nomenclature which are addressed below. In this chapter, the ecology of phytotoxins, including several glycosides, non-protein amino acids, furanocoumarins, condensed tannins, gossypol and specific anti-nutritional proteins, is considered. A number of these compounds occur exclusively in tropical plants while others are more universally distributed but with particular relevance in temperate environments. The aforementioned phytotoxins have been selected for diverse manifestations and implications in toxicology. Effects in mammals include digestive dysfunction caused by anti-nutritional proteins in legume seeds; irreversible spastic paralysis and cognition defects induced by cyanogenic glycosides in cassava; goitrogenic activity precipitated by *Brassica* glucosinolate breakdown products; favism associated with pyrimidine glycosides in faba beans; cardiotoxicity and reproductive abnormalities caused by gossypol in cottonseed; phytodermatitis following contact with furanocoumarins in celery and other plants; and cancer induced by ptaquiloside in bracken fern.

Lower organisms possess variable mechanisms for metabolizing several of these phytotoxins. Nevertheless, it is consistently maintained that phytotoxins serve in a protective role in plants against invertebrate herbivores and fungal pathogens. It is concluded that a defence system based on protein phytotoxins may be relatively more robust than those involving secondary compounds. Furthermore, it is suggested that these heat-labile protein phytotoxins should form part of a plant breeding programme to enhance pest and pathogen resistance without compromising food safety. However, the challenge for the future is to exploit the wide array of the other phytotoxins as environmentally friendly protectants for food crops. Structure-activity and other functional relationships have been established for a limited number of phytotoxins, thus improving the prospects for the development of more effective bio-pesticides.

## 1.2 Introduction

A diverse array of toxic compounds occurs in higher plants as a consequence of endogenous

* E-mail address: jpfdmello@hotmail.co.uk

synthesis, involving intricate and extensive pathways of secondary metabolism (D'Mello, 1997). These products differ widely in terms of molecular weights, structure and biological activities, eliciting responses independently of each other or in synergy with other molecules and perhaps even implying a role for specific receptors. The term 'phytotoxins' has been adopted in this chapter, but there is a plethora of synonyms in the extensive literature, justifying further clarification as detailed below. Although these substances are considered to be constitutive in origin, there are also significant mechanisms for induced synthesis in response to biotic and environmental stresses. In general, phytotoxins are of low to medium molecular weights, but more complex compounds regularly occur, particularly in seeds and certain roots. Toxicity is directed towards, and readily expressed in, a wide range of organisms, irrespective of taxonomic classification, contributing to a proposed constitutive and induced mechanism of defence in higher plants. Thus, other plant genera may succumb to adverse effects just as well as microbial pathogens, insect herbivores and vertebrates.

The study of phytotoxins is relevant to this volume for several reasons, in addition to enhancing our knowledge of fundamental and applied toxicology. It highlights issues in chemical ecology, particularly with regard to the differences between temperate and tropical plant species and the interaction with biotic and environmental factors, thus assisting in our understanding of niche biodiversity. An emerging issue is the putative role of these toxins in the elucidation of host–pathogen and host–parasite interactions and the expression of innate immunity in plants. A logical progression in this aspect is the exploitation of plant toxins for the development of bio-pesticides as environmentally friendly alternatives to persistent chemical protectants currently in use.

An important exclusion clause applies here to further define the terms of reference. Phytotoxins with established pharmacological activities and those with medical uses fall outside the scope of this review, due to limited environmental implications. Nevertheless, there is a need for flexibility in this restrictive approach, since several phytotoxins may potentially serve as novel pharmaceuticals partly designed to alleviate drug pollution, an emerging problem highlighted by Larsson (2014).

## 1.3 Terminology

It is essential at the outset to clarify several issues concerning nomenclature. It is generally accepted that phytotoxins comprise those secondary compounds of plants that are toxic to other living organisms. This definition, by convention, extends to certain protein molecules that clearly do not arise as a result of secondary metabolism and may serve additional functions in the plant. Over the course of time, additional terms have come into common usage to emphasize distinct functional characteristics of certain phytotoxins. These include allelochemicals, anti-metabolites and particularly phytoalexins. In addition, secondary substances may be classified as carcinogens, phytoestrogens and anti-nutritive factors.

According to recognized nomenclature, phytoalexins refer specifically to those secondary compounds synthesized by plants to inhibit infection and proliferation of pathogenic fungi. It is important to note overlap in common usage in that several phytotoxins are also classified as phytoalexins. The term 'death acids' has also been used with reference to cytotoxic phytoalexins in maize, *Zea mais* (Christensen *et al.*, 2015).

It is instructive to consider the broad context of the term 'phytotoxins'. For example, Dadler (2014) referred to a bacterial phytotoxin in the inhibition of eukaryotic proteasomes, and Bignell *et al.* (2014) reported the synthesis of phytotoxins by plant pathogenic *Streptomyces* species, while Cimmino *et al.* (2015) indicated the existence of a fungal phytotoxin with herbicidal potential. Furthermore, Chang *et al.* (2016) identified multiple phytotoxins as well as phytotoxic effector in *Fusarium virguliforme*. This chapter is concerned exclusively with toxins produced by plants.

As will be seen later, the concept of secondary metabolism is now under renewed scrutiny since it implies that the derived molecules exist on the periphery of plant survival, growth and reproduction, surplus to requirements. There is now overwhelming evidence of important signalling and defence roles for a wide range of these compounds. Examination of metabolic pathways indicates that primary and secondary processes are not parallel but integrated with respect to utilization of common substrates, cofactors and shared bioenergetics. Thus, the distinction between primary and secondary metabolism appears to be largely arbitrary in the light of present-day evidence.

## 1.4 Objectives

It is not intended to provide an exhaustive catalogue of phytotoxins but rather to focus on key compounds that illustrate how molecular and structural diversity affect toxicology. In determining this selection, the aim has been to indicate the disparate range of manifestations attributed to the different phytotoxins. For example, in vertebrate animals, including humans, deleterious effects range from inhibition of proteolytic enzymes in the gut to organ damage and dysfunction, photosensitization reactions, genetic damage and carcinogenesis. The selection of plant sources of these toxins will reflect species occurring in natural and managed ecosystems. An additional consideration includes the putative role of phytotoxins and other secondary compounds in defence mechanisms of plants specifically directed at insect pests and fungal pathogens. Attempts to elucidate the detoxification mechanisms in target species are also relevant in this chapter, as efforts continue to identify potential targets for the development of effective bio-pesticides.

## 1.5 Distribution and Ecology

The phytotoxins under consideration here have been selected from the important primary groups in the conventional classification of secondary compounds, namely glycosides, non-protein amino acids, phenolic compounds and derivatives, alkaloids and proteins. The occurrence of these compounds in plants of economic and ecological importance is presented in Table 1.1. The diversity of individual compounds present as glycosides is immediately apparent. It will be noted that the cyanogens occur in tubers as well as in seeds. The distribution of several phytotoxins is characteristically associated with tropical species, for example mimosine in *Leucaena leucephala* (ipil-ipil) and gossypol in *Gossypium hirsutum* and *G. barbadense* (cotton). In contrast, canavanine occurs more widely, including both temperate and tropical plants. Legumes tend to contain a greater diversity of phytotoxins, with the seed being the most concentrated source. The co-occurrence of *S*-methylcysteine sulfoxide (SMCO) with glucosinolates in *Brassica* species is worth noting. Condensed tannins also invariably occur with other phytotoxins in legume plants. Proteinase inhibitors, lectins and certain enzymes complete the spectrum of large molecular-weight phytotoxins.

## 1.6 Adverse Effects in Vertebrate Animals

### 1.6.1 Toxic glycosides

It is instructive to commence with the glycosides, due to the diverse range in molecular complexity and mode of deployment of the toxic principle. The effects of the glycosides listed in Table 1.2 are only expressed after completion of an enzyme-dependent reaction releasing the deleterious component from its precursor. This enzyme reaction is triggered by tissue damage to the plant, as for example after insect herbivory or fungal penetration and infection.

The cyanogens exist as a distinctive class of glycosides in the foliage and seeds of different species of plants (Davis, 1991). The classical example is amygdalin, present in bitter almonds, while another well-known cyanogen, linamarin, occurs in cassava. In sorghum plants, the principal cyanogen is dhurrin. The structures and a comprehensive list of cyanogens are presented by Davis (1991). Following tissue damage, hydrogen cyanide (HCN) is released from the cyanogen, causing dose-related toxicity and lethality to insect herbivores and vertebrate animals. Extensive human health issues continue to be reported in sub-Saharan communities dependent upon cyanogenic cassava even to the present time (Tshala-Katumbay *et al.*, 2016). Sub-lethal blood HCN concentrations of up to 80 μmol $l^{-1}$ are regularly observed, leading to symptoms of acute toxicity. Sub-lethal effects include distinct and irreversible spastic paralysis, known in sub-Saharan African regions as 'konzo', and cognition deficits. However, the issue may be confounded by other factors such as age, gender, dietary protein insufficiency and even micronutrient deficiency. Health disorders may also result from thiocyanate, a detoxication product of HCN. Thiocyanate increases the metabolic demand for iodine and goitre may occur in cassava-reliant populations if the intake of the essential element is inadequate (Davis, 1991).

**Table 1.1.** Classification and distribution of important phytotoxins in higher plants. This table is not designed to be exhaustive but illustrative of the biophysical features of the selected phytotoxins. The compounds below are arranged in order of increasing complexity of the bioactive groups. Of particular note is the occurrence of phytotoxins as glycosides.[a] For comparison of chemical structures, refer to *Toxic Substances in Crop Plants* (D'Mello *et al.*,1991) and *Handbook of Plant and Fungal Toxicants* (D'Mello, 1997).

| Classification | Phytotoxins (precursors/ active compounds) | Occurrence |
|---|---|---|
| Cyanogens[a] | Linamarin | Cassava (*Manihot esculenta*) |
| | Amygdalin | Bitter almonds (*Prunus dulcis*), and seeds of peaches (*Prunus persica*) and apricots (*Prunus armeniaca*) |
| | Dhurrin | Sorghum (*Sorghum vulgare*) |
| Glycoalkaloids[a] | Solanine and chaconine | Potato (*Solanum tuberosum*) |
| Glucosinolates[a] | Nitriles, isothiocyanates and other derivatives and breakdown products | Cabbage, cauliflower, kale and other *Brassica* species |
| Saponins[a] | Triterpenoid and steroid aglycones | Wide distribution in plants destined for consumption by humans and farm animals |
| Pyrimidine glycosides[a] | Vicine and convicine | Faba beans (*Vicia faba*) |
| Flavones[a] | Quercetin, myricetin, rutin, etc. | Constituents of fruits, vegetables, cereals, tea and cocoa |
| Isoflavonoids[a] | Daidzin, glycitin, genistin, etc. | Wide distribution in plants, especially legumes |
| Illudane glycoside (bracken toxin) | Ptaquiloside[a] | Bracken fern (*Pteridium* species) |
| Non-protein amino acids | *S*-Methylcysteine sulfoxide | *Brassica* species |
| | Mimosine | Ipil ipil (*Leucaena leucocephala*) |
| | Canavanine | Alfalfa (*Medicago sativa*), jackbean (*Canavalia ensiformis*) and certain other tropical legume seeds |
| | Selenoamino acids | *Brassica* vegetables |
| Linear furanocoumarins | Psoralen, bergapten, xanthotoxin, isopimpinellin | Distributed in at least 15 plant families |
| Polyphenolic compounds | Condensed tannins | Distributed in cereal grains, legume seeds and forages |
| | Gossypol | Cotton plant (*Gossypium* spp.) |
| Proteins | Proteinase inhibitors | Ubiquitous in plants, especially legume seeds, e.g. soyabeans (*Glycine max*) |
| | Lectins | Concentrated in legume seeds, e.g. jackbean (*Canavalia ensiformis*), kidney beans (*Phaseolus vulgaris*) |
| | Threonine dehydratase | Tomato (*Solanum lycopersicum*) |

Common manifestations of glycoalkaloid poisoning in human subjects include headache, vomiting, diarrhoea, abdominal pain and neurological disorders (Percival and Dixon, 1997). Potato glycoalkaloids are consumed by significant numbers of the global population over their lifespans, but there appears to be a dearth of data on potential long-term effects.

Histological lesions in the liver and kidney of rats have been attributed to breakdown products of glucosinolates (Duncan, 1991). Nitriles, isothiocyanates and 5-vinyl oxazolidinethione (5-OZT) are the principal degradation products under regular investigation. The goitrogenic effects of 5-OZT are attributed to the organic iodination of thyroxine. Due to intermittent intakes,

**Table 1.2.** Contrasting effects of phytotoxins in mammals and other vertebrates.

| Phytotoxins | Effects | References |
|---|---|---|
| Cyanogens | Irreversible spastic paralysis in humans (konzo); cognition deficits; goitre | Tshala-Katumbay *et al.* (2016) |
| Glycoalkaloids | Headache, vomiting, diarrhoea, abdominal pain and neurological disorders in humans | Percival and Dixon (1997) |
| Glucosinolates | Histological lesions in the liver and kidney of rats attributed to breakdown products of glucosinolates; goitrogenic effects | Duncan (1991) |
| Saponins | Interaction with biological membranes; haemolytic and cytotoxic effects | Tong *et al.* (2017) |
| Pyrimidine glycosides | Favism in humans with a genetic deficiency of glucose-6-phosphate dehydrogenase in erythrocytes; precipitated by ingestion of faba beans or by certain drugs or infection; acute haemolytic anaemia | Reading *et al.* (2016) |
| Flavones | Moderate to low cytotoxicity in human cell lung carcinoma and colorectal cancer cell lines | Gaydou (1997) |
| Isoflavonoids | Biliary atresia in neonates exposed to biliatresone from *Dysphania* species | Lorent *et al.* (2015) |
| Ptaquiloside | Major carcinogen of bracken fern | Virgilio *et al.* (2015) |
| Non-protein amino acids | *S*-Methylcysteine sulfoxide: haemolytic anaemia in ruminants; mimosine: disruption of reproductive processes, teratogenic effects and loss of hair and wool; canavanine: antagonist of arginine in avian species; selenoamino acids: anti-metabolite activity towards sulfur amino acids | D'Mello (2015a) |
| Furanocoumarins | Phytophotodermatitis, mutagenic and carcinogenic properties | Diawara and Trumble (1997) |
| Condensed tannins | Anti-nutritional effects in ruminants, including reduced feed intake and digestibility of nutrients | Kumar and D'Mello (1995) |
| Gossypol | Cardiotoxic effects; reproductive dysfunction | Santana *et al.* (2015) |
| Proteinase inhibitors | Depressed dietary protein digestibility; pancreatic hypertrophy | Norton (1991) |
| Lectins | Reduced appetite and severe digestive disorders | D'Mello (1995) |

glucosinolates and associated breakdown products are not risk factors in human health; on the contrary, several of the sulfur-containing secondary compounds in *Brassica* vegetables are attributed with beneficial properties as potential anti-cancer agents (see below).

Although saponins are known for interactions with biological membranes and haemolytic and cytotoxic effects under experimental conditions (Tong *et al.*, 2017), there is no evidence of adverse implications for human health with common foods containing these secondary compounds. Indeed, the hypocholesterolaemic effect of saponins has generated considerable clinical interest.

The pyrimidine glycosides are associated with favism in humans with a genetic deficiency of glucose-6-phosphate dehydrogenase (G6PD) in erythrocytes, precipitated by ingestion of faba beans or by certain drugs or infection; acute haemolytic anaemia is another feature. Reading *et al.* (2016) observed that the most common form of severe haemolytic anaemia in Palestinian children varies in severity with three different variants of G6PD deficiency within the same community.

Flavones in general are attributed with moderate to low cytotoxicity, as shown by studies with human cell lung carcinoma and colorectal cancer cell lines (Gaydou, 1997). However, in studies with Chinese hamster ovary cells, quercetin aglycone induced the highest degree of toxicity of all the flavonoids tested by Engen *et al.* (2015). With isoflavonoids, biliary atresia in neonates exposed to biliatresone from

*Dysphania* species has emerged in a recent study (Lorent *et al.*, 2015).

In contrast, ptaquiloside, the major toxin in bracken fern, is positively associated with carcinogenic properties towards livestock and possibly humans through contact or consumption. Virgilio *et al.* (2015) reported ptaquiloside in the pooled raw milk of healthy sheep and goats, representing an underestimated, global concern for food safety.

### 1.6.2 Non-protein amino acids

Non-protein amino acids precipitate a diverse array of toxic effects in mammals (see D'Mello, 2015a). SMCO causes haemolytic anaemia in ruminants fed largely or exclusively on *Brassica* forage. Adverse effects of SMCO develop after its metabolism by anaerobic rumen bacteria to dimethyl disulfide, a reactive metabolite. SMCO in *Brassica* vegetables is safe for human consumption due to the absence of significant microbial activity in the gut. Mimosine causes disruption of reproductive processes, teratogenic effects and loss of hair and wool in cattle and sheep grazing *Leucaena* pastures. The toxicity of *Leucaena* is determined by geographical differences in rumen ecology. In regions where *Leucaena* is indigenous (Central America) or is naturalized (Hawaii and Indonesia), ruminants possess the full complement of bacteria to completely degrade mimosine, which accounts for the absence of *Leucaena* toxicity in these countries. However, in Australia, the USA and Kenya, ruminants lack the requisite bacteria required for the complete degradation of mimosine and succumb to the goitrogenic effects of its intermediate compound, 3-hydroxy-4(1*H*)-pyridone (see D'Mello, 2015a). Anecdotal evidence indicates that consumption of *Leucaena* as a vegetable may cause hair loss in humans. In avian species, canavanine acts as an antagonist of arginine, an essential amino acid for these animals. The selenoamino acids also display anti-metabolite activity, but the target is the sulfur amino acids.

### 1.6.3 Linear furanocoumarins

Manifestations of mammalian toxicity of linear furanocoumarins include phytophotodermatitis, mutagenesis and carcinogenesis (Diawara and Trumble, 1997). The linear furanocoumarins are photoactivated by ultraviolet A radiation in the wavelength range 320–400 nm. The epidermal symptoms include bullous eruptions, pigmentation, erythema and potential vesicle formation. These changes may appear at points of contact with plants containing linear furanocoumarins, or over the entire body if exposure occurs via ingestion.

### 1.6.4 Condensed tannins

With condensed tannins, the primary observations relate to anti-nutritional effects in ruminants, including reduced feed intake and digestibility of nutrients. However, these effects may vary according to the ability of certain species to produce and secrete salivary proline-rich proteins (PRPs). It is considered that PRPs form the first line of defence against ingested tannins and that deer and possibly goats produce copious quantities of PRPs, whereas they are absent in the salivary secretions of cattle and sheep (Kumar and D'Mello, 1995).

### 1.6.5 Gossypol

Gossypol toxicity is characterized by two features: cardiotoxic effects, for example in pigs, lambs and calves consuming cottonseed or its by-products (Risco and Chase, 1997); and reproductive dysfunction. The latter issue continues to attract attention, with Santana *et al.* (2015), for example, investigating the mechanisms involved in reproductive damage caused by gossypol in rats and the protective effects of vitamin E.

### 1.6.6 Protein phytotoxins

The protein phytotoxins exert moderate to powerful anti-nutritional effects in vertebrate animals. Proteinase inhibitors depress dietary protein digestibility in a wide range of animal species and also cause pancreatic hypertrophy (Norton,1991). In contrast, lectins are associated with rapid and wasteful growth of the small intestine and pancreas in laboratory animals, causing, in addition, a disruption of the absorptive epithelium and reduced nutrient availability for peripheral organs. Mortality in these animals is attributed to severe depletion of body reserves (Grant, 1991).

## 1.7 Role as Plant Defence Compounds

The role of phytotoxins as plant defence compounds has been a major justification for continuing research (see D'Mello *et al.*, 1991). In recent years, a high degree of sophistication has emerged in the description of innate immunity in higher plants (comparable to that in mammals), involving signal transduction and deployment of secondary compounds in response to insect herbivory or fungal infection (D'Mello, 2015b). Signal transduction in plants is critically dependent upon the metabolism and utilization of amino acids. Specific amino acids may directly contribute to signal transduction, as in the case of glutamate. For example, $Ca^{2+}$ conduction by an amino acid-gated channel has been attributed to the existence of glutamate-like receptor homologues of mammalian ionotropic glutamate receptors. The role of D-serine in a novel plant signalling mechanism has also been demonstrated, reflecting a similar system in mammals. Contemporary evidence and opinions implicate a diverse array of phytohormones and other bioactive compounds in complex signal transduction networks integrated into a complex defence system. Examples frequently cited include abscisic acid, salicylic acid, jasmonate, ethylene, nitric oxide and indole acetic acid. The context for signal transduction in plants is generally associated with stress and defence metabolism and the initiation of induced resistance, though signalling compounds rarely operate in isolation. The emerging evidence confirms complex interactions or crosstalk in communication networks of plants exposed to biotic or environmental challenges. Specific signalling compounds have been associated with crosstalk functions. It is generally accepted that salicylic acid signalling mediates defences towards biotrophic pathogens, whereas jasmonate is effective against insect herbivores and necrotrophic microorganisms. However, much still needs to be uncovered about salicylic acid/jasmonate crosstalk at the molecular and ecological levels (see D'Mello, 2015b). Insect herbivory activates several endogenous signals from the damaged tissues, resulting in the deployment of plant defence compounds. Collectively, these processes comprise the 'wound response' of plants to insect herbivory. Molecular analysis of the wound response indicates that systemin and prosystemin are upstream components of a jasmonate-dependent cascade, though a jasmonate-independent pathway may also operate. Others implicate oligosaccharides, ethylene and abscisic acid as additional contributors to the wound response.

The foregoing account amply demonstrates the extensive and in-depth advances relating to the characterization of signalling and defence compounds in plants. It is also evident that the role of signalling molecules involved in the initiation of defence reactions in plants applies to both constitutive and inducible phytotoxins. However, the integration of these components into a working model to explain plant immune responses towards insect herbivores or fungal pathogens remains largely elusive. This relative lack of progress may be attributed, in part, to the unpredictable efficacy of certain phytotoxins in plant responses to biotic stress.

### 1.7.1 Effects on insect herbivores

Although considerable attention has been given to the activity of constitutive phytotoxins, there is evidence that several of these defence compounds may also be induced during and after insect predation.

Several phytotoxins exert their defence functions towards predatory insects by reducing feeding activity. These include glycoalkaloids, saponins, furanocoumarins and condensed tannins (Table 1.3).

On the basis of mammalian evidence, it would be logical to assume that plant cyanogens might serve as highly effective defence compounds against insect herbivores. Any such perceptions are not entirely supported by experimental observations. The widespread herbivory associated with cyanogen-containing plants argues against any protective functions for this phytotoxin. Davis (1991) suggested that cyanogens may serve in a defensive role arising from deterrence to feeding rather than from outright toxicity. Differences may also be apparent

**Table 1.3.** Diverse effects of phytotoxins on insects of economic importance, including herbivores and stored-product pests: a selection of recent data.

| Phytotoxins | Effects | References |
|---|---|---|
| Cyanogens | Insect herbivore employs multiple strategies to overcome plant cyanogens | Pentzold *et al.* (2014) |
| Glycoalkaloids (GAs) | Feeding preferences affected by concentrations and profile of GAs in specialist and generalist insect herbivores | Altesor *et al.* (2014) |
| Glucosinolates (GSLs) | Insects are capable of diverting hydrolysis of GSLs to less toxic compounds or desulfate parent GSLs to prevent hydrolysis by plant myrosinases; insects may also sequester ingested GSLs for use as defence molecules | Jeschke *et al.* (2015) |
| Saponins | In combinations with the flavonoid apigenin, saponins reduce both the number and duration of feeding probes in aphids; responses depend upon the concentration and proportions of compounds in the mixture | Goławska *et al.* (2014) |
| Flavones | Survival, food utilization and cell organelles of a generalist insect herbivore adversely affected by quercetin | Selin-Rani *et al.* (2016) |
| Non-protein amino acids | Mimosine inhibitory to growth and development of insect larvae; insecticidal activity of canavanine attributed to its mimicry of arginine | D'Mello (2015a) |
| Furanocoumarins | Insecticidal and anti-feeding effects observed with extracts containing furanocoumarins | Pavela and Vrchotova (2013) |
| Condensed tannins | Anti-feeding activity of condensed tannins to various non-adapted insect herbivores | Griffiths (1991) |
| Gossypol | Aldehyde groups of gossypol are critical for its toxicity to cotton-feeding insect larvae; up-regulation of cytochrome P450 enzyme attributed to a general stress response to plant toxins and not just gossypol | Krempl *et al.* (2016) |
| Proteinase inhibitors | Trypsin inhibitor causes high mortality in the raspberry weevil | Medel *et al.* (2015) |
| Lectins | Anti-nutritional effects of lectins affecting larval weight, fecundity, pupation and survival | Macedo *et al.* (2015) |
| Threonine dehydratase | Depletion of threonine (an essential amino acid) in insect gut | D'Mello (2015b) |

between specialist and generalist insect predators. The southern armyworm (*Spodoptera eridania*) exhibits preference for the linamarin-containing lima bean (*Phaseolus lunatus*) and it has been demonstrated that cyanide acts as a feeding attractant. In choice-feeding studies, larvae of a Southern African butterfly (*Acraea horta*) preferred leaves of cyanogen-containing plant species. Davis (1991) concluded that the role of cyanogens was 'neither clear nor consistent'. As shown in Table 1.3, multiple strategies may exist in *Zygaena filipendulae* larvae to overcome plant cyanogen defence. These mechanisms include behavioural, morphological, physiological and biochemical adaptations at different phases during feeding and digestion to avoid toxic breakdown of cyanogens (Pentzold *et al.*, 2014; Fig.1.1). Despite the lack of consistency on the efficacy of cyanogens as defence against insect herbivores, others point to the prophylactic and curative properties of extracts from cassava biowaste on the pseudostem weevil (*Odoiporous longicollis* Oliver) (Coleoptera) in banana (Krishnan *et al.*, 2015). The practical implications of this

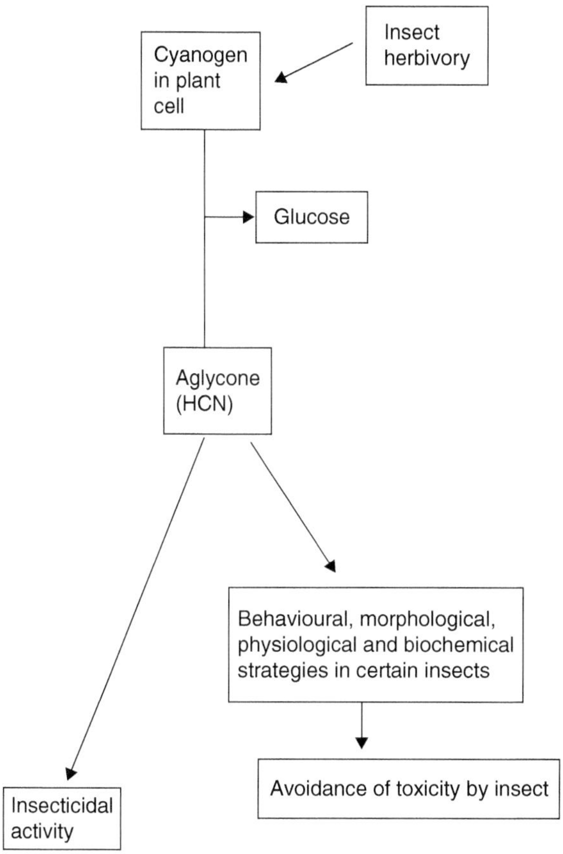


**Fig. 1.1.** Fate of plant cyanogens in insect herbivores.

approach for pest control in low-input tropical agriculture is all too clear.

The review by Percival and Dixon (1997) indicated that glycoalkaloids are characterized by repellent properties towards the Colorado potato beetle (*Leptinotarsa decemlineata*) causing effects in both larval and adult beetles. The type of glycoalkaloid present in plant foliage may be as important as total glycoalkaloid concentrations in conferring resistance to insect predation. In wild *Solanum* species, glycoalkaloid concentrations were inversely correlated with nymphal infestation by the potato leaf hopper (*Empoasca fabae* Harris) and positively affected nymphal survival and feeding behaviour. It was concluded that foliar concentrations of glycoalkaloids may markedly assist in the defence of wild potato towards this pest. However, glycoalkaloid composition and concentrations of foliage were not associated with resistance to the potato aphid (*Myzus persicae*). In recent studies (Table 1.3), it has emerged that different glycoalkaloid profiles affect feeding preferences of insect herbivores and that domestication has affected the defensive potential of *Solanum tuberosum*. Nevertheless, the wild potato plant is susceptible to predation by specialist insect herbivores guided by glycoalkaloids (Altesor *et al.*, 2014).

According to Duncan (1991), the rapid turnover of glucosinolates reflects an adaptive function, possibly related to a defensive role in plants. The glucosinolate content of various plants has been shown to affect larval development, pupation and the extent of predation by insect herbivores. Some insect species have been designated '*Brassica* specialists', being endowed with biochemical and other mechanisms of detoxification or avoidance of adverse effects of glucosinolates. One such pathway may involve glucosinolate desulfation, as in the phloem-feeding insect *Bemisia tabaci*. Insect herbivores are capable of diverting hydrolysis of glucosinolates to less toxic compounds or desulfate parent glucosinolates to prevent hydrolysis by plant myrosinases; insects may also sequester ingested glucsinolates for use as defence molecules (Jeschke *et al.*, 2015; Table 1.3). Adult flea beetles (*Phyllotreta striolata*) selectively acquire intact glucosinolates from their host plant and also produce their own myrosinase enzyme to enable utilization of the glucosinolate breakdown products for their own purposes (Beran *et al.*, 2014). An intriguing aspect relates to the attractant–defence axis, in that young leaves high in glucosinolates stimulate oviposition by a specialist herbivore despite reduced survival of larvae due to high saponin content (Badenes-Perez *et al.*, 2014).

Marked insecticidal activity of non-protein amino acids has been known for several decades, with the effects of canavanine being most consistent (D'Mello, 2015a). The efficacy of canavanine as a plant defence molecule is based on its structural analogy to arginine, an essential amino acid for insect herbivores. Substitution of canavanine for arginine results in the synthesis of aberrant dysfunctional canavanyl peptides and proteins, thus compromising insect survival.

It is assumed that the cotton plant is well protected by its primary defence compound, gossypol. It is now known that the aldehyde groups of gossypol are critical for its toxicity to cotton-feeding insect larvae. Furthermore, upregulation of cytochrome P450 enzyme was attributed to a general stress response to plant toxins and not just gossypol (Krempl *et al.*, 2016). Nevertheless, gossypol may be partially detoxified via glycosylation, thus providing a mechanism even for generalist insect herbivores to utilize cotton as a host plant.

In contrast, protein phytotoxins may provide a more robust system of defence against insect pests. Norton (1991) noted that proteinase inhibitors conferred protection to various parts of the plant against insects. This conclusion was based on the observation that wounding the leaves of potato and tomato plants by larvae or adult Colorado beetle induced a rapid accumulation of proteinase inhibitors throughout the aerial parts of the plant. It was further stated that high levels of proteinase inhibitors in the seed of a cultivar of cowpea (*Vigna unguiculata*) provided resistance against the bruchid beetle larva (*Callobruchus maculatus*). The insecticidal properties of a proteinase inhibitor from *Clitoria fairchildiana* seeds has also been investigated (Dantzger *et al.*, 2015).

An elegant model involving the activity of a plant enzyme, threonine dehydratase, in response to insect herbivory has been reviewed by D'Mello (2015b). Following initiation of damage to tissues, a cascade of molecular events ensues that vary according to plant species and insect herbivore. Several signals emanate from injured tissues, including calcium ion fluxes, phosphorylation reactions and jasmonic acid signalling. In plants, threonine dehydratase catalyses the dehydration/deamination of threonine. In the tomato plant, this enzyme is encoded by duplicate genes. One acts in the housekeeping mode, allowing the synthesis of isoleucine. The other (*TD2*) acts in a defence mode against insect herbivores. Following insect-provoked damage to the plant, jasmonic acid signal transduction induces *TD2* and the resulting enzyme causes depletion of threonine in the insect gut, thereby depriving the predator of an essential amino acid. The enhanced survival and growth rates of larvae on transgenic tomato lines silenced for *TD2* expression support the hypothesis that threonine dehydratase serves in defence towards insect herbivores.

### 1.7.2 Defence against nematodes

There is good evidence that phytotoxins may confer resistance towards nematode damage to crop plants and glucosinolate breakdown products are frequently implicated in this role, due to the volatile nature of these metabolites. For example, benzyl isothiocyanates exert such protection to the plant by affecting development, hatching and reproduction of the soybean cyst nematode (*Heterodera glycines*) (Wu *et al.*, 2014). Allylisothiocyanate in extracts from horseradish (*Armoracia rusticana*) were also effective against the nematode *Meloidogyne incognita* (Aissani *et al.*, 2013).

Similarly, flavonoids, total phenolics and gossypol have been correlated with resistance of the cotton plant to the root-knot nematode *M. incognita* (Alves *et al.*, 2016).

Proteinases in plants may also confer resistance to nematodes. For example, cysteine proteinases are lethal to parasitic nematodes *in vitro* and exhibit anti-helminthic activity *in vivo* (Phiri *et al.*, 2014). However, susceptibility of nematodes to these proteinases may be modulated by the production of inhibitors such as cystatins.

The complex hormone signalling operating in plants has been extended to the effects of the root-knot nematode *M. incognita* in tomato (Martinez-Medina *et al.*, 2016). It is known that beneficial root endophytes such as *Trichoderma* species provide protection to the host plant by stimulating endogenous defences. In this study, it emerged that *Trichoderma* protects tomato against the root-knot nematode by shifting from priming of salicylic acid-regulated to jasmonate-regulated defences.

Thus, it is clear that a number of phytotoxins have the capacity to respond to nematode infestation in plants and may participate in a complex defence system under hormonal control.

### 1.7.3 Defence against microbial pathogens

Following initial fungal infection, plants generally exhibit enhanced resistance to subsequent challenge by that particular pathogen (see D'Mello, 2015b). This systemic acquired resistance (SAR) is widely recognized and constitutes a major and distinctive feature of defence mechanisms in plants. Signal transduction and synthesis of pathogenesis-related proteins and defence compounds (including phytoalexins) are major manifestations of the establishment and efficacy of SAR. Recent research by Seneviratne *et al.* (2015) provided an example of defence responses and non-host disease resistance. Following exposure to *Fusarium solani*, pea (*Pisum sativum*) pod tissue undergoes an inducible transcriptional activation of pathogenesis-related genes and also synthesizes (+)-pisatin, its principal phytoalexin. These changes are consistent with the widely accepted concept that plant–pathogen interactions are characterized by modulations in plant metabolism involving the activation of defence responses. The major phytoalexins produced, including camalexin, glucosinolates, phenylpropanoids, terpenes and fatty acid derivatives, exert defensive functions (Konig *et al.*, 2014). Gossypol (together with brassinosteroids and jasmonic acid) has also been implicated in a role as phytoalexin in the resistance of cotton to *Verticillium* wilt (Gao *et al.*, 2013).

The precise nature of pathogenesis-related proteins needs to be addressed in order to define SAR more effectively. It is likely that these proteins will include enzymes necessary for the synthesis of signalling molecules and phytoalexins, but the need for structural components and specific inhibitory compounds may also contribute to plant immunity. It may be relevant that lectin-like proteins promote systemic rather than local immunity, possibly in conjunction with salicylic acid (Breitenbach *et al.*, 2014). Van Holle *et al.* (2016) distinguished between the classical lectins localized in the vacuole of plants and a group of inducible cytoplasmic/nuclear lectins conferring tolerance towards microbial infection and other stressors.

The outcome of plant–pathogen interaction may partly depend upon the ability of microbes to detoxify phytoalexins. Jeandet (2015), for example, pointed to the detoxification of brassinin, the indole phytoalexin occurring in *Brassica* plants.

### 1.7.4 Herbicidal potential

There is increasing evidence that deleterious properties of phytotoxins towards a wide range of plants may form the basis for the development of novel bio-herbicides. Several non-protein amino acids are attributed with this type of activity (see D'Mello, 2015a).

Mimosine is unequivocally toxic to plants, inhibiting seed germination and seedling growth in studies with mung beans, rice, wheat, perennial ryegrass and cocksfoot. Mimosine also inhibits root growth in soybean. Consistent with these and other observations is the emerging consensus that mimosine may serve as a model or parent compound for the development of potent bio-herbicides.

Limited data suggests that canavanine and canaline are inhibitory towards organogenesis in leafy spurge, specifically with respect to root and shoot growth, and that arginine is partially effective in restoring root development. A canavanine–arginine antagonism was also implied in rice shoots treated with canavanine.

It is consistently maintained that selenoamino acids are toxic to plants that are deemed to be 'non-accumulators' with respect to the Se element. There are indications that selenoamino acids interfere with cellular biochemical processes and that methionine status is compromised by selenomethionine in an anti-metabolite mechanism.

## 1.8 Clinical Applications

Pharmaceutical pollution represents an emerging risk factor for the environment particularly affecting water quality and ecotoxicity, though efforts are currently being directed at remediation technologies (Marquez Brazon *et al.*, 2016). It is possible that mitigation of potential risks may be achieved by the use of alternative medicinal compounds based on biodegradable phytochemicals with relatively short half-lives.

Despite the overriding theme of negativity normally associated with particular secondary compounds of plants, there is a favourable perspective deserving attention. Plants synthesize a diverse array of beneficial secondary compounds that are expressed in the edible parts of vegetables and fruit. The health attributes of plant antioxidants are extensively publicized, but phytotoxins are increasingly perceived as preventive and curative measures for a variety of human diseases. The term 'nutraceuticals' is often used as a collective term in this context.

The anti-cancer properties of *Brassica* secondary compounds are under active investigation, particularly with respect to glucosinolates, SMCO and certain selenoamino acids. For example, Awasthi and Saraswathi (2016) identified sinigrin as a potent anti-cancer glucosinolate in cruciferous vegetable, in efforts to elucidate its molecular mechanism of action. Efforts to conserve glucosinolate breakdown products during processing are now under way (Park *et al.*, 2013). The efficacy of SMCO and its metabolite methyl methane thiosulfinate has been investigated, with the conclusion that these two compounds may contribute to the anticarcinogenic effects of *Brassica* vegetables. However, epidemiological evidence only supports the case for the prevention of gastric and lung cancers through the consumption of these vegetables (see D'Mello, 2015a). The anti-cancer role of selenoamino acids is reflected in the activity of *Se*-methylselenocysteine, which is effective against mammary tumours. In addition, this amino acid is highly effective in potentiating the activity of anti-cancer drugs and in protecting against drug-induced toxicity. Indeed, it has been suggested that the efficacy of methylselenocysteine in enhancing the therapeutic index of the anti-cancer drug irinotecan was dependent upon the dose of the amino acid. Other studies, however, indicate that *Se*-allylselenocysteine is more effective than a number of other selenoamino acids in the chemoprevention of mammary cancer in a rat methylnitrosourea model (see D'Mello, 2015a).

In addition, gossypol has been attributed with a wide range of therapeutic properties, including anti-fertility, antioxidant, antimicrobial and anti-cancer activities (Keshmiri and Goliaei, 2014). The preparation of novel gossypol nanoparticles for therapeutic use in prostate cancer has been reported (Jin *et al.*, 2015). The potential role of gossypol as a contraceptive is also being considered (Lopez-Charcas *et al.*, 2016).

## 1.9 Conclusions

The distribution of the major phytotoxins has been reviewed in this chapter. The cyanogens, glycoalkaloids, glucosinolates, non-protein amino acids, furanocoumarins, gossypol and bioactive proteins have been selected for detailed study due to diverse toxicology towards vertebrate animals, insects, nematodes and fungal pathogens. Mammals display maximal manifestations of toxicity to these compounds, whereas lower organisms possess variable detoxification mechanisms. Nevertheless, it is consistently maintained that phytotoxins serve in a defence role in plants against invertebrate herbivores and fungal pathogens.

The resilience of a defence system based on protein phytotoxins is strongly implied as a result of recent evidence reviewed in this chapter. Whereas, for example, cyanogens and gossypol are readily metabolized and detoxified by specialist insect herbivores, proteinase inhibitors and at least one enzyme, threonine dehydratase (TD), offer greater resistance in this respect. Following insect-provoked damage to the plant, jasmonic acid signal transduction induces the *TD2*, the defence gene and the resulting enzyme causes depletion of threonine in the insect gut, thereby depriving the predator of an essential amino acid. It is suggested, therefore, that these heat-labile protein phytotoxins should form part of a plant breeding programme to enhance pest and pathogen resistance without compromising food safety; it is widely acknowledged that protein phytotoxins are readily denatured during cooking. However, attempts to use transgenics for this purpose would be controversial, due to continuing consumer disquiet over the ethics and acceptability of genetically modified foods.

The lack of a working model for fungal pathogens that encompasses the major components of SAR in plants should not deter future researchers, despite the daunting nature of the work involved. It will require the intellectual resources of molecular biologists, biochemists and geneticists to unravel the complex interactions associated with the synthesis of pathogenesis-related proteins, signalling compounds and phytoalexins. It will be necessary to dissect out the components of the said proteins into enzymes, receptors and inhibitory factors, but the large number of differentially expressed proteins adds to the difficulties in functional analysis. Crosstalk between salicylic acid and other signalling networks following fungal infection is another essential step towards elucidating SAR as well as the role of individual phytotoxins operating in defence mode (as phytoalexins).

It is clear, therefore, that considerable fundamental work still needs to be undertaken in order to develop the aforementioned phytotoxins into practical bio-pesticides. In addition, it should be acknowledged that other secondary compounds may emerge as preferred candidates for this role. Researchers in this field will not be short of encouragement: Matthews (2017) states that 'bio-pesticides have a bright future, but more attention is needed on their application'. Furthermore, European pesticide regulations favour the use of reduced-risk substitutes in integrated pest management protocols. At the same time, it is important to recognize that future work should be underpinned by a substantial element of studies at the molecular level if current understanding of host–parasite interactions is to evolve into innovative solutions.

## References

Aissani, N., Tedeschi, P., Maietti, A., Brandolini, V., Garau, V.L. and Caboni, P. (2013) Nematicidal activity of allylisothiocyanate from horseradish (*Armoracia rusticana*) roots against *Meloidogyne incognita*). *Journal of Agricultural and Food Chemistry* 61, 4723–4727.

Altesor, P., Garcia, A., Font, E., Oesterheld, M., Soler, R. and Gonzalez, A. (2014) Glycoalkaloids of wild and cultivated *Solanum*: effects on specialist and generalist insect herbivores. *Journal of Chemical Ecology* 40, 599–608.

Alves, G.C.S., Ferri, P.H., Seraphin, J.C., Fortes, G.A.C., Rocha, M.R. and Santos, S.C. (2016) Principal response curves analysis of polyphenol variation in resistant and susceptible cotton after infection by a root-knot nematode (RKN). *Physiological and Molecular Plant Pathology* 96, 19–28.

Awasthi, S. and Saraswathi, N.T. (2016) Elucidating the molecular interaction of sinigrin, a potent anticancer glucosinolate from cruciferous vegetables with bovine serum albumin: effect of methylglyoxal modification. *Journal of Biomolecular Structure and Dynamics* 34(10), 2224–2232.

Badenes-Perez, F.R., Gershenzon, J. and Heckel, D.G. (2014) Insect attraction versus plant defense: young leaves high in glucosinolates stimulate oviposition by a specialist herbivore despite poor larval survival due to high saponin content. *PLoS ONE* 9(4): e95766. doi: 10.1371/journal.pone.0095766.

Beran, F., Pauchet, Y., Kunert, G., Reichelt, M., Vogel, H., Gershenzon, J. and Heckel, D.G. (2014) *Phyllotreta striolata* flea beetles use host plant defense compounds to create their own glucosinolate-myrosinase system. *Proceedings of the National Academy of Sciences* 111, 7349–7354.

Bignell, D.R.D., Fyans, J.K. and Cheng, Z. (2014) Phytotoxins produced by plant pathogenic *Streptomyces* species. *Journal of Applied Microbiology* 116, 223–235.

Breitenbach, H.H., Wenig, M., Witek, F., Jordan, L., Colby, T. *et al.* (2014) Contrasting roles of the apoplastic aspartyl protease apoplastic, enhanced disease susceptibility1-dependent1 and legume lectin-like protein1 in *Arabidopsis* systemic acquired resistance. *Plant Physiology* 165, 791–809.

Chang, H-X., Domier, L.L., Radwan, O., Yendrek, C.R., Hudson, M.E. and Hartman, G.L. (2016) Identification of multiple phytotoxins produced by *Fusarium virguliforme* including a phytotoxic effector (Fv NIS 1) associated with sudden death syndrome foliar symptoms. *Molecular Plant–Microbe Interactions* 29, 96–108.

Christensen, S.A., Huffaker, A. and Kaplan, F. (2015) Maize death acids, 9-lipoxygenase-derived cyclopente (a) nones, display activity as cytotoxic phytoalexins and transcriptional mediators. *Proceedings of the National Academy of Sciences of the United States of America* 112(26),11407–11412.
Cimmino, A., Masi, M., Evidente, M., Superchi, S. and Evidente, A. (2015) Fungal phytotoxins with potential herbicidal activity: chemical and biological characterization. *Natural Product Reports* 32, 1629–1653.
Dadler, R. (2014) The role of bacterial phytotoxins in inhibiting eukaryotic proteasomes. *Trends in Microbiology* 22, 28–35.
Dantzger, M., Vasconcelos, I.M., Scorsato, V., Aparicio, R. and Macedo, M.L.R. (2015) Bowman-Birk proteinase inhibitor from *Clitoria fairchildiana* seeds: isolation, biochemical properties and insecticidal potential. *Phytochemistry* 118, 224–235.
Davis, R.H. (1991) Cyanogens. In: D'Mello, J.P.F., Duffus, C.M. and Duffus, J.H. (eds) *Toxic Substances in Crop Plants*. The Royal Society of Chemistry, Cambridge, UK, pp. 202–225.
Diawara, M.M. and Trumble, J.T. (1997) *Handbook of Plant and Fungal Toxins*, 1st edn. CRC Press, Boca Raton, Florida, pp.175–189.
D'Mello, J.P.F. (1995) Anti-nutritional substances in legume seeds. In: D'Mello, J.P.F. and Devendra, C. (eds) *Tropical Legumes in Animal Nutrition*. CAB International, Wallingford, UK, pp.135–172.
D'Mello, J.P.F. (1997) *Handbook of Plant and Fungal Toxins*, 1st edn. CRC Press, Boca Raton, Florida.
D'Mello, J.P.F. (2015a) Toxicology of non-protein amino acids. In: D'Mello, J.P.F. (ed.) *Amino Acids in Higher Plants*. CAB International, Wallingford, UK, pp. 507–537.
D'Mello, J.P.F. (2015b) Delivering innovative solutions and paradigms for a changing environment. In: D'Mello, J.P.F. (ed.) *Amino Acids in Higher Plants*. CAB International, Wallingford, UK, pp. 538–583.
D'Mello, J.P.F., Duffus, C.M. and Duffus, J.H. (1991) *Toxic Substances in Crop Plants*. The Royal Society of Chemistry, Cambridge, UK.
Duncan, A. (1991) Glucosinolates. In: D'Mello, J.P.F., Duffus, C.M. and Duffus, J.H. (eds) *Toxic Substances in Crop Plants*. The Royal Society of Chemistry, Cambridge, UK, pp. 126–147.
Engen, A., Maeda, J., Wozniak, D.E., Brents, C.A, and Kato, T.A. (2015) Induction of cytotoxic and genotoxic responses by natural and novel quercetin glycosides. *Mutation Research/Genetic Toxicology and Environmental Mutagenesis* 784, 15–22.
Gao, W., Long, L., Zhu, L.-F., Xu, L., Sun, L.-Q., Liu, L.-L. and Zhang, X.-L. (2013) Proteomic and virus-induced gene silencing (VIGS) analyses reveal that gossypol, brassinosteroids and jasmonic acid contribute to the resistance of cotton to *Verticillium dahliae*. *Molecular & Cellular Proteomics* 12, 3690–3703.
Gaydou, E.M. (1997) Flavones. In: D'Mello, J.P.F. (ed.) *Handbook of Plant and Fungal Toxicants*. CRC Press, Boca Raton, Florida, pp. 99–115.
Golawska, S., Sprawka, I. and Lukasik, I. (2014) Effect of saponins and apigenin mixtures on feeding behaviour of the pea aphid, *Acyrthosiphon pisum* Harris. *Biochemical Systematics and Ecology* 55, 137–144.
Grant, G. (1991) Lectins. In: D'Mello, J.P.F., Duffus, C.M. and Duffus, J.H. (eds) *Toxic Substances in Crop Plants*. The Royal Society of Chemistry, Cambridge, UK, pp. 49–67.
Griffiths, D.W. (1991) Condensed tannins. In: D'Mello, J.P.F., Duffus, C.M. and Duffus, J.H. (eds) *Toxic Substances in Crop Plants*. The Royal Society of Chemistry, Cambridge, UK, pp. 180–201.
Jeandet, P. (2015) Phytoalexins: current progress and future prospects. *Molecules* 20, 2770–2774.
Jeschke, V., Gershenzon, J. and Vassao, D.G. (2015) Metabolism of glucosinolates and their hydrolysis products in insect herbivores. *Recent Advances in Phytochemistry* 45, 163–194.
Jin, C.-L., Chen, M.-L., Wang, Y., Kang, X.-C., Han, G.-Y. and Xu, S.-L. (2015) Preparation of novel (-)-gossypol nanoparticles and the effect on growth inhibition in human prostate cancer PC-3 cells *in vitro*. *Experimental & Therapeutic Medicine* 9, 675–678.
Keshmiri-Neghab, H. and Goliaei, B. (2014) Therapeutic potential of gossypol: an overview. *Pharmaceutical Biology* 52, 124–128.
Konig, S., Feussner, K., Kaever, A., Thurow, C., Karlovsky, P., Gatz, C., Polle, A. and Feussner, I. (2014) Soluble phenylpropanoids are involved in the defense response of *Arabidopsis* against *Verticillium longisporum*. *New Phytologist* 202, 823–837.
Krempl, C., Heidel-Fischer, H.M., Jimenez-Aleman, G.H., Reichelt, M., Menezes, R.C., Boland, W. and Vogel, H. (2016) Gossypol toxicity and detoxification in *Helicoverpa armigera* and *Heliothis virescens*. *Insect Biochemistry and Molecular Biology* 78, 69–77.
Krishnan, J.U., Jayaprakas, C.A., Leskshmi, N.R., Rajeshwari, L.S. and Leena, S. (2015) Toxicity of insecticidal principles from cassava (*Manihot esculenta* Crantz) on pseudostem weevil (*Odoiporous longicollis* Oliver) (Coleoptera: Curculionidae) in banana. *Journal of Root Crops* 41(2), 55–61.

Kumar, R. and D'Mello, J.P.F. (1995) Anti-nutritional factors in forage legumes. In: D'Mello, J.P.F. and Devendra, C.(eds) *Tropical Legumes in Animal Nutrition*. CAB International, Wallingford, UK, pp. 95–133.
Larsson, D.G.J. (2014) Pollution from drug manufacturing: a review and perspectives. *Philosophical Transactions of the Royal Society B* 369, 1656. doi: 10.1098/rstb.2013.0571.
Lopez-Charcas, O., Herrera-Carrillo, Z., Montiel-Reyes, L.E. and Gomora, J.C. (2016) Block of recombinant T-type calcium channels by gossypol, a potential contraceptive. *Biophysical Journal* 110, 440a.
Lorent, K., Gong, W., Koo, K.A., Zhao, X. and Sealy, I. (2015) Identification of a plant isoflavonoid that causes biliary atresia. *Science Translational Medicine* 7(286), 286ra67. doi: 10.1126/scitranslmed.aaa1652.
Macedo, M.L.R., Oliveira, C.F.R. and Oliveira, C.T. (2015) Insecticidal activity of plant lectins and potential application in crop protection. *Molecules* 20, 2014–2033.
Marquez Brazon, E., Piccirillo, C., Moreira, I.S. and Castro, P.M.L. (2016) Photodegradation of pharmaceutical persistent pollutants using hydroxyapatite-based materials. *Journal of Environmental Management* 182, 486–495.
Martinez-Medina, A., Fernandez, I., Lok, G.B., Pozo, M.J., Pieterse, C.M.J. and Van Wees, S.C.M. (2016) Shifting from priming of salicylic acid- to jasmonic acid-regulated defences by *Trichoderma* protects tomato against the root knot nematode *Meloidogyne incognita*. *New Phytologist* 213, 1363–1377.
Matthews, G. (2017) Biopesticides have a bright future, but more attention is needed on their application. *International Pest Control* 59, 48–49.
Medel, V., Palma, R., Mercado, D., Rebolledo, R., Quiroz, A. and Mutis, A. (2015) The effect of protease inhibitors on digestive proteolytic activity in the raspberry weevil, *Aegorhinus superciliosus* (Guerin) (Coleoptera: Curculionidae). *Neotropical Entomology* 44, 77–83.
Norton, G. (1991) Proteinase inhibitors. In: D'Mello, J.P.F., Duffus, C.M. and Duffus, J.H. (eds) *Toxic Substances in Crop Plants*. The Royal Society of Chemistry, Cambridge, UK, pp. 68–106.
Park, M.H., Arasu, M.V., Park, N.Y. and Choi, Y.J (2013) Variation of glucoraphanin and glucobrassicin: anticancer components in *Brassica* during processing. *Food Science and Technology (Campinas)* 33(4), 624–631. doi: 10.1590/S0101-20612013000400005.
Pavela, R. and Vrchotova, N. (2013) Insecticidal effect of furanocoumarins from fruits of *Angelica archangelica* L. against larvae *Spodoptera littoralis* Boisd. *Industrial Crops and Products* 43, 33–39.
Pentzold, S., Zagrobelny, M., Roelsgaard, P.S., Moller, B.L. and Bak, S. (2014) The multiple strategies of an insect herbivore to overcome plant cyanogenic glucoside defence. *PLoS ONE* 9(3), e91337. doi: 10.1371/journal.pone.0091337.
Percival, G.C. and Dixon, G.R. (1997) Glycoalkaloids. In: D'Mello, J.P.F. (ed.) *Handbook of Plant and Fungal Toxicants*. CRC Press, Boca Raton, Florida, pp.19–35.
Phiri, A.M., De Pomerai, D., Buttle, D.J. and Behnke, M.B. (2014) Developing a rapid throughput screen for detection of nematicidal activity of plant cysteine proteinases: the role of *Caenorhabditis elegans* cystatins. *Parasitology* 141, 164–180.
Reading, N.S., Sirdah, M.M., Shubair, M.E., Nelson, B.E., Al-Tayeb, J.M. and Prchal, J.T. (2016) Favism, the commonest form of severe haemolytic anemia in Palestinian children, varies in severity with three different variants of G6PD deficiency within the same community. *Blood Cells, Molecules and Diseases* 60, 58–64.
Risco, C.A. and Chase, C.C. (1997) Gossypol. In: D'Mello,J.P.F. (ed.) *Handbook of Plant and Fungal Toxicants*. CRC Press, Boca Raton, Florida, pp.87–98.
Santana, A.T., Guelfi, M., Medeiros, H.C.D., Tavares, M.A., Bizerra, P.F.V. and Mingatto, F.E. (2015) Mechanisms involved in reproductive damage caused by gossypol in rats and protective effects of vitamin E. *Biological Research* 48, 43. doi: 10.1186/s40659-015-0026-7.
Selin-Rani, S., Senthil-Nathan, S., Thanigaivel, A., Edwin, E-S., Ponsankar, A. *et al*. (2016) Toxicity and physiological effect of quercetin on generalist herbivore, *Spodoptera litura* Fab. and a non-target earthworm *Eisenia fetida* Savigny. *Chemosphere* 165, 257–267.
Seneviratne, H.K., Dalisay, D.S., Kim, K.-W., Yang, H., Hartshorn, C.M., Davin, L.B. and Lewis, N.G. (2015). Non-host disease resistance response in pea (*Pisum sativum*) pods: biochemical function of DRR206 and phytoalexin pathway localization. *Phytochemistry* 113, 140–148.
Tong, X., Han, L., Duan, H., Cui, Y., Feng, Y. and Yang, S. (2017) The derivatives of Pulsatilla saponin A, a bioactive compound from *Pulsatilla chinensis*: their synthesis, cytotoxicity, haemolytic toxicity and mechanism of action. *European Journal of Medicinal Chemistry* 129, 325–336. doi:10.1016/j.ejmech.2917.02.025.
Tshala-Katumbay, D.D., Ngombe, N.N., Okitundu, D., David, L., Westaway, S.K. and Banea, J.-P. (2016) Cyanide and the human brain: perspectives from a model of food (cassava) poisoning. *Annals of the New York Academy of Sciences* 1378, 50–57.

Van Holle, S., Smagghe, G. and Van Damme, E.J.M. (2016) Overexpression of *Nictaba*-like lectin genes from *Glycine max* confers tolerance toward *Pseudomonas syringae* infection, aphid infestation and salt stress in transgenic *Arabidopsis* plants. *Frontiers in Plant Science* 7, 1590.

Virgilio, A., Sinisi, A., Russo, V., Gerardo,S., Santoro, A. and Roperto, F. (2015) Ptaquiloside, the major carcinogen of bracken fern, in the pooled raw milk of healthy sheep and goats: an underestimated, global concern of food safety. *Journal of Agricultural and Food Chemistry* 63, 4886–4892.

Wu, H., Masler, E.P., Rogers, S.T., Chen, C. and Chitwood, D.J. (2014) Benzyl isothiocyanate affects development, hatching and reproduction of the soybean cyst nematode *Heterodera glycines*. *Nematology* 16, 495–504.

# 2 Mycotoxins

**J.P.F. D'Mello***
*Formerly of SAC, University of Edinburgh King's Buildings Campus, West Mains Road, Edinburgh, UK*

## 2.1 Abstract

Mycotoxins are a diverse and ubiquitous group of fungal metabolites characterized by the propensity to induce adverse effects in humans and other vertebrate animals. The production of these compounds is determined by ecological and environmental factors, particularly temperature and humidity or substrate water activity. The preponderance of specialized fungi in particular niches determines the type and range of mycotoxins that may contaminate food or the indoor environment. *Claviceps*, *Fusarium* and *Alternaria* species are classical exponents of plant pathogens with toxigenic potential. *Claviceps purpurea* produces the ergot alkaloids, while the principal *Fusarium* mycotoxins include the trichothecenes, zearalenone and fumonisins. *Alternaria alternata* synthesizes tenuazonic acid, alternariol and altenuene. *Aspergillus* and *Penicillium* species exemplify food-spoilage fungi, associated with particular conditions in post-harvest ecology. *Aspergillus flavus* and *A. parasiticus* produce the aflatoxins $B_1$, $B_2$, $G_1$ and $G_2$ ($AFB_1$, $AFB_2$, $AFG_1$ and $AFG_2$, respectively), but *A. ochraceus* together with *Penicillium viridicatum* and *P. cyclopium* synthesize ochratoxin A (OTA). *Penicillium citrinum* and *P. expansum* are principal sources of citrinin, with the latter also producing patulin.

In temperate countries, mycotoxin residues in cereal grains are largely the result of fungal disease of standing crops in the field. Consequently, host–pathogen interactions are important components in mycotoxin production. In warm humid tropical regions, fungal proliferation generally arises during post-harvest storage, particularly if the products have been inadequately dried, but the inoculum for these microorganisms may originate from field sources such as plant debris and soil.

Current surveillance indicates widespread mycotoxin contamination of primary and processed plant products with global implications for human health. Concentrations of aflatoxins in maize and peanuts regularly exceed safety threshold limits. OTA, certain trichothecenes and zearalenone are ubiquitous, occurring primarily in cereal grains and derived products. In addition, OTA may occur in dried vine fruits and green coffee beans. Of considerable concern is the widespread contamination of maize and associated products with fumonisins. The use of contaminated feedstocks in livestock nutrition may result in the transfer of mycotoxins to animal products, particularly milk and offal. Consequently, humans may be exposed to combinations of different foodborne mycotoxins.

Although mycotoxins may be graded according to acute lethality tests, the major concerns in

* E-mail address: jpfdmello@hotmail.co.uk

human health relate to epidemiological evidence. A broad spectrum of adverse outcomes has been associated with chronic exposure, including carcinogenesis, hepatitis, nephrotoxicity and endocrine disruption. Furthermore, mycotoxins may compromise health by modulating other disorders. For example, foodborne aflatoxins may enhance the carcinogenic potential of hepatitis B virus. In addition, it has been proposed that kwashiorkor in African children may be a manifestation of aflatoxicosis. Nevertheless, in toxicological classification, $AFB_1$ has been designated as a group 1 carcinogen, specifically implicated in liver, lung and gallbladder malignancy. Epidemiological evidence also links human oesophageal cancer in South Africa with dietary exposure to the fumonisins. In addition, this group has been cited as a possible contributory risk factor in primary hepatocellular cancer in China. More controversially, OTA has been linked with the incidence of Balkan (and possible Tunisian) endemic nephropathy, but the co-occurrence of OTA with citrinin suggests an interaction between the two mycotoxins.

Current studies focus on molecular and biochemical dimensions, particularly in the context of mycotoxin-induced carcinogenesis. In the case of $AFB_1$, this work includes nucleotide excision repair, reduction of DNA adducts, modulation of cellular gene expression, activation of signalling pathways, mutational spectra, expression of microRNA (miRNA), interferon anti-cancer pathway and factors in cancer cell migration. It is envisaged that molecular and biochemical investigations should resolve cause-and-effect issues raised by epidemiological evidence and assist in evaluating alternative mechanisms in the aetiology of hepatocellular carcinoma. In addition, there may be scope for the characterization of improved biomarkers for the assessment of mycotoxin-induced malignancy in humans.

Despite enhanced awareness of health risks and the adoption of legal or advisory guidelines, human exposure to foodborne mycotoxins continues unabated and on a global scale. The evidence for residues of aflatoxins and OTA is particularly striking as demonstrated by analysis of body fluids, mother's milk and tissue specimens. Further studies are being undertaken to establish or confirm the link between mutational fingerprints and mycotoxin exposure.

Of considerable concern in environmental toxicology is the inefficacy of fungicides to control fungal diseases of plants and, therefore, mycotoxin contamination of harvested grain. The development of fungicide resistance in these fungal phytopathogens is an added risk.

It is concluded that factors such as ecology, environmental temperature and humidity or substrate water activity predispose to production of mycotoxins by plant pathogenic and saprophytic fungi, resulting in worldwide contamination of staple foods. These compounds constitute a continuing hazard to human health following acute and/or chronic exposure. Measures to mitigate risk, such as the use of fungicides and preservatives, are of limited efficacy. Indeed, the use of sub-lethal doses or the development of fungicide resistance may exacerbate potential hazards.

## 2.2 Introduction

Secondary metabolism in certain fungi is characterized by the synthesis of a diverse and ubiquitous group of bioactive compounds. Those metabolites that are associated with pathological conditions in humans and other vertebrates are termed mycotoxins. The contamination of cereal grains, nuts, fruit and green coffee beans with mycotoxins represents a major worldwide food safety issue (D'Mello, 2003). If farm livestock are offered feeds containing mycotoxins, then associated residues and metabolites may appear in animal products. The mycotoxins of particular relevance in human health arise from the secondary metabolism of *Claviceps*, *Aspergillus*, *Penicillium*, *Fusarium* and *Alternaria* species. Mycotoxins are classified, and some named, according to their fungal origin. In addition, mycotoxins may be categorized on the basis of their biosynthetic origin from mainstream intermediates. For example, the polyketide mycotoxins are derived from acetyl coenzyme A (CoA), while the terpene mycotoxins originate from mevalonic acid. Amino acids are incorporated in the synthesis of a third group of mycotoxins comprising cyclic polypeptides and their derivatives.

The issue of mycotoxins is relevant in this volume for several reasons. Apart from enhancing overall knowledge of toxicology, the study of mycotoxins exemplifies how ecological and environmental factors combine to affect food safety and human health on a global scale. The impact of ambient temperature and humidity (Smith, 1997; Wu *et al.*, 2016) is well acknowledged, but questions remain regarding other factors such as ultraviolet (UV) radiation (Garcia-Cela *et al.*, 2015) and drought conditions (Rheeder *et al.*, 2016). There is also speculation about the likely impact of global climate change on mycotoxin contamination of foods (Russell *et al.*, 2010). Battilani *et al.* (2016) were more forthright in predicting that mycotoxin contamination in maize is likely to increase in Europe as a result of climate change. In addition, the development of fungicide resistance among plant pathogens is a major issue (Hahn, 2014). It is possible that mycotoxin production may be affected in chemically stressed fungal pathogens. It is, therefore, pertinent to review recent evidence at both gross and gene levels (Becher *et al.*, 2010; Popiel *et al.*, 2017, respectively).

There is also renewed interest in the use of fungal entomopathogens in biological control of insect pests as an alternative to chemical agents (Vega *et al.*, 2009). The potential for fungal biopesticides as substitutes for insecticide-resistant compounds in malaria parasite control has also been advanced (Heinig *et al.*, 2015). Such an approach should, however, recognize any potential safety issues concerning the dissemination of mycotoxins and associated particulate hazards posed by spores and mycelia.

Within the indoor environment, there is evidence of exposure to mycotoxins in water-damaged buildings, referred to as 'sick buildings syndrome' (SBS) (Brewer *et al.*, 2013). Associated with SBS is the question whether some fungal volatile organic compounds may be classified as mycotoxins (Bennett and Inamdar, 2015).

## 2.3 Ecology

Mycotoxin contamination of plant products is determined primarily, but not exclusively, by fungal ecology. It is conventional to subdivide toxigenic fungi into plant-pathogenic ('field') and food-spoilage ('saprophytic' or 'storage') species. *Claviceps*, *Fusarium* and *Alternaria* are classical exponents of toxigenic plant pathogens, while *Aspergillus* and *Penicillium* exemplify food-spoilage fungi, reflecting post-harvest ecology.

The distinction between field and storage fungi is largely academic since the inoculum for post-harvest spoilage of grain and fruit, for example, frequently originates from field sources such as soil or plant debris.

Ecological features are exemplified in the distribution of *A. flavus* and *A. parasiticus*, which occur widely and frequently in tropical and sub-tropical countries with hot and humid climates (Smith, 1997). Contamination of food crops with these fungi may occur in the field or during post-harvest storage as a result of inadequate drying. Although these particular conditions may well apply in tropical environments, in the USA *Aspergillus* contamination of maize occurs predominantly in the field due to insect or bird damage. The ecology of mycotoxins in peanuts and its relationship with climatic conditions in China is also emphasized by Wu *et al.* (2016).

According to Vega *et al.* (2009), the ecological role of entomopathogenic fungi in the environment is not fully appreciated and restricts exploitation of these organisms in insect pest management. Further research is advocated on the basis of existing knowledge that these fungi act as endophytes, antagonists of plant pathogens, growth-promoting agents and enhancers within the rhizosphere.

The association of SBS with water-damaged buildings suggests a particular, if not unique, ecology of fungi and work continues to unravel the implications for human exposure to particulates (e.g. spores and hyphae) and mycotoxins. The putative role of fungal volatile organic compounds as mycotoxins in this context provides an additional dimension in risk assessment (Bennett and Inamdar, 2015).

## 2.4 Pathways in Mycotoxin Contamination of Foods

Environmental factors and fungal ecology combine to determine the extent and type of mycotoxins that may contaminate basic foods such as cereal grains, nuts and other plant products. In temperate countries, mycotoxin residues in cereal grains are largely the result of fungal disease of standing crops in the field. In warm humid tropical regions, fungal colonization generally

**Table 2.1.** Principal food-borne mycotoxins of confirmed or potential significance in human health (D'Mello, 2003).

| Mycotoxins | Fungal species | Foods |
|---|---|---|
| Ergot alkaloids | *Claviceps purpurea* | Cereal grains |
| Aflatoxins | *Aspergillus flavus*; *A. parasiticus* | Nuts; maize kernels; dried fruits |
| Cyclopiazonic acid | *A. flavus* | Nuts |
| Ochratoxin A | *A. ochraceus*; *Penicillium viridicatum*; *P. cyclopium* | Cereal grains and products; pig products; raw coffee |
| Citrinin | *P. citrinum*; *P. expansum* | Cereal grains |
| Patulin | *P. expansum* | Apple products |
| Citreoviridin | *P. citreo-viride* | Rice |
| T-2 toxin (type A trichothecene) | *Fusarium sporotrichioides*; *F. poae* | Cereal grains |
| Diacetoxyscirpenol (type A trichothecene) | *F. sporotrichioides*; *F. poae* | Cereal grains |
| Deoxynivalenol (type B trichothecene) | *F. culmorum*; *F. graminearum* | Cereal grains |
| Zearalenone | *F. culmorum*; *F.graminearum*; *F. sporotrichioides* | Cereal grains |
| Fumonisins; moniliformin; fusaric acid | *F. moniliforme* | Maize kernels |
| Tenuazonic acid; alternariol; alternariol methyl ether; altenuene | *Alternaria alternata* | Fruit; vegetables; cereal grains |

arises during post-harvest storage, particularly if the products have been inadequately dried. These differences in ecology result in contrasting features of secondary metabolism of plant-pathogenic compared to food-spoilage fungi and the formation of characteristic mycotoxins that contaminate the final products. The biosynthesis of mycotoxins occurs via the well-established polyketide and isoprenoid pathways using, respectively, acetyl CoA and mevalonic acid as the primary starting substrates.

The mycotoxins most commonly associated with human health disorders are listed in Table 2.1, which also indicates the fungal origin and occurrence of these compounds as food contaminants. Details of the chemistry of the important mycotoxins are presented by D'Mello (2003).

### 2.4.1 Ergot alkaloids

The major ergot alkaloids synthesized by *Claviceps purpurea* include the lysergic acid derivatives ergocristine and ergotamine. However, ecological differences mean that ergot alkaloids not related to *Claviceps* are produced in endophyte-infected grasses (Porter, 1997).

### 2.4.2 Aflatoxins and cyclopiazonic acid

Aflatoxins $B_1$, $B_2$, $G_1$ and $G_2$ ($AFB_1$, $AFB_2$, $AFG_1$ and $AFG_2$, respectively) are synthesized by *A. flavus* and *A. parasiticus* (Smith, 1997). In addition, aflatoxin $M_1$ ($AFM_1$) may appear in the milk of dairy cows and women consuming and metabolizing $AFB_1$ from contaminated diets. *A. flavus* only produces $AFB_1$ but is also capable of synthesizing cyclopiazonic acid, a mycotoxin recently confirmed as a co-contaminant in the batch of peanuts associated with mass mortality in turkey poults in 1960. In contrast, *A. parasiticus* regularly synthesizes all four aflatoxins. The two *Aspergillus* species grow and produce aflatoxins when temperature and humidity/water activity conditions are favourable. In the case of *A. parasiticus*, temperatures of 25–30°C are optimal for maximizing aflatoxin synthesis. However, both temperature and water activity may interact in aflatoxin synthesis; consequently the risk of contamination is considerably enhanced in foods produced in warm and humid countries (Smith, 1997). Global aflatoxin monitoring of peanuts in relation to climatic conditions/agroecology continues to be an active area of research, for example in Democratic Republic of Congo, South Africa (Kamika *et al.*, 2014),

China (Wu *et al.*, 2016) and recently in Zambia (Kachapulula *et al.*, 2017). However, as indicated below, surveillance has been extended to other foods and products (D'Mello, 2003). It is now clear that diverse foods contain levels of aflatoxin that exceed international statutory limits. An outstanding feature in recent surveillance has been the high level of $AFB_1$ contamination of Indonesian maize (at 428 μg $kg^{-1}$). Of equal concern is the relatively high concentrations of total aflatoxins (up to 20 μg $kg^{-1}$) in maize-based gruels used as weaning food for children in Nigeria. Following an outbreak of human aflatoxicosis in Kenya, 55% of maize products were found with aflatoxin levels exceeding the local regulatory directive of 20 μg $kg^{-1}$, and 35% containing concentrations above 100 μg $kg^{-1}$ (Lewis *et al.*, 2005). The focus on aflatoxins in maize is likely to continue in view of predictions of contamination in home-grown sources in Europe due to global warming (Battilani *et al.*, 2016).

In the UK, samples of peanut butter analysed in 1986 and 1991 showed that 'crunchy' types continued to contain more aflatoxin than 'smooth' varieties. The maximum concentrations of total aflatoxin found in the two surveys were similar (53 μg $kg^{-1}$) in a sample of crunchy peanut butter obtained in 1986 and 51 μg $kg^{-1}$ in a smooth sample collected in1991. The UK Committee on Toxicity of Chemicals in Food, Consumer Products and the Environment (COT) expressed concern that there had been no decrease in contamination since the previous report, but anticipated reductions after implementation of updated regulations. Contamination of peanut butter appears to be a persistent problem, for example in Haiti with total aflatoxin concentrations of up to 2720 μg $kg^{-1}$ being recorded (Schwartzbord and Brown, 2015).

Aflatoxin contamination of imported pistachio nuts has also been a source of concern in the UK and elsewhere. UK surveillance conducted between 1990 and 1992 indicated that 28–52% of samples exceeded the 4 μg $kg^{-1}$ statutory limit for total aflatoxins in finished products. In addition, 25–38% of samples exceeded the 10 μg $kg^{-1}$ limit in products destined for further processing. Concentrations of $AFB_1$ in Californian pistachio nuts of up to 149 μg $kg^{-1}$ have been recorded, while in the Netherlands levels of up to 165 μg $kg^{-1}$ were published (see D'Mello, 2003). Although Dani *et al.* (2013) suggested that $AFB_1$ levels in Iranian pistachio nuts during the period 2009–2011 represented a satisfactory improvement over a previous surveillance, it should be noted that contamination was found in 23% of samples.

Whole dried figs may also be contaminated, with UK surveillance data in 1993, for example, indicating a decline in the percentage contaminated with total aflatoxins at levels above 4 μg $kg^{-1}$, but samples containing up to 427 μg $kg^{-1}$ were found. The incidence of aflatoxins above 4 μg $kg^{-1}$ in fig paste samples was also at a reduced level in the UK data with a concomitant decline in total aflatoxin concentrations from 165 to 15 μg $kg^{-1}$. The overall results attracted comment by COT, who were clearly concerned about the high levels of aflatoxin contamination but were satisfied that consignments exceeding the 10 μg $kg^{-1}$ limit were refused entry by the UK port health authorities.

Aflatoxin contamination of milk is a continuing global risk due to the ubiquitous contamination of animal feeds. For example, a recent study in Ethiopia indicated $AFM_1$ concentrations of up to 4.98 μg $l^{-1}$ in milk samples which was attributed to contamination of noug cake fed to dairy cows (Gizachew *et al.*, 2016).

### 2.4.3 Ochratoxins and citrinin

*Aspergillus ochraceus* and two important species of *Penicillium* are associated with the synthesis of OTA and ochratoxin B (OTB) (Table 2.1). However, OTA is more ubiquitous, often occurring with another mycotoxin, citrinin, in cereal grains, dried vine fruits and green coffee. Relatively high values of OTA in Bulgarian cereals (up to 140 μg $kg^{-1}$) were associated with grain samples taken from villages with the incidence of Balkan endemic nephropathy (BEN) (see D'Mello, 2003). Observations in France indicated consistent contamination of cereals and oilseeds with OTA, the values ranging from 0.6 to 12.8 μg $kg^{-1}$ in positive samples. In the UK, OTA levels in dried vine fruit originating from Greece and other countries indicated that 88% were contaminated, with values in the range 0.2–53.6 μg $kg^{-1}$. OTA results for green coffee beans were in the range 4.1–22.1 μg $kg^{-1}$, but a maximum value of 360 μg $kg^{-1}$ was also recorded (see

D'Mello, 2003). Use of contaminated cereals in brewing and as animal feed regularly results in OTA contamination of the final products. For example, of 104 samples of porcine kidney, 12% contained OTA at 1–5 μg $kg^{-1}$, while 3% had concentrations of up to 10 μg $kg^{-1}$. Of black pudding samples examined, 13% were contaminated with OTA in the range 1–5 μg $kg^{-1}$. Citrinin often co-occurs with OTA; in Bulgarian wheat, citrinin levels of up to 420 μg $kg^{-1}$ have been found (see D'Mello, 2003).

### 2.4.4 Patulin and citreoviridin

A number of *Penicillium* species are capable of synthesizing patulin and *P. expansum* is of particular relevance due to its association with storage rot of apples and a wide variety of other fruits (D'Mello, 2003). The presence of patulin in apple juice, attributed to the use of mouldy fruit prior to processing, has been a cause of concern in the UK and elsewhere in Europe. This contamination is associated with the production of cloudy apple juice, which requires fewer processing steps compared with the procedure for the production of clear juices. In UK surveillance, the incidence of patulin contamination was higher in cloudy juices, with a median value of 28 μg $kg^{-1}$, compared with 0–10 μg $kg^{-1}$ for clear juices. In a more extensive study in Spain, patulin was detected in 82% of commercial samples of apple juice (see D'Mello, 2003). Surveillance continues to the present time; for example, Rahimi and Jeiran (2015) indicated that only 2.5% of fruit juices in Iran exceeded the maximum set by EU regulations for patulin (50 μg $kg^{-1}$). In contrast, Zouaoui *et al.* (2015) concluded that patulin is a problem in fruit juices available in Tunisia, with 22% of samples exceeding the EU limit and values averaging at 89 μg $l^{-1}$ in positive samples.

In contrast, *P. citreo-viride* produces citreoviridin, a contaminant of rice and other cereal grains (Wang *et al.*, 2015).

### 2.4.5 *Fusarium* mycotoxins

The principal *Fusarium* mycotoxins considered in this chapter include the trichothecenes, zearalenone and fumonisins. Diversity in this group is exemplified by the general association of trichothecenes with cereal grains in cooler latitudes whereas the fumonisins are common contaminants of maize kernels originating from tropical regions (see D'Mello, 2003).

The trichothecenes include T-2 toxin, HT-2 toxin, diacetoxyscirpenol (DAS), deoxynivalenol (DON) and nivalenol (NIV). A regular feature of mycotoxin production is the biosynthesis of zearalenone (ZEN). Data compiled by D'Mello (2003) demonstrated the widespread occurrence of trichothecenes and ZEN as contaminants in cereal grains. The levels of DON in Polish wheat and maize and in Japanese barley were striking, but it was noted that some US samples exceeded advisory limits. Within the USA, highest levels of DON were observed in Missouri, North Dakota and Tennessee for the 1991 harvest. In 1993, 86% of samples from Minnesota and up to 78% of samples from North and South Dakota had levels in excess of 2 mg $kg^{-1}$. Canadian data indicated consistently higher levels of DON in maize compared with soft wheat over a 15-year period. On the basis of this and other evidence, D'Mello (2003) concluded that DON was a regular contaminant of Canadian cereals.

An outstanding feature of ZEN contamination of cereal grains is its co-occurrence with other *Fusarium* mycotoxins, particularly several trichothecenes. This observation is entirely consistent with the confirmed production of ZEN by virtually all toxigenic *Fusarium* phytopathogens (D'Mello *et al.*, 1997). The highest values for ZEN (11–15 mg $kg^{-1}$) were recorded for two barley samples in Japan which were also contaminated with DON (61–71 mg $kg^{-1}$) and NIV (14–26 mg $kg^{-1}$) (D'Mello, 2003).

The fumonisins (FBs) are relatively recent additions to the list of carcinogenic mycotoxins. Several structurally related forms of FBs have been characterized, with $FB_1$, $FB_2$ and $FB_3$ occurring regularly in maize. In addition, $FB_1$ is structurally similar to sphinganine and sphingosine, intermediates in the biosynthesis and degradation of sphingolipids. The widespread contamination of maize with FBs is now an unequivocal fact (D'Mello, 2003). In most instances the predominant form is $FB_1$. Outstanding features noted by D'Mello (2003) included relatively high $FB_1$ concentrations in samples from China (up to

26 mg $kg^{-1}$), Thailand (up to 19 mg $kg^{-1}$), South Africa (up to 47 mg $kg^{-1}$) and Kenya (up to 12 mg $kg^{-1}$). Highest levels of $FB_2$ were recorded for Argentinian and South African samples of maize kernels. In the Philippines, Thailand and Indonesia, $FB_1$ and $FB_2$ occurred in over 50% of maize samples, while incidence rates of 82–100% were noted for samples from Italy, Portugal, Zambia and Benin.

## 2.5 Toxicology

### 2.5.1 Lethality

Insight into the acute effects of mycotoxins may be gauged by comparisons of lethal dose ($LD_{50}$) data (D'Mello, 2003). These values are subject to wide variation due, for example, to age, sex and species of test animals as well as the mycotoxin in question. Thus, day-old ducklings are more susceptible to $AFB_1$ than laboratory animals, whereas ZEN is associated more with endocrine disruption rather than with overt manifestations such as lethality. Although acute toxicity measures are often viewed as academic assessments, in at least two cases human aflatoxicosis has been associated with high mortality preceded by hepatitis (Krishnamachari *et al.*, 1975; Ngindu *et al.*, 1982). Hepatitis was accompanied by jaundice, rapidly developing ascites and portal hypertension, while at necropsy, hepatic lesions included bile duct abnormalities.

### 2.5.2 Mycotoxicosis: Case reports

Over several decades, considerable epidemiological evidence of adverse effects has accumulated in case studies of chronic exposure associating specific disease outbreaks in particular geographical locations with mycotoxin contamination of staple foods (Table 2.2). One of the ancient episodes of mycotoxicosis relates to ergotism (St Anthony's Fire) caused by the bioactive alkaloids produced in the sclerotia of *C. purpurea*. These compounds cause constriction of peripheral blood capillaries, leading to oxygen depletion and gangrene of the limbs (Flannigan, 1991).

The enduring example of adverse effects in humans is represented by episodes of aflatoxicosis in the tropics. For example, in 1974, an outbreak of liver disease in India was attributed to the consumption of mouldy grain contaminated with aflatoxins (Krishnamachari *et al.*, 1975). Principal pathological lesions in hepatic tissues included destruction of centrilobular zones, thickening of central veins and cirrhosis (see D'Mello, 2003). An outbreak of acute hepatitis in Kenya in 1981 was also linked with aflatoxin poisoning (Ngindu *et al.*, 1982). Subsequently, Lewis *et al.* (2005) reported aflatoxin contamination of commercial maize associated with an outbreak of acute aflatoxicosis in Eastern and Central Kenya.

In the case of OTA, there have been consistent links with Balkan endemic nephropathy (BEN), a chronic disease occurring among rural populations of Bulgaria, Romania and the former state

**Table 2.2.** Mycotoxins implicated in specific human diseases (based on D'Mello, 2003, updated in text of this chapter).

| Mycotoxin | Disease | Food | Geographical location |
|---|---|---|---|
| Ergot alkaloids | Ergotism | Rye | Europe |
| Aflatoxins | Mortality; hepatic and gallbladder cancers; kwashiorkor; cirrhosis; acute hepatitis; Reye's syndrome | Peanuts; maize | Sub-Saharan Africa; India; South-east Asia; South America |
| Cyclopiazonic acid | 'Kodua poisoning' | Millet | India |
| Ochratoxins | Balkan (and possible Tunisian) endemic nephropathy | Cereal grains | Balkan states; Tunisia |
| Citreoviridin | 'Shoshin-kakke' | Rice | Japan |
| T-2 toxin | Alimentary toxic aleukia | Cereal grains | Former USSR |
| Fumonisins | Oesophageal cancer; primary liver cancer | Maize | South Africa; China |
| Moniliformin | 'Keshan disease' | Maize | China |

of Yugoslavia (see D'Mello, 2003). In affected individuals, the kidneys are markedly reduced in size and, histologically, the condition is characterized by tubular degeneration and dysfunction, interstitial fibrosis and glomerular defects. However, the co-occurrence of OTA with citrinin in cereals consumed locally implies an interaction between the two mycotoxins in the aetiology of BEN. The role of joint mycotoxin exposure in BEN is supported by the conclusions of Stoev (2017). Others dispute a mycotoxin connection altogether, citing aristolochic acid contamination of wheat in the Balkans as the causative agent (Bui-Klimke and Wu, 2014). It may be relevant that a possible endemic ochratoxin-related nephropathy has also been suggested to occur in Tunisia. Affected patients were classified into groups with chronic forms of interstitial nephropathy, glomerular nephropathy and vascular nephropathy (see D'Mello, 2003).

An outbreak of trichothecene mycotoxicosis associated with consumption of mouldy wheat products in Kashmir Valley in India was reported by Bhat *et al.* (1989). The condition was characterized by gastrointestinal symptoms which were reproduced in dogs fed extracts of contaminated samples.

### 2.5.3 Carcinogenesis

The pre-eminent public health issue is, arguably, the association of mycotoxins with carcinogenesis. In particular, there is good epidemiological and molecular evidence linking aflatoxin exposure with the incidence of liver cancer. However, other factors such as chronic malnutrition and disease may contribute to the incidence of hepatocellular cancer (HCC). In addition, aflatoxin exposure may enhance the carcinogenic potential of hepatitis B virus. In toxicological classification, $AFB_1$ has been designated as a Group 1 carcinogen (i.e. sufficient evidence in humans for carcinogenicity), whereas its product in milk ($AFM_1$) is classified in the Group 2B category (i.e. probable human carcinogen).

Epidemiological evidence also links aflatoxin exposure with other forms of cancer. Nogueira *et al.* (2015) observed a statistically significant association between dietary aflatoxin and the incidence of gallbladder cancer in Chile, but stressed the need for identifying specific mutations precipitated by $AFB_1$. High gallbladder cancer incidence rates in Bolivia and Peru had previously been tentatively linked with aflatoxin contamination of red chilli peppers (Asai *et al.*, 2012).

Recently, however, there has been a strategic change in mycotoxin research in favour of a systematic approach to probe the underlying biochemical and molecular aetiology. This shift in emphasis is clearly seen in the case of emerging themes in $AFB_1$ toxicology (Table 2.3). Initial investigations centred on the metabolic activation of $AFB_1$ and the formation of DNA adducts (Smith, 1997). Recent studies include understanding of nucleotide excision repair processes, $AFB_1$-DNA adducts, gene expression, mutational spectra and genetic features, miRNA expression, type 1 interferon anti-cancer response pathways and sarcoma (Src) phosphorylation. The need for biomarkers appears to be a consistent underlying objective in several investigations to enable early detection of mycotoxin exposure and cancer development. Although emphasis continues in respect of HCC, there is accumulating evidence linking aflatoxin exposure to other forms of malignancy, for example, lung cancer and its underlying biochemistry (Table 2.3) (Cui *et al.*, 2015). Identification of specific *TP53* mutations has been advanced as a means of further clarification of the epidemiological evidence for gallbladder cancer in Chile (Nogueira *et al.*, 2015).

Other epidemiological evidence indicates a link between dietary fumonisin exposure and human oesophageal cancer episodes in South Africa, China and Iran. $FB_1$ is classified as a Group 2B carcinogen (possibly carcinogenic to humans). It is suggested that $FB_1$ stimulates proliferation of oesophageal cells by modulating the cell cycle and apoptosis (Wang *et al.*, 2014). Attenuation of DNA methylation might also contribute to the mechanism $FB_1$ toxicity (Demirel *et al.*, 2015). In China, it has been suggested that fumonisins may promote primary liver cancer initiated by $AFB_1$ and/or hepatitis B virus.

Emerging observations point to the carcinogenic properties of OTA. In particular, Ibrahim *et al.* (2013) implicated OTA in HCC, based on case report evidence. Mantle (2016) proposed OTA as a potent naturally occurring kidney carcinogen. Structural characterization of

**Table 2.3.** Molecular aspects of $AFB_1$-induced carcinogenesis and associated interactions.

| Observations | Reference |
|---|---|
| $AFB_1$-up-regulation of nucleotide excision repair in lung and liver is dependent on *p53* genotype | Mulder *et al.* (2014) |
| Role of hepatic glutathione *S*-transferases in reducing levels of $AFB_1$-DNA adducts | Techapiesancharoenkij *et al.* (2015) |
| $AFB_1$ treatment modulates cellular gene expression, activates signalling pathways and stimulates Epstein-Barr virus-induced B-cell transformations | Accardi *et al.* (2015) |
| Mutational spectra as potential biomarkers in early detection of $AFB_1$-induced hepatocarcinogenesis in humans | Chawanthayatham *et al.* (2017) |
| Identification of specific genetic and mutation features of hepatocellular carcinomas associated with aflatoxin exposure | Zhang *et al.* (2017) |
| Expression of miRNA (post-transcriptional gene expression regulators) altered by $AFB_1$-induced cancer | Livingstone *et al.* (2017) |
| $AFB_1$ inhibits the type 1 interferon anti-cancer response pathway, implying a novel mechanism for induction of hepatocellular carcinoma | Narkwa *et al.* (2017) |
| $AFB_1$ induces Src phosphorylation and stimulates lung cancer cell migration | Cui *et al.* (2015) |

the major adduct formed from OTA has also been established (Sharma *et al.*, 2014).

The co-occurrence of several mycotoxins in the same sample of cereal grains continues to provoke toxicological concerns. Kim *et al.* (2016) provided a molecular explanation on the potential risk of crosstalk between carcinogenic mycotoxins. On the basis of studies elucidating an $AFB_1$–OTA interaction, it was suggested that OTA may contribute to the survival of tumour cells with deleterious mutations by genotoxic mycotoxins, potentially increasing the risk of carcinogenesis.

### 2.5.4 Systemic dysfunction

An emerging feature in environmental toxicology concerns the identification and characterization of endocrine disruptors (Chapter 15). Whereas ZEN has long been linked with reproductive disorders in experimental models (see D'Mello, 2003), emphasis is now shifting to the potential of $AFB_1$ as an antagonist of the androgen biosynthetic pathway (Adedara *et al.*, 2014). Furthermore, it is conceivable that the extra-hepatic effects of mycotoxins may extend to other aspects of human health such as immunity and susceptibility to bacterial and viral diseases.

## 2.6 Risk Assessment

Analysis of foods provides a valuable initial assessment of potential risks of mycotoxin contamination. However, analysis of blood, breast milk, organ samples, urine and stools gives an in-depth picture of exposure. The results continue to cause concern among toxicologists and health personnel, particularly as there may be interactions with other conditions. For example, analysis of urine and stools in Kenya indicated that following feeding of an aflatoxin-free diet, children with kwashiorkor continued to excrete aflatoxins in urine for 2 days, but those with marasmus excreted aflatoxins for up to 4 days. Differences were also seen in the type of aflatoxins discharged in faeces (see D'Mello, 2003). Another investigation demonstrated widespread fetal exposure to aflatoxins in East and West Africa following analysis of cord and maternal blood samples. Aflatoxins were also detected in breast milk samples of mothers. D'Mello (2003) concluded that there was widespread prenatal and postnatal exposure of infants to aflatoxins which might predispose children to infection. The current situation indicates that contamination of foods and, therefore, human exposure to mycotoxins continues unabated. For example, Castelino *et al.* (2014) reported seasonal and gestation-stage differences in aflatoxin exposure in Gambian women, while Obade *et al.* (2015) confirmed similar aflatoxin risks to mothers and infants in Kenya.

D'Mello (2003) noted that human OTA exposure was also widespread, based on analyses of physiological fluids, with geographical and regional differences in risk. In Croatia, the highest blood OTA levels were observed in subjects living in areas known for the incidence of endemic nephropathy. In Tunisia, an endemic OTA-related nephropathy is believed to occur with similarities to the Balkan syndrome. OTA exposure appears to be a persistent problem: Malir *et al.* (2013) noted that in Czech women in early pregnancy, 96% of serum samples were positive for OTA.

It is therefore clear that, despite enhanced awareness and the promulgation of advisory and statutory directives, human exposure to mycotoxins continues and not just in developing countries (Solfrizzo *et al.*, 2014). Adi *et al.* (2016) pointed to greater frequency and levels of DON in urine samples of German compared with Bangladeshi adults despite long-standing regulatory limits in EU countries.

## 2.7 Mitigation of Risk

As previously indicated, the overriding aims in food safety should focus on prevention of contamination, as corrective methods may be of limited efficacy (D'Mello, 2003). In theory, effective use of fungicides against diseases such as *Fusarium* head blight (FHB) of cereal crops should result in reduced mycotoxin contamination of harvested grain. However, it is generally accepted that fungicide control is only partially effective. Moreover, the potential exists for the development of resistance to fungicides by certain phytopathogens, which may possibly be linked to stimulation of mycotoxin production. Hahn (2014), using *Botrytis* as a case study, cautioned on the rising threat of fungicide resistance in plant pathogenic fungi. Of equal concern is the increasing evidence of ineffectiveness of certain fungicides in the control of mycotoxin synthesis in plant pathogens. For example, Schmidt-Heydt *et al.* (2013) evaluated seven different fungicides for the ability to inhibit growth of *Penicillium verrucosum*. Fungicides belonging to a class of imidazoles decreased growth but stimulated ochratoxin and citrinin biosynthesis. Terra *et al.* (2016), in studies with wine grapes, demonstrated that OTA production was enhanced when growth of *Aspergillus carbonarius* was not inhibited by sub-optimal doses of fungicide. Similarly, fungicide control of FHB in the field was not accompanied by satisfactory reduction of DON contamination in harvested wheat (Marques *et al.*, 2017). Overall these studies are consistent with the classification of fungicide efficacy for mycotoxin control as advanced by D'Mello (2003). Five classes of fungicides were proposed, including: Class I (effective); Class IIA (partially effective, mycotoxin residues possible); Class IIB (partially effective, direct inhibition of mycotoxin synthesis but disease/infection/fungal growth possible); Class IIIA (ineffective); and Class IIIB (stimulatory and/or inducing resistance). The relationship between fungicide resistance and mycotoxin production in phytopathogens, however, still remains to be ascertained.

In view of the foregoing evidence, there is growing interest in the potential of plant selection and breeding as environmentally friendly alternatives to fungicide use (see D'Mello, 2003). For example, experimental studies show that breeding maize plants that are resistant to infection and ear rot caused by *A. flavus* generally results in reduced $AFB_1$ contamination. Similarly, exploitation of genetic resistance to FHB in wheat has been used successfully to reduce DON adulteration of harvested grain. In addition, selection of FHB-resistant Chinese cultivars of wheat in preference to Canadian varieties can also reduce DON contamination of kernels.

Adequate storage of harvested grain, nuts and fruit constitutes a crucial element in the prevention of mycotoxin adulteration, particularly from spoilage fungi. Grain moisture content (expressed as water activity) and environmental temperature are critical factors during storage and distribution. In addition, insect and rodent invasion should be minimized, as these pests adversely affect the microclimate within grain silos and also act as significant vectors in transmission of fungal inoculum (D'Mello, 2003).

The malting process provides the potential for mould and mycotoxin contamination. Oliveira *et al.* (2015) suggested that lactic acid bacteria bio-protection applied at the malting stage can ensure microbial stability and improve safety in terms of mycotoxin (DON) status. Other fermentation situations, such as silage preparation, also present hazards regarding mycotoxin synthesis. The occurrence of pre- and post-harvest mycotoxins and related fungal metabolites in maize silage was stressed by Drejer Storm *et al.* (2014) with the emphasis on carry-over into edible products from animals fed contaminated silage. In the case of moist cereal grains destined for animal feeds, addition of organic acids appears to be an effective treatment for reducing both fungal proliferation and mycotoxin production. Alam *et al.* (2014) demonstrated the efficacy of calcium propionate in controlling aflatoxin production in broiler feeds, with interactions caused by factors such as water activity and duration of storage, but the effects of ambient temperature were not examined in this study.

Once mycotoxins have been detected in a particular matrix, it is possible to perform decontamination technologies, for example by hydrothermal treatment in the presence of sodium metabisulfite, methylamine and calcium hydroxide (Rempe *et al.*, 2013). It was concluded that even relatively high concentrations of DON and ZEN may be decontaminated by this method. Alternatively, mycotoxin degradation might be implemented by biocontrol processes involving, for example, yeast fermentation. Ianiri *et al.* (2017) examined the efficacy of *Sporobolomyces* species for the breakdown of patulin.

## 2.8 Conclusions

The prevailing consensus reflects widespread and continuing mycotoxin contamination of basic food staples, with global implications for human health. Concentrations of mycotoxins in cereals and nuts regularly exceed safety threshold values. At particular risk are consumers in warm and humid latitudes where exposure to aflatoxins and fumonisins may occur simultaneously via contamination of maize and peanut kernels.

In general, however, mycotoxin contamination of primary staples continues unabated even in specific regions associated epidemiologically with the incidence of hepatocellular and oesophageal cancers and nephropathy due to consumption of, respectively, aflatoxins, fumonisins and ochratoxins. In addition there may be interactions with other human diseases. For example, foodborne aflatoxins can enhance the carcinogenic potential of hepatitis B virus. A link between aflatoxicosis and kwashiorkor in African children has also been suggested. Moreover, it is possible that prenatal exposure to combinations of aflatoxins and ochratoxins may predispose to low birth weights and premature mortality of infants, as implied by studies in Sierra Leone and elsewhere. Prenatal risks are inevitable in view of contamination of breast milk and weaning foods with combinations of aflatoxins and ochratoxins. Looking forward, there is a need to acquire direct assessments of human exposure to combinations of aflatoxins and fumonisins in regions where peanuts and maize constitute the staple foods. Although acute risks are considered to be relatively low in Europe, there is consistent evidence of chronic exposure to certain mycotoxins. The detection of specific aflatoxin-albumin adducts in the serum of UK individuals implies problems with sampling and monitoring of regulated imported foodstuffs. The occurrence of ochratoxin A in the blood and breast milk of donors in several European countries emphasizes the need for concerted action to institute or enforce preventive and remedial measures.

Furthermore, the spectre of volatile mycotoxins adds another dimension in ensuring public health safety, particularly in water-damaged buildings and the indoor environment.

Looking ahead, there may be some merit in current attempts to probe the underlying biochemical and molecular aetiology of mycotoxicosis. This approach would be of assistance in resolving persistent doubts regarding epidemiological evidence. Establishing cause-and-effect issues is a fundamental requirement in any strategy to minimize health risks associated with mycotoxins. In unravelling possible underlying mechanisms, there may be benefits for the development of biomarkers to assess mycotoxin exposure in humans. It is conceivable that enhanced understanding of the interaction between mycotoxin-induced conditions and the carcinogenic activity of hepatitis B virus or kwashiorkor may also emerge in future research. Finally, the role of mycotoxins as endocrine disruptors deserves in-depth elucidation in view of current evidence implicating aflatoxins and zearalenone in antagonistic activity towards androgens and oestrogens, respectively.

## References

Accardi, R., Gruffat, H., Sirand, C., Gheit, T., Cosset, F.-L. and Manet, E. (2015) The mycotoxin aflatoxin $B_1$ stimulates Epstein-Barr virus-induced B-cell transformation in *in vitro* and *in vivo* experimental models. *Carcinogenesis* 36, 1440–1451.

Adedara, I.A., Nanjappa, M.K., Farombi, E.O. and Akingbemi, B.T. (2014) Aflatoxin $B_1$ disrupts the androgen biosynthetic pathway in rat Leydig cells. *Food and Chemical Toxicology* 65, 252–259.

Adi, N., Blaszkewicz, M. and Degen, G.H. (2016) Assessment of deoxynivalenol exposure among Bangladeshi and German adults by a biomarker-based approach. *Toxicology Letters* 258, 20–28.

Alam, S., Shah, H.U., Afzal, M. and Magan, N. (2014) Influence of calcium propionate, water activity and storage time on mold incidence and aflatoxin production in broiler starter feed. *Animal Feed Science and Technology* 188, 137–144.

Asai, T., Tsuchiya, Y., Nishi, C.Y., Ikoma, T., Oyama, T., Ikegami, K. and Yamamoto, M. (2012) Aflatoxin contamination of red chili peppers from Bolivia and Peru, countries with high gallbladder cancer incidence rates. *Asian Pacific Journal of Cancer Prevention* 13, 5167–5170.

Battilani, P., Toscano, P., Moretti, A., Brera,C., and Robinson, T. (2016) Aflatoxin $B_1$ contamination in maize in Europe increases due to climate change. *Scientific Reports (Nature Publishing Group)* 6, 24328. doi:10.1038/srep 24328.

Becher, R., Hettwer, U., Karlovsky, P., Deising, H.B. and Wirsel, S.G.R. (2010) Adaptation of *Fusarium graminearum* to tebuconazole yielded descendants diverging for levels of fitness, fungicide resistance, virulence and mycotoxin production. *Phytopathology* 100, 444–453.

Bennett, J.W. and Inamdar, A.A. (2015) Are some fungal volatile organic compounds (VOCs) mycotoxins? *Toxins* 7, 3785–3804.

Bhat, R.V., Ramakrishna, Y., Beedu, S.R. and Munshi, K.L. (1989) Outbreak of trichothecene mycotoxicosis associated with consumption of mould-damaged wheat products in Kashmir Valley, India. *The Lancet* 333, 35–37.

Brewer, J.H., Thrasher, J.D., Straus, D.C., Madison, R.A. and Hooper, D. (2013) Detection of mycotoxins in patients with chronic fatigue syndrome. *Toxins* 5, 605–617.

Bui-Klimke, T. and Wu, F. (2014) Evaluating weight of evidence in the mystery of Balkan endemic nephropathy. *Risk Analysis* 34, 1688–1705.

Castelino, J.M., Dominguez-Salas, P., Routledge, M.N., Prentice, A.M., Moore, S.E., Wild, C.P. and Gong, Y.Y. (2014) Seasonal and gestation stage associated differences in aflatoxin exposure in pregnant Gambian women. *Tropical Medicine & International Health* 19, 348–354.

Chawanthayatham, S., Valentine, C.C., Fedeles, B.I., Fox, E.J., Levine, S.S. *et al.* (2017) Mutational spectra of aflatoxin $B_1$ *in vivo* establish biomarkers of exposure for human hepatocellular carcinoma. *Proceedings of the National Academy of Sciences of the United States of America* 114, E3101–E3109.

Cui, A., Hua, H., Shao, T., Song, P., Kong, Q. and Jiang, Y. (2015) Aflatoxin $B_1$ induces Src phosphorylation and stimulates lung cancer cell migration. *Tumor Biology* 36, 6507–6513.

Dani, A., Khazaeli, P., Madadiou, A., Doraki, N., Farrokhi, H., Moradi, H. and Khodadadi, E. (2013) Aflatoxin concentration level in Iran's pistachio nuts during year 2009–2011. *Food Control* 30, 540–544.

Demirel, G., Alpertunga, B. and Ozden, S. (2015) Role of fumonisin $B_1$ on DNA methylation changes in rat kidney and liver cells. *Pharmaceutical Biology* 53, 1302–1310.

D'Mello, J.P.F. (2003) Mycotoxins in cereal grains, nuts and other plant products. In: D'Mello, J.P.F. (ed.) *Food Safety: Contaminants and Toxins*. CAB Publishing, Wallingford, UK, pp. 65–90.

D'Mello, J.P.F., Porter, J.K., Macdonald, A.M.C. and Placinta, C. M. (1997) *Fusarium* mycotoxins. In: D'Mello, J.P.F. (ed.) *Handbook of Plant and Fungal Toxicants*. CRC Press, Boca Raton, Florida, pp. 287–301.

Drejer Storm, I.M.L., Rasmussen, R.R. and Rasmussen, P.H. (2014) Occurrence of pre- and post-harvest mycotoxins and other secondary metabolites in Danish maize silage. *Toxins* 6, 2256–2269.

Flannigan, B. (1991) Mycotoxins. In: D'Mello, J.P.F. (ed.) *Toxic Substances in Crop Plants*. The Royal Society of Chemistry, Cambridge, UK, pp. 226–257.

Garcia-Cela, E., Marin, S., Sanchis, V., Crespo-Sempere, A. and Ramos, A.J. (2015) Effect of ultraviolet radiation A and B on growth and mycotoxin production by *Aspergillus carbonarius* and *Aspergillus parasiticus* in grape and pistachio media. *Fungal Biology* 119, 67–78.

Gizachew, D., Szonyi, B., Tegegne, A., Hanson, J. and Grace, D. (2016) Aflatoxin contamination of milk and dairy feeds in the Greater Addis Ababa milk shed, Ethiopia. *Food Control* 59, 773–779.

Hahn, M. (2014) The rising threat of fungicide resistance in plant pathogenic fungi: *Botrytis* as a case study. *Journal of Chemical Biology* 7, 133–141.
Heinig, R.L., Paaijmans, K.P., Hancock, P.A. and Thomas, M.B. (2015) The potential for fungal biopesticides to reduce malaria transmission under diverse environmental conditions. *Journal of Applied Ecology* 52, 1558–1566.
Ianiri, G., Pinedo, C., Fratianni, A., Panfili, G. and Castoria, R. (2017) Patulin degradation by the biocontrol yeast *Sporobolomyces* species is an inducible process. *Toxins* 9, 61. doi:10.3390/toxins 9020061.
Ibrahim, A.S., Zaghloul, H. and Badria, F.A. (2013) Case report evidence of relationships between hepatocellular carcinoma and ochratoxicosis. *PLoS ONE* 8 (8): e71423. doi: 10.1371/journal.pone.0071423.
Kachapulula, P.W., Akello, J., Bandyopadhyay, R. and Cotty, P.J. (2017) Aflatoxin contamination of groundnut and maize in Zambia: observed and potential concentrations. *Journal of Applied Microbiology* 122(6), 1471–1482. doi: 10.1111/jam.13448.
Kamika, L., Mngqawa, P., Rheeder, J.P., Teffo, S.L. and Katerere, D.R. (2014) Mycological and aflatoxin contamination of peanuts sold at markets in Kinshasa, Democratic Republic of Congo, and Pretoria, South Africa. *Food Additives & Contaminants: Part B. Surveillance* 7(2), 120–126.
Kim, J., Park, S.-H., Do, K.H., Kim, D. and Moon, Y. (2016) Interference with mutagenic aflatoxin $B_1$-induced checkpoints through antagonistic action of ochratoxin A in intestinal cancer cells: a molecular explanation on potential risk of crosstalk between carcinogens. *Cancer* 20, 21.
Krishnamachari, K.A.V.R., Nagarajan, V., Bhat, R.V. and Tilak, T.B.G. (1975) Hepatitis due to aflatoxicosis: an outbreak in Western India. *The Lancet* 305, 1061–1063.
Lewis, L., Onsongo, M., Njapau, H., Luber, G., Nyamongo, J. *et al.* (2005) Aflatoxin contamination of commercial maize products during an outbreak of acute aflatoxicosis in Eastern and Central Kenya. *Environmental Health Perspectives* 113, 1763–1767.
Livingstone, M.C., Johnson, N.M., Roebuck, B.D., Kensler, T.W. and Groopman, J.D. (2017) Profound changes in miRNA expression during cancer initiation by aflatoxin $B_1$ and their abrogation by the chemopreventive triterpenoid CDDO-Im. *Molecular Carcinogenesis* 56(11), 2382–2390. doi: 10.1002/mc.22635.
Malir, F., Ostry, V., Roubal, T., Dvorak, V. and Dohnal, V. (2013) Ochratoxin A levels in blood serum of Czech women in the first trimester of pregnancy and its correspondence with dietary intake of the mycotoxin contaminant. *Biomarkers* 18(8), 673–678.
Mantle, P. (2016) Rat kidney cancers determined by dietary ochratoxin A in the first year of life. *Journal of Kidney Cancer and VHL* 3, 1–10.
Marques, L.N., Pizzutti, I.R., Balardin, R.S., Dos Santos, I.D., Dias, J.V. and Serafini, P.T. (2017) Occurrence of mycotoxins in wheat grains exposed to fungicides on *Fusarium* control in Brazil. *Journal of Environmental Science and Health* 12, 1–7.
Mulder, J.E., Bardy, C.S., Mehta, R. and Massey, T.E. (2014) Up-regulation of nucleotide excision repair in mouse lung and liver following chronic exposure to aflatoxin $B_1$ and its dependence on *p53* genotype. *Toxicology and Applied Pharmacology* 275, 96–103.
Narkwa, P.W., Blackbourn, D.J. and Mutocheluh, M. (2017) Aflatoxin $B_1$ inhibits the type 1 interferon response pathway via STAT1 suggesting another mechanism of hepatocellular carcinoma. *Infectious Agents and Cancer* 12, 17. doi: 10.1186/s1 3027-017-0127-8.
Ngindu, A., Kenya, P.R., Ocheng, D., Omondi, T.N., Johnson, B.K., Jansen, A.J. and Siongok, T.A. (1982) Outbreak of acute hepatitis caused by aflatoxin poisoning in Kenya. *The Lancet* 319, 1346–1348.
Nogueira, L., Foerster, C. and Groopman, J. (2015) Association of aflatoxin with gallbladder cancer in Chile. *Journal of the American Medical Association* 313, 2075–2077.
Obade, M., Andango, P., Obonyo, C. and Lusweti, F. (2015) Aflatoxin exposure in pregnant women in Kisumu County, Kenya. *Current Research in Nutrition and Food Science* 3 (2), 42–50. doi: 10.12944/CRNFSJ.3.2.06.
Oliveira, P., Brosnan, B., Jacob, F., Furey, A., Coffey, A. and Arendt, E.K. (2015) Lactic acid bacteria bioprotection applied to the malting process. Part II: substrate impact and mycotoxin reduction. *Food Control* 51, 444–452.
Popiel, D., Dawidziuk, A., Koczyk, G., Mackowiak, A. and Marcinkowska, K. (2017) Multiple facets of response to fungicides – the influence of azole treatment on expression of key mycotoxin biosynthetic genes and candidate resistance factors in the control of resistant *Fusarium* strains. *European Journal of Plant Pathology* 147, 773–785.
Porter, J.K. (1997) Endophyte alkaloids. In: D'Mello, J.P.F. (ed.) *Handbook of Plant and Fungal Toxins*, 1st edn. CRC Press, Boca Raton, Florida, pp. 51–62.

Rahimi, E. and Jeiran, M.R. (2015) Patulin and its dietary intake by fruit juice consumption in Iran. *Food Additives & Contaminants: Part B Surveillance* 8, 40–43.
Rempe, I., Kersten, S., Valenta, H. and Danicke, S. (2013) Hydrothermal treatment of naturally contaminated maize in the presence of sodium metabisulfite, methylamine and calcium hydroxide: effects on the concentration of zearalenone and deoxynivalenol. *Mycotoxin Research* 29, 169–175.
Rheeder, J.P., Van de Westhuizen, Imrie, G. and Shephard, G.S. (2016) *Fusarium* species and fumonisins in subsistence maize in the former Transkei region, South Africa: a multi-year study in rural villages. *Food Additives & Contaminants: Part B. Surveillance* 9, 176–184.
Russell, R., Paterson, M. and Lima, N. (2010) How will climate change affect mycotoxins in food? *Food Research International* 43, 1902–1914.
Schmidt-Heydt, M., Stoll, D. and Geisen, R. (2013) Fungicides effectively used for growth inhibition of several fungi could induce mycotoxin biosynthesis in toxigenic species. *International Journal of Food Microbiology* 166, 407–412.
Schwartzbord, J.R. and Brown, D.L. (2015) Aflatoxin contamination in Haitian peanut products and maize and the safety of oil processed from contaminated peanuts. *Food Control* 56, 114–118.
Sharma, P., Manderville, R.A. and Wetmore, S.D. (2014) Structural and energetic characterization of the major DNA adduct formed from the food mutagen ochratoxin A in the *Nar* I hotspot sequence: influence of adduct ionization on the conformational preferences and implications for the NER propensity. *Nucleic Acids Research* 42, 11831–11845.
Smith, J.E. (1997) Aflatoxins. In: D'Mello, J.P.F. (ed.) *Handbook of Plant and Fungal Toxins*, 1$^{st}$ edn. CRC Press, Boca Raton, Florida, pp. 269–285.
Solfrizzo, M., Gambacorta, L. and Visconti, A. (2014) Assessment of multi-mycotoxin exposure in Southern Italy by urinary multi-biomarker determination. *Toxins* 6, 523–538.
Stoev, S. D. (2017) Balkan endemic nephropathy – still continuing enigma, risk assessment and underestimated hazard of joint mycotoxin exposure of animals and humans. *Chemico-Biological Interactions* 261, 63–79.
Techapiesancharoenkij, N., Fiala, J.L.A., Navasumrit, P., Croy, R.G., Wogan, G.N., Groopman, J.D. and Essigmann, J.M. (2015) Sulforaphane, a cancer chemopreventive agent, induces pathways associated with membrane biosynthesis in response to tissue damage by aflatoxin $B_1$. *Toxicology and Applied Pharmacology* 282, 52–60.
Terra, M.F., Lira, N. deA., Passamani, F.R.F., Santiago, W.D., das Gracas Cardoso, M. and Batista, L.R. (2016) Effect of fungicides on growth and ochratoxin A production by *Aspergillus carbonarius* from Brazilian wine grapes. *Journal of Food Protection* 79, 1508–1516.
Vega, F.E., Goettel, M.S., Blackwell, M., Chandler, D., Pell, J.K., Rangel, D.E.N. and Roy, H.E. (2009) Fungal entomopathogens: new insights on their ecology. *Fungal Ecology* 2, 149–159.
Wang, S.-K., Wang, T.-T., Huang, G.-L., Shi, R.-F. and Sun, G.-J. (2014) Stimulation of the proliferation of human normal esophageal epithelial cells by fumonisin $B_1$ and its mechanism. *Experimental and Therapeutic Medicine* 7, 55–60.
Wang, Y., Liu, Y., Liu, X., Jiang, L., Li, Q., Yao, X. and Chen, M. (2015) Citreoviridin induces autophagy-dependent apoptosis through lysosomal-mitochondrial axis in human liver HepG2 cells. *Toxins* 7, 3030–3044.
Wu, L.X., Ding, X.X., Li, P.W., Du, X.H. and Zhang, L.X. (2016) Aflatoxin contamination of peanuts at harvest in China from 2010 to 2013 and its relationship with climatic conditions. *Food Control* 60, 117–123.
Zhang, W., He, H., Zang, M., Wu, Q., Zhao, H., Lu, L.-L., Zeng, Y.-X. and Jiao, Y. (2017) Genetic features of aflatoxin-associated hepatocellular carcinomas. *Gastroenterology* 153(1), 249–262. doi: 10.1053/j.gastro. 2017.03.024.
Zouaoui, N., Sbaii, N., Bacha, H. and Abid-Essefi, S. (2015) Occurrence of patulin in various fruit juices marketed in Tunisia. *Food Control* 51, 356–360.

# 3 Cyanobacterial Toxins

**J.S. Metcalf* and N.R. Souza**
*Brain Chemistry Labs, Institute for Ethnomedicine, Jackson, Wyoming, USA*

## 3.1 Abstract

Cyanobacteria are capable of producing a wide range of low-molecular-weight toxic compounds, largely identified as a result of poisoning incidents and through meticulous screening of cyanobacterial strains and blooms. Their molecular modes of action vary and can include hepatotoxic and neurotoxic effects in acute doses. Furthermore, long-term exposure to cyanobacterial toxins has been implicated in a number of human health conditions, from primary liver cancer to human neurodegenerative disease. Through various routes and media, exposure to cyanobacterial toxins can occur and toxicological and quantitative assessment of cyanobacterial blooms should be performed to help minimize adverse effects on human and animal health. Management and treatment of cyanobacterial blooms can, in the long term, prevent periodic exposure to these toxic compounds.

## 3.2 Introduction

Cyanobacteria, also known as blue-green algae, are a group of Gram-negative photosynthetic bacteria that also have the ability to grow heterotrophically under certain conditions (Fogg *et al.*, 1973). They inhabit a wide range of environments from freshwater to oceans, and are also found in hot springs, on snow and in deserts. With the discovery of fossils comprising cyanobacteria that dated to over 3.5 billion years ago (Schopf, 2000), this group of organisms is considered to have contributed to the Great Oxidation Event, which resulted in the oxygenation of the Earth's atmosphere allowing life to exist in its current state (Knoll, 2003).

Most frequently observed in aquatic environments, cyanobacteria present as blooms and scums within and on the surface of water, often as macroscopic structures visible to the naked eye. Blooms of cyanobacteria have been reported in many countries around the world, mostly in lakes, rivers, estuaries and reservoirs used for the preparation of drinking water, recreation and fisheries (Metcalf and Codd, 2012).

Cyanobacteria are important primary producers establishing the basis of food chains and webs within aquatic environments. The presence of high concentrations of nitrogen and phosphorus in waterbodies can result in mass populations of cyanobacteria that can be observed by satellite using techniques such as remote sensing (Hunter *et al.*, 2017). In addition, climate change and global warming may contribute to increases in bloom formation and frequency, along with geographical expansion of certain species of cyanobacteria (Paerl and Huisman, 2008, 2009; Paerl and Paul, 2012).

* E-mail address: james@ethnomedicine.org

The phenomenon of bloom formation has, increasingly in recent times, gained attention due to potential adverse effects on economies and the health of humans and wildlife. These effects can range from aesthetic (visual and odour) to possible intoxications and deaths of animals and humans caused by exposure to cyanobacterial low-molecular-weight metabolites: the cyanotoxins.

## 3.3 Classes of Cyanobacterial Toxins

The identification of cyanobacterial toxins has manifested through a variety of means including, but not limited to: animal deaths; toxicological screening of cyanobacterial strains; human health illness investigations; and retrospective toxicological and analytical assessment of bloom material. All have proven useful for the identification and elucidation of toxins produced by cyanobacteria. The cyanotoxins can be divided into classes, largely based upon their mode of action.

### 3.3.1 Hepatotoxins

Considered to be one of the most common cyanobacterial toxins, the hepatotoxic microcystins (Fig. 3.1) and related nodularins (Fig. 3.2) are cyclic peptides produced by a number of cyanobacterial genera (Metcalf and Codd, 2012; Catherine *et al.*, 2017). They act as potent protein phosphatase inhibitors and, in high doses, can result in destruction of the liver cytoskeleton and macrostructure, with death possible due to hypovolaemic shock (Carmichael, 1994). Repeated low-dose exposure to microcystins and nodularins is considered to cause tumour promotion and carcinogenicity, respectively (Nishiwaki-Matsushima *et al.*, 1992; Ohta *et al.*, 1994). At present, at least 200 individual microcystin and ten nodularin variants are known, each with differing toxicities (Spoof and Catherine, 2017).

### 3.3.2 Cytotoxins

Cylindrospemopsin (Fig 3.3), also hepatotoxic, is a commonly found low-molecular-weight genotoxic alkaloid produced by many Australian strains of *Cylindrospermopsis raciborskii* (Hawkins *et al.*, 1985), in addition to other cyanobacterial genera (Metcalf and Codd, 2012). Cylindrospermopsin inhibits protein synthesis in plants (Metcalf *et al.*, 2004) and animals (Terao *et al.*, 1994) with the liver and kidney being the main organs affected, but the lungs and intestines may also be susceptible (Seawright *et al.*, 1999). Further assessment of cyanobacteria led to the isolation of the cylindrospermopsin variants 7-epicylindrospermopsin, 7-deoxy-desulfo-cylindrospermopsin, 7-deoxy-desulfo-12-acetylcylindrospermopsin and deoxycylindrospermopsin, with the latter considered to be non-toxic (Banker *et al.*, 1997, 2001; Norris *et al.*, 1999; Wimmer *et al.*, 2014).

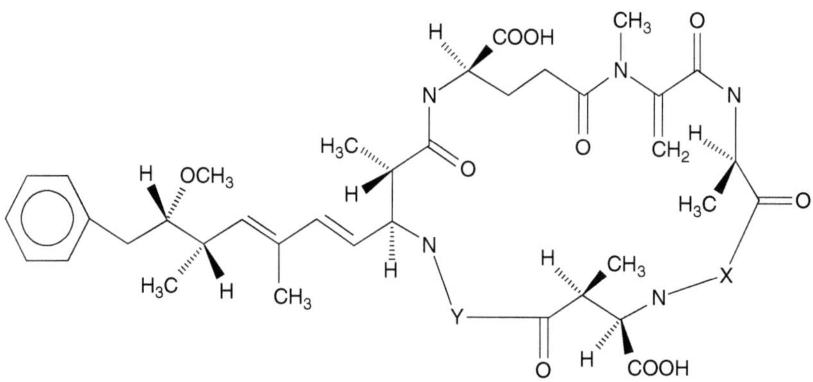


**Fig. 3.1.** The generic structure of microcystins where X and Y represent variable amino acids.

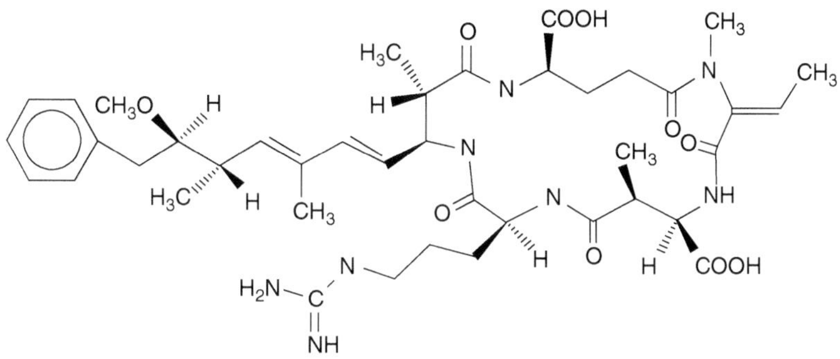


**Fig. 3.2.** Nodularin.

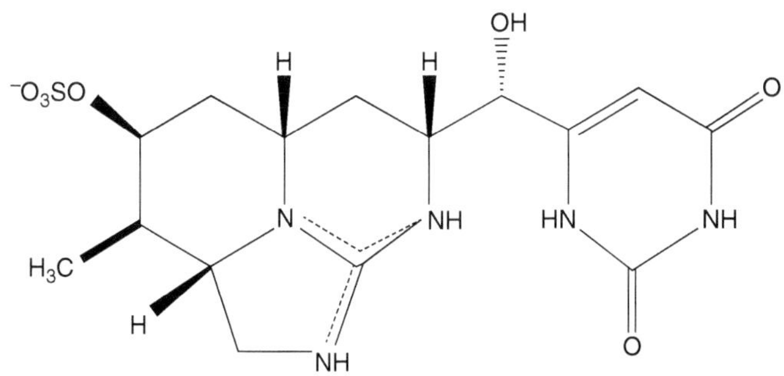


**Fig. 3.3.** Cylindrospermopsin.

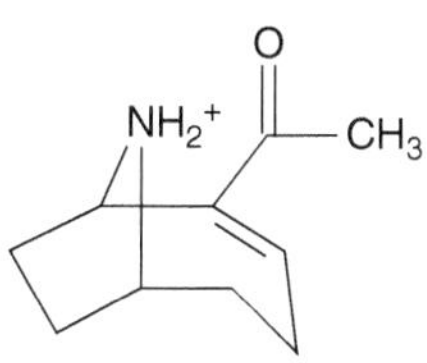


**Fig. 3.4.** Anatoxin-a.

### 3.3.3 Neurotoxins

#### *3.3.3.1 Anatoxin-a*

An acetylcholine mimic, anatoxin-a (Fig. 3.4) was first identified in *Dolichospermum* (formerly *Anabaena;* Carmichael *et al.*, 1975). Anatoxin-a and its congener homoanatoxin-a target the cholinergic synapse and act as a potent postsynaptic depolarizing neuromuscular blocking agent, binding to the nicotinic acetylcholine receptor at the neuromuscular junction, causing persistent stimulation and consequently blocking further electrical transmission. In high enough doses anatoxin-a exposure may lead to paralysis, asphyxiation and death (Devlin *et al.*, 1977; Carmichael *et al.*, 1979; Carmichael, 1994).

#### *3.3.3.2 Anatoxin-a*(S)

Although not as commonly found as other cyanobacterial toxins, anatoxin-a(*S*) (Fig. 3.5) is structurally unrelated to its similarly named counterpart anatoxin-a. A naturally occurring organophosphate, it has more toxicological similarity to synthetic organophosphate pesticides and insecticides (Mahmood and Carmichael, 1986). Consequently, if present in sufficient doses, anatoxin-a(*S*) is able to inhibit acetylcholine esterases and can result in paralysis and death (Mahmood and Carmichael, 1986; Cook *et al.*, 1989; Carmichael, 1994). Largely produced by *Dolichospermum*, anatoxin-a(*S*) has been responsible for bird deaths on lakes supporting this cyanobacterium (Henriksen *et al.*, 1997). Although primarily an aquatic cyanotoxin, using acetylcholine esterase (AChE) inhibition assays its presence has been inferred in terrestrial desert environments where cyanobacteria comprise significant amounts of

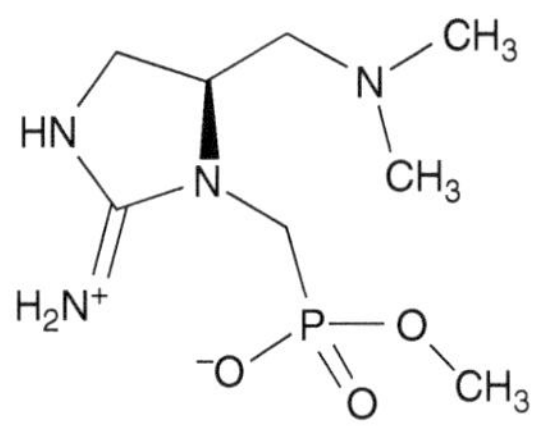


**Fig. 3.5.** Anatoxin-a(*S*).

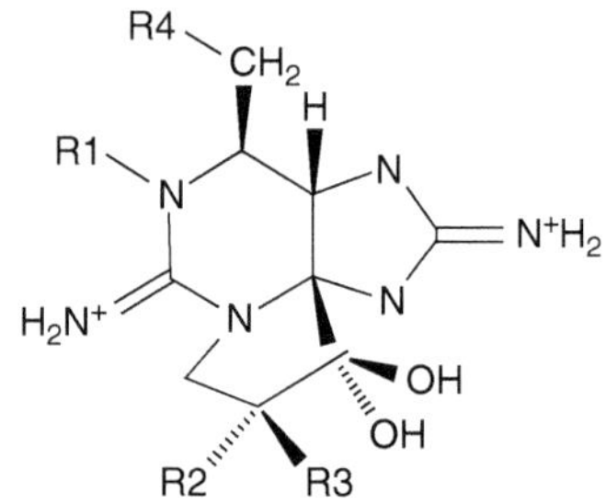


**Fig 3.6.** The generic structure of the saxitoxin molecule where R represents variable groups.

the desert surface as cryptogamic crusts (Metcalf *et al.*, 2012; Richer *et al.*, 2012). Genetically engineered AChEs are now able to distinguish anatoxin-a(*S*) from synthetic organophosphorus compounds (Devic *et al.*, 2002).

#### 3.3.3.3 Saxitoxins

As products of dinoflagellates and through contamination of shellfish with the resulting paralytic shellfish toxins (PST), saxitoxins have caused significant impacts on marine shellfish fisheries (Ballot *et al.*, 2017). Also produced by cyanobacteria, these toxins have been documented in strains of *Aphanizomenon*, *Dolichospermum* and *Cylindrospermopsis* (Metcalf and Codd, 2012) as examples. There are currently over 20 different variants of the saxitoxins known (Fig. 3.6), each with differing toxicities and structures (Ballot *et al.*, 2017). In mammalian systems, these alkaloid toxins (Schantz *et al.*, 1957; Bordner *et al.*, 1975) are known sodium channel blocking compounds and, consequently, in high enough doses can result in paralysis and death (Carmichael, 1994; Ballot *et al.*, 2017).

### 3.3.4 Lipopolysaccharide

As cyanobacteria are Gram-negative bacteria, they have the potential to produce lipopolysaccharide (LPS) endotoxins. These complex, outer components of the Gram-negative bacterial cell wall (Drews and Weckesser, 1982) are known to be toxic to mammals, often causing gastrointestinal discomfort (Metcalf and Codd, 2012). Although large amounts of cyanobacterial LPS can be present in blooms (Bláhová *et al.*, 2013), cyanobacterial LPS is considered to be of lower toxicity than heterotrophic bacterial LPS from *Salmonella typhimurium* and *Escherichia coli* as examples (Monteiro *et al.*, 2017).

### 3.3.5 Dermatotoxins

Aplysiatoxin, debromoaplysiatoxin and lyngbyatoxins are among the common cyanotoxins showing dermatotoxic effects and are also considered to be skin tumour promoters (Fujiki *et al.*, 1981, 1983). Aplysiatoxin and debromoaplysiatoxin are phenolic bislactones causing skin irritation, rashes and blistering (Mynderse *et al.*, 1977). Lyngbyatoxins (with variants A, B and C) are indole alkaloids that can be produced by genera of benthic cyanobacteria (e.g *Moorea producens*) and have been implicated in dermatitis and inflammation of oral and gastrointestinal tissues (Cardellina *et al.*, 1979; Aimi *et al.*, 1990).

### 3.3.6 Neurotoxic amino acids

β-*N*-Methylamino-L-alanine (BMAA) (Fig. 3.7), one of many non-protein amino acids, was identified in the environment (Vega and Bell, 1967) and in the diet of Chamorro villagers who died of amyotrophic lateral sclerosis/Parkinsonism dementia complex (ALS/PDC) on Guam (Murch *et al.*, 2004). Cycads were identified as sources of BMAA, as ultimately were cyanobacteria of the genus *Nostoc*, present within specialized coralloid roots of the cycad (Murch *et al.*, 2004). Since then, analysis of cyanobacterial blooms and strains from outside of Guam has shown BMAA production to be widespread (Cox *et al.*, 2005, 2009; Metcalf *et al.*, 2008). Dosing of vervets with BMAA showed neurofibrillary tangles and amyloid plaques, consistent with human ALS/PDC neuropathology within the brain (Cox *et al.*, 2016). Within the mammalian cell, BMAA has a number of molecular effects, including binding

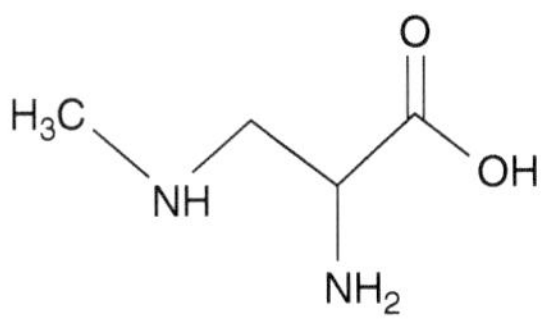

**Fig. 3.7.** β-*N*-Methylamino-L-alanine (BMAA).

to glutamate receptors and incorporation into proteins in place of L-serine (Dunlop *et al.*, 2013). Furthermore, BMAA can be metabolized and converted from L- to D-BMAA in the central nervous system, with the different enantiomers showing toxicity at different receptors (Metcalf *et al.*, 2017).

## 3.4 Exposure Routes

To better understand, mitigate and assess the adverse effects of cyanotoxins in humans and animals, the various potential exposure routes and media require identification and evaluation. The following contains common routes of exposure to cyanobacteria and their toxins.

### 3.4.1 Water

After a die-off of livestock at Lake Alexandrina in South Australia, which supported a bloom of *Nodularia spumigena*, Francis (1878) conducted an investigation and analysed the stomach contents of the dead animals, finding the presence of filamentous cyanobacteria. Considering that this filamentous material was the cause of death, sheep were dosed with the *Nodularia* scum to reproduce the same signs of poisoning observed in the livestock die-off, providing one of the first scientific reports documenting the adverse effects of exposure to cyanobacteria.

Wild and domestic animals (Sivonen and Jones 1999), including sheep, cattle (Mez *et al.*, 1997; Saker *et al.*, 1999), horses, pigs, bats (Pybus *et al.*, 1986), fish (Rodger *et al.*, 1994), birds (Henriksen *et al.*, 1997; Codd *et al.*, 2005; Metcalf and Codd, 2012) and dogs (Mahmood *et al.*, 1988; Codd *et al.*, 1992; Edwards *et al.*, 1992; Wood *et al.*, 2007; Chatziefthimiou *et al.*, 2014), have succumbed to the effects of accidental exposure to cyanotoxins.

In 1985, approximately 1000 bats and 24 ducks (mallard and American wigeon) were reported to have died in Cross Lake Provincial Park (Alberta, Canada). They were found floating in the lake, which had a bloom of cyanobacteria that was later identified as *Anabaena*. Carcasses of seven bats and two mallard ducks, covered in 'green slime', were sent for analysis. The animals were in good health overall and death had occurred rapidly. The analysis of this material indicated the presence of a toxic cyanobacterial alkaloid (Pybus *et al.*, 1986).

Fish kills due to the direct action of cyanobacterial toxins have been documented, such as that observed with microcystins (Rodger *et al.*, 1994). Indirectly, once environmental conditions change and cyanobacterial blooms senesce, the decay of such large amounts of organic matter can result in decreases in dissolved oxygen within the water column. This oxygen-consuming bacterial decay of the cyanobacterial bloom can place a strain on the aquatic organisms that are present. Under such conditions, it is not uncommon for large numbers of fish to die as a result of a lack of oxygen.

Bird deaths caused by cyanobacterial toxins are well known (Metcalf and Codd, 2012). Although lesser flamingos (*Phoeniconaias minor*) consume microalgae and cyanobacteria (*Arthrospira*) as their food source, they undergo periodic mass mortalities which have been attributed to many causes, including cyanotoxins (Krienitz *et al.*, 2003; Metcalf and Codd 2012). Mass mortalities of greater flamingo chicks (*Phoenicopterus ruber*) from a national park in Spain and of Chilean flamingos (*Phoenicopterus chlilensis*) from Sea World were also attributed to the ingestion of microcystins (Codd *et al.*, 2003).

Of domestic animals, intoxications from cyanobacterial toxins are frequently reported in dogs. The odour and taste compounds that cyanobacterial biomass can produce (e.g. geosmin, methylisoborneol) can be attractive to dogs (Codd *et al.*, 1992). They regularly ingest mats of cyanobacteria (e.g. *Phormidium*) that are toxic, resulting in intoxication and/or death, in addition to drinking water and scums containing cyanobacteria, and self-cleaning their fur, resulting in secondary intoxication (Codd *et al.*, 1992; Edwards *et al.*, 1992; Wood *et al.*, 2007). According to Backer *et al.* (2013), from the late 1920s to mid-2012 a total of 115 events was identified from media, state, federal agencies, scientific and medical communities, involving the poisoning of 260 dogs in which 83% of the dogs died and 17% became ill but recovered. These events involved several genera of cyanobacteria, including *Dolichospermum*, *Microcystis*, *Lyngbya* and *Aphanizomenon*, and cyanotoxins including anatoxin-a and microcystins. In Qatar, from November 2013 to March 2014, four cases of dog poisoning were associated with neurotoxins from cyanobacteria. Two of the dogs died despite veterinary intervention and care. The symptoms followed recreational activities of the dogs in the desert and consumption of rainwater that had been trapped in natural depressions (Chatziefthimiou *et al.*, 2014). The cyanobacteria in Qatar are present as crusts in the depressions covering up to 87% of the land surface, which could potentially lead to further intoxications (Cox *et al.*, 2009; Metcalf *et al.*, 2012; Richer *et al.*, 2012). In the case of the desert dog poisonings, crust material from the site of the intoxication was positive for acetylcholine esterase inhibition, indicating the presence of anatoxin-a(*S*) (Chatziefthimiou *et al.*, 2014). Dermatitis in dogs has also been reported and attributed to cyanobacteria (Puschner *et al.*, 2017).

Warm temperatures, especially during the summer season, can increase the frequency and number of outdoor recreation activities in many rivers, lakes and beaches. This is also the time when cyanobacterial bloom events mostly occur and intensify. Water sports and recreational activities frequently involve unintentional swallowing of water and may include long periods of skin exposure.

Cases of human dermatitis, eye irritation, diarrhoea and vomiting with allergic-like symptoms following contact with water have been associated, reported and attributed to cyanotoxins and cyanobacteria (Pilotto *et al.*, 1997; Osborn and Shaw, 2008). In 2016, four Florida counties declared a state of emergency due to a cyanobacterial bloom that caused flu-like symptoms, respiratory issues, rashes, burning eyes and headaches in people who came into contact with the mist/aerosol from the bloom (Lantigua, 2017). A severe case that led to a 19-year-old man being hospitalized for 20 days, presenting flu-like symptoms, abdominal pain, hepatotoxicosis and pneumonia, was also reported after immersion in a scum of *Microcystis aeruginosa* (Giannuzzi *et al.*, 2011).

Human intoxication and poisonings, in some cases resulting in gastroenteritis, human liver failure and death, have been reported (Teixeira *et al.*, 1993; Annadotter *et al.*, 2001; Azevedo *et al.*, 2002). In Palm Island, Australia, in 1979, more than 100 people were hospitalized with symptoms of vomiting, headaches, hepatomegaly, bloody diarrhoea and dehydration, associated with consumption of water (Byth, 1980). Follow-up investigations at the source of the drinking water isolated a strain of *C. raciborskii* and identified the production of cylindrospermopsin (Hawkins *et al.*, 1985). In 2015, a family of four in Uruguay were exposed to cyanotoxins after recreational use of water. The three adults presented gastrointestinal symptoms while the 20-month-old baby presented severe gastrointestinal conditions resulting in liver failure, requiring a liver transplant, with the removed liver showing the presence of microcystin-LR (MC-LR) and a D-leucine variant (Vidal *et al.*, 2017). Accidental ingestion of cyanobacteria has also occurred, as demonstrated by the hospitalization of army recruits with pneumonia-like symptoms after 'barrel rolling' in kayaks in a bloom of *Microcystis* at a UK waterbody (Turner *et al.*, 1990).

Cyanotoxins have been reported in several drinking-water supplies from different countries (Molica *et al.*, 2005; Mhlanga *et al.*, 2006; Chatziefthimiou *et al.*, 2016; Gaget *et al.*, 2017) with closures and restrictions occurring, as evidenced in 2014 in Toledo (Pelley, 2016). Due to a large cyanobacterial bloom containing microcystins, which subsequently entered the water treatment plant, drinking-water supplies to residents were affected, with alternative drinking-water provision and toxin analysis and risk assessment being carried out (Pelley, 2016).

The highest-profile case of poisoning and death caused by cyanotoxins occurred in Caruaru, Northeast Brazil, in 1996. One hundred patients developed acute liver failure and 52 of these patients died after treatment with haemodialysis. The clinic used water taken from a reservoir that supported a cyanobacterial bloom before being inadequately treated at the clinic. Subsequent toxin analysis of the water showed the presence of microcystins and cylindrospermopsin, but microcystins were considered to be the major contributor to the dialysis patients' death (Jochimsen *et al.*, 1998; Pouria *et al.*, 1998; Carmichael *et al.*, 2001; Azevedo *et al.*, 2002). This led to the identification of medicinal water, especially that which is administered intravenously, as another possible exposure route for cyanotoxins.

### 3.4.2 Food

The presence of cyanotoxins, including microcystins, cylindrospermopsins, saxitoxins, BMAA and anatoxin-a in various trophic levels, have been documented, including aquatic plants, molluscs and fish, with many used as food (Magalhaes *et al.*, 2001; Sipiä *et al.*, 2007a; Berry *et al.*, 2011; Al-Sammak *et al.*, 2014; see Testai *et al.*, 2016, for an extensive review on cyanotoxins in food). Codd *et al.* (1999) found *Microcystis aeruginosa* and microcystins in salad lettuce that was spray-irrigated with water containing cyanobacteria. Kittler *et al.* (2012) also showed the accumulation of cyanotoxins, in this case cylindrospermopsin, in tissues of *Brassica* crop plants, including the edible parts that were irrigated with contaminated water. Saqrane and Oudra (2009) highlighted the need for surveillance and monitoring of irrigation water as cyanotoxin-contaminated water can negatively affect plant crops and pose a potential risk for human and animal health due to possible accumulation in edible plants.

Dietary supplements containing cyanobacteria are also possible sources of cyanotoxin exposure and poisoning, with microcystins, anatoxin-a and BMAA detected in commercial cyanobacterial food supplements (Glover *et al.*, 2015; Roy-Lachapelle *et al.*, 2017). Bautista *et al.* (2015) reported a case of microcystin poisoning in a dog after 3½ weeks of use of a dietary supplement containing 100% certified organic *Aphanizomenon flos-aquae* which had been cultured and harvested outdoors. The presence of cyanotoxins in cyanobacterial supplements has also helped with the introduction of legislation in certain places, such as the state of Oregon, which has introduced microcystin guidelines of 1 mg $g^{-1}$ in food supplements due to this state being a location of cyanobacterial harvesting for supplement production (Dietrich and Hoeger, 2005).

### 3.4.3 Aerosols and airborne cyanotoxins

During cyanobacterial bloom formation, the concentration of cyanobacterial cells can be enriched as a consequence of environmental conditions. In addition to still, warm weather which increases the potential for blooms, scum formation can be encouraged through wind-driven processes. Largely, this is seen in the accumulation of scums in bays and on shorelines on the leeward side of a lake after the cyanobacterial cells have floated to the surface and are then pushed by the action of the wind. Although light wind leads to the formation of scums and large concentration factors in cell number can be achieved, little is known about the movement of cyanobacteria produced as a result of sprays and aerosols generated through wind action. With regards to BMAA, Caller *et al.* (2009) mapped the addresses of ALS patients, finding clusters around lakes with cyanobacterial blooms. BMAA and isomers have also been detected in air samples taken at lakes with cyanobacterial blooms (Banack *et al.*, 2015). Microcystins have been detected in aerosols generated either in the laboratory or at the lakeside (Cheng *et al.*, 2007; Backer *et al.*, 2008, 2010), indicating that exposure can occur through inhalation. Toxicity assessment of cyanobacterial toxins in laboratory animals has shown that adverse effects can occur, either from individual cyanotoxins or in combination (Fitzgeorge *et al.*, 1994).

Chatziefthimiou *et al.* (2015) highlighted the potential for aerosol exposure to cyanobacteria and cyanotoxins as a natural phenomenon of dust storms in desert environments. Calculations on the amount of microcystins present in desert dust, when compared with the amount of dust potentially inhaled during a dust storm, suggested that this exposure route may be a cause for concern (Metcalf *et al.*, 2012). Certainly, the inhalation of cyanotoxins is an understudied route of exposure and toxicity, and subsequently guidelines may be altered due to different exposure routes. As cyanobacteria can be significant biological components of desert crust, the finding that veterans deployed to the Persian Gulf during Operation Desert Storm had an increased risk of developing ALS compared with those who underwent the same training and were not deployed (Horner *et al.*, 2003) suggests that inhalation exposure to substances such as BMAA, naturally present within the desert environment, may be significant (Cox *et al.*, 2009).

## 3.5 Toxicological Assessment

As guidelines are introduced for permissible concentrations of cyanobacterial toxins in various media, there is an increasing need for their assessment. This can be carried out by various means, depending on the subsequent information required by the end user, and largely falls into three categories depending on the potential for production, toxicity assessment and quantitative assessment of cyanotoxins.

The potential for production of cyanotoxins is largely achieved through analysis of genes associated with their production. Largely they are not produced by the traditional protein synthetic machinery but are synthesized by enzymes that are transcribed and translated from the cyanobacterial genome. Polymerase chain reaction (PCR) methods, both standard and real-time (quantitative) PCR (qPCR) have been used successfully to detect the presence of these genes in cyanobacteria for microcystins (Rantala *et al.*, 2004), cylindrospermopsin (Schembri *et al.*, 2001), anatoxin-a (Legrand *et al.*, 2016) and saxitoxins (Kellman *et al.*, 2008). Although these cannot accurately quantify the amount of toxin present in the water, they do show the potential for what cyanotoxins may be produced. Furthermore, when cyanobacteria present as a mixed bloom containing cyanotoxins, the isolation of colonies for PCR or the use of nucleotide-labelled probes can indicate which species or genera of cyanobacteria may be producing cyanotoxins (Metcalf *et al.*, 2009). Consequently, once the toxigenic species or genera are known, then microscopy may aid in determining whether the concentration of cyanotoxins is likely to increase or decrease, depending on population dynamics with regular toxin verification by analytical methods.

For toxicological assessment, an appropriate organism is required to verify and assess the concentration of cyanobacterial toxins that may be present within a sample. Traditionally, this has been achieved using the mouse bioassay but, due to ethical concerns and often the need to concentrate samples to show toxicity, alternatives have been sought. Many fish and invertebrate species have been assessed for various cyanotoxins and, in the main, all show an ability to assess toxicity of cyanobacterial material (Blaha *et al.*, 2017). A further issue with toxicity testing is that extraction needs to be performed so that minimal interference from the vehicle is obtained during toxicity assessment. This may manifest as simple extraction methods or through drying down and re-suspending extracts in more suitable solvents.

If specific cyanotoxin assessment is required, then analytical methods are available. These range from simple methods such as enzyme-linked immunosorbent assays (ELISA) which are amenable to field use (Metcalf and Codd, 2017), through to enzyme assays such as the protein phosphatase inhibition assay for microcystins (Metcalf *et al.*, 2001), the rabbit reticulocyte lysate assay for cylindrospermopsin (Froscio *et al.*, 2001) or the acetylcholine esterase inhibition assay for anatoxin-a(S) (Mahmood and Carmichael, 1986), as examples. More complex methods require the use of liquid chromatography, such as high-performance liquid chromatography (HPLC) and ultra-performance liquid chromatography (UPLC) with a potential range of detectors from UV methods such as photodiode array (PDA) detection to mass spectrometric methods, including single quadrupole spectrometers (QMS) and triple quadrupole mass spectrometers (MS/MS) (Codd *et al.*, 2001). With more complex methods, the degree of specificity, selectivity and sensitivity is increased and, depending on the event being investigated, may require multiple methods to confirm the presence of toxins during, for example, human illness investigations.

## 3.6 Prevention, Treatment and Remediation

### 3.6.1 Prevention

With an increasing human population, the need for resources such as freshwater and food will result in increased demand and pressure. Anthropogenic effects combined with climate change can add further stresses and strains to environments, with the potential for a greater frequency of algal blooms (Paerl and Paul, 2012).

Management strategies and mitigation actions need to be researched and implemented to minimize these risks. Waterbodies used for recreation, drinking-water preparation and irrigation purposes, as examples, should be monitored frequently so that dynamics of cyanobacterial populations in the area can be understood and early signs of blooms detected. Appropriate warning notices and information in cases of blooms should also be provided to alert the population.

Hilborn and Beasley (2015) concluded that rapid communication between the public and appropriate authorities concerning animal illnesses or death and the potential human health effects related to harmful algal blooms is important to reduce risks. They also emphasized the importance of monitoring programmes, incident reports and protocols, and public outreach as tools to reduce risks associated with cyanobacterial exposure. Their use of 'One Health', involving multidisciplinary professionals and recognizing the interdependence of human, animal, plant, microbial and ecosystem health, is an efficient approach to managing environmental risks. According to Vidal *et al.* (2017) it is important to bring awareness to medical staff about cyanotoxin-related hepatotoxicoses for correct differential diagnosis, especially for at-risk populations in areas where cyanobacteria blooms are known to occur.

Another tool that could aid in responses to blooms and their remediation is the use of remote sensing. Wynne and Stumpf (2015) recommended the use of maps generated using satellite imagery to assist managers as they implement mitigation protocols for blooms, such as avoiding certain intakes where blooms are likely to be present. This tool can also aid in sampling strategies, as well as aiding the public in planning their recreational use of waters, with the ultimate goal of reducing the risk of exposure. Remote-sensing with process-based models to forecast algal blooms requires active research and large databases of *in situ* measurements to provide the necessary predictive power (Recknagel *et al.*, 2018), in addition to ground-truthing to validate such maps and models.

### 3.6.2 Treatment

When a cyanobacterial bloom occurs, if the waterbody is not used for any human or animal activities, then the level of sampling, analysis and risk assessment can be reduced. However, when waterbodies have active human and animal uses, then some degree of treatment may be required. With waterbodies, this may include such uses as the introduction of barley straw, which has shown some efficacy in bloom management (Barrett *et al.*, 1996), through to biological control, such as with the introduction of planktivorous fish species to consume cyanobacteria (Peczula, 2012) or through physical manipulation with methods such as aeration to destratify waterbodies (Simmons, 1998) or ultrasonication to lyse potential cyanobacterial bloom-forming species (Peczula, 2012). Although manipulation of the waterbody may have some efficacy in preventing cyanobacterial blooms, it is likely that at some point a cyanobacterial bloom with toxigenic potential or confirmed cyanotoxins may be present. Therefore, methods may be required to eliminate the cyanotoxins or prevent exposure. In the case of waterbodies that are used to allow animals to drink, then this may be as simple as restricting access until the waterbloom has passed. However, in the case of waterbodies used for the preparation of drinking water, steps may be required to monitor and remove cyanotoxins. This will largely occur during drinking-water treatment and monitoring of the final finished water, using many treatment processes that are known to remove cyanotoxins (He *et al.*, 2016).

### 3.6.3 Biological processing

With acute and chronic cyanotoxin poisoning, illness and death can result. However, depending on the health status of the person and/or animal being exposed to the cyanobacterial toxins, low-dose exposure may be manageable to some degree. Some insight has been obtained into how cyanotoxin metabolism can occur. At low doses, microcystins are able to be metabolized by the glutathione-*S*-transferase detoxication system (Pflugmacher *et al.*, 1998) with the transformed products having lower toxicity (Metcalf *et al.*, 2000). Detoxication products by the cytochrome $P_{450}$ mechanism have also been demonstrated with cylindrospermopsin (Norris *et al.*, 2002). Therefore, biological processing may reduce the risk of adverse effects to some degree. Other methods include deposition into biological tissues or fluids, with excretion through urine or faeces, and potential deposition into keratinous tissues. Cyanotoxin deposition into feathers has been shown for lesser flamingos exposed to dietary cyanotoxins (Metcalf *et al.*, 2006) and also for eider ducks that feed on *Mytilus* mussels containing nodularin from the Baltic Sea (Sipiä *et al.*, 2007b).

### 3.6.4 Remediation

The increased prevalence of cyanobacterial blooms with the potential to produce toxins may be influenced by a number of factors, including population pressures on water and climate change. However, ultimately cyanobacteria require nutrients, such as nitrogen and phosphorus, to proliferate. Efforts to manage diffuse and point-source nutrients such as fertilizers, sewage or animal wastes, for example, are required. Therefore, measures that minimize the introduction of these nutrients may prove useful. Furthermore, even when nutrients are present in a waterbody, there are activities that can be performed such as the use of a nutrient-locking compounds (for example, phoslock) or through dredging and removing sediments to prevent release of nutrients into the water column when destratification occurs and waterbodies are turned over. Ultimately, preventing nutrient use by cyanobacteria may be the most effective method to prevent cyanobacterial blooms (Metcalf and Codd, 2012).

## 3.7 Conclusions and Future Directions

Cyanobacteria are capable of producing a plethora of toxic compounds with known acute and chronic effects. Through their identification and elucidation, an understanding of their impact on human and animal health has grown. Methods for their detection, removal and processing have been developed and implemented. Although acute exposure to cyanotoxins is a well known and recognized phenomenon, further work is required to understand the links to, and effects of, chronic low-dose exposure to cyanotoxins such as the microcystins and BMAA. Ultimately, efforts to control and remove cyanotoxins are required to protect human and animal health and to preserve water supplies that are under increasing pressure.

## References

Aimi, N., Odaka, H., Sakai, S., Fujiki, H., Suganuma, M., Moore, R.E. and Patterson, G.M. (1990) Lyngbyatoxins B and C, two new irritants from *Lyngbya majuscula*. *Journal of Natural Products* 53, 1593–1596.

Al-Sammak, M.A, Hoagland, K.D., Cassada, D. and Snow, D.D. (2014) Co-occurrence of the cyanotoxins BMAA, DABA and anatoxin-*a* in Nebraska reservoirs, fish, and aquatic plants. *Toxins* 6, 488–508.

Annadotter, H., Cronberg, G., Lawton, L.A., Hansson, H.B., Gothe, U. and Skulberg, O.M. (2001) An extensive outbreak of gastroenteritis associated with the toxic cyanobacterium *Planktothrix agardhii* (Oscillatoriales, Cyanophyceae) in Scania, South Sweden. In: Chorus, I. (ed.) *Cyanotoxins – Occurrence, Causes, Consequences*. Springer, Berlin, pp. 200–208.

Azevedo, S.M.F.O., Carmichael, W.W., Jochimsen, E.M., Rinehard, K.L., Lau, S., Shaw, G.R. and Eaglesham, G.K. (2002) Human intoxication by microcystin during renal dialysis treatment in Caruaru – Brazil. *Toxicology* 181–182, 441–446.

Backer, L.C., Carmichael, W., Kirkpatrick, B., Williams, C., Irvin, M. *et al.* (2008) Recreational exposure to low concentrations of microcystins during an algal bloom in a small lake. *Marine Drugs* 6, 389–406.

Backer, L.C., McNeel, S.V., Barber, T., Kirkpatrick, B., Williams, C. *et al.* (2010) Recreational exposure to microcystins during algal blooms in two California lakes. *Toxicon* 55, 909–921.

Backer, L.C., Landsberg, J.H., Miller, M., Keel K. and Taylor, T.K. (2013). Canine cyanotoxin poisonings in the United States (1920s–2012). Review of suspected and confirmed cases from three data sources. *Toxins* 5, 1597–1628.

Banack, S.A., Caller, T., Henegan, P., Haney, J., Murby, A., Metcalf, J.S., Powell, J., Cox, P.A. and Stommel, E. (2015) Detection of cyanotoxins, β-*N*-methylamino-L-alanine and microcystins, from a lake surrounded by cases of amyotrophic lateral sclerosis. *Toxins* 7(2), 322–36.

Ballot, A., Bernard, C. and Fastner, J. (2017) Saxitoxins and analogues. In: Meriluoto, J., Spoof, L. and Codd, G.A. (eds) *Handbook of Cyanobacterial Monitoring and Cyanotoxin Analysis*. John Wiley & Sons, Chichester, UK, pp. 148–154.

Banker, R., Carmeli, S., Hadas, O., Teltsch, B., Porat, R. and Sukenik, A. (1997) Identification of cylindrospermopsin in *Aphanizomenon ovalisporum* (Cyanophyceae) isolated from Lake Kinneret, Israel. *Journal of Phycology* 33, 613–616.
Banker, R., Carmeli, S., Werman, M., Teltsch, B., Porat, R. and Sukenik, A. (2001) Uracil moiety is required for toxicity of the cyanobacterial hepatotoxin cylindrospermopsin. *Journal of Toxicology and Environmental Health, Part A* 62, 281–288.
Barrett, P.R.F., Curnow, J.C. and Littlejohn, J.W. (1996) The control of diatom and cyanobacterial blooms in reservoirs using barley straw. *Hydrobiologia* 340, 307–311.
Bautista, A.C., Moore, C.E., Lin, Y., Cline, M.G., Benitah, N. and Puschner, B. (2015) Hepatopathy following consumption of a commercially available blue-green algae dietary supplement in a dog. *BMC Veterinary Research* 11, 136.
Berry, J.P., Lee, E., Walton, K., Wilson, A.E. and Bernal-Brooks, F. (2011) Bioaccumulation of microcystins by fish associated with a persistent cyanobacterial bloom in Lago de Patzcuaro (Michoacan, Mexico). *Environmental Toxicology and Chemistry* 30, 1621–1628.
Blaha, L., Camean, A.M., Fessard, V., Gutierrez-Praena, D., Jos, A. *et al.* (2017) Bioassay use in the field of toxic cyanobacteria. In: Meriluoto, J., Spoof, L. and Codd, G.A. (eds) *Handbook of Cyanobacterial Monitoring and Cyanotoxin Analysis*. John Wiley & Sons, Chichester, UK, pp. 272–279.
Bláhová, L., Adamovský, O., Kubala, L., Švihálková Šindlerová, L., Zounková, R. and Bláha, L. (2013) The isolation and characterization of lipopolysaccharides from *Microcystis aeruginosa*, a prominent toxic water bloom forming cyanobacteria. *Toxicon* 76, 187–196.
Bordner, J., Thiessen, W.E., Bates, H.A. and Rapoport, H. (1975) Structure of a crystalline derivative of saxitoxin. Structure of saxitoxin. *Journal of the American Chemical Society* 97, 6008–6012.
Byth, S. (1980) Palm Island mystery disease. *Medical Journal of Australia* 2, 40–42.
Caller, T.A., Doolin, J.W., Haney, J.F., Murby, A.J., West, K.G. *et al.* (2009) A cluster of amyotrophic lateral sclerosis in New Hampshire: a possible role for toxic cyanobacteria blooms. *Amyotrophic Lateral Sclerosis* 10 (Suppl. 2), 101–108.
Cardellina, J.H., Marner, F.-J. and Moore, R.E. (1979) Seaweed dermatitis: structure of lyngbyatoxin A. *Science* 204, 193–195.
Carmichael, W.W. (1994) The toxins of cyanobacteria. *Scientific American* 270, 78–86.
Carmichael, W.W., Biggs, D.F. and Gorham, P.R. (1975) Toxicology and pharmacological action of *Anabaena flos-aquae* toxin. *Science* 187, 542–544.
Carmichael, W.W., Biggs, D.F. and Peterson, M.A. (1979) Pharmacology of anatoxin-a, produced by the freshwater cyanophyte *Anabaena flos-aquae* NRC-44-1. *Toxicon* 17, 229–236.
Carmichael, W.W., Azevedo, S.M.F.O., An, J.S., Molica, R.J.R., Jochimsen, E.M. *et al.* (2001) Human fatalities from cyanobacteria: chemical and biological evidence for cyanotoxins. *Environmental Health Perspectives* 109(7), 663–668.
Catherine, A., Bernard, C., Spoof, L. and Bruno, M. (2017) Microcystins and nodularins. In: Meriluoto, J., Spoof, L. and Codd, G.A. (eds) *Handbook of Cyanobacterial Monitoring and Cyanotoxin Analysis*. John Wiley & Sons, Chichester, UK, 109-126.
Chatziefthimiou, A.D., Richer, R., Rowles, H., Powell, J.T. and Metcalf, J.S. (2014) Cyanotoxins as a potential cause of dog poisonings in desert environments. *Veterinary Record* 174(19), 484–485.
Chatziefthimiou, A.D., Metcalf, J.S., Powell, J.T., Glover, W.B., Bannack, S., Cox, P. and Richer, R. (2015) One Health: the case of human exposure to cyanobacteria toxins in natural and built environments. *QScience Proceedings* 2015, Qatar.
Chatziefthimiou, A.D., Metcalf, J.S., Glover, W.B., Banack, A.A., Dargham, S.R. and Richer, R.A. (2016) Cyanobacteria and cyanotoxins are present in drinking water impoundments and groundwater wells in desert environments. *Toxicon* 114, 75–84.
Cheng, Y.S., Zhou, Y., Irvin, C.M., Kirkpatrick, B. and Backer, L.C. (2007) Characterization of aerosols containing microcystin. *Marine Drugs* 5, 136–150.
Codd, G.A., Edwards, C. and Beattie, K.A. (1992) Fatal attraction to cyanobacteria? *Nature* 359, 110–111.
Codd, G.A, Metcalf, J.S. and Beattie, K.A. (1999) Retention of *Microcystis aeruginosa* and microcystin by salad lettuce (*Lactuca sativa*) after spray irrigation with water containing cyanobacteria. *Toxicon* 37, 1181–1185.
Codd, G.A., Metcalf, J.S., Ward, C.J., Beattie, K.A., Kaya, K. and Poon, G.K. (2001) Analysis of cyanobacterial toxins by physicochemical and biochemical methods. *Journal of AOAC International* 84, 1626–1635.
Codd, G.A., Metcalf, J.S., Morrison, L.F., Krienitz, L., Ballot, A., Pflugmacher, S., Wiegand, C. and Kotut, K. (2003) Susceptibility of flamingos to cyanobacterial toxins via feeding. *Veterinary Record* 152, 722–723.

Codd, G.A., Lindsay, J., Young, F.M., Morrison, L.F. and Metcalf, J.S. (2005) Harmful cyanobacteria: from mass mortalities to management measures. Huisman, J., Matthijs, H.C.P. and Visser, P.M. (eds.) *Harmful Cyanobacteria*. Springer, Dordrecht, pp. 1–23.

Cook, W.O., Beasley, V.R., Lovell, R.A., Dahlem, A.M., Hooser, S.B., Mahmood, N.A. and Carmichael, W.W. (1989) Consistent inhibition of peripheral cholinesterases by neurotoxins from the freshwater cyanobacterium *Anabaena flos–aquae*: studies of ducks, swine, mice, and a steer. *Environmental Toxicology and Chemistry* 8(10), 915–922.

Cox, P.A., Banack, S.A., Murch, S.J., Rasmussen, U., Tien, G. *et al.* (2005) Diverse taxa of cyanobacteria produce β-*N*-methylamino-L-alanine, a neurotoxic amino acid. *Proceedings of the National Academy of Sciences of the United States of America* 102. 5074–5078.

Cox, P.A., Richer, R., Metcalf, J.S., Banack, S.A., Codd, G.A. and Bradley, W.G. (2009) Cyanobacteria and BMAA exposure from desert dust: a possible link to sporadic ALS among Gulf War veterans. *Amyotrophic Lateral Sclerosis* 10 (Suppl. 2), 109–117.

Cox, P.A., Davis, D.A., Mash, D.C., Metcalf, J.S. and Banack, S.A. (2016) Dietary exposure to an environmental toxin triggers neurofibrillary tangles and amyloid deposits in the brain. *Proceedings of the Royal Society of London, Part B* 283, 2015–2397.

Devic, E., Li, D., Dauta, A., Henriksen, P., Codd, G.A., Marty, J.-L. and Fournier, D. (2002) Detection of anatoxin-a(S) in environmental samples by using a biosensor with engineered acetylcholinesterases. *Applied and Environmental Microbiology* 68, 4102–4106.

Devlin, J.P., Edwards, O.E., Gorham, P.R., Hunter, N.R., Pike, R.K. and Stavric, B. (1977) Anatoxin-a, a toxic alkaloid from *Anabaena flos-aquae* NRC-44H. *Canadian Journal of Chemistry* 55, 1367–1371.

Dietrich, D. and Hoeger, S. (2005) Guidance values for microcystins in water and cyanobacterial supplement products (blue-green algal supplements): a reasonable or misguided approach? *Toxicology and Applied Pharmacology* 203, 273–289.

Drews, G. and Weckesser, J. (1982). Function, structure and composition of cell wall and external layers. In: Carr, N.G. and Whitton, B.A. (eds) *The Biology of Cyanobacteria*. Blackwell Scientific Publications, Oxford, UK. pp. 33–357.

Dunlop. R.A., Cox, P.A., Banack, S.A. and Rodgers, K.J. (2013) The non-protein amino acid BMAA is misincorporated into human proteins in place of L-serine causing protein misfolding and aggregation. *PLoS ONE* 8(9), e75376. doi:10.1371/journal.pone.0075376.

Edwards, C., Beattie, K.A., Scrimgeour, C.M. and Codd, G.A. (1992) Identification of anatoxin-a in benthic cyanobacteria (blue-green algae) and in associated dog poisonings at Loch Insh, Scotland. *Toxicon* 30, 1165–1175.

Fitzgeorge, R., Clark, S. and Keevil, C. (1994) Routes of intoxication. In: Codd, G.A., Jeffries, T.M., Keevil, C.W. and Potter, P. (eds) *Detection Methods for Cyanobacterial Toxins*. Royal Society of Chemistry, Cambridge, UK, pp. 69–74.

Fogg, G., Stewart, W.D.P., Fay, P. and Walsby, A.E. (1973) *The Blue-Green Algae*. Academic Press, London.

Francis, G. (1878). Poisonous Australian lake. *Nature* 18, 11–12.

Froscio, S.M., Humpage, A.R., Burcham, P.C. and Falconer, I.R. (2001) Cell free protein synthesis inhibition assay for the cyanobacterial toxin cylindrospermopsin. *Environmental Toxicology* 16, 408–412.

Fujiki, H., Mori, M., Nakayasu, M., Terada, M., Sugimura, T. and Moore, R.E. (1981) Indole alkaloids: dihydroteleocidin B, teleocidin, and lyngbyatoxin A as members of a new class of tumor promoters. *Proceedings of the National Academy of Sciences of the United States of America* 78, 3872–3876.

Fujiki, H., Mori, M., Sugimura, T. and Moore, R.E. (1983) New classes of environmental tumor promoters: indole alkaloids and polyacetates. *Environmental Health Perspectives* 50, 85–90.

Gaget, V., Humpage, A.R., Huang, Q., Monis, P. and Brookes, J.D. (2017) Benthic cyanobacteria: a source of cylindrospermopsin and microcystin in Australian drinking water reservoirs. *Water Research* 124, 454–464.

Giannuzzi, L., Sedan, D., Echenique, R. and Andrinolo, D. (2011) An acute case of intoxication with cyanobacteria and cyanotoxins in recreational water in Salto Grande Dam, Argentina. *Marine Drugs* 9, 2164–2175.

Glover, W.B., Baker, T.C., Murch, S.J., Brown, P.N. (2015) Determination of β-*N*-methylamino-L-alanine, *N*-(2-aminoethyl)glycine, and 2,4-diaminobutyric acid in food products containing cyanobacteria by ultra-performance liquid chromatography and tandem mass spectrometry: single-laboratory validation. *Journal of AOAC International* 98, 1559–1565.

Hawkins, P., Runnegar, M.C., Jackson, A.B. and Falconer, I.R. (1985) Severe hepatotoxicity caused by the tropical cyanobacterium (blue-green alga) *Cylindrospermopsis raciborskii* (Woloszynska) Seenaya and Subba Raju isolated from a domestic water supply reservoir. *Applied and Environmental Microbiology* 50(5), 1292–1295.

He, X., Liu, Y-L., Conklin, A., Westwrick, J., Weavers, L. *et al.* (2016) Toxic cyanobacteria and drinking water: impact, detection, and treatment. *Harmful Algae* 54, 174–193.
Henriksen, P., Carmichael, W.W., An, J.S. and Moestrup, O. (1997) Detection of an anatoxin-a(s)-like anticholinesterase in natural blooms and cultures of cyanobacteria/blue–green algae from Danish lakes and in the stomach contents of poisoned birds. *Toxicon* 35(6), 901–913.
Hilborn, E.D. and Beasley, V.R. (2015) One Health and cyanobacteria in freshwater systems: animal illnesses and deaths are sentinel events for human health risks. *Toxins* 7, 1374–1395.
Horner, R.D., Kamins, K.G., Feussner, J.R., Grambow, S.C., Hoff-Lundquist, J. *et al.* (2003) Occurrence of amyotrophic lateral sclerosis among Gulf War veterans. *Neurology* 61, 742–749.
Hunter, P.D., Matthews, M.W., Kutser, T. and Tyler, A.N. (2017) Remote sensing of cyanobacterial bloom in inland, coastal, and ocean waters. In: Meriluoto, J., Spoof, L. and Codd, G.A. (eds) *Handbook of Cyanobacterial Monitoring and Cyanotoxin Analysis*. John Wiley & Sons, Chichester, UK, pp. 89–99.
Jochimsen, E.M., Carmichael, W.W., An, J.S. and Cardo, D.M. (1998) Liver failure and death after exposure to microcystins at a hemodialysis center in Brazil. *New England Journal of Medicine* 338, 873–878.
Kellman, R., Michali, T.K. and Neilan, B.A. (2008) Identification of a saxitoxin biosynthesis gene with a history of frequent horizontal gene transfers. *Journal of Molecular Evolution* 67, 526–538.
Kittler, K., Schreiner, M., Krumnain, A., Manzei, S., Koch, M., Rohn, S. and Maul, R. (2012) Uptake of the cyanobacterial toxin cylindrospermopsin in Brassica vegetables. *Food Chemistry* 133: 875–879. doi: 10.1016/j.foodchem.2012.01.107.
Knoll, A.H. (2003) *Life on a Young Planet: The First Three Billion Years of Evolution on Earth*. Princeton University Press, Princeton, New Jersey.
Krienitz, L., Ballot, A., Kotut, K., Wiegand, C., Pütz, S., Metcalf, J.S., Codd, G.A. and Pflugmacher, S. (2003) Contribution of hot spring cyanobacteria to the mysterious deaths of Lesser Flamingos at Lake Bogoria, Kenya. *FEMS Microbiology Ecology* 43, 141–148.
Lantigua, J. (2017) *Tainted Waters – Threats to Public Health and the People's Right to Know*. American Civil Liberties Union (ACLU), Miami, Florida.
Legrand, B., Lesobre, J., Colombet, J., Latour, D. and Sabart, M. (2016) Molecular tools to detect anatoxin-a genes in aquatic ecosystems. Toward a new nested PCR-based method. *Harmful Algae* 58, 16–22.
Magalhaes, V.F., Azevedo, S.M.O. and Soares, R.M. (2001) Microcystin contamination in fish from the Jacarepagua Lagoon (Rio de Janeiro, Brazil): ecological implication and human health risk. *Toxicon* 39(7), 1077–1085.
Mahmood, N.A. and Carmichael, W.W. (1986) The pharmacology of anatoxin-a(s), a neurotoxin produced by the freshwater cyanobacterium *Anabaena flos-aquae*. *Toxicon* 24, 425–434.
Mahmood, N.A., Charmichael, W.W. and Pfahler, D. (1988) Anticholinesterase poisonings in dogs from cyanobacterial (blue-green algae) bloom dominated by *Anabaena flos-aquae*. *American Journal of Veterinary Research* 49(4), 500–5003.
Metcalf, J.S. and Codd, G.A. (2012) Cyanotoxins. In: Whitton, B.A. (ed.) *Ecology of Cyanobacteria II: Their Diversity in Space and Time*. Springer, Berlin, pp. 651–676. doi: 10.1007/978-94-007-3855_3224.
Metcalf, J.S. and Codd, G.A. (2017) Immunoassays and other antibody applications. In: Meriluoto, J., Spoof, L. and Codd, G.A. (eds) *Handbook of Cyanobacterial Monitoring and Cyanotoxin Analysis*. John Wiley & Sons, Chichester, UK, pp. 263–266.
Metcalf, J.S., Beattie, K.A., Pflugmacher, S. and Codd, G.A. (2000) Immuno-crossreactivity and toxicity assessment of conjugation products of the cyanobacterial toxin, microcystin-LR. *FEMS Microbiology Letters* 189, 155–158.
Metcalf, J.S., Bell, S.G. and Codd, G.A. (2001) Colorimetric immuno-protein phosphatase inhibition assay for specific detection of microcystins and nodularins of cyanobacteria. *Applied Environmental Microbiology* 67, 904–909.
Metcalf, J.S., Barakate, A. and Codd, G.A. (2004) Inhibition of plant protein synthesis by the cyanobacterial hepatotoxin cylindrospermopsin. *FEMS Microbiology Letters* 235, 125–129.
Metcalf, J.S., Morrison, L.F., Krienitz, L., Ballot, A., Krause, E. *et al.* (2006) Analysis of the cyanotoxins anatoxin-a and microcystins in Lesser Flamingo feathers. *Toxicology and Environmental Chemistry* 88, 159–167.
Metcalf, J.S., Banack, S.A., Lindsay, J., Morrison, L.F. and Cox, P.A. (2008) Co-occurrence of beta-*N*-methylamino-L-alanine, a neurotoxic amino acid with other cyanobacterial toxins in British waterbodies, 1990–2004. *Environmental Microbiology* 10, 702–708.

Metcalf, J.S., Reilly, M., Young, F.M. and Codd, G.A. (2009) Localisation of microcystin synthetase genes in colonies of the cyanobacterium *Microcystis* using fluorescence in situ hybridization. *Journal of Phycology* 45, 1400–1404.
Metcalf, J.S., Richer, R., Cox, P.A. and Codd, G.A. (2012) Cyanotoxins in desert environments may present a risk to human health. *Science of the Total Environment* 421–422, 118–123.
Metcalf, J.S., Lobner, D., Banack, S.A., Cox, G., Nunn, P.B., Wyatt, P.B. and Cox, P.A. (2017) Analysis of BMAA enantiomers in cycads, cyanobacteria, and mammals: *in vivo* formation and toxicity of D-BMAA. *Amino Acids* 49, 1427–1439.
Mez, K., Beattie, K., Codd, G., Hanselmann, K., Hauser, B., Naegeli, H. and Preisig, H. (1997) Identification of a microcystin in benthic cyanobacteria linked to cattle deaths on alpine pastures in Switzerland. *European Journal of Phycology* 32(2), 111–117. doi: 10.1080/09670269710001737029
Mhlanga, L., Day, J., Cronberg, G., Chimbari, M., Siziba, N. and Annadotter, H. (2006) Cyanobacteria and cyanotoxins in the source water from Lake Chivero, Harare, Zimbabwe, and the presence of cyanotoxins in drinking water. *African Journal of Aquatic Science* 31(2), 165–173.
Molica, R.J. R., Oliveira, E.J.A., Carvalho, P.V.C., Costa, A.N.S.F., Cunha, C.C., Melo, G. and Azevedo, S.M. (2005) Occurrence of saxitoxins and an anatoxin-a(s)-like anticholinesterase in a Brazilian drinking water supply. *Harmful Algae* 4, 743–753.
Monteiro, S., Santos, R., Blaha, L. and Codd, G.A. (2017) Lipopolysaccharide endotoxins. In: Meriluoto, J., Spoof, L. and Codd, G.A. (eds) *Handbook of Cyanobacterial Monitoring and Cyanotoxin Analysis*. John Wiley & Sons, Chichester, UK, pp. 109–126.
Murch, S.J., Cox, P.A. and Banack, S.A. (2004) A mechanism for slow release of biomagnified cyanobacterial neurotoxin and neurodegenerative disease in Guam. *Proceedings of the National Academy of Sciences of the United States of America* 101, 12228–12231.
Mynderse, J.S., Moore, R.E., Kashiwagi, M. and Norton T.R. (1977) Antileukemia activity in the Oscillatoriaceae: isolation of debromoaplysiatoxin from *Lyngbya*. *Science* 196, 538–540.
Nishiwaki-Matsushima, R., Ohta, T., Nishiwaki, S., Suganuma, M., Kohyama, K. *et al.* (1992) Liver tumor protomotion by the cyanobacterial cyclic peptide toxin microcystin-LR. *Journal of Cancer Research and Clinical Oncology* 118, 420–424.
Norris, R.L., Eagelsham, G.K., Pierans, G., Shaw, G.R., Smith, M.J. *et al.* (1999) Deoxycylindrospermopsin, an analog of cylindrospermopsin from *Cylindrospermopsis raciborskii*. *Environmental Toxicology* 14, 163–165.
Norris, R.L., Seawright, A.A., Shaw, G.R., Senogles, P., Eaglesham, G.K. *et al.* (2002) Hepatic xenobiotic metabolism of cylindrospermopsin in vivo in the mouse. *Toxicon* 40, 471–476.
Ohta, T., Sueoka, E., Lida, N., Komori, A., Suganuma, M. *et al.* (1994) Nodularin, a potent inhibitor of protein phosphatases 1 and 2A, is a new environmental carcinogen in male F344 rat liver. *Cancer Research* 54, 6402–6406.
Osborne, N.J. and Shaw, G.R. (2008) Dermatitis associated with exposure to a marine cyanobacterium during recreational water exposure. *BMC Dermatology* 8, 5.
Paerl, H.W. and Huisman, J. (2008) Blooms like it hot. *Science* 320, 57–58.
Paerl, H.W. and Huisman, J. (2009) Climate change: a catalyst for global expansion of harmful cyanobacterial blooms. *Environmental Microbiology Reports* 1(1), 27–37.
Paerl, H.W. and Paul, V.J. (2012) Climate change: links to global expansion of harmful cyanobacteria. *Water Research* 46, 1349–1363.
Peczula, W. (2012) Methods applied in cyanobacterial bloom control in shallow lakes and reservoirs. *Ecological Chemistry and Engineering A* 19, 795–806.
Pelley, J., (2016) Taming toxic algae blooms. *American Chemical Society Central Science* 2(5), 270–273.
Pflugmacher, S., Wiegand, C., Oberemm, A., Beattie, K.A., Krause, E., Codd, G.A. and Steinberg, C.E.W. (1998) Identification of an enzymatically formed glutathione conjugate of the cyanobacterial hepatotoxin microcystin-LR: the first step of detoxication. *Biochimica et Biophysica Acta* 1425, 527–533.
Pilotto, L., Douglas, R., Burch, M., Cameron, S., Beers, M. *et al.* (1997) Health effects of exposure to cyanobacteria (blue-green algae) during recreational water-related activities. *Australia and New Zealand Journal of Public Health* 21, 562–566.
Pouria, S., de Andrade, A., Barbosa, J., Cavalcanti, R.L., Barreto, V.T.S. *et al.* (1998) Fatal microcystin intoxication in haemodialysis unit in Caruaru, Brazil. *The Lancet* 352 (9121), 21–26.
Puschner, B., Bautista, A. and Wong, C. (2017) Debromoaplysiatoxin as the causative agent of dermatitis in a dog after exposure to freshwater in California. *Frontiers in Veterinary Science* 4, 50.

Pybus, M.J., Hobron, D.P. and Onderka, D.K. (1986) Mass mortality of bats due to probable blue-green algal toxicity. *Journal of Wildlife Disease* 22(3), 449–450.

Rantala, A., Fewer, D.P., Hisbergues, M., Rouhiaianen, L., Vaitomaa, J., Börner, T. and Sivonen, K. (2004) Phylogenetic evidence for the early evolution of microcystin synthesis. *Proceedings of the National Academy of Sciences of the United States of America* 101, 568–573.

Recknagel, F., Orr, P., Swanepoel, A., Joehnk, K. and Anstee, J. (2018) Operational forecasting in ecology by inferential models and remote sensing. In: Recknagel, F. and Michener, W.K. (eds) *Ecological Informatics. Data Management and Knowledge Discovery*, 3rd edn. Springer International Publishing, Cham, Switzerland, pp. 319–339. doi: 10.1007/978-3-319-59928-1_15.

Richer, R., Anchassi, D., El-Assaad, I., El-Matbouly,M., Ali, F., Makki, I. and Metcalf, J.S. (2012) Variation in the coverage of biological soil crusts in the State of Qatar. *Journal of Arid Environments* 78, 187–190.

Rodger, H.D., Turnbull, T., Edwards, C. and Codd, G.A. (1994) Cyanobacterial (blue-green algal) bloom associated pathology in brown trout, *Salmo trutta* L., in Loch Leven, Scotland. *Journal of Fish Diseases* 17, 177–181.

Roy-Lachapelle, A., Solliec, M., Bouchard, M.F. and Sauvé, S. (2017) Detection of cyanotoxins in algae dietary supplements. *Toxins* 9, 76.

Saker, M.L., Thomas, A.D. and Norton, J.H. (1999) Cattle mortality attributed to the toxic cyanobacterium *Cylindrospermopsis raciborskii* in an outback region of North Queensland. *Environmental Toxicology*,14(1), 179–182.

Saqrane, S. and Oudra, B. (2009) CyanoHAB occurrence and water irrigation cyanotoxin contamination: ecological impacts and potential health risks. *Toxins* 1, 113–122.

Schantz, E.J., Mold, J., Stanger, D., Shavel, J., Riel, F. *et al.* (1957) Paralytic shellfish poison VI. A procedure for the isolation and purification of the poison from toxic clams and mussel tissues. *Journal of the American Chemical Society* 79, 5230–5235.

Schembri, M.A., Neilan, B.A. and Saint, C.P. (2001) Identification of genes implicated in toxin production in the cyanobacterium *Cylindrospermopsis raciborskii*. *Environmental Toxicology* 16, 413–421.

Schopf, J.W. (2000) The fossil record: tracing the roots of the cyanobacterial lineage. In: Whitton, B.A. and Potts, M. (eds) *The Ecology of Cyanobacteria*. Kluwer Academic Publishers, Dordrecht, pp. 13–35.

Seawright, A.A., Nolan, C.C., Shaw, G.R., Chiswell, R.K., Norris, R.L., Moore, M.R. and Smith, M.J. (1999) The oral toxicity for mice of the tropical cyanobacterium *Cylindrospermopsis raciborskii* (Woloszynska). *Environmental Toxicology* 14, 135–142.

Simmons, J. (1998) Algal control and destratification at Hanningfield Reservoir. *Water Science & Technology* 37, 309–316.

Sipiä, V., Kankaanpää, H., Peltonen, H., Vinni, M. and Meriluoto, J. (2007a) Transfer of nodularin to three-spined stickleback (*Gasterosteus aculeatus* L.), herring (*Clupea harengus* L.), and salmon (*Salmo salar* L.) in the northern Baltic Sea. *Ecotoxicology and Environmental Safety* 66, 421–425. doi: 10.1016/j.ecoenv.2006.02.006.

Sipiä, V., Neffling, M.-L., Metcalf, J.S., Nybom, S.M.K., Meriluoto, J. and Codd, G.A. (2007b) Nodularin in feathers and livers of eiders (*Somateria mollissima*) caught from the western Gulf of Finland in June–September 2005. *Harmful Algae* 7, 99–105.

Sivonen, K. and Jones, G. (1999) Cyanobacterial toxins. In: Chorus, I. and Bartram, J. (eds) *Toxic Cyanobacteria in Water: A Guide to Public Health Significance, Monitoring and Management*. E&FN Spon, London, pp. 41–111.

Spoof, L. and Catherine, A. (2017) Appendix 3: Tables of microcystins and nodularins. In: Meriluoto, J., Spoof, L. and Codd, G.A. (eds) *Handbook of Cyanobacterial Monitoring and Cyanotoxin Analysis*. John Wiley & Sons, Chichester, UK, pp. 526–537.

Teixeira, M.G.L.C., Costa, M.C.N., Carvalho, V.L.P., Peireira, M.S. and Hagner, E. (1993) Gastroenteritis epidemic in the area of the Itaparica Dam, Bahia, Brazil. *Bulletin of the Pan American Health Organization* 27(3).

Terao, K., Ohmori, S., Igarashi, K., Ohtani, I., Watanabe, M.F. *et al.* (1994) Electron microscopic studies on experimental poisoning in mice induced by cylindrospermopsin isolated from blue-green alga *Umezakia natans*. *Toxicon* 32, 833–843.

Testai, E., Buratti, F.M., Funari, E., Manganelli, M., Vichi, S. *et al.* (2016) *Review and Analysis of Occurrence, Exposure and Toxicity of Cyanobacteria Toxins in Food*. EFSA supporting publication 2016: EN-998, 309 pp. European Food Safety Authority, Parma, Italy.

Turner, P.C., Gammie, A.J., Hollinrake, K. and Codd, G.A. (1990) Pneumonia associated with contact with cyanobacteria. *British Medical Journal* 300, 1440–1441.

Vega, A. and Bell, E.A. (1967). α-Amino-β-methylaminoproprionic acid, a new amino acid from seeds of *Cycas circinalis*. *Phytochemistry* 6, 759–762.

Vidal, F., Sedan, D., D'Agostino, D., Cavalieri, M.L., Mullen, E. *et al.* (2017). Recreational exposure during algal bloom in Carrasco beach, Uruguay: a liver failure case report. *Toxins* 9, 267. doi: 10.3390/toxins9090267.

Wimmer, K.M., Strangman, W.K. and Wright, J.L.C. (2014) 7-Deoxy-desulfo-cylindrospermopsin and 7-deoxy-desulfo-12-acetylcylindrospermopsin: two new cylindrospermopsin isolated from a Thai strain of *Cylindrospermopsis raciborskii*. *Harmful Algae* 37, 203–206.

Wood, S.A., Selwood, A.I., Rueckert, A., Holland, P.T., Milne, J.R. *et al.* (2007) First report of homoanatoxin-a and associated dog neurotoxicosis in New Zealand. *Toxicon* 50, 292–301.

Wynne, T.T. and Stumpf, R.P. (2015) Spatial and temporal patterns in the seasonal distribution of toxic cyanobacteria in western Lake Erie from 2002–2014. *Toxins* 7, 1649–1663.

# 4 Amino Acids and Peptides as Mediators of Abiotic Stress Tolerance in Higher Plants

**J.P.F. D'Mello***

*Formerly of SAC, University of Edinburgh King's Buildings Campus, West Mains Road, Edinburgh, UK*

## 4.1 Abstract

The obligatory demand for amino acids in the primary metabolism of plants is no longer a matter of debate but, nevertheless, continues to be the focus of current research for a variety of strategic reasons. In particular, considerable attention is currently being devoted to the role of amino acids in abiotic stress responses in plants, affecting signal transduction and crosstalk with biotic pathways. Common stressors under investigation include salinity, ambient temperature, drought, acid rain, anoxia and heavy-metal contamination. Certain amino acids operate directly as signalling molecules or perform defence and adaptive functions in response to internal and external stimuli. Glutamate is the prime example with transduction and stress-related functions in higher plants. Of equal significance is the indirect role of amino acids as precursors, activators or structural components of key plant hormones. For example, arginine is the source of nitric oxide, methionine is the precursor of ethylene, while isoleucine is an activator of jasmonate. Among the aromatic amino acids, phenylalanine is, at least in part, the source of salicylic acid and tryptophan is the precursor of the auxin, indole-3-acetic acid. Interactions with other networks involving abscisic acid and the role of leucine zipper proteins, particularly under stress conditions, add to the complexity of signal transduction and stress responses in plants. A diverse array of defence molecules are also synthesized during the course of secondary metabolism of amino acids. While a number of these compounds represent the constitutive framework for plant protection, others are induced in response to specific biotic or environmental challenges involving up-regulation of metabolic pathways. Thus, the catecholamine biosynthesis route has been implicated in plant responses to stress and similar comments apply to the secondary metabolism of phenylalanine and tryptophan. The intersection between primary and secondary metabolism provides the potential for regulation at the gene and enzyme level and the opportunity for metabolic intervention. In a number of instances, amino acids serve as precursors of both signal transduction molecules and secondary compounds. Thus methionine is the precursor of ethylene and aliphatic glucosinolates, while tryptophan metabolism results in the synthesis of indole acetic acid, camalexin and indole glucosinolates. In addition, phenylalanine is the substrate for salicylic acid as well as flavonoids, coumarins and

* E-mail address: jpfdmello@hotmail.co.uk

 

stilbenes. The effects of environmental stressors such as salinity, anoxia, drought and exposure to heavy metals on specific aspects of amino acid metabolism in plants should be considered against this background.

Review of the adverse effects of abiotic stress is relevant in this volume due to distinct manifestations of toxicity in higher plants. These effects include: excess ammonia ($NH_3$) production as a consequence of enhanced endogenous proteolysis; intrinsic toxicity of certain amino acids, for example, proline; generation of reactive oxygen species (ROS) and oxidative stress; and exposure to heavy metals. A number of mechanisms exist in plants to avoid or alleviate potential toxicity. Thus, activation and attenuation are key components of homeostatic mechanisms in plants, as in other living organisms. Under conditions such as salinity, increased proteolytic activity creates increased intracellular hyper-ammonia and potential metabolic toxicity if homeostasis is not restored. Ammonium ions are therefore incorporated into glutamine and glutamate by glutamine synthetase/glutamate synthase, thus directing the pathway towards proline synthesis. It is well known that proline accumulates in higher plants in response to a variety of environmental factors such as wilting, drought, salinity, temperature, anoxia, exposure to heavy metals and oxidative stress. Concentrations of $\gamma$-amino butyric acid, alanine, polyamines and glycine betaine also increase in response to abiotic stress.

On exposure of plants to toxic heavy metals, the direction of metabolism of glutathione is altered in favour of the biosynthesis of phytochelatins. These peptides are endowed with properties of chelation, transport and vacuolar sequestration of toxic metals within plant cells. However, the lack of stoichiometric relationships between phytochelatin expression and, for example, arsenic (As), zinc (Zn) and lead (Pb) concentrations and the possible chelation of valuable nutrients such as copper (Cu) and sulfur (S) illustrate the complexity of metal detoxification and establishment of homeostasis in plants.

The pathways involved in plant responses to abiotic stress are highly complex, entailing initial perception, signal transduction and defence gene expression, culminating in the establishment of tolerance or programmed cell death. The potential significance of crosstalk between biotic and abiotic defence pathways is only just unfolding but needs to be incorporated into future models. In addition, signal transduction, whether effected directly or via associated metabolites of amino acids, represents another significant example transcending traditional concepts of plant biochemistry. Amino acid accumulation and the phytochelatin-induced transport and sequestration of potentially toxic metals will depend on the relative activities of signalling compounds such as ethylene (ET), salicyclic acid (SA), auxins, jasmonate (JA) and abscisic acid (ABA), but research is still at a preliminary stage and any conclusions would be premature.

It is readily admitted that other aspects of abiotic stress responses need to be addressed in order to obtain a more comprehensive perspective of metabolism in higher plants. For example, the contribution of proteomics in the understanding of stress responses should be considered and incorporated within a general model of plant toxicology.

Despite significant developments in recent decades, immense challenges lie ahead, not least at the fundamental level in discerning the genes encoding enzymes of the major pathways of metabolism, particularly insofar as they impact on metabolic regulation and abiotic stress responses of plants. Future advances will inevitably depend on the application of molecular techniques and the development of transgenic lines and mutants, particularly in elucidating regulatory mechanisms. Concurrently, the complexities and significance of additive, synergistic and antagonistic interactions of signal transduction in plant immunity require elucidation in the context of environmental pressures. These aspects will be of crucial value if we are to breed lines more tolerant to abiotic stresses. This area is now set to provide us with valuable data for a number of pressing questions on the relationship between plant metabolism and acclimation to abiotic stress.

On the basis of current evidence, it is concluded that abiotic stress in plants is profoundly associated with aspects of cellular toxicology. Manifestations are reflected in alterations in the metabolism of amino acids and certain peptides. The mechanisms underlying these changes remain largely speculative, but may involve amino acid-derived and other signalling compounds in complex crosstalk between abiotic and biotic pathways.

## 4.2 Introduction

The metabolic versatility of plants is such that all 21 canonical amino acids are synthesized *de novo* from basic substrates linked to the glycolytic and tricarboxylic acid (TCA) pathways (Fig. 4.1). Simpler amino acids may also be used, or provide active groups, for the biosynthesis of more complex amino acids, as for example in the formation of histidine, which requires inputs of both glutamine and glutamate in separate stages of the pathway (Ingle, 2015). Methionine acts as a donor of sulfur and methyl groups for the synthesis of a diverse range of compounds. In addition, the interconversion between glycine and serine is well recognized as an important feature of amino acid metabolism. Glycine is a major substrate available for light-induced oxidation in plant leaf mitochondria (Oliver *et al.*, 1990).

In steady-state and minimal-stress circumstances, amino acids are directed towards primary metabolism resulting in the synthesis of enzymes, peptides and storage proteins. Protein synthesis is under genetic control, only allowing the 21 canonical amino acids to participate in a

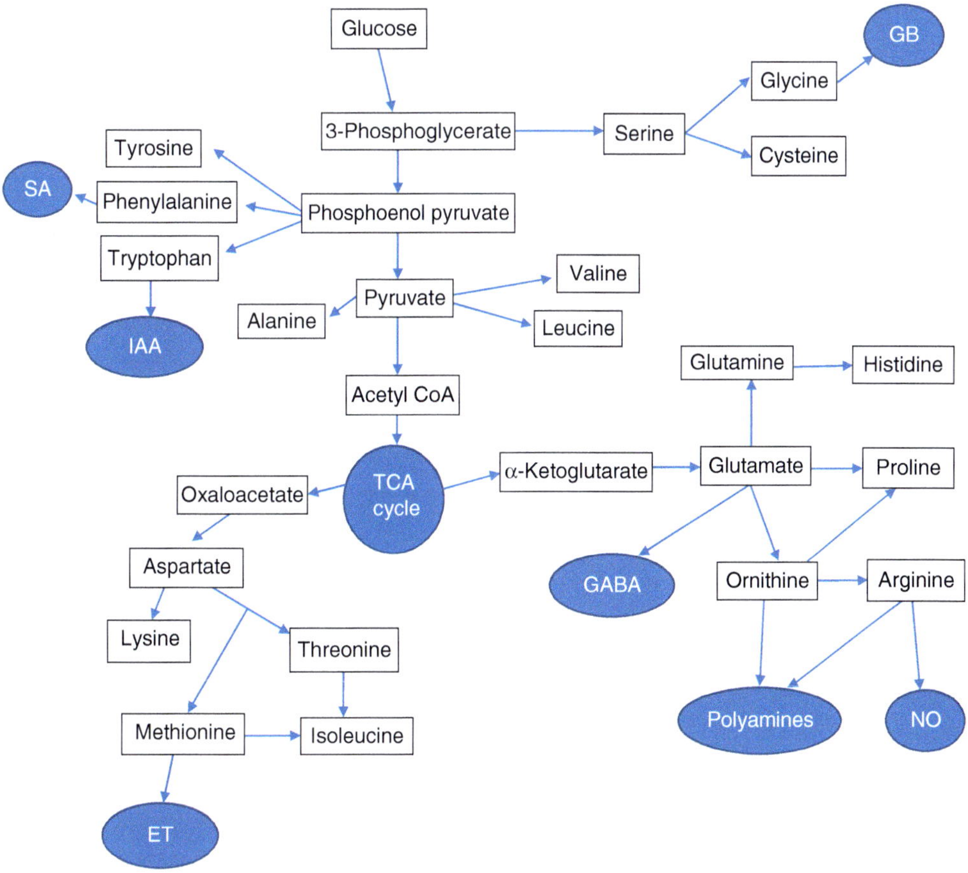


**Fig. 4.1.** Simplified scheme depicting the biosynthesis and utilization of amino acids in higher plants. The central role of the tricarboxylic acid (TCA) cycle in linking carbon and amino acid metabolism is emphasized. The sources of glycine betaine (GB) and signalling compounds salicylic acid (SA), indole-3-acetic acid (IAA), γ-aminobutyric acid (GABA), nitric oxide (NO) and ethylene (ET) are shown. IAA normally exists in free form, but amide conjugates with alanine, leucine, phenylalanine, aspartate, glutamate and tryptophan also occur naturally in plants. In addition, isoleucine activates jasmonate and, separately, leucine-rich receptor kinases and leucine zipper transcription factors are implicated in signal transduction.

diverse array of anabolic pathways. However, the secondary metabolism of amino acids is equally significant, since it provides for the synthesis of signalling and defence compounds (including phytoalexins) which plants require in order to react to pathogenic and abiotic pressures. Certain plant species also contain non-protein amino acids that are endowed with toxic and anti-predatory properties of potential use in medicine and plant protection. Of these, mimosine occurring in the tropical forage legume *Leucaena leucocephala* (Acamovic and D'Mello, 1981; Acamovic *et al.*, 1982) is arguably the best example of a candidate pharmaceutical, bio-pesticide and putative metal detoxification amino acid.

It is now generally acknowledged that amino acid metabolism in higher plants is profoundly affected by abiotic stress, but may also provide potential mechanisms for restoration of homeostasis and amelioration of adverse effects. The impact of environmental stressors such as salinity, anoxia, drought, acid rain and exposure to heavy metals on specific aspects of amino acid metabolism and peptide functions in plants is currently under active investigation. The updated evidence is reviewed in this chapter for each respective amino acid and selected peptides. In addition, it is instructive to summarize the major routes available for the synthesis and utilization of amino acids in plants (Fig. 4.1).

## 4.3 Pathways of Amino Acid Metabolism

In general, amino acid metabolism is driven by a diverse range of enzymes belonging to major classes including aminotransferases, dehydrogenases, decarboxylases, synthetases and racemases (D'Mello, 2012). The universal features of amino acid metabolism have long been recognized, particularly with respect to the reactions catalysed by glutamate dehydrogenase (GDH), aminotransferases and enzymes associated with polyamine biosynthesis. However, there is now an emerging consensus of a much greater degree of metabolic compatibility, specifically between plants and animals, although significant differences are also apparent. In higher plants, as in animals, a number of enzymes are subject to or respond to stress stimuli, with the requirement to generate defence reactions and products, including a wide range of signalling compounds and antioxidants. For example, the tripeptide glutathione (GSH) acts in a protective role in both plants and animals (Wu *et al.*, 2004).

The secondary metabolism of plants yields a diverse spectrum of active groups, as well as signalling and other compounds, with specific amino acids acting as major precursors (Fig. 4.1). For example, methionine metabolism results in the biosynthesis of *S*-adenosylmethionine (SAM; also known as AdoMet). It is established that SAM is essential as a methyl donor and as an activator of threonine synthase. In addition, SAM contributes to the biosynthesis of polyamines and ethylene (ET), a significant signal transduction molecule in plants (Ravanel *et al.*, 1998). Current investigations focus on the synthesis of the intermediate 1-amino-cyclopropane-1-carboxylic acid (ACC) and the associated enzyme ACC synthase (Huang *et al.*, 2013). This enzyme is encoded by a number of genes, some of which are inducible by hypoxia, while others are involved in *Bortytis cinerea*-induced ET production in *Arabidopsis*. Although ET was initially associated with fruit ripening, it is now also implicated in seed germination, root development, abscission and flower senescence (Wang *et al.*, 2002). However, interactions with other hormones, particularly under abiotic stress conditions, ensure that ET will continue to be the focus of considerable attention (Kazan, 2015; Lestari *et al.*, 2018).

The secondary metabolism of phenylalanine and tryptophan in higher plants results in the formation of two significant signalling compounds, namely salicylic acid (SA) and indole-3-acetic acid (IAA). Opinion is divided over the mechanisms of SA biosynthesis in higher plants. Initial observations indicated that the production of SA precursors was a major function of phenylalanine ammonia-lyase (PAL) in the resistance of *Arabidopsis* to *Peronospora parasitica* (Mauch-Mani and Slusarenko, 1996). However, others claim that utilization of phenylalanine or chorismate as precursors for SA synthesis depends upon plant species, developmental stage and growth conditions (Tounekti *et al.*, 2013). It is widely accepted that SA is a critical signal transduction molecule for the induction of immune responses in plants towards biotic stress and particularly in the establishment of localized and systemic acquired resistance (SAR) (He *et al.*, 2002; Moreau *et al.*, 2012). In addition, SA has been linked with plant responses to abiotic stress by modulating terpenoid and flavonoid metabolism (Tounekti *et al.*, 2013). Khan *et al.* (2015) implicated SA in tolerance to salinity, ozone, heat and drought. Another aspect relates to the occurrence and activity of the conjugate SA-aspartate (SA-Asp), which functions as a mobile molecule, possibly in the induction of pathogenesis-related gene expression (Chen *et al.*, 2013). It is thus clear that SA exerts its defence functions in two ways: through the induction of SAR (Maier *et al.*, 2011; Moreau *et al.*, 2012) and by stimulating the synthesis of key secondary metabolites such as terpenoids and flavonoids (Tounekti *et al.*, 2013).

The chorismate–anthranilate pathway is a well-recognized route for the synthesis of tryptophan, occurring exclusively in plants (Zhao and Last, 1996) and microorganisms. Tryptophan is a rich source of physiologically active metabolites in both plants and animals. The synthesis of IAA, serotonin, melatonin and indole glucosinolates illustrates the diversity of pathways of tryptophan metabolism continuing to attract attention (see D'Mello, 2015a). For example, the question of the role of IAA conjugates has recently emerged as a factor in its regulation and homeostasis (Ludwig-Muller, 2011). IAA normally exists in free form, but its amide conjugates with alanine, leucine, phenylalanine, aspartate, glutamate and tryptophan also occur naturally in plants.

It is also now clear that the induction of SA and defence molecules such as indole alkaloids, flavanones, isoflavanones and phytoalexins is accompanied by up-regulation of associated biosynthetic enzymes. Huang *et al.* (2010) examined the role of the *Arabidopsis PAL* gene family on the response to abiotic stress, while Ishihara *et al.* (2008) implicated the tryptophan pathway in defence responses to pathogens via serotonin production.

Other features of amino acid utilization in plants relate to the synthesis of the signalling molecules γ-aminobutyric acid (GABA) from glutamate, nitric oxide (NO) from arginine and polyamines from ornithine (Fig. 4.1).

## 4.4 Specific Amino Acids Associated with Abiotic Stress Responses of Higher Plants

Abiotic factors may impact on amino acid metabolism in plants via effects on enzymes and signalling compounds. It is instructive to consider effects for the major participants in abiotic stress responses. In several instances individual amino acids accumulate in response to environmental stress factors. However, it is not clear whether these accumulations are manifestations of stress symptoms as a result of reduced growth or an acclimation response. Furthermore, the role of amino acid-derived signalling compounds in the instigation of plant responses to abiotic stress remains largely unexplored. In contrast, there is substantial evidence of the involvement of both signalling compounds and phytoalexins derived from amino acids in the response of higher plants to fungal pathogens and insect herbivory.

### 4.4.1 Glutamate and γ-aminobutyric acid (GABA)

All living organisms synthesize and utilize glutamate in diverse metabolic reactions. Although glutamate is a major constituent of proteins, its ubiquitous distribution as an unconjugated molecule and its signal transduction roles place it in a unique position in the biochemistry of amino acids. Indeed, glutamate has been referred to as an amino acid of 'particular distinction' (Young and Ajami, 2000). Glutamate dehydrogenase is a key enzyme in amino acid metabolism of higher plants and its role in abiotic stress responses is now established. According to Skopelitis *et al.* (2006), abiotic stress generates ROS, which signals the expression of GDH. Under conditions such as salinity, enhanced proteolytic activity creates intracellular hyper-ammonia and potential metabolic toxicity if homeostasis is not restored. Ammonium ions are therefore incorporated into glutamine and glutamate by glutamine synthetase/glutamate synthase (GS/GOGAT), thus directing the pathway towards proline biosynthesis.

Glutamate is considered to be the major precursor of GABA via the action of glutamate decarboxylase (GAD), though polyamines and particularly putrescine may also contribute to GABA synthesis. It is hypothesized that the catabolism of putrescine to GABA is regulated by a combination of gene-dependent and gene-independent mechanisms (Shelp *et al.*, 2012). GABA has been associated with stress responses with rapid accumulation to high levels following exposure of plants to conditions such as anoxia, low pH, drought, salinity, heat shock, low temperature and osmotic and mechanical manipulation (Planchet and Limami, 2015). The stress-induced accumulation of GABA varies with developmental age and organ, as two distinct GAD isoforms have been characterized and expressed in different tissues. Under stress conditions, GABA accumulation is accompanied by increased calcium (Ca) levels and a decrease in cytosolic pH and is also dependent upon activation of enzymes involved in GABA metabolism. The impairment of GABA transaminase, for example, leads to reduced salt tolerance in a loss-of-function mutant of *Arabidopsis*, indicating the importance of GABA degradation in stress responses. A functional GABA shunt may be important in stress responses, but the mechanism of action of the amino acid still remains obscure. There is speculation that GABA may function in glutamate homeostasis, nitrate uptake, osmoregulation, pH regulation, antioxidant scavenging, TCA cycle bypassing and cell signalling (Planchet and Limami, 2015).

### 4.4.2 Proline

As a derivative of glutamate, proline is ubiquitous in living organisms and is now emerging as a key multifunctional amino acid in primary metabolism and in the response of plants to environmental stress (Sharma and Dietz, 2006). Although GDH is the pre-eminent exponent of its class, it should be recognized that other dehydrogenase enzymes participate actively in N transactions and stress responses of plants. For example, two proline dehydrogenase isoforms have been identified in *Arabidopsis* differing in spatial, developmental and salinity stress regulation of expression (Funck *et al.*, 2010). It is well known that proline accumulates in higher plants in response to a variety of environmental factors such as wilting, drought, salinity, temperature, anoxia, exposure to heavy metals and oxidative stress (Planchet and Limami, 2015). This accumulation varies with plant species and has been linked with improved plant performance under diverse abiotic stress conditions.

Proline is associated with specific attributes, including regulation of cellular homeostasis, as a signal transduction molecule modulating mitochondrial function and in gene expression, particularly under adverse conditions (Szabados and Savoure, 2010). In addition, Lehmann *et al.* (2010) emphasized the physical properties of proline arising from its cyclic structure, which restricts conformational flexibility as well as its activity as a compatible solute in the protection of cellular components. In its physical role, proline protects the cellular structure of proteins and membranes during dehydration. It is believed that, in its functions as an osmoprotectant, proline accumulates in the cytosol and chloroplasts, as opposed to other solutes such as organic acids and sugars which are concentrated in the vacuole. Proline may also act as a regulator of cellular redox potential via mechanisms for counteracting cytosolic acidosis and maintenance of nicotinamide adenine dinucleotide phosphate ($NADP^+$/NADPH) balance. In addition, proline is associated with the property to scavenge free radicals (ROS), thereby affording cells with the mechanism to avoid oxidative stress (Planchet and Limami, 2015). Proline may also protect the activity of enzymes such as nitrate reductase and glucose-6-phosphate dehydrogenase (G6PD) against toxic heavy metals through the formation of complexes.

The signalling role of proline is an emerging issue, as is the case with a number of other amino acids and associated metabolites (Planchet and Limami, 2015). Consistent with this theme is the observation that proline induces the expression of certain saline-stress response genes to enhance salt tolerance via up-regulation of stress-protective proteins, reminiscent of the generation of pathogenesis-related proteins in fungal diseases of higher plants.

Despite the forgoing evidence, it is salutary to note that proline accumulation following heat stress results in increased sensitivity of affected plants rather than tolerance. Significantly, drought-induced accumulation of proline is abolished during a combination of drought and heat stresses (Planchet and Limami, 2015).

It has also been reported that proline accumulation in barley was not linked with salt tolerance, illustrating the complexity underlying proline-stress relationships in higher plants.

According to the comprehensive review of Planchet and Limami (2015), the glutamate pathway is the primary route for proline synthesis and accumulation in plants exposed to osmotic stress. The reactions involved include: (i) the reduction of glutamate to glutamate-semialdehyde (GSA), catalysed by the bifunctional enzyme $\Delta^1$-pyrroline-5-carboxylate synthetase (P5CS); (ii) the spontaneous conversion of GSA to pyrroline-5-carboxylate (P5C); and (iii) the reduction of P5C to proline by the action of P5C reductase (P5CR). The reduction of glutamate may be differentially regulated, depending upon plant organ and imposition of osmotic stress. The aforementioned reactions occur in the cytosol or chloroplasts, whereas proline catabolism is undertaken by the mitochondria. Proline degradation represents essentially the reversal of the synthetic pathway except for the involvement of two dehydrogenase reactions. Proline accumulation during abiotic stress is the result of a trade-off between biosynthetic and catabolic reactions. It is envisaged that, under abiotic stress conditions, proline biosynthesis is up-regulated whereas its catabolism is down-regulated. Under these conditions, proline concentrations are correlated with the expression of the *P5CS* gene, but not the *P5CR* gene, and with enhanced stress tolerance. In transgenic plants, overexpression of the *P5CS* gene results in increased proline biosynthesis and improved tolerance to high salinity, freezing and osmotic stress. Consistent with this theme is the observation that, in *Arabidopsis P5CS1* knockout plants, proline biosynthesis was reduced with increased sensitivity to salinity. Proline accumulation may also be the result of reduced utilization for protein synthesis and a decrease in its catabolism to glutamate. In line with this supposition, it has been demonstrated that suppression of proline catabolism in *Arabidopsis* plants results in accumulation of proline and improved tolerance to freezing and salinity stress. It is emphasized that proline degradation is enhanced during recovery from stress to provide reducing equivalents for mitochondrial oxidative phosphorylation in the generation of ATP to repair damage caused by adverse environmental factors.

The ornithine/arginine pathway represents an alternative route for proline biosynthesis in higher plants (Planchet and Limami, 2015). In mitochondria, ornithine, a common intermediate in both biosynthesis and degradation of proline, can be converted to P5C via the action of ornithine-δ-aminotransferase (OAT). However, the involvement of OAT in stress-induced accumulation of proline is unclear. While transgenic *Arabidopsis* plants overexpressing the *OAT* gene in tobacco and rice accumulated a higher level of proline and had improved stress tolerance, *Arabidopsis* mutants deficient in this gene accumulated proline at a similar level and exhibited tolerance to salinity stress. On the basis of these observations, it was suggested that the biosynthesis of proline under abiotic stress conditions occurs directly from glutamate, independently of the ornithine/arginine pathway.

It has been suggested that efforts to manipulate proline metabolism might assist in unlocking further potential to enhance abiotic stress tolerance in plants. However, even with proline, metabolic conditions can be devised to demonstrate toxicity (Nanjo *et al.*, 2003). Indeed, a nuclear gene encoding mitochondrial $\Delta^1$–pyrroline-5-carboxylate dehydrogenase and its potential role in protection of plants from proline toxicity has already been intimated (Deuschle *et al.*, 2001).

### 4.4.3 Arginine and ornithine: Functional metabolism

As in other organisms, arginine is metabolized by arginase (arginine amidohydrolase, ARGAH) to ornithine and urea. Arginine is also the substrate for nitric oxide synthase (NOS) resulting in the production of NO and citrulline. A third pathway involves the action of arginine decarboxylase leading to the synthesis of putrescine, a significant member of the polyamine family (Kim *et al.*, 2013).

The urea cycle, its enzymes and substrates are normally associated with the disposal of excess N in mammals. However, the key enzyme, arginase, is widely distributed in nature, even occurring in diverse species of higher plants. In studies on the regulation of plant arginase by wounding, jasmonate and the phytotoxin coronatine, Chen *et al.* (2004) reported the characterization of two genes *LeARG1* and *LeARG2* encoding arginase in tomato. Marked similarities in the regulation of arginase in plants and animals were noted. In particular, *LeARG2* expression and arginase activity were induced by stress factors. The overexpression of the two genes (designated *ARGAH1* and *ARGAH2*) in *Arabidopsis thaliana* has also been reported (Brauc *et al*., 2012). Previously, Brownfield *et al.* (2008) had proposed multiple functions for arginase genes in *Arabidopsis*. Modulation of arginase activity is a significant emerging theme emphasized in studies with chilling tolerance in tomato fruits (Zhang *et al*., 2010) and other forms of abiotic stress (Shi *et al*., 2013).

Ornithine-δ-aminotransferase catalyses the transamination of ornithine to glutamate-γ-semialdehyde, but the potential significance of OAT may reside in stress-induced proline synthesis (Stranska *et al*., 2008, 2010). The concept that proline is toxic to plants under non-stress conditions was alluded to by Kalamaki *et al.* (2009a, b) with the suggestion that overproduction of its precursor, ornithine, may offer the mechanism to provide plants with an amino acid that can readily be used for proline production under stress conditions. It was proposed that genetic manipulation of plants to accumulate a non-toxic metabolite that can rapidly be transformed into a protective molecule might present an alternative method for modulating stress tolerance. The efficacy of ornithine is further underlined in exogenous feeding studies in salt-stressed cashew leaves. Under these conditions, proline concentrations were several times higher than glutamate or salt-treated plants (da Rocha *et al*., 2012). A significant contribution of the OAT pathway was implied, but others may disagree (see Planchet and Limami, 2015).

#### *4.4.3.1 Polyamines*

Polyamines are a class of ubiquitous biogenic amines with wide-ranging functions in biological processes, including cellular growth and cell division, embryogenesis, organ development, apoptosis and adaptation/tolerance to biotic and abiotic stressors (Mattoo *et al*., 2015). In higher plants, both arginine and ornithine normally serve as precursors of polyamines in distinct pathways, initiated respectively by arginine decarboxylase (ADC) and ornithine decarboxylase (ODC). The primary product of these two pathways is the diamine putrescine. The triamine spermidine is derived from putrescine by the action of spermidine synthase (SPDS), while the tetraamine spermine arises through the action of spermine synthase (SPMS) on spermidine. However, the synthesis of spermidine and spermine is dependent on the activity of *S*-adenosylmethionine decarboxylase (SAMDC). Polyamines have been associated with plant responses to abiotic stress for over six decades (Mattoo *et al*., 2015). It is now clear that polyamines accumulate in response to a variety of adverse environmental conditions, including exposure to drought, salinity, chilling, heat, hypoxia, ozone, UV radiation, heavy metals, mechanical wounding and herbicides. Transgenic lines of *Arabidopsis*, rice, tobacco and tomato employed in recent studies have led to the hypothesis that polyamines may exert an ameliorative role in plants exposed to a variety of abiotic stressors (Table 4.1). For example, in transgenic plants, overexpression of *ADC* and *ODC* conferred tolerance to salt, drought, freezing and multiple stressors. Similarly, overexpression of *SAMDC* in transgenic rice, tobacco or tomato enhanced tolerance to specific and multiple stressors.

The development of polyamine-deficient mutants has contributed to improved understanding of abiotic stress responses. For example, *Arabidopsis* mutants deficient in polyamine accumulation showed impaired tolerance to salinity, while an *ADC2*-knockout mutant had increased susceptibility to salt stress which was reversed by exogenous putrescine administration. Another *Arabidopsis* mutant deficient in spermine synthesis exhibited sensitivity to drought and salinity which could be reversed by external supplementation of the polyamine. These results underline the roles of specific polyamines in abiotic stress tolerance (see also Mattoo *et al*., 2015). In *Arabidopsis*, the *ADC1* gene is up-regulated by cold whereas the *ADC2* gene is stimulated by a variety of abiotic stressors. *SPMS*, *SPDS1* and *SPDS2* genes are upregulated in response to dehydration and high salinity; *SAMDC1* is induced by cold and *SAMDC2* is additionally up-regulated by salt. Thus, the accumulation of polyamines is supported by up-regulation of key enzyme genes (Mattoo *et al*., 2015). It has been further suggested that exogenous supplementation of polyamines might offer a credible option for enhancing abiotic stress tolerance in plants (Gill and Tuteja, 2010).

**Table 4.1.** Abiotic stress tolerance in transgenic plants expressing genes of the polyamine biosynthesis pathway (adapted from Mattoo *et al.*, 2015).

| Gene[a] | Gene source | Transfer to | Type | Overproduction | Tolerance to |
|---|---|---|---|---|---|
| ADC | Oat | Rice | Inducible | Putrescine | Salt |
| ADC | *Datura stramonium* | Rice | Inducible | Spermidine and spermine | Drought |
| ADC1 | *Arabidopsis* | *Arabidopsis* | Constitutive | Putrescine | Freezing |
| ADC2 | *Arabidopsis* | *Arabidopsis* | Constitutive | Putrescine | Drought |
| ADC | *Arabidopsis* | Aubergine | Constitutive | Putrescine, spermidine and spermine | Multiple stresses |
| ODC | Mouse | Tobacco | Constitutive | Putrescine | Salt |
| SAMDC | *Tritordeum* | Rice | Inducible | Spermidine and spermine | Salt |
| SAMDC | Human | Tobacco | Constitutive | Putrescine and spermidine | Salt and osmotic pressure |
| SAMDC | Carnation | Tobacco | Constitutive | Putrescine, spermidine and spermine | Multiple stresses |
| SAMDC1 | *Arabidopsis* | *Arabidopsis* | Constitutive | Spermine | Multiple stresses |
| SAMDC | Yeast | Tomato | Constitutive | Spermidine and spermine | Heat |
| SPDS | *Cucurbita ficifolia* | *Arabidopsis* | Constitutive | Spermidine | Multiple stresses |
| SPDS | *Cucurbita ficifolia* | Sweet potato | Constitutive | Spermidine | Drought and salt |
| SPDS | Apple | Pear | Constitutive | Spermidine | Multiple stresses |

[a]Abbreviations: *ADC*, arginine decarboxylase; *ODC*, ornithine decarboxylase; *SAMDC*, *S*-adenosylmethionine decarboxylase; *SPDS*, spermidine synthase.

As might be predicted, the relationship between abiotic stress and polyamine accumulation is integrated into phytohormone networks, particularly involving abscisic acid (ABA). In addition, it should be noted that polyamines may be associated with regulatory functions in abiotic stress tolerance via other processes including ROS and NO signalling, modulation of ion channel activation and $Ca^{2+}$ homeostasis (Alcazar *et al.*, 2010). The mediation of spermidine-derived hydrogen peroxide ($H_2O_2$) in the induction of tolerance to salinity has also been proposed by Moschou *et al.* (2008).

### 4.4.4 Citrulline

Yokota *et al.* (2002) observed that under drought conditions in the presence of strong sunlight, high concentrations of citrulline, arginine and glutamate may accumulate in the leaves of wild watermelon. It was suggested that citrulline and arginine synthesis may be related to the induction of DRIP-1, a homologue of ArgE in *Escherichia coli* in which it functions to incorporate the C-skeleton of glutamate into the urea cycle. Salt treatment of wild melon increased concentrations of γ-aminobutyrate, glutamine and alanine in addition to smaller quantities of citrulline. The relationship between citrulline accumulation and salt tolerance during vegetative growth of melon has been examined by Dasgan *et al.* (2009) who concluded that citrulline may protect green tissues from secondary oxidative stress. Alterations in the molar ratios of citrulline to free proline suggested that salt tolerance might be linked with high values for this ratio and the converse for sensitivity. The potential use of citrulline as a biochemical marker for salt tolerance in melon genotypes is now an emerging issue.

### 4.4.5 Alanine

A consistent feature of plant responses to adverse environmental conditions is the accumulation of alanine and the regulation of the activity of alanine aminotransferase (AlaAT) (Raychaudhuri, 2015). It is assumed that, due to its role as a compatible solute, alanine accumulation to relatively high levels should cause no toxic effects in abiotic stress situations such as flooding, water deficit and low temperatures. It is envisaged that alanine acts by: (i) stabilizing the quaternary structure of cellular proteins and membranes; (ii) regulation of pH balance during anoxia; and (iii) mitigation of some of the biochemical aberrations associated with hypoxia. Flooding induces a rapid expression of the gene for AlaAT with a concomitant increase in enzyme activity. However, it appears that alanine accumulation is not solely dependent upon the activity of AlaAT but also on an alternative pathway catalysed by GABA transaminase.

### 4.4.6 Glycine betaine

Glycine and choline are precursors of glycine betaine (GB), a low-molecular-mass molecule which may confer protection to plants exposed to stress (Holmstrom *et al.*, 2000; Sakamoto and Murata, 2002). BeGora *et al.* (2010) commented on the dearth of information concerning the enzymes or the factors regulating the biosynthesis of choline and phosphatidylcholine. However, evidence was presented to demonstrate that the enzyme phosphomethylethanolamine *N*-methyltransferase catalyses three sequential *N*-methylations of phosphoethanolamine to produce phosphocholine, with SAM providing the methyl groups. It was further claimed that the activity of this enzyme was not the rate-limiting factor and that choline transport into chloroplasts and insufficiency of other precursors should be considered as potential constraints in the supply of GB. This issue is relevant in the light of evidence indicating that the GB accumulation is more effective in the chloroplasts than in the cytosol for abiotic stress protection in plants (Park *et al.*, 2007).

GB is considered to act as a compatible solute exerting a role in response to abiotic stressors such as high salinity and low environmental temperature. Initial work emphasized the role of GB in the maintenance of cellular osmotic potential in plants, but current studies demonstrate the protective functions of GB against a variety of abiotic stress factors. At the concentrations present, the effects of GB are attributed more significantly to stress tolerance than to the maintenance of cellular osmotic potential (Chen and Murata, 2011). GB accumulation is more effective in chloroplasts than in the cytosol for protection against abiotic stress, judging by responses in transgenic tomato plants (Park *et al.*, 2007). In addition, GB accumulation varies, being relatively high in the chloroplasts of spinach and barley compared with other species such as *Arabidopsis* and tobacco. Transgenic tobacco lines accumulating GB show increased tolerance to salinity and low temperatures (Holmstrom *et al.*, 2000). It may be argued that GB accumulation is a product of stress and not an adaptive response. However, exogenous applications of GB may elicit significant increases in growth and crop yield under environmental stress conditions, though species differences in responses are apparent (Ashraf and Foolad, 2007).

### 4.4.7 Branched-chain amino acids

A significant aspect relates to the interaction between regulatory proteins and DNA, exemplified by leucine zipper transcription factors. The orientation of leucine residues along the segments of the protein double-helix promotes binding to DNA. Certain members of this class of proteins are attributed with mediating the response of plants to a range of processes, including ABA-dependent signal transduction under drought and high-salinity regimes (Uno *et al.*, 2000). ABA exerts a pivotal role in stress responses in plants and Kang *et al.* (2002) emphasized that leucine zipper proteins may be integrated into this pathway.

### 4.4.8 β-Aminobutyrate

The xenobiotic non-protein amino acid β-aminobutyrate (BABA) is worthy of consideration at this juncture. BABA acts as a glycine receptor agonist in animals but is also effective in priming plants against pathogens and osmotic stress, while significantly enhancing acquired thermotolerance. As such, there is accumulating recognition of the role of BABA in a general role of potentiating the stress resistance of plants. For example, Zimmerli *et al.* (2008) noted that BABA increased thermotolerance in *Arabidopsis*, implicating crosstalk with ABA signalling cascades as part of the mechanism. In addition, proteomic analysis indicated large-scale changes in primary metabolism in response to BABA application and on exposure of *Arabidopsis* to simulated acid rain. Liu *et al.* (2011) identified 175 proteins responding to BABA in the absence and presence of simulated acid rain. It appears that such changes are complementary to the activation of antioxidant systems and SA, JA and ABA signalling pathways. The priming effect of BABA on gene expression in *Arabidopsis* subjected to abiotic stress is relevant at this point in that pretreated plants show earlier and elevated expression of certain SA-dependent and ABA-dependent genes following salt and drought stress (Jakab *et al.*, 2005). It has been suggested that priming with BABA is a cost-effective method of inducing abiotic stress tolerance, based on studies with sodium chloride/polyethylene glycol (NaCl/PEG) exposure (Jisha and Puthur, 2016). However, any implementation would be subject to satisfactory environmental and toxicological evaluation. It should be noted that repeated treatment with high doses of BABA induces female sterility in *Arabidopsis* (see D'Mello, 2015b). The observation that BABA is an agonist at the glycine receptor is also of relevance here and warrants further investigation in any assessment of product safety. Nevertheless, the role of BABA in plant protection continues along distinct lines outlined above in a scenario that is still evolving and that may yield dividends in our understanding of plant responses to abiotic stress.

## 4.5 Secondary Metabolism

It is now acknowledged that the term 'secondary' in this particular context is a misnomer, as it implies functional passivity in the biochemical response of plants to endogenous and environmental stimuli. However, there is almost universal agreement and ample evidence that secondary compounds actively support ecological competitiveness, survival and growth of plants. Amino acids serve as sources of a diverse range of secondary compounds, including signal transduction molecules and defence agents.

Conventional accounts of secondary metabolism tend to focus on the biosynthesis of low-molecular-weight phytochemicals, particularly in response to stress challenges such as biotic or physical pressures. It is generally assumed that these compounds support the constitutive defences in the plant. In addition, the appearance of the oligopeptide, systemin, has been associated with stress responses in plants (Leon *et al.*, 2001).

Amino acids in higher plants are a prolific source of secondary compounds, including auxins, serotonin, melatonin, glucosinolates, natural phytotoxins, alkaloids, condensed tannins, cyanogens, indole phytoalexins and flavonoids. Plants also contain a diverse array of non-protein amino acids. Although a number of secondary compounds form part of the constitutive defence system, others are synthesized as a result of up-regulation of specific biosynthetic genes and pathways in response to biotic or environmental elicitors.

## 4.6 Signal Transduction and Regulation in Stress Responses

Current evidence and opinion implicate a diverse array of phytohormones and other bioactive compounds in complex signal transduction networks, contributing significantly to defence reactions following abiotic challenge and/or microbial pathogenesis. Examples frequently cited are ABA (Ishitani *et al.*, 1995), SA (Pal *et al.*, 2013), JA (Pedranzani *et al.*, 2016), ET (Fujita *et al.*, 2006; Wang *et al.*, 2010), NO (Neill *et al.*, 2008; Fan and Du, 2012) and IAA (Zhang *et al.*, 2009; Tiwari *et al.*, 2012). Hormone signals are relevant here for several reasons. Importantly, in their respective biosynthetic pathways, ET, SA, NO and IAA can all be traced back to respective precursor amino acids (Fig.4.1). Subsequent signalling activity may be modulated by conjugation with various amino acids, as in the cases of SA, JA and IAA. Thus, SA–Asp is the endogenous conjugate which accumulates after pathogen infection (Chen *et al.*, 2013) and possibly abiotic stressors. Jasmonate–isoleucine conjugate (JA–Ile) is the classical example of a conjugate required for optimal signalling in *Arabidopsis* and modulated by antagonistic interactions with SA (Kazan and Manners, 2008). Uniquely, IAA–trytophan conjugate (IAA–Trp) is generally implicated in the regulatory inhibition of auxin activity. Other examples of the indirect role of amino acids include leucine-rich repeat receptor kinases in brassinosteroid (BR) signalling, effective through serine/threonine phosphorylation and tyrosine protein nitration in peroxynitrite mediation of defence responses.

The context for signal transduction is generally associated with stress and defence metabolism and the initiation of induced resistance. For example, in the case of proline, stress-induced accumulation is considered to be modulated by ABA and certain regulatory macromolecules such as protein kinases (Planchet and Limami, 2015). However, signalling compounds rarely operate in isolation and the emerging evidence confirms complex interactions or crosstalk in communication networks of plants exposed to biotic and environmental challenges (Fujita *et al.*, 2006). Indeed, Igarashi *et al.* (2012) maintained that activation of the immune system by plant pathogens carries cost in the form of reduced growth and the capacity to respond to abiotic stresses. Selected evidence either implied or explicitly stated for crosstalk is summarized in Table 4.2. The activity of a peptide growth factor, phytosulfokine, in the attenuation of pattern-triggered immunity contributes to complex homeostatic mechanisms. Similarly, the ethylene response factor (ERF) and the ethylene responsive-element binding protein (EREBP) act as transcriptional regulators, exerting a key role in the adaptation of plants to various abiotic and biotic stresses (Tiwari *et al.*, 2012). Further evidence of overlapping response pathways has emerged in evaluating the role of the *Arabidopsis* leucine-rich repeat receptor kinase (LRRK) in abiotic and biotic responses (Van der Does *et al.*, 2017). It is important to emphasize that the final outcome will be determined by the extent of interactions among signalling compounds both within and between abiotic and biotic cascades. For example, although signalling compounds have been considered separately as derivatives of specific amino acids, the concept of crosstalk in hormonal activity is gaining momentum. Robert-Seilaniantz *et al.* (2011) indicated that such interactions were 'more than just JA–SA antagonism' and that other components such as ET, auxin, ABA and BR are important regulators of plant disease and defence. Thus, elements of synergism and antagonism contribute to the eventual outcome of biotic and abiotic elicitors.

## 4.7 Mechanisms of Metal Stress Tolerance in Higher Plants

Two significant mechanisms exist in plants to reduce the impact of deleterious metal ions. One involves the mediation of specific amino acids and the other relies on the synthesis of peptides with particular origins and properties. Sources of toxic metals include agricultural applications, sewage sludge, disposal of mine wastes and emissions from industrial smelters.

### 4.7.1 Amino acids

The notion of metal–amino acid linkages in living organisms is well established. When selenium (Se) replaces sulfur (S) in methionine, cysteine and cystathionine in certain leguminous and brassica plants, striking structural analogues are produced. Indeed, selenomethionine has been declared a 'canonical' amino acid and now formally included in the list of 21 amino acids that are required for protein biosynthesis. However, a number of other amino acids are credited with a protective role in higher plants by virtue of specific biophysical attributes.

Proline is, arguably, a prominent exponent of a number of amino acids with metal-complexing activities. Proline accumulation has been associated with protective properties towards specific enzymes, for example, nitrate reductase and glucose-6-phosphate dehydrogenase, and against heavy metals by forming proline–metal complexes. The physical properties of proline arising from its cyclic structure may confer particular attributes in complex formation.

Ingle (2015) referred to the role of histidine in nickel (Ni) hyperaccumulation in certain plants of the genus *Alyssum*. Coordination of Ni by histidine has been demonstrated *in vivo* in the xylem, root and shoot tissues of these

**Table 4.2.** Stress responses in higher plants: selected evidence of crosstalk in abiotic and biotic pathways involving signalling molecules derived from, or physically associated with, amino acids. Note that methionine is the precursor of ethylene, while isoleucine is an activator of jasmonate. Among the aromatic amino acids, phenylalanine is, at least in part, the source of salicylic acid and tryptophan is the precursor of the auxin, indole acetic acid and arginine is the source of nitric oxide. Interactions with other networks involving abscisic acid and the role of leucine zipper proteins, particularly under stress conditions, add to the complexity of signal transduction in plants. Leucine-rich repeat receptor kinases constitute a substantive sub-family of transmembrane receptor-like proteins in plants. The role of these signalling molecules is well established for plants subjected to fungal infection and/or insect herbivory (see D'Mello, 2015a).

| Source | Criteria/methodologies | Outcomes/results[a] |
|---|---|---|
| Tiwari *et al.* (2012) | Transcription studies | ERF/EREBP family of transcriptional regulators plays a key role in adaptation to various abiotic and biotic stressors |
| Wei *et al.* (2015) | Salt, polyethylene glycol and cold stress | Ethylene is involved in abiotic stress in cucumber (*Cucumis sativus* L.) seedlings |
| Alonso *et al.* (1999) | Molecular studies in mutant plants exposed to paraquat-induced oxygen radicals | EIN2 constitutively activates ethylene responses and restores responsiveness to jasmonate; EIN2 recognized as the molecular link between previously perceived distant hormone response pathways |
| Kazan (2015) | Review | During cold stress, ethylene and jasmonates differentially regulate signalling pathways |
| Lestari *et al.* (2018) | Overexpression studies in *Hevea brasiliensis* | ERF1 enhances tolerance to abiotic stress and is an essential integrator of the ethylene and jasmonate signalling pathways |
| Creelman and Mullet (1995) | Dehydration in soybean leaves | Jasmonate levels increased fivefold following dehydration |
| Nguyen *et al.* (2016) | Drought stress and insect herbivory: hypothesis | Synergistic interaction between jasmonate and abscisic acid |
| Pedranzani *et al.* (2016) | Arbuscular mycorrhizal symbiosis in *Digitaria eriantha* plants | Jasmonate increased in inoculated plants subjected to drought |
| Khan *et al.* (2015) | Review | Salicylic acid induced tolerance to salinity, ozone and drought |
| Jain and Khurana (2009) | Transcript profiling | Auxin-responsive genes exert diverse effects in abiotic stress signalling in rice |
| Zhang *et al.* (2009) | Physiology and gene expression/activation | Altered architecture and enhanced drought tolerance in rice via downregulation of auxin |
| Asgher *et al.* (2017) | Review | Nitric oxide implicated in crosstalk with other plant growth regulators in abiotic stress tolerance |
| Van der Does *et al.* (2017) | Reverse-genetic approach | The *Arabidopsis* leucine-rich repeat receptor kinase connects cell wall integrity sensing, root growth and response to abiotic and biotic stresses |

[a]Abbreviations: ERF, ethylene response factor; EREBP, ethylene responsive-element binding protein; EIN2, ethylene insensitive2.

hyper-accumulators and, in addition, exogenous histidine supplementation to a non-accumulator species of *Alyssum* reduced the toxic effects of Ni. Steady-state measurements indicate that hyper-accumulators have constitutively elevated concentrations of histidine in root tissues. It is suggested that histidine serves as a ligand for Ni and promotes the xylem loading of this metal ion in Ni hyper-accumulators.

Mimosine, occurring in the tropical legume *Leucaena leucocephala*, is also attributed with metal-complexing properties, for example towards iron salts such as $FeSO_4$, and aluminium (Al) as $Al_2(SO_4)_3$ (D'Mello and Acamovic, 1982). In human breast cancer cells, mimosine blocks cell cycle progression by chelating iron (Kulp and Vulliet, 1996), while others suggest that mimosine attenuates serine hydroxymethyltransferase transcription by chelating Zn (see D'Mello, 2015b).

### 4.7.2 Phytochelatins

A wide array of peptides synthesized in plants are associated with complex mechanisms to promote primary metabolic activities and to confer stress tolerance towards potentially toxic metals. The phytochelatins are the most prominent members of this group, attracting considerable research to the present day. In order to circumvent the deleterious effects of metal stressors, relatively efficient but complex mechanisms exist in plants for chelation, transport and vacuolar sequestration within plant cells through the action of phytochelatins. The phytochelatins are a group of Cys-rich peptides synthesized exclusively from reduced glutathione in transpeptidation reactions catalysed by phytochelatin synthase. Data exemplifying a broad spectrum of functional attributes of phytochelatins in homeostatic regulation and metal stress tolerance in plants are presented in Table 4.3.

It is emphasized that a greater definition of specified metal/abiotic stressors needs to be considered for different plant species before firm and unequivocal conclusions can be recommended. The potential for the development of biomarkers and the implications for phytoremediation are now emerging. However, contrasting effects have been observed regarding tolerance and disruption of homeostasis of essential mineral elements, implying that responses may depend on the metal stressor in question and the affected plant species. The role of high-affinity phosphate transporters adenosine triphosphate (ATP)-binding cassette (ABC) transporters ABCC1 and ABCC2, conjugation with phytochelatin and enzymes involved in As tolerance in plants, as summarized by Zhang *et al.* (2017), is depicted in Fig. 4.2.

It would be premature to adopt this model for general application for all toxic metals. Thus, Negrin *et al.* (2017) comment on the lack of stoichiometric relationships between phytochelatin expression and As, Zn and Pb concentrations in salt marsh plants, with others indicating roles for phytochelatins in cell wall re-modelling, metabolism of phenylpropanoids and even modulation of auxin biosynthesis. Another factor for consideration is the possible chelation of valuable nutrients such as Cu and S which might compromise optimum enzyme function and the synthesis of glucosinolates, *S*-methylcysteine sulfoxide and other S-containing compounds serving defence roles towards fungal pathogens and insect herbivores. These observations are probably first indications of active crosstalk in abiotic–biotic networks associated with metal tolerance in higher plants (Table 4.3).

## 4.8 Implications and Future Directions

Research into the role of amino acids and certain peptides in abiotic stress tolerance is destined to continue as the impact of climate change on plant survival and productivity unravels. Furthermore, the application of transgenic methodologies to underpin future investigations has only just commenced and should play a major role in elucidating the dynamics of amino acid metabolism following exposure to abiotic stressors such as salinity, acid rain, drought and heavy metals. The critical role of amino acids in the primary metabolism of higher plants remains undisputed throughout several decades of sustained research. Thus, the contribution of amino acids to structural and catalytic functions of proteins and other macromolecules is well recognized and continues to offer new opportunities for potential applications. However, other aspects have recently emerged to add to the diversity of objectives; thus, signal transduction and secondary metabolism of amino acids are increasingly integrated within plant defence networks, for example during cold, drought and salinity stress, but the mechanisms remain elusive (Xiong *et al.*, 2002).

In a number of instances, amino acids serve as sources of both signal transduction molecules and secondary compounds regularly implicated in plant defence systems. Thus methionine is the precursor of ET and aliphatic glucosinolates, while tryptophan metabolism results in the synthesis of auxin, camalexin and indole glucosinolates. In addition, phenylalanine is the substrate for SA as well as flavonoids, coumarins and stilbenes (Huang *et al.*, 2010).

The indirect role of amino acids in plants (as in other living organisms) is exemplified by

**Table 4.3.** Phytochelatins: data exemplifying a broad spectrum of functional attributes in homeostatic regulation and metal stress tolerance in plants. It is emphasized that a greater definition of specified metal/abiotic stressors needs to be considered for different plant species before firm conclusions and practical implications can be recommended in unequivocal terms. The potential for the development of biomarkers and the implications for phytoremediation are now emerging. However, contrasting effects have been observed regarding tolerance and disruption of homeostasis of essential mineral elements, implying that responses may depend on the metal stressor in question and plant species.

| Metal stressor/ conditions | Evidence summarizing main effects, mechanisms and implications | Reference |
|---|---|---|
| None | Phytochelatin synthase plays a role in the homeostatic control of Fe (II)/(III), based on studies with *Nitella mucronata* (charophyta) | Fontanini *et al.* (2018) |
| Salinity | Phytochelatins positively correlated with As, Zn and Pb in salt marsh plants; complex mechanism implied | Negrin *et al.* (2017) |
| Zn-contaminated soils | Phytochelatin synthesis in *Arabidopsis thaliana* is essential for survival in Zn-contaminated soils | Kuhnlenz *et al.* (2016) |
| Heavy metals | Glutathione and phytochelatins mediate redox homeostasis and stress signal transduction in plants (presented in an overview) | Singh *et al.* (2016) |
| Hg-rich industrial waste | Phytochelatin synthesis in macrophytes may serve as a marker of bioavailable Hg in aquatic environments | Turull *et al.* (2017) |
| *Arabidopsis thaliana* mutants | The knockout mutant for phytochelatin synthase is defective in callose deposition, bacterial pathogen defence and auxin production; putative link to glucosinolate biosynthesis | De Benedictis *et al.* (2018) |
| Cd | Genetic manipulation of plants using a phytochelatin synthase variant associated with improved Cd tolerance | Cahoon *et al.* (2015) |
| Cd | Cd-inducible expression of the ABC-type transporter increases phytochelatin-mediated Cd tolerance in *A. thaliana* | Brunetti *et al.* (2015) |
| Cd and Cu | Specific mechanisms of tolerance to Cd and Cu compromised by a limited concentration of glutathione in alfalfa plants; Cd promoted phytochelatin synthesis but Cu caused GSH diminution | Flores-Caceres *et al.* (2015) |
| Cd and Cu | Cd may increase phytochelatin levels which complex both Cd and Cu, potentially resulting in Cu deficiency; Cd exposure can, therefore disrupt homeostasis of essential inorganic elements | Gielen *et al.* (2017) |
| Cd and As accumulation | The existence of at least partially distinct phytochelatin synthase-dependent pathways for As and Cd accumulation in rice grains | Uraguchi *et al.* (2017) |
| As | Cytokinin determines thiol-mediated As tolerance and accumulation in *A. thaliana*; thiol groups of GSH and phytochelatins are essential for As sequestration | Mohan *et al.* (2016) |
| As | Identification of amino acid residues important for the As resistance function of *A. thaliana* ABCC1 transporter | Zhang *et al.* (2017) |
| As | Enhanced As sensitivity with excess phytochelatin accumulation in shoots of a SULTR1;2 knockout mutant of *A. thaliana*; implies requirement for S nutrient in phytochelatin synthetic pathway (and S defence compounds, e.g. glucosinolates?) | Nishida *et al.* (2016) |
| Heavy metals | Phytoremediation capacity of aquatic plants associated with the degree of phytochelatin polymerization; metal-induced phytochelatin synthesis and degree of polymerization highest in *Lemna minor* compared with *Salvinia natans* or *Elodea canadensis* | Torok *et al.* (2015) |

complex structural–functional relationships which are evident in virtually all aspects of metabolism. For example, basic leucine zipper proteins mediate ABA-dependent signal transduction in drought and salinity stress, reflecting protein–DNA interactions (Uno *et al.*, 2000; Kang *et al.*, 2002). Leucine-rich repeat receptor kinases are transmembrane structures also regulating a diverse range of developmental and defence-related processes, but via protein–protein interactions (Kobe and Kajava, 2001).

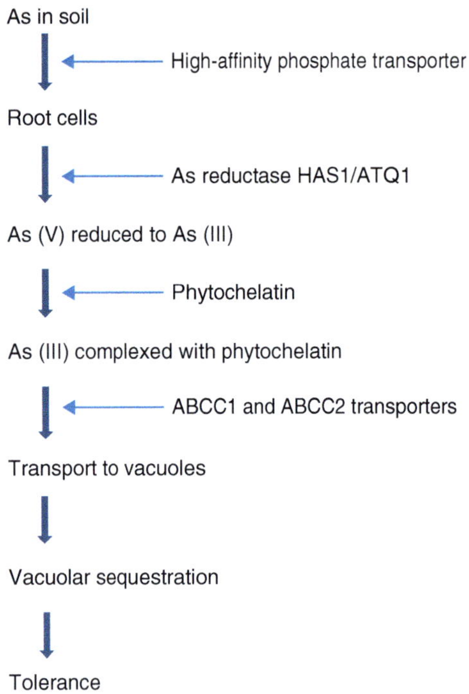


**Fig. 4.2.** A model for phytochelatin-mediated arsenic (As) tolerance. Based upon studies with the *Arabidopsis* ATP-binding cassette (ABC) transporter (Zhang *et al.*, 2017).

Reviewing data relating to the impact of osmotic and acid rain effects, Horvath *et al.* (2007) concluded that abiotic stress tolerance may be induced by SA. Yasuda *et al.* (2008) extended this role with observations of antagonistic crosstalk at multiple steps between SA and ABA following NaCl treatment. Thus, SA appears to exert a major influence in both biotic and environmental stresses and, in addition, is involved in the induction of low-molecular-weight defence secondary compounds (Tounekti *et al.*, 2013) However, the role of JA and other signalling compounds in defence reactions and the implications of complex phytohormone interactions should not be underestimated Thus, synergy between ET and JA is essential for the induction of the osmotin gene (Xu *et al.*, 1994) and the *Arabidopsis* defensin gene (*PDF 1.2*) (Penninckx *et al.*, 1998).

Light regulates diverse developmental and morphological changes in plants and effects on amino acid metabolism are only to be expected. A role for BR in light development of higher plants has been proposed (Li *et al.*, 1996). In addition, evidence for the involvement of glutamate receptors in light-signal transduction was advanced by Lam *et al.* (1998). Another theme relates to the functional association of glycine decarboxylase with the control of carbon flow through both photosynthesis and photorespiration (Hylton *et al.*, 1988; Timm *et al.*, 2012) and the sensitivity of this enzyme to oxidative stress (Taylor *et al.*, 2002). Several defence reactions, including upregulation of PAL, SA synthesis and expression of pathogenesis-related protein PR-1, appear to be light-dependent, whereas pathogen-induced production of JA, camalexin and myrosinase are enhanced in the dark (Zeier *et al.*, 2004). Of particular note is the effect of light on the promotion of JA-Ile conjugation via two distinct mechanisms (Svyatyna *et al.*, 2013).

In abiotic stress responses, antagonisms regulated by ABA may also provide the basis of further investigation (Moons *et al.*, 1997; Yasuda *et al.*, 2008). An integrated view of the complex interactions involving phytohormones, mediators and transcription factors has been attempted but considerable uncertainty still exists (Kazan and Manners, 2008).

Proline and citrulline accumulate in plants exposed to abiotic stress (Sharma and Dietz, 2006; Dasgan *et al.*, 2009; Lehmann *et al.*, 2010; Szabados and Savoure, 2010) but the sequence of metabolic events from signalling to protective functions remains obscure, despite extensive speculation as to the mechanisms involved in osmolyte functions. Thus, Szabados and Savoure (2010) invoked effects on homeostasis, mitochondrial function, cellular proliferation, apoptosis and gene expression as potential areas for research. In other abiotic stress studies, ABA (Gomez *et al.*, 1988) and SA signalling (Horvath *et al.* 2007) have been implicated. Specifically, the role of SA in modulating terpenoid and flavonoid metabolism in plant responses to abiotic stress has been investigated (Tounekti *et al.*, 2013).

In the course of recent research, complex interactions have emerged with respect to signal transduction involving, in particular, phytohormones derived from amino acids or those in structural associations with isoleucine, leucine or tryptophan. The identification of additive, synergistic and antagonistic effects and the influence of biotic and abiotic stresses on the outcomes of these interactions underline the need for continuing investigations. The task is daunting, given the complexity of interactions and persistent evidence of crosstalk between biotic and abiotic pathways (Divi *et al.*, 2010) (Table 4.2).

Convergence of abiotic and biotic stress pathways is supported by studies implicating the signal peptide systemin. Orsini *et al.* (2010) proposed that the systemin-dependent salinity tolerance in tomato reflected this overlapping feature via interactions with JA. This theme had previously been developed by Meier *et al.* (2008), who assigned a role for plant natriuretic peptides, a class of systemically mobile molecules, in abiotic and biotic stress responses.

The metal-complexing properties of mimosine that are present naturally in *Leucaena* and selenoamino acids accumulating in certain other legumes and in *Brassica* plants opens up the prospects for phytoremediation in agricultural soils, particularly as the selenoamino acids are attributed with beneficial properties for human health. However, the major contributors to homeostasis and metal stress tolerance in higher plants are represented by the phytochelatin peptides. It is emphasized that a greater definition of specified metal/abiotic stressors needs to be considered for different plant species before firm and unequivocal conclusions can be made for phytochelatins. The potential for the development of biomarkers and the implications for phytoremediation are now emerging. However, contrasting effects have been observed regarding toxic metal tolerance and disruption of homeostasis of essential mineral elements, implying that responses may depend on the metal stressor in question and the affected plant species. Furthermore, as part of a comprehensive strategy, the role of specific enzymes in conferring tolerance during metal stress in plants is worthy of investigation (Sanjaya *et al.*, 2008).

## 4.9 Conclusions

Abiotic stresses impinge profoundly on manifestations of toxicity in higher plants and are therefore relevant to the general theme of this volume. Adverse effects include $NH_3$ production as a consequence of endogenous proteolysis, intrinsic toxicity of certain amino acids (for example, proline), generation of ROS and exposure to heavy metals. A number of mechanisms exist in plants to avoid or alleviate potential toxicity. Under conditions such as salinity, increased proteolytic activity creates increased intracellular hyper-ammonia and potential metabolic toxicity if homeostasis is not restored. Ammonium ions are therefore incorporated into glutamine and glutamate by glutamine synthetase/glutamate synthase (GS/GOGAT), thus directing the pathway towards proline synthesis. It is well known that proline accumulates in higher plants in response to a variety of environmental factors such as wilting, drought, salinity, temperature, anoxia, exposure to heavy metals and oxidative stress. This accumulation varies with plant species and has been linked with improved plant performance under diverse abiotic stress conditions. Salt treatment of wild melon increased concentrations of γ-aminobutyrate, glutamine and alanine in addition to smaller quantities of citrulline. It has been concluded that citrulline may protect green tissues from secondary oxidative stress. The potential use of citrulline as a biochemical marker for salt tolerance in plants has been proposed. It appears that the amino acid metabolism of plants in response to abiotic stress is up-regulated not only to neutralize potential toxicity but also to provide for the synthesis of protectants such as proline, ornithine, alanine, glycine betaine and polyamines. At the same time, mechanisms exist to equilibrate accumulations for protection of toxicity arising from specific amino acids such as proline. It is also clear that polyamines, derived from arginine and/or ornithine, accumulate in response to a variety of adverse environmental conditions, including exposure to drought, salinity, chilling, heat, hypoxia, ozone, UV radiation, heavy metals, mechanical wounding and herbicides.

Similarly, the direction of metabolism of the tripeptide, GSH, is altered in favour of the biosynthesis of phytochelatins during exposure of plants to toxic heavy metals. In order to circumvent the deleterious effects of metal stressors, relatively efficient but complex mechanisms exist in plants for chelation, transport and vacuolar sequestration within plant cells. These processes involve the intervention of phytochelatins, but the lack of stoichiometric relationships between phytochelatin expression and As, Zn and Pb concentrations and the possible chelation of valuable nutrients such as Cu and S might compromise optimum enzyme function and the synthesis of glucosinolates, *S*-methylcysteine sulfoxide and other *S*-containing compounds serving defence roles towards pathogens and insect pests.

The regulatory pathways involved in the responses of plants to abiotic stress remain essentially unexplored. Amino acid accumulation and the phytochelatin-induced transport and sequestration of potentially toxic metals will depend on the relative activities of signalling compounds such as ET, SA, auxins, JA and ABA, but research is still at a preliminary stage and any conclusions would be premature.

It is readily admitted that other aspects of abiotic stress responses need to be addressed in order to obtain a more comprehensive perspective of metabolism in higher plants. For example, the contribution of proteomics in the understanding of stress responses should be considered and there may be general implications here for plant toxicology.

Future advances will inevitably depend on the application of molecular techniques and the development of transgenic lines and mutants, particularly in elucidating regulatory mechanisms. These aspects will be of crucial value if we are to breed lines more tolerant to abiotic stresses. This area is now set to provide us with valuable data for a number of pressing questions on the relationship between plant metabolism and acclimation to adverse environmental conditions.

## References

Acamovic, T.A. and D'Mello, J.P.F. (1981) Determination of mimosine by ion-exchange chromatography. *Journal of Chromatography* 206, 416–420.

Acamovic, T.A., D'Mello, J.P.F. and Fraser, K.W. (1982) Determination of mimosine and 3-hydroxy-4(1H)-pyridone in *Leucaena*, avian excreta and serum using reversed-phase high-performance liquid chromatography. *Journal of Chromatography* 236, 169–179.

Alcazar, R., Altabella, T., Marco, F., Bortolotti, C., Reymond, M., Carrasco, P. and Tiburcio, A.F. (2010) Polyamines: molecules with regulatory functions in plant abiotic stress tolerance. *Planta* 231, 1237–1249.

Alonso, J.M., Hirayama, T., Roman, G., Nourizadeh, S., Ecker, J.R. *et al.* (1999) EIN2, a bifunctional transducer of ethylene and stress responses in *Arabidopsis*. *Science* 284, 2148–2152.

Asgher, M., Per, T.S., Masood, A., Fatma, M., Freschi, L., Corpas, F.J. and Khan, N.A. (2017) Nitric oxide signaling and its crosstalk with other plant growth regulators in plant responses to abiotic stress. *Environmental Science and Pollution Research* 24, 2273–2285.

Ashraf, M. and Foolad, M.R. (2007) Roles of glycine betaine and proline in improving plant abiotic stress resistance. *Environmental and Experimental Botany* 59, 206–216.

BeGora, M.D., Macleod, M.J.R., Summers, P.S. and Weretilnyk, E.A. (2010) Identification of phosphomethylethanolamine *N*-methyltransferase from *Arabidopsis* and its role in choline and phospholipid metabolism. *Journal of Biological Chemistry* 285, 29147–29155.

Brauc, S., De Vooght, E., Claeys, M., Hoffe, M. and Angenon, G. (2012) Over-expression of arginase in *Arabidopsis thaliana* influences defence responses against *Botrytis cinerea*. *Plant Biology (Stuttgart, Germany)* 1, 39–45.

Brownfield, D.L., Todd, C.D. and Deyholos, M.K. (2008) Analysis of *Arabidopsis* arginase gene transcription patterns indicates specific biological functions for recently diverged paralogs. *Plant Molecular Biology* 67, 429–440.

Brunetti, P., Zanella, L., DePaolis, A., Cecchetti, V., Falasca, G. *et al.* (2015) Cadmium-inducible expression of the ABC-type transporter *AtBCC3* increases phytochelatin-mediated cadmium tolerance in *Arabidopsis*. *Journal of Experimental Botany* 66, 3815–3829.

Cahoon, R.E., Lutke, W.K., Cameron, J.C., Chen, S., Lee, S.G. *et al.* (2015) Adaptive engineering of phytochelatin-based heavy metal tolerance. *Journal of Biological Chemistry* 290, 17321–17330.

Chen, T.H.H. and Murata, N. (2011) Glycine betaine protects plants against abiotic stress: mechanisms and biotechnological applications. *Plant, Cell and Environment* 34, 1–20.

Chen, H., McCaig, B.C., Melotto, H., He, S.Y. and Howe, G.A. (2004) Regulation of plant arginase by wounding, jasmonate and the phytotoxin coronatine. *Journal of Biological Chemistry* 279, 45998–46007.

Chen, Y., Shen, H., Wang, M. and He, Z. (2013) Salicyloyl-aspartate synthesized by the acetyl-amido synthetase GH 3.5 is a potential activator of plant immunity in *Arabidopsis*. *Acta Biochimica et Biophysica Sinica* 45, 827–836.

Creelman, R.A. and Mullet, J.E. (1995) Jasmonic acid distribution and action in plants: regulation during development and response to biotic and abiotic stress. *Proceedings of the National Academy of Sciences of the United States of America* 92, 4114–4119.

da Rocha, I.M., Vitorello, V.A., Silva, J.S. and Silveira, J.A. (2012) Exogenous ornithine is an effective precursor and the δ-ornithine aminotransferase pathway contributes to proline accumulation under high N recycling in salt-stressed cashew leaves. *Journal of Plant Physiology* 169, 41–49.

Dasgan, H.Y., Kusvuran, S., Abak, K. and Bouchereau, A. (2009) The relationship between citrulline accumulation and salt tolerance during the vegetative growth of melon (*Cucumis melo* L.). *Plant Soil Environment* 55, 51–57.

De Benedictis, M., Brunetti, C., Brauer, E.K., Andreucci, A., Popescu, S.C., Commisso, M., Guzzo, F., Solo, A., Castiglione, M.R. and di Toppi, L.S. (2018) The *Arabidopsis thaliana* knockout mutant *for phytochelatin synthase 1 (cad 1-3)* is defective in callose deposition, bacterial pathogen defense and auxin content, but shows an increased stem lignification. *Frontiers in Plant Science* 9, 19; published online 22 January 2018. doi: 10.3389/fpts.2018.00019.

Deuschle, K., Funck, D., Hellmann, H., Dsachner, K., Binder, S. and Frommer, W.B. (2001) A nuclear gene encoding mitochondrial $\Delta^1$–pyrroline-5-carboxylate dehydrogenase and its potential role in protection from proline toxicity. *The Plant Journal* 27, 345–356.

Divi, U.K., Rahman, T. and Krishna, P. (2010) Brassinosteroid-mediated stress tolerance in *Arabidopsis* shows interactions with abscisic acid, ethylene and salicylic acid pathways. *BMC Plant Biology* 10, 151. doi: 10.1186/1471-2229-10-151.

D'Mello, J.P.F. (2012) Emergence of a new momentum. In: D'Mello, J.P.F. (ed.) *Amino Acids in Human Nutrition and Health*. CAB International, Wallingford, UK, pp. 479–524.

D'Mello, J.P.F. (2015a) Delivering innovative solutions and paradigms for a changing environment. In: D'Mello, J.P.F. (ed.) *Amino Acids in Higher Plants*. CAB International, Wallingford, UK, pp. 538–583.

D'Mello, J.P.F. (2015b) Toxicology of non-protein amino acids. In: D'Mello, J.P.F. (ed.) *Amino Acids in Higher Plants*. CAB International, Wallingford, UK, pp. 507–537.

D'Mello, J.P.F. and Acamovic, T. (1982) Growth performance of, and mimosine excretion by, young chicks fed on *Leucaena leucocephala*. *Animal Feed Science and Technology* 7, 247–255.

Fan, H-F. and Du, C-X. (2012) Effect of nitric oxide on proline metabolism in cucumber seedlings under salinity stress. *Journal of the American Society for Horticultural Science* 137, 127–133.

Flores-Caceres, M.L., Hattab, S., Boussetta, H., Banni, M. and Hernandez, L.E. (2015) Specific mechanisms of tolerance to copper and cadmium are compromised by a limited concentration of glutathione in alfalfa plants. *Plant Science* 233, 165–173.

Fontanini, D., Andreucci, A., Castiglione, M.F., Basile, A., Degola, F., bellini, E., Bruno, L. and Varotto, C. (2018) The phytochelatin synthase from *Nitella mucronata* (Charophyta) plays a role in the homeostatic control of iron (II)/(III). *Plant Physiology and Biochemistry* 127, 88–96.

Fujita, M., Fujita, Y., Narusaka, Y. and Shinozaki, K. (2006) Crosstalk between abiotic and biotic stress responses: a current view from the points of convergence in the stress signaling networks. *Current Opinion in Plant Biology* 9, 436–442.

Funck, D., Eckard, S. and Muller, G. (2010) Non-redundant functions of two proline dehydrogenase isoforms in *Arabidopsis*. *BMC Plant Biology* 10, 70. doi: 10.1186/1471-2229-10-70.

Gielen, H., Vangronsveld, J. and Cuypers, A. (2017) Cd-induced Cu deficiency responses in *Arabidopsis thaliana*: are phytochelatins involved? *Plant, Cell & Environment* 40, 390–400.

Gill, S.S. and Tuteja, N. (2010) polyamines and abiotic stress tolerance in plants. *Plant Signaling & Behavior* 5, 26–33.

Gomez, J., Sanchez-Martinez, M., Stiefel, V. and Pages, M. (1988) A gene induced by the plant hormone abscisic acid in response to water stress encodes a glycine-rich protein. *Nature* 334, 262–264.

He, C.Y., Zhang, J.-S. and Chen, S.Y. (2002) A soybean gene encoding a proline-rich protein is regulated by salicylic acid, endogenous circadian rhythm and by various stresses. *Theoretical and Applied Genetics* 104, 1125–1131.

Holmstrom, K.-O., Somersalo, S., Mandal, A. and Welin, B. (2000) Improved tolerance to salinity and low temperature in transgenic tobacco producing glycine betaine. *Journal of Experimental Botany* 51, 177–185.

Horvath, E., Szalai, G. and Janda, T. (2007) Induction of abiotic stress tolerance by salicylic acid signaling. *Journal of Plant Growth Regulation* 26, 290–300.

Huang, J., Gu, M., Lai, Z., Zhou, Y.-H. and Chen, Z. (2010) Functional analysis of the *Arabidopsis PAL* gene family in plant growth, development and response to environmental stress. *Plant Physiology* 153, 1526–1538.

Huang, S.-J., Chang, C.-L., Wang, P.-H. and Chang, I.-F. (2013) A type III ACC synthase, ACS7, is involved in root gravitropism in *Arabidopsis thaliana*. *Journal of Experimental Botany* 64, 4343–4360.

Hylton, C.M., Rawsthorne, S., Smith, A.M., Jones, D.A. and Woolhouse, H.W. (1988) Glycine decarboxylase is confined to the bundle-sheath cells of leaves of $C_3$- $C_4$ intermediate species. *Planta* 175, 452–459.

Igarashi, D., Tsuda, K. and Katagiri, F. (2012) The peptide growth factor, phytosulfokine, attenuates pattern-triggered immunity. *The Plant Journal* 71, 194–204.
Ingle, R.A. (2015) Histidine. In: D'Mello, J.P.F. (ed.) *Amino Acids in Higher Plants*. CAB International, Wallingford, UK, pp. 251–261.
Ishihara, A., Hashimoto, Y., Tanaka, C. and Wakasa, K. (2008) The tryptophan pathway is involved in the defense responses of rice against pathogenic infection via serotonin production. *The Plant Journal* 54, 481–495.
Ishitani, M., Nakamura, T., Han, S.Y. and Takabe, T. (1995) Expression of the betaine aldehyde dehydrogenase gene in barley in response to osmotic stress and abscisic acid. *Plant Molecular Biology* 27, 307–315.
Jain, M. and Khurana, J.P. (2009) Transcript profiling reveals diverse roles of auxin-responsive genes during reproductive development and abiotic stress in rice. *FEBS Journal* 276, 3148–3162.
Jakab, G., Ton, J., Flors, V., Zimmerli, L. and Metraux, J.P. (2005) Enhancing *Arabidopsis* salt and drought stress tolerance by chemical priming for its drought stress tolerance by chemical priming for its abscisic acid responses. *Plant Physiology* 139(1), 267–274. doi: 10.1104/pp.105.065698.
Jisha, K.C. and Puthur, J.T. (2016) Seed priming with BABA (γ-aminobutyric acid): a cost-effective method of abiotic stress tolerance in *Vigna radiata* (L.) Wilczek. *Protoplasma* 253, 277–289.
Kalamaki, M.S., Merkouropoulos, G. and Kanellis, A.K. (2009a) Can ornithine accumulation modulate stress tolerance in *Arabidopsis*? *Plant Signaling & Behaviour* 4, 1099–1101.
Kalamaki, M.S., Alexandrou, D.A., Lazari, D., Pateraki, I. and Kanellis, A.K. (2009b) Over-expression of a tomato *N*-acetyl-L-glutamate synthase gene (*SlNAG S1*) in *Arabidopsis thaliana* results in high ornithine levels and increased tolerance in salt and drought stresses. *Journal of Experimental Botany* 60, 1859–1871.
Kang, J.-Y., Choi, H., Im, M. and Kim, S.Y. (2002) *Arabidopsis* basic leucine zipper proteins that mediate stress-responsive abscisic acid signaling. *The Plant Cell* 14, 343–357.
Kazan, K. (2015) Diverse roles of jasmonates and ethylene in abiotic stress tolerance. *Trends in Plant Science* 20, 219–229.
Kazan, K. and Manners, J.M. (2008) Jasmonate signaling: towards an integrated view. *Plant Physiology* 146, 1459–1468.
Khan, M.I.R., Fatma, M., Per, T.S., Anjum, N.A. and Khan, N.A. (2015) Salicylic acid-induced abiotic stress tolerance and underlying mechanism in plants. *Frontiers in Plant Science* 6, 462. doi: 10.3389/fpls.2015.00462.
Kim, N.H., Kim, B.S. and Hwang, B.K. (2013) Pepper arginine decarboxylase is required for polyamine and γ-aminobutyric acid signaling in cell death and defense response. *Plant Physiology* 162, 2067–2083.
Kobe, B. and Kajava, A.V. (2001) The leucine-rich repeat as a protein recognition motif. *Current Opinion in Structural Biology* 11, 725–732.
Kuhnlenz, T., Hofmann, C., Uraguchi, S., Weber, M., Lahner, B., Salt, D.E. and Clemens, S. (2016) Phytochelatin synthesis promotes leaf Zn accumulation of *Arabidopsis thaliana* plants grown in soil with adequate Zn supply and is essential for survival in Zn-contaminated soil. *Plant and Cell Physiology* 57, 2342–2352.
Kulp, K.S. and Vulliet, P.R. (1996) Mimosine blocks cell cycle progression by chelating iron in asynchronous human breast cancer cells. *Toxicology and Applied Pharmacology* 139, 356–364.
Lam, H.-M., Chiu, J., Hsieh, M.-H., Oliveira, I.C. and Coruzzi, G. (1998) Glutamate receptor genes in plants. *Nature* 396, 125–126.
Lehmann, S., Funck, D., Szabados, I. and Rentsch, D. (2010) Proline metabolism and transport in plant development. *Amino Acids* 39, 949–962.
Leon, J., Rojo, E. and Sanchez-Serrano, J.J. (2001) Wound signaling in plants. *Journal of Experimental Botany* 52, 1–9.
Lestari, R., Rio, M., Martin, F., Leclercq, J., Roques, S. *et al.* (2018) Overexpression of *Hevea brasiliensis* ethylene response factor *HbERF-1Xe5* enhances growth and tolerance to abiotic stress and affects laticifer differentiation. *Plant Biotechnology Journal* 16, 322–336.
Li, J., Nagpal, P., Vitart, V., McMorris, T.C. and Chory, J. (1996) A role for brassinosteroids in light-dependent development of *Arabidopsis*. *Science* 272, 398–401.
Liu, T., Jiang, X., Shi, W., Chen, J., Pei, Z. and Zheng, H. (2011) Comparative proteomic analysis of differentially expressed proteins in β-aminobutyric acid-enhanced *Arabidopsis thaliana* tolerance to simulated acid rain. *Proteomics* 11, 2079–2094.
Ludwig-Muller, J. (2011) Auxin conjugates: their role for plant development and in the evolution of land plants. *Journal of Experimental Botany* 62, 1757–1773.

Maier, F., Zwicker, S., Funck, J. and Pfitzner, U.M. (2011) Nonexpressor of pathogenesis-related proteins 1 (NPR1) and some NPR1-related proteins are sensitive to salicylic acid. *Molecular Plant Pathology* 12, 73–91.

Mattoo, A.K., Fatima, T., Upadhyay, R.K. and Handa, A.K. (2015) Polyamines in plants: biosynthesis from arginine, and metabolic, physiological and stress-response roles. In: D'Mello, J.P.F. (ed.) *Amino Acids in Higher Plants*. CAB International, Wallingford, UK, pp. 177–194.

Mauch-Mani, B. and Slusarenko, A.J. (1996) Production of salicylic acid precursors is a major function of phenylalanine ammonia-lyase in the resistance of *Arabidopsis* to *Paronospora parasitica*. *The Plant Cell* 8, 203–212.

Meier, S., Bastian, R., Donaldson, L., Murray, S., Bajic, V. and Gehring, C. (2008) Co-expression and promoter content analyses assign a role in biotic and abiotic stress responses to plant natriuretic peptides. *BMC Plant Biology* 8, 24. doi: 10.1186/1471-2229-8-24.

Mohan, T.C., Castrillo, G., Navarro, C., Zarco-Fernandez, S., Mateo, C. *et al.* (2016) Cytokinin determines thiol-mediated arsenic tolerance and accumulation in *Arabidopsis thaliana*. *Plant Physiology* 171(2), 1418–1426. doi: 10.1104/pp.16.00372.

Moons, A., Prinsen, E., Bauw, G. and Van Montagu, M. (1997) Antagonistic effects of abscisic acid and jasmonates on salt stress-inducible transcripts in rice shoots. *The Plant Cell* 9, 2243–2259.

Moreau, M., Tian, M. and Klessig, D.F. (2012) Salicylic acid binds NPR3 and NPR4 to regulate NPR1-dependent defense responses. *Cell Research* 22, 1631–1633.

Moschou, P.N., Paschalidis, K.A., Dells, I.D., Andriopoulou, A.H. and Roubelakis-Angelakis, K.A. (2008) Spermidine exodus and oxidation in the apoplast induced by abiotic stress is responsible for $H_2O_2$ signatures that direct tolerance responses in tobacco. *The Plant Cell* 20, 1708–1724.

Nanjo, T., Fujita, M., Seki, M., Kato, T., Tabata, S. and Shinozaki, K. (2003) Toxicity of free proline in an *Arabidopsis* T-DNA-tagged mutant deficient in proline dehydrogenase. *Plant & Cell Physiology* 44, 541–548.

Negrin, V.L., Teixeira, B., Godinho, R.M., Mendes, R. and Vale, C. (2017) Phytochelatins and monothiols in salt marsh plants and their relation with metal tolerance. *Marine Pollution Bulletin* 121, 78–84.

Neill, S., Barros, R., Bright, J., Hancock, J. and Wilson, I. (2008) Nitric oxide, stomatal closure and abiotic stress. *Journal of Experimental Botany* 59, 165–176.

Nguyen, D., Rieu, I., Mariani, C. and Van Dam, N.M. (2016) How plants handle multiple stresses: hormonal interactions underlying responses to abiotic stress and insect herbivory. *Plant Molecular Biology* 91, 727–740.

Nishida, S., Duan, G., Ohkama-Ohtsu, N., Uraguchi, S. and Fujiwara, T. (2016) Enhanced arsenic sensitivity with excess phytochelatin accumulation in shoots of a *SULTR1;2* knockout mutant of *Arabidopsis thaliana* (L.) Heynh. *Soil Science and Plant Nutrition* 62(4), 367–372.

Oliver, D.J., Neuburger, M., Bourguignon, J. and Douce, R. (1990) Glycine metabolism by plant mitochondria. *Physiologia Plantarum* 80, 487–491.

Orsini, F., Cascone, P., De Pascale, S., Barbieri, G., Corrado, G., Rao, R. and Maggio, A. (2010) Systemin-dependent salinity tolerance in tomato: evidence of specific convergence of abiotic and biotic stress responses. *Physiologia Plantarum* 138, 10–21.

Pal, M., Kovacs, V., Soos, V., Liu, H. and Janda, T. (2013) Salicylic acid and abiotic stress responses in rice. *Journal of Agronomy and Crop Science* 200(1), 1–11. doi: 10.1111/jac.12037.

Park, E.J., Jenkic, Z., Pino, M.T., Murata, N. and Chen, T.H. (2007) Glycinebetaine accumulation is more effective in chloroplasts than in the cytosol for protecting transgenic tomato plants against abiotic stress. *Plant, Cell & Environment* 30, 994–1005.

Pedranzani, H., Rodriguez-Rivera, M., Gutierrez, M., Porcel, R., Hause, B. and Ruiz-Lozano, J.M. (2016) Arbuscular mycorrhizal symbiosis regulates physiology and performance of *Digitaria eriantha* plants subjected to abiotic stresses by modulating antioxidant and jasmonate levels. *Mycorrhiza* 26, 141–152.

Penninckx, I.A.M.A., Thomma, B.P.H.J., Buchala, A. and Broekaert, W.F. (1998) Concomitant activation of jasmonate and ethylene response pathways is required for induction of a plant defensin gene in *Arabidopsis*. *The Plant Cell* 10, 2103–2113.

Planchet, E. and Limami, A.M. (2015) Amino acid synthesis under abiotic stress. In: D'Mello, J.P.F. (ed.) *Amino Acids in Higher Plants*. CAB International, Wallingford, UK, pp. 262–276.

Ravanel, S., Gakiere, B., Job, D. and Douce, R. (1998) The specific features of methionine biosynthesis and metabolism in plants. *Proceedings of the National Academy of Sciences of the United States of America* 95, 7805–7812.

Raychaudhuri, A. (2015) Alanine aminotransferase: amino acid metabolism in higher plants. In: D'Mello, J.P.F. (ed.) *Amino Acids in Higher Plants*. CAB International, Wallingford, UK, pp. 30–56.
Robert-Seilaniantz, A., Grant, M. and Jones, J.D.G. (2011) Hormone crosstalk in plant disease and defense: more than just jasmonate-salicylate antagonism. *Annual Review of Phytopathology* 49, 317–343.
Sakamoto, A. and Murata, N. (2002) The role of glycine betaine in the protection of plants from stress: clues from transgenic plants. *Plant, Cell & Environment* 25, 163–171.
Sanjaya, H.P.-Y., Su, R.-C., Ko, S.-S., Tong, C.-G. and Chan, M.-J. (2008) Overexpression of *Arabidopsis thaliana* tryptophan synthase beta1 (AtTSB1) in *Arabidopsis* and tomato confers tolerance to cadmium stress. *Plant, Cell & Environment* 31, 1074–1085.
Sharma, S.S. and Dietz, K.-J. (2006) The significance of amino acids and amino acid-derived molecules in plant response and adaptation to heavy metal stress. *Journal of Experimental Botany* 57, 711–726.
Shelp, B.J., Bozzo, C.G., Trzobacher, C.P., Zarei, A., Deyman, K.L. and Brikis, C.J. (2012) Hypothesis/review: contribution of putrescine to 4-aminobutyrate (GABA) production in response to abiotic stress. *Plant Science* 194, 130–135.
Shi, H., Ye, S., Wang, Y., Zhang, Y. and Chen, Z. (2013) Manipulation of arginase expression modulates abiotic stress tolerance in *Arabidopsis*: effect on arginine metabolism and ROS accumulation. *Journal of Experimental Botany* 64, 1367–1379. doi: 10.1093/jxb/ers400.
Singh, S., Tripathi, D.K., Chauhan, D.K. and Dubey, N.K. (2016) Glutathione and phytochelatins mediated redox homeostasis and stress signal transduction in plants: an integrated overview. In: Ahmad, P. (ed.) *Plant Metal Interaction: Emerging Remediation Techniques*. Elsevier, Amsterdam, pp. 285-310. doi: 10.1016/B978-0-12-803158-2.00011-4.
Skopelitis, D.S., Paranychianakis, N.V., Paschalidis, K.A., Kouvarakis, K.A. and Roubelakis-Angelakis, K.A. (2006) Abiotic stress generates ROS that signal expression of anionic glutamate dehydrogenase to form glutamate for proline synthesis in tobacco and grapevine. *The Plant Cell* 18, 2767–2781.
Stranska, J., Kopecny, D., Snegaroff, J. and Sebela, M. (2008) Ornithine δ-aminotransferase: an enzyme implicated in salt tolerance in higher plants. *Plant Signalling & Behavior* 3, 929–935.
Stranska, J., Tylichova, M., Snegaroff, J. and Sebela, M. (2010) Biochemical characterization of pea ornithine-delta-aminotransferase: substrate specificity and inhibition by di- and polyamines. *Biochimie* 92, 940–948.
Svyatyna, K., Brendel, R., Takano, M., Nick, P. and Riemann, M. (2013) Light induces jasmonate-isoleucine conjugation via OsJAR 1-dependent and -independent pathways in rice. *Plant, Cell & Environment* 37, 827–839. doi: 10.1111/pce.12201.
Szabados, L. and Savoure, A. (2010) Proline: a multifunctional amino acid. *Trends in Plant Science* 15, 89–97.
Taylor, N.L., Day, D.A. and Millar, A.H. (2002) Environmental stress causes oxidative damage to plant mitochondria leading to inhibition of glycine decarboxylase. *Journal of Biological Chemistry* 277, 42663–42668.
Timm, S., Florian, A., Stitt, M., Frenie, A.R. and Bauwe, H. (2012) Glycine decarboxylase controls photosynthesis and plant growth. *FEBS Letters* 586, 3692–3697.
Tiwari, S.B., Belachew, A., Ma, S.F., Young, M., Ade, J. *et al.* (2012) The EDLL motif: a potent plant transcriptional activation domain from AP2/ERF transcription factors. *The Plant Journal* 70, 855–865.
Torok, A., Gulyas, Z., Szalai, G., Kocsy, G. and Majdik, C. (2015) Phytoremediation capacity of aquatic plants is associated with the degree of phytochelatin polymerization. *Journal of Hazardous Materials* 299, 371–378.
Tounekti, T., Hernandez, I., and Munne-Bosch, S. (2013) Salicylic acid biosynthesis and role in modulating terpenoid and flavonoid metabolism in plant responses to abiotic stress. In: Hayat, S., Ahmad, A. and Alyemeni, M.N. (eds) *Salicylic Acid*. Springer Science & Business Media, Dordrecht, pp. 141–160.
Turull, M., Grmanova, G., Dago, A., Arino, C., Diez-Cruz, J.M. and Esteban, M. (2017) Phytochelatin synthesis in response to Hg uptake in aquatic plants near a chlor-alkali factory. *Chemosphere* 176, 74–80.
Uno, Y., Furihata, T., Abe, H., Yoshida, R. and Yamaguchi-Shinozaki, K. (2000) *Arabidopsis* basic leucine zipper transcription factors involved in an abscisic acid-dependent signal transduction pathway under drought and high-salinity conditions. *Proceedings of the National Academy of Sciences of the United States of America* 97, 11632–11637.
Uraguchi, S., Tanaka, N., Hofmann, C., Abiko, K., Weber, M. *et al.* (2017) Phytochelatin synthase has contrasting effects on cadmium and arsenic accumulation in rice grains. *Plant & Cell Physiology* 58, 1730–1742.
Van der Does, D., Boutrot, F., Engelsdork, T., Rhodes, J., Mckenna, J.F. *et al.* (2017) The *Arabidopsis* leucine-rich repeat receptor kinase MIK2/LRR-KISS connects cell wall integrity sensing, root growth

and response to abiotic and biotic stresses. *PLoS Genetics* 13(6), e1006832. doi: 10.1371/journal.pgen.1006832.
Wang, H., Liang, X., Huang, J., Liu, Z. and Bi, Y. (2010) Involvement of ethylene and hydrogen peroxide in induction of alternative respiratory pathway in salt-treated *Arabidopsis* calluses. *Plant & Cell Physiology* 51, 1754–1765.
Wang, K.L.-C., Li, H. and Ecker, J.R. (2002) Ethylene biosynthesis and signaling networks. *The Plant Cell* 14, S131–S151.
Wei, L.-J., Deng, X.-G., Zhu, T., Zheng, T., Li, P.-X., Wu, J.-Q., Zhang, D.-W. and Lin, H.-H. (2015) Ethylene is involved in brassinosteroids induced alternative respiratory pathway in cucumber (*Cucumis sativus* L.) seedlings response to abiotic stress. *Frontiers in Plant Science* 6, 982. doi: 10.3389/fpls.2015.00982.
Wu, G., Fang, Y.-Z., Yang, S., Lupton, J.R. and Turner, N.D. (2004) Glutathione metabolism and its implications for health. *Journal of Nutrition* 134, 489–492.
Xiong, L., Schumaker, K.S. and Zhu, J-K. (2002) Cell signaling during cold, drought and salt stress. *The Plant Cell* 14, S165–S183.
Xu, Y., Chang, D., Liu, M.L., Hasegawa, P.M. and Bressan, R.A. (1994) Plant defense genes are synergistically induced by ethylene and methyl jasmonate. *The Plant Cell* 6, 1077–1085.
Yasuda, M., Ishikawa, A., Seki, M., Shinozaki, K., Yoshida, S. and Nakashita, H. (2008) Antagonistic interaction between systemic acquired resistance and the abscisic acid-mediated abiotic stress response in *Arabidopsis*. *The Plant Cell* 20, 1678–1692.
Yokota, A., Kawasaki, M., Iwano, M. and Akashi, K. (2002) Citrulline and DRIP-1 protein (ArgE homologue) in drought tolerance of wild watermelon. *Annals of Botany* 89, 825–832.
Young, V.R and Ajami, A.M (2000) Glutamate: an amino acid of particular distinction. *Journal of Nutrition* 130, 892S–900S.
Zeier, J., Pink, B., Mueller, M.J. and Berger, S. (2004) Light conditions influence specific defence responses in incompatible plant-pathogen interactions: uncoupling systemic resistance from salicylic acid and PR-1 accumulation. *Planta* 219, 673–683.
Zhang, S.-W., Li, C.-H., Cao, J., Zhang, Y.-C., Zhang, S.-Q., Xia, Y.-F., Sun, D.-Y. and Sun, Y. (2009) Altered architecture and enhanced drought tolerance in rice via the down-regulation of indole-3-acetic acid by *TLD1/OsGH3.13* activation. *Plant Physiology* 151, 1889–1901.
Zhang, X., Shen, L., Li, F., Zhang, Y., Mang, D. and Sheng, J. (2010) Up-regulating arginase contributes to amelioration of chilling stress and the antioxidant system in cherry tomato fruits. *Journal of the Science of Food and Agriculture* 90, 2195-2202.
Zhang, J., Hwang, J.-U., Song, W.-Y., Martinola, E. and Lee, Y. (2017) Identification of amino acid residues important for the arsenic resistance function of *Arabidopsis* ABCC1. *FEBS Letters* 591, 656–666.
Zhao, J. and Last, R.L. (1996) Coordinate regulation of the tryptophan biosynthetic pathway and indolic phytoalexin accumulation in *Arabidopsis*. *The Plant Cell* 8, 2235–2244.
Zimmerli, L., Hou, B.H., Tsai, C.H., Jakab, G., Mauch-Mani, B. and Somerville, S. (2008) The xenobiotic beta-aminobutyric acid enhances *Arabidopsis* thermotolerance. *The Plant Journal for Cell and Molecular Biology* 54, 144–151.

# Part II

# Ambient Gases Affecting Human Health and Adaptation in Higher Plants

# 5 Ozone I. Human Disorders: an Overview

**P. Silveyra,* N. Fuentes and L. Rivera**
*The University of North Carolina at Chapel Hill, School of Nursing, Chapel Hill, North Carolina, USA*

## 5.1 Abstract

Ground-level ozone is one of the most widespread pollutants in the world. Millions of people are exposed to unhealthy levels of ozone air pollution on a daily basis. Often called 'smog', ozone is harmful to breathe and is associated with the development and exacerbation of pulmonary and cardiovascular disease. Scientists have studied the effects of ozone exposure on human health for decades. Hundreds of research studies have confirmed that ozone harms the human body when inhaled at levels currently found in many cities and countries. Ozone is a gas molecule composed of three oxygen atoms that is formed by a reaction between atmospheric oxygen and organic volatile compounds from vehicle and industrial emissions in the presence of sunlight. When inhaled, ozone aggressively attacks lung tissue and triggers systemic inflammation that can result in a series of health complications, ranging from inflammation and injury of the airways to alterations of vital functions at the cardiovascular, immune, endocrine and neurological levels and even death. In women, exposure to ozone can also affect fertility, cause pregnancy complications and alter fetal development. It has also been noted that susceptibility to ozone varies with age, gender and health status and studies have indicated that sex hormones can contribute to the damaging effects of ozone in the lung. Patients suffering from inflammatory lung disease, including asthma, bronchitis, chronic obstructive pulmonary disease and lung infection, are at significantly higher risk for negative health effects associated with ozone exposure. Numerous studies in cells and animal models have helped to elucidate some of the mechanisms of ozone toxicity and have identified cell receptors and inflammatory cytokines that activate the inflammatory response upon ozone exposure. These have also provided information for environmental agencies to control emissions and warn the population about unhealthy levels. In this chapter, we summarize the known effects of ozone exposure on human health, as well as the main results from clinical and animal studies testing ozone exposure at different doses. We also discuss current policies and regulations aimed to control emissions and potential harm to various populations. A better understanding of the negative effects of ozone exposure on human health and of the mechanisms associated with its toxicity will likely inform policy makers to introduce new guidelines to control environmental emissions and will allow for the development of strategies for symptom management

* E-mail address: patry@email.unc.edu

and potential therapeutics for different individuals exposed to unhealthy ozone levels.

## 5.2 Introduction

In the past decade, air pollution has become the world's single biggest environmental health risk, causing about 7 million deaths worldwide annually, or nearly one in every eight deaths (Soriano *et al.*, 2017). Exposure to air pollutants, including ground-level ozone, biomass fuels, particulate matter, sulfur dioxide and nitric dioxide, has been strongly associated with increased mortality from cardiovascular and lung disease (Ostro *et al.*, 2010; Kurt *et al.*, 2016). These pollutants are generally present in the environment as a mixture of gases and particles that are products of combustion of fossil fuels, diesel traffic, wood smoke, industrial processes and some sources of domestic energy (Bernstein *et al.*, 2008; Kampa and Castanas, 2008).

Epidemiological studies have shown that the onset and clinical course of lung diseases are strongly influenced by poor air quality (Kurt *et al.*, 2016; Perng and Chen, 2017). This represents a significant healthcare and economic burden, especially since low-income and underserved communities tend to live in areas with higher air pollution levels (Forno and Celedon, 2009; Miranda *et al.*, 2011; Collins *et al.*, 2017). Exposure to ozone has been associated with decreases in lung function and immunity, increased lung injury and higher lung disease exacerbations and hospitalization rates (Ghio *et al.*, 2000; Goss *et al.*, 2004; Hollingsworth *et al.*, 2007; Mikerov *et al.*, 2008; Durrani *et al.*, 2011; Kim *et al.*, 2011; Rice *et al.*, 2013; Sesé *et al.*, 2017). While the mechanisms by which ozone affects human health remain poorly understood, a great deal of its negative effects can relate to its ability to cause oxidative stress in various cell types. To this end, studies in animal models have identified some of the key players of this response and its regulation (Ciencewicki *et al.*, 2008).

Different from the 'good' ozone present in the upper atmospheric layer, ground-level (tropospheric) ozone is a reactive oxidant gas formed by the photochemical reactions of carbon monoxide, nitric dioxide and volatile organic compounds, which are found at higher concentrations in major cities (Jariyasopit *et al.*, 2014; Bromberg, 2016). Ozone together with particulate matter represent the two dominant air pollutants worldwide. The 'State of the Air 2017' from the American Lung Association (ALA) recently reported that almost 40% of Americans live in areas with unhealthy levels of ozone and about 116.5 million people (36% of the US population) live in areas that have earned an F (lowest grade) for ozone pollution in 2016 (American Lung Association, 2017). Either alone or in combination with airborne particulate matter, ozone increases risk of death from respiratory causes and correlates with shortened life expectancy (Romieu *et al.*, 2012; Chen *et al.*, 2017; Ebenstein *et al.*, 2017). Furthermore, extensive epidemiological evidence has demonstrated age, gender and inter-individual differences in susceptibility to environmental exposures, with specific genetic polymorphisms associated with negative health effects (Silveyra and Floros, 2012). In this regard, the negative effects of ozone are markedly higher in women than in men (Yoda *et al.*, 2014; Fernandez *et al.*, 2015; Coogan *et al.*, 2017). Studies in animal models have demonstrated that the inflammatory response to ozone varies with the sex and hormonal status of the animal (Cabello *et al.*, 2015; Mishra *et al.*, 2016). It has also been postulated that since women have more relative fat mass than men, they possess a larger distribution volume for highly lipophilic substances such as chemical particles present in the environment. Importantly, these substances are also metabolized faster in women's cells, resulting in higher toxicity (Ben-Zaken Cohen *et al.*, 2007).

Scientists have studied the effects of ozone exposure on human health for decades. The results from these investigations have influenced policy and regulations aimed to improve air quality. Our understanding of the disease processes has also allowed for the development of guidelines to help to prevent disease exacerbations in vulnerable populations. While experts in the field have concluded that ozone pollution poses multiple serious threats to health, the specific mechanisms associated with these effects remain poorly described. In this chapter, we summarize the major health complications arising from exposure to ground-level ozone in various organs and tissues, and the influences of factors such as age, gender, or ethnicity in the severity of these effects. We also provide an overview of the animal models used to study the mechanisms of ozone toxicity and the major findings that lead to changes in environmental policy worldwide.

## 5.3 Ambient Ozone

Ozone develops in the atmosphere from gases that come out of tailpipes, smokestacks and many other sources. When these gases come into contact with sunlight, they react and form ozone smog. The main substrates for ozone production are nitrogen oxides (NOx), volatile organic compounds (VOCs) and carbon monoxide (CO) (Fig. 5.1). These gases are produced primarily by combustion of fossil fuels such as gasoline, oil and coal. NOx is emitted from power plants, motor vehicles and other sources of high heat combustion, especially in big cities. CO is primarily emitted from motor vehicles and VOCs are emitted from motor vehicles, chemical plants, refineries, factories and other sources. Because ultraviolet (UV) radiation from sunlight affects the rate by which ozone is produces, its accumulation at different latitudes, seasons and time of the day also varies.

As mentioned earlier, ozone air pollution at ground level causes serious health problems. This represents a significant environmental health problem in highly industrialized countries like the USA, where nearly 40% of the population live in areas that have unhealthy levels of ozone, and in many cities in countries including China, India, Mexico, Turkey, Iran and Saudi Arabia, which are particularly known for having some of the worst ozone pollution levels in history and where ozone continues to reach dangerous levels to this day (Anonymous, 2014; Guarnieri and Balmes, 2014). As a consequence, several countries have established regulations aimed to control ozone levels by setting standards and penalties for different types of emissions.

In the USA, the Clean Air Act was established by the US Environmental Protection Agency (EPA) in 1965 and represents one of the first and most influential environmental and air quality laws in the world. In 1999, Canada adopted the Canadian Environmental Protection Act, which established comprehensive regulatory policies for the country. China has recently established an Airborne Pollution Prevention and Control Action Plan (2012), aimed to reduce over a quarter of emissions in the space of 5 years. New Zealand established a Clean Air Act in 1972 that was later amended by the Resource Management Plan of 1991. Many of these regulations resulted from seriously adverse events associated with high levels of air pollution. For example, the UK responded to the Great Smog of 1952, a historical high-pollution event in London responsible for over 6000 deaths and over 25,000 health-related events, by implementing its Clean Air Acts of 1956 and 1968, later updated in 1993.

International agencies have also established agreements to monitor and control transnational air quality. Among these, the Convention on Long-Range Transboundary Air Pollution (directed by the United Nations Economic Commission for Europe (UNECE) and implemented by the European Monitoring and Evaluation Programme (EMEP)) has established protocols to reduce levels of various pollutants, including ozone (under the

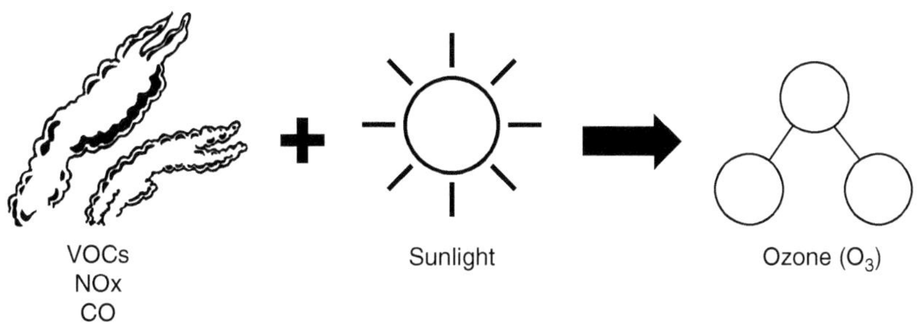


**Fig. 5.1.** Tropospheric ozone formation. At ground level, ozone is formed by chemical reactions between volatile organic compounds (VOCs), nitrogen oxides (NOx) and carbon monoxide (CO) from anthropogenic sources, in the presence of sunlight.

multi-effect protocol), and the United Nations Framework Convention on Climate Change (UNFCCC) (which includes the Kyoto protocol and the Paris Agreement), aims to reduce emissions and stop the increase in global average temperatures. Thanks to the provisions in these policies, many counties have continued to reduce ozone pollution and other pollutants for decades, despite increases in population, transportation, industrialization and miles driven per capita. However, oil and gas production wells, processing plants, transmission pipelines and storage units continue to emit harmful gases, contributing to unhealthy ozone levels, and millions of people still live in areas where monitors show unhealthy levels of ozone. Additional policies, including the US EPA's Clean Power Plan, have aimed to reduce carbon pollution from power plants. This type of pollution is responsible for climate warming, which enhances conditions for ozone formation. It is estimated that a reduction in carbon pollution from electricity generation will also reduce ozone pollution from these plants and will prevent up to 3600 premature deaths and 90,000 asthma attacks in children in 2030 in the USA.

In 2015, the US EPA began to implement additional protective limits on ozone pollution. These are commonly known as National Ambient Air Quality Standards, and aim to regulate the clean-up of ozone pollution nationwide. The Clean Air Act requires that these standards are reviewed every 5 years to make sure that they protect the health of the public. Based on the monitored levels of ozone in the environment over a period of 8 h, the Air Quality Index assigns a daily value ranging from 0 to 500 for a particular area. This level is associated with a colour that represents its health concern. These data can be accessed in real time (http://airnow.gov) and are summarized in Table 5.1.

The current national ambient air quality standard for ozone is 70 parts per billion over 8 h. Levels above this threshold are considered unhealthy. The ALA assigns different 'grades', ranging from A (best) to F (worst), to cities and counties on a yearly basis and summarizes these results in its 'State of the Air' reports (American Lung Association, 2017). An A or B grade will be earned by having less than 2 orange days and no red days, and a grade F will be assigned to a region exceeding 9 days over the standard,

**Table 5.1.** Air quality index and health concerns associated with environmental ozone levels.

| 8-h Ozone concentration (ppb) | Air Quality Index (AQI) values | Colour | Level of health concern |
|---|---|---|---|
| 0–54 | 0–50 | Green | **Good**<br>Air quality is considered satisfactory, and air pollution poses little or no risk |
| 55–70 | 51–100 | Yellow | **Moderate**<br>Air quality is acceptable; however, for some pollutants there may be a moderate health concern for a very small number of people who are unusually sensitive to air pollution |
| 71–85 | 101–150 | Orange | **Unhealthy for Sensitive Groups**<br>Members of sensitive groups may experience health effects. The general public is not likely to be affected |
| 86–105 | 151–200 | Red | **Unhealthy**<br>Everyone may begin to experience health effects; members of sensitive groups may experience more serious health effects |
| 106–200 | 201–300 | Purple | **Very Unhealthy**<br>Health alert: everyone may experience more serious health effects |
| > 201 | 301–500 | Maroon | **Hazardous**<br>Health warnings of emergency conditions. The entire population is more likely to be affected |

10 orange days, or 9 total days including at least 1 or more red, purple, or maroon days. This information has been crucial to understand the effects of air pollution on human health and raise awareness about the dangers of breathing ozone. Aggregated data from the 2017 report show clear associations of lung and cardiovascular disease incidence over a period of 2 years in counties that earned an F grade. For example, the number of adults with asthma in areas earning F grades was over 7.6 million, while it reached only 2.1 million in areas earning A grades. The differences for paediatric asthma were even more marked, with over 2.3 million in areas earning F versus only half a million in areas receiving A grades. Moreover, chronic obstructive pulmonary disease (COPD) was about three times more prevalent in F areas (4.7 million) than in A areas (1.6 million). Similarly, cardiovascular disease was over three times more frequent in F areas (6.58 million) than in A areas (2.1 million). Health disparities were also evident from this report. Not only did most of the populations studied live in areas with F grades (116 million people) versus A grades (only 31.5 million), but also populations living in poverty were four times more likely to live in areas earning F versus A grades (16.5 million versus 4.5 million, respectively). The most vulnerable populations (children under 18 years of age and elders over 65 years of age) were also found to be more likely to live in areas earning F grades (American Lung Association, 2017).

## 5.4 Ozone Effects on Human Health

Ozone is a gas with a dual action for health. While it has a protective function blocking UV radiation from the sun in the stratosphere or ozonosphere (Xia *et al.*, 2010), it has severe negative effects when inhaled at ground level, where it increases morbidity and mortality from pulmonary and cardiovascular disease (Bell *et al.*, 2004). Additionally, ozone possesses therapeutic properties and has been extensively used as a therapy agent in patients with cardiovascular disease and in many branches of medicine and medical specialties. Its clinical uses include cardiovascular disease treatment, peripheral vascular disease, neurological disease, activation of the immune and neuro-endocrine systems, orthopaedic treatment, gastrointestinal and genito-urinary applications, cosmetic use and to improve oxygen transport to ischaemic tissue. These applications have been widely supported by the medical community due to its safety and effectiveness in multi-international studies (Bocci *et al.*, 2011; Smith *et al.*, 2017).

At ground (tropospheric) level, ozone exposure can range from mildly dangerous to deadly (Table 5.2). A variety of factors influence the effects of ozone inhalation on human health. These include pre-existing diseases and conditions, age and gender of the individual, duration and concentration of exposure, tidal volume and respiratory rate (e.g. during exercise), as well as other anatomical and physiological factors (Fig. 5.2).

When inhaled, ozone damages the nasal and upper respiratory tracts as it travels to the alveoli, where most of the damage is exerted and where it can activate alveolar macrophages and epithelial cells to produce inflammatory cytokines. The effects vary with exposure time, with shorter exposures leading to decreased pulmonary function, airway inflammation and increased bronchial reactivity, which are often reversible. The effects of longer exposures, on the other hand, often result in permanent changes to lung

**Table 5.2.** Main effects of ozone exposure in human health.

| Short-term exposure | Long-term exposure |
|---|---|
| Wheezing | Premature death |
| Coughing | Developmental harm |
| Shortness of breath | Reproductive harm |
| Lung tissue redness and swelling | Lung cancer |
| Asthma attack | Asthma (new onset) |
| Exacerbation of pre-existing lung disease | Cardiovascular harm |
| | Increased susceptibility to infections |

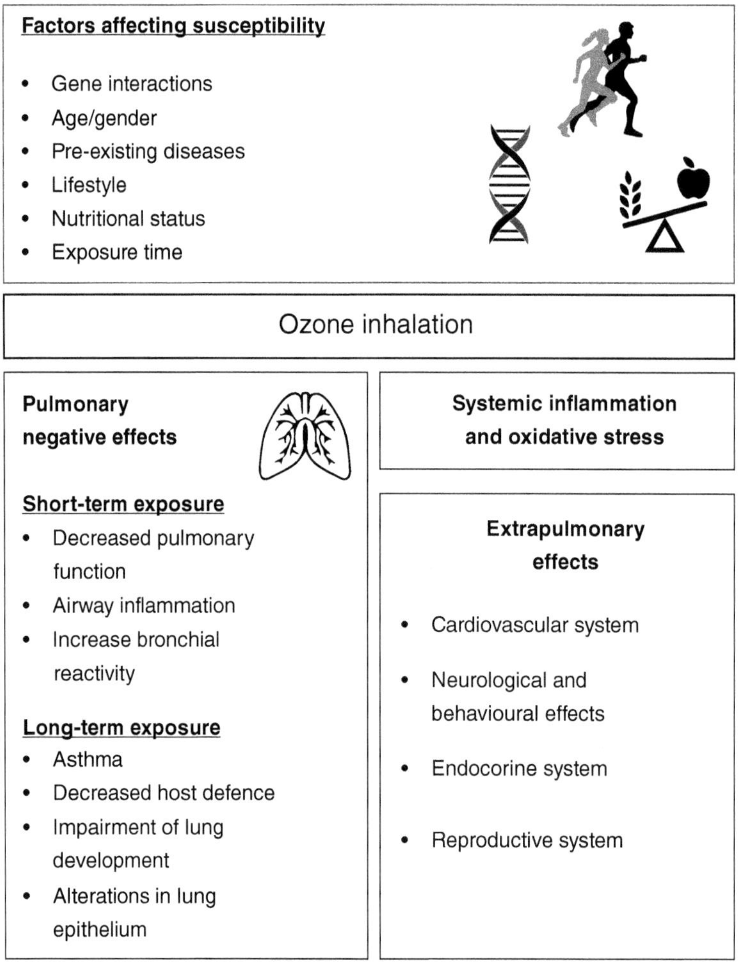


**Fig. 5.2.** The toxic pathway of ozone starts in the respiratory tract and is influenced by pre-existing conditions and factors. Short-time exposure to ozone results in acute and reversible effects, whereas long-term exposure leads to structural changes and extrapulmonary effects affecting the important systems of the body.

architecture via airway remodelling and triggering of oxidative stress that results in systemic inflammation and extrapulmonary damage (Fig. 5.2). Moreover, exposures to high ozone concentrations are associated with increased hospital admissions for lung disease, including pneumonia, asthma, allergic rhinitis and lung infection, as well as with premature death (Ito *et al.*, 2005). Healthy people also exhibit a small but significant decrease in lung function following a prolonged exposure to ozone levels as low as 60 ppb during mild exercise (Kim *et al.*, 2010).

For years, it has been noted that children and older adults have an increased risk for negative health effects associated with ozone exposure. Investigators have postulated that this is partially due to these groups spending more time outdoors than other age groups. In addition, children are particularly vulnerable to ozone exposure because their lungs are still growing and developing until 18–20 years of age and because they have rapid breathing rates that increase their exposure to the inhaled pollutant (Sarangapani *et al.*, 2003; Vinikoor-Imler *et al.*, 2014). This is particularly true for children who play sports in communities with high daytime ozone concentrations, in which a study showed that children who played sports were more likely to develop asthma than children who played no sports (McConnell *et al.*, 2002). A separate study in children showed that ozone exposure results in decreased lung volumes, forced vital capacity (FVC) and forced expiratory volume in 1 second (FEV1). The same study reported sex differences in ozone susceptibility, with boys being more affected than girls (Hwang et al., 2015). Moreover, genetic factors have also been proposed to affect oxidative metabolism and predisposition to negative health effects triggered by ozone (Thurston *et al.*, 1997; Vinikoor-Imler *et al.*, 2014). In the case of elderly adults, the mechanisms of increased susceptibility and risk of ozone-related hospitalization or death have been often associated with pre-existing lung conditions and a physiological decrease in immune function associated with age (Delfino *et al.*, 1998).

Regarding nutritional contributions, deficiency in vitamins C and E have also been show to predispose patients to ozone damage. In children with asthma, supplementation with these two vitamins has been successful in preventing ozone-mediated inflammation (Romieu *et al.*, 2009). Additionally, obese individuals are highly susceptible to ozone, partially due to associated co-morbidities leading to decreases in lung function that could worsen the effects of ozone inhalation (Alexeeff *et al.*, 2007).

Not surprisingly, patients suffering from lung disease are particularly susceptible to ozone exposure. Specifically, people with asthma have greater bronchoconstriction after 2 h of ozone exposure at 400 ppb (Kreit *et al.*, 1989). Smokers and patients with COPD have decreased spirometry values following ozone exposure, when compared with healthy patients exposed to similar ozone levels. In respiratory diseases disproportionately affecting women (e.g. asthma, COPD), women appear to be more susceptible to the damaging effects of ozone air pollution than men, but the mechanisms associated with these disparities remain unclear (Chauhan and Johnston, 2003; Butter, 2006; Martinez *et al.*, 2007; Makri and Stilianakis, 2008; Ostro *et al.*, 2010; Sorheim *et al.*, 2010; Negrisoli and Nascimento 2013; Wang and Chau, 2013; Pinkerton *et al.*, 2015; Zhao *et al.*, 2017). In this regard, pregnant women are highly susceptible to air pollution, which is responsible for premature delivery and complications of pregnancy, including low birthweight and health issues (Salam *et al.*, 2005; Ha *et al.*, 2018). This is important, considering that each year 15 million babies die due to complications from pre-term birth. Some studies have also suggested increased risk of stillbirth with ozone exposure, especially in women in the third trimester of pregnancy (Brown *et al.*, 2015; Green *et al.*, 2015; Arroyo *et al.*, 2016; Yang *et al.*, 2018). Interestingly, female newborns appear to be more susceptible to these effects than males (Fernando Costa Nascimento *et al.*, 2017). Studies in children and teenagers reported that children growing up in areas with higher ozone pollution face increased risk of having underdeveloped lungs, and girls with asthma and boys who spend more time outdoors show significant reductions in lung function (Galizia and Kinney, 1999; Peters *et al.*, 1999).

Breathing ozone can also significantly shorten life expectancy. Strong evidence exists of the deadly impact of ozone in large studies conducted in cities around the world. The risk of premature death increases with higher levels of ozone, either alone or in combination with other air pollutants. Patients with pre-existing conditions or from vulnerable populations are at higher risk. The next sections discuss population studies addressing the specific effects of ozone exposure in various organs and systems, as well as in healthy and vulnerable populations.

### 5.4.1 Respiratory effects

Several studies have indicated a direct relationship of ozone environmental levels with increases in population risk for respiratory infection and inflammatory lung disease. The main cause associated with these effects is the impairment of pulmonary host immune defence mechanisms that lead to an increased risk for infectious disease. Exposure to ozone also results in increased airway inflammation and injury, as demonstrated by increases in bronchoalveolar lavage fluid inflammatory cells, including neutrophils and eosinophils, as well as increased levels of pro-inflammatory cytokines (Arjomandi *et al.*, 2015). Interestingly, this inflammatory response can persist for longer than 18 h. Furthermore, short-term exposure to ozone results in shortness of breath, wheezing and coughing, as well as decreased pulmonary function including changes in lung volumes and increased airway resistance (Brown *et al.*, 2008). In patients with asthma, ozone exposure results in an increase of the severity of symptoms such as coughing, wheezing and shortness of breath, as well as increases in asthma attacks, medication use and hospitalization rates (McDonnell and Smith, 1994; Delfino *et al.*, 1997; Devlin *et al.*, 2012). These effects are markedly higher during the summer months, which is the high ozone season, where asthmatic patients showed increases in the frequency of respiratory symptoms, eosinophilia and exhaled NO levels (Khatri *et al.*, 2009). In fact, many areas in the USA produce enough ozone during the summer months to cause health problems that can be felt immediately. While short-term ozone exposures have been associated with increases in asthma exacerbations, long-term exposures for periods longer than 8 h (including days, months, or years), have been linked with new onset of asthma or early death, especially in patients carrying specific genetic variants (Fig. 5.2). On the other hand, studies in patients suffering from COPD indicate that ozone exposure may induce an increase in cardiovascular acute events, rather than respiratory events (Peel *et al.*, 2007).

### 5.4.2 Cardiovascular effects

For years, clinical studies have examined acute cardiovascular effects of air pollution exposure (Dugas, 2018). Collectively, these studies have found that exposure to ozone increases the risk for acute cardiovascular events, via mechanisms that involve increased myocardial work, and impairment of pulmonary gas exchange that can affect pre-existing cardiovascular conditions. Several studies around the world have reported increased risk of hospital admissions or emergency department visits for cardiovascular disease (Wong *et al.*, 1999; Middleton *et al.*, 2008; Linares and Diaz, 2010; Azevedo *et al.*, 2012), increased risk of heart attacks in middle-aged adults without heart disease (Ruidavets *et al.*, 2005) and elevated risk of cardiac arrhythmias associated with premature death and stroke (Rich *et al.*, 2006). Moreover, long-term exposure to ozone has been associated with systemic oxidative stress leading to endothelial cell activation and pro-coagulation effects, stimulating thrombosis and triggering myocardial infarction, stroke and cardiac arrhythmias. Recently, studies have also found specific associations between ozone exposure and the development of cardiovascular disease, even at ozone concentrations below the EPA recommended levels (Mirowsky *et al.*, 2017). One of the principal mechanisms for this increased risk is the alteration of the fibrinolysis pathway, which induces an increase in tissue-type plasminogen activator (tPA) and plasminogen activator inhibitor-1 (PAI-1). Both tPA and plasminogen form plasmin, which is the component that degrades fibrin and prevents coagulation (Angles-Cano, 1994). However, in ozone-exposed patients, plasma concentrations of tPA are lower than those of PAI-1, resulting in decreased fibrinolysis leading to an impairment of the coagulation process (Mirowsky *et al.*, 2017). In addition, thrombotic effects were found in a study of young healthy volunteers exposed to ozone concentrations as low as 0.03 ppm, in which vascular markers of thrombosis in blood showed a decrease in PAI-1 and an increase in tPA (Devlin *et al.*, 2012). This study also reported increased levels of inflammatory markers such as interleukin (IL)-1 β, IL-8 and tumor necrosis factor (TNF) in blood, and alterations in autonomic nervous system as measured by electrographic changes.

Vascular changes have also been associated with ozone exposure. In particular, vasoconstriction of brachial artery diameter in humans following exposure to ozone has been related to coronary vasoconstriction and increased risk of acute coronary events (Brook *et al.*, 2002; Mirowsky *et al.*, 2017). Autonomic dysfunction also has been reported, through increase of sympathetic activity, and these changes persisted for up to 24 h following ozone exposure (Arjomandi *et al.*, 2015). This autonomic dysfunction appears to be associated with activation of neural reflexes through participation of bronchial C fibres, as indicated in studies conducted in dogs (Schelegle *et al.*, 1993). In these, neural reflexes are activated leading to stimulation of the autonomic nervous system and subsequent electrophysiological responses in the heart.

### 5.4.3 Endocrine effects

Exposure to air pollution has recently been associated with increased incidence of metabolic diseases but the mechanisms responsible for these effects remain unknown. Experimental and epidemiological evidence suggests that exposure to air pollution, and specifically ozone, can affect metabolic and endocrine responses (Thomson *et al.*, 2018). For example, a study in healthy children from Mexico City who had been exposed to high levels of ozone and particulate matter during their prenatal stages found that they had increased serum levels of circulating leptin, a hormone involved in the regulation of appetite, but also a pro-inflammatory adipokine, which plays important roles in brain development, diabetes, obesity, cardiovascular disease and insulin resistance (Steppan and Swick 1999). In the same study, leptin increase was higher in females than in males and children also showed alterations in other hormones such as glucagon-like peptide 1(GLP-1), glucagon and ghrelin, which are important regulators of feeding behaviour (Calderon-Garciduenas *et al.*, 2015). From these and similar studies, investigators have concluded that ozone acts through both glucocorticoid-dependent and independent pathways (Thomson *et al.*, 2016) in combination with oxidative stress mechanisms to mediate adverse metabolic effects. Given the interrelated nature of metabolic, inflammatory and stress pathways, continued characterization of the collective responses to air pollution exposure in a variety of models should yield a better understanding of pollutant effects that underlie endocrine and metabolic disease processes.

### 5.4.4 Neurological effects

Recently, the negative effects of ozone exposure have been associated with diseases of the central nervous system, including stroke, Alzheimer's disease, Parkinson's disease and neurodevelopmental disorders (Genc *et al.*, 2012). In these, associations of ozone exposure and cognitive impairment in patients with Alzheimer's disease have been reported (Hwang *et al.*, 2015; Cleary *et al.*, 2018). A proposed mechanism involves oxidative stress and brain lipid peroxidation leading to memory deficiency, and alteration of neural function (Jung *et al.*, 2015). In patients with Parkinson's disease, alteration of dopaminergic neurons has also been found in patients exposed to ozone (Kirrane *et al.*, 2015). Additionally, the risk of some neurological and behavioural disorders as is the case of autism has also been linked to air pollution (Sheng *et al.*, 2010; Volk *et al.*, 2014). Finally, the association of ozone exposure and psychiatric disorders has also been evaluated. In these studies, increases in the number of emergency visits due to psychiatric causes and suicide risk have been found during the warmer seasons, when ozone levels are higher (Kim *et al.*, 2015; Oudin *et al.*, 2018).

### 5.4.5 Vulnerable populations

According to the American Lung Association, anyone who spends time outdoors where ozone pollution levels are high is at risk for its negative effects (American Lung Association, 2017). Five groups of people have been identified as especially vulnerable to the effects of breathing ozone: (i) children and teenagers; (ii) elderly individuals of age 65 years and older; (iii) individuals who work or exercise outdoors; (iv) people with existing inflammatory lung disease; and (v) individuals with existing cardiovascular disease. The effects observed in these individuals range from reduced lung function to airway obstruction that persists for several days, to reduced life expectancy. In addition, some evidence suggests that other groups, including women and obese individuals, as well as patients with cancer are also at higher risk for ozone negative effects. Breathing ozone may also increase the response to allergens in people with allergies.

Low-income populations, as well as some minority groups, also suffer greater impacts when exposed to ozone pollution, partially due to their limited access to healthcare resources. To this end, socio-economic status is an important determinant of differences in asthma prevalence and severity among ethnic minorities in the USA (Forno and Celedon, 2009). Younger children and children from low-income families are also more likely than other children to need hospital admissions even during the same time periods. Furthermore, children from low-income and Spanish-speaking families appear to be at particularly high risk for inadequate asthma therapy and use of inhalers (Wilson *et al.*, 2012).

Additionally, communities of colour, include non-Hispanic people of colour and Hispanics, are more likely to live in counties with higher levels of ozone pollution. More research is needed to determine whether genetic factors play a role in these health disparities.

## 5.5 Animal Models

The use of animal models for scientific purposes is a longstanding practice in biological research and medicine. The remarkable anatomical and physiological resemblances between humans and other animals, particularly mammals, have driven researchers to examine a large range of mechanisms and assess novel treatments in animal models before applying their discoveries to humans. Because they are affected by similar diseases, animals can act as models for the study of human illnesses. From such models, investigators have learnt how disorders affect the body and what are the physiological responses associated with the various illnesses. Once researchers learn more about a particular disease, animals can be used to develop and test potential therapies as part of the applied research process. These models are an essential part of applying biological research to real medical problems, allowing new targets for disease intervention to be identified.

### 5.5.1 Mice

Knowledge of differences among strains of animals in disease models can provide ideal tools for the discovery of mechanisms of disease development. Genetic variability is becoming gradually known as a substantial host factor in environmental disease predisposition. In both human and animal models, evidence supporting a genetic basis for susceptibility to the toxic effects of inhaled pollutants has been well established and animal studies show that ozone toxicity is species and strain dependent (Vancza *et al.*, 2009). In addition, a recently developed murine strain, collaborative cross, is derived from an eight-way cross using several founder strains (Churchill *et al.*, 2004).

There are several mouse strains, such as the C3H/HEJ, that are resistant to ozone exposure and have been used to study potential genes associated with ozone exposure susceptibility. In these studies, differential ozone responsiveness in resistant and susceptible mice was associated with a toll-like receptor 4 polymorphism that was previously identified as an important determinant of endotoxin susceptibility in mice (Kleeberger *et al.*, 2000). Similarly, different strains of mice have been found to vary in their acute inflammatory and pulmonary function responses to single or short-term exposure to ozone. For example, studies have shown that exposures to ozone causes type 2 immune responses with associated eosinophilic inflammation and epithelial mucous cell metaplasia in both C57BL/6 and BALB/c mice (Harkema *et al.*, 2017). However, the magnitude of these ozone-induced changes was consistently greater in C57BL/6 compared with BALB/c mice. These strain differences in airway responses to repeated ozone exposures indicate that genotype is an important factor in this murine model of non-atopic (i.e. without allergic sensitization) asthma and rhinitis. This further suggests that an individual's genetic background may determine the clinical manifestation of ozone-induced new-onset asthma (and rhinitis) in non-atopic children or adults (Harkema *et al.*, 2017).

Several studies have investigated the effect of ozone on allergen sensitization and challenge using the mouse ovalbumin model. This model has been commonly used to study sub-phenotypes of allergic asthma. Interestingly, ozone appears to antagonize the specific inflammatory effects of ovalbumin exposure, especially when given before or during exposure to ovalbumin in a murine model. To determine if timing of ozone and ovalbumin challenge affects the potential to exacerbate an allergic response, pre-sensitized BALB/c mice were exposed to 0.2 or 0.5 ppm ozone either before, after, continuously, or intermittently during ovalbumin aerosol challenge (Backus-Hazzard *et al.*, 2004; Last *et al.*, 2004). When ozone was administered prior to ovalbumin challenge, the response to allergen was similar to responses observed following ozone exposure (i.e. increased neutrophils and less eosinophilia). In contrast, when ozone was administered daily after or intermittently during ovalbumin challenge, the response resembled an allergic response (increased immunoglobulin E and eosinophilia) (Last *et al.*, 2004).

The house dust mite (HDM)-induced asthma model is used to study the *in vivo* efficacy of anti-asthma drugs and the development of diseases. This model features many similarities to human allergic asthma, including the presence of eosinophilic lung inflammation and the release of inflammatory mediators and cytokines primarily associated with Th2-type inflammation (Blacquière *et al.*, 2010; Draijer *et al.*, 2013). Exposure to HDM causes significant airway eosinophilia, mucous goblet cell metaplasia and airway hyper-responsiveness in mice. Subsequent exposure to ozone in HDM-treated mice decreased static lung compliance, respiratory frequency, tidal volume and eosinophils and increased lung resistance and eosinophil cell death. In non-human primates, however, ozone exposure resulted in airway eosinophilia mediated by different mechanisms than those triggered by HDM exposure, indicating that the interaction of ozone effects on allergen-sensitized lungs results in a distinct profile of eosinophil trafficking in both peripheral blood and lung compared with those triggered by ozone alone (Chou *et al.*, 2011).

Ozone affects and contributes to the development of chronic pulmonary diseases such as pneumonia (Silveyra *et al.*, 2017). Animal models have been widely used in this research area and have often provided insight into the physiological processes associated with the disease. A study showed that mice infected with *Klebsiella pneumoniae* following exposure to 2 ppm of ozone decreased the ability of mice to clear bacteria from the lungs, and that females were more affected and had higher mortality than males (Mikerov *et al.*, 2008, 2011). Contrarily, in the absence of ozone-induced oxidative stress, males were more prompted to have a higher level of propagation of infection compared with females. Studies using this model also suggest that a differential regulation of the lung immune response could be implicated in the observed increased susceptibility to adverse health effects from ozone observed in women versus men (Cabello *et al.*, 2015).

Besides the well-described inflammatory and dysfunction effects on the respiratory tract, accumulating evidence indicates that ozone exposure also affects the central nervous system and cardiovascular functions. Studies using Wistar rats support the conclusion that ozone causes a specific activation of vagal lung afferents rather than non-specific overall brain alterations mediated by blood-borne agents (Chounlamountry *et al.*, 2015). Moreover, studies using murine models have shown ozone as an accelerator of Alzheimer's disease pathogenesis in genetically predisposed populations (Akhter *et al.*, 2015). It has also been revealed that ozone exposure induces lipid peroxidation in the hippocampus and cortex in rats (Dorado-Martínez *et al.*, 2001; Martínez-Canabal *et al.*, 2008). In an animal model of glucose homeostasis, acute ozone exposure induced marked systemic metabolic impairments in Brown Norway rats of all ages (Bass *et al.*, 2013). On the other hand, repetitive pre-treatment with small doses of ozone has been demonstrated to protect against myocardial ischaemia-reperfusion injury in rat models of renal, hepatic and cardiac ischemia (Merin *et al.*, 2007; Chen *et al.*, 2008; Leon Fernandez *et al.*, 2008). Similarly, a decrease in morphometric changes as irregularity of the elastic lamina, disruption of the endothelial cells, vacuolization and haemorrhage was observed in a femoral artery vasospasm rat model (Orakdogen *et al.*, 2016). However, studies using murine models are limited, due to observed variation in susceptibility to ozone-induced health effects in these models (Hatch *et al.*, 2015; Ward and Kodavanti, 2015).

### 5.5.2 Ferrets

Ferrets share many anatomical, metabolic and physiological features with humans, which has promoted their use as an animal model. Ferrets are used in biomedical research in a wide variety of studies, including cardiopulmonary, neurological, reproductive, gastrointestinal and environmental research. Ozone-exposed ferret lungs display a severe, acute infiltration of neutrophils in regions with necrotic epithelial cells, especially in the centriacinar region. This response is more severe in ferrets than in other animal models (Sterner-Kock *et al.*, 2000). Findings also indicate that ozone exposure increases airway inflammation and IL-1 release (Wu *et al.*, 2008). Oxidant injury caused by ozone not only increases basal secretion of respiratory glycoconjugates but also increases tracheal gland sensitivity in ferrets (McBride *et al.*, 1991).

Therefore, ozone exposure in ferrets induces severe epithelial necrosis and inflammation, resulting in similar epithelial injury compared with monkeys and humans (Sterner-Kock *et al.*, 2000).

### 5.5.3 Non-human primates

Non-human primate models provide a crucial and unique opportunity to study airway disease in association with ozone exposure. The rhesus monkey (*Macaca mulatta*) is a useful model, as its immune system is similar to that of humans. Likewise, the intrapulmonary conducting airway components that are affected in asthmatic patients are present in the respiratory system of the rhesus monkey and appear to be affected in a similar way (Backus-Hazzard *et al.*, 2004).

There is a lack of data concerning the effect of ozone on lung development of young children who are in danger of atopic airway disease onset. In a model of childhood asthma using sensitized rhesus monkeys, episodic exposure to HDM or combined with ozone inhalation during the first 6 months of life resulted in a condition with many of the hallmarks of asthma, such as altered development of the distal airways and parenchyma (Avdalovic *et al.*, 2012; Plopper *et al.*, 2012). It was also discovered that combined ozone exposure and HDM amplified synergistic allergic and structural remodelling airway response in sensitized 30-day-old rhesus monkeys (Schelegle *et al.*, 2003). Other studies indicate that early life exposure to HDM and/or ozone alters the development process in the lung alveoli and the anatomical distribution of T cells throughout the proximal and distal airways (Miller *et al.*, 2009; Herring *et al.*, 2015). Atypical tracheal basement membrane development and the changes in plasticity of nucleus tractus solitarius neurons regulating respiratory motor responses to episodic ozone exposure were also found in this model (Chen *et al.*, 2003; Evans *et al.*, 2003). Even though monkeys are phylogenetically close to the human species, the disadvantages associated with their use as models is that they are only useful for a limited number of studies and they are expensive to house and feed, slow to breed and genetically diverse.

## 5.6 Conclusions

Exposure to ozone is a serious threat for human health. It is clear from the multiple epidemiological, clinical and animal studies that exposure to ozone poses serious negative health effects that are affected by genetic, physiological and even socio-economical factors. The negative effects of ozone exposure range from respiratory harm, including worsening of existing lung disease and inflammation, to increased susceptibility to premature death at both short- and long-term exposures. Moreover, ozone exposure is highly likely to cause cardiovascular harm, including heart attacks, strokes, heart disease and congestive heart failure, as well as harm to the central nervous system, cognitive, endocrine and reproductive function, and metabolic and developmental functions. While animal models have provided information on some of the inflammatory mechanisms activated during ozone exposure, as well as potential genes involved in the observed increased susceptibility of some individuals, the concise mechanisms associated with ozone toxicity are complex and have not been fully elucidated. More research is needed to understand and develop proper treatment for these conditions, as well as to implement adequate public health policies aimed to prevent more serious health complications, especially in young and vulnerable populations.

## References

Akhter, H., Ballinger, C., Liu, N., Van Groen, T., Postlethwait, E.M. and Liu, R.M. (2015) Cyclic ozone exposure induces gender-dependent neuropathology and memory decline in an animal model of Alzheimer's disease. *Toxicological Science* 147, 222–234.

Alexeeff, S.E., Litonjua, A.A., Suh, H., Sparrow, D., Vokonas, P.S. and Schwartz, J. (2007) Ozone exposure and lung function: effect modified by obesity and airways hyperresponsiveness in the VA normative aging study. *Chest* 132, 1890–1897.

American Lung Association (2017) State of the Air report. Available at: www.lung.org, accessed February 2018.
Angles-Cano, E. (1994) Overview on fibrinolysis: plasminogen activation pathways on fibrin and cell surfaces. *Chemistry and Physics of Lipids* 67-68, 353–362.
Anonymous (2014) Editorial: (Barely) living in smog: China and air pollution. *The Lancet* 383(9920), 845.
Arjomandi, M., Wong, H., Donde, A., Frelinger, J., Dalton, S. *et al.* (2015) Exposure to medium and high ambient levels of ozone causes adverse systemic inflammatory and cardiac autonomic effects. *American Journal of Physiology – Heart and Circulatory Physiology* 308, H1499–H1509.
Arroyo, V., Diaz, J., Carmona, R., Ortiz, C. and Linares, C. (2016) Impact of air pollution and temperature on adverse birth outcomes: Madrid, 2001–2009. *Environmental Pollution* 218, 1154–1161.
Avdalovic, M.V., Tyler, N.K., Putney, L., Nishio, S.J., Quesenberry, S. *et al.* (2012) Ozone exposure during the early postnatal period alters the timing and pattern of alveolar growth and development in nonhuman primates. *Anatomical Record (Hoboken)* 295(10), 1707–1716.
Azevedo, J.M., Gonçalves, F.L. and de Fátima Andrade, M. (2012) Long-range ozone transport and its impact on respiratory and cardiovascular health in the north of Portugal. *International Journal of Biometeorology* 55, 187–202.
Backus-Hazzard, G.S., Howden, R. and Kleeberger, S.R. (2004) Genetic susceptibility to ozone-induced lung inflammation in animal models of asthma. *Current Opinion in Allergy and Clinical Immunology* 4(5), 349–353.
Bass, V., Gordon, C.J., Jarema, K.A., MacPhail, R.C., Cascio, W.E. *et al.* (2013) Ozone induces glucose intolerance and systemic metabolic effects in young and aged Brown Norway rats. *Toxicology and Applied Pharmacology* 273(3), 551–560.
Bell, M.L., McDermott, A., Zeger, S.L., Samet, J.M. and Dominici, F. (2004) Ozone and short-term mortality in 95 US urban communities, 1987–2000. *Journal of the American Medical Association* 292(19), 2372–2378.
Ben-Zaken Cohen, S., Paré, P.D., Man, S.F. and Sin, D.D. (2007) The growing burden of chronic obstructive pulmonary disease and lung cancer in women: examining sex differences in cigarette smoke metabolism. *American Journal of Respiratory and Critical Care Medicine* 176(2), 113–120.
Bernstein, J.A., Alexis, N., Bacchus, H., Bernstein, I.L., Fritz, P. *et al.* (2008) The health effects of non-industrial indoor air pollution. *Journal of Allergy and Clinical Immunology* 121(3), 585–591.
Blacquière, M.J., Hylkema, M.N., Postma, D.S., Geerlings, M., Timens, W. and Melgert, B.N. (2010) Airway inflammation and remodeling in two mouse models of asthma: comparison of males and females. *International Archives of Allergy and Immunology* 153(2), 173–181.
Bocci, V.A., Zanardi, I. and Travagli, V. (2011) Ozone acting on human blood yields a hormetic dose-response relationship. *Journal of Translational Medicine* 9, 66.
Bromberg, P.A. (2016) Mechanisms of the acute effects of inhaled ozone in humans. *Biochimica et Biophysica Acta* 1860(12), 2771–2781.
Brook, R.D., Brook, J.R., Urch, B., Vincent, R., Rajagopalan, S. and Silverman, F. (2002) Inhalation of fine particulate air pollution and ozone causes acute arterial vasoconstriction in healthy adults. *Circulation* 105(13), 1534–1536.
Brown, J.S., Bateson, T.F. and McDonnell, W.F. (2008) Effects of exposure to 0.06 ppm ozone on FEV1 in humans: a secondary analysis of existing data. *Environmental Health Perspectives* 116(8), 1023–1026.
Brown, J.M., Harris, G., Pantea, C., Hwang, S.A. and Talbot, T.O. (2015) Linking air pollution data and adverse birth outcomes: environmental public health tracking in New York State. *Journal of Public Health Management and Practice* 21 (Suppl. 2), S68–74.
Butter, M.E. (2006) Are women more vulnerable to environmental pollution? *Journal of Human Ecology* 20(3), 221–336.
Cabello, N., Mishra, V., Sinha, U., DiAngelo, S.L., Chroneos, Z.C. *et al.* (2015) Sex differences in the expression of lung inflammatory mediators in response to ozone. *American Journal of Physiology. Lung Cellular and Molecular Physiology* 309(10), L1150–1163. doi: 10.1152/ajplung.00018.2015.
Calderon-Garciduenas, L., Franco-Lira, M., D'Angiulli, A., Rodriguez-Diaz, J., Blaurock-Busch, E. *et al.* (2015) Mexico City normal weight children exposed to high concentrations of ambient PM2.5 show high blood leptin and endothelin-1, vitamin D deficiency, and food reward hormone dysregulation versus low pollution controls. Relevance for obesity and Alzheimer disease. *Environmental Research* 140, 579–592.
Chauhan, A.J. and Johnston, S.L. (2003) Air pollution and infection in respiratory illness. *British Medical Bulletin* 68, 95–112.

Chen, C.Y., Bonham, A.C., Plopper, C.G. and Joad, J.P. (2003) Neuroplasticity in nucleus tractus solitarius neurons after episodic ozone exposure in infant primates. *Journal of Applied Physiology* 94(2), 819–827.

Chen, H., Xing, B., Liu, X., Zhan, B., Zhou, J., Zhu, H. and Chen, Z. (2008) Ozone oxidative preconditioning inhibits inflammation and apoptosis in a rat model of renal ischemia/reperfusion injury. *European Journal of Pharmacology* 581, 306–314

Chen, R., Yin, P., Meng, X., Liu, C., Wang, L. *et al.* (2017) Fine particulate air pollution and daily mortality. A nationwide analysis in 272 Chinese cities. *American Journal of Respiratory and Critical Care Medicine* 196(1), 73–81.

Chou, D.L., Gerriets, J.E., Schelegle, E.S., Hyde, D.M. and Miller, L.A. (2011) Increased CCL24/eotaxin-2 with postnatal ozone exposure in allergen-sensitized infant monkeys is not associated with recruitment of eosinophils to airway mucosa. *Toxicology and Applied Pharmacology* 257(3), 309–318.

Chounlamountry, K., Boyer, B., Penalba, V., François-Bellan, A.M., Bosler, O., Kessler, J.P. and Strube, C. (2015) Remodeling of glial coverage of glutamatergic synapses in the rat nucleus tractus solitarii after ozone inhalation. *Journal of Neurochemistry* 134(5), 857–864.

Churchill, G.A., Airey, D.C., Allayee, H., Angel, J.M., Attie, A.D. *et al.* (2004) The Collaborative Cross, a community resource for the genetic analysis of complex traits. *Nature Genetics* 36, 1133–1137.

Ciencewicki, J., Trivedi, S. and Kleeberger, S.R. (2008) Oxidants and the pathogenesis of lung diseases. *Journal of Allergy and Clinical Immunology* 122(3), 456–68; quiz 469–470.

Cleary, E.G., Cifuentes, M., Grinstein, G., Brugge, D. and Shea, T.B. (2018) Association of low-level ozone with cognitive decline in older adults. *Journal of Alzheimer's Disease* 61(1), 67–78.

Collins, T.W., Grineski, S.E. and Morales, D.X. (2017) Environmental injustice and sexual minority health disparities: a national study of inequitable health risks from air pollution among same-sex partners. *Social Science & Medicine* 191, 38–47.

Coogan, P.F., White, L.F., Yu, J., Brook, R.D., Burnett, R.T. *et al.* (2017) Long-term exposure to $NO_2$ and ozone and hypertension incidence in the Black Women's Health Study. *American Journal of Hypertension* 30(4), 367–372.

Delfino, R.J., Zeiger, R.S., Seltzer, J.M., Street, D.H., Matteucci, R.M., Anderson, P.R. and Koutrakis, P. (1997) The effect of outdoor fungal spore concentrations on daily asthma severity. *Environmental Health Perspectives* 105(6), 622–635.

Delfino R.J., Murphy-Moulton, A.M. and Becklake, M.R. (1998) Emergency room visits for respiratory illnesses among the elderly in Montreal: association with low level ozone exposure. *Environmental Research* 76(2), 67–77.

Devlin, R.B., Duncan, K.E., Jardim, M., Schmitt, M.T., Rappold, A.G. and Diaz-Sanchez, D. (2012) Controlled exposure of healthy young volunteers to ozone causes cardiovascular effects. *Circulation* 126(1), 104–111.

Dorado-Martínez, C., Paredes-Carbajal, C., Mascher, D., Borgonio-Pérez, G. and Rivas-Arancibia, S. (2001) Effects of different ozone doses on memory, motor activity and lipid peroxidation levels, in rats. *International Journal of Neuroscience* 108(3–4), 149–161.

Draijer, C., Robbe, P., Boorsma, C.E., Hylkema, M.N. and Melgert, B.N. (2013) Characterization of macrophage phenotypes in three murine models of house-dust-mite-induced asthma. *Mediators of Inflammation* 2013, 632049.

Dugas, T.R. (2018) Unraveling mechanisms of toxicant-induced oxidative stress in cardiovascular disease. *Current Opinions in Toxicology* 7, 1–8. doi: 10.1016/j.cotox.2017.10.007.

Durrani, F., Phelps, D.S., Weisz, J., Silveyra, P., Hu, S., Mikerov, A.N. and Floros, J. (2011) Gonadal hormones and oxidative stress interaction differentially affects survival of male and female mice after lung *Klebsiella pneumoniae* infection. *Experimental Lung Research* 38(4), 165–172.

Ebenstein, A., Fan, M., Greenstone, M., He, G. and Zhou, M. (2017) New evidence on the impact of sustained exposure to air pollution on life expectancy from China's Huai River Policy. *Proceeds of the National Academy of Sciences of the United States of America* 114(39), 10384–10389.

Evans, M.J., Fanucchi, M.V., Baker, G.L., Van Winkle, L.S., Pantle, L.M. *et al.* (2003) Atypical development of the tracheal basement membrane zone of infant rhesus monkeys exposed to ozone and allergen. *American Journal of Physiology – Lung Cellular and Molecular Physiology* 285(4), L931–939.

Fernandez, R., Ariza, M., Iscar, M., Martinez, C., Rubinos, G. *et al.* (2015) Impact of environmental air pollutants on disease control in asthmatic patients. *Lung* 193(2), 195–198.

Fernando Costa Nascimento, L., Blanco Machin, A. and Antonio Almeida Dos Santos, D. (2017) Are there differences in birth weight according to sex and associations with maternal exposure to air pollutants? A cohort study. *Sao Paulo Medical Journal* 135(4), 347–354.

Forno, E. and Celedon, J.C. (2009) Asthma and ethnic minorities: socioeconomic status and beyond. *Current Opinions in Allergy and Clinical Immunology* 9(2), 154–160.

Galizia, A. and Kinney, P.L. (1999) Year-round residence in areas of high ozone: association with respiratory health in a nationwide sample of nonsmoking young adults. *Environmental Health Perspectives* (107), 675–679.

Genc, S., Zadeoglulari, Z., Fuss, S.H. and Genc, K. (2012) The adverse effects of air pollution on the nervous system. *Journal of Toxicology* 2012, Article ID 782462. doi: 10.1155/2012/782462.

Ghio, A.J., Kim, C. and Devlin, R.B. (2000) Concentrated ambient air particles induce mild pulmonary inflammation in healthy human volunteers. *American Journal of Respiratory and Critical Care Medicine* 162(3), 981–988.

Goss, C.H., Newsom, S.A., Schildcrout, J.S., Sheppard, L. and Kaufman, J.D. (2004) Effect of ambient air pollution on pulmonary exacerbations and lung function in cystic fibrosis. *American Journal of Respiratory and Critical Care Medicine* 169(7), 816–821.

Green, R., Sarovar, V., Malig, B, and Basu, R. (2015) Association of stillbirth with ambient air pollution in a California cohort study. *American Journal of Epidemiology* 181(11), 874–882.

Guarnieri, M. and Balmes, J.R. (2014) Outdoor air pollution and asthma. *Lancet* 383(9928), 1581–1592.

Ha, S., Sundaram, R., Buck Louis, G.M., Nobles, C., Seeni, I., Sherman, S. and Mendola, P. (2018) Ambient air pollution and the risk of pregnancy loss: a prospective cohort study. *Fertility and Sterility* 109(1), 148–153.

Harkema, J.R., Hotchkiss, L.A., Vetter, N.A., Jackson-Humbles, D.N., Lewandowski, R.P. and Wagner, J.G. (2017) Strain differences in a murine model of air pollutant-induced nonatopic asthma and rhinitis. *Toxicologic Pathology* 45(1), 161–171.

Hatch, G.E., Crissman, K., Schmid, J., Richards, J.E., Ward, W.O. *et al.* (2015) Strain differences in antioxidants in rat models of cardiovascular disease exposed to ozone. *Inhalation Toxicology* 27 (Suppl. 1), 54–62

Herring, M.J., Putney, L.F., St George, J.A., Avdalovic, M.V., Schelegle, E.S., Miller, L.A. and Hyde, D.M. (2015) Early life exposure to allergen and ozone results in altered development in adolescent rhesus macaque lungs. *Toxicology and Applied Pharmacology* 283, 35–41.

Hollingsworth, J.W., Kleeberger, S.R. and Foster, W.M. (2007) Ozone and pulmonary innate immunity. *Proceedings of the American Thoracic Society* 4(3), 240–246.

Hwang, B.F., Chen, Y.H., Lin, Y.T., Wu, X.T. and Leo Lee, Y. (2015) Relationship between exposure to fine particulates and ozone and reduced lung function in children. *Environmental Research* 137, 382–390.

Ito, K.S., De Leon, S.F. and Lippman, M. (2005) Associations between ozone and daily mortality: analysis and meta-analysis. *Epidemiology* 16(4), 446–457.

Jariyasopit, N., Zimmermann, K., Schrlau, J., Arey, J., Atkinson, R. *et al.* (2014) Heterogeneous reactions of particulate matter-bound PAHs and NPAHs with $NO_3/N_2O_5$, OH radicals, and $O_3$ under simulated long-range atmospheric transport conditions: reactivity and mutagenicity. *Environmental Science and Technology* 48(17), 10155–10164.

Jung, C.R., Lin, Y.T. and Hwang, B.F. (2015) Ozone, particulate matter, and newly diagnosed Alzheimer's disease: a population-based cohort study in Taiwan. *Journal of Alzheimer's Disease* 44(2), 573–584.

Kampa, M. and Castanas, E. (2008) Human health effects of air pollution. *Environmental Pollution* 151(2), 362–367.

Khatri, S.B., Holguin, F.C., Ryan, P.B., Mannino, D., Erzurum, S.C. and Teague, W.G. (2009) Association of ambient ozone exposure with airway inflammation and allergy in adults with asthma. *Journal of Asthma* 46(8), 777–785.

Kim, C.S., Kehrl, H., Hazucha, M.J., Rappold, A., Brown, J., Devlin, R,. and Diaz-Sanchez, D. (2010) Pulmonary responses in healthy young adults exposed to low concentration of ozone for 6.6 hours with mild exercise. *American Journal of Respiratory and Critical Care Medicine* 181, A1728.

Kim, C.S., Alexis, N.E., Rappold, A.G., Kehrl, H., Hazucha, M.J. *et al.* (2011) Lung function and inflammatory responses in healthy young adults exposed to 0.06 ppm ozone for 6.6 hours. *American Journal of Respiratory and Critical Care Medicine* 183(9), 1215–1221.

Kim, Y., Myung, W., Won, H.-H., Shim, S., Jeon, H.J. *et al.* (2015) Association between air pollution and suicide in South Korea: a nationwide wtudy. *PLoS ONE* 10(2), e0117929.

Kirrane, E.F., Bowman, C., Davis, J.A., Hoppin, J.A., Blair, A. *et al.* (2015) Associations of ozone and $PM_{2.5}$ concentrations with Parkinson's disease among participants in the Agricultural Health Study. *Journal of Occupational and Environmental Medicine* 57(5), 509–517.

Kleeberger, S.R., Reddy, S., Zhang, L.Y. and Jedlicka, A.E. (2000) Genetic susceptibility to ozone-induced lung hyperpermeability: role of toll-like receptor 4. *American Journal of Respiratory and Critical Care Medicine* 22(5), 620–627.

Kreit, J.W., Gross, K.B., Moore, T.B., Lorenzen, T.J., D'Arcy, J. and Eschenbacher, W.L. (1989) Ozone-induced changes in pulmonary function and bronchial responsiveness in asthmatics. *Journal of Applied Physiology* 66(1), 217–222.
Kurt, O.K., Zhang, J. and Pinkerton, K.E. (2016) Pulmonary health effects of air pollution. *Current Opinion in Pulmonary Medicine* 22(2), 138–143.
Last, J.A., Ward, R., Temple, L. and Kenyon, N.J. (2004) Ovalbumin-induced airway inflammation and fibrosis in mice also exposed to ozone. *Inhalation Toxicology* 16(1),33–43.
Leon Fernandez, O.S., Ajamieh, H.H., Berlanga, J., Menendez, S., Viebahn-Hansler, R., Re, L. and Carmona, A.M. (2008) Ozone oxidative preconditioning is mediated by A1 adenosine receptors in a rat model of liver ischemia/reperfusion. *Transplant International* 21:, 39–48
Linares, C. and Diaz, J. (2010) Short-term effect of concentrations of fine particulate matter on hospital admissions due to cardiovascular and respiratory causes among the over-75 age group in Madrid, Spain. *Public Health* 124, 28–36.
Makri, A. and Stilianakis, N.I. (2008) Vulnerability to air pollution health effects. *International Journal of Hygiene and Environmental Health* 211(3–4), 326–336.
Martinez, F.J., Curtis, J.L., Sciurba, F., Mumford, J., Giardino, N.D. *et al.* (2007) Sex differences in severe pulmonary emphysema. *American Journal of Respiratory and Critical Care Medicine* 176(3), 243–252.
Martínez-Canabal, A., Angoa-Pérez, M., Rugerio-Vargas, C., Borgonio-Perez, G. and Rivas-Arancibia, S. (2008) Effect of growth hormone on cyclooxygenase-2 expression in the hippocampus of rats chronically exposed to ozone. *International Journal of Neuroscience* 118(3), 455–469.
McBride, R.K., Oberdoerster, G. and Marin, M.G. (1991) Effects of ozone on the cholinergic secretory responsiveness of ferret tracheal glands. *Environmental Research* 55(1), 79–90.
McConnell, R., Berhane, K., Gilliland, F., London, S.J., Islam, T. *et al.* (2002) Asthma in exercising children exposed to ozone: a cohort study. *Lancet* 359(9304),386–391.
McDonnell, W.F. and Smith, M.V. (1994) Description of acute ozone response as a function of exposure rate and total inhaled dose. *Journal of Applied Physiology* 76(6), 2776–2784.
Merin, O., Attias, E., Elstein, D., Schwalb, H., Bitran, D., Zimran, A. and Silberman, S. (2007) Ozone administration reduces reperfusion injury in an isolated rat heart model. *Journal of Cardiac Surgery* 22, 339–342
Middleton, N., Yiallouros, P., Kleanthous, S., Kolokotroni, O., Schwartz J. *et al.* (2008) A 10-year time-series analysis of respiratory and cardiovascular morbidity in Nicosia, Cyprus: the effect of short-term changes in air pollution and dust storms. *Environmental Health* 7, 39.
Mikerov, A.N., Gan, X., Umstead, T.M., Miller, L., Chinchilli, V.M., Phelps, D.S. and Floros, J. (2008) Sex differences in the impact of ozone on survival and alveolar macrophage function of mice after *Klebsiella pneumoniae* infection. *Respiratory Research* 9, 24.
Mikerov A.N., Cooper T.K., Wang, G., Hu, S., Umstead, T.M., Phelps, D.S. and Floros, J. (2011) Histopathologic evaluation of lung and extrapulmonary tissues show sex differences in *Klebsiella pneumoniae*-infected mice under different exposure conditions. *International Journal of Physiology, Pathophysiology and Pharmacology* 3, 176-–190.
Miller, L.A., Gerriets, J.E., Tyler, N.K., Abe,I K., Schelegle, E.S., Plopper, C.G. and Hyde DM. (2009) Ozone and allergen exposure during postnatal development alters the frequency and airway distribution of CD25+ cells in infant rhesus monkeys. *Toxicology and Applied Pharmacology* 236(1), 39–48.
Miranda, M.L., Edwards, S.E., Keating, M.H. and Paul, C.J. (2011) Making the environmental justice grade: the relative burden of air pollution exposure in the United States. *International Journal of Environmental Research and Public Health* 8(6), 1755–1771.
Mirowsky, J.E., Carraway, M.S., Dhingra, R., Tong, H., Neas, L. *et al.* (2017) Ozone exposure is associated with acute changes in inflammation, fibrinolysis, and endothelial cell function in coronary artery disease patients. *Environmental Health* 16(1), 126.
Mishra, V., DiAngelo, S.L. and Silveyra, P. (2016) Sex-specific IL-6-associated signaling activation in ozone-induced lung inflammation. *Biology of Sex Differences* 7,16. doi: 10.1186/s13293-016-0069-7.
Negrisoli, J. and Nascimento, L.F. (2013) Atmospheric pollutants and hospital admissions due to pneumonia in children. *Revista Paulista de Pediatria* 31(4), 501–506.
Orakdogen, M., Uslu, S., Emon, S.T., Somay, H., Meric, Z.C. and Hakan, T. (2016) The effect of ozone therapy on experimental vasospasm in the rat femoral artery. *Turkish Neurosurgery* 26(6), 860–865.
Ostro, B., Lipsett, M., Reynolds, P., Goldberg, D., Hertz,.A. *et al.* (2010) Long-term exposure to constituents of fine particulate air pollution and mortality: results from the California Teachers Study. *Environmental Health Perspectives* 118(3), 363–369.

Oudin, A., Åström, D.O., Asplund, P., Steingrimsson, S., Szabo, Z. and Carlsen, H.K. (2018) The association between daily concentrations of air pollution and visits to a psychiatric emergency unit: a case-crossover study. *Environmental Health* 17,4.
Peel, J.L., Metzger, K.B., Klein, M., Flanders, W.D., Mulholland, J.A. and Tolbert, P.E. (2007) Ambient air pollution and cardiovascular emergency department visits in potentially sensitive groups. *American Journal of Epidemiology* 165(6), 625–633.
Perng D.W. and Chen, P.K. (2017) The relationship between airway inflammation and exacerbation in chronic obstructive pulmonary disease. *Tuberculosis and Respiratory Diseases* (Seoul) 80(4), 325–335.
Peters, J.M., Avol, E., Gauderman, W.J., Linn, W.S., Navidi, W. *et al.* (1999) A study of twelve southern California communities with differing levels and types of air pollution. II. Effects on pulmonary function. *American Journal of Respiratory and Critical Care Medicine* 159, 768–775
Pinkerton, K.E., Harbaugh, M., Han, M.K., Jourdan Le Saux, C., Van Winkle, L.S. *et al.* (2015) Women and lung disease. Sex differences and global health disparities. *American Journal of Respiratory and Critical Care Medicine* 192(1), 11–16.
Plopper, C.G., Joad, J.P., Miller, L.A., Schelegle, E.S., Fanucchi, M.V. *et al.* (2012) Lung effects of inhaled corticosteroids in a rhesus monkey model of childhood asthma. *Clinical & Experimental Allergy* 42(7), 1104–1118.
Rice, M.B., Ljungman, P.L., Wilker, E.H., Gold, D.R., Schwartz, J.D. *et al.* (2013) Short-term exposure to air pollution and lung function in the Framingham Heart Study. *American Journal of Respiratory and Critical Care Medicine* 188(11), 1351–1357.
Rich, D.Q., Mittleman, M.A., Link, M.S., Schwartz, J., Luttmann-Gibson, H. *et al.* (2006) Increased risk of paroxysmal atrial fibrillation episodes associated with acute increases in ambient air pollution. *Environmental Health Perspectives* 114, 120–123.
Romieu, I., Barraza-Villarreal, A., Escamilla-Nunez, C., Texcalac-Sangrador, J.L., Hernandez-Cadena, L. *et al.* (2009) Dietary intake, lung function and airway inflammation in Mexico City school children exposed to air pollutants. *Respiration Research* 10, 122.
Romieu, I., Gouveia, N., Cifuentes, L.A., de Leon, A.P., Junger, W. *et al.* (2012) Multicity study of air pollution and mortality in Latin America (the ESCALA study). *Research report Health Effects Institute* 171, 5–86.
Ruidavets, J.B., Cournot, M., Cassadou, S., Giroux, M., Meybeck, M. and Ferrières, J. (2005) Ozone air pollution is associated with acute myocardial infarction. *Circulation* 111, 563–569.
Salam, M.T., Millstein, J., Li ,Y.F., Lurmann, F.W., Margolis, H.G. and Gilliland, F. (2005) Birth outcomes and prenatal exposure to ozone, carbon monoxide and particulate matter: results from the children's health study. *Environmental Health Perspectives* 113(11), 1638–1644.
Sarangapani, R., Gentry, P.R., Covington, T.R., Teeguarden, J.G. and Clewell, H.J. 3rd (2003) Evaluation of the potential impact of age- and gender-specific lung morphology and ventilation rate on the dosimetry of vapors. *Inhalation Toxicology* 15(10), 987–1016.
Schelegle, E.S., Carl, M.L., Coleridge, H.M., Coleridge, J.C. and Green, J.F. (1993) Contribution of vagal afferents to respiratory reflexes evoked by acute inhalation of ozone in dogs. *Journal of Applied Physiology* 74, 2338–2344.
Schelegle, E.S., Miller, L.A., Gershwin, L.J., Fanucchi, M.V., Van Winkle, L.S. *et al.* (2003) Repeated episodes of ozone inhalation amplifies the effects of allergen sensitization and inhalation on airway immune and structural development in Rhesus monkeys. *Toxicology and Applied Pharmacology* 191(1), 74–85.
Sesé, L., Nunes, H., Cottin, V., Sanyal, S., Didier, M. *et al.* (2017) Role of atmospheric pollution on the natural history of idiopathic pulmonary fibrosis. *Thorax* 73(2), 145–150.
Sheng, L., Ding, X., Ferguson, M., McCallister, M., Rhoades, R. *et al.* (2010) Prenatal polycyclic aromatic hydrocarbon exposure leads to behavioral deficits and downregulation of receptor tyrosine kinase, MET. *Toxicological Science* 118(2), 625–634.
Silveyra, P. and Floros, J. (2012) Air pollution and epigenetics: effects on SP-A and innate host defence in the lung. *Swiss Medical Weekly* 142, w13579.
Silveyra, P., Rivera, L. and Fuentes, N. (2017) Understanding the intersection of environmental pollution, pneumonia, and inflammation: does gender play a role? In: Chroneos, Z. (ed.) *Contemporary Topics of Pneumonia*. InTechOpen Books, Croatia, pp. 3–34.
Smith, N.L., Wilson, A.L., Gandhi, J., Vatsia, S. and Khan, S.A. (2017) Ozone therapy: an overview of pharmacodynamics, current research, and clinical utility. *Medical Gas Research* 7(3), 212–219.

Sorheim, I.C., Johannessen, A., Gulsvik, A., Bakke, P.S., Silverman, E.K. and DeMeo, D.L. (2010) Gender differences in COPD: are women more susceptible to smoking effects than men? *Thorax* 65(6), 480–485.

Soriano, J.B., Abajobir, A.A., Abate, K.H., Abera, S.F., Agrawal, A. *et al.* (2017) Global, regional, and national deaths, prevalence, disability-adjusted life years, and years lived with disability for chronic obstructive pulmonary disease and asthma, 1990-2015: a systematic analysis for the Global Burden of Disease Study 2015. *Lancet Respiratory Medicine* 5(9), 691–706.

Steppan, C.M. and Swick, A.G. (1999) A role for leptin in brain development. *Biochemical and Biophysical Research & Communications* 256(3), 600–602.

Sterner-Kock, A., Kock, M., Braun, R. and Hyde, D.M. (2000) Ozone-induced epithelial injury in the ferret is similar to nonhuman primates. *American Journal of Respiratory and Critical Care Medicine* 162(3 Pt 1), 1152–1156.

Thomson, E., Pal, J., Guenette, J., Wade, M.G., Atlas, A.C., *et al.* (2016) Ozone inhalation provokes glucocorticoid-dependent and -independent effects on inflammatory and metabolic pathways. *Toxicological Sciences* 152 (1), 17–28.

Thomson, E., Pilon, S., Guenette, J., Williams, A. and Holloway, A.C. (2018) Ozone modifies the metabolic and endocrine response to glucose: reproduction of effects with the stress hormone corticosterone. *Toxicology and Applied Pharmacology* 342, 31–38.

Thurston, G.D., Lippmann, M., Scott, M.B. and Fine, J.M. (1997) Summertime haze air pollution and children with asthma. *American Journal of Respiratory and Critical Care Medicine* 155(2), 654–660.

Vancza, E.M., Galdanes, K., Gunnison, A., Hatch, G. and Gordon, T. (2009) Age, strain, and gender as factors for increased sensitivity of the mouse lung to inhaled ozone. *Toxicological Sciences* 107(2), 535–543.

Vinikoor-Imler, L.C., Owens, E.O., Nichols, J.L., Ross, M., Brown, J.S. and Sacks, J.D. (2014) Evaluating potential response-modifying factors for associations between ozone and health outcomes: a weight-of-evidence approach. *Environmental Health Perspectives* 122(11), 1166–1176.

Volk, H.E., Kerin, T., Lurmann, F., Hertz-Picciotto, I., McConnell, R. and Campbell, D.B. (2014) Autism spectrum disorder: interaction of air pollution with the MET receptor tyrosine kinase gene. *Epidemiology* 25(1), 44–47.

Wang, K.Y. and Chau, T.T. (2013) An association between air pollution and daily outpatient visits for respiratory disease in a heavy industry area. *PLoS ONE* 8(10), e75220.

Ward, W.O. and Kodavanti, U.P. (2015) Pulmonary transcriptional response to ozone in healthy and cardiovascular compromised rat models. *Inhalation Toxicology* 27 (Suppl. 1), 93–104

Wilson, S.R., Rand, C.S., Cabana, M.D., Foggs, M.B., Halterman, J.S. *et al.* (2012) Asthma outcomes: quality of life. *Journal of Allergy and Clinical Immunology* 129 (3 Suppl.), S88–123.

Wong, T.W., Lau, T.S., Yu, T.S., Neller, A., Wong, S.L., Tam, W. and Pang, S.W. (1999) Air pollution and hospital admissions for respiratory and cardiovascular diseases in Hong Kong. *Occupational and Environmental Medicine* 56, 679–683.

Wu, Z.X., Barker, J.S., Batchelor, T.P. and Dey, R.D. (2008) Interleukin (IL)-1 regulates ozone-enhanced tracheal smooth muscle responsiveness by increasing substance P (SP) production in intrinsic airway neurons of ferret. *Respiratory Physiology & Neurobiology* 164(3), 300–311.

Xia, L., Lenaghan, S.C., Zhang, M., Zhang, Z. and Li, Q. (2010) Naturally occurring nanoparticles from English ivy: an alternative to metal-based nanoparticles for UV protection. *Journal of Nanobiotechnology* 8, 12–12.

Yang, S., Tan, Y., Mei, H., Wang, F., Li, N. *et al.* (2018) Ambient air pollution the risk of stillbirth: a prospective birth cohort study in Wuhan, China. *International Journal of Hygiene and Environmental Health* S1438–4639.

Yoda, Y., Otani, N., Sakurai, S. and Shima, M. (2014) Acute effects of summer air pollution on pulmonary function and airway inflammation in healthy young women. *Journal of Epidemiology* 24,312–320.

Zhao, D., Zhou, Y., Jiang, C., Zhao, Z., He, F. and Ran, P. (2017) Small airway disease: A different phenotype of early stage COPD associated with biomass smoke exposure. *Respirology* 23(2), 198–205.

# 6 Ozone II. Biophysical Observations

**K.C. Thompson***
*Department of Biological Sciences and Institute of Structural and Molecular Biology, Birkbeck College, University of London, UK*

## 6.1 Abstract

Exposure to the ubiquitous pollutant gas, ozone, is linked to a range of respiratory problems. Inhaled ozone that reaches the alveoli will first encounter the fluid lining these, known as lung surfactant. Lung surfactant is a complex mixture of lipids and proteins and is essential as it lowers the surface tension of the air–liquid interface of the lung, preventing alveolar collapse. In this chapter possible changes to lung surfactant following exposure to low levels of gas-phase ozone, as are typically found in polluted air, are presented and how these changes may affect the biophysical properties of the lung surfactant, and hence its ability to function correctly. It is shown that a number of components in lung surfactant are expected to react with ozone and whilst some oxidation products will leave the interfacial area, other species, such as damaged lipids with truncated tails and surfactant proteins with oxidized amino acid residues, are expected to remain in the lung surfactant film, impairing its ability to function.

## 6.2 Introduction

### 6.2.1 Lung surfactant composition

The outer surface of the lung is lined with a complex fluid known as lung surfactant. Lung surfactant self-assembles such as to form a layer of amphipathic molecules at the air–water interface, with the hydrophobic portion of the molecules in contact with the air whilst the hydrophilic portion is embedded in the water surface, where it disrupts the hydrogen bonding between the water molecules. The presence of lung surfactant is crucial for the proper functioning of the lungs, providing not just a barrier against infection, but also in reducing the surface tension of the air–water interface of the lung, thus reducing the work required to expand and contract the lungs during the breathing cycle and reducing the risk of alveolar collapse after expiration when the surface area is at a minimum. A comprehensive review of the biophysical properties of lung surfactant can be found in the article by Parra and Pérez-Gil (2015).

* Email address: k.thompson@mail.cryst.bbk.ac.uk

Lung surfactant is composed of around 10% proteins and 90% lipids by mass. A wide range of lipids are present but most (> 90%) are phospholipids (Creuwels *et al.*, 1997). The exact composition of the phospholipids present varies slightly between species but in humans it is essentially around 80% phosphocholine lipids and 20% anionic lipids such as phosphoglycerols and phosphatidylinositols, though other minor lipids are also present (Postle *et al.*, 2001). Of the phosphocholines, the main lipid is dipalmitoyl-*sn*-glycero-3-phosphtidylcholine (DPPC), which makes up about 54% of the phosphocholine in humans; other disaturated phospholipids comprise a further 10% of the phosphocholine species. The lipid DPPC itself is capable of forming a monolayer at the air–water interface, with the phosphocholine head group embedded in the water and the acyl tails pointing out into the air, which when compressed achieves very low surface tensions. However, treating neonatal babies who have a surfactant deficiency with DPPC only does not lead to an improvement in respiratory function. It is thought that unsaturated lipids, which form more fluid monolayers, are an essential component of lung surfactant (Fujiwara *et al.*, 1980). Around 36% of the phosphocholine content of lung surfactant are unsaturated lipids, predominantly 1-palmitoyl-2-oleoyl-*sn*-glycero-3-phosphatidylcholine (POPC) and 1-palmitoyl-2-palmitoleoyl-*sn*-glycero-3-phosphatidylcholine, with smaller amounts of other lipids (Postle *et al.*, 2001). Of the anionic phosphoglycerol lipids, the main components are all at least partially unsaturated species (i.e. at least one of the lipid tails contains a carbon–carbon double bond), the most prevalent being 1-palmitoyl-2-oleoyl-*sn*-glycero-3-phosphatidylglycerol (POPG), 1-stearoyl-2-oleoyl-*sn*-glycero-3-phosphatidylglycerol and dioleoyl-*sn*-glycero-3-phosphatidylcholine (Postle *et al.*, 2001). Similarly the phosphatidylinositol lipids are predominantly unsaturated lipids species such as 1-stearoyl-2-oleoyl-*sn*-glycero-3-phospho-(1′-myo-inositol-4′-phosphate). In addition to phospholipids, lung surfactant contains a small amount (approximately 8% by mass of the total surfactant components) of cholesterol, an unsaturated sterol. Thus of all the lipids present in lung surfactant, around half contain at least one carbon–carbon double bond, which, as will be shown later, makes them susceptible to oxidation upon exposure to the pollutant gas, ozone. The chemical structures of cholesterol, DPPC, POPC and POPG are shown in Fig. 6.1.

Of the four surfactant proteins, surfactant protein A (SP-A) and surfactant protein D (SP-D) are hydrophilic proteins that are found in the bulk surfactant fluid. Two of the surfactant proteins, however, are hydrophobic and are found in lipid bilayers structures, or in the lipid monolayer which forms at the air-water interface of the lung. The hydrophobic surfactant proteins are known as surfactant protein B (SP-B) and surfactant protein C (SP-C). The protein SP-B is known to be essential for life and genetic defects or lack of this protein can lead to death soon after birth (Nogee *et al.*, 1994; Clark *et al.*, 1995). The protein SP-C does not appear to be vital, but lack of SP-C is associated with respiratory problems (Glasser *et al.*, 2001). The primary structures of SP-B and SP-C are shown in Fig. 6.2.

The role of the proteins SP-B and SP-C is not completely understood but both interact with lipids and are believed to be involved in maintaining the correct lipid composition in the crucial interfacial layer during the breathing cycle, linking the surface monolayer of lipid material to bilayer structures below it (Parra and Pérez-Gil, 2015). At physiological pH, SP-B carries a net positive charge of around plus 7. It is thought that it interacts strongly with the anionic lipids, which are predominantly unsaturated, perhaps ensuring that they return to the interface as the lungs expand after expiration. SP-C has two palmitoyl groups bonded, via a thio-ester bond, to the two cysteine residues near the N terminus. These palmitoyl residues may be inserted into lipid monolayers and bilayers, acting in effect as an anchor for the protein. Like SP-B, SP-C carries a net positive charge at physiological pH, which may help it to interact with anionic lipids.

This chapter considers changes that occur to lung surfactant, and individual components of lung surfactant, upon exposure to gas-phase ozone at levels encountered in polluted ambient air. It will be shown that, at least for some components, much is known about how individual species react with ozone but far less is understood as to how these changes influence the overall properties of the lung surfactant.

### 6.2.2 Ozone in ambient air

Ozone is present as a secondary pollutant in ambient air. It is formed following the photolysis of nitrogen dioxide ($NO_2$):

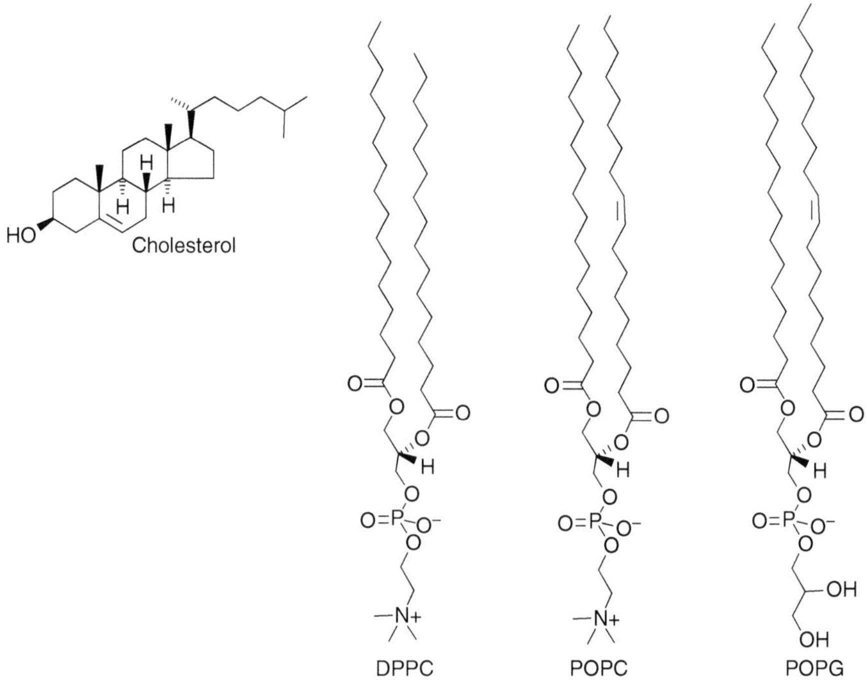


**Fig. 6.1.** Chemical structures of selected lipids found in lung surfactant: cholesterol and the phospholipids DPPC[a], POPC[b] and POPG[c]. Cholesterol, POPC and POPG each contain a carbon-carbon double bond, making them susceptible to attack by ozone. ([a]Dipalmitoyl-*sn*-glycero-3-phosphatidylcholine; [b]1-palmitoyl-2-oleoyl-*sn*-glycero-3-phosphatidylcholine; [c]1-palmitoyl-2-oleoyl-*sn*-glycero-3-phosphatidylglycerol.)

SP-B (Uniprot ID P07988):
FPIPLPYCWLCRALIKRIQAMIPKGALAVAVAQVCRVVPLVAGGICQCLAERYSVILLDTLLGRMLPQLVCRLVLRCSM

SP-C (Uniprot ID P11686):
FGIPCCPVHLKRLLIVVVVVVLIVVVIVGALLMGL

**Fig. 6.2.** Primary structure of the human lung surfactant proteins SP-B and SP-C. All the cysteine residues, C, in SP-B are involved in either intramolecular or intermolecular disulfide bonds. The two cysteine residues in SP-C are covalently link to two palmitoyl chains. Both SP-B and SP-C contain unsaturated amino acid residues, such as tryptophan (W) and tyrosine (Y), found in SP-B, and histidine (H), found in SP-C, making them potentially susceptible to attack by ozone. In addition, methionine (M), present in both proteins, is highly susceptible to oxidation, though the mechanism is different to that of ozone reacting with a double bond.

$$NO_2 + hv \rightarrow NO + O \qquad \text{(Eqn 6.1)}$$

$$O + O_2 \rightarrow O_3 \qquad \text{(Eqn 6.2)}$$

The nitric oxide (NO) itself can be oxidized back to $NO_2$ in the atmosphere, in a pathway facilitated by the presence of volatile organic compounds (VOCs), thus providing a route for the build-up of ozone in ambient air parcels containing oxides of nitrogen, VOCs and exposed to sunlight. The current UN World Health Organization (WHO) guidelines for ozone recommend that any 8 h period should not exceed an average of 100 μg ozone $m^{-3}$ (approximately 50 ppb). This level is regularly exceeded in many locations around the world, in both developing and developed countries, during the summer months, with London providing a good example of this. Many studies have shown that exposure to increased levels of ozone in ambient air is linked to respiratory problems, increased hospital admissions and increased mortality; for examples see Cohen *et al.* (2017) and Jerrett *et al.* (2009).

## 6.3 Reaction Pathways

When ambient air is inhaled into the lungs the air comes into contact with the lung surfactant lining the alveoli. Any inhaled ozone may be exhaled without reaction, react with one or more components in the lung surfactant, or cross the interfacial layer into the bulk surfactant fluid where it may react with other species. The primary component of lung surfactant present at the air–water interface is the saturated lipid DPPC. Ozone, at least at the levels found in ambient air, will not react with saturated hydrocarbons and the reaction of ozone with DPPC is not therefore expected to occur. However, a significant portion of the lipids in lung surfactant are unsaturated and ozone has long been known to react with unsaturated organic compounds. The initial step is the formation of a primary ozonide, which decomposes to yield a carbonyl-containing product and a Criegee bi-radical, as shown in Fig. 6.3.

Examining the scheme shown in Fig. 6.3, it is clear that the products of the reaction of ozone with unsaturated phospholipids are expected to be lipids with a tail truncated at the previous position of the double bond, a new oxygen-containing functional group at this point, plus a short-chain aldehyde, or other short-chain oxygen-containing species. As an example, the reaction of POPC with ozone, based on the scheme shown in Fig. 6.3, is shown in Fig. 6.4. It is expected that products from both potential reaction paths, I and II, will be found, i.e. a mixture of nonanal and Criegee I, and the oxidized lipid POnPC and Criegee II.

As can be seen in Fig. 6.4, the newly formed truncated lipid tail is terminated with an oxygen-containing group, such as an aldehyde, which for POPC oxidation yields the lipid 1-palmitoyl-2-(9′-oxo-nonanoyl)-*sn*-glycero-3-phosphocholine, referred to here at POnPC. The equivalent oxidized lipid with a carboxylic acid group in place of the new aldehyde group would be 1-palmitoyl-2-azelaoyl-*sn*-glycero-3-phosphocholine, referred to here as PAzPC. In addition to phospholipids, the cholesterol present in lung surfactant contains a carbon–carbon double bond (see Fig. 6.1) and is susceptible to oxidation by ozone by the same chemistry as shown in Fig. 6.3, the reaction leading to ring opening and an oxygenated product.

In addition to potentially reacting with unsaturated lipids, ozone could react, via the same mechanism as shown in Fig. 6.3, with unsaturated amino acid residues in SP-B and SP-C. SP-B contains the unsaturated residues tryptophan (W), tyrosine (Y) and phenylalanine (F). Tryptophan is known to react readily with ozone to yield kynurenine, shown in Fig. 6.5, and *N*-formyl kynurenine, whilst tyrosine reacts with ozone to yield a mixture of hydroxy-containing products (Mudd *et al.*, 1969).

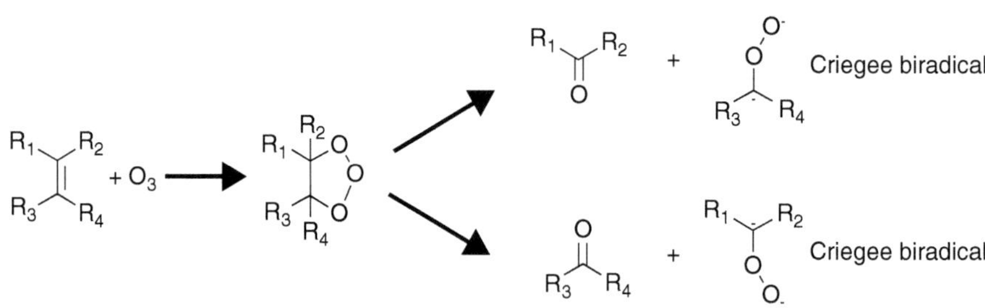


**Fig. 6.3.** General reaction scheme for the reaction of ozone with unsaturated organic compounds, showing the formation of a carbonyl containing compound and a Criegee bi-radical. The Criegee bi-radical shown will go on to react further, the pathway in the liquid phase being different to that in the gas phase. The pathway followed by these Criegee bi-radicals, and products formed, at the air–water interface is not clear.

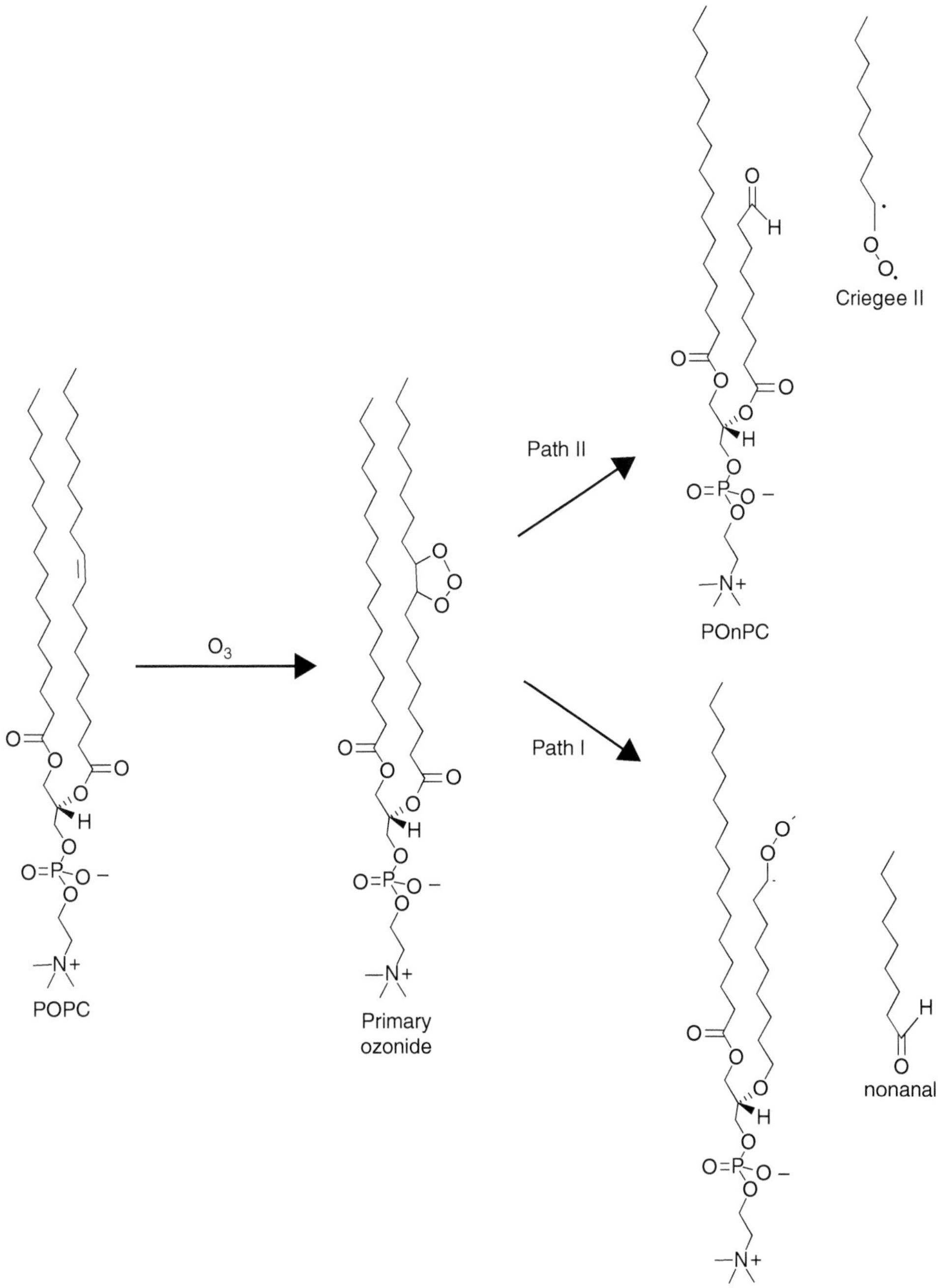


**Fig. 6.4.** Initial mechanism of the reaction of ozone with the unsaturated lipid POPC. It is expected that products from both paths I and II will be found. The Criegee bi-radicals formed will react further and are expected to yield a range of oxygenated products.

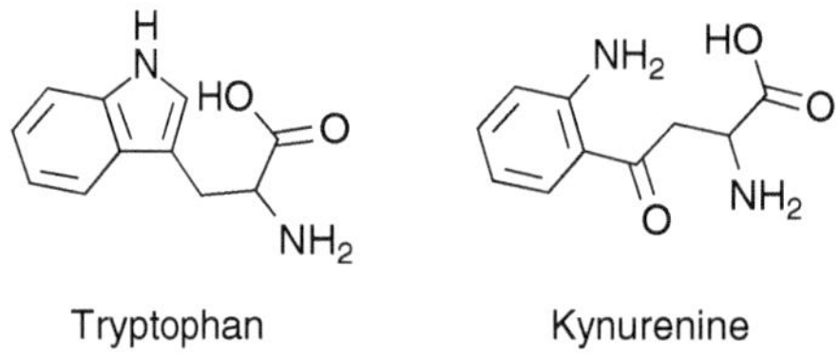


**Fig. 6.5.** Structure of the amino acid tryptophan and kynurenine, an expected product of the reaction of ozone with tryptophan.

Reaction of phenylalanine with ozone is also possible, though this reaction is slower than the reaction with tryptophan or tyrosine. SP-C contains the unsaturated residues histidine (H) and phenylalanine. Histidine reacts rapidly with ozone and is suggested to yield ammonia and aspartic acid (Berlett *et al.*, 1996). Thus, if attack of unsaturated amino acid residues of both SP-B and SP-C by ozone occurred, the product would be oxidized amino acids residues, more polar than the original residues, and a protein with very different biophysical properties. The reaction of ozone with proteins is not limited to attack of carbon–carbon double bonds. Ozone may react with sulfur in low oxidation states, and thus could potentially react with the amino acid residues containing sulfur: cysteine (C) and methionine (M). The reaction of methionine, present in both SP-B and SP-C, with ozone leads to the formation of methionine sulfoxide (Mudd *et al.*, 1969). Cysteine oxidation by ozone has been reported to yield a variety of products, including cysteine sulfenate, cysteine sulfinate, cysteine sulfonate, cysteic acid and cystine (Mudd *et al.*, 1969; Enami *et al.*, 2009), but the original cysteines in SP-B are all thought to be involved in disulfide bonds, to yield cystine, which is significantly less reactive towards ozone. In SP-C the cysteine residues are link through thioester bonds to palmitoyl residues, again making them less susceptible to attack by ozone.

It is clear that both a significant fraction of the lung surfactant lipids and the hydrophobic surfactant proteins SP-B and SP-C present at the air–water interface of the lung are susceptible to direct attack by inhaled ozone. In addition to this, the reaction of ozone in aqueous solution may lead to the production of secondary reactive species, including highly reactive radicals such as the hydroxyl radical ·OH, that are more reactive than ozone and can attack species not reactive to ozone itself. Whether any reaction between ozone and components of lung surfactant occurs at the air–water interface of the lung, and what the products are, is considered below.

## 6.4 Inhalation of Ozone: Evidence of Damage to Lung Surfactant

### 6.4.1 *In vitro* product studies

Analysis of lung surfactant recovered from calves, spread at the air–water interface and exposed to levels of ozone found in pollution events (125 ppb) reveals the presence of the oxidized lipid POnPC, an expected product of the reaction of POPC with ozone, as shown in Fig. 6.4 (Uhlson *et al.*, 2002). Almstrand *et al.* (2015) exposed the fluid recovered from human bronchoalveolar lavage to 60 ppb of ozone. The authors examined the lipid content of the lavage fluid before and after exposure and found that, following exposure, there were 13 different phosphocholines and six different phosphoglycerol lipids formed from oxidation, the usual species being a tail-shortened aldehyde product such as POnPC.

### 6.4.2 *In vivo* product studies

Although the *in vitro* experiments show that ozone reacts with unsaturated lipids present in lung surfactant when spread at the air–water interface, this does not imply that inhaled ozone will react with components of lung surfactant lining the alveoli. The inhaled ozone could potentially react in the upper respiratory tract, cross the surfactant layer and react with species deeper inside the lungs, or react with antioxidants present in lung surfactant. There are numerous *in vivo* studies looking at the effects of ozone on lung surfactant. Mudway and Kelly (2000) reviewed the available literature and concluded that ozone did penetrate as far as the alveoli, where some reacted with antioxidants but some reacted with lipids and/or proteins present in the lung surfactant fluid. Putman *et al.* (1997) demonstrated that the lowest surface tension attained by lung surfactant recovered from rats that had been exposed to ~800 ppb of ozone over several hours was higher than that attained by surfactant recovered from rats without exposure. If inhaled ozone were to react with any of the unsaturated lipids present in lung surfactant, then aldehydes such as the C9 aldehyde nonanal, as shown in Fig. 6.4, and phospholipids containing shorten tails truncated with a carbonyl group at the position of the former double bond, such as POnPC, would be expected to be formed. Pryor *et al.* (1996) reported that a range of aldehydes (C6, C7, C8 and C9) were detected in the bronchoalveolar lavage of rats after they had been exposed to 500 ppb of ozone. Frampton *et al.* (1999) exposed healthy humans to 220 ppb of ozone and found that this led to an increase in C9 aldehydes in lung lavage fluid retrieved from the subjects after the exposure. The *in vivo* studies therefore indicate that exposure to ozone leads to products consistent with the reaction of ozone with unsaturated lipids.

The *in vitro* and *in vivo* experiments both suggest that inhaled ozone causes the unsaturated lipids in lung surfactant to be oxidized. The short-chain aldehyde species produced, such as nonanal, can leave the surfactant layer, either into the gas phase and being exhaled (and thus can be detected), or dissolved into the bulk subphase, potentially causing a biological response when they interact with the cells lining the lung. The larger products, such as the lipids with truncated tails, could remain in the surfactant layer, affecting its efficiency of action. The review by Müller *et al.* (1998) concluded that lung surfactant is structurally impaired by exposure to ozone and other oxidants and that this will affect its biophysical properties *in vivo*.

### 6.4.3 Biophysical studies on model lipid membranes

A number of different biophysical experiments have been performed on lung surfactant, or lung surfactant mimics, either before and after exposure to gas-phase ozone, or continuously during exposure. Lai *et al.* (1994) exposed monolayers of POPC to levels of gas-phase ozone between 0.3 ppm and 30 ppm. Plots of surface pressure, $\Pi$, versus area per molecule were recorded before and after exposure to the ozone gas. A lower surface pressure equates to a higher surface tension (see Eqn 6.3):

$$\Pi = \gamma_0 - \gamma \qquad \text{(Eqn 6.3)}$$

where $\gamma_0$ is the surface tension recorded with no lipid monolayer and $\gamma$ is the surface tension recorded with the lipid monolayer. The results revealed that significantly lower surface pressures, i.e. higher surface tensions, were achieved upon compression of the film after exposure to ozone. The authors also noted that carbonyl lipid products were detected after exposure to ozone, and that more extensive reaction, at least at these high concentrations of ozone, led to further oxidation of the carbonyl species yielding the corresponding carboxylic acid. Wadia *et al.* (2000) exposed monolayers of OPPC, an isomer of POPC, to low levels of ozone at the air–water interface and, by following the rate of nonanal production, concluded that the reaction at the air–water interface was kinetically enhanced compared with that in the gas phase. In addition they found no evidence for the production of the secondary oxidant ·OH radicals, as would be expected if the reaction occurred in the gas phase. Thompson *et al.* (2010) followed changes in the surface pressure when a monolayer of POPC, spread at the air–water interface, was exposed to fairly high levels of ozone (~ 2 ppm) and noted that the oxidation led to a rapid increase in surface pressure (reduction in surface tension) followed by a slow decrease in surface pressure, i.e. an increase in surface tension. In a later paper, Thompson *et al.* (2013) found that a similar trend was observed when much lower levels of ozone were used (~100 ppb), as encountered in polluted air. The authors used neutron reflection and selective deuteration of the POPC to show that the increase in surface pressure occurred as the oleoyl chain was cleaved and the nine-carbon terminal portion left the interface, presumably as nonanal, indicating that the remaining lipid species (e.g. POnPC) underwent a rearrangement, such that the truncated tail, with the now more polar termination, reoriented itself and associated with the water. The very much slower decrease in surface pressure corresponded with the damaged lipid being lost from the interface. Qiao *et al.* (2015) also exposed monolayers of POPC and POPC/DPPC spread at the air–water interface to very low levels of ozone gas (~20 ppb). The authors used surface pressure measurements combined with vibrational spectroscopy and atomic force microscopy imaging to follow the reaction. The result also provided evidence, even at these very low ozone levels, of rapid POPC oxidation and that the product POnPC, or the carbocyclic acid equivalent, PAzPC, reoriented the newly truncated tail and thus disrupted the hydrated head region, changing the surface tension of the monolayer. There is therefore clear evidence that exposure to gas-phase ozone, at levels found in polluted air, causes direct damage to the unsaturated lipids in the films. A short-chain carbonyl-containing product is released from the interfacial layer, a fraction of which may dissolve in the subphase, leading to potential affects further down. The remainder of the lipid continues to reside, at least initially, at the air–water interface but in an altered conformation, changing the properties of the film; ultimately these damaged lipids solubilize into the subphase and/or are targeted by phospholipases. It should be noted that, although cholesterol is present in lung surfactant and is a likely target of ozone attack owing to its double bond, there are no studies in the literature on the reaction of ozone specifically with cholesterol in lipid monolayers at the air–water interface.

A monolayer of pure DPPC undergoes a number of phase transitions as it is compressed from large to small surface area. A region exists where domains of liquid condensed DPPC molecules are found surrounded by DPPC in a less ordered liquid expanded form. Natural lung surfactant is a more complicated fluid than pure DPPC but domain formation, perhaps with a biological role, is at least a possibility. Alonso *et al.* (2004) reported the presence of liquid ordered domains when films of the commercial lung surfactant replacement fluid, Survanta, were spread at the air–water interface, but Survanta is unlike natural lung surfactant in that it does not contain any cholesterol. Bernardino de la Serna *et al.* (2004) reported domains in bilayer structures of natural procine lung surfactant, suggesting that domains would also be found at equivalent molecular areas in surfactant monolayers spread at the air–water interface. In addition to changing the structure of the interfacial layer directly, the presence of oxidized lipids could potentially disrupt the organization of lipids at the air–water interface. To investigate this Sabatini *et al.* (2006) used fluorescence microscopy, with a small amount of fluorescently labelled DPPC, to show that the introduction of POnPC, and also the acid equivalent PAzPC, disrupted the formation of the liquid condensed DPPC domains but also that the domains, once formed, persisted at higher surface pressures than those of pure DPPC monolayers. Volinsky *et al.* (2012) looked at the ternary lipid system POPC, sphingomyelin and cholesterol, and the 'oxidized' system where part of the POPC was replaced with the oxidized lipid PAzPC, using the label-free method, Brewster Angle Microscopy (BAM). The results indicated that the presence of the oxidized lipid led to the persistence of liquid ordered domains at higher surface pressures than found in the unoxidized system. Although sphingomyelin is not a significant component of lung surfactant, the results are consistent with the DPPC/PAzPC system: the presence of oxidized lipids in monolayers at the air–water interface enhances lipid segregation between liquid condensed and liquid expanded domains. Thus the presence of oxidized lipids in lung surfactant monolayers not only leads to increased hydration of the head group area and thus surface tension changes, but also may change the in-plane ordering of the lipids. Whether changes to the formation and persistence of domains would impair the functioning of the surfactant monolayer has not yet been explored fully. It is worth noting that cholesterol is well known to influence lipid domain formation, but little is known of how the presence of cholesterol oxidation products might alter this in a lung surfactant monolayer.

### 6.4.4 Biophysical studies on surfactant proteins and peptide mimics

No studies have looked at the reaction of ozone with full length SP-B at the air–water interface. This is undoubtedly at least partly due to difficulties working with this protein: a correctly folded full-length version has not been chemically synthesized and a folded recombinant version is also difficult to obtain. A few studies have looked at the more general oxidation of SP-B by reactive oxygen species (ROS) (e.g. Manzanares *et al.*, 2007). Manzanares *et al.* (2007) found that the surfactant film's ability to re-spread after compression, and achieve low surface tensions during compression, was impaired following exposure to ROS. However, reaction with ROS in general will give different products to those expected from reaction with ozone and components of ROS can oxidize amino acid residues that would not react directly with ozone. Sarker *et al.* (2011) looked at the structure of two peptides: SP-B(8–25), containing residues 8 to 25 of full-length SP-B, and an 'oxidized' version of this peptide where the tryptophan at position 2 was replaced with kynurenine, a known product of the reaction of tryptophan with ozone. The structures of the peptides were studied by circular dichroism spectroscopy, nuclear magnetic resonance and molecular dynamics simulations in systems designed to mimic the lipid environment of lung surfactant: detergent and lipid micelles, lipid bilayers and, for the molecular dynamics, lipid monolayers at the air–water interface. The authors found that replacement of tryptophan with kynurenine had a dramatic effect on the properties of the peptide. The oxidized peptide had significantly reduced helical content compared with the original peptide and positioned itself almost in plane with the lipid bilayer and lipid monolayer, rather than almost perpendicular to it, as the original peptide did, indicating that the kynurenine-containing peptide interacted very differently with the lipids than the tryptophan-containing peptide. The tryptophan residue of the original peptide was directed towards the acyl chains of the lipids, whereas the more polar kynurenine was directed towards the water subphase. Oxidation of just one amino acid thus dramatically alters the structure, positioning and interactions of the peptide with the lipid membrane. Sarker's study involved only a subset of the amino acids present in full-length SP-B; it is not known whether the full-length SP-B protein is more resistant to structural changes upon tryptophan oxidation.

A few authors have looked directly at the ozone-initiated oxidation of peptide mimics of full-length SP-B at the air–water interface, either alone or together with phospholipids. Kim *et al.* (2010) exposed SP-B(1–25), a peptide composed of the first 25 amino acids of full-length SP-B, spread at the air–water interface, to high levels of gas-phase ozone (30 ppm). They found that whereas when oxidized in bulk solution a large number of residues were damaged, oxidation at the air–water interface led to damage but to a smaller number of residues, suggested to be tryptophan and possibly cysteine, presumably as other residues were protected from the ozone in some way, such as by being embedded in the water subphase. In addition to films of pure SP-B(1–25) at the interface, Kim *et al.* (2010) looked at mixed films of SP-B(1–25) and the lipid 1-palmitoyl-2-oleoyl-*sn*-glycerol (POG). POG was used as a structurally related mimic of POPC and POPG, which were not suitable for the experimental detection system used. The authors reported that the oxidation followed a similar trend when the SP-B(1–25) was in the presence of the lipid: a small number of residues suffered oxidative damage upon exposed to ozone. Hemming *et al.* (2015) used a range of techniques to follow the reaction of SP-B(1-25) and the larger peptide SP-B(1–25,63–78), known as SMB, which contains both the first 25 and the final 16 amino acid residues of SP-B and is thought to better mimic the behaviour of the full-length protein. Like full-length SP-B, all the cysteine residues in SMB are involved in disulfide bonds, reducing the likelihood of further oxidation by ozone. First, Hemming *et al.* (2015) showed that monolayer films of SP-B(1–25) and SMB, spread at the air–water interface, showed a change in surface pressure when exposed to gas-phase ozone, indicating that a reaction had occured. Exposure to ozone led to a rapid drop in surface pressure, an increase in surface tension, which did not recover. This drop in surface pressure was also observed when the SP-B peptides where mixed with the lipid DPPC at the air–water interface. Second, Hemming *et al.* (2015) recorded the intrinsic fluorescence from the one tryptophan residue in SMB, when SMB was spread at the air–water interface and exposed to ozone. Exposure caused a very rapid loss in the fluorescence signal, indicating that the tryptophan residue reacts readily with ozone, in agreement with the work of Kim *et al.* (2010), who used an order-of-magnitude higher ozone concentration. Third, both X-ray and neutron reflectivity measurements were performed. The reflectivity of a surface to X-rays and neutrons can tell you about the amount of material present at the air–water interface and the thickness of the film. No measurable change was observed when either SP-B(1–25) or SMB, spread at the air–water interface, was exposed to gas-phase ozone. The surface pressure and fluorescence data clearly showed that a reaction was occurring, and the reflectivity measurements that the reaction products, damaged peptides, remained at the air–water interface. Finally, Hemming *et al.* (2015) recovered the product material from the interface and analysed it by sodium dodecyl sulfate-polyacrylamide gel electrophoresis (SDS-PAGE) and reverse-phase HPLC. They found clear evidence for a more hydrophobic reaction product being formed, attributed to oxidation leading to unfolding of the peptides, consistent with the work of Sarker *et al.* (2011).

There are no reports in the literature on changes to SP-C when exposed to gas-phase ozone at the air–water interface. The cysteine residues of SP-C are unlikely to be oxidized further, as they are bonded to palmitoyl chains in the functional form of the protein. However, there is a histidine residue on SP-C, located just at the start of the hydrophobic region, that is expected to be embedded in the lipid tails and thus potentially exposed to gas-phase ozone. It would be interesting to discover if this histidine is oxidized by ozone when present in a lipid layer at the air–water interface and, if so, if the oxidation leads to a significant structural change.

In summary, experimental evidence exists to support the idea that SP-B in lung surfactant is likely to be damaged by exposure to ozone. This damage is likely to change the structure of the protein and its interaction with the lipid components of lung surfactant. The degree of the disruption caused by this on the functioning of lung surfactant is something on which we can only speculate.

### 6.4.5 Molecular dynamics simulations

Molecular dynamics simulations of lipid monolayers at the air–water interface are not routine and are much less commonly undertaken than lipid bilayer studies. However, a number of researchers have used molecular dynamics simulations to assess the impact of the presence of oxidized lipids on the physical properties of lipid bilayers and monolayer. Wong-ekkabut *et al.* (2007) performed molecular dynamics simulations on lipid bilayers composed of 1-palmitoyl-2-linoleoyl-*sn*-glycero-3-phosphatidylcholine (PLPC) where some of the PLPC was replaced with lipid oxidation products. They found that the oxidation products, which had more polar groups at the truncated terminal of the *sn*-2 chain, tended to orientate themselves towards the lipid head groups and water, leading to a thinner bilayer that was potentially more permeable. Khabiri *et al.* (2012) used molecular dynamics simulations to investigate changes to monolayers of the lipid dioleoyl-*sn*-glycero-3-phosphtidylcholine when lipid oxidation products are present. Similarly to what had been found by Wong-ekkabut *et al.* (2007) in lipid bilayers, in lipid monolayers the presence of new more polar groups at the truncated terminal ends of the lipid tails led to a rearrangement of the lipid, with the new polar group orientating itself towards the aqueous subphase. The short-chain oxidation product nonanal was predicted to be ultimately lost from the interface, in agreement with experiment. Khandelia and Mouritsen (2009) looked at lipid bilayers of POPC with varying levels of the oxidation products PAzPC and POnPC. They found that in the case of PAzPc the azelaoyl chain entered the water layer and in the case of POnPC the 9′-oxo-nonanoyl orientated itself towards the water layer. Siani *et al.* (2016) used molecular dynamics to explore how a range of hydroperoxidized lipids would behave in lipid bilayers and lipid monolayers. Again, these damaged lipids, with more polar residues at mid-points in the formerly hydrophobic lipids tail, gave increased hydration of the lipid head region and required an increased area to attain the same surface pressure.

Molecular dynamics simulations to look at the effects of oxidation on the properties of monolayers involving the surfactant proteins SP-B and SP-C are less common. Studies involving the full-length SP-B are rare, at least in part because there is no experimental structure of full-length SP-B to use as a starting point for the simulation. As mentioned previously, Sarker *et al.* (2011) did use molecular dynamics to reveal substantial changes in the behaviour of SP-B(8–25), when the tryptophan residue was replaced with an oxidized version. No other studies involving oxidized proteins are available. There is also a lack of simulations looking at how oxidized lipids interact with the lung surfactant proteins.

## 6.5 Conclusions

It is clear that inhaled ozone reacts with lipids, and probably proteins, present in lung surfactant at the air–water interface. Some of the shorter-chain aldehyde products have been detected in exhaled air but others will be solubilized in the epithelial lining fluid, potentially triggering a biological response. The major lipid oxidation product, a lipid with one undamaged saturated tail, and a second truncated tail with a new polar terminating group, will, initially at least, remain at the air–water interface but in an altered conformation, causing substantial disruption and changes to the biophysical properties of the surfactant monolayer. Changes to the ordering of the lipids in the surfactant film are also likely. The vital surfactant protein B is likely to undergo at least tryptophan oxidation when exposed to low levels of ozone at the air–water interface, potentially causing significant structural changes. More work is needed to determine how the presence of both oxidized lipids and oxidized protein changes the nature of the protein/lipid surfactant film.

## References

Almstrand, A.-C., Voelker, D. and Murphy, R.C. (2015) Identification of oxidized phospholipids in bronchoalveolar lavage exposed to low ozone levels using multivariate analysis. *Analytical Biochemistry* 474, 50–58.

Alonso, C., Alig, T., Yoon, J., Bringezu, F., Warriner, H. and Zasadzinski, J.A. (2004) More than a monolayer: relating lung surfactant structure and mechanics to composition. *Biophysical Journal* 87, 4188–4202.

Berlett, B.S, Levine, R.L. and Stadtman, E.R. (1996) Comparison of the effects of ozone on the modification of amino acid residues in glutamine synthetase and bovine serum albumin. *Journal of Biological Chemistry* 271, 4177–4182.

Bernardino de la Serna, J., Pérez-Gil, J., Simonsen, A.C. and Bagatolli, L.A. (2004) Cholesterol rules: direct observation of the coexistence of two fluid phases in native pulmonary surfactant membranes at physiological temperatures. *Journal of Biological Chemistry* 279, 40715–40722.

Clark, J.C., Wert, S.E., Bachurski, C.J., Stahlman, M.T., Stripp, B.R., Weaver, T.E. and Whitsett, J.A. (1995) Targeted disruption of the surfactant protein B gene disrupts surfactant homeostasis, causing respiratory failure in newborn mice. *Proceedings of the National Academy of Science of the United States of America* 92, 7794–7798.

Cohen, A.L., Brauer, M., Burnett, R., Anderson, H.R., Frostad, J. *et al.* (2017) Estimates and 25-year trends of the global burden of disease attributable to ambient air pollution: an analysis of data from the Global Burden of Diseases Study 2015. *Lancet* 389, 1907–1918.

Creuwels, L.A., van Golde, L.M., Haagsman, H.P. (1997) The pulmonary surfactant system: biochemical and clinical aspects. *Lung* 175, 1–39.

Enami, S., Hoffmann, M.R. and Colussi, A.J. (2009) Simultaneous detection of cysteine sulfenate, sulfinate, and sulfonate during cysteine interfacial ozonolysis. *Journal of Physical Chemistry B* 113, 9356–9358.

Frampton, M.W., Pryor, W.A., Cueto, R., Cox, C., Morrow, P.E. and Utell, M.J. (1999) Ozone exposure increases aldehydes in epithelial lining fluid in human lung. *American Journal of Respiratory and Critical Care Medicine* 159, 1134–1137.

Fujiwara, T., Maeta, H., Chida, S., Morita, T., Watabe, Y. and Abe, T. (1980) Artificial surfactant therapy in hyaline-membrane disease. *Lancet* 1, 55–59.

Glasser, S.W., Burhans, M.S., Korfhagen, T.R., Na, C.L., Sly, P.D. *et al.* (2001) Altered stability of pulmonary surfactant in SP-C-deficient mice. *Proceedings of the National Academy of Science of the United States of America* 98, 6366–6371.

Hemming, J.M., Hughes, B.R., Rennie, A.R., Tomas, S., Campbell, R.A. *et al.* (2015) Environmental pollutant ozone causes damage to lung surfactant protein B (SP-B). *Biochemistry* 54, 5185–5197.

Jerrett, M., Burnett, R.T., Pope, C.A., Ito, K., Thurston, G. *et al.* (2009) Long-term ozone exposure and mortality. *New England Journal of Medicine* 360, 1085–1095.

Khabiri, M., Roeselova, M. and Cwiklik, L. (2012) Properties of oxidized phospholipid monolayers: an atomistic molecular dynamics study. *Chemical Physics Letters* 519, 93–99.

Khandelia, H. and Mouritsen, O.G. (2009) Lipid gymnastics: evidence of complete acyl chain reversal in oxidized phospholipids from molecular simulations. *Biophysical Journal* 96, 2734–2743.

Kim, H.I., Kim, H., Shin, Y.S., Beegle, L.W., Jang, S.S. *et al.* (2010) Interfacial reactions of ozone with surfactant protein B in a model lung surfactant system. *Journal of the American Chemical Society* 132, 2254–2263.

Lai, C.C., Yang, S.H. and Finlayson-Pitts, B.J. (1994) Interactions of monolayers of unsaturated phosphocholines with ozone at the air–water interface. *Langmuir* 10, 4637–4644.

Manzanares, D., Rodriguez-Capote, K., Liu, S., Haines, T., Ramos, Y. *et al.* (2007) Modification of tryptophan and methionine residues is implicated in the oxidative inactivation of surfactant protein B. *Biochemistry*, 46, 5604–5615.

Mudd, J.B., Leavitt, R., Ongun, A. and McManus, T.T. (1969) Reaction of ozone with amino acids and proteins. *Atmospheric Environment* 3, 669–682.

Mudway, I.S. and Kelly, F.J. (2000) Ozone and the lung: a sensitive issue. *Molecular Aspects of Medicine* 21, 1–48.

Müller, B., Seifart, C. and Barth, P.J. (1998) Effect of air pollutants on the pulmonary surfactant system. *European Journal of Clinical Investigation* 28, 762–777.

Nogee, L.M., Garnier, G., Dietz, H.C., Singer, L., Murphy, A.M., deMello, D.E. and Colten, H.R. (1994) A mutation in the surfactant protein B gene responsible for fatal neonatal respiratory disease in multiple kindreds. *Journal of Clinical Investigation* 93, 1860–1863.

Parra, E. and Pérez-Gil, J. (2015) Composition, structure and mechanical properties define performance of pulmonary surfactant membranes and films. *Chemistry and Physics of Lipids* 185, 153–175.

Postle, A.D., Heeley, E.L. and Wilton, D.C. (2001) A comparison of the molecular species compositions of mammalian lung surfactant phospholipids. *Comparative Biochemistry and Physiology Part A: Molecular & Integrative Physiology* 129, 65–73.

Pryor, W.A., Bermúdez, E., Cueto, R. and Squadrito, G.L. (1996) Detection of aldehydes in bronchoalveolar lavage of rats exposed to ozone. *Fundamental and Applied Toxicology* 34, 148–156.

Putman, E., Liese, W., Voorhout, W.F., van Bree, L., van Golde, L.M. and Haagsman, H.P. (1997) Short-term ozone exposure affects the surface activity of pulmonary surfactant. *Toxicology and Applied Pharmacology* 142, 288–296.

Qiao, L., Ge, A., Liang, Y. and Ye, S. (2015) Oxidative degradation of the monolayer of 1-palmitoyl-2-oleoyl-*sn*-glycero-3-phosphocholine (POPC) in low-level ozone. *Journal of Physical Chemistry B*, 119, 14188–14199.

Sabatini, K., Mattila, J.-P., Megli, F.M. and Kinnunen, P.K.J. (2006) Characterization of two oxidatively modified phospholipids in mixed monolayers with DPPC. *Biophysical Journal* 90, 4488–4499.

Sarker, M., Rose, J., McDonald, M., Morrow, M.R. and Booth, V. (2011) Modifications to surfactant protein B structure and lipid interactions under respiratory distress conditions: consequences of tryptophan oxidation. *Biochemistry* 50, 25–36.

Siani, P., de Souza, R.M., Dias, L.G., Itri, R. and Khandelia, H. (2016) An overview of molecular dynamics simulations of oxidized lipid systems, with a comparison of ELBA and MARTINI force fields for coarse grained lipid simulations. *Biochimica Biophysica Acta* 1858, 2498–2511.

Thompson, K.C., Rennie, A.R., King, M.D., Hardman, S.J.O., Lucas, C.O.M. *et al.* (2010) Reaction of a phospholipid monolayer with gas-phase ozone at the air–water interface: measurement of surface excess and surface pressure in real time. *Langmuir* 26, 17295–17303.

Thompson, K.C., Jones, S.H., Rennie, A.R., King, M.D., Ward, A.D. *et al.* (2013) Degradation and rearrangement of a lung surfactant lipid at the air–water interface during exposure to the pollutant gas ozone. *Langmuir* 29, 4594–4602.

Uhlson, C., Harrison, K., Allen, C.B., Ahmad, S., White, C.W. and Murphy, R.C. (2002) Oxidized phospholipids derived from ozone-treated lung surfactant extract reduce macrophage and epithelial cell viability. *Chemical Research in Toxicology* 15, 896–906.

Volinksy, R., Paananen, R. and Kinnunen, P.K.J. (2012) Oxidized phosphatidylcholines promote phase separation of cholesterol–sphingomyelin domains. *Biophysical Journal* 103, 247–254.

Wadia, Y., Tobias, D.J., Stafford, R. and Finlayson-Pitts, B.J. (2000) Real-time monitoring of the kinetics and gas-phase products of the reaction of ozone with an unsaturated phospholipid at the air–water interface. *Langmuir* 16, 9321–9330.

Wong-ekkabut, J., Xu, Z., Triampo, W., Tang, I.-M., Peter Tieleman, D. and Monticelli, L. (2007) Effect of lipid peroxidation on the properties of lipid bilayers: a molecular dynamics study. *Biophysical Journal* 93, 4225–4236.

# 7 Nitrogen Dioxide: Ambient Exposure in Human Disorders

**Y.-C.T. Huang* and J.L. Tucker**
*Division of Pulmonary, Allergy and Critical Care Medicine, Duke University Medical Center, Durham, North Carolina, USA*

## 7.1 Abstract

Nitrogen dioxide ($NO_2$) is produced in certain industrial processes. Point exposure to high concentration of $NO_2$ (> 100 ppm), such as in silo filler's disease and during chemical warfare, produces acute lung injury, and death in the worst cases. Acute exposure to lower concentration of $NO_2$ (0.5–15 ppm) may increase airway responsiveness, especially in patients with obstructive airway diseases. The effects of chronic exposure to ambient $NO_2$ (< 0.5 ppm) are much harder to quantify, in part due to the low concentration and the existence of other co-pollutants. Because $NO_2$ is a reactive gas, it remains possible that exposure to ambient concentration of $NO_2$ for a prolonged period of time may result in cellular and biochemical changes that ultimately cause pulmonary dysfunction.

## 7.2 Introduction

Nitrogen dioxide ($NO_2$) is one of several nitrogen oxides existing in gaseous form at room temperatures. It is a reddish-brown gas with a pungent, acrid odor, and is produced in certain industrial processes. Workers in industries where $NO_2$ is used may be exposed and are at risk for occupational lung diseases. The US National Institute for Occupational Safety and Health (NIOSH) has set exposure limits and safety standards. The recommended exposure limit (REL) is 1 ppm as a 15-minute short-term limit. $NO_2$ is also a ubiquitous atmospheric pollutant and one of the six criteria air pollutants that the US Environmental Protection Agency (EPA) regulates as directed by the Clean Air Act for the purposes of protecting health and the public welfare (EPA, 1993). The current National Ambient Air Quality Standard (NAAQS) for nitrogen oxides includes a 1-hour (1-h) standard of 100 ppb based on the 3-year average of 98th percentile of the yearly distribution of 1-h daily maximum concentrations, and an annual standard of 53 ppb. Environmental exposure to $NO_2$ has been associated with various human disorders. The aim of this chapter is to describe the environmental toxicology of $NO_2$ and its role in human respiratory tract disease.

## 7.3 Environmental Sources of Nitrogen Dioxide

### 7.3.1 Nitrogen dioxide in outdoor air

Natural sources of $NO_2$ include intrusion of stratospheric deposits of nitrogen oxides, volcanic

* E-mail: huang002@mc.duke.edu

 *A Handbook of Environmental Toxicology: Human Disorders and Ecotoxicology*. J.P.F. D'Mello (ed.)

activity, environmental bacterial activity, forest fires and lightning. These natural emissions are dispersed over the entire earth, resulting in relatively small background ambient concentrations that are typically less than 0.0001 ppm (Atwell *et al.*, 1995; EPA, 2008). The major source of anthropogenic emissions of $NO_2$ is fossil fuel combustion from both stationary and mobile sources (heating, cooking fuel, power generation, motor vehicles, etc.) In urban settings the compound is often used as a sentinel for traffic-based air pollution, which is a complex mixture of particulate matter and various carcinogens, including volatile organic compounds, metals and carbonyls. Annually, mean ambient $NO_2$ levels in the USA are well below the current NAAQS of 0.053 ppm. There has been a general decrease in the nationwide average atmospheric $NO_2$ concentration; levels in 2006 were decreased from levels in 1980 by 41% (EPA, 2008). There are few recent data describing global atmospheric $NO_2$ concentrations, but a World Health Organization (WHO) report in the early 1990s established trends in $NO_2$ concentration for six selected cities from various regions of the world: Bombay, Mexico City, London, Sao Paulo, Moscow and Tokyo. In general, the overall trends appeared to be relatively stable for most of the cities (WHO, 1997).

### 7.3.2 Nitrogen dioxide in indoor air

Considerable indoor $NO_2$ exposure may also result from a variety of common household sources. Indoor sources of $NO_2$ include tobacco burning, wood-burning stoves, candles and the use of gas-fired appliances and oil stoves. Indoor concentrations may sometimes exceed those outdoors, especially with the use of unvented combustion appliances. The average concentration over a period of several days may exceed 0.1 ppm when unvented gas stoves are used for cooking, supplementary heating, or clothes drying. Maximum 1-h concentrations in a kitchen or with a room heater may be in the range of 0.4–1.5 ppm and can reach concentrations well above those measured in ambient outdoor locations. Therefore, aside from ambient outdoor exposures to $NO_2$, total personal $NO_2$ exposure may be highly impacted by personal activities, which in fact may dominate short-term exposures (WHO, 2006; EPA, 2017). Nitrogen oxides including $NO_2$ are also produced commercially, usually as the first step in the production of nitric acid, either by the oxidation of atmospheric nitrogen via electric arc (Birkeland-Eyder process) or by catalytic oxidation of anhydrous ammonia (Oswald process). Trace-metal impurities also cause nitrogen oxides to form in nitric acid and its solutions. Nitrogen oxides are intermediates in the production of lacquers, dyes and other chemicals as well. Another important but less common indoor source of $NO_2$ is inherent in the storage of corn, oats and other agricultural products in silos. The gas that accumulates just above the silage product in a recently filled silo contains a mixture of nitrogen oxides due to the bacterial conversion of plant nitrates into nitric oxide (NO), which readily reacts with the ambient air to form $NO_2$. This exposure is the source of lung toxicity in silo filler disease, discussed later (Epler, 1989).

## 7.4 Toxicology of Nitrogen Dioxide

$NO_2$ is a reddish-brown gas above 21.2°C with an acrid odour and becomes a yellowish-brown liquid below this temperature. It typically arises via the oxidation of nitric oxide by oxygen in air. $NO_2$ is poorly soluble in water but is hydrolysed to yield nitrous acid ($HNO_2$), nitric acid ($HNO_3$) and NO (Elsayed, 1994; Huie, 1994). It is capable of producing a diversity of nitro-, nitrate or nitroso- addition products, as well as other nitrogen oxides or acids. $NO_2$ is thus quite a reactive compound that can initiate free-radical reactions, react with unsaturated fatty acids, induce auto-oxidation of organic compounds and is itself a nitrogen-centred free radical. Most of the toxic effects of $NO_2$ are related to its oxidative properties (Utell and Frampton, 1997; Searl, 2004; ATSDR, 2014).

### 7.4.1 Uptake, metabolism and oxidative damage

As mentioned previously, $NO_2$ has a limited solubility in water, thus uptake of inhaled gas is governed by reactions with constituents of pulmonary surfactant rather than physical dissolution of the gas itself. The abstraction of hydrogen ions from organic molecules leads to the formation of nitrous acid and organic radicals (Postlethwait and Bidani, 1994). $NO_2$ is only a moderately potent oxidant and whereas compounds with high reactivity, such as ozone ($O_3$), react in the upper respiratory tract, compounds with lower oxidative reactivity, such as $NO_2$, will act on the level of gas exchange at the alveoli. This is where typical $NO_2$-induced lesions have been observed in several species following high levels of exposure. Experimental studies have shown that $NO_2$ or its chemical products can remain in the lung for prolonged periods. Once absorbed in lung fluids, $NO_2$ dissolves to form nitrous and nitric acids and both nitrite and nitrate anions are translocated in the bloodstream (Elsayed, 1994; Januszkiewicz and Mayorga, 1994; Mayorga, 1994). The oxidant properties of $NO_2$ also induce the peroxide detoxification pathway of glutathione peroxidase, glutathione reductase and glucose-6-phosphate dehydrogenase. Such free radical-mediated reactions are the mechanism by which $NO_2$ exerts direct toxicity on lung cells. This mechanism of action is supported by animal studies showing the importance of lung antioxidant defences in protecting against the effects of $NO_2$ (Elsayed, 1994; Januszkiewicz and Mayorga, 1994; Utell and Frampton, 1997; WHO, 1997). Phospholipid peroxidation plays a chief role in the pulmonary epithelial toxicity of $NO_2$. The effects of lipid peroxidation include changes in pulmonary surfactant and membrane fluidity. Oxidation of the fatty acids of surfactant phospholipids occurs at relatively low levels of exposures and impairs their adsorption to the air–liquid interface. Oxidative degradation of surfactant also results in reduced pulmonary clearance rates and increased retention of inhaled particles. The products of oxidation may promote airways inflammation and pulmonary oedema (Putman *et al.*, 1997; WHO, 1997).

### 7.4.2 Inflammatory response and host defences

Host defence responses to $NO_2$ exposure have been observed in several laboratory animal species, resulting in the conclusion that these effects could occur in humans. In addition, mathematical dosimetry models suggest that the greatest dose of $NO_2$ is delivered to the same region in both animal and human lungs. Thus, the responses of laboratory animals can be qualitatively extrapolated to humans. Acute exposure to $NO_2$ is associated with lung inflammation characterized by infiltration of serum inflammatory cells and hyperplasia of type 2 respiratory epithelial cells. Transcription and release of pro-inflammatory cytokines including interleukin IL-6 and tumour necrosis factor-α (TNF-α) follow, with consequent activation of pulmonary macrophages and lymphocyte proliferation. The majority of available studies demonstrating an inflammatory esponse to $NO_2$ involved concentrations that greatly exceed those in ambient air, however, and data regarding the inflammatory response to chronic low-level exposure to $NO_2$ are limited.

Several studies have examined the cellular and biochemical changes present in sputum, nasal lavage or broncheoalveolar lavage (BAL) fluid associated with short-term $NO_2$ exposure in concentrations in healthy human subjects. These studies have generally reported either small or statistically non-significant effects, but observations have included elevated levels of natural killer (NK) cells and neutrophils, as well as the pro-inflammatory mediators thromboxane B2, IL-6 and IL-8 (Hesterberg, 2009). Overall the evidence is inconsistent for markers of pulmonary inflammation from short-term $NO_2$ inhalation exposures at concentrations below 1 ppm among healthy volunteers but is more consistent at concentrations up to 2 ppm.

In terms of mechanical host defences, exposure to $NO_2$ can cause loss of cilia and ciliated epithelial cells. Such changes are reflected in the functional impairment of mucociliary clearance at high levels of $NO_2$ in animal models (> 9400 $\mu g\ m^{-3}$, 5.0 ppm). The same effect has not been demonstrated in animal models at lower levels of exposure. Inhalation of $NO_2$ leads to mucus hypersecretion and altered mucus composition, affecting lung clearance and respiration (Samet and Chang, 1994). Animal investigations of the effects of $NO_2$ on particle clearance found that all epithelial defence functions, including ciliary activity, mucociliary transport velocity and epithelial permeability were significantly inferior compared with subjects not exposed to $NO_2$ (Januszkiewicz and Mayorga, 1994; Samet and Chang, 1994; WHO, 1997). $NO_2$ also increases susceptibility to bacterial and viral pulmonary infections in animal models. Reduced phagocytic activity and reduced mobility were observed in alveolar macrophages in those models; and $NO_2$ exposure has been shown to cause a clear dose-related decrease in pulmonary antibacterial defences.

### 7.4.3 Effect on respiratory mechanics

In many human clinical studies of $NO_2$ exposure, both pulmonary function and airway responsiveness to bronchoconstrictors have been measured. $NO_2$ causes decrements in lung function, particularly increased airway resistance in resting healthy subjects at 2 h concentrations as low as 2.5ppm. $NO_2$ exposure results in increased airway responsiveness to broncoconstrictive agents in exercising healthy, non-smoking subjects exposed to concentrations as low as 1.5 ppm for exposure durations of 1 h or longer (Abe, 1967; Von Nieding *et al.*, 1973; WHO, 1997). Further, exposure of asthmatics to $NO_2$ has been associated with increased airway responsiveness to a variety of provocative mediators, including cholinergic and histaminergic chemicals, sulfur dioxide and cold air. These responses may begin at concentrations as low as 0.20 ppm. Modest increases in airway resistance may occur in patients with chronic obstructive pulmonary disease (COPD) from brief exposure (15–60 min) to concentrations of $NO_2$ as low as 1.5 ppm and decrements in spirometric measures of lung function (3–8%) change in forced expiratory volume in 1 second (FEV1) may also be observed with longer exposures to concentrations as low as 0.3 ppm (Von Nieding *et al.*, 1979; WHO, 1997).

## 7.5 Disease Associated with Acute Point-Source Exposure

### 7.5.1 Chemical weapon exposure

One of the oldest sources of historical data about the role of nitrogen oxides including $NO_2$ and human disease comes from military research involving chemical weapons of warfare and terrorism. A number of chemical agents have been used or have potential use for injuring the upper respiratory tract and/or the lung parenchyma itself. $NO_2$ is not itself a discrete weapon of chemical warfare, but high concentrations of the gas are found in the deployment of weapons such as nitrogen mustards and phosgene. Phosgene is a colourless gas and was the most lethal chemical warfare agent used in World War I. Because of its low water solubility, phosgene causes extensive lung parenchymal damage with little irritation to the upper airway. Massive exposure to agents such as phosgene may cause immediate destruction of the alveolar epithelial cells and adjacent capillary endothelial cells, leading to death from acute respiratory failure. More common is the delayed onset of acute lung injury, potentially progressing to the acute respiratory distress syndrome (ARDS). Depending on the extent of injury, clinical manifestations may vary from mild acute lung injury to severe ARDS. A variety of chronic respiratory diseases have been attributed to exposure to chemical pulmonary agents, including chronic bronchitis, bronchiectasis, interstitial pulmonary fibrosis, airway hyperreactivity and large airway obstruction. Bronchiolitis obliterans and bronchiolitis obliterans organizing pneumonia (BOOP) are two distinct patterns of disease that have been described as late sequelae from injury to small distal airways after exposure to a variety of pulmonary agents, including those described above (Mayorga, 1994; Greenfield *et al.*, 2002).

### 7.5.2 Silo filler disease

Silo filler disease is an acute lung injury caused by inhalation of toxic gases, including $NO_2$, in or near an agricultural silo (Fig. 7.1). The entity represents an occupational hazard associated specifically with the ensiled crops, most commonly corn, oats, alfalfa, and hay.

Silage is a high-moisture product usually made from grass crops, sorghum or corn by using the entire green plant in the fermentation process by silo storage. Silage is then used to feed cattle or sheep. After a silo is stocked with silage (usually September and October), $NO_2$ is released into the air, with maximum concentrations being reached in 1–2 days. Laboratory studies have shown that toxic levels of nitrogen oxides are regularly produced by the silage. Levels of 200–2000 ppm are common, and levels of 50–100 ppm are capable of producing pulmonary disease (Scott and Hunt, 1973). $NO_2$ is heavier than air and may concentrate along the top of the silage

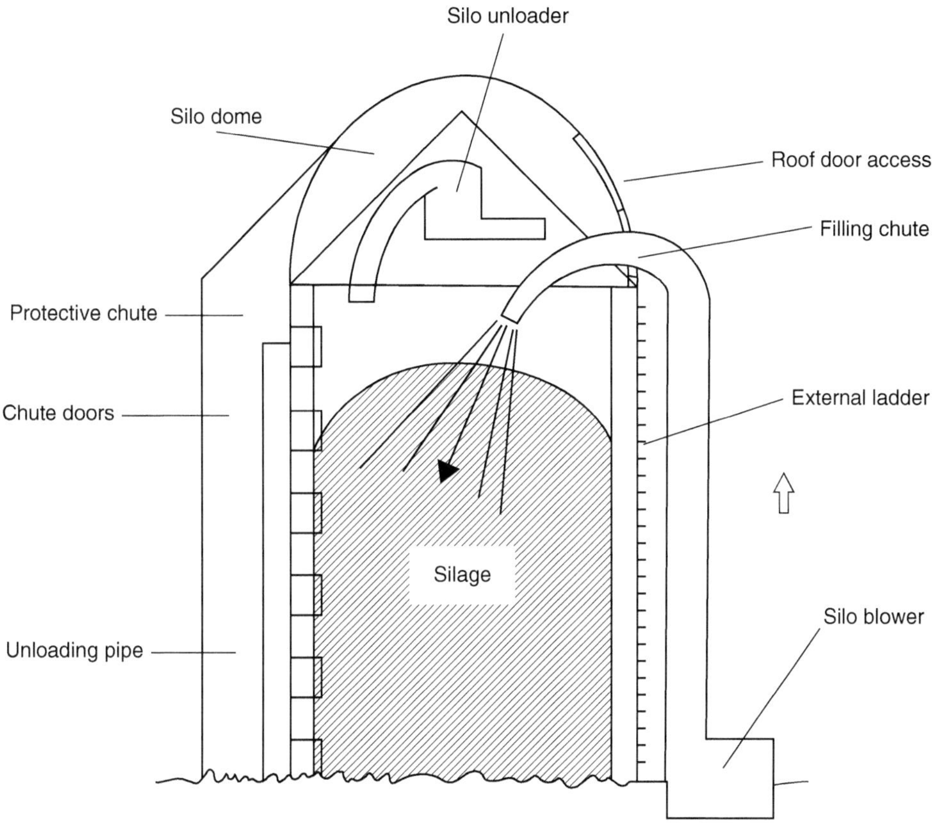


**Fig. 7.1.** Typical grain silo. Modified from Douglas *et al.* (1989).

or the floor of the silo entrance. Exposed individuals usually experience minor symptoms, such as eye irritation, cough and nausea. More serious consequences may also occur, such as pulmonary oedema, bronchiolitis obliterans, rapid asphyxiation, respiratory collapse and death within minutes when exposed to high concentrations. Less frequent symptoms include myalgias, palpitations, haemoptysis and cyanosis. These symptoms usually indicate acute exposure when the history is consistent; however, persistent or delayed symptoms can also occur for days or even weeks after a single exposure (Leavey *et al.*, 2004). Examination of the patient with silo filler disease may uncover predominantly pulmonary abnormalities, such as rales, rhonchi, wheezes and decreased breath sounds. Hypotension may be present either as a result of NO formation in the vascular system, which reduces systemic vascular resistance and causes vasodilation, or acute right ventricular failure due to hypoxic pulmonary vasoconstriction. Cyanosis may rarely be present and indicates either methemoglobinaemia or severe hypoxaemia. Haemoptysis if present indicates severe inflammatory reaction within the respiratory tract. Pulmonary function studies performed acutely have shown both restrictive and obstructive defects, with hypoxaemia and a reduction in carbon monoxide diffusion. Supportive care including the use of nebulized bronchodilators, corticosteroids, supplemental oxygen and pulmonary toilet is the mainstay of management of patients with exposures consistent with silo filler disease. Respiratory support with non-invasive or invasive mechanical ventilation is often indicated for the patient with severe respiratory symptoms. The diagnosis of silo filler disease must be considered as soon as possible so that corticosteroid therapy can be given early to prevent a fulminant course and death. Corticosteroids seem to have a substantial influence on prognosis, as a thorough review of reported cases indicates that most patients have a complete recovery when full supportive treatment is administered (CDC, 1982; Douglas *et al.*, 1989; Epler, 1989; Auerbach *et al.*, 2016).

## 7.6 Respiratory Disease Linked to Short-term and Long-term Ambient Exposure

### 7.6.1 Lung function and $NO_2$ exposure

Most experimental studies of $NO_2$ exposure in humans show no adverse lung function effects at exposure concentrations of up to 14 ppm in healthy volunteers, though some have found effects at concentrations less than 0.3 ppm. Studies have thus shown inconsistent results and some findings have not been reproducible in subsequent studies by the same laboratories. A greater proportion of experiments with asthmatics has found adverse effects on lung function at concentrations of less than 14 ppm than those with healthy subjects. One meta analysis of asthma patients showed no effects at $NO_2$ concentrations of 0.2–1.2 ppm. Small effects were seen in some studies, however, at concentrations of 0.5 ppm (Berglund *et al.*, 1993). Overall, there is no clear evidence that adverse lung function effects increase with concentration or duration of exposure, or that superimposition of peak exposures on a background level of exposure greatly alters observed effects. There is also no clear evidence that exercise clearly increases effects in healthy volunteers, but there is some limited evidence that concurrent exercise may increase effects on lung function in asthmatic volunteers. The results of individual human studies are unfortunately highly inconsistent and it is impossible to determine exposure–response relationships. Some studies of patients with COPD have found effects on airways resistance at 0.9 ppm and lung function at 0.5 ppm, but no information is available about the presence or absence of effects at lower levels of exposure (WHO, 1997). Other studies have reported no effects associated with 1-h exposures to 3.7 ppm (Berglund *et al.*, 1993) and one found a mild decrease in FEV1 in COPD patients exposed to 0.5 ppm for 2 h, but similar effects were not seen in healthy or asthmatic volunteers (Vagaggini *et al.*, 1996). Although individuals with COPD may be expected to be more sensitive to $NO_2$ than healthy subjects, they may have limited capacity to respond to $NO_2$ due to poor physiological reserve.

### 7.6.2 $NO_2$ and airway responsiveness and allergy

Airways responsiveness has been studied in both healthy and asthmatic volunteers. A variety of challenges have been used to assess airways responsiveness so that the results of different studies are not always directly comparable. Further, the results of individual studies have not always been internally consistent. Some studies suggest that effects in both healthy and asthmatic subjects may be related to both concentration and duration of exposure, though the evidence is extremely limited. A number of studies of the effects of $NO_2$ on the allergic response of asthmatics have been undertaken; in general the results of these studies indicate that exposure to concentrations of $NO_2$ of more than 0.5 ppm increases the asthmatic response to a range of allergens. There is limited evidence of possible effects at concentrations of only 0.3 ppm. Effects are observed on both immediate and delayed responses to allergen exposure. Note that the concentrations used in the studies were much higher than EPA $NO_2$ exposure standard and unlikely to be encountered in ambient environments. Interestingly, there may be biomarkers of $NO_2$-related disease observable in the respiratory secretions of susceptible individuals such as asthmatics. In asthmatic subjects, reactive oxygen species such as superoxide are elevated and correlate negatively with FEV1; the reaction of superoxide with nitric oxide (elevated in the setting of exposure to $NO_2$) yields the powerful oxidant peroxynitrite, which is considered largely responsible for the adverse effects of nitric oxide generation in asthmatics. 3-Nitrotyrosine (3-NT) may represent a biomarker for the generation of reactive nitrogen intermediates such as peroxynitrite *in vivo*. One study demonstrated that the median 3-NT concentration was fivefold higher in asthmatic children than in healthy subjects (Baraldi *et al.*, 2006). Another study showed 3-NT concentration in the epithelial lining fluids in asthma children was three- to fivefold that of control children (Fitzpatrick *et al.*, 2009).

### 7.6.3 Epidemiological studies

#### *7.6.3.1 Short-term health effects of $NO_2$*

Results of numerous epidemiological studies of the short-term health effects of exposure to $NO_2$ are inconsistent and there is considerable uncertainty about the relationship between personal exposure concentrations of $NO_2$ and the varied environmental measurements used in epidemiological analyses. Many studies have failed to find statistically significant associations between $NO_2$ and a range of health endpoints. The health endpoints for which there is the most consistent evidence of a causal relationship specifically with $NO_2$ include daily mortality (Chen *et al.*, 2012; Faustini *et al.*, 2014), emergency hospital visits for asthma and use of primary health care facilities (Mills *et al.*, 2015, 2016). The apparent adverse health effects of $NO_2$ often disappear in multi-pollutant models that include particulate matter and/or ozone. $NO_2$ is strongly associated with traffic pollution and it is possible that effects attributed to $NO_2$ are actually due to some other component of traffic pollution, such as fine particles. $NO_2$ acts simply as a signal of high levels of particulate pollution. That said, significant associations have been found between $NO_2$ and various health end points, even in the absence of a strong correlation between concentrations of $NO_2$ and fine particulate matter. Overall, the evidence points towards a small independent role for $NO_2$ in the association between air quality and health that is separate from that associated with fine particles. Uncertainties in exposure estimation and the strong correlations that exist between concentrations of different pollutants have tended to obscure the influence of $NO_2$ on health overall (Searl, 2004).

#### *7.6.3.2 Long-term health effects of $NO_2$*

The difficulty in determining personal exposures to $NO_2$ over prolonged timescales has made the determination of the independent effects of long-term exposure to $NO_2$ in epidemiological studies relatively difficult. There are few studies of the long-term effects of $NO_2$ compared with the effects of day-to-day fluctuations of ambient concentrations. Further (and similar to epidemiological studies of short-term exposures and health outcomes), there is a poor correlation between personal exposure concentrations of $NO_2$ and concentrations measured at central monitoring sites. The presence of a high proportion of smokers in the population who are exposed to more $NO_2$ in cigarette smoke than is present in ambient air may also have contributed to weakening the power of epidemiological studies to detect long-term exposure effects. The results of the limited number of studies available suggest that long-term exposure to $NO_2$ may lead to slightly increased risks of respiratory symptoms, including bronchitis and a small reduction in lung function growth in children and lung function in adults. Long-term exposure to $NO_2$ may also lead to a small loss of life expectancy but the evidence for this is weak. It is difficult to be certain to what extent apparent effects are due to $NO_2$ and not concurrent exposure to fine particulate matter (Searl, 2004).

## 7.7 Summary and Conclusions

Exposure to high concentration of $NO_2$ may cause acute lung injury or silo filler disease, a condition that still occurs sporadically in farming communities. The effects of exposure to ambient $NO_2$ are much harder to quantify, in part due to the low concentration and the existence of other co-pollutants. Because $NO_2$ is a reactive gas, it remains possible that exposure to low concentration of $NO_2$ for a prolonged period of time may result in cellular and biochemical changes that ultimately cause pulmonary dysfunction.

## References

Abe, M. (1967) Effects of mixed $NO_2$-$SO_2$ gas on human pulmonary functions. Effects of air pollution on the human body. *Bulletin of Tokyo Medical and Dental University* 14(4), 415–433.

ATSDR (2014) *Medical Management Guidelines for Nitrogen Oxides*. Agency for Toxic Substances and Disease Registry Atlanta, Georgia

Atwell, V., Edwards, C. and McKee, D. (1995) *Review of the national ambient air quality standards for nitrogen dioxide assessment of scientific and technical information*. Staff paper (Final). Office of Air

Quality Planning and Standards, US Environmental Protection Agency, Research Triangle Park, Durham, North Carolina.
Auerbach, P.S., Cushing, T.A. and Harris, N.S. (2016) *Auerbach's Wilderness Medicine E-Book*: Elsevier Health Sciences.
Baraldi, E., Giordano, G., Pasquale, M.F., Carraro, S., Mardegan, A. *et al.* (2006) 3-Nitrotyrosine, a marker of nitrosative stress, is increased in breath condensate of allergic asthmatic children. *Allergy* 61, 90–96.
Berglund, M., Boström, C.-E., Bylin, G., Ewetz, L., Gustafsson, L. *et al.* (1993) Health risk evaluation of nitrogen oxides. *Scandinavian Journal of Work, Environment & Health*, 19 (Suppl. 2) 1–72.
CDC (1982) Silo-filler's disease in rural New York. *Morbidity and Mortality Weekly Report* 31, 389–391.
Chen, R., Samoli, E., Wong, C.-M., Huang, W., Wang, Z. *et al.* (2012) Associations between short-term exposure to nitrogen dioxide and mortality in 17 Chinese cities: the China Air Pollution and Health Effects Study (CAPES). *Environment International* 45, 32–38.
Douglas, W.W., Hepper, N.G. and Colby, T.V. (1989) Silo-filler's disease. *Mayo Clinic Proceedings* 64, 291–304.
Elsayed, N.M. (1994) Toxicity of nitrogen dioxide: an introduction. *Toxicology* 89, 161–174.
EPA (1993) *Air Quality Criteria for Oxides of Nitrogen (Final Report, 1993)*. US Environmental Protection Agency, Washington, DC.
EPA (2008) Integrated science assessment for oxides of nitrogen: Assessment of scientific and technical information. Office of Air Quality Planning and Standards, US Environmental Protection Agency, Research Triangle Park, Durham, North Carolina.
EPA (2017) *Nitrogen Dioxide's Impact on Indoor Air Quality*. US EPA/Office of Radiation and Indoor Air, Washington, DC
Epler, G.R. (1989) Silo-filler's disease: a new perspective. *Mayo Clinic Proceedings* 64, 368–370.
Faustini, A., Rapp, R. and Forastiere, F. (2014) Nitrogen dioxide and mortality: review and meta-analysis of long-term studies. *European Respiratory Journal* 44, 744–753.
Fitzpatrick, A.M., Brown, L.A., Holguin, F., Teague, W.G. and National Institutes of Health/National Heart, Lung, and Blood Institute Severe Asthma Research Program (2009) Levels of nitric oxide oxidation products are increased in the epithelial lining fluid of children with persistent asthma. *Journal of Allergy and Clinical Immunology* 124(5), 990–996, e991–999.
Greenfield, R.A., Slater, L.N., Bronze, M.S., Brown, B.R., Jackson, R., Iandolo, J.J. and Hutchins, J.B. (2002) Microbiological, biological, and chemical weapons of warfare and terrorism. *American Journal of Medical Sciences* 323, 326–340.
Hesterberg, T.W., Bunn, W.B., McClellan, R.O., Hamade, A.K., Long, C.M. and Valberg, P.A. (2009) Critical review of the human data on short-term nitrogen dioxide ($NO_2$) exposures: evidence for $NO_2$ no-effect levels. *Critical Reviews in Toxicology* 39(9), 743–781.
Huie, R.E. (1994) The reaction kinetics of $NO_2$. *Toxicology* 89, 193–216.
Januszkiewicz, A.J. and Mayorga, M.A. (1994) Nitrogen dioxide-induced acute lung injury in sheep. *Toxicology* 89(3), 279–300.
Leavey, J.F., Dubin, R.L., Singh, N. and Kaminsky, D.A. (2004) Silo-filler's disease, the acute respiratory distress syndrome, and oxides of nitrogen. *Annals of Internal Medicine* 141(5), 410.
Mayorga, M.A. (1994) Overview of nitrogen dioxide effects on the lung with emphasis on military relevance. *Toxicology* 89(3), 175–192.
Mills, I., Atkinson, R., Anderson, H., Maynard, R. and Strachan, D. (2016) Distinguishing the associations between daily mortality and hospital admissions and nitrogen dioxide from those of particulate matter: a systematic review and meta-analysis. *BMJ Open* 6(7), e010751.
Mills, I., Atkinson, R., Kang, S., Walton, H. and Anderson, H. (2015) Quantitative systematic review of the associations between short-term exposure to nitrogen dioxide and mortality and hospital admissions. *BMJ Open* 5(5), e006946.
Postlethwait, E.M. and Bidani, A. (1994) Mechanisms of pulmonary $NO_2$ absorption. *Toxicology* 89(3), 217–237.
Putman, E., Van Golde, L. and Haagsman, H. (1997) Toxic oxidant species and their impact on the pulmonary surfactant system. *Lung* 175, 75–103.
Samet, J.M. and Cheng, P.-W. (1994) The role of airway mucus in pulmonary toxicology. *Environmental Health Perspectives* 102 (Suppl. 2), 89.
Scott, E.G. and Hunt, W. (1973) Silo filler's disease. *Chest* 63, 701–706.
Searl, A. (2004) *A Review of the Acute and Long-term Impacts of Exposure to Nitrogen Dioxide in the United Kingdom*. Institute of Occupational Medicine, Edinburgh.

Utell, M.J. and Frampton, M.W. (1997) Oxides of nitrogen. In: Sipes, I.G., McQueen, C.A. and Gandolfi, A.J. (eds) *Toxicology of the Respiratory System*. Elsevier, Oxford, UK, pp. 303–312.

Vagaggini, B., Paggiaro, P., Giannini, D., Franco, A. D., Cianchetti, S., Carnevali, S., Taccola, M., Bacci, E., Bancalari, L. and Dente, F. (1996) Effect of short-term $NO_2$ exposure on induced sputum in normal, asthmatic and COPD subjects. *European Respiratory Journal* 9, 1852–1857.

Von Nieding, G., Krekeler, H., Fuchs, R., Wagner, M. and Koppenhagen, K. (1973) Studies of the acute effects of $NO_2$ on lung function: influence on diffusion, perfusion and ventilation in the lungs. *Internationales Archiv für Arbeitsmedizin* 31, 61–72.

Von Nieding, G., Wagner, H., Krekeler, H., Loellgen, H., Fries, W. and Beuthan, A. (1979) Controlled studies of human exposure to single and combined action of $NO_2$, $O_3$, and $SO_2$. *International Archives of Occupational and Environmental Health* 43, 195–210.

WHO (1997) *Environmental Health Criteria 188: Nitrogen Oxides*. International Programme on Chemical Safety, World Health Organization, Geneva.

WHO (2006) *Air Quality Guidelines: Global Update 2005*. World Health Organization, Copenhagen.

# 8 Sulfur Dioxide and Human Disorders

**S. Ahmad,*[,1] A. Ahmad[2] and A. Ahmad[1]**
[1]*Department of Anesthesiology and Perioperative Medicine, School of Medicine, University of Alabama at Birmingham, USA;* [2]*Department of Material Science Engineering, Purdue University, West Lafayette, Indiana, USA*

## 8.1 Abstract

Large amounts of sulfur dioxide are released into the atmosphere from the burning of fossil fuels, by power plants and industrial facilities that utilize it for making chemicals like sulfuric acid, pesticides, preservatives, wine and other beverages. Atmospheric sulfur dioxide is itself an enormous health concern and other oxides produced upon its reaction with environmental compounds further increase the human health risks. Moreover, high concentrations of sulfur dioxide also damages trees, plants and vegetation. Sulfur dioxide in humid atmosphere causes acid rain and leads to destruction of sensitive ecosystems. Environmental protection agencies throughout the world monitor atmospheric sulfur dioxide and have defined rules to reduce its emissions. However, industrial workers and the general public are at risk of occupational or accidental sulfur dioxide exposure due to its use or transport. In this chapter we discuss the human health risks and adverse effects of acute accidental exposure to sulfur dioxide and its increased environmental presence.

## 8.2 Introduction

Sulfur dioxide ($SO_2$) is a highly reactive, toxic, colourless gas with a very pungent odour. It is primarily used in industry as a food preservative, antioxidant and fungicide. It is also an environmental pollutant and often causes acid rains that result from its high solubility in water (Li *et al.*, 2013). Sources of atmospheric $SO_2$ include areas with volcanic activity as well as combustion of common fossil fuels such as coal and petroleum. In addition to this, $SO_2$ is formed and released by metal factories where it is a by-product of the heating/smelting of metals. $SO_2$ is characterized as a non-metal oxide, which in the presence of water displays acidic properties. In the presence of oxygen, water and other commonly found chemicals in the atmosphere, $SO_2$ reacts to form sulfuric acid. This is also the basis of the acidification of water bodies and soil causing serious damage to trees and plants.

$$SO_2\ (\text{gas}) + H_2O\ (\text{liquid}) \rightarrow H_2SO_3\ (\text{aqueous}) \qquad \text{(Eqn 8.1)}$$

$$H_2SO_3\ (\text{aqueous}) + O_2\ (\text{gas}) \rightarrow H_2SO_4\ (\text{aqueous}) \qquad \text{(Eqn 8.2)}$$

Exposure to air contaminated with pollutants such as $SO_2$ causes disease and disability and, in the case of the US state of Michigan alone, has led to a health cost impact of US$6.5 billion (Martenies *et al.*, 2017). Control of $SO_2$ along with particulate carbon emissions improved mortality, provided significant health

* E-mail address: shamaahmad@uabmc.edu

benefit and reduced air quality management costs (Liao *et al.*, 2016). The largest world contributors of atmospheric $SO_2$ are China and India, where it is primarily produced by the burning of coal for electricity generation. While China has taken strong measures and reduced emission of $SO_2$ by 75%, India has increased it by 50% (Li *et al.*, 2017b). The increase in $SO_2$ emission in East Asia is a matter of great scientific interest and political concern (Qu *et al.*, 2016). Industrial use of $SO_2$ also includes the risk of acute accidental exposures via spillage of leaky gas tanks or via transportation accidents. $SO_2$ is also used as a pesticide, preservative or fumigant in an agricultural setting, where workers are routinely exposed to its higher concentrations. Although there are extensive studies performed to understand the effect of environmental $SO_2$, there are clear gaps in knowledge about its toxicity and the mechanisms of action.

## 8.3 General Toxicity of Atmospheric Sulfur Dioxide

The worst air pollution event in the history of Europe occurred in London in 1952 where about 4000 people died and 150,000 were hospitalized. It was recently revealed that the killer fog that enveloped London may have consisted of sulfates and sulfuric acid produced by $SO_2$ formed from burning of coal (Wang *et al.*, 2016; Guo *et al.*, 2017). A similar incident was recorded in 1930 over the Meuse Valley where 63 people died and several were hospitalized following a thick fog that lasted 6 days (Setterstrom, 1940). $SO_2$ was suspected then but there was lack of evidence. However, the New York Academy of Medicine's committee on public health relations, writing on effects of air pollution on health, reported: 'It is probable that sulphurous fumes are the most deadly of all the gaseous constituents of smoke' (Anonymous, 1931). A Clean Air Act was passed by the British parliament in 1956 to regulate $SO_2$ and other pollutants but poor air quality still persists in parts of the world. Increased atmospheric $SO_2$ has been associated with heart failures globally (Shah *et al.*, 2013, Zhang *et al.*, 2017). Serious cardiopulmonary morbidities resulting in lost working days and increased hospitalizations are also a major concern (Bernard *et al.*, 2001; Wang *et al.*, 2013). Although epidemiological studies have clearly demonstrated a link between increased atmospheric $SO_2$ and increase in hospital admission for cardiovascular diseases, the underlying mechanisms remain unknown. It has been proposed that inhaled $SO_2$ reacts with moist airway epithelium to form sulfuric acid. Sulfuric acid and its bisulfite and sulfite derivatives that are readily absorbed in blood contribute to human/biological toxicity. $SO_2$ can also be endogenously generated by metabolism of sulfur-containing amino acids such as cysteine. At neutral pH $SO_2$ is broken into bisulfite and sulfite. The sulfites are also the physiological form of endogenous $SO_2$ (Lester, 1995). Sulfites are further oxidized to sulfur trioxide or sulfates and the enzymes sulfite oxidase or peroxidase catalyses the reaction. There is a large population that is sensitive to sulfites owing to deficiency of sulfite oxidase or other mitochondrial enzymes (Lester, 1995). $SO_2$ forms adducts with biomolecules such as nucleotides, flavins, DNA or proteins (Shapiro and Gazit, 1977; Xie *et al.*, 2007; Sang *et al.*, 2009). High concentrations of $SO_2$ forms photolabile complexes with vitamins such as $B_1$ and $B_{12}$ and destroys them (Steel, 1997; Dereven'kov *et al.*, 2017). Although a direct role of $SO_2$ was not shown, it was suggested that smog kills the beneficial mutualistic microbiota of human skin and mucosal surfaces may cause additional health risks, especially in children (Wong, 2017) (Fig. 8.1).

## 8.4 Acute Pulmonary Effects and Exacerbation of Pre-existing Respiratory Diseases

Sulfur dioxide inhalation causes irritation to the nose, throat and the lungs. The typical symptoms are sore throat, runny nose, cough and difficulty in breathing. Reactive airway dysfunction syndrome (RADS), non-specific bronchial hyperreactivity and pulmonary oedema can develop after $SO_2$ inhalation. People with asthma and chronic obstructive pulmonary disease (COPD) may experience bronchospasms upon $SO_2$ inhalation. Inhaled $SO_2$ reacts with the respiratory mucosa to form sulfuric/sulfurous acid and reactants such as sulfites and sulfite ions. These reactants cause allergic

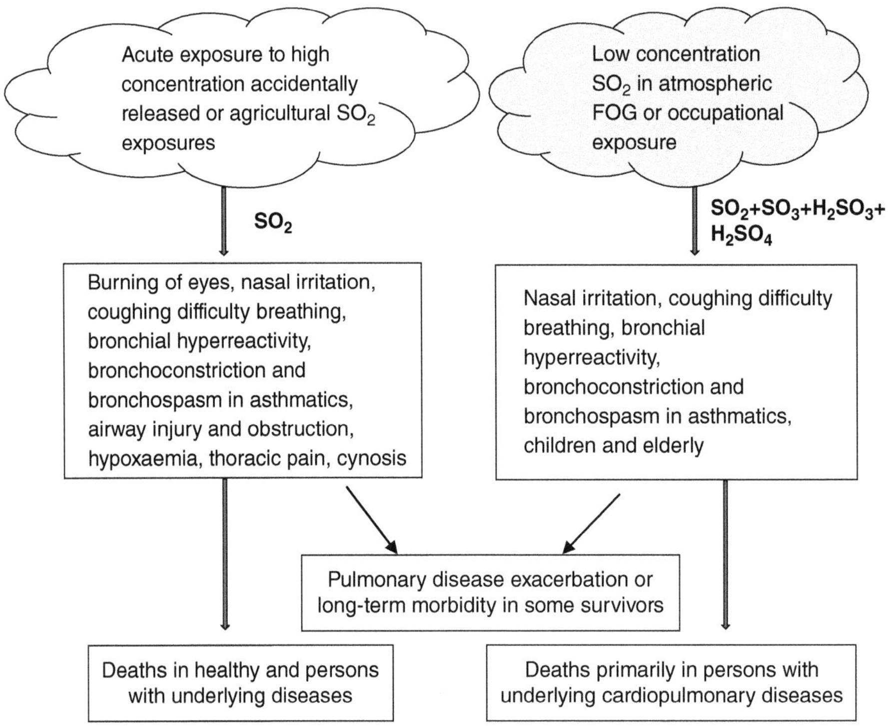


**Fig. 8.1.** Effect of acute accidental or environmental $SO_2$ on human health.

reactions in the airways and are reabsorbed into the bloodstream to affect other organs in the body. Accidental inhalation of high concentrations of $SO_2$ can be fatal. Two of five previously healthy paper-mill workers died from accidental exposure to $SO_2$ (Charan *et al.*, 1979). The autopsy revealed extensive sloughing of the airways and haemorrhagic alveolar oedema. One of the survivors developed severe airway obstruction that was refractive to bronchodilators. A similar accident caused severe airway obstruction, hypoxaemia, reduced tolerance to exercise, ventilation–perfusion mismatch and inflammation (Rabinovitch *et al.*, 1989). There was another incident where nine miners were exposed to high concentrations of $SO_2$ after being trapped in a mine (Piirila *et al.*, 1996). One of them died and the survivors experienced thoracic pain, cyanosis and difficulty in breathing. These also developed permanent bronchial hyperreactivity and had obstructive impairment of ventilator function. These conditions partly improved with time, suggesting long-term pulmonary morbidity as a consequence of $SO_2$ inhalation incidents (Piirila *et al.*, 1996; Henneberger *et al.*, 1993; Rabinovitch *et al.*, 1989). Pulmonary oedema followed by subsequent irreversible obstructive syndrome occurred in an individual exposed to high concentration of $SO_2$ in an enclosed space (Woodford *et al.*, 1979). Exposures caused by agricultural use of $SO_2$, such as those of apricot sulfurization, cause bronchoconstriction and asthma-like syndrome in the workers (Yildirim *et al.*, 2005). However, in another study such exposures did not cause chronic pulmonary damage in the workers when observed for extended periods of time (Ermis *et al.*, 2010). Other studies have indicated that workplace $SO_2$ exposure may cause chronic bronchitis or recurrent wheezing attacks that were not due to asthma (Henneberger *et al.*, 2005).

Incidents of increased mortality caused by deadly historical pollution episodes such as the

London fog illuminated the role of environmental $SO_2$ levels in causing serious adverse health effects in humans, especially those with underlying cardiopulmonary co-morbidities. The industrial threshold limit $SO_2$ values may not be tolerated by individuals with existing pulmonary diseases such as asthma and COPD (Koenig *et al.*, 1992; Sandstrom, 1995; Piirila *et al.*, 1996; Winterton *et al.*, 2001). Several studies have shown that the levels of atmospheric $SO_2$ tolerated by a normal individual can cause severe bronchoconstriction in asthmatic patients (Sheppard, 1988; Koenig *et al.*, 1992; Sandstrom, 1995; Piirila *et al.*, 1996; Winterton *et al.*, 2001). $SO_2$ in presence of other pollutants such as nitrogen dioxide ($NO_2$) may enhance the sensitivity to subsequent allergen challenge in asthmatics (Peden, 1997). A synergistic response of declined lung function was observed in smokers who were exposed to a high $SO_2$ environment (Xu and Wang, 1998). Therefore, the detrimental effects of $SO_2$ may be exaggerated by presence of other environmental contaminants such as particulates and gases like $NO_2$. Pre-existing morbidities such as asthma may further enhance the susceptibility to $SO_2$.

Understanding of the signalling mechanisms of $SO_2$ may clarify the pathophysiological effects that ensue. However, mechanistic studies on the effects of $SO_2$ are still few. Acute exposure to high concentrations of $SO_2$ injures the lung and can be fatal. Severe respiratory distress was observed in rats exposed to 2200 ppm of $SO_2$ for 10 min (Wigenstam *et al.*, 2016). There was acute inflammation with neutrophils and macrophages in the airways observed 5 h after exposure in this study. The mucosal epithelial lining was damaged in the upper airways and large bronchi. Within 24 h the rats developed airway hyperreactivity and the inflammation persisted for up to 14 days post exposure. There was $T_h1$ and $T_h2$ T helper cell activation along with M1 and M2 macrophages. The pro-fibrotic cytokine, transforming growth factor-beta (TGF-β1), was detected in airways after 24 h of exposure and its levels remained increased up to 28 days. At that late time point, collagen deposition and fibrosis were observed in the lungs (Wigenstam *et al.*, 2016). Repeated acute high-concentration exposure (1000 ppm, 3 h per day for 4 days) caused inflammation, remodelling and goblet cell metaphasia in the airways of guinea pigs exposed to $SO_2$ (McLeod *et al.*, 2007). There was increased transient receptor potential vanilloid 1 (TRPV1) receptor function and increased intracellular $Ca^{2+}$ in the ganglia of $SO_2$-exposed animals (McLeod *et al.*, 2007). In another study, exposure to $SO_2$ (500 ppm, 3 h per day for 3 days a week for 12 weeks) caused pulmonary oedema, inflammatory cell infiltration in the airways and lesions in the tracheobronchial mucosa (Miller *et al.*, 1985). Similar $SO_2$ exposures for 6 months in beagle dogs caused tracheobronchial inflammation and increased mucus production (Malo *et al.*, 1983). These beagles had hyperactive airways as demonstrated by enhanced sensitivity to prostaglandin F2 alpha (PGF2 alpha) (Malo *et al.*, 1983). Another study reported that chronic exposure of dogs to $SO_2$ caused chronic bronchitis, increased pulmonary resistance and persistent lung inflammation (Shore *et al.*, 1987). In this study there was decreased airway hyperresponsiveness, which was attributed to some inhibitory influence on the muco-epithelial barrier (Shore *et al.*, 1987). Exposure of rats to 400 ppm of $SO_2$ for up to 7 weeks resulted in progressive yet reversible hypertrophy and hyperplasia of the submucosal glands and flattening of the epithelium (Clark *et al.*, 1980). Studies on rabbit tracheal epithelium $SO_2$ treatment (10–30 ppm for 1 h) reduced ciliary beat frequency (Sakai *et al.*, 1993; Blanquart *et al.*, 1995). These effects were, however, reversed after 24 h (Blanquart *et al.*, 1995). Similarly, the ciliary beat frequency of human nasal epithelial cells was decreased upon $SO_2$ treatment (Kienast *et al.*, 1993, 1996). Mucociliary activity of tracheal epithelium *ex vivo* was also affected by $SO_2$ (Riechelmann *et al.*, 1995). $SO_2$ also reduced the chemotactic response of alveolar macrophages and blood monocytes (Knorst *et al.*, 1996). These data indicate that pulmonary clearance could be affected by $SO_2$ inhalation and that it may enhance infections (Ferin and Leach, 1973). It has been demonstrated that sulfite oxidase plays an important role in the defence against $SO_2$-mediated organ injury (Gunnison *et al.*, 1987). In sulfite oxidase-sufficient rats the injury was limited to lung and trachea but in deficient animals the injury was found in distant organs (Gunnison *et al.*, 1987). Therefore, sulfite oxidase deficiency may be a high-risk factor for $SO_2$ inhalation toxicity (Calabrese *et al.*, 1981; Reno *et al.*, 2015). $SO_2$ inhalation decreased cytochrome P450 enzymes CYP1A1 and CYP1A2 in the lungs of rats (Qin and Meng, 2005). This reduction in CYP1A1 and CYP1A2 was proposed to be an adaptive response to minimize cell damage (Qin and Meng, 2005). Other forms of pulmonary CYP are also affected by $SO_2$ inhalation (Qin and Meng, 2005).

$SO_2$ inhalation also decreased pulmonary mitochondrial cytochrome c oxidase (COX) activity, expression of complex IV and V subunits and expression of regulatory peroxisome proliferator-activated receptor gamma coactivator 1a (PGC-1a), nuclear respiratory factor 1 (NRF1) and mitochondrial transcription factor A (TFAM) (Qin *et al.*, 2017). There was loss of mitochondrial membrane potential as well suggesting mitochondrial dysfunction. $SO_2$ (10–30 ppm for 1 h) treatment caused mitochondrial damage and reduced total cell adenosine triphosphate (ATP) in rabbit tracheal epithelium (Sakai *et al.*, 1993; Blanquart *et al.*, 1995). Such impairment can cause cell dysfunction, death and eventually lead to pulmonary disease. Accordingly, it was shown that inhalation of $SO_2$ increased apoptosis-related genes in the lungs (Qin *et al.*, 2017). Inhalation of $SO_2$ increased caspase-3 activity and proapoptotic proteins such as p53 and Bax and decreased Bcl-2 protein in the lungs of rats (Meng *et al.*, 2004; Bai and Meng, 2005, 2010b). There are no known receptors through which $SO_2$ would cause the biological effects it does. Exposure to sulfur dioxide with fine particulate matter induces inflammatory responses via toll-like receptor 4 (TLR4) activation (Li *et al.*, 2017c). The biological effects and the signalling mechanisms of endogenously produced $SO_2$ are reviewed elsewhere and are beyond the scope of this chapter (Wang *et al.*, 2014, 2015).

## 8.5 Cardiovascular Effects

Exposure to atmospheric pollutants, especially particulates, are known to exacerbate pre-existing heart disease and cause severe cardiovascular effects (Hoek *et al.*, 2001; Shah *et al.*, 2013; Chung *et al.*, 2017). Heart failure is an important sequella of inhalation of gases such as carbon monoxide. $SO_2$ is known to cause detrimental cardiovascular events such as heart failure, ischaemic heart disease and cardiac arrhythmia (Hoek *et al.*, 2001; Routledge *et al.*, 2006). Increased susceptibility of the heart to oxidative stress due to low levels of antioxidants may play a role. High levels of reactive oxygen species produced due to the high metabolic rate in cardiac muscle are also important in rendering this susceptibility. Inhalation of high concentrations of oxidant halogen gases cause significant cardiac injury, dysfunction and failure that can be critical in causing mortalities following exposures (Ahmad *et al.*, 2014, 2015; Zaky *et al.*, 2015a, b). It was also demonstrated that the cardiac dysfunction was independent of coexisting hypoxia, since it is not fully reversed by oxygen supplementation (Zaky *et al.*, 2015b). Therefore, studies on offsite organ effects of inhaled toxic gases are important and can impact development of treatment strategies upon accidental or deliberate exposures to these agents.

Inhaled $SO_2$ causes systemic oxidative damage. Exposure of mice to low concentrations (22, 56 or 112 ppb) of sulfur dioxide for 6 h per day for 7 days increased lipid peroxidation and reduced the antioxidants and antioxidant enzyme content in the heart (Meng, 2003; Meng *et al.*, 2003; Wu and Meng, 2003). Similar low-concentration inhalations for 6 h per day for 7 days caused increased expression of proinflammatory and proapoptotic genes in the hearts of exposed rats (Yun *et al.*, 2011). $SO_2$ may also cause protein oxidation, DNA protein crosslinking and apoptotic cell death and subsequent cerebral ischaemic stroke in rats (Xie *et al.*, 2007; Yun *et al.*, 2010; Sang *et al.*, 2009, 2010). Mice inhaling $SO_2$ (56 ppb, 4 h per day for 7 days) had significant ultrastructural cardiac damage (Meng and Liu, 2007). There was extensive mitochondrial swelling and loss of crista. Disruption of myocardium, dissociation of intercalated disc and endothelial oedema occurred. Myocardial membrane breach and inflammatory cell infiltration was also observed (Meng and Liu, 2007). Long-term low-dose exposure to $SO_2$ (3.5, 7, or 14 ppm for 30 days) in rats caused ultrastructural damage to the myocardium as visualized by interspersed vacuoles, myofibrillar lattice disarray with clusters of swollen mitochondria with low amounts of cristae (Qin *et al.*, 2016). Mitochondrial dysfunction was indicated by loss of mitochondrial membrane potential (MMP) and decreased ATP content (Qin *et al.*, 2016). The mitochondrial dysfunction was also associated with cardiac dysfunction demonstrated by decreased ejection fraction and decreased fractional shortening and increase in left ventricular (LV) end systolic volume (Qin *et al.*, 2016). There was decrease in coactivator of peroxisome proliferator-activated receptor gamma (PGC-1a), nuclear respiratory factor 1 and TFAM mRNA and protein. Overexpression of TFAM and treatment with antioxidant *N*-acetyl-l-cysteine (NALC) reversed the mitochondrial dysfunction (Qin et al., 2016). $SO_2$ also induces vasorelaxant effect on the aorta by altering the expression of several potassium and calcium channel subunits that may play a role in pathogenesis of $SO_2$-associated cardiovascular diseases (Zhang *et al.*, 2014, 2016b). Adverse cardiorespiratory heart effects were further observed in rodent neonates that were exposed to $SO_2$ before birth and postnatally. The heart rate was altered in these neonatal rats and it was attributed to the loss of parasympathetic control of brainstem (Woerman and Mendelowitz, 2013a, b).

## 8.6 Effects on Skin, Eyes and Brain

Dermal exposures of $SO_2$ are rare. However, direct contact with liquefied compressed gas may cause frostbite and irritation as seen in the case of a factory worker who experienced freezing of skin and cornea after accidental exposure to liquid sulfur dioxide (Kennon, 1927; Grant, 1947). Some of these cases had corneal opacification and loss of vision after direct ocular contact with liquid $SO_2$ (Grant, 1947). The severe irritant effect of $SO_2$ vapours produces redness and blistering on skin and burns on the cornea. $SO_2$-containing volcanic fog caused a variety of ocular symptoms, including itching, tearing, burning and foreign body sensation (Camara and Lagunzad, 2011). There was eyelid swelling, oedema and chemosis along with mucous discharge (Camara and Lagunzad, 2011). Dry eye and non-specific conjunctivitis are also associated with exposure to air containing $SO_2$ (Chang *et al.*, 2012). Increased lipid peroxidation, decreased antioxidant enzymes and decreased visual evoked potential upon inhalation of $SO_2$ (10 ppm for 1 h per day for 7 days for 6 weeks) was observed in older rats as compared with the young ones (Kilic, 2003). Low doses of $SO_2$ (0.5–2 ppm) did not affect the eyelid function and did not cause ocular irritation in the subjects of an experimental $SO_2$ exposure study (van Thriel *et al.*, 2010). Irritation of the eye was also observed in pigs and guinea pigs that were exposed to 5–40 ppm of $SO_2$ (D'Donoghue and Graesser, 1962).

Loss of parasympathetic control of brainstem was observed in neonatal rats that were exposed to $SO_2$ (Woerman and Mendelowitz, 2013a, b). Exposure of rats to 10 ppm $SO_2$ for 1 h per day for 30 days depleted total lipids and caused extensive lipid peroxidation in all brain regions (Haider *et al.*, 1981, 1982). Oxidative damage to brains of mice exposed to $SO_2$ (22–112 ppm for 6 h per day for 7 days) has also been described (Meng and Zhang, 2003). Increased lipid peroxidation and reduced antioxidants are also a feature of pathogenesis of brain diseases such as Alzheimer's and it seems pertinent to consider environmental impact on the brains of the elderly in the progression of such diseases (Marcus *et al.*, 1998). $SO_2$ inhalation causes an imbalance of pro- and anti-inflammatory cytokines and leads to accumulation of Aβ42 peptide in the cerebral cortex and hippocampus of rats (Yang *et al.*, 2017). Increased proinflammatory cytokines were observed in the brains of mice inhaling $SO_2$ (Li *et al.*, 2017a). It was also suggested that $SO_2$ inhalation may cause brain injury similar to cerebral ischaemia and that atmospheric $SO_2$ may contribute to the progression of ischaemic stroke (Sang *et al.*, 2010; Szyszkowicz *et al.*, 2012).

## 8.7 Carcinogenicity and Teratogenicity

Most of the information about carcinogenicity of $SO_2$ comes from the observations in factory workers where $SO_2$ is produced or utilized for various purposes. It is important to note that in these work environments exposures to other metals, non-metals and smoke also occurs. Other genetic predispositions of these workers were also not excluded. Several studies have determined that exposure to $SO_2$ itself does not cause cancer (Lubin *et al.*, 1981; Enterline *et al.*,

1987; Ades and Kazantzis, 1988). Although no evidence of lung cancer was observed in the workers exposed to $SO_2$ in sulfuric acid-producing factories, significant increases in incidences of bladder cancer was observed (Englander *et al.*, 1988). Association between long-term exposures to increased $SO_2$ and increases in lung cancer was established in a large cohort of subjects in Tianjin region of China (Yue *et al.*, 2017). Positive correlation exists between prolonged exposure to atmospheric $SO_2$ and lung cancer prevalence (Chen *et al.*, 2016; Yang *et al.*, 2016). Increased deaths due to lymphosarcoma, rectal and pancreatic cancer and Hodgkin's disease were observed in the sulfite and sulfate process workers of pulp and paper industries (Milham and Demers, 1984). Gastric tumours were reported in the workers in sulfite mills, whereas lymphosarcomas and reticulosarcomas were found in the workers in sulfate mills (Milham and Demers, 1984; Robinson *et al.*, 1986). In animal studies, long-term exposure to inhaled sulfur dioxide alone did not lead to tumour formation; however, it potentiated tumour-promoting effects of other known carcinogens such as benzo[a] pyrene (Laskin *et al.*, 1980; Pauluhn *et al.*, 1985). These studies suggest that sulfur dioxide may increase carcinogenicity of other carcinogens. On the other hand, few studies have demonstrated that endogenously produced sulfur dioxide may have a tumour-suppressing role (Bai and Meng, 2010a; Qin and Meng, 2010; Cui Y. *et al.*, 2017). The teratogenic effects of $SO_2$ are unclear in humans.

## 8.8 Decontamination and Antidotes for Acute Accidental Exposures

$SO_2$ is a colourless gas with strong pungent odour that can be easily smelled at concentrations of less than 1 ppm. Several methods to adsorb and neutralize $SO_2$ have been developed to decrease $SO_2$ and other gases from factory exhausts and reduce environmental pollution (Rubio and Izquierdo, 2010; Plens *et al.*, 2015; Zhang *et al.*, 2016a). However, these processes are hampered by reduced efficiency of $SO_2$ removal, higher costs, accumulation of by-products such as sulfuric acid and increased energy consumption.

During acute accidental exposures, there is no secondary contamination risk from persons exposed to only sulfur dioxide gas; however, individuals exposed to liquid $SO_2$ do pose a secondary contamination risk to rescuers via direct contact or inhalation of vapours. The frostbite caused by liquid $SO_2$ can be relieved by washing with warm water. Washing and irrigation of eyes with copious amounts of water are recommended decontamination procedures, with appropriate measures to ensure that no hypothermia occurs. There are no specific countermeasures to treat victims of accidental $SO_2$ inhalation. The Agency for Toxic Substances and Disease Registry (ATSDR) lists several supportive medical management strategies. There is a need for appropriate evaluation and support of cardiopulmonary function. Supplemental oxygen is provided if cardiopulmonary dysfunction is suspected. Endotracheal intubation may be considered in case of respiratory compromise. Aerosolized/ nebulized bronchodilators (β2 agonists) may be used for patients who have bronchospasm. When using the bronchodilators it is critical to ensure the health of the myocardium. Mechanical ventilation with positive end-expiratory pressure can be undertaken. Racaemic epinephrine aerosols are recommended for children who develop stridor. $SO_2$ poisoning is not known to cause additional risk during the use of bronchodilators or cardiac sensitizing agents. However, if there are multiple chemical exposures, extreme caution is required in determining the therapeutic strategies.

Few studies on animal models have been performed to evaluate therapeutic efficacies against damaging effects of exposures to $SO_2$. $SO_2$ oxidizes lipids in various tissues and organs. Treatment with antioxidants such as salicylic acid and vitamin C are protective against such oxidative stress caused by $SO_2$ inhalation in mice (Zhao *et al.*, 2008; Poljsak and Fink, 2014). Similarly, vitamin E and C administration was found to be effective in reducing $SO_2$ inhalation-induced red blood cell damage in guinea pigs (Etlik *et al.*, 1995). $SO_2$-induced LV dysfunction in rats was reversed by administration of NALC (Qin *et al.*, 2016). Recently, the role of endocannabinoids in protecting against brain inflammation caused by $SO_2$ inhalation in mice was demonstrated (Li *et al.*, 2015, 2017a).

Sulfur dioxide is one of the most abundantly used chemicals in industry. Its atmospheric concentrations are increasing as well along with its use. However, studies on acute exposure to increased environmental concentrations of $SO_2$ are limited. There seems to be a gap in understanding of its mechanism of actions and filling this gap is necessary for the development of rescue therapies.

## Acknowledgements

Shama Ahmad is supported by intramural funds from the Department of Anesthesiology and Perioperative Medicine (UAB), a Bridge fund from the Dean's office, Department of Medicine (UAB) and CounterACT Program, National Institutes of Health Office of the Director (NIH OD), the National Institute of Environmental Health Sciences (NIEHS) Grant Number U01ES028182 (SA). Aftab Ahmad is supported by the CounterACT Program, National Institutes of Health Office of the Director (NIH OD), the National Institute of Environmental Health Sciences (NIEHS), and the National Heart Lung and Blood Institute (NHLBI) Grant Numbers U01ES025069 and R01HL114933.

## References

Ades, A.E. and Kazantzis, G. (1988) Lung cancer in a non-ferrous smelter: the role of cadmium. *British Journal of Industrial Medicine* 45(7), 435–442.

Ahmad, S., Ahmad, A., Hendry-Hofer, T.B., Loader, J.E., Claycomb, W.C. *et al.* (2015) Sarcoendoplasmic reticulum Ca(2+) ATPase. A critical target in chlorine inhalation-induced cardiotoxicity. *American Journal of Respiratory Cell and Molecular Biology* 52(4), 492–502.

Ahmad, S., Ahmad, A., Neeves, K.B., Hendry-Hofer, T., Loader, J.E., White, C.W. and Veress, L. (2014) In vitro cell culture model for toxic inhaled chemical testing. *Journal of Visualized Experiments* 87, e51539

Anonymous (1931) Effect of Air Pollution on Health: Report of the Committee on Public Health Relations of The New York Academy of Medicine. *Bulletin of the New York Academy of Medicine* 7(9), 751–775.

Bai, J. and Meng, Z. (2005) Effects of sulfur dioxide on apoptosis-related gene expressions in lungs from rats. *Regulatory Toxicology and Pharmacology*, 43(3), 272–279.

Bai, J. and Meng, Z. (2010a) Effect of sulfur dioxide on expression of proto-oncogenes and tumor suppressor genes from rats. *Environmental Toxicology*, 25(3), 272–283.

Bai, J. and Meng, Z. (2010b) Expression of caspase and apoptotic signal pathway induced by sulfur dioxide. *Environmental and Molecular Mutagenesis*, 51(2), 112–122.

Bernard, S.M., Samet, J.M., Grambsch, A., Ebi, K.L. and Romieu, I. (2001) The potential impacts of climate variability and change on air pollution-related health effects in the United States. *Environmental Health Perspectives*, 109 (Suppl. 2), 199–209.

Blanquart, C., Giuliani, I., Houcine, O., Jeulin, C., Guennou, C. and Marano, F. (1995) In vitro exposure of rabbit tracheal epithelium to SO(2): Effects on morphology and ciliary beating. *Toxicology In Vitro*, 9(2), 123–132.

Calabrese, E., Sacco, C., Moore, G. and DiNardi, S. (1981) Sulfite oxidase deficiency: a high risk factor in SO2, sulfite, and bisulfite toxicity? *Medical Hypotheses*, 7(2), 133–145.

Camara, J.G. and Lagunzad, J.K. (2011) Ocular findings in volcanic fog induced conjunctivitis. *Hawaii Medical Journal* 70(12), 262–265.

Chang, C.J., Yang, H.H., Chang, C.A. and Tsai, H.Y. (2012) Relationship between air pollution and outpatient visits for nonspecific conjunctivitis. *Investigative Ophthalmology and Visual Science*, 53(1), 429–433.

Charan, N.B., Myers, C.G., Lakshminarayan, S. and Spencer, T.M. (1979) Pulmonary injuries associated with acute sulfur dioxide inhalation. *American Review of Respiratory Disease* 119(4), 555–560.

Chen, X., Zhang, L.W., Huang, J.J., Song, F.J., Zhang, L.P. *et al.* (2016) Long-term exposure to urban air pollution and lung cancer mortality: A 12-year cohort study in Northern China. *Science of the Total Environment*, 571, 855–861.

Chung, J.W., Bang, O.Y., Ahn, K., Park, S.S., Park, T.H. *et al.* (2017) Air pollution is associated with ischemic stroke via cardiogenic embolism. *Stroke* 48(1), 17–23.

Clark, J. N., Dalbey, W. E. and Stephenson, K. B. (1980) 'Effect of sulfur dioxide on the morphology and mucin biosynthesis by the rat trachea', *J Environ Pathol Toxicol*, 4, pp. 197-207.
Cui Y., H.S., Yang, Y., Wang, L., Sheng, J., Fan, Y. *et al.* (2017) Endogenous sulfur dioxide, a novel gasotransmitter, plays tumor suppressor role in prostate cancer. *Journal of Urology*, 197(4), e1114.
D'Donoghue, J.G. and Graesser, F.E. (1962) Effects of sulphur dioxide on guinea pigs and swine. *Canadian Journal of Comparative Medicine and Veterinary Science* 26(11), 255–263.
Dereven'kov, I.A., Ivlev, P.A., Bischin, C., Salnikov, D.S., Silaghi-Dumitrescu, R., Makarov, S.V. and Koifman, O.I. (2017) Comparative studies of reaction of cobalamin (II) and cobinamide (II) with sulfur dioxide. *Journal of Biological Inorgic Chemistry* 22(6), 969–975.
Englander, V., Sjoberg, A., Hagmar, L., Attewell, R., Schutz, A., Moller, T. and Skerfving, S. (1988) Mortality and cancer morbidity in workers exposed to sulphur dioxide in a sulphuric acid plant. *International Archives of Occupational and Environmental Health* 61(3), 157–162.
Enterline, P.E., Marsh, G.M., Esmen, N.A., Henderson, V.L., Callahan, C.M. and Paik, M. (1987) Some effects of cigarette smoking, arsenic, and SO2 on mortality among US copper smelter workers. *Journal of Occupational Medicine* 29(10), 831–838.
Ermis, H., Gokirmak, M., Yildirim, Z., Yologlu, S. and Ankarali, H. (2010) Exposure to SO2 does not have a chronic effect on pulmonary functions of apricot workers. *Inhalation Toxicology*, 22(3), 219–223.
Etlik, O., Tomur, A., Kutman, M.N., Yorukan, S. and Duman, O. (1995) The effects of sulfur dioxide inhalation and antioxidant vitamins on red blood cell lipoperoxidation. *Environmental Research*, 71(1), 25–28.
Ferin, J. and Leach, L.J. (1973) The effect of SO2 on lung clearance of TiO2 particles in rats. *American Industrial Hygiene Association Journal* 34(6), 260–263.
Grant, W.M. (1947) Ocular injury due to sulfur dioxide; experimental study and comparison with ocular effects of freezing. *Archives of Ophthalmology*, 38(6), 762–774.
Gunnison, A.F., Sellakumar, A., Currie, D. and Snyder, E.A. (1987) Distribution, metabolism and toxicity of inhaled sulfur dioxide and endogenously generated sulfite in the respiratory tract of normal and sulfite oxidase-deficient rats. *Journal of Toxicology and Environmental Health* 21(1–2), 141–162.
Guo, H., Weber, R.J. and Nenes, A. (2017) High levels of ammonia do not raise fine particle pH sufficiently to yield nitrogen oxide-dominated sulfate production. *Scientific Reports* 7(1), 12109.
Haider, S.S., Hasan, M., Hasan, S.N., Khan, S.R. and Ali, S.F. (1981) Regional effects of sulfur dioxide exposure on the guinea pig brain lipids, lipid peroxidation and lipase activity. *Neurotoxicology* 2(3), 443–450.
Haider, S.S., Hasan, M. and Khan, N.H. (1982) Air pollutant sulfur dioxide-induced alterations on the levels of lipids, lipid peroxidation and lipase activity in various regions of the rat brain. *Acta Pharmacologica et Toxicologica (Copenhagen)* 51(1), 45–50.
Henneberger, P.K., Ferris, B.G. Jr. and Sheehe, P.R. (1993) Accidental gassing incidents and the pulmonary function of pulp mill workers. *American Review of Respiratory Disease* 148(1), 63–67.
Henneberger, P.K., Olin, A.C., Andersson, E., Hagberg, S. and Toren, K. (2005) The incidence of respiratory symptoms and diseases among pulp mill workers with peak exposures to ozone and other irritant gases. *Chest* 128(4), 3028–3037.
Hoek, G., Brunekreef, B., Fischer, P. and van Wijnen, J. (2001) The association between air pollution and heart failure, arrhythmia, embolism, thrombosis, and other cardiovascular causes of death in a time series study. *Epidemiology* 12(3), 355–357.
Kennon, B.R. (1927) Report of a case of injury to the skin and eyes by liquid sulphur dioxide. *Journal of Industrial Hygiene* 9, 486–487.
Kienast, K., Knorst, M., Riechelmann, H., Schellenberg, J., Muller-Quernheim, J. and Ferlinz, R. (1993) [In vitro studies of the beat frequency of ciliary cell cultures after short-term exposure to $SO_2$ and $NO_2$]. *Medizinische Klinik (Munich)* 88(9), 520–524.
Kienast, K., Riechelmann, H., Knorst, M., Haffner, B., Muller-Quernheim, J., Schellenberg, J. and Ferlinz, R. (1996) Combined exposures of human ciliated cells to different concentrations of sulfur dioxide and nitrogen dioxide. *European Journal of Medical Res*, 1(11), 533–536.
Kilic, D. (2003) The effects of ageing and sulfur dioxide inhalation exposure on visual-evoked potentials, antioxidant enzyme systems, and lipid-peroxidation levels of the brain and eye. *Neurotoxicology and Teratology* 25(5), 587–598.
Knorst, M.M., Kienast, K., Gross, S., Fries, B., Muller-Quernheim, J. and Ferlinz, R. (1996) Chemotactic response of human alveolar macrophages and blood monocytes elicited by exposure to sulfur dioxide. *Research in Experimental Medicine (Berlin)*, 196, 127–135.

Koenig, J.Q., Dumler, K., Rebolledo, V., Williams, P.V. and Pierson, W.E. (1992) Theophylline mitigates the bronchoconstrictor effects of sulfur dioxide in subjects with asthma. *Journal of Allergy and Clinical Immunology* 89(4), 789–794.

Laskin, S., Sellakumar, A.R., Kuschner, M., Nelson, N., La Mendola, S. *et al.* (1980) Inhalation carcinogenicity of epichlorohydrin in noninbred Sprague-Dawley rats. *Journal of the National Cancer Institute* 65(4), 751–757.

Lester, M.R. (1995) Sulfite sensitivity: significance in human health. *Journal of the American College of Nutrition* 14(3), 229–232.

Li, Q., Zhang, J., Li, L., He, Z., Yang, X. *et al.* (2013) Solubility properties and spectral investigation of dilute SO2 in a triethylene glycol + water + La3+ system. *Journal of Physical Chemistry B*, 117(18), 5633–5646.

Li, B., Chen, M., Guo, L., Yun, Y., Li, G. and Sang, N. (2015) Endogenous 2-arachidonoylglycerol alleviates cyclooxygenases-2 elevation-mediated neuronal injury from SO2 inhalation via PPARgamma pathway. *Toxicological Sciences* 147(2), 535–548.

Li, B., Chen, M., Guo, L., Yun, Y., Li, G. and Sang, N. (2017a) Endocannabinoid 2-arachidonoylglycerol protects inflammatory insults from sulfur dioxide inhalation via cannabinoid receptors in the brain. *Journal of Environmental Sciences (China)* 51, 265–274.

Li, C., McLinden, C., Fioletov, V., Krotkov, N., Carn, S. *et al.* (2017b) India is overtaking China as the world's largest emitter of anthropogenic sulfur dioxide. *Scientific Reports* 7(1), 14304.

Li, R., Zhao, L., Tong, J., Yan, Y. and Xu, C. (2017c) Fine particulate matter and sulfur dioxide coexposures induce rat lung pathological injury and inflammatory responses via TLR4/p38/NF-kappaB pathway', *International Journal of Toxicology* 36(2), 165–173.

Liao, K.J., Hou, X. and Strickland, M.J. (2016) Resource allocation for mitigating regional air pollution-related mortality: a summertime case study for five cities in the United States. *Journal of Air and Waste Management Association* 66(8), 748–757.

Lubin, J.H., Pottern, L.M., Blot, W.J., Tokudome, S., Stone, B.J. and Fraumeni, J.F. Jr (1981) Respiratory cancer among copper smelter workers: recent mortality statistics. *Journal of Occupatinal Medicine*, 23(11), 779–784.

Malo, P.E., Wasserman, M.A., Griffin, R.L., Leong, B.K. and Powell, D.J. (1983) Enhanced bronchoconstriction responses to prostaglandin F2 alpha following inhalation of sulfur dioxide. *Prostaglandins* 25(2), 179–192.

Marcus, D.L., Thomas, C., Rodriguez, C., Simberkoff, K., Tsai, J.S., Strafaci, J.A. and Freedman, M.L. (1998) Increased peroxidation and reduced antioxidant enzyme activity in Alzheimer's disease. *Experimental Neurology* 150(1), 40–44.

Martenies, S.E., Milando, C.W., Williams, G.O. and Batterman, S.A. (2017) Disease and health inequalities attributable to air pollutant exposure in Detroit, Michigan. *International Journal of Environmental Research Public Health*, 14(10) e1243.

McLeod, R.L., Jia, Y., McHugh, N.A., Fernandez, X., Mingo, G.G. *et al.* 2007) Sulfur-dioxide exposure increases TRPV1-mediated responses in nodose ganglia cells and augments cough in guinea pigs. *Pulmonary Pharmacology and Therapeutics* 20(6), 750–757.

Meng, Z. (2003) Oxidative damage of sulfur dioxide on various organs of mice: sulfur dioxide is a systemic oxidative damage agent. *Inhalation Toxicology* 15(2), 181–195.

Meng, Z. and Liu, Y. (2007) Cell morphological ultrastructural changes in various organs from mice exposed by inhalation to sulfur dioxide. *Inhalation Toxicology* 19(6–7), 543–551.

Meng, Z. and Zhang, B. (2003) Oxidative damage of sulfur dioxide inhalation on brains and livers of mice. *Environmental Toxicology and Pharmacology* 13(1), 1–8.

Meng, Z., Qin, G., Zhang, B., Geng, H., Bai, Q., Bai, W. and Liu, C. (2003) Oxidative damage of sulfur dioxide inhalation on lungs and hearts of mice. *Environmental Research* 93(3), 285–292.

Meng, Z., Qin, G., Zhang, B. and Bai, J. (2004) DNA damaging effects of sulfur dioxide derivatives in cells from various organs of mice. *Mutagenesis* 19(6), 465–468.

Milham, S. Jr and Demers, R.Y. (1984) Mortality among pulp and paper workers. *Journal of Occupational Medicine* 26(11), 844–846.

Miller, M.L., Andringa, A., Rafales, L. and Vinegar, A. (1985) Effect of exposure to 500 ppm sulfur dioxide on the lungs of the ferret. *Respiration* 48(4), 346–354.

Pauluhn, J., Thyssen, J., Althoff, J., Kimmerle, G. and Mohr, U. (1985) Long-term inhalation study with benzo(a)pyrene and SO2 in Syrian golden hamsters. *Experimental Pathology* 28(1), 31.

Peden, D.B. (1997) Mechanisms of pollution-induced airway disease: in vivo studies. *Allergy* 52(38 Suppl.), 37–44; discussion 57–58.

Piirila, P.L., Nordman, H., Korhonen, O.S. and Winblad, I. (1996) A thirteen-year follow-up of respiratory effects of acute exposure to sulfur dioxide. *Scandinavian Journal of Work, Environment and Health* 22(3), 191–196.
Plens, A.C., Monaro, D.L. and Coutinho, A.R. (2015) Adsorption of SOx and NOx in activated viscose fibers. *Anais da Academia Brasileira de Ciências* 87, 1149–1160.
Poljsak, B. and Fink, R. (2014) The protective role of antioxidants in the defence against ROS/RNS-mediated environmental pollution. *Oxidative Medicine and Cellular Longevity* 2014, 671539.
Qin, G. and Meng, Z. (2005) Effect of sulfur dioxide inhalation on CYP1A1 and CYP1A2 in rat liver and lung. *Toxicology Letters* 160(1), 34–42.
Qin, G. and Meng, Z. (2010) Expression of oncogenes and tumor suppressor genes in lungs of rats exposed to sulfur dioxide and benzo(a)pyrene. *Inhalation Toxicology* 22(4), 322–329.
Qin, G., Wu, M., Wang, J., Xu, Z., Xia, J. and Sang, N. (2016) Sulfur dioxide contributes to the cardiac and mitochondrial dysfunction in rats. *Toxicological Sciences* 151(2), 334–346.
Qin, G., Wang, J. and Sang, N. (2017) Sulfur dioxide inhibits expression of mitochondrial oxidative phosphorylation genes encoded by both nuclear DNA and mitochondrial DNA in rat lungs. *Environmental Science and Pollution Research International* 24(3), 2527–2534.
Qu, Y., An, J., He, Y. and Zheng, J. (2016) An overview of emissions of SO2 and NOx and the long-range transport of oxidized sulfur and nitrogen pollutants in East Asia. *Journal of Environmental Sciences (China)* 44, 13–25.
Rabinovitch, S., Greyson, N.D., Weiser, W. and Hoffstein, V. (1989) Clinical and laboratory features of acute sulfur dioxide inhalation poisoning: two-year follow-up. *American Review of Respiratory Disease* 139(2), 556–558.
Reno, A.L., Brooks, E.G. and Ameredes, B.T. (2015) Mechanisms of heightened airway sensitivity and responses to inhaled SO2 in asthmatics. *Environmental Health Insights*, 9(Suppl. 1), 13–25.
Riechelmann, H., Maurer, J., Kienast, K., Hafner, B. and Mann, W.J. (1995) Respiratory epithelium exposed to sulfur dioxide--functional and ultrastructural alterations. *Laryngoscope* 105(3 Pt 1), 295–299.
Robinson, C.F., Waxweiler, R.J. and Fowler, D.P. (1986) Mortality among production workers in pulp and paper mills. *Scandinavian Journal of Work, Environment and Health* 12(6), 552–560.
Routledge, H.C., Manney, S., Harrison, R.M., Ayres, J.G. and Townend, J.N. (2006) Effect of inhaled sulphur dioxide and carbon particles on heart rate variability and markers of inflammation and coagulation in human subjects. *Heart* 92(2), 220–227.
Rubio, B. and Izquierdo, M.T. (2010) Coal fly ash based carbons for SO2 removal from flue gases. *Waste Management*, 30(7), 1341–1347.
Sakai, N., Tamaoki, J., Chiyotani, A., Takeyama, K. and Konno, K. (1993) [Inhibitory effect on sulfur dioxide on ciliary motility in rabbit tracheal epithelium and its prevention by intracellular cyclic AMP]. *Nihon Kyobu Shikkan Gakkai Zasshi* 31(6), 733–737.
Sandstrom, T. (1995) Respiratory effects of air pollutants: experimental studies in humans. *European Respiratory Journal* 8(6), 976–995.
Sang, N., Hou, L., Yun, Y. and Li, G. (2009) SO(2) inhalation induces protein oxidation, DNA-protein crosslinks and apoptosis in rat hippocampus. *Ecotoxicology and Environmental Safety* 72(3), 879–884.
Sang, N., Yun, Y., Li, H., Hou, L., Han, M. and Li, G. (2010) SO2 inhalation contributes to the development and progression of ischemic stroke in the brain. *Toxicological Sciences* 114(2), 226–236.
Setterstrom, C. (1940) Effects of sulfur dioxide on plants and animals. *Industrial and Engineering Chemistry*, 32(4), 473–479.
Shah, A.S., Langrish, J.P., Nair, H., McAllister, D.A., Hunter, A.L. *et al.* (2013) Global association of air pollution and heart failure: a systematic review and meta-analysis. *Lancet* 382(9897), 1039–1048.
Shapiro, R. and Gazit, A. (1977) Crosslinking of nucleic acids and proteins by bisulfite. *Advances in Experimental Medicine and Biology* 86A, 633–640.
Sheppard, D. (1988) Sulfur dioxide and asthma – a double-edged sword? *Journal of Allergy and Clinical Immunology* 82(6), 961–964.
Shore, S.A., Kariya, S.T., Anderson, K., Skornik, W., Feldman, H.A. *et al.* 1987) Sulfur-dioxide-induced bronchitis in dogs. Effects on airway responsiveness to inhaled and intravenously administered methacholine. *American Review of Respiratory Disease* 135(4), 840–847.
Steel, R.J. (1997) Thiamine deficiency in a cat associated with the preservation of 'pet meat' with sulphur dioxide. *Australian Veterinary Journal* 75(10), 719–721.
Szyszkowicz, M., Porada, E., Tremblay, N. and Grafstein, E. (2012) Sulfur dioxide and emergency department visits for stroke and seizure. *Stroke Research and Treatment* 2012, 824724.

van Thriel, C., Schaper, M., Kleinbeck, S., Kiesswetter, E., Blaszkewicz, M. *et al.* (2010) Sensory and pulmonary effects of acute exposure to sulfur dioxide (SO2). *Toxicology Letters* 196(1), 42–50.

Wang, D.Z., Jiang, G.H., Zhang, H., Song, G.D. and Zhang, Y. (2013) [Effect of air pollution on coronary heart disease mortality in Tianjin, 2001-2009: a time-series study]. *Zhonghua Liu Xing Bing Xue Za Zhi* 34(5), 478–483.

Wang, X.B., Du, J.B. and Cui, H. (2014) Sulfur dioxide, a double-faced molecule in mammals. *Life Sciences* 98(2), 63–67.

Wang, X.B., Du, J.B. and Cui, H. (2015) Signal pathways involved in the biological effects of sulfur dioxide. *European Journal of Pharmacology* 764, 94–99.

Wang, G., Zhang, R., Gomez, M.E., Yang, L., Levy Zamora, M. *et al.* (2016) Persistent sulfate formation from London Fog to Chinese haze. *Proceedings of the National Academy of Sciences of the United States of America* 113, 13630–13635.

Wigenstam, E., Elfsmark, L., Bucht, A. and Jonasson, S. (2016) Inhaled sulfur dioxide causes pulmonary and systemic inflammation leading to fibrotic respiratory disease in a rat model of chemical-induced lung injury. *Toxicology* 368–369, 28–36.

Winterton, D.L., Kaufman, J., Keener, C.V., Quigley, S., Farin, F.M., Williams, P.V. and Koenig, J.Q. (2001) Genetic polymorphisms as biomarkers of sensitivity to inhaled sulfur dioxide in subjects with asthma. *Annals of Allergy, Asthma and Immunology* 86(2), 232–238.

Woerman, A.L. and Mendelowitz, D. (2013a) Perinatal sulfur dioxide exposure alters brainstem parasympathetic control of heart rate. *Cardiovascular Research* 99(1), 16–23.

Woerman, A.L. and Mendelowitz, D. (2013b) Postnatal sulfur dioxide exposure reversibly alters parasympathetic regulation of heart rate. *Hypertension* 62(2), 274–280.

Wong, T.Y. (2017) Smog induces oxidative stress and microbiota disruption. *Journal of Food and Drug Analysis* 25(2), 235–244.

Woodford, D.M., Coutu, R.E. and Gaensler, E.A. (1979) Obstructive lung disease from acute sulfur dioxide exposure. *Respiration* 38(4), 238–245.

Wu, D. and Meng, Z. (2003) Effect of sulfur dioxide inhalation on the glutathione redox system in mice and protective role of sea buckthorn seed oil. *Archives of Environtal Contamination and Toxicology* 45(3), 423–428.

Xie, J., Fan, R. and Meng, Z. (2007) Protein oxidation and DNA-protein crosslink induced by sulfur dioxide in lungs, livers, and hearts from mice. *Inhalation Toxicology* 19(9), 759–765.

Xu, X. and Wang, L. (1998) Synergistic effects of air pollution and personal smoking on adult pulmonary function. *Archives of Environmental Health* 53(1), 44–53.

Yang, W.S., Zhao, H., Wang, X., Deng, Q., Fan, W.Y. and Wang, L. (2016) An evidence-based assessment for the association between long-term exposure to outdoor air pollution and the risk of lung cancer. *European Journal of Cancer Prevention* 25(3), 163–172.

Yang, Z., Chen, Y., Zhang, Y., Li, R. and Dong, C. (2017) The role of pro-/anti-inflammation imbalance in Abeta42 accumulation of rat brain co-exposed to fine particle matter and sulfur dioxide. *Toxicology Mechanisms and Methods* 27(8), 568–574.

Yildirim, Z., Kilic, T., Koksal, N. and Kotuk, M. (2005) Protective effect of ipratropium bromide on bronchoconstriction induced by sulfur dioxide exposure during apricot sufurization processes that causes asthma-like syndrome in agricultural environment. *Pharmacological Research* 51(5), 479–482.

Yue, S., Wang, Y., Wang, J. and Chen, J. (2017) Relationships between lung cancer incidences and air pollutants. *Technology and Health Care* 25(S1), 411–422.

Yun, Y., Li, H., Li, G. and Sang, N. (2010) SO2 inhalation modulates the expression of apoptosis-related genes in rat hippocampus via its derivatives in vivo. *Inhalation Toxicology* 22(11), 919–929.

Yun, Y., Hou, L. and Sang, N. (2011) SO(2) inhalation modulates the expression of pro-inflammatory and pro-apoptotic genes in rat heart and lung. *Journal of Hazardous Material* 185(1), 482–488.

Zaky, A., Ahmad, A., Dell'Italia, L.J., Jahromi, L., Reisenberg, L.A., Matalon, S. and Ahmad, S. (2015a) Inhaled matters of the heart. *Cardiovascular Regenerative Medicine* 2, e997.

Zaky, A., Bradley, W.E., Lazrak, A., Zafar, I., Doran, S. *et al.* (2015b) Chlorine inhalation-induced myocardial depression and failure. *Physiological Reports* 3(6), e12439.

Zhang, Q., Tian, J., Bai, Y., Lei, X., Li, M., Yang, Z. and Meng, Z. (2014) Effects of gaseous sulfur dioxide and its derivatives on the expression of KATP, BKCa and L-Ca(2+) channels in rat aortas in vitro. *European Journal of Pharmacology* 742, 31–41.

Zhang, C., Yang, D., Jiang, X. and Jiang, W. (2016a) Desulphurization performance of TiO2-modified activated carbon by a one-step carbonization-activation method. *Environmental Technology* 37(15), 1895–1905.

Zhang, Q., Bai, Y., Yang, Z., Tian, J. and Meng, Z. (2016b) The molecular mechanism of the effect of sulfur dioxide inhalation on the potassium and calcium ion channels in rat aortas. *Human and Experimental Toxicology* 35(4), 418–427.

Zhang, J., Liu, Y., Cui, L. L., Liu, S. Q., Yin, X. X. and Li, H. C. (2017) Ambient air pollution, smog episodes and mortality in Jinan, China. *Scientific Reports* 7(1), 11209.

Zhao, H., Xu, X., Na, J., Hao, L., Huang, L., Li, G. and Xu, Q. (2008) Protective effects of salicylic acid and vitamin C on sulfur dioxide-induced lipid peroxidation in mice. *Inhalation Toxicology* 20, 865–871.

# 9 Plant Response to Acid Rain Stress

**C. Liang***
*Jiangsu Key Laboratory of Anaerobic Biotechnology, School of Environmental and Civil Engineering, Jiangnan University, Wuxi, China*

## 9.1 Abstract

Acid rain is known as the 'air killer' because of its devastating damage to the ecosystem. As the major factor in the terrestrial ecosystem, the toxic effect of acid rain on plants has attracted more attention of researchers. The loss in agriculture and forestry caused by acid rain aggravates the challenge for achieving sustainable food production to feed the world's increasing population. This chapter reviews the history of acid rain and current situation, and then discusses the toxic effects of acid rain on plants, including morphology and growth of plant leaf and root, photosynthesis, nutrient uptake, plasma membrane and reactive oxygen species and its scavenging, as well as mechanisms on the combined effects of acid rain and other abiotic stress factors on plants. Based on these, the countermeasures to reduce the damage caused by acid rain to plants are considered.

## 9.2 Introduction

Acid rain is one of the most widespread sources of polluton, severely inhibiting plant growth and productivity. Acid rain, also termed acid deposition, refers to atmospheric deposition of acidic constituents in the form of rain, snow, hail or fog and has a pH lower than 5.6, the value expected on the equilibrium of pure water and atmospheric $CO_2$ (Seinfeld and Pandis, 1998). Acid rain was first described by the English chemist Robert Angus Smith (Smith, 1852), whose pioneering studies linked the sources to industrial emissions and reported many of its potentially harmful effects (Smith, 1872). Our modern understanding of acid rain as an environmental problem caused largely by regional emissions of sulfur dioxide ($SO_2$) and nitrogen oxides (NOx) developed in the 1960s and early 1970s. Since the 1970s, acid rain has remained in the public spotlight in both Europe and the USA and recently has emerged as an important problem in other regions such as South-east Asia, particularly China (Menz and Seip, 2004). In most regions its pH values range from 4.4 to 2.3 (Shan *et al.*, 1996). In 1998–2000, acid rain in the Adirondack region of New York had an average pH of 4.5 and in the Catskill region an average pH of 4.4. These values are about 10 times more acidic than background conditions (Driscoll *et al.*, 2003). Gravano *et al.* (1999) reported pH values as low as 1.4 in San Rossore Estate (Pisa, Italy). The extreme acidity of rainfall reached pH 2.54 in China in 2012 (Wang *et al.*, 2014). To control acid rain pollution and to protect the

* E-mail address: cjliang78@yahoo.com; liangchanjuan@jiangnan.edu.cn

ecological environment, governments and environmental protection departments have formulated laws and regulations to inhibit further expansion of harm by acid rain. However, the area affected by acid rain has not obviously decreased. Acid rain still receives worldwide attention because acidification damage is often the result of atmospheric transport of sulfur and nitrogen emissions across state and/or national boundaries. In Europe, acidic deposition crosses national boundaries, with Scandinavian countries (principally) concerned about acidification damages resulting from emissions coming from the UK and the central and eastern European continents (Menz and Seip, 2004). Because of its transboundary nature, controlling acid rain is very difficult politically.

Acid rain affects human life in a variety of ways. It can acidify soil and surface waters, bringing about a series of ecological changes such as hampering the growth of forests and agricultural crops and threatening animal species. For example, freshwater shrimp cannot survive at a pH level of 6.0 or below. At a pH level of 5.5, bottom-dwelling bacterial decomposers begin to die, causing non-decomposed leaf litter and other organic debris to lie on the bottom and depriving plankton of a food supply. At a pH level of 4.5 or below, all fish and most frogs and insects die (Nagase and Silva, 2007). In addition, acid rain damages buildings and cultural heritage (particularly outdoor marble and bronze sculpture), leads to the release of harmful chemicals, such as aluminum, from rocks and soils into drinking-water sources and corrodes lead and copper piping (Nagase and Silva, 2007).

## 9.3 Toxic Effects of Acid Rain on Plants

Achieving sustainable food production to feed the increasing population of the problematic lands of the world is an enormous challenge. As the main producer of the terrestrial ecosystem, toxic effects of acid rain on plants have attracted more attention in the world today. Inhibition on plants by acid rain includes direct and indirect effects. The direct effect is damaging the foliage. Acid rain directly causes morphological alterations of plant foliage, such as decrease in thickness of cuticle, reduction in leaf area, discoloration and occurrence of necrotic spots. The physiological injury caused by acid rain in plant foliage includes reduction in photosynthetic rate, variation in stomatal conductance, decrease in chlorophyll content, destruction of membrane integrity, disorder of intracellular homeostasis and accumulation of reactive oxygen species (ROS) (Chen *et al.*, 2013; Yi *et al.*, 2014; Ramlall *et al.*, 2015;). One the other hand, acid rain indirectly affects plant growth by influencing soil properties, leaching of soil nutrient cations, increasing the solubility and bioavailability of toxic metals, altering microbial community composition and function and depressing activities of soil animal (Bäck *et al.*, 1995; Larssen and Carmichael, 2000; Wei *et al.*, 2017). Such damage caused by acid rain is even worse when other stress factors such as heavy metals, ozone and enhanced ultraviolet-B radiation exist simultaneously (Momen and Helms, 1996; Tarhanen *et al.*, 1999; Liang *et al.*, 2005; Liang and Wang, 2013; Wang *et al.*, 2013; Zhang *et al.*, 2014).

### 9.3.1 Morphological characteristics and growth

Acid rain firstly destroys the wax and cuticle of leaf surfaces, damages epidermal structure of leaves, and then acids diffuse into the cells through stomata or epidermis. Leaf surface characteristics are important in determining the leaf water-holding capacity, foliar permeability and penetration, and rates of exchange of water and dissolved substances between plant and atmosphere (Fernández and Eichert, 2009). Therefore, changes in leaf surface characteristics caused by acid rain can affect the sensitivity of the plant. Pathogens and other pollutants can more easily invade and reduce plant resistance as acid rain leads to cell damage, stomatal opening, water loss and leaf wilt (Haines *et al.*, 1980; Sant'Anna-Santos *et al.*, 2006; Singh and Agrawal, 2008). The damage caused by acid rain to leaves is related to plant species, growth stages, exposure time and acidity. Based on recent reviews of acid rain exposure experiments on Asian trees, crops and garden plants (Kohno, 2017; Matsumura and Izuta, 2017), it can be concluded that visible leaf injury indicators, such as necrotic spots, defoliation and discoloration, are not affected by acid rain up to a level of pH 4.0. Haines *et al.*

(1980) treated eight plant species (*Erechtites*, *Robinia*, *Pinus*, *Quercus*, *Carya*, *Liriodendron*, *Acer* and *Cornus*) with artificial acid rains of pH 2.5, 2.0, 1.5, 1.0 and 0.5 to determine the threshold and symptoms of damage. They found that droplets of pH 2.0 produced brown necrotic spots on all species except *Pinus*, while droplets of pH 1.0 produced necroses on leaves of all species examined. The size of necrotic spots increased with increasing acidity. However, woody-plant studies that have examined acute foliar damage by acidic deposition are typically performed under controlled environmental conditions. Most data have been produced from experiments conducted in growth chambers or greenhouses which typically have higher temperatures and lower light than ambient conditions (Fan and Wang, 2000).

Acid rain also can affect plant roots by changing the pH value and other properties of soil. Root growth and development directly affect the growth of a plant's aboveground parts by absorbing nutrients and water and synthesizing substances physiologically (Miller, 2011). Root morphology (root length, surface area, volume and number of root tips, etc.) and biomass can be used to reflect root growth conditions because they are sensitive to environmental factors. Zhang *et al.* (2016a) reported that acid rain (pH 5.0 or 3.5) increased the density of root hair and root volume by increasing concentrations of $K^+$, $Na^+$ and $Ca^{2+}$ in rice roots, and the root dry weight was increased. However, strong acid rain (pH 2.5) decreased the root length, surface area, volume and number of root tips by decreasing the concentrations of $K^+$, $Na^+$ and $Mg^{2+}$ in rice root, and fresh and dry weight were both decreased. Similarly, the morphology of soybean root showed no obvious change under acid rain at pH 4.5, while acid rain at pH 3.0 caused the total root length, root surface area, root volume and root tip number in soybean seedlings to decrease (Sun *et al.*, 2013). The deleterious effects of acid rain on root morphology partly result from the decreased macro-element contents and increased micro-element contents in roots (Fageria and Moreira, 2011).

### 9.3.2 Photosynthesis

Chlorophyll is essential in photosynthesis to support plant growth and can be a sensitive indicator of the damaging effect of various environmental stresses on leaf function and health (Carter and Knapp, 2001). Actually, leaf chlorophyll content can directly reflect foliage damage induced by acid rain and is strongly related to inhibition on plant productivity by acid rain. Morrison (1984) pointed out that chlorophyll formation may be decreased by acid rain due to foliar leaching of nutrient elements, especially magnesium (Mg), which is one of the major components of chlorophyll. The degradation of chlorophyll to pheophytin has also been proved, because the absorption spectrum of chlorophyll extracted from leaves exposed to acid rain showed peaks at 665, 606 and 535 nm, characteristic of pheophytin a (Shan, 1998). In general, leaf chlorophyll content showed a linear reduction in response to acid rain exposure (across pH ranges from 2.0 to 5.6) and acid rain substantially reduced leaf chlorophyll content by 6.71% per pH unit across the recorded plant species (Du *et al.*, 2017). In particular, leaf chlorophyll content of deciduous species is more sensitive to acid rain in comparison with evergreen species. Moreover, vegetables and fruit trees are more sensitive to acid rain than other economically useful plants. Some reports show that chlorophyll b may be more sensitive to acid rain than chlorophyll a, because chlorophyll a content is not affected by acid rain whereas chlorophyll b level is reduced at lower pH values (Abouguendia and Baschak, 1987; Westman and Temple, 1989). However, Fan and Wang (2000) found that chlorophyll a may be more sensitive and be reduced by acid rain due to foliar leaching, especially of magnesium, a major chlorophyll component. Neves *et al.* (2009) reported that the contents of chlorophyll a and b are both significantly reduced by acid rain at pH 2.0, while the ratio of these two pigments is not affected. Differences in plant species and rain acidity may be partially responsible for these variations.

Photosynthetic apparatus is one the most stress-sensitive physiological systems. With acid rain, a major impact site is the chloroplast. Acid rain at pH 4.5, pH 3.5 or pH 3.0 causes the granum thylakoids to thin, and the lamellar structure of thylakoid to become loose (Wen *et al.*, 2011; Sun *et al.*, 2012). In the chloroplast, the structure of granum thylakoids affects the absorption of light energy because chlorophyll is distributed on the granum thylakoids (Jagendorf, 1982). The lamellar structure of thylakoid and the chloroplast membrane depend on the Hill reaction rate and the capacity of electron transport (Guo *et al.*, 2005; Neves *et al.*, 2009). Thus the destruction of the structure of chloroplasts led to a decrease in the content of chlorophyll, Hill reaction rate, the activity of $Mg^{2+}$-ATPase and the chlorophyll fluorescence parameters (Mathobo *et al.*, 2017). Gabara *et al.* (2003) also found that the main ultrastructural changes in leaf mesophyll of *Lycopersicon esculentum* following spraying with acid rain (pH 1.8) are located in chloroplasts and mitochondria. In chloroplasts from *L. esculentum* leaves, wave-like oriented thylakoids, often a little swollen and occupying the central part of these organelles, are seen. Similar changes in chloroplasts accompanied by large grana composed of about 40–50 thylakoids are reported in *Phaseolus vulgaris* after acid rain spraying (Stoyanova and Velikova, 1997). All these malformations are non-specific for acid rain since they are observed in a variety of stress situations, such as metal toxicity, acidity, ion deficiency and oxygen stress.

Usually, photosynthesis in leaves of plant has already been inhibited by acid rain before the presence of visible injuries, even before the decrease in chlorophyll content. Simulated acid rain (pH 3.0–2.3) increases the chlorophyll contents in *Pinus armandy* but decreases photosynthetic rate (Shan *et al.*, 1995, 1996). This may be because acid rain reduces the efficiency of the use of chlorophyll in photosynthesis and this reduced efficiency may be linked to the increase in the rate of degradation of chlorophyll to pheophytin. Besides, acid rain (pH 4.0 or less) decreases leaf water content, destroys chloroplast structure, decreases chloroplast ATP synthase activity and its expression at transcriptional level, and finally inhibits photosynthesis (Sun *et al.*, 2016). The increased acidity in chloroplasts can cause injury of chlorophyll proteins, especially including a decrease in PSII activity (whole electron transport chain activity) and inhibition of the rate of Calvin cycle enzymatic reactions (Woodrow *et al.*, 1984; Mutchuchelian *et al.*, 1995). Hence, the inhibition of photosynthesis in plants treated with acid rain not only results from reduced efficiency of chlorophyll and injury to photosynthetic apparatus, but is also due to a reduced photochemical activity and lowered activity of Calvin cycle enzymes, as well as stomata limitation.

### 9.3.3 Nutrient uptake

Hydrogen ions ($H^+$) in acid rain are potentially disruptive of mineral element cycling because of their ability to displace mineral elements out of plant leaves, to inhibit element uptake by plant roots and to leach nutrients from soils (Haines *et al.*, 1980). Nutrient elements are essential to maintain physiological metabolisms for plant growth and development. Hence, the negative effects of acid rain on morphology and growth of plant roots can be one of main reasons for inhibiting plant growth, biomass accumulation and final yield by disturbing water metabolism, nutrient uptake and hormone synthesis in plants (Russell, 1979; Ericsson, 1995; Zheng *et al.*, 2016). When soil pH is lowered to 4, most of nutritional elements such as Mg, calcium (Ca), phosphorus (P) and soluble nitrogen in plant roots are decreased, resulting in nutrient deficiency and abnormal growth of plants (Sun *et al.*, 2013; Du *et al.*, 2014). Zhang *et al.* (2016a) found the decrease in $K^+$, $Na^+$, $Mg^{2+}$, nitrogen and phosphorus contents in rice roots exposed to strong acid rain (pH 2.5) is related to the decrease in the activity of plasma membrane $H^+$-ATPase. The plasma membrane $H^+$-ATPase can provide an energy source for nutrient transport into the cell and extrusion of positive charges ($H^+$) thus forming membrane potential (Palmgren, 2001). Acid rain (pH 2.5) decreases plasma membrane $H^+$-ATPase activity and its expression at transcriptional levels (*OSA1-OSA10*) (Zhang *et al.*, 2017a). Besides, acid rain increases extracellular concentration of $H^+$ and can be a threat to membrane structure by exchanging with positive ions (which is most important in calcium) in the membrane, resulting in increased membrane permeability and leaking out of intracellular substance. Therefore, the decrease in absorbability of plant roots and the increase in membrane leakage are two direct causes for deficiency of nutritional elements in plants (DeHayes *et al.*, 1999; Tarhanen *et al.*, 1999).

The nutrients uptake of plant is also limited by the loss of nutrients in soil. There is increasing evidence that the loss of soil fertility as a consequence of acid rain is distressingly widespread and represents a serious blow to our environment (Bäck *et al.*, 1995). Acid rain can cause soil acidification and then influences the solubility of nutrient ions, such as calcium ($Ca^{2+}$), iron

($Fe^{2+}/Fe^{3+}$), potassium ($K^+$) and magnesium ($Mg^{2+}$) ions, with the result that water removes these nutrients more rapidly from the soil into streams and lakes (Driscoll *et al.*, 2003). Since the $H^+$ ions have a stronger affinity for negative soil particles than other ions (with the exception of $Al^{3+}$), the $H^+$ in acid rain displaces $Ca^{2+}$, $Mg^{2+}$, $K^+$ and $Na^+$ and so these are not available in soil to plants. Thus, the loss of nutrients in soils affected by acidic deposition provides poorer growing conditions for vegetation and extends the time needed for terrestrial and aquatic ecosystems to recover from acidic deposition.

### 9.3.4 Plasma membrane

The plasma membrane of the plant cell acts as an important barrier that separates and shields the cell from its environment, especially the first target affected by abiotic stress, including acid rain (Lüthje *et al.*, 2013). The main components of the plasma membrane are lipids, proteins and carbohydrates (Larsson and Møller, 1990). The lipid composition, including sterols, phospholipids and glycolipids, provides an optimal environment for membrane function, i.e. permeability and enzyme activity (Cooke *et al.*, 1994). There is increasing evidence that acid rain causes elevated levels of ROS, which induces peroxidation of the cell membrane and an increase in permeability and contributes significantly to damage on plants. Appearance of such changes depends on pH of acid rain and plant species (Tarhanen *et al.*, 1999; Liang *et al.*, 2005; Liang and Wang, 2013). Velikova *et al.* (2000) reported that acid rain causes a decrease in unsaturated fatty acid and an increase in lipid peroxidation by inducing excessive accumulation of ROS. The ratio of unsaturated to saturated fatty acids in rice leaves treated with acid rain is decreased by 7.6% (pH 3.5) and 19.5% (pH 2.5), because acid rain reduces unsaturated fatty acid in lipid composition (Su *et al.*, 2016). The changes in membrane lipid composition, such as the degree of saturation or unsaturation of the fatty acid, may affect membrane fluidity, which can collectively influence enzyme activity and ion permeability (Kasamo, 2003). Many studies show that strong acid rain causes obvious increases in membrane permeability, resulting in outflow of intercellular substance and imbalance of intercellular ions (Tarhanen *et al.*, 1999; Liang *et al.*, 2005; Kubis, 2006). The loss of membrane integrity in plants under acid rain stress could be one of the important causes for destroying balance of intracellular ions and disturbing normal metabolism in cells. Besides, high concentrations of $H^+$ can inhibit the activities of functional groups in enzymes and transport systems such as $H^+$-ATPase and $Ca^{2+}$-ATPase, leading to increased membrane permeability and leakage of intracellular solutes, e.g. $K^+$, $Mg^{2+}$ (Wei and Liang, 2014; Bu *et al.*, 2015). Plasma membrane $H^+$-ATPase plays a role in stabilizing intracellular pH because of its high capacity to remove $H^+$ out of cells, and also can provide an energy source for transport of nutrients into the cell by extruding positive charges ($H^+$) and thus forming a membrane potential (Zhang *et al.*, 2011; Zeng *et al.*, 2015). Acid rain with moderate acidity (pH 3.5) increases the hydrolytic activity of plasma membrane $H^+$-ATPase for alleviating membrane damage and stabilizing intercellular $H^+$, whereas acid rain with high acidity inhibits the hydrolytic activity of plasma membrane $H^+$-ATPase, resulting in membrane damage and destabilizing intercellular $H^+$, then reducing photosynthetic efficiency and relative growth rate (Liang *et al.*, 2015). Meanwhile, the decrease of $H^+$-ATPase activity tends to aggravate the degree of peroxidation of plasma membrane, because the increase in intracellular pH causes excessive ROS accumulation (Lu and Liang, 2013).

### 9.3.5 Reactive oxygen species and its scavenging

The increase in hydrogen ion ($H^+$) uptake caused by acid rain would not only influence metabolic processes but may also affect the detrimental oxidative processes in the tissue (Jagels *et al.*, 2002). Oxidative stress is a serious imbalance between ROS production and antioxidant defence. Under acid rain stress, significant accumulation of ROS and increased degree of lipid peroxidation have been observed in previous studies (Scalet *et al.*, 1995; Koricheva *et al.*, 1997; Wyrwicka and Skłodowska, 2006; Neves *et al.*, 2009). ROS are a main part of free radicals, e.g. superoxide radicals ($O_2^- \cdot$), singlet oxygen ($^1O_2$), hydrogen peroxide ($H_2O_2$) and hydroxyl radical ($OH \cdot$).

These ROS can react with lipids, proteins, pigments and nucleic acid and cause lipid peroxidation, membrane damage and inactivation of enzymes, thus affecting cell viability and reasulting in a reduction of plant growth and development (Dixit *et al.*, 2001). On the other hand, an antioxidative defence system has been developed in plant cells to fight against toxic free radicals in order to protect themselves from oxidative stress, including that induced by acid rain (Mittler, 2002; Gabara *et al.*, 2003). The antioxidative defence system comprises both enzymatic and non-enzymatic antioxidants. Antioxidative enzymes include superoxide dismutase (SOD), catalase (CAT), peroxidases (POD), guaiacol peroxidase (GPX), ascorbate peroxidase (APX) and glutathione reductase (GR). Non-enzymatic antioxidants involve lipid-soluble membrane-associated antioxidants (e.g. α-tocopherol and β-carotene) and water-soluble reductants (e.g. ascorbic acid and glutathione). The alteration of antioxidant enzyme activities and non-enzymatic antioxidant contents is an important metabolic process of plants (Du *et al.*, 2013). Wyrwicka and Skłodowska (2006) reported that activities of APX and GST are enhanced for contributing to the scavenging of ROS and detoxification of potentially toxic products originating from acid rain stress. However, antioxidant enzymatic response to acid rain stress is quite sufficient in leaf tissue exposed to acid rain at pH 4.4 but was not effective at pH below 3.0. Ren *et al.* (2018) also found that the regulating effect of antioxidant systems on acid tolerance of plants is limited by the intensity of acid rain. They observed that acid rain at pH higher than 4.0 has no obvious effect on MDA content, concentrations of $O_2^{-}\cdot$ and $H_2O_2$ and antioxidant enzymes (SOD, CAT, POD) activities in rice leaves, while acid rain at pH below 3.0 causes the increase in MDA and membrane permeability by increasing concentrations of $O_2^{-}\cdot$ and $H_2O_2$ which exceeded the scavenging ability of antioxidant enzymes in rice leaves. The increase in activities of antioxidant enzymes and contents of non-enzymatic antioxidants are beneficial to maintained normal ROS levels, which are related to the tolerance of the plant to an acid environment. When they are not enough to avoid cellular damage, membrane lipids are injured and necrosis occurs (Liu and Liu, 2011; Ju *et al.*, 2017).

## 9.4 Combined Effects of Acid Rain and Other Abiotic Stress on Plants

Acid rain can activate $Al^{3+}$ in soil, and hence result in secondary aluminium (Al) toxic problems (von Uexküll and Mutert, 1995). First, $Al^{3+}$ toxicity severely affects crop production in acid soil due to rapid inhibition of root growth and changes in other metabolic activities of plant cells (Horst *et al.*, 2010). After $Al^{3+}$ has entered the plant root, multiple target sites of $Al^{3+}$ include root apex, cell wall, plasma membrane, cytoskeleton, signal transduction, nucleus and oxidative stress (Aggarwal *et al.*, 2015). Second, activated $Al^{3+}$ can reduce bacteria in soil (i.e. kills bacteria) and then decrease ammonification, resulting in loss of ammonia in soil. Third, the combined pollution of acid rain and activated $Al^{3+}$ also affects growth of soil animals (Meng *et al.*, 2011). The mortality rate of the tested soil animals is increased with the decrease of pH value and the rise of $Al^{3+}$ concentration. All these changes can aggravate adverse impacts of acid rain on agricultural and forest health.

Acid rain has been shown to increase the mobility of heavy metals in the soil and their uptake by plants. Therefore, it is expected that the combination of acid rain and heavy metals would cause more pronounced changes in plants than acid rain and heavy metals applied separately. Liao *et al.* (2005) studied complex toxic effects of $Cd^{2+}$, $Zn^{2+}$ and acid rain on the growth of kidney bean (*Phaseolus vulgaris* L) in a pot experiment. They found that all treatments of $Cd^{2+}$, $Zn^{2+}$ and/or acid rain significantly decrease fresh weights of kidney bean and caused toxic effects on growth of the plants, especially higher amounts of $Cd^{2+}$ and $Zn^{2+}$ and higher acidity of acid rain. Combination of these three pollutant factors results in more serious toxic effects than any single pollutant and than combinations of any two pollutants. Sun *et al.* (2012) found that the combined pollution of $Cd^{2+}$ and acid rain aggravates the toxic effect of the single pollution of $Cd^{2+}$ or acid rain on the photosynthetic parameters, due to the serious damage to the chloroplast structure. However, significant enzyme responses to mixed pollutants were usually additive or antagonistic, while no synergistic responses were detected. Koricheva *et al.*

(1997) measured the responses of different antioxidants in 2-year-old birch (*Betula pendula* Roth) seedlings subjected to simulated acid rain (pH 4.0) and heavy metals (Cu/Ni), applied alone or in combination for 2 months. The results show that the effects of mixed pollutants on enzyme activities were usually less than the summed effects of individual pollutants, and enzyme responses also depended on concentrations of pollutants and where the pollutants were applied (root or leaf).

Rare-earth elements (REE) are a series of elements with similar physical and chemical properties that are widely used in the pharmaceutical, chemical, agricultural, electronics, new materials and aerospace industries (Tyler, 2004). In many agricultural areas, acid rain and REE pollution exist simultaneously and the combination of these two environmental factors can affect plant growth and development, ultimately impacting the safety of agricultural ecosystems. The combined effects of acid rain and lanthanum ($La^{3+}$) on plants have been studied for several aspects, such as root phenotype, nitrogen assimilation, chloroplast ultrastructure, photoysnthesis and antioxdant enzymes activity (Liang *et al.*, 2010, 2017; Wen *et al.*, 2011; Liang and Wang, 2013; Sun *et al.*, 2013; Hu *et al.*, 2016; Zhang *et al.*, 2016b, 2017b; Xia *et al.*, 2017). It is testified that the combined effects of $La^{3+}$ and acid rain on growth, physiological and biochemical parameters depend on the growth stage of plant, concentrations of $La^{3+}$ and pH of acid rain. An antagonistic interaction of acid rain and $La^{3+}$ can be observed when the concentration of La is low. However, the combined effect of acid rain and $La^{3+}$ at high concentrations is obviously heavier than the single acid rain or $La^{3+}$ treatment.

Ozone ($O_3$), produced by photochemical reactions, is a phytotoxic air pollutant with the dramatic increase of traffic and growth of the petrochemical industry, and its current ambient and elevated levels are able to induce severe damage to plant growth (Takemoto *et al.*, 1988). Because of the potential for coincident stress by ozone and acid rain on agriculture and forestry, some researchers have been investigating their effects on plants singly or combined. Significant interactive effects of $O_3$ and simulated acid rain on net photosynthetic rate were observed when the seedlings of *Pinus armandi* Franch. were exposed to $O_3$ at 300 ppb for 8 h per day, 6 days a week, and simulated acid rain of pH 3.0 or 2.3, six times a week, alone or in combination, for 14 weeks (Shan *et al.*, 1996). The significant interactive effects of $O_3$ and acid rain have also been observed on alfalfa (Takemoto *et al.*, 1988) and soybean (Troiano *et al.*, 1983). However, no significant interactive effects of $O_3$ and simulated acid rain were observed on chlorophyll contents, carbon allocation and biomass accumulation of the seedlings of armand pine (Shan *et al.*, 1995). The interactive effects of $O_3$ and simulated acid rain on green pepper (Bytnerowicz and Olszyk, 1988), loblolly pine (Edwards *et al.*, 1992), radish (Johnston *et al.*, 1986) sugar maple and northern red oak (Reich *et al.*, 1986) were not significant either. These varying results might be due to the difference in species, environment, dose or process of exposure and may also be related to the determined parameters.

## 9.5 Conclusions and Future Perspectives

With the ever-increasing demands of agricultural production, food security is imminently threatened by global climate change and environmental pollution. Acid rain severely limits the productivity of agricultural crops, most of which are sensitive to the presence of the low pH value of acid rain. Despite substantial research, we are still far from a deep understanding of molecular mechanisms of acid rain inducing intracellular production of reactive oxygen species, inhibition on light and dark reaction sites in photosynthesis, and imbalance of the energy metabolism of cells, etc. There is a need to create an interacting network of genes, proteins and metabolisms participating in stress-regulated biological processes. To clarify the toxic effect of acid rain on plants, finding possible ways to alleviate such damage is also essential. Limiting emission amount of $SO_2$ and NOx is at the first strategy for controlling acid rain pollution and protecting the ecological environment. According to the concept of sustainable development, some measures must be taken, such as enhancing environmental management, electing practical energy technologies of clean coal, developing other energy sources to replace coal, etc. Second, to decrease agricultural and forestry losses, acid-resistant types of crops or trees can be selected and cultured in areas where acid rain happens. Researchers need to explore the resistance difference in species, and improve plant acid resistance by modern biotechnology or conventional selection and breeding. Third, planting green manure crops is also a strategy, by using organic fertilizers and liming. This method can improve acidification of soil by increasing buffer capacities of the soil in the short term. Last but not least, finding the appropriate method of chemical control can be one effective way to alleviate the negative effects of acid rain on plants, especially considering people's acceptance of genetically modified crops. Some plant growth regulators such as calcium (Zhou *et al.*, 1999), abscisic acid (Wu and Liang, 2017), mefluidide [*N*-(2,4-dimethyl-5-(((trifluoromethyl)sulfony) amino)phenyl)acetamide] (Bi and Zhang, 1993) and light rare-earth elements (Yan *et al.*, 1999) have been applied for alleviating the toxic effects of acid rain on some plants under laboratory conditions. Further study is still needed, especially under farm field conditions.

## Acknowledgments

The author is grateful for the financial support from the National Natural Science Foundation of China (31000245, 31370517) and the Natural Science Foundation of Jiangsu Province (No.BK20161131).

## References

Abouguendia, Z.M. and Baschak, L.A. (1987) Response of two western Canadian conifers to simulated acidic precipitation. *Water Air and Soil Pollution* 33, 15–22.

Aggarwal, A., Ezaki, B., Munjal, A. and Tripathi, B.N. (2015) Stress responses in plants. In: Tripathi, B.N. and Müller, M. (eds.) *Stress Responses in Plants: Mechanisms of Toxicity and Tolerance*. Springer International Publishing, Basel, pp. 35–58.

Bäck, J., Huttunen, S., Turunen, M. and Lamppu, J. (1995) Effects of acid rain on growth and nutrient concentrations in Scots pine and Norway spruce seedlings grown in a nutrient-rich soil. *Environmental Pollution* 89, 177–187.

Bi, Y. and Zhang, C. (1993) Damage of modeled acid rain on *Phaseolus* leaves and the protect effects of Mefluidide. *Acta Scientiae Circumstantiae* 13, 379–384.

Bu, J., Su, L., Lv, X. and Liang, C.J. (2015) Effect of simulated acid rain on plasma membrane $H^+$-ATPase activity and mineral elements contents in rice leaves. *Acta Scientiae Circumstantiae* 35, 3020–3024.

Bytnerowicz, A. and Olszyk, D.M. (1988) Depression of photosynthesis, growth, and yield in field-grown green pepper (*Capsicum annuum* L.) exposed to acidic fog and ambient ozone. *Plant Physiology* 88, 477–482.

Carter, G.A. and Knapp, A.K. (2001) Leaf optical properties in higher plants: linking spectral characteristicsto stress and chlorophyll concentration. *American Journal of Botany* 88, 677–684.

Chen, J., Wang, W., Liu, T., Wu, F. and Zheng, H. (2013) Photosynthetic and antioxidant responses of *Liquidambar formosana* and *Schima superba* seedlings to sulfuric-rich and nitric-rich simulated acid rain. *Plant Physiology and Biochemistry* 64, 41–51.

Cooke, D.T., Burden, R.S., James, C.S., Seco, T. and Sierra, B. (1994) Influence of sterols on plasma membrane proton-pumping ATPase activity and membrane fluidity in oat shoots. *Plant Physiology and Biochemistry* 32, 769–773.

DeHayes, D.H., Schaberg, P.G., Hawley, G.J. and Strimbeck, G.R. (1999) Acid rain impacts on calcium nutrition and forest health – alteration of membrane-associated calcium leads to membrane destabilization and foliar injury in red spruce. *Bioscience* 49, 789–800.

Dixit, V., Pandey, V. and Shyam, R. (2001) Differential antioxidative responses to cadmium in roots and leaves of pea (*Pisum sativum* L. cv. Azad ). *Journal of Experimental Botany* 52, 1101.

Driscoll, C.T., Driscoll, K.M., Mitchell, M.J. and Raynal, D.J. (2003) Effects of acidic deposition on forest and aquatic ecosystems in New York State. *Environmental Pollution* 123, 327–336.

Du, H., Zhou, P. and Huang, B. (2013) Antioxidant enzymatic activities and gene expression associated with heat tolerance in a cool-season perennial grass species. *Environmental and Experimental Botany* 87, 159–166.

Du, Y., Wei, M., Reddy, K.R., Liu, Z. and Jin, F. (2014) Effect of acid rain pH on leaching behavior of cement stabilized lead-contaminated soil. *Journal of Hazardous Materials* 271, 131–140.
Du, E., Dong, D., Zeng, X., Sun, Z., Jiang, X. and de Vries, W. (2017) Direct effect of acid rain on leaf chlorophyll content of terrestrial plants in China. *Science of the Total Environment* 605, 764–769.
Edwards, N.T., Edwards, G.L., Kelly, J.M. and Taylor, G.E. (1992) Three-year growth responses of *Pinus taeda* L. to simulated rain chemistry, soil magnesium status, and ozone. *Water Air and Soil Pollution* 63, 105–118.
Ericsson, T. (1995) Growth and shoot: root ratio of seedlings in relation to nutrient availability. *Plant and Soil* 168–169, 205–214.
Fageria, N.K. and Moreira, A. (2011) The role of mineral nutrition on root growth of crop plants. *Advances in Agronomy* 110, 251–331.
Fan, H.B and Wang, Y.H. (2000) Effects of simulated acid rain on germination, foliar damage, chlorophyll contents and seedling growth of five hardwood species growing in China. *Forest Ecology and Management* 126, 321–329.
Fernández, V. and Eichert, T. (2009) Uptake of hydrophilic solutes through plant leaves: current state of knowledge and perspectives of foliar fertilization. *Critical Reviews in Plant Sciences* 28, 36–68.
Gabara, B., Skłodowska, M., Wyrwicka, A., Glińska, S. and Gapińska, M. (2003) Changes in the ultrastructure of chloroplasts and mitochondria and antioxidant enzyme activity in *Lycopersicon esculentum* Mill. leaves sprayed with acid rain. *Plant Science* 164, 507–516.
Gravano, E., Ferretti, M., Bussotti, F. and Grossoni, P. (1999) Foliar symptoms and growth reduction of *Ailanthus altissima* Desf. in an area with high ozone and acidic deposition in Italy. *Water, Air, and Soil Pollution* 116, 267–272.
Guo, D.P., Guo, Y.P., Zhao, J.P., Liu, H., Peng, Y., Wang, Q.M., Chen, J.S. and Rao, G.Z. (2005) Photosynthetic rate and chlorophyll fluorescence in leaves of stem mustard (*Brassica juncea* var. tsatsai) after turnip mosaic virus infection. *Plant Science* 168, 57–63.
Haines, B., Stefani, M. and Hendrix, F. (1980) Acid rain: threshold of leaf damage in eight plant species from a Southern Appalachian forest succession. *Water Air and Soil Pollution* 14, 403–407.
Horst, W.J., Wang, Y. and Eticha, D. (2010) The role of the root apoplast in aluminium-induced inhibition of root elongation and in aluminium resistance of plants: a review. *Annals of Botany* 106, 185–197.
Hu, H., Wang, L., Zhou, Q. and Huang, X. (2016) Combined effects of simulated acid rain and lanthanum chloride on chloroplast structure and functional elements in rice. *Environmental Science and Pollution Research* 23, 8902–8916.
Jagels, R., Jiang, M., Marden, S. and Carlisle, J. (2002) Red spruce canopy response to acid fog exposure. *Atmospheric Research* 64, 169–178.
Jagendorf, A.T. (1982) Oligomycin effects on ATPase and photophosphorylation of pea chloroplast thylakoid membranes. *Plant Physiology* 69, 888–896.
Johnston, J.W. Jr, Shriner, D.S. and Kinerley, C.K. (1986) The combined effects of simulated acid rain and ozone on injury, chlorophyll, and growth of radish. *Environmental and Experimental Botany* 26, 107–113.
Ju, S., Yin, N., Wang, L., Zhang, C. and Wang, Y. (2017) Effects of silicon on *Oryza sativa* L. seedling roots under simulated acid rain stress. *PLoS ONE* 12(3), e0173378.
Kasamo, K. (2003) Regulation of plasma membrane H+-ATPase activity by the membrane environment. *Journal of Plant Research* 116, 517–523.
Kohno, Y. (2017) Effects of simulated acid rain on asian crops and garden plants. In: Izuta, T. (ed.) *Air Pollution Impacts on Plants in East Asia*. Springer Japan, Tokyo, pp. 223–235.
Koricheva, J., Roy, S., Vranjic, J.A., Haukioja, E., Hughes, P.R. and Hänninen, O. (1997) Antioxidant responses to simulated acid rain and heavy metal deposition in birch seedlings. *Environmental Pollution* 95, 249–258.
Kubis, J. (2006) Exogenous spermidine alters in different way membrane permeability and lipid peroxidation in water stressed barley leaves. *Acta Physiologiae Plantarum* 28, 27–33.
Lüthje, S., Möller, B., Perrineau, F.C. and Wöltje, K. (2013) Plasma membrane electron pathways and oxidative stress. *Antioxidants and Redox Signaling* 18, 2163–2183.
Larssen, T. and Carmichael, G.R. (2000) Acid rain and acidification in China: the importance of base cation deposition. *Environmental Pollution* 110, 89–102.
Larsson, C. and Møller, I.M. (1990) The plant plasma membrane. Structure, function and molecular biology. *Physiological and Molecular Plant Pathology* 37, 492–493.
Liang, C.J. and Wang, W. (2013) Antioxidant response of soybean seedlings to joint stress of lanthanum and acid rain. *Environmental Science and Pollution Research* 20, 8182–8191.

Liang, C.J., Huang, X., Tao, W. and Zhou, Q. (2005) Responses of antioxidant enzyme and photosynthesis in rape seedling to the combined stresses of acid rain and ultraviolet-B radiation. *Journal of Environmental Sciences* 17, 1038–1041.

Liang, C.J., Pan, D., Xu, Q. and Zhou, Q. (2010) Combined injured effects of acid rain and lanthanum on growth of soybean seedling. *Chinese Journal of Environmental Science* 31, 1652–1656.

Liang, C.J., Ge, Y., Su, L. and Bu, J. (2015) Response of plasma membrane $H^+$-ATPase in rice (*Oryza sativa*) seedlings to simulated acid rain. *Environmental Science and Pollution Research* 22, 535–545.

Liang, C.J., Li, L. and Su, L. (2017) Effect of lanthanum on plasma membrane $H^+$-ATPase in rice (*Oryza sativa*) under acid rain stress. *Journal of Plant Growth Regulation* 37(2), 380–390. doi: 10.1007/s00344-017-9740-4.

Liao, B., Liu, H., Zeng, Q., Yu, P., Probst, A. and Probst, J.L. (2005) Complex toxic effects of $Cd^{2+}$, $Zn^{2+}$, and acid rain on growth of kidney bean (*Phaseolus vulgaris* L). *Environment International* 31, 891–895.

Liu, E. and Liu, C.P. (2011) Effects of simulated acid rain on the antioxidative system in *Cinnamomum philippinense* seedlings. *Water Air and Soil Pollution* 215, 127–135.

Lu, X. and Liang, C.J. (2013) Influence of simulated acid rain on the plasma membrane $H^+$-ATPase and peroxidation of plasma membrane in rice leaves. *Journal of Safety and Environment* 13, 1–4.

Mathobo, R., Marais, D. and Steyn, J.M. (2017) The effect of drought stress on yield, leaf gaseous exchange and chlorophyll fluorescence of dry beans (*Phaseolus vulgaris* L.). *Agricultural Water Management* 180, 118–125.

Matsumura, H. and Izuta, T. (2017) Effects of simulated acid rain on Asian trees. In: Izuta, T. (ed.) *Air Pollution Impacts on Plants in East Asia*. Springer Japan, Tokyo, pp. 237–247.

Meng, L., Zhang, J., Xu, H., Yu, J., Qin, Z., Xie, J. and Quan, G. (2011) Killing effects of compound pollution from the simulated acid rain and its activated $Al^{3+}$ on small and medium-sized soil animals. *Ecology and Environment* 20, 1491–1495.

Menz, F.C. and Seip, H.M. (2004) Acid rain in Europe and the United States: an update. *Environmental Science & Policy* 7, 253–265.

Miller, D.M. (2011) Studies of root function in *Zea mays* I. Apparatus and methods. *Canadian Journal of Botany* 58, 351–360.

Mittler, R. (2002) Oxidative stress, antioxidants and stress tolerance. *Trends in Plant Science* 7, 405–410.

Momen, B. and Helms, J.A. (1996) Effects of simulated acid rain and ozone on foliar chemistry of field-grown *Pinus ponderosa* seedlings and mature trees. *Environmental Pollution* 91, 105–111.

Morrison, I.K. (1984) Acid rain, forests and forestry. In: Stone, E.L. (ed.) *Proceedings of the 6th North American Forest Soils Conference, University of Tennessee, Knoxville*, pp. 209–219.

Mutchuchelian, K., Murugan, C., Harigovindan, R. and Nedunchezhian, N. (1995) Growth, $^{14}CO_2$ fixation, activities of photosystems, ribulose 1,5-bisphosphate carboxylase and nitrate reductase in trees as affected by simulated acid rain. *Biologia Plantarum* 37, 355–362.

Nagase, Y. and Silva, E.C.D. (2007) Acid rain in China and Japan: a game-theoretic analysis. *Regional Science and Urban Economics* 37, 100–120.

Neves, N.R., Oliva, M.A., da Cruz Centeno, D., Costa, A.C., Ribas, R.F. and Pereira, E.G. (2009) Photosynthesis and oxidative stress in the restinga plant species *Eugenia uniflora* L. exposed to simulated acid rain and iron ore dust deposition: Potential use in environmental risk assessment. *Science of the Total Environment* 407, 3740–3745.

Palmgren, M.G. (2001) Plant plasma membrane $H^+$-ATPases: powerhouses for nutrient uptake. *Annual Review of Plant Physiology and Plant Molecular Biology* 52, 817–845.

Ramlall, C., Varghese, B., Ramdhani, S., Pammenter, N., Bhatt, A. and Berjak, P., Sershen (2015) Effects of simulated acid rain on germination, seedling growth and oxidative metabolism of recalcitrant-seeded *Trichilia dregeana* grown in its natural seed bank. *Physiologia Plantarum* 153, 149–160.

Reich, P.B., Schoettle, A.W. and Amundson, R.G. (1986) Effects of $O_3$ and acidic rain on photosynthesis and growth in sugar maple and northern red oak seedlings. *Environmental Pollution* 40, 1–15.

Ren, X., Zhu, J., Liu, H., Xu, X. and Liang, C. (2018) Response of antioxidative system in rice (*Oryza sativa*) leaves to simulated acid rain stress. *Ecotoxicology and Environmental Safety* 148, 851–856.

Russell, R.S. (1979) Plant root systems: their function and interaction with the soil. *Field Crops Research* 2, 177–179.

Sant'Anna-Santos, B.F., Silva, L.C.D., Azevedo, A.A., Araújo, J.M.D., Alves, E.F., Silva, E.A.M.D. and Aguiar, R. (2006) Effects of simulated acid rain on the foliar micromorphology and anatomy of tree tropical species. *Environmental and Experimental Botany* 58, 158–168.

Scalet, M., Federico, R., Guido, M.C. and Manes, F. (1995) Peroxidase activity and polyamine changes in response to ozone and simulated acid rain in Aleppo pine needles. *Environmental and Experimental Botany* 35, 417–425.
Seinfeld, J.H. and Pandis, S.N. (1998) *Atmospheric Chemistry and Physics: from Air Pollution to Climate Change*. Wiley, New York.
Shan, Y. (1998) Effects of simulated acid rain on *Pinus densiflora*: inhibition of net photosynthesis by the pheophytization of chlorophyll. *Water, Air, and Soil Pollution* 103, 121–127.
Shan, Y., Feng, Z., Izuta, T., Aoki, M. and Totsuka, T. (1995) The individual and combined effects of ozone and simulated acid rain on chlorophyll contents, carbon allocation and biomass accumulation of armand pine seedlings. *Water, Air, and Soil Pollution* 85, 1399–1404.
Shan, Y., Feng, Z., Izuta, T., Aoki, M. and Totsuka, T. (1996) The individual and combined effects of ozone and simulated acid rain on growth, gas exchange rate and water-use efficiency of *Pinus armandi* Franch. *Environmental Pollution* 91, 355-361.
Singh, A. and Agrawal, M. (2008) Acid rain and its ecological consequences. *Journal of Environmental Biology* 29, 15–24.
Smith, R.A. (1852) On the air and rain of Manchester. *Memoirs and Proceedings of the Manchester Literary & Philosophical Society* 10, 207–217.
Smith, R.A. (1872) *Air and Rain. The Beginnings of a Chemical Climatology*. Long-mans & Green, London.
Stoyanova, D. and Velikova, V. (1997) Effects of simulated acid rain on chloroplast ultrastructure of primary leaves of *Phaseolus Vulgaris*. *Biologia Plantarum* 40, 581–588.
Su, L., Wu, X. and Liang, C.J. (2016) Effect of lanthanum on plasma membrane $H^+$-ATPase activity in rice leaves under acid rain. *Acta Scientiae Circumstantiae* 36, 353–358.
Sun, Z., Wang, L., Chen, M., Wang, L., Liang, C., Zhou, Q. and Huang, X. (2012) Interactive effects of cadmium and acid rain on photosynthetic light reaction in soybean seedlings. *Ecotoxicology and Environmental Safety* 79, 62–68.
Sun, Z., Wang, L., Zhou, Q. and Huang, X. (2013) Effects and mechanisms of the combined pollution of lanthanum and acid rain on the root phenotype of soybean seedlings. *Chemosphere* 93, 344–352.
Sun, J., Hu, H., Li, Y., Wang, L., Zhou, Q. and Huang, X. (2016) Effects and mechanism of acid rain on plant chloroplast ATP synthase. *Environmental Science and Pollution Research* 23, 18296–18306.
Takemoto, B.K., Hutton, W.J. and Olszyk, D.M. (1988) Responses of field-grown *Medicago sativa* L. to acidic fog and ambient ozone. *Environmental Pollution* 54, 97–107.
Tarhanen, S., Metsärinne, S., Holopainen, T. and Oksanen, J. (1999) Membrane permeability response of lichen *Bryoria fuscescens* to wet deposited heavy metals and acid rain. *Environmental Pollution* 104, 121–129.
Troiano, J., Colavito, L., Heller, L., Hccune, D.C. and Jacobson, J.S. (1983) Effects of acidity of simulated rain and its joint action with ambient ozone on measures of biomass and yield in soybean. *Environmental and Experimental Botany* 23, 113–119.
Tyler, G. (2004) Rare earth elements in soil and plant systems – a review. *Plant and Soil* 267, 191–206.
Velikova, V., Yordanov, I. and Edreva, A. (2000) Oxidative stress and some antioxidant systems in acid rain-treated bean plants: protective role of exogenous polyamines. *Plant Science* 151, 59–66.
von Uexküll, H.R. and Mutert, E. (1995) Global extent, development and economic impact of acid soils. *Plant and Soil* 171, 1-15.
Wang, S., Wang, L., Zhou, Q. and Huang, X. (2013) Combined effect and mechanism of acidity and lead ion on soybean biomass. *Biological Trace Element Research*, 298–307.
Wang, L., Wang, W., Zhou, Q. and Huang, X. (2014) Combined effects of lanthanum (III) chloride and acid rain on photosynthetic parameters in rice. *Chemosphere* 112, 355–361.
Wei, J. and Liang, C.J. (2014) Effect of simulated acid rain on plasma membrane $H^+$-ATPase activity and intracellular $Ca^{2+}$ concentration in rice leaves. *Acta Scientiae Circumstantiae* 34, 532–536.
Wei, H., Liu, W., Zhang, J. and Qin, Z. (2017) Effects of simulated acid rain on soil fauna community composition and their ecological niches. *Environmental Pollution* 220, 460–468.
Wen, K., Liang, C.J., Wang, L., Hu, G. and Zhou, Q. (2011) Combined effects of lanthanumion and acid rain on growth, photosynthesis and chloroplast ultrastructure in soybean seedlings. *Chemosphere*, 601–608.
Westman, W.E. and Temple, P.J. (1989) Acid mist and ozone effects on the leaf chemistry of two Western Conifer species. *Environmental Pollution* 57, 9–26.
Woodrow, I.E., Murphy, D.J. and Latzko, E. (1984) Regulation of stromal sedoheptulose 1,7-bisphosphatase activity by pH and $Mg^{2+}$ concentration. *Journal of Biological Chemistry* 259, 3791.

Wu, X. and Liang, C. (2017) Enhancing tolerance of rice (*Oryza sativa*) to simulated acid rain by exogenous abscisic acid. *Environmental Science and Pollution Research* 24, 4860–4870.

Wyrwicka, A. and Skłodowska, M. (2006) Influence of repeated acid rain treatment on antioxidative enzyme activities and on lipid peroxidation in cucumber leaves. *Environmental and Experimental Botany* 56, 198–204.

Xia, B., Sun, Z., Wang, L., Zhou, Q. and Huang, X. (2017) Analysis of the combined effects of lanthanum and acid rain, and their mechanisms, on nitrate reductase transcription in plants. *Ecotoxicology and Environmental Safety* 138, 170–178.

Yan, C., Hong, Y., Lin, P., Yang, X., Fu, S., Wu, S. and Zhu, K. (1999) Effects of rare-earth elements on physiological and biochemical responses of wheat under acid rain stress. *Progress in Natural Science:Materials International*, 929–933.

Yi, L., Liu, M., Yu, S., Yu, F. and Yin, X. (2014) Effects of simulated acid rain stress on chlorophyll fluorescence characteristics and growth in leaves of *Lithocarpus glaber* and *Schima superba* seedlings. *Asian Journal of Chemistry* 26, 4619–4622.

Zeng, H., Di, T., Zhu, Y. and Subbarao, G.V. (2015) Transcriptional response of plasma membrane H+-ATPase genes to ammonium nutrition and its functional link to the release of biological nitrification inhibitors from sorghum roots. *Plant and Soil* 398, 301–312.

Zhang, R., Liu, G., Wu, N., Gu, M., Zeng, H., Zhu, Y. and Xu, G. (2011) Adaptation of plasma membrane H+ ATPase and H+ pump to P deficiency in rice roots. *Plant and Soil* 349, 3–11.

Zhang, J.E., Yu, J., Ouyang, Y. and Xu, H. (2014) Impact of simulated acid rain on trace metals and aluminum leaching in Latosol from Guangdong Province, China. *Soil and Sediment Contamination: An International Journal* 23, 725–735.

Zhang, B., Bu, J. and Liang, C. (2016a) Root morphology and growth regulated by mineral nutrient absorption in rice roots exposed to simulated acid rain. *Water Air and Soil Pollution* 227, 457.

Zhang, X., Wang, L., Zhou, A., Zhou, Q. and Huang, X. (2016b) Alterations in cytosol free calcium in horseradish roots simultaneously exposed to lanthanum(III) and acid rain. *Ecotoxicology and Environmental Safety* 126, 62–70.

Zhang, B., Bu, J. and Liang, C. (2017a) Regulation of nitrogen and phosphorus absorption by plasma membrane $H^+$-ATPase in rice roots under simulated acid rain. *International Journal of Environmental Science and Technology* 14, 101–112.

Zhang, F., Cheng, M., Sun, Z., Wang, L., Zhou, Q. and Huang, X. (2017b) Combined acid rain and lanthanum pollution and its potential ecological risk for nitrogen assimilation in soybean seedling roots. *Environmental Pollution* 231, 524–532.

Zheng, Y., Zhang, Y. and Wu, J. (2016) Yield and quality of *Moringa oleifera* under different planting densities and cutting heights in southwest China. *Industrial Crops and Products* 91, 88–96.

Zhou, Q., Huang, X., Wang, D., Cao, Y., Ning, Y. and Jin, J. (1999) Effect of calcicum on muskmelon seedling harmed by acid rain. *Chinese Journal of Plant Ecology* 23, 186–191.

# Part III

# Persistent Organic Pollutants

# 10 Polycyclic Aromatic Hydrocarbons: Ecotoxicity in the Aquatic Environment and Implications for Human Health

**D.M. Pampanin*[,1] and D. Schlenk[2]**
[1]*University of Stavanger and International Research Institute of Stavanger, Stavanger, Norway;* [2]*University of California Riverside, Riverside, California, USA*

## 10.1 Abstract

Polycyclic aromatic hydrocarbons (PAHs) are global environmental pollutants of profound concern. Their presence is ubiquitous in water, air and soil. Their ecotoxicity is continuously under assessment and researchers are engaged linking their presence with risks for human health and wildlife.

It is generally recognized that PAHs cause adverse effects on aquatic organisms at all life stages and these effects include growth reduction, DNA damage, cytotoxicity, endocrine disruption and immunosuppression, as well as malformations of embryos and larvae.

Many research articles and reviews have been published and this chapter will help to guide the reader in finding answers regarding the linkage of PAH ecotoxicity and human health.

The major route of exposure to PAHs in the general population is from contaminated air. With regard to human exposure from the aquatic environment, PAH exposure occurs primarily via ingestion (i.e. consumption of contaminated food or product cooked at high temperature).

Although PAHs have been under evaluation for about 50 years, some knowledge gaps have been identified in this chapter and suggestions regarding future research focus have been provided.

For effective management of PAH contamination, the development of new analytical tools is necessary (e.g. new cost-efficient bioassays with higher throughput, preferably capable of providing highly sensitive and specific data in real time).

The development of adverse outcome pathways to identify biomarkers for exposure, for both individual PAHs and complex mixtures containing PAHs, is needed since the aquatic environment is often the final and major receiver of many contaminants. Analyses of PAH mixtures with fingerprinting, using rapidly developing mass spectrometry technology, will help in the identification of contamination sources for consequent removal or reduction.

This will provide the required knowledge for human health and environmental risk assessments as well as lead to enhanced management of aquatic resources.

## 10.2 Introduction

As documented in a large number of scientific articles and books, polycyclic aromatic hydrocarbons (PAHs) are global environmental pollutants of high concern. Their presence is recognized as

* E-mail: daniela.m.pampanin@uis.no

ubiquitous in water, air and soil. Their ecotoxicity, as single compounds and as mixtures, is continuously under assessment and researchers are engaged linking their presence with risks for human and environmental health. Since water resources tend to receive most contaminants, the study and assessment of PAHs in the aquatic environment has been a priority since the 1970s (Hahn and Stegeman, 1994; EPA, 2008). Google Scholar reports 22,300 results when searching for PAHs and ecotoxicity in the aquatic environment, and over 6810 are from 2018.

General information regarding PAH toxicity and carcinogenic potential can be found in reports from the US Environmental Protection Agency (EPA) and IARC (International Agency for Research on Cancer) (EPA, 2008). The IARC classifies seven PAHs as possibly or probably carcinogenic to humans (Group 1, 2A and 2B): benzo[*a*]pyrene, dibenzo[*a,h*]anthracene, benzo[*a*] anthracene, benzo[*b*]fluoranthene, benzo[*k*]fluoranthene, indeno[1,2,3-cd]pyrene and chrysene. In addition, these seven are well-known genotoxic compounds, i.e. able to damage the genetic information within a cell. The EPA has also established toxicity equivalent factors (TEFs) for quantifying mixtures of individual compounds, with the highest level (1) assigned to benzo[*a*]pyrene (Adeniji *et al.*, 2018). The European Commission Joint Research Centre and the Institute for Reference Materials and Measurements provide PAH Factsheets, which contain information regarding standards and guidelines for toxicity measurements (Lerda, 2011). Recently, a book has been published documenting PAH, analytical methods for detection and toxicity as well as an evaluation of their environmental impact in the aquatic environment (Pampanin and Sydnes, 2017). This is considered a good starting point for the reader who wishes to gain knowledge of PAH ecotoxicity, as important basic information is reported.

In the late 1980s, one of the first comprehensive books discussing the fate of PAHs in the aquatic environment was published (Varanasi, 1989). Since then, our knowledge regarding the fate and effects of PAHs has increased exponentially, aiding regulators in the management of contaminated environments. Many research articles and reviews have been published in the literature, but the focus of this chapter will be to evaluate linkages between PAH ecotoxicity and human health.

The most studied PAHs are benzo[*a*]pyrene and dimethylbenzo anthracene, while the most commonly analysed in environmental monitoring are 16 PAHs recommended by the EPA. Their physiochemical properties are tightly linked to their toxicity and their adverse effects on biota (Alegbeleye *et al.*, 2017). In general, the 16 are characterized by high melting point, low vapour pressure and low aqueous solubility (high hydrophobicity). The last two properties tend to decrease with an increase in the PAH molecular weight. Behera *et al.* (2018) summarized the physiochemical and toxicological parameters of EPA listed PAHs, including key references.

It is generally recognized that PAHs cause adverse effects on aquatic organisms at all life stages. These effects include growth reduction, DNA damage, cytotoxicity, endocrine disruption, immunosuppression and carcinogenesis, as well as malformations of embryos and larvae (Abdel-Shafy and Mansour, 2016; Alegbeleye *et al.*, 2017; Behera *et al.*, 2018). Table 10.1 reports some of the toxic effects of PAHs in fish. For example, embryonic exposure to PAHs can cause sub-lethal effects such as yolk sac and pericardial oedema, disruption of cardiac function, cardiac deformities, neuronal cell death and impaired swimming (White *et al.*, 1999; Incardona and Scholz, 2016; Cherr *et al.*, 2017; Gao *et al.*, 2018).

Many PAHs have mutagenic and/or carcinogenic properties; and, being highly lipid soluble, they can easily be absorbed from the gastrointestinal tract of organisms, including mammals. They are also rapidly distributed in tissues with a tendency to be stored in the body fat. Metabolism of PAHs occurs in most organisms via the cytochrome P450 monooxygenase family of enzymes with oxidation or hydroxylation as the first step (Walker *et al.*, 2012). Due to their genotoxicity and ubiquitous presence in the environment, there is a continuous necessity for monitoring PAH concentrations and effects on aquatic life.

## 10.3 Sources of PAH Contamination

PAH contamination can be found in all environmental compartments and their concentrations vary widely (Abdel-Shafy and Mansour, 2016). PAHs are generally divided into two groups: pyrogenic and petrogenic (Pampanin and Sydnes,

**Table 10.1.** Examples of toxic effect of polycyclic aromatic hydrocarbons in fish (modified from Behera *et al.*, 2018), including IARC group classification (1 = carcinogenic to humans, 2A = probably carcinogenic to humans, 2B = possibly carcinogenic to humans, 3 = unclassifiable as to carcinogenicity to humans).

| PAH compound | No. rings | Fish species | Toxic effect | IARC group |
|---|---|---|---|---|
| Naphthalene | 2 | Fathead minnows | Growth reduction | 2B |
| | | Rainbow trout | Growth reduction | |
| | | Pink salmon | Growth reduction | |
| | | European eel | Growth reduction | |
| Acenaphthene | 3 | Fathead minnow | Lethargic behaviour, $LC_{50}$ 1600 µg $l^{-1}$ | 3 |
| | | Channel catfish | $LC_{50}$ 1720 µg $l^{-1}$ | |
| | | Rainbow trout | $LC_{50}$ 670 µg $l^{-1}$ | |
| | | Brown trout | $LC_{50}$ 580 µg $l^{-1}$ | |
| Fluorene | 3 | Bluegill sunfish | Abnormal swimming | 3 |
| | | English sole | Liver lesions | |
| | | Zebrafish | Defect cardiac function | |
| Phenanthrene | 3 | Zebrafish | Defect cardiac function | 3 |
| | | Goldfish | Oxidative stress | |
| | | Golden grey mullet | Oxidative damage | |
| | | Japanese medaka | Prolonged hatching duration | |
| Anthracene | 3 | Bluegill sunfish | Hypoxia, $LC_{50}$ 11.9 µg $l^{-1}$ | 3 |
| | | Fathead minnow | Hatching reduction | |
| | | Milkfish | Oxidative stress, neurotoxicity | |
| Fluoranthene | 4 | Fathead minnow | $LC_{50}$ 12.2 µg $l^{-1}$ | 3 |
| | | Rainbow trout | $LC_{50}$ 7.7 µg $l^{-1}$ | |
| | | Bluegill sunfish | $LC_{50}$ 12.3 µg $l^{-1}$ | |
| | | Sheephead minnow | $LC_{50}$ 159 µg $l^{-1}$ | |
| | | Inland silverside | $LC_{50}$ 30 µg $l^{-1}$ | |
| | | Winter flounder | $LC_{50}$ 0.1 µg $l^{-1}$ | |
| Pyrene | 4 | Zebrafish | Defect cardiac function | 3 |
| | | Fathead minnow | Photoinduced toxicity | |
| | | Common goby | Inhibition of AChE activity | |
| Benzo[*a*]pyrene | 5 | Milkfish | Increase in lipid peroxidation and catalase activity (oxidative stress, inhibition of AChE activity) | 1 |
| | | Zebrafish | Reduction in egg count, increase P450 expression | |
| | | Japanese medaka | Liver cell lesions, carcinogenic effects | |
| | | Killifish | Hepatocellular carcinomas | |
| Benzo[*k*] fluoranthene | 5 | Northern pike | DNA adduct formation | 2B |
| | | Zebrafish | Induction of P450 | |
| | | Crucian carp | Increased EROD activity | |

2013). Pyrogenic PAHs are formed by the incomplete combustion of organic material, while petrogenic PAHs are present in fossil fuel products. Sources of pyrogenic PAHs are forest fires, incomplete combustion of fossil fuels and tobacco smoke (Lang *et al.*, 1962, 1964; Wakeham *et al.*, 1980). Petrogenic PAHs are present in crude oil (Lauhglin and Neff, 1979; Harvey, 1996; Feng *et al.*, 2009) and coal (Harvey, 1996). In coastal areas, PAHs enter the water primarily from sewage, runoff from roads (Durand *et al.*, 2004), the smelter industry (Beyer *et al.*, 1998; Næs and Oug, 1998) and oil spills (Redondo and Platonov, 2009; Mascarelli, 2010), while offshore PAHs enter the water through oil seeps (Tedesco, 1985), oil spills (Mascarelli, 2010) and produced water discharge from offshore oil installations (Røe Utvik, 1999). PAHs are not normally chemically synthesized for industrial purposes; however, numerous intermediates are derived from fossil fuels and used in numerous domestic products as well as pharmaceutical and agricultural agents (Kaminski *et al.*, 2008).

Anthropogenic sources of PAHs are of concern since they can be accidentally released into the environment and cause significant damage. A recent case study is the acute PAH contamination of the Gulf of Mexico, caused by the *Deep Water Horizon* oil spill in May 2010. This oil spill covered more than 11,200 $km^2$ on the ocean surface, contaminating more than 2000 km of shoreline in five US states. Beyer *et al.* (2016) recently reviewed the environmental effects of this event, focusing on biological effects on various animals and on seafood contamination. The oil caused a number of adverse effects in a wide range of organisms (e.g. invertebrates, fish, birds and sea mammals), impairing organism growth, reproduction and physiological health and in some cases causing death. The enormous amount of data that has been collected since the accident occurred is helping the scientific community and authorities to better understand the short- and long-term effects of PAH contamination, as well as the efficiency of current techniques for removal of PAHs and the possibilities for remediation. Other aspects of this and the *Exxon Valdez* oil spill are reviewed in Chapters 20 to 22 of this volume.

Levels of PAHs in sediment and water have been quantified around the world (Adeniji *et al.*, 2018). In general, PAH pollution seems more pronounced in Asia and Africa, with concentrations higher in the sediment, intermediate in biota and very low in the water column (CCME, 2008). Their fate in sediments is due to their tendency to adsorb organic components of particulate matter and precipitate into benthic environments. The Mussel Watch Program set up in 1986 in the USS by the National Oceanic and Atmospheric Administration (NOAA) has been used for decades to assess exposure in biota. An example where water concentrations may be relatively high is specific point-source discharges such as those observed in produced water from an oil platform. PAH concentrations around the discharge point in the water column is sufficiently high to be detected in oceanic regions (Hylland *et al.*, 2008; Bakke *et al.*, 2013).

There are many individual regulations aimed at limiting the occurrence of PAHs in aquatic environments and there is also extensive regulation when it comes to standards in the working environment. However, limited regulation exists for oceanic concentrations of PAHs. Difficulties in quantifying exposure and source of contamination significantly impair risk analyses in environmental media. Although the EPA has established ambient water quality criteria to provide effect thresholds to protect aquatic organisms, the overall exposure through aquatic media is difficult to measure and predict. To effectively assess the risks of PAH contamination in aquatic ecosystems, the development of analytical tools and technology for rapid, ultrasensitive, highly specific and low-cost detection of PAH mixtures is essential (Behera *et al.*, 2018).

## 10.4 Human Exposure to PAHs

The major route of PH exposure to the general human population is from breathing ambient and indoor air, eating food containing PAHs, smoking cigarettes, or breathing smoke from open fireplaces (ACGIH, 2005). With regard to human exposures from the aquatic environment, PAH exposure occurs primarily via ingestion (i.e. consumption of contaminated food) and through dermal contact (Beyer *et al.*, 2010; Sun *et al.*, 2016; Puri *et al.*, 2017).

Contaminated food from the aquatic environment and some products such as meats cooked at high temperature are probably the major sources of human exposures (Chen *et al.*, 1997; Menichini and Bocca, 2003). The average levels in non-cooked products range between 0.01 and 1 $\mu g/kg^{-1}$. However, in smoked or processed foods, levels as high as 200 $\mu g/kg^{-1}$ have been observed (Standing Committee on Foodstuffs, 2001). Some aquatic organisms are particularly prone to bioaccumulation of PAHs and also serve as vectors for exposure to humans. Filter-feeding bivalves, such as mussels and oysters, may accumulate PAHs due to their feeding strategy and their low capability in metabolizing these compounds. Their consumption as food represents a potentially higher risk in comparison with other marine products.

Fish is an important source of protein for the human diet, providing about 20% of the global intake of animal protein and almost 7% of all protein (FAO, 2016). Thus, contaminated fish is an important source for human exposure to PAHs. In addition to basic bioaccumulation within some species, biomagnification of PAHs through the aquatic food chain needs to be taken into account with regard to human exposure and has started to attract the attention of authorities responsible for risk assessment of PAHs in humans.

As in other organisms, the effects in humans depend on many factors: duration and rate of exposure, concentration, and toxicity of the specific compound or mixture. Other aspects influencing the potential effects of PAHs include pre-exposure health status, age, presence of other pollutants and social habits (ATSDR, 1995). It is difficult to perform epidemiological studies related to exposure from the aquatic environment, therefore most of the toxicological information and risk evaluations are performed using extrapolation models to estimate risk factors.

A good approach to evaluate the risks of PAHs associated with food consumption is to use per capita consumption rates and compare the measured concentrations with the maximum allowable level of benzo[a]pyrene (BaP) in fish and mussels, with subsequent application of TEFs for the measured compounds besides BaP (EPA, 1996; WHO, 1998; European Commission, 2005). Using this approach, risks of carcinogenicity can be quantified (EPA, 2001). Barhoumi *et al.* (2016) used this approach for the first time to evaluate the risk of seafood consumption from Tunisian coastal waters, concluding that the risk related to consumption of mussels and fish from the area was of very little concern. A similar approach was used for risk estimates for cancer from the consumption of fish from a lake in China, using TEFs for PAHs (Zhao *et al.*, 2014).

Regarding model organisms for monitoring of PAH contamination in the aquatic environment in relation to human health, Table 10.1 reports numerous species that have been used as model organisms for laboratory exposures and also field studies. In addition to fish, bivalves are considered highly relevant due to their important role in the food chain. In particular, mussels have been used for environmental monitoring worldwide and they can serve both purposes: monitoring of human health exposure (in term of food contamination) and environmental health (being used as bioindicators) (Beyer *et al.*, 2017).

### 10.4.1 New methods of PAH measurement

Methods for the determination of PAH concentrations in the environment have undergone significant advances, particularly regarding mass spectroscopy (MS) instrumentation. Reviews are available providing information regarding the most commonly used approaches (Kumar *et al.*, 2017; Adeniji *et al.*, 2018). High-performance liquid chromatography (HPLC) coupled with UV/fluorescence detector or photodiode-array and gas chromatography (GC) combined with MS or flame ionization have been the preferred methods for years. They allow the quantification of PAHs present in biological material with high sensitivity. PAHs and their metabolites have very characteristic ultraviolet (UV) absorbance and most of them are fluorescent (i.e. emitting specific wavelengths of light when they are excited). These properties are commonly used for screening PAH metabolites in fish bile collected in field surveys (van der Oost *et al.*, 2003).

The development of MS techniques and the increased sensitivity of instruments is making possible the determination of other 'unknown' compounds per analysis, without the use of standards. This approach allows for tracking of contamination sources, a very important step in monitoring and managing aquatic resources (Pampanin *et al.*, 2014; Enerstvedt *et al.*, 2018). PAH fingerprint determination is regarded as the best available method for this purpose and is becoming more popular in research. For example, the analytical field of conducting oil fingerprinting is now referred to as petroleomics (Sydnes, 2017).

In general, the presented methods are time consuming, the cost per analysis is quite high, and in some cases the use of instrumentations requires highly trained personnel. These practical limitations have encouraged researchers to develop new analytical techniques for real-time evaluation of PAHs. Biosensors, in particular, are considered a very good alternative to the traditional methods (Behera *et al.*, 2018). PAH biosensors have been developed using recognition element enzyme-linked immunosorbent assay (ELISA) (Scharnweber *et al.*, 2001; Moore *et al.*, 2004; Matschulat *et al.*, 2005; Meng *et al.*, 2015), electrochemical and surface plasmon resonance (SPR) biosensors (Gobi *et al.*, 2003; Lin *et al.*,

2012; Li *et al.*, 2016), fluorescence polarization immunoassays (FPIA) (Yu Goryacheva *et al.*, 2007; Meimaridou *et al.*, 2010), whole cell bacterial sensors (Chang *et al.*, 2004; Cho *et al.*, 2014) and DNA-based biosensors (Doong *et al.*, 2005; Ni *et al.*, 2014). The use of these new sensors still needs validation and intercalibration with traditional methods before applying these techniques in environmental monitoring or evaluation of food contamination.

One example of a PAH biosensor used in PAH monitoring is the ELISA base immunoassay PAH RISc® (EnSys, Inc,), a commercially available tool used for the quantification of these contaminants in soil samples. The EPA has noted that this assay is suitable for monitoring PAHs in sediment samples (Behera *et al.*, 2018).

## 10.5 Bioassays

The presence of xenobiotic compounds in an aquatic ecosystem does not, by itself, imply harmful effects. Therefore, connections need to be established between contaminant levels in the environment, levels of tissue contamination in organisms and biologically adverse effects. A bioassay is usually defined as an estimation of concentration or potency of chemical(s) by means of measuring the magnitude of the response over a suitable biological system under standard set of conditions. Bioassays are used predict the adverse biological effects of PAHs by using a representative organism to estimate effects in the field, and short-term acute toxicity tests are the most commonly used (Blasco and Pico, 2009). Somewhat labour-intensive aquatic toxicity bioassays are usually conducted over a relatively short duration but tend to be less expensive than other approaches. Hader and Erzinger (2018) published a book entitled *Bioassays: Advanced Methods and Applications*, which provides a thorough understanding of the applications of bioassays in monitoring toxicity in aquatic ecosystems. Herein, the newest tests and applications in discovering compounds and toxins in the aquatic environment were reviewed, covering all suitable organisms from bacteria to vertebrates.

Overviews of other commercially available bioassays for PAH assessment in aqueous samples are also available in the literature (Kokkali and van Delft, 2014; Azizullah and Hader, 2018). The choice of the most appropriate bioassay depends on the user needs and the available facilities. The Organization for Economic Cooperation and Development (OECD) also provides guidelines for testing of chemicals, which are periodically reviewed in the light of scientific progress. Established OECD bioassay toxicity tests are usually performed for both ecotoxicological evaluation of chemicals and risk assessment.

In this context, the use of fish embryos as an alternative to adult animal testing has been highly promoted. In 2008, the International Life Science Institute, the Health and Environmental Science Institute and the European Centre for Ecotoxicology and Toxicology of Chemicals held a workshop about the application of the fish embryo test as an animal alternative method in hazard and risk assessment and scientific research. The workshop included scientists and authority representatives from North America, Europe and Asia and the outcome of this activity was in general agreement about the development and validation of fish embryo toxicity tests (Embry *et al.*, 2010).

Zooplankton and phytoplankton tend to be sensitive to PAH toxicity with thresholds in the micrograms per litre ($\mu g\ l^{-1}$) range (Othman *et al.*, 2018). Benzo[a]pyrene, for example, has $LC_{50}$ values (lethal concentration that kills 50% of the individuals) of 5 $\mu g\ l^{-1}$ and 58 $\mu g\ l^{-1}$ for *Daphnia pulex* and *Eurytemora affinis*, respectively (Ikenaka *et al.*, 2013). A summary of the toxicity in zooplankton and phytoplankton species can be found in the review of Behera *et al.* (2018). Some of these organisms are also used in standard toxicity tests to assess the discharge of PAHs in oil and gas offshore activities.

Smaller organisms are frequently used, due to their short life cycle and ease of manipulation in the laboratory. Copepods are aquatic invertebrates that have been successfully applied as test organisms for various chemicals (Raisuddin *et al.*, 2007; Dahms *et al.*, 2016). They are key elements of the food web and take part in the transfer of chemicals in the trophic chain, which makes them highly ecologically relevant species. Both whole organism bioassays and molecular markers (e.g. gene expression studies) have demonstrated excellent capabilities to evaluate the impact of PAHs (Won *et al.*, 2018). Several bioassays have been developed to evaluate PAH toxicity in organisms at multiple trophic levels, e.g. algae (producers), invertebrates and vertebrates (consumers), and bacteria (decomposers). Bacterial assays tend to be simple, fast and relatively inexpensive and have shown good correlations with other toxicity tests. For example, the bioluminescent *Vibrio fisheri* assay has been effectively used to estimate the toxicity of sediment samples, both as solid matter and as elutriates (Jarque *et al.*, 2016).

However, bioassays are not always sensitive, or inexpensive. Consequently, other approaches have become preferable (Fent, 2003; van der Oost, 2003). Moreover, bioassays should be used in combination with other tools (e.g. biomarkers) for specific focus on PAH sub-lethal effects.

## 10.6 Biomarkers

Exposure levels, fate and effects of PAHs in the aquatic ecosystem have been extensively studied by ecotoxicologists. Due to their lipophilicity, PAH bioavailability after entering an organism is substantial. PAHs are prone to multi-step metabolic activation, which can occur through specific enzymatic reactions. Following epoxidation by cytochrome P450, PAH epoxides can then be conjugated with glutathione or glucuronides, which are regarded as detoxification metabolites. The epoxides that are not conjugated can react with molecules like proteins and DNA, causing severe adverse effects. Most metabolites are, however, excreted in faeces and urine, allowing the organism to survive (Walker *et al.*, 2012). Short- and long-term health effects together with a detailed description of the major biological effects have been reviewed by many (see, for example, the review of Abdel-Shafy and Mansour, 2016). Most of the sub-lethal problems related to exposure to PAHs is caused by the oxidized metabolites that organisms generate *in vivo* in order to excrete the unwanted compounds. The metabolites are much more reactive and therefore also more toxic than their parent compounds; however, few studies have characterized the oxygenated compounds in the environment or biota.

Biomarkers of PAH contamination have been developed at different levels of biological organization (Franco *et al.*, 2008; Pampanin, 2017). The most common include: P450 induction (both at gene or protein levels) primarily in vertebrate species; oxidative stress parameters (e.g. oxidative stress enzymes like catalase, production of ROS); lipid peroxidation (e.g. malondyaldehyde assay); DNA damage (e.g. DNA adducts); and tissue damage (e.g. liver or gill histopathological conditions). Examples of specific methods are as follows:

- PAH metabolites in bile/urine, which represent the internal dose within an organism after short-term exposure (maximum few days) and can be evaluated using screening methods such fluorescence wavelength analysis, or more sensitive methods like GC–MS or LC MS/MS (Beyer *et al.*, 2010).
- DNA adducts, which also represent an effective dose of PAH capable of reacting with genetic material, causing genotoxicity and potentially carcinogenicity. Adducts are measured using highly sensitive radioactive techniques (i.e. $^{32}P$-postlabelling) but more recently by the use of tandem MS (MS/MS) applications (Pampanin *et al.*, 2017).
- Cytogenetic alterations (e.g. chromosomal aberration, micronuclei induction), which are biomarkers of early effects correlated with cancer risk in epidemiological studies and exposure studies using marine organisms (Yang *et al.*, 2010).
- Histopathological alterations, which can be linked with carcinogenicity (Loughery *et al.*, 2018a, b; Wolf and Wheeler, 2018).

## 10.7 Species Sensitivity Distribution and Biomarker Bridge

The species sensitivity distribution (SSD) model has been widely used by decision makers to derive thresholds for ecological risk assessment and water quality criteria (Posthuma *et al.*, 2002; He *et al.*, 2015; Del Signore *et al.*, 2016). A recent trilogy of articles reported the use of biomarkers as environmental risk indicators for PAH contamination. Sanni *et al.* (2016a, 2016b, 2017) established a link between biomonitoring and risk assessment procedures by using biomarker species sensitivity distributions. The authors have been bridging the gap between biomarker and whole-organism responses related to oil-based offshore discharges. Biomarker-based species sensitivity distributions (SSD biomarkers) have been constructed for relevant groups of biomarkers based on laboratory data from oil exposures. SSD curves express the fraction of species responding to different types of biomarkers. They have been connected to SSDs for whole-organism responses (WORs), in order to relate the SSD biomarkers to animal fitness parameters that are commonly used in environmental risk assessment. The resulting SSD curves show that biomarkers and WORs can be linked, enhancing the capability to monitor field parameters with better correlation to impact and risk assessment criteria and providing improved chemical/biological integration. The environmental risk can therefore be expressed in terms of biomarker responses and *vice versa*. The quantitative biomarker information can form the statistical basis for bridging diagnosis with prognosis, as well as monitoring with risk prediction.

## 10.8 Adverse Outcome Pathway

Another method used to link sub-lethal effects to apical end-points useful for threshold determination in biological systems is the adverse outcome pathway (AOP) paradigm. The term adverse outcome pathway is defined as 'the sequence of events from the exposure of individual/population to a chemical(s) through a final adverse toxic effect at individual level (from a human health prospective) or population level (from an environmental prospective)' (Fig. 10.1) (Ankley *et al.*, 2010).

This concept has been introduced to overcome some of the limitations of biomarkers, like the low relevance to adverse outcomes used in threshold assessments such as reproduction and survival. General guidelines have been proposed and vetted through the EPA, as well as OECD (Villaneuve *et al.*, 2014). AOPs utilize a host of conceptual methods such as Bradford–Hill plausibility coupled with real-time updates and database additions through an AOPwiki. AOPs begin with a molecular initiating event (MIE) which is subsequently linked to key events (KE) through KE relationships (KERs). Many of the MIEs can be used as biomarkers for KE thresholds for risk assessments. Moreover, the paradigm may also be used quantitatively to predict impacts of chemicals through KERs. KERs may also be established using high-throughput *in silico* and *in vitro* methods. In addition to their use in the identification of data gaps for linkages between biomarkers and apical effects in the problem formulation stage of risk assessments, AOPs may also be used in weight of evidence (WoE) discussions in risk characterization stages to help to determine causal relationships between stressors and effects as well as uncertainties associated with risk estimates. The utility of the AOPs in biomarker-based environmental risk assessment has been demonstrated (Lee *et al.*, 2015) and provides a way forward in the evaluation of ecotoxicity of PAH as single compounds, mixtures of PAHs or cocktails of multiple classes of contaminants.

## 10.9 Ecological Risk Assessment of PAHs

Ecological risk assessment of contaminants can be divided in four components: (i) problem formulation; (ii) exposure assessment; (iii) effects assessment (using biomarkers); and (iv) risk characterization (uncertainty analyses with

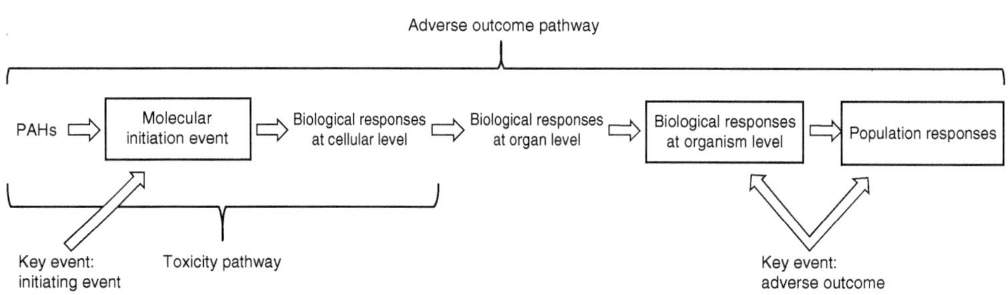


**Fig. 10.1.** Adverse outcome pathway scheme. Horizontal arrows represent key event relationships, modified from Ankley *et al.* (2010). PAHs, polycyclic aromatic hydrocarbons.

WoE analyses). When using ecotoxicological tests, the risk is expressed as no observed effect concentration (NOEC) values and no effect concentration (NEC). When mathematical models are used to predict these values, the threshold concentration is expressed as a predicted no-effect concentration (PNEC) and the exposure of the compound expressed as a predicted environmental concentration (PEC) (Riva *et al.*, 2019). In recent years, additional efforts have been directed towards improving risk assessment methodologies aiming to consider mechanisms of toxicity. In the field of human risk assessment, the EPA and the World Health Organization (WHO) have been focusing on the mode of action (MoA) of chemicals in WoE analyses after hazard has been identified. The MoA framework promotes the evaluation of both toxico kinetics (TK) and toxico dynamic (TD) processes at various levels of biological organization (EFSA, 2014). In environmental risk assessment, various MoA classifications have been developed and more details can be found in the work of Kienzler *et al.* (2017). Recently, the European Food Safety Authority (EFSA) published a review on methods for human hazard assessment of chemicals focusing on TK and TD process data.

The main scientific association concerning this area is the Society of Environmental Toxicology and Chemistry (SETAC), which promotes scientific research related to contaminants in the environment and the use of the scientific information in environmental policy and decision making (https://www.setac.org/).

In general, the combination of chemical analysis of PAHs in water/sediment/biota and biological analysis of PAH effects is the only way to assess the ecological risk posed by PAHs. This type of approach is considered an essential tool for the complete implementation of efficient regulations and environmental monitoring, like the Water Framework Directive of the European Union (European Commission, 2000; Lepom *et al.*, 2009).

## 10.10 PAH Interactions with Other Stressors (Microplastics)

Due to co-occurrence of multiple chemical stressors in the aquatic environment, research is moving towards new strategies for evaluating the effects of mixtures (Beyer *et al.*, 2014). Assessing one single chemical at a time might underestimate the environmental impact of the substance on the single organisms, the population and possibly the ecosystem. Two approaches are currently in use to overcome this challenge: (i) toxicity can be measured as the whole mixture (which has the disadvantage of not determining the stressor(s) required for monitoring); or (ii) component-based (which can use effects coupled with chromatography to identify the cause).

Co-exposure with other chemicals can influence the toxicity of PAHs (Billiard *et al.*, 2008). Non-additive toxicity, with 'more-than' or 'less-than' additive effects, may occur (Gauthier *et al.*, 2014). Due to similarities in the toxic mechanism of PAHs, 'more-than' addictive effects have been hypothesized (with about 45% of the studied cases confirming the hypothesis) as the major mechanisms involving ion regulatory dysfunction and reactive oxygen species (ROS) imbalance. However, due to the limited available knowledge, the co-toxicity evaluation requires case-by-case evaluations.

Microplastics (MPs) and plastic debris have been receiving great attention in recent years, due to a general concern about their effects in aquatic ecosystems, mainly as vectors/transporters for hydrophobic contaminants. Plastic debris in the oceans contained measurable concentrations of multiple organic pollutants (Frias *et al.*, 2010; Hirai *et al.*, 2011; Fisner *et al.*, 2013). However, the interaction between MPs/plastic debris and organic pollutants has not been adequately evaluated (Batel *et al.*, 2018).

In one study, the partitioning of phenanthrene within three difference plastic particles (polyethylene, polystyrene and polyvinylchloride) was compared with natural sediment (used as a control). MPs had higher capacities for phenanthrene adsorption and released larger amounts of the sorbed phenanthrene into the water compared with natural sediment (Wang and Wang, 2018). Other studies have shown that MPs reduce the effects of PAHs, by sequestering them and making them less bioavailable (Diepens and Koelmans, 2018; Kleinteich *et al.*, 2018). Clearly, additional study is needed to better understand the interactions of MP with PAHs.

## 10.11 Conclusions and Recommendations

Even though the fate and effects of PAHs have been evaluated for more than 50 years, some knowledge gaps have been identified in this chapter and suggestions regarding future research focus have been provided. For effective management of PAH contamination, the development of new analytical tools, preferably capable of providing highly sensitive and specific data in real time, is necessary. Bioassays provide useful information regarding the ecotoxicity of PAHs but more cost-efficient methods with higher throughput are desirable. The development of AOPs, identifying biomarkers for predictive assessments of risk, for both individual PAHs and complex mixtures is needed, since the aquatic environment tends to be the final and major receiver of contaminants. Using this approach, uncertainties may be diminished for environmental risk assessment and management of aquatic resources.

Finally, the development in MS instrumentation will allows PAH fingerprinting with the potential for identification of unknown compounds as well as the potential sources of contamination.

## References

Abdel-Shafy, H.I. and Mansour, M.S.M. (2016) A review on polycyclic aromatic hydrocarbons: source, environmental impact, effect on human health and remediation. *Egyptian Journal of Petroleum* 25, 107–123.

ACGIH (2005) *Polycyclic Aromatic Hydrocarbons (PAHs) Biologic Exposure Indices (BEI)*. American Conference of Governmental Industrial Hygienists, Cincinnati, Ohio.

Adeniji, A.O., Okon, O.O. and Okoh, A.I. (2018) Analytical methods for polycyclic hydrocarbons and their global trend of distribution in water and sediment: a review. In: Zoveidavianpoor, M. (ed.) *Recent Insights in Petroleum Science and Engineering*. InTech, Rijeka, Croatia, pp.393–426.

Alegbeleye, O.O., Opeolu, B.O. and Jackson, V.A. (2017) Polycyclic aromatic hydrocarbons: a critical review of environmental occurrence and bioremediation. *Environmental Management* 60, 758–783.

Ankley, G.T., Bennett, R.S., Erickson, R.J., Hoff, D.J., Hornung, M.W. and Johnson, R.D. (2010) Adverse outcome pathways: a conceptual framework to support ecotoxicology research and risk assessment. *Environmental Toxicology and Chemistry* 29, 730–741.

ATSDR (1995) *Toxic Substances –Polycyclic Aromatic Hydrocarbons (PAHs)*. Agency for Toxic Substances and Disease Registry, Atlanta, Georgia.

Azizullah, A. and Hader, D.P. (2018) A comparison of commonly used and commercially available bioassays for aquatic ecosystems. In: Hader, D.P. and Erzinger, G.S. (eds) *Bioassay*. Elsevier Inc., Atlanta, Georgia, pp. 347–368

Bakke, T., Klungsøyr, J. and Sanni, S. (2013) Environmental impacts of produced water and drilling waste discharges from the Norwegian offshore petroleum industry. *Marine Environmental Research* 92, 154–169.

Barhoumi, B., Megdiche, Y.E., Clérandeau, C., Ameur, W.B., Mekni, S. *et al*. (2016) Occurrence of polycyclic aromatic hydrocarbons (PAHs) in mussel (*Mytilus galloprovincialis*) and eel (*Anguilla anguilla*) from Bizerte lagoon, Tunisia, and associated human health risk assessment. *Continental Shelf Research* 124, 104–116.

Batel, A., Borchert, F., Reinwald, H., Erdinger, L. and Braunbeck, T. (2018) Microplastic accumulation patterns and transfer of benzo[a]pyrene to adult zebrafish (*Danio rerio*) gills and zebrafish embryos. *Environmental Pollution* 235, 918–930.

Behera, B.K., Das, A., Sarkar, D.J., Weeranthunge, P., Parida, P.K. *et al*. (2018) Polycyclic aromatic hydrocarbons (PAHs) in inland aquatic ecosystems: perils and remedies through biosensors and bioremediation. *Environmental Pollution* 241, 212–233.

Beyer, J., Aas, E., Borgenvik, H.K. and Ravn, P. (1998) Bioavailability of PAH in effluent water from an aluminium works evaluated by transplant caging and biliary fluorescence measurements of Atlantic cod (*Gadus morhua* L.). *Marine Environmental Research* 46, 233–236.

Beyer, J., Jonsson, G., Porte, C., Krahn, M.M. and Ariese, F. (2010) Analytical methods for determining metabolites of polycyclic aromatic hydrocarbon (PAH) pollutants in fish bile: a review. *Environmental Toxicology and Pharmacology* 30, 224–244.

Beyer, J., Peterson, K., Song, Y., Ruus, A., Grung, M., Bakke, T. and Tollefsen, K.E. (2014) Environmental risk assessment of combine effects in aquatic ecotoxicology: a discussion paper. *Marine Environmental Research* 96, 81–91.

Beyer, J., Trannum, H.C., Bakke, T., Hodson, P.V. and Collier, T. (2016) Environmental effects of the Deep Horizon oil spill: a review. *Marine Pollution Bulletin* 110, 28–51.

Beyer, J., Green, N.W., Brooks, S., Allan, I.J., Ruus, A. *et al.* (2017) Blue mussels (*Mytilus edulis* spp.) as sentinel organisms in coastal pollution monitoring: a review. *Marine Environmental Research* 130, 338–365.

Billiard, S., Meyer, J.N., Wassenberg, D.M., Hodson, P.V. and Di Giulio, R.T (2008) Nonoadditive effects of PAHs on early vertebrate development: mechanisms and implications for risk assessment. *Toxicology Review* 105, 5–23.

Blasco, C. and Pico, Y. (2009) Prospects for combining chemical and biological methods for integrated environmental assessment. *Trends in Analytical Chemistry* 28, 745–757.

CCME (2008) Canadian soil quality guidelines: carcinogenic and other polycyclic aromatic hydrocarbons (PAHs) (environmental and human health effects). Canadian Council of Ministers of the Environment, Ottawa, Canada.

Chang, S.T., Lee, H.J. and Gu, M.B. (2004) Enhancement in the sensitivity of an immobilized cell-based soil biosensor for monitoring PAH toxicity. *Sensor Actuator B Chemistry* 97, 272–276.

Chen, B.H. and Lin, Y.S. (1997) Formation of polycyclic aromatic hydrocarbons during processing of duck meat. *Journal of Agriculture Food Chemistry* 45, 1394–1403.

Cherr, G.N., Fairbairn, E. and Whitehead, A. (2017) Impacts of petroleum-derived pollutants on fish development. *Annual Review in Animal Bioscience* 5, 185–203.

Cho, J.H., Lee, D.Y., Lim, W.K. and Shin, H.J. (2014) A recombinant *Escherichia coli* biosensor for detecting polycyclic aromatic hydrocarbons in gas and aqueous phases. *Preparation in Biochemistry and Biotechnology* 44, 849–860.

Dahms, H.U., Won, E.J., Kim, H.S., Han, J., Park, H.G. *et al.* (2016) Potential of the small cyclopoid copepod *Paracyclopina nana* as an invertebrate model for ecotoxicity testing. *Aquatic Toxicology* 180, 282–294.

Del Signore, A., Hendriks, A.J., Lenders, H.J.R., Leuven, R.S.E.W. and Breure, A.M. (2016) Development and application of the SSD approach in scientific case studies for ecological risk assessment. *Environmental Toxicology and Chemistry* 35, 2149–2161.

Diepens, N.J. and Koelmans, A.A. (2018) Accumulation of plastic debris and associated contaminants in aquatic food webs. *Environmental Science and Technology* 52, 8510–8520.

Doong, R.A., Shih, H.M. and Lee, S.H. (2005) Solegel-derived array DNA biosensor for the detection of polycyclic aromatic hydrocarbons in water and biological samples. *Sensor Actuator B Chemistry* 111–112, 323–330.

Durand, C., Ruban, V., Amblès, A. and Oudot, J. (2004) Characterization of the organic matter of sludge: determination of lipids, hydrocarbons and PAHs from road retention/infiltration ponds in France. *Environmental Pollution* 132, 375–384.

EFSA (2014) Modern Methodologies and Tools for Human Hazard Assessment of Chemicals. 12. *EFSA Journal* 12(4), 3638.

Embry, M.R., Belanger, S.E., Braunbeck, T.A., Galay-Burgos, M. and Halder, M. (2010) The fish embryo toxicity test as an animal alternative method in hazard and risk assessment and scientific research. *Aquatic Toxicology* 97, 79–87.

Enerstvedt, K.S., Sydnes, M.O. and Pampanin, D.M. (2018) Screening for protein adducts of naphthalene and chrysene in plasma of exposed Atlantic cod (*Gadus morhua*). *Chemosphere* 200, 67–79.

EPA (1996) *Proposed Guidelines for Carcinogenic Risk Assessment*. EPA/600/P-92/003C. US Environment Protection Agency, Washington, DC.

EPA (2001) *Integrated Risk Information System (IRIS)*. National Center for Environmental Assessment, Office of Research and Development, Washington, DC. (Office of Solid Waste and Emergency Response: http://www.epa.gov/iris/)

EPA (2008) *Polycyclic Aromatic Hydrocarbons (PAHs) – EPA Fact Sheet*. National Center for Environmental Assessment, Office of Research and Development, Washington, DC.

European Commission (2000) Directive 2000/60/EC of the European Parliament and of the Council establishing a framework for the Community action in the field of water policy. *Official Journal of the European Community* L 327.

European Commission (2005) Commission regulation (EC) No. 208/2005 of 4 February 2005 amending Regulation (EC) No 466/2001 as regards polycyclic aromatic hydrocarbons. *Official Journal of the European Union* 34, 3–5.

FAO (2016) *The State of World Fisheries and Aquaculture 2016*. FAO Fisheries & Aquaculture Department, Rome.
Feng, X., Pisula, W. and Müllen, K. (2009) Large polycyclic aromatic hydrocarbons: Synthesis and discotic organization. *Pure Applied Chemistry* 81, 2203–2224.
Fent, K. (2003) Ecotoxicological problems associated with contaminated sites. *Toxicology Letters* 140–141, 353–365.
Fisner, M., Taniguchi, S., Majer, A.P., Bícego, M.C. and Turra, A. (2013) Concentration and composition of polycyclic aromatic hydrocarbons (PAHs) in plastic pellets: implications for small-scale diagnostic and environmental monitoring. *Marine Pollution Bulletin* 76, 349–354.
Franco, S.S., Nardocci, A.C. and Gunther, W.M.R (2008) PAH biomarkers for human health risk assessment: a review of the state-of-the-art. *Cadernos de Saúde Pública* 24, 569–580.
Frias, J.P.G.L., Sobral, P. and Ferreira, A.M. (2010) Organic pollutants in microplastics from two beaches of the Portuguese coast. *Marine Pollution Bulletin* 60, 1988–1992.
Gao, D., Lin, J., Ou, K., Chen, Y., Li, H., Dai, Q., Yu, Z., Zuo, Z. and Wang, C. (2018) Embryonic exposure to benzo(a)pyrene inhibits reproductive capability in adult female zebrafish and correlation with DNA methylation. *Environmental Pollution* 240, 403–411.
Gauthier, P.T., Norwood, W.P., Prepas, E.E. and Pyle, G.G. (2014) Metal–PAH mixture in the aquatic environment: a review of co-toxic mechanisms leading to more-than-additive outcomes. *Aquatic Toxicology* 154, 253–269.
Gobi, K.V., Sasaki, M., Shoyama, Y. and Miura, N. (2003) Highly sensitive detection of polycyclic aromatic hydrocarbons (PAHs) and association constants of the interaction between PAHs and antibodies using surface plasmon resonance immunosensor. *Sensor Actuator B Chemistry* 89, 137–143.
Hader, D.P. and Erzinger, G.S. (eds) (2018) *Bioassays*. Elsevier Inc., Atlanta, Georgia.
Hahn, M.E. and Stegeman, J.J. (1994) Regulation of cytochrome P4501A1 in teleosts: sustained induction of CYP1A1 mRNA, protein, and catalytic activity by 2,3,7,8-tetrachlorodibenzofuran in the marine fish *Stenotomus chrysops*. *Toxicology and Applied Pharmacology* 127, 187–198.
Harvey, R.G. (1996) *Polycyclic Aromatic Hydrocarbons*. Wiley-VCH, New York.
He, W., Xu, F.L., Qin, N. and Kong, X.Z. (2015) Development of species sensitivity distribution (SSD) models for setting up the management priority with water quality criteria of toxic chemicals. *Development in Environmental Modelling* 27, 163–187.
Hirai, H., Takada, H., Ogata, Y., Yamashita, R., Mizukawa, K. *et al.* (2011) Organic micropollutants in marine plastics debris from the open ocean and remote and urban beaches. *Marine Pollution Bulletin* 62, 1683–1692.
Hylland, K., Tollefsen, K.E., Ruus, A., Jonsson, G., Sundt, R.C. *et al.* (2008) Water column monitoring near oil installations in the North Sea 2001–2004. *Marine Pollution Bulletin* 56, 414–429.
Ikenaka, Y., Sakamoto, M., Nagata, T., Takahashi, H., Miyabara, Y. *et al.* (2013) Effects of polycyclic aromatic hydrocarbons (PAHs) on an aquatic ecosystem: acute toxicity and community-level toxic impact tests of benzo [a] pyrene using lake zooplankton community. *Journal of Toxicology Sciences* 38, 131–136.
Incardona, J.P. and Scholz, N.L. (2016) The influence of heart developmental anatomy on cardiotoxicity-based adverse outcome pathways in fish. *Aquatic Toxicology* 177, 515–525.
Jarque, S., Masner, P., Klanova, J., Prokes, R. and Blaha, L. (2016) Bioluminescent *Vibrio fisheri* assays in the assessment of seasonal and spatial patterns in toxicity of contaminated river sediments. *Frontier in Microbiology* 7, article 1738.
Kaminski, N.E., Faubert Kaplan, B.L. and Holsapple, M.P. (2008) Toxic responses of the immune system. In: Klaassen, C.D. (ed) Casarett & Doull's *Toxicology: The Basic Science of Poisons*. Mc-Graw Hill, Inc, New York, pp. 485–555.
Kienzler, A., Barron, M.G., Belanger, S.E., Beasley, A. and Embry, M.R. (2017). Mode of action (MoA) assignment classifications for ecotoxicology: an evaluation of approaches. *Environmental Science and Technology* 51, 10203–10211.
Kleinteich, J., Seidensticker, S., Marggrander, N. and Zarfl, C. (2018) Microplastics reduce short-term effects of environmental contaminants. Part II: polyethylene particles decrease the effect of polycyclic aromatic hydrocarbons on microorganisms. *International Journal of Environmental Research and Public Health* 15, 287.
Kokkali, V. and van Delft, W. (2014) Overview of commercially available bioassays for assessing chemical toxicity in aqueous samples. *Trends in Analytical Chemistry* 61, 133–155.
Kumar, S., Negi, S. and Maiti, P. (2017) Biological and analytical techniques used for detection of polyaromatic hydrocarbons. *Environmental Science and Pollution Research* 24, 25810–25827.

Lang, K.F., Buffleb, H. and Kalowy, J. (1962) 2-Phenyl-phenanthren und binaphthyl-(2,2') aussteinkohlenteer. *Chemische Berichte* 95, 1052–1053.
Lang, K.F., Buffleb, H. and Kalowy, J. (1964) Fulminen (1,2-benzo-picen) im steinkohlenteer. *Chemische Berichte* 97, 494–497.
Laughlin, R.B. and Neff, J.M. (1979) Interactive effects of salinity, temperature and polycyclic aromatic hydrocarbons on the survival and development rate of larvae of the mud crab *Rhithropanopeus harrisii*. *Marine Biology* 53, 281–291.
Lee, J.W., Won, E.J, Raisuddin, S. and Lee, J.S. (2015) Significance of advance outcome pathways in biomarker-based environmental risk assessment in aquatic organisms. *Journal of Environmental Science* 35, 115–127.
Lepom, P., Brown, B., Hanke, G., Loos., R., Quevauviller, P. and Wollgast, J. (2009) Needs for reliable analytical methods for monitoring chemical pollutants in surface water under the European Framework Directive. *Journal of Chromatography A* 1216, 302–315.
Lerda, D. (2011) *Polycyclic Aromatic Hydrocarbons (PAHs) Factsheet*. JRC Technical Notes 66955. Joint Research Centre, European Commission, Geel, Belgium.
Li, X., Kaattari, S.L., Vogelbein, M.A., Vadas, G.G. and Unger, M.A. (2016) A highly sensitive monoclonal antibody based biosensor for quantifying 3-5 ring polycyclicaromatic hydrocarbons (PAHs) in aqueous environmental samples. *Sensor and Biosensing Research* 7, 115–120.
Lin, M., Liu, Y., Sun, Z., Zhang, S., Yang, Z. and Ni, C. (2012) Electrochemical immunoassay of benzo [a] pyrene based on dual amplification strategy of electron-accelerated Fe 3 O 4/polyaniline platform and multi-enzyme-functionalized carbon sphere label. *Analytica Chimica Acta* 722, 100–106.
Loughery, J.R., Kidd, K.A., Mercer, A. and Martyniuk, C.J. (2018a) Part A: Temporal and dose-dependent transcriptional responses in the liver of fathead minnows following short term exposure to the polycyclic aromatic hydrocarbon phenanthrene. *Aquatic Toxicology* 199, 90–102.
Loughery, J.R., Kidd, K.A., Mercer, A. and Martyniuk, C.J. (2018b) Part B: Morphometric and transcriptomic responses to sub-chronic exposure to the polycyclic aromatic hydrocarbon phenanthrene in the fathead minnow (*Pimephales promelas*). *Aquatic Toxicology* 199, 77–89.
Mascarelli, A (2010) After the oil. *Nature* 467, 22–24.
Matschulat, D., Deng, A., Niessner, R. and Knopp, D. (2005) Development of a highly sensitive monoclonal antibody based ELISA for detection of benzo [a] pyrene in potable water. *Analyst* 130, 1078–1086.
Meimaridou, A., Haasnoot, W., Noteboom, L., Mintzas, D., Pulkrabova, J., Hajslová, J. and Nielen, M.W. (2010) Color encoded microbeads-based flow cytometric immunoassay for polycyclic aromatic hydrocarbons in food. *Analytica Chimica Acta* 672, 9–14.
Meng, X.Y., Li, Y.S., Zhou, Y., Zhang, Y.Y., Yang, L., Qiao, B., Wang, N.N., Hu, P., Lu, S.Y., Ren, H.L., Liu, Z.S., Zhang, J.H. and Wang, X.R. (2015) An enzyme-linked immunosorbent assay for detection of pyrene and related polycyclic aromatic hydrocarbons. *Analytical Biochemistry* 473, 1–6.
Menichini, E. and Bocca, B. (2003) Polycyclic aromatic hydrocarbons: an overview. In: Caballero, B., Trugo, L.C. and Finglas, P.M. (eds) *Encyclopaedia of Food Sciences and Nutrition*. Academic Press, Amsterdam.
Moore, E.J., Kreuzer, M.P., Pravda, M. and Guilbault, G.G. (2004) Development of a rapid single-drop analysis biosensor for screening of phenanthrene in water samples. *Electroanalysis* 16, 1653–1659.
Næs, K. and Oug, E. (1998) The distribution and environmental relationships of polycyclic aromatic hydrocarbons (PAHs) in sediments from Norwegian smelter-affected fjords. *Chemosphere* 36, 561–576.
Ni, Y., Wang, P., Song, H., Lin, X. and Kokot, S. (2014) Electrochemical detection of benzo(a)pyrene and related DNA damage using DNA/hemin/nafion-graphene biosensor. *Analytica Chimica Acta* 22, 821.
Othman, H.B., Lanougu, E., Got, P., Hlaili, A.S. and Leboulanger, C. (2018) Structural and functional responses of coastal marine phytoplankton communities to PAH mixtures. *Chemosphere* 209, 908–919.
Pampanin, D.M. (2017) The presence of petrogenic PAHs in the aquatic environment – monitoring studies. In: Pampanin, D.M. and Sydnes, M.O. (eds) *Petrogenic Polycyclic Aromatic Hydrocarbons in the Aquatic Environment: Analysis, Synthesis, Toxicity and Environmental Impact*. Betham Science Publishers, Sharjah, UAR, pp. 18–49.
Pampanin, D.M. and Sydnes, M.O. (2013) Polycyclic aromatic hydrocarbons a constituent of petroleum: presence and influence in the aquatic environment. In: Kutcherov, V. and Kilesnikove, A. (eds) *Hydrocarbons*. InTech, Rijeka, Croatia, pp. 83–118.
Pampanin, D.M. and Sydnes, M.O. (eds) (2017) *Petrogenic Polycyclic Aromatic Hydrocarbons in the Aquatic Environment: Analysis, Synthesis, Toxicity and Environmental Impact*. Betham Science Publishers, Sharjah, UAR.

Pampanin, D.M., Larssen, E., Øysæd, K.B., Sundt, R.C. and Sydnes, M.O. (2014) Study of the bile proteome of Atlantic cod (*Gadus morhua*): multi biological markers of exposure to polycyclic aromatic hydrocarbons. *Marine Environmental Research* 101, 161–168.
Pampanin, D.M., Brooks, S.J., Grøsvik, B.E., Le Goff, J., Meier, S. and Sydnes, M.O. (2017) DNA adducts in marine fish as biological marker of genotoxicity in environmental monitoring: the way forward. *Marine Environmental Research* 125, 49–62.
Posthuma, L., Suter II, G.W. and Traas, T.P. (2002) *Environmental and Ecological Risk Assessment: Species Sensitivity Distributions in Ecotoxicology*. Lewis Publishers, Washington, DC.
Puri, P., Nandar, S.K., Kathuria, S. and Remesh, V. (2017) Effects of air pollution on the skin: a review. *Indian Journal of Dermatology, Venereology and Leprology* 83, 415–423.
Raisuddin, S., Kwok, K.W.H., Leung, K.M.Y., Schlenk, D. and Lee, J.S. (2007) The copepod *Tigriopus*: a promising marine model organism for ecotoxicology and environmental genomics. *Aquatic Toxicology* 83, 161–173.
Redondo, J. and Platonov, A.K. (2009) Self-similar distribution of oil spills in European coastal waters. *Environmental Research Letters* 4, 014008.
Riva, F., Zuccato, E., Davoli, E., Fattore, E. and Castiglioni, S. (2019) Risk assessment of a mixture of emerging contaminants in surface water in a highly urbanized area in Italy. *Journal of Hazardous Material* 361, 103–110.
Røe Utvik, T. (1999) Chemical characterization of produced water from four offshore oil production platforms in the North Sea. *Chemosphere* 39, 2593–2606.
Sanni, S., Björkblom, C., Jonsson, H., Godal, B.F., Liewenborg, B., Lyng, E. and Pampanin, D.M. (2016a) I: Biomarker quantification in fish exposed to crude oil as input to species sensitivity distributions and threshold values for environmental monitoring. *Marine Environmental Research* 125, 10–24.
Sanni, S., Lyng, E., Pampanin, D.M. and Smit, M.G.D. (2016b) II: Species sensitivity distributions based on biomarkers and whole organism responses for integrated impact and risk assessment criteria. *Marine Environmental Research* 127, 11–23.
Sanni, S., Lyng, E. and Pampanin, D.M. (2017) III: Use of biomarkers as Risk Indicators in Environmental Risk Assessment of oil-based discharges offshore. *Marine Environmental Research* 127, 1–10.
Scharnweber, T., Fisher, M., Suchanek, M., Knopp, D. and Niessner, R. (2001) Monoclonal antibody to polycyclic aromatic hydrocarbons based on a new benzo [a] pyrene immunogen. Fresenius. *Journal of Analytical Chemistry* 371, 578–585.
Standing Committee on Foodstuffs (2001) *Outcome of the expert group meeting on 3 October on ways to prevent contamination of olive residue oil and other oils with polycyclic aromatic hydrocarbons (PAH)*. Summary record of the 85th meeting of the Standing Committee on Foodstuffs, 25th October 2001, agenda item 9. Available at: http://europa.eu.int/comm/food/fs/rc/scfs/rap09_en.pdf (accessed October 2015).
Sun, R.X., Lin, Q., Ke, C.L., Du, F.Y., Gu, Y.G., Cao, K., Xiao, J.L. and Mai, B.X. (2016) Polycyclic aromatic hydrocarbons in surface sediments and marine organisms from the Daya Bay, South China. *Marine Pollution Bulletin* 103, 325–332.
Sydnes, M.O. (2017) Oil spill fingerprinting – lientification of crude oil source of contamination. In: Pampanin, D.M. and Sydnes, M.O. (eds) *Petrogenice Polycyclic Aromatic Hydrocarbons in the Aquatic Environment: Analysis, Synthesis, Toxicity and Environmental Impact*. Betham Science Publishers, Sharjah, UAR, pp. 50–64.
Tedesco, S.A. (1985) *Surface Geochemistry in Petroleum Exploration*. Chapman & Hall, New York.
van der Oost, R., Beyer, J. and Vermeulen, N.P.E. (2003) Fish bioaccumulation and biomarkers in environmental risk assessment: a review. *Environmental Toxicology and Pharmacology* 13, 57–149.
Varanasi, U. (1989) *Metabolism of Polycyclic Aromatic Hydrocarbons in the Aquatic Environment*. CRC Press, Boca Raton, Florida.
Villeneuve, D., Crump, D., Garcia-Reyero, N., Hecker, M., Hutchinson, T.H. *et al.* (2014) Adverse outcome pathway (AOP) development I: Strategies and principles. *Toxicological Sciences* 142, 312–320.
Wakeham, S.G., Schaffner, C. and Giger, W. (1980) Polycyclic aromatic hydrocarbons in recent lake sediments – I. Compounds having anthropogenic origins. *Geochimica Cosmo Acta* 44, 403–413.
Walker, C.H., Sibly, R.M., Hopkin, S.P. and Peakall D.B. (2012) *Principles of Ecotoxicology*, 4th edn. CRC Press, Boca Raton, Florida.
Wang, W. and Wang, J. (2018) Different proportion in polycyclic aromatic hydrocarbons on environmental particulates in freshwater: microplastics in comparison to natural sediment. *Ecotoxicology and Environmental Safety* 147, 648–655.

White, P.A., Robitaille, S. and Rasmussen, J.B. (1999) Heritable reproductive effects of benzo(a)pyrene on the fathead minnow (*Pimephales promelas*). *Environmental Toxicology and Chemistry* 18, 1843–1847.
WHO (1998) *Environmental Health Criteria 202: Selected Non-Heterocyclic Polycyclic Aromatic Hydrocarbons*. International Programme on Chemical Safety, World Health Organization, Geneva.
Wolf, J.C. and Wheeler, J. R. (2018) A critical review of histopathological findings associated with endocrine and non-endocrine hepatic toxicity in fish models. *Aquatic Toxicology* 197, 60–78.
Won, E.J., Lee, Y., Gang, Y., Kim, M.S., Kim, C.J. *et al.* (2018) Chronic adverse effects of oil dispersed sediments on growth, hatching, and reproduction of benthic copepods: indirect exposure for long-term tests. *Marine Environmental Research* 137, 225–233.
Yang, F., Zhang, Q., Guo, H. and Zhang, S. (2010) Evaluation of cytotoxicity, genotoxicity and teratogenicity of marine sediments from Qingdao coastal areas using in vitro fish cell assay, comet assay and zebrafish embryo test. *Toxicology In Vitro* 24, 2003–2011.
Yu Goryacheva, I., Eremin, S.A., Shutaleva, E.A., Suchanek, M., Niessner, R. and Knopp, D. (2007) Development of a fluorescence polarization immunoassay for polycyclic aromatic hydrocarbons. *Analytical Letters* 40, 1445–1460.
Zhao, A., Zhang, L., Cai, Y. and Chen. Y. (2014) Distribution of polycyclic aromatic hydrocarbon (PAH) residues in several tissues of edible fishes from the largest freshwater lake in China, Poyang Lake, and associated human health risk assessment. *Ecotoxicology and Environmental Safety* 104, 323–331.

# 11 The Developmental Neurotoxicity of Polychlorinated Biphenyls: a Continuing Environmental Health Concern

**S. Sethi and P.J. Lein***
*Department of Molecular Biosciences, University of California Davis, California, USA*

## 11.1 Abstract

Polychlorinated biphenyls (PCBs) are synthetic chemicals widely used in diverse industrial applications and commercial products until their production was banned in the 1970s due to concerns regarding their environmental persistence and human health risks. Despite the ban, environmental levels of PCBs have not declined significantly over the past two decades, and more recently, contemporary PCBs not synthesized in the original industrial mixtures have been identified as inadvertent by-products of modern pigment manufacturing processes. PCBs remain contaminants of considerable concern because of their demonstrated adverse effects on neurodevelopment in human and animal models. This chapter briefly summarizes human literature and animal data documenting the developmental neurotoxicity (DNT) of PCBs and then focuses on the mechanisms by which PCBs are postulated to interfere with neurodevelopment, a question that remains a subject of much debate. The predominant mechanistic hypotheses of PCB DNT include disruption of thyroid hormone signalling, altered neurotransmitter signalling, perturbation of calcium homeostasis and the induction of oxidative stress. Of these, calcium signalling is the most sensitive and experimental data support a causal link between PCB effects on calcium signalling and the perturbation of critical neurodevelopmental processes, specifically dendritic arborization and apoptosis. Nonetheless, it cannot be ruled out that more than one mechanism contributes to PCB DNT, and it seems likely that the various molecular effects of PCBs may be interrelated. This chapter also identifies key data gaps regarding PCB DNT and describes strategies for addressing these gaps to develop a better understanding of the threat PCBs pose to the developing brain.

## 11.2 Introduction

Polychlorinated biphenyls (PCBs) are a structurally related family of synthetic compounds characterized chemically as biphenyls with variable chlorine substitutions for the hydrogen atoms in the benzene rings (Lein, 2017). General Electric originally discovered that these compounds have ideal heat dissipating and flame-retardant properties, which led to their worldwide production in large quantities from the 1920s to the late

* E-mail address: pjlein@ucdavis.edu

1970s. PCBs were synthesized and marketed in the USA and the UK as Aroclor® mixtures whose degree of chlorination was identified by a four-digit designation (e.g. 1248, 1254, 1260), with the first two digits indicating the number of carbon atoms in the biphenyl backbone and the last two digits indicating the percentage of chlorine by mass in the mixture (Lein, 2017). Similar PCB mixtures were synthesized in other countries under varying trade names such as Kanechlor® (Japan), Phenclor® (France) and Clophen® (Germany). Because of their chemical stability, electrical insulating properties and low flammability, PCBs were widely used as coolants and lubricants in electrical transformers, capacitors, fluorescent light ballasts and hydraulic equipment (Erickson and Kaley, 2011). PCBs were also broadly incorporated into diverse commercial products such as paints, varnishes, adhesives, cements, caulking compounds, thermal insulation material, pesticides, carbonless copy paper and newsprint.

The same physicochemical properties that made PCBs desirable for industrial and commercial applications also made them highly resistant to degradation in the environment. In addition, many PCBs are highly lipophilic and thus bioaccumulate up the food chain. PCBs were first recognized as pervasive pollutants in the 1960s (Jensen, 1972), and it was this recognition, coupled with growing concerns regarding potential human health risks, that led the US Congress to ban PCB production in 1979 and the Stockholm Convention on Persistent Organic Pollutants to issue a similar ban in 2001 (Carpenter, 2006; White and Birnbaum, 2009). While mean levels of PCBs in the environment and in human tissues steadily declined in the USA during the first two decades after the 1979 production ban, there is evidence that environmental levels of PCBs stabilized after the late 1990s and, more recently, may actually be increasing in some regions (Hopf *et al.*, 2009; Consonni *et al.*, 2012). These latter trends likely reflect the accelerated release of legacy PCBs from ageing products. For example, the release of PCBs from paints and caulking materials in ageing municipal buildings built during the era when PCBs were commonly incorporated into construction materials likely contributes to the higher than expected levels of airborne PCBs in Chicago (Herrick *et al.*, 2004, 2007; Marek *et al.*, 2017; Herkert *et al.*, 2018). Similarly, elevated levels of PCBs in the indoor air of elementary schools in the USA have been attributed to release of legacy PCBs from paints and caulking materials (Thomas *et al.*, 2012).

Research from the past decade also suggests a shift in the PCB congener profile in multiple environmental matrices, including human tissues. There are 209 possible PCB compounds – each of which is referred to as a congener – that vary according to the number and position of chlorine substitutions (Lein, 2017). Based on their structure, these 209 congeners are broadly divided into two categories: dioxin-like (DL) or non-dioxin-like (NDL) (Fig. 11.1). Early studies indicated that DL PCBs predominated in environmental samples, consistent with the predominance of DL congeners in the technical Aroclor mixtures. However, more recent data suggests the predominance of NDL PCBs in environmental matrices, including human tissues (DeCaprio *et al.*, 2005). Additionally, congeners not detected in the commercial mixtures synthesized prior to the bans on PCB production are emerging as ubiquitous contaminants in the human chemosphere (Marek *et al.*, 2013; Koh *et al.*, 2015). In contrast to the heavily chlorinated legacy PCB congeners, most of these 'contemporary' PCBs are lightly chlorinated. Produced as unintended by-products of modern pigment manufacturing processes, these contemporary PCBs have been documented in air and water worldwide (Choi *et al.*, 2008; Du *et al.*, 2008, 2009; Hu *et al.*, 2008; Basu *et al.*, 2009; Rodenburg *et al.*, 2010; Guo *et al.*, 2014). More recently, one of these lightly chlorinated congeners, PCB 11, has been reported at detectable levels in cow's milk in California (Chen *et al.*, 2017) and in the serum of pregnant women living in northern California (Sethi *et al.*, 2017) as well as women and adolescent children living in the greater Chicago area and rural Iowa (Koh *et al.*, 2015).

Humans are predominantly exposed to PCBs via dietary intake and inhalation (Ampleman *et al.*, 2015; Chen *et al.*, 2017). Once PCBs enter the human body they tend to deposit in fatty tissues, such as adipose, liver and brain, but they are also present at quantifiable levels in blood (Marek *et al.*, 2013). Until recently, most research on adverse human health effects of PCBs focused on the DL PCBs. The DL PCBs resemble dioxin structurally and they bind to the

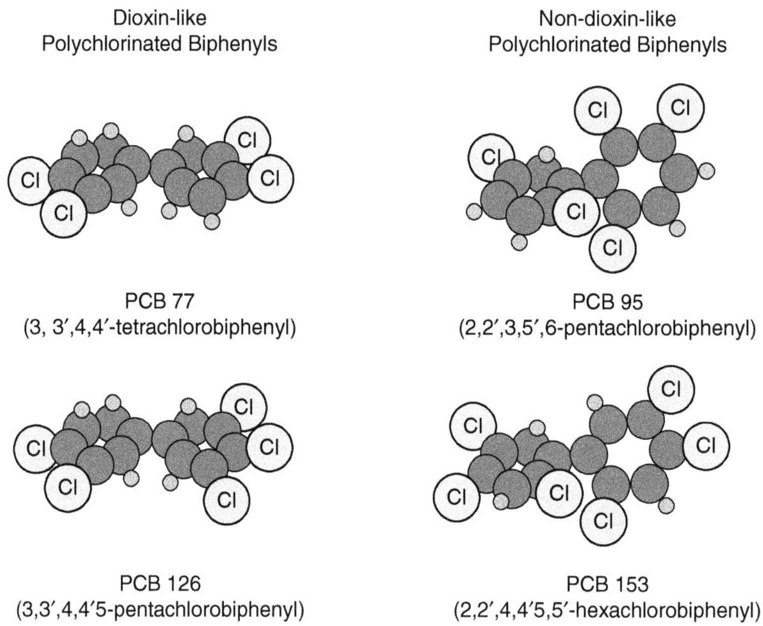


**Fig. 11.1.** Structures of dioxin-like (DL) and non-dioxin-like (NDL) PCBs. DL PCB congeners typically have $\leq$1 chlorine at the *ortho* positions of the carbon rings and are at their lowest energy state when lying in a co-planar configuration, similar to dioxin. In contrast, the NDL PCBs have more than one *ortho*-substituted chlorine, which creates steric hindrance, thereby preventing these congeners from assuming a co-planar configuration. Thus, NDL PCBs are typically non-coplanar.

arylhydrocarbon receptor (AhR), which is the major receptor that mediates the toxic actions of dioxin (Denison *et al.*, 2011). Thus, the toxicity profiles of DL PCBs resemble that of dioxin, with high-level exposures causing chloracne and liver damage and low-level chronic exposures linked to immune dysfunction and cancer (Carpenter, 2006; White and Birnbaum, 2009). Based on epidemiological, animal and mechanistic data, PCBs are classified as human carcinogens (Group 1) by the International Agency for Research on Cancer (IARC) and probable human carcinogens by the US Environmental Protection Agency (EPA) and current regulations of PCBs are based on carcinogenic risks.

In contrast, the NDL PCB congeners have negligible AhR binding activity. Thus, it was long assumed that NDL PCBs were toxicologically inert. However, in the late 1980s and early 1990s, pioneering research conducted at the EPA and at the Wadsworth Center in the New York State Department of Health suggested that this assumption was wrong. Specifically, NDL PCBs, but not DL PCBs, were reported to alter intracellular calcium dynamics in primary neurons (Kodavanti *et al.*, 1993, 1998; Kodavanti and Tilson, 2000) and to deplete dopamine in a neuronal cell line (Seegal *et al.*, 1994, 1997). These initial experimental observations indicating that NDL PCBs are not toxicologically inert coincided with emerging epidemiological and preclinical studies suggesting that PCBs are neurotoxic and that the developing nervous system is particularly vulnerable to the neurotoxic effects of PCBs (Berghuis *et al.*, 2015; Ulbrich and Stahlmann, 2004). The initial concerns generated by these reports have been heightened by subsequent data from the National Health and Nutrition Examination Survey (NHANES) confirming widespread PCB exposures among women of childbearing age living in the USA (Thompson and Boekelheide, 2013) and evidence that PCBs cross the placenta (Lanting *et al.*, 1998; Soechitram *et al.*, 2004) and accumulate in breast milk (Jorissen, 2007). Below, we summarize the evidence indicating that PCBs are developmental neurotoxicants and discuss postulated mechanisms by which PCBs interfere with normal neurodevelopment.

## 11.3 Human Evidence of PCB Developmental Neurotoxicity

PCBs first gained attention as developmental neurotoxicants during two accidental poisoning incidents: the Yusho incident in Japan in 1968 (Mitoma *et al.*, 2015) and the Yu-Cheng incident in Taiwan in 1979 (Hsu *et al.*, 1985). In both cases, liquids contaminated with PCBs were mistakenly mixed with cooking oil that was then consumed by local residents. Cognitive assessments of Yu-Cheng children born shortly after the PCB poisoning incident revealed cognitive deficits at 4–5 and 6–7 years of age compared with age-matched controls and with older siblings born before the incident (Chen *et al.*, 1992). While these incidents involved exposures to high levels of PCBs, numerous epidemiological studies have subsequently focused on the potential developmental neurotoxicity (DNT) risk of environmentally relevant levels of PCBs. The majority of these studies have reported a significant association between prenatal PCB exposure and neuropsychological deficits in children. Several critical reviews of the epidemiological literature (Schantz *et al.*, 2003; Boucher *et al.*, 2009; Berghuis *et al.*, 2015) have concluded that the weight of evidence indicates that impaired executive function is consistently associated with prenatal PCB exposure. As summarized in Table 11.1, subsequent studies generally support this conclusion, though there are several discordant studies (addressed below). In addition, more recent epidemiological studies have linked developmental

**Table 11.1.** Summary of PCB exposure and neurodevelopment.

| PCBs measured | Exposure assessment | Geographical location | Outcome | Testing method | Reference |
|---|---|---|---|---|---|
| PCB 207 | Cord blood | Japan | Poorer cognitive function (boys only) | Kaufman Assessment Battery for Children | Tatsuta *et al.*, 2014 |
| PCB 118, 138, 153, 156, 170 180 | Maternal blood | Greece | Poorer cognitive function | McCarthy Scales | Kyriklaki *et al.*, 2016 |
| PCB 153 | Cord blood | Canada | Increased inattention | Video recordings | Verner *et al.*, 2015 |
| PCB 118, 138, 153, 156, 170 180 | Maternal blood | Finland | Increased ASD risk | ASD diagnosis | Cheslack-Postava *et al.*, 2013 |
| PCB 138/158, 153 | Maternal blood | California | Increased ASD risk | ASD diagnosis | Lyall *et al.*, 2016 |
| Dioxin-like PCBs | Maternal blood | Germany | Lower social scores | Social Responsiveness Scale | Nowack *et al.*, 2015 |
| PCB 118, 138, 153, 180 | Cord blood | Massachusetts | Higher risk for attention deficit hyperactivity disorder | Conners' Rating Scale for Teachers | Sagiv *et al.*, 2010 |
| PCB 153 | Maternal blood | Greenland and Ukraine | Higher prevalence of hyperactivity | Strengths and Difficulties Questionnaire | Rosenquist *et al.*, 2017 |
| PCB 118, 138, 153, 180 | Cord blood | Massachusetts | No effect | Wide Range Assessment of Memory and Learning | Orenstein *et al.*, 2014 |
| PCB 118, 138, 153, 180 | Maternal blood | Ohio | No effect | Wechsler Intelligence Scales for Children IQ | Zhang *et al.*, 2016 |
| Dioxin-like PCBs | Maternal blood | Japan | Negative in boys Positive in girls | Kaufman Assessment Battery for Children | Ikeno *et al.*, 2017 |

PCB exposures to increased risk for neurodevelopmental disorders, specifically, attention deficit/hyperactivity disorder (ADHD) and autism spectrum disorders (ASD).

While the majority of human studies report evidence of PCB developmental neurotoxicity, a few found either no effect or, surprisingly, a positive effect of developmental PCB exposure on neuropsychological outcomes. One explanation of these discrepancies is the diversity of analytical methods used to quantify PCB exposures, which has made it difficult to compare across studies because of significant variation with respect to the limits of detection, the number and type of congeners analysed and data handling. A comparative study of PCB levels reported in epidemiological studies of cohorts from around the world concluded that, with the exception of communities whose main food source was marine mammals, PCB levels were rather comparable across cohorts and that PCB 153 is generally a good biomarker of cumulative PCB exposure (Longnecker *et al.*, 2003). If this conclusion is valid, it suggests that differences in congener profiles between cohorts may contribute to the divergent findings regarding PCB DNT reported in the human literature. In other words, variation in the number and type of congeners analysed in different cohorts could result in different conclusions across studies. An additional problem is that many human studies have focused on DL-congeners, despite evidence from animal studies suggesting that DL-PCBs are likely not responsible for many of the cognitive and behavioural abnormalities observed in humans (Bernhoft *et al.*, 1994; Schantz *et al.*, 1996; Bushnell and Rice, 1999). Thus, in future human studies, it would be informative to include an increased number and wider variety of PCB congeners in the exposure assessment to gain a more comprehensive understanding of the congener(s) posing the greatest risk to the developing human brain.

Another factor that likely contributes to inconsistencies across human studies is the type of cognitive domain assessed and the task administered to assess cognitive ability. For example, some studies only tested visual–spatial abilities, which have been shown to be not very sensitive to PCBs, where other studies tested IQ or response inhibition, which appear to be more sensitive to PCBs (Boucher *et al.*, 2009). Socio-economic status or home environment of the child, co-exposure to other neurotoxic compounds (e.g. methylmercury (MeHg)) and the age at which the child is tested may also contribute to variable results in human studies. A recent review summarizing the neuropsychological tests that have been used in human studies of PCB DNT provided a detailed discussion of tests that may be the most beneficial to perform and proposed standardization of test protocols across studies with the goal of enabling more rigorous comparisons across studies (Boucher *et al.*, 2009). While discrepancies between studies with respect to the spectrum and persistence of adverse neurobehavioral outcomes, confounding co-exposures and differences in congener profiles that comprise the exposure have raised questions concerning the causative role of PCBs in human developmental neurotoxicity (Winneke, 2011), extensive experimental findings in non-human primate and rodent models confirm that developmental PCB exposure causes deficits in learning and memory (Ulbrich and Stahlmann, 2004; Sable and Schantz, 2006). Perinatal exposure to PCBs has recently been reported to also alter social behaviours in rats (Jolous-Jamshidi *et al.*, 2010).

## 11.4 Mechanisms of PCB DNT

Epidemiological evidence of PCB DNT has largely been corroborated in non-human primate and rodent models (Ulbrich and Stahlmann, 2004; Sable and Schantz, 2006). But the mechanism(s) by which PCBs interfere with neurodevelopment, and whether all PCB congeners act as developmental neurotoxicants, remain key questions in the field. Prevailing hypotheses for how PCBs derail normal neurodevelopment include disruption of thyroid hormone signalling, altered neurotransmitter signalling, perturbation of calcium homeostasis and oxidative stress.

### 11.4.1 Evidence for and against thyroid hormone signalling in PCB DNT

Disruption of thyroid hormone signalling is widely debated as a mechanism of PCB DNT. Complex mixtures of PCBs have consistently been shown to cause dose-dependent decreases in serum levels of thyroxine (T4) in humans and in animal models (Goldey and Crofton, 1998; Zoeller *et al.*, 2000; Kato *et al.*, 2003) and maternal or infant hypothyroidism is linked to significant neurodevelopmental defects, including mental retardation (Oppenheimer and Schwartz, 1997; Williams, 2008). These observations led to the hypothesis that decreased thyroid hormone levels are responsible for the neurobehavioural changes observed in children exposed developmentally to PCBs (Winneke *et al.*, 2002). In early support of this hypothesis, developmental exposure of rats to Aroclor 1254 was shown to cause hypothyroxinaemia (a decrease in T4 levels) coincident with hearing loss, and supplementation with T4 throughout development attenuated PCB-induced auditory deficits (Goldey and Crofton, 1998). However, other neurobehavioural effects of PCBs, such as learning and memory, do not appear to be mediated by thyroid hormone-dependent mechanisms. In rats developmentally exposed to Aroclor 1254 at levels that significantly decreased their serum T4 levels, learning and memory were not affected as assessed using the T-maze or the Morris water maze (Zahalka *et al.*, 2001). Conversely, rats developmentally exposed to Aroclor 1254 at 1 mg $kg^{-1}$ exhibited impaired performance in the Morris water maze but no significant alterations in serum triiodothyronine (T3) or T4 levels (Yang *et al.*, 2009). Other studies with individual PCB congeners also provide little support for the hypothesis that thyroid hormone deficits mediate the cognitive effects of PCBs. For example, developmental exposure to the NDL PCBs 28, 118 or 153 produced similar deficits on a T-maze delayed spatial alternation task; however, their effects on serum T4 levels varied from a marked reduction to no effect (Ness *et al.*, 1993; Schantz *et al.*, 1995). Another complication with the thyroid hormone disruption hypothesis is that PCBs have been shown to have both antagonistic and agonistic effects at the thyroid hormone receptor, depending on the congener and on the dose or concentration (Goldey *et al.*, 1995; Zoeller *et al.*, 2000; Fritsche *et al.*, 2005). For example, PCB 118 has been shown to mimic the effects of T3 by inducing the differentiation of human neural progenitor cells into oligodendrocytes, but this effect was not seen with PCB 126 (Fritsche *et al.*, 2005). In rats, developmental exposure to Aroclor 1254 decreased T4 levels as expected, but surprisingly increased the expression of two thyroid hormone responsive genes, suggesting that PCBs do not follow a binary mechanism of inducing either hypo- or hyperthyroidism (Zoeller *et al.*, 2000).

Human studies also reveal conflicting results, inconsistent correlations and lack of a strong dose-response relationship (Hagmar, 2003; Dallaire *et al.*, 2009). For example, in one study, a negative association was observed between perinatal PCB exposure and psychomotor function, but not between serum levels of thyroid hormone and neurodevelopment (Koopman-Esseboom *et al.*, 1996). The general trend is for PCBs to dose-dependently decrease circulating T4 levels (Maervoet *et al.*, 2007) but one recent study reported a positive association between serum levels of hydroxylated PCBs and T3 (Dallaire *et al.*, 2009). The discrepancies across both human and rodent studies do not support a consistent association between serum thyroid hormone levels and adverse neurodevelopmental outcomes. This may be due to multiple factors, including species, the PCBs used in exposure (mixtures versus single congeners, DL versus NDL congeners), the period of PCB exposure and the time points assessed in the offspring; however, collectively, the data do not provide strong conclusive evidence that perturbation of thyroid hormone signalling is the major mechanism of PCB DNT.

### 11.4.2 Altered neurotransmitter levels as a mechanism of PCB DNT

Experimental studies have demonstrated that PCBs can alter neurotransmitter levels, particularly dopamine. Much of this work, however, has focused on adult exposures to PCBs and, as described in a comprehensive review of the early studies of rodent and non-human primates (Fonnum and Mariussen, 2009), PCB exposures generally tend to decrease brain dopamine levels. Mechanistic studies indicate that decreased dopamine levels are likely due to inhibition of the vesicular monoamine transporter (VMAT) or the plasma membrane dopamine transporter (DAT) (Mariussen and Fonnum, 2001; Seegal *et al.*, 2002; Caudle *et al.*, 2006). Decreased levels of dopamine as a consequence of PCB exposure have been proposed as a mechanism to explain epidemiological evidence of a positive association between PCB exposures and increased risk of Parkinson's disease (Seegal *et al.*, 1990; Goodwill *et al.*, 2007).

Some of the first studies to examine the effects of developmental PCB exposure on dopamine-related behaviours reported that adult mice exposed prenatally to the DL congener PCB 77 in the maternal diet were hyperactive and that brain levels of dopamine and dopamine receptor binding sites were decreased in hyperactive animals compared with animals with normal activity levels (Chou *et al.*, 1979; Agrawal *et al.*, 1981). In contrast, studies in which animals were exposed perinatally to technical mixtures of PCBs (Aroclor 1016 or 1254) observed no changes in dopamine levels in the adult brain (Morse *et al.*, 1996; Zahalka *et al.*, 2001). A more recent study using single NDL congeners revealed either no change or depression of activity levels amongst adult rats exposed to the PCB 52, 138 or 180 in the maternal diet throughout gestation and lactation (Boix *et al.*, 2011). PCB effects on motor activity were sex specific and depended on the degree of chlorination in the PCB congener: PCB 52 had no effect on motor activity; PCB 138 decreased activity in both males and females; and PCB 180 reduced activity in males but not females. The effects of developmental exposures to these NDL PCBs on dopamine levels also varied. Neither PCB 52 nor 138 altered dopamine levels in the nucleus accumbens (NAcc), whereas PCB 180 increased NAcc dopamine levels in both males and females. Interestingly, glutamate levels in the NAcc were significantly reduced by all three NDL PCB congeners. Conversely, developmental exposure to Aroclor 1254 was observed to increase activity levels in females, and this effect coincided with decreased *N*-methyl-D-aspartic acid receptor (NMDAR) binding in multiple brain regions in the absence of any changes in dopamine transporters or receptors (Tian *et al.*, 2011).

In summary, no consistent pattern of PCB effects on neurotransmitter levels or consistent correlations between changes in neurotransmitter levels and behavioural effects have emerged from the published studies, suggesting that PCB effects on neurotransmitter levels are not the sole, or perhaps even a primary, mechanism by which PCBs alter behaviour. Also arguing against PCB effects on neurotransmitter levels as a predominant mechanism contributing to PCB DNT is the fact that most, if not all, of the developmental PCB studies mentioned here used doses of PCBs that are at the high end or even above the range of PCB levels found in humans. Thus, the relevance of these studies to the human situation is questionable. These studies do, however, underscore the fact that different congeners can elicit differing profiles of developmental neurotoxicity.

### 11.4.3 Altered calcium signalling as a mechanism of PCB DNT

Calcium signalling regulates critical processes of neurodevelopment as well as synaptic plasticity associated with many behaviours (Konur and Ghosh, 2005; Berridge, 2006; Brini *et al.*, 2014). Thus, experimental evidence indicating that NDL, but not DL, PCBs increased intracellular $Ca^{2+}$ levels in cultured neuronal cells (Kodavanti *et al.*, 1993; Inglefield and Shafer, 2000; Wayman *et al.*, 2012a) prompted research efforts to identify functional consequences of this PCB effect. A well-documented consequence of increased intracellular $Ca^{2+}$ is translocation of protein kinase C (PKC) to the membrane where it may be activated (Trilivas and Brown, 1989). PKC is a key signalling molecule in learning and memory processes (Lee, 2006; Purkayastha *et al.*, 2009), raising the possibility that developmental exposure to PCBs causes cognitive deficits by disrupting PKC signalling. In support of this hypothesis, NDL PCBs increased the translocation of PKC to the membrane in cultured cerebellar neurons (Kodavanti and Tilson, 2000; Yang and Kodavanti, 2001), whereas DL PCBs had no effect on PKC translocation (Do and Lee, 2012). *In vivo*, the effects of developmental exposure to Aroclor 1254 on subcellular distribution of PKC isoforms in the brain were more complicated, with region-dependent increases and decreases in membrane localization observed (Yang *et al.*, 2003). The question of whether PCB effects on PKC activity are causally linked to the effects of developmental PCB exposures on behaviour remains unanswered.

Alternatively, there is strong evidence linking PCB effects on intracellular $Ca^{2+}$ to changes in neurodevelopmental processes that shape neuronal connectivity in the developing brain.

This research was prompted by efforts to identify the molecular mechanisms by which NDL PCBs increase intracellular $Ca^{2+}$. Numerous mechanisms have been identified, including disruption of membrane properties (Kodavanti *et al.*, 1996), influx of extracellular $Ca^{2+}$ through L-type voltage-sensitive $Ca^{2+}$ channels or the NMDA receptor (Mundy *et al.*, 1999; Inglefield and Shafer, 2000), or release of $Ca^{2+}$ from intracellular stores via activation of ryanodine receptors (RyR) (Wong *et al.*, 1997a) or $IP_3$ receptors (Inglefield *et al.*, 2001). Of these, RyR activation is the most sensitive identified to date (Pessah *et al.*, 2010). RyRs are $Ca^{2+}$-induced $Ca^{2+}$ ion channels localized to the endoplasmic reticulum that function to regulate intracellular $Ca^{2+}$ storage and release. Nanomolar concentrations of PCBs interact with RyRs to dramatically sensitize their activation by nM $Ca^{2+}$ and attenuate their inhibition by mM $Ca^{2+}$ and $Mg^{2+}$, thereby stabilizing these channels in their open configuration. A stringent structure–activity relationship has been characterized for this effect with the most potent and efficacious PCBs possessing two or three chlorine substitutions in the ortho positions (Pessah *et al.*, 2010).

PCB sensitization of RyRs increases intracellular $Ca^{2+}$ levels in neurons (Wong *et al.*, 1997b) and, consistent with this observation, picomolar to nanomolar concentrations of NDL PCB 95 activate two $Ca^{2+}$-dependent signalling pathways in cultured rat hippocampal neurons: (i) sequential activation of $Ca^{2+}$/calmodulin-dependent protein kinase kinase (CaMKK), $Ca^{2+}$/calmodulin kinase-I (CaMKIα/γ) and extracellular signal-regulated kinase kinase (MEK/ERK) and cyclic AMP response element binding protein (CREB), which increases transcription of Wnt2 (Wayman *et al.*, 2012a); and (ii) CREB-mediated miR132 up-regulation, which suppresses the translation of p250GAP (Lesiak *et al.*, 2014). In cultured rat hippocampal neurons, the former signalling pathway mediates PCB 95-induced dendritic growth (Wayman *et al.*, 2012a), whereas the latter mediates PCB 95-induced synaptogenesis, which is evident as increased spine density and increased frequency of miniature excitatory post-synaptic currents (Lesiak *et al.*, 2014). Several lines of evidence establish that PCB effects on RyR activity are causally linked to dendritic growth and spine formation: (i) RyR-active PCB congeners, such as PCB 95 and PCB 136, but not PCB congeners that lack activity at the RyR, such as PCB 66, enhance dendritic growth (Wayman *et al.*, 2012b) and promote spine formation (Lesiak *et al.*, 2014) in cultured hippocampal and cortical neurons; and (ii) pharmacological blockade or small interfering RNA (siRNA) knockdown of RyRs inhibits the dendrite-promoting activity of these NDL PCBs (Yang *et al.*, 2009, 2014). An interesting observation from these studies is that PCB-induced dendritic growth exhibits a non-monotonic concentration–effect relationship, with effects observed in the pico- and nanomolar range but not at femto- and micromolar concentrations. Elucidating the mechanisms responsible for this non-monotonic concentration–effect relationship remains a critical data gap in the field.

Dendritic arborization is a critical determinant of neuronal connectivity (Kennedy, 2000; Matus, 2000), and altered neuronal connectivity, including hyperconnectivity, is associated with neurodevelopmental disorders, including ASD (Supekar *et al.*, 2013; Alaerts *et al.*, 2016; Copf, 2016). These observations suggest that PCB effects on dendritic growth contribute to PCB-induced behavioural deficits. In support of this possibility, developmental exposure of rats to Aroclor 1254 at 1 or 6 mg $kg^{-1}$ in the maternal diet was found to cause learning and memory deficits in the Morris water maze at doses that modulated RyR activity and dendritic arborization in brain regions that subserve Morris water maze behaviour (Yang *et al.*, 2009). Several interesting findings emerged from this study. First, developmental PCB exposure promoted dendritic growth in cerebellar Purkinje cells and neocortical pyramidal neurons among untrained animals but attenuated or reversed experience-dependent dendritic growth among Morris water maze-trained littermates. Second, deficits in learning and memory were only seen in the 1 mg $kg^{-1}$ group and the most robust changes in dendritic arborization were also observed in this dose group, replicating the non-monotonic dose–response relationship observed in the *in vitro* studies. In a separate study, a similar non-monotonic dose–response relationship was observed for dendritic arborization of CA1 pyramidal neurons in the hippocampus of rats exposed to PCB 95 in the maternal diet throughout gestation and lactation (Wayman *et al.*, 2012b). Third, the Aroclor 1254 study demonstrated that the behavioural deficits correlated better with changes in RyR expression and activity than with changes in serum levels of thyroid hormone or sex hormones. While these data strongly support the hypothesis that PCB effects on RyR activity contribute to behavioural deficits via modulation of calcium-dependent signalling pathways that regulate dendritic plasticity, experimental evidence demonstrating that the molecular and cellular effects are causally linked to behavioural deficits are lacking. Testing this hypothesis using RyR knock-out animals is not feasible, because of embryo lethality, but alternative approaches using knock-in animals expressing human mutations that increase RyR sensitization by halogenated hydrocarbons (Yuen *et al.*, 2012) may increase confidence in causal inferences.

### 11.4.4 Oxidative stress as a mechanism of PCB DNT

Oxidative stress occurs when the production of reactive oxygen species (ROS) and reactive nitrogen species (RNS) exceeds the antioxidant capacity of the system (Betteridge, 2000). This imbalance can cause oxidative and nitrative damage to macromolecules, which in turn can cause cell damage or even cell death. Conversely, at physiological levels, ROS are important signalling molecules that regulate neuronal cell fate and development (Kennedy *et al.*, 2012; Olguin-Albuerne and Moran, 2015). This is the classic goldilocks effect, where just the right amount of ROS is critical for normal neurodevelopment, whereas too much can be detrimental. PCBs have been shown to tamper with this balance, causing oxidative stress both *in vitro* (Mariussen *et al.*, 2002; Howard *et al.*, 2003; Do and Lee, 2012) and *in vivo* (Yang and Lein, 2010). In cultured cerebellar granule neurons, exposure to the NDL congeners PCB 4 or 153, or to Aroclor 1254, increased cellular ROS levels, which caused concentration-dependent cell death (Mariussen *et al.*, 2002; Do and Lee, 2012). Similarly, exposure of primary hippocampal neurons to Aroclor 1254 or the NDL PCB 47 triggered neuronal cell apoptosis that was blocked by co-exposure to the antioxidant alpha-tocopherol (Howard *et al.*, 2003), suggesting that increased ROS mediated PCB-induced apoptosis. In all three *in vitro* studies a DL PCB congener was also tested and found to have no effect on ROS or on apoptosis.

In rats exposed throughout gestation and lactation to Aroclor 1254 at 1 mg $kg^{-1}$ but not 0.1 mg $kg^{-1}$ in the maternal diet, apoptosis was increased in multiple brain regions at PND 3, as indicated using terminal deoxynucleotidyl transferase dUTP nick-end labelling (TUNEL) staining and caspase activity assays, and these effects coincided with increased expression of oxidative stress biomarkers (Yang and Lein, 2010). In contrast, a study of mice exposed to a mixture of six NDL PCBs via lactation failed to find any evidence of oxidative damage in the brain (Elnar *et al.*, 2016). There are multiple reasons for these discrepant findings, including differences in model species, timing and length of exposure, doses and composition of PCB mixtures to which animals were exposed, and different oxidative stress biomarkers. Based on these two studies, it is difficult to determine whether oxidative stress plays a predominant role in PCB DNT. More studies assessing oxidative stress in preclinical models of PCB DNT are needed to evaluate the importance of oxidative stress as a mechanism of PCB-induced behavioural deficits.

### 11.4.5 Interactions between the proposed mechanisms

As described above, multiple biological activities have been ascribed to NDL PCBs, including disruption of thyroid hormone signalling (Zoeller, 2007; Crofton, 2008), altered levels of dopamine (Fonnum *et al.*, 2006; Mariussen and Fonnum, 2006), increased calcium signalling (Pessah *et al.*, 2010) and increased intracellular levels of ROS (Fonnum and Mariussen, 2009). Are these biological activities causally related to PCB developmental neurotoxicity, and if so, do they represent divergent or convergent mechanisms of PCB DNT? It is likely that PCBs interact with multiple targets in the developing brain and that these processes have numerous convergence points (Fig. 11.2). Thus, it is the cumulative effect of these multiple molecular 'hits' that mediate PCB-induced behavioural deficits. For example, while PCB sensitization of RyRs stimulates calcium-dependent signalling pathways that promote dendritic growth (Wayman *et al.*,

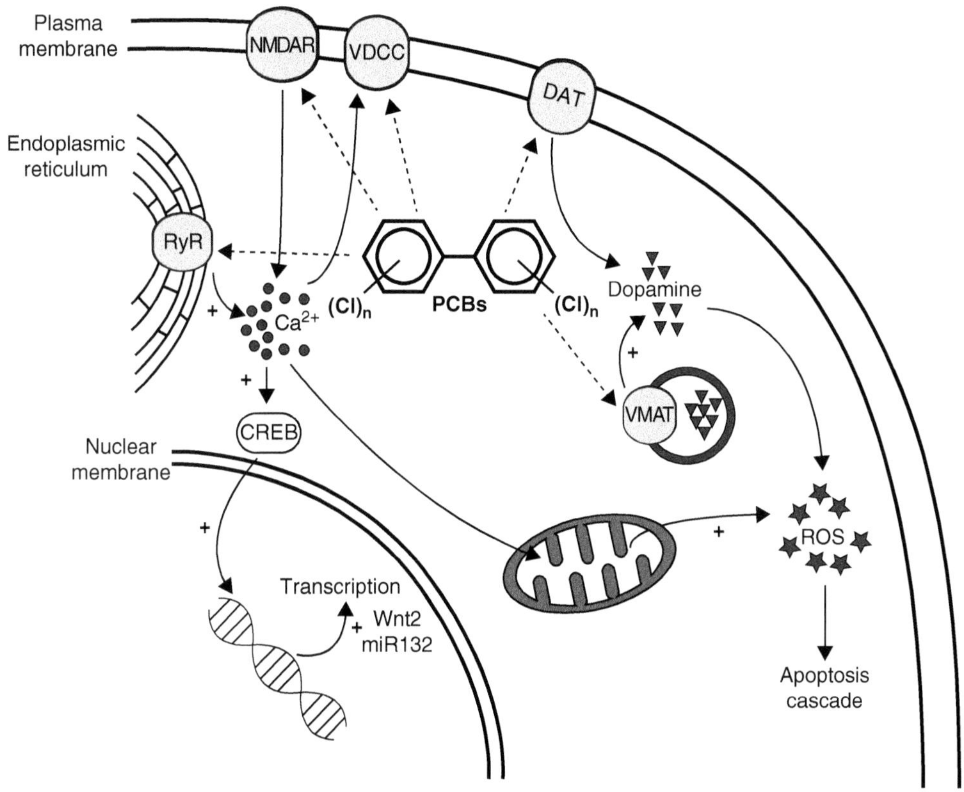


**Fig. 11.2.** Schematic illustrating mechanisms of PCB developmental neurotoxicity. Dashed arrows represent identified molecular targets of PCBs; + signs indicate an increase due to PCB exposure. CREB, cAMP response element binding protein; DAT, plasma membrane dopamine transporter; NMDAR, *N*-methyl-D-aspartate receptor; PCBs, polychlorinated biphenyls; ROS, reactive oxygen species; RyR, ryanodine receptor; VDCC, voltage-dependent calcium channel; VMAT, vesicular monoamine transporter.

2012a) and spine formation (Lesiak *et al.*, 2014), the increased calcium resulting from the initial molecular effect likely also increases ROS to induce neuronal apoptosis (Howard *et al.*, 2003; Yang and Lein, 2010). Consistent with this, multiple *in vitro* studies have shown that blocking calcium channels such as the RyR (Howard *et al.*, 2003), NMDAR (Mariussen *et al.*, 2002), or both separately in the same cell preparation (Kang *et al.*, 2004) prevents PCB-induced oxidative stress. *In vitro* studies indicate that PCB effects on dopamine can also increase oxidative stress. In a co-culture of developing rat striatum and ventral mesencephalon, PCBs altered dopamine levels coincident with increased ROS (Lee *et al.*, 2012). Co-exposure to the antioxidant, *N*-acetyl-cysteine, reduced ROS levels but did not block PCB effects on dopamine; conversely, depletion of dopamine prevented oxidative stress upon exposure to PCBs (Lee *et al.*, 2012). However, there is also evidence that these molecular targets may indeed be independent. Blocking calcium influx did not change PCB effects on intracellular dopamine levels in a catecholaminergic cell line, suggesting that PCBs likely affect dopamine homeostasis independent of calcium signalling (Kang *et al.*, 2004).

### 11.4.6 Data gaps and approaches for addressing them

The most critical data gaps in the field of PCB DNT include identification of PCB congeners that pose the greatest risk to the developing brain, and lack of experimental evidence addressing causal relationships between known molecular effects of PCBs and behavioural deficits. To address these data gaps, there are a number of approaches that should be undertaken.

**1.** Epidemiological studies should increase the profile of congeners analysed, and standardize behavioural tests to enable rigorous comparisons across studies.
**2.** Human studies should leverage mechanistic data from experimental studies to stratify populations according to genetic mutations and polymorphisms in well established molecular targets of PCB DNT. This is important for identifying gene × environment interactions that influence individual risk for PCB DNT, but will be increasingly important for studies focused on determining whether PCBs are risk factors for neurodevelopmental disorders. These types of studies will also provide insight as to the importance of differences in genetic backgrounds as factors contributing to discrepant findings between cohorts.
**3.** Preclinical studies should shift away from doses that are too high to be considered physiologically relevant and focus on dose ranges reflective of current environmental levels.
**4.** Preclinical studies should stop using industrial mixtures, since these do not model congener profiles in current human samples (Frame *et al.*, 1996; Longnecker *et al.*, 2003). Rather, researchers should use PCB mixtures that mimic the PCB composition in human serum or placenta since these likely better represent the gestational environment. This includes the lightly chlorinated contemporary PCB congeners for which there is growing evidence of human exposure (Koh *et al.*, 2015; Chen *et al.*, 2017), but negligible data regarding their DNT potential.
**5.** Preclinical studies should be designed to include not only behavioural outcomes but also mechanistically relevant end points and should leverage contemporary gene editing techniques to generate the data needed to understand causal relationships between the molecular, cellular and behavioural effects of PCBs.
**6.** Advances in the use of *in vitro* and alternative models for toxicity testing should be leveraged to expand the database regarding the relative DNT potency of individual legacy and contemporary PCB congeners. For example, an *in vitro* screen of PCB congeners for effects on structural parameters of neuronal connectivity in primary neuron–glia co-cultures identified a contemporary PCB congener, PCB 11, and two of its known biological metabolites as potential developmental neurotoxicants. Specifically, these compounds were found to increase both dendritic and axonal growth in primary hippocampal and cortical neurons (Sethi *et al.*, 2017). Results from such screens will be useful for prioritizing PCB congeners to test in preclinical models, and will inform interpretation of results generated using PCB mixtures *in vivo*.

## 11.5 Conclusions

PCBs remain a continuing environmental health problem. The cumulative evidence from the past 10 years of PCB research suggests that there is a need to revise the conventional understanding of the environmental health risks associated with PCBs (Lein, 2017). For example, while PCBs remain persistent in the environment, the congener profile of PCBs is changing such that NDL congeners predominate. Risk assessors also need to reconsider dietary exposures as the primary route of human exposure to PCBs in light of emerging evidence of widespread contamination of indoor air with PCBs. Understanding the impact of these changing human exposures to PCBs on toxic outcomes is a critically important research need.

Evidence from the past 10 years has also built a convincing case for NDL PCB congeners as significant environmental health risks and for developmental neurotoxicity as a significant end point of concern (Lein, 2017). Epidemiological data demonstrating a positive association between developmental PCB exposure and adverse neurodevelopmental outcomes has been largely confirmed by preclinical studies. However, inconsistent findings in the human data have been the basis for some to question whether PCBs are developmental neurotoxicants in humans. There is an urgent need to conduct a systematic review of the human, animal and mechanistic literature to determine whether current regulatory guidelines should be reconsidered to address developmental neurotoxicity as an end point of concern (Lein, 2017). Another pressing research need is to identify which PCB congeners are developmental neurotoxicants and to elucidate the mechanism(s) by which they interfere with neurodevelopment (Lein, 2017). Answers to these questions will inform rigorous approaches for assessing risks to the developing brain associated with exposures to complex PCB mixtures (Lein, 2017).

## References


Agrawal, A.K., Tilson, H.A. and Bondy, S.C. (1981) 3,4,3′,4′-tetrachlorobiphenyl given to mice prenatally produces long-term decreases in striatal dopamine and receptor binding sites in the caudate nucleus. *Toxicology Letters* 7, 417–424.

Alaerts, K., Swinnen, S. and Wenderoth, N. (2016) Sex differences in autism: A resting-state fmri investigation of functional brain connectivity in males and females. *Social Cognitive and Affective Neuroscience* 11(6), 1002–1016.

Ampleman, M.D., Martinez, A., Dewall, J., Rawn, D.F., Hornbuckle, K.C. and Thorne, P.S. (2015) Inhalation and dietary exposure to PCBs in urban and rural cohorts via congener-specific measurements. *Environmental Science Technology* 49, 1156–1164.

Basu, I., Arnold, K.A., Venier, M. and Hites, R.A. (2009) Partial pressures of PCB-11 in air from several great lakes sites. *Environmental Science Technology* 43, 6488–6492.

Berghuis, S.A., Bos, A.F., Sauer, P.J. and Roze, E. (2015) Developmental neurotoxicity of persistent organic pollutants: An update on childhood outcome. *Archives of Toxicology* 89, 687–709.

Bernhoft, A., Nafstad, I., Engen, P. and Skaare, J.U. (1994) Effects of prenatal and postnatal exposure to 3,3′,4,4′,5-pentachlorobiphenyl on physical development, neurobehavior and xenobiotic-metabolizing enzymes in rats. *Environmental Toxicology and Chemistry* 13, 1589–1597.

Berridge, M.J. (2006) Calcium microdomains: organization and function. *Cell Calcium* 40, 405–12.

Betteridge, D.J. (2000) What is oxidative stress? *Metabolism* 49, 3–8.

Boix, J., Cauli, O., Leslie, H. and Felipo, V. (2011) Differential long-term effects of developmental exposure to polychlorinated biphenyls 52, 138 or 180 on motor activity and neurotransmission. Gender dependence and mechanisms involved. *Neurochemistry International* 58, 69–77.

Boucher, O., Muckle, G. and Bastien, C.H. (2009) Prenatal exposure to polychlorinated biphenyls: A neuropsychologic analysis. *Environmental Health Perspectives* 117, 7–16.

Brini, M., Cali, T., Ottolini, D. and Carafoli, E. (2014) Neuronal calcium signaling: function and dysfunction. *Cellular and Molecular Life Sciences* 71, 2787–2814.

Bushnell, P.J. and Rice, D.C. (1999) Behavioral assessments of learning and attention in rats exposed perinatally to 3,3′,4,4′,5-pentachlorobiphenyl (PCB 126). *Neurotoxicology and Teratology* 21, 381–392.

Carpenter, D.O. (2006) Polychlorinated biphenyls (pcbs): routes of exposure and effects on human health. *Reviews on Environmental Health* 21, 1–23.

Caudle, W.M., Richardson, J.R., Delea, K.C., Guillot, T.S., Wang, M., Pennell, K.D. and Miller, G.W. (2006) Polychlorinated biphenyl-induced reduction of dopamine transporter expression as a precursor to Parkinson's disease-associated dopamine toxicity. *Toxicological Sciences* 92, 490–499.

Chen, X., Lin, Y., Dang, K. and Puschner, B. (2017) Quantification of polychlorinated biphenyls and polybrominated diphenyl ethers in commercial cows' milk from California by gas chromatography-triple quadruple mass spectrometry. *PLoS ONE* 12, e0170129.

Chen, Y.C., Guo, Y.L., Hsu, C.C. and Rogan, W.J. (1992) Cognitive development of yu-cheng ('oil disease') children prenatally exposed to heat-degraded pcbs. *Journal of the American Medical Association* 268, 3213–3218.

Cheslack-Postava, K., Rantakokko, P.V., Hinkka-Yli-Salomaki, S., Surcel, H.M., Mckeague, I.W., Kiviranta, H.A., Sourander, A. and Brown, A.S. (2013) Maternal serum persistent organic pollutants in the finnish prenatal study of autism: a pilot study. *Neurotoxicology and Teratology* 38, 1–5.

Choi, S.D., Baek, S.Y., Chang, Y.S., Wania, F., Ikonomou, M.G. *et al.* (2008) Passive air sampling of polychlorinated biphenyls and organochlorine pesticides at the Korean arctic and antarctic research stations: implications for long-range transport and local pollution. *Environmental Science Technology* 42, 7125–7131.

Chou, S.M., Miike, T., Payne, W.M. and Davis, G.J. (1979) Neuropathology of 'Spinning syndrome' induced by prenatal intoxication with a PCB in mice. *Annals of the New York Academy of Sciences* 320, 373–395.

Consonni, D., Sindaco, R. and Bertazzi, P.A. (2012) Blood levels of dioxins, furans, dioxin-like PCBs, and TEQs in general populations: a review, 1989–2010. *Environment International* 44, 151–162.

Copf, T. (2016) Impairments in dendrite morphogenesis as etiology for neurodevelopmental disorders and implications for therapeutic treatments. *Neuroscience & Biobehavioral Reviews* 68, 946–978.
Crofton, K.M. (2008) Thyroid disrupting chemicals: mechanisms and mixtures. *International Journal of Andrology* 31, 209–223.
Dallaire, R., Muckle, G., Dewailly, E., Jacobson, S.W., Jacobson, J.L. *et al.* (2009) Thyroid hormone levels of pregnant Inuit women and their infants exposed to environmental contaminants. *Environmental Health Perspectives* 117, 1014–1020.
DeCaprio, A.P., Johnson, G.W., Tarbell, A.M., Carpenter, D.O., Chiarenzelli, J.R. *et al.* (2005) Polychlorinated biphenyl (PCB) exposure assessment by multivariate statistical analysis of serum congener profiles in an adult native american population. *Environmental Research* 98, 284–302.
Denison, M.S., Soshilov, A.A., He, G., Degroot, D.E. and Zhao, B. (2011) Exactly the same but different: promiscuity and diversity in the molecular mechanisms of action of the aryl hydrocarbon (dioxin) receptor. *Toxicological Sciences* 124, 1–22.
Do, Y. and Lee, D.K. (2012) Effects of polychlorinated biphenyls on the development of neuronal cells in growth period; structure-activity relationship. *Experimental Neurobiology* 21, 30–36.
Du, S., Belton, T.J. and Rodenburg, L.A. (2008) Source apportionment of polychlorinated biphenyls in the tidal Delaware river. *Environmental Science & Technology* 42, 4044–4051.
Du, S., Wall, S.I., Cacia, D. and Rodenburg, L.A. (2009) Passive air sampling for polychlorinated biphenyls in the Philadelphia metropolitan area. *Environmental Science & Technology* 43, 1287–1292.
Elnar, A.A., Desor, F., Legay, S., Nemos, C., Yen, F.T., Oster, T., Bohn, T. and Soulimani, R. (2016) No evidence for oxidative stress in the cerebellar tissues or cells of juvenile male mice exposed via lactation to the 6 non-dioxin-like PCBs at levels below the regulatory safe limits for humans. *Toxicology Letters* 245, 7–14.
Erickson, M.D. and Kaley, R.G. 2nd (2011) Applications of polychlorinated biphenyls. *Environmental Science and Pollution Research Intermatopma;* 18, 135–151.
Fonnum, F., Mariussen, E. and Reistad, T. (2006) Molecular mechanisms involved in the toxic effects of polychlorinated biphenyls (PCBs) and brominated flame retardants (BFRs). *Journal of Toxicology and Environmental Health A* 69, 21–35.
Fonnum, F. and Mariussen, E. (2009) Mechanisms involved in the neurotoxic effects of environmental toxicants such as polychlorinated biphenyls and brominated flame retardants. *Journal of Neurochemistry* 111, 1327–1347.
Frame, G.M., Cochran, J.W. and Bowadt, S.S. (1996) Complete pcb congener distributions for 17 aroclor mixtures determined by 3 hrgc systems optimized for comprehensive, quantitative, congener-specific analysis. *Journal of Separation Science* 19, 657–668.
Fritsche, E., Cline, J.E., Nguyen, N.H., Scanlan, T.S. and Abel, J. (2005) Polychlorinated biphenyls disturb differentiation of normal human neural progenitor cells: clue for involvement of thyroid hormone receptors. *Environmental Health Perspectives* 113, 871–876.
Goldey, E.S. and Crofton, K.M. (1998) Thyroxine replacement attenuates hypothyroxinemia, hearing loss, and motor deficits following developmental exposure to Aroclor 1254 in rats. *Toxicological Sciences* 45, 94–105.
Goldey, E.S., Kehn, L.S., Lau, C., Rehnberg, G.L. and Crofton, K.M. (1995) Developmental exposure to polychlorinated biphenyls (Aroclor 1254) reduces circulating thyroid hormone concentrations and causes hearing deficits in rats. *Toxicology and Applied Pharmacology* 135, 77–88.
Goodwill, M.H., Lawrence, D.A. and Seegal, R.F. (2007) Polychlorinated biphenyls induce proinflammatory cytokine release and dopaminergic dysfunction: protection in interleukin-6 knockout mice. *Journal of Neuroimmunology* 183, 125–132.
Guo, J., Capozzi, S.L., Kraeutler, T.M. and Rodenburg, L.A. (2014) Global distribution and local impacts of inadvertently generated polychlorinated biphenyls in pigments. *Environmental Science & Technology* 48, 8573–8580.
Hagmar, L. (2003) Polychlorinated biphenyls and thyroid status in humans: a review. *Thyroid* 13, 1021–1028.
Herkert, N.J., Jahnke, J.C. and Hornbuckle, K.C. (2018) Emissions of tetrachlorobiphenyls (PCBs 47, 51, and 68) from polymer resin on kitchen cabinets as a non-aroclor source to residential air. *Environmental Science & Technology* 52(9), 5154–5160.
Herrick, R.F., McClean, M.D., Meeker, J.D., Baxter, L.K. and Weymouth, G.A. (2004) An unrecognized source of PCB contamination in schools and other buildings. *Environmental Health Perspectives* 112, 1051–1053.
Herrick, R.F., Lefkowitz, D.J. and Weymouth, G.A. (2007) Soil contamination from PCB-containing buildings. *Environmental Health Perspectives* 115, 173–175.
Hopf, N.B., Ruder, A.M. and Succop, P. (2009) Background levels of polychlorinated biphenyls in the u.S. Population. *Science of the Total Environment* 407, 6109–6119.

Howard, A.S., Fitzpatrick, R., Pessah, I., Kostyniak, P. and Lein, P.J. (2003) Polychlorinated biphenyls induce caspase-dependent cell death in cultured embryonic rat hippocampal but not cortical neurons via activation of the ryanodine receptor. *Toxicology and Applied Pharmacology* 190, 72–86.

Hsu, S.T., Ma, C.I., Hsu, S.K., Wu, S.S., Hsu, N.H., Yeh, C.C. and Wu, S.B. (1985) Discovery and epidemiology of PCB poisoning in taiwan: a four-year followup. *Environmental Health Perspectives* 59, 5–10.

Hu, D., Martinez, A. and Hornbuckle, K.C. (2008) Discovery of non-Aroclor PCB (3,3′-dichlorobiphenyl) in Chicago air. *Environmental Science & Technology* 42, 7873–7877.

Ikeno, T., Miyashita, C., Nakajima, S., Kobayashi, S., Yamazaki, K. *et al.* (2017) Effects of low-level prenatal exposure to dioxins on cognitive development in Japanese children at 42 months. *Science of the Total Environment* 618, 1423–1430.

Inglefield, J.R. and Shafer, T.J. (2000) Polychlorinated biphenyl-stimulation of Ca(2+) oscillations in developing neocortical cells: a role for excitatory transmitters and l-type voltage-sensitive Ca(2+) channels. *Journal of Pharmacology and Experimental Therapy* 295, 105–113.

Inglefield, J.R., Mundy, W.R. and Shafer, T.J. (2001) Inositol 1,4,5-triphosphate receptor-sensitive Ca(2+) release, store-operated Ca(2+) entry, and cAMP responsive element binding protein phosphorylation in developing cortical cells following exposure to polychlorinated biphenyls. *Journal of Pharmacology and Experimental Therapy* 297, 762–773.

Jensen, S. (1972) The PCB story. *Ambio* 1, 123–131.

Jolous-Jamshidi, B., Cromwell, H.C., Mcfarland, A.M. and Meserve, L.A. (2010) Perinatal exposure to polychlorinated biphenyls alters social behaviors in rats. *Toxicology Letters* 199, 136–143.

Jorissen, J. (2007) Literature review. Outcomes associated with postnatal exposure to polychlorinated biphenyls (pcbs) via breast milk. *Advances in Neonatal Care* 7, 230–237.

Kang, J.H., Park, I.S., Oh, W.Y., Lim, H.K., Wang, S.Y. *et al.* (2004) Inhibition of aroclor 1254-induced depletion of stored calcium prevents the cell death in catecholaminergic cells. *Toxicology* 200, 93–101.

Kato, Y., Haraguchi, K., Yamazaki, T., Ito, Y., Miyajima, S. *et al.* (2003) Effects of polychlorinated biphenyls, kanechlor-500, on serum thyroid hormone levels in rats and mice. *Toxicological Sciences* 72, 235–241.

Kennedy, K.A., Sandiford, S.D., Skerjanc, I.S. and Li, S.S. (2012) Reactive oxygen species and the neuronal fate. *Cellular and Molecular Life Sciences* 69, 215–221.

Kennedy, M.B. (2000) Signal-processing machines at the postsynaptic density. *Science* 290, 750–754.

Kodavanti, P.R. and Tilson, H.A. (2000) Neurochemical effects of environmental chemicals: in vitro and in vivo correlations on second messenger pathways. *Annals of the New York Academy of Sciences* 919, 97–105.

Kodavanti, P.R., Shin, D.S., Tilson, H.A. and Harry, G.J. (1993) Comparative effects of two polychlorinated biphenyl congeners on calcium homeostasis in rat cerebellar granule cells. *Toxicology and Applied Pharmacology* 123, 97–106.

Kodavanti, P.R., Ward, T.R., Mckinney, J.D. and Tilson, H.A. (1996) Inhibition of microsomal and mitochondrial Ca2+-sequestration in rat cerebellum by polychlorinated biphenyl mixtures and congeners. Structure-activity relationships. *Archives of Toxicology* 70, 150–157.

Kodavanti, P.R., Derr-Yellin, E.C., Mundy, W.R., Shafer, T.J., Herr, D.W. *et al.* (1998) Repeated exposure of adult rats to Aroclor 1254 causes brain region-specific changes in intracellular ca2+ buffering and protein kinase c activity in the absence of changes in tyrosine hydroxylase. *Toxicology and Applied Pharmacology* 153, 186–198.

Koh, W.X., Hornbuckle, K.C. and Thorne, P.S. (2015) Human serum from urban and rural adolescents and their mothers shows exposure to polychlorinated biphenyls not found in commercial mixtures. *Environmental Science & Technology* 49, 8105–8112.

Konur, S. and Ghosh, A. (2005) Calcium signaling and the control of dendritic development. *Neuron* 46, 401–405.

Koopman-Esseboom, C., Weisglas-Kuperus, N., De Ridder, M.A., Van Der Paauw, C.G., Tuinstra, L.G. andSauer, P.J. (1996) Effects of polychlorinated biphenyl/dioxin exposure and feeding type on infants' mental and psychomotor development. *Pediatrics* 97, 700–706.

Kyriklaki, A., Vafeiadi, M., Kampouri, M., Koutra, K., Roumeliotaki, T. *et al.* (2016) Prenatal exposure to persistent organic pollutants in association with offspring neuropsychological development at 4years of age: The rhea mother-child cohort, crete, greece. *Environmental International* 97, 204–211.

Lanting, C.I., Huisman, M., Muskiet, F.A., Van Der Paauw, C.G., Essed, C.E. and Boersma, E.R. (1998) Polychlorinated biphenyls in adipose tissue, liver, and brain from nine stillborns of varying gestational ages. *Pediatric Research* 44, 222–225.

Lee, H.K. (2006) Synaptic plasticity and phosphorylation. *Pharmacology & Therapeutics* 112, 810–832.

Lee, D.W., Notter, S.A., Thiruchelvam, M., Dever, D.P., Fitzpatrick, R. *et al.* (2012) Subchronic polychlorinated biphenyl (aroclor 1254) exposure produces oxidative damage and neuronal death of ventral midbrain dopaminergic systems. *Toxicological Sciences* 125, 496–508.

Lein, P.J. (2017) Polychlorinated biphenyls (PCBs): a continuing environmental health concern. UC Davis Veterinary Medicine E-book, available at: https://www.openaccessgovernment.org/polychlorinated-biphenyls-pcbs-a-continuing-environmental-health-concern/40570/ (accessed 2 May 2019).

Lesiak, A., Zhu, M., Chen, H., Appleyard, S.M., Impey, S., Lein, P.J. and Wayman, G.A. (2014) The environmental neurotoxicant pcb 95 promotes synaptogenesis via ryanodine receptor-dependent mir132 upregulation. *Journal of Neuroscience* 34, 717–725.

Longnecker, M.P., Wolff, M.S., Gladen, B.C., Brock, J.W., Grandjean, P. *et al.* (2003) Comparison of polychlorinated biphenyl levels across studies of human neurodevelopment. *Environmental Health Perspectives* 111, 65–70.

Lyall, K., Croen, L.A., Sjodin, A., Yoshida, C.K., Zerbo, O., Kharrazi, M. and Windham, G.C. (2016) Polychlorinated biphenyl and organochlorine pesticide concentrations in maternal mid-pregnancy serum samples: Association with autism spectrum disorder and intellectual disability. *Environmental Health Perspectives* 125(3), 474–480.

Maervoet, J., Vermeir, G., Covaci, A., Van Larebeke, N., Koppen, G. *et al.* (2007) Association of thyroid hormone concentrations with levels of organochlorine compounds in cord blood of neonates. *Environmental Health Perspectives* 115, 1780–1786.

Marek, R.F., Thorne, P.S., Wang, K., Dewall, J. and Hornbuckle, K.C. (2013) Pcbs and oh-pcbs in serum from children and mothers in urban and rural U.S. communities. *Environmental Science & Technology* 47, 3353–3361.

Marek, R.F., Thorne, P.S., Herkert, N.J., Awad, A.M. and Hornbuckle, K.C. (2017) Airborne PCBs and OH-PCBs inside and outside urban and rural U.S. schools. *Environmental Science & Technology* 51, 7853–7860.

Mariussen, E. and Fonnum, F. (2001) The effect of polychlorinated biphenyls on the high affinity uptake of the neurotransmitters, dopamine, serotonin, glutamate and gaba, into rat brain synaptosomes. *Toxicology* 159, 11–21.

Mariussen, E. and Fonnum, F. (2006) Neurochemical targets and behavioral effects of organohalogen compounds: an update. *Critical Reviews in Toxicology* 36, 253–289.

Mariussen, E., Myhre, O., Reistad, T. and Fonnum, F. (2002) The polychlorinated biphenyl mixture aroclor 1254 induces death of rat cerebellar granule cells: the involvement of the n-methyl-d-aspartate receptor and reactive oxygen species. *Toxicology and Applied Pharmacology* 179, 137–144.

Matus, A. (2000) Actin-based plasticity in dendritic spines. *Science* 290, 754–758.

Mitoma, C., Uchi, H., Tsukimori, K., Yamada, H., Akahane, M., *et al.* (2015) Yusho and its latest findings – a review in studies conducted by the Yusho group. *Environment International* 82, 41–48.

Morse, D.C., Seegal, R.F., Borsch, K.O. and Brouwer, A. (1996) Long-term alterations in regional brain serotonin metabolism following maternal polychlorinated biphenyl exposure in the rat. *Neurotoxicology* 17, 631–638.

Mundy, W.R., Shafer, T.J., Tilson, H.A. and Kodavanti, P.R. (1999) Extracellular calcium is required for the polychlorinated biphenyl-induced increase of intracellular free calcium levels in cerebellar granule cell culture. *Toxicology* 136, 27–39.

Ness, D.K., Schantz, S.L., Moshtaghian, J. and Hansen, L.G. (1993) Effects of perinatal exposure to specific PCB congeners on thyroid hormone concentrations and thyroid histology in the rat. *Toxicology Letters* 68, 311–323.

Nowack, N., Wittsiepe, J., Kasper-Sonnenberg, M., Wilhelm, M. and Scholmerich, A. (2015) Influence of low-level prenatal exposure to PCDD/Fs and PCBs on empathizing, systemizing and autistic traits: Results from the duisburg birth cohort study. *PLoS ONE* 10, e0129906.

Olguín-Albuerne, M. and Morán, J. (2015) ROS produced by NOX2 controls in vitro development of cerebellar granule neurons development. *ASN Neuro* 7, published online 8 April 2015. doi: 10.1177/1759091415578712.

Oppenheimer, J.H. and Schwartz, H.L. (1997) Molecular basis of thyroid hormone-dependent brain development. *Endocrine Reviews* 18, 462–475.

Orenstein, S.T., Thurston, S.W., Bellinger, D.C., Schwartz, J.D., Amarasiriwardena, C.J., Altshul, L.M. and Korrick, S.A. (2014) Prenatal organochlorine and methylmercury exposure and memory and learning in school-age children in communities near the new Bedford Harbor Superfund site, Massachusetts. *Environmental Health Perspectives* 122, 1253–1259.

Pessah, I.N., Cherednichenko, G. and Lein, P.J. (2010) Minding the calcium store: ryanodine receptor activation as a convergent mechanism of pcb toxicity. *Pharmacology and Therapeutics* 125, 260–285.

Purkayastha, S., Fernando, S.S., Diallo, S., Cohen, L., Ranasinghe, B., Levano, K. and Banerjee, P. (2009) Regulation of protein kinase c isozymes during early postnatal hippocampal development. *Brain Research* 1288, 29–41.

Rodenburg, L.A., Guo, J., Du, S. and Cavallo, G.J. (2010) Evidence for unique and ubiquitous environmental sources of 3,3′-dichlorobiphenyl (PCB 11). *Environmental Science & Technology* 44, 2816–2821.
Rosenquist, A.H., Hoyer, B.B., Julvez, J., Sunyer, J., Pedersen, H.S., Lenters, V., Jonsson, B.a.G., Bonde, J.P. and Toft, G. (2017) Prenatal and postnatal PCB-153 and p,p′-DDE exposures and behavior scores at 5–9 years of age among children in Greenland and Ukraine. *Environmental Health Perspectives* 125, 107002.
Sable, H.J.K. and Schantz, S.L. (2006) Executive function following developmental exposure to polychlorinated biphenyls (PCBs): what animal models have told us. In: Levin, E.D. and Buccafusco, J.J. (eds) *Animal Models of Cognitive Impairment*. CRC Press, Boca Raton, Florida, pp. 147–168.
Sagiv, S.K., Thurston, S.W., Bellinger, D.C., Tolbert, P.E., Altshul, L.M. and Korrick, S.A. (2010) Prenatal organochlorine exposure and behaviors associated with attention deficit hyperactivity disorder in school-aged children. *American Journal of Epidemiology* 171, 593–601.
Schantz, S.L., Moshtaghian, J. and Ness, D.K. (1995) Spatial learning deficits in adult rats exposed to ortho-substituted PCB congeners during gestation and lactation. *Fundamental Applied Toxicology* 26, 117–126.
Schantz, S.L., Seo, B.W., Moshtaghian, J., Peterson, R.E. and Moore, R.W. (1996) Effects of gestational and lactational exposure to TCDD or coplanar PCBs on spatial learning. *Neurotoxicology and Teratology* 18, 305–313.
Schantz, S.L., Widholm, J.J. and Rice, D.C. (2003) Effects of PCB exposure on neuropsychological function in children. *Environmental Health Perspectives* 111, 357–576.
Seegal, R.F., Bush, B. and Shain, W. (1990) Lightly chlorinated ortho-substituted PCB congeners decrease dopamine in nonhuman primate brain and in tissue culture. *Toxicology and Applied Pharmacology* 106, 136–144.
Seegal, R.F., Bush, B. and Brosch, K.O. (1994) Decreases in dopamine concentrations in adult, non-human primate brain persist following removal from polychlorinated biphenyls. *Toxicology* 86, 71–87.
Seegal, R.F., Brosch, K.O. and Okoniewski, R.J. (1997) Effects of in utero and lactational exposure of the laboratory rat to 2,4,2′,4′- and 3,4,3′,4′-tetrachlorobiphenyl on dopamine function. *Toxicology and Applied Pharmacology* 146, 95–103.
Seegal, R.F., Okoniewski, R.J., Brosch, K.O. and Bemis, J.C. (2002) Polychlorinated biphenyls alter extraneuronal but not tissue dopamine concentrations in adult rat striatum: An in vivo microdialysis study. *Environmental Health Perspectives* 110, 1113–1117.
Sethi, S., Keil, K.P., Chen, H., Hayakawa, K., Li, X. *et al.* (2017) Detection of 3,3′-dichlorobiphenyl in human maternal plasma and its effects on axonal and dendritic growth in primary rat neurons. *Toxicological Sciences* 158(2) 409–411.
Soechitram, S.D., Athanasiadou, M., Hovander, L., Bergman, A. and Sauer, P.J. (2004) Fetal exposure to PCBs and their hydroxylated metabolites in a dutch cohort. *Environmental Health Perspectives* 112, 1208–1212.
Supekar, K., Uddin, L.Q., Khouzam, A., Phillips, J., Gaillard, W.D. *et al.* (2013) Brain hyperconnectivity in children with autism and its links to social deficits. *Cell Reports* 5, 738–747.
Tatsuta, N., Nakai, K., Murata, K., Suzuki, K., Iwai-Shimada, M. *et al.* (2014) Impacts of prenatal exposures to polychlorinated biphenyls, methylmercury, and lead on intellectual ability of 42-month-old children in japan. *Environmental Research* 133, 321–326.
Thomas, K., Xue, J., Williams, R., Jones, P., and Whitaker, D. (2012) *Polychlorinated Biphenyls (PCBs) in School Buildings: Sources, Environmental Levels, and Exposures*. EPA/600/R-12/051. US Environmental Protection Agency: Office of Research and Development, National Exposure Research Laboratory.
Thompson, M.R. and Boekelheide, K. (2013) Multiple environmental chemical exposures to lead, mercury and polychlorinated biphenyls among childbearing-aged women (NHANES 1999–2004): body burden and risk factors. *Environmental Research* 121, 23–30.
Tian, Y.H., Hwan Kim, S., Lee, S.Y. and Jang, C.G. (2011) Lactational and postnatal exposure to polychlorinated biphenyls induces sex-specific anxiolytic behavior and cognitive deficit in mice offspring. *Synapse* 65, 1032–1041.
Trilivas, I. and Brown, J.H. (1989) Increases in intracellular Ca2+ regulate the binding of [3h]phorbol 12,13-dibutyrate to intact 1321n1 astrocytoma cells. *Journal of Biological Chemistry* 264, 3102–3107.
Ulbrich, B. and Stahlmann, R. (2004) Developmental toxicity of polychlorinated biphenyls (PCBs): a systematic review of experimental data. *Archives of Toxicology* 78, 252–268.
Verner, M.A., Plusquellec, P., Desjardins, J.L., Cartier, C., Haddad, S. *et al.* (2015) Prenatal and early-life polychlorinated biphenyl (PCB) levels and behavior in inuit preschoolers. *Environment International* 78, 90–94.

Wayman, G.A., Bose, D.D., Yang, D., Lesiak, A., Bruun, D. *et al.* (2012a) PCB-95 modulates the calcium-dependent signaling pathway responsible for activity-dependent dendritic growth. *Environmental Health Perspectives* 120, 1003–1009.

Wayman, G.A., Yang, D., Bose, D.D., Lesiak, A., Ledoux, V. *et al.* (2012b) PCB-95 promotes dendritic growth via ryanodine receptor-dependent mechanisms. *Environmental Health Perspectives* 120, 997–1002.

White, S.S. and Birnbaum, L.S. (2009) An overview of the effects of dioxins and dioxin-like compounds on vertebrates, as documented in human and ecological epidemiology. *Journal of Environmental Science and Health, Part C: Environmental Carcinogenesis and Ecotoxicology Reviews* 27, 197–211.

Williams, G.R. (2008) Neurodevelopmental and neurophysiological actions of thyroid hormone. *Journal of Neuroendocrinology* 20, 784–794.

Winneke, G. (2011) Developmental aspects of environmental neurotoxicology: lessons from lead and polychlorinated biphenyls. *Journal of Neurological Sciences* 308, 9–15.

Winneke, G., Walkowiak, J. and Lilienthal, H. (2002) PCB-induced neurodevelopmental toxicity in human infants and its potential mediation by endocrine dysfunction. *Toxicology* 181–182, 161–165.

Wong, P.W., Brackney, W.R. and Pessah, I.N. (1997a) Ortho-substituted polychlorinated biphenyls alter microsomal calcium transport by direct interaction with ryanodine receptors of mammalian brain. *Journal of Biological Chemistry* 272, 15145–15153.

Wong, P.W., Joy, R.M., Albertson, T.E., Schantz, S.L. and Pessah, I.N. (1997b) Ortho-substituted 2,2′,3,5′,6-pentachlorobiphenyl (PCB 95) alters rat hippocampal ryanodine receptors and neuroplasticity in vitro: evidence for altered hippocampal function. *Neurotoxicology* 18, 443–456.

Yang, D. and Lein, P.J. (2010) Polychlorinated biphenyls increase apoptosis in the developing rat brain. *Current Neurobiology* 1, 70–76.

Yang, J.H. and Kodavanti, P.R. (2001) Possible molecular targets of halogenated aromatic hydrocarbons in neuronal cells. *Biochemical and Biophysical Research Communications* 280, 1372–1377.

Yang, J.H., Derr-Yellin, E.C. and Kodavanti, P.R. (2003) Alterations in brain protein kinase c isoforms following developmental exposure to a polychlorinated biphenyl mixture. *Brain Research. Molecular Brain Research* 111, 123–135.

Yang, D., Kim, K.H., Phimister, A., Bachstetter, A.D., Ward, T.R. *et al.* (2009) Developmental exposure to polychlorinated biphenyls interferes with experience-dependent dendritic plasticity and ryanodine receptor expression in weanling rats. *Environmental Health Perspectives* 117, 426–435.

Yang, D., Kania-Korwel, I., Ghogha, A., Chen, H., Stamou, M. *et al.* (2014) PCB 136 atropselectively alters morphometric and functional parameters of neuronal connectivity in cultured rat hippocampal neurons via ryanodine receptor-dependent mechanisms. *Toxicological Sciences* 138, 379–392.

Yuen, B., Boncompagni, S., Feng, W., Yang, T., Lopez, J.R. *et al.* (2012) Mice expressing t4826i-ryr1 are viable but exhibit sex- and genotype-dependent susceptibility to malignant hyperthermia and muscle damage. *FASEB Journal* 26, 1311–1322.

Zahalka, E.A., Ellis, D.H., Goldey, E.S., Stanton, M.E. and Lau, C. (2001) Perinatal exposure to polychlorinated biphenyls aroclor 1016 or 1254 did not alter brain catecholamines nor delayed alternation performance in long-evans rats. *Brain Research Bulletin* 55, 487–500.

Zhang, H., Yolton, K., Webster, G.M., Sjodin, A., Calafat, A.M. *et al.* (2016) Prenatal PBDE and PCB exposures and reading, cognition, and externalizing behavior in children. *Environmental Health Perspectives* 125(4) 746–752.

Zoeller, R.T. (2007) Environmental chemicals impacting the thyroid: targets and consequences. *Thyroid* 17, 811–817.

Zoeller, R.T., Dowling, A.L. and Vas, A.A. (2000) Developmental exposure to polychlorinated biphenyls exerts thyroid hormone-like effects on the expression of rc3/neurogranin and myelin basic protein messenger ribonucleic acids in the developing rat brain. *Endocrinology* 141, 181–189.

# 12 Dioxins I. Dynamics and Legal Directives in Europe

**M. Dopico and A. Gómez***
*Universidad de Oviedo, Gijón, Asturias, Spain*

## 12.1 Abstract

Polychlorinated dibenzo-*p*-dioxins and polychlorinated dibenzofurans are a group of aromatic hydrocarbons, emitted mostly from anthropogenic sources, that have negative effects on human health. Chlorine and carbon compounds, in combination with temperature and combustion processes, are the main factors in the production of these products that, depending on their structural configuration and formation conditions, may have higher or lower impact in the environment, along the different phases of emission, ambient distribution, final deposition and absorption.

Given their potential negative effect on human health, the importance of gaining better knowledge and analysing emission patterns of dioxins from different sources is critical in order to observe the actual trend of transmission of dioxins around the world and take measures to reduce their emission and minimize their impact in the atmosphere.

For these reasons, the main objective of this chapter is to provide a basic outline about the dioxin situation, going briefly through the historical evolution of dioxin emissions in order to understand the change in perception towards this chemical agent, followed by a description of the main sources of these pollutants, as well as the typical emission patterns and the most important factors that can affect ambient distribution. The production and formation of these compounds has been followed by important research on different minimization and abatement technologies, to reduce the formation and emission of dioxins to the atmosphere, in order to meet legal emission standards that become more restrictive over the years.

## 12.2 Introduction

Polychlorinated dibenzo-*p*-dioxins (PCDDs) and polychlorinated dibenzofurans (PCDFs) are a group of aromatic hydrocarbons formed by a structure of two benzene rings interconnected by a third oxygenated ring. Whether this oxygenated ring has one or two oxygen atoms establishes the difference between dioxins (two oxygen atoms in the intermediate ring) and furans (one oxygen atom in the intermediate ring). Chlorine is an important component of dioxins and furans. In fact, depending on the position of chlorine atoms within the structural configuration, different forms and congeners of these compounds can be formed. Theoretically, 75 PCDD and 135 PCDF congeners are possible, their physical and chemical properties being determined by the number of chlorine atoms and their respective position in the structure.

* E-mail address: albertogomez@uniovi.es

The major issue with dioxins is that, even though they are basically introduced in the environment via airborne sources after combustion processes, they are persistent pollutants that can become attached to body fat and remain there, entering the food chain and ending up being ingested by humans. For that reason, considering that the intake of dioxins and their consequences in human health is generally caused by this chain of events, the impact and risk of these compounds have followed a progressive tendency in which higher protection measures were implemented as more information was discovered.

One of the reasons for the late reaction in considering the high toxicity and carcinogenic effect of dioxins in industrial environments is that these products have never been produced for the market. They were the consequence of the production of other marketable products that involved combustion processes of organic compounds and chlorine. Formation of dioxins can take place under natural circumstances as well, so this secondary effect of industrial production, considering the difficulties in characterizing and identifying each possible congener, was neglected at first, since the effects on human health were not yet proven.

Industrial sources of dioxins have been present since the birth of the chemical industry in the early 1900s, especially in chlorine industries such as Leblanc-process soda, leading to a progressive increment in dioxin production and emission to the atmosphere that increased swiftly from the 1930s, moving to a higher production rate that was probably motivated by two important factors: first the industrial impact of World War II on manufacturing rates and processes, and then the introduction of products such as plastics as an important part of chemical industries. The combustion of these products pushed the generation of dioxins until maximum levels were reached in the 1970s, when the trend started to decrease and the emission of dioxins from industrial sources started to diminish, motivated by incipient social awareness, restrictive legal policies and the consequent and progressive installation of remediation and minimization technologies.

As science and research kept progressing and more was learnt about the effects of dioxins, the change in public and social perception was also motivated by particular incidents, such as: (i) human exposure in a herbicide plant of Monsanto in the USA in 1949, causing skin lesions in hundreds of workers; (ii) the release of a great amount of dioxins to nearby communities in Germany after an accident in 1953 at a BASF plant; (iii) the contamination in 1957 of millions of chickens in the USA due to dioxin inclusion in the fatty acids fed to the animals; (iv) the employment of Agent Orange during the Vietnam war (1962–1970) that contained 2378-TCDD, affecting millions of people; (v) the exposure to contaminated waste oils for a whole town in Times Beach, USA, in the 1970s; (vi) the explosion in 1976 of a chemical plant in Seveso, Italy, exposing thousands of people to a toxic cloud containing dioxins (Hites, 2010); or (vii) the evacuation of Love Canal in New York state in 1978; until finally scientific publications started to link 2378-TCDD with cancer rates in animals in the late 1970s and early 1980s (Kociba *et al.*, 1978), and consequent modification in legal regulations started to take place. But still, there were episodes in the following years that kept manifesting the risks of dioxins, like the crisis in Belgium in 1999 in which millions of chickens and thousands of other farm animals were contaminated with dioxin-like compounds.

All these events and increasing knowledge about dioxins motivated a tendency towards change in industrial activities, imposed by legal regulations. However, in the case of non-industrial sources a different trend could be observed. Nowadays, non-industrial sources of dioxins account for similar total values as the industrial sources because they have not experienced the same drastic reduction that industrial sources needed to achieve in order to fulfil legal policies. There are two basic reasons for this: (i) non-industrial sources comprise a great number of particular and small production sites, which are very difficult to identify and thus to regulate; and (ii) because social pressure is still not enough to change practices such as backyard burning.

## 12.3 Emission Patterns

The main source of dioxins and dioxin-like products is anthropogenic. Even though there is also evidence of their formation in nature, such as in forest fires and volcanic eruptions, it can be said that dioxin formation is directly associated with human interaction in combustion-related processes that involve the presence of chlorine and volatile organic compounds within the feed materials.

Given the nature of the transformations and the variability of the different sources, the predominant mechanism or pathway for the formation of dioxins can vary significantly from process to process. The reason for this is that there is no universal controlling factor during these transformations. For example, although carbon is a basic element for the formation of dioxins, there is a point where a precursor or a chemical reaction must push this carbon to assume an aromatic structure in order to continue and lead to the formation of dioxins. Globally speaking, there are two basic routes for the formation of dioxins: (i) from precursors such as chlorinated phenols; and (ii) from carbonaceous structures via de novo synthesis.

The first mechanism involves a gas-phase transformation in which different precursors react to form dioxins after several cyclization and chlorination reactions; while de novo synthesis generally involves a gas–solid mechanism, usually in the presence of catalysts, involving the breakdown of a carbon matrix to aromatic compounds in a series of oxidation and chlorination reactions. From these possibilities, de novo synthesis is usually indicated as the most plausible mechanism for the formation of dioxins in industrial environments (Buekens *et al.*, 2001), even though other authors point out that many studies have been developed without being able to determine the leading mechanism in their formation (Tuppurainen *et al.*, 1998).

Obviously, these two pathways are influenced by many other variables in the process, such as the following.

- The efficiency of combustion. Poor combustion conditions may enhance the formation of dioxins. These poor conditions are characterized by low oxygen levels, high moisture, low flame temperature, low residence time, or disturbances.
- Temperature of the process. Higher formation rates are found between 200°C and 450°C for de novo synthesis.
- Presence of catalysers, such as different metals like copper, iron, zinc, aluminium, chromium or manganese, or other compounds like polychlorinated biphenyls.
- Presence of inhibitors, like sulfur and nitrogen, even though they can improve the formation of other undesired by-products (UNEP, 2005).
- Surface area of particles and of course the presence and concentration of feed materials like oxygen and chlorine, independently if it is presented as HCl or $Cl_2$ (Addink *et al.*, 1995).

Besides the great number of different combustion processes that can yield these transformations, one of the key reasons for the variability of possible formation mechanisms is based on the great number of different congeners that can subsequently be formed after these transformations take place. Because dioxins and furans as a group include 210 possible congeners, minor modifications in the raw materials, combustion conditions, or operational variables can determine the formation of one congener or another, hence it is extremely difficult to characterize a formation pathway.

The configuration of different congeners is basically structured depending on the number of chlorine atoms within the chain and the position in which they are inserted. In the case of dioxins, congener configurations can reach a total of 75 possible compounds. In summary, Table 12.1 represents this distribution (Shields *et al.*, 2010).

**Table 12.1.** Congener distribution and nomenclature.

| No. of chlorine atoms | Group | Possible congeners |
|---|---|---|
| 1 | Monochlorodibenzo-*p*-dioxin (MCDD) | 2 |
| 2 | Dichlorodibenzo-*p*-dioxin (DCDD) | 10 |
| 3 | Trichlorodibenzo-*p*-dioxin (TrCDD) | 14 |
| 4 | Tetrachlorodibenzo-*p*-dioxin (TCDD) | 22 |
| 5 | Pentachlorodibenzo-*p*-dioxin (PeCDD) | 14 |
| 6 | Hexachlorodibenzo-*p*-dioxin (HxCDD) | 10 |
| 7 | Heptachlorodibenzo-*p*-dioxin (HpCDD) | 2 |
| 8 | Octachlorodibenzo-*p*-dioxin (OCDD) | 1 |

Concerning the toxicity of the congener, a typical rule of thumb is to consider that the most toxic compounds are those in which the chlorine atom is found in positions 2, 3, 7 and 8. For that reason, the most toxic dioxin congener has being identified to be 2,3,7,8-tetrachlorodibenzo-*p*-dioxin, or 2378-TCDD. In fact, this is the only toxic considered congener from the group of four chlorine atoms (out of 22 possible congeners). Regarding other dioxin compounds with the same chlorine pattern, there is a total of seven congeners out of the possible 75 that show chlorine in these positions, thus these are the most important ones from a toxicity point of view.

In the case of furans, an analogous characterization can be established in which, depending on the number of chlorine atoms inserted in the structure (also from one to eight), different congeners are classified until reaching a total of 135, including tetrachlorodibenzofurans (four chlorine atoms), the species that allow for a greater number of congeners (38 in this case). As in the case of dioxins, most toxic congeners are found when chlorine is in positions 2, 3, 7, and 8. From the possible total of 135 congeners, ten of them have this chlorine distribution.

As each congener has a different degree of associated toxicity, characterization cannot be done simply by raw quantification, but by analysing also the emission rate or emission factor that considers as well the toxicity and therefore the total risk associated to each source. Considering 2378-TCDD as the most toxic of all congeners, it is possible to associate a toxicity equivalent factor (TEF) with all the series in relation to 2378-TCDD (TEF = 1). This classification, based on real samples analysis, is continuously evolving as more accurate evaluations can keep adjusting these factors. For example, according to the most recent World Health Organization (WHO) classification, congeners like 12378-PeCDD are considered as dangerous as 2378-TCDD (TEF = 1), while others like 123678-HxCDD or OCDD have lower qualifications of TEF = 0.1 or TEF = 0.0003, respectively (Van der Berg *et al.*, 2006).

Once all congeners have been allocated their own TEF, in order to characterize the toxicity of the whole mix of the stream another concept can be used: denominated toxicity equivalency (TEQ). The main objective of this unit is to give a homogeneous idea of the toxicity of the sample by considering that all the congeners would behave as if they all were 2378-TCDD. In order to do so, it is necessary to multiply the concentration of each congener by its TEF and then add all the values to reach a final result that expresses toxicity as if it was all referring to 2378-TCDD, the most toxic congener. Since different organizations can establish different TEF rates for each congener, it is possible that different final values for TEQ of the mix could be obtained depending on the bibliographical source. Nevertheless, in order to keep a certain degree of concordance with other classifications, those differences should never be higher than 28% (Ren and Zheng, 2009).

In the end, given the difficulty in characterizing each fingerprint source, it is necessary to combine TEF and TEQ values from a concrete source in a concrete time, along with raw emission values of total mass per unit of time, in order to give sense to the interpretation of results, because the most important congeners from a mass point of view will not always be the same as the most important congeners from a TEQ perspective for the same considered source.

For that reason, it is complicated for typical inventories of dioxin emission on a worldwide scale to give an exact description of the current situation, not only because of the difficult quantification in mass per unit of time of diffuse non-industrial sources, but also because, even when referring to industrial sources, the same processes can lead to different TEQ patterns depending on raw materials and operational values. That is why TEQ information should be complemented with data from the process like feed flow from the plant, fuel consumption or the definition of raw materials and process stages.

This kind of endeavour is only possible for big organizations like the Environment Protection Agency (EPA) in the USA (EPA, 2006), which can measure the evolution in emissions: for example, the reduction of emissions from municipal waste incineration (going from 9500 g TEQ in year 1987 to 77 g TEQ in year 2000); or how, as time goes by, industrial emissions decrease their emissions while non-industrial emissions remain constant: thus in 2000, from the top three generators of dioxins, forest fires and backyard burning led the classification with 600–700 g TEQ.

## 12.4 Main Sources of Dioxins

It is known that the formation of dioxins is characterized by conditions in which organic compounds react with chlorine species during combustion processes. This definition can be extended to include situations of high temperature after combustion in which fly ash or unburnt carbon still has the capacity to transform into cyclic species that can lead to the formation of dioxins. For that reason, the group of industrial processes that can provide the conditions for the formation of dioxins constitutes quite a wide range of possibilities, to which it is important to add as well those non-industrial sources, smaller in magnitude but higher in frequency, that also contribute in great measure to the emissions into the atmosphere.

Basically, the most important industrial sources of dioxins are waste combustion, metal production, coke making and sintering, pulp and paper mills, wastewater treatment, cement production, coal power plants, chemical plants and crematory incineration. Waste combustion and metal production are currently the main sources of dioxins in industrial environments.

On the other hand, dioxin emissions from non-industrial sources can proceed from many controlled and uncontrolled combustion sources, the most important contributors being accidental fires, diesel engines, release of PCDD/F from pentachlorophenol-treated wood products, illegal incineration of household wastes and domestic heating (Quass *et al.*, 2004).

### 12.4.1 Industrial sources

Concerning industrial waste combustion, special attention is needed for all possible alternatives, whether referring to municipal solid waste incinerators (MSWI), hazardous waste incinerators (HWI), medical waste incinerators (MWI) or industrial waste incinerators (IWI), since the formation of different congeners can be expected. Nevertheless, even though HWI or IWI might yield emissions with higher toxicity levels of dioxins, the number of MSWI installed over the world are significantly higher than other types of waste incinerators and therefore emission from MSWI is the most important in absolute values, serving as a good example that, when referring to dioxins, emission factors and total production must be pondered when the emissions are quantified (Lee *et al.*, 2003).

The formation of dioxins from metal production is to be considered not only because process conditions promote the formation of dioxins, but for the presence as well of metals that can act as catalysts in these transformations. Within this group, zinc, copper, lead and aluminium metallurgies can be highlighted, especially in the case of zinc manufacturing, in which the content of 2378-TCDD can reach concentrations as high as 6% in the stack gas (Ba *et al.*, 2009a, b).

Steel manufacturing from coal and iron is also a concern regarding dioxin emissions, but with one particularity in relation to previous metal-smelting industries. In the case of steel production, the most sensitive stages in the process are not related to the core transformations in steel shops, but in the preceding steps of coke formation and sinter production (Aries *et al.*, 2006). Sinter and coke constitute basic raw materials for the production of hot metal that will later be transformed into steel, and it is precisely the composition of these raw materials of coal and iron that promotes the conditions for dioxin formation and emission.

Usually, the emission of dioxins to the atmosphere from industrial sources is associated with air emission, since it is the most common event. However, wastewater treatment plants are the perfect example of dioxin transference to water and soils (Wcgiel *et al.*, 2014). The operation in these installations is associated with the formation of by-products like sludges, generating a residue with similar fingerprints to the original water, and that should be dealt with prior to definitive disposal.

It has already been said that the source and origin of raw materials can influence final production of dioxins. Industries such as paper mills, in which chlorine can be directly a feed flow in the process, are one example that highlights the importance of the operational configuration in relation with dioxin formation and emission (Zheng *et al.*, 2001). Another example is given by cement plants, which are commonly used to recycle different types of residues such as used tyres, fly ash from power plants, blast furnace slag, animal residues etc., which modifies input concentration; therefore, depending on the operation regimen, different congeners and total volume of dioxins can be formed (Rivera-Austrui *et al.*, 2014).

Finally, there is another category of industrial sources that when analysed individually do not possess a great capacity for dioxin generation, but when considering the total amount of these type of installations around the world, global emissions gain importance. This is the case with coal power plants (Lin *et al.*, 2010) and crematoria (Chiu *et al.*, 2011).

### 12.4.2 Non-industrial sources

The advantage of industrial sources is that they refer to specific locations, thus measurement, control and legislation are more easily applied, and this is the key parameter that allowed for the reducing tendency that has been observed over the years. Emission limits can be imposed, and different industries must meet them in order to avoid legal and economic consequences. On the other hand, when referring to non-industrial sources with multiple diffuse origins in uncontrolled combustion processes, it is more difficult to implement restrictions since the emission of dioxins can depend on: (i) accidental variables, as in the case of forest fires; (ii) geographical and cultural variables, such as in domestic heating systems and wood combustion; (iii) life habits, like selection of use of diesel vehicles; (iv) particular country legislation regarding backyard burning; and(v) in the end, the degree of social awareness of the population.

For that reason, in the case of non-industrial sources, it is even more complicated to develop a full characterization of the emission sources, so fewer conclusions can be drawn. For example, it is known that the application of conservative treatments to wood furniture like varnishing generally involves the application of chlorine compounds. So, in the case of burning these products for heating purposes, a higher amount of dioxins will be generated (Tame *et al.*, 2007). In fact, nowadays, open waste burning, backyard barrel burning and biomass waste burning in domestic environments are already showing higher emission factors than those from modern industrial waste incinerators, because poor and uncontrolled combustion in these domestic environments is also aggravated by the variability in feed composition and lack of proper remediation technologies (Solorzano-Ochoa *et al.*, 2012; Wu *et al.*, 2014).

In the case of diesel engines, even though their individual contribution might not be especially important and with congeners of low toxicity, the great numbers of vehicles in circulation in the world are translated into a total dioxin emission that must be considered (Chang *et al.*, 2014). However, this tendency is being changed by new legislation in Europe regarding diesel vehicle manufacturing and use, as well as the promotion and encouragement of electric cars, in which diesel is no longer needed.

### 12.4.3 Congener profiles of different sources

There are many studies that over the years have analysed the congener profile and emission factors of many industrial and non-industrial sources. However, this type of comparison must not only contemplate specific process variables or feed materials, depending on the industrial goal, but it is important to consider as well the date when the study took place, because given the evolution in the policies to restrict dioxin emissions, studies might have taken place in a period of time that no longer represents the current disposition; thus the congener profile might not be viable for comparison against other installations.

Nevertheless, as long as these issues are kept in mind, it is possible to try to summarize the information from published reviews (Dopico and Gomez, 2015) in order to determine what are the main typical congeners, in other words, the most abundant components from the total fingerprint that are expected to be formed depending on the source. Table 12.2 attempts to summarize this tendency.

The first immediate conclusion that can be drawn from Table 12.2 is that the most dangerous compound, 2378-TCDD, is not present among the most abundant congeners in any of the typical sources. On the other hand, OCDD and 1234678-HpCDF are common important congeners among the fingerprints from different sources.

**Table 12.2.** Typical dioxin predominant congeners from different sources[a]

| | 2378-TeCDF | 23478-PeCDD | 23478-PeCDF | 123478-HxCDF | 1234678-HpCDD | 1234678-HpCDF | OCDD | OCDF |
|---|---|---|---|---|---|---|---|---|
| MSWI | | | X | X | X | X | X | |
| MWI/HWI/IWI | | X | X | | | X | X | |
| Zn/Cu/Al/Pb | X | | | | | | | X |
| Sintering plant | X | | X | | | X | | X |
| Coke plant | X | | | | | X | X | X |
| Wastewater | X | | | | X | X | X | X |
| Paper mill | | X | | | | | X | |
| Cement | X | | X | | | X | X | X |
| Coal power | | | X | | | X | X | |
| Crematory | | | | | X | X | X | |
| Wood comb. | X | | | X | X | X | | |
| Diesel engine | | | | | X | X | X | X |

[a]Distribution indicated by X.

Table 12.1 referred to the total amount generated of each congener, but, as said before, when depicting a congener profile it is important to consider not only raw mass formation, but final TEQ in the mix. This is more dependent on the particular situation in each source, since higher toxic congeners are generally presented in the lower concentration intervals, thus minor changes in the fingerprint can yield significant changes in TEQ expression.

## 12.5 Factors Affecting Ambient Distributions

Once dioxins are formed and emitted to the atmosphere, different interactions can be expected in ambient air, water and soils, depending in different conditions. Deposition from air to soil, vegetation and water can occur by wet and dry processes or in mist. According to dispersion models, from the annual emission rate (approximately 287 kg TEQ), it has been predicted that 3% remains in the air, 57% falls to land areas, while the remaining 40% is received by ocean waters (Booth *et al.*, 2013). Later on, dioxins are going to interact in the environment: there is the possibility of resuspension or revolatilization of the less chlorinated congeners back to air if certain conditions are met; dioxins can move from land to water via runoffs; dioxins in water can absorb organic matter and settle out in sediments, persisting for years in a slow degradation, etc.

Before that, there are already several variables in the point of origin that will determine ambient distribution. In addition to process variables like feed material or the efficiency of abatement technologies, influencing parameters such as geographical, demographical, economical or climatological conditions in the context of the origin of the emission, whether referring to industrial or non-industrial sources, can play a key role in the subsequent distribution of dioxins. For example, noting the seasonal evolution, higher concentrations are found during winter months, when more domestic heating devices are working. Other factors such as meteorological conditions and height affect air movement and, consequently, final dispersion, concentration and dilution mechanisms (Gunes and Saral, 2014).

In relation to demographic variables, tradition and cultural issues play an important role too, which is subsequently translated in different ambient distribution of dioxins. This is manifested in the divergences between Asian and European industries (Zheng *et al.*, 2001). For example, in Europe paper mills generally use wood or recycled paper for fibre, while in China non-wood plant fibres are employed as raw material (mainly cereals, rice and reeds). Congener profiles are then different and their posterior interaction in the atmosphere is expected to differ as well. Another example of these cultural divergences is found at open burning of biomass, where rice straw is an extremely common raw material in Asian countries, while in Europe other feeding materials with their associated congener fingerprint are used (Shih *et al.*, 2008).

In addition, since dioxins fall under the category of persistent organic pollutants (POPs), a certain 'memory effect' should be considered, opening the opportunity for continuous interaction with the atmosphere long after its direct emission (Trivedi and Majumbar, 2013). This can lead to a 'secondary emission', that is, the presence of dioxins that are not related to a direct emission, but as a result of different transformations and transportation mechanisms that can diffuse the presence of dioxins to locations far away from the original source, yielding the possibility of a concrete area in which the dioxin accumulation rate is higher than the dioxin production rate (Booth *et al.*, 2013). This phenomenon explains why remote regions without local sources, like the Arctic or Antarctic continent, show certain levels of dioxin concentration in the air due to long-range transportation mechanisms.

These particularities, as well as traffic and the combination of deposition mechanisms from different sources, make it near impossible to associate pollution levels with a concrete source even though it is known that industrial sources are critical origin points of dioxin emission. For that reason, with the influence of a diffuse network of many sources, both industrial and non-industrial, which combine and deposit in the atmosphere, the gap between urban and rural ambient concentrations is not as significant as should be expected (Wohrnschimmel and Yao, 2014). Nevertheless, there is still a higher influence of industrial sites over urban and remote ones.

One key parameter that influences transportation mechanisms is that dioxins, as semivolatile agents, can exist in gaseous form or bound to particles, depending on formation and ambient conditions such as temperature or initial particle concentration in the air. This repartition is also related to seasonal parameters: in summer, higher temperatures are translated into higher concentration in the gas phase of the less chlorinated congeners, while in winter the split is more balanced. This separation between phases is a determinant for ambient distribution, since dioxins in vapour phase can undergo photochemical transformations to less toxic compounds, while particles are more resistant to this degradation (Watterson *et al.*, 1999).

In the end, the behaviour of dioxins is determined by physical and chemical properties like their low vapour pressure, low solubility in water, solubility in organic bodies and binding properties to organic matter in soil.

In summary, time interval, type of source, geography, meteorology and cultural issues are the main parameters that can determine from the origin how the intensity and homogeneity of dioxin distribution in the atmosphere can take place later on, as well as other important variables like releasing height, temperature and particle size that have an important role. After that, persistent properties of dioxins allow for further transportation and transformations that can lead to a secondary type of emission. Hence, in the end, given the numerous sources of dioxins and their persistent qualities, it is extremely difficult to associate ambient concentrations to a particular source: it is more accurate to state that it is a result of a mix of many points of origin and transportation mechanisms.

## 12.6 Remediation and Other Methodologies

In order to reduce dioxin emissions from a concrete source, first it is important to determine what congeners are being formed in order to select the most adequate control and removal technology. However, given the great number of possible congeners that can be formed and the small presence of many of them in a typical sample, the analysis and characterization of the exact fingerprint of each process is not an easy task. Nevertheless, there has been a great effort over the past decades to improve characterization techniques with more precise methodologies, given their important role and interest.

Basically, as happens in the field of chemical analysis, a good preparation and cleaning of the sample is one of the most important steps for the characterization of the dioxins, including chemical treatments and other low-pressure flow methods for conditioning the sample and removing any interference (method EN1948 is recommended by the European Committee of Standardization). Later, the combination of high-resolution gas chromatography and high-resolution mass spectrometry technologies seems to be the optimal strategy to gain proper sensitivity in order to detect different congeners (Shields *et al.*, 2010).

Remediation technologies are based on the minimization of the formation and emission of dioxins. In the field of non-industrial sources, remediation techniques are generally reduced to energy efficiency measures (such as upgrading heating equipment, insulation and monitoring devices), banning backyard burning, promotion of district heating in households and replacement of fuel by renewable energy sources, awareness raising, or favouring subsidy measures. On the other hand, in industrial environments there is more room for actuation. If the destruction of the formed dioxins is sought, methods with high temperature are required since dioxins are stable up to 800°C and not totally destroyed until 1300°C (BiPRO, 2009). Nevertheless, this chapter will focus on the existing mechanisms to control and reduce dioxins in industrial environments during the formation stage and the avoidance of their release into the environment, which can be classified into primary and secondary methods, respectively. While primary methods usually deal with modifications in the industrial activity towards preferable conditions that minimize formation, secondary methods involve the addition of chemical agents to capture the dioxins that have already been formed. In general, even though primary methods are effective in reducing the formation of dioxins, it is generally not sufficient and some secondary methods should be used as well to improve mitigation performance.

Primary methods pay attention to the agents and the conditions involved during the formation stage, and try to alter them without compromising the industrial operation. Under this assumption, there are three basic mitigation alternatives:

- reduction of concentration of chemical agents that form dioxins;
- employment of inhibitors that decelerate the formation mechanisms; and
- modification of the operational configuration in a way that avoids the conditions that promotes the formation of dioxins.

For example, recirculation of the off-gas to reduce molecular oxygen (Buekens *et al.*, 2001) or application of feed pre-treatments to reduce the presence of chlorine, are strategies that follow the premise of reducing agents that promote dioxin formation. On occasions when it is not possible to minimize in origin the presence of dioxins reactants, primary methods must use different inhibitors that reduce and interrupt the formation mechanisms of dioxins. Interesting inhibitors in this field are agents such as triethanolamine, $SO_2$ (high-sulfur coal), CaO or nitrogen compounds (Buekens *et al.*, 2001; Wielgosinski, 2011).

One final group of measures that fall under the category of primary methods refer to the modification in the design of the equipment, acting in important conditionings such as combustion efficiency or post-combustion temperature, like minimizing the formation of dust, or improving the cooling of the off-gas section to reach safer temperatures quickly from a dioxin formation point of view (Zeng *et al.*, 2011).

Typical secondary methods are designed to capture dioxins that have been formed but without destroying them, so that dioxins are simply transferred to a different medium which should be dealt with later. Most of these methods are based on the following technologies:

- absorption/adsorption technologies focused in gas streams;
- filtration technologies focused in particle dust; or
- newer methods involving plasma, electron irradiation, oxidation catalysis etc. directed to the degradation of dioxins as well.

In the first case, the selection of the capture agent is the key parameter for an efficient operation. The broad range of possible adsorbents can go from typical active carbon to more elaborate solutions including polymers like polypropylene (Lindgren and Andersson, 2008), carbon nanotubes (Wielgosinski, 2011), or ionic liquids (Pan *et al.*, 2013).

On the other hand, filtration technologies are useful for scenarios in which dioxins are concentrated in the surface of particle dust. Filters or electrostatic precipitators, alone or in combination with scrubbers and absorption agents, are typical configurations when referring to dioxin capture and mitigation (Buekens *et al.*, 2001).

## 12.7 Legal Directives in Europe

Since dioxins are a global issue with worldwide implications, joint efforts are required between nations in order to address the health and environmental concerns due to these pollutants. For that reason the European Commission (EC) adopted a communication to the European Parliament in 2001 that achieved as a result the setting of a Community strategy for dioxins, furans and PCBs (Dioxin Strategy) (EC, 2010). The Dioxin Strategy consists of two parts, addressing the reduction of the presence of dioxins both in the atmosphere, and in feed and food.

Regarding the activities focused on reducing the presence of dioxins in the environment, it was agreed to contribute and implement many multilateral environmental agreements such as the Regulation (EC) 850/2004 in order to implement the Stockholm Convention (which aims to reduce and ultimately cease the manufacture, use and emission of POPs) and the 1998 Aarhus Protocol on POPs under the convention on Long-Range Transboundary Air Pollution within the European Union (EU). Besides the obligation of the member states to set up emission inventories of dioxins, this regulation also enforces measures regarding production, use of chemicals, waste management and actions to reduce unintentional releases of dioxins and other POPs. POPs protocol was amended in 2009, including further requirements for reduction of dioxins, and setting emission limit values for several industrial facilities such as waste incinerators or sinter plants, which are already covered by existing EU legislation.

Regulations are not only focused on air emissions, but also include a number of specific EU environmental quality and emission standards for POPs in coastal and inland waters or drinking waters, focusing on the required quality of the receiving waters, such as Directives 76/464/EC and 86/280/EC on dangerous substances, and the Water Framework Directive (2000/98/EC) (EEA, 2017b). Regarding soil contamination, according to the Sixth Environment Action Programme in 2006, it is also a proposal to identify and ensure the remediation of contaminated sites (EC, 2010).

In addition the EC gives financial support to the secretariat of the Stockholm Convention of POPs for elaboration and updating the *Standardized Toolkit for Identification and Quantification of Dioxin and Furan Releases*, which includes expected emission factors from different sources and classification of their potential release routes.

One important part of the Dioxin Strategy resides in the preparation of emission inventories that allow measuring the evolution over time in different areas. EU Member States report their emissions of different pollutants under Directive (EU) 2016/2284 on the reduction of certain atmospheric pollutants, which in the case of dioxins apply on annual basis. Each member state is responsible for reporting this inventory in order for the EC to prepare and develop a final EU inventory. According to the annual EU emission inventory report for 1990–2015 (EEA, 2017a), emission of the main dioxins and furans have dropped substantially since 1990, by around 85% (1848 g I-TEQ in 2015 for EU-28). Currently, members that are reporting higher contributions are Poland, Italy and the UK, while other countries like Greece have not reported any data; thus this total value has been underestimated. This significant reduction is the combination of greater industrial control and abatement techniques encouraged by more restrictive legislation. As a result, nowadays, around 40% of the total share of dioxin emission in Europe is predicted to be from commercial, institutional and households (non-industrial sources); followed by 16% from energy use in industry, 15% from industrial processes sector and 14% from waste sector. However, these reports are based on the recompilation of available data; thus real data from non-industrial sources are more difficult to clarify. Nevertheless, although emission trends from non-industrial sources, have experienced a reduction as well, this has been done with lower intensity (reduction of 60% since 1990).

Referring now to the regulation of legal limits associated to each industry, this is a more complicated issue because, even though there are examples like waste incinerators that have indeed established a limit of 0.1 ng TEQ/$m^{-3}$ according with directive 2000/76/EC, there is the case of countries or industries in which a limit value has not been fixed, given the variation in processes that make it difficult to establish a standardized limit. In those cases, the rule of thumb regarding reduction of industrial emissions is the implementation and enforcement of the IPPC Directive, which is envisaged by the EU Dioxin Strategy and can be considered the most important instrument to reduce industrial dioxin emissions, since installations covered by the directive must reach levels of Best Available Technologies (BAT), given that the emission values presented in BAT will be mandatory for installations under BAT conclusions within 4 years from publication (Paradiz and Dilara, 2003). Assumption of BAT implies the application of the most effective and advanced stage for providing the basis for emission limit values designed to reduce emissions and their impact.

Within the scope of dioxin emission from industrial sources, the following is a list of regulations at European level that complement Dioxin Strategy (BiPRO, 2009; SEPA, 2013).

- Directive 2010/75/EU (IED). Covering the majority of large industries in the EU, this is a compilation revision of the IPPC directive and six sectorial EU directives, to include a total of around 52,000 installations. It also includes some emission limits, such as 0.3 ng/l for discharges of wastewater from the cleaning of waste gases from co-incineration plants. This regulation has replaced older regulations such as Directive 2008/1/EC (concerning integrated pollution prevention and control), Directive 2001/80/EC (concerning the limitation of emissions of pollutants from large combustors) and Directive 2000/76/EC (regarding waste incineration).
- Regulation EC 850/2004 on persistent organic pollutants. The objective of this regulation is to protect human health and the environment by reducing and finally prohibiting the release of POPs. Each party to the Stockholm Convention has to establish an implementation plan to show the concrete action that will be taken against the POPs listed in the convention.
- Waste framework directive 2008/98/EC. This establishes the basic definitions regarding waste management and how to distinguish between waste and by-product.
- Council directive 1999/31/EC. This states the measures that must be taken to minimize risk and hazards coming from landfill among others fires.
- Regulation EC 1069/2009. This relates to the procedures for the regulation of disposal of animal by-products not intended for human consumption, via methods such as incineration of the corpses.

Finally, regulations affecting the domestic sector and thus non-industrial sources relate to general policies such as Directive 2008/50/EC or Directive 2001/81/EC in relation to air quality and emission ceilings, respectively; or Directive 2005/32/EC, in relation to the eco-design of energy-using products and their requirements regarding environmentally relevant product characteristics, including solid-fuel boilers and domestic heaters.

## 12.8 Conclusions

The heterogeneity of the different forms and congeners of dioxins and furans is so broad that, depending on the formation source, different emission patterns are expected. But even when referring to the same source (whether it is an industrial process or a forest fire), variations in the raw materials, process variables, or even the geographical or demographic regions around the world can lead to significant differences in the fingerprint and behaviour of dioxins. Being persistent organic pollutants, dioxins can experience transformations after their emission, depending on different physical and chemical variables; thus their quantification is a complex task in which determining the traceability of ambient concentration is almost impossible, it being preferable to observe general tendencies of measurements in origin. For example, over the past decades, industrial sources of dioxins have experienced a great reduction thanks to the improvements in abatement methods and stricter legislation. On the other hand, among the challenges to face in the future there is a need to increase social awareness against non-industrial sources, harder to control and measure, which has not experienced the same trend and includes nowadays some of the highest contributors.

Regulation is always easier to implement on industrial sources. In fact, right now the main EU source-oriented dioxin policy instruments still do not target small sources, which added to the current lack of capacities in the field of monitoring of the emissions could be identified as a barrier for effective implementation of EU legislation in some countries. For that reason, national approaches are demanded to improve this situation.

## References

Addink, R., Bakker, W. C. and Olie, K. (1995) Influence of HCl and $Cl_2$ on the formation of polychlorinated dibenzo-*p*-dioxins/dibenzofurans in a carbon/fly ash mixture. *Environmental Science and Technology* 29, 2055–2058. doi: 10.1021/es00008a026.

Aries, E., Anderson, D.R., Fisher, R., Fray, T.A. and Hemfrey, D. (2006) PCDD/F and 'Dioxin-like' PCB emissions from iron ore sintering plants in the UK. *Chemosphere* 65(9), 1470–1480. doi: 10.1016/j.chemosphere.2006.04.020.

Ba, T., Zheng, M., Zhang, B., Liu, W., Su, G. and Xiao, K. (2009a) Estimation and characterization of PCDD/Fs and dioxin-like PCB emission from secondary zinc and lead metallurgies in China. *Journal of Environmental Monitoring* 11, 867–872. doi: 10.1039/b818555g.

Ba, T., Zheng, M., Zhang, B., Liu, W., Xiao, K., & Zhang, L. (2009b) Estimation and characterization of PCDD/Fs and dioxin-like PCBs from secondary copper and aluminum metallurgies in China. *Chemosphere* 75, 1173–1178. doi: 10.1016/j.chemosphere.2009.02.052.

BiPRO (2009) *Information Exchange on Reduction of Dioxin Emissions from Domestic Sources*. Beratungsgesellschaft für integrierte Problemlösungen, Munich, for European Commission.

Booth, S., Hui, J., Alojado, Z., Lam, V., Cheung, W. *et al.* (2013) Global deposition of airborne dioxin. *Marine Pollution Bulletin* 75, 182–186. doi: 10.1016/j.marpolbul.2013.07.041.

Buekens, A., Stieglitz, L., Hell, K., Huang, H. and Segers, P. (2001). Dioxins from thermal and metallurgical processes: recent studies for the iron and steel industry. *Chemosphere* 42, 729-735. doi: 10.1016/S0045-6535(00)00247-2.

Chang, Y., Lee, W., Wang, L., Yang, H., Cheng, M. *et al.* (2014) Effects of waste cooking oil-based biodiesel on the toxic organic pollutant emissions from a diesel engine. *Applied Energy* 113, 631–638. doi: 10.1016/j.apenergy.2013.08.005.

Chiu, J.-C., Shen, Y.-H., Li, H.-W., Lin, L.-F., Wang, L.-C. *et al.* (2011) Emissions of polychlorinated dibenzo-*p*-dioxins and dibenzofurans from an electric arc furnace, secondary aluminum smelter, crematory and joss paper incinerators. *Aerosol and Air Quality Research* 11, 13–20. doi: 10.4209/aaqr.2010.06.0051.

Dopico, M. and Gomez, A. (2015) Review of the current state and main sources of dioxins around the world. *Journal of the Air and Waste Management Association* 65(9), 1033–1049. doi: 10.1080/10962247.2015.1058869.

EC (2010) *Communication from the Commission to the Council, the European Parliament, the European Economic and Social Committee. On the implementation of the Community Strategy for dioxins, furans, and polychlorinated biphenyls (COM(2001)593) – Third progress report*. European Commission, Brussels.

EEA (2017a) *European Union Emission Inventory Report 1990–2015 under the UNECE Convention on Long-range Transboundary Air Pollution (LRTAP)*. EEA Report 9/2017. European Environment Agency, Copenhagen.

EEA (2017b) *Persistent Organic Pollutant Emissions*. European Environment Agency, Copenhagen.

EPA (2006) *An Inventory of Sources and Environmental Releases of Dioxin-Like Compounds in the United States for the Years 1987, 1995, and 2000*. EPA/600/P-03/002F. National Center for Environmental Assessment, Office of Research and Development, US Environmental Protection Agency, Washington, DC.

Gunes, G. and Saral, A. (2014) Seasonal variation of PCDD/Fs in the metropolis of Istanbul, Turkey. *Environmental Science and Pollution Research* 1-12. doi: 10.1007/s11356-014-2798-7.

Hites, R. A. (2010 Dioxins: an overview and history. *Environmental Science and Technology* 45, 16–20. doi: 10.1021/es1013664.

Kociba, R., Keyes, D., Beyer, J., Carreon, R., Wade, C. *et al.* (1978) Results of a two-year chronic toxicity and oncogenicity study of 2,3,7,8-tetrachlorodibenzo-p-dioxin in rats. *Toxicology and Applied Pharmacology* 46, 279–303. doi: 10.1016/0041-008X(78)90075-3.

Lee, W.-S., Chang-Chien, G.-P., Wang, L.-C., Lee, W.-J., Tsai, P.-J. and Chen, C.-K. (2003) Emissions of polychlorinated dibenzo-*p*-dioxins and dibenzofurans from the incinerations of both medical and municipal solid wastes. *Aerosol and Air Quality Research* 3, 1–6. doi: 10.4209/aaqr.2003.06.0001.

Lin, W., Wu, Y., Tu, L., Wang, L. and Lu, X. (2010) The emission and distribution of PCDD/Fs in municipal solid waste incinerators and coal-fired power plant. *Aerosol and Air Quality Research* 10, 519–532. doi: 10.4209/aaqr.2010.03.0017.

Lindgren, P. and Andersson, S. (2008) Adiox (R) for dioxin removal in wet scrubbers and semi-wet or dry absorbers. *WIT Transactions on Ecology and the Environment* 109, 569–577. doi: 10.2495/WM080581.

Pan, W., Qi, Y., Wang, R., Han, Z., Zhang, D. and Zhan, J. (2013) Adsorption of TCDD with 1-butyl-3-methylimidazolium dicyanamide ionic liquid: a combined molecular dynamics simulation and quantum chemistry study. *Chemosphere* 91(2), 157–164. doi: 10.1016/j.chemosphere.2012.12.021.

Paradiz, B. and Dilara, P. (2003) *Dioxin Emissions in the Candidate Countries: sources, emission inventories, reduction policies and measures*. EUR 20779 EN. European Commission Institute for Environment and Sustainability Emissions and Health Unit, Ispra, Italy.

Quass, U., Fermann, M. and Broker, G. (2004) The European dioxin air emission inventory project – final results. *Chemosphere* 54, 1319–1327. doi: 10.1016/S0045-6535(03)00251-0.

Ren, Z. and Zheng, M. (2009) Impacts of human activities on dioxins emissions at national scale. *Chemosphere* 76, 853–859. doi: 10.1016/j.chemosphere.2009.03.070.

Rivera-Austrui, J., Martinez, K., Marco-Almagro, L., Abalos, M. and Abad, E. (2014) Long-term sampling of dioxin-like substances from a clinker kiln stack using alternative fuels. *Science of the Total Environment* 485, 528–533. doi: 10.1016/j.scitotenv.2014.03.021.

SEPA (2013) Rules, regulations and recommendations on combustion and incineration related to the emission of dioxins. (Summary produced by the Swedish Environmental Protection Agency to be used as background material for the dioxin network meeting in Stockholm 27 May.) Swedish Environmental Protection Agency, Stockholm.

Shields, W.J., Tondeur, Y., Benton, L. and Edwards, M.R. (2010) 14 Dioxins and Furans. In: Morrison, R.D. and Murphy, B.L. (eds) *Environmental Forensics: Contaminant Specific Guide*. Academic Press, New York, pp. 293–312.

Shih, S., Lee, W., Lin, L., Huang, J., Su, J. and Chang-Chien, G. (2008) Significance of biomass open burning on the levels of polychlorinated dibenzo-*p*-dioxins and dibenzofurans in the ambient air. *Journal of Hazardous Materials* 153, 276–284. doi: doi.org/10.1016/j.jhazmat.2007.08.048.

Solorzano-Ochoa, G., de la Rosa, D., Maiz-Larralde, P., Gullett, B., Tabor, D. *et al.* (2012) Open burning of household waste: effect of experimental condition on combustion quality and emission of PCDD, PCDF and PCB. *Chemosphere* 87, 1003–1008. doi: 10.1016/j.chemosphere.2011.11.038.

Tame, N.W., Dlugogorski, B.Z. and Kennedy, E.M. (2007) Formation of dioxins and furans during combustion of treated wood. *Progress in Energy and Combustion Science* 33, 384–408. doi: 10.1016/j.pecs.2007.01.001.

Trivedi, J. and Majumdar, D. (2013) Memory effect driven emissions of persistent organic pollutants from industrial thermal processes, their implications and management: a review. *Journal of Environmental Management* 119, 111–120. doi: 10.1016/j.jenvman.2013.01.026.

Tuppurainen, K., Halonen, I., Ruokojarvi, P., Tarhanen, J. and Ruuskanen, J. (1998) Formation of PCDDs and PCDFs in municipal waste incineration and its inhibition mechanisms: a review. *Chemosphere* 36, 1493–1511. doi: 10.1016/S0045-6535(97)10048-0.

UNEP (2005) *Draft guidelines on best available techniques and provisional guidance on best environmental practices relevant to Article 5 and Annex C***. Conference of the Parties of the Stockholm Convention on Persistent Organic Pollutants First meeting. Punta del Este, Uruguay, 2–6 May 2005.

Van den Berg, M., Birnbaum, L., Denison, M., De Vito, M., Farland, W. *et al.* (2006) The 2005 World Health Organization reevaluation of human and mammalian toxic equivalency factors for dioxins and dioxin-like compounds. *Toxicological Sciences* 93, 223–241. doi: 10.1093/toxsci/kfl055.

Watterson, J., Buckley-Golder, D. and Woodfield, M. (1999) *Compilation of EU Dioxin Exposure and Health Data. Task 3 – Environmental fate and transport.* Report produced for European Commission DG Environment, and UK Department of Environment, Transport and the Regions (DETR). AEA Technology, Harwell, UK.

Wcgiel, M., Chrzaszcz, R., Maslanka, A. and Grochowalski, A. (2014) Study on the impact of industrial flue gases on the PCDD/Fs congener profile in ambient air. *Chemosphere* 114, 76–83. doi: 10.1016/j.chemosphere.2014.03.104.

Wielgosinski, G. (2011) The reduction of dioxin emissions from the processes of heat and power generation. *Journal of the Air and Waste Management Association* 61, 511–526. doi: doi.org/10.3155/1047-3289.61.5.511.

Wohrnschimmel, H. and Yao, Y. (2014) *Assessing Comparability of Atmospheric PCDD, PCDF and Coplanar PCB Data from North American Ambient Air Monitoring Networks*. Commission for Environmental Cooperation, Montreal.

Wu, J.-L., Lin, T.-C., Wang, Y.-F., Wang, J.-W., Wang, C.-T. and Kuo, Y.-M. (2014) Polychlorinated dibenzo-*p*-dioxin and dibenzofuran (PCDD/F) emission behavior during incineration of laboratory waste. Part 1: Emission profiles obtained using chemical assay and bioassay. *Aerosol and Air Quality Research* 14, 1199–1205. doi: 10.4209/aaqr.2013.05.0140.

Zeng, X., Ren, J., Pan, W., Zong, T. and Li, Y. (2011) Research on the formation and control measures of dioxin emissions from Municipal Solid Waste incinerator. In: *Proceedings 2011 International Conference on Electrical and Control Engineering (ICECE), Sep 16–18, 2011, Yichang, China*. Institute of Electrical and Electronics Engineers, Piscataway, New Jersey, pp. 1707–1710. doi: 10.1109/ICECENG.2011.6058442.

Zheng, M.-H., Bao, Z.-C., Zhang, B. and Xu, X.-B. (2001) Polychlorinated dibenzo-*p*-dioxins and dibenzofurans in paper making from a pulp mill in China. *Chemosphere* 44, 1335–1337. doi: 10.1016/S0045-6535(00)00488-4.

# 13 Dioxins II. Human Exposure and Health Risks

**J. Tuomisto* and M. Viluksela**
*National Institute for Health and Welfare, Kuopio, and the University of Eastern Finland, Kuopio, Finland*

## 13.1 Abstract

There are two very different facets of dioxins. On the one hand they are feared environmental 'superpoisons'; on the other hand they are highly interesting tools for studying the mechanisms of intracellular receptors, gene expression, growth and development of organs, metabolism of chemicals in the body, carcinogenesis, food intake and hunger, as well as interactions of chemicals, microbes and immunological systems. The aryl hydrocarbon receptor (AHR, AH receptor), through which most effects of dioxins are conveyed, is now seen as an important physiological actor in the body, similar to other intracellular receptors such as steroid or thyroid receptors, not only as a route to toxic effects. This also takes us back to the ultimate principle of Paracelsus: '*Alle Dinge sind Gift, und nichts ist ohne Gift, allein die Dosis macht dass ein Ding kein Gift ist*' ('All things are poison, and nothing is without poison, the dosage alone makes it so that a thing is not a poison'). AH receptors are essential for life, their activation at certain levels regulates our wellbeing, but their inappropriate activation leads to multiple forms of toxicity.

The chapter describes the chemistry and sources of dioxins, their behaviour in nature and in the body, their mechanisms of action, high but variable acute toxicity in animals and controversial delayed toxicity, carcinogenicity, and developmental toxicity – mostly from human viewpoint but in many instances based on animal experiments. Some interesting physiological roles of AH receptors, as revealed by dioxin studies, are briefly mentioned.

## 13.2 Introduction

'Dioxins' is a loose and inaccurate term for related groups of chemicals including polychlorinated dibenzo-*p*-dioxins (PCDDs) and polychlorinated dibenzofurans (PCDFs), and usually also dioxin-like polychlorinated biphenyls (dl-PCBs). The common toxicity is based on a common mechanism of action: inappropriate stimulation of AH receptor ('dioxin receptor'). Some compounds may, however, cause toxic effects that are not specific to the whole group. This is more likely if the affinity to AH receptor, and thus dioxin-like toxicity, is low.

By far the best studied compound is 2,3,7,8-tetrachlorodibenzo-*p*-dioxin (TCDD), which is the most toxic. The toxicity of others is compared with this prototype of the group. TCDD is given a toxicity equivalence factor (TEF) of 1, and each other compound their own TEF that

* E-mail address: j.tuomisto@dnainternet.net

may vary from 1 to 0.000 03. This indicates that the potency and toxicokinetics of various compounds are highly different. TEF can be used to calculate toxic equivalency (TEQ) in assessing the toxicity of mixtures (see below).

Dioxins are metabolized and excreted very slowly, and are also persistent and accumulating in biosphere. This makes them potential sources of delayed toxicity. However, there have been a few dramatic accidental or deliberate cases of acute poisoning. In 1998 two women were poisoned at their workplace in Vienna by huge doses of TCDD. In one of them, dioxin concentrations were the highest ever measured in humans: 144,000 pg $g^{-1}$ in serum fat, implying a dose of about 25 μg $kg^{-1}$ (presently the concentrations in young people are 5–10 pg $g^{-1}$ fat, and in older people around 50 pg $g^{-1}$ fat, and present daily intakes are about 1–2 pg $g^{-1}$TEQ). She survived, but had a severe chloracne lasting for years. Surprisingly, other symptoms were few: mild gastrointestinal symptoms and amenorrhea. Even laboratory findings were few (Geusau *et al.*, 2001). In 2004 the then presidential candidate of Ukraine, Victor Yushchenko, was deliberately poisoned with a huge dose of TCDD. TCDD concentration in fat was 108,000 pg $g^{-1}$. He suffered from severe chloracne, but after initial malaise and stomach pain other symptoms were unremarkable (Sorg *et al.*, 2009).

The most publicized dioxin accident happened in Seveso, Italy, in 1976. A tank containing 2,4,5-trichlorophenol contaminated by TCDD released its contents to air and most of the town was polluted with TCDD. The highest concentrations (up to 56,000 pg $g^{-1}$ in serum lipid) were found in children, probably from eating local food and playing in the contaminated environment. Acute effects were limited to about 200 cases of chloracne (Mocarelli *et al.*, 1991) but over the years a slight excess of cancer was found (Warner *et al.*, 2011). After heavy industrial exposures chloracne has been a typical finding, as well as a slight increase in cancer of all sites at the highest exposures (Flesch-Janys *et al.*, 1995; Ott and Zober, 1996; Steenland *et al.*, 1999).

## 13.3 Chemistry

There are 75 possible congeners of PCDDs and 135 possible congeners PCDFs. Compounds with so-called lateral chlorine substitutions at positions 2, 3, 7 and 8 (Fig. 13.1) bind to the AH receptor with high affinity and are very slowly metabolized. Therefore, they are specifically toxic. TEF values have been estimated for 17 of them (seven dibenzo-*p*-dioxins and ten dibenzofurans) having four to eight chlorine substitutes. Each chlorine substitute in excess of the four (2, 3, 7 and 8) decreases the potency (exception 2,3,4,7, 8-PCDF) but the toxic effects remain mainly the same (van den Berg *et al.*, 2006).

There are 209 PCB-compounds, and by dioxin-like activities they can be divided into three groups (Fig. 13.2). The four non-*ortho* compounds having no chlorine substitute in any *o*-position to the inter-ring C-C bridge (2, 2′, 6 or 6′) are the most potent. Among them 3,3′,4,4′,5-penta-CB (PCB126) is almost as toxic as the most potent dioxins (van den Berg *et al.*, 2006). The eight mono-*ortho* PCBs have some activity. All other PCBs are devoid of noticeable dioxin-like effects. This is because only compounds able to assume a planar (flat) position are able to bind

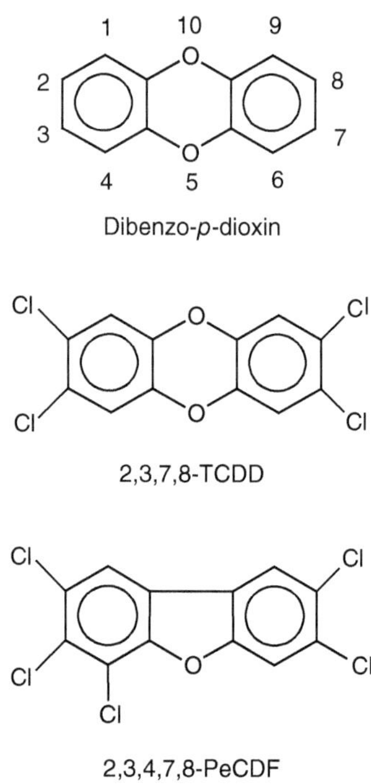


**Fig. 13.1.** Structures of dibenzo-*p*-dioxin, 2,3,7,8-tetrachlorodibenzo-*p*-dioxin and 2,3,4,7,8-pentachlorodibenzofurane (after Tuomisto *et al.*, 2011).

to the AH receptor. Only non-*ortho* compounds are freely rotating along the C-C bridge between the rings, and each *o*-chlorine makes it more difficult for the molecule to assume a planar conformation (Fig. 13.2).

Brominated dioxins, furans and biphenyls seem to share the toxicity and the ability to bind to AH receptor. They might deserve TEF values as well, but lack sufficient data (van den Berg *et al.*, 2006). Many other compounds bind to AH receptors, such as polyaromatic hydrocarbons and polychlorinated azoxy-benzenes and naphthalenes (Poland and Knutson, 1982).

Surprisingly, a number of natural compounds have been identified that may have very high affinity to AH receptors. These include indoles, flavones, benzoflavones, imidazoles and pyridines, among others (for review, see DeGroot *et al.*, 2012). They are usually metabolized fast, but due to continuous intake from food, especially vegetables, it has been claimed that they cause receptor activation at the same level as or higher than the present background concentrations of contaminant dioxins (Connor *et al.*, 2008).

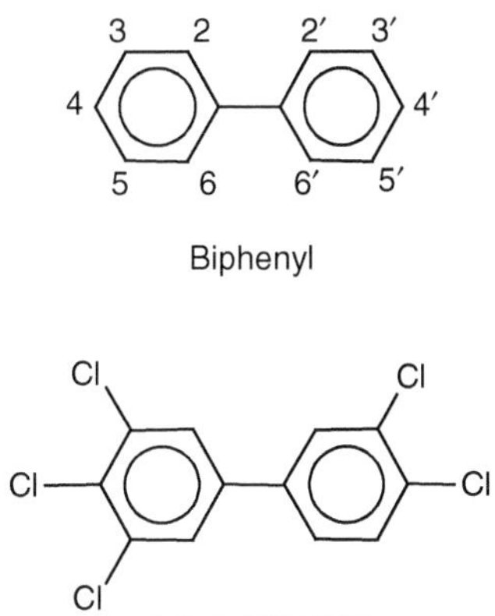


**Fig. 13.2.** Structures of biphenyl and 3, 3′, 4, 4′, 5-pentachlorobiphenyl (after Tuomisto *et al.*, 2011).

## 13.4 Mechanism of Action: the Aryl Hydrocarbon Receptor

Most actions of dioxins, particularly of the most potent ones such as TCDD, are transmitted by AHR (Fig. 13.3). This is an ancient receptor, an over 500-million-year-old protein occurring in all vertebrates. Ancient homologues of the AHR have even been discovered in invertebrates and insects. The primitive AHR forms do not bind dioxin or other ligands of the mammalian AHR, but they seem to play important developmental roles in neuronal differentiation and regulation of feeding-related aggregation behaviour or in regulation of normal morphogenesis (see Lindén *et al.*, 2010; Hahn and Karchner, 2012).

The AHR belongs to basic helix–loop–helix/PAS (bHLH/PAS) proteins, which have important roles in, for example, regulation of neural development, in generation and maintenance of circadian rhythms and as transcriptional partners and co-activators. Functionally (though not structurally) the AHR resembles the nuclear receptors such as steroid receptors, acting as a ligand-activated transcription factor.

The AHR exists in the cytosol in a protein complex including heat shock protein 90 (HSP90),

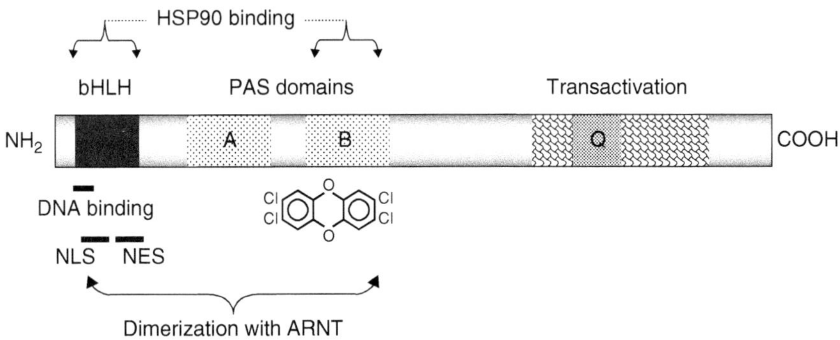


**Fig. 13.3.** The structure of AHR. The approximate sites for DNA binding, nuclear translocation (NLS) and export (NES), ligand binding (TCDD molecule), HSP90 binding, heterodimerization, and transactivation are shown (adapted from Lindén *et al.*, 2010, with permission of the authors and Elsevier).

AHR-associated protein-9 (ARA9 or XAP2 or AIP) and p23 (Fig. 13.4). These chaperone proteins keep the AHR in a conformation that is able to bind a ligand but unable to enter the nucleus. After ligand binding, the nuclear translocation site (NLS) of the AHR (Fig. 13.3) is exposed and recognized by importin-β, and the protein complex translocates into the nucleus. There the AHR dissociates from the chaperones and heterodimerizes with another bHLH/PAS protein, ARNT (AHR nuclear translocator). The AHR/ARNT dimer binds to the DNA within the major groove of the DNA helix at specific sites, AHR elements also known as dioxin response elements (DREs).

In addition to this canonical pathway, some actions of dioxins and AHR are mediated via non-canonical pathways. These may be involved in, for example, interactions with other receptors, such as oestrogen receptor (cf. Lindén *et al.*, 2010).

In response to activation by dioxins, the AHR signalling pathway modifies the expression levels of numerous genes. The best characterized of these at the molecular level is the induction of the gene for a phase I drug-metabolizing cytochrome P450 enzyme, CYP1A1 (Okey *et al.*, 2005, Ma, 2012).

Dioxin-activated AHR induces other phase I and II drug-metabolizing enzymes in liver including CYP1A2, CYP1B1, CYP2S1, CYP2A5, ALDH3, GSTA1, UGT1A1, UGT1A6, UGT1A7 and NQO1. In addition to drug-metabolizing enzymes, TCDD exposure modifies the expression of a large number of other genes, presumably by a similar mechanism. For example, in adult

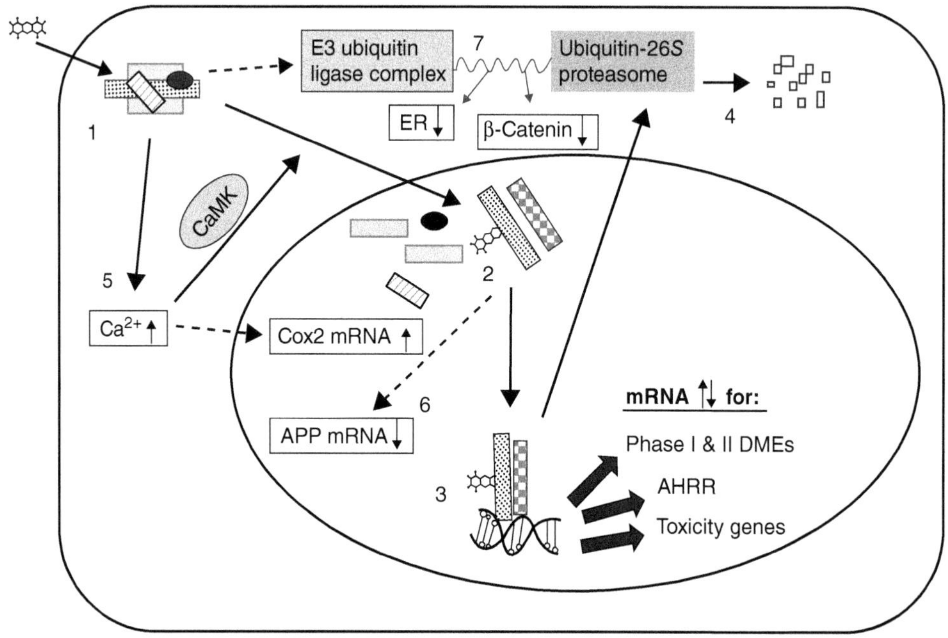


**Fig. 13.4.** A schematic diagram of some AHR signalling pathways. The canonical pathway is depicted with solid arrows, alternative pathways with dashed arrows, and an intersection of these two with a solid thin arrow. The dark-spotted bars represent the AHR, white-spotted bars ARNT, striped bars ARA9, grey bars HSP90 and the ovals p23. Dioxin binding to the AHR (1) leads to its translocation into the nucleus (2), heterodimerization with ARNT and binding to the DNA at DREs, modulating expression levels of target genes (3). One of the gene products elevated by this mechanism is AHRR, which forms a feedback loop by inhibiting AHR action. The AHR is finally degraded by the ubiquitin–proteasome system (4). AHR activation can also rapidly increase intracellular $Ca^{2+}$ concentration (5) which in turn may ultimately result in augmented Cox2 gene expression. Elevation of $Ca^{2+}$ activates CaMKs, which appear to have a critical role in the translocation of the AHR. Other examples of effects mediated by the AHR via non-canonical pathways are suppression of acute-phase proteins (6), which does not involve DNA binding, and degradation of e.g. ER and β-catenin by acting as an atypical E3 ubiquitin ligase (adapted from Lindén *et al.*, 2010, with permission).

mouse or rat liver, hundreds or even thousands of genes are affected (see Lindén *et al.*, 2010). It is still unclear which genes are the most important for the main toxic effects such as lethality, anorexia and wasting syndrome and various hyperplastic and atrophic tissue changes (Okey, 2007).

The role of AH receptor as an inducer of metabolic enzymes to protect us from xenobiotics is rapidly changing. AH receptors participate in many regulatory functions in the body but this is outside the scope of this chapter. The reader is referred to recent reviews ((Okey, 2007); Lindén *et al.*, 2010; Casado *et al.*, 2010; Pohjanvirta, 2012; Van Voorhis *et al.*, 2012; Esser and Rannug, 2015; Sibilano *et al.*, 2015; Kolluri *et al.*, 2017).

## 13.5 Toxicity Equivalents

Different congeners have different potencies and different toxicokinetics. Therefore the toxicity of a mixture cannot be simply calculated by adding up the amounts or concentrations of all chemicals in the mixture. However, if the toxicity of a congener is standardized to the equivalent amount of TCDD, chemicals with different potencies can be summed up and this equivalent quantity is very useful for regulatory and even some scientific purposes (Tuomisto, 2012).

Because toxicities vary by a factor of 30,000, and TCDD is given the TEF of 1, other chemicals are given TEF values of 1 to 0.000 03 (Table 13.1). The amount or concentration of a given compound is multiplied by its TEF, resulting in the amount or concentration toxicologically equivalent to that of TCDD. These partial equivalent amounts of congeners are then added up to make the sum toxic equivalent (TEQ) of the mixture. This can be used as a proxy of the total dose of dioxin-like compounds, while appreciating that it is a consensus value based on several assumptions (van den Berg *et al.*, 2006). PCDD/F congeners usually seem to act additively, which justifies the use of TEFs. With less potent compounds, though, partial antagonism might be expected (Safe, 1998; Peters *et al.*, 2006; Howard *et al.*, 2010). This may lead to overestimation of the total toxicity (Howard *et al.*, 2010).

**Table 13.1.** Toxic equivalency factors for PCDD/Fs and PCBs. Other congeners are not assumed to have dioxin-like effects. IUPAC numbers for PCBs are given in parenthesis (van den Berg *et al.*, 2006).

| Congener | WHO-TEF |
|---|---|
| PCDDs | |
| 2,3,7,8-TCDD | 1 |
| 1,2,3,7,8-PeCDD | 1 |
| 1,2,3,4,7,8-HxCDD | 0.1 |
| 1,2,3,6,7,8-HxCDD | 0.1 |
| 1,2,3,7,8,9-HxCDD | 0.1 |
| 1,2,3,4,6,7,8-HpCDD | 0.01 |
| OCDD | 0.0003 |
| PCDFs | |
| 2,3,7,8-TCDF | 0.1 |
| 1,2,3,7,8-PeCDF | 0.03 |
| 2,3,4,7,8-PeCDF | 0.3 |
| 1,2,3,4,7,8-HxCDF | 0.1 |
| 1,2,3,6,7,8-HxCDF | 0.1 |
| 1,2,3,7,8,9-HxCDF | 0.1 |
| 2,3,4,6,7,8-HxCDF | 0.1 |
| 1,2,3,4,6,7,8-HpCDF | 0.01 |
| 1,2,3,4,7,8,9-HpCDF | 0.01 |
| OCDF | 0.0003 |
| Non-*ortho*-PCBs | |
| 3,3′,4,4′-TCB (77) | 0.0001 |
| 3,4,4′,5-TCB (81) | 0.0003 |
| 3,3′,4,4′,5-PeCB (126) | 0.1 |
| 3,3′,4,4′,5,5′-HxCB (169) | 0.03 |
| Mono-*ortho*-PCBs | |
| 2,3,3′,4,4′-PeCB (105) | 0.00003 |
| 2,3,4,4′,5-PeCB (114) | 0.00003 |
| 2,3′,4,4′,5-PeCB (118) | 0.00003 |
| 2′,3,4,4′,5-PeCB (123) | 0.00003 |
| 2,3,3′,4,4′,5-HxCB (156) | 0.00003 |
| 2,3,3′,4,4′,5′-HxCB (157) | 0.00003 |
| 2,3′,4,4′,5,5′-HxCB (167) | 0.00003 |
| 2,3,3′,4,4′,5,5′-HpCB (189) | 0.00003 |

TEF values are based on toxicity studies when available, or on *in vitro* information. Most studies are based on oral intake, so the values best correlate with oral toxicity. Different end points of toxicity may result in different TEF values; hence the values are always balanced compromises. Because of varying toxicokinetics, an internal TEF based on target concentrations would be more reliable (and perhaps closer to AH receptor affinity), but there is not yet enough data to compile internal TEF values (van den Berg *et al.*, 2006, 2013).

## 13.6 Toxicokinetics: Absorption, Distribution and Elimination

The main source of dioxins is food. Oral absorption of dioxins depends on the carrier. Dioxins in the fat of fish or meat are well absorbed, but those in soils, for example, are poorly absorbed. After absorption they are distributed mostly to adipose tissue and to some extent to the liver.

Elimination of dioxins is slow, because they are not easily metabolized and urinary excretion is negligible. Elimination is mainly via faeces after slow metabolism in the liver, and biliary excretion. Elimination half-lives may vary tenfold (Table 13.2). Very high concentrations seem to induce metabolizing enzymes and shorten the half-lives (Aylward *et al.*, 2005).

Nursing mothers excrete dioxins in milk fat at approximately the same concentrations as their own level in body fat. This means that maternal dioxin levels decrease during the lactation period (even by 20%) (Vartiainen *et al.*, 1998). Also placental PCDD/F concentrations are in the same range as in maternal body or breast milk (as pg $g^{-1}$ fat) (Virtanen *et al.*, 2012), and placental transfer to the fetus occurs (cf. Feeley and Brouwer, 2000). Each delivery and lactation decreases the mother's body burden by 25–30%. In newborn babies the elimination is much faster than later, with a half-life of months rather than years (Kreuzer *et al.*, 1997).

**Table 13.2.** Elimination half-lives of some PCDD/Fs (Milbrath *et al.*, 2009).

| Congener | Half-life (years) |
|---|---|
| 2,3,7,8-TCDD | 7.2 |
| 1,2,3,7,8-PeCDD | 11.2 |
| 1,2,3,4,7,8-HxCDD | 9.8 |
| 1,2,3,6,7,8-HxCDD | 13.1 |
| 1,2,3,7,8,9-HxCDD | 5.1 |
| 1,2,3,4,6,7,8-HpCDD | 4.9 |
| OCDD | 6.7 |
| 2,3,7,8-TCDF | 2.1 |
| 1,2,3,7,8-PeCDF | 3.5 |
| 2,3,4,7,8-PeCDF | 7.0 |
| 1,2,3,4,7,8-HxCDF | 6.4 |
| 1,2,3,6,7,8-HxCDF | 7.2 |
| 1,2,3,7,8,9-HxCDF | 7.2 |
| 2,3,4,6,7,8-HxCDF | 2.8 |
| 1,2,3,4,6,7,8-HpCDF | 3.1 |
| 1,2,3,4,7,8,9-HpCDF | 4.6 |
| OCDF | 1.4 |

## 13.7 Sources of Dioxins

Different classes of dioxin-like chemicals come from different sources. PCDD/F compounds were never synthesized on purpose except for scientific research. They are unwanted side products in burning processes, as well as in the synthesis of PCBs, chlorophenol fungicides and phenoxy acid herbicides. Due to control measures, main sources are very different today than they were 30 or 40 years ago.

Any burning will produce PCDD/Fs, if chlorine (and metal catalysts) is available. Therefore poorly controlled urban waste incineration used to be one of the most important sources. This can be technically solved by ensuring high incineration temperature (1000°C or higher), long enough burning time, and effective flue-gas filtration. In modern good-quality incinerators PCDD/Fs are not a problem (Zhang *et al.*, 2017). Accidental dumpsite fires and backyard burning of waste are much more problematic and poorly controlled, because in poor burning conditions the production of PCDD/Fs can be high (Dopico and Gomez, 2015; Zhang *et al.*, 2017).

Many important sources of PCDD/Fs have disappeared or are duly under control (chlorine bleaching of pulp, syntheses of PCBs, chlorophenols and phenoxy acids, etc.). Metal industries and local burning of solid fuels remain as sources (Zhang *et al.*, 2017). The decline in ambient air dioxin emissions in different regions is depicted in Fig. 13.5. In Europe, emissions decreased between 1985 and 2004 by about 80% (from 14 kg per year I-TEQ to 2–4 kg) (Quass *et al.*, 2004) and in the USA between 1987 and 2000 even more (from 14 kg to 1.4 kg) (EPA 2006). (Note that I-TEQ, i.e. international TEQ for PCDD/Fs, was used before present TEQs were agreed under the auspices of the World Health Organization, but the differences are minor.) In the USA the top three sources of dioxin emissions to air are forest fires, backyard burning of trash, and medical waste incinerators (EPA, 2013). PCDD/F emissions have not decreased in all countries, however (Momeniha *et al.*, 2017). PCDD/Fs are also known to be produced naturally (Jin and Chen, 2017).

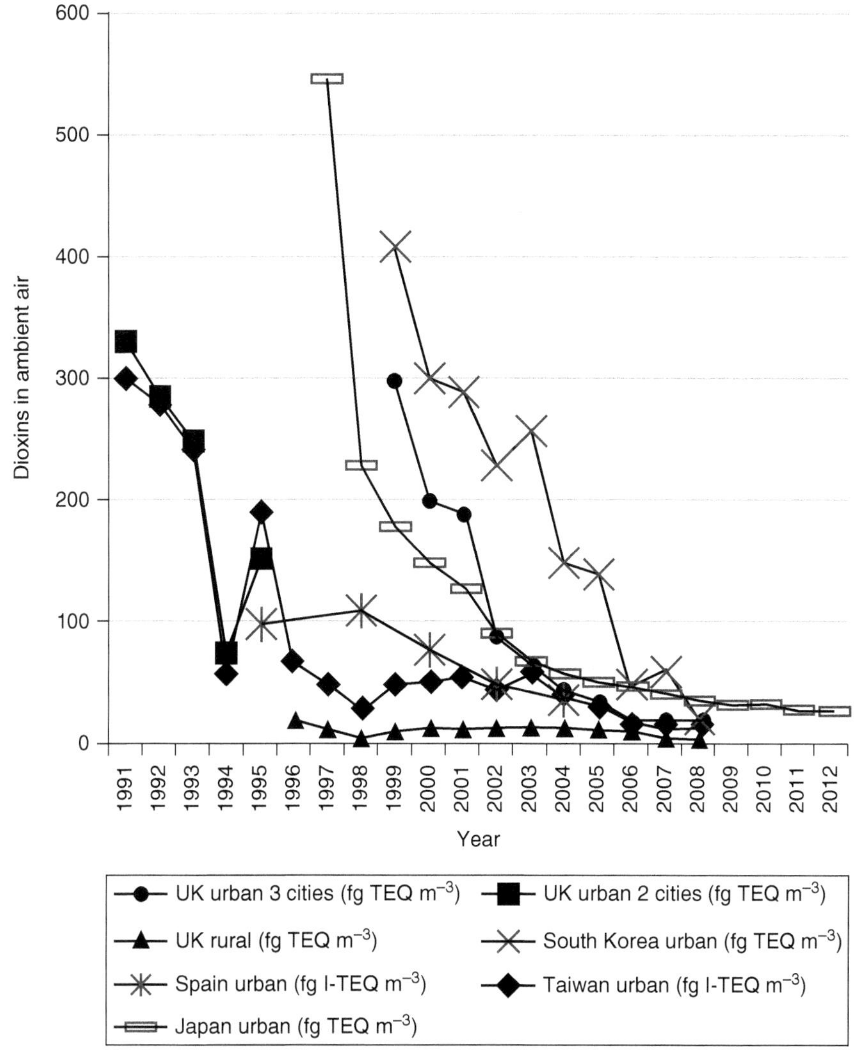


**Fig. 13.5.** Decrease of dioxins in ambient air in different regions (after Dopico and Gomez, 2015, with permission of the authors and Taylor & Francis).

PCB compounds (see Chapter 11) were used widely from 1930s to 1980s for multiple purposes as technically excellent oils, resistant to pressure, chemically resistant, non-flammable, and not conducting electricity. Even after their production was discontinued in most countries in the 1980s, these compounds linger in many products such as electrical transformers and plastic materials. Some of it ends up to the general environment. Only a minor part of PCBs are dioxin-like, depending on the matrix, non-*ortho* of the order of 1:1000, mono-*ortho* 1:10 (Kiviranta *et al.*, 2004).

## 13.8 Environmental Fate

Because dioxins are persistent and not easily degraded by environmental microbes, they tend to accumulate in the environment. In addition, their lipid/water distribution coefficient is very high and they tend to accumulate in, for example, phytoplankton, and then their concentration tends to magnify at each trophic level (biomagnification). This leads to high concentrations at the highest trophic levels, e.g. seals and predatory birds. Humans are also at a high trophic level but, due to a very mixed diet, human concentrations are not nearly as high as in the most endangered wild species. However, this has led to concerns about the safety of wild fish in the human diet (see below).

## 13.9 Human Intake and Concentrations

By far the most important source of dioxins is food, especially animal-source food (Liem *et al.*, 2000). Food items vary from country to country: in many countries meat and milk products dominate, in others fish. In all of these foods the concentrations have decreased in the Western countries during the past 30–40 years, and the present daily intake is 1–2 pg $kg^{-1}$ body weight (TEQ).

Because the half-lives of these compounds are very long, they accumulate during the whole lifetime. Therefore PCDD/F concentrations in young people are 5–10 pg $g^{-1}$ TEQ in fat, but 40–100 pg $g^{-1}$ in older generations (Kiviranta *et al.*, 2005). In older generations there seems to be also carrying over from earlier decades when the intake was 5–10 times higher than at present (Tuomisto *et al.*, 2016). For this reason any concentration data without information on age and the year of sampling are of little value.

Dioxin concentrations (but not all PCBs) have been decreasing for over 30 years that they have been actively monitored, in line with decreasing environmental levels (Consonni *et al.*, 2012). The World Health Organization (WHO) has organized dioxin follow-up measurements in breast milk since 1987. In more recent surveys PCBs and some other persistent chlorinated compounds have also been measured (van den Berg *et al.*, 2017).

Breast milk concentrations were very high in 1970s, about 50 pg $g^{-1}$ for PCDD/Fs and 50 pg $g^{-1}$ for dl-PCBs (TEQ in fat), but there are few data (Norén and Meironyté, 2000). During the first systematic round of breast milk measurements in 1987, PCDD/F concentrations in many countries were between 30 and 40 pg $g^{-1}$ TEQ in milk fat (WHO, 1989) and during the last round in 2005–2010 they were between 5 and 10 pg $g^{-1}$ in the European countries, but high in India, for example, and low in many African countries (van den Berg *et al.*, 2017). Thus the concentrations have decreased by 80–90% in many but not all countries.

## 13.10 Toxic Effects in Humans

### 13.10.1 Accidents, contamination episodes and occupational risks

Apart from acute accidents, there have been several food contamination episodes. In Japan (Yusho incident, 1968) and in Taiwan (Yucheng incident, 1979) PCB oil used in heat exchangers leaked to rice bran oil, and more than 2000 people (Kashima *et al.*, 2015) and about 2000 people (Tsai *et al.*, 2007), respectively, were poisoned by consuming contaminated oil. Most of the toxic effects have been attributed to PCDFs and dl-PCBs. The most dramatic health effects were caused by developmental toxicity during pregnancy.

In the Yusho incident, the average daily intake was 154,000 pg I-TEQ $kg^{-1}$, or 100,000 times higher than average background intake at present. In the Yu-cheng incident they were roughly similar and the concentrations were still over 1300 pg I-TEQ $g^{-1}$ fat about 15 years later (Hsu *et al.*, 2005). Many skin problems were seen: hypersecretion of Meibomian glands in the eyes, swelling of eyelids, abnormal pigmentation of skin, hyperkeratosis and chloracne. Babies born to Yusho and Yu-cheng mothers were smaller than normal, with dark brown pigmentation, gingival hyperplasia, and sometimes dentition at birth or other tooth deformities. There were fetal deaths and miscarriages. Increase in cancer was initially seen in Yusho males, but not in females (Onozuka *et al.*, 2009), nor in Yucheng victims in spite of the heavy exposure (Tsai *et al.*, 2007). Later, a pooled analysis indicated increased mortality from all causes, all cancers, lung cancer and heart disease in men, and liver cancer in women (Li *et al.*, 2015).

Phenoxy acid herbicides (Agent Orange, contaminated by dioxins, especially TCDD) were used in large quantities during the Vietnam War.

The veterans have been thoroughly studied, but variable levels complicate assessments. There is evidence of increased cancer, diabetes (Michalek and Pavuk, 2008) and hypertension (Cypel *et al.*, 2016) in the highly exposed groups.

In the Seveso accident (1976), described above, the most conspicuous finding was chloracne. Cancer studies have suggested a slightly increased number of haematopoietic and lymphatic tissue malignancies (Consonni *et al.*, 2008; Pesatori *et al.*, 2009). In a cohort of women with measured individual TCDD levels, an increased risk of all cancers was found (Warner *et al.*, 2011).

Several developmental consequences were also detected. Dental aberrations associated with TCDD levels were found 25 years after the accident in persons who had been less than 5 years old at the time of the accident (Alaluusua *et al.*, 2004). Lowered male/female sex ratios were found in the offspring of males exposed to high concentrations of TCDD (Mocarelli *et al.*, 2000). Decreased sperm quality was observed in young men exposed to TCDD *in utero* and lactationally or during infancy or prepuberty (Mocarelli *et al.*, 2008, 2011). Slightly increased risk of endometriosis (Eskenazi *et al.*, 2002) as well as a dose-dependently increased time to pregnancy and infertility was found among the most heavily exposed women (Eskenazi *et al.*, 2010). However, in 30 years of follow-up no association between TCDD exposure and adverse pregnancy outcomes were detected except for a non-significant decrease in birthweight (Wesselink *et al.*, 2014). Some metabolic and endocrine effects were seen for a limited time period (Sweeney and Mocarelli, 2000).

In Belgium a large tank of recycled fats was contaminated with 40–50 kg PCB oil in 1999 and used for animal feed. Low fertility of chickens and deformed chicks were noted. About 1 g of dioxins and 2 g dl-PCBs (TEQ) were involved (Debacker *et al.*, 2007). This caused a major dioxin alarm in Europe, and the European Union (EU) set very strict limits for dioxins in food. Due to fairly rapid intervention, total dioxin concentrations in the population did not increase even in Belgium: 23.1 versus 22.9 pg $g^{-1}$ TEQ in fat (Debacker *et al.*, 2007). No health effects have been noted. The incidence shows that strict food controls are necessary, but no individual health measures are rational, because the impact on human levels is very slow.

Several industrial settings have caused high exposures to dioxins when synthesizing chlorophenols or phenoxy acid herbicides (Flesch-Janys *et al.*, 1995; Ott and Zober, 1996; Steenland *et al.*, 1999; Boers *et al.*, 2010). Chloracne was a hallmark characteristic at the higher end of exposure levels. Occupational cancer studies have been pooled in a large international combined cohort, suggesting an increased risk of all cancers and of soft-tissue sarcoma (Kogevinas *et al.*, 1997; Kogevinas, 2000). The problem lies in indirectly assessed and variable exposure levels, i.e. very high industrial levels and marginally increased levels in workers spraying phenoxy herbicides (cf. Tuomisto and Tuomisto, 2012). The study was crucial for International Agency for Research on Cancer (IARC) evaluations (IARC, 1997, 2012), which have also been criticized (Yamaguchi, 1999; Cole *et al.*, 2003; Boffetta *et al.*, 2011). The evidence on soft-tissue sarcoma is especially weak and based on very few cases (Tuomisto and Tuomisto, 2012) but a slight increase of all cancers is likely to be real, considering recent new evidence on the Yusho, Yu-cheng and Seveso accidents.

A review of high-exposure studies suggested that dioxin exposure is associated with mortality from cardiovascular disease and especially ischaemic heart disease (Humblet *et al.*, 2008).

### 13.10.2 Risks connected with exposures of general population

Population risks have been assessed at several occasions. To give guidance for assessing tolerable daily intake (TDI) values, an international panel organized by the WHO and International Programme on Chemical Safety met in 1998 (Van Leeuwen and Younes, 2000). Animal body burdens were compared with estimated human intakes, and on that basis the following effects were considered most relevant: sperm count, immune suppression, genital malformations, and neurobehavioural effects in offspring and endometriosis in adults (van den Berg *et al.*, 2000). This means that the safety margins are lowest for different developmental effects.

One of the few plausible developmental effects in the general population is tooth deformities (Alaluusua *et al.*, 1996, 1999), based on mean dioxin concentrations of 48.8 (range 7.7–258) pg $g^{-1}$ TEQ in milk fat. When dioxin levels in milk decreased over the years, the effects were no more seen. No association of cryptorchidism and placental levels of dioxins and PCBs was seen (Virtanen *et al.*, 2012) but adipose tissue levels at time of operation may give some support to an association (Koskenniemi *et al.*, 2015). Sperm counts at age 18–19 years were inversely associated with dioxin levels at age 8–9 years in a cohort of Russian boys (Mínguez-Alarcón *et al.*, 2017). The range of PCDD/F+PCB TEQ was 4.88–107 pg $g^{-1}$ lipid, or relatively high for age.

Cancer risk of dioxins has been hotly debated. IARC (1997, 2012) deemed TCDD and 2,3,4,7,8-TCDF as carcinogenic to humans (class 1). However, the assessments were based on animal experiments and high accidental or occupational exposures (Schrenk and Chopra, 2012). Because IARC only assesses hazard and not risk, it remains unclear what is the risk for general population. The high-exposure populations (Kogevinas, 2000) were exposed to 100–1000 or more times higher levels than the general population. Moreover, the assessment has been challenged in several papers on various grounds (Yamaguchi, 1999; Cole *et al.*, 2003; Boffetta *et al.*, 2011; Tuomisto and Tuomisto, 2012). At this point it may be safely concluded that dioxins are carcinogenic in animals, and probably carcinogenic at high dose levels in humans. However, there is no good evidence that there would be any cancer risk at present general population levels. The WHO consultation group (van den Berg *et al.*, 2000) concluded that the possible cancer risk is taken care of if TDI is determined on the basis of developmental effects.

Several arguments speak against a population risk in humans. Dioxins are not genotoxic (IARC, 2012). Therefore linear extrapolation is not likely to be valid, and one could rely on safety margins as in other forms of toxicity. An important physiological role of the AH receptor means that a certain level of receptor activation is beneficial, and only inappropriate stimulation is harmful, as for other receptors such as steroid receptors (Tuomisto and Tuomisto, 2012). Positive findings are based on case-control studies relying on exposure assessment by questionnaires after diagnosing cancer. These are notoriously unreliable because of recall bias (Tuomisto *et al.*, 2017). Cohort studies give equivocal results (Tuomisto and Tuomisto, 2012). A specific cancer associated with dioxins is soft-tissue sarcoma. In a large case-control study with individual measured concentration data, no positive association was found between soft-tissue sarcoma and TEQs or individual dioxins or PCBs (Tuomisto *et al.*, 2004a). Rather there was a trend of decreasing risk at higher exposure groups, suggesting a hormetic effect (Tuomisto *et al.*, 2005). The other side of the coin may even be that AH receptor agonists could be used in the search for drugs in treating cancer (Kolluri *et al.*, 2017).

In conclusion as to contemporary human risks, the safety margins seem to be lowest for developmental effects. Sex ratio changes were seen at levels about 20 times the present levels (Mocarelli *et al.*, 2000), and for enamel defects in teeth and sperm quality the margin may be slightly lower (Alaluusua *et al.*, 1996, 2004; Mínguez-Alarcón *et al.*, 2017; Pilsner *et al.*, 2017). These are in line with the assessment by the WHO panel (van den Berg *et al.*, 2000).

The WHO panel based their assessment on the exposure of child-bearing women, who excrete much of their body burden to the child during pregnancy and lactation. The panel concluded that even if the safety margin concerning the child is fairly narrow, the benefits of breast feeding clearly exceed the risks. Similarly, the health benefits of fish consumption clearly exceed the risks of dioxins or other persistent organic compounds (Tuomisto *et al.*, 2004b).

## 13.11 Animal Toxicity and its Relevance in Assessing Human Risks

Many effects have been described in animals and they can be broadly divided into clearly toxic effects (such as lethality, wasting syndrome, liver injury) and metabolic effects that often can be classified as adaptive responses (such as induction of enzymes metabolizing xenobiotic chemicals). Only the most pertinent features will be described here and the reader is referred to several reviews (Pohjanvirta and Tuomisto, 1994; Birnbaum and Tuomisto, 2000; Tuomisto, 2005; White and Birnbaum, 2009; Lindén *et al.*, 2010).

### 13.11.1 The most conspicuous acute toxic effects in adult animals

Acute toxicity of dioxins is highly variable among species: the $LD_{50}$ of TCDD in guinea pigs is in the order of 1 μg $kg^{-1}$, while in hamsters it is more than 1000-fold higher. The differences are sometimes due to different ligand binding affinities (e.g. C57BL/6 mice and ten times more resistant DBA2/2J mice) and sometimes the structure of the transactivation domain of the receptor (such as a 1000-fold difference between Long-Evans and Han/Wistar/Kuo rats, and possibly between guinea pig and hamster). What is spectacular in acute toxicity is that the animals do not die immediately, but after a reduced feed intake and wasting in 2–3 weeks (Pohjanvirta and Tuomisto, 1994). The wasting syndrome is associated with decreased appetite and food intake but the exact mechanism is not clear (Lindén *et al.*, 2010). Even at low doses there is clear aversion response to novel foods, which may not be related to the fatal wasting syndrome (Lensu *et al.*, 2011a, b).

Because the changes in the transactivation domain of AH receptor drastically influence the wasting syndrome and lethality, but very little enzyme induction, two types of dioxin effects have been proposed. Type I responses include developmental effects (see below), aversion to novel foods and the typical induction of CYP1A1 and other oxidative enzymes that occur similarly regardless of the structure of the AHR. Type II responses include several high-dose effects such as wasting syndrome, lethality and liver toxicity (Tuomisto *et al.*, 1999; Simanainen *et al.*, 2003). There is some evidence that tumour promotion might belong to type II responses (Viluksela *et al.*, 2000). This difference also implies that type I effects are relatively similar among species, but type II effects cannot be reliably predicted over species. It is of interest that many of the type I responses associate with defences towards noxious chemicals via the AH receptor (induction of metabolism, aversion to toxic foods) and can therefore be considered adaptive and protective.

Various pleiotropic effects are typical of dioxin toxicity. There may be both proliferative responses and atrophic responses. Thymic atrophy is a consistent finding, as well as some immunological effects. Liver toxicity is variable; it is typical in rabbits but seen to a variable extent in other species, e.g. disturbances of porphyrin metabolism, oxidative damage and fatty infiltration. There are also multiple high-dose effects on the nervous system, such as tryptophan metabolism or neuropathies. Generally speaking, adverse effects at low doses in adult animals are few.

### 13.11.2 Developmental effects

Experimental studies in multiple animal species have indicated that developmental effects are the most sensitive adverse effects of TCDD and that some of them are observed at exposure levels close to the human background exposure (Birnbaum and Tuomisto, 2000). These findings have promoted studies focusing on developmental end points in both experimental models and dioxin-exposed populations. The sensitive targets include developing male and female reproductive system, immune system, nervous system and teeth.

Several studies from different laboratories have indicated a variety of adverse effects on the male reproductive system after *in utero* and lactational exposure of rats to low doses of TCDD. These include reduction of cauda epididymal sperm counts, daily sperm production and weight of accessory sex organs as well as increased proportion of abnormal sperm and delayed puberty (reviewed by Bell *et al.*, 2010). Although there is variability among different studies, the delay in developmental milestones for male reproductive end points proved to be the most consistent and sensitive finding. Also decreased male/female sex ratios were reported in the offspring of male mice exposed to TCDD for 12 weeks prior to mating (Ishihara *et al.*, 2007). However, maternal exposure did not affect the sex ratio of rat offspring (Bell *et al.*, 2007).

Recent studies reported epigenetically mediated transgenerational effects of TCDD in rats and mice, some of which were paternally mediated or resulted in adult-onset disease states. Toxic effects are considered transgenerational if neither the parent nor the offspring is directly exposed. In these studies pregnant females were exposed to low doses of TCDD and the outcome of several end points of toxicity were monitored in F1–F3 (or F4) generations. In rats, primordial follicle loss, polycystic ovaries and early onset of puberty were observed in female F1 and F3 offspring, and histopathological alterations of testis and kidney abnormalities in male F1 and F3 offspring (Manikkam *et al.*, 2012). These changes were associated with differentially DNA methylated regions in F3 generation sperm epigenome.

In the offspring of TCDD-exposed mice reduced fertility, increased incidence of premature birth and increased uterine sensitivity to inflammation were found in F1–F4 generations (Bruner-Tran *et al.*, 2011). Interestingly, infertility and increased incidence of premature birth were also found in unexposed female mice mated with males exposed to TCDD *in utero* (Ding *et al.*, 2011). Premature birth was associated with reduced progesterone receptor expression and inflammation of placenta. In male mice, infertility and increased premature births in unexposed mating partners that persisted to F2 and F3 generations were associated with testicular inflammation and apoptosis of developing spermatocytes (Bruner-Tran *et al.*, 2014). The role of paternal exposure was also studied in male rat offspring (F1) exposed *in utero* and lactationally to low doses of TCDD and mated with unexposed females to obtain the F2 generation and further the F3 generation (Sanabria *et al.*, 2016). The proportion of implantations per corpus luteum was significantly decreased in all three generations. These findings emphasize the significance of paternal preconception exposure as a determinant of pregnancy outcome in mice.

In addition to rats and mice, transgenerationally inherited dioxin-induced effects have been studied in the zebrafish model (reviewed by Baker *et al.*, 2014a, b). In zebrafish, TCDD-induced transgenerational and partly paternally mediated effects include reproductive dysfunction, reduced fertility, skeletal malformations and lowered male/female sex ratio. These effects seem to be phenotypically very similar across these vertebrate classes.

Development of teeth and the skeleton have been identified as highly sensitive targets of dioxin toxicity in several vertebrate species (reviewed by Viluksela *et al.*, 2012). Teeth are useful indicators of developmental toxicity, because unlike bone they do not undergo continuous remodelling after mineralization. Remodelling may repair mineralization defects of bones. Developmental defects of teeth can therefore be detected later in life, as in the case of the Seveso accident, when dental defects were observed 25 years after the accident (Alaluusua *et al.*, 2004).

*In utero* and lactational exposure to TCDD was shown to result in a wide range of alterations in rats and mice. They included smaller molar size, delayed eruption, increased susceptibility to caries, altered mineral composition of enamel, increased fluctuating asymmetry of molars and complete arrest of development of the third molars (Kattainen *et al.*, 2001; Miettinen *et al.*, 2002, 2006; Keller *et al.*, 2007a, b). In addition to rodents, sensitivity of tooth development to TCDD was shown in rhesus monkeys, mink, rainbow trout and zebrafish (Hornung *et al.*, 1999; Render *et al.*, 2001; Yasuda *et al.*, 2005; Planchart and Mattingly, 2010). In tooth development (as well as in the development of several other organs), the target of toxicity seems to be the developing epithelium. Developmental defects are the consequence of impaired epithelial–mesenchymal signalling, and AHR, epidermal growth factor (EGF) and transforming growth factor α (TGFα) are involved in mediating the effects (Partanen *et al.*, 1998; Abbott *et al.*, 2003, Alaluusua and Lukinmaa, 2006; Viluksela *et al.*, 2012).

Cleft palate is the best-known skeletal effect of dioxins at relatively high maternal doses (Birnbaum, 1995). *In utero* and lactational exposure to lower doses of TCDD was shown to affect long bones of rats, mice and rhesus monkeys by inducing altered bone geometry, decreased bone mineral density and biomechanical strength and retardation of bone matrix maturation (Miettinen *et al.*, 2005; Hermsen *et al.*, 2008; Nishimura *et al.*, 2009; Finnilä *et al.*, 2010). Further studies indicated that differentiation of bone marrow stem cells to bone-forming osteoblasts and bone-resorbing osteoclasts is disrupted by TCDD in AHR-dependent manner (Korkalainen *et al.*, 2009).

### 13.11.3 Cancer in animals

Dioxins are clear multi-site carcinogens in animal studies, but not genotoxic as indicated by both mutagenicity assays and tumour promotion studies. Also the ability of TCDD to inhibit apoptosis and enhance proliferation supports a non-mutagenic mechanism of carcinogenicity.

Much of the cancer risk assessment has been based on an early rat study (Kociba *et al.*,

1978), demonstrating liver tumours in female rats at low doses (TCDD 10 ng $kg^{-1}$ per day for 2 years). Other studies have confirmed multi-site carcinogenicity in several species but the doses have usually been higher. Toxic hepatitis has also been found in animals with tumours. Non-genotoxic or promoting mechanisms are favoured (Dragan and Schrenk, 2000). When differently sensitive Long-Evans (Turku/AB) (L-E) and H/W rat sub-strains were compared in a 3-month tumour promotion study, there was a difference of almost two orders of magnitude, and in both strains tumour promotion was associated with signs of liver toxicity (Viluksela *et al.*, 2000). Such findings suggest that carcinogenicity may be secondary to toxicity.

## 13.12 Conclusions

Dioxins are a group of related, persistent, bioaccumulating environmental poisons that act via aryl hydrocarbon receptor. AHR is an intracellular receptor that has important physiological functions. Hence a certain level of activity is necessary for life, but inappropriate activation leads to a number of deleterious effects. The most sensitive adverse effects of dioxins seem to involve several aspects of development, from teeth and bones to sexual organs. At present environmental levels the safety margins are only an order of magnitude or a little more, and possibly the safe level was slightly exceeded in 1970s and 1980s. This means that strict environmental controls of dioxin emissions are important, but on the other hand one should avoid measures that would increase competing risks. The benefits of, say, breast feeding are estimated as clearly greater than possible risks of contaminants, and the nutritional benefits of fish consumption outweigh toxic effects, if any. In particular, carcinogenicity has caused confusion, because it is probably true at high industrial or accidental exposure levels, but dioxins are not genotoxic and there is neither good evidence nor logical reason that they would cause cancer at levels below those causing developmental effects.

## References


Abbott, B.D., Buckalew, A.R., DeVito, M.J., Ross, D., Bryant, P.L. and Schmid, J.E. (2003) EGF and TGF-alpha expression influence the developmental toxicity of TCDD: dose response and AhR phenotype in EGF, TGF-alpha, and EGF + TGF-alpha knockout mice. *Toxicological Sciences* 71, 84–95.

Alaluusua, S. and Lukinmaa, P.L. (2006) Developmental dental toxicity of dioxin and related compounds – a review. *International Dental Journal* 56, 323–331.

Alaluusua, S., Lukinmaa, P.-L., Vartiainen, T., Partanen, M., Torppa, J. and Tuomisto, J. (1996) Polychlorinated dibenzo-*p*-dioxins and dibenzofurans via mother's milk may cause developmental defects in the child's teeth. *Environmental Toxicology and Pharmacology* 1, 193–197.

Alaluusua, S., Lukinmaa, P.-L., Torppa, J., Tuomisto, J. and Vartiainen, T. (1999) Developing teeth as biomarker of dioxin exposure. *The Lancet* 353, 206.

Alaluusua, S., Calderara, P., Gerthoux, P.M., Lukinmaa, P.-L., Kovero, O. *et al.* (2004) Developmental dental aberrations after the dioxin accident in Seveso. *Environmental Health Perspectives* 112, 1313–1318.

Aylward, L.L., Brunet, R.C., Carrier, G., Hays, S.M., Cushing, C.A. *et al.* (2005) Concentration-dependent TCDD elimination kinetics in humans: toxicokinetic modeling for moderately to highly exposed adults from Seveso, Italy, and Vienna, Austria, and impact on dose estimates for the NIOSH cohort. *Journal of Exposure Analysis and Environmental Epidemiology* 15, 51–65.

Baker, T.R., King-Heiden, T.C., Peterson, R.E. and Heideman, W. (2014a) Dioxin induction of transgenerational inheritance of disease in zebrafish. *Molecular and Cellular Endocrinology* 398, 36–41.

Baker, T.R., Peterson, R.E. and Heideman, W. (2014b) Using zebrafish as a model system for studying the transgenerational effects of dioxin. *Toxicological Sciences* 138, 403–411.

Bell, D.R., Clode, S., Fan, M.Q., Fernandes, A., Foster, P.M. *et al.* (2007) Toxicity of 2,3,7,8-tetrachlorodibenzo-*p*-dioxin in the developing male Wistar(Han) rat. II: Chronic dosing causes developmental delay. *Toxicological Sciences* 99, 224–233.

Bell, D.R., Clode, S., Fan, M.Q., Fernandes, A., Foster, P.M. *et al.* (2010) Interpretation of studies on the developmental reproductive toxicology of 2,3,7,8-tetrachlorodibenzo-*p*-dioxin in male offspring. *Food and Chemical Toxicology* 48, 1439–1447.

Birnbaum, L.S. (1995) Developmental effects of dioxins. *Environmental Health Perspectives* 103 (Suppl. 7), 89–94.

Birnbaum, L.S. and Tuomisto, J. (2000) Non-carcinogenic effects of TCDD in animals. *Food Additives and Contaminants* 17, 275–288.

Boers, D., Portengen, L., Bueno-de-Mesquita, H.B., Heederik, D. and Vermeulen, R. (2010) Cause-specific mortality of Dutch chlorophenoxy herbicide manufacturing workers. *Occupational and Environmental Medicine* 67, 24–31.

Boffetta, P., Mundt, K.A., Adami, H.O., Cole, P. and Mandel, J.S. (2011) TCDD and cancer: a critical review of epidemiologic studies. *Critical Reviews in Toxicology* 41, 622–636.

Bruner-Tran, K.L. and Osteen, K.G. (2011) Developmental exposure to TCDD reduces fertility and negatively affects pregnancy outcomes across multiple generations. *Reproductive Toxicology* 31, 44–50.

Bruner-Tran, K.L., Ding, T., Yeoman, K.B., Archibong, A., Arosh, J.A. and Osteen, K.G. (2014) Developmental exposure of mice to dioxin promotes transgenerational testicular inflammation and an increased risk of preterm birth in unexposed mating partners. *PLoS ONE* 15, e105084.

Casado, F.L., Singh, K.P. and Gasiewicz, T.A. (2010) The aryl hydrocarbon receptor: regulation of hematopoiesis and involvement in the progression of blood diseases. *Blood Cells, Molecules and Diseases* 44, 199–206.

Cole, P., Trichopoulos, D., Pastides, H., Starr, T. and Mandel, J.S. (2003) Dioxin and cancer: a critical review. *Regulatory Toxicology and Pharmacology* 38, 378–388.

Connor, K.T., Harris, M.A., Edwards, M.R., Budinsky, R.A., Clark, G.C. *et al.* (2008) AH receptor agonist activity in human blood measured with a cell-based bioassay: Evidence for naturally occurring AH receptor ligands in vivo. *Journal of Exposure Science & Environmental Epidemiology* 18, 369–380..

Consonni, D., Pesatori, A.C., Zocchetti, C., Sindaco, R., D'Oro, L.C., Rubagotti, M. and Bertazzi, P.A. (2008) Mortality in a population exposed to dioxin after the Seveso, Italy, accident in 1976: 25 years of follow-up. *American Journal of Epidemiology* 167, 847–858.

Consonni, D., Sindaco, R. and Bertazzi, P.A. (2012) Blood levels of dioxins, furans, dioxin-like PCBs, and TEQs in general populations: a review, 1989–2010. *Environment International* 44, 151–162.

Cypel, Y.S., Kress, A.M., Eber, S.M., Schneiderman, A.I. and Davey, V.J, (2016) Herbicide exposure, Vietnam service, and hypertension risk in Army Chemical Corps veterans. *Journal of Occupational and Environmental Health* 58, 1127–1136.

Debacker, N., Sasse, A., van Wouwe, N., Goeyens, L., Sartor, F. and van Oyen, H. (2007) PCDD/F levels in plasma of a Belgian population before and after the 1999 Belgian PCB/DIOXIN incident. *Chemosphere* 67, S217–S223.

DeGroot, D., He, G., Fraccalvieri, D., Bonati, L., Pandini, A., and Denison, M.S. (2012) AHR ligands: promiscuity in binding and diversity in response. In: Pohjanvirta, R. (ed.) *The AH receptor in Biology and Toxicology*. John Wiley & Sons, Hoboken, New Jersey, pp. 63–79.

Ding, T., McConaha, M., Boyd K.L., Osteen, K.G. and Bruner-Tran, K.L. (2011) Developmental dioxin exposure of either parent is associated with an increased risk of preterm birth in adult mice. *Reproductive Toxicology* 31, 351–358.

Dopico, M. and Gomez, A. (2015) Review of the current state and main sources of dioxins around the world. *Journal of the Air & Waste Management Association* 65, 1033–1049.

Dragan, Y.P. and Schrenk, D. (2000) Animal studies addressing the carcinogenicity of TCDD (or related compounds) with an emphasis on tumour promotion. *Food Additives and Contaminants* 17, 289–302.

EPA (2006) An Inventory of Sources and Environmental Releases of Dioxin-like Compounds in the United States for the Years 1987, 1995, and 2000 (Final, Nov. 2006). Available at: https://cfpub.epa.gov/ncea/dioxin/recordisplay.cfm?deim=159286.

EPA (2013) Update to An Inventory of Sources and Environmental Releases of Dioxin-Like Compounds in the United States for the Years 1987, 1995, and 2000 (2013, External Review Draft). Available at: https://cfpub.epa.gov/ncea/dioxin/recordisplay.cfm?deid=235432.

Eskenazi, B., Mocarelli, P., Warner, M., Samuels, S., Vercellini, P. *et al.* (2002) Serum dioxin concentrations and endometriosis: a cohort study in Seveso, Italy. *Environmental Health Perspectives* 110, 629–634

Eskenazi, B., Warner, M., Marks, A.R., Samuels, S., Needham, L., Brambilla, P. and Mocarelli, P. (2010) Serum dioxin concentrations and time to pregnancy. *Epidemiology* 21, 224–231.

Esser, C. and Rannug, A. (2015) The aryl hydrocarbon receptor in barrier organ physiology, immunology, and toxicology. *Pharmacological Reviews* 67, 259–279.

Feeley, M. and Brouwer, A. (2000) Health risks to infants from exposure to PCBs, PCDDs and PCDFs. *Food Additives and Contaminants* 17, 325–333.

Finnilä , M.A., Zioupos, P., Herlin, M., Miettinen, H.M., Simanainen, U., Håkansson, H., Tuukkanen, J., Viluksela, M. and Jämsä, T. (2010) Effects of 2,3,7,8-tetrachlorodibenzo-*p*-dioxin exposure on bone material properties. *Journal of Biomechanics* 43, 1097–1103.

Flesch-Janys, D., Berger, J., Gurn, P., Manz, A., Nagel, S., Waltsgott, H. and Dwyer, J.H. (1995) Exposure to polychlorinated dioxins and furans (PCDD/F) and mortality in a cohort of workers from a herbicide-producing plant in Hamburg, Federal Republic of Germany, *American Journal of Epidemiology* 142, 1165–1175.

Geusau, A., Abraham, K., Geissler, K., Sator, M.O., Stingl, G. and Tschachler, E. (2001) Severe 2,3,7,8-tetrachlorodibenzo-*p*-dioxin (TCDD) intoxication: clinical and laboratory effects. *Environmental Health Perspectives* 109, 865–869.

Hahn, M.E. and Karchner, S.I. (2012) Structural and functional diversification of AHRs during metazoan evolution. In: Pohjanvirta, R. (ed.) *The AH Receptor in Biology and Toxicology*. John Wiley & Sons, Hoboken, New Jersey, pp. 389–403.

Hermsen, S.A., Larsson, S., Arima, A., Muneoka, A., Ihara, T. *et al.* (2008) In utero and lactational exposure to 2,3,7,8-tetrachlorodibenzo-*p*-dioxin (TCDD) affects bone tissue in rhesus monkeys. *Toxicology* 253, 147–152.

Hornung, M.W., Spitsbergen, J.M., Peterson, R.E. and Jan, T.S. (1999) 2,3,7,8-Tetrachlorodibenzo-*p*-dioxin alters cardiovascular and craniofacial development and function in sac fry of rainbow trout (*Oncorhynchus mykiss*). *Toxicological Sciences* 47, 40–51.

Howard, G.J., Schlezinger, J.J., Hahn, M.E. and Webster, T.F. (2010) Generalized concentration addition predicts joint effects of aryl hydrocarbon receptor agonists with partial agonists and competitive antagonists. *Environmental Health Perspectives* 118, 666–672.

Hsu, J.-F., Guo, Y.-L., Yang, S.-Y. and Liao, P.-C. (2005) Congener profiles of PCBs and PCDD/Fs in Yucheng victims fifteen years after exposure to toxic rice-bran oils and their implications for epidemiologic studies. *Chemosphere* 61, 1231–1243.

Humblet, O., Birnbaum, L., Rimm, E., Mittleman, M.A. and Hauser, R. (2008) Dioxins and cardiovascular disease mortality. *Environmental Health Perspectives* 116, 1443–1448.

IARC (1997) *Polychlorinated Dibenzo-*para*-dioxins and Polychlorinated Dibenzofurans*. IARC Monographs on the Evaluation of Carcinogenic Risks to Humans, Vol. 69. International Agency for Research on Cancer, Lyon, France.

IARC (2012) 2,3,7,8-Tetrachlorodibenzopara-dioxin, 2,3,4,7,8-pentachlorodibenzofuran, and 3,3′,4,4′,5-pentachlorobiphenyl. In: *Chemical Agents and Related Occupations*. IARC Monographs on the Evaluation of Carcinogenic Risks to Humans, Vol. 100F, pp. 339–378. International Agency for Research on Cancer, Lyon, France.

Ishihara, K., Warita, K., Tanida, T., Sugawara, T., Kitagawa, H. and Hoshi, N. (2007) Does paternal exposure to 2,3,7,8-tetrachlorodibenzo-*p*-dioxin (TCDD) affect the sex ratio of offspring? *Journal of Veterinary Medical Science* 69, 347–352.

Jin L.J. and Chen B.L. (2017) Natural origins, concentration levels, and formation mechanisms of organohalogens in the environment. *Progress in Chemistry (China)* 29, 1093–1114.

Kashima, S., Yorifuji, T., Tsuda, T. and Eboshida, A. (2015) Cancer and non-cancer excess mortality resulting from mixed exposure to polychlorinated biphenyls and polychlorinated dibenzofurans from contaminated rice oil: 'Yusho'. *International Archives of Occupational and Environmental Health* 88, 419–430.

Kattainen, H., Tuukkanen, J., Simanainen, U., Tuomisto, J.T., Kovero, O. *et al.* (2001) In utero / lactational TCDD exposure impairs molar tooth development in rats. *Toxicology and Applied Pharmacology* 174, 216–224.

Keller, J.M., Allen, D.E., Davis, C.R. and Leamy, L.J. (2007a) 2,3,7,8-Tetrachlorodibenzo-*p*-dioxin affects fluctuating asymmetry of molar shape in mice, and an epistatic interaction of two genes for molar size. *Heredity* 98, 259–267.

Keller, J.M., Huet-Hudson, Y.M. and Leamy, L.J. (2007b) Qualitative effects of dioxin on molars vary among inbred mouse strains. *Archives of Oral Biology* 52, 450–454.

Kiviranta, H., Ovaskainen, M.-L. and Vartiainen T. (2004) Market basket study on dietary intake of PCDD/Fs, PCBs and PCDEs in Finland. *Environment International* 30, 923–932.

Kiviranta, H., Tuomisto, J.T., Tuomisto, J., Tukiainen, E. and Vartiainen, T. (2005) Polychlorinated dibenzo-*p*-dioxins, dibenzofurans, and biphenyls in the general population in Finland. *Chemosphere* 60, 854–869.

Kociba, R.J., Keyes, D.G., Beyer, J.E., Carreon, R.M., Wade, C.E. *et al.* (1978) Results of a 2-year chronic toxicity and oncogenicity study of 2,3,7,8-tetrachlorodibenzo-p-dioxin in rats. *Toxicology and Applied Pharmacology* 46, 279–303.

Kogevinas, M. (2000) Studies of cancer in humans. *Food Additives and Contaminants* 17, 317–324.
Kogevinas, M., Becher, H. Benn, T., Bertazzi, P.A. *et al.* (1997) Cancer mortality in workers exposed to phenoxy herbicides, chlorophenols and dioxins. *American Journal of Epidemiology* 145, 1061–1075.
Kolluri, S.K, Jin, U.-H. and Safe, S. (2017) Role of the aryl hydrocarbon receptor in carcinogenesis and potential as an anti-cancer drug target. *Archives of Toxicology* 91, 2497–2513.
Korkalainen, M., Kallio, E., Olkku, A., Nelo, K., Ilvesaro, J. *et al.* (2009) Dioxins interfere with differentiation of osteoblasts and osteoclasts. *Bone* 44, 1134–1142.
Koskenniemi, J.J., Virtanen, H.E., Kiviranta, H., Damgaard, I.N., Matomäki, J. *et al.* (2015) Association between levels of persistent organic pollutants in adipose tissue and cryptorchidism in early childhood: a case-control study. *Environmental Health* 14, 78.
Kreuzer, P.E., Csanády, G.A., Baur, C., Kessler, W., Päpke, O., Greim, H. and Filser, J.G. (1997) 2,3,7,8-Tetrachlorodibenzo-*p*-dioxin (TCDD) and congeners in infants. A toxicokinetic model of human lifetime body burden by TCDD with special emphasis on its uptake by nutrition. *Archives of Toxicology* 71, 383–400.
Lensu, S., Tuomisto, J.T., Tuomisto, J. and Pohjanvirta, R. (2011a) Characterization of the 2,3,7,8-tetrachlorodibenzo-*p*-dioxin (TCDD)-provoked strong and rapid aversion to unfamiliar foodstuffs in rats. *Toxicology* 283, 140–150.
Lensu, S., Tuomisto, J.T., Tuomisto, J., Viluksela, M., Niittynen, M. and Pohjanvirta, R. (2011b) Immediate and highly sensitive aversion response to a novel food item linked to AH receptor stimulation. *Toxicology Letters* 203, 252–257
Li, M.-C., Chen, P.-C., Tsai, P.-C., Furue, D., Onozuka, D. *et al.* (2015) Mortality after exposure to polychlorinated biphenyls and polychlorinated dibenzofurans: a meta-analysis of two highly exposed cohorts. *International Journal of Cancer* 137, 1427–1432.
Liem, A.K.D., Fürst, P. and Rappe, C. (2000) Exposure of populations to dioxins and related compounds. *Food Additives and Contaminants* 17, 241–259.
Lindén, J., Lensu, S., Tuomisto, J. and Pohjanvirta, R. (2010) Dioxins, the aryl hydrocarbon receptor and the central regulation of energy balance. *Frontiers in Neuroendocrinology* 31, 452–478.
Ma, Q. (2012) Overview of AHR functional domains and the classical AHR signalling pathway: induction of drug metabolizing enzymes. In: Pohjanvirta, R. (ed.) *The AH Receptor in Biology and Toxicology*. John Wiley & Sons, Hoboken, New Jersey, pp. 35–45.
Manikkam, M., Tracey, R., Guerrero-Bosagna, C. and Skinner, M.K. (2012) Dioxin (TCDD) induces epigenetic transgenerational inheritance of adult onset disease and sperm epimutations. *PLoS ONE* 7, e46249.
Michalek, J.E. and Pavuk, M. (2008) Diabetes and cancer in veterans of Operation Ranch Hand after adjustment for calendar period, days of spraying, and time spent in Southeast Asia. *Journal of Occupational and Environmental Medicine* 50, 330–340.
Miettinen, H.M., Alaluusua, S., Tuomisto, J. and Viluksela, M. (2002) Effect of in utero and lactational TCDD exposure on rat molar development: the role of exposure time. *Toxicology and Applied Pharmacology* 184, 57–66
Miettinen, H.M., Pulkkinen, P., Jämsä, T., Koistinen, J., Simanainen, U. *et al.* (2005) Effects of in utero and lactational TCDD exposure on bone development in differentially sensitive rat lines. *Toxicological Sciences* 85, 1003–1012.
Miettinen, H.M., Sorvari, R., Alaluusua, S., Murtomaa, M., Tuukkanen, J. and Viluksela, M. (2006) The effect of perinatal TCDD exposure on caries susceptibility in rats. *Toxicological Sciences*, 91, 568–575.
Milbrath, M.O., Wenger Y., Chang, C.-W., Emond, C., Garabrant, D., Gillespie, B.W. and Jolliet, O. (2009) Apparent half-lives of dioxins, furans, and polychlorinated biphenyls as a function of age, body fat, smoking status, and breast-feeding. *Environmental Health Perspectives* 117, 417–425.
Mínguez-Alarcón, L., Sergeyev, O., Burns, J.S., Williams, P.L., Lee, M.M., Korrick, S.A., Smigulina, L., Revich, B. and Hauser, R. (2017) A longitudinal study of peripubertal serum organochlorine concentrations and semen parameters in young men: the Russian Children's Study. *Environmental Health Perspectives* 125, 460–466.
Mocarelli P, Needham, L.L., Marocchi, A., Patterson, D.G. Jr, Brambilla, P. *et al.* (1991) Serum concentrations of 2,3,7,8-tetrachlorodibenzo-*p*-dioxin and test results from selected residents of Seveso, Italy. *Journal of Toxicology and Environmental Health* 32, 357–366.
Mocarelli, P., Gerthoux, P.M., Ferrari, E., Patterson, D.G. Jr, Kieszak, S.M. *et al.* (2000) Paternal concentrations of dioxin and sex ratio of offspring. *The Lancet* 355, 1858–1863.

Mocarelli, P., Gerthoux, P.M., Patterson, D.G. Jr, Milani, S., Limonta, G. *et al.* (2008) Dioxin exposure, from infancy through puberty, produces endocrine disruption and affects human semen quality. *Environmental Health Perspectives* 116, 70–77.

Mocarelli, P., Gerthoux, P.M., Needham, L.L., Patterson, D.G. Jr, Limonta, G. *et al.* (2011) Perinatal exposure to low doses of dioxin can permanently impair human semen quality. *Environmental Health Perspectives* 119, 713–718.

Momeniha, F., Faridi, S., Amini, H., Shamsipour, M., Kazem Naddafi, K. *et al.* (2017) Estimating national dioxins and furans emissions, major sources, intake doses, and temporal trends in Iran from 1990–2010. *Journal of Environmental Health Science and Engineering* 15, 20.

Nishimura, N., Nishimura, H., Ito, T., Miyata, C., Izumi, K., Fujimaki, H. and Matsumura, F. (2009) Dioxin-induced up-regulation of the active form of vitamin D is the main cause for its inhibitory action on osteoblast activities, leading to developmental bone toxicity. *Toxicology and Applied Pharmacology* 236, 301–309.

Norén, K. and Meironyté, D. (2000) Certain organochloride and organobromine contaminants in Swedish human milk in perspective of past 20–30 years. *Chemosphere* 40, 1111–1123.

Okey, A.B. (2007) An aryl hydrocarbon receptor odyssey to the shores of toxicology: the Deichmann Lecture, International Congress of Toxicology XI. *Toxicological Sciences* 98, 5–38.

Okey, A.B., Franc, M.A., Moffat, I.D., Tijet, N., Boutros, P.C. *et al.* (2005) Toxicological implications of polymorphisms in receptors for xenobiotic chemicals: the case of the aryl hydrocarbon receptor. *Toxicology and Applied Pharmacology* 207, 43–51.

Onozuka, D., Yoshimura, T., Kaneko, S. and Furue, M. (2009) Mortality after exposure to polychlorinated biphenyls and polychlorinated dibenzofurans: a 40-year follow-up study of Yusho patients. *American Journal of Epidemiology* 169, 86–95.

Ott, M.G. and Zober, A. (1996) Cause specific mortality and cancer incidence among employees exposed to 2,3,7,8-TCDD after a 1953 reactor accident. *Occupational and Environmental Medicine* 53, 606–612.

Partanen, A.M., Alaluusua, S., Miettinen, P.J., Thesleff, I., Tuomisto, J., Pohjanvirta, R. and Lukinmaa, P.-L. (1998) Epidermal growth factor receptor as a mediator of developmental toxicity of dioxin in mouse embryonic teeth. *Laboratory Investigation* 78, 1473–1481.

Pesatori, A.C., Consonni, D., Rubagotti, M., Grillo, P. and Bertazzi, P.A. (2009) Cancer incidence in the population exposed to dioxin after the 'Seveso accident': twenty years of follow-up. *Environmental Health* 8, 39.

Peters, A.K., Leonards, P.E., Zhao, B., Bergman, A., Denison, M.S. and van den Berg, M. (2006) Determination of in vitro relative potency (REP) values for mono-ortho polychlorinated biphenyls after purification with active charcoal. *Toxicological Letters* 165, 230–241.

Pilsner, J.R., Parker, M., Sergeyev, O. and Suvorova, A. (2017) Spermatogenesis disruption by dioxins: epigenetic reprogramming and windows of susceptibility. *Reproductive Toxicology* 69, 221–229.

Planchart, A. and Mattingly, C.J. (2010) 2,3,7,8-Tetrachlorodibenzo-*p*-dioxin upregulates FoxQ1b in zebrafish jaw primordium. *Chemical Research in Toxicology* 23, 480–487.

Pohjanvirta, R. (ed.) (2012) *The AH Receptor in Biology and Toxicology*. John Wiley & Sons, Hoboken, New Jersey.

Pohjanvirta, R. and Tuomisto, J. (1994) Short-term toxicity of 2,3,7,8-tetrachlorodibenzo-*p*-dioxin in laboratory animals: effects, mechanisms, and animal models. *Pharmacological Reviews* 46, 483–549.

Poland, A. and Knutson, J.C. (1982) 2,3,7,8-Tetrachlorodibenzo-*p*-dioxin and related halogenated aromatic hydrocarbons: examination of the mechanisms of toxicity. *Annual Review of Pharmacology and Toxicology* 22, 517–554.

Quass, U., Fermann, M. and Bröker, G. (2004) The European Dioxin Air Emission Inventory Project – Final Results. *Chemosphere* 54, 1319–1327.

Render, J.A., Bursian, S.J., Rosenstein, D.S. and Aulerich, R.J. (2001). Squamous epithelial proliferation in the jaws of mink fed diets containing 3,3′,4,4′,5-pentachlorobiphenyl (PCB 126) or 2,3,7,8-tetrachlorodibenzo-*p*-dioxin (TCDD). *Veterinary and HumanToxicology* 43, 22–26.

Safe, S.H. (1998) Hazard and risk assessment of chemical mixtures using the toxic equivalency factor approach. *Environmental Health Perspectives* 106 (Suppl. 4), 1051–1058.

Sanabria, M., Cucielo, M.S., Guerra, M.T., Dos Santos Borges, C., Banzato, T.P. *et al.* (2016) Sperm quality and fertility in rats after prenatal exposure to low doses of TCDD: a three-generation study. *Reproductive Toxicology* 65, 29–38.

Schrenk, D. and Chopra, M. (2012) Dioxin activated AHR and cancer in laboratory animals. In: Pohjanvirta, R. (ed.) *The AH Receptor in Biology and Toxicology*. John Wiley & Sons, Hoboken, New Jersey, pp. 245–256.

Sibilano, R., Pucillo, C.E. and Gri, G. (2015) Allergic responses and aryl hydrocarbon receptor novel pathway of mast cell activation. *Molecular Immunology* 63, 69–73.

Simanainen, U., Tuomisto, J.T., Tuomisto, J. and Viluksela, M. (2003) Dose-response analysis of short-term effects of 2,3,7,8-tetrachlorodibenzo-*p*-dioxin in three differentially susceptible rat lines. *Toxicology and Applied Pharmacology* 187, 128–136.

Sorg, O., Zennegg, M., Schmid, P., Fedosyuk, R., Valikhnovskyi, R. *et al.* (2009) 2,3,7,8-Tetrachlorodibenzo-p-dioxin (TCDD) poisoning in Victor Yushchenko: identification and measurement of TCDD metabolites. *Lancet* 374, 1179–1185.

Steenland, K., Piacitelli, L., Deddens, J., Fingerhut, M. and Chang, L.I. (1999) Cancer, heart disease, and diabetes in workers exposed to 2,3,7,8-tetrachlorodibenzo-*p*-dioxin. *Journal of the National Cancer Institute* 91, 779–786.

Sweeney, M.H. and Mocarelli, P. (2000) Human health effects after the exposure to 2,3,7,8-TCDD. *Food Additives and Contaminants* 17, 303–316.

Tsai, P.C., Ko, Y.C., Huang, W.Y., Liu, H.S. and Guo, Y.L. (2007). Increased liver and lupus mortalities in 24-year follow-up of the Taiwanese people highly exposed to polychlorinated biphenyls and dibenzofurans. *Science of the Total Environment* 374, 216–222.

Tuomisto, J. (2005) Does mechanistic understanding help in risk assessment – the example of dioxins. *Toxicology and Applied Pharmacology* 207 (Suppl. 1), 2–10.

Tuomisto, J. (2012) The toxic equivalency principle and its application in dioxin risk assessment. In: Pohjanvirta, R. (ed.) *The AH Receptor in Biology and Toxicology*. John Wiley & Sons, Hoboken, New Jersey, pp. 317–330.

Tuomisto, J. and Tuomisto, J.T. (2012) Is the fear of dioxin cancer more harmful than dioxin? *Toxicology Letters* 210, 338–344.

Tuomisto, J.T., Viluksela, M., Pohjanvirta, R. and Tuomisto, J. (1999) The AH receptor and a novel gene determine acute toxic responses to TCDD: segregation of the resistant alleles to different rat lines. *Toxicology and Applied Pharmacology* 155, 71–81.

Tuomisto, J.T., Pekkanen, J., Kiviranta, H., Tukiainen, E., Vartiainen, T. and Tuomisto, J. (2004a) Soft tissue sarcoma and dioxins – a case control study. *International Journal of Cancer* 108, 893–900.

Tuomisto, J.T., Tuomisto, J., Tainio, M., Niittynen, M., Verkasalo, P. *et al.* (2004b) Risk-benefit analysis of eating farmed salmon. *Science* 305, 476.

Tuomisto, J., Pekkanen, J., Kiviranta, H., Tukiainen, E., Vartiainen, T., Viluksela, M. and Tuomisto, J.T. (2005) Dioxin cancer risk – example of hormesis? *Dose-Response* 3, 332–341.

Tuomisto, J., Vartiainen, T. and Tuomisto, J.T. (2011) *Synopsis on Dioxins and PCBs*. Report of the National Institute for Health and Welfare, Helsinki, Finland. Available at: www.thl.fi/dioxin.

Tuomisto, J., Airaksinen, R., Kiviranta, H., Tukiainen, E., Pekkanen, J. and Tuomisto, J.T. (2016) A pharmacokinetic analysis and dietary information are necessary to confirm or reject the hypothesis on persistent organic pollutants causing type 2 diabetes. *Toxicology Letters* 261, 41–48.

Tuomisto, J., Airaksinen, R., Pekkanen, J., Tukiainen, E., Kiviranta, H. and Tuomisto, J.T. (2017) Comparison of questionnaire data and analyzed dioxin concentrations as a measure of exposure in soft-tissue sarcoma studies. *Toxicology Letters* 270, 8–11.

van den Berg, M. with a group of 30 temporary advisors (2000) Consultation on assessment of the health risk of dioxins; re-evaluation of the tolerable daily intake (TDI): Executive summary. *Food Additives and Contaminants* 17, 223–240.

van den Berg, M., Birnbaum, L.S., Denison, M., De Vito, M., Farland, W. *et al.* (2006) The 2005 World Health Organization reevaluation of human and mammalian toxic equivalency factors for dioxins and dioxin-like compounds, *Toxicological Sciences* 93, 223–241.

van den Berg, M., Denison, M.S., Birnbaum, L.S., DeVito, M.J., Fiedler, H., Falandysz, J., Rose, M., Schrenk, D., Safe, S., Tohyama, C., Tritscher, A., Tysklind, M. and Peterson, R.E. (2013) Polybrominated dibenzo-p-dioxins, dibenzofurans and biphenyls: inclusion in the toxicity equivalency factor concept for dioxin-like compounds. *Toxological Sciences* 133, 197–208.

van den Berg, M., Kypke, K., Kotz, A., Tritscher, A., Lee, S.Y. *et al.* (2017) WHO/UNEP global surveys of PCDDs, PCDFs, PCBs and DDTs in human milk and benefit–risk evaluation of breastfeeding. *Archives of Toxicology* 91, 83–96.

Van Leeuwen, F.X.R. and Younes, M.M. (2000) Assessment of health risks of dioxins: re-evaluation of the tolerable daily intake (TDI). *Food Additives and Contaminants* 17(4), 223–369.

Van Voorhis, M., Fechner, J.H., Zhang, X. and Mezrich, J.D. (2012) The aryl hydrocarbon receptor: a novel target for immunomodulation in organ transplantation. *Transplantation* 95, 983–990.

Vartiainen, T., Jaakkola, J.J.K., Saarikoski, S. and Tuomisto, J. (1998) Birth weight and sex of children and the correlation to the body burden of PCDDs/PCDFs and PCBs of the mother. *Environmental Health Perspectives* 106, 61–66.

Viluksela, M., Bager, Y., Tuomisto, J.T., Scheu, G., Unkila, M., *et al.* (2000) Liver tumor-promoting activity of 2,3,7,8.tetrachlodibenzo-*p*-dioxin (TCDD) in TCDD-sensitive and TCDD-resistant rat strains. *Cancer Research* 60, 6911–6920.

Viluksela, M., Miettinen, H.M. and Korkalainen, M. (2012) Effects of dioxins on teeth and bone: the role of AHR. In: Pohjanvirta, R. (ed.) *The AH Receptor in Biology and Toxicology*. John Wiley & Sons, Hoboken, New Jersey, pp. 285–297.

Virtanen, H.E., Koskenniemi, J.J., Sundqvist, E., Main, K.M., Kiviranta, H. *et al.* (2012) Associations between congenital cryptorchidism in newborn boys and levels of dioxins and PCBs in placenta. *International Journal of Andrology* 35, 283–293.

Warner, M., Mocarelli, P., Samuels, S., Needham, L., Brambilla, P. and Eskenazi, B. (2011) Dioxin exposure and cancer risk in the Seveso Women's Health Study. *Environmental Health Perspectives* 119, 1700–1705.

Wesselink, A., Warner, M., Samuels, S., Parigi, A., Brambilla, P., Mocarelli, P. and Eskenazi, B. (2014) Maternal dioxin exposure and pregnancy outcomes over 30 years of follow-up in Seveso. *Environment International* 63, 143–148.

White, S.S. and Birnbaum, L.S. (2009) An overview of the effects of dioxins and dioxinlike compounds on vertebrates, as documented in human and ecological epidemiology. *Journal of Environmental Science and Health, Part C: Environmental Carcinogenesis and Ecotoxicology Reviews* 27, 197–211.

WHO (1989) *Levels of PCBs, PCDDs, and PCDFs in Breast Milk*. Environmental Health Series 34, World Health Organization, Copenhagen.

Yamaguchi, N. (1999) Uncertainty in risk characterization of weak carcinogens. In: Bailer, A.J., Maltoni, C., Bailar, J.C. *et al.* (eds) *Uncertainty in the Risk Assessment of Environmental and Occupational Hazards*. New York Academy of Sciences, New York, pp. 338–347.

Yasuda, I., Yasuda, M., Sumida, H., Tsusaki, H., Arima, A. *et al.* (2005). In utero and lactational exposure to 2,3,7,8-tetrachlorodibenzo-*p*-dioxin (TCDD) affects tooth development in rhesus monkeys. *Reproductive Toxicology* 20, 21–30.

Zhang, M., Buekens, A. and Li, X. (2017) Open burning as a source of dioxins. *Critical Reviews in Environmental Science and Technology* 47, 543–620

# 14 Dioxins III. Relationship to Pre-Diabetes, Diabetes and Diabetic Nephropathy

**C.J. Everett***
*Medical University of South Carolina, Charleston, USA*

## 14.1 Abstract

Toxic equivalency factors (TEFs) have been defined for seven polychlorinated dibenzo-*p*-dioxins, ten polychlorinated dibenzofurans and 12 polychlorinated biphenyls (van den Berg *et al.*, 2006). Twenty-three of the 29 dioxin-like chemicals were measured in human blood during the 1999–2004 National Health and Nutrition Examination Survey (NHANES) conducted in the USA (Everett and Thompson, 2012, 2014). Multiplying the concentration by the TEF for these 23 dioxin-like chemicals, and summing the products, yielded a measure called toxic equivalency ($TEQ_{23}$). In these investigations, pre-diabetes was defined as glycohaemoglobin (type A1c) 5.7–6.4%, diabetes was defined as diagnosed or A1c ≥ 6.5%, and nephropathy (kidney disease) as urinary albumin to creatinine ratio > 30 mg $g^{-1}$ (microalbuminuria or macroalbuminuria). Expressed as a continuous variable, logarithm-transformed toxic equivalency ($\ln(TEQ_{23}+1)$) was associated with 'high' pre-diabetes (in the range of A1c 5.9–6.4%) with an odds ratio of 1.33 (95% confidence interval (CI): 1.03–1.72). Logarithm-transformed $TEQ_{23}$ was also associated with diabetes (odds ratio 1.60, 95% CI: 1.22–2.09), diabetes without nephropathy (odds ratio 1.44, 95% CI: 1.11–1.87) and diabetic nephropathy (odds ratio 2.35, 95% CI: 1.57–3.52). Eight of the dioxin-like chemicals included in $TEQ_{23}$ were most important in terms of the proportion of persons, included in the NHANES 1999–2004, with detectable values (> 25% above the maximum limit of detection). Of these eight dioxin-like chemicals, six were associated with diabetes, five associated with diabetes without nephropathy and seven associated with diabetic nephropathy. While 1,2,3,4,6,7,8-heptachlorodibenzo-*p*-dioxin and PCB 126 were not associated with diabetes, they were associated with diabetes without nephropathy. The association of logarithm-transformed $TEQ_{23}$ with diabetic nephropathy appears to be a case of reverse causality, or perhaps both due to reserve causality and a risk factor for the disease.

## 14.2 Introduction

Diabetes is a growing worldwide health problem. In the USA, the prevalence of diabetes was 12.3% (95% CI: 10.8–14.1%) in 2011–2012 (Menke *et al.*, 2015) and is projected to be 14.5% in 2031 (Mainous *et al.*, 2007). A diabetes risk score developed by Schmidt *et al.* (2005) utilized data from the Atherosclerosis Risk in Communities (ARIC) cohort. One point each was assigned

* E-mail address: everettc@musc.edu

for high waist circumference (women > 88 cm, men > 102 cm), high blood pressure (> 130/85 mmHg or using antihypertensive medication), low high-density lipoprotein (HDL) cholesterol (men < 1.03 mmol $l^{-1}$, women < 1.29 mmol $l^{-1}$), high triglycerides (> 1.7 mmol $l^{-1}$) and obesity (body mass index ≥ 30 kg $m^{-2}$); in addition, two points were assigned for fasting glucose ≥ 5.6 mmol $l^{-1}$ or five points for fasting glucose ≥ 6.1 mmol $l^{-1}$. A score of 4 or more indicated a high risk of developing diabetes (sensitivity 68% and specificity 75%) with 32% of the ARIC sample being at high risk. While such a diabetes risk score is useful there is no consideration of the effect of environmental pollutants such as dioxin-like chemicals, which are the subject of this chapter.

Releases of dioxins and furans from industrial sources in the USA decreased by approximately 80% from the 1980s to 2005. Currently, release of these chemicals is due to open burning of household and municipal trash, landfill fires and agricultural and forest fires. Polychlorinated biphenyls (PCBs) were once used for electrical insulation and heat-exchange fluids. Their production peaked in the early 1970s and was banned in the USA after 1979 (CDC, 2005). Co-planar and mono-*ortho*-substituted PCBs are considered dioxin-like.

## 14.3 Toxic Equivalency

The history of the toxic equivalency factor approach to assessing chlorinated dioxin-like chemicals was given by Haws *et al.* (2006) and covered the period 1984–2004. Haws *et al.* (2006) summarized relative effect potency (REP) data for 17 laterally substituted (2,3,7,8-substituted) polychlorinated dibenzo-*p*-dioxins (PCDDs) and polychlorinated dibenzofurans (PDCFs), and 12 PCBs). This database was focused on biological effects mediated by binding to and activating the aryl hydrocarbon (Ah) receptor in mammals. In humans, aryl hydrocarbon receptor binding has been hypothesized to antagonize the peroxisome proliferator-activated receptor (PPAR) and contribute to the pathophysiology of diabetes (Remillard and Bunce, 2002). The numbers of studies included in the REP database were 48 studies with *in vivo* data and 37 studies with *in vitro* data, the total being 83 studies. These investigations yielded 383 *in vivo* REPs and 251 *in vitro* REPs. Sixteen of the 29 dioxin-like chemicals in the database had ten or more REPs each, with PCB 126 (115 REPs) and 2,3,4,7,8-pentachlorodibenzofuran (99 REPs) being the best represented.

The work of Haws *et al.* (2006) was followed by a public session of the World Health Organization (WHO), working through the International Programme on Chemical Safety (IPCS), held in Geneva in June 2005 (van den Berg *et al.*, 2006). The purpose of the meeting was to re-evaluate toxic equivalency factors (TEFs) for the 29 dioxin-like chemicals. In this assessment, 2,3,7,8-tetrachlorodibenzo-*p*-dioxin (TCDD) was assigned a TEF of 1. It was decided to assign TEFs using half-order-of-magnitude increments on a logarithmic scale of 0.01, 0.03, 0.1, 0.3, etc. and these are referred to as WHO 2005 TEFs. It is important to note that van den Berg *et al.* (2006) recommended caution when applying WHO 2005 TEFs to human tissue samples, because the concept is primarily designed for intake situations.

## 14.4 National Health and Nutrition Examination Survey

The National Health and Nutrition Examination Survey (NHANES) is a programme of studies conducted in the USA by the Centers for Disease Control and Prevention (CDC). The NHANES has been conducted on a continuing basis since 1999. Fifteen communities are visited each year and data released every 2 years. In the 1999–2004 NHANES dioxins, furans and PCBs in human blood were measured in a one-third subsample of participants. Twenty-three of the 29 dioxin-like chemicals having WHO 2005 TEFs were included in all three data releases of interest (1999–2000, 2001–2002 and 2003–2004). The concentrations and WHO 2005 TEFs of these 23 chemicals were used to calculate toxic equivalency by Everett and Thompson (2012, 2014) and are referred to here as $TEQ_{23}$ (Table 14.1).

Whether or not a measurement was detectable was dependent on the concentration of each dioxin-like chemical, the amount of blood available for analysis and the survey year. Organochlorine pesticides were measured along with

**Table 14.1.** Twenty-nine dioxins and dioxin-like compounds having toxic equivalency factors (WHO 2005 TEF) and those included in $TEQ_{23}$ and $TEQ_8$ measures of toxic equivalency. Adapted from Everett and Thompson (2014) with permission. See van den Berg *et al.* (2006) for WHO 2005 TEF.

| | $TEQ_{23}$ | $TEQ_8$ | Toxic equivalency at maximum limit of detection (TEQ fg $g^{-1}$) |
|---|---|---|---|
| **Dioxins** | | | |
| 2,3,7,8-Tetrachlorodibenzo-*p*-dioxin (TCDD) | * | | 47.73 |
| 1,2,3,7,8-Pentachlorodibenzo-*p*-dioxin (PeCDD) | * | | 53.81 |
| 1,2,3,4,7,8- Hexachlorodibenzo-*p*-dioxin (HxCDD) | | | |
| 1,2,3,6,7,8-Hexachlorodibenzo-*p*-dioxin (HxCDD) | * | * | 8.86 |
| 1,2,3,7,8,9- Hexachlorodibenzo-*p*-dioxin (HxCDD) | * | | 8.15 |
| 1,2,3,4,6,7,8-Heptachlorodibenzo-*p*-dioxin (HpCDD) | * | * | 1.57 |
| 1,2,3,4,6,7,8,9-Octachlorodibenzo-*p*-dioxin (OCDD) | * | * | 0.58 |
| **Furans** | | | |
| 2,3,7,8-Tetrachlorodibenzofuran (TCDF) | * | | 5.19 |
| 1,2,3,7,8-Pentachlorodibenzofuran (PeCDF) | * | | 1.44 |
| 2,3,4,7,8-Pentachlorodibenzofuran (PeCDF) | * | * | 15.34 |
| 1,2,3,4,7,8-Hexachlorodibenzofuran (HxCDF) | * | | 4.73 |
| 1,2,3,6,7,8-Hexachlorodibenzofuran (HxCDF) | * | | 4.93 |
| 1,2,3,7,8,9-Hexachlorodibenzofuran (HxCDF) | * | | 4.72 |
| 2,3,4,6,7,8-Hexachlorodibenzofuran (HxCDF) | * | | 5.52 |
| 1,2,3,4,6,7,8-Heptachlorodibenzofuran (HpCDF) | * | | 0.72 |
| 1,2,3,4,7,8,9- Heptachlorodibenzofuran (HpCDF) | | | |
| 1,2,3,4,6,7,8,9-Octachlorodibenzofuran (OCDF) | * | | 0.050 |
| **PCBs** | | | |
| 3,3′,4,4′-Tetrachlorobiphenyl (PCB 77) | | | |
| 3,4,4′,5-Tetrachlorobiphenyl (PCB 81) | * | | 0.082 |
| 3,3′,4,4′,5-Pentachlorobiphenyl (PCB 126) | * | * | 8.65 |
| 3,3′,4,4′,5,5′-Hexachlorobiphenyl (PCB 169) | * | * | 3.45 |
| 2,3,3′,4,4′-Pentachlorobiphenyl (PCB 105) | * | | 0.0014 |
| 2,3,4,4′,5-Pentachlorobiphenyl (PCB 114) | | | |
| 2,3′,4,4′,5-Pentachlorobiphenyl (PCB 118) | * | * | 0.0012 |
| 2′,3,4,4′,5-Pentachlorobiphenyl (PCB 123) | | | |
| 2,3,3′,4,4′,5-Hexachlorobiphenyl (PCB 156) | * | * | 0.0014 |
| 2,3,3′,4,4′,5′-Hexachlorobiphenyl (PCB 157) | * | | 0.0014 |
| 2,3′,4,4′,5,5′-Hexachlorobiphenyl (PCB 167) | * | | 0.0014 |
| 2,3,3′,4,4′,5,5′-Heptachlorobiphenyl (PCB 189) | | | |

dioxins, furans and PCBs in the 1999–2000 and 2001–2002 survey years, but in a separate subsample in 2003–2004. Therefore there was more blood available for analysis in the 2003–2004 survey years. This resulted in a concentration range for a chemical where some measurements were detectable and some were not. To deal with this ambiguity the highest concentration that was not detectable was used as the limit of detection and is referred to as the maximum limit of detection (MLOD). The product of the WHO 2005 TEF multiplied by the MLOD is shown in Table 14.1. The toxic equivalency at the MLOD is high for 2,3,7,8-TCDD and 1,2,3,7,8-PeCDD because the WHO 2005 TEF has a value of 1. Similarly, the toxic equivalency at the MLOD is low for PCB 105, PCB 118, PCB 156, PCB 157 and PCB 167 because the WHO 2005 TEF has a value of 0.00003. In between these extremes are cases where the toxic equivalencies at the MLOD, and the WHO 2005 TEF, have intermediate values. One example is PCB 126, which has a WHO 2005 TEF of 0.1 and toxic equivalency at the MLOD of 8.65. When all 23 dioxin-like

**Table 14.2.** Diabetes categories, number of compounds elevated and distribution of toxic equivalency ($TEQ_{23}$) in a representative sample of the US population, 1999–2004. Adapted from Everett and Thomson (2014), with permission.

| | Unweighted *N* | Population estimate | Proportion (%) |
|---|---|---|---|
| **Diabetes Category** | | | |
| Normal A1c (< 5.7%) | | | |
| - Without nephropathy | 1788 | 89,282,573 | 75.9 |
| - With nephropathy | 156 | 6,497,735 | 5.5 |
| Pre-diabetes (A1c 5.7–6.4%) | | | |
| - Without nephropathy | 312 | 11,463,344 | 9.7 |
| - With nephropathy | 50 | 1,585,421 | 1.3 |
| Pre-diabetes | | | |
| - A1c 5.7–5.8% | 191 | 6,914,475 | 5.9 |
| - A1c 5.9–6.4% | 171 | 6,134,289 | 5.2 |
| Total diabetes (Diagnosed or A1c ≥ 6.5%) | | | |
| - Without nephropathy | 184 | 6,143,212 | 5.2 |
| - With nephropathy | 98 | 2,686,073 | 2.3 |
| **Number of compounds elevated**[a] | | | |
| 0 | 946 | 47,558,488 | 40.4 |
| 1 | 335 | 16,809,962 | 14.3 |
| 2 | 235 | 11,035,270 | 9.4 |
| 3 | 163 | 7,372,429 | 6.3 |
| 4–8 | 519 | 20,513,596 | 17.4 |
| 9–13 | 317 | 11,828,906 | 10.0 |
| ≥ 14 | 73 | 2,539,706 | 2.2 |
| **Toxic equivalency** ($TEQ_{23}$ fg $g^{-1}$) | | | |
| < 13.87 | 593 | 29,382,702 | 25.0 |
| 13.87–34.53 | 585 | 29,417,231 | 25.0 |
| 34.54–81.65 | 615 | 29,400,551 | 25.0 |
| ≥ 81.66 | 795 | 29,457,873 | 25.0 |
| **Total sample** | 2588 | 117,658,357 | 100 |

[a]Only 11.9% of the sample (unweighted *N* = 284) had no compounds above the maximum limit of detection. The number of compounds elevated does not include values above the maximum limit of detection, and below the 75th percentile, for the eight chemicals included in $TEQ_8$. For the other 15 chemicals, values above the maximum limits of detection are included.

chemicals for an individual are summed, the resulting $TEQ_{23}$ is a function of both the concentrations and the sensitivity of the measurements as indicated by the toxic equivalency at the MLOD. When the toxic equivalency at the MLOD is high, the measurement is not particularly sensitive, but it does substantially influence the final summed $TEQ_{23}$ result if the concentration of the chemical is above the MLOD.

Only 11.9% of the US population had none of the 23 dioxin-like chemicals above their respective MLODs and 59.6% had one or more of the compounds substantially elevated (Table 14.2). In terms of $TEQ_{23}$, 25% of the US population had summed $TEQ_{23}$ < 13.87 fg $g^{-1}$ and 25% had summed $TEQ_{23}$ ≥ 81.66 fg $g^{-1}$, making the interquartile range 67.79 fg $g^{-1}$. The proportion of the US population having diabetes was 7.5% and the proportion having nephropathy (kidney disease) was 9.1% (Table 14.2). In Everett and Thompson (2014), diabetes was defined as diagnosed or glycohaemoglobin (A1c) ≥ 6.5%, and nephropathy defined as urinary albumin to creatinine ratio > 30 mg $g^{-1}$ (microalbuminuria or macroalbuminuria). In addition, 11% of the US population had pre-diabetes, defined by

Everett and Thompson (2012, 2014) as A1c 5.7–6.4% (Table 14.2).

There are eight dioxin-like chemicals in human blood that are most important in terms of the proportion of persons, included in the NHANES 1999–2004, with detectable values (> 25% above the MLOD). Three are dioxins, one is a furan and four are PCBs and are referred to as $TEQ_8$ (Table 14.1). The relationship between the summed $TEQ_{23}$ and the summed $TEQ_8$ is given by the following equation:

$$TEQ_{23} = 1.56 * TEQ_8 - 11.86 \quad r^2 = 0.92 \qquad \text{(Eqn 14.1)}$$

The 75th percentile of $TEQ_{23}$ reported in Everett and Thompson (2014) was 81.66 fg $g^{-1}$. Using the above equation, the comparable level of $TEQ_8$ is 59.95 fg $g^{-1}$, which is 73.4% of the $TEQ_{23}$ value (Everett and Thompson, 2016). Odds ratios for these eight dioxin-like chemicals and diabetes with or without nephropathy (kidney disease) are shown in Fig. 14.1. When considering concentrations above the 75th percentile, 12 of 16 of the relationships have odds ratios significantly greater than 1.00 (Everett and Thompson, 2014).

## 14.5 Pre-Diabetes

Pre-diabetes is a state that precedes diabetes. What is not clear is if a single dioxin-like chemical being elevated is most important for the relationship with pre-diabetes, or if the toxic equivalency is most important for the relationship. There is evidence to support both lines of reasoning, but only 14.3% of the US population have a single dioxin-like chemical substantially elevated and 45.3% have two or more elevated. What is meant by substantially elevated is that the concentration of one of the chemicals included in $TEQ_8$ is above the 75th percentile, or the concentration of one of the other 15 chemicals included in $TEQ_{23}$ is above the MLOD (Table 14.2). When one or more of the 23 dioxin-like chemicals is substantially elevated, the odds ratio for pre-diabetes with nephropathy is 4.70 (95% CI: 1.17–18.92) and the odds ratio for pre-diabetes without nephropathy is 1.54 (95% CI: 1.04–2.26) (Everett, 2014). The proportion of the US population having pre-diabetes with nephropathy is only 1.3%, while the proportion having pre-diabetes without nephropathy is 9.7% (Table 14.2).

Considering the relationship of toxic equivalency with pre-diabetes, one has to divide the

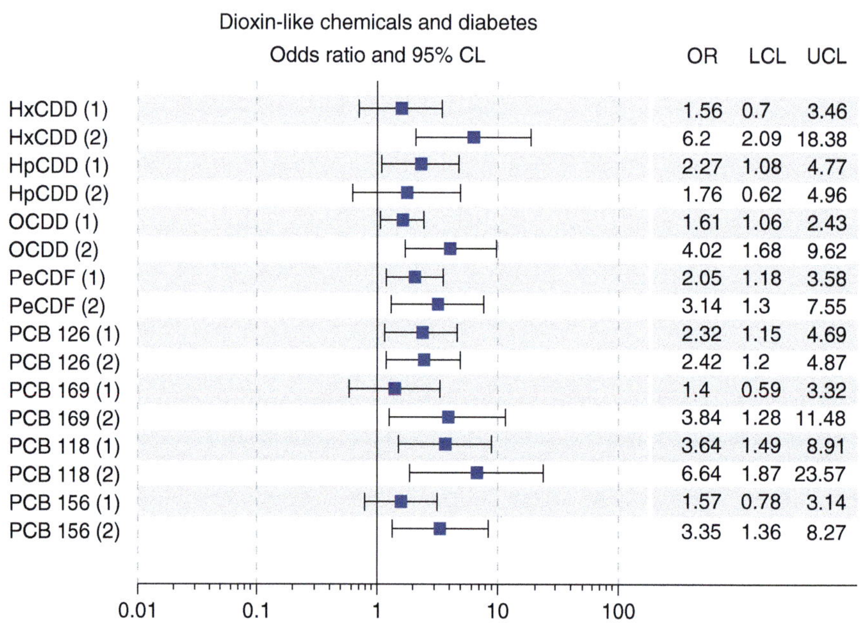


**Fig. 14.1.** Association of eight dioxin-like chemicals, used in $TEQ_8$, with (1) diabetes without nephropathy and (2) diabetic nephropathy. Adapted from Everett and Thompson (2014) with permission.

data based on glycohaemoglobin (A1c) concentration in blood (Table 14.2). Approximately half of those with pre-diabetes had A1c 5.7–5.8% (5.9% of the US population) and half had A1c 5.9–6.4% (5.2% of the US population). In Everett and Thompson (2012), the relationship of toxic equivalency with pre-diabetes was investigated using quartiles of $TEQ_{23}$; however, here we report relationships with continuous logarithm-transformed toxic equivalency ($\ln(TEQ_{23}+1)$). Those persons having A1c 5.7–5.8% do not have an association of logarithm-transformed $TEQ_{23}$ with pre-diabetes (odds ratio: 1.06; 95% CI: 0.87–1.30), while those having A1c 5.9–6.4% do have an association of logarithm-transformed $TEQ_{23}$ with pre-diabetes (odds ratio: 1.33; 95% CI: 1.03–1.72). The reference category for these logistic regressions was the group with normal A1c (< 5.7%) without nephropathy (Table 14.2).

It is possible to convert the odds ratio ($OR_1$) for continuous logarithm-transformed toxic equivalency to a more easily understood odds ratio ($OR_2$) comparing the 75th percentile of $TEQ_{23}$ to the 25th percentile, i.e. over the interquartile range (IQR). This relationship can be expressed by the following equation:

$$OR_2 = \left(\left(OR_1 - 1\right) * \ln\left(IQR\right)\right) + 1 \qquad \text{(Eqn 14.2)}$$

Given $OR_1$ = 1.33, IQR = 67.79 fg $g^{-1}$ and ln(IQR) = 4.216, then $OR_2$ = 2.39. Similarly, the 95% confidence limits can be calculated. In this example, the association of A1c 5.9–6.4% with toxic equivalency over the interquartile range has an odds ratio of 2.39 (95% CI: 1.13–4.04). While this odds ratio applies to persons 20 years of age and older (Everett and Thompson, 2014), there is no evidence of a relationship of pre-diabetes with dioxin-like chemicals before 30 years of age (Everett and Thompson, 2016).

## 14.6 Diabetes

There are a few published meta-analyses on the association of diabetes with dioxins, furans and PCBs (Henley *et al.*, 2012; Wu *et al.*, 2013; Tang *et al.*, 2014; Song *et al.*, 2016). A meta-analysis of four studies on occupational exposures to 2,3,7,8-tetrachlorodibenzo-*p*-dioxin (TCDD) and diabetes had a summary odds ratio of 1.48 (95% CI: 1.10–1.98) for a comparison of the highest category with the lowest category (Henley *et al.*, 2012). Two studies of exposure to TCDD due to an industrial accident had a summary odds ratio of 0.45 (95% CI: 0.39–0.52), indicating a lower prevalence of diabetes. Hence, repeated exposure to TCDD appears to be more important when considering associations with diabetes. A meta-analysis of six studies on dioxins (and furans) and diabetes produced a summary relative risk of 1.91 (95% CI: 1.44–2.54) for the highest category compared with the lowest category (Song *et al.*, 2016).

There are several meta-analyses on PCBs, but most are focused on non-dioxin-like PCBs, or include both dioxin-like PCBs and non-dioxin-like PCBs. Thirteen cross-sectional studies of PCBs and diabetes had a summary relative risk of 2.90 (95% CI: 2.14–3.92) and eight prospective studies produced a summary relative risk of 1.65 (95% CI: 1.16–2.34) (Song *et al.*, 2016). Tang *et al.* (2014) used ten studies (eight cross-sectional and two case-control) and also found an association of PCBs with diabetes. The summary odds ratio for the meta-analysis was 2.36 (95% CI: 1.64–3.41). Looking specifically at dioxin-like PCB 118, Wu *et al.* (2013) conducted a meta-analysis of four prospective studies that yielded a summary odds ratio of 1.20 (95% CI: 0.73–1.96). It is common in the literature on persistent organic pollutants for prospective studies to have lower odds ratios, or relative risks, than cross-sectional studies. This is likely due to the inclusion of persons having complications of diabetes in cross-sectional studies. In prospective studies, new cases of diabetes are identified and while some cases are complex, there are fewer such instances.

When considering associations between dioxin-like chemicals and diabetes, there are two opposing factors. The first is the size of the group that has diabetes, which in the case of Everett and Thompson (2014) would have been 7.5% of the US population (Table 14.2). The second factor is how well-defined subgroups of diabetes are. Looking at diabetes with and without nephropathy gives a more detailed view of the associations with dioxin-like chemicals but requires analysis of smaller groups. Therefore the logistic regressions for these subgroups may have wider confidence intervals solely due to the smaller size of the groups involved. As shown in Table 14.2, diabetes without nephropathy was 5.2% of the US population and diabetic nephropathy was 2.3%. It is appropriate to consider both approaches to the problem.

In Everett and Thompson (2012), the fourth quartile of $TEQ_{23}$ compared with the first quartile had an odds ratio of 3.08 (95% CI: 1.20–7.90) for diabetes, and six of the eight dioxin-like chemicals included in $TEQ_8$ showed associations with diabetes. The two that were not associated were 1,2,3,4,6,7,8-heptachlorodibenzo-*p*-dioxin and PCB 126. However, these two dioxin-like chemicals had significant associations in Everett and Thompson (2014) for diabetes without nephropathy, indicating that the better-defined subgroup was more appropriate in the case of these two chemicals.

Similarly, the relationship between diabetes and continuous logarithm-transformed toxic equivalency ($\ln(TEQ_{23}+1)$) can be looked at in two ways. For diabetes and logarithm-transformed $TEQ_{23}$, the odds ratio is 1.60 (95% CI: 1.22–2.09); and for diabetes without nephropathy and logarithm-transformed $TEQ_{23}$, the odds ratio is 1.44 (95% CI: 1.11–1.87). The reference category for both of these logistic regressions was the normal A1c (< 5.7%) without nephropathy group (Table 14.2). As the diabetes without nephropathy category has a more conservative odds ratio, the association for that group is considered the best measure of the effect. Expressed in another way, the odds ratio for diabetes without nephropathy and toxic equivalency over the interquartile range is 2.86 (95% CI: 1.46–4.67).

## 14.7 Diabetic Nephropathy

Seven of the eight dioxin-like chemicals included in $TEQ_8$ are associated with diabetic nephropathy in spite of the fact that only 2.3% of the US population is involved (Everett and Thompson, 2014). Logarithm-transformed toxic equivalency is also associated with diabetic nephropathy, having an odds ratio of 2.35 (95% CI: 1.57–3.52). The odds ratio for diabetic nephropathy and toxic equivalency over the interquartile range is 6.69 (95% CI: 3.40–11.62). Given this very high odds ratio, the relationship between toxic equivalency and diabetic nephropathy appears to be a case of reverse causality. Reverse causality, in this case, is when the disease precedes the rise in the chemical concentration in blood.

However, the relationship between $TEQ_{23}$ and diabetic nephropathy may be more complicated. High levels of dioxin-like chemicals may be both a case of reverse causality and a risk factor for diabetic nephropathy. A negative feedback loop may exist, with a rise in dioxin-like chemicals in blood causing the beginnings of diabetic nephropathy, followed by a build-up of more dioxin-like chemicals as the kidneys become less efficient at removing toxins from the blood (Everett and Thompson, 2014). Notably, dioxin-like chemicals cannot be detected in urine, which suggests another organ is responsible for metabolism of these compounds (D.O. Carpenter, New York, 2015, personal communication). Lee *et al.* (2006) found fatty liver and gamma-glutamyltransferase (GGT) to be associated with elevated levels of dioxins and furans, which suggests liver function may be involved.

## 14.8 Conclusions

TEFs have been defined for seven polychlorinated dibenzo-*p*-dioxins, ten polychlorinated dibenzofurans and 12 polychlorinated biphenyls (van den Berg *et al.*, 2006). Concentrations of these 29 dioxin-like chemicals can be multiplied by their respective TEFs and summed to yield a measure called toxic equivalency (TEQ). Toxic equivalency provides a way to quantify the overall toxicity of this class of chemicals. The NHANES 1999–2004, conducted in the USA, has been used to assess toxic equivalency in human blood. Two summary measures have been defined. The first uses 23 of the 29 dioxin-like chemicals and is referred to as $TEQ_{23}$. The second measure, $TEQ_8$, uses the eight dioxin-like chemicals that have the highest proportion of persons with detectable levels of the compounds. $TEQ_8$ includes three dioxins, one furan and four PCBs.

There are significant associations of toxic equivalency with pre-diabetes, diabetes and diabetic nephropathy in the NHANES 1999–2004. Expressed as a continuous variable, logarithm-transformed toxic equivalency ($\ln(TEQ_{23}+1)$) was associated with 'high' pre-diabetes (in the range of A1c 5.9–6.4%) with an odds ratio of 1.33 (95% CI: 1.03–1.72). Logarithm-transformed $TEQ_{23}$ was also associated with diabetes (odds ratio: 1.60; 95% CI: 1.22–2.09), diabetes without nephropathy (odds ratio 1.44; 95% CI 1.11–1.87) and diabetic nephropathy (odds ratio: 2.35; 95% CI: 1.57–3.52). In cross-sectional studies it is more appropriate to exclude persons having diabetic nephropathy when considering the association with diabetes. Therefore, the best estimate of the effect size is the odds ratio for diabetes without nephropathy. The association of logarithm-transformed $TEQ_{23}$ with diabetic nephropathy appears to be a case of reverse causality, or perhaps both due to reserve causality and a risk factor for the disease. The odds ratio of 2.86 (95% CI: 1.46–4.67) for diabetes without nephropathy and toxic equivalency over the interquartile range can be used for comparisons to other studies on dioxin-like chemicals and diabetes.

## References


CDC (2005) *Third National Report on Human Exposure to Environmental Chemicals*. National Center for Environmental Health Pub. No. 05-0570. Centers for Disease Control and Prevention, Atlanta, Georgia.

Everett, C.J. (2014) Commentary on nephropathy and longitudinal studies of diabetes and dioxins, furans, and dioxin-like PCBs. *Environmental Research* 134, 8.

Everett, C.J. and Thompson, O.M. (2012) Associations of dioxins, furans and dioxin-like PCBs with diabetes and pre-diabetes: is the toxic equivalency approach useful? *Environmental Research* 118, 107–111.

Everett, C.J. and Thompson, O.M. (2014) Dioxins, furans and dioxin-like PCBs in human blood: causes or consequences of diabetic nephropathy? *Environmental Research* 132, 126–131.

Everett, C.J. and Thompson, O.M. (2016) Association of dioxins, furans and dioxin-like PCBs in human blood with nephropathy among US teens and young adults. *Reviews on Environmental Health* 31, 195–201.

Haws, L.C., Su, S.H., Harris, M., DeVito, M.J., Walker, N.J. *et al.* (2006) Development of a refined database of mammalian relative potency estimates for dioxin-like compounds. *Toxicological Sciences* 89, 4–30.

Henley, P., Hill, J., Moretti, M.E., Jahedmotlagh, Z., Schoeman, K., Koren, G. and Bend, J.R. (2012) Relationships between exposure to polyhalogenated aromatic hydrocarbons and organochlorine pesticides and the risk for developing type 2 diabetes: a systematic review and a meta-analysis of exposures to 2,3,7,8-tetrachlorodibenzo-*p*-dioxin (TCDD). *Toxicology and Environmental Chemistry* 94, 814–845.

Lee, C.C., Yao, Y.J., Chen, H.L., Guo, Y.L. and Su, H.J. (2006) Fatty liver and hepatic function for residents with markedly high serum PCDD/Fs levels in Taiwan. *Journal of Toxicology and Environmental Health, Part A* 69, 367–380.

Mainous, A.G. 3rd, Baker, R., Koopman, R.J., Saxena, S., Diaz, V.A., Everett, C.J. and Majeed, A. (2007) Impact of population at risk of diabetes on projections of diabetes burden in the United States: an epidemic on the way. *Diabetologia* 50, 934–940.

Menke, A., Casagrande, S., Geiss, L. and Cowie, C.C. (2015) Prevalence of and trends in diabetes among adults in the United States, 1988–2012. *Journal of the American Medical Association* 314, 1021–1029.

Remillard, R.B. and Bunce, N.J. (2002) Linking dioxins to diabetes: epidemiology and biologic plausibility. *Environmental Health Perspectives* 110, 853–858.

Schmidt, M.I., Duncan, B.B., Bang, H., Pankow, J.S., Ballantyne, C.M. *et al.* (2005) Identifying individuals at high risk for diabetes. *Diabetes Care* 28, 2013–2018.

Song, Y., Chou, E.L., Baecker, A., You, N.C.Y., Song, Y., Sun, Q. and Liu, S. (2016) Endocrine-disrupting chemicals, risk of type 2 diabetes, and diabetes-related metabolic traits: a systematic review and meta-analysis. *Journal of Diabetes* 8, 516–532.

Tang, M., Chen, K., Yang, F. and Liu, W. (2014) Exposure to organochlorine pollutants and type 2 diabetes: a systematic review and meta-analysis. *PLoS ONE* 9, 10.

van den Berg, M., Birnbaum, L.S., Denison, M., De Vito, M., Farland, W. *et al.* (2006) The 2005 World Health Organization reevaluation of human and mammalian toxic equivalency factors for dioxins and dioxin-like compounds. *Toxicological Sciences* 93, 223–241.

Wu, H., Bertrand, K.A., Choi, A.L., Hu, F.B., Laden, F., Grandjean, P. and Sun, Q. (2013) Persistent organic pollutants and type 2 diabetes: a prospective analysis in the nurses' health study and meta-analysis. *Environmental Health Perspectives* 121, 153–161.

# 15 Environmental Endocrine-Disrupting Chemicals and Human Health

**P.D. Darbre***

*School of Biological Sciences, University of Reading, Reading, UK*

## 15.1 Abstract

Many environmental pollutant chemicals have been now found to interfere in the actions of hormones and have been termed endocrine-disrupting chemicals (EDCs). They may act by altering hormone synthesis in the endocrine gland, by altering transport of the hormone to the target site or by interfering in cellular responses in the target cells. This may lead to adverse physiological consequences, which may even become manifest at a population level. Some EDCs are found as natural components of microbes (mycoestrogens) or plants (phytoestrogens) but most EDCs are man-made chemicals that have been released into the environment from human activity. Such chemicals include pesticides, herbicides, industrial chemicals such as polychlorinated biphenyls, products of incineration (polychlorinated dibenzo-dioxins and -furans), components of plastics (bisphenol A, phthalates), surfactants (alkyl phenols), flame retardants (polybrominated diphenyl ethers), stain-resistance coatings (perfluoro compounds) and a range of components used in personal care products as antimicrobials (parabens, triclosan), antiperspirants (aluminium salts), conditioning agents (cyclic volatile methylsiloxanes), UV filters (benzophenones) and for fragrance (butylphenylmethylproprianol, polycyclic musks, nitromusks). This chapter discusses the extent to which EDCs may enter human tissues from environmental exposures and the mechanisms by which EDCs disrupt hormone action. Evidence is discussed linking EDC exposure with adverse effects on human health, including on reproductive health, thyroid health and energy metabolism. Concerns are discussed for long-term effects of fetal exposure to EDCs on adverse outcomes for adult endocrine health.

## 15.2 Introduction: What are Endocrine-Disrupting Chemicals?

> An endocrine disrupting chemical has been defined as 'an exogenous substance that causes adverse health effects in an intact organism, and/or its progeny, consequent to changes in endocrine function'. (EC, 1996).

Human health is dependent on an intact endocrine system in which hormones are secreted from glands across the body to be carried by the blood to act as chemical messengers at distant target sites in order to coordinate and regulate physiological functions. Over recent years, it has become apparent that many environmental chemicals have the ability to interfere in the actions of hormones and have been termed endocrine-disrupting chemicals

* E-mail address: p.d.darbre@reading.ac.uk

(EDCs) (Darbre, 2015). Such compounds may interfere in the action of hormones in many different ways, as illustrated in Fig. 15.1. They may act by altering hormone synthesis in the endocrine gland, by altering transport of the hormone to the target site or by interfering in cellular responses in the target cells. This can lead to physiological changes, which may then expand to consequences at a population level.

Although it was only in the 1990s that endocrine disruption received formal recognition, the phenomenon had already been known about for many decades. In the 1920s, pig farmers in the USA reported problems with infertility in their herds which was found to be linked to feeding the animals mouldy grain (McNutt, Purwin and Murray, 1928). In the 1940s, this was followed by reports from sheep farmers in Western Australia of infertility in their sheep when grazing on certain clover fields (Bennets, Underwood and Shier, 1946). More recent research has identified the underlying reasons as consumption of specific oestrogenic compounds contained in the mould (mycoestrogens) or in the clover (phytoestrogens) that disrupted fertility due to their potent oestrogenic activity. In the 1950s and 1960s, it became increasingly apparent that widespread release of synthetic chemicals into the environment, including pesticides and herbicides, was causing loss of wildlife populations. In 1962, the book published by Rachel Carson entitled *Silent Spring* brought the problems to public attention (Carson, 1962). One of the early meetings to discuss the issues was the World Wildlife Fund Wingspread Conference in the USA in 1991 and it was at this meeting that the term endocrine disruptor was first proposed (Colborn and Clement, 1992).

Over the past 50 years, cases of endocrine disruption in aquatic and terrestrial wildlife have been increasingly documented and linked to specific chemical exposures. One of the early studies linked loss of bivalves and gastropods in harbour waters to exposure to tributyltin (TBT) escaping into the aquatic environment from its use in antifouling paints applied to the underside of ships. Further research has shown that TBT causes imposex in molluscs, a condition in which female snails develop male sex organs, including penis and vas deferens, and this has been shown to lead to reproductive failure in over 150 species worldwide (Horiguchi, 2006). Another study identified loss of the alligator population in Lake Apopka in Florida, USA, following a spill of dicofol in 1980. Genital abnormalities were found in both male and female alligators and female alligators had abnormal ovarian morphology and raised plasma oestradiol levels (Guillette and Gunderson, 2001). In the UK, extensive studies of the feminization of

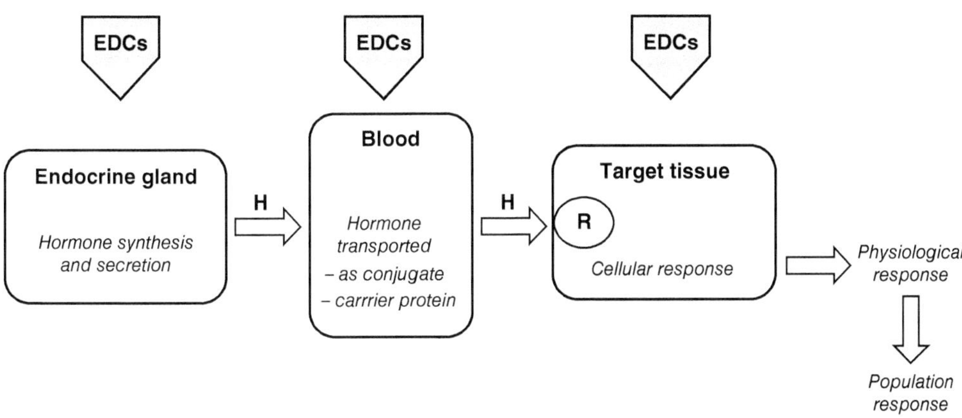


**Fig. 15.1.** Mechanisms of action of endocrine disrupting chemicals (EDCs). EDCs may act at the endocrine gland by altering hormone (H) synthesis and/or secretion. EDCs may alter transport and bioavailability of hormone by altering hormone conjugation and/or binding to carrier proteins in the blood. EDCs may alter response at the target tissue by competing for binding to receptor (R), modifying receptor levels and/or altering hormone metabolism and excretion. Responses at a cellular level may result in physiological consequences for individuals or at a population level.

male fish in river waters have highlighted the existence of oestrogenic components in sewage effluent (Jobling *et al.*, 1998). The much debated question now continues as to the extent to which the effects reported in wildlife might also occur in the human population. This chapter discusses the sources of human exposure to EDCs, the extent to which EDCs may enter human tissues from environmental exposures, the mechanisms by which EDCs may disrupt hormone action and evidence linking EDC exposure with adverse outcomes for human health.

## 15.3 Sources of Human Exposure to Endocrine-Disrupting Chemicals

Some endocrine disruptors are natural components of biological material but the majority are man-made chemicals which have been released into the environment from the activities of humans often without any prior knowledge of their effects on ecosystems, animal welfare or human health. Although some exposure may be through specific occupational circumstances, the population is now widely exposed on a daily basis to EDCs in the domestic and workplace environment.

The naturally occurring EDCs are found in microbial and plant material where some potent oestrogenic compounds have been identified and termed mycoestrogens or phytoestrogens (Woods, 2003), respectively. Man-made EDCs are contained within many agricultural, industrial and consumer products that have become ubiquitous environmental pollutants due to their widespread use (Darbre, 2015). This includes many organic compounds, often halogenated, which, due to their resistance to breakdown and lipid solubility, have become persistent in the environment and thus have been termed persistent organic pollutants (POPs). Some specific examples are:

- pesticides (including dichlorodiphenyl trichloroethane (DDT), dieldrin, lindane, endosulfan) and herbicides (including atrazine, glyphosate) used both in an agricultural setting and in urban environments;
- industrial chemicals (e.g. polychlorinated biphenyls (PCBs));
- by-products of combustion (polychlorinated dibenzodioxins (PCDDs), polychlorinated dibenzofurans (PCDFs)), including from vehicles, ships and aircraft;
- plastics (bisphenols, phthalates) which are used widely in building materials, food containers and water bottles;
- surfactants (alkyl phenols) used in detergents for cleaning in both industrial and domestic applications;
- flame retardants (polybrominated diphenyl ethers (PBDEs)) used in soft furnishings;
- stain-resistance coatings (perfluoro compounds);
- components of personal care products such as chemicals used for preservation (alkyl esters of *p*-hydroxybenzoic acid (parabens)), as deodorant (e.g. triclosan), as antiperspirant (aluminium salts), for conditioning (cyclic volatile methylsiloxanes), for absorption of ultraviolet light (e.g. benzophenones), as anti-ageing agents (e.g. synthetic oestrogens) and as fragrance/fragrance fixers (e.g. butylphenylmethylpropional, polycyclic musks, nitromusks); and
- some pharmaceutical (paracetamol) and nutraceutical products.

## 15.4 Entry of Endocrine-Disrupting Chemicals into Human Tissues

The first step in assessing the potential for EDCs to exert any functional role in the human body is to determine the extent to which they can enter and/or bioaccumulate in human tissues from environmental exposures. A main exposure route is oral, through dietary intake of water and food. However, inhalation of polluted air and dermal absorption of components of personal care products are also important routes of exposure. Due to the widespread presence of EDCs in the ecosystem and the use of individual EDCs in multiple consumer goods, exposure rarely occurs from a single source or a single product, or even from a single exposure route. For example, exposure to pesticides and herbicides may result from inhalation during agricultural spraying, but these products are also now used widely in urban gardens and in the indoor environment, and furthermore residues may enter through the oral route from remnants in water and food or through the dermal route from skin contamination. Another example is personal care products where usual application is dermal, but inhalation can be a route of exposure from products in aerosol format and cosmetics such as lipsticks may enter via the oral route. Furthermore, it is also now known that pollutants in air may enter not only by inhalation but also through the dermal route (Weschler and Nazaroff, 2014).

The extent to which EDCs are retained in tissues will depend on the metabolic capacity to remove them through detoxification reactions involving cytochrome P450 enzymes and/or through conjugation systems. Some EDCs such as phytoestrogens may be consumed in relatively large amounts in diets rich in plant materials but most are removed from the body within hours (Woods, 2003). However, other EDCs, most notably the POPs, may be taken into the body in only small amounts but, due to their resistance to clearance mechanisms, they may bioaccumulate in body tissues, especially fatty tissues, over the years. In general, inhalation and oral entry of compounds results in greater uptake than from dermal application. However, the low level of dermal absorption may be offset by the long-term exposure from products applied and left on the skin without washing off; and furthermore, the local absorption through skin will result in avoidance of detoxification mechanisms in the liver to which chemicals absorbed systemically into the blood from oral or inhaled routes would be subjected.

Many hundreds of EDCs have now been measured as present in human urine, blood and/or breast milk, because these are the most easily accessible tissues to sample. Blood serum samples from adults provide a measure of circulating EDCs, and since all organs are perfused with blood, this provides an indication of overall body burden. However, blood levels can fluctuate according to levels of exposure at the time of sampling, and whilst some dermally applied chemicals may increase rapidly within hours, they can also decrease within hours after cessation of exposure (Janjua *et al.*, 2007, 2008). Urine samples are most easily obtained because sampling is non-invasive and this offers the option for large sample numbers with statistical power. However, such measurements provide a measure of clearance rather than homeostatic body burdens, which is particularly pertinent to lipophilic EDCs that are not easily cleared from the body but bioaccumulate over decades in fatty tissues. More targeted measurements have been made using cord blood or placenta to determine the potential for EDCs to cross from mother to baby during pregnancy.

Biomonitoring of levels of EDCs is now increasingly being performed and one extensive study is from the National Health and Nutrition Examination Survey (NHANES) of the general population of the USA. Due to the multiplicity of sources of EDCs, most tissue measurements cannot identify the origin of the chemicals measured, but surveys such as from NHANES have greater potential to uncover information on sources because the blood and urine levels are linked to interviews and physical examinations. The NHANES studies have shown that many EDCs are now found ubiquitously in the samples, thus demonstrating the widespread contamination of the human population with compounds such as PCBs, DDT and PBDEs (Sjodin *et al.*, 2014), phthalates (Silva *et al.*, 2004), bisphenol A and octylphenol (Calafat *et al.*, 2008a), triclosan (Calafat *et al.*, 2008b) and parabens (Calafat *et al.*, 2010).

## 15.5 Mechanisms by Which Endocrine-Disrupting Chemicals Interfere in Hormone Action

Endocrine disruptors may interfere in the action of hormones in many different ways (Fig. 15.1). First, they may act by altering hormone synthesis in the endocrine gland, including through altering levels and/or activities of key enzymes. For example, oestrogens are derived from androstenedione and testosterone by the enzyme aromatase, which is a cytochrome P450 enzyme (19A1), and some EDCs can alter biosynthesis of endogenous oestrogens either through altering expression of the aromatase gene or through binding to the aromatase enzyme and so altering its enzymatic activity (Whitehead and Rice, 2006). Many EDCs are now known also to alter synthesis of thyroid hormones in the thyroid gland, including through inhibiting the sodium–iodide transporter responsible for iodine uptake from the blood or through inhibiting the key enzyme thyroperoxidase responsible for synthesis of thyroid hormones (Odum, 2015).

Alternatively, EDCs may act through altering transport of the hormone to the target organ. Most hormones are carried from the organ of synthesis to the distant target organ by the blood either in a conjugated form or bound to carrier proteins. The unbound, unconjugated or 'free' fraction is only a small proportion of the total (usually 1–2%) and is considered to be the biologically available hormone that is free to enter target cells. EDCs may act by interfering with activity of conjugation enzymes (Waring *et al.*, 2008), by altering levels of carrier binding proteins (Pino *et al.*, 2000) and/or by competing for binding to carrier proteins (Dechaud *et al.*, 1999). In so doing, the EDCs interfere in the biological availability of hormones.

At the target cells, hormones act through binding to specific cellular receptors and EDCs can interfere by altering levels of these receptors and/or competing for binding to one or more of these receptors, thereby giving rise to inappropriate agonist and/or antagonist actions. Binding of hormone to receptor sends signals into the cell and intracellular signalling through either genomic or non-genomic mechanisms may be disturbed by EDCs. By the genomic mechanism, hormone binds to receptor causing displacement of receptor-associated chaperone proteins and enabling dimerization of receptors. The receptor dimers then act by binding to specific 'response element' nucleotide sequences in the DNA to cause alteration to gene expression. Through an oestrogen receptor-mediated mechanism, the physiological oestrogen 17β-oestradiol is known to alter the expression level of hundreds of genes, and in breast cancer cells the majority of alterations are down-regulation rather than up-regulation (Frasor *et al.*, 2003). EDCs with oestrogenic activity also alter global gene expression profiles but not always in an identical manner to that of oestradiol or even in similar ways to each other (Pugazhendhi *et al.*, 2007). By the non-genomic mechanisms, hormone may bind to cell surface receptors, triggering intracellular signalling pathways which lead to cellular responses, and EDCs can also interfere in these cell signalling pathways (Bulayeva *et al.*, 2004).

### 15.5.1 Effect of mixtures of EDCs

The ability of many different EDCs to act by a common mechanism, such as through binding to a specific hormone receptor, implies that an end-point response could be achievable by mixing several chemicals. Such additive effects enable individual chemicals to act in combination at lower concentrations than would be needed for each chemical alone (Rajapakse *et al.*, 2002; Kortenkamp, 2007). The ubiquitous measurement of so many EDCs in human blood or urine demonstrates that the human tissues are not exposed to one chemical at a time but rather to complex mixtures of chemicals that have entered the human body according to individual lifestyle choices. Therefore it is highly likely that end-point responses are the result of long-term exposure to low doses of mixtures of chemicals. A further implication of chemical mixtures is that if EDCs act by a common mechanism, then the same end response could be generated by different combinations of EDCs. This poses a challenge to epidemiological studies of toxicity where evidence of causality has classically relied on measuring different levels of a single chemical in cases of adversity. A new paradigm is therefore now needed to enable the analysis of different mixtures of chemicals to be assessed for a common end point. This is supported by a study of measurements of 19 individual POPs showing different levels of each POP in different people but overall showing one or more POPs were measurable at a high level (Porta *et al.*, 2012). Analogous results have been reported for parabens in that five paraben esters were measured in 158/160 human breast tissue samples from women with primary breast cancers but samples with high levels of one ester did not necessarily have high levels of another ester (Barr *et al.*, 2012). Further analysis of functionality showed that levels of parabens were at sufficient concentrations in some of these breast tissue samples to enable adverse cellular responses *in vitro* but it was different combinations of esters that enabled the end-point response (Charles and Darbre, 2013).

### 15.5.2 Dose–response considerations

Another feature of the actions of EDCs is that their dose-responses may not always be monotonic (Vandenberg *et al.*, 2012) and therefore not always in line with the classical toxicology premise of 'the dose makes the poison'. This concept was first introduced by the Swiss chemist Paracelsus (1493–1541) holds much truth, because in a linear manner some toxic substances can become harmless at low doses whilst even some harmless substances can become toxic at high doses. However, hormones act with a sigmoidal dose–response curve rather than linear, because hormone receptors are limited in number in a cell and responses are mediated by a 1:1 ratio of one molecule of hormone binding to one molecule of receptor. This means that there comes a point of saturation where the response cannot be increased, however much more hormone is added, because every molecule of receptor has hormone bound to it. The ability of EDCs to act through hormone receptor-mediated mechanisms allows them to work at very low doses but also changes the nature of their dose–responses from linear to sigmoidal. Furthermore, EDCs may also act through multiple receptors and/or multiple other non-receptor mechanisms, which can result in non-monotonic responses where the sign of the slope (positive or negative) can change at some point across the dose range, giving a range of shapes from U-shaped to inverted U-shaped or even more complex shapes such as M-shaped or W-shaped (Vandenberg *et al.*, 2012). For these reasons, it may not always be possible to predict effects at low doses from responses measured at high doses and vice versa. As an example, the phytoestrogen genistein can increase growth of oestrogen-responsive cells at low doses by an oestrogen receptor-mediated mechanism but at high doses it inhibits cell growth by a different mechanism that does not involve oestrogen receptor (Matsumura *et al.*, 2005).

### 15.5.3 Effect of metabolism on activity of EDCs

Metabolic capabilities using cytochrome P450 enzymes have evolved to remove unwanted foreign compounds and any harmful compounds produced during the normal course of metabolism. However, in the process of detoxifying foreign compounds such as pharmaceuticals and environmental chemicals, these metabolic reactions can also act on occasions to inadvertently turn a harmless parent foreign compound into a compound with unwanted activity. In this way some environmental compounds that of themselves possess only weak or no endocrine disrupting properties can be converted into compounds with greater endocrine disrupting activity. For example, environmental contamination of animal tissues with DDT includes not only the parent compound but also a mixture of related compounds generated from breakdown and metabolism, and o,p′-DDT has been reported to possess more potent oestrogenic activity than the parent DDT compound (Bitman *et al.*, 1968). Some PCB congeners bind only weakly to oestrogen receptors and their more potent oestrogenic activity has been related to metabolic conversion to more reactive intermediates through hydroxylation (Nesaretnam *et al.*, 1996) or generation of catechol metabolites (Garner *et al.*, 1999). On the other hand, some foreign compounds may have their endocrine-disrupting activity reduced by endogenous metabolism. For example, the oestrogenic activity of the alkyl esters of *p*-hydroxybenzoic acid (parabens) increases with the linear length of the alkyl chain from methylparaben to *n*-butylparaben (Byford *et al.*, 2002) and with branching in the alkyl chain from *n*-butylparaben to isobutylparaben (Darbre *et al.*, 2002) but all the parabens are subject to endogenous esterase activity that converts them into the common metabolite *p*-hydroxybenzoic acid with lower oestrogenic activity (Pugazhendhi *et al.*, 2005). The presence of intact paraben esters in human tissues is indicative that they have escaped metabolism by esterases (Barr *et al.*, 2012).

### 15.5.4 Variations between tissues and between individuals

In assessing the outcome of exposure to EDCs, it must also be taken into account that there will be differences in response between different tissues and between different individuals. Since the actions of EDCs are dependent on the presence of specific receptors, effects can only be expected in cells that possess those specific receptors, and responses may even vary between tissues possessing those receptors if the levels of receptors differ or other cellular components essential for the actions of those receptors also vary. For example, although breast and uterus both possess oestrogen receptors, it has long been known that these tissues can display opposing responses to oestrogen, and in particular the use of tamoxifen as an anti-oestrogen to inhibit oestrogen receptor-mediated proliferation of breast cancer cells can result in unwanted agonist responses in the uterus (Lonning, 2004). Furthermore, actions of EDCs can also be expected to vary between individuals. Consequences of exposure to EDCs depend on a complex array of interactions between the individual genetic background and the lifestyle factors and choices of that individual. Genetic factors play a major role in regulating susceptibility to any disease but the speed of increase in many endocrine-related disorders over recent decades rules out genetic factors as a sole determinant. For example, many of the children born to women who were prescribed the synthetic oestrogen diethylstilboestrol (DES) to prevent miscarriage in pregnancy have grown up to suffer reproductive abnormalities and increased rates of some cancers in endocrine sensitive tissues, but although many did, not every child had these associated health effects (Harris and Waring, 2012). Developing an understanding of the mechanisms underlying susceptibility to adverse effects of EDCs will need to be a research priority over coming years.

### 15.5.5 Effect of timing of exposure

Timing of exposure is another important determinant of the action of EDCs. Hormones are released at regulated levels in the body with highly specific timings, but EDCs do not enter the human body in any physiologically regulated manner and so may impact through disrupting normal hormonal patterns. For example, the female is exposed to the highest levels of oestrogen only between puberty and menopause; furthermore, even during those times, the levels of oestrogen cycle during the normal monthly menstrual cycle. However, exposure to EDCs with oestrogenic activity does not occur in synchrony with physiological oestrogen exposure and may occur even inappropriately prior to puberty and after menopause. One critical window of susceptibility to disruption by EDCs is prior to birth (Darbre, 2015). This was first highlighted by the developmental disorders in offspring of mothers who were prescribed DES in pregnancy (Harris and Waring, 2012). Development of the human embryo/fetus is tightly regulated by the endocrine system and small changes to hormone levels through interference from EDCs have now been shown to alter developmental programming, leading to alterations to tissue and organ development that become permanent into adult life and with consequences for adult health, most notably in terms of reproductive functions, obesity and neurodevelopment (Darbre, 2015). Of even greater consequence is that changes during this critical window of susceptibility are now known to produce long-lasting effects that are even passed on to future generations without need for any further exposure. Such transgenerational effects were first reported for the pesticide vincolzolin which, when administered to developing rodents, produced adverse effects on the developing testis and the adverse effects were passed on for three generations (Skinner *et al.*, 2011).

## 15.6 Endocrine-Disrupting Chemicals and Human Health

Hormones of the endocrine system are ultimately responsible not only for regulating major physiological processes of development and reproduction, but also for maintenance of all the tissues and organs of the body and for enabling adaptations to environmental changes. It is not surprising, therefore, that EDCs are now being found to have very wide ranging consequences for human health (Fig. 15.2) (Darbre, 2015).

Much of the reported disruptive activity has been in relation to the action of oestrogens and androgens, and since these steroid hormones are the main regulators of reproductive functions, many of the reported effects of exposure to EDCs have been on adverse consequences for reproductive health. However, physiological consequences have also been widely demonstrated as resulting from disruption to thyroid function and consequent alterations to thyroid hormone levels, which has implications for regulation of metabolism in all tissues. More widely, adverse effects have also been reported as resulting from alterations to adrenocortical function and impairment of the immune system. Exposure to EDC can also cause loss of control on energy metabolism, with consequent implications for development of obesity, diabetes and cardiovascular disease.

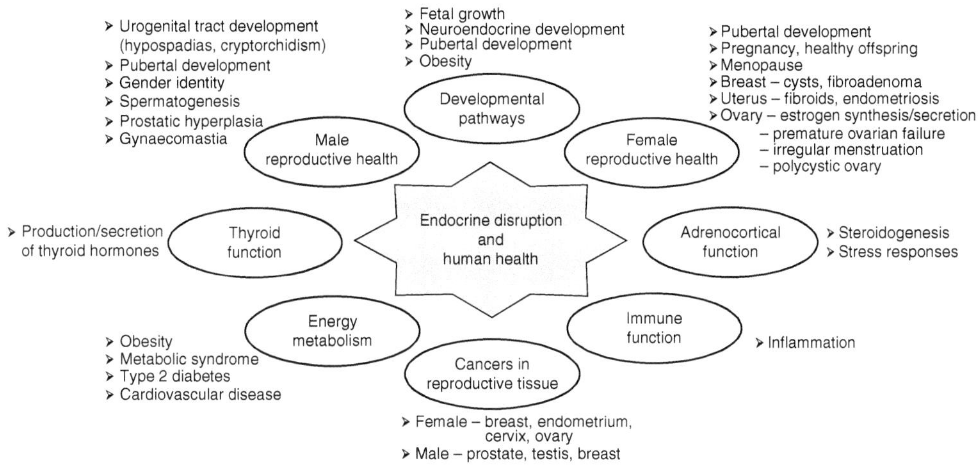


**Fig. 15.2.** Outline of the influences of endocrine disrupting chemicals (EDCs) on human health. Oval boxes indicate broad health issues affected; bullet arrows indicate specific effects linked to EDCs.

Embryonic and fetal exposure to EDCs may result in impaired reproductive abilities, aberrant brain function, reduced immunity and altered metabolic programming in adult life. The effects of EDCs on neurodevelopment have implications for many of the neurological conditions in later life, such as behavioural abnormalities, Alzheimer's disease, and Parkinson's disease.

### 15.6.1 Female reproductive health

The strongest evidence for adverse effects on female reproductive health following exposure to an EDC comes from studies of the outcomes for women who were prescribed DES to prevent the threat of miscarriage during the first trimester of pregnancy over the years 1940–1971 (Smith, 1948). In 1971, exposure to this synthetic non-steroidal oestrogen was reported to cause a rare vaginal cancer in daughters who were born to these women and who had therefore been exposed *in utero* to DES (Herbst *et al.*, 1971). Further prescription ceased, but long-term follow-up studies of these daughters born to DES-treated mothers have shown associations with many adverse reproductive health outcomes (Hoover *et al.*, 2011). Animal studies have confirmed that DES causes tumours in oestrogen-sensitive tissues in several animal species, that it can cause tumours in adult animals following prenatal exposure and furthermore that effects of DES exposure *in utero* can pass down to the F2 and F3 generations (Harris and Waring, 2012). Although it is too early to ascertain any F2 or F3 transgenerational effects in the DES-exposed daughters, the consequences documented in the F1 generation demonstrate that exposure to a synthetic oestrogen in the fetal life of a woman can have consequences for her health in later adult life. However, the fact that not every daughter born to a DES-exposed mother had adverse symptoms illustrates the diversity in response between individuals and highlights the difficulty in identifying causative agents where varied susceptibility influences outcomes. The ability more widely of environmental EDCs to cause also adverse effects on female reproductive health remains to be fully ascertained, but studies now indicate effects on the timing of puberty and on the functions of the ovary, uterus and breast.

Puberty is the period during late childhood when secondary sexual characteristics develop and reproductive capacity is attained. The changes during puberty are endocrine-regulated and thus susceptible to disruption by EDCs (Bourguignon and Parent, 2012). One extreme illustration is provided by the case report of a 3-year-old girl who presented with vaginal bleeding, uterus enlargement and thelarche (breast development) following exposure to ethinyl oestradiol contained in her mother's hair lotion (Guaneri *et al.*, 2008). The child was in the habit of playing with her mother's hair, her mother's combs and the empty lotion bottles, and analysis of her long scalp hairs revealed ethinyl oestradiol concentrations of 10.6 μg g$^{-1}$ in her hair. Six months after ceasing exposure to the hair lotion, the girl's hyperoestrogenic symptoms resolved. The reversibility of this case demonstrates a causative link between exposure to oestrogenic components of a personal care product and early development of puberty. Epidemiological studies demonstrate that puberty is occurring generally at an earlier age in girls today in the western world compared with a century ago (Parent *et al.*, 2003) and environmental exposures to EDCs are thought to be a contributing factor (WHO, 2012; Diamanti-Kandarakis and Gore, 2012). Some studies suggest an association with raised serum levels of phthalates (Colon *et al.*, 2000) and other studies to raised levels of POPs or bisphenol A (WHO, 2012), but no simple picture has yet emerged.

The ovary is not only an endocrine organ responsible for synthesis of steroid hormones but is also itself subject to endocrine regulation. Disruption of ovarian function by EDCs can lead to a range of adverse outcomes such as anovulation, infertility, oestrogen deficiency, premature ovarian failure and ovarian cyst development (Craig *et al.*, 2011). Premature ovarian failure occurs in about 1% of women under the age of 40 and leads to reproductive disorders and early menopause (Nelson, 2009). A potential precursor is the development of multi-oocyte follicles and exposure to some EDCs has been shown to cause this in animal models. Neonatal exposure to DES has been shown to cause multi-oocyte follicles in mice (Kirigaya *et al.*, 2009). Early postnatal exposure to parabens has been shown to inhibit the early phase of folliculogenesis in rats (Ahn *et al.*, 2012). Exposure of adult mice and rats to methoxychlor can cause ovarian atrophy also due to inhibition of folliculogenesis (Martinez and Swartz, 1991). Polycystic ovary syndrome is one of the most common female endocrine disorders, affecting 5–10% of women of reproductive age, and is a leading cause of female subfertility. Although there is a genetic component, EDCs are also thought to be contributing to the aetiology of this condition, most notably bisphenol A (Rutkowska and Rachon, 2014). Animal models have shown that exposure to bisphenol A during the prenatal period in mice (Newbold *et al.*, 2009) or neonatal period in rats (Fernández *et al.*, 2010) can disrupt ovarian function as the animals reach adulthood, and these studies noted specifically the development of a large number of ovarian cysts. In the human population, serum levels of bisphenol A have been reported to be increased in women with polycystic ovary syndrome (Kandari *et al.*, 2011).

The uterus is another endocrine-sensitive organ and EDCs are thought to be contributing to common benign uterine disorders such as uterine fibroids and endometriosis. Uterine fibroids are common benign tumours of uterine smooth muscle tissue termed leiomyomas and occur mainly between puberty and menopause. Fibroid tissue is regulated by oestrogen and progesterone and EDCs are thought to contribute to the development of these tumours (McLachlan *et al.*, 2006). Raised levels of some polychlorinated biphenyl congeners have been found in the abdominal fat of women with fibroids (Trabert *et al.*, 2014), and raised levels of bisphenol A have been measured in the urine of women with fibroids (Shen *et al.*, 2013). *In vitro* studies have demonstrated that bisphenol A can increase the growth of uterine leiomyoma cells (Shen *et al.*, 2014) and *in vivo* studies have reported that neonatal exposure of mice to bisphenol A results in leiomyomas in adulthood (Newbold *et al.*, 2007). Endometriosis is another common abnormality of the uterus where cells from the lining of the uterus (endometrium) grow outside the uterine cavity and this may cause pelvic pain and lead to infertility. Although there are genetic determinants, endometrial tissue is strongly influenced by oestrogens,which is again suggestive of an involvement of EDCs, particularly organochlorine pollutants (Porpora *et al.*, 2013). Using animal models, exposure to polychlorinated dibenzo-*p*-dioxins, dibenzofurans and biphenyls has been shown to result in development of endometriosis in monkeys (Rier *et al.*, 1993) and in mice (Johnson *et al.*, 1997), as has also bisphenol A in mice (Signorile *et al.*, 2010). Follow-up of the DES-exposed daughters has suggested that they also have a higher incidence of endometriosis (Missmer *et al.*, 2004).

The breast is yet another endocrine-regulated organ and animal models have demonstrated that EDCs can disrupt normal mammary gland development in rodents (Fenton *et al.*, 2012).

EDCs are thought to contribute to benign breast abnormalities such as cysts and fibroadenomas. Gross cystic breast disease is the most common benign breast disorder and results from dilation and/or obstruction of terminal lobular units, which is associated with retention of fluid and secretory material from the lining epithelial cells (Mannello *et al.*, 2006). Fibroadenomas are benign tumours which also arise in terminal duct lobular units. Both these conditions occur most frequently before menopause, indicating their hormonal dependence, but also arise at a disproportionately high rate in the upper outer quadrant of the breast (Rimsten, 1976), which has been pointed out to be the site of application of EDCs in underarm personal care products (Darbre, 2001, 2015).

### 15.6.2 Male reproductive health

Although male sexuality is initially determined by the presence of a Y chromosome (Gilbert, 2000), development and maintenance of the male phenotype (masculinization) is dependent on a complex network of endocrine signals, most notably testicular hormones (Sharpe, 2006). Individuals who lack any gonad are phenotypically female (Gilbert, 2000) and to become a male requires endocrine intervention to modify the default female phenotype (Sharpe, 2006). This makes both development and maintenance of masculinization vulnerable to endocrine disrupting influences at all developmental stages from the early embryo to adulthood. Furthermore, disruption of early embryonic developmental processes has inevitable consequences for male reproductive health in adult life (Sharpe, 2006). One important determinant is that of the ratio of androgen to oestrogen in male life, which makes exposure to EDCs with oestrogenic or anti-androgenic activity especially pertinent. Evidence for consequences to male reproductive health following fetal exposure to excess oestrogen is demonstrated from the adverse reproductive health outcomes for sons whose mothers were treated with DES during pregnancy (Palmer *et al.*, 2009; Harris and Waring, 2012). Since it is difficult to relate specific chemical exposures of women in early pregnancy to outcomes in their sons, especially if outcomes are not manifested until into adulthood, the inadvertent exposure of so many boys to DES *in utero* provides a unique view of the effects in the human population on male reproductive health following early embryonic exposure to a potent oestrogen. Evidence that exposure to exogenous oestrogens can impact also in adult life is provided by the reported cases of gynaecomastia (inappropriate breast growth) following exposure to oestrogenic chemicals in dermally applied creams. The first classical report, entitled the 'mortician's mystery', was of an embalmer who presented with hypogonadotropic hypogonadism (diminished function of gonads) together with gynaecomastia caused by exposure to embalming creams which he was applying to corpses without wearing gloves (Finkelstein *et al.*, 1988). By 1 year after cessation of exposure to the creams, simply by wearing gloves, it is reported that the symptoms reversed, including increase in size of his testes (Finkelstein *et al.*, 1988). This demonstrates a clear case where long-term topical exposure to oestrogenic components of embalming creams (sometimes termed 'the embalmer's curse') can result in endocrine disruption to the adult male. Development of breast tissue is inhibited by testosterone but stimulated by oestrogen, and therefore the condition of gynaecomastia is usually assumed to be caused by excess exposure to oestrogen. Cases such as this aid understanding of endocrine-disrupting influences, because the reversibility of symptoms in adult life enables identification of the specific causative environmental factors.

Studies of the rising numbers of hypospadias and cryptorchidism (urogenital tract malformations), abnormalities of spermatogenesis (declining sperm counts and sperm quality) and testicular cancer in westernized countries have shown that these adverse trends in male reproductive health may be linked and may be arising from a common origin during fetal testicular development. Accordingly, this collection of conditions has been termed testicular dysgenesis syndrome with disturbances in early fetal life to Leydig cell function leading to hypospadias and cryptorchidism after birth and to Sertoli cell function leading to reduced semen quality and testicular cancer in adult life (Skakkebaek *et al.*, 2001). Whilst there are genetic determinants, it is thought that the increasing coincident trends may indicate common environmental factors and especially those that lead to androgen insufficiency or disturbance to the androgen/oestrogen balance in fetal life (Skakkebaek *et al.*, 2001).

Hypospadias result from abnormal development of the urethra, and cryptorchidism is characterized by maldescent of the testes, resulting in the absence of one or both testes from the scrotum. The incidence of both conditions has been reported to be increasing over recent decades in the Western world, with a correlation between them (Weidner *et al.*, 1999) and with a rapidity suggesting environmental rather than genetic determinants (Virtanen and Adamsson, 2012). In the human population, the risk of cryptorchidism has been found to be increased in the sons exposed to DES *in utero* (Palmer *et al.*, 2009), which demonstrates the effect of exposure to a potent oestrogen in fetal life. Animal models have demonstrated that rats exposed *in utero* to dibutylphthalate develop hypospadias and cryptorchidism (Fisher *et al.*, 2003), which further shows that EDCs can also cause these malformations of the urogenital system. Environmental epidemiological studies have shown that living near hazardous-waste landfill sites is correlated in five European countries with risk of hypospadias (Dolk *et al.*, 1998). Cryptorchidism has been correlated in Spain with living close to intensive farming where pesticides are used (García-Rodriquez *et al.*, 1996), in Denmark with mothers working in gardens (Weidner *et al.*, 1998) and in the Netherlands with paternal pesticide exposure (Pierik *et al.*, 2004). In Denmark, maternal use of mild analgesics, especially during the second trimester of pregnancy, has been dose-dependently correlated with cryptorchidism (Kristensen *et al.*, 2011).

Anogenital distance is the distance from the anus to the base of the penis and reduction in length is another indicator of abnormal male reproductive tract masculinization and considered to be influenced by EDCs. In the rodent model, perineal growth is dependent on dihydrotestosterone and males with a shortened anogenital distance have increased risk of hypospadias and cryptorchidism (Welsh *et al.*, 2008). Prenatal exposure of male rodents to phthalates (Mylchreest *et al.*, 2000), *n*-butylparaben (Zhang *et al.*, 2014) or bisphenol A (Christiansen *et al.*, 2014) has been reported to cause shortening of anogenital distance. In human epidemiological studies of mothers during pregnancy, shortening of anogenital distance in their sons has been associated with raised levels of urinary phthalates (Swan *et al.*, 2005) or with raised blood levels of the pesticide metabolite dichlorodiphenyldichloroethylene (DDE) (Torres-Sanchez *et al.*, 2008), bisphenol A (Miao *et al.*, 2011) or dioxins (Vafeiadi *et al.*, 2013).

Over the past 50 years, there have also been reports of a decline in sperm counts and sperm quality across the western world (Carlsen *et al.*, 1992; Swan *et al.*, 2000) and EDCs are suspected of being a contributory factor (Darbre, 2015). For animal models it has been reported that early life exposure to some esters of *p*-hydroxybenzoic acid (parabens) leads to reduced sperm number and decreased sperm motility (Kang *et al.*, 2002). In order to investigate the potential for effects in the human population, urinary levels of EDCs have been measured in men attending infertility clinics, and sperm DNA damage has been reported to be associated with higher urinary levels of butylparaben (Meeker *et al.*, 2010a) and bisphenol A (Meeker *et al.*, 2010b).

EDCs are also increasingly implicated in other male reproductive health issues, including alterations to puberty, development of prostatic hyperplasia and gender identity. Animal models and human epidemiological studies have implicated several EDCs, including PCBs, dioxins and some pesticides, in altering the timing of onset and attainment of pubertal milestones (Wu *et al.*, 2012). Benign prostatic hyperplasia is an overproliferative enlargement of the prostate gland common in older men that is associated with an altered balance of androgen to oestrogen (Nicholson and Ricke, 2011). Animal models have demonstrated that low-level exposure to oestrogens (vom Saal *et al.*, 1997) or bisphenol A (Nagel *et al.*, 1997) in early life can lead to enhanced prostate growth in later life, and further work showing that bisphenol A can alter growth properties of human prostate cells (Prins *et al.*, 2014) has opened up questions about a role for environmental exposures to EDCs. Gender identity concerns how a man feels about himself regardless of physical characteristics, and the many reports in wildlife of hermaphroditism and sexual abnormalities in behaviour now pose serious questions about the potential for EDCs to have the same effects in humans (Hood, 2005).

### 15.6.3 Cancer of reproductive tissues

Development of cancer is a lengthy and multi-step process involving both genetic and environmental components but the unprecedented worldwide rise in incidence of cancers in endocrine-sensitive tissues has raised questions concerning exposure to EDCs as a contributory factor (Darbre, 2015). Breast cancer has now become the major cancer of women and prostate cancer the major cancer of men in many westernized countries (Ferlay *et al.*, 2010). Although life expectancy may be greater in modern times than in ancient days and most cancer incidence tends to increase with age, some cancers with rising incidence such as testicular cancer affect young people (Shanmugalingam *et al.*, 2013). Although in the UK the majority of breast cancer occurs over the age of 50 years, the 20% proportion of breast cancers under the age of 50 has remained constant despite a near doubling of incidence over the past 30 years, which demonstrates that numbers of young-age breast cancers are also rising (ONS, 2014).

Cancer results from a loss of normal cell growth control, starting from multiple rounds of initiation (involving DNA damage) and promotion (enabling of the damaged cells to grow), and followed by tumour growth (uncontrolled growth of the damaged cells), tumour progression (loss of genetic and phenotypic stability) and metastasis (spread of tumour cells to distant sites around the body) (Weinberg, 2013). Hanahan and Weinberg (2011) provided a framework for understanding the complexity of these stages in cancer development by defining a set of hallmarks and enabling characteristics. The six basic hallmarks are: sustained proliferative signalling; evasion of growth suppressors; resistance to cell death; replicative immortality; induction of angiogenesis; and activation of invasion and metastasis. Two underlying enabling characteristics are genome instability and inflammation. Two emerging hallmarks are evasion of immune destruction and reprogramming of energy metabolism. In endocrine-sensitive tissues, where cell growth is under endocrine control, all of these steps of deregulation could be influenced by EDCs through their ability to disrupt normal endocrine controls. Therefore, EDCs may influence the development of female cancers of breast, endometrium, ovary and cervix and of male cancers of prostate, testis and breast (Darbre, 2015).

### 15.6.4 Thyroid health

The main function of the thyroid gland is to secrete thyroid hormones (thyroxine (T4), triiodothyronine (T3)), which have wide-ranging effects on intermediary metabolism, the cardiovascular system, energy intake and thermogenesis, and which, during early life development, are critical for neurodevelopment and skeletal development. The thyroid is in turn itself under neuroendocrine control, such that thyrotropin-releasing hormone produced by the hypothalamus then stimulates production/secretion of thyroid-stimulating hormone (TSH) by the pituitary gland, which in turn stimulates production/secretion of T4/T3 by the thyroid. Thyroid dysfunction often results in loss of T4 and T3 secretion, which then results in a compensatory increase in levels of TSH. The main cause of such hypothyroidism is iodine deficiency and one-third of the world population lives in iodine-deficient regions (Vanderpump, 2011). However, the rising incidence of hypothyroidism over recent years even in iodine-replete regions suggests alternative environmental factors (Vanderpump, 2011; Leese *et al.*, 2008), and there is now evidence for thyroid-disrupting effects of exposure to polychlorinated biphenyls, phthalates, bisphenol A, brominated flame retardants and perfluorinated chemicals (Boas *et al.*, 2012). Thyroid cancer is also on the increase and environmental factors are suspected (Davies and Welch, 2006). Since hypothyroidism results in elevated TSH and elevated TSH is thought to play a central role in development of thyroid carcinomas (Boelaert, 2009), it is plausible that EDCs such as PCBs, which act to elevate TSH, could be contributory factors.

### 15.6.5 Energy metabolism

Energy metabolism is under strict endocrine control in order to maintain a homeostatic balance between the supply and the usage of fuels, and disruption to these processes can cause disordered metabolism leading to development of conditions such as obesity, metabolic syndrome, type 2 diabetes and cardiovascular disease. The incidence of obesity, as defined by a body mass index greater than 30 kg $m^{-2}$, has risen sharply over recent decades (OECD, 2014). Although genetic predisposition, excess food intake and lack of exercise contribute to these trends, some EDCs are now known to possess specific obesogenic properties whereby they can disrupt lipid homeostasis, promoting adipogenesis and lipid accumulation, and so lead to weight gain (Darbre, 2017). An especially sensitive time frame for exposure is either prior to birth or during the perinatal period, because developmental programming of adipose tissue during this time becomes irreversible in adult life. This has been verified in animal models where exposure of pregnant mice to tributyltin has been shown to give rise to offspring that are heavier than those not exposed (Grün *et al.*, 2006), and where exposure of neonatal mice to DES has also been reported to lead to increased body weight (Newbold *et al.*, 2005). Obesity is not just a matter of carrying excess weight through life but has been established as an underlying risk factor for many diseases, including metabolic syndrome (NCEP, 2002), diabetes (Ross *et al.*, 2011), cardiovascular disease (Bastien *et al.*, 2014) and cancer (Berger, 2014), which have themselves also been linked to exposure to EDCs (Darbre, 2015). In this context, it is noteworthy that many EDCs are lipophilic, and POPs in particular are known to bioaccumulate in body fat over the years. It is therefore plausible that the link of obesity to disease relates not just simply to the deposition of fat but to the increased retention of lipophilic pollutants (including those with or without endocrine-disrupting properties) in the greater volume of fat.

## 15.7 Conclusions and Regulatory Needs

Evidence is pointing towards serious long-term consequences if nothing is done to reduce exposure of the human population to EDCs but this has resulted in considerable controversy, not about what the science shows but about the interpretation by those with differing points of view. On the one hand, there are those who would like to ban all EDCs; on the other hand, there are those for whom economics of the status quo is the overriding factor. The majority of the population live somewhere in the middle, acknowledging that there may be issues but feeling unable to stop consumer-driven chemical usage. It is said that 'the road to hell is paved with good intentions' and many environmental chemicals have been brought into existence for valid reasons. The problem is how to respond when chemicals brought in for a good initial purpose are then found to have endocrine-disrupting properties. There is undoubtedly a need for governments to regulate use of EDCs at national and international level. However, there is also a need for producers to be willing to make adjustments for the public good and for every consumer to start to understand the implications of endocrine disruption and to act responsibly.

Risk assessment is a prerequisite for any regulatory action by government. It requires identification of the hazard, assessment of the likely exposure and consideration of the susceptibility of a person or population. A hazard is posed by any EDC (or mixture of EDCs) that has the potential to cause an adverse health effect. Evidence of hazard may be based on human epidemiology, clinical experience, animal models or laboratory studies, and involves consideration of both toxicokinetics (how the body absorbs, distributes, metabolizes and eliminates EDCs) and toxicodynamics (mechanisms of action of EDCs at a cellular and physiological level). Exposure assessment requires knowledge of the magnitude and duration of human exposure to an EDC, which may use measurements of concentrations of EDCs in body tissues but is most often based on an estimate from human intake over time. Susceptibility of individuals may be influenced by age, gender, genetic background and presence of other co-acting risk factors. An emerging influence is the stage of life at which the exposure occurs, with greatest susceptibility arising from early life exposures.

Government regulation occurs primarily at a national level and thus varies in different countries. However, international cooperation, trade and travel have created a global economy and this has served to turn EDC exposure from a national issue to one of international concern, particularly for those EDCs that are produced in high volume and which can bioaccumulate. The Rotterdam Convention (www.pic.int) was established to promote shared responsibility and co-operative efforts in the international trade of certain hazardous chemicals in order to protect human health and the environment from potential harm. The text of the Convention was adopted in 1998 at a Conference in Rotterdam in the Netherlands, and entered into force in 2004. The Stockholm Convention (www.pops.int) is a global treaty to protect human health and the environment from POPs which entered into force in 2004, which was signed by over 150 countries and which continues to add compounds on a rolling basis. Within the European Union, EDCs remain under surveillance and regulation through the 'REACH' (Registration, Evaluation, Authorisation and restriction of CHemicals) legislation. Under this legislation, which came into effect in 2007, member states can nominate substances of very high concern to a candidate list for regulation.

In addition to these initiatives, the role of non-governmental organizations (NGOs) needs also to be acknowledged, since several NGOs have championed the issues of endocrine disruption from an early stage and have played a significant role in disseminating information, establishing international awareness and campaigning for regulatory actions on EDCs. The mass media have also contributed to the issue of EDCs through broadcast media, print media and the worldwide web. On the negative side, mass advertising has served to encourage increasing consumption of a range of consumer goods containing EDCs and has fuelled the widespread usage of chemicals by portraying them as essential for daily life. However, on the positive side, reports and documentaries have served to bring awareness of the underlying problems to the general public. Many initiatives to remove EDCs from consumer products are now resulting from media coverage as much as from more traditional routes of scientific output and risk assessment. Perhaps the most controversial resolution to EDCs lies in the use of the 'Precautionary Principle', which reiterates the old saying of 'better safe than sorry' and states that 'if an action or policy has a suspected risk of causing harm to the public or to the environment, in the absence of scientific consensus that the action or policy is not harmful, the burden of proof that it is not harmful falls on those taking an action'. Use of the precautionary principle will always cause strong controversy among those with differing viewpoints, but there is a growing international awareness that exposure to EDCs needs to be reduced, and with the challenge that EDCs pose to existing regulatory frameworks, a precautionary approach may become a necessary part of the route forward.

## References


Ahn, H.J., An, B.S., Jung, E.M., Yang, H., Choi, K.C. and Jeung, E.B. (2012) Parabens inhibit the early phase of folliculogenesis and steroidogenesis in the ovaries of neonatal rats. *Molecular Reproduction and Development* 79, 626–636.

Barr, L., Metaxas, G., Harbach, C.A.J., Savoy, L.A. and Darbre, P.D. (2012) Measurement of paraben concentrations in human breast tissue at serial locations across the breast from axilla to sternum. *Journal of Applied Toxicology* 32, 219–232.

Bastien, M., Poirier, P., Lemieux, I. and Després, J.P. (2014) Overview of epidemiology and contribution of obesity to cardiovascular disease. *Progress in Cardiovascular Diseases* 56, 369–381.

Bennets, H., Underwood, E.J. and Shier, F.L. (1946) A specific breeding problem of sheep on subterranean clover pasture in Western Australia. *Australian Veterinary Journal* 22, 2–12.

Berger, N.A. (2014) Obesity and cancer pathogenesis. *Annals of the New York Academy of Sciences* 131, 57–76.

Bitman, J., Cecil, H.C., Harris, S.J. and Fries, G.F. (1968) Estrogenic activity of o,p'-DDT in the mammalian uterus and avian oviduct. *Science* 162, 371–372.

Boas, M., Feldt-Rasmussen, U. and Main, K.M. (2012) Thyroid effects of endocrine disrupting chemicals. *Molecular and Cellular Endocrinology* 355, 240–248.

Boelaert, K. (2009) The association between serum TSH concentration and thyroid cancer. *Endocrine-Related Cancer* 16, 1065–1072.

Bourguignon, J.P. and Parent, A.S. (2012) The impact of endocrine disruptors on female pubertal timing. In: Diamanti-Kandarakis, E. and Gore, A.C. (eds) *Endocrine Disruptors and Puberty*. Humana Press, Springer, New York, pp. 325–337.

Bulayeva, N.N. and Watson, C.S. (2004) Xenoestrogen-induced ERK-1 and ERK-2 activation via multiple membrane-initiated signalling pathways. *Environmental Health Perspectives* 112, 1481–1487.

Byford, J.R., Shaw, L.E., Drew, M.G., Pope, G.S., Sauer, M.J. and Darbre, P.D. (2002) Oestrogenic activity of parabens in MCF7 human breast cancer cells. *Journal of Steroid Biochemistry and Molecular Biology* 80, 49–60.

Calafat, A.M., Ye, X., Wong, L.Y., Reidy, J.A. and Needham, L.L. (2008a) Exposure of the US population to bisphenol A and 4-tertiary-octylphenol: 2003–2004. *Environmental Health Perspectives* 116, 39–44.

Calafat, A.M., Ye, X., Wong, L.Y., Reidy, J.A. and Needham, L.L. (2008b) Urinary concentrations of triclosan in the US population: 2003–2004. *Environmental Health Perspectives* 116, 303–307.

Calafat, A.M., Ye, X., Wong, L.Y., Bishop, A.M. and Needham, L.L. (2010) Urinary concentrations of four parabens in the US population: NHANES 2005–2006. *Environmental Health Perspectives* 118, 679–685.

Carlsen, E., Giwercman, A., Keiding, N. and Skakkebaek, N.E. (1992) Evidence for decreasing quality of semen during past 50 years. *British Medical Journal* 305, 609–613.

Carson, R. (1962) *Silent Spring*. Houghton Mifflin, Boston, Massachusetts.

Charles, A.K. and Darbre, P.D. (2013) Combinations of parabens at concentrations measured in human breast tissue can increase proliferation of MCF-7 human breast cancer cells. *Journal of Applied Toxicology* 33, 390–398.

Christiansen, S., Axelstad, M., Boberg, J., Vinggaard, A.M., Pedersen, G.A. and Hass, U. (2014) Low-dose effects of bisphenol A on early sexual development in male and female rats. *Reproduction* 147, 477–487.

Colborn, T. and Clement, C. (1992) *Chemically-induced Alterations in Sexual and Functional Development: the Wildlife/Human Connection*. Princeton Scientific Publishing, Princeton, New Jersey.

Colon, I., Caro, D., Bourdony, C.J. and Rosario, O. (2000) Identification of phthalate esters in the serum of young Puerto Rican girls with premature breast development. *Environmental Health Perspectives* 108, 895–900.

Craig, Z.R., Wang, W. and Flaws, J.A. (2011) Endocrine-disrupting chemicals in ovarian function: effects on steroidogenesis, metabolism and nuclear signalling. *Reproduction* 142, 633–646.

Darbre, P.D. (2001) Hypothesis: Underarm cosmetics are a cause of breast cancer. *European Journal of Cancer Prevention* 10, 389–393.

Darbre, P.D. (2015) *Endocrine Disruption and Human Health*. Elsevier/Academic Press, New York.

Darbre, P.D. (2017) Endocrine disruptors and obesity. *Current Obesity Reports* 6, 18–27.

Darbre, P.D., Byford, J.R., Shaw, L.E., Horton, R.A., Pope, G.S. and Sauer, M.J. (2002) Oestrogenic activity of isobutylparaben in vitro and in vivo. *Journal of Applied Toxicology* 22, 219–226.

Davies, L. and Welch, H.G. (2006) Increasing incidence of thyroid cancer in the United States, 1973–2002. *Journal of the American Medical Association* 295, 2164–2167.

Dechaud, H., Ravard, C., Claustrat, F., de la Perriere, A.B. and Pugeat, M. (1999) Xenoestrogen interaction with human sex hormone-binding globulin (hSHBG). *Steroids* 64, 328–334.

Diamanti-Kandarakis, E. and Gore, A.C. (eds) (2012) *Endocrine Disruptors and Puberty*. Humana Press, Springer, New York.

Dolk, H., Vrijheid, M., Armstrong, B., Abramsky, L., Bianchi, F. *et al.* (1998) Risk of congenital anomalies near hazardous-waste landfill sites in Europe: the EUROHAZCON study. *Lancet* 352, 423–427.

EC (1996) *Report of the Proceedings of the European workshop on the impact of endocrine disrupters on human health and wildlife. Weybridge, UK.* Report EUR17549 of the Environment and Climate Change Research Programme of DGXII of the European Commission. European Commission DGXII, Brussels.

Fenton, S.E., Beck, L.M., Borde, A.R. and Rayner, J.L. (2012) Developmental exposure to environmental endocrine disruptors and adverse effects on mammary gland development. In: Diamanti-Kandarakis, E. and Gore, A.C. (eds) *Endocrine Disruptors and Puberty*. Humana Press, Springer, New York, pp. 201–224.

Ferlay, J., Shin, H.R., Bray, F., Forman, D., Mathers, C. and Parkin, D.M. (2010) Estimates of worldwide burden of cancer in 2008: GLOBOCAN 2008. *International Journal of Cancer* 127, 2893–2917.

Fernández, M., Bourguignon, N., Lux-Lantos, V. and Libertun, C. (2010) Neonatal exposure to bisphenol a and reproductive and endocrine alterations resembling the polycystic ovarian syndrome in adult rats. *Environmental Health Perspectives* 118, 1217–1222.

Finkelstein, J.S., McCully, W.F., MacLaughlin, D.T., Godine, J.E. and Crowley, W.F. (1988) The Mortician's Mystery. *New England Journal of Medicine* 318, 961–965.
Fisher, J.S., Macpherson, S., Marchetti, N. and Sharpe, R.M. (2003) Human 'testicular dysgenesis syndrome': a possible model using in-utero exposure of the rat to dibutylphthalate. *Human Reproduction* 18, 1383–1394.
Frasor, J., Danes, J.M., Komm, B., Chang, K.C.N., Lyttle, C.R. and Katzenellenbogen, B.S. (2003) Profiling of estrogen up- and down-regulated gene expression in human breast cancer cells: insights into gene networks and pathways underlying estrogenic control of proliferation and cell phenotype. *Endocrinology* 144, 4562–4574.
García-Rodríguez, J., García-Martín, M., Nogueras-Ocaña, M., de Dios Luna-del-Castillo, J., Espigares García, M., Olea, N. and Lardelli-Claret, P. (1996) Exposure to pesticides and cryptorchidism: geographical evidence of a possible association. *Environmental Health Perspectives* 104, 1090–1095.
Garner, C.E.I., Jefferson, W.N., Burka, L.T., Matthews, H.B. and Newbold, R.R. (1999) In vitro estrogenicity of the catechol metabolites of selected polychlorinated biphenyls. *Toxicology and Applied Pharmacology* 154, 188–197.
Gilbert, S.F. (2000) Chromosomal sex determination in mammals. Chapter 17 in: Gilbert, S.F. *Developmental Biology*, 6th edn. Sinauer Associates, Sunderland, Massachusetts, pp. 547–574.
Grün, F., Watanabe, H., Zamanian, Z., Maeda, L., Arima, K. *et al.* (2006) Endocrine-disrupting organotin compounds are potent inducers of adipogenesis in vertebrates. *Molecular Endocrinology* 20, 2141–2155.
Guaneri, M.P., Brambilla, G., Loizzo, A., Colombo, I. and Chiumello, G. (2008) Estrogen exposure in a child from hair lotion used by her mother: clinical and hair analysis data. *Clinical Toxicology* 46, 762–764.
Guillette, L.J. Jr and Gunderson, M.P. (2001) Alterations in development of reproductive and endocrine systems of wildlife populations exposed to endocrine-disrupting contaminants. *Reproduction* 122, 857–864.
Hanahan, D. and Weinberg, R.A. (2011) Hallmarks of cancer: the next generation. *Cell* 144, 646–674.
Harris, R.M. and Waring, R.H. (2012) Diethylstilboestrol – a long-term legacy. *Maturitas* 72, 108–112.
Herbst, A.L., Ulfelder, H. and Poskanzer, D.C. (1971) Adenocarcinoma of the vagina: association of maternal stilboestrol therapy with tumor appearance in young women. *New England Journal of Medicine* 284, 878–881.
Hood, E. (2005) Are EDCs blurring issues of gender? *Environmental Health Perspectives* 113, A670–A677.
Hoover, R.N., Hyer, M., Pfeiffer, R.M., Adam, E., Bond, B. *et al.* (2011) Adverse health outcomes in women exposed in utero to diethylstilboestrol. *New England Journal of Medicine* 365, 1304–1314.
Horiguchi, T. (2006) Masculinization of female gastropod mollusks induced by organotin compounds, focusing on mechanism of actions of tributyltin and triphenyltin for development of imposex. *Environmental Science* 13, 77–87.
Janjua, N.R., Mortensen, G.K., Andersson, A.M., Kongshoj, B., Skakkebaek, N.E. and Wulf, H.C. (2007) Systemic uptake of diethyl phthalate, dibutyl phthalate, and butyl paraben following whole-body topical application and reproductive and thyroid hormone levels in humans. *Environmental Science and Technology* 41, 5564–5570.
Janjua, N.R., Kongshoj, B., Andersson, A.M. and Wulf, H.C. (2008) Sunscreens in human plasma and urine after repeated whole-body topical application. *Journal of the European Academy of Dermatology and Venereology* 22, 456–461.
Jobling, S., Nolan, M., Tyler, C.R., Brighty, G. and Sumpter, J.P. (1998) Widespread sexual disruption in wild fish. *Environmental Science and Technology* 32, 2498–2506.
Johnson, K.L., Cummings, A.M. and Birnbaum, L.S. (1997) Promotion of endometriosis in mice by polychlorinated dibenzo-*p*-dioxins, dibenzofurans, and biphenyls. *Environmental Health Perspectives* 105, 750–755.
Kandari, E., Chatzigeorgiou, A., Livadas, S., Palioura, E., Economou, F. *et al.* (2011) Endocrine disruptors and polycystic ovary syndrome (PCOS): elevated serum levels of bisphenol A in women with PCOS. *Journal of Clinical Endocrinology and Metabolism* 96, E480–484.
Kang, K.S., Che, J.H., Ryu, D.Y., Kim, T.W., Li, G.X. and Lee, Y.S. (2002) Decreased sperm number and motile activity on the F1 offspring maternally exposed to butyl *p*-hydroxybenzoic acid (butyl paraben). *Journal of Veterinary Medical Science* 64, 227–235.
Kirigaya, A., Kim, H., Hayashi, S., Chambon, P., Watanabe, H., Lquchi, T. and Sato, T. (2009) Involvement of estrogen receptor beta in the induction of polyovular follicles in mouse ovaries exposed neonatally to diethylstilbestrol. *Zoological Science* 26, 704–712.
Kortenkamp, A. (2007) Ten years of mixing cocktails: a review of combination effects of endocrine-disrupting chemicals. *Environmental Health Perspectives* 115 (Suppl. 1), 98–105.

Kristensen, D.M., Hass, U., Lesné, L., Lottrup, G., Jacobsen, P.R. *et al.* (2011) Intrauterine exposure to mild analgesics is a risk factor for development of male reproductive disorders in human and rat. *Human Reproduction* 26, 235–244.

Leese, G.P., Flynn, R.V., Jung, R.T., MacDonald, T.M., Murphy, M.J. and Morris, A.D. (2008) Increasing prevalence and incidence of thyroid disease in Tayside, Scotland: the Thyroid Epidemiology, Audit and Research Study (TEARS). *Clinical Endocrinology (Oxford)* 68, 311–316.

Lonning, P.E. (ed.) (2004) Endocrinology and treatment of breast cancer. *Best Practice & Research: Clinical Endocrinology & Metabolism* 18, 1–130.

Mannello, F., Tonti, G.A.M. and Papa, S. (2006) Human gross cyst breast disease and cystic fluid: biomolecular, morphological, and clinical studies. *Breast Cancer Research and Treatment* 97, 115–129.

Martinez, E.M. and Swartz, W.J. (1991) Effects of methoxychlor on the reproductive system of the adult female mouse. Gross and histologic observations. *Reproductive Toxicology* 5, 139–147.

Matsumura, A., Ghosh, A., Pope, G.S. and Darbre, P.D. (2005) Comparative study of oestrogenic properties of eight phytoestrogens in MCF7 human breast cancer cells. *Journal of Steroid Biochemistry and Molecular Biology* 94, 431–443.

McLachlan, J.A., Simpson, E. and Martin, M. (2006) Endocrine disrupters and female reproductive health. *Best Practice & Research in Clinical Endocrinology & Metabolism* 20, 63–75.

McNutt, S.H., Purwin, P. and Murray, C. (1928) Vulvo-vaginitis in swine: preliminary report. *Journal of the American Veterinary and Medical Association* 73, 484.

Meeker, J.D., Yang, T., Ye, X., Calafat, A.M. and Hauser, R. (2010a) Urinary concentrations of parabens and serum hormone levels, semen quality parameters, and sperm DNA damage. *Environmental Health Perspectives* 119, 252–257.

Meeker, J.D., Ehrlich, S., Toth, T.L., Wright, D.L., Calafat, A.M. *et al.* 2010b) Semen quality and sperm DNA damage in relation to urinary bisphenol A in men from an infertility clinic. *Reproductive Toxicology* 30, 532–539.

Miao, M., Yuan, W., He, Y., Zhou, Z., Wang, J. *et al.* (2011) In utero exposure to bisphenol-A and anogenital distance of male offspring. *Birth Defects Research Part A: Clinical and Molecular Teratology* 91, 867–872.

Missmer, S.A., Hankinson, S.E., Spiegelman, D., Barberi, R.L., Michels, K.B. and Hunter, D.J. (2004) In utero exposures and the incidence of endometriosis. *Fertility and Sterility* 82, 1501–1508.

Mylchreest, E., Wallace, D.G., Cattley, R.C. and Foster, P.M. (2000) Dose-dependent alterations in androgen-regulated male reproductive development in rats exposed to di(n-butyl) phthalate during late gestation. *Toxicological Sciences* 55, 143–151.

Nagel, S.C., vom Saal, F.S., Thayer, K.A., Dhar, M.G., Boechler, M. and Welshons, W.V. (1997) Relative binding affinity-serum modified access (RBA-SMA) assay predicts the relative in vivo bioactivity of the xenoestrogens bisphenol A and octylphenol. *Environmental Health Perspectives* 105, 70–76.

NCEP (2002) Third report of the National Cholesterol Education Program (NCEP) Expert Panel on Detection, Evaluation, and Treatment of High Blood Cholesterol in Adults (Adult Treatment Panel III) final report. *Circulation* 106, 3143–3421.

Nelson, L.M. (2009) Clinical Practice. Primary ovarian insufficiency. *New England Journal of Medicine* 360, 606–614.

Nesaretnam, K., Corcoran, D., Dils, R.R. and Darbre, P. (1996) 3,4,3',4'-tetrachlorobiphenyl acts as an estrogen in vitro and in vivo. *Molecular Endocrinology* 10, 923–936.

Newbold, R.R., Padilla-Banks, E., Snyder, R.J. and Jefferson, W.N. (2005) Developmental exposure to estrogenic compounds and obesity. *Birth Defects Research Part A: Clinical and Molecular Teratology* 73, 478–480.

Newbold, R.R., Jefferson, W.N. and Padlila-Banks, E. (2007) Long-term adverse effects of neonatal exposure to bisphenol A on the murine female reproductive tract. *Reproductive Toxicology* 24, 253–258.

Newbold, R.R., Jefferson, W.N. and Padilla-Banks, E. (2009) Prenatal exposure to bisphenol A at environmentally relevant doses adversely affects the murine female reproductive tract later in life. *Environmental Health Perspectives* 117, 879–885.

Nicholson, T.M. and Ricke, W.A. (2011) Androgens and estrogens in benign prostatic hyperplasia: past, present and future. *Differentiation* 82, 184–199.

Odum, J. (2015) Disrupters of thyroid hormone action and synthesis. In: Darbre, P. (ed.) *Endocrine Disruption and Human Health*. Elsevier/Academic Press, New York, pp. 91–109.

OECD (2014) Obesity Update. June 2014. Organisation for Economic Co-operation and Development. Available at: www.oecd.org/health/obesity-update-2014.pdf , accessed 28 April 2019

ONS (2014) *Series MB1 from 1979 to 2014 (MB series 1 numbered to 43)*. Office of National Statistics, Newport, Wales.
Palmer, J.R., Herbst, A.L., Noller, K.L., Boggs, D.A., Troisi, R. *et al.* (2009) Urogenital abnormalities in men exposed to diethylstilboestrol in utero: a cohort study. *Environmental Health* 8, 37, 1–6.
Parent, A.S., Teilman, G., Juul, A., Skakkebaek, N.E., Toppari, J. and Bourguignon, J.P. (2003) The timing of normal puberty and the age limits of sexual precocity: variations around the world, secular trends, and changes after migration. *Endocrine Reviews* 24, 668–693.
Pierik, F.H., Burdorf, A., Deddens, J.A., Juttman, R.E. and Weber, R.F. (2004) Maternal and paternal risk factors for cryptorchidism and hypospadias: a case-control study in newborn boys. *Environmental Health Perspectives* 112, 1570–1576.
Pino, A.M., Valladares, L.E., Palma, M.A., Mancilla, A.M., Yanez, M. and Albala, C. (2000) Dietary isoflavones affect sex-hormone-binding globulin levels in postmenopausal women. *Journal of Clinical Endocrinology and Metabolism* 85, 2797–2800.
Porpora, M.G., Resta, S., Fuggetta, E., Storelli, P., Megiorni, F., Manganaro, L. and De Felip, E. (2013) Role of environmental organochlorinated pollutants in the development of endometriosis *Clinical and Experimental Obstetrics and Gynecology* 40, 565–567.
Porta, M., Pumarega, J. and Gasull, M. (2012) Number of persistent organic pollutants detected at high concentrations in a general population. *Environment International* 44, 106–111.
Prins, G.S., Hu, W.Y., Shi, G.B., Hu, D.P., Majumdar, S. *et al.* (2014) Bisphenol A promotes human prostate stem-progenitor cell self-renewal and increases in vivo carcinogenesis in human prostate epithelium. *Endocrinology* 155, 805–817.
Pugazhendhi, D., Pope, G.S. and Darbre, P.D. (2005) Oestrogenic activity of *p*-hydroxybenzoic acid (common metabolite of paraben esters) and methylparaben in human breast cancer cell lines. *Journal of Applied Toxicology* 25, 301–309.
Pugazhendhi, D., Sadler, A.J. and Darbre, P.D. (2007) Comparison of the global gene expression profiles produced by methylparaben, n-butylparaben and 17β-oestradiol in MCF7 human breast cancer cells. *Journal of Applied Toxicology* 27, 67–77.
Rajapakse, N., Silva, E. and Kortenkamp, A. (2002) Combining xenoestrogens at levels below individual no-observed-effect concentrations dramatically enhances steroid hormone action. *Environmental Health Perspectives* 110, 917–921.
Rier, S.E., Martin, D.C., Bowman, R.E., Dmowski, W.P. and Becker, J.L. (1993) Endometriosis in rhesus monkeys (*Macaca mulatta*) following chronic exposure to 2,3,7,8-tetrachlorodibenzo-*p*-dioxin. *Fundamental and Applied Toxicology* 21, 433–441.
Rimsten, A. (1976) Symptoms and signs in benign and malignant tumours of the breast. *Upsala Journal of Medical Sciences* 81, 54–60.
Ross, S.A., Dzida G, Vora J, Khunti, K., Kaiser, M. and Ligthelm, R.J. (2011) Impact of weight gain on outcomes in type 2 diabetes. *Current Medical Research and Opinion* 27, 1431–1438.
Rutkowska, A. and Rachon, D. (2014) Bisphenol A (BPA) and its potential role in the pathogenesis of the polycystic ovary syndrome (PCOS). *Gynecological Endocrinology* 30, 260–265.
Shanmugalingam, T., Soultati, A., Chowdhury, S., Rudman, S. and Hemelrijck, M.V. (2013) Global incidence and outcome of testicular cancer. *Clinical Epidemiology* 5, 417–427.
Sharpe, R.M. (2006) Pathways of endocrine disruption during male sexual differentiation and masculinisation. *Best Practice & Research in Clinical Endocrinology & Metabolism* 20, 91–110.
Shen, Y., Xu, Q., Ren, M., Feng, X., Cai, Y. and Gao, Y. (2013) Measurement of phenolic environmental estrogens in women with uterine leiomyoma. *PLoS ONE* 8, e796838, 1–5.
Shen, Y., Shen, Y., Ren, M.L., Feng, X., Cai, Y.L., Gao, Y.X. and Xu, Q. (2014) An evidence in vitro for the influence of bisphenol A on uterine leiomyoma. *European Journal of Obstetrics & Gynecology and Reproductive Biology* 178, 80–83.
Signorile, P.G., Spugnini, E.P., Mita, L., Mellone, P., D'Avino, A. *et al.* (2010) Pre-natal exposure of mice to bisphenol A elicits an endometriosis-like phenotype in female offspring. *General and Comparative Endocrinology* 168, 318–325.
Silva, M.J., Barr, D.B., Reidy, J.A., Malek, N.A., Hodge, C.C. *et al.* (2004) Urinary levels of seven phthalate metabolites in the US population from the National Health and Nutrition Examination Survey (NHANES) 1999–2000. *Environmental Health Perspectives* 112, 331–338.
Sjödin, A., Jones, R.S., Caudill, S.P., Wong, L.Y., Turner, W.E. and Calafat, A.M. (2014) Polybrominated diphenyl ethers, polychlorinated biphenyls, and persistent pesticides in serum from the National Health and Nutrition Examination Survey: 2003–2008. *Environmental Science and Technology* 48, 753–760.

Skakkebaek, N.E., Rajpert-De Meyts, E. and Main, K.M. (2001) Testicular dysgenesis syndrome: an increasingly common developmental disorder with environmental aspects. *Human Reproduction* 16, 972–978.
Skinner, M.K., Manikkam, M. and Guerrero-Bosagna, C. (2011) Epigenetic transgenerational actions of endocrine disruptors. *Reproductive Toxicology* 31, 337–343.
Smith, O.W. (1948) Diethylstilboestrol in the prevention and treatment of complications of pregnancy. *American Journal of Obstetrics and Gynecology* 56, 821–834.
Swan, S.H., Elkin, E.P. and Fenster, L. (2000) The question of declining sperm density revisited: an analysis of 101 studies published 1934–1996. *Environmental Health Perspectives* 108, 961–966.
Swan, S.H., Main, K.M., Liu, F., Stewart, S.L., Kruse, R.L. *et al.* (2005) Decrease in anogenital distance among male infants with prenatal phthalate exposure. *Environmental Health Perspectives* 113, 1056–1061.
Torres-Sanchez, L., Zepeda, M., Cebrian, M.E., Belkind-Gerson, J., Garcia-Hernandez, R.M., Belkind-Valdovinos, U. and Lopez-Carrillo, L. (2008) Dichlorodiphenyldichloroethylene exposure during the first trimester of pregnancy alters the anal position in male infants. *Annals of the New York Academy of Sciences* 1140, 155–162.
Trabert, B., Chen, Z., Kannan, K., Peterson, C.M., Pollack, A.Z., Sun, L. and Buck Louis, G.M. (2014) Persistent organic pollutants (POPs) and fibroids: results from the ENDO study. *Journal of Exposure Science & Environmental Epidemiology* 25, 278–285.
Vafeiadi, M., Agramunt, S., Papadopoulou, E., Besselink, H., Mathianaki, K. *et al.* (2013) In utero exposure to dioxins and dioxin-like compounds and anogenital distance in newborns and infants. *Environmental Health Perspectives* 121, 125–130.
Vandenberg, L.N., Colborn, T., Hayes, T.B., Heindel, J.J., Jacobs, D.R. Jr *et al.* (2012) Hormones and endocrine-disrupting chemicals: low-dose effects and nonmonotonic dose responses. *Endocrine Reviews* 33, 378–455.
Vanderpump, M.P.J. (2011) The epidemiology of thyroid disease. *British Medical Bulletin* 99, 39–51.
Virtanen, H.E. and Adamsson, A. (2012) Cryptorchidism and endocrine disrupting chemicals. *Molecular and Cellular Endocrinology* 355, 208–220.
vom Saal, F.S., Timms, B.G., Montano, M.M., Palanza, P., Thayer, K.A. *et al.* (1997) Prostate enlargement in mice due to fetal exposure to low doses of estradiol or diethylstilbestrol and opposite effects at high doses. *Proceedings of the National Academy of Sciences of the USA* 94, 2056–2061.
Waring, R.H., Ayers, S., Gescher, A.J., Glatt, H.R., Meinl, W. *et al.* (2008) Phytoestrogens and xenoestrogens: the contribution of diet and environment to endocrine disruption. *Journal of Steroid Biochemistry and Molecular Biology* 108, 213–220.
Weidner, I.S., Moller, H., Jensen, T.K. and Skakkebaek, N.E. (1998) Cryptorchidism and hypospadias in sons of gardeners and farmers. *Environmental Health Perspectives* 106, 793–796.
Weidner, I.S., Moller, H., Jensen, T.K. and Skakkebaek, N.E. (1999) Risk factors for cryptorchidism and hypospadias. *Journal of Urology* 161, 1606–1609.
Weinberg, R.A. (2013) *The Biology of Cancer*, 2nd edn. Garland Science, New York
Welsh, M., Saunders, P.T.K., Fisken, M., Scott, L.B. and Sharpe, R.M. (2008) Identification in rats of a programming window for reproductive tract masculinisation, disruption of which leads to hypospadias and cryptorchidism. *Journal of Clinical Investigation* 118, 1479–1490.
Weschler, C.J. and Nazaroff, W.W. (2014) Dermal uptake of organic vapors commonly found in indoor air. *Environmental Science and Technology* 48, 1230–1237.
Whitehead, S.A. and Rice, S. (2006) Endocrine-disrupting chemicals as modulators of sex steroid synthesis. *Best Practice & Research Clinical Endocrinology & Metabolism* 20, 45–61.
Woods, H.F. (2003) *Phytoestrogens and Health*. COT report: Committee on Toxicity of Chemicals in Food, Consumers Products and the Environment. Food Standards Agency, London.
WHO (2012) *Endocrine Disrupters and Child Health: Possible Developmental Early Effects of Endocrine Disrupters on Child Health*. World Health Organization, Geneva.
Wu, X., Zhang, N. and Lee, M.M. (2012) The influence of endocrine disruptors on male pubertal timing. In: Diamanti-Kandarakis, E. and Gore, A.C. (eds) *Endocrine Disruptors and Puberty*. Humana Press, Springer, New York.
Zhang, L., Dong, L., Ding, S., Qiao, P., Wang, C. *et al.* (2014) Effects of n-butylparaben on steroidogenesis and spermatogenesis through changed E2 levels in male offspring. *Environmental Toxicology and Pharmacology* 37, 705–717.

# 16 Organochlorine Insecticides: Neurotoxicity

**W.M. Caudle***
*Department of Environmental Health, Emory University, Atlanta, Georgia, USA*

## 16.1 Abstract

Organochlorine insecticides have an extensive history in the mitigation and control of insect populations in agricultural as well as residential and personal settings. Their utility was defined by unique characteristics that rendered these compounds highly toxic to insects, coupled with a persistence and chemical stability that allowed them long-lasting effects after application. Unfortunately, these same traits that made them so powerful led to their eventual demise as the health and well-being of the ecosystem, especially the human population, became a critical consequence of organochlorine insecticides. In the past several decades, these health concerns have been repeatedly addressed through population-based as well as laboratory studies that have focused on the neurological effects of chronic exposure. These evaluations have been extended to include pre- and postnatal exposures as these represent critical periods of children's neurodevelopment. The persistence of these chemicals ensures their continued legacy in our environment and our bodies, necessitating a concerted effort to better understand the neurological impact of these exposures.

## 16.2 Introduction

Over the past 80 years, organochlorine insecticides have been a major player in the complex relationship that exists between the environment and human health. Developed at a critical point in our history, organochlorines helped to solidify human dominance over nature. By harnessing chemicals, not only could humans regulate the insect population that served as a vector for disease, but also their dominance extended to the agricultural realm and the preservation of crop yield and a robust agricultural output. However, with time, the benefits of organochlorines began to be questioned, as the health effects at various levels of the ecosystem, especially in the human population, began to command attention. Even as action was taken to manage these results through the regulation of use, the legacy of these compounds persisted, both in the environment and human bodies, initiating a new phase of study of the organochlorine compounds and the health effects that follow continued exposure.

Organochlorine insecticides represent an extensive class of compounds that distinguish themselves from other insecticides by their high degree of chlorine groups attached to the main

* E-mail address: william.m.caudle@emory.edu

chemical backbone (Coats, 1990). These chlorine substitutions also allow for the further stratification of organochlorines into more specific chemical classes, including the chlorinated ethanes (DDT), cyclodienes (endrin, dieldrin, chlordane, heptachlor, aldrin, endosulfan and toxaphene), hexachlorocyclohexanes (lindane) and the organochlorines with caged structures (mirex and chlordecone). In addition to their chemical structure, organochlorine compounds possess an array of other physical features that gives them value as insecticides (Coats, 1990). For example, they are highly lipophilic, which allows them to easily penetrate biological membranes, such as skin and cells in the body. Their affinity for lipids also allows them to migrate to and deposit in higher concentrations in fatty tissues, such as the liver, adipose tissue and the brain (Fleming *et al*., 1994; Corrigan *et al*., 2000). Organochlorine compounds are also very resistant to breakdown or metabolism, both in the environment and in the human body. Taken together, these chemical characteristics of organochlorine insecticides allow them to persist in the environment and human tissue for many years, even after manufacture and use have ceased. Indeed, although the production and use of organochlorine insecticides has been dormant for decades in the USA, we are still able to measure and quantify concentrations of these chemicals in human blood and breast milk (LaKind *et al*., 2004; Shen *et al*., 2007).

In addition to their physiochemical properties, organochlorines are unified by their generalized mechanisms of action on the peripheral and central nervous systems of invertebrates and vertebrates alike, leading to neurotoxicity and death (Casida, 2009). As the normal function of these neuronal circuits is mediated by the integration of electrical signals maintained by the highly regulated flow of ions across the neuronal membrane and chemical signalling at the neuronal synapse, organochlorine insecticides have been explicitly developed to target and disrupt these processes. Consequences of these alterations can manifest in a variety of sensory, motor and cognitive difficulties. The aim of this chapter is to discuss the general characteristics and neurotoxic properties of several organochlorine insecticides. Within this discussion, a greater focus is given to those compounds that have been more extensively studied over the past several decades, providing a much richer appreciation of their neurotoxicological effects in the human population as well as the cellular and molecular pathologies that may underlie these deficits.

## 16.3 Chlorinated Ethane: 1,1,1-trichloro-2,2-*bis* (4-chlorophenyl) ethane (DDT)

By far, the most recognized organochlorine insecticide is DDT. Initially developed in 1874, its utility as an insecticide was not recognized until 1939 when the Swiss scientist, Paul Muller, noted DDT's extraordinary ability to kill insects over an extended period of time following application. These properties thrust DDT to the front lines during World War II, proving to be one of the military's most powerful weapons against insect-borne diseases, such as typhus and malaria, which were decimating troops in Europe and Africa. The popularity of DDT continued after the war, as a means to control mosquito populations as well as protect America's expanding food supply in agricultural settings. Over time, the impact of DDT on the environment and its potential for harm to human health gained greater attention. Rachel Carson's book, *Silent Spring*, highlighted the extensive harmful effects on every level of the ecosystem, leading to the eventual banning of DDT in 1972 for use in the USA (Carson, 1962).

The success of DDT was partly due to its relative lack of acute toxicity to mammalian species, compared with its chemical predecessors (Smith, 2000; Gaines, 1969) (Table 16.1). Indeed, evidence for these claims emerged through a variety of early studies on human subjects, who were exposed to DDT via different routes (dermal, inhalational, oral) for specified periods of time, followed by a battery of neurological assessments (Cameron and Burgess, 1945). For the most part, dermal exposure was very well tolerated and did not result in an adverse neurological effect. In contrast, acute oral exposures were most notably accompanied by hypersensitivity of the mouth and tongue, followed by spontaneous motor movements, muscular hyperexcitability, tremors and general confusion. These neurological signs and symptoms have been attributed to DDT's well-defined mechanism of action that

**Table 16.1.** Acute toxicity (oral $LD_{50}$ in rat) or organochlorine insecticides.

| Organochlorine | Male | Female |
|---|---|---|
| **Chlorinated ethane** | | |
| DDT | 217 | N/A |
| **Cyclodiene** | | |
| Aldrin | 39 | 60 |
| Chlordane | 335 | 460 |
| Dieldrin | 46 | 46 |
| Endosulfan | 43 | 18 |
| Endrin | 18 | 7.5 |
| Heptachlor | 100 | 162 |
| Toxaphene | 90 | 80 |
| **Hexachlorcyclohexane** | | |
| Lindane | 80 | 91 |
| **Caged Structures** | | |
| Chlordecone (Kepone) | 125 | 125 |
| Mirex | 740 | 600 |

specifically targets the function of the sodium channels that reside on neurons. As neuronal communication is defined by the generation of electrical signals that propagate the length of the axon, the highly orchestrated movement of sodium ions through voltage-gated sodium channels embedded in the neuronal membrane underlies the action potential. As sodium channels open and sodium rushes into the axon, a dramatic change in voltage across the membrane forces the sodium channels to close and an opposing opening of potassium channels to release potassium to the extracellular space, effectively re-equilibrating the neuronal environment. Exposure to DDT binds to sodium channels, increasing the time it takes for them to close, which prolongs the activation of the neuron, leading to neuronal hyperexcitability. In addition to sodium channels, DDT also affects the movement of other ions into and out of the neuron through their interaction with $Na^+$, $K^+$, and $Ca^{2+}$-ATPases (Casida, 2009). As these enzymes participate in maintaining the ionic balance across the neuronal membrane, inhibition of these processes by DDT can also contribute to the neuronal excitation and toxicological effects. In addition to ionic disruption, DDT has been shown to affect neurotransmitter signalling. Briefly, DDT exposure has caused elevations in the concentration of glutamate in the brain, as well as alteration to the levels and function of the neurotransmitters serotonin, norepinephrine, and dopamine in different brain regions (Hong *et al.*, 1986). More recent work has focused on elucidating the possible cellular mechanisms that underlie alteration to the dopamine system, following DDT exposure.

The dopamine circuit serves a variety of critical functions in the brain, including aspects of learning and memory and emotion, as well as being a major target for more pathological situations, such as addiction and neurodegenerative diseases, including Parkinson's disease. The neuropathology of Parkinson's disease is well defined, involving a loss of dopamine-producing neurons in the *substantia nigra pars compacta* (SNpc) and a concomitant reduction of dopamine in the striatum, which is heavily innervated by the dopamine projections from the SNpc (Obeso *et al.*, 2017). These losses are accompanied by aggregates of the intracellular protein, alpha-synuclein, which has been highlighted as a premier pathological feature of the disease, following the formation of insoluble fibrils. Loss of dopamine in the nigrostriatal pathway results in hallmark behavioural features, including difficulty walking, deficits in cognitive abilities, depression, as well as other symptoms such as constipation and trouble sleeping.

Experiments pairing an *in vitro* and *in vivo* model system found exposure to DDT, as well as its major metabolites, dichlorodiphenyldichloroethylene (DDE) and dichlorodiphenyldichloroethane (DDD), caused a significant deficit in dopamine handling by the vesicular monoamine transporter 2 (VMAT2) and the dopamine transporter (DAT) (Hatcher *et al.*, 2008) (Table 16.2). Deficits in these functions have been shown to elicit the pathological accumulation of dopamine in the cytosol of the dopamine neurons, leading to oxidation to neurotoxic metabolites and cell death (Caudle *et al.*, 2008). In sum, these findings are of interest as DDT and DDE have been identified in postmortem brain samples and found to be associated with neurodegenerative diseases, such as Parkinson's and Alzheimer's disease (Fleming *et al.*, 1994; Corrigan *et al.*, 2000; Richardson *et al.*, 2014).

The success of DDT was also highlighted by its chemical stability and persistence in the environment, ensuring its continued effectiveness with less reapplication. This continued environmental presence directly translated to a defined persistence within mammalian species

**Table 16.2.** Effects of organochlorine insecticides on dopamine signallling.

| Organochlorine | *In vitro* | *In vivo* | | References |
|---|---|---|---|---|
| | | Adult | Gestational/lactational | |
| DDT | Reduced DAT uptake[a]<br>Reduced VMAT2 uptake[a]<br>Reduced dopamine release[a] | No change in expression of striatal DAT, TH, VMAT2[a]<br>No change in striatal alpha-synuclein[a]<br>No change in striatal dopamine levels[a] | N/A | Hatcher *et al*. (2007)[a] |
| Dieldrin | Increased oxidative stress[d]<br>Increased mitochondrial dysfunction[d]<br>Increased apoptotic mechanisms[d]<br>Increased alpha-synuclein aggregation[d] | Increased oxidative stress in striatum[b]<br>Increased striatal alpha-synuclein expression[b]<br>Reduced striatal DAT expression and function[b]<br>Reduced striatal dopamine metabolites[b]<br>Reductions in dopamine[c,f] | Increased striatal DAT expression in offspring[e]<br>Increased striatal VMAT2 expression in offspring[e]<br>Increased vulnerability to MPTP in male offspring[e] | Hatcher *et al*. (2007)[b]<br>Heinz *et al*. (1980)[c]<br>Kanthasamy *et al*. (2005)[d]<br>Richardson *et al*. (2006)[e]<br>Sharma *et al*. (1976)[f] |
| Endosulfan | Reduction in cell viability[i]<br>Reduction in dopaminergic neuronal morphology[i] | No changes in striatal DAT expression[i]<br>No change in striatal TH expression[i] | Reduced striatal DAT expression in offspring[i]<br>Reduced striatal TH expression in offspring[i]<br>Increased vulnerability to MPTP in offspring[i]<br>Alteration in dopamine metabolism in offspring[h]<br>Reduced cortical DAT, TH, VMAT2 expression in offspring[j]<br>Increased cortical $D_2$ receptor expression in offspring[j]<br>Alteration in dopamine metabolism in offspring[g] | Cabaleiro *et al*. (2008)[g]<br>Lafuente and Pereiro (2013)[h]<br>Wilson *et al*. (2014)[i]<br>Wilson *et al*. (2014)[j] |
| Heptachlor | Increased dopamine release[m]<br>Reduced VMAT2 uptake[n] | Reduced SNpc TH+ neurons[l]<br>Reduced striatal TH expression[l]<br>Increased striatal gliosis[l] | Increased striatal DAT, TH, VMAT2 expression in offspring[k,o]<br>Increased vulnerability to MPTP in offspring[o]<br>No change in striatal dopamine levels in offspring[k] | Caudle *et al*. (2005)[k]<br>Hong *et al*. (2014)[l]<br>Kirby *et al*. (2002)[m]<br>Miller *et al*. (1999)[n]<br>Richardson *et al*. (2008)[o] |

[a, b, c, d, e, f, g, h, i, j, k, l, m, n] References relevant to each statement.

Abbreviations: DAT, dopamine transpsorter; DDT, 1,1,1-trichloro-2,2-*bis*(4-chlorophenyl) ethane; TH, tyrosine hydroxylase; VMAT2, vesicular monoamine transporter 2.

that were exposed to DDT, which further facilitated accumulation of this compound and an elevated body burden, over time. Thus, while the initial exposure to DDT may not have been acutely toxic, the risk for toxicity or an adverse biological response increased with body burden and time. These processes are important to keep in mind, as acute exposures to high concentrations of DDT are rare, especially as the use of DDT has been phased out in many parts of the world. Instead, the focus should be directed towards a situation of a more chronic exposure to low levels of DDT, due to the persistence of this compound in the environment and our bodies. Related to this concern is the relative ease with which DDT can be transported around the body, especially from the mother to the developing fetus, as well as through breast milk. Although these concentrations of DDT are still relatively low, fetal development marks a critical period of neurological development and highlights the vulnerability of the fetus to neurotoxic agents, like DDT (Grandjean and Landrigan, 2014).

More recently epidemiological studies have been conducted in order to assess the impact of exposure to organochlorine insecticides during periods of child development that encompass *in utero* and lactational exposure, including other routes of exposure as infants and children. Many of these studies have considered an occupational exposure scenario, such as a farm worker or pesticide applicator, while others have examined non-occupational or residential exposure routes to the mother and child. These types of non-occupational exposure may represent off-target effects of pesticides, as they drift via air movement from their site of application into the surrounding community. Additionally, these exposures could arise from the transfer or transport of pesticides on agricultural workers' clothing or shoes into the residence. In general, a major focus has been directed towards effects of DDT and some of its metabolites, such as DDE, on neurodevelopment following pre- and postnatal exposure to these compounds. The overall findings from these studies are still controversial, with many finding associations between DDT and DDE concentrations and impairment in multiple neurological end points, including abnormal reflexes, impairments in memory, executive function, and social and attentional processes (Ribas-Fito *et al.*, 2003, 2006; Eskenazi *et al.*, 2006; Engel *et al.*, 2007; Fenster *et al.*, 2007; Torres-Sanchez *et al.*, 2007). For example, a study by Eskenazi *et al.* (2006) measured mean concentrations of DDT and DDE, 22.0 ppb and 1436.9 ppb, respectively, in the serum of pregnant women living in an agricultural setting in California. Alterations in psychomotor and mental development were then assessed in offspring at 6, 12 and 24 months of age. Interestingly, a study in the same cohort did not find an association between DDT or DDE levels and neurodevelopmental effects when children were assessed at 2 months of age (Fenster *et al.*, 2007). A similar longitudinal effect was observed in a cohort of children developmentally exposed to DDT and DDE, in which a negative association between umbilical cord levels of DDE and cognitive function was seen when the children were 12 months old. These neurodevelopmental deficits persisted when the children were evaluated at 48 and 60 months of age (Torres-Sanchez *et al.*, 2013).

These studies are of considerable interest as they represent neurodevelopmental findings following pre- and postnatal exposure to DDT in North America, where the manufacture and use has been banned for many years. This suggests that these findings are related to exposure to residual levels of DDT that persist in the environment, as well as our food chain, presenting a scenario of chronic or continual exposure to DDT. However, in many countries outside the USA, DDT is still readily used to mitigate mosquito-borne illnesses, such as malaria. With this approach, indoor residual spraying (IRS) is employed, which applies DDT to walls and other surfaces inside homes and buildings, killing mosquitoes when they come into contact with these surfaces (Mabaso *et al.*, 2004). While this situation represents a current or recent DDT exposure scenario, very little is known about the potential neurological health effects of this procedure. Recent reports have highlighted the serum levels of DDT and DDE in residents who are exposed to DDT through IRS practices. For example, pregnant women in South Africa who are participating in IRS had serum levels of DDT and DDE measured at 575 ng $g^{-1}$ and 1850 ng $g^{-1}$, respectively (Gaspar *et al.*, 2017). In contrast, pregnant women who were not participating in the IRS programme had levels of DDT and DDE measured at 40 ng $g^{-1}$ and 180 ng $g^{-1}$, respectively. While similar levels have been reported by other studies, concomitant neurological health effects have not been studied. This paucity of neurological data is critical, as maternal levels of DDT and DDE reported to be associated with neurodevelopmental deficits in the USA are lower or within the range of levels observed in African populations participating in IRS.

## 16.4 Cyclodienes and Hexachlorocyclohexanes

In addition to DDT, a variety of other organochlorine insecticides have been produced and used extensively over the past several decades. Although broadly classified as organochlorines, cyclodienes, hexachlorocyclohexanes and organochlorines with caged structures can be differentiated from DDT based upon their specific chemical structures, as well as their mechanisms of neurotoxicity. Of these compounds, the most well studied and extensively used are the cyclodienes, comprising the insecticides aldrin, endrin, dieldrin, chlordane, heptachlor, endosulfan and toxaphene, as well as the hexachlorocyclohexanes, including a variety of isomers such as lindane and β-hexachlorocyclohexane (β-HCH). Like DDT, these insecticides were used extensively in agricultural, residential and personal settings to mitigate pests. In many cases their toxicity was greater than or equal to that of DDT, making them extremely effective in most pest control situations. However, their toxicity, coupled with their persistence in the environment and ability to accumulate, eventually led to their being phased out and banned in the USA. Even though, in most instances, decades have passed since these compounds were widely used, their stability has ensured the persistence of measurable levels of these compounds in the environment and human tissue.

Similar to DDT, cyclodienes and hexachlorocyclohexanes primarily target the central and peripheral nervous systems, inducing a sustained neuronal hyperexcitation that results in a quick onset of convulsions and seizures. These effects are attributed to the ability of cyclodienes to inhibit ion transport by the $Na^+$, $K^+$, $Ca^{2+}$ and $Mg^{2+}$-ATPases, in addition to interfering with the signalling of the main inhibitory neurotransmitter, γ-aminobutryric acid (GABA), in the central and peripheral nervous system. These compounds block the activity of GABA by blocking the influx of chloride through the $GABA_A$ receptor–ionophore complex, effectively removing the inhibitory signalling critical to modulating neuronal activity, resulting in hyperexcitation (Casida, 2009).

Our understanding of the neurological effects of exposure to cyclodienes and hexachlorocyclohexanes continues to evolve. Although an extensive library of epidemiological studies has been conducted on these compounds, the focus has been directed towards their role as risk factors for specific neurodegenerative diseases, such as Parkinson's disease, with only a few recent studies highlighting the potential neurodevelopmental impacts of exposure to these chemicals. Of these compounds, occupational exposures to dieldrin and β-HCH have received the most attention as possible risk factors for the disease (Kamel, 2013). Similar elevations in serum levels of β-HCH have been found in Parkinson's disease patients compared with control as well as individuals diagnosed with Alzheimer's disease (Richardson *et al*., 2009, 2011), demonstrating a link between exposure to these compounds and risk of neurodegenerative disease. More recent work has found exposure to cyclodiene insecticides as risk factors for neurodevelopmental deficits, including autism spectrum disorder (ASD), though these findings remain controversial (Roberts *et al*., 2007).

Given the epidemiological associations between cyclodiene exposure and risk for Parkinson's disease, the majority of laboratory-based studies have focused on the potential cellular and molecular mechanisms by which cyclodienes disrupt the dopamine system, contributing to Parkinson's disease pathogenesis. Furthermore, a hallmark feature of organochlorine insecticides, cyclodienes included, is the ability to transport easily across the placenta, creating an *in utero* exposure model that could potentially impact a variety of neurodevelopmental processes (Polishuk *et al*., 1977). Although Parkinson's disease is generally considered a disease of ageing, efforts have been made to determine the potential role of prenatal exposure on the development and function of the dopamine system. It has been hypothesized that such an exposure may initiate a pathological cascade within the dopamine circuit that is exacerbated over time or is more vulnerable to future insults, chemical or otherwise, that may increase the risk for Parkinson's disease.

### 16.4.1 Dieldrin neurotoxicity

From the 1950s until 1970, dieldrin was extensively used as an insecticide on crops such as corn and cotton. However, due to ecological and human health concerns, the use of dieldrin was phased out in 1970, but re-approved in 1972 for use in residential settings to control termites, where it was used until the manufacturer cancelled the registration in 1987. The neurological effects of dieldrin mirror many of the symptoms reported following exposure to DDT, including rapid onset of convulsions following an acute exposure. Even longer-term exposures to lower concentrations of dieldrin affect the central nervous system, causing headaches, dizziness, muscle twitching and hyperirritability, most likely attributed to dieldrin's disruption of ionic homeostasis and GABAergic signalling (Bloomquist and Soderlund, 1985).

Although dieldrin has been reported to affect different neurotransmitter systems, the effects on the dopamine circuit have received the most attention, in laboratory-based as well as population-based studies. Initially, exposure of mallard ducks to low levels of dieldrin was found to significantly reduce the brain levels of dopamine, as well as serotonin and norepinephrine, but not GABA (Sharma *et al.*, 1976). Similar reductions in dopamine and norepinephrine were reported in ring doves fed low levels of dieldrin (Heinz *et al.*, 1980). As dopamine, norepinephrine and serotonin are part of the monoamine family of neurotransmitters, these effects suggest that dieldrin may be acting on a shared intracellular pathway involved in monoamine transmitter signalling. These findings align with several epidemiological studies that have identified an association between levels of dieldrin found in postmortem brain samples and Parkinson's disease. Indeed, elevated levels of dieldrin were found in specific brain regions, including those uniquely damaged in Parkinson's disease, compared with control patients (Fleming *et al.*, 1994; Corrigan *et al.*, 2000). While measurable levels of DDT and DDE were also found in these samples, dieldrin was the only compound that was significantly correlated with disease.

While these findings provide strong evidence for dieldrin exposure as a risk factor for dopaminergic damage and Parkinson's disease pathology, the specific mechanisms that underlie damage to the dopamine system are currently under investigation and have provided several potential intracellular targets. Work from Hatcher *et al.* (2007) utilized an *in vivo* model system to directly evaluate the effect of dieldrin exposure on the nigrostriatal dopamine circuit. While dopamine levels in the striatum remained unchanged and the integrity of the dopamine neurons in the SNpc was maintained, dieldrin was found to elicit a significant dampening of antioxidant capacity and a concomitant increase in oxidative stress in dopaminergic regions on the brain. These findings are of interest, as elevations in oxidative stress and damage in the nigrostriatal dopamine circuit are a hallmark pathological feature of Parkinson's disease. These findings, as well as a much richer understanding of the intracellular mechanisms involved in dieldrin-induced dopaminergic damage, have been reported by the Kanthasamy research group, as well as others. In addition to demonstrating similar alterations to the redox status of dopaminergic neurons, a series of elegant studies have been performed that significantly extended these findings to implicate deficits in mitochondrial function, alpha-synuclein fibril formation and apoptotic signalling, as well as epigenetic modifications as critical contributors to dieldrin-induced pathogenesis in Parkinson's disease (Kanthasamy *et al.*, 2005).

As discussed above, Parkinson's disease is considered a disease of ageing, generally manifesting in the sixth decade of life, and increasing in incidence as we age. However, little is known about the pathological cascade or timeline leading up to the loss of dopamine and the first clinical indications of the disease, whether it is initiated just prior to disease onset or whether an insult that occurred much earlier in life could initiate a pathological cascade, making the dopamine system vulnerable to future insults. To address this question, female mice were exposed to dieldrin at 0.3 ppm, 1.0 ppm, or 3.0 ppm throughout gestation and lactation. Evaluation of male and female offspring for specific dopaminergic proteins found significant elevations in the expression of the DAT and VMAT2 in the striatum of offspring. These elevations appeared to be aligned with similar increases in the transcription factors Nurr1 and Pitx3, which are critically involved in the development of DAT and VMAT2 proteins (Richardson *et al.*, 2006). The molecular mechanisms that underlie these alterations are unclear; however, previous work has shown that GABA serves as a trophic signal during development of neurotransmitter systems, especially the dopamine system, which could be disrupted by inhibition of GABAergic signalling by dieldrin, as well as other cyclodienes (Liu *et al.*, 1997).

Extending the initial findings, when a subset of animals developmentally exposed to dieldrin were challenged with the dopaminergic neurotoxin 1-methyl-4-phenyl-1,2,3,6-tetrahydropyridine (MPTP), male mice demonstrated more robust damage to the dopamine system, compared with female offspring. These findings suggest that exposure to dieldrin during critical periods of neurodevelopment can have detrimental effects on the developing dopamine system and render it more vulnerable to additional insults that may occur later in life. Indeed, the potential role of early life exposures to various environmental toxicants as a contributing factor in Parkinson's disease pathogenesis has been previously shown (Ling *et al.*, 2002; Thiruchelvam *et al.*, 2002; Barlow *et al.*, 2004; Richardson *et al.*, 2006, 2008). The implications of these findings are important considering that we are continually exposed to a number of chemical toxicants, in addition to organochlorine pesticides, that could cause further damage to the dopamine system and increase the risk for developing neurological diseases such as Parkinson's disease.

### 16.4.2 Heptachlor neurotoxicity

From the 1950s until it was banned in 1974, exposure to the cyclodiene heptachlor occurred most often through its use in agricultural settings to control insects on seed grains and crops, as well as through residential uses to kill termites and fire ants. Although heptachlor is highly neurotoxic, it is rapidly metabolized via an epoxidation reaction to form heptachlor epoxide, which has been shown to be more toxic than the parent compound (Tashiro and Matsumura, 1978). Similar to other organochlorines, exposure to elevated levels of heptachlor manifests in several central nervous system symptoms related to muscle control, such as muscle tremors, convulsions and irritability (Dadey and Kammer, 1953). Aside from these findings following acute exposure, laboratory-based studies have provided some insight into the potential implications of heptachlor exposure on dopaminergic function. A study by Hong *et al.* (2014) administered heptachlor at 7 mg $kg^{-1}$ twice a week for 8 weeks to adult male mice. With this exposure paradigm, they measured a reduction in the number of dopamine-producing neurons in the SNpc as well as a concurrent reduction in striatal tyrosine hydroxylase (TH), which is involved in dopamine synthesis. These alterations appeared to be selective for the dopamine systems, as heptachlor exposure did not disrupt markers of GABAergic neurons in the same brain regions. These findings were accompanied by elevations in markers of astrogliosis and microgliosis in the SNpc, similar to that seen in postmortem Parkinson's disease brains.

The mechanisms that underlie these pathological responses are still unclear. However, focused investigation of heptachlor-induced disruption of specific processes involved in regulating dopamine handling and homeostasis has been assessed using *in vitro* and *in vivo* model systems. Studies of heptachlor as well as heptachlor epoxide have shown that heptachlor selectively releases dopamine, but not serotonin, from isolated synaptosomes. In addition to synaptosomal release, treatment of mice with heptachlor resulted in an increase in the expression of DAT and VMAT2 in the striatum. Although these proteins were elevated, *in vivo* treatment of cells stably expressing DAT or VMAT2 with heptachlor epoxide resulted in inhibition of VMAT2 function, while leaving DAT function relatively spared (Miller *et al.*, 1999; Kirby *et al.*, 2001, 2002). As these transporters are intimately involved in regulating dopamine handling and homeostasis in the intracellular environment, disruption in their function or expression could have significant consequences for the accumulation of neurotoxic species that can damage the dopamine neuron (Caudle *et al.*, 2008).

Although explicit epidemiological data documenting the potential neurotoxicological effects of heptachlor is not widespread, situations of elevated exposure to heptachlor have been documented, raising the concern for potential health effects of these exposures. In the early 1980s, residents of Hawaii were exposed to elevated levels of heptachlor in their milk supply after dairy cattle were fed pineapple green chop that had been contaminated with heptachlor (Baker *et al.*, 1991). This incident raised concern for the developmental consequences of this exposure in the offspring of mothers, who were exposed *in utero* and during breast feeding, and in children who consumed the contaminated milk. Following a similar protocol to that used with dieldrin, female mice were treated with heptachlor at 3 ppm through gestation and lactation. This treatment protocol resulted in an increase in specific dopaminergic proteins DAT and VMAT2 in both male and female offspring at 3 months of age (Caudle *et al.*, 2005). Similar elevations in DAT were observed by other investigators who administered heptachlor at 4.2 mg $kg^{-1}$ and 8.4 mg $kg^{-1}$ during development (Purkerson-Parker *et al.*, 2001). Elaborating upon these data, alterations to these proteins appeared to be occurring at the transcriptional level, as elevations in transcription factors Nurr1 and Pitx3 were also observed (Richardson *et al.*, 2008). These elevations translated to an increased vulnerability of the dopamine circuit following treatment with MPTP, in male offspring developmentally exposed to heptachlor.

### 16.4.3 Endosulfan neurotoxicity

Of all of the organochlorine insecticides, endosulfan was the most recent to have its manufacture and use phased out in 2016, as a result of its acute toxicity and persistence in the environment and human tissue. The extensive duration of use of endosulfan, relative to other organochlorine insecticides, raises important questions regarding the potential current health effects of this compound, as well as longitudinal health effects from persistent exposures to elevated levels in the environment. Similar to other cyclodienes, endosulfan targets the nervous system through disruption of ion transport and GABA signalling, resulting in neuronal hyperexcitation, hyperactivity, convulsions and seizures (Cole *et al.*, 1984).

Of particular concern regarding endosulfan is the contribution of this compound to neurodevelopmental deficits in children whose mothers have been exposed. Work from Moreno Frias *et al.* (2004) and Jimenez Torres *et al.* (2006) found elevated levels of endosulfan in the cord blood and breast milk of pregnant women. These data become more considerable when taken in the context of recent epidemiological studies that have reported associations between endosulfan exposure and Parkinson's disease, as well as incidence of ASD (Roberts *et al.*, 2007; Rhodes *et al.*, 2013). The underlying cellular deficits associated with these disorders are still being evaluated. However, animal studies focused on the neurodevelopmental outcomes of pre- and postnatal endosulfan exposure have identified alterations in the levels of neurotransmitters such as norephinephrine, serotonin and dopamine in specific regions of the brain (Lakshmana and Raju, 1994; Cabaleiro *et al.*, 2008; Lafuente and Pereiro, 2013).

Two recent studies have aimed to further elucidate the cellular processes that may be disrupted following pre- and postnatal exposure to endosulfan, as they relate to Parkinson's disease as well as other neurodevelopmental disorders that may demonstrate neuropathological processes in the frontal cortex, including ASD. In each of these studies, female mice were exposed to 500 ppb endosulfan every other day, beginning prior to pregnancy and continuing throughout gestation and lactation. At 3–4 months of age, male offspring demonstrated significant reductions in the DAT and TH in the striatum, and these reductions were exacerbated when a subset of mice were challenged with the dopaminergic neurotoxin MPTP. Even more intriguing was the finding that exposure to endosulfan during neurodevelopment appeared to be the more important exposure scenario. Indeed, exposure of adult mice to similar levels of endosulfan for 30 days did not cause an alteration to the dopamine neurons. Furthermore, when these animals were challenged with MPTP, they did not show an elevated vulnerability to the toxin (Wilson *et al.*, 2014b). Interestingly, assessment of GABAergic and glutamatergic markers in the striatum did not show any effect following developmental or adult exposure to endosulfan, demonstrating a selective effect on the dopaminergic circuit in this region. As previously seen with other cyclodienes, these findings suggest that developmental exposure to endosulfan could represent the initiation of a pathological cascade that begins during development and persists throughout adulthood. This damage could predispose the dopamine neurons to further damage following exposure to additional chemical insults that may occur in the lifetime.

Using this same exposure scenario, the effects of neurodevelopmental exposure to endosulfan on specific neuronal populations in the frontal cortex of the mouse brain was also explored (Wilson *et al.*, 2014a). This brain region is important as it is highly involved in facilitating learning and memory, aspects of emotion and response to stressful situations, and has been shown to be damaged in diseases such as ASD and schizophrenia, and in general deficits in learning and memory. Indeed, assessment of the dopamine pathway as well as the GABA and glutamate system in this region demonstrated alterations to each of these neurotransmitter circuits, showing similar deficits as those observed in many of the neurological diseases listed above (Seamans and Yang, 2004; Floresco and Magyar, 2006; Gonzalez-Burgos and Lewis, 2012). These findings are important, as many neurological diseases do not have a clearly identified cause. As research progresses, we gain a richer appreciation for the contribution of the environment and how exposure to environmental chemicals such as organochlorines can impact normal neurological function and the risk for disease.

## 16.5 Conclusions and Future Directions

The development of organochlorine insecticides sits at the intersection of industry and environmental health. Their benefit to insect control as a means to mitigate the spread of disease and to protect our agricultural presence is without question. Yet, the irony still exists that the very characteristics that contributed to the utility of these compounds, including their explicit neurotoxic actions and chemical persistence, eventually led to health concerns that would undermine their use. We are still disentangling the health effects of these compounds, with an especially strong focus on defining the neurological impacts. Although most of these chemicals have not been used for almost 50 years, their continued presence in our environment and our bodies ensures that we will carry their effects into the future. With this in mind, a clear path of investigation begins to take shape, as a way of truly understanding the neurotoxicological effects of organochlorine insecticides on the human population. From a population-based perspective, there is a need for the merging of epidemiological and exposure science approaches to cast a wider neurological net, in order to identify other neurological diseases and disorders that have organochlorine exposure as a risk factor. Within this approach, a clearer understanding of the most vulnerable populations would come into focus, allowing for a more defined intervention scenario. These findings can be seamlessly blended with a deeper understanding of the cellular and molecular pathways that are disrupted and underlie neurological disease. Pairing high-throughput '-omics' (transcriptomics, metabolomics, proteomics, epigenomics) platforms as well as leveraging cutting-edge stem cell techniques, such as inducible pluripotent stem cells, to delineate the specific pathways and explore the interaction of genetic predisposition and organochlorine neurotoxicity. The blending of each of these approaches into a multidisciplinary platform would significantly enrich our appreciation of the neurological effects of organochlorine insecticides and allow for additional points of input to resolve their neurotoxicity in the human population.

## References

Baker, D.B., Loo, S. and Barker, J. (1991) Evaluation of human exposure to the heptachlor epoxide contamination of milk in Hawaii. *Hawaii Medical Journal*, 50, 108–112, 118.

Barlow, B.K., Richfield, E.K., Cory-Slechta, D.A. and Thiruchelvam, M. (2004) A fetal risk factor for Parkinson's disease. *Developmental Neuroscience* 26, 11–23.

Bloomquist, J.R. and Soderlund, D.M. (1985) Neurotoxic insecticides inhibit GABA-dependent chloride uptake by mouse brain vesicles. *Biochemical and Biophysical Research Communications* 133, 37–43.
Cabaleiro, T., Caride, A., Romero, A. and Lafuente, A. (2008) Effects of in utero and lactational exposure to endosulfan in prefrontal cortex of male rats. *Toxicology Letters*, 176, 58–67.
Cameron, G.R. and Burgess, F. (1945) The toxicity of DDT. *British Medical Journal*, 1, 865–871.
Carson, R. (1962) *Silent Spring*. Houghton Mifflin Harcourt, New York.
Casida, J.E. (2009) Pest toxicology: the primary mechanisms of pesticide action. *Chemical Research in Toxicology* 22, 609–619.
Caudle, W. M., Richardson, J. R., Wang, M. and Miller, G.W. (2005) Perinatal heptachlor exposure increases expression of presynaptic dopaminergic markers in mouse striatum. *Neurotoxicology* 26, 721–728.
Caudle, W.M., Colebrooke, R.E., Emson, P.C. and Miller, G.W. (2008) Altered vesicular dopamine storage in Parkinson's disease: a premature demise. *Trends in Neuroscience*, 31, 303–308.
Coats, J.R. (1990) Mechanisms of toxic action and structure-activity relationships for organochlorine and synthetic pyrethroid insecticides. *Environmental Health Perspectives* 87, 255–262.
Cole, L.M., Lawrence, L.J. and Casida, J.E. (1984) Similar properties of [35S]t-butylbicyclophosphorothionate receptor and coupled components of the GABA receptor-ionophore complex in brains of human, cow, rat, chicken and fish. *Life Sciences*, 35, 1755–1762.
Corrigan, F.M., Wienburg, C.L., Shore, R.F., Daniel, S.E. and Mann, D. (2000) Organochlorine insecticides in substantia nigra in Parkinson's disease. *Journal of Toxicology and Environmental Health Part A*, 59, 229–234.
Dadey, J.L. and Kammer, A.G. (1953) Chlordane intoxication; report of a case. *Journal of the American Medical Association* 153, 723–725.
Engel, S.M., Berkowitz, G.S., Barr, D.B., Teitelbaum, S.L., Siskind, J. *et al.* (2007) Prenatal organophosphate metabolite and organochlorine levels and performance on the Brazelton Neonatal Behavioral Assessment Scale in a multiethnic pregnancy cohort. *American Journal of Epidemiology* 165, 1397–1404.
Eskenazi, B., Marks, A.R., Bradman, A., Fenster, L., Johnson, C., Barr, D.B. and Jewell, N.P. (2006) In utero exposure to dichlorodiphenyltrichloroethane (DDT) and dichlorodiphenyldichloroethylene (DDE) and neurodevelopment among young Mexican American children. *Pediatrics* 118, 233–241.
Fenster, L., Eskenazi, B., Anderson, M., Bradman, A., Hubbard, A. and Barr, D.B. (2007) In utero exposure to DDT and performance on the Brazelton neonatal behavioral assessment scale. *Neurotoxicology* 28, 471–477.
Fleming, L., Mann, J. B., Bean, J., Briggle, T. and Sanchez-Ramos, J.R. (1994) Parkinson's disease and brain levels of organochlorine pesticides. *Annals of Neurology* 36, 100–103.
Floresco, S.B. and Magyar, O. (2006) Mesocortical dopamine modulation of executive functions: beyond working memory. *Psychopharmacology (Berlin)*, 188, 567–585.
Gaines, T.B. (1969) Acute toxicity of pesticides. *Toxicology and Applied Pharmacology* 14, 515–534.
Gaspar, F.W., Chevrier, J., Quiros-Alcala, L., Lipsitt, J. M., Barr, D.B. *et al.* (2017) Levels and determinants of DDT and DDE exposure in the VHEMBE Cohort. *Environmental Health Perspectives* 125, 077006.
Gonzalez-Burgos, G. and Lewis, D.A. (2012) NMDA receptor hypofunction, parvalbumin-positive neurons, and cortical gamma oscillations in schizophrenia. *Schizophrenia Bulletin* 38, 950–957.
Grandjean, P. and Landrigan, P.J. (2014) Neurobehavioural effects of developmental toxicity. *Lancet Neurology*, 13, 330–338.
Hatcher, J.M., Delea, K.C., Richardson, J.R., Pennell, K.D. and Miller, G.W. (2008) Disruption of dopamine transport by DDT and its metabolites. *Neurotoxicology* 29, 682–690.
Hatcher, J.M., Richardson, J.R., Guillot, T.S., McCormack, A.L., Di Monte, D.A. *et al.* (2007) Dieldrin exposure induces oxidative damage in the mouse nigrostriatal dopamine system. *Experimental Neurology* 204, 619–630.
Heinz, G.H., Hill, E.F. and Contrera, J.F. (1980) Dopamine and norepinephrine depletion in ring doves fed DDE, dieldrin, and Aroclor 1254. *Toxicology and Applied Pharmacology* 53, 75–82.
Hong, J.S., Herr, D.W., Hudson, P.M. and Tilson, H.A. (1986) Neurochemical effects of DDT in rat brain in vivo. *Archives of Toxicology Supplement* 9, 14–26.
Hong, S., Hwang, J., Kim, J.Y., Shin, K.S. and Kang, S.J. (2014) Heptachlor induced nigral dopaminergic neuronal loss and Parkinsonism-like movement deficits in mice. *Experimental Molecular Medicine* 46, e80.
Jimenez Torres, M., Campoy Folgoso, C., Canabate Reche, F., Rivas Velasco, A., Cerrillo Garcia, I., Mariscal Arcas, M. and Olea-Serrano, F. (2006) Organochlorine pesticides in serum and adipose tissue of pregnant women in Southern Spain giving birth by cesarean section. *Science of the Total Environment* 372, 32–38.

Kamel, F. (2013) Epidemiology. Paths from pesticides to Parkinson's. *Science* 341, 722–723.
Kanthasamy, A.G., Kitazawa, M., Kanthasamy, A. and Anantharam, V. (2005) Dieldrin-induced neurotoxicity: relevance to Parkinson's disease pathogenesis. *Neurotoxicology* 26, 701–719.
Kirby, M.L., Barlow, R.L. and Bloomquist, J.R. (2001) Neurotoxicity of the organochlorine insecticide heptachlor to murine striatal dopaminergic pathways. *Toxicological Sciences* 61, 100–106.
Kirby, M.L., Barlow, R.L. and Bloomquist, J.R. (2002) Selective effects of cyclodiene insecticides on dopamine release in mammalian synaptosomes. *Toxicology and Applied Pharmacology* 181, 89–92.
Lafuente, A. and Pereiro, N. (2013) Neurotoxic effects induced by endosulfan exposure during pregnancy and lactation in female and male rat striatum. *Toxicology* 311, 35–40.
LaKind, J.S., Amina Wilkins, A. and Berlin, C.M. Jr (2004) Environmental chemicals in human milk: a review of levels, infant exposures and health, and guidance for future research. *Toxicology and Applied Pharmacology* 198, 184–208.
Lakshmana, M.K. and Raju, T.R. (1994) Endosulfan induces small but significant changes in the levels of noradrenaline, dopamine and serotonin in the developing rat brain and deficits in the operant learning performance. *Toxicology* 91, 139–150.
Ling, Z., Gayle, D.A., Ma, S.Y., Lipton, J.W., Tong, C.W., Hong, J.S. and Carvey, P.M. (2002) In utero bacterial endotoxin exposure causes loss of tyrosine hydroxylase neurons in the postnatal rat midbrain. *Movement Disorders* 17, 116–124.
Liu, J., Morrow, A.L., Devaud, L., Grayson, D.R. and Lauder, J.M. (1997) GABAA receptors mediate trophic effects of GABA on embryonic brainstem monoamine neurons in vitro. *Journal of Neuroscience* 17, 2420–2428.
Mabaso, M.L., Sharp, B. and Lengeler, C. (2004) Historical review of malarial control in southern African with emphasis on the use of indoor residual house-spraying. *Tropical Medicine and International Health* 9, 846–856.
Miller, G.W., Kirby, M.L., Levey, A.I. and Bloomquist, J.R. (1999) Hepachlor alters expression and function of dopamine transporters. *Neurotoxicology* 20, 631–637.
Moreno Frias, M., Jimenez Torres, M., Garrido Frenich, A., Martinez Vidal, J.L., Olea-Serrano, F. and Olea, N. (2004) Determination of organochlorine compounds in human biological samples by GC-MS/MS. *Biomedical Chromatography* 18, 102–111.
Obeso, J. A., Stamelou, M., Goetz, C. G., Poewe, W., Lang, A.E. *et al.* (2017) Past, present, and future of Parkinson's disease: a special essay on the 200th Anniversary of the Shaking Palsy. *Movement Disorders* 32, 1264–1310.
Polishuk, Z.W., Wassermann, D., Wassermann, M., Cucos, S. and Ron, M. (1977) Organochlorine compounds in mother and fetus during labor. *Environtal Research* 13, 278–284.
Purkerson-Parker, S., McDaniel, K.L. and Moser, V.C. (2001) Dopamine transporter binding in the rat striatum is increased by gestational, perinatal, and adolescent exposure to heptachlor. *Toxicological Sciences* 64, 216–223.
Rhodes, S.L., Fitzmaurice, A.G., Cockburn, M., Bronstein, J.M., Sinsheimer, J.S. and Ritz, B. (2013) Pesticides that inhibit the ubiquitin-proteasome system: effect measure modification by genetic variation in SKP1 in Parkinsons disease. *Environmental Research*, 126, 1–8.
Ribas-Fito, N., Cardo, E., Sala, M., Eulalia De Muga, M., Mazon, C., Verdu, A. *et al.* (2003) Breastfeeding, exposure to organochlorine compounds, and neurodevelopment in infants. *Pediatrics* 111, e580–585.
Ribas-Fito, N., Torrent, M., Carrizo, D., Munoz-Ortiz, L., Julvez, J., Grimalt, J.O. and Sunyer, J. (2006) In utero exposure to background concentrations of DDT and cognitive functioning among preschoolers. *American Journal of Epidemiology* 164, 955–962.
Richardson, J.R., Caudle, W.M., Wang, M., Dean, E.D., Pennell, K.D. and Miller, G.W. (2006) Developmental exposure to the pesticide dieldrin alters the dopamine system and increases neurotoxicity in an animal model of Parkinson's disease. *FASEB Journal* 20, 1695–1697.
Richardson, J.R., Caudle, W.M., Wang, M.Z., Dean, E.D., Pennell, K.D. and Miller, G.W. (2008) Developmental heptachlor exposure increases susceptibility of dopamine neurons to N-methyl-4-phenyl-1,2,3,6-tetrahydropyridine (MPTP) in a gender-specific manner. *Neurotoxicology* 29, 855–863.
Richardson, J.R., Shalat, S.L., Buckley, B., Winnik, B., O'Suilleabhain, P. *et al.* (2009) Elevated serum pesticide levels and risk of Parkinson disease. *Archives of Neurology* 66, 870–875.
Richardson, J.R., Roy, A., Shalat, S.L., Buckley, B., Winnik, B. *et al.* (2011) beta-Hexachlorocyclohexane levels in serum and risk of Parkinson's disease. *Neurotoxicology* 32, 640–645.
Richardson, J.R., Roy, A., Shalat, S.L., Von Stein, R.T., Hossain, M.M. *et al.* (2014) Elevated serum pesticide levels and risk for Alzheimer disease. *JAMA Neurology* 71, 284–290.

Roberts, E.M., English, P.B., Grether, J.K., Windham, G.C., Somberg, L. and Wolff, C. (2007) Maternal residence near agricultural pesticide applications and autism spectrum disorders among children in the California Central Valley. *Environmental Health Perspectives* 115, 1482–1489.
Seamans, J.K. and Yang, C.R. (2004) The principal features and mechanisms of dopamine modulation in the prefrontal cortex. *Progress in Neurobiology* 74, 1–58.
Sharma, R.P., Winn, D.S. and Low, J.B. (1976) Toxic, neurochemical and behavioral effects of dieldrin exposure in mallard ducks. *Archives of Environmental Contamination and Toxicology* 5, 43–53.
Shen, H., Main, K.M., Virtanen, H.E., Damggard, I.N., Haavisto, A.M. *et al.* (2007) From mother to child: investigation of prenatal and postnatal exposure to persistent bioaccumulating toxicants using breast milk and placenta biomonitoring. *Chemosphere* 67, S256–262.
Smith, A.G. (2000) How toxic is DDT? *The Lancet* 356, 267–268.
Tashiro, S. and Matsumura, F. (1978) Metabolism of trans-nonachlor and related chlordane components in rat and man. *Archives of Environmental Contamination and Toxicology* 7, 113–127.
Thiruchelvam, M., Richfield, E.K., Goodman, B.M., Baggs, R.B. and Cory-Slechta, D.A. (2002) Developmental exposure to the pesticides paraquat and maneb and the Parkinson's disease phenotype. *Neurotoxicology* 23, 621–633.
Torres-Sanchez, L., Rothenberg, S. J., Schnaas, L., Cebrian, M.E., Osorio, E. *et al.* (2007) In utero p,p'-DDE exposure and infant neurodevelopment: a perinatal cohort in Mexico. *Environmental Health Perspectives* 115, 435–439.
Torres-Sanchez, L., Schnaas, L., Rothenberg, S.J., Cebrian, M.E., Osorio-Valencia, E. *et al.* (2013) Prenatal p,p -DDE exposure and neurodevelopment among children 3.5–5 years of age. *Environmental Health Perspectives* 121, 263–268.
Wilson, W.W., Onyenwe, W., Bradner, J.M., Nennig, S.E. and Caudle, W.M. (2014a) Developmental exposure to the organochlorine insecticide endosulfan alters expression of proteins associated with neurotransmission in the frontal cortex. *Synapse* 68, 485–497.
Wilson, W.W., Shapiro, L.P., Bradner, J.M. and Caudle, W.M. (2014b) Developmental exposure to the organochlorine insecticide endosulfan damages the nigrostriatal dopamine system in male offspring. *Neurotoxicology* 44, 279–287.

# 17 Organophosphates I. Human Health Effects and Implications for the Environment: an Overview

**T. Wille,* H. Thiermann and F. Worek**
*Bundeswehr Institute of Pharmacology and Toxicology, Munich, Germany*

## 17.1 Abstract

Organophosphorus (OP) compounds were developed as pesticides in the 1930s, including the discovery of the even more toxic compounds later called nerve agents. Newspapers in the Western world focus on the individual dissemination of nerve agents during the Iraq or Syrian war. In contrast, more than 100,000 fatalities are due to deliberate intake of or accidental contamination by OP pesticides in developing countries every year, underlining this large public health problem. This chapter gives an overview on human toxicokinetics, toxicodynamics and clinical signs, established and experimental therapeutic approaches and implications for the environment.

## 17.2 Introduction

### 17.2.1 History

Despite first esterification of alcohols resulting in synthesis of phosphoric acid and the synthesis of organophosphorus (OP) compounds in the 19th century (De Clermont, 1855), systematic research on the toxicity of OP started in the mid 1930s with the German scientist Gerhard Schrader being successful in the synthesis of the structurally related OPs paraoxon, parathion, schradan, tabun and sarin (Holmstedt, 1959; Chambers, *et al.*, 2010). Following World War II, a series of aminoethanthiol-derived organophosphates with exceptionally high toxicity were developed and termed V-type agents with VX being discovered in the 1950s followed by VX analogues developed by Russia (VR) and China (CVX) (Table 17.1).

### 17.2.2 Application and field of use

OP pesticides are used in large quantities for pest control and to improve harvest yield worldwide. Table 17.1 shows selected OP pesticides and their WHO classification according to hazard classes. The total number of pesticide poisoning is believed to be > 3 million cases per year that require medical treatment (Jeyaratnam, 1990). The most life-threatening and fatal cases are mainly due to deliberate drinking of large amounts of pesticide solution (Thiermann *et al.*, 1997; Eyer, 2003). Whereas these self-poisoning cases are scarce in Europe (Hrabetz *et al.*, 2013) they accounted for a conservatively estimated 260,000 cases of fatal pesticide poisoning per year worldwide in the period of 1990–2007 and decreased

* E-mail address: TimoWille@Bundeswehr.org

**Table 17.1.** Important OP pesticides according to the WHO recommended classification of pesticides by hazard and nerve agents sorted according to their NATO codes (increasing toxicity of compounds from left to right).

| Pesticides (WHO classification by hazard) | | | | Nerve agents | |
|---|---|---|---|---|---|
| Slightly (III) | Moderately (II) | Highly (Ib) | Extremely (Ia) | G-type | V-type |
| Malathion | Diazinon | Fenamiphos | Parathion | Tabun (GA) | VX |
| Temephos | Chlorpyrifos | Methamidophos | Phorate | Sarin (GB) | VR |
| Fosamine | Dimethoate | Chlorfenvinphos | Mevinphos | Soman (GD) | CVX |
| | Profenofos | Omethoate | Terbufos | Cyclosarin (GF) | |

to 168,000 cases per year in the period of 2006–2015 (Gunnell *et al.*, 2007; Mew *et al.*, 2017). This means that 20–30 % of the world's suicides are due to pesticide poisoning, with a proportion of 50% in Southeast Asia. After deliberate self-poisoning, even in well equipped hospitals the fatality rate of up to 15% is exceptionally high, thereby exceeding the common 1% fatality rate in acute deliberate self-poisoning with drugs in the industrialized world (Hrabetz *et al.*, 2013).

Although insecticides with decreased human toxicity, e.g. neonicotinoides, have been developed in the meantime, OP pesticides remain an important class for pest control and are extensively used in poor and developing countries. As a general rule it can be assumed that the higher the socio-economic status of a country, the higher are the safety precautions regarding distribution and wearing personal protective equipment during application of OP, resulting in low numbers of OP poisoning. However, in developing and emerging countries access to these potentially lethal agents is not well controlled and results in poisoning of agricultural workers, manufacturing workers and children due to inappropriate storage, dissemination and/or insufficient personal protective equipment (O'Malley, 1997). This may result in accidental poisoning mainly via percutaneous exposure during preparation, loading and distribution or after direct skin contact with contaminated areas (Knaak *et al.*, 1993) which mostly results, similar to inhalational uptake after occupational exposure, in mild to moderate poisoning but is rarely life threatening (Cocker *et al.*, 2002).

In the public perception, the military aspect of OP nerve agents is a major focus. This is particularly noticeable when nerve agents were used against the civilian population: the use of sarin by the troops of Saddam Hussein against the Kurdish minority in the Iran–Iraq war in the 1980s (Macilwain, 1993) and the use of self-synthesized sarin by the Aum Shinrikyo cult in Matsumoto and the Tokyo subway in the 1990s (Morita *et al.*, 1995). Recently, more than 1000 civilians died in the Syrian civil war after dissemination of sarin in 2013 and 2017 (Rosman *et al.*, 2014; Dewan and Alkashli, 2017; OPCW, 2017). Quite recently, Novichok nerve agent was used against individuals in the United Kingdom (Vale *et al.*, 2018).

### 17.2.3 Chemistry and nomenclature

The chemical structure of OP pesticides is very heterogeneous. The general formula was first proposed by Schrader in 1937 (Fig. 17.1) and comprises: (i) a P=O group or a P=S group: (ii) a

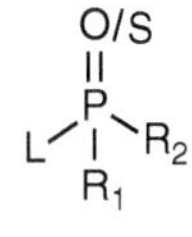


**Fig. 17.1.** Generic structure of OP pesticides and nerve agents. L denotes the leaving group, the most reactive and variable part of OP that is released after phosphylating the target enzyme acetylcholinesterase (EC 3.1.1.7). The residues $R_1$ and $R_2$ are less reactive alkyl, alkoxy or amino groups and usually linked to the central phosphorus atom via an oxygen or a sulfur atom. In phosphates the central phosphorus atom is surrounded by four oxygen atoms; in phosphonates by three oxygen atoms and one carbon atom; and in phosphinates by two oxygen and two carbon atoms (Chambers *et al.*, 2010).

leaving group (L), susceptible to hydrolysis and easily interchangeable by nucleophilic reactants; and (iii) two organic residues ($R_1$ and $R_2$) (Eto, 1977; Schrader, 1950).

## 17.3 Toxicokinetics of Organophosphorus Compounds

The fate of OP within an organism is determined by its absorption, distribution, metabolism and excretion (ADME). OP can be absorbed into the body via inhalation, ingestion or dermal absorption through the skin (Kwong, 2002). Inhalational uptake is usually the most rapid absorption and depends on the volatility of the OP pesticide or nerve agent. Oral ingestion with OP pesticides is mostly accidental with children but usually due to self-poisoning with suicidal intention in adults. Dermal absorption depends mainly on the contact time of the agent, its lipophilicity and the possible presence of a solvent, which can contribute additional toxicity (Vale, 1998; Eddleston *et al.*, 2012). OP rapidly distributes in the body with peak plasma levels after oral intake within hours. Due to its lipophilicity, it has a large volume of distribution and can penetrate the blood–brain barrier. Accumulation typically occurs in fatty tissue and the kidney, liver, lung and brain (Vale, 1998; Timchalk, 2010). Regarding toxicokinetics, the differentiation of OP pesticides into phosphorothioates (P=S) and phosphates (P=O) as shown in Fig. 17.1 is important. Phosphorothioates are more lipophilic than their respective phosphates/oxon metabolites. For parathion this results in an approximately tenfold higher affinity to fat compared with paraoxon (Timchalk, 2010). This may lead to an extensive accumulation of the phosphorothioates in fatty tissue, with prolonged release and repetitive clinical relapses. Major determinants for persistence in the body are the incorporated dose of the OP, which might be $> 100 \times LD_{50}$ in oral mega-dose poisoning after drinking a pesticide solution and the lipophilicity of the pesticide. OP pesticides show a high binding to plasma and tissue proteins ranging from 89–99% in different species with parathion, diazinon and chlorpyrifos (Iverson *et al.*, 1975; Nielsen *et al.*, 1991; Brimer *et al.*, 1994; Wu *et al.*, 1996; Timchalk, 2010).

OP pesticides are subject to extensive metabolism influencing the toxicity in both directions – bioactivation and detoxification. One of the most important reactions of bioactivation is the oxidative desulfuration reaction of phosphorothioates to generate the respective oxon analogues that are inhibitors of human AChE (Forsyth and Chambers, 1989; Vale, 1998; Kwong, 2002; Timchalk, 2010). The reaction requires the most important class of xenobiotic metabolizing enzymes: the cytochrome P450 enzyme family (Brown *et al.*, 2008). These monooxygenases are haemoproteins and catalyse the addition of oxygen with electrons transferred from nicotinamide adenine dinucleotide phosphate (NADPH) thereby releasing the sulfur. Cytochrome P450 enzymes also degrade OP pesticides via a mechanism called dearylation, which is similar to hydrolysis but involving the enzyme, oxygen and NADPH (Chambers *et al.*, 2010). Further oxidizing enzymes are the flavin-containing monooxygenases (Levi and Hodgson, 1992). An additional phase I enzyme involved in the detoxification of OP pesticides is paraoxonase 1 (PON1, A-esterases; EC 3.1.8.1) (Aldridge and Reiner, 1972). PON1 is produced in the liver, attached to high-density lipoproteins, and plays a role in protection against cardiovascular diseases (Mackness *et al.*, 1997). The affinity of PON1 to OP pesticides varies and prefers the oxon analogues of parathion and chlorpyrifos, but the potential implications of wild-type PON1 activity in OP poisoning have not been completely elucidated (Chambers *et al.*, 2010; Androutsopoulos *et al.*, 2011). However, the mutated PON IIG1 variant proved to be beneficial in an *in vivo* model for G-type nerve agent poisoning (Worek *et al.*, 2014).

B-esterases, including the enzymes carboxylesterase (EC 3.1.1.1), acetylcholinesterase (AChE; EC 3.1.1.7) and butyrylcholinesterase (BChE; EC 3.1.1.8) are inhibited by OP and do not significantly contribute to hydrolyzation of these agents (Aldridge, 1953; Ross *et al.*, 2010). As an endogenous stoichiometric bioscavenger without known physiological function, BChE is found in the whole body and with a concentration of ~50 nM in the plasma providing protection for only very low doses of OP (Schallreuter *et al.*, 2007; Masson and Rochu, 2009). As OP binds to albumin – the most abundant protein in blood – a role for this molecule in sequestering

OP is under debate (Masson and Rochu, 2009; Wille *et al.*, 2014).

## 17.4 Toxicodynamics of OP Pesticides and Nerve Agents

After uptake, OP pesticides and nerve agents are distributed within the blood compartment. Here, OP is bound to AChE expressed on the membranes of erythrocytes, plasma BChE and other proteins before OP can reach the target tissues and organs. However, the scavenged amount is in the nanomolar range and too low to effectively counteract distribution. PON1 can attenuate toxicity by hydrolysis of particular pesticides (e.g. paraoxon and chlorpyrifos-oxon) before reaching its target – the AChE in neural tissue. The inhibition of BChE and binding to albumin and other proteins does not result in an additional toxicity. Quite the contrary, concepts to use BChE and A-esterases as therapeutic bioscavengers are under research but have not yet been approved for clinical use (Worek *et al.*, 2014; Masson and Nachon, 2017). Under physiological conditions the neurotransmitter acetylcholine (ACh) binds to the active serine residue 203 of AChE and forms an enzyme intermediate. Subsequently, this intermediate breaks down and releases acetate and choline and thus terminates its action as neurotransmitter (Massoulié *et al.*, 1993). In the presence of OP pesticides and nerve agents, however, the intermediate of enzyme and OP is built and by cleaving the leaving group this active OH-group is phosphylated (i.e. phosphonylation, phosphorylation and phosphinylation), rendering the enzyme in a more stable but inactive form (Fig. 17.2) (Silman *et al.*, 1999). The inhibitory potency (inhibition rate constant $k_i$) of nerve agents is higher than that of pesticides, resulting in per se lower lethal doses and lower toxic concentrations in the body. OP pesticides are categorized into three major hazard classes according to the WHO classification from extremely toxic (e.g. parathion: Ia) to slightly hazardous compounds (malathion: III) (Table 17.1). Additionally, the co-formulated solvents in commercially available pesticides seem to play a role in inducing additional toxicity (Eddleston *et al.*, 2012).

The covalent binding of OP to the serine residue is stable, resulting in a persistent enzyme inhibition, and regeneration of the inhibited enzyme in the absence of free OP may take up to 100 days in the case of AChE attached to red blood cells. Depending on the OP residue covalently bound to AChE, several secondary reactions are possible (Fig. 17.2) (Worek *et al.*, 2004): (i) spontaneous reactivation and regeneration of active AChE, being relevant only in poisoning by low-dose dimethoxy OP pesticides but not in nerve agent poisoning (Worek *et al.*, 2004); (ii) a dealkylation reaction with half-lives ranging from a few seconds for crotylsarin to 40 h with the nerve agent VX, resulting in a further stabilization of the OP-inhibited AChE (so-called 'ageing'); this process prevents the reactivation of the inhibited AChE by therapeutic nucleophiles (Busker *et al.*, 1991; Worek *et al.*, 2004); and (iii) as therapeutic approach nucleophiles, mostly oximes are used to initiate an accelerated enzyme reactivation (Worek and Thiermann, 2013).

## 17.5 Clinical Signs and (Laboratory) Diagnosis of OP Pesticide and Nerve Agent Poisoning

OP pesticides exert their toxic effects by irreversibly binding to the pivotal enzyme AChE. Failure of inhibited AChE to hydrolyse ACh results in an endogenous overflow at muscarinic and nicotinic synapses comprising the central and vegetative nervous system and the neuromuscular junctions in the peripheral nerval system (Kwong, 2002). In an early phase the nicotinic ACh receptors in the sympathetic gangliae are stimulated, which results in a release of catecholamines and subsequent tachycardia and increase of the blood pressure. Thereafter, the muscarinic overstimulation overrides the sympathetic overstimulation, resulting in parasympathetic signs at the target organs. This vagal overstimulation leads to cholinergic crisis with hypersecretion of glands (Salivation, Lacrimation), smooth muscle contraction (miosis, bronchoconstriction), Urination, Diarrhoea, Gastrointestinal signs like abdominal cramps and Emesis: acronym SLUDGE (Table 17.2). Finally, cardiovascular signs such as bradycardia, arrhythmia and hypotension occur. Overstimulation of the perspiratory glands is mediated by the sympathetic nervous system and often observed in

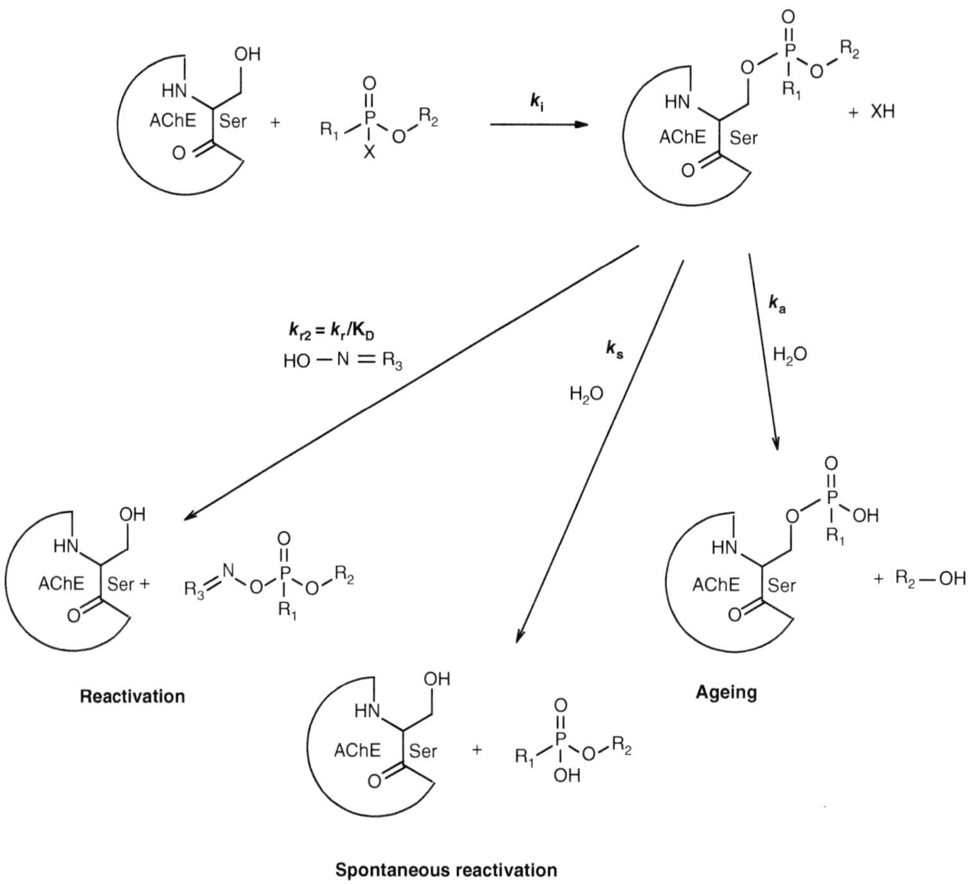


**Fig. 17.2.** Possible reactions of the serine residue (Ser) in the active centre of AChE with organophosphonates, modified according to Millard *et al.* (1999), Worek *et al.* (2004) and Wille (2015). AChE-OH corresponds to the active AChE, HO-N = $R_3$ to the nucleophilic reactivator (oxime). The constant $k_i$ describes the inhibition of AChE by the organophosphate; $k_{r2}$ represents the second order reactivation rate constant by an oxime and is calculated from the quotient of the reactivity constant ($k_r$) and the dissociation constant ($K_D$). Spontaneous reactivation is described with the spontaneous reactivation constant ($k_s$) and the formation of aged enzyme (dealkylation reaction) by the ageing constant ($k_a$).

OP poisoning. Nicotinic signs present at the neuromuscular endplates of skeletal muscles as muscle fasciculations, twitching, cramps and severe muscle dysfunction are due to a depolarization block. Respiratory failure and finally death may occur due to strong secretions in the respiratory system, paralysis of the diaphragm and intercostal muscles and central depression of the respiratory center (Hulse *et al.*, 2014). The onset of signs is dependent on the dose and toxicity of OP, route of exposure and stability within the body. The exposure of mucous membranes by, for example, droplets or volatile exposure will result in a rapid onset of local effects (rhinorrhoea, miosis and eye pain) followed by systemic toxicity. A more delayed onset of clinical signs is expected after percutaneous uptake, but the time interval between first signs and respiratory distress might be short, as first toxic signs occur at 50% inhibition of the target enzyme AChE and life-threatening signs occur after inhibition of > 80% AChE (Kwong, 2002).

This necessitates the identification of early markers or physiological parameters prior to the onset of toxic signs to identify cases of OP poisoning in an early stage and start treatment

**Table 17.2.** Clinical effects of OP exposure. Occurrence of clinical signs is faster after vapour exposure (few minutes) compared with dermal exposure (up to 6 h). Low-dose vapour exposure might result in direct effects of eyes and airways. Bronchial hypersecretion, bronchoconstriction, muscle flaccidity and central respiratory disturbance (*in italic letters*) may ultimately lead to death by hypoxia.

| Muscarinic effects | Nicotinic effects | CNS effects |
|---|---|---|
| Hypersecretion of glands (salivary, lacrimal, nasal, *bronchial*) | Muscle fasciculations, twitching | Seizures |
| Smooth muscle contraction (miosis, *bronchoconstriction*, urination, diarrhoea, emesis) | Muscle weakness | *Central respiratory disturbance* |
| Cardiovascular (bradycardia, arrhythmia, hypotension) | *Muscle flaccidity* | |

before the first clinical signs develop, the so-called 'triggers to treat'. For clinical routine and on-site testing blood samples are a meaningful tool. As the cholinesterases are the primary OP targets, analysis of AChE and BChE activity can be used to support clinical diagnosis, provide evidence in case of untypical signs, exclude exposure to OP pesticides or nerve agents and finally optimize antidotal treatment. This indirect detection of OP exposure via inhibition of ChE activities is far more rapid than direct determination of free OP, its metabolites in blood or plasma, and the respective phosphyl–protein complexes, with these being more suitable for the forensic verification of exposure.

Human AChE is expressed not only in the neural system, but also on the membranes of erythrocytes. AChE is coded by a single gene (Massoulié *et al.*, 1993) and several studies have demonstrated comparable kinetic properties of neural and erythrocyte AChE, indicating that erythrocyte AChE is a reliable surrogate parameter for neural AChE (Eckert *et al.*, 2008; Thiermann *et al.*, 2009). BChE differs from AChE regarding kinetic interactions with OP and oximes. Moreover, BChE activities underlie large inter- and intra-individual variations, which reduces the value of BChE activity determinations in plasma as a diagnostic marker for OP poisoning or therapeutic monitoring (Augustinsson, 1955; Kwong, 2002; Aurbek *et al.*, 2009; Worek *et al.*, 2016). However, numerous test kits for the determination of BChE activity are commercially available and this parameter is widely used in clinical chemistry departments as a global marker for liver function and therefore commonly used for laboratory diagnosis of OP poisoning (Eyer and Worek, 2009). Contrarily, liver diseases and malnutrition might also result in BChE activity reduction in the absence of OP poisoning (Kwong, 2002). Unfortunately, determination of the more valuable AChE activity is not a routine parameter in clinical laboratories. The determination of ChE activity including both the AChE and BChE is based on the Ellman method in most laboratories (Ellman *et al.*, 1961; Worek *et al.*, 1999). Here, acetylthiocholine and butyrylthiocholine are used as substrates for AChE and BChE and the thiocholines react with the dye 5,5′-dithio-*bis*-2-nitrobenzoic acid (DTNB) to yield a yellow product thionitrobenzoate ($TNB^-$) that is monitored photometrically. Inhibition of BChE by ethopropazine or AS1397 allows a highly selective determination of AChE in whole blood. As there are large inter-individual differences in AChE and especially BChE activity, single values can (if not completely out of range) neither confirm nor exclude OP poisoning and advocate the repetitive determination of enzyme activities (Kwong, 2002; Worek *et al.*, 2016). For personnel with an elevated risk of OP exposure, such as OP pesticide applicators or military and NGO personnel deployed to areas with a potential risk of OP exposure, occupational monitoring should be conducted, including determination of basal values (Worek *et al.*, 2016). Meanwhile mobile test kits that allow on-site determination of AChE or BChE activity are commercially available and allow cholinesterase activity determination on site without additional infrastructure. In the case of clear cholinergic signs, therapy should be started immediately and not be delayed by determination of BChE and AChE.

## 17.6 Treatment of OP Pesticide Poisoning

Before entering a medical treatment facility, potentially contaminated clothes should be removed rapidly and sealed in bags. This is a simple but highly effective method for removing external OP (Chilcott, 2014). The skin should then be decontaminated with water and detergents and/or other excipients to reduce OP by physical means or detoxification and to avoid potential cross-contamination of medical personnel and infrastructure. For spot decontamination the Food and Drug Administration (FDA)-approved and European Union (EU) CE-certified reactive skin decontamination lotion (RSDL) is available covering both OP nerve agents and pesticides.

Clinical data on nerve agent-poisoned patients are scarce and the approved treatment regimens are mainly based on animal experiments with a broad range of species. Due to the high number of poisoned patients in developing countries, a large cohort of humans poisoned with pesticides exists. Whereas clinical parameters are well described, in-depth information on the type of pesticide, incorporated dose, starting point of first treatment and scheme of the first emergency treatment is most frequently unknown or not well described. This results in an ongoing debate on the efficacy of the standard treatment protocol, especially the administration of reactivators (oximes). As mentioned above, the clear signs of OP poisoning justify a start of treatment without laboratory confirmation. As in many cases the OP compound is unknown, the initial therapy is standardized and comprises three groups of drugs: (i) the competitive muscarinic receptor antagonist atropine; (ii) an oxime to reactivate the inhibited AChE; and (iii) a benzodiazepine, e.g. diazepam.

Atropine is a competitive antagonist at muscarinic receptors and represents a cornerstone in the successful treatment of life-threatening muscarinic signs. In clinical routine it is mostly used in cardiology to treat bradycardia with a dosing of 0.5–1 mg. Substantially higher doses are necessary for treatment of the cholinergic crisis in OP poisoning and need to be administered rapidly after onset of signs (Grob, 1956; Connors *et al.*, 2014). Therefore, atropine autoinjectors for intramuscular administration have been developed for nerve agent poisoning in military scenarios to allow self and buddy aid in the field. For practical reasons, different formulations with higher atropine concentrations are used in OP poisoning by medical personnel. The regimen established by Eddleston appears currently as the most appropriate one and is based on a dose doubling if signs do not improve (Eddleston *et al.*, 2004). Starting with 2 mg, 4–8–16–32 mg are administered intravenously every 5–10 min. Adequate atropinization is rapidly achieved if the following end points are attained: clear chest during auscultation, heart rate > 80 beats/min, a systolic blood pressure > 80 mm Hg, no pinpoint pupils (miosis), dry axilla (Eddleston *et al.*, 2004). Maintenance dose of atropine in the post-acute phase is usually in the range of 0.5–1 mg $h^{-1}$ and has to be adjusted according to clinical signs. Atropine dosing requires a balance between the desired beneficial and potential side effects known as anticholinergic syndrome.

Whereas atropine is a symptomatic treatment for muscarinic signs, causal treatment of nicotinic signs requires reactivation of OP-inhibited AChE to remove the phosphyl residue from the active serine residue of the AChE (Fig. 17.2). This is mostly achieved by nucleophilic oximes (Fig. 17.2). These antidotes break the covalent bond between the phosphyl residue and AChE, rendering the enzyme in an active form by nucleophilic attack (Eyer, 2003).

Currently four oximes are marketed worldwide. US and British scientists independently developed pralidoxime in the 1950s and it is still the standard antidote in military and civilian institutions in the USA, UK, France and many other countries (Childs *et al.*, 1955; Ginsburg and Wilson, 1957). Trimedoxime (TMB-4) was synthesized in 1958 but appears to cause dangerous adverse side effects and is only in use in Israel and some Eastern European countries (Poziomek *et al.*, 1958; Xue *et al.*, 1985). Obidoxime was developed in the early 1960s by Lüttringhaus and Hagedorn and is the standard therapy for OP poisoning in several European countries (Lüttringhaus and Hagedorn, 1964). Obidoxime shows higher reactivating potency *in vitro* compared to the most extensively used pralidoxime in reactivation of AChE inhibited by various pesticides and nerve agents (Worek *et al.*, 2004). In 1968, HI-6 was developed by Stark and Hagedorn and used in OP pesticide poisoning in former Yugoslavia (Stark, 1968; Kusić *et al.*, 1991). Intensive research on HI-6 was initiated after the discovery of weaponized cyclosarin during Gulf War I, as obidoxime and pralidoxime revealed a weak reactivation potency of cyclosarin-inhibited AChE. Although it is currently stockpiled in autoinjectors for military forces in Canada, Sweden and Czech Republic and several European countries are having licensing programs, no HI-6 is at present commercially available. In the past 60 years several thousand oximes were synthesized by various groups. However, no oxime with a broad spectrum covering structurally different OP with superior efficacy in *in vivo* experiments compared to obidoxime and HI-6 was discovered (Worek and Thiermann, 2013).

The reactivation process depends on two reactions that can be quantified by kinetic constants. The affinity of an oxime for a particular OP-inhibited AChE is described by the dissociation constant $K_D$ (e.g. 32 μM for paraoxon-ethyl inhibited AChE with obidoxime but 187 μM with pralidoxime) and the release of active AChE after reactivation is described by the reactivity constant $k_r$ (e.g. 0.81 $min^{-1}$ for paraoxon-ethyl inhibited AChE with obidoxime but 0.17 $min^{-1}$ with pralidoxime). These constants allow calculation of oxime concentrations necessary to reactivate a certain fraction of inhibited AChE. The reactivating potency of an oxime for a particular OP is described by the second-order reactivation constant $k_{r2} = k_r / K_D$ (Fig. 17.2). Model calculations indicate that minimal requirements for a successful oxime therapy comprise a $K_D < 100$ μM and a $k_r > 0.1$ $min^{-1}$ (Worek *et al.*, 2011). Despite fulfilment of these theoretic requirements for pralidoxime in paraoxon-ethyl poisoning that advocate a therapeutic value, a large meta-analysis revealed inefficacy of pralidoxime in pesticide poisoning (Buckley *et al.*, 2011). Albeit its low reactivating potency against the majority of pesticides, especially if compared to obidoxime, this observation may be biased by application of low pralidoxime doses or delayed onset of treatment (Silva *et al.*, 1992; Sungur and Güven, 2001). It has to be stated that obidoxime is a much more potent reactivator in the case of VX, sarin and common pesticides such as dimethoxy- or diethoxy-OP (Worek *et al.*, 2004). Further, obidoxime and trimedoxime are the only oximes that show at least a partial reactivation of tabun-inhibited AChE (Worek *et al.*, 2007; Kassa *et al.*, 2008). The outcome of an oxime therapy is dependent on several factors. First of all, the reactivation of a particular OP with the oxime of choice must be per se possible, i.e. no premature ageing of OP-inhibited AChE. In addition, the body load of the OP is a major determinant, as a high persisting OP concentration results in re-inhibition of initially reactivated AChE, thereby preventing effective net-reactivation. However, ongoing reactivation of (re-)inhibited AChE might prevent AChE from ageing (Fig. 17.2) and could keep the therapeutic window open. Administration of oximes should be maintained for a sufficient period of time, i.e. presence of OP in the body in an effective concentration. As mentioned above, devices for diagnosis and therapeutic monitoring of OP poisoning are commercially available.

As previously proposed protocols for pralidoxime administration were inefficient, current protocols for pralidoxime administration, mostly as its chloride salt, comprise an intravenous 2 g pralidoxime loading dose followed by 0.5 g $h^{-1}$ for a maximum of 6 days (Eddleston *et al.*, 2009). However, although a reactivation of AChE was recorded in dimethyl- and diethyl-OP poisoning in this study, this was not correlated with increased survival compared with the placebo group. In contrast, a study comparing different protocols of pralidoxime administration (Pawar *et al.*, 2006) or a pralidoxime scheme according to plasma cholinesterase showed lower morbidity and mortality (Due, 2014). For obidoxime the data are less controversial and the established therapeutic regime consists of an initial intravenous administration of 250 mg followed by a continuous infusion of 750 mg per 24 h. However, obidoxime is the standard oxime in countries where OP pesticide poisoning is scarce and not used in regions with highest incidence of poisoning.

Another cornerstone in treatment of OP poisoning next to atropine and an oxime is the administration of benzodiazepines. Cholinergic over-excitation results in central seizures (electroencephalographic (EEG) findings) and convulsions (muscle cramps). Benzodiazepines reduce neural damage in nerve agent-poisoned animals (Murphy *et al.*, 1993) and diazepam was used after nerve agent poisoning in the Iran–Iraq war and the sarin attack by the Aum Shinrikyo cult in Matsumoto 1994 and Tokyo 1995 (Newmark, 2004; Yanagisawa *et al.*, 2006). Similarly, patients with OP pesticide poisoning benefited from administration of the benzodiazepines diazepam, lorazepam and midazolam in several case studies and are therefore recommended (Blain, 2011). As most experiences are with diazepam, its intravenous administration of doses ranging from 5 mg to 10 mg in the absence of convulsions, or up to 30–40 mg in the presence of convulsions, is recommended.

New experimental approaches in therapy of OP poisoning focus on the development of (bio-)scavengers and modulation of the nicotinic receptor but are still in the early stage of preclinical studies (Elsinghorst *et al.*, 2013; Nachon *et al.*, 2013; Goldsmith *et al.*, 2016; Tattersall, 2016).

## 17.7 Implications for the Environment

### 17.7.1 General aspects on stability of organophosphates in the environment

Long-lasting pesticides were preferred in the past and allowed pest control for a long time with a single treatment. However, some of these chlorinated hydrocarbon insecticides (e.g. DDT) persisted in the environment and accumulated through food chains (Eto, 1977). OP pesticides are not as stable in the environment and are generally acknowledged as short-lived in the environment and living organisms. Nevertheless, there is a broad range of persistence of OP in nature, dependent on chemical structure and its physicochemical properties, and this may reach from days to a few weeks. Under specific conditions, in cold and slightly acidic groundwater after application on soil, some types of OP are believed to be stable for more than a year (Murty and Ramani, 1992; Ragnarsdottir, 2000). This is of special importance; as although new technologies try to keep pesticide drift to a minimum, substantial losses by wind and airblasts have to be taken into account (Hall and Fox, 1996). Typical reactions to degrade OP are hydrolysis, oxidation and reduction similar to metabolism and excretion in the human or animal body (see Section 17.3). Additionally, photochemical degradation by sunlight and thermal degradation by heat may occur. In general, biodegradability of OP in the environment (soil and water) is correlated with its hydrophilicity. The pH value plays a major role in hydrolysis velocity of OP, i.e. an increase of the pH results in an increased rate of hydrolysis (Akhtar, 1977). Consequently, alkaline solutions such as sodium hydroxide or sodium hypochlorite are used for decontamination purposes.

### 17.7.2 Stability of organophosphates in soil and mammalian and avian toxicity

Stability in soil is, next to moisture, temperature and pH, dependent on the soil type (Eto, 1977). Additionally, type and content of metal ions and microflora catalysing the degradation of OP play a major role. A simple experiment showed that parathion is more active in dry sand compared with wet sand (Harris, 1964). This result may be explained by a higher adsorptive capacity of dry sand. However, in organic moist muck compared with moist sand the inactivation of parathion was even enhanced and proportional to the organic content. In the same year first observations of a microbial breakdown of parathion suggesting a catalytic mechanism were made in agricultural soil samples (Lichtenstein and Schulz, 1964; Munnecke and Hsieh, 1974), and finally led to the identification of the first parathion hydrolase (Serdar *et al.*, 1982; Serdar and Gibson, 1985). Consequently, soil sterilization resulted in a decreased breakdown of OP. Microorganisms in soil can be trained by repetitive challenge with OP resulting in a faster biodegradation, attributed to a competitive advantage over other microorganisms regarding energy sources (Eto, 1977; Singh and Walker, 2006). Until now, numerous OP-hydrolysing enzymes from various species have been identified (Gupta *et al.*, 2011; Nachon *et al.*, 2013; Jacquet *et al.*, 2016). Originally thought of as instruments for OP decontamination purposes in the field (bioremediation) (Singh and Walker, 2006), some of the promising candidate enzymes were developed further and successfully tested as a treatment option in OP poisoning (Worek *et al.*, 2014).

In a study in northern Greece, specimens of dead wild animals were analysed and in 44% the OP compounds methamidophos, parathion-methyl, phorate or fenthion were found (Samouris *et al.*, 2007; Guitart *et al.*, 2010). Other compounds found in dead animal bodies included carbamates, cyanides and rodenticides. In additional studies OP compounds were found in birds of prey (Toutoudaki *et al.*, 2006; Redig and Arent, 2008). In a large study in the USA with 35,000 avian necropsies, 335 cases showed substantial cholinesterase inhibition, with famphur ($n = 59$), diazinon ($n = 40$) and fenthion ($n = 17$) being the most frequent OP compounds (Fleischli *et al.*, 2004).

### 17.7.3 Stability of organophosphates in water and aquatic toxicity

If not degraded by soil, leached OP and runoff from cropland may enter the aquatic environment. Similar to degradation in soil, OP degrades by chemical and microbial pathways. In water the hydroxide ion initiates a nucleophilic attack at the phosphorus, resulting in hydrolysis of the OP. Hydrolysis of OP into less toxic and water-soluble products is the most important pathway in OP degradation. Persistence of OP in water is comparatively short. Nevertheless, the high acute toxicity poses a serious danger for aquatic wildlife and some insects, fish, mussels and snails are used to monitor environmental pollution of water by OP (Van Scoy *et al.*, 2016). Similar to humans, an acute and chronic type of injury are described. Acute poisoning mostly results in inhibition of AChE first described by Weiss (1958). Long-term effects impair physical activity to search for food, elude predators, or chances to find a sexual partner and finally result in mortality (Murty and Ramani, 1992). *Daphnia magna*, a small planktonic crustacean, is a standard model in ecotoxicology and was successfully used as sensor for both OP nerve agents and pesticides (Green *et al.*, 2003; Pérez *et al.*, 2015). In the past few years the zebrafish (*Danio rerio*) has been repetitively used as standard model for OP poisoning and the reader is referred to a recent review by Koenig *et al.* (2016).

## 17.8 Summary and Outlook

OP pesticides still pose a major threat, accounting for more than 100,000 deaths per year, mostly by oral intake in suicidal intention, especially in developing countries. Nerve agents, such as sarin and VX, are highly toxic OP compounds and were repeatedly used during wars and for assassinations. OP poisoning results in inhibition of AChE, with subsequent overstimulation of nicotinic and muscarinic receptors by acetylcholine accumulation, and may finally lead to death by respiratory failure. If typical clinical signs can be observed, medical treatment should be started immediately after decontamination of the patient to avoid cross-contamination of medical personnel. For delayed onset of OP poisoning, mobile test kits are available and can confirm suspected OP poisoning before life-threatening signs present. Unfortunately, the mortality rate of 15% is still high, despite optimized treatment with oxime, atropine, benzodiazepines and supportive care. Novel therapeutic approaches focus on scavenging of OPs before they reach target tissue, i.e. neural AChE, or on attenuation of cholinergic response by direct interaction with nicotinic receptor with antagonists. Although originally developed to replace the environmentally stable organochlorines, OP may also be stable for weeks in water and soil, presenting a danger for birds, mammals, and aquatic organisms.

## References

Akhtar, M.H. (1977) Degradation of tetrachlorvinphos and its major metabolite 2,4,5-trichlorophenacyl chloride in aqueous media. *Journal of Agricultural and Food Chemistry* 25(4), 848–851.

Aldridge, W.N. (1953) Serum esterases. I. Two types of esterase (A and B) hydrolysing p-nitrophenyl acetate, propionate and butyrate, and a method for their determination. *Biochemistry Journal* 53(1), 110–117.

Aldridge, W.N. and Reiner, E. (1972) *Enzyme Inhibitors as Substrates: Interactions of Esterases with Ethers of Organophosphorus and Carbonic Acids*, 1st edn. North-Holland Publishing, Amsterdam.
Androutsopoulos, V.P., Kanavouras, K. and Tsatsakis, A.M. (2011) Role of paraoxonase 1 (PON1) in organophosphate metabolism: implications in neurodegenerative diseases. *Toxicology and Applied Pharmacology* 256(3), 418–424.
Augustinsson, K.B. (1955) The normal variation of human blood cholinesterase activity. *Acta Physiologica Scandinavica* 35(1), 40–52.
Aurbek, N., Thiermann, H., Eyer, F., Eyer, P. and Worek, F. (2009) Suitability of human butyrylcholinesterase as therapeutic marker and pseudo catalytic scavenger in organophosphate poisoning: a kinetic analysis. *Toxicology* 259(3), 133–139.
Blain, P.G. (2011) Organophosphorus poisoning (acute). *BMJ Clinical Evidence* pii:2102.
Brimer, L., Gyrd-Hansen, N. and Rasmussen, F. (1994) Disposition of parathion after dermal application in pigs. *Journal of Veterinary Pharmacology and Therapeutics* 17(4), 304–308.
Brown, C.M., Reisfeld, B. and Mayeno, A.N. (2008) Cytochromes P450: a structure-based summary of biotransformations using representative substrates. *Drug Metabolism Reviews* 40(1), 1–100.
Buckley, N.A., Eddleston, M., Li, Y., Bevan, M. and Robertson, J. (2011) Oximes for acute organophosphate pesticide poisoning. *Cochrane Database Systemic Reviews* 2, p. CD005085.
Busker, R.W., Zijlstra, J.J., van der Wiel, H.J., Melchers, B.P.C. and van Helden, H.P.M. (1991) Organophosphate poisoning: a method to test therapeutic effects of oximes other than acetylcholinesterase activation in the rat. *Toxicology* 69(3), 331–344.
Chambers, H.W., Meek, E.C. and Chambers, J.E. (2010) The metabolism of organophosphorus insecticides. In: Krieger, R. *et al.* (ed.) *Hayes´ Handbook of Pesticide Toxicology*, 3rd edn. Elsevier, Boston, Massachusetts, pp. 1399–1407.
Chilcott, R.P. (2014) Managing mass casualties and decontamination. *Environment International* 72, 37–45.
Childs, A.F., Davies, D.R., Green, A.L. and Rutland, J.P. (1955) The reactivation by oximes and hydroxamic acids of cholinesterase inhibited by organo-phosphorus compounds. *British Journal of Pharmacology and Chemotherapy* 10(4), 462–465.
Cocker, J., Mason, H.J., Garfitt, S.J. and Jones, K. (2002) Biological monitoring of exposure to organophosphate pesticides. *Toxicology Letters* 134(1–3), 97–103.
Connors, N.J., Harnett, Z.H. amd Hoffman, R.S. (2014) Comparison of current recommended regimens of atropinization in organophosphate poisoning. *Journal of Medical Toxicology* 10(2), 143–147.
De Clermont, P. (1855) Mémoire sur les éthers phosphoriques. *Annales de Chimie et de Physique* 44, 330–336.
de Silva, H.J., Wijewickrema, R. and Senanayake, N. (1992) Does pralidoxime affect outcome of management in acute organophosphorus poisoning? *Lancet*, 339(8802), 1136–1138.
Dewan, A. and Alkashli, H. (2017) *Syria Chemical Attack: Authority Finds 'Incontrovertible' Evidence of Sarin*. Available from: http://edition.cnn.com/2017/04/20/middleeast/syria-chemical-attack-sarin-opcw/index.html (accessed 28 June 2019).
Due, P. (2014) Effectiveness of high dose pralidoxime for treatment of organophosphate poisoning. *Asia Pacific Journal of Medical Toxicology*, 3 (3), 97–103.
Eckert, S., Eyer, P., Herkert, N., Bumm, R., Weber, G., Thiermann, H. and Worek, F. (2008) Comparison of the oxime-induced reactivation of erythrocyte and muscle acetylcholinesterase following inhibition by sarin or paraoxon, using a perfusion model for the real-time determination of membrane-bound acetylcholinesterase activity. *Biochemical Pharmacology* 7 (3), 698–703.
Eddleston, M., Buckley, N.A., Checketts, H., Senarathna, L., Mohamed, F., Sheriff, M.H.R. and Dawson, A. (2004) Speed of initial atropinisation in significant organophosphorus pesticide poisoning - a systematic comparison of recommended regimens. *Journal of Toxicology: Clinical Toxicology* 42(6), 865–875.
Eddleston, M., Eyer, P., Worek, F., Juszczak, E., Alder, N. *et al.* (2009) Pralidoxime in acute organophosphorus insecticide poisoning – a randomised controlled trial. *PLoS Medicine*, 6(6), e1000104.
Eddleston, M., Street, J.M., Self, I., Thompson, A., King, T. *et al.* (2012) A role for solvents in the toxicity of agricultural organophosphorus pesticides. *Toxicology* 294, 94–103.
Ellman, G.L., Courtney, K.D., Andres, V. and Feather-Stone, R.M. (1961) A new and rapid colorimetric determination of acetylcholinesterase activity. *Biochemical Pharmacology* 7, 88–95.
Elsinghorst, P.W., Worek, F., Thiermann, H. and Wille, T. (2013) Drug development for the management of organophosphorus poisoning. *Expert Opinion on Drug Discovery* 8(12), 1467–1477.

Eto, M. (1977) Introduction. In: Eto, M. (ed.) *Organophosphorus Pesticides: Organic and Biological Chemistry*. CRC Press, Cleveland, Ohio, pp. 1–15.
Eyer, P. (2003) The role of oximes in the management of organophosphorus pesticide poisoning. *Toxicology Review* 22(3), 165–190.
Eyer, P. and Worek, F. (2009) Cholinesterases. In: *Clinical Toxicological Analysis*. Wiley-VCH, Weinheim, pp. 755–774.
Fleischli, M.A., Franson, J.C., Thomas, N.J., Finley, D.L. and Riley, W. (2004) Avian mortality events in the United States caused by anticholinesterase pesticides: a retrospective summary of National Wildlife Health Center records from 1980 to 2000. *Archives of Environmental Contamination and Toxicology* 46(4), 452–550.
Forsyth, C.S. and Chambers, J.E. (1989) Activation and degradation of the phosphorothionate insecticides parathion and EPN by rat brain. *Biochemical Pharmacology* 38(10), 1597–1603.
Ginsburg, S. and Wilson, I.B. (1957) Oximes of the pyridine series. *Journal of the American Chemical Society* 79(2), 481–485.
Goldsmith, M., Ashani, Y., Margalit, R., Nyska, A., Mirelman, D. and Tawfik, D.S. (2016) A new post-intoxication treatment of paraoxon and parathion poisonings using an evolved PON1 variant and recombinant GOT1. *Chemico-Biological Interactions* 259 (Pt B), 242–251.
Green, U., Kremer, J.H., Zillmer, M. and Moldaenke, C. (2003) Detection of chemical threat agents in drinking water by an early warning real-time biomonitor. *Environmental Toxicology* 18(6), 368–374.
Grob, D. (1956) The manifestations and treatment of poisoning due to nerve gas and other organic phosphate anticholinesterase compounds. *AMA Archives of Internal Medicine* 98(2), 221–239.
Guitart, R., Sachana, M., Caloni, F., Croubels, S., Vandenbroucke, V. and Berny, P. (2010) Animal poisoning in Europe. Part 3: Wildlife. *Veterinary Journal* 183(3), 260–265.
Gunnell, D., Eddleston, M., Phillips, M.R. and Konradsen, F. (2007) The global distribution of fatal pesticide self-poisoning: systematic review. *BMC Public Health* 7, 357.
Gupta, R.D., Goldsmith, M., Ashani, Y., Simo, Y., Mullokandov, G. *et al.* (2011) Directed evolution of hydrolases for prevention of G-type nerve agent intoxication. *Nature Chemical Biology* 7(2), 120–125.
Hall, F.R. and Fox, R.D. (1996) The reduction of pesticide drift. In: Foy, C.L. and Pritchard, D.W. (eds) *Pesticide Formulation and Adjuvant Technology*. CRC Press, Boca Raton, Florida, pp. 209–239.
Harris, C.R. (1964) Influence of soil type and soil moisture on the toxicity of insecticides in soils to insects. *Nature* 202(4933), 724–724.
Holmstedt, B. (1959) Pharmacology of organophosphorus cholinesterase inhibitors. *Pharmacological Reviews* 11, 567–688.
Hrabetz, H., Thiermann, H., Felgenhauer, N., Zilker, T., Haller, B. *et al.* (2013) Organophosphate poisoning in the developed world – a single centre experience from here to the millennium. *Chemico-Biolical Interactions* 206(3), 561–568.
Hulse, E.J., Davies, J.O.J., Simpson, A.J., Sciuto, A.M. and Eddleston, M. (2014) Respiratory complications of organophosphorus nerve agent and insecticide poisoning. Implications for respiratory and critical care. *American Journal of Respiratory and Critical Care Medicine* 190(12), 1342–1354.
Iverson, F., Grant, D.L. and Lacroix, J. (1975) Diazinon metabolism in the dog. *Bulletin of Environmental Contamination and Toxicology* 13(5), 611–618.
Jacquet, P., Daudé, D., Bzdrenga, J., Masson, P., Elias, M. and Chabrière, E. (2016) Current and emerging strategies for organophosphate decontamination: special focus on hyperstable enzymes. *Environmental Science and Pollution Research International* 23(9), 8200–8218.
Jeyaratnam, J. (1990) Acute pesticide poisoning: a major global health problem. *World Health Statistics Quarterly* 43(3), 139–144.
Kassa, J., Karasova, J., Musilek, K. and Kuca, K. (2008) An evaluation of therapeutic and reactivating effects of newly developed oximes (K156, K203) and commonly used oximes (obidoxime, trimedoxime, HI-6) in tabun-poisoned rats and mice. *Toxicology* 243(3), 311–316.
Knaak, J.B., Al-Bayati, M. and Raabe, O.G. (1993) Physiologically based pharmacokinetic modeling to predict tissue dose and cholinesterase inhibition in workers exposed to organphosphorus and carbamate pesticides. In: Wang, R.G.M., Knaak, J.B. and Maibach, H.I. (eds) *Health Risk Assessment: Dermal and Inhalation Exposure and Absorption of Toxicants*. CRC Press, Boca Raton, Florida, pp. 3–29.
Koenig, J.A., Dao, T.L., Kan, R.K. and Shih, T.-M. (2016) Zebrafish as a model for acetylcholinesterase-inhibiting organophosphorus agent exposure and oxime reactivation: zebrafish model of OP exposure and AChE reactivation. *Annals of the New York Academy of Sciences* 1374(1), 68–77.

Kusić, R., Jovanović, D., Randjelović, S., Joksović, D., Todorovic, V. *et al.* (1991) HI-6 in man: efficacy of the oxime in poisoning by organophosphorus insecticides. *Human & Experimental Toxicology* 10(2), 113–118.

Kwong, T.C. (2002) Organophosphate pesticides: biochemistry and clinical toxicology. *Therapeutic Drug Monitoring* 24(1), 144–149.

Levi, P.E. and Hodgson, E. (1992) Metabolism of organophosphorus compounds by flavin-containing monooxygenase. In: Chambers, J.E. and Levi, P.E. (eds) *Organophosphates: Chemistry, Fate and Effects*. Academic Press, San Diego, pp. 141–154.

Lichtenstein, E.P. and Schulz, K.R. (1964) The effects of moisture and microorganisms on the persistence and metabolism of some organophosphorous insecticides in soils, with special emphasis on parathion. *Journal of Economic Entomology* 57(5), 618–627.

Lüttringhaus, A. and Hagedorn, I. (1964) Quartäre Hydroxyiminomethylpyridinium Salze. *Drug Research* 14, 1–5.

Macilwain, C. (1993) Study proves Iraq used nerve gas. *Nature*, 363 (6424), 3.

Mackness, M.I., Arrol, S., Mackness, B. and Durrington, P.N. (1997) Alloenzymes of paraoxonase and effectiveness of high-density lipoproteins in protecting low-density lipoprotein against lipid peroxidation. *Lancet*, 349(9055), 851–852.

Masson, P. and Nachon, F. (2017) Cholinesterase reactivators and bioscavengers for pre- and post-exposure treatments of organophosphorus poisoning. *Journal of Neurochemistry* 142 (Suppl. 2), 26–40.

Masson, P. and Rochu, D. (2009) Catalytic bioscavengers against toxic esters, an alternative approach for prophylaxis and treatments of poisonings. *Acta Naturae*, 1(1), 68–79.

Massoulié, J., Pezzementi, L., Bon, S., Krejci, E. and Vallette, F.M. (1993) Molecular and cellular biology of cholinesterases. *Progress in Neurobiology* 41(1), 31–91.

Mew, E.J., Padmanathan, P., Konradsen, F., Eddleston, M., Chang, S.-S., Phillips, M.R. and Gunnell, D. (2017) The global burden of fatal self-poisoning with pesticides 2006–15: systematic review. *Journal of Affective Disorders* 219, 93–104.

Millard, C.B., Kryger, G., Ordentlich, A., Greenblatt, H.M., Harel, M. and Raves, M.L. (1999) Crystal structures of aged phosphonylated acetylcholinesterase: nerve agent reaction products at the atomic level. *Biochemistry* 38, 7032–7039.

Morita, H., Yanagisawa, N., Nakajima, T., Shimizu, M., Hirabayashi, H. *et al.* (1995) Sarin poisoning in Matsumoto, Japan. *Lancet*, 346(8970), 290–293.

Munnecke, D.M. and Hsieh, D.P. (1974) Microbial decontamination of parathion and p-nitrophenol in aqueous media. *Applied Microbiology* 28(2), 212–217.

Murphy, M.R., Blick, D.W., Dunn, M.A., Fanton, J.W. and Hartgraves, S.L. (1993) Diazepam as a treatment for nerve agent poisoning in primates. *Aviation, Space, and Environmental Medicine* 64(2), 110–115.

Murty, A.S. and Ramani, A.V. (1992) Toxicity of anticholinesterases to aquatic organisms. In: Ballantyne, B. and Marrs, T.C. (eds) *Clinical and Experimental Toxicology of Organophosphates and Carbamates*. Butterworth-Heinemann, Oxford, UK, pp. 305–320.

Nachon, F., Brazzolotto, X., Trovaslet, M. and Masson, P. (2013) Progress in the development of enzyme-based nerve agent bioscavengers. *Chemico-Biological Interactions* 206(3), 536–544.

Newmark, J. (2004) The birth of nerve agent warfare: lessons from Syed Abbas Foroutan. *Neurology* 62(9), 1590–1596.

Nielsen, P., Friis, C., Gyrd-Hansen, N. and Kraul, I. (1991) Disposition of parathion in neonatal and young pigs. *Pharmacology & Toxicology* 69(4), 233–237.

O'Malley, M. (1997) Clinical evaluation of pesticide exposure and poisonings. *Lancet* 349(9059), 1161–1166.

OPCW (2017) *Report of the OPCW Fact-Finding Mission in Syria Regarding an Alleged Incident in Khan Shaykhun, Syrian Arab Republic*. Organisation for the Prohibition of Chemical Weapons Technical Secretariat, The Hague, Netherlands.

Pawar, K.S., Bhoite, R.R., Pillay, C.P., Chavan, S.C., Malshikare, D.S. and Garad, S.G. (2006) Continuous pralidoxime infusion versus repeated bolus injection to treat organophosphorus pesticide poisoning: a randomised controlled trial. *Lancet* 368(9553), 2136–2141.

Pérez, S., Rial, D. and Beiras, R. (2015) Acute toxicity of selected organic pollutants to saltwater (mysid *Siriella armata*) and freshwater (cladoceran *Daphnia magna*) ecotoxicological models. *Ecotoxicology* 24(6), 1229–1238.

Poziomek, E.J., Hackley, B.E. Jr and Steinberg, G.M. (1958) Pyridinium aldoximes. *Journal of Organic Chemistry* 23(5), 714–717.

Ragnarsdottir, K.V. (2000) Environmental fate and toxicology of organophosphate pesticides. *Journal of the Geological Society* 157(4), 859–876.
Redig, P.T. and Arent, L.R. (2008) Raptor toxicology. *Veterinary Clinics of North America: Exotic Animal Practice* 11(2), 261–282, vi.
Rosman, Y., Eisenkraft, A., Milk, N., Shiyovich, A., Ophir, N., Shrot, S., Kreiss, Y. and Kassirer, M. (2014) Lessons learned from the Syrian sarin attack: evaluation of a clinical syndrome through social media. *Annals of Internal Medicine* 160(9), 644–648.
Ross, M.K., Streit, T.M., Herring, K.L. and Xie, S. (2010) Carboxylesterases: dual roles in lipid and pesticide metabolism. *Journal of Pesticide Science* 35(3), 257–264.
Samouris, G., Antoniou, V., Zantopoulos, N. and Ioannidou, M. (2007) Impact of toxic substances on animals of wild fauna in northern Greece. *Journal of Environmental Protection and Ecology* 8(2), 287–291.
Schallreuter, K.U., Gibbons, N.C.J., Elwary, S.M., Parkin, S.M. and Wood, J.M. (2007) Calcium-activated butyrylcholinesterase in human skin protects acetylcholinesterase against suicide inhibition by neurotoxic organophosphates. *Biochemical and Biophysical Research Communications* 355(4), 1069–1074.
Schrader, G. (1950) Organische Phosphor-Verbindungen als neuartige Insektizide (Auszug). *Angewandte Chemie* 62(20), 471–473.
Serdar, C.M. and Gibson, D.T. (1985) Enzymatic hydrolysis of organophosphates: cloning and expression of a parathion hydrolase gene from *Pseudomonas diminuta*. *Nature Biotechnology* 3(6), 567–571.
Serdar, C.M., Gibson, D.T., Munnecke, D.M. and Lancaster, J.H. (1982) Plasmid involvement in parathion hydrolysis by *Pseudomonas diminuta*. *Applied and Environmental Microbiology* 44(1), 246–249.
Silman, I., Millard, C.B., Ordentlich, A., Greenblatt, H.M., Harel, M. *et al.* (1999) A preliminary comparison of structural models for catalytic intermediates of acetylcholinesterase. *Chemico-Biological Interactions* 119–120, 43–52.
Singh, B.K. and Walker, A. (2006) Microbial degradation of organophosphorus compounds. *FEMS Microbiology Reviews* 30(3), 428–471.
Stark, I. (1968) *Versuche zur Darstellung eines LüH6 (Toxogonin) überlegenen Acetylcholinesterasereaktivators*. Albert Ludwigs Universität, Freiburg, Germany
Sungur, M. and Güven, M. (2001) Intensive care management of organophosphate insecticide poisoning. *Critical Care*, 5(4), 211–215.
Tattersall, J.E.H. (2016) Nicotinic receptors as targets for nerve agent therapy. In: Worek, F., Jenner, J. and Thiermann, H. (eds) *Chemical Warfare Toxicology. Volume 2: Management of Poisoning*, 1st edn. Royal Society of Chemistry, Cambridge, UK, pp. 82–119.
Thiermann, H., Mast, U., Klimmek, R., Eyer, P., Hibler, A. *et al.* (1997) Cholinesterase status, pharmacokinetics and laboratory findings during obidoxime therapy in organophosphate poisoned patients. *Human and Experimental Toxicology* 16(8), 473–480.
Thiermann, H., Worek, F., Eyer, P., Eyer, F., Felgenhauer, N. and Zilker, T. (2009) Obidoxime in acute organophosphate poisoning: 2 – PK/PD relationships. *Clinical Toxicology* 47(8), 807–813.
Timchalk, C. (2010) Organophosphorus insecticides pharmakokinetics. In: Krieger, R. *et al.* (ed.) *Hayes' Handbook of Pesticide Toxicology*. 3rd edn. Elsevier Inc., Boston, Massachusetts, pp. 1409–1433.
Toutoudaki, M., Tzatzarakis, M., Christakis, M., Margariti, M., Xirouchakis *et al.* (2006) Detection of organophosphorous pesticides in poisoned birds of prey. *Toxicology Letters* 164, S247–S248.
Vale, J.A. (1998) Toxicokinetic and toxicodynamic aspects of organophosphorus (OP) insecticide poisoning. *Toxicology Letters* 102–103, 649–652.
Vale, J.A., Marrs, T.C. and Maynard, R.L. (2018) Novichok. A murderous nerve agent attack in the UK. *Clinical Toxicology* 56, 1093–1097.
Van Scoy, A., Pennell, A. and Zhang, X. (2016) Environmental fate and toxicology of dimethoate. *Reviews of Environmental Contamination and Toxicology* 237, 53–70.
Weiss, C.M. (1958) The determination of cholinesterase in the brain tissue of three species of freshwater fish and its inactivation in vivo. *Ecology* 39, 194–198.
Wille, T. (2015) Arzneimittelforschung zur Therapie von Organophophatvergiftungen. Habilitation thesis, Technical University, Munich.
Wille, T., Thiermann, H. and Worek, F. (2014) In vitro kinetics of nerve agent degradation by fresh frozen plasma (FFP). *Archives of Toxicology* 88(2), 301–307.
Worek, F. and Thiermann, H. (2013) The value of novel oximes for treatment of poisoning by organophosphorus compounds. *Pharmacology and Therapeutics* 139(2), 249–259.

Worek, F., Mast, U., Kiderlen, D., Diepold, C. and Eyer, P. (1999) Improved determination of acetylcholinesterase activity in human whole blood. *Clinica Chimica Acta*, 288 (1–2), 73–90.

Worek, F., Thiermann, H., Szinicz, L. and Eyer, P. (2004) Kinetic analysis of interactions between human acetylcholinesterase, structurally different organophosphorus compounds and oximes. *Biochemical Pharmacology* 68(11), 2237–2248.

Worek, F., Aurbek, N., Koller, M., Becker, C., Eyer, P. and Thiermann, H. (2007) Kinetic analysis of reactivation and aging of human acetylcholinesterase inhibited by different phosphoramidates. *Biochemical Pharmacology* 73(11), 1807–1817.

Worek, F., Aurbek, N., Wille, T., Eyer, P. and Thiermann, H. (2011) Kinetic prerequisites of oximes as effective reactivators of organophosphate-inhibited acetylcholinesterase: a theoretical approach. *Journal of Enzyme Inhibition and Medicinal Chemistry* 26(3), 303–308.

Worek, F., Seeger, T., Goldsmith, M., Ashani, Y., Leader, H. *et al.* (2014) Efficacy of the rePON1 mutant IIG1 to prevent cyclosarin toxicity in vivo and to detoxify structurally different nerve agents in vitro. *Archives of Toxicology* 88(6), 1257–1266.

Worek, F., Schilha, M., Neumaier, K., Aurbek, N., Wille, T., Thiermann, H. and Kehe, K. (2016) On-site analysis of acetylcholinesterase and butyrylcholinesterase activity with the ChE Check Mobile Test Kit – determination of reference values and their relevance for diagnosis of exposure to organophosphorus compounds. *Toxicology Letters* 249, 22–28.

Wu, H.X., Evreux-Gros, C. and Descotes, J. (1996) Diazinon toxicokinetics, tissue distribution and anticholinesterase activity in the rat. *Biomedical and Environmental Science* 9(4), 359–369.

Xue, S.Z., Ding, X.J. and Ding, Y. (1985) Clinical observation and comparison of the effectiveness of several oxime cholinesterase reactivators. *Scandinavian Journal of Work, Environment & Health* 11 (Suppl 4), 46–48.

Yanagisawa, N., Morita, H. and Nakajima, T. (2006) Sarin experiences in Japan: acute toxicity and long-term effects. *Journal of Neurological Sciences* 249(1), 76–85.

# 18 Organophosphates II. Neurobehavioural Problems Following Low-Level Exposure: Methodological Considerations for Future Research

**S.J. Mackenzie Ross*,[1] and V. Harrison[2]**

[1]*University College London, London, UK;* [2]*The Open University, Milton Keynes, UK*

## 18.1 Abstract

Organophosphate (OP) pesticides are one of the most widely used insecticides in the world and concern about the effects of OPs on human health has been growing as they are increasingly used throughout the world. The neurotoxic effects of high-level acute poisoning are well established, but the possibility that long-term low-level exposure to OPs in doses below that causing acute toxicity causes ill health is controversial. In 2013 we published a systematic review of the literature regarding the neurotoxicity of long-term low-level exposure to organophosphate pesticides in adult populations. In this chapter we summarize and update our previous review, discuss possible reasons for divergent evaluations of the same literature and highlight unanswered questions and methodological considerations that future researchers need to address before firm conclusions can be drawn.

## 18.2 Introduction

Pesticides are widely used in agriculture throughout the world to ensure sufficient food supplies by eliminating a variety of pests, weeds and pathogens that can spread diseases, damage crops and reduce productivity. The US Environmental Protection Agency (EPA) estimates that over 5.2 billion pounds (2.4 million tonnes) of pesticides were used worldwide in 2006 and 2007 and this figure has been steadily increasing since then, exceeding 6 billion pounds (2.7 million tonnes) in 2011 (EPA, 2011, 2017).

Organophosphate (OP) pesticides are one of the most widely used insecticides in the world and are considered by the World Health Organization (WHO) to be one of the most hazardous pesticides to vertebrate animals. They are responsible for poisoning a large number of people worldwide, particularly in developing countries where adequate protective measures may be lacking (De Silva *et al*., 2006; WHO, 1990). OP pesticides kill insects by interfering with nervous system function and are potentially harmful to humans. Concern about the effects of OPs on human health has been growing as they are increasingly used throughout the world for a variety of agricultural, domestic and industrial purposes.

The neurotoxic effects of acute OP poisoning are well established and involve inhibition of the enzyme acetylcholinesterase (AChE) causing

* E-mail address: s.mackenzie-ross@ucl.ac.uk

changes in peripheral, autonomic and central nervous system function (cholinergic crisis). It is estimated that approximately 3–5 million pesticide poisonings occur annually worldwide and pesticide poisoning is around 13 times more likely in developing countries than in highly industrialized ones (WHO, 1990; Rosenstock *et al.*, 1991; Karalliede *et al.*, 2001). This, coupled with the fact that pesticide ingestion is one of the leading causes of suicide, has meant that a significant amount of research has focused on the relationship between high levels of exposure and acute health effects; and there is a general consensus in the literature about its negative effects on physical, neurological and psychiatric functioning.

However, the impact of long-term low-level exposure to OPs on human health remains unclear, as the findings from previous research are inconsistent. Since many more individuals are likely to be at risk this type of exposure, rather than acute poisoning, establishing whether low-level exposure to OPs causes ill health is of the utmost importance. In this chapter we review the available evidence concerning the neurotoxicity of long-term low-level exposure to OPs and highlight the methodological issues that need to be addressed before firm conclusions can be drawn. This chapter provides a brief overview of some of the key studies that have investigated this issue, before considering the findings from recent systematic reviews of the literature.

## 18.3 Overview of Individual Studies

This review will focus on studies that: (i) investigate the neurotoxicity of OP pesticides; (ii) use objective measures of exposure (e.g. biomarkers and/or questionnaires about exposure/work history); (iii) use sensitive, neurobehavioural outcome measures; and (iv) are observational group studies of human adults. Studies concerning children and adolescents do not form part of this review, as developmental issues may complicate interpretation of neurobehavioural data. Studies that include participants with a history of acute poisoning and papers that were not written in English are also excluded.

A variety of different methodologies have been used to investigate whether low-level exposure to OPs has an adverse effect on neurobehavioural function, but most studies utilize one of three approaches: (i) cross-sectional epidemiological studies comparing exposed participants with non-exposed groups; (ii) studies evaluating neurobehavioural function before and/or after a brief period of exposure; and (iii) correlational studies examining the relationship between exposure metrics and health outcomes in a particular occupational groups exposed to OPs. The latter are not included in this review, because correlational studies tell us little about cause-and-effect. Cohort studies, which compare the neuropsychological abilities of exposed and non-exposed groups, are more appropriate for identifying the neurotoxic effects of long-term low-level exposure and are therefore the main focus of this review. Relevant studies published since 1960 were identified using computerized databases, including Medline, Embase and Psychinfo, and findings are summarized in Tables 18.1 and 18.2.

### 18.3.1 Cognitive functioning

As Tables 18.1 illustrates, the literature concerning the neurotoxicity of low-level exposure to OPs is equivocal. Of the 22 studies that met our inclusion/exclusion criteria, 17 found evidence of cognitive deficits associated with chronic OP exposure, while five did not. Studies that evaluated participants before and after a brief period of exposure were less likely to report adverse effects than cohort studies that compared exposed and unexposed groups, which suggests that long-term cumulative exposure may be more harmful than brief periods of exposure. Indeed, two studies claimed that they found no relationship between exposure and cognitive performance using a pre/post design; however, when their exposed participants were compared with controls, they found evidence of cognitive impairment (Maizlish *et al.*, 1987; Daniell *et al.*, 1992).

However, of the studies that found poorer performance in exposed groups, the picture was not entirely clear, as there was considerable variation in the number and type of deficits identified. Some studies found evidence of subtle deficits in only one or two cognitive domains (e.g. Fiedler *et al.*, 1997; London *et al.*, 1997; Steenland *et al.*, 2000; Stephens and Sreenivasan, 2004), whilst others reported more profound deficits across

**Table 18.1.** Summary of studies investigating neurobehavioural function following low-level exposure to organophosphates.

| Author | Study number | Research question | Design | Participants (exposed/ referent) | Job title | Developed/ developing | Exposure measures | Sig. effect? | Summary of findings |
|---|---|---|---|---|---|---|---|---|---|
| Rodnitzky *et al*. (1975) | 1 | NB changes following chronic exposure to OPs | Group comparisons | 23 (12 farmers; 11 applicators)/ 23 not exposed in last 2 weeks only | Pesticide applicators | Developed (USA) | EHQ, AChE | No | No significant differences between groups on cognitive tests; AChE within normal limits, but slightly lower in applicators than controls |
| Maizlish *et al*. (1987) | 2 | NB effects of chronic exposure over a single work shift | Pre/post and group comparisons | 46/56 | Pesticide applicators | Developed (USA) | Urinary metab- olites | No/yes | No negative exposure-related changes in pre/post performance on NB tests. But performance on Symbol-Digit was poorer in exposed group overall |
| Daniell *et al*. (1992) | 3 | NB effects of chronic exposure over a season of spraying (~6 months) | Pre/post and group comparisons | 49/40 | Fruit tree sprayers | Developed (USA) | EHQ, AChE | No/yes | Authors claim no significant decrements in performance were found. But, significantly poorer performance found in applicators on Symbol-Digit when baseline scores taken into account |

*Continued*

**Table 18.1.** Continued.

| Author | Study number | Research question | Design | Participants (exposed/ referent) | Job title | Developed/ developing | Exposure measures | Sig. effect? | Summary of findings |
|---|---|---|---|---|---|---|---|---|---|
| Ames *et al*. (1995) | 4 | LTLL exposure to OPs and NB function; does prevention of acute poisoning prevent chronic ill health | Group comparisons | 45/90 | Pesticide applicators | Developed (USA) | Clinical records | No | No group differences. Authors conclude preventing acute poisoning prevents chronic sequelae |
| Stephens *et al*. (1995)[a] | 5 | LTLL exposure to OPs and NB function | Group comparisons | 146/143 | Sheep dippers | Developed (UK) | EHQ | Yes | Farmers slower than controls on all timed tests, impaired attention but memory intact. Farmers with the highest exposure level performed worst on syntactic reasoning (even after controlling for covariates). Farmers were 50% more vulnerable to psychiatric disorder |
| Fiedler *et al*. (1997)[a] | 6 | Effect of LTLL exposure to OPs on NB function | Group comparisons | 27/42 | Fruit tree sprayers | Developed (USA) | EHQ | Yes: cog No: mood | Fruit farmers had slower simple RT than controls. Within farmers higher related to slower RT. No differences on mood |

| | | | | | | | | | |
|---|---|---|---|---|---|---|---|---|---|
| Cole *et al*. (1997)[a] | 7 | Compared NB performance of farm and non-farm members | Group comparisons | Farm members: 23 consumers, 28 exposed, 123 applicators/72 | Farm members | Developing (Equador) | EHQ, AChE | Yes: cog No: mood | Evidence of detrimental performance on visual-spatial tasks, language and attention tasks. No differences on mood scores |
| London *et al*. (1997) | 8 | Associations between LTLL exposure and adverse effects on vibration sense and NB function | Group comparisons and correlational | 163/84 | Fruit tree sprayers | Developing (South Africa) | EHQ, BuChE | Yes | Small associations found between exposure and Pursuit-Aiming and the Santa Ana (nondominant-hand) subtest |
| Bazylewicz-Walczak *et al*. (1999)[a] | 9 | Behavioural effects of chronic exposure to OPs | Group comparisons and pre/post | 26/25 | Greehouse workers | Developed (Poland) | Air and clothing concentrations | Yes | No change in performance on NB tests pre/post season, but exposure group showed impaired perceptuomotor function and increased anxiety, depression, irritability, fatigue and memory problems on both occasions suggesting cumulative exposure affects NB function more than a single exposure episode |

*Continued*

**Table 18.1.** Continued.

| Author | Study number | Research question | Design | Participants (exposed/ referent) | Job title | Developed/ developing | Exposure measures | Sig. effect? | Summary of findings |
|---|---|---|---|---|---|---|---|---|---|
| Steenland *et al.* (2000)[a] | 10 | Chronic neurological effects of OP exposure | Group comparisons | 191/189 | Pest control | Developed (USA) | EHQ, PON1, Urinary metab-olites | No: cog Yes: mood | Exposed group reported more problems with memory, anxiety, fatigue and strength but few differences on cognitive tests. Exposed group showed impairment on pegboard turning and some postural sway tests but were similar to controls on the other measures. 8 subjects who were acutely exposed had impaired reaction time and continuous performance |
| Srivastava *et al.* (2000) | 11 | Health risks associated with the manufacture of OP | Group comparisons | 59/17 | Manufactur-ers | Developing (India) | EHQ, AChE | Yes | Similar AChE levels in both groups, but exposed had altered reflexes and neurobehavioural deficits, i.e. lower scores on digit span, digit symbol and vigilance task |

| | | | | | | | | | |
|---|---|---|---|---|---|---|---|---|---|
| Salvi *et al.* (2003)[b] | 12 | NB outcomes after 3 months with and without OP exposure | Pre/Post | 37/25 | Tobacco workers | Developing (Brazil) | EHQ, AChE | No: cog yes: mood | AChE levels normal for exposed cohort. No difference between groups on cognitive tests. But higher anxiety and depression rates following recent exposure |
| Stephens & Sreenivasan (2004) | 13 | Effect of LTLL exposure to OPs on NB function | Group comparisons | 37 orchard sprayers/26 pig farmers/31 construction workers | Fruit tree sprayers | Developed (UK) | EHQ | Yes | Orchard workers slower on syntactic reasoning than controls but no relationship with exposure index |
| Roldan-Tapia *et al.* (2005)[a] | 14 | Continuous exposure to OPs (subsymtomatic) and NB effects | Group comparisons | 40/26 | Greehouse workers | Developed (Spain) | BuChE, EHQ | Yes | Association between cumulative exposure and lower performance on verbal memory, visual memory and increased anxiety. Those exposed for more than 10 years also have lower scores on tests of visuo-spatial ability |
| Farahat *et al.* (2003) | 15 | NB effects of pesticide exposure | Group comparisons | 52/50 | Pesticide applicators (PA) | Developing (Egypt) | EHQ, AChE | Yes | PA performed worse than controls on similarities, tests of attention, visual memory and RT, but this did not correlate with AChE levels (so not due to recent exposure) but did correlate with lifetime exposure |

*Continued*

**Table 18.1.** Continued.

| Author | Study number | Research question | Design | Participants (exposed/ referent) | Job title | Developed/ developing | Exposure measures | Sig. effect? | Summary of findings |
|---|---|---|---|---|---|---|---|---|---|
| Roldan-Tapia *et al*. (2006)[a] | 16 | Association between different levels of exposure to OPs and NB function | Group comparisons | 24 acute/40 chronic/26 controls | Greenhouse workers | Developed (Spain) | BuChE, EHQ | Yes | Exposed had reduced visuo-motor, perceptual and constructive abilities, verbal learning, processing speed and increased anxiety. Acutely exposed and those exposed for > 10 yrs had similar profile of deficits. Those exposed for <10 yrs had similar profiles to controls |
| Mackenzie Ross *et al*. (2007)[a] | 17 | Nature and extent of NB problems in farmers who report chronic ill health | Group comparisons | 25/22 | Sheep dippers | Developed (UK) | EHQ | Yes | Exposed had lower scores on tests of mental flexibility, verbal memory; 72% reported anxiety and 76% depression (significantly more than the controls) |
| Mackenzie Ross *et al*. (2010)[a] | 18 | Does LTLL exposure to OPs cause ill health (NB problems) in sheep farmers | Group comparisons | 127/78 | Sheep dippers | Developed (UK) | EHQ, PON1 | Yes | Exposure group performed poorer on tests of response speed, working, verbal and visual memory, mental flexibility and fine motor control; and reported higher levels of anxiety and depression |

| | | | | | | | | | |
|---|---|---|---|---|---|---|---|---|---|
| Rothlein *et al*. (2006) | 19 | Relationship between LL exposure and NB performance | Group comparisons and correlational | 92/45 | Farm workers | Developed (USA) | EHQ, Urinary metab-olites, dust samples | Yes | Higher metabolites associated with poorer performance on digit symbol, a selective attention task, finger tapping and continuous performance. Lower backward digit span in exposed group, and finger tapping for females |
| Starks *et al*. (2012) | 20 | Relationship between high pesticide exposure events (without acute toxicity) and NB outcomes | Group comparisons and correlational | 156 HPEE/537 No HPEE | Pesticide applicators | Developed (USA) | EHQ | Yes | History of HPEE (without acute toxicity) was associated with poorer performance on digit-symbol substitution and Sequences A. No effects were observed on any of the other neuropsychological tests. NB - all participants in the study were pesticide applicators, so all have exposure |

*Continued*

**Table 18.1.** Continued.

| Author | Study number | Research question | Design | Participants (exposed/ referent) | Job title | Developed/ developing | Exposure measures | Sig. effect? | Summary of findings |
|---|---|---|---|---|---|---|---|---|---|
| Malekirad *et al.* (2013)[a] | 21 | OP effects on neurocognitive impairment and health status | Group comparisons | 187/187 | Horticultural workers | Developing (Iran) | EHQ | Yes | Exposed group showed significantly poorer psychomotor speed, attention, verbal memory, nonverbal memory, prospective memory, spatial functioning, and initiative/energy; and had higher anxiety and depression scores |
| Berent *et al.* (2014)[a] | 22 | Compare NB function of exposed and non-exposed workers over 1 year period | Group comparisons and longitudinal | 53/60 | Chemical factory workers | Developed (USA) | BuCHE, AChE and Urinary | No | Exposed group performed better than controls on verbal memory. No other differences found |

Abbreviations: AChE, acetylcholinesterase; BuCHE, butyrylcholinesterase; cog-cognitive tests; HPEE, high pesticide exposure events; LTLL, long-term low-level; EHQ, Exposure history questionnaire; PON1, serum paraoxonase/arylesterase; NB, neurobehavioural; RT, response time.

[a]Included self-report measures of psychiatric function (most commonly mood, anxiety or depression).

[b]Investigated mood disorder using clinical diagnostic interviews.

**Table 18.2.** Summary of the study measures used to assess the different cognitive domains.

| Cognitive domain | Neuropsychological measures | Study number |
|---|---|---|
| Attention | Continuous Performance | 2,3,8,10,20 |
| | Digit Vigilance | 7 |
| | Rapid Visual Information Processing | 22 |
| | Selective Attention Trials | 19 |
| | Sustained Attention | 4 |
| | Vigilance Task | 11 |
| Executive function/mental flexibility/inhibition | CALCAP (Choice) | 18 |
| | Category Search | 5,13 |
| | Sequences B | 20 |
| | Similarities | 7,15,17,18 |
| | Stockings of Cambridge | 22 |
| | Stroop | 6,17,18 |
| | Syntactic Reasoning | 5,13 |
| | Trails B | 6,7,15,17,18 |
| Working memory | Arithmetic (WAIS) | 17 |
| | Digit Span (Backwards) | 4–9,11,14–19 |
| | Letter Number Sequencing | 17 |
| | Manipulating Numbers | 8 |
| | Paced Auditory Serial Addition Test | 15 |
| | WMS (Working Memory Subtests) | 18 |
| Processing speed | CALCAP (Simple) | 18 |
| | Digit Symbol/Symbol digit | 2–11,14–18,20 |
| | Inspection Time | 8 |
| | Letter Cancel | 15 |
| | Reaction Time (Simple) | 1,4–10,13,14,16,19,22 |
| | Sequences A | 20 |
| | Trails A | 6,7,15,17,18 |
| Verbal memory | Associate Learning and Recall | 10 |
| | Auditory Verbal Learning Test | 20 |
| | California Verbal Learning Test | 6 |
| | CogniSyst Story Recall Test | 22 |
| | Digit Span (Forward) | 4–9,11,14–19 |
| | Sentence Repetition | 1 |
| | Serial Digit | 10,19 |
| | Short-Term Memory Scanning | 8 |
| | Story Recall | 15 |
| | Verbal Recall | 1 |
| | WMS (Verbal Subtests) | 18 |
| | Word Learning | 5,13 |
| | Word Span | 12 |
| Verbal abilities/verbal reasoning/language skills | Comprehension (WAIS) | 17,18 |
| | Graded Naming | 6,18 |
| | Information (WAIS) | 6,7,17,18 |
| | National Adult Reading Test | 17 |
| | Picture Completion | 17 |
| | Pointing and Speaking Arrows | 8 |
| | Semantic Fluency | 6 |
| | Token Test | 6 |
| | Verbal Fluency | 17 |
| | Vocabulary (WAIS) | 7,17,18 |

*Continued*

**Table 18.2.** Continued.

| Cognitive domain | Neuropsychological measures | Study number |
|---|---|---|
| Visual memory | AMIPB | 17 |
| | Benton Visual Retention Test | 7–10,14,16 |
| | Face Recognition | 17 |
| | Matching to Sample | 22 |
| | Pattern Memory | 2–4,10 |
| | Visual memory | 5,13 |
| | WMS Visual Subtests | 6,18 |
| Visuo-motor abilities/ coordination/fine motor control | Finger Tapping | 3,4,10,19,20 |
| | Grooved Pegboard | 6,18,20 |
| | Hand–eye Coordination Task | 2–4,6 |
| | Motor Performance Series | 22 |
| | Proprioception | 1 |
| | Pursuit Aiming | 4,7–9,14,16 |
| | Santa Ana Pegboard | 4,7–9,14,16 |
| Visuoperception/visual spatial skills | Benton Visual Form Discrimination | 15 |
| | Block Design (WAIS) | 7,15,17,18 |
| | Line Orientation | 17 |
| | Pattern Comparison | 2,10 |
| | Symbol Search | 17 |
| General IQ or aptitude | Armed Forces Qualifying Test | 2 |
| | Matrix Reasoning | 17 |
| | MMSE | 12 |
| | WAIS-III | 18 |
| | WAIS-R | 17 |
| | Wide Range Achievement Test | 6,22 |
| Other, subjective measures | Subjective Neurocognition Inventory | 21 |
| | Symptoms Questionniare | 9,14,16 |
| Mood | Profile of Mood States | 7,9,10,14,16 |
| | Hospital Anxiety & Depression Scale | 17,18 |
| | Neuropsychiatric Interview | 12 |
| | MMPI-2 | 6 |
| | General Health Questionnaire | 5,21 |
| | Brief Symptom Inventory | 22 |

Abbreviations: CALCAP-California computerized assessment package; MMPI-2 Minnesota multiphasic personality inventory-2; MMSE,mini mental state examiniation; WAIS Wechsler adult intelligence scale; WMS Wechsler memory scale. For further information about the psychometric tests in this table see Lezak *et al.* (2012) and Strauss *et al.* (2006).

several cognitive domains (e.g. Srivastava *et al.*, 2000; Farahat *et al.*, 2003; Roldan-Tapia *et al.*, 2005, 2006; Rothlein *et al.*, 2006; Mackenzie Ross *et al.*, 2010). Furthermore, there is some variability in the cognitive domains reported to be impaired, with seven studies identifying deficits in memory, six finding problems with attention, five with motor skills and four identifying issues with executive functioning and spatial processing. Despite this, some consistency in study findings is also evident, as few studies identified verbal/language impairments or global intellectual decline. A commonly reported finding (in ten of the 22 studies) was a relationship between long-term low-level exposure to OPs and reduced processing speed.

### 18.3.2 Psychiatric symptoms

Psychiatric symptoms have been reported following acute OP poisoning, particularly depression, and this is biologically plausible because animal studies have shown that OPs disrupt neurotransmission in a number of neural pathways involved in mood regulation (e.g. serotoninergic, dopaminergic and noradrenergic pathways) (for review, see Stallones and Beseler, 2016). However, it is unclear whether individuals with a history of low-level exposure to OPs are at increased risk of psychiatric illness, as some studies find evidence of mood disorder following low-level exposure to OPs whilst others do not. For example, 11 of the studies outlined in Table 18.1 (those flagged with 'a') used different self-report and standardized scales to evaluate mood with varying sensitivity and specificity (see Table 18.2). Four found evidence of an association between low-level exposure to OPs and elevated levels of anxiety and depression (Bazylewicz-Walczak *et al.*, 1999; Mackenzie Ross *et al.*, 2007, 2010; Malekriard *et al.*, 2013). Three found this association only existed for anxiety (Roldan-Tapia *et al.*, 2005, 2006; Steenland *et al.*, 2000) and Stephens *et al.* (1995) found that OP-exposed farmers showed greater vulnerability to psychiatric disorder, but did not elaborate further. In contrast, Berent *et al.* (2014), Fiedler *et al.* (1997) and Cole *et al.* (1997) found no mood-related differences between exposed and non-exposed cohorts.

While self-report mood scales are often used in clinical research (due to their ease of use and cost-effectiveness), they are only designed to screen for emotional distress, not to diagnose psychiatric disorder. To date, only three researchers have used strict diagnostic criteria, considered to be the gold standard of psychiatric diagnoses. Salvi *et al.* (2003) assessed 37 tobacco workers immediately after OP use and then again, 3 months following cessation. Diagnoses were made in accordance with the Diagnostic and Statistical Manual of Psychiatric Disorders (DSM-IV). Almost half of their sample met the criteria for a psychiatric disorder (35% anxiety and 21% major depression) when interviewed soon after exposure; but this prevalence declined once exposure ceased, suggesting a strong association between poor mental health and *recent* exposure to pesticides. However, the impact of *long-term* low-level exposure to OPs cannot be determined from this study.

Amr *et al.* (1997) assessed 208 Egyptian pesticide formulators, 172 pesticide applicators and 223 control subjects using DSM-III-R criteria. Psychiatric disorders were more common in exposed subjects, particularly depression and dysthymic disorder, but this study failed to provide enough detailed information about exposure history to confirm the exclusion of individuals with a history of acute poisoning.

In 2016 we investigated whether UK sheep farmers with a history of low-level exposure to OPs were more vulnerable to psychiatric symptoms than unexposed controls (Harrison and Mackenzie Ross, 2016). In-depth exposure information was collected from all participants and any with a history of acute exposure were excluded from the study. Several measures of emotional well-being were used, including self-report measures and structured clinical interviews (based on the DSM-IV). When using self-report scales, a significant effect of exposure group was found, even after controlling for age and psychosocial risk factors such as ill health, pain and stressful life events. However, when mood was evaluated using more stringent criteria, this relationship was only found for anxiety disorder and not depression. This suggests self-report measures may overestimate depressive symptoms in OP-exposed cohorts, so findings from studies that only utilize these measures should be interpreted with caution.

The individual studies described in this chapter differ in terms of methodological quality and study populations, and these factors may explain the variability in study findings. Major differences in study design are apparent, such as examination of different occupational groups with different levels and routes of exposure, from different cultural backgrounds, examined over different time periods (e.g. following a single episode of exposure, several years of exposure or over a lifetime), using a variety of different measures that vary in terms of sensitivity and specificity. To complicate matters further, the sample sizes examined were often relatively small and may have lacked power to detect what might be quite subtle effects of low-level exposure to OPs on neurobehavioural function. Although the majority of studies found an association between long-term low-level exposure and impaired neurobehavioural function, it is unclear which findings are most reliable and valid and should be relied upon.

Systematic reviews (SRs) of the literature (including meta-analyses) are often used to resolve ongoing controversies such as this, because they enable researchers to address some of the methodological problems identified above. SRs allow researchers to explore the similarities and differences between studies and the possible reasons for variation in study findings; they give researchers criteria to rate the methodological quality of individual studies, thus enabling more weight to be given to the findings from studies of higher quality.

Additionally, meta-analysis allows researchers to summarize, quantify and combine the results of different studies, thereby increasing the number of participants, reducing random error, narrowing confidence intervals and increasing statistical power to detect small effects that may be missed by individual studies that are too small to yield valid conclusions (CRD, 2009). It represents each study's findings in the form of effect sizes and gives a more reliable estimate of whether a significant association exists between specified variables than one study alone. Meta-analysis moves discussion away from individual studies towards an overview of a body of literature and is considered to be the method of choice in situations where research findings may be used to inform public policy (CRD, 2009).

## 18.4 Narrative versus Systematic Reviews of the Literature

SRs use explicit protocols to identify, select, analyse and appraise research findings, to reduce bias, increase transparency and ensure that scientifically robust interpretations are made of the available evidence. SR methodology is routinely used in healthcare and social sciences to evaluate study findings that may be used to determine treatment protocols, health and public policy. However, only recently have SR methods been identified as having a potentially useful role to play in chemical risk assessment (CRA). Prior to this narrative reviews (NRs) were used in CRA and range from reflecting the opinion of a single individual, to the consensus view of a small committee or large organization. For example, in 1998, the UK Department of Health commissioned an independent scientific committee (Committee on Toxicity of Chemicals in Food, Consumer Products and the Environment (COT)) to review the available scientific evidence concerning the impact of prolonged or repeated low-level exposure to OPs on human health. COT formed a small working group to undertake this work, but was unable to reach a consensus regarding this issue. COT identified several gaps in knowledge that might be contributing to this uncertainty and recommended that the UK government should commission further research in this area (COT, 1999).

In 2014, COT updated its review, incorporating findings from research studies undertaken in the interim (COT, 2014). This time it concluded that: (i) although there is an excess of neuropsychiatric symptoms in people who have been exposed to low levels of OPs, it is unclear whether this is a consequence of chemical toxicity or other, unrelated psychological mechanisms; and (ii) overall, there is no consistent evidence that low-level exposure to OPs has adverse effects on neuropsychological functioning, but if OPs do cause long-term neuropsychological impairment in the absence of overt poisoning, then the effects, in most cases, must be minor and subtle.

Whilst COT's conclusions appear reassuring, it is difficult to know how much confidence to have in its findings because of the review methodology employed by the working group. COT undertook a narrative review and relied upon a small panel of experts to form an opinion. However, in the past decade the validity and credibility of narrative reviews has been questioned (Whaley *et al.*, 2016; Mackenzie Ross *et al.*, 2016). Many factors can influence the conclusions drawn by experts in these circumstances, such as personal biases and subjectivity in the interpretation of scientific evidence, the political context and agenda within which a review is undertaken, and the influence of dominant personalities on the committee. Additionally, the limits of human memory and decision making can lead some to use heuristics (either consciously or unconsciously) to reduce the effort associated with reviewing large amounts of information, resulting in poorer integration of information, inconsistent treatment of material, and/or examination of fewer alternatives (Shah and Oppenheimer, 2008). These factors can lead to biased interpretations that do not adequately reflect the evidence base. Worryingly, the conclusions reached by COT were the complete opposite of those drawn by several independent research groups who used SR methods to evaluate the same body of literature.

Muñoz-Quezada *et al.* (2016) reviewed 33 English and Spanish language studies and found that over two-thirds showed evidence of a link between chronic occupational exposure to OP pesticides and low neuropsychological performance. While the authors made no attempt to group the results of the studies statistically, they put more weight on the outcomes of studies using objective measures of exposure (such as biomarkers and environmental arrays) and those they classified as 'high quality' using a standardized review protocol. From this, they concluded that increased exposure to OPs (and the associated low-levels of AChE) was likely to result in poorer performance in the following domains: motor speed; motor coordination; visuospatial coordination; memory; processing speed and attention.

A further three papers have been published that review the literature using SR techniques incorporating meta-analysis. The first was conducted by Ismail *et al.* (2012) and included 17 studies (and 21 cohorts). They found that exposed participants showed consistent, significant decrements in neurobehavioural performance across several cognitive domains (attention, visuomotor integration, verbal abstraction and perception). Significant exposure-related decrements were also seen in one of three memory tests, two of five tests of sustained attention and four of seven tests of motor speed. While this review suggests that a range of neurobehavioural deficits are associated with long-term low-level exposure to OPs, there are some problems with the way it was carried out that limit the conclusions that can be drawn. For example, the papers included in the analysis contained both adult and adolescent samples, with differing exposure histories. This has serious implications for the reliability and validity of the fixed-effects model that they used to analyse the data, as this assumes that all included studies have exactly the same underlying 'true' effect. Given the heterogeneity of the study participants (not to mention other differing factors, such as exposure, measures, methodologies etc.), this is unlikely to be the case. Thus, it is difficult to draw firm conclusions from this analysis.

In 2013, we published our own meta-analysis using a random-effects model to assimilate the data from 14 studies investigating the association between exposure to OPs and neurobehavioural impairment (Mackenzie Ross *et al.*, 2013). Only studies using adult populations were considered, and data from more than 1600 participants were aggregated. The meta-analysis showed a small but significant association between exposure to low levels of OP and decrements in cognitive function. Working memory/attention, visual memory, psychomotor speed, executive function and visuospatial ability were compromised, while other cognitive domains, such as language and general knowledge, appeared intact. However, the statistical approach used in this review has limitations. Effect sizes were averaged across different neuropsychological tests, considered to assess the same cognitive domain. It may have been more appropriate to investigate each test separately, as they may differ in terms of sensitivity and specificity. In addition, dose–response relationships were not analysed beyond the exploration of broad differences between exposed and unexposed cohorts.

A recent meta-analysis by Meyer-Baron *et al.* (2015) explicitly sought to address the shortcomings of Ismail *et al.* (2012) and Mackenzie Ross *et al.* (2013). They reviewed 22 studies (including 32 samples; 1758 exposed and 1260 unexposed participants) and looked at the effect of exposure on participants' performance in individual tests. While they included both adult and adolescent samples, the analyses for the different populations were carried out separately. Little consistency was found between the outcomes of adolescent studies, although a negative association was found between exposure and a measure of working memory (digit span backwards). In contrast, chronically exposed adults showed a relatively consistent pattern of significantly poorer performance on tests of memory and attention; and somewhat smaller effects (with few non-significant outcomes) on tests of psychomotor speed. Analysis of dose–response relationships suggested that lower performance was directly related to cumulative exposure.

Despite the different approaches taken by the four research groups employing SR techniques, similar conclusions were reached regarding the neurotoxicity of long-term low-level exposure OPs. All of the reviews conclude that a significant association exists between long-term low-level exposure to OPs and impairments in neurobehavioural function with memory, attention and psychomotor speed appearing uniquely susceptible to neurotoxic damage. This is in complete opposition to the conclusion drawn by COT, who did not utilize SR methodology and failed to incorporate the findings from existing SRs into its review. The next section will explore the methodological issues associated with research in the area, which may explain some of the variation seen in different study findings.

## 18.5 Methodological Issues

### 18.5.1 Exposure assessment

One of the most difficult challenges faced by researchers investigating the relationship between OP exposure and ill health is establishing the most reliable measures of exposure. Tables 18.1 gives an overview of the different approaches used by the studies in this review. In an ideal world, biological monitoring techniques (urine and blood analysis) that can detect the presence of OP markers would provide researchers with an objective measure of exposure. However, OPs are metabolized quickly by the body and so biological monitoring is of limited value in retrospective studies of long-term health effects, as they only provide a measure of recent exposure. They may be useful in prospective studies but unfortunately the costs involved in running longitudinal projects mean that few are commissioned.

Often the most that can be achieved in retrospective studies is a rough estimate of lifetime exposure based on proxy measures (for example, where someone lives, or their occupation) and/or an individual's testimony regarding their exposure history (for example, using exposure history questionnaires (EHQs)). However, given the limits of human memory, information collected in this way may be unreliable and critical exposure data may be missed. Furthermore, it remains unclear what the critical aspects of exposure actually are (i.e. whether it is dose, frequency, intensity or duration of exposure, the particular OP compound an individual is exposed to, or the route of exposure, such as oral/dermal/inhalation) and so critical measurements may be missed in some studies. There is also little agreement in the literature about whether these dimensions of exposure should be treated separately, or be amalgamated into a single metric that captures multiple aspects of exposure simultaneously. Where exposure metrics have been used, they vary enormously in terms of complexity, the variables included and their weightings. While metrics are often considered to be an improvement over simple measures of exposure (for example, ever/never been exposed, or exposure duration), their validity and reliability remain unclear.

An additional complicating factor is the lack of any agreed definition of low-level exposure, beyond classifying it as 'that which does not provoke symptoms of acute toxicity which require medical intervention'. This rather crude definition is likely to encompass a wide range of different populations with different durations, frequencies and modes of exposure, with some occupational groups at one end of the continuum reporting daily exposure for prolonged periods of time (e.g. chemical plant manufacturers and pesticide applicators) and others at the opposite end of the continuum reporting infrequent exposure, maybe as little as twice a year for a couple of days (e.g. many sheep farmers). Furthermore, this definition assumes that individuals can determine whether their symptoms reflect pesticide poisoning; and it does not address the fact that variables other than severity of illness determine whether individuals consult physicians (Pitts and Phillips, 1991).

Finally, few retrospective studies investigate the potential synergistic effects of exposure to more than one chemical substance at a time, a common occurrence in some professions, such as farming. Metabolism of OPs involves a number of enzyme systems, but several of the enzymes involved in metabolizing OPs are involved in metabolizing other substances, which means that their ability to metabolize OPs may be altered if an individual is exposed to compounds that share the same metabolic pathway, such as certain prescribed medicines, or exposure to other industrial chemicals at the same time as OPs (Abou-Donia *et al.*, 1996; Costa and Furlong, 2002; van Himbergen *et al.*, 2008).

### 18.5.2 Vulnerable sub-groups

In the past few decades, sub-groups of individuals at increased risk of neurobehavioural impairment following exposure to OPs have been identified. Clearly individuals in particular occupations such as farming, or those involved in applying, manufacturing or transporting OP pesticides, may be at greater risk than others; but it has become increasingly apparent that children, adolescents and the elderly may be at greater risk of sustaining neurotoxic damage than adults, because the organs involved in metabolizing and excreting toxins (e.g. liver and kidneys) may not be fully developed or may be compromised by ageing and the coexistence of other health conditions. Many adolescents in developing countries work as pesticide applicators; many migrant workers who move to developed countries such as the USA are adolescents who may end up working in the agricultural industry (Rohlman *et al.*, 2007; Meyer-Baron *et al.*, 2015); and some researchers attribute the rise in neurodevelopmental and neurodegenerative disorders to increased exposure to industrial and environmental chemicals (Grandjean and Landrigan, 2014).

Studies have also identified individuals who are at greater risk of developing ill health following exposure to toxic substances because of inter-individual differences in the capacity to metabolize and detoxify certain chemicals; hence an individual's response to exposure may be affected by polymorphisms in genes involved in pesticide metabolism (Cherry *et al.*, 2002; Mackness *et al.*, 2003), which means that level of exposure is not the only biologically critical variable.

### 18.5.3 Developing versus developed countries

Many of the populations included in this review differ in terms of their country of origin. Some of the largest exposure effects were found in studies from developing nations (Amr *et al.*, 1997; Srivastava *et al.*, 2000; Farahat *et al.*, 2010) where not only is daily exposure more frequent and intense, but also workers may not receive pesticide safety training or wear suitable protective clothing (Farahat *et al.*, 2010). Heat and humidity may alter the characteristics and toxicity of chemical products and influence decisions regarding the use of personnel protective clothing. Linguistic differences and possible illiteracy may mean that instructions for use, storage and other health and safety advice are not followed, and economic factors may mean that products that have been banned from other countries due to health and safety concerns may still be in use. This is particularly concerning, because chemical manufacturing is expected to grow fastest in developing countries over the next 5 years, making it imperative that workers, employers, government officials and policy makers are educated about the potential risks involved in working with industrial chemicals.

### 18.5.4 Outcome measures

Although most studies included in this review find that individuals with a history of low-level exposure to OPs show evidence of cognitive impairment, there is considerable variation in the number and type of deficits identified. This may reflect the wide variety of test batteries administered (see Table 18.2), which makes direct comparison of study findings from across the world challenging. Even when similar tests have been used across studies, researchers may have different opinions regarding the cognitive domain that they represent; for example, Digit Symbol tests have been considered a measure of processing speed, memory and executive function. What is needed is an internationally agreed consensus on which tests or test batteries to use to enable data to be amalgamated and compared and common effects identified. The Neurobehavioural Core Test Battery has been proposed by some researchers but may not detect neurotoxic effects among people with low levels of education or among diverse cultures (Anger, 1994).

The variety of different methods used to evaluate mental health may also explain the variance in study findings. Many researchers use self-report symptom questionnaires, with differing degrees of sensitivity and specificity. Whilst they may be cost-effective for screening purposes in large populations, they appear to overestimate the prevalence of psychiatric conditions in comparison with clinical interviews undertaken by mental health professionals using internationally agreed diagnostic criteria for mental disorders. Furthermore, many of these measures will misclassify people with physical health complaints as having a mental health condition, because they incorporate cognitive and somatic symptoms in addition to the psychic manifestations of emotional distress. For example, symptom checklists for depression often include irritability, difficulty concentrating, indecision, sleep disturbance, lack of energy and fatigue; while anxiety measures include references to numbness, tingling, dizziness, palpitations and breathing difficulties. These symptoms are often reported by individuals who have been exposed to OPs (e.g. Ahmed and Davies, 1997; Tahmaz *et al.*, 2003; Mackenzie Ross *et al.*, 2007), making it difficult to know whether they are truly psychiatric in nature, or whether they are non-psychiatric effects of exposure. Future research exploring this issue should aim to use more objective diagnostic measures; and modern imaging techniques such as positron emission tomography (PET) scans and magnetic resonance imaging (MRI) diffuse tensor imaging (DTI) and functional MRI (fMRI) may be a useful addition to future study designs.

### 18.5.5 Confounds

Most of the studies included in this review compared the performance of exposed and unexposed individuals on neurobehavioural measures and did not always take account of all of the factors that might affect the outcomes of both cognitive testing and mood measures, such as age, gender, years of education, alcohol and drug consumption, and stressful life events; so unless these factors are controlled for (either by matching the exposed and non-exposed groups, or by factoring out their effects as part of the analyses, e.g. as covariates), firm conclusions regarding causation cannot be drawn. Many studies failed to consider the possible confounding effects of mood disorder or other health-related factors (e.g. pain and fatigue) known to affect performance on cognitive tests. Those that did (e.g. Mackenzie Ross *et al.*, 2010) found evidence of cognitive impairment in exposed cohorts, even after controlling for the effects of mood. Researchers also need to ensure that they do not control for too many potentially confounding variables at once, as some may be inextricably linked to exposure variables (e.g. age) and statistical control of multiple variables may reduce the likelihood of finding meaningful associations between exposure metrics and neurobehavioural test performance.

## 18.6 Conclusions

The literature regarding the impact of long-term low-level exposure to OPs on neurobehavioural functioning in adult populations is often described as equivocal, but recent systematic reviews of the literature find the majority of well-designed studies report a significant association between long-term low-level exposure to OP pesticides and impaired neurobehavioural function. Cognitive functions such as memory, attention and psychomotor speed appear to be particularly vulnerable to the neurotoxic effects of OPs; and individuals with a history of low-level exposure appear to be at increased risk of developing a mental health condition (particularly anxiety). Thus, it seems reasonable to conclude that long-term low-level exposure to OPs can have an adverse effect on neurobehavioural function. However, research in this area faces a number of methodological challenges, and several questions remain unanswered. Future research will need to address these issues, including the following points.

- The critical exposure variables need to be identified, and more precise information obtained about dose–response relationships and the time course over which neurobehavioural problems may develop. Level of exposure is frequently assumed to be the only biologically critical variable, but this review has identified several other variables that may influence toxicity, which need to be evaluated.
- More rigorous selection of study participants is needed, to ensure that the human health risks of exposure to OPs are not overestimated by accidental inclusion of individuals with a history of acute poisoning.
- The profile of cognitive and emotional sequelae needs further elucidation and could be expedited if researchers could agree on which tests to use; and modern brain imaging techniques may be a useful addition to future study designs.
- When reviewing the literature on the neurotoxicity of OPs, researchers need to group studies according to the populations studied, their countries of origin and the age of participants, as some populations

appear more vulnerable to the neurotoxic effects of OPs.

- Finally, researchers need to investigate the long-term prognosis of those who report ill health following exposure to OPs and determine whether there are any treatment protocols that could ameliorate their symptoms.

## References

Abou-Donia, M., Wilmarth, K.R., Abdel-Rahman, A., Jensen, K.F., Oeheme, F.W. and Kurt, T.L. (1996). Increased neurotoxicity following concurrent exposure to pyridostigmine bromide, DEET, and chlorpyrifos. *Fundamental Applied Toxicology* 34, 201–222.

Ahmed, G.M. and Davies, D.R. (1997) Chronic organophosphate exposure: towards the definition of a neuropsychiatric syndrome. *Journal of Nutritional and Environmental Medicine* 7, 169–176.

Ames, R.G., Steenland, K., Jenkins, B., Chrislip, D. and Russo, J. (1995) Chronic neurologic sequelae to cholinesterase inhibition among agricultural pesticide applicators. *Archives of Environmental Health* 50(6), 440–444.

Amr, M.M., Halim, Z.S. and Moussa, S.S. (1997) Psychiatric disorders among Egyptian pesticide applicators and formulators. *Environmental Research* 73, 193–199.

Anger, W.K., Moody, L., Burg, J., Brightwell, W.S., Taylor, B.J. and Russo, J.M. (1986) Neurobehavioral evaluation of soil and structural fumigators using methyl-bromide and sulfuryl fluoride. *NeuroToxicology* 7, 137–156.

Anger, W.K., Letz, R., Chrislip, D.W., Frumkin, H., Hudnell, K., Russo, J.M., Chappell, W. and Hutchinson, L. (1994) Neurobehavioral test methods for environmental health studies of adults. *Neurotoxicology and Teratology* 16(5), 489–497.

Bazylewicz-Walczak, B., Majczakowa, W. and Szymczak, M. (1999) Behavioural effects of occupational exposure to organophosphorous pesticides in female greenhouse planting workers. *NeuroToxicology* 20(5), 819–826.

Berent, S., Giordani, B., Albers, J.W., Garabrant, D.H., Cohen, S.S. and Garrison, R.P. (2014) Effects of occupational exposure to chlorpyrifos on neuropsychological function: a prospective longitudinal study. *NeuroToxicology* 41, 44–53.

CRD (2009) *Systematic Reviews: CRD's guidance for undertaking reviews in health care*. Centre for Reviews and Dissemination, University of York, UK.

Cherry, N., Mackness, M., Durrington, P., Povey, A., Dippnall, M., Smith, T. and Mackness B. (2002) Paraoxonase (PON1) polymorphisms in farmers attributing ill health to sheep dip. *The Lancet* 359,763–764.

Cole, D.C., Carpio, F., Julian, J., Leon, N., Carbottes, R. and de Almeida, H (1997) Neurobehavioural outcomes among farm and nonfarm rural Ecuadorians. *Neurotoxicology & Teratology* 19(4), 277–286.

Costa, L.G. and Furlong, C.E. (eds) (2002) *Paraoxonase (PON1) in Health and Disease: Basic and Clinical Aspects*. Kluwer Academic Publishers Group, Norwell, Massachusetts.

COT (1999) *Organophosphates*. Committee on Toxicity of Chemicals in Food, Consumer Products and the Environment, Food Standards Agency, London.

COT (2014) *Statement on Long Term Neurological, Neuropsychological and Psychiatric Effects of Low Level Exposure to Organophosphates in Adults*. Committee on Toxicity of Chemicals in Food, Consumer Products and the Environment, Food Standards Agency, London. Available at: https://cot.food.gov.uk/sites/default/files/cot/cotstate.pdf (accessed 1 February 2018).

Daniell, W., Barnhart, S., Demers, P. Costa, L.G., Eaton, D.L., Miller, M. and Rosenstock, L (1992) Neuropsychological performance among agricultural pesticide applicators. *Environmental Research* 59, 217–228.

De Silva, H.J., Samarawickrema, N.A. and Wickremasinghe, A.R. (2006) Toxicity due to organophosphorous compounds: what about chronic exposure? *Transactions of the Royal Society of Tropical Medicine and Hygiene* 100(9), 803–806.

EPA (2011) *Pesticides Industry Sales and Usage 2006 and 2007 Market Estimates*. US Environmental Protection Agency, Washington, DC.

EPA (2017) *Pesticides Industry Sales and Usage 2008–2012 Market Estimates*. US Environmental Protection Agency, Washington, DC.

European Commission (2018) Eurostat data – Pesticide Sales Statistics. Available at: http://ec.europa.eu/eurostat/statistics-explained/index.php/Pesticide_sales_statistics (accessed 1 February 2018).

Farahat, T.M., Abdelrasoul, G.M., Amr, M.M., Shebl, M.M., Farahat, F.M. and Anger, W.K. (2003) Neurobehavioural effects among workers occupationally exposed to organophosphorus pesticides. *Occupational and Environmental Medicine* 60, 279–286.
Farahat, F.M., Fenske, R.A., Olson, J.R., Galvin, K., Bonner, M.R. *et al.* (2010) Chlorpyrifos exposures in Egyptian cotton field workers. *NeuroToxicology* 31(3), 297–304.
Fiedler, N., Kipen, H., Kelly-McNeil, K. and Fenske, R. (1997) Long term use of organophosphates and neuropsychological performance. *American Journal of Industrial Medicine* 32, 487–496.
Grandjean, P. and Landrigan, P.J. (2014) Neurobehavioural effects of developmental toxicity. *The Lancet Neurology* 13(3), 330–338.
Harrison V. and Mackenzie Ross, S.J. (2016) Anxiety and depression following cumulative low-level exposure to organophosphate pesticides. *Environmental Research* 151, 528–536.
Ismail, A.A., Bodner, T.E. and Rohlman, D.S. (2012) Neurobehavioral performance among agricultural workers and pesticide applicators: a meta-analytic study. *Occupational Environmental Medicine* 69(7), 457–464.
Karalliedde, L., Feldman, S., Henry, J. and Marrs, T (eds) (2001) *Organophosphates and Health*. Imperial College Press, London.
Lezak, M.D., Howieson, D.B., Bigler, E.D. and Tranel, D. (eds) (2012) *Neuropsychological Assessment*, 5th edn. Oxford University Press, Oxford, UK.
London, L., Myers, J.E., Nell, V., Taylor, T. and Thompson, M.L (1997) An investigation into neurologic and neurobeavioural effects of long-term agricultural use among deciduous fruit farm workers in the Western Cape, South Africa. *Environmental Research* 73,132–145.
Mackenzie Ross, S.J., Clark, J.S., Harrison, V. and Abraham, K.M. (2007) Cognitive impairment following exposure to organophosphate pesticides: a pilot study. *Journal of Occupational Health & Safety: Australia & New Zealand* 23(2),133–142.
Mackenzie Ross, S.J., Brewin, C.R., Curran, H.V., Furlong, C.E., Abraham-Smith, K.M. and Harrison, V. (2010) Neuropsychological and psychiatric functioning in sheep farmers exposed to low levels of organophosphate pesticides. *Neurotoxicology & Teratology* 32(4), 452–459.
Mackenzie Ross, S.J., McManus, I.C., Harrison, V. and Mason, O. (2013) Neurobehavioural problems following low level exposure to organophosphate pesticides: a systematic and meta-analytic review. *Critical Reviews in Toxicology* 43,21–44.
Mackenzie Ross, S.J., McManus, I.C., Harrison, V. and Mason, O. (2016) Reflections on the process of using systematic review techniques to evaluate the literature regarding the neurotoxicity of low level exposure to organophosphate pesticides. *Environment International* 92, 569–573.
Mackness, B., Durrington, P., Povey, A., Thomson, S., Dippnall, M. *et al.* (2003) Paraoxonase and susceptibility to organophosphorus poisoning in farmers dipping sheep. *Pharmacogenetics* 13(2), 81–88.
Maizlish, N., Schenker, M., Weisskopf , C., Seiber, J. and Samuels, S.A (1987). Behavioural evaluation of pest control workers with short-term, low-level exposure to the organophosphate diazinon. *American Journal of Industrial Medicine* 12, 153–172.
Malekirad, A.A., Faghih, M., Mirabdollahi, M., Kiani ,M., Fathi, A. and Abdollahi, M. (2013) Neurocognitive, mental health, and glucose disorders in farmers exposed to organophosphorus pesticides. *Archives of Industrial Hygiene and Toxicology* 64(1), 1–8.
Meyer-Baron, M., Knapp, G., Schäper, M. and van Thriel, C. (2015) Meta-analysis on occupational exposure to pesticides – neurobehavioral impact and dose–response relationships. *Environmental Research* 136, 234–245.
Muñoz-Quezada, M.T., Lucero, B.A., Iglesias, V.P., Muñoz, M.P., Cornejo, C.A., Achu, E. and Villalobos, M. (2016) Chronic exposure to organophosphate (OP) pesticides and neuropsychological functioning in farm workers: a review. *International Journal of Occupational and Environmental Health* 22(1), 68–79.
Pitts, M. and Phillips, K. (eds) (1991) *The Psychology of Health: an Introduction*. Routledge, London.
Rohlman, D.S., Lasarev, M., Anger, W.K., Scherer, J., Stupfel, J. and McCauley, L (2007) Neurobehavioural performance of adult and adolescent agricultural workers. *NeuroToxicology* 28, 374–380.
Rodnitzky, R., Levin, H.S. and Mick, D.L. (1975) Occupational exposure to organophosphate pesticides: a neurobehavioral study. *Archives of Environmental Health* 30, 98–103.
Roldan-Tapia, L., Parron, T. and Sanchez Santed, F. (2005) Neuropsychological effects of long-term exposure to organophosphate pesticides. *Neurotoxicology & Teratology* 27(2), 259–266.
Roldan-Tapia, L., Nieto-Escamez, F.A., del Aguila, E.M., Laynez, F., Parron, T. and Sanchez-Santed, F. (2006) Neuropsychological sequelae from acute poisoning and long-term exposure to carbamate and organophosphate pesticides. *Neurotoxicology & Teratology* 28, 694–703.
Rosenstock, L., Keifer, M., Daniell, W.E., McConnell, R. and Claypoole, K. (1991) Chronic central nervous system effects of acute organophosphate pesticide intoxication. *The Lancet* 338, 223–227.

Rothlein, J., Rohlman, D., Lasarev, M., Phillips, J., Muniz, J. and McCauley, L. (2006) Organophosphate pesticide expossure and neurobehavioral performance in agricultural and non-agricultural Hispanic workers. *Environmental Health Perspectives* 114(5), 691–696.
Salvi, R.M., Lara, D.R., Ghisolfi, E.S., Portela, L.V., Dias, R.D. and Souza, D.O. (2003) Neuropsychiatric evaluation in subjects chronically exposed to organophosphate pesticides. *Toxicological Sciences* 72, 267–271.
Shah, A.K. and Oppenheimer, D.M. (2008) Heuristics made easy: an effort-reduction framework. *Psychological Bulletin* 134(2), 207.
Srivastava, A.K., Gupta, B.N., Bihar, V., Mathur, N., Srivastava, L.P. *et al.* (2000) Clinical, biochemical and neurobehavioural studies in workers engaged in the manufacture of quinalphos. *Food and Chemical Toxicology* 38, 65–69.
Stallones, L. and Beseler, C. (2016) Assessing the connection between organophosphate pesticide poisoning and mental health: a comparison of neuropsychological symptoms from clinical observations, animal models and epidemiological studies. *Cortex* 74, 405–416.
Starks, S.E., Gerr, F., Kamel, F., Lynch, C.F., Alavanja ,M.C. *et al.* (2012) High pesticide exposure events and central nervous system function among pesticide applicators in the agricultural health study. *International Archives of Occupational and Environmental Health* 85, 505–515.
Steenland, K., Dick, R.B, Howell, R.J., Chrislip, D.W., Hines, C.J. *et al.* (2000). Neurologic function among termiticide applicators exposed to chlorpyrifos. *Environmental Health Perspectives* 108, 293–300.
Stephens, R., Spurgeon, A., Calvert, I.A., Beach, J., Levy, L.S., Berry, H. and Harrington, J.M. (1995) Neuropsychological effects of long-term exposure to organophosphate in sheep dip. *The Lancet* 345(6), 1135–1139.
Stephens, R. and Sreenivasan, B. (2004). Neuropsychological effects in long-term low level organophosphate exposure in orchard sprayers in England. *Archives of Environmental Health* 59(11), 566–574.
Strauss, E., Sherman, E.M.S. and Spreen, O. (eds) (2006) *A Compendium of Neuropsychological Tests: Administration, Norms and Commentary*, 3rd edn. Oxford University Press, Oxford, UK.
Tahmaz, N., Soutar, A. and Cherrie, J.W. (2003) Chronic fatigue and organophosphate pesticides in sheep farming: a retrospective study amongst people reporting to a UK pharmacovigilance scheme. *Annals of Occupational Hygiene* 47, 261–267.
van Himbergen, T.M., van der Schouw, Y.T., Vorbij, H.A.M., van Tits, L.J.H., Stalenhoef, A.F.H., Peeters, P.H.M. and Roest, M. (2008) Paraoxonase (PON1) and the risk for coronary heart disease and myocardial inafarction in a general population of Dutch women. *Atheroscelrosis* 199, 408–414.
Whaley, P., Halsall, C., Ågerstrand, M., Aiassa, E., Benford, D., Bilotta, G. and FitzGerald, R. (2016) Implementing systematic review techniques in chemical risk assessment: challenges, opportunities and recommendations. *Environment International* 92, 556–564.
WHO (1990) *Public Health Impact of Pesticides Used in Agriculture*. World Health Organisation/UNEP Working Group. Geneva.

# 19 Glyphosate as a Glycine Analogue

**S. Seneff***
*Computer Science and Artificial Intelligence Laboratory, MIT, Cambridge, Massachusetts, USA*

## 19.1 Abstract

Glyphosate is the most heavily used herbicide on the planet, due to its ability to kill all types of weeds and its perceived non-toxicity to humans. Its usage on core crops in the USA has increased dramatically in the past two decades, in step with an alarming increase in a long list of debilitating diseases and conditions, including autism, diabetes, obesity, Alzheimer's disease, pancreatic cancer, thyroid cancer, inflammatory bowel disease, coeliac disease and many others. Most have dismissed these strong correlations because there is disbelief that a chemical deemed so safe could cause so many diseases. However, no other toxic chemical being used in agriculture matches the time trends nearly as well. If a biological mechanism could be found to explain how glyphosate could cause these diseases, it would greatly enhance the case for a causal relationship. In this chapter, it is proposed that glyphosate has an insidious cumulative mechanism of toxicity due to its ability to substitute for the coding amino acid glycine during protein synthesis. Multiple naturally produced amino acid analogues exhibit this kind of toxicity through erroneous insertion into proteins, and they cause neurological diseases. Glyphosate substitution for glycine at certain specific highly conserved residues in specific proteins would disable the protein's function and produce misfolded proteins, leading to amyloidoses. Alarming increases in many modern diseases can be explained by specific proteins whose dysfunction or misfolding is linked to those diseases. Monsanto's early studies on bluegill sunfish exposed to radiolabelled glyphosate came close to proving that glyphosate is getting incorporated into proteins. There is now an urgent need to reconsider the safety of glyphosate not only with regard to its proposed action as a glycine analogue but also in the context of ongoing and emerging issues of concern. In particular, the current debate centres on the classification of glyphosate as a 'probable human carcinogen', its effects on developmental neurotoxicity in animal models, its putative role as an endocrine disruptor and potential adverse impacts on biodiversity in agroecology and in aquatic ecosystems.

## 19.2 Introduction

Glyphosate is the active ingredient in the pervasive herbicide Roundup. It is by far the most used herbicide on the planet and its usage rate has gone up exponentially over the past two decades, in step with the widespread introduction of

* E-mail address: Seneff@csail.mit.edu

genetically modified glyphosate-resistant crops. The USA consumes more glyphosate per capita than any other nation and it also has the highest healthcare costs, two facts that are probably directly related. Glyphosate has been found as a contaminant in many common foods, as well as in cotton products such as tampons and sterile gauze, and in vaccines (Samsel and Seneff, 2017). Analyses of disease trends have shown that multiple debilitating diseases and conditions have increased dramatically in the past two decades in the USA and that the rate of increase of glyphosate usage on corn and soya crops exhibits a remarkably similar trend over time (Swanson *et al.*, 2014). These diseases include diabetes, obesity, Alzheimer's disease, Parkinson's disease, autism, inflammatory bowel disease, liver cancer, kidney failure, pancreatic cancer, thyroid cancer, bladder cancer and many others. Autism rates in the USA have risen almost exactly in step with glyphosate usage on core crops, as shown in Fig. 19.1.

It is accepted that autism is a complex multifactorial disorder and that the underlying mechanisms cannot be elucidated just by statistical correlations. However, Shaw (2017) and Argou-Cardozo and Zeidan-Chulia (2018) proposed a gut–brain hypothesis for the effects of glyphosate in autistic children. In addition, Good (2018) announced that 'the US autism epidemic initiated by acetaminophen (*Tylenol*) is aggravated by oral antibiotic amoxicillin/clavulanate (*Augmentin*) and now exponentially by herbicide glyphosate (*Roundup*)'. Others (Cattani *et al.*, 2017) claimed a potential relationship between developmental exposure to glyphosate-based herbicides and depressive-like behaviour in adult offspring, based on evidence with a rat model,

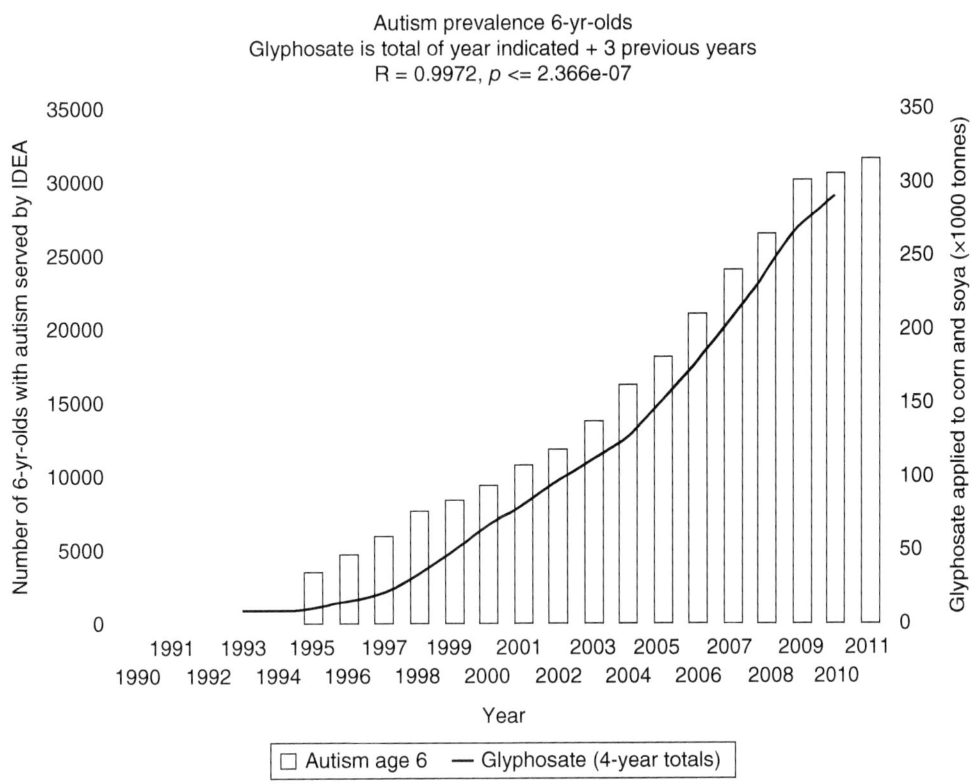


**Fig. 19.1.** Graph of US rates of autism prevalence in first grade served under Individuals with Disabilities Education Act (IDEA) compared with glyphosate usage on corn and soy crops over the previous 4 years (for details, see Seneff *et al.*, 2015). Data in this figure do not constitute proof of a causal relationship between autism and glyphosate usage; see text for further comments.

while Ji *et al.* (2018) referred to reports of glyphosate neurotoxicity in human and animal models.

In a separate development, a relatively large group of experts (Myers *et al.*, 2016) issued a 'statement of concern' over the safety of glyphosate-based herbicides to the effect that: (i) these formulations often contaminate drinking water, precipitation and air, especially in agricultural regions; (ii) the half-life of glyphosate in water and soil has hitherto been underestimated; (iii) the herbicide and its metabolites widely contaminate the global soybean supply; (iv) regulatory estimates of tolerable daily intakes in the USA and European Union (EU) are based on outdated science; and (v) glyphosate is authoritatively classified as a 'probable human carcinogen'. It should be noted, however, that a joint FAO/WHO group (JMPR, 2016) declared that there was no risk of malignancy associated with glyphosate residues in food, but the impact of other routes of exposure need to be considered in view of the foregoing comments of Myers *et al.* (2016) and of the substantial damages in a US court cancer trial judgement awarded to an operative who regularly worked with glyphosate formulations. There are, therefore, grounds to evaluate the underlying risks and mechanisms that may be involved in the potential toxicity of glyphosate with respect to neuropathology, endocrine disruption and carcinogenicity. Furthermore, there are questions regarding the ecological implications of glyphosate-resistant weeds and the impact of this herbicide on aquatic biodiversity (Myers *et al.*, 2016).

Glyphosate is believed to be relatively non-toxic to humans, because its main mechanism of toxicity to weeds involves a blockage in the shikimate pathway, which is essential for plant survival but not found in human cells. However, disruption of the shikimate pathway in our microbiome can result in deficiencies in the essential aromatic amino acids produced through this pathway. Furthermore, glyphosate has been shown to disrupt the balance of gut microbes favouring pathogenic overgrowth (Shehata *et al.*, 2013), leading to inflammatory bowel disease and leaky gut syndrome (Gildea *et al.*, 2017).

It is not disputed that glyphosate is an amino acid analogue of glycine, and much of its toxicity is likely linked to interference with glycine's many roles in the body. Glyphosate is a complete glycine molecule except that a hydrogen atom normally attached to the nitrogen atom has been swapped out and replaced by a much bulkier and negatively charged methylphosphonyl group ($CH_2PO_3^{-2}$). Glyphosate has been shown to excite *N*-methyl-D-aspartate (NMDA) receptors in neurons, leading to neuroexcitotoxicity, and this is attributed in part to its binding to the receptor in place of glycine (Cattani *et al.*, 2014; Beecham and Seneff, 2015). Glyphosate interferes with the synthesis of the porphyrin ring, a central building block of both chlorophyll in plants and haem in animals (Kitchen *et al.*, 1981). Suppression of chlorophyll synthesis is a contributory factor in its toxicity to plants. Its interference is likely due in part to disruption of glycine's access as a substrate to δ-aminolevulinic acid (ALA) synthase. Experiments on mice show that glyphosate is actively taken up along L-type amino acid transporters and transported directly to the brain following intranasal exposure (Xu *et al.*, 2016). That glyphosate reaches the brain through nasal exposure was verified in another experiment on mice exposed intranasally, which demonstrated significant neurological impairment, including memory loss, increased anxiety and reduced motility (Baier *et al.*, 2017).

Despite the very low *p*-value associated with correlation studies between glyphosate usage and the alarming rise in a long list of debilitating diseases, most people are sceptical that this correlation implies a causal role. Part of the scepticism stems from a false belief that glyphosate is non-toxic to humans, and furthermore it seems hard to explain how a single chemical could have so many distinct adverse effects. The answer can come from a relatively simple idea that glyphosate may be erroneously incorporated into proteins during protein synthesis, in place of the coding amino acid, glycine.

There is a broad research literature on the many roles of glycine residues in various proteins and on the sometimes crippling effects of even simple mutations swapping out a specific glycine residue for alanine, the amino acid with the greatest similarity to glycine. Glycine is the smallest amino acid and it has no side chain. Alanine is the next smallest, with a single methyl group as a side chain. Other amino acids that are bulkier and negatively charged, such as glutamate (and glyphosate), would have a much greater effect on protein function if substituted for glycine. Glyphosate is negatively charged and bulky because of its methyl-phosphonyl side chain on the nitrogen atom.

What is perhaps even more remarkable than the idea that glyphosate may be substituting for glycine during protein synthesis is the fact that this idea has been so successfully repressed from the research agenda of toxicologists and biochemists for so many years, which has allowed this insidiously and cumulatively toxic synthetic chemical to remain widely used over four decades, broadly threatening ecosystems. While it is not completely straightforward to design an experiment that definitively proves this idea, Monsanto researchers came tantalizingly close to proving it in 1989 when they exposed bluegill sunfish to $^{14}C$-radiolabelled glyphosate and found significant radiolabel in tissue samples that was undetected as glyphosate by standard chromatography tests (Ridley and Chott, 1989; Samsel and Seneff, 2017). Extensive exposure of the tissue to the digestive enzyme proteinase K brought the yield from 17–20% up to 57–70%, thus closing but not eliminating the gap. Presciently, these authors wrote: 'Proteinase K hydrolyses proteins to amino acids and small oligopeptides, suggesting that a significant portion of the $^{14}C$ activity residing in the bluegill sunfish tissue was tightly associated with or *incorporated into protein*' (emphasis added) (Ridley and Chott, 1989).

Logically, following the evidence from this experiment, glyphosate contamination in food proteins will lead to autoimmune disease due to immune cell exposure to proteolysis-resistant glyphosate-contaminated food peptides. We have witnessed a huge increase in the past two decades in allergies to proteins in foods, which is correlated with adoption of a Western diet (Prescott and Allen, 2011). The most prominent allergens are from proteins that can be expected to be contaminated with glyphosate, such as soya, casein, gluten and peanuts.

Perhaps the most compelling evidence that glyphosate is substituting for glycine during protein synthesis comes from the enzyme that is disrupted in the shikimate pathway in plants that is considered to be its main toxic effect. This enzyme, called 5-enolpyruvylshikimate-3-phosphate (EPSP) synthase, has a highly conserved glycine at the active site where the substrate, phosphoenolpyruvate (PEP), fits snugly into a pocket. It has been established that glyphosate displaces PEP from its pocket but there is controversy over exactly how it accomplishes this feat. Remarkably, multiple species of plants and multiple species of microbes have independently developed a mutated form of EPSP synthase that has an alanine residue in place of the highly conserved glycine, and this change results in a complete insensitivity to glyphosate even at very high concentrations (Eschenburg *et al.*, 2002; Seneff and Orlando, 2018). The most plausible explanation is that glyphosate gets inserted by mistake into the peptide sequence in place of this highly conserved glycine residue and then protrudes into the pocket, preventing PEP from gaining access.

It might be thought that the side chain on the nitrogen atom would interfere with glyphosate's ability to incorporate into the amino acid chain. However, proline, one of the basic coding amino acids, also has a side chain on its nitrogen atom and this does not preclude its integration into the chain. Proline, however, is resistant to proteolysis and it requires a specialized enzyme to detach it from the chain, called prolyl aminopeptidase. This implies that glyphosate would also resist proteolysis, which makes a glyphosate contaminated protein more likely to cause an autoimmune reaction. Glyphosate substitution will interfere with protein folding and cause native proteins to be recognized as foreign by the immune cells. Thus, the epidemic we are witnessing today in autoimmune diseases (Ray *et al.*, 2012) could well be explained by glyphosate contamination in various proteins.

## 19.3 Amino Acid Analogue Incorporation into Proteins

Most people are unaware of the fact that protein synthesis is inherently an errorful process. While DNA synthesis has multiple strategies to edit and repair mistakes immediately as they occur, such is not the case for protein synthesis. Instead, a protein is fully assembled, carelessly, and then allowed to fold into its native state before a decision is made as to whether it is defective. Many proteins are assembled with errors inserted into the peptide chain, where either the wrong coding amino acid is selected, or even a non-coding amino acid analogue of a coding amino acid is inserted by mistake. If a mistake is catastrophic, the protein is marked for clearance through a process called ubiquitination and then is disassembled and fully reconstructed. This process may seem inefficient, but often mistakes are tolerable and there is no need to repair them.

There are only 22 coding amino acids, associated with the three-letter code encapsulated in the four nucleotides of DNA. However, there are hundreds of amino acid analogues, many of which occur naturally (see Chapter 1) and some of which are associated with severe disease (Rodgers and Shiozawa, 2008). Amino acid analogue misincorporation is associated with autoimmune disease, because the defective protein is sometimes perceived as foreign by the immune system. Examples abound of severe disease, often involving neurological damage, linked to naturally produced amino acid analogues. An epidemic in an amyotrophic lateral sclerosis (ALS)-like disease in Guam was traced to β-methylamino-L-alanine (BMAA), an amino acid analogue of serine produced by cyanobacteria resident as root symbionts in native cycad trees (Spencer *et al.*, 1987; Murch *et al.*, 2004). Glufosinate is an amino acid analogue of glutamate synthesized by *Streptomyces* soil bacteria and it is commonly used as a herbicide (Hoerlein, 1994). Exposure to low-dose glufosinate pre- and postnatally induces autism-like behaviours in mice (Laugeray *et al.*, 2014). Azetidine-2-carboxylic acid (Aze), an analogue of proline, is produced by sugarbeet in response to stress and it has been linked to multiple sclerosis (Rubenstein, 2008). L-Canavanine, an amino acid analogue of arginine, is believed to be the cause of death for William McCandless, whose fated wilderness experience in Alaska was described in the book *Into the Wild* (Rosenthal, 1990; Krakauer *et al.*, 2015).

An excellent review article of the role of misincorporation of BMAA in place of serine in disease was published by Murch *et al.* (2004), who pointed out that, although BMAA is not likely to accumulate in fat tissues because it is not lipophilic, it can instead be expected to accumulate within proteins and be slowly released as a neurotoxin over time, when those proteins are metabolized. They also suggested – and this too would be expected to be a property of glyphosate as well – that BMAA would alter the folding properties of affected proteins, and would disrupt the bioavailability of metal ions by binding to them. It could also induce tauopathies by causing protein synthesis to abort prematurely, leaving behind a truncated defective version of the protein. Its insidious effects would result in latency and a slow progressive disease, as opposed to an acute toxic effect. All of these ideas would easily carry over to glyphosate incorporation into proteins.

BMAA is taken up by cells along amino acid transport channels and it has been shown in multiple studies to be neurotoxic, in part through exciting NMDA and glutamate receptors (Chiu *et al.*, 2011). Ince and Codd (2005) found in a study on BMAA-contaminated tissues that it was essential to subject the sample to proteolysis to retrieve protein-bound BMAA. These authors wrote (p. 348): ‘When the insoluble, protein-containing fraction following TCA [trichloroacetic acid] extraction is further hydrolysed to release BMAA from protein, there is a further pool of protein-bound BMAA that is present in a ratio of between 60:1 and 120:1 compared with the pool of free BMAA.’ This parallels Monsanto researchers’ findings regarding bound glyphosate in tissue samples from bluegill sunfish. This also implies that testing for glyphosate in protein-containing foods is likely to miss most of the glyphosate if proper prior proteolysis is omitted.

Rubenstein (2008) explicitly considered the probable effects of Aze substitution for proline in proteins where proline is highly conserved. He pointed to both myelin basic protein (MBP) and collagen, as well as ion channels, among others, as likely candidates for disturbance by Aze substitution. Fifteen per cent of the amino acid residues in collagen are prolines. Numerous studies have shown that Aze causes severe impairment of collagen formation in chicks, mice, pigs and human cells *in vitro* (Rubenstein, 2000). The links between Aze and multiple sclerosis can be explained through disturbances in MBP, which contains a conserved hexapeptide sequence, PRTPPP, and an α helix bounded by prolyls.

## 19.4 Highly Conserved Glycine Residues and Disease

Glycine, the smallest amino acid, is highly conserved in certain residue sites in diverse protein types, including enzymes, transporters, receptors, contractile proteins, ion channels and pumps, regulatory proteins, structural proteins, proteins involved in DNA repair and many others. Over 25% of proteins in cell physiology are membrane proteins. An important role they play is the diffusion of both small and large molecules across membranes. The broad α-helical class of membrane proteins includes receptors, ion channels, transporters and redox proteins. Many form either homodimers or helical bundles in association with other transmembrane helices (Senes *et al.*, 2004). Glycine residues play crucial roles in stabilizing α helices in transmembrane proteins. As we shall expand upon later, a GxxxG motif is a common pattern in multiple transmembrane proteins that are associated with amyloidoses, including Alzheimer's disease and Parkinson's disease (Brosig and Langosch, 1998; Munter *et al.*, 2007).

Glycine substitutions can lead to a major change in the way a protein folds, causing a transformation of a normally transmembrane protein into a soluble protein that remains in the cytoplasm. Glycine mutations can prevent receptors from going to the membrane, completely blocking receptor activity. Glycine residues often control a hinge region in transmembrane transporter proteins. A terminal glycine is often essential for attachment to cytoskeletal proteins or the plasma membrane. A glycine doublet terminates ubiquitin and it is essential for the ubiquitination process involved in initiating clearance of a defective protein or a protein whose activity needs to be terminated (Xu *et al.*, 2009). Genetic mutations where specific glycines are replaced by other amino acids lead in many cases to severe disease.

Samsel and Seneff (2016) published the first paper on the idea that glyphosate might be substituting for glycine during protein synthesis and causing major impairment in select proteins as a consequence. In that paper, multiple correlations between the rise in glyphosate usage on core crops and the alarming rise in various diseases were explained through specific enzymes that would be adversely affected by loss of certain critical and highly conserved glycine residues. They wrote in the abstract:

> Glyphosate substitution for conserved glycines can easily explain a link with diabetes, obesity, asthma, chronic obstructive pulmonary disease (COPD), pulmonary edema, adrenal insufficiency, hypothyroidism, Alzheimer's disease, ALS, Parkinson's disease, prion diseases, lupus, mitochondrial disease, non-Hodgkin's lymphoma, neural tube defects, infertility, hypertension, glaucoma, osteoporosis, fatty liver disease and kidney failure.

While there is no space here to go into the details of these various glycine dependencies, the interested reader can browse this paper and its many references to gain a further understanding of the mechanisms of protein dysfunction in the specific diseases. Here, we will just provide a few examples of specific protein dependencies on glycine residues and the severe disease that develops when these proteins are mutated. Of course, glyphosate substitution would likely only affect a small percentage of the affected protein and so the disease manifestation would be considerably milder than what is experienced with a genetic mutation, where 100% of the protein molecules are affected.

The insulin receptor contains at least eight repeats of a glycine-centred motif that is essential for the protein to fold into the correct secondary structure. Glycine mutations disrupt protein folding and lead to an impaired ability to migrate to the membrane and form a transmembrane dimer, thus blocking insulin signalling and leading to type 2 diabetes (Wertheimer *et al.*, 1994). Similarly, the high-density lipoprotein (HDL) receptor in the liver depends critically on highly conserved glycines in a GxxxG motif to form transmembrane dimers that are essential for the uptake of cholesterol from HDL particles (Gaidukov *et al.*, 2011). Impairment in the return of cholesterol to the liver leads to high serum cholesterol associated with increased risk to cardiovascular disease. Myosin, the most common protein in muscle cells, contains an absolutely essential glycine at residue 699. If this residue is replaced with alanine, the protein's ability to contract is reduced to only 1% of the original ability (Kinose *et al.*, 1996). Glyphosate substituting for glycine in myosin could be a key factor in chronic fatigue syndrome, a new disease now affecting up to 15% of the US population that did not exist before glyphosate became widespread.

Glyphosate has been shown to severely suppress cytochrome P450 enzymes (CYPs) in rat liver (Syed and Mashele, 2014). A comparative study on CYP enzymes across multiple species identified a characteristic glycine-rich CxG motif, FxxGxRxCxG, containing two and often three glycine residues, that is essential for their proper function (Hietanen *et al.*, 1983). CYP enzymes play multiple important roles, including bile acid synthesis, detoxification of xenobiotics and activation of vitamin D.

Sulfite oxidase is a mitochondrial enzyme that oxidizes sulfite to sulfate. Sulfite is extremely toxic and sulfate plays essential roles in the extracellular matrix of most cells. Sulfite oxidase deficiency is a rare but fatal disease that causes neurological disorders, mental retardation, physical deformities and early death. A mutation involving a substitution of bulkier and negatively charged aspartate for the highly conserved glycine residue at G473 in sulfite oxidase results in impaired dimer formation and severe dysfunction (Kisker *et al.*, 1997). Glyphosate can be expected to have a similar effect. Sulfite oxidase also contains haem, whose synthesis could be compromised by glyphosate (Kitchen *et al.*, 1981), and it depends on molybdenum as a cofactor, which as a +2 cation could be chelated by glyphosate, making it unavailable. Defective sulfite oxidase could account for sulfur sensitivities in foods, since sulfite is highly toxic. Excess sulfite in the gut will induce excessive growth of sulfur-reducing microbes such as *Bilophilia wadsworthia* and *Desulfovibrio*, whose overgrowth has been linked to regressive autism (Finegold, 2011).

Heparan sulfate is a long, highly diverse carbohydrate polymer that links covalently to glycoproteins at serine residues to form heparan sulfate proteoglycans (HSPGs). Heparan sulfate assists in diverse biological processes including cell–cell and cell–matrix adhesion, apoptosis, growth factor regulation, proteolysis and angiogenesis (Langford *et al.*, 1998). Syndecans are the predominant HSPGs at cell surfaces, and they play a crucial role in neurodevelopment. Heparan sulfate deficiency in the brain is linked to autism in both humans (Pérez *et al.*, 2016) and mouse models (Irie *et al.*, 2012) and is associated with impaired maturation of neuronal dendrites (Pérez *et al.*, 2016). A structural motif that is shared by many heparan sulfate proteoglycans, including the syndecans, consists of multiple acidic amino acids closely followed by the tetrapeptide SGXG. This pattern is necessary for enzymatic action by a xylosyltransferase that initiates the assembly of glycosaminoglycan chains attached to the serine residue. HSPGs also play an important role in invasive tumours. Heparan sulfate binding to all three highly conserved serine residues within this motif in syndecan-1 is necessary to effectively inhibit invasiveness of tumour cells into collagen gels (Vekemans and David, 1999).

Collagen is the most common protein in the body, representing 25% of the protein mass. Collagen folds into a triple helix structure that is crucial for its tensile strength, flexibility and ability to hold water. The tripeptide sequence G-X-Y repeats in a strong pattern in the triple helix structural segment of collagen molecules, where X and Y are frequently proline or hydroxyproline. Over 25% of the amino acids in a typical collagen molecule are glycine residues (Eastoe, 1955). Glyphosate has been found as a contaminant in bovine bone, gelatin and collagen (Samsel and Seneff, 2017). It is possible that the epidemic we are seeing today in the USA in bone and joint pain leading to an opioid drug overdose crisis is related to glyphosate damaging the crystalline structure of collagen.

There is epidemiological evidence that glyphosate causes neurodevelopmental issues, and Seneff and Nigh (2017) have described how multiple proteins that are critical for neurodevelopment would be disrupted if glyphosate were to substitute for certain essential glycine residues. Some of these proteins are listed in Table 19.1, indicating the specific glycine residue(s) that are vulnerable. Details can be found in the original paper. Mesoamerican nephropathy is a new disease ravaging young sugarcane workers in Nicaragua and El Salvador. Seneff and Orlando (2018) have elaborated on how glyphosate substitution for glycine in multiple proteins related to the pathology can explain this disease.

## 19.5 GxxxG Motif, Alzheimer's Disease and Amyloidoses

Alzheimer's disease is a debilitating neurological disease that is very costly to society, because patients in advanced stages require around-the-clock care. While significant funding has gone into developing drugs to treat Alzheimer's, results of clinical trials have generally been disappointing and most of these potential drugs cost

**Table 19.1.** Various proteins with highly conserved glycines whose substitution by glyphosate would disrupt protein function, leading to neurodevelopmental disorders. Details can be found in Seneff and Nigh (2017).

| Protein | Conserved G | Effect of glycine substitution |
|---|---|---|
| Insulin receptor | **G**-centered motif | Impaired membrane transport<br>Diabetes |
| Folate receptor | **G**137 | Impaired folate binding |
| LDL receptor | **G**34 | Fetal DHEA-sulfate deficiency |
| CDK1 | **GEG**TY**G** motif | Cell cycle delay; inhibited mitosis |
| Kinases | **G**x**G**xx**G** | Enhanced activity<br>Impaired timing in development |
| Tyrosine phosphatase | **G**127 | Apoptosis in brain and spinal cord |
| ACTH | Terminal **G** | Impaired activation |
| Metallothionein | MSCC**GG**NC**G**CS motif | Impaired antioxidant capacity |
| Serine protease | **G**193 | Impaired neural tube closure |
| Myosin | **G**699 | Holoprosencephaly<br>Impaired folate uptake |

the pharmaceutical industry millions of dollars with ultimately nothing to show for it.

The accumulation of amyloid beta (Aβ) plaque is a hallmark of Alzheimer's, but much controversy surrounds exactly which form of Aβ is toxic. Recently, there has been considerable convergence on the idea that the toxic form is an intermediate-sized soluble oligomer and it has even been argued that the formation of the precipitated aggregates may be protective (Caughey and Lansbury, 2003; Goure *et al.*, 2014). Unfortunately, these toxic soluble oligomers make up a very small percentage of the total Aβ burden and the difficulty then is to design antibodies that would be specific just to this form (Goure *et al.*, 2014).

The precursor protein to Aβ, termed amyloid precursor protein (APP), is a transmembrane protein whose exact role is unclear. APP contains three copies of a highly conserved glycine-containing motif (GxxxG) that is also found in several other transmembrane proteins that are linked to amyloidoses. Specifically, the transmembrane portion of APP contains a GxxxGxxxGxxxG motif, not known to be present in any other protein (Munter *et al.*, 2007). The intervening wildcard amino acids are usually hydrophobic, supporting membrane penetration. The transmembrane portion of APP takes the form of a strongly lipophilic α helix, and two such helices in two molecules of APP bind together to form a dimer within the membrane. This motif appears to be important for stabilizing the alpha helix (α helix) dimer through hydrogen bonding on the nitrogen atoms in the conserved glycine residues.

The amino acid proline is known as a 'helix buster', because the absence of a hydrogen atom bound to its nitrogen atom prevents α helix stabilization. Glyphosate, like proline, does not have a free hydrogen atom bound to its nitrogen, and so it too can be expected to be a helix buster. Glyphosate's side chain, the methyl phosphonyl group, is attached to its nitrogen atom rather than the normal carbon-atom attachment of a side chain. This side chain will greatly disrupt the α helix configuration. Furthermore, the negative charge on glyphosate can be expected to increase solubility of the peptide. All of these alterations can result in the protein folding into a water-soluble beta sheet instead of the fat-soluble α helix. The toxic soluble oligomer form of Aβ configures as a beta sheet. An increase in beta sheet structure is a typical feature of amyloidogenic proteins and of prions (Pan *et al.*, 1993) and it reduces the sensitivity towards proteolytic degradation (von Bergen *et al.*, 2005). Tau protein, another Alzheimer's-linked protein, also forms aggregates after transitioning to a beta sheet structure (von Bergen *et al.*, 2005).

Aggregation-prone proteins accumulate in part because of an impaired ability to clear misfolded proteins. This is often due to impaired function of the ubiquitin–proteasome system. Ubiquitin plays an important role in protein clearance by binding to proteins as a complex poly-ubiquitin chain (McKinnon *et al.*, 2016). Ubiquitin contains a highly conserved C-terminal glycine–glycine doublet that is essential for chain formation (Xu *et al.*, 2016). Hence, substitution of glyphosate for glycine in ubiquitin can be expected to disrupt signalling for subsequent degradation.

Another way to clear the toxic soluble form of Aβ is to excrete it from the cell using a transporter. A protein that has been shown to export soluble Aβ is P glycoprotein, which is a general-purpose transporter for multiple xenobiotics as well (Wang *et al.*, 2016). P glycoprotein critically depends upon glycine residues in a hinge region within the transmembrane portion of the protein, which provides the flexibility to orchestrate the complex transformations necessary to open and close the transport channel (Wen *et al.*, 2013).

Another highly conserved glycine-containing sequence that has been carefully studied due to its link to prion diseases is the palindromic sequence AGAAAAGA found in prion proteins. The particular small fibrillogenic peptide isolated from positions 106 to 126 of the prion protein contains the above motif and maintains the neurotoxicity associated with the full prion peptide. It is often used in isolation to elucidate the toxic properties of prion peptides (Florio *et al.*, 2003). This study from 2003 showed that substitution of the two glycines with alanine resulted in a peptide that tended to form soluble oligomers that were highly toxic to neuroblastoma cells. This was associated with a transformation into a beta sheet crystalline structure, and the peptide resisted proteolysis. They argued that the glycine residues protect from beta sheet formation. Glyphosate substitution can be expected to be even more disruptive than alanine substitution.

Amyotrophic lateral sclerosis (ALS) is a fatal neurodegenerative disease where 90–95% of the cases are idiopathic. However, for the 5–10% of cases of familial ALS, it is remarkable how many of the genetic mutations linked to ALS involve either substitutions of other amino acids for critical glycine residues or mutations within glycine-rich regions of target proteins (Seneff *et al.*, 2016). One of the susceptibility proteins is transactive response DNA-binding protein 43 (TDP-43), and a large number of mutations within its glycine-rich region are linked to early-onset ALS. This region contains multiple putative GxxxG motifs (Dewey *et al.*, 2011). In both sporadic and familial ALS, TDP-43 is found in aggregates, along with ubiquitin, in inclusion bodies within affected neurons (Banks *et al.*, 2008).

The GxxxG motif appears in multiple other transmembrane proteins. Gly338 and Gly342 of the serotonin transporter form a GxxxG motif that is essential for dimerization and for uptake of serotonin by the protein (Horschitz *et al.*, 2008). Serotonin dysregulation is associated with autism, obesity and violent behaviours. As mentioned previously, the HDL receptor contains this motif. The syndecans, crucial for neurodevelopment, also contain a GxxxG motif (Brosig and Langosch, 1998).

Alzheimer's is also associated with neuronal oxidative stress and mitochondrial dysfunction. *In vitro* studies have shown that glyphosate-based herbicides cause mitochondrial damage by disrupting oxidative phosphorylation (Peixoto, 2005; Kim *et al.*, 2013). One possible mechanism is through impairment of the function of cytochrome c oxidase, a crucial enzyme in the electron transport chain. This enzyme has two highly conserved glycine residues at positions 219 and 226, in close proximity to the copper-binding site (Holm *et al.*, 1987). Displacement of either of these glycines by glyphosate would interfere with the catalytic action of copper.

## 19.6 Glyphosate and Aluminium

The neurotoxic metal aluminium has been strongly implicated as a risk factor in Alzheimer's, and it is well established that aluminium as a contaminant in dialysis fluid can lead to dialysis-related dementia (Wills and Savory, 1985). Because of its property as an extremely high charge density cation, aluminium induces amyloidogenesis by cross-linking amino acids in peptides to form clumps consisting of neurotoxic oligomers. Aluminium also impairs the cellular machinery involved with phagocytosis and clearance of Aβ aggregates (Kawahara and Kato-Negishi, 2011; Zhao *et al.*, 2014). Aluminium is present as a contaminant in Aβ plaque and Alzheimer's patients have been shown to have excessively high levels of aluminium in their brains (Mirza *et al.*, 2017). Elevated aluminium was also found recently in postmortem brains of people diagnosed with autism (Exley, 2014).

Theoretical analysis has shown that two glyphosate molecules can form a cage around an aluminium atom, hiding its +3 charge and producing a small neutral molecule that can bridge barriers much more easily than free aluminium (Purgel *et al.*, 2009). This directly follows the model of citrate binding to aluminium which has been shown to greatly enhance aluminium uptake across the gut barrier (Coburn *et al.*, 1991). It stands to reason that two glyphosate molecules embedded within two monomers of APP might similarly be able to bind to a bridging aluminium atom, with aluminium then seeding dimer formation and subsequent oligomerization.

Aluminium is taken up by cells along L-type calcium channels (Seneff *et al.*, 2015), which are overexpressed in association with glyphosate exposure (de Liz Oliveira *et al.*, 2013). The calcium channel itself has multiple highly conserved glycine residues within its transmembrane portion and mutations of certain of these glycines lead to excessive calcium uptake (Splawski *et al.*, 2004; Teng *et al.*, 2010). Glyphosate likely promotes aluminium uptake from the gut through its chelating effects and escorts it to the brainstem nuclei, freeing up the aluminium ion once the glyphosate–aluminium–glyphosate complex reaches the acidic terminal watershed region in the brainstem, as proposed in Seneff *et al.* (2015).

Another way in which glyphosate would enhance the bioavailability of aluminium in the brain is through impairment of the proton-dependent transport protein, monocarboxylate transporter 1 (MCT1), which exports aluminium citrate out of the extracellular fluid in the brain (Ackley and Yokel, 1998; Yokel *et al.*, 2002). MCT1 has an essential glycine residue at location 153. A mutation where this glycine, situated in the middle of helix 5 of the protein, is replaced by valine results in an inactive protein that is completely impaired in its ability to enter the membrane (Galić *et al.*, 2003). This strikingly resembles the pattern of disrupting α helices in amyloidogenic proteins.

## 19.7 Conclusions

Glyphosate is by far the most used herbicide on the planet. Its insidious cumulative mechanism of toxicity has been overlooked for more than four decades while it continues to steadily accumulate in soil, water, food, cotton products, drugs and vaccines. There are strong correlations between the alarming rise in glyphosate usage on core crops in the USA and the similar rise in frequency of a long list of debilitating autoimmune and neurological diseases. While correlation does not necessarily mean causation, correlation data become much more compelling when there is a plausible mechanism by which the toxic substance could cause the disease. In this chapter, we have provided strong evidence that glyphosate may be getting inserted into proteins by mistake in place of the coding amino acid glycine. We have provided plausible mechanisms by which such corruption of certain specific proteins can account for the observed symptoms of many of the diseases and conditions that are on the rise. If this unusual mechanism of toxicity can be validated, then there would be profound implications for the safety assessment of glyphosate, requiring immediate regulatory interventions to protect human health and biodiversity in different ecosystems.

## References

Ackley, D.C. and Yokel, R.A. (1998) Aluminum transport out of brain extracellular fluid is proton dependent and inhibited by mersalyl acid, suggesting mediation by the monocarboxylate transporter (MCT1). *Toxicology* 127(1–3), 59–67. doi: 10.1016/S0300-483X(98)00037-7.

Argou-Cardozo, I. and Zeidan-Chulia, F. (2018) Clostridia bacteria and autism spectrum conditions: a systematic review and hypothetical contribution of environmental glyphosate levels. *Medical Sciences* 6, 29. doi: 10.3390/medsci6020029.

Baier, C.J., Gallegos, C.E., Raisman-Vozarid, R. and Minetti, A. (2017) Behavioral impairments following repeated intranasal glyphosate-based herbicide administration in mice. *Neurotoxicology and Teratology* 64, 63–72. doi: 10.1016/j.ntt.2017.10.004

Banks, G.T., Kuta, A., Isaacs, A.M. and Fisher, E.M.C. (2008) TDP-43 is a culprit in human neurodegeneration, and not just an innocent bystander. *Mammalian Genome* 19(5), 299–305. doi: 10.1007/s00335-008-9117-x.
Beecham, J.E. and Seneff, S. (2015) The possible link between autism and glyphosate acting as glycine mimetic – a review of evidence from the literature with analysis. *Journal of Molecular Genetics and Medicine* 9, 4. doi: 10.4172/1747-0862.1000187.
Brosig, B. and Langosch, D. (1998) The dimerization motif of the glycophorin A transmembrane segment in membranes: importance of glycine residues. *Protein Science* 7, 1052–1056. doi: 10.1002/pro.5560070423.
Cattani, D., de Liz Oliveira Cavalli, V.L., Rieg, C.E.H., Domingues, J.T., Dal-Cim, T. *et al.* (2014) Mechanisms underlying the neurotoxicity induced by glyphosate-based herbicide in immature rat hippocampus: involvement of glutamate excitotoxicity. *Toxicology* 320, 340–345. doi: 10.1016/j.tox.2014.03.001
Cattani, D., Cesconetto, P.A. Tavares, M.K., Parisotto, E.B., De Oliveira, P.A. *et al.* (2017) Developmental exposure to glyphosate-based herbicide and depressive-like behaviour in adult offspring: implications of glutamate excitotoxicity and oxidative stress. *Toxicology* 387, 67–80.
Caughey, B. and Lansbury, P.T. (2003) Protofibrils, pores, fibrils, and neurodegeneration: separating the responsible protein aggregates from the innocent bystanders. *Annual Review of Neuroscience* 26, 267–298. doi: 10.1146/annurev.neuro.26.010302.081142
Chiu, A.S., Gehringer, M.M., Welch, J.H. and Neilan, B.A. (2011) Does α-amino-β-methylaminopropionic acid (BMAA) play a role in neurodegeneration? *International Journal of Environmental Research and Public Health* 8(9), 3728–3746. doi: 10.3390/ijerph8093728.
Coburn, J.W., Mischel, M.G., Goodman, W.G. and Salusky, I.B. (1991) Calcium citrate markedly enhances aluminum absorption from aluminum hydroxide. *American Journal of Kidney Diseases* 17, 708–711. doi: 10.1016/S0272-6386(12)80356-8.
de Liz Oliveira Cavalli, V.L., Cattani, D., Rieg, C.E.H., Pierozan, P., Zanatta, L. *et al.* (2013) Roundup disrupts male reproductive functions by triggering calcium-mediated cell death in rat testis and Sertoli cells. *Free Radical Biology and Medicine* 65, 335–346. doi: 10.1016/j.freeradbiomed.2013.06.043
Dewey, C.M., Cenik, B., Sephton, C.F., Dries, D.R., Mayer, P. 3rd *et al.* (2011) TDP-43 is directed to stress granules by sorbitol, a novel physiological osmotic and oxidative stressor. *Molecular Cell Biology* 31(5), 1098–1108. doi: 10.1128/MCB.01279-10.
Eastoe, J.E. (1955) The amino acid composition of mammalian collagen and gelatin. *Biochemistry Journal* 61, 589–600.
Eschenburg, S., Healy, M.L., Priestman, M.A., Lushington, G.H. and Schonbrunn, E. (2002) How the mutation glycine 96 to alanine confers glyphosate insensitivity to 5-enolpyruvyl shikimate-3-phosphate synthase from Escherichia coli. *Planta* 216, 129–135. doi: 10.1007/s00425-002-0908-0.
Exley, C. (2014) Why industry propaganda and political interference cannot disguise the inevitable role played by human exposure to aluminum in neurodegenerative diseases, including Alzheimer's disease. *Frontiers in Neurology* 5, 212. doi: 10.3389/fneur.2014.00212.
Finegold, S.M. (2011) Desulfovibrio species are potentially important in regressive autism. *Medical Hypotheses* 77(2), 270–274. doi: 10.1016/j.mehy.2011.04.032.
Florio, T., Paludi, D., Villa, V., Principe, D.R., Corsaro, A. *et al.* (2003) Contribution of two conserved glycine residues to fibrillogenesis of the 106-126 prion protein fragment. Evidence that a soluble variant of the 106-126 peptide is neurotoxic. *Journal of Neurochemistry* 85(1), 62–72. doi: 10.1046/j.1471-4159.2003.01664.x
Gaidukov, L., Nager, A.R., Xu, S., Penman, M. and Krieger, M. (2011) Glycine dimerization motif in the N-terminal transmembrane domain of the high density lipoprotein receptor SR-BI required for normal receptor oligomerization and lipid transport. *Journal of Biological Chemistry* 86(21), 18452–18464. doi: 10.1074/jbc.M111.229872.
Galić, S., Schneider, H.P., Bröer, A., Deitmer, J.W. and Bröer, S. (2003) The loop between helix 4 and helix 5 in the monocarboxylate transporter MCT1 is important for substrate selection and protein stability. *Biochemistry Journal* 376(Pt 2), 413–422. doi: 10.1042/BJ20030799.
Gildea, J.J., Roberts, D.A. and Bush, Z.M. (2017) Protective effects of lignite extract supplement on intestinal barrier function in glyphosate-mediated tight junction injury. *Journal of Clinical and Nutrition and Dietetics* 3, 1. doi: 10.4172/2472-1921.100035.
Good, P. (2018) Evidence that the US autism epidemic initiated by acetaminophen (*Tylenol*) is aggravated by oral antibiotic amoxicillin/clavulanate (*Augmentin*) and now exponentially by herbicide glyphosate (*Roundup*). *Clinical Nutrition ESPEN* 23, 171–183.
Goure, W.F., Krafft, G.A., Jerecic, J. and Franz Hefti, F. (2014) Targeting the proper amyloid-beta neuronal toxins: a path forward for Alzheimer's disease immunotherapeutics. *Alzheimer's Research & Therapy* 6, 42. doi: 10.1186/alzrt272

Hietanen, E., Linnainmaa, K. and Vainio, H. (1983) Effects of phenoxyherbicides and glyphosate on the hepatic and intestinal biotransformation activities in the rat. *Acta Pharmacolica et Toxicologica (Copenhagen)* 53(2), 103–112. doi: 10.1111/j.1600-0773.1983.tb01876.x
Hoerlein, G. (1994) Glufosinate (phosphinothricin), a natural amino acid with unexpected herbicidal properties. *Reviews of Environmental Contamination and Toxicology* 138, 73–145. doi: 10.1007/978-1-4612-2672-7_4
Holm, L., Saraste, M. and Wikstrom, M. (1987) Structural models of the redox centres in cytochrome oxidase. *EMBO Journal* 6, 2819–2823.
Horschitz, S., Lau, T. and Schloss, P. (2008) Glycine residues G338 and G342 are important determinants for serotonin transporter dimerisation and cell surface expression. *Neurochemistry International* 52(4–5), 770–775. doi: 10.1016/j.neuint.2007.09.005.
Ince, P.G. and Codd, G.A. (2005) Return of the cycad hypothesis: does the amyotrophic lateral sclerosis/parkinsonism dementia complex (ALS/PDC) of Guam have new implications for global health? *Neuropathology and Applied Neurobiology* 31, 345–353. doi: 10.1111/j.1365-2990.2005.00686.x.
Irie, F., Badie-Mahdavi, H. and Yamaguchi, Y. (2012) Autism-like socio-communicative deficits and stereotypies in mice lacking heparan sulfate. *Proceedings of the National Academy of Sciences of the United States of America* 109(13), 5052–5056. doi: 10.1073/pnas.1117881109.
Ji, H., Xu, L., Wang, Z, Fan, X. and Wu, L. (2018) Differential microRNA expression in the prefrontal cortex of mouse offspring induced by glyphosate exposure during pregnancy. *Experimental and Therapeutic Medicine* 15, 2457–2467.
JMPR (2016) Joint FAO/WHO Meeting on Pesticide Residues. Available at: https://www.who.int/foodsafety/faq
Kawahara, M. and Kato-Negishi, M. (2011) Link between aluminum and the pathogenesis of Alzheimer's disease: the integration of the aluminum and amyloid cascade hypotheses. *International Journal of Alzheimer's Disease* 2011, article 276393. doi: 10.4061/2011/276393.
Kim, Y.-H., Hong, J.-R., Gil, H.-W., Song, H.-Y. and Hong, S.-Y. (2013) Mixtures of glyphosate and surfactant TN20 accelerate cell death via mitochondrial damage-induced apoptosis and necrosis. *Toxicology in Vitro* 27, 191–197. doi: 10.1016/j.tiv.2012.09.021.
Kinose, F., Wang, S.X., Kidambi, U.S., Moncman, C.L. and Winkelmann, D.A. (1996) Glycine 699 is pivotal for the motor activity of skeletal muscle myosin. *Journal of Cell Biology* 134(4), 895–909.
Kisker, C., Schindelin, H., Pacheco, A., Wehbi, W.A., Garrett, R.M. *et al.* (1997) Molecular basis of sulfite oxidase deficiency from the structure of sulfite oxidase. *Cell* 91(7), 973–983. doi: 10.1016/S0092-8674(00)80488-2.
Kitchen, L.M., Witt, W.M. and Rieck C.E. (1981) Inhibition of δ-aminolevulinic acid synthesis by glyphosate. *Weed Science* 29(5), 571–577. doi: 10.1017/S004317450006375X.
Krakauer, J., Long Kolbert, A., Thanedar, S. and Southard, J. (2015) Presence of L-canavanine in *Hedysarum alpinum* seeds and its potential role in the death of Chris McCandless. *Wilderness & Environmental Medicine* 26, 36–42. doi: 10.1016/j.wem.2014.08.014.
Langford, J.K., Stanley, M.J., Cao, D. and Sanderson, R.D. (1998) Multiple heparan sulfate chains are required for optimal syndecan-1 function. *Journal of Biological Chemistry* 273(45), 29965–29971.
Laugeray, A., Herzine, A., Perche, O., Hébert, B., Aguillon-Naury, M. *et al.* (2014) Pre- and postnatal exposure to low dose glufosinate ammonium induces autism-like phenotypes in mice. *Frontiers in Behavioral Neuroscience* 8, 390. doi: 10.3389/fnbeh.2014.00390.
McKinnon, C., Goold, R., Andre, R., Devoy, A., Ortega, Z. *et al.* (2016) Prion-mediated neurodegeneration is associated with early impairment of the ubiquitin-proteasome system. *Acta Neuropathology* 131(3), 411–425. doi: 10.1007/s00401-015-1508-y.
Mirza, A., King, A., Troakes, C. and Exley, C. (2017) Aluminium in brain tissue in familial Alzheimer's disease. *Journal of Trace Elements in Medicine and Biology* 40, 30–36. doi: 10.1016/j.jtemb.2016.12.001.
Munter, L.-M., Voigt, P., Harmeier, A., Kaden, D., Gottschalk, K.E. *et al.* (2007) GxxxG motifs within the amyloid precursor protein transmembrane sequence are critical for the etiology of Ab42. *The EMBO Journal* 26, 1702–1712. doi: 10.1038/sj.emboj.7601616.
Murch, S.J., Cox, P.A. and Banack, S.A. (2004) A mechanism for slow release of biomagnified cyanobacterial neurotoxins and neurodegenerative disease in Guam. *Proceedings of the National Academy of Sciences USA* 101, 12228–12231. doi: 10.1073/pnas.0404926101.
Myers, J.P., Antoniou, M.J., Blumberg, B., Carroll, L., Colborn, T. *et al.* (2016) Concerns over the use of glyphosate-based herbicides and risks associated with exposures: a consensus statement. *Environmental Health* 15, 19. doi: 10.1186/s12940-016-0117-0.

Pan, K.M., Baldwin, M., Nguyen, J., Gasset, M., Serban, A. *et al.* (1993) Conversion of alpha-helices into beta-sheets features in the formation of the scrapie prion proteins. *Proceedings of the National Academy of Sciences of the United States of America* 90(23), 10962–10966.

Peixoto, F. (2005) Comparative effects of the Roundup and glyphosate on mitochondrial oxidative phosphorylation. *Chemosphere* 61, 1115–1122. doi: 10.1016/j.chemosphere.2005.03.044.

Pérez, C., Sawmiller, D. and Tan, J. (2016) The role of heparan sulfate deficiency in autistic phenotype: potential involvement of Slit/Robo/srGAPs-mediated dendritic spine formation. *Neural Development* 11, 11. doi: 10.1186/s13064-016-0066-x.

Prescott, S. and Allen, K.J. (2011) Food allergy: riding the second wave of the allergy epidemic. *Pediatric Allergy and Immunology* 22(2), 155–160. doi: 10.1111/j.1399-3038.2011.01145.x.

Purgel, M., Takács, Z., Jonsson, C.M., Nagy, L., Andersson, I. *et al.* (2009) Glyphosate complexation to aluminium(III). An equilibrium and structural study in solution using potentiometry, multinuclear NMR, ATR-FTIR, ESI-MS and DFT calculations. *Journal of Inorganic Biochemistry* 103, 1426–1438. doi: 10.1016/j.jinorgbio.2009.06.011.

Ray, S., Sonthalia, N., Kundu, S. and Ganguly, S. (2012) Autoimmune disorders: an overview of molecular and cellular basis in today's perspective. *Journal of Clinical & Cellular Immunology* S10, 003. doi: 10.4172/2155-9899.S10-003.

Ridley, W.P. and Chott, K.A. (1989) Uptake, depuration and bioconcentration of C-14 glyphosate to bluegill sunfish (*Lepomis machrochirus*). Part II: Characterization and quantitation of glyphosate and its metabolites. Monsanto Agricultural Company, St Louis, Missouri (unpublished study).

Rodgers, K.J. and Shiozawa, N. (2008) Misincorporation of amino acid analogues into proteins by biosynthesis. *International Journal of biochemistry and Cell Biology* 40(8), 1452–1466. doi: 10.1016/j.biocel.2008.01.009.

Rosenthal, G.A. (1990) The biochemical basis for the deleterious effects of L-canavanine. *Phytochemistry* 30, 1055–1058.

Rubenstein, E. (2000) Biologic effects of and clinical disorders caused by nonprotein amino acids. *Medicine (Baltimore)* 79, 80–89.

Rubenstein, E. (2008) Misincorporation of the proline analog azetidine-2-carboxylic acid in the pathogenesis of multiple sclerosis: a hypothesis. *Journal of Neuropathology and Experimental Neurology* 67(11), 1035–1040. doi: 10.1097/NEN.0b013e31818add4a.

Samsel, A. and Seneff, S. (2016) Glyphosate, pathways to modern diseases V: Amino acid analogue of glycine in diverse proteins. *Journal of Biological Physics and Chemistry* 16, 9–46. doi: 10.4024/03SA16A.jbpc.16.01.

Samsel, A. and Seneff, S. (2017) Glyphosate pathways to modern diseases VI: Prions, amyloidoses and autoimmune neurological diseases. *Journal of Biological Physics and Chemistry* 17, 8–32. doi: 10.4024/25SA16A.jbpc.17.01.

Seneff, S. and Nigh, G.L. (2017) Glyphosate and anencephaly: death by a thousand cuts. *Journal of Neurology and Neurobiology* 3(2), 1–15. doi: 10.16966/2379-7150.140.

Seneff, S. and Orlando, L.F. (2018) Glyphosate substitution for glycine during protein synthesis as a causal factor in Mesoamerican nephropathy. *Journal of Environmental and Analytical Toxicology* 8, 1. doi: 10.4172/2161-0525.1000541.

Seneff, S., Swanson, N. and Li, C. (2015) Aluminum and glyphosate can synergistically induce pineal gland pathology: connection to gut dysbiosis and neurological disease. *Agricultural Sciences* 6, 42–70. doi: 10.4236/as.2015.61005.

Seneff, S., Morley, W.A., Hadden, M.J. and Michener, M.C. (2016) Does glyphosate acting as a glycine analogue contribute to ALS? *Journal of Bioinformatics and Proteomics Review* 2, 3. doi: 10.15436/2381-0793.16.1173.

Senes, A., Engel, D.E. and DeGrado, W.F. (2004) Folding of helical membrane proteins: the role of polar, GxxxG-like and proline motifs. *Current Opinion in Structural Biology* 14(4), 465–479. doi: 10.1016/j.sbi.2004.07.007.

Shaw, W. (2017) Elevated urinary glyphosate and Clostridia metabolites with altered dopamine metabolism in triplets with autistic spectrum disorder or suspected seizure disorder: a case study. *Integrative Medicine: A Clinician's Journal* 16, 50–57.

Shehata, A.A., Schrödl, W., Aldin, A.A., Hafez, H.M. and Krüger, M. (2013) The effect of glyphosate on potential pathogens and beneficial members of poultry microbiota in vitro. *Current Microbiology* 66, 350–358. doi: 10.1007/s00284-012-0277-2.

Spencer, P.S., Nunn, P.B., Hugon, J., Ludolph, A.C., Ross, S.M., Roy, D.N. and Robertson, R.C. (1987) Guam amyotrophic lateral sclerosis-parkinsonism-dementia linked to a plant excitant neurotoxin. *Science* 237(4814), 517–522.

Splawski, I., Timothy, K.W., Sharpe, L.M., Decher, N., Kumar, P. *et al.* (2004) Ca(V)1.2 calcium channel dysfunction causes a multisystem disorder including arrhythmia and autism. *Cell* 119(1), 19–31. doi: 10.1016/j.cell.2004.09.011.

Swanson, N.L., Leu, A., Abrahamson, J. and Wallet, B. (2014) Genetically engineered crops, glyphosate and the deterioration of health in the United States of America. *Journal of Organic Systems* 9, 6–37.

Syed, K. and Mashele, S.S. (2014) Comparative analysis of P450 signature motifs EXXR and CXG in the large and diverse kingdom of fungi: identification of evolutionarily conserved amino acid patterns characteristic of P450 family. *PLoS ONE* 9(4), e95616. doi: 10.1371/journal.pone.0095616.

Teng, J., Iida, K., Ito, M., Izumi-Nakaseko, H., Kojima, I., Adachi-Akahane, S. and Iida, H. (2010) Role of glycine residues highly conserved in the S2-S3 linkers of domains I and II of voltage-gated calcium channel alpha(1) subunits. *Biochimica et Biophysica Acta* 1798(5), 966–974. doi: 10.1016/j.bbamem.2010.01.004.

Vekemans, S. and David, G. (1999) Molecular polymorphism of the syndecans. Identification of a hypoglycanated murine syndecan-1 splice variant. *Journal of Biological Chemistry* 274(26), 18667–18674. doi: 10.1074/jbc.274.26.18667.

von Bergen, M., Barghorn, S., Biernat, J., Mandelkow, E.M. and Mandelkow, E. (2005) Tau aggregation is driven by a transition from random coil to beta sheet structure. *Biochimica et Biophysica Acta* 1739(2–3), 158–166. doi: 10.1016/j.bbadis.2004.09.010.

Wang, W., Bodles-Brakhop, A.M. and Barger, SW. (2016) A role for p-glycoprotein in clearance of Alzheimer amyloid β-peptide from the brain. *Current Alzheimer Research* 13(6), 615–620. doi: 10.2174/156720 5013666160314151012.

Wen, P.C., Verhalen, B., Wilkens, S., Mchaourab, H.S. and Tajkhorshid, E. (2013) On the origin of large flexibility of P-glycoprotein in the inward-facing state. *Journal of Biological Chemistry* 288(26), 19211–19220. doi: 10.1074/jbc.M113.450114.

Wertheimer, E., Barbetti, F., Muggeo, M., Roth, J. and Taylor, S.I. (1994) Two mutations in a conserved structural motif in the insulin receptor inhibit normal folding and intracellular transport of the receptor. *Journal of Biological Chemistry* 269, 7587–7592.

Wills, M.R., and Savory, J. (1985) Water content of aluminum, dialysis dementia, and osteomalacia. *Environmental Health Perspectives* 63, 141–147.

Xu, P., Duong, D.M., Seyfried, N.T., Cheng, D., Xie, Y. *et al.* (2009) Quantitative proteomics reveals the function of unconventional ubiquitin chains in proteasomal degradation. *Cell* 137, 133–145. doi: 10.1016/j.cell.2009.01.041.

Xu, J., Li, G., Wang, Z., Si, L., He, S., Cai, J., Huang, J. and Donovan, M.D. (2016) The role of L-type amino acid transporters in the uptake of glyphosate across mammalian epithelial tissues. *Chemosphere* 145, 487–494. doi: 10.1016/j.chemosphere.2015.11.062.

Yokel, R.A., Wilson, M., Harris, W.R. and Halestrap, A.P. (2002) Aluminum citrate uptake by immortalized brain endothelial cells: implications for its blood–brain barrier transport. *Brain Research* 930(1–2), 101–110. doi: 10.1016/S0006-8993(02)02234-5.

Zhao, Y., Hill, J.M., Bhattacharjee, S., Percy, M.E., Pogue, A.I.D. and Lukiw, W.J. (2014) Aluminum-induced amyloidogenesis and impairment in the clearance of amyloid peptides from the central nervous system in Alzheimer's disease. *Frontiers in Neurology* 5, 167. doi: 10.3389/fneur.2014.00167.

# Part IV

# Petroleum Pollution

# 20 Crude Oil Pollution I. *Deepwater Horizon* Contamination: Human Health Effects and Health Risk Assessments, a Case Study

**M.J. Wilson***

*Department of Global Environmental Health Sciences, Tulane University School of Public Health and Tropical Medicine, New Orleans, Louisiana, USA*

## 20.1 Abstract

On 20 April 2010 the *Deepwater Horizon* mobile oil exploration platform experienced a devastating accident that led to the largest marine oil spill in history. This event led to concern about health risk due to oil-related chemical exposures in oil spill response personnel, including workers and volunteers, and community members in Texas, Louisiana, Mississippi, Alabama and Florida. Thousands of individuals were involved in clean-up operations and millions were potentially at risk due to contamination of seafood. Occupational exposure monitoring and sensory and chemical analysis of seafood were conducted in order to determine the potential for exposure to workers and community members via inhalation or dietary exposure routes. The US Environmental Protection Agency (EPA) employed air monitoring to determine if community members were likely to be exposed to oil-associated volatile organic compounds. The US Food and Drug Administration (FDA) and National Oceanic and Atmospheric Administration (NOAA) developed fishery closure and re-opening guidelines designed to minimize potential dietary exposures. The occupational monitoring determined that clean-up workers experienced exposures that were below accepted occupational exposure limits and there was minimal exposure potential to community members associated with volatile compound that were released from the crude oil. The potential for dietary exposure to oil-related chemicals via seafood consumption was minimal. The *Deepwater Horizon* event provided an opportunity to evaluate the current regulatory and health risk assessment framework, which is primarily designed to deal with individual chemicals, as it applies to a complex scenario that involved mixtures of chemicals, non-chemical agents (such as psychological stress) and physical agents (such as heat). Reconciling the complexity of this event with the current regulatory health risk assessment process proved to be a difficult endeavour and highlighted many of the limitations of the current approach.

## 20.2 Introduction

Paracelsus tells us that it is the dose that makes the poison. This maxim lays the foundation for

* E-mail address: mwilson9@tulane.edu

the field of toxicology. It is well established that chemical, physical and biological hazards exist in our environment and upon exposure of sufficient magnitude, duration and frequency some hazards may cause deleterious effects on human health. A simple conceptual model represented by the idea of the completed exposure pathway can summarize the relationship between health risks as they relate to both exposures and hazards. The conceptual model is:

Hazard × Exposure = Health Risk

When chemically mediated health risks are evaluated under this conceptual model, it is apparent that risk (the probability of a negative outcome) is a function of the presence of an environmental hazard coupled with exposure to that hazard. That is to say, in the absence of either hazard or exposure there is no risk of adverse health effects. In addition to providing a framework to evaluate the probability of negative outcomes, this conceptual model allows risk managers, with the input of toxicologists and health risk assessors, to determine levels of concern (LOCs) or occupational exposure limits (OELs) for either hazards or exposures that correspond to acceptable levels of health risk. Typically environmental hazards are quantified by measuring the concentration of the hazard in some environmental media, such as air, water, food, soil, etc., and these measured levels may then be directly compared to LOCs or OELs to determine whether there is a possibility for health risks to be present. The exposure component of the conceptual model not only deals with the magnitude, frequency and duration of contact with the contaminated environmental media but also should consider the bioavailability and toxicokinetics of chemical hazards.

## 20.3 Health Risk Assessment

This simple exposure pathway-based conceptual model is an appropriate framework within which to evaluate both the human health risk assessment process and epidemiological investigations that seek to link chemical exposures to adverse health effects.

The overlap of the exposure pathway-based conceptual model and the four distinct steps of the health risk assessment process is shown in Fig. 20.1.

### 20.3.1 Hazard identification

Step one of the health risk assessment process is hazard identification. Here the particular chemical (or class of chemicals), physical or biological

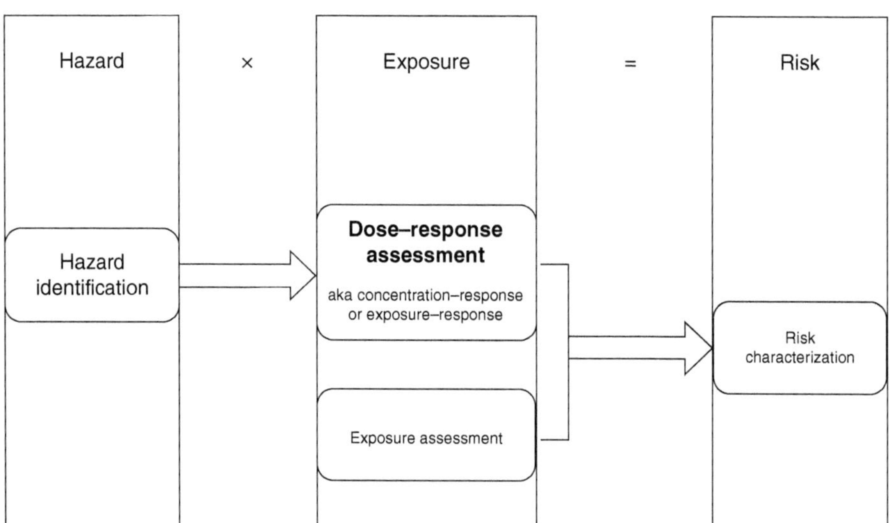


**Fig. 20.1.** Conceptual model and regulatory human health risk assessment framework.

agent of concern is selected. Based on the weight of evidence associated with the clinical, epidemiological and toxicological literature base, likely health outcomes related to exposure to the hazard are also identified. Following identification, hazards may be quantified via sampling and analysis of various environmental media such as air, water, food and soil.

### 20.3.2 Dose–response assessment

Step two of the health risk assessment process involves the evaluation of the dose–response relationship between the hazard and the health outcome of interest. In general it is anticipated that the severity of effect will increase with a greater magnitude of exposure. Typically, dose response is evaluated by one of two models based on the mode of action for toxicity associated with the specific hazard and related health outcome. The first dose or exposure response model to consider is a threshold-based non-linear model where there is a defined dose or exposure range that corresponds to no detectable effect. The top of this range is often referred to as the No Observable Effect Level (NOEL) or No Observable Adverse Effect Level (NOAEL). The highest data point of the NOAEL region is used in tandem with uncertainty factors to derive Reference Doses (RfDs) and Reference Concentrations (RfCs), which represent lifetime daily doses or exposures for individual chemicals that are not anticipated to be associated with negative health effects (EPA, 2002).

A linear model is used to evaluate the probability of negative health effects for chemicals and health outcomes that are not anticipated to have thresholds. Cancer risk, for example, is evaluated with a non-threshold linear model. It is assumed that no exposure to a carcinogen is without an incremental increase in risk (EPA, 2005). The potency of the carcinogen is equal to the slope of the linear dose–response model and is referred to as the cancer slope factor or cancer potency factor.

### 20.3.3 Exposure assessment

The third step of the health risk assessment process is exposure assessment. During this phase the magnitude, duration and frequency of exposure to the hazard of interest is determined. Robust exposure assessments should not only identify if an exposure has occurred or is likely to occur but also it should identify the route (inhalation, dermal, ingestion, etc.) by which the exposure occurred or is likely to occur.

Often in community-based risk assessments the exposure assessment process relies on assumptions related to environmental media contact rates, such as from the EPA exposure factors handbook (EPA, 2011). These contact rates allow for the estimation of mass of chemical ingested when coupled with chemical analysis data from environmental media samples. Often occupational exposure assessments, for example with airborne exposures, will involve placing samplers on individual workers. This type of approach allows for the integration of time spent in an area with actual environmental conditions. While it is possible to determine the average concentration of a pollutant in the breathing zone, this approach does not account for the use of personal protective equipment (PPE) and unless PPE use is accounted for, such as with the application of a respiratory protection factor, the actual exposure may be significantly overestimated.

Frequently it is not feasible to sample each individual in a given population and so sampling data from individuals or even data from area sampling is extrapolated to groups of people who are likely to have experienced a similar exposure. The development of a job-exposure matrix is an example of this approach. Biomonitoring also provides an avenue to determine if exposure has occurred. Biomonitoring for either the chemical of interest or its metabolites provides unequivocal evidence of whether an exposure has occurred or not. One of the limitations of this approach is that it cannot identify the route of exposure, only that an exposure has occurred. This is problematic for the application of biomonitoring strategies for chemicals that are frequently encountered in daily life. This also limits the utility of biomonitoring in efforts to make definitive statements regarding the causality of specific exposure scenarios as they relate to chemically mediated health effects.

In epidemiological evaluations, exposure assessment is often based on self-reported exposures or proximity to an environmental hazard (Niehoff *et al.*, 2016; Tsuda *et al.*, 2016). The reliance on self-reported exposures can be problematic as they are at best qualitative. This makes reconciling exposure metrics with dose–response information difficult. Furthermore, recall bias can be a significant confounding issue when relying on self-reported exposures or effects (Deziel *et al.*, 2015). Each of the exposure assessment strategies mentioned has its own strengths and weaknesses. The most robust exposure estimates may employ more than one strategy, such as personal or environmental sampling, exposure matrices and biomonitoring, in tandem (Friesen *et al.*, 2012; Ji *et al.*, 2012). Exposure assessment is often the weak link in both health risk assessments and environmental epidemiological evaluations, yet it is a critical aspect of determining whether a chemically mediated effect has or is likely to occur. Care must be taken when evaluating the toxicological and epidemiological literature to determine the overall confidence in the exposure assessment metrics reported, based on the strategies employed in individual studies.

### 20.3.4 Risk characterization

The fourth and final step of the health risk assessment process is risk characterization. If the mode of action for a specific chemical hazard is based on a non-linear threshold model, then the estimated dose or exposure can be compared with an RfD, RfC or OEL by setting up a simple ratio, often referred to as a hazard quotient, with the estimated dose divided by the RfD, RfC or OEL. If the value from the ratio exceeds 1, it indicates the presence of health risk associated with a given exposure scenario. In other words, a hazard quotient that is greater than 1 indicates that the threshold for toxicity based on the non-linear dose–response model has been exceeded. If a non-threshold linear model is used, then the slope of the line is multiplied by the estimated daily dose of the hazard, resulting in a probability of an outcome. In regulatory cancer health risk assessment, the acceptable range for risks assessed with linear models is from $1 \times 10^{-4}$ to $1 \times 10^{-6}$. For cancer risk assessments these values correspond to the number of cancers, above the background in a population, specifically related to a given exposure scenario. For example, the interpretation of a risk level of $1 \times 10^{-4}$ would be an expectation of one excess case of cancer per 10,000 people that is specifically related to the exposure of interest. It is important to note that a risk level that exceeds $1 \times 10^{-4}$ would be considered unacceptable, while a risk level that is less than $1 \times 10^{-6}$, such as $1 \times 10^{-8}$, would be acceptable. The ultimate decision on what constitutes an acceptable risk level is up to the individual risk assessor but, in general, acceptable risk levels will not exceed $1 \times 10^{-4}$. The units of dose are typically expressed as mass of chemical per unit body weight per unit time. Common units of dose are mg $kg^{-1}$ per day. Slope or potency factors have units of (mg $kg^{-1}$ per day)$^{-1}$. When estimated daily doses are multiplied by a slope factor, the result is a unitless probability that indicates the likelihood of a negative health outcome occurring. If this probability exceeds the acceptable risk level, then it indicates that an exposure has occurred that may increase the risk of cancer to an unacceptable level in a given population.

## 20.4 Risk in Epidemiology

In environmental epidemiology studies the concept of risk is often presented as a relative risk, prevalence ratio, or odds ratio. This metric measures the strength of association between an exposure and health outcome and may be interpreted in a similar fashion to a hazard quotient. Relative risks and odds ratios that exceed 1, with confidence intervals that do not contain 1, represent a statistically significant positive association between a given exposure scenario and health outcome. This method is fundamentally different from how risk is conceived in the regulatory health risk assessment process but still provides valuable information related to how real-world environmental exposures may impact rates of adverse health outcomes. The health risk assessment process uses either acceptable risk ranges or lifetime average daily doses compared with acceptable exposures as benchmarks for decisions about whether environmental exposures may pose health risks. Case-control and cohort epidemiological studies identify the strength of association between a given exposure and health outcome based on the number of people with the health outcome in the exposed group compared with the number of people with the health outcome in the unexposed control group. The risk metrics from epidemiology often inform the health risk assessment process and are representative of potential exposure scenarios in the real world. However, direct comparison of an odds ratio or a risk ratio to hazard quotients or acceptable ranges of risk is not possible, because they represent wholly different conceptual approaches.

## 20.5 *Deepwater Horizon* Oil Spill: Background

The *Deepwater Horizon* oil spill in the Gulf of Mexico was the largest event of its kind and is the only oil spill to be officially declared a 'Spill of National Significance' under the National Contingency Plan of the Oil Pollution Act by the US government (Lubchenco *et al.*, 2012). The blowout of the Macondo wellhead and subsequent explosion and destruction of the *Deepwater Horizon* mobile oil-drilling platform occurred on 20 April 2010. The event led to 11 fatalities and 17 critical injuries among the crew of the *Deepwater Horizon* oil-drilling platform (CSHIB, 2014). This was followed by an 87-day period during which an estimated 4.9 million gallons (18.5 million litres) of crude oil spilled out into the Gulf of Mexico (McNutt *et al.*, 2012). Crude oil contaminated approximately 1700 miles of shorelines and beaches throughout all states within the Gulf South region (Michel *et al.*, 2013).

In response to this massive environmental catastrophe, thousands of workers and volunteers mobilized for onshore and offshore operations to contain and mitigate oil migration and to clean up coastal areas and wildlife that were already directly impacted (DHSS, 2011). Clean-up activities included booming and skimming of surface waters, *in situ* burning of surface oil, removal of oil and tarballs from beaches, wildlife and shorelines, cleaning vessels that came into contact with surface oil, and the application of subsurface dispersants (Kwok *et al.*, 2017a; Stewart *et al.*, 2017). In addition approximately 1.8 million gallons (6.8 million litres) of chemical dispersants were applied to surface and subsurface oil (McGowan *et al.*, 2017).

## 20.6 Chemical Hazards

During the oil spill event and during response and clean-up operations there was a potential for exposure to chemical constituents of crude oil, dispersed crude oil and combustion products associated with burning oil among clean-up workers, volunteers and community members (Goldstein *et al.*, 2011; Middlebrook *et al.*, 2012; Reddy *et al.*, 2012; Black *et al.*, 2016 ). Crude oil is a complex chemical mixture and some of the chemical constituents represent chemical hazards. These hazards include aliphatic, cyclic and aromatic hydrocarbons that are relevant to adverse human health outcomes. The crude oil associated with the *Deepwater Horizon* event originated from a well within the Macondo Prospect oil field designated Mississippi Canyon Block 252 (MS252) (NOAA, 2010). One class of chemicals of concern in crude oil included volatile organic compounds (VOCs), especially benzene, toulene, ethylbenzene and xylene (BTEX). VOCs have significant vapour pressure and are likely to represent airborne inhalation hazards. The Macondo well is located approximately 50 miles offshore of Louisiana. Due to the distance of the well from the shore and environmental conditions, such as heat and sunlight, the chemical characteristics of oil that reached the shorelines were different in composition than fresh crude oil leaking directly from the well. The change in composition is due to a process known as weathering. Weathered oil tends to have a lower percentage composition of volatile compounds relative to fresh crude oil, due to the loss of VOCs to the atmosphere during the weathering process. Weathered oil is not only different in chemical composition but also differs in terms of its toxicological properties compared with fresh crude oil. Air monitoring conducted by the EPA from 28 April to 28 August 2010 demonstrated that the VOC levels, including BTEX, were low in the weathered MC252 oil (EPA, 2010).

Another class of chemicals of concern associated with the *Deepwater Horizon* event was polycyclic aromatic hydrocarbons (PAHs). These compounds belong to a large chemical family of planar aromatic hydrocarbons, some of which are known carcinogens (Nisbet and LaGoy, 1992). Compared with VOCs, PAHs are much heavier, they do not have significant vapour pressure and are unlikely to be inhalation hazards. All crude oil contains some amount of PAHs but light crude tends to contain less PAHs, including carcinogenic PAHs, compared with heavy crude oil (NRC TRB, 2003). PAHs are also associated with combustion of organic material. As such they are frequently found in the environment in the absence of oil contamination. PAHs were thought to be the primary toxicological risk associated with contact with weathered oil.

## 20.7 Exposure Routes

There are two primary routes of exposure that were a concern for both clean-up workers and community members during and after the *Deepwater Horizon* oil spill. Inhalation exposures were of concern in particular for clean-up workers. This group was the most likely to come into contact with fresh or lightly weathered crude oil that still had a significant VOC emission potential. Inhalation exposure to VOCs was a lesser concern for community members, due to the fact that oil had weathered significantly before it reached land. The other primary exposure route of concern was ingestion related to the consumption of contaminated seafood. Due to the physical properties of VOCs, they are not likely to contaminate seafood, but the heavier hydrocarbons associated with weathered oil may. Coastal populations tend to consume seafood and so this was the primary exposure scenario that was relevant for residents of the affected areas. If this is framed as a health risk assessment issue, two distinct classes of chemical hazards, VOCs and PAHs, were identified. VOCs, in particular BTEX, and PAHs have been associated with both acute and chronic health effects.

## 20.8 Oil Spill Response Worker Exposure

There was extensive air monitoring carried out to assess exposure potential in clean-up workers. The National Institute of Occupational Safety and Health (NIOSH) conducted a formal health hazard evaluation (HHE) focused on oil spill response-worker safety (NIOSH, 2011). From an exposure standpoint this group, which consisted of onshore and offshore personnel, represented the people most at risk for inhalation and dermal chemical exposures. The increased exposure potential via these routes of exposure was a function of the activities the workers were involved in and their proximity to fresh and lightly weathered crude oil.

There is a wealth of toxicological, epidemiological and occupational studies that inform the understanding of the dose–response relationship of these hazards as they relate to health outcomes (ATSDR, 1995, 2004). VOC exposures were assessed and evaluated using a threshold based approach where OELs, such as Occupational Safety and Health Administration (OSHA) permissible exposure limits (PELs) or American Conference of Governmental Industrial Hygienists (ACGIH) threshold limit values (TLVs), were used as benchmarks for comparison with personal and area air sampling data. None of the air sampling data exceeded these OELs. Additional secondary analysis of publicly available air monitoring data collected by British Petroleum (BP) indicated that the largest contributor to BTEX exposure among response workers was not VOCs originating from the oil but from diesel exhaust emissions from ships (Avens *et al.*, 2011). When the exposure data related to VOCs, and BTEX in particular, are evaluated it is clear that no excess risk should be expected in this group as their exposure levels are below OELs that are designed to protect worker health. The primary hazards that were identified in the NIOSH HHE were associated with physical and ergonomic hazards. The self-reported symptom surveys conducted as part of the HHE identified headache, respiratory irritation and nausea as common symptoms among response workers. Several of the response workers who reported symptoms had signs of dehydration or a diagnosis of heat exhaustion. Given this information, it is difficult to ascribe symptoms specifically to chemical exposures for two reasons. The first reason is the low VOC exposure levels, which were based on air sampling data that should be representative of the environmental exposure potential. The second reason is that the symptom surveys and review of clinical medical records cannot parse out the contribution of heat stress versus chemical exposures, as they relate to the symptoms reported by the personnel.

In addition to the HHE, which focused on monitoring for acute health effects and exposures, there are two large cohort studies that are focused on respiratory and general health evaluations of spill response workers. The first is the coast guard cohort study, which evaluated the association between self-reported oil exposure and respiratory symptoms in over 8500 US Coast Guard (USCG) service members who were deployed as part of the spill response operation (Alexander *et al.*, 2018). Findings from this study indicate that oil exposure is correlated with increased reports of adverse respiratory symptoms. This finding is in line with other epidemiological studies focused on communities and workers who have been impacted by previous oil spills (reviewed in Aguilera *et al.*, 2010; Laffon *et al.*, 2016). The exposure metric in this study is based on a five-point Likert scale answer to the survey question, 'How often were you exposed to crude oil/oily water via inhalation/direct skin contact/ingestion/submersion?' This information was then distilled into three categorical exposure groups (no exposure, low exposure, high exposure) for statistical analysis. Reconciling these categorical exposure groups with dose–response models is difficult and making sense of the apparent chemically mediated self-reported effects and exposures with the air monitoring data collected during the event is challenging as well. This cohort study also examined the effect of exposure, again self-reported, to not only crude oil but to dispersed oil as well. The study reported a clear increase in the prevalence rate for acute respiratory effects among workers exposed to an oil/dispersant mixture. This effect, seen in a cross-sectional epidemiological evaluation, can clearly inform the health risk assessment process as it identifies additional chemical hazards, dispersants and dispersed oil, that should be included, in addition to VOCs and PAHs, for hazard evaluations in future oil spills.

The second large cohort study is focused on long-term health outcomes for oil spill clean-up workers and volunteers. It is sponsored by the National Institutes of Health (NIH) through the National Institute of Environmental Health Sciences (NIEHS). This study is designed to determine how participation in different types of oil spill response activities impact current and future health (Kwok *et al.*, 2017a). This study, known as the GuLF study, enrolled approximately 33,000 participants between 2011 and 2013. Of this group over 11,000 individuals who either worked or were trained to work in clean-up operations completed home survey examinations and supplied biospecimens for future analysis (Engel *et al.*, 2017). A subset of study participants completed a clinical evaluation that included pulmonary function testing, neurological function assessment, physiological measurements and a mental health evaluation. The exposure assessment in this study is based on a job exposure matrix that links job titles with environmental sampling, resulting in an extrapolated exposure value based on the amount of time individual study participants were engaged in specific activities (Stewart *et al.*, 2017). The exposure metrics indicate that none of the study participants experienced exposures that exceeded OELs. Recent publications related to the evaluation of health outcome data from this study have not found long-term respiratory effects among individuals who were involved in oil spill response and clean-up operations (Gam *et al.* 2018a, b). BTEX biomonitoring was also conducted on blood samples from a subset ($n = 849$) of study participants. The values were compared with the National Health and Nutrition Examination Survey (NHANES) database and the primary predictor that was associated with BTEX detection in blood was determined to be smoking status (Werder *et al.*, 2017). A primary limitation of this study is that the biospecimen collection occurred 2–3 years after the *Deepwater Horizon* event. The half-lives of BTEX compounds range from hours to days so it would not be possible to link these biomonitoring data to the oil spill in any sort of temporally logical way. This points out the need to develop biomonitoring sampling strategies that focus on collecting samples at relevant times to augment exposure assessments that rely on self-reported exposures and/or environmental monitoring. If this approach is taken in the future, then it will be possible to reconstruct doses related to specific events, such as exposures related to environmental disasters, and reconcile the exposure levels with dose–response models in meaningful ways.

## 20.9 Coastal Community Member Exposure

Exposure of oil spill chemicals to coastal community members was also a concern. One study evaluated the BTEX levels in a pooled blood sample and determined that the detection of BTEX in blood samples was evidence of community-wide exposures to BTEX (Sammarco *et al.*, 2016). This finding is in opposition to the EPA's VOC air monitoring data, which showed weathered oil contained low levels of VOCs including BTEX compounds (EPA, 2010). The study also does not attempt to account for other environmental sources of BTEX, such as cigarette smoke or consumer products. A search of the consumer products database for benzene, toluene, ethylbenzene and xylene returns over 1500 individual consumer products across ten product classes (Fig. 20.2) (DHSS, 2017). While detection of BTEX in blood samples demonstrates that an exposure has occurred, it does not provide direct evidence that the exposure should be attributed to a specific event.

Contamination of seafood was a significant concern related to the potential exposure to PAHs from crude oil. The US Food and Drug Administration (FDA) worked with the National Oceanographic and Atmospheric Administration (NOAA) to make regulatory decisions related to fishery closures and to develop a specific protocol for fishery reopening (FDA, 2010). The closure of a fishery assumed a worst-case scenario and was intended to protect seafood consumers. There were two chemical hazard scenarios that were evaluated during the reopening protocol. The first was the sensory evaluation, aka 'the sniff test' for crude oil taint on seafood. Basically, if seafood smelled like oil it was deemed unfit for consumption and the fishery remained closed. The second chemical hazard of concern was PAH contamination of the edible portion of seafood. These levels were assessed by mass spectroscopy chemical analysis. A level of concern (LOC) was established that corresponded to the acceptable cancer risk level of $1 \times 10^{-5}$ for carcinogenic PAHs and levels corresponding to average daily doses that would not exceed RfDs for non-carcinogenic PAHs (Dickey, 2012). These LOCs were based on an assumed bodyweight of 80 kg, a seafood consumption rate corresponding the NHANES 90th percentile for shrimp, crab, oysters and finfish with an assumed exposure duration of 5 years averaged over a 78-year lifetime. No seafood samples exceeding the LOCs for either cancer or non-cancer health outcomes have been reported (Gohlke *et al.*, 2011; Xia *et al.*, 2012). It is worth mentioning that smoked and chargrilled foods

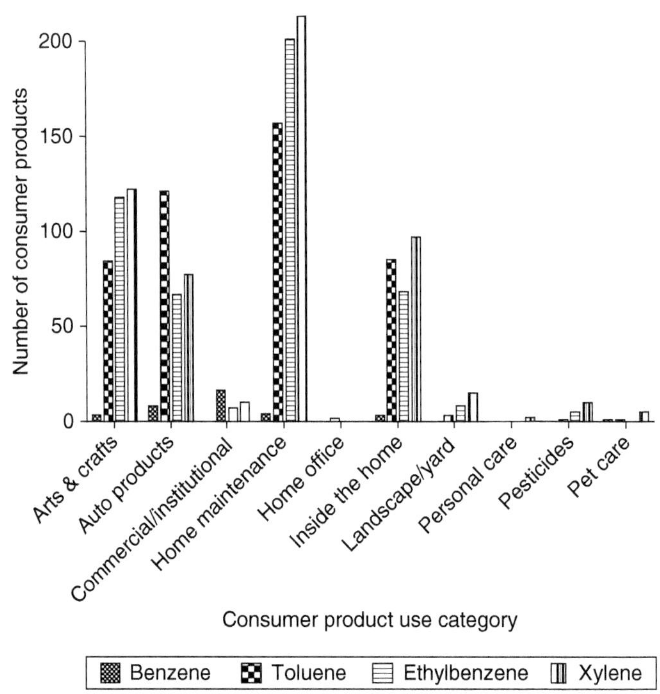


**Fig. 20.2.** Consumer products containing benzene, toluene, ethylbenzene and xylene, by category.

represent a much larger exposure potential to PAHs than would be likely from seafood consumption (Xia *et al.*, 2012). The assumptions, which were the basis of the LOC determination, were called into question in terms of their ability to protect the health of high-end seafood consumers. The rationale for the selection of $1 \times 10^{-5}$, as opposed to the more health conservative $1 \times 10^{-6}$, as the acceptable cancer risk level was also called into question (Rotkin-Ellman *et al.*, 2012). Additional seafood risk assessment investigations have determined that Gulf Coast residents' seafood consumption rates were underestimated by using the 90th percentile NHANES consumption estimates (Wilson *et al.*, 2015). Even when these consumption estimates were re-evaluated to represent actual coastal population seafood consumption rates, there were no unacceptable risks detected due to consumption of PAHs in seafood. The lack of risk was a function of the low levels of PAHs detected in the edible portion of seafood (Gohlke *et al.*, 2011). Had PAH contamination been present at the LOC, based on the 90th percentile NHANES consumption estimate, then high-end seafood consumers would likely exceed the acceptable risk levels.

## 20.10 Non-Chemical Stressors

Workers and community members experienced stress associated with experiencing the *Deepwater Horizon* oil spill as an event (Osofsky *et al.*, 2011; Kwok *et al.*, 2017b). The effects of event-associated stress were not limited to people involved in clean-up operations or even to adults in coastal communities (Rung *et al.*, 2015; Osofsky *et al.*, 2016). Many of the self-reported symptoms among Gulf South community members were potentially due to being involved in a stressful event but were ascribed to a self-reported chemical exposure that likely did not occur (Peres *et al.*, 2016). That does not mean that affected community members are not experiencing the self-reported symptoms, but it does point out a limitation related to the chemically focused health risk assessment process. This is another example of where observational epidemiology can provide insight into gaps in the current regulatory health risk assessment framework.

## 20.11 Limitations of the Dietary Health Risk Assessment Process

One primary issue with the regulatory health risk assessment process, in particular for cancer health risk assessment, is that the reference PAHs are associated with incomplete combustion of organic material. For cancer health risk assessment purposes, benzo[a]pyrene is used as an index chemical for a relative potency factor-based approach (EPA, 1993). The problem with this approach for assessment of dietary health risk associated with seafood potentially contaminated with crude oil is that these combustion-associated chemicals were not frequently detected in the MS252 crude oil. PAHs were present, but they were primarily alkylated chrysenes and phenanthrenes (NOAA, 2010). There is very limited toxicological data for the alkylated PAHs and using unsubstituted combustion-associated PAHs as proxies relies on the assumption that the toxicological properties of the petrogenic alkylated PAHs are similar to those of the unsubstituted pyrogenic PAHs. This assumption may or may not be true. Furthermore, it is difficult to apply relative potency factors to chemicals that have not been well evaluated from a toxicological or regulatory point of view (Wickliffe *et al.*, 2014). The field of risk assessment is struggling to include variables, such as non-chemical stressors and/or mixtures of chemicals, into a holistic or cumulative risk assessment framework (Sexton, 2012). There is an emerging regulatory framework for this type of approach but there are still many gaps and unknowns associated with its application in a real-world setting (EPA, 2007). Evaluations of end points that do not focus on a single chemical as the only potential exposure for health risk assessments are a step in the right direction but there is clearly much room for improvement, not only in terms of considering chemical mixtures and interactions, but also the effect that non-chemical stressors may have on human health outcomes.

## 20.12 Conclusions

The *Deepwater Horizon* event began with an explosion on an oil-drilling and exploration platform that resulted in deaths and injuries among the crew. Oil spill response operations began quickly and involved many thousands of workers and volunteers. These response workers had the highest potential for exposure to inhalation hazards. Personnel sampling data and job-exposure matrix exposure estimates indicate that clean-up workers were not exposed to volatile oil chemicals at levels that exceeded established OELs. Heat stress was determined to be a very significant physical hazard associated with clean-up operations. Many of the symptom surveys among workers who sought medical attention had signs of heat stress and/or dehydration. There have been two large cohort studies tailored to evaluate the impact of being involved in oil spill clean-up operations on health outcomes. Mental health issues related to stress are widely reported among members of the GuLF study cohort and among community members in affected communities. Chemical exposures directly related to the oil spill in community members are unlikely, based on evaluation of air monitoring data and chemical analysis of seafood. Exposure to hydrocarbon chemicals may be occurring but is likely originating from other environmental sources, such as cigarette smoke and consumer products.

## References


Aguilera, F., Mendez, J., Pasaro, E. and Laffon, B. (2010) Review on the effects of exposure to spilled oils on human health. *Journal of Applied Toxicology* 30, 291–301.

Alexander, M., Engel, L.S., Olaiya, N., Wang, L., Barrett, J. *et al.* (2018) The Deepwater Horizon oil spill coast guard cohort study: a cross-sectional study of acute respiratory health symptoms. *Environmental Research* 162, 196–202.

ATSDR (1995) *Toxicological Profile for Polycyclic Aromatic Hydrocarbons (PAHs).* Agency for Toxic Substances and Disease Registery, Atlanta, Georgia. Available at: https://www.atsdr.cdc.gov/toxprofiles/tp.asp?id=122&tid=25 (accessed 20 February 2018).

ATSDR (2004) *Interacton Profile for Benzene, Toluene, Ethylbenzne,and Xylenes (BTEX).* Agency for Toxic Substances and Disease Registry, Atlanta, Georgia. Available at: https://www.atsdr.cdc.gov/interactionprofiles/ip05.html (accessed 20 February 2018).

Avens, H.J., Unice, K.M., Sahmel, J., Gross, S.A., Keenan, J.J. and Paustenbach, D.J. (2011) Analysis and modeling of airborne BTEX concentrations from the Deepwater Horizon oil spill. *Environmental Science & Technology* 45, 7372–7379.

Black, J.C., Welday, J.N., Buckley, B., Ferguson, A., Gurian, P.L. *et al.* (2016) Risk assessment for children exposed to beach sands impacted by oil spill chemicals. *International Journal of Environmental Research and Public Health* 13.

CSHIB (2014) *Investigation Report Explosion and Fire at the Macondo Well.* 2010-10-I-OS. US Chemical Safety and Hazard Investigation Board, Washington, D.C. Available at: http://www.csb.gov/macondo-blowout-and-explosion/ (accessed 20 Febuary 2018).

Deziel, N.C., Colt, J.S., Kent, E.E., Gunier, R.B., Reynolds, P. *et al.* (2015) Associations between self-reported pest treatments and pesticide concentrations in carpet dust. *Environmental Health* 14, 27.

DHHS (2011) *Lessons Learned for the Deepwater Horizon Response*, 2012–117. US Department of Health and Human Services. National Institute for Occupational Safety and Health, Atlanta, Georgia. Available at: http://www.cdc.gov/niosh/docs/2012-117/pdfs/2012-117.pdf (accessed 20 February 2018).

DHHS (2017) Household products database. US Department of Health and Human Services, Atlanta, Georgia. Available at: https://householdproducts.nlm.nih.gov/index.htm (accessed 20 February 2018).

Dickey, R.W. (2012) FDA risk assessment of seafood contamination after the BP oil spill. *Environmental Health Perspectives* 120, a54–55; author reply a55–56.

Engel, L.S., Kwok, R.K., Miller, A.K., Blair, A., Curry, M.D., McGrath, J.A. and Sandler, D.P. (2017) The Gulf long-term follow-up study (Gulf Study): biospecimen collection at enrollment. *Journal of Toxicology and Environmental Health Part A* 80, 218–229.

EPA (1993) *Provisional Guidance for Quantitative Risk Assessment of Polycyclic Aromatic Hydrocarbons.* EPA/600/R-93/089, US Environmental Protection Agency, Washington, DC. Available at: http://ofmpub.epa.gov/eims/eimscomm.getfile?p_download_id=466885 (accessed 20 February 2018).

EPA (2002) *A Review of the Reference Dose and Reference Concentration Processes.* EPA/630/P-02/002. US Environmental Protection Agency, Washington, DC. Available at: http://www.epa.gov/sites/production/files/2014-12/documents/rfd-final.pdf (accessed 20 February 2018).

EPA (2005) *Guidelines for Carcinogen Risk Assessment*. EPA/630/P-03/001F. Risk Assessment Forum, US Environmental Protection Agency, Washington, DC. Available at: https://www.epa.gov/sites/production/files/2013-09/documents/cancer_guidelines_final_3-25-05.pdf (accessed 20 February 2018).

EPA (2007) *Concepts, Methods and Data Sources for Cumulative Health Risk Assessment of Multiple Chemicals, Exposures and Effects: a Resource Document*. EPA/600/R-06/013F. National Center for Environmental Assessment, Office of Research and Development, US Environmental Protection Agency, Cincinnati, Ohio. Available at: http://ofmpub.epa.gov/eims/eimscomm.getfile?p_download_id=474337 (accessed 20 February 2018).

EPA (2010) *Air Monitoring Data for BP Spill/Deepwater Horizon*. US Environmental Protection Agency, Washington, DC. Available at: https://edg.epa.gov/data/Public/OLEM/DWH_Air_Monitoring.zip (accessed, 2/20/2018]

EPA (2011) *Exposure Factors Handbook, 2011 edn*. EPA/600/R-09/052F. US Environmental Protection Agency, Washington, DC. Available at: https://cfpub.epa.gov/ncea/risk/recordisplay.cfm?deid=236252 (accessed 20 February 2018).

FDA (2010) *Protocol for interpretation and use of sensory testing and analytical chemistry results for re-opening oil-impacted areas closed to seafood harvesting due to the Deepwater Horizon oil spill*. US Food and Drug Administration, Silver Spring, Maryland. Available at: https://www.fda.gov/downloads/Food/RecallsOutbreaksEmergencies/Emergencies/UCM233818.pdf (accessed 20 February 2018).

Friesen, M.C., Coble, J.B., Lu, W., Shu, X.O., Ji, B.T. *et al*. (2012) Combining a job-exposure matrix with exposure measurements to assess occupational exposure to benzene in a population cohort in Shanghai, China. *The Annals of Occupational Hygiene* 56, 80–91.

Gam, K.B., Kwok, R.K., Engel, L.S., Curry, M.D., Stewart, P.A. *et al*. (2018a) Lung function in oil spill response workers 1–3 years after the Deepwater Horizon disaster. *Epidemiology* 29(3), 1.

Gam, K.B., Kwok, R.K., Engel, L.S., Curry, M.D., Stewart, P.A. *et al*. (2018b) Exposure to oil spill chemicals and lung function in Deepwater Horizon disaster response workers. *Journal of Occupational and Environmental Medicine* 60(6), 1.

Gohlke, J.M., Doke, D., Tipre, M., Leader, M. and Fitzgerald, T. (2011) A review of seafood safety after the Deepwater Horizon blowout. *Environmental Health Perspectives* 119,1062–1069.

Goldstein, B.D., Osofsky, H.J. and Lichtveld, M.Y. (2011) The Gulf oil spill. *New England Journal of Medicine* 364, 1334–1348.

Ji, B.T., Gao, Y.T., Shu, X.O., Yang, G., Yu, K. *et al*. (2012) Nightshift work job exposure matrices and urinary 6-sulfatoxymelatonin levels among healthy Chinese women. *Scandinavian Journal of Work, Environment & Health* 38, 553–559.

Kwok, R.K., Engel, L.S., Miller, A.K., Blair, A., Curry, M.D. *et al*. (2017a) The Gulf Study: a prospective study of persons involved in the Deepwater Horizon oil spill response and clean-up. *Environmental Health Perspectives* 125, 570–578.

Kwok, R.K., McGrath, J.A., Lowe, S.R., Engel, L.S., Jackson, W.B.N. *et al*. (2017b) Mental health indicators associated with oil spill response and clean-up: cross-sectional analysis of the Gulf Study cohort. *The Lancet Public Health* 2, e560–e567.

Laffon, B., Pasaro, E. and Valdiglesias, V. (2016) Effects of exposure to oil spills on human health: updated review. *Journal of Toxicology and Environmental Health Part B, Critical Reviews* 19, 105–128.

Lubchenco, J., McNutt, M.K., Dreyfus, G., Murawski, S.A., Kennedy, D.M. *et al*. (2012) Science in support of the Deepwater Horizon response. *Proceedings of the National Academy of Sciences of the United States of America* 109, 20212–20221.

McGowan, C.J., Kwok, R.K., Engel, L.S., Stenzel, M.R., Stewart, P.A. and Sandler, D.P. (2017) Respiratory, dermal, and eye irritation symptoms associated with Corexit EC9527a/EC9500a following the Deepwater Horizon oil spill: findings from the Gulf Study. *Environmental Health Perspectives* 125, 097015.

McNutt, M.K., Camilli, R., Crone, T.J., Guthrie, G.D., Hsieh, P.A. *et al*. (2012) Review of flow rate estimates of the Deepwater Horizon oil spill. *Proceedings of the National Academy of Sciences of the United States of America* 109, 20260–20267.

Michel, J., Owens, E.H., Zengel, S., Graham, A., Nixon, Z. *et al*. (2013) Extent and degree of shoreline oiling: Deepwater Horizon oil spill, Gulf of Mexico, USA. *PloS ONE* 8, e65087.

Middlebrook, A.M., Murphy, D.M., Ahmadov, R., Atlas, E.L., Bahreini, R. *et al*. (2012) Air quality implications of the Deepwater Horizon oil spill. *Proceedings of the National Academy of Sciences of the United States of America* 109, 20280–20285.

Niehoff, N.M., Nichols, H.B., White, A.J., Parks, C.G., D'Aloisio, A.A. and Sandler, D.P. (2016) Childhood and adolescent pesticide exposure and breast cancer risk. *Epidemiology* 27, 326–333.

NIOSH (2011) *Health Hazard Evaluation of Deepwater Horizon Response Workers*. HETA 2010-0115 & 2010-0129-3138. National Institute of Occupational Safety and Health, Atlanta, Georgia. Available at: http://www.cdc.gov/niosh/hhe/reports/pdfs/2010-0115-0129-3138.pdf (accessed 20 February 2018).
Nisbet, I.C.T. and LaGoy, P.K. (1992) Toxic equivalency factors (TEFs) for polycyclic aromatic hydrocarbons (PAHs). *Regulatory Toxicology and Pharmacology* 16, 290–300.
NOAA (2010) *Deepwater Horizon Oil: Characteristics and Concerns*. National Oceanic and Atmospheric Administration , Silver Spring, Maryland. Available at: http://sero.nmfs.noaa.gov/deepwater_horizon/documents/pdfs/fact_sheets/oil_characteristics.pdf (accessed 20 February 2018).
NRC TRB (2003) *Oil in the Sea III: Inputs, Fates, and Effects*. Transportation Research Board, National Research Council. The National Academies Press Washington, DC.
Osofsky, H.J., Osofsky, J.D. and Hansel, T.C. (2011) Deepwater Horizon oil spill: mental health effects on residents in heavily affected areas. *Disaster Medicine and Public Health Preparedness* 5, 280–286.
Osofsky, J.D., Osofsky, H.J., Weems, C.F., Hansel, T.C. and King, L.S. (2016) Effects of stress related to the Gulf oil spill on child and adolescent mental health. *Journal of Pediatric Psychology* 41, 65–72.
Peres, L.C., Trapido, E., Rung, A.L., Harrington, D.J., Oral, E. *et al.* (2016) The Deepwater Horizon oil spill and physical health among adult women in southern Louisiana: the women and their children's health (watch) study. *Environmental Health Perspectives* 124, 1208–1213.
Reddy, C.M., Arey, J.S., Seewald, J.S., Sylva, S.P., Lemkau, K.L. *et al.* (2012) Composition and fate of gas and oil released to the water column during the Deepwater Horizon oil spill. *Proceedings of the National Academy of Sciences of the United States of America* 109, 20229–20234.
Rotkin-Ellman, M., Wong, K.K. and Solomon, G.M. (2012) Seafood contamination after the BP Gulf oil spill and risks to vulnerable populations: a critique of the FDA risk assessment. *Environmental Health Perspectives* 120, 157–161.
Rung, A.L., Oral, E., Fontham, E., Harrington, D.J., Trapido, E.J. and Peters, E.S. (2015) Mental health impact of the Deepwater Horizon oil spill among wives of clean-up workers. *Epidemiology* 26, e44–46.
Sammarco, P.W., Kolian, S.R., Warby, R.A., Bouldin, J.L., Subra, W.A. and Porter, S.A. (2016) Concentrations in human blood of petroleum hydrocarbons associated with the BP/Deepwater Horizon oil spill, Gulf of Mexico. *Archives of Toxicology* 90, 829–837.
Sexton, K. (2012) Cumulative risk assessment: an overview of methodological approaches for evaluating combined health effects from exposure to multiple environmental stressors. *International Journal of Environmental Research and Public Health* 9, 370–390.
Stewart, P.A., Stenzel, M.R., Ramachandran, G., Banerjee, S., Huynh, T.B. *et al.* (2017) Development of a total hydrocarbon ordinal job-exposure matrix for workers responding to the Deepwater Horizon disaster: The Gulf Study. *Journal of Exposure Science & Environmental Epidemiology* 28(3), 223–230.
Tsuda, T., Tokinobu, A., Yamamoto, E. and Suzuki, E. (2016) Thyroid cancer detection by ultrasound among residents ages 18 years and younger in Fukushima, Japan: 2011 to 2014. *Epidemiology* 27, 316–322.
Werder, E.J., Gam, K.B., Engel, L.S., Kwok, R.K., Ekenga, C.C. *et al.* (2017) Predictors of blood volatile organic compound levels in Gulf Coast residents. *Journal of Exposure Science & Environmental Epidemiology* 28(4), 358–370.
Wickliffe, J., Overton, E., Frickel, S., Howard, J., Wilson, M. *et al.* (2014) Evaluation of polycyclic aromatic hydrocarbons using analytical methods, toxicology, and risk assessment research: seafood safety after a petroleum spill as an example. *Environmental Health Perspectives* 122, 6–9.
Wilson, M.J., Frickel, S., Nguyen, D., Bui, T., Echsner, S. *et al.* (2015) A targeted health risk assessment following the Deepwater Horizon oil spill: polycyclic aromatic hydrocarbon exposure in Vietnamese-American shrimp consumers. *Environmental Health Perspectives* 123,152–159.
Xia, K., Hagood, G., Childers, C., Atkins, J., Rogers, B. *et al.* (2012) Polycyclic aromatic hydrocarbons (PAHs) in Mississippi seafood from areas affected by the Deepwater Horizon oil spill. *Environmental Science & Technology* 46, 5310–5318.

# 21 Crude Oil Pollution II. Effects of the *Deepwater Horizon* Contamination on Sediment Toxicity in the Gulf of Mexico

**P.A. Montagna*,[1] and S.S. Arismendez[2]**
[1]*Texas A&M University-Corpus Christi, Harte Research Institute for Gulf of Mexico Studies, Corpus Christi, Texas, USA;* [2]*Texas Commission on Environmental Quality, Austin, Texas, USA*

## 21.1 Abstract

The *Deepwater Horizon* accident and subsequent oil spill released between 3.2 and 6.2 million barrels (509–986 million litres) of oil. Following capping of the well site, a series of cruises was undertaken from September through October 2010 to determine if oil had reached the bottom and if sediment toxicity was occurring. A sediment quality triad approach was used where chemical contaminants were measured to indicate a dose effect, *in situ* bacterial toxicity was measured to indicate a biological effect and macroinfauna community structure was measured to indicate an ecological effect. Oil and toxic responses were found as far away as 35 km from the spill site and there were direct correlations between the presence of oil and biological and ecological effects of reduced macrofauna abundance and diversity. The dispersants were also toxic, so a combination of dispersants and oil is responsible for the biological and ecological responses.

## 21.2 Introduction

The *Deepwater Horizon* (DWH) accident was caused by a blowout on 20 April 2010 in a water depth of 1525 m. There are considerable differences in estimates of the amount of oil released. The US District Court for the Eastern District of Louisiana (2015) found that the accident released 3.19 million barrels (134 million gallons, 507 million litres), but other credible estimates are as high as 6.2 million barrels (Griffith, 2012). The oil was released under great pressure and with methane gas, so that it formed very small droplets that diffused into a subsurface plume (Aman *et al.*, 2015). Because of the plumes, the oil was transported over great distances (Peterson *et al.*, 2012; Ryerson *et al.*, 2012). In addition, 2.1 million gallons (7.95 million litres) of Corexit dispersant was applied to disperse the oil and enable people at the surface to work on clean-up and capping the well; and some of the dispersant was also incorporated into the deep-sea plumes (Kujawinski *et al.*, 2011).

The oil in the deep-sea plume and surface waters was deposited to the seafloor by a large marine oil snow sedimentation and flocculent accumulation (MOSSFA) event. During the MOSSFA event the oil was incorporated into faecal pellets (Passow, 2014), plankton and microbial mucus, which then rapidly sank to the bottom as marine snow (Daly *et al.*, 2016). Advective transport of the deep-sea plume (adsorption on

* E-mail address: Paul.Montagna@tamucc.edu

to particles or incorporation in) and direct contact of the plume along the continental slope at depths between 1000m and 2000 m were also likely mechanisms for hydrocarbon deposition (Romero *et al.*, 2015). Other possible mechanisms for deep-sea deposition include onshore–offshore transport, or sinking of heavy by-products after burning (UAC, 2010).

Once the DWH well was capped on 15 July 2010, it was then possible to assess the effects to the deep-sea. Prior to that date, the area was quarantined so that responders could work on clean-up and capping the wellhead. At the time, there was knowledge of the deep-sea plumes but not of the MOSSFA event, so there was some controversy over whether there was oil on the bottom. Many contended that oil floats and could not possibly be in deep-sea sediments, while others pointed out the presence of plumes and possibility of sedimentation events. A series of cruises was commissioned by British Petroleum (BP) to determine the presence or absence of Macondo oil and dispersants within surficial sediments and supernatant water and its environmental effects. The sampling protocol was developed to employ a sediment quality triad (SQT) approach (Long and Chapman, 1985). The SQT components are: (i) measures of chemical contaminants to indicate dose; (ii) sediment toxicity to indicate biological effects; and (iii) benthic community structure to indicate ecological effects. Data from the three components of the triad can be used in an assessment process to form a weight of evidence to compare and rank the relative quality of sediment samples and regions of a study area (Chapman *et al.*, 1987; Long, 1989).

## 21.3 Methods

The oil spill occurred in the lease block named Macondo 252 (MC-252). Field missions were conducted on the R/V *Gyre* (16 September to 19 October 2010), R/V *Ocean Veritas* (24 September to 30 October 2010) (Montagna *et al.*, 2013; Baguley *et al.*, 2015; Washburn *et al.*, 2016) and M/V *Ryan Chouest* (an offshore support ship converted to collect samples). While 227 stations were sampled at water depths ranging from 10 m to 2767 m, only 179 had contaminants and toxicity measured, and 111 had contaminants, toxicity and macrofauna measured. Stations were located along a suspected gradient of contaminant effects where 16 of the stations were arranged in a 'bull's-eye' design. This survey design was used because transects extending in radial patterns from the source of contamination and the statistical analysis of such designs is well known (Kennicutt *et al.*, 1996). Sediment samples were collected using a Bowers and Connelly MAXI corer manufactured by Ocean Scientific International Ltd (OSIL). The cores are 10 cm inner diameter and 60 cm in length. The corer collects 12 simultaneous cores from a single deployment at each station but only five were allocated for the work described here. Three cores were set aside for benthic macrofauna, one core was used for sediment toxicity and one core was used for measuring oil and other drilling-related contaminants. Details of the macrofauna processing and chemical contaminant measurements are provided in Montagna *et al.* (2013). The chemical contaminant data was part of the initial Operational Science Advisory Team (OSAT, 2010) report. Samples of two dispersants, Corexit® 9527 and Corexit® 9500, were purchased and tested for toxicity.

Toxicity was measured using the Microtox® system. Microtox uses the fluorescence of the bioluminescent bacterium, *Vibrio fischeri* NRRL B-11177, to assess toxicity of samples. Reductions in bioluminescent activity indicate corresponding increases in toxicity. The bioluminescence of *V. fischeri* is sensitive to most toxicants, such as pesticides, phenolic compounds and metals (Ramaiah and Chandramohan, 1993). The Microtox method yields results comparable to standard amphipod and urchin toxicity tests for water and sediment samples (De Zwart and Sloof, 1983). Results are reported as effect concentration 50 ($EC_{50}$) values (mg $l^{-1}$) required to elicit a negative 50% effect relative to a control sample. Therefore, low $EC_{50}$ values denote a large toxic effect. It is much less labour intensive than other amphipod or sea urchin exposure tests and utilizes bulk sediment rather than pore water or preservation of bulk sediment to expose animals in arrays of dilutions.

Three sediment sub-cores were taken from one core using a 10 cc (= 10 ml) syringe from the undisturbed surface. The three sub-cores were composited into one 30 cc sediment sample for toxicity analysis. Some sediments were analysed immediately on shipboard but other sediments were preserved at –20°C and analysed onshore. Samples measured on shipboard, were held at 4°C in the dark. The bacteria preparation is added to bulk sediment samples and mixed thoroughly, then a dilution series is created and the samples are exposed for 30 min (Morehead *et al.*, 2008). The bacteria and sediment are pelleted and fluorescence measurements are taken to determine the concentration that provides an $EC_{50}$ response.

## 21.4 Results

Typically, the toxic components of total petroleum hydrocarbons (TPHs) are the polycyclic aromatic hydrocarbons (PAHs). For analysis here, the PAH values were summed and the total PAH values were used in the analyses. The range of PAH concentrations was large, from near zero to near 20,000 μg $kg^{-1}$ (Fig. 21.1). The toxicity response is labelled $EC_{50}$ and ranged from near zero to 100,000 mg $l^{-1}$ (Fig. 21.1). Low $EC_{50}$ values mean high toxicity. On a linear scale, the shape of the curve resembles a hockey stick, where there is a large range of $EC_{50}$ values at low PAH concentrations and a low range of $EC_{50}$ values over the wider range of higher PAH values (Fig. 21.1A).

The average PAH concentration was 140 μg $kg^{-1}$ during a survey of sediment chemistry between 2000 and 2002 throughout the northern Gulf of Mexico (Wade *et al.*, 2008) in the Deep Gulf of Mexico Benthos (DGOMB) Program (Rowe and Kennicutt, 2008). This average was used as a cut-off point to calculate a linear regression between PAH and $EC_{50}$, which yielded a regression equation of $EC_{50} = 6016 - 0.3351 \times$ PAH (Fig. 21.1a). This implies that the toxic effects are restricted to $EC_{50}$ values below 6000 mg $l^{-1}$. A total of 24 of the 179 stations had an $EC_{50}$ below this threshold and a PAH concentration above this threshold (Table 21.1). A total of 77 stations presented no evidence of toxicity nor high PAH values. There were only nine stations where PAH concentrations were high, but there was no toxic response. However, there were 69 stations with false positives, that is, there was a toxic response but the PAH concentrations were lower than those expected to provoke a toxic response.

The false positives can be easily visualized when the data is plotted on log–log scales (Fig. 21.1B). The log–log plot has the general trend line expected, that is, decreasing $EC_{50}$ with increasing PAH concentration. The area of the curve where false positives reside is bounded by $EC_{50}$ values below 6000 mg $l^{-1}$ and PAH concentrations below 140 μg $kg^{-1}$ (Fig. 21.1B). Except

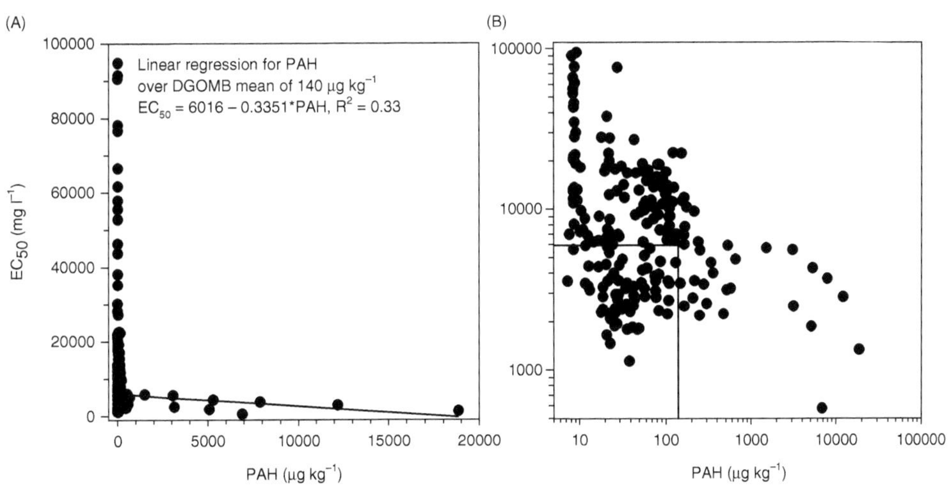


**Fig. 21.1.** Relationship between total polycylic aromatic hydrocarbons (PAH) and $EC_{50}$ from microtoxicity tests in bulk sediments. (A) Microtoxicity and PAH plotted on linear scales with a regression through values over 140 μg $kg^{-1}$. (B) Microtoxicity and PAH plotted on a $\log_{10}$ scales with drop lines at PAH values of 140 μg $kg^{-1}$ and $EC_{50}$ values of 6000 mg $l^{-1}$.

for the lowest $EC_{50}$ at a PAH concentration of about 6000 μg $kg^{-1}$, there were seven very low $EC_{50}$ values, indicating sediments that are highly toxic without the presence of PAH.

The spatial distribution of the toxic samples (EC50 < 6000) with high PAH (> 140 μg $kg^{-1}$) was within 25 km of the DWH accident site, and mainly towards the south-west (Fig. 21.2). All of the samples that had low PAH and no toxicity were in shallow water closer to shore. The distribution of the stations with false positives was mainly among the deeper locations but extended to the north-east of the DWH accident site.

Over all samples, the PAH concentrations were not significantly correlated with $EC_{50}$ ($p$ = 0.78); however, TPH values were significantly inversely correlated, as would be expected ($p$ = 0.0004) (Table 21.1). Neither PAH nor TPH was significantly correlated with either macrofauna abundance or macrofauna species richness. Macrofauna abundance was significantly correlated with $EC_{50}$ ($p$ = 0.0005), which indicates that there were more macrofauna present in samples that were less toxic. As expected, both PAH and TPH were highly correlated with one another ($p$ < 0.0001) and macrofauna abundance was highly correlated with species richness ($p$ < 0.0001).

Samples of the same dispersants that were used for dispersing the spill were tested for toxicity and were highly toxic even when diluted 10,000 times (Fig. 21.3). The two dispersants

**Table 21.1.** Classification of the stations based on $EC_{50}$ toxicity response threshold values of < 6000 mg $l^{-1}$ and polycyclic aromatic hydrocarbon (PAH) threshold of > 140 μg $kg^{-1}$.

| | PAH contaminated | |
|---|---|---|
| Toxicity response | No | Yes |
| **No** | 77 | 9 |
| **Yes** | 69 | 24 |

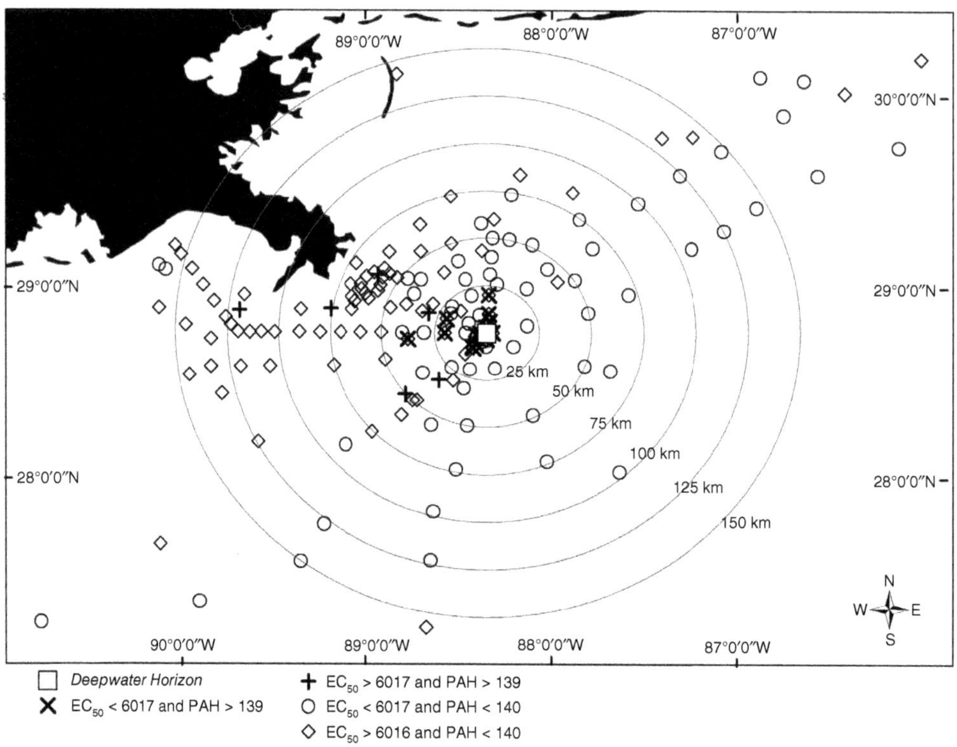


**Fig. 21.2.** Map of study area with classification of stations based on exceeding the $EC_{50}$ ($EC_{50}$) and polycyclic aromatic hydrocarbon (PAH) thresholds from Fig. 21.1.

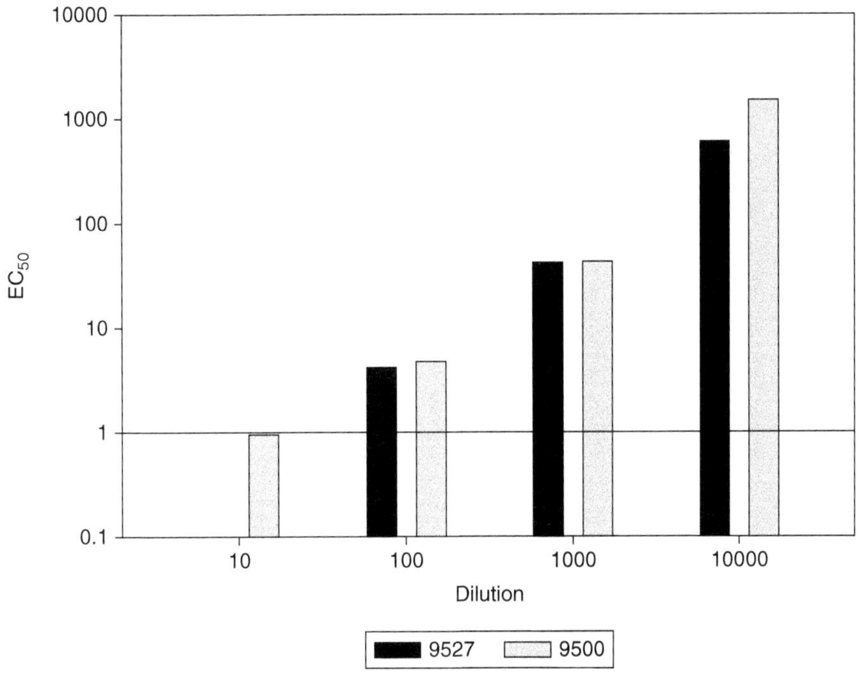


**Fig. 21.3.** Toxicity of two dispersants, Corexit® 9527 and Corexit® 9500.

had similar toxicity. There was a log–log linear relationship between toxicity and dilution strength.

## 21.5 Discussion

The SQT approach is a powerful weight-of-evidence approach (Long, 1989). In the current study, there were inverse correlations with TPH and PAH and toxicity (Fig. 21.1) and inverse correlations between toxicity and macrofauna abundance and species richness (Table 21.1). These correlations indicate a toxic response to the DWH oil spill, but it was limited to the deep-sea (Fig. 21.2 and Table 21.2). Interestingly, the dispersants themselves are highly toxic (Fig. 21.3). Therefore, it is likely that both oil and dispersants were contributing to the toxicity at the spill site.

Sediment quality benchmarks using logistic regression models were previously developed for the deep-sea sediments of the DWH site (Balthis *et al.*, 2017). These models were based on some of the TPH and macrofauna taxa richness data from 37 of the stations presented in the current study. Likelihood of impacts to benthic macrofauna and meiofauna communities is low (< 20%) at total PAH concentrations of < 4000 $\mu g\ kg^{-1}$ for both macrofauna and meiofauna and high (> 80%) at concentrations greater than 24,000 $\mu g\ kg^{-1}$ for macrofauna and 25,000 $\mu g\ kg^{-1}$ for meiofauna. These concentrations align with the highest ones reported here (Fig. 21.1).

Dispersants are toxic as well (Fig. 21.3) (Almeda *et al.*, 2014). In fact, the dispersants can be as toxic to invertebrates as the PAHs (Barron *et al.*, 2013). Dispersants were used during the DWH Response operations, including subsurface injections of Corexit 9500 at the wellhead (DWH NRDA, 2016). A principal compound in Corexit 9500 is dioctyl sulfosuccinate (DOSS). DOSS measured in the upper 1 cm of sediment in 96 replicate samples from 32 stations sampled in May to June 2011 (Montagna *et al.*, 2017) ranged from 0 to 84 ppb, but 35% of the measurements were below detection limits (Balthis *et al.*, 2017). However, DOSS was not measured in 2010 when the current study was performed, and it could have degraded in the intervening months since September to October 2010 when the samples reported on here were taken. The dispersants and oil can have synergistic effects.

**Table 21.2.** Spearman correlations among sediment quality triad (SQT) variables.

| | TPH | $EC_{50}$ | Abundance | Richness | Statistic[a] |
|---|---|---|---|---|---|
| **PAH** (polycyclic aromatic | 0.67 | −0.02 | 0.09 | −0.09 | *r* |
| hydrocarbons | < 0.0001 | 0.7849 | 0.3260 | 0.3585 | *p* |
| | 181 | 179 | 110 | 110 | *n* |
| **TPH** (total petroleum | | −0.27 | −0.16 | −0.17 | *r* |
| hydrocarbons) | | 0.0004 | 0.1052 | 0.0792 | *p* |
| | | 175 | 106 | 106 | *n* |
| **$EC_{50}$** microtoxicity | | | 0.32 | 0.16 | *r* |
| | | | 0.0005 | 0.1025 | *p* |
| | | | 111 | 111 | *n* |
| **Abundance** | | | | 0.61 | *r* |
| | | | | < 0.0001 | *p* |
| | | | | 112 | *n* |

[a]*r* = correlation coefficient; *p* = probability that *r* = 0; *n* = number of sample pairs.

In toxicity tests using blue crab *Callinectes sapidus* megalopae, the addition of Corexit 9500 and Corexit 9527 had higher toxicity of the dispersed oil relative to the dispersant alone (Fern *et al.*, 2015). This is probably because dispersed oil increases the water-accommodated fraction of phenathrenrene sorption by as much as 22% (Gong *et al.*, 2014). It is likely that dispersants added to the toxicity found in the bulk sediment samples reported on in the current study and may have contributed to the high number of false positives.

As much as 35% of the oil was entrained in the deep-sea plumes (Ryerson *et al.*, 2012). From 4% to 31% of the oil was trapped in the deep-sea and contamination from the fall-out plume was extensive, covering an area of 3200 km$^2$ (Valentine *et al.*, 2014). With oil covering this large an area, it is not surprising that toxicity was found to cover a large area as well. A high toxic response was found as far away as 35 km from the well-head to the west, but most of the highly toxic responses were within 25 km away from the accident site (Fig. 21.2). There were two sites exhibiting highly toxic responses 125 km north of the well-head but there are two explanations for this finding: (i) it could have been oil sinking from surface slicks that did move directly north to the Mississippi coastline; or (ii) it could be locally derived (for example, from another well, runoff from land, or from a previous accident or spill). Other studies have also reported high PAH concentrations of the Mississippi coastline (Adhikari *et al.*, 2016).

One odd trend in the distribution of the Microtox toxicity is the large number of apparent false positives, i.e. stations where there was toxicity but the PAH concentrations were lower than those expected to be responsible for the toxic response (i.e. the circles in Fig. 21.2). What is interesting about all of these stations is that, except for nine, they are in deeper water between 1000 m nd 2000 m depth toward the south-west, or in the DeSoto Canyon to the north-east. We know the deep-sea plume moved along the 1000 m and 2000 m depth contours (Weisberg *et al.*, 2011), which helps to explain those false positives, because there could be undiscovered oil there. Less obvious is the trend to the north-east. However, since both trends, and most of the false positives, occur in deeper water it is likely that there is something else in the deep-sea sediments that the luminescent bacteria are sensitive to.

The wide area exhibiting biological toxicity is much further than previously acknowledged for ecological responses. In past studies, the most severe impacts on macrofaunal abundance and diversity extended to 3 km from the wellhead in all directions, and moderate impacts were observed up to 17 km towards the south-west and 8.5 km towards the north-east (Montagna *et al.*, 2013). However, for meiofauna abundance and diversity, moderate impacts extended up to 60 km from the wellhead site (Baguley *et al.*, 2015).

While sediment toxicity is reported here, no toxicity was found in overlying water samples from the current study. Other studies report limited or no toxicity in water-column exposure samples following the DWH incident. It is reported that Corexit dispersant did not change microbial community structure in water-column samples (Chakraborty *et al.*, 2012). The mysid *Americamysis bahia*, silverside fish *Menidia beryllina* and bacterium *Vibrio fischeri* (i.e. the Microtox assay) were exposed to Gulf of Mexico water samples collected from May to December 2010, and Echols *et al.* (2015) concluded that 'the potential for acute toxicity from water exposures to MC-252 were also limited'. They found acute toxicity to *A. bahia* and *M. beryllina* in only two samples collected before the well was capped in July 2010.

## 21.6 Conclusions

Oil and toxic responses were found in sediments as far away as 35 km from the spill site, but were common within 25 km of the spill site, and there were direct correlations between the presence of oil and biological and ecological effects. The oil and toxic effects were not found symmetrically around the site, but skewed to the south-west in the direction of the deep-water plumes. The dispersants were also toxic. A combination of the deep-water plumes, circulation patterns and toxicity from oil and dispersants is responsible for the distribution patterns of reduced benthic macrofauna abundance and diversity that occurred following the blow out.

## Acknowledgements

Funding of the sample collection and analysis was provided by British Petroleum (BP) and National Oceanic and Atmospheric Administration (NOAA) as part of the Response Phase of the *Deepwater Horizon* oil spill incident. These activities occurred prior to the initiation of the Natural Resource Damage Assessment (NRDA) Phase. The scientific results and conclusion of this publication, as well as any views or opinions expressed herein, are those of the authors and do not necessarily represent the view of NOAA or any other natural resource Trustee for the BP/*Deepwater Horizon* NRDA. Any use of trade, firm, or product names is for descriptive purposes only and does not imply endorsement by the US Government. Funding for the data analysis and writing was provided by the Harte Research Institute. Data are publically available through the DIVER (Data Integration Visualization Exploration and Reporting) website (https://www.diver.orr.noaa.gov/deepwater-horizon-nrda-data).

## References

Adhikari, P.L., Maiti, M., Overton, E.B., Rosenheim, B.E. and Marx, B.E. (2016) Distributions and accumulation rates of polycyclic aromatic hydrocarbons in the northern Gulf of Mexico sediments. *Environmental Pollution* 212, 413–423.

Almeda, R., Bona, S., Foster, C.R. and Buskey, E.J. (2014) Dispersant Corexit 9500A and chemically dispersed crude oil decreases the growth rates of meroplanktonic barnacle nauplii (*Amphibalanus improvisus*) and tornaria larvae (*Schizocardium* sp.). *Marine Environmental Research* 99, 212–217.

Aman, Z.M., Paris, C.B., May, E.F., Johns, M.L and Lindo-Atichati, D. (2015) High-pressure visual experimental studies of oil-in-water dispersion droplet size. *Chemical Engineering Science* 127, 392–400.

Baguley, J.G., Montagna, P.A., Cooksey, C., Hyland, J.L., Bang, H.W. *et al.* (2015) Community response of deep-sea soft-sediment metazoan meiofauna to the Deepwater Horizon blowout and oil spill. *Marine Ecology Progress Series* 528, 127–140.

Balthis, W.L., Hyland, J.L., Cooksey, C., Montagna, P.A., Baguley, J.G., Ricker, R.W. and Lewis, C. (2017) Sediment quality benchmarks for assessing oil-related impacts to the deep-sea benthos. *Integrated Environmental Assessment and Management* 13, 840–851.

Barron, M.G., Hemmer, M.J. and Jackon, C.R. (2013) Development of aquatic toxicity benchmarks for oil products using species sensitivity distributions. *Integrated Environmental Assessment and Management* 9, 610–615.

Chakraborty, R., Borglin, S.E., Dubinsky, E.A., Andersen, G.L. and Hazen, T.C. (2012) Microbial response to the MC-252 oil and Corexit 9500 in the Gulf of Mexico. *Frontiers in Microbiology* 3, 357. doi: 10.3389/fmicb.2012.00357.

Chapman, P.M., Dexter, R.N. and Long, E.R. (1987) Synoptic measures of sediment contamination, toxicity and infaunal community composition (the Sediment Quality Triad) in San Francisco Bay. *Marine Ecology Progress Series* 37, 75–96.

Daly, K.L., Passow, U., Chanton, J. and Hollander, D. (2016) Assessing the impacts of oil-associated marine snow formation and sedimentation during and after the Deepwater Horizon oil spill. *Anthropocene* 13, 18–33. doi 10.1016/j.ancene.2016.01.006.

De Zwart, D. and Slooff, W. (1983) The Microtox as an alternative assay in the acute toxicity assessment of water pollutants. *Aquatic Toxicology* 4, 129–138.

DWH NRDA (2016) *Deepwater Horizon oil spill: Final Programmatic Damage Assessment and Restoration Plan and Final Programmatic Environmental Impact Statement*. Deepwater Horizon Natural Resource Damage Assessment Trustees. National Ocean and Atmospheric Administration (NOAA), Silver Spring, Maryland. Available at: http://www.gulfspillrestoration.noaa.gov/restoration-planning/gulf-plan (accessed 30 April 2019).

Echols, B.S., Smith, A.J., Gardinali, P.R. and Rand, G.M. (2015) Acute aquatic toxicity studies of Gulf of Mexico water samples collected following the Deepwater Horizon incident (May 12, 2010 to December 11, 2010). *Chemosphere* 120, 131–137.

Fern, R., Wither, K., Zimba, P., Wood, T. and Shoech, L. (2015) Toxicity of three dispersants alone and in combination with crude oil on blue crab *Callinectes sapidus* megalopae. *Southeastern Naturalist* 14, G82–G92.

Gong, Y., Zhoa, X., O'Reilly, S.E., Qian, T. and Zhoa, D. (2014) Effects of oil dispersant and oil on sorption and desorption of phenanthrene with Gulf Coast marine sediments. *Environmental Pollution* 185, 240–249.

Griffiths, S.K. (2012) Oil release from Macondo Well MC-252 following the Deepwater Horizon Accident. *Environmental Science & Technology* 46, 5616–5622.

Kennicutt, M.C. II, Green, R.H., Montagna, P.A. and Roscigno, P.F. (1996) Gulf of Mexico Offshore Operations Experiment (GOOMEX) Phase I: sublethal responses to contaminant exposure introduction and overview. *Canadian Journal of Fisheries and Aquatic Sciences* 53, 2540–2553.

Kujawinski, E.B., Kido Soule, M.C., Valentine, D.L., Boysen, A.K., Longnecker, K. and Redmond, M.C. (2011) Fate of dispersants associated with the Deepwater Horizon oil spill. *Environmental Science & Technology* 45, 1298–1306.

Long, E.R. (1989) Use of the sediment quality triad in classification of sediment contamination. In: *Contaminated Marine Sediments – Assessment and Remediation*. National Research Council, Marine Board. National Academies Press, Washington, DC, pp. 78–99.

Long, E.R. and Chapman, P.M. (1985) A sediment quality triad: measures of sediment contamination, toxicity and in faunal community composition in Puget Sound. *Marine Pollution Bulletin* 16, 405–415.

Montagna, P.A., Baguley, J.G., Cooksey, C., Hartwell, I., Hyde, L.J., Hyland, J.L., Kalke, R.D., Kracker, L.M., Reuscher, M. and Rhodes, A.C.E. (2013) Deep-sea benthic footprint of the Deepwater Horizon blowout. *PloS ONE* 8, e70540.

Montagna, P.A., Baguley, J.G., Cooksey, C. and. Hyland, J.L (2017). Persistent impacts to the deep soft-bottom benthos one year after the Deepwater Horizon event. *Integrated Environmental Assessment and Management* 13, 342–351.

Morehead, S., Montagna, P.A. and Kennicutt, M.C. II. (2008) Comparing fixed-point and probabilistic sampling designs for monitoring the marine ecosystem near McMurdo Station, Antarctica. *Antarctic Science* 20, 471–484.

OSAT (2010) *Summary Report for the Sub-sea and Sub-Surface Oil and Dispersant Detection: Sampling and Monitoring*. Operational Science Advisory Team. Available at: https://www.restorethegulf.gov/release/2015/07/01/osat-summary-report-sub-sea-and-sub-surface-oil-and-dispersant-detection (accessed 2 July 2018).

Passow, U. (2014) Formation of rapidly-sinking, oil-associated marine snow. *Deep-Sea Research II* 129, 232–240.

Peterson, C.H., Anderson, S.S., Cherr, G.N., Ambrose, R.F., Anghera, S. *et al*. (2012) A tale of two spills: novel science and policy implications of an emerging new oil spill model. *BioSience* 62, 461–469.

Ramaiah, N. and Chandramohan, D. (1993) Ecological and laboratory studies on the role of luminous bacteria and their luminescence in coastal pollution surveillance. *Marine Pollution Bulletin* 26, 190–201.

Romero, I.C., Schwing, P.T., Brooks, G.R., Larson, R.A., Hastings, D.W. *et al*. (2015) Hydrocarbons in deep-sea sediments following the 2010 Deepwater Horizon blowout in the Northeast Gulf of Mexico. *PLoS ONE* 10, e0128371.

Rowe, G.T. and Kennicutt, M.C. (2008) Introduction to the Deep Gulf of Mexico Benthos Program. *Deep-Sea Research II* 55, 2536–2540.
Ryerson, T.B., Camilli, R., Kessler, J.D., Kujawinski, E.B., Reddy, C.M. *et al.* (2012) Chemical data quantify Deepwater Horizon hydrocarbon flow rate and environmental distribution. *Proceedings of the National Academy of Sciences* 109, 20246–20253.
UAC (2010) *Deepwater Horizon MC 252 response. Strategic Plan for Sub-sea and Sub-surface Oil and Dispersant Detection, Sampling, and Monitoring. November 13, 2010*. Unified Area Command, US Coast Guard and BP Exploration and Production, Inc., New Orleans, Louisiana.
US District Court for the Eastern District of Louisiana (2015) *Oil spill by the oil rig 'Deepwater Horizon' in the Gulf of Mexico, on April 20, 2010*. MDL No. 2179, Section J. New Orleans, Louisiana.
Valentine, D.L., Fisher, G.B., Bagby, S.C., Nelson, R.K., Reddy, C.M., Sylva, S.P. and Woo, M.A. (2014) Fallout plume of submerged oil from Deepwater Horizon. *Proceedings of the National Academy of Sciences* 111, 15906–15911.
Wade, T.L., Soliman, Y., Sweet, S.T., Wolff, G.A., and Presley, B.J. (2008). Trace elements and polycyclic aromatic hydrocarbons (PAHs) concentrations in deep Gulf of Mexico sediments. *Deep-Sea Research II* 55, 2585–2593.
Washburn, T., Rhodes, A.C.E. and Montagna, P.A. (2016) Benthic taxa as potential indicators of a deep-sea oil spill. *Ecological Indicators* 71, 587–597.
Weisberg, R.H., Zheng, L. and Liu, Y. (2011) Tracking subsurface oil in the aftermath of the Deepwater Horizon well blowout. In: *Monitoring and Modeling the Deepwater Horizon Oil Spill: A Record-Breaking Enterprise*. Geophysical Monograph Series 195, 205–215. American Geophysical Union, Washington, DC.

# 22 Crude Oil Pollution III. *Exxon Valdez* Contamination: Ecological Recovery, a Case Study

S. Haycox*
*University of Alaska Anchorage, Alaska, USA*

## 22.1 Abstract

The *Exxon Valdez* oil spill in Prince William Sound, Alaska, in March 1989 was considered at the time the worst environmental disaster in North America. Though Exxon Corporation expended $2 billion over 4 years on clean-up operations, oil remained on the beaches contaminated by the spill. The US court civil settlement of charges against the corporation included $900 million to fund a high-level joint federal–state body, the Exxon Valdez Oil Spill Trustee Council, assigned to undertake restoration and preservation of the affected environment. The council is advised by scientists and by members of the affected communities. It has been exemplary in objective assessment and remediation of the lingering impact of the spilled oil. It has to date identified biological species and human services as recovered, recovering, or not recovered, based on scientific analysis and social science study.

## 22.2 Introduction

When the Exxon Shipping Company supertanker *Exxon Valdez* ran on to Bligh Reef in Prince William Sound, Alaska, at 12:04am on 24 March 1989, spilling between 11 and 32 million gallons (50–145 million litres) of crude oil into pristine northern waters, analysts regarded the disaster as the worst environmental catastrophe in North American history (Haycox, 2012). Because programmed spill response capability had been severely compromised by poor planning, indecision and inept preparation, little of the spilled oil was contained and after 4 days a strong Pacific storm with winds of 70 mph drove the oil on to rocks and into sand and intertidal beaches in the Sound, and then south-west along Alaska's outer coast on the Gulf of Alaska, eventually reaching 800 km from the spill site and into more than 1500 km of irregular shoreline. In 1991 the US National Transportation Safety Board, charged with investigating the accident, determined that its probable causes included the following: (i) the failure to manoeuvre the vessel properly, possibly due to fatigue and excessive workload; (ii) failure of the master to provide a proper navigation watch, possibly due to impairment from alcohol; (iii) failure of Exxon Shipping Company to properly supervise the master and to provide a rested and sufficient crew for the vessel; (iv) failure of the US Coast Guard to provide an effective vessel traffic system; and (v) lack of effective pilot and escort services (NTSB, 1990). The causes of the accident are not the focus of this

* E-mail address: swhaycox@alaska.edu

chapter, which will be directed toward the toxicity generated by the spill to the shoreland and waters of the Sound and the Gulf of Alaska, and to human and wildlife, but a brief discussion of the general impact of the spill is provided here to establish context.

## 22.3 Context

The Exxon Valdez Oil Spill Trustee Council (EVOSTC) noted that the areas of Alaska impacted by the oil spill supported a large diverse ecosystem that was valued by large numbers of the American public who did not visit the area (EVOSTC, 2018a). The spill killed substantial numbers of different bird species and marine mammals as well as oiling much of the coastline in the impacted areas. The spill also had substantial effects on the fish, bird and wildlife populations. While some of these effects may be of relatively short duration, others such as recovery of various bird populations and the herring fishery are likely to take decades. Subsistence harvest of fish and wildlife in 11 of 15 villages affected by the spill declined by 4–77% in 1989 when compared with pre-spill levels. At least four of the 11 villages showed continued lower-than-average levels of use in the period 1990–1991; this decline is particularly noticeable in the Prince William Sound villages of Chenega and Tatitlek (Miraglia, 2002). In 1989–1991, chemical analysis indicated that most resources tested, including fish, marine mammals, deer and ducks, were safe to eat. Starting in 1989, health advisories were issued indicating that shellfish from oiled beaches should not be eaten. During 1989, emergency commercial fishery closures were ordered in Prince William Sound, Cook Inlet, Kodiak and the Alaska Peninsula. This affected salmon, herring, crab, shrimp, rockfish and sablefish. The 1989 closures resulted in sockeye over-escapement in the Kenai River and portions of Kodiak Island. In 1990, portions of Prince William Sound were closed to shrimp and salmon fishing. Recreation use of Prince William Sound and other areas of Alaska impacted declined significantly in 1989. Between 1989 and 1990, a decline in sport fishing (number of anglers, fishing trips and fishing days) were recorded for Prince William Sound, Cook Inlet and the Kenai Peninsula. Some commercial recreation and tourism businesses were injured by reduction in the number of visitors and visitor spending. The oil spill caused injury to the way people perceive recreation opportunities in the spill area. The location of recreation use was altered by changed use patterns and displaced use. A few recreation facilities were impacted by the spill, most from overuse or misuse during 1989 and 1990.

Assessing the impact of the spill, scientists estimated mass mortality among sea otters, harbour seals and seabirds. The oil itself, and subsequent pressure washing, caused decimation of microalgae and benthic invertebrates. The emotional, psychological and economic cost among Native Alaskans living in the two villages whose residents depended most heavily on harvesting marine species, and on fishers and others in the towns of Valdez and Cordova, was substantial. All of these consequences of the spill linger 30 years after the spill. Though many affected species of wildlife have recovered, others have not and oil remains on beaches of the Sound. The spill constitutes a continuing human and biological tragedy.

## 22.4 Clean-up

Exxon Corporation, the State of Alaska and agencies of the federal government spent four summers working to clean up the spilled oil, paid for by the Exxon Corporation, which asserts it spent $2.1 billion. At its peak the effort involved about 10,000 workers, 1000 boats and perhaps 100 airplanes and helicopters. The principal method utilized high-pressure hot and cold water spraying of oiled areas, with the washed-away oil collected in booms at the shoreline (Shinnefield, 2018). Some lightly oiled beaches were treated with bioremediation but most were not. Some beaches were tilled with mechanical equipment. Few chemical agents were used. Exxon ceased clean-up activities after 4 years. Subsequent surveys have found oil on many beaches in the Sound.

## 22.5 Settlement

On 9 October 1991, the US District Court (Alaska) approved a settlement of the various state and federal government charges against Exxon

(US Alaska District Court, 1991a, b). The court fined Exxon Corporation $150 million as penalty for the spill, the largest civil penalty for environmental despoliation in US history. Of this sum, $125 million was forgiven in recognition of Exxon's work on cleaning the beaches. The remaining $25 million went to the North American Wetlands Conservation Fund and a national Victims of Crime Fund. Exxon paid an additional $100 million in civil penalty for restitution for injury to fish, wildlife and land, which was shared by the state and federal governments.

The court also imposed a civil settlement on the corporation of $900 million. A central element of the settlement was creation of a joint federal–state trust council, established to expend this award: the Exxon Valdez Oil Spill Trustee Council (EVOSTC, 2018b). Intended to remediate the environmental impact of the spill over time, the funds are to be used for restoration of the ecosystem. The EVOSTC is advised both by members of the general public and by the scientific community. Two vital components of the Trustee Council's restoration effort are the research, monitoring and general restoration programme and the habitat protection and acquisition programme. Extensive work has been done to restore and monitor resources and communicate these findings to the public. The research, monitoring and general restoration programme is funded each year through the annual work plan, which documents the projects that are currently funded to implement restoration activities for injured resources and services. This includes two long-term monitoring programmes. The habitat protection programme preserves habitat important to injured resources through the acquisition of land or interests in land. As of 2006, the Council had protected more than 630,000 acres (255,000 ha) of habitat, including more than 1400 miles (2250 km) of coastline and over 300 streams valuable for salmon spawning and rearing. Other public information efforts in which the Council is currently engaged include its website (EVOSTC, 2018a), which offers detailed information regarding past, current and future restoration efforts; additionally, the Council prepares a number of documents for distribution to the public. These include: (i) invitation for proposals, issued at 5-year intervals, which solicit restoration project ideas from the scientific community and the public for the Council's restoration activities, including two long-term monitoring programmes; (ii) the annual work plan (described above); (iii) updates to the restoration plan which periodically provide new information on the recovery status of injured resources and services; and (iv) project final reports, which are available to the public at the Council's website, through the Alaska Resource Library and Information Services (ARLIS) in Anchorage as well as at several other libraries in the State, at the Library of Congress, and through the National Technical Information Service. In addition, the Council supports researchers in publishing their project results in peer-reviewed scientific literature, which expands their audience well beyond Alaska. The Public Advisory Committee (PAC) is an important means of keeping stakeholders and others informed of the progress of restoration and providing the public's opinions to the EVOSTC as they make decisions. The Council has commissioned and cooperated with a variety of surveys and remediation actions in its 30 years. It has compiled the most reliable data on initial devastation and on the status of lands and waters impacted by the spill, and on the health of Prince William Sound. It has become the most widely respected and utilized resource for study of the spill impact. Much of the information presented here relies on studies conducted or commissioned by the Council.

In addition to civil and criminal prosecutions, individual personal suits for damages continued in the courts for many years following the spill (Lebedo, 1997; Chambers, 2003; US Ninth Circuit Court, 2004)

The EVOSTC estimates that initially the spill resulted in the deaths of 250,000 seabirds, 2800 sea otters, 300 harbour seals, 250 bald eagles, as many as 22 orcas, and billions of salmon and herring eggs. Several surveys over the 30 years since the spill, the most recent in 2016, show that toxic oil remains in areas of Prince William Sound.

## 22.6 Lingering Oil

The extensive clean-up activities and natural processes did remove much of the stranded oil over the 4 years immediately following the spill and scientists expected that the remaining oil would be reduced to minimal amounts. A 1994 study concluded that 20% of the oil had evaporated, 14% biodegraded and 14% was cleaned. An estimated 13% remained in sub-tidal sediments, 2% remained on shorelines and less than 1% remained in the water (EVOSTC, 2018c).

However, a survey 8 years after the spill found that oil remained in many intertidal sediments. Such lingering oil has been the subject of numerous subsequent studies to determine its extent, its effect om biota and the best methods of continuing remediation. A survey 10 years after the spill found that most beaches in the Sound appeared clean on the surface. Such subsurface oil as remained was expected to dissipate over time. However, continuing studies have found oil in numerous locations in discontinuous patches at depths of 12–18 cm. It is estimated that today, 23,000 gallons (105,000 l) of oil remain deposited. Comparative surveys in 2004 and 2015 found little change in the amount of oil remaining. The distribution of remaining oil seems to be influenced by several factors. First is the degree of the initial oiling and the permeability of the landscape oiled. More oiling at the outset results in more lingering oil, as do beach soils that are more permeable and rocky. In addition, low wave action results in more lingering oil. Moreover, protected beaches and those without freshets running from inland preserved more oil than those less protected and with flowing fresh waters. Beaches with larger gravels held more oil than those with smaller gravel and finer sediments. The researchers also suggested that there are in fact numerous areas in the Sound that have not been surveyed for lingering oil. The 2015 study indicated that in many of the still-oiled areas, low oxygen levels in the water in the intertidal zone will be inadequate to dissipate the oil by natural washing. In 2016 a team commissioned by the EVOSTC sampled 400 dug pits in six areas of the Sound that had been examined in 2015 and in 2004, as representative of oiled areas. The team reported no significant difference in the average subsurface oil per unit of the sites measured in 2004.

The team had developed spatially explicit machine learning models for oiled areas in both Prince William Sound and along the Gulf of Alaska. The models were based on data collected at 314 shoreline segments that had been surveyed in 2001 and 2008. The models allowed identification of the factors noted that help to explain the continued presence and lack of dissipation of the oil, and to suggest areas not yet fully surveyed where oil may be present. The implications of lingering oil are discussed below.

## 22.7 Recovery

To fully appraise the continuing impact and lasting toxicity of the oil spilled from the *Exxon Valdez* on the environment in Prince William Sound and Alaska, it is useful to examine the initial injury and rate of recovery of the individual species affected. The creation of the EVOSTC with its mandate for preservation and restoration of the affected environment, and its extraordinary funding to carry out that mission, has produced the most complete study and remediation efforts of any environmental catastrophe and should be studied for its role in assessing damage and directing repair following the spill. Below are summaries of restoration work on the affected species, and their current status, drawn from the Council's 2014 update summarizing studies to that date, the most complete summary available at present (EVOSTC, 2018d).

### 22.7.1 Mammals

#### *22.7.1.1 Harbour seals*

Harbour seal numbers were declining in the Gulf of Alaska, including in Prince William Sound, before the oil spill. *Exxon Valdez* oil affected their habitat. Estimated mortality as a direct result of the oil spill was about 300. In some parts of the Sound, 80% of the seals had oil on them in May 1989 and remained oiled until their moult in August. Some of the haul-out sites were oiled through the pupping season, and many pups became oiled shortly after birth. Based on aerial surveys conducted at trend-count haul-out sites, seals in oiled areas declined by 43%, compared with 11% in unoiled areas. Based on annual counts from haul-outs concentrated in the south-central region of the Sound, seal numbers stabilized from 1996 to 2005 and likely increased between 2001 and 2005. Harbour seals are considered 'Recovered' due to collective evidence from the past 10 years indicating that harbour seal population numbers are stabilizing or increasing.

#### 22.7.1.2 *Killer whale (orcas)*

More than 160 killer whales in eight resident (fish-eating) pods regularly use Prince William Sound/Kenai Fjords as part of their ranges. Transient (marine mammal-eating) groups are observed in the Sound less frequently but some use the Sound year-round. After the Spill, the loss of individual whales from the resident AB pod was of particular concern. At the time of the Spill, this group numbered 36 animals, and 14 whales disappeared in 1989–1990. During that time no young were recruited into the population. Members of the transient AT1 population were also observed in the area of the Spill and adjacent to the tanker, as it was leaking oil. Whales were observed surfacing in *Exxon Valdez* oil slicks following the Spill in 1989 and nearly all of the deaths occurred at the time of the Spill or the following winter. It is likely they inhaled petroleum or petroleum vapours, and it is also possible that they ate contaminated fish. The mortality rate for the AB pod was 19% in 1989 and 21% in 1990, compared with an expected natural mortality rate of 2.2% or less. The AT1 numbers were stable at 22 regularly observed individuals, but in a retrospective analysis it was determined that nine whales disappeared shortly after the Spill. In the subsequent 20 years researchers have not seen these individuals. Since 1989, a total of 15 of 22 whales have gone missing from the AT1 group and are now presumed dead (five of the carcasses were found on beaches). During that same period there has been no recruitment of calves into this genetically unique group of transients.

From 1990 to1995 seven calves were born within the AB pod; however, additional mortalities occurred and by 2005 the number of whales was only 28. The AB pod continued a slow recovery and in 1990 numbered 30 individuals, though the pod has now split and travels as two distinct units. Killer whales are long-lived and slow to reproduce. A disproportionate number of females were lost at the time of the Spill and population modelling has demonstrated that the Spill impacted the AB pod primarily through the loss of young and reproductive females.

The AB pod is the only tracked pod that has experienced a decline following the Spill. Other pods have increased at an average rate of 3% per year. Since 1990, the AB pod females that survived have produced nearly as many calves as would be expected based on the number of females and their ages. The lack of recovery of AB pod, thus, can be largely attributed to the loss of young adult females, which reduced the number of reproductive females by half, and by the loss of juveniles. The annual birth rate in AB pod since the spill has been about 70% of the birth rate observed in other resident pods, which was significantly lower than expected. Nonetheless, this pod is considered to be 'Recovering'. The AT1 transient population of killer whales has remained stable at seven individuals with no recorded births or deaths since 2010 and is considered 'Not Recovering'. Progress toward recovery appears unlikely, as key breeding females have been lost and no new recruitment observed.

#### 22.7.1.3 *River otters*

River otters have a low population density in Prince William Sound. Twelve river otter carcasses were found following the Spill, but the actual total mortality is not known. Although some of the differences (e.g. values of blood characteristics) between river otters in oiled and unoiled areas in Prince William Sound were apparent through 1996, they did not persist. Thus, the Trustee Council considered river otters to be 'Recovered', because indications of possible lingering injury from the Spill were not present.

#### 22.7.1.4 *Sea otters*

By the late 19th century, sea otters had been eliminated from most of their range due to overharvest by fur traders. Sea otters came under international protection in 1911 and since then their numbers have rebounded. Today, sea otters can only be harvested for subsistence purposes. More than 1000 otters became coated with oil in the days following the Spill and 871 carcasses were collected throughout the Spill area. Estimates of the total number of sea otters lost to acute mortality vary, but range as high as 40% (2650) of the approximately 6500 sea otters inhabiting the western areas of the Sound. In 1990 and 1991, higher than expected proportions of prime-age adult sea otters were found dead in western Prince William Sound. Higher mortality of recently weaned juveniles in oiled areas was documented through 1993. Continuing studies suggest that relatively poor survival of otters in the oiled area persisted for well over a decade. From 1993 to 2000, the number of otters increased by 600 animals which represents an annual growth rate of 4%. However, in areas that were heavily oiled, sea otter populations remained well below pre-Spill numbers and population trends continued to decline through 2010. Aerial surveys in 2013 indicated that population abundance in Prince William Sound has converged in oiled and unoiled areas. The estimated number of sea otters more than doubled relative to the 1993 estimate and the increase over that time frame was greater to or similar to estimates of sea otters that died within the first years of the Spill. Starting in 2011, there was a distinct change in the age-class proportions of dying sea otters, with a return to the pre-Spill pattern of predominantly young and older sea otters recovered as carcasses. This pattern continued in 2012 and 2013, which may be interpreted as evidence that from 2011 to 2013 chronic exposure to lingering oil and/or chronic effects due to previous exposure abated to the point where they are no longer factors constraining survival. The return to pre-Spill numbers and mortality patterns suggests a gradual dissipation of exposure to lingering oil over the past two decades. Therefore, sea otters are considered to be 'Recovered'.

### 22.7.2 Birds

#### *22.7.2.1 Bald eagles*

Prince William Sound provides year-round and seasonal habitat for about 6000 bald eagles and within the Sound it is estimated that about 250 bald eagles died as a result of the Spill. In addition, productivity was reduced in oiled areas. Productivity (or reproductive success, as measured by chicks per nest) was back to pre-Spill levels in 1990 and 1991 and an aerial survey of adults in 1995 indicated that the population had returned to or exceeded its pre-Spill level in the Sound. In September 1996, the Trustee Council classified the bald eagle as 'Recovered'.

#### *22.7.2.2 Barrow's goldeneyes*

Barrow's goldeneyes are sea ducks that winter in protected near-shore marine waters in Prince William Sound and feed in the intertidal zone, consuming primarily mussels. Sea ducks, generally, were vulnerable to acute mortality and constituted approximately 25% of the carcasses recovered in Prince William Sound. Given the number of Barrow's goldeneyes present at the time of the Spill, acute mortality was probably in the low thousands. Of more concern are longer-term effects due to either chronic exposure to lingering oil or indirect effects of trophic web disruption. Because Barrow's goldeneyes occur exclusively in intertidal and shallow sub-tidal habitats, they are particularly vulnerable to lingering oil exposure and the potential for physiological effects. Barrow's goldeneyes were shown to have higher levels of induction of cytochrome P4501A (CYP1A) in oiled areas in 1996, 1997 and 2005. However, in March 2009, average CYP1A was similar between areas, suggesting that exposure to residual oil had abated by that time. Prince William Sound supports between 20,000 and 50,000 wintering individuals. US Fish and Wildlife Service surveys through April 2012 show that population growth rates were the same between oiled and unoiled sites and remained relatively unchanged between 1998 and 2012. There is no evidence that Barrow's goldeneyes are currently being exposed to lingering oil in the intertidal habitat. Lack of elevated CYP1A in oiled relative to unoiled areas suggests that exposure to lingering oil has ceased in the Barrow's goldeneyes. Barrow's goldeneyes are considered to be 'Recovered' from the effects of the Spill.

#### *22.7.2.3 Black oystercatchers*

Black oystercatchers spend their entire lives in or near intertidal habitats and are highly vulnerable to oil pollution. They are fully dependent on the nearshore environment and forage exclusively on invertebrate species along shorelines. It was estimated at the time of the Spill that 1500–2000 oystercatchers bred in south-central Alaska. In addition to direct mortalities, breeding activities were disrupted by the oil and clean-up activities. Black oystercatchers are long-lived (15+ years) and territorial. In the early 1990s, elevated hydrocarbons in faeces were measured in chicks living on oiled shorelines. Deleterious behavioural and physiological changes, including lower body weights of females and chicks, were also recorded. Because foraging areas are limited to a few kilometres around a nest, contamination of mussel beds in the local vicinity was thought to provide a source of exposure. In 1998, data indicated that oystercatchers had fully reoccupied and were nesting at oiled sites in the Sound. The breeding phenology of nesting birds was relatively synchronous in oiled and unoiled areas and no oil-related differences in clutch size, egg volume, or chick growth rates were detected. Based on this study and 1 year of boat-based surveys (2000) of marine birds in Prince William Sound indicating that there were increases in numbers of oystercatchers in both the oiled and unoiled areas for that year, the black oystercatcher was identified as recovered. However, a long-term (1989–2007) evaluation of marine bird population trends suggested that populations of black oystercatchers in the Sound may not have recovered to pre-Spill levels, making the recovered designation premature, and their status was changed to recovering in the 2010 update. Surveys through 2012 have shown a stable population trend. Recent studies show no evidence of change in black oystercatcher abundance in oiled areas and no evidence that trends differ between oiled and unoiled areas. Therefore, black oystercatchers are regarded as 'Very Likely Recovered'.

#### 22.7.2.4 Common loons

Common loons in the Spill area may number only a few thousand, including only hundreds in Prince William Sound. Common loons injured by the Spill probably included a mixture of wintering and migrating birds. Common loons have an intrinsically low population growth rate and relatively large numbers of carcasses were recovered after the Spill. Yet post-Spill winter population counts of common loons have met or exceeded available pre-Spill counts for all years measured since the Spill, except 1993. Given the long-term positive changes in winter population information, common loons are considered 'Recovered'.

#### 22.7.2.5 Common murres

About 30,000 carcasses of oiled birds were picked up in the first 4 months following the Spill, and 74% of them were common and thick-billed murres. Many more probably died than actually were recovered. Based on surveys of index breeding colonies, the Spill area populations may have declined by about 40% following the Spill. There is evidence that the timing of reproduction was disrupted and productivity decreased. Post-Spill monitoring at the breeding colonies in the Barren Islands indicated that productive success was within normal bounds by 1993, and it has stayed within these bounds each breeding season since then. Although counts were low in 1996, the counts in 1997 at the index site brought the colony size to pre-Spill levels. The population size coupled with normal reproductive success (productivity) indicates that recovery has been achieved for common murres.

#### 22.7.2.6 Cormorants

Cormorants are large fish-eating birds that spend much of their time on the water or perched on rocks near the water. Direct Spill-related mortality was estimated at between 2900 and 8800 deaths. In 1996, the US Fish and Wildlife Service Alaska Seabird Colony Catalog, however, listed counts of 7161 pelagic cormorants, 8967 red-faced cormorants and 1558 double-crested cormorants in the Spill area. These are direct counts at colonies, not overall population estimates, but they suggest that population sizes are small. In this context, it appears that injury to all three cormorant species was significant. Marine bird surveys were conducted during ten of the 16 years during1989–2005. For cormorants, trends for both summer and winter populations were increasing in the oiled area of Prince William Sound. Moreover, population estimates for cormorants in summer 2004 ranged from 9000 to 11,000 birds, which falls within the range of 10,000–30,000 estimated in 1972. Therefore, although population estimates of cormorants are highly variable throughout their range, cormorants are considered to be 'Recovered'.

#### 22.7.2.7 Harlequin ducks

Harlequin ducks spend most of their time in intertidal and shallow sub-tidal habitats where much of the oil was initially stranded. An estimated 1000 harlequins were killed by the initial oiling event, which represented about 7% of the wintering population. In addition to acute effects, harlequin ducks were one of the few species for which chronic injury related to long-term exposure to lingering oil was documented. Winter populations of harlequin ducks in Prince William Sound have been as high as 19,000. The 2010 estimate of wintering harlequin ducks in the Sound was approximately 15,425. Several post-Spill studies were designed. Through 1998, Spill effects were still evident. Supporting studies provided evidence of continuing injury to harlequins. Lingering oil appeared to remain in habitats used by harlequins, thereby maintaining the possibility of chronic effects related to continued exposure. But in 2013 studies, hepatic CYP1A levels in harlequin ducks, based on EROD (7-ethoxyresorufin-O-deethylase) activity, were similar between areas oiled during the Spill and in nearby unoiled areas. This constitutes the first time since initiation of harlequin duck CYP1A sampling in 1998 that EROD activity has not been higher in oiled areas than in unoiled areas of Prince William Sound. This would indicate that harlequin ducks are no longer exposed to residual oil from the Spill. Harlequin ducks are considered to be 'Recovered'.

#### 22.7.2.8 *Kittlitz's murrelets*

The Kittlitz's murrelet is found only in Alaska and the Russian Far East. A large percentage of the world population, which may number only a few tens of thousands, breed in Prince William Sound. Estimates of the total number of Kittlitz's murrelets that died as a result of the Spill vary from 255 to 2000; it has been suggested that this represents 5–10% of the world's population. Knowledge of the ecology and trends of these birds is limited. They lay a single egg and have an intrinsically low population growth rate, thus recovery from an acute loss is likely to be slow. What little evidence is available from studies in Alaska suggest possible predator limitation in some areas, and lack of productivity due to food availability and chick predation. A 2014 study also found that paralytic shellfish poisoning had contributed to chick mortality at nest sites on Kodiak Island. While the population decline appears to have abated, current recovery of Kittlitz's murrelets remains 'Unknown'.

#### 22.7.2.9 *Marbled murrelets*

Marbled murrelets are found throughout the northern Gulf of Alaska and are known to concentrate in Prince William Sound. Carcasses of nearly 1100 *Brachyramphus* murrelets were found after the Spill, and about 90% of the murrelets that could be identified to the species level were marbled murrelets. Since they are a small bird and not easily seen, many more murrelets probably were killed as a result of the oil than were recovered. Estimates vary but between 2900 and 14,800 individuals were killed by the initial oiling and this represented 6–12% of the marbled murrelets in the Spill area. It is estimated that marbled murrelets declined at a rate of 5% per annum from 1989 to 2012, a cumulative population loss of –69%. It is listed as a threatened species in Washington, Oregon, California and British Columbia. Marbled murrelets have low intrinsic productivity and a slow population growth rate. Therefore, recovery from an acute loss will likely take many years. They are considered to be 'Not Recovering'.

#### 22.7.2.10 *Pigeon guillemots*

An estimated 2000–6000 guillemots, representing 10–15% of the Spill area population, died from acute oiling. Additionally, an increase in nest predation of pigeon guillemot chicks and incubating adult birds occurred in the Sound after the Spill. Researchers speculated that immediately after the Spill, predators such as river otters and minks preyed more heavily on nesting guillemots due to heavy oiling and subsequent reduction of their customary shellfish prey. Summer surveys along both oiled and unoiled shorelines of the Sound have indicated that numbers of guillemots continued to decline through 2005. March surveys reveal no significant trends in abundance, though the data appear to suggest a decline at this time of year as well. Reduction in forage fish, specifically herring and sand lance, has been implicated in declines of pigeon guillemots. Nest predation is a potential source of mortality that may be limiting recovery in some areas, implying that predator removals could prove an effective restoration option. More data on productivity levels is needed to determine if the recovery objective of increasing abundance and productivity has been met.

Pigeon guillemots are considered to be 'Not Recovered'.

### 22.7.3 Shellfish

#### *22.7.3.1 Clams*

Clams are widely distributed throughout the Spill area. They are important prey for various fish and wildlife resources, including sea otters and some seabirds. Clean-up technologies, including hot-water high-pressure washing, manual and mechanical scrubbing and physical removal of oiled sediments, were detrimental to clam populations. Hot-water washing caused thermal stress, oil dispersal into the water column, animal displacement and burial, and the transportation of fine-grain sediment from the upper intertidal into the lower intertidal zone. Early assessments reported that clean-up activities resulted in reductions in clam abundance and distribution on treated (oiled-but-treated) beaches up to 3 years after the Spill. Studies have indicated that abundances of some species of clam were lower on treated beaches through 1996. Densities of littleneck and butter clams were depressed through 1997 on cleaned mixed-sedimentary shores where fine sediments had been washed down the beach during pressured water treatments. As of 2005, several wildlife species that use the intertidal zone and feed on bivalves (e.g. harlequin ducks and sea otters) were still being exposed to oil. Between 2002 and 2010, bivalve assemblages declined substantially. Recent (2012) studies indicate that the decline is in response to changes in regional conditions rather than the Spill or subsequent clean-up activities. Clams are considered 'Recovered'.

#### *22.7.3.2 Mussels*

Mussels are a keystone species in the nearshore environment throughout the Spill area and are locally important for subsistence users. They provide prey for harlequin ducks, black oystercatchers, juvenile sea otters, river otters and many other species. Mussel beds are also important components of intertidal habitats, because they provide physical stability and habitat for other organisms in the intertidal zone. Although mussels were coated with oil from the *Exxon Valdez*, dense mussel beds were purposely not disturbed during clean-up operations so that the stability and habitat they provided would be preserved. After the Spill, concentrations of oil in mussel tissue from the oiled area increased rapidly. These concentrations were typically far higher than in mussels from unoiled areas Long-term mussel contamination occurred where substantial amounts of oil were trapped in sediment, primarily within coarse-textured habitats, including heavily oiled beaches exposed to considerable wave and storm energy. The primary route by which mussels accumulate oil is through ingestion of petroleum hydrocarbons in the water. The current distribution of oil within a mussel bed is determined by water flow, amount of oil present, sediment grain size and disturbance history. Data from 2012 indicated that hydrocarbon concentrations in mussels, even on armoured beaches where elimination has been slow, are no longer different from background. Thus, mussels are considered 'Recovered'.

### 22.7.4 Fish

#### *22.7.4.1 Cutthroat trout*

Researchers documented that cutthroat trout emigrating within the oiled areas in 1989–1990 grew more slowly than those in the unoiled areas. When trout leave their freshwater spawning areas they feed primarily in the nearshore environment, thus it is likely that cutthroats were exposed to oil in this environment. The difference in growth rates between trout in oiled versus unoiled streams persisted through 1991. Populations in the Sound are small and geographically isolated from each other. Given the ecological similarities in summer diet and foraging ecology along shorelines between cutthroat trout, juvenile pink salmon and Dolly Varden, and the absence of ongoing injury to those other two species, further research would be very unlikely to demonstrate any evidence of continuing differences between oiled and unoiled areas due to the Spill. Thus, funding the additional research necessary to provide current growth rate and abundance data for this species is not a cost-effective scientific priority. Cutthroat trout are 'Very Likely Recovered'.

##### *22.7.4.2 Dolly Varden*

Dolly Varden migration into the marine environment occurs in the summer where the fish spend time feeding in nearshore waters. Many fish were in freshwater when the Spill occurred but emigrated in and out of the Spill area later in the season. Concentrations of hydrocarbons in the bile of Dolly Varden were some of the highest of any fish sampled in 1989. The growth differences between Dolly Varden in oiled and unoiled streams did not persist into the 1990–1991 winter, but no growth data have been gathered since 1991. Current exposure to lingering oil is unlikely, because most of the bioavailable oil is confined to subsurface intertidal areas and not dissolved in the water column. Given the available evidence, Dolly Varden are considered to be 'Recovered'.

##### *22.7.4.3 Pacific herring*

Pacific herring are an ecologically and commercially important species in the Prince William Sound ecosystem. They are central to the marine food web, providing food to marine mammals, birds, invertebrates and other fish. Herring are also commercially fished for food, bait, sac-roe and spawn on kelp. Pacific herring spawned in intertidal and subtidal habitats in Prince William Sound shortly after the Spill. All age classes were contaminated by oil. Lesions and elevated hydrocarbon levels were documented in some adult Pacific herring from the oiled areas. Laboratory studies showed abnormalities and possible depressed immune functions. Four years after the Spill a dramatic collapse of the fishery occurred, and the herring population has never rebounded. The herring fishery in the Sound has been closed for 19 of the 25 years since the Spill. The population began increasing again in 1997 and the fishery was opened briefly in 1997 and 1998. However, the population increase stalled in 1999, and research suggests that the opening of the fishery in 1997 and 1998 stressed an already weakened population and may have contributed to the 1999 decline. The fishery has been closed since then and no trend suggesting healthy recovery has occurred. Pacific herring are considered 'Not Recovering', eliminating a once-lucrative commercial fishery.

##### *22.7.4.4 Pink salmon*

Up to 75% of wild pink salmon in Prince William Sound spawn in the intertidal portions of streams. Eggs deposited in gravel and developing embryos were chronically exposed to hydrocarbon contamination from the water column and from leaching oil deposits on adjacent beaches. Juvenile salmon entering seawater from both wild and hatchery sources were likely exposed to oil as they swam through contaminated waters and fed along oiled beaches. In 1999, dissolved oil was measured in six pink salmon streams that had been oiled in 1989. Only one of the six streams had detectable concentrations of oil, and they were about 1000 times lower than concentrations reported as toxic to developing pink salmon embryos. Based on these results, continuing exposure of pink salmon embryos to lingering oil is negligible and unlikely to limit pink salmon populations. Thus, pink salmon were considered 'Recovered' in 1999.

##### *22.7.4.5 Rockfish*

Dead rockfish were observed throughout the Sound immediately following the Spill, but an absolute count was never documented. From 1989 to 1991, higher petroleum hydrocarbon concentrations were measured in rockfish from oiled areas when compared with unoiled areas. It is unlikely that rockfish are currently being exposed to lingering oil, because known pockets of lingering oil rarely occur in their preferred habitat. There have been no demonstrated differences in population or breeding success between oiled and unoiled areas. The EVOSTC considers rockfish to be 'Very Likely Recovered'.

##### *22.7.4.6 Sockeye salmon*

Commercial salmon fishing was closed in Prince William Sound and in portions of Cook Inlet and near Kodiak in 1989 to avoid the possibility of contaminated salmon being sold at market. As a result, there were higher-than-desirable numbers (i.e. 'over-escapement') of spawning sockeye salmon entering the Kenai River and on Kodiak Island. Initially, these high escapements produced an overabundance of juvenile sockeye that overgrazed the zooplankton and altered planktonic food webs in the nursery lakes. As a result, growth rates were reduced during the freshwater stage of the salmon's life cycle, which led to a decline in returns of spawning adults. The net result was an initial loss of sockeye production. Starting in 1993, however, the production of smolts per adult increased sharply and the smolt sizes and age composition suggested that rearing conditions had improved. On the basis of catch data through 2001 and in view of recent analyses of return-per-spawner estimates presented to the Alaska Board of Fisheries in 2001, the return-per-spawner in the Kenai River system is within historical bounds. In 2002, this species was considered to be 'Recovered'.

### 22.7.5 Coastline

#### *22.7.5.1 Intertidal communities*

Over 1400 miles of coastline were oiled by the Spill in Prince William Sound, on the Kenai and Alaska peninsulas, and in the Kodiak Archipelago. Heavy oiling affected approximately 220 miles of this shoreline. It is estimated that 40–45% of the 11 million gallons (50 million litres) of crude oil spilled by the *Exxon Valdez* washed ashore in the intertidal zone. For months after the Spill in 1989, and again in 1990 and 1991, both oil and intensive clean-up activities had significant impacts on the flora and fauna of this environment. Direct assessment of the Spill effects included sediment toxicity testing, documenting abundance and distribution of intertidal organisms and sampling ecological parameters of community structure. Dominant species of algae and invertebrates directly affected by the Spill included common rockweed, speckled limpet, several barnacle species, blue mussels, periwinkles and oligochaete worms. At lower elevations on gravel and mixed sand/gravel beaches, the abundance of sediment organisms and densities of clams declined. Large numbers of dead and moribund clams were documented on treated beaches, likely due to a combination of oil toxicity and hot-water washing.

By 1991, in the lower and middle intertidal zones, algal coverage and invertebrate abundances on oiled rocky shores had returned to conditions similar to those observed in unoiled areas. The brown algae canopy was initially eliminated in most of the areas that underwent extensive cleaning. Full recovery of brown algae is crucial for the recovery of intertidal communities at oiled sites, because many intertidal organisms depend on the shelter this seaweed provides. Lingering oil is still present in some intertidal areas within the Spill zone. Recent studies indicate that at beaches with pockets of buried lingering oil, high amphipod mortality is associated with elevated hydrocarbon concentrations. Nonetheless, intertidal communities are considered to be 'Recovering'.

#### *22.7.5.2 Sediments*

Intertidal shorelines captured approximately 40–45% of the spilled oil and up to 13% of the oil settled in sub-tidal habitats. Manual removal eliminated some of the oil from the intertidal zone early in the response phase, and within a few months of the Spill, 89% of the moderately to heavily oiled beaches had been treated. By 1992, approximately 10 km of the original estimated 583 km of beaches with surface oiling remained uncleaned. Approximately 10 acres (4 ha) of *Exxon Valdez* oil remains in surface sediments, primarily in the form of highly weathered, asphalt-like or tar deposits. In 2003, it was estimated that 20 acres (8 ha) of unweathered lingering oil may still be present in subsurface, which could represent up to 100 t of remaining oil. Most of this oil is found in protected, unexposed bays and beaches. Subsurface oil was not subjected to the original clean-up activities; and because this oil is trapped beneath a matrix of cobbles, gravel and finer sediments, it is not easily exposed to natural weathering processes. Through 2012, surface oil was not frequently observed in these areas and subsurface oil was present as mostly unweathered mousse. It is likely that oil in sub-tidal sediments has decreased substantially since the Spill. Yet, 25 years after the Spill, lingering oil persists in the intertidal zones of Prince William Sound and on north-west shorelines of the Spill area. The presence of subsurface oil continues to compromise wilderness and recreational values, expose and potentially harm living organisms, and offend visitors and residents, especially those who engage in subsistence activities along still-oiled shorelines. Although much of the oil has diminished over time, pockets of unweathered oil exist and natural degradation of this oil is very slow.

Therefore, sediments are still considered to be 'Recovering'.

#### 22.7.5.3 Sub-tidal communities

Sub-tidal habitats encompass all of the seafloor below the mean lower low-water tide line to about 800 m. The impacted sub-tidal zone generally ranges from the lower intertidal zone to a depth of about 20 m. Communities in the near sub-tidal areas are typically characterized by dense stands of kelp or eelgrass and comprise various invertebrate species, such as amphipods, polychaete worms, snails, clams, sea urchins and crabs. Sub-tidal habitats provide shelter and food for an array of nearshore fish, birds and marine mammals. It is estimated that up to 13% of the oil that was spilled was deposited in the sub-tidal zones, which caused changes in the abundance and species composition of plant and animal populations below lower tides.

Invertebrate assemblages within eelgrass beds and adjacent areas of soft sediment were compared at oiled and unoiled sites from 1990 to 1995. In the early 1990s, several benthic organisms using the sub-tidal zones showed trends towards recovery and hydrocarbon concentrations had declined in many areas. However, consistent systematic surveys have not been conducted for many species. Nonetheless, sub-tidal communities are 'Very Likely Recovered'.

### 22.7.6 Human services

The social consequences of the Spill reflect the toxic character of the disaster and are briefly noted here.

#### 22.7.6.1 Commercial fishing

The Trustee Council specifically recognized the declines in pink salmon and Pacific herring populations and considered the reduction in these two fisheries as the biggest contributors to injury of the commercial fishing service in the Spill area. Therefore, many restoration activities were focused towards these resources. By 2002, the Trustee Council considered pink salmon and sockeye salmon to be recovered from the Spill. However, recovery was not considered complete for Pacific herring and the recovery status of this resource remains 'Not Recovering'. Income from commercial fishing dramatically declined immediately after the Spill and, for a variety of reasons, disruptions to income from commercial fishing continue today, as evidenced by changes in average earnings, ex-vessel prices and limited entry permit values. The Prince William Sound herring fishery has been closed for 19 of the 25 years since the Spill and herring are still considered to be recovering. Commercial fishing is considered to be 'Recovering' from the effects of the Spill.

#### 22.7.6.2 Passive use

Passive use is the service provided by natural resources to people who will probably not visit, contact, or otherwise use the resource. Thus, injuries to passive use are tied to public perceptions of injured resources. Passive use is also the appreciation of the aesthetic and intrinsic values of undisturbed areas and the value derived from simply knowing that a resource exists. The loss to passive use following the Spill was estimated by the State of Alaska at \$2.8 billion. Until the public no longer perceives that lingering oil is adversely affecting the aesthetics and intrinsic value of the Spill area, it cannot be considered 'Recovered'.

#### 22.7.6.3 Recreation and tourism

Recreation and tourism in the Spill area dramatically declined in 1989 in Prince William Sound, Cook Inlet and the Kenai Peninsula. Injuries to natural resources led resource managers to limit access to hunting and fishing areas. Recreation was also affected by changes in human use in response to the Spill, because areas that were unoiled become more heavily used as activity was displaced from the oiled areas. Recreation and tourism accounted for 26,000 jobs, generated \$2.4 billion in gross sales and contributed \$1.5 billion to Alaska's economy in 2003. The number of visitors to Alaska has increased in the years since the Spill and it is expected that the recreation and tourism industry in south-central Alaska will grow approximately 28% per year through 2020. Lingering oil remains on beaches and this remains a concern for users in some localized areas. Moreover, some of the natural resources upon which recreation and tourism rely have not recovered from the effects of the

Spill. Therefore, the Trustee Council finds recreation and tourism to be 'Recovering', but not yet 'Recovered'.

#### *22.7.6.4 Subsistence*

Fifteen predominantly Alaskan Native communities (with a total population of about 2200 people) in the oil Spill area rely heavily on harvests of subsistence resources, such as fish, shellfish, seals, deer and waterfowl. Oil from the Spill disrupted subsistence activities for the people of these villages and approximately 13,000 other subsistence permit holders in the area. Harvest levels have generally increased in many communities since the Spill, but results of harvest surveys have been variable. Even though the harvest levels in Sound communities were not as high as pre-Spill estimates, they were within the range of other Alaska rural communities. Harvest composition was also altered by the Spill. In the first few years following the Spill, people harvested more fish and shellfish than marine mammals because of the reduced number of marine mammals and the perception that these resources were contaminated and unsafe to eat. From 1989 to 1994, subsistence foods were tested for evidence of hydrocarbon contamination, with no or very low concentrations of petroleum hydrocarbons found in most subsistence foods. However, concerns about oil contamination remained and there was a belief that the increase in paralytic shellfish poisoning was linked with *Exxon Valdez* oil. By 2006, most subsistence users expressed confidence in foods such as seals, finfish and chitons. However, the safety of certain shellfish, such as clams, was still met with scepticism. But harvest levels from villages in the Spill area are comparable to other Alaskan communities. For these reasons, subsistence is considered to be 'Recovering' (Picou and Martin, 2007).

## 22.8 Conclusions

In summary, 30 years after the Spill, of the species affected, based on extensive scientific study and assessment, the Exxon Valdez Oil Spill Trustee Council, charged with preservation and restoration of the spill environment, finds 14 species 'Recovered', three species and the sub-tidal communities 'Likely Recovered', one pod of orcas and the intertidal communities and sediments 'Recovering', four species (including one pod of orcas) 'Not Recovering', and the status of Kittlitz's Murrelets 'Unknown'. Commercial fishing has not fully recovered; nor has passive use (the perception of the spill damage among the general public); recreation and tourism and subsistence continue to be affected. The herring fishery, once a central economic and community element of the Spill area, has been effectively eliminated.

The comprehensive and continuing assessment by the Trustee Council stands as a model of evaluation of environmental catastrophe, worthy of admiration and emulation.

## References


Chambers, J.H. (2003) *In re Exxon Valdez:* Application of Due Process Constraints on Punitive Damage Awards. *Alaska Law Review* 20, 195–212.

EVOSTC (2018a) Website for Exxon Valdez Oil Spill Trustee Council, Anchorage, Alaska. Available at: www.evostc.state.ak.us (accessed 1 May 2019).

EVOSTC (2018b) A State and Federal Partnership. Available at: http://www.evostc.state.ak.us/people/tc.cfm (accessed 1 May 2019).

EVOSTC (2018c) Oil Spill Facts. Available at: http://www.evostc.state.ak.us/index.cfm?FA=facts.home (accessed 1 May 2019).

EVOSTC (2018d) Exxon Valdez Oil Spill Restoration Plan, 2014 Update, Injured Resources and Services. Available at: http://www.evostc.state.ak.us/static/PDFs/2014IRSUpdate.pdf. (accessed 1 May 2019).

Haycox, S.W. (2012) 'Fetched up': unlearned lessons from the Exxon Valdez. *Journal of American History* 99, 219–228.

Lebedo, D. (1997) *Cleaning Up: The Story behind the Biggest Legal Bonanza of Our Time.* Simon & Schuster, New York.

Miraglia, R.A. (2002) The cultural and behavioral impact of the *Exxon Valdez* oil spill on the native peoples of Prince William Sound, Alaska. *Spill Science and Technology Bulletin* 7, 75–87.

NTSB (1990) *Marine Accident Report: Grounding of the US Tankship Exxon Valdez on Bligh Reef, Prince William Sound near Valdez, Alaska, March 24, 1989*. Report No. NTSB/Mar-90/04. National Transportation Safety Board, Washington, DC.
Picou, J. S. and Martin, C.G. (2007) *Long Term Community Impacts of the Exxon Valdez Oil Spill: Patterns of Social Disruption and Psychological Stress Seventeen Years After the Disaster*. Report to the National Science Foundation, Office of Polar Research, Washington, DC. Available at: http://www.arlis.org/docs/vol1/B/243478793.pdf (accessed 1 May 2019).
Shinnefield, P. (2018) Clean Up of the Exxon Valdez Spill. Available at: https://web.stanford.edu/class/e297c/trade_environment/energy/hexxon.html (accessed 1 May 2019).
US District Court of Alaska (1991a) *United States v. Exxon Corp.,* No. A91-082-CV. Available at: http://www.arlis.org/docs/vol1/A/42964164.pdf (accessed 1 May 2019).
US District Court of Alaska (1991b) *State of Alaska v. Exxon Corp. and Exxon Shipping Co.,* No. A91-083-CV. Available at: http://www.arlis.org/docs/vol1/A/294858686.pdf (accessed 1 May 2019).
US Ninth Circuit Court (2004) *In re Exxon Valdez*, No. A89-0095-CV.

# 23 Review of Studies of Composition, Toxicology and Human Health Impacts of Wastewater from Unconventional Oil and Gas Development from Shale

**L.M. Crosby* and W.H. Orem**
*US Geological Survey, Reston, Virginia, USA*

## 23.1 Abstract

Unconventional oil and gas (UOG) extraction has produced large economic benefits. However, prudent management of UOG wastes necessitates a thorough understanding of the complex composition, fate and potential impacts of wastewater releases. UOG production results in large volumes of wastewater. Despite limited reuse of the wastewater, the majority needs to be disposed of, usually by underground injection. The wastewater contains myriad organic, inorganic and radioactive substances from hydraulic fracturing and production activities or from the (typically shale) formation. Many substances in this wastewater are either proprietary, or are known or potential toxicants. Limited toxicological studies of these mixtures suggest that some of the components are highly toxic. Thus, any releases of untreated wastewater may represent a threat to environmental integrity and human health.

## 23.2 Background and Introduction

The oil and gas industry in the USA and worldwide has changed dramatically in the past few decades through the development of technologies such as directional drilling and hydraulic fracturing (Gandossi and Von Estorff, 2015). These new Unconventional oil and gas (UOG) technologies have allowed development of resources from previously inaccessible sources, especially shales. UOG production (also called continuous oil and gas) has made large plays, such as the Marcellus Shale in the eastern USA and the Wolfcamp Shale in the Permian Basin, Texas, available for exploitation (Gaswirth, 2017; USEIA, 2017). This revolution in the oil/gas industry has dramatically altered the international market (Deutch, 2011), creating jobs in regions of UOG production (Weber, 2012; Lee, 2015; Munasib and Rickman, 2015). However, as pointed out by Jacquet (2014), risks to communities near UOG development include overburdened municipal services, upended social and cultural patterns, and volatile economic growth. In communities invested with continuing resource extraction (including UOG), declines in population, development and financial investment may ensue (Jacquet, 2014). Powers (2015) found that local populations were concerned about threats to water, population growth and its implications, and changes to the rural landscape.

UOG technology used in the development of shale has been described (Short, 1993; Arthur

* E-mail address: Lynn.crosby@fda.hhs.gov

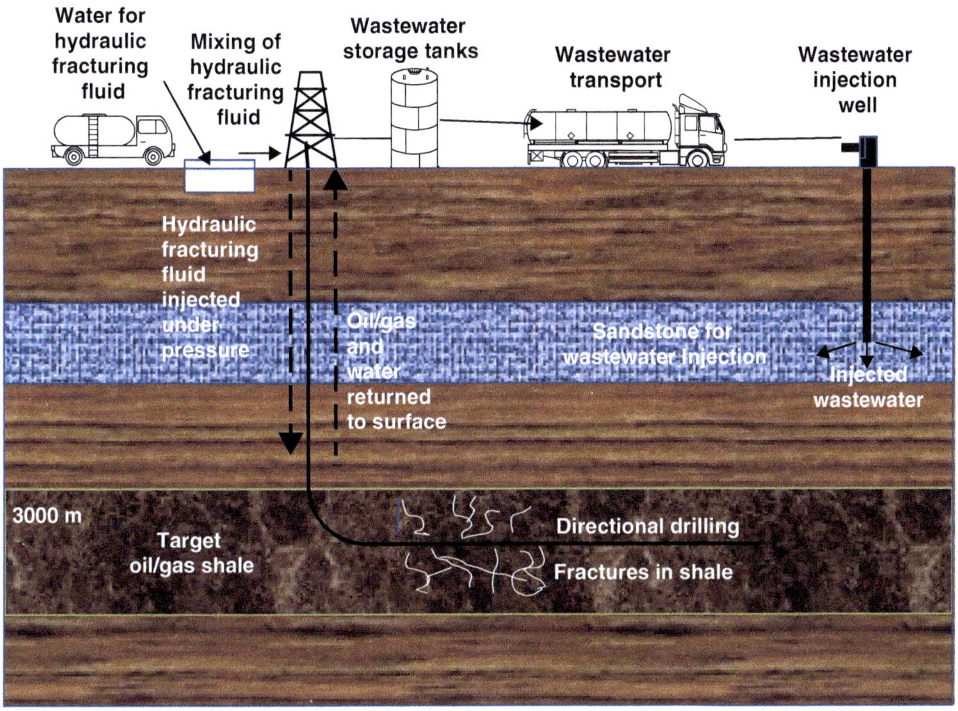


**Fig. 23.1.** Major steps in unconventional oil and gas development in shale, including directional drilling and hydraulic fracturing.

*et al.*, 2009). Briefly, after site selection and preparation, directional drilling of the targeted shale proceeds by vertical drilling to a kick-off point, where a gradual turn to horizontal drilling through the shale begins. The average depth of shales targeted in the USA is 8300 ft (2530 m) but can vary from < 1000 ft to > 10,000 ft (305–3050 m) (Jackson *et al.*, 2015). Wells are cased with steel and concrete to return all fluids and gas to the surface and minimize leaks. The horizontal well casing is perforated at regular intervals with an explosive charge using a perforation gun. Hydraulic fracturing fluids, which are often unidentified, and proppant, usually silica or sand, are injected under high pressure into the well, and the resulting force through the perforations in the casing fractures the shale. The required pressure varies depending on the shale characteristics but can be as high as 15,000 psi. The injected sand 'props' the shale fractures open after the pressure has been removed. Gas and fluids then flow from the shale to the well to be captured, separated and transported for use or disposal. Major steps in UOG production are depicted in Fig. 23.1. Equipment used in gas production from a site in the Marcellus Shale is identified in Fig. 23.2. The hydraulic fracturing fluid contains water, sand and ~1% chemicals that are added for purposes such as the modification of viscosity, biocidal activity and corrosion inhibition (Stringfellow *et al.*, 2014). A recent review (Luek and Gonsior, 2017) highlighted that BTEX (benzene, toluene, ethylbenzene and xylene), acetate, acetone and halogenated organics are among the most frequently found organic compounds in hydraulic fracturing fluids. Fracfocus (website: https://fracfocus.org) is a good source of public information on chemicals used in hydraulic fracturing, but specific information on individual wells is often hard to obtain due to the proprietary nature of many of the mixtures used by individual operators.

Water use ranges from 13.7 to 23.8 million litres per well for UOG shale gas, and 1.3–15.1 million litres per well for UOG shale oil (Kondash and Vengosh, 2015). More than 21 billion barrels of UOG wastewater from all sources were generated in the USA in 2012 (Groundwater

**Fig. 23.2.** View of an unconventional oil and gas drill site in the Marcellus Shale during preparations for hydraulic fracturing. (Photo courtesy of Matthew Varonka, USGS.)

Protection Council & Veil Environmental LLC, 2015; Chittick and Srebotnjak, 2017) and average water use per well in the USA was 9.2 million litres (Jackson *et al.*, 2015). A range of 10–20% of the injected water is returned to the surface in the flowback period (discussed in the next section), which is typically the first several weeks after hydraulic fracturing. The fate of the remaining 80–90% of the injected water is unknown. Additional water is returned to the surface during gas production ('produced water'), much of which is formation water. Both flowback and produced water are increasingly reused for additional hydraulic fracturing (Lutz *et al.*, 2013). However, all water returned to the surface ultimately becomes wastewater for disposal, usually by underground injection (Gregory *et al.*, 2011). Underground injection of UOG wastewater has been shown to be associated with induced seismicity (Ellsworth, 2013; White *et al.*, 2014; Bommer *et al.*, 2015; Bao and Eaton, 2016; Lei *et al.*, 2017, Scanlon *et al.*, 2017; Kozłowska *et al.*, 2018; Schultz *et al.*, 2018). Further, municipal wastewater treatment plants lack the capacity to deal with the volumes and composition of the wastewater (Rozell and Reaven, 2012), which includes organic substances, metals, radioactive substances and, often, high salt concentrations (Gregory *et al.*, 2011). The sources of the chemicals are: (i) production chemicals; and (ii) shale and formation water.

Although many benefits may be derived from UOG production in shales, there are potential costs to human health and environmental integrity, as depicted in Fig. 23.3. Air quality issues, such as emissions of volatile organic substances or air particulates containing toxic metals or organics, may be of concern near production wells but this topic has not been extensively studied and the impacts are mostly undefined (McKenzie *et al.* 2012, 2017; Pacsi *et al.*, 2013; Field *et al.*, 2014; Ahmadi and John, 2015; McCawley, 2015). The biggest public health concern may be the contamination of surface and groundwater drinking-water supplies from

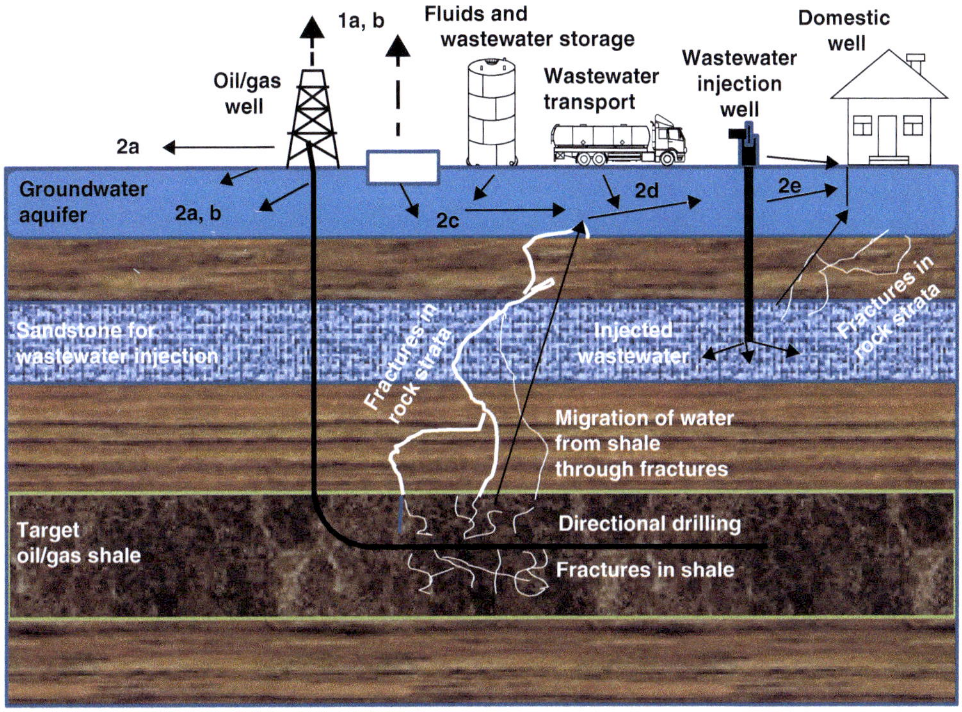


**Fig. 23.3.** Potential contamination routes and exposure pathways from unconventional oil and gas production and related activities, including disposal of wastewater. These include: (1a) releases to the air of gas and particulates from the well site; (1b) releases of gases from open storage ponds; (2a) liquid spills at the well site on the surface; (2b) releases of liquids to groundwater from infiltration of surface spills, or direct releases from cracked well casings; (2c) releases of liquids from storage pond/tank leaks; (2d) releases of liquids during transport; (2e) releases of liquids from injection well sites from surface spills or leaks/migration from the wells. Transport to household wells where exposure may occur may be through groundwater or surface water flow. Exposure from air emissions may also occur.

UOG wastewater releases (Gross *et al.*, 2013). Reuse of UOG wastewater containing naturally occurring radioactive materials (NORM) was demonstrated to enrich $^{226}Ra$ (Zhang *et al.*, 2015) in centralized storage impoundments, demonstrating the build-up of radioactive components over time. Radioactivity is not removed through conventional wastewater treatment. A risk assessment study determined that a significant contamination risk may result from disposal of hydraulic fracturing wastewater on site (Rozell and Reaven, 2012). In this analysis, the risk of migration of contaminated fluid to drinking-water supplies through fractures in rock layers overlying shale carried a high degree of uncertainty, mainly due to a dearth of information. The depth of most shale deposits being exploited (on average about 2500 m in the USA) may mitigate the risk of contaminant migration from the zone of hydraulic fracturing (Jackson *et al.*, 2015). The highest risk of wastewater release into the environment was considered to be greatest for spills during transport, and leaks from well casings or storage tanks (Rozell and Reaven, 2012).

The following sections review the chemical composition, toxicology and known impacts of wastewater from UOG production, beginning with some important working definitions.

## 23.3 Definitions

Definitions shown here are adapted from those given in Engle *et al.* (2014).

### 23.3.1 Hydraulic fracturing fluid

This fluid is the mixture of water and chemical additives that is added to the well with proppant (typically sand) for the purpose of fracturing the reservoir rock (shale). This produces fractures and increases rock porosity and permeability that allows gas to flow. The fluid is frequently referred to as a 'slickwater' type fluid and the base is often a mixture of freshwater and reused wastewater. The base water and additives (often proprietary) can vary significantly from well to well and between plays.

### 23.3.2 Flowback

During UOG development, hydraulic fracturing fluid and proppant (usually sand) are injected into the reservoir under high pressure. The early period of time immediately following hydraulic fracturing when water production volumes are highest is referred to as 'flowback'. Flowback water consists primarily of injected water and may contain both chemicals added by the producer and those extracted from the shale. During this period, water needs to be continuously trucked away for disposal. Flowback ends after flow rates have decreased enough so that the water can be piped through the separator into an on-site tank, generally in the first few days to weeks following hydraulic fracturing.

### 23.3.3 Produced water

Any water produced from a hydrocarbon well is referred to as produced water. It includes flowback water, formation water, injected fluids, water condensing from the gas phase, and mixtures thereof. It is sometimes referred to as co-produced water, as it is co-produced with the oil or gas. Some substances used in UOG production and recovered in produced water are listed in Table 23.1. The produced water is often stored on site in holding tanks. These are periodically emptied and removed to a disposal site (injection well) as wastewater, or reused for additional hydraulic fracturing. Figure 23.4 shows holding tanks and the physical appearance of produced water from a well.

**Table 23.1.** Common organic substances in wastewater from unconventional oil and gas production in shale and sources of these substances.

| Group | Sources |
|---|---|
| Hydrofracking fluids | Viscosity modifiers (e.g. guar gum, which allows sand to be injected as a proppant downhole)<br>Solvents for mixing base fluid (alcohols, diesel fuel) |
| Production chemicals | Friction reducers (e.g. ethylene glycol; other glycol ethers used as friction reducers and cross-linkers)<br>Corrosion inhibitors (e.g. alkyldimethylbenzylammonium chloride; 2,2-dibromo-3-nitriloproprionamide; prop-2-enal; 2-bromo-2-nitropropane-1,3-diol; f 3,5-dimethyl-1,3,5-thiadiazine-2-thione<br>Removal of gas hydrate blockages (e.g. methanol). Biocides (e.g. glutaraldehyde; didecyl dimethyl ammonium chloride; 2,2-dibromo-3-nitriloproprionamide; prop-2-enal; 2-bromo-2-nitropropane-1, 3-diol; f 3,5-dimethyl-1,3,5-thiadiazine-2-thione |
| Sourced from the formation | Petroleum hydrocarbons (aliphatic and aromatic)<br>Polycyclic aromatic hydrocarbons<br>Organic acids (microbial decomposition products)<br>Phenols and heterocyclic compounds |

### 23.3.4 Formation water/brine/fluid

This fluid is naturally occurring water that is found in oil and gas reservoirs in association with the source rock (shale) and prior to UOG production. Formation water in most oil and gas reservoirs is typically of brackish to brine salinity, but formation water in many coalbed methane reservoirs is relatively fresh. The formation brines from the Marcellus Shale in the Appalachian Basin are believed to be primarily highly evaporated palaeo-seawater (Haluszczak *et al.*, 2013) and are up to six times saltier than seawater (210,000 mg $l^{-1}$; Ohio Department of Natural Resources, 2017).

## 23.4 Wastewater Composition

Many different chemical substances are added to water used in UOG production from shale

**Fig. 23.4.** Picture of bottles containing produced water from an unconventional natural gas production well in the Marcellus Shale (eastern USA). The cloudy, orange colour is from oxidation of $Fe^{2+}$ to $Fe^{3+}$ in the samples that produces a precipitate of iron oxyhydroxides. Inset: tanks at a gas well site where wastewater is stored. Note containment wall around the storage tanks to contain any leaks or spills of liquids. (Photos courtesy of Matthew Varonka and Jeanne Jaeschke, USGS.).

(Table 23.1). These lubricate horizontal drilling operations, facilitate hydraulic fracturing and prevent corrosion and build-up of microbial populations. Hydraulic fracturing fluid is typically 98–99% water and sand by volume (Zoback, 2010). The remaining 1–2% is composed of chemical additives specifically designed for drilling at a particular site, thus the composition and amounts may vary. Examples of chemicals added to injection water include: glutaraldehyde (biocide), *N,N*-dimethylformamide (corrosion inhibitor), polyacrylamide and petroleum distillate (friction reducer), borate salts (cross-linking) and ethylene glycol (scale inhibitor) (Arthur *et al.*, 2009; Groundwater Protection Council and ALL Consulting, 2009; Colburn *et al.*, 2011; Luek and Gonsior, 2017). In addition, brines (hypersaline saltwater) and organic, inorganic and radioactive substances from the formation water and shale are mobilized and may be returned in the wastewater (Rich and Crosby, 2013).

Wastewater from many UOG wells is hypersaline, especially after the flowback period. As previously mentioned, these hypersaline brines can reach levels of up to six times the salinity of seawater in the Marcellus Shale, with total dissolved solids (TDS) as high as 210,000 mg $l^{-1}$ (TDS in the range of 100,000–150,000 mg $l^{-1}$ is more typical of the Marcellus Shale brines) (Kondash *et al.*, 2017). These brines are primarily sodium and calcium chlorides but with a host of other minor elements (Dresel and Rose, 2010). UOG production in the Bakken Shale (Williston Basin, North Dakota) and Wolfcamp Shale (Permian Basin, Texas) also results in wastewater with high salinities (TDS of about 105,000–123,000 in the Wolfcamp and 194,000–293,000 in the Bakken) (Engle *et al.*, 2016). The composition of the brines in the Bakken and Wolfcamp Shales is also primarily sodium and calcium chlorides. Sulfate levels are relatively low in wastewater from all such shales, indicating that sulfate reduction occurred in the formation. Iron, present mostly as $Fe^{2+}$ in the formation water, ranges from roughly 20 mg $l^{-1}$ to 200 mg $l^{-1}$ (Engle *et al.*, 2016). $Fe^{2+}$ rapidly oxidizes to $Fe^{3+}$ in the wastewater and forms a precipitate of iron oxide (Fig. 23.4). Increased bromide and iodide originating from the shale and formation water and high levels of ammonium (> 100 mg $l^{-1}$) from the denitrification of organic matter in the shale are also observed in the wastewater (Harkness *et al.*, 2015). Notably, characterization of Marcellus Shale UOG wastewater revealed that most samples exceeded relevant water quality standards for barium (Ba), strontium (Sr), benzene, copper (Cu), lead (Pb), chromium (Cr), mercury (Hg), zinc (Zn), cadmium (Cd) and arsenic (As) by as much as 1000-fold (Shih *et al.*, 2015).

Organic substances present in wastewater from UOG extraction include compounds used in hydraulic fracturing, production chemicals (e.g. biocides, corrosion inhibitors, friction reducers) and compounds from the shale, including petroleum hydrocarbons in some instances (EPA, 2004; Arthur *et al.*, 2009; Waxman *et al.*, 2011; Orem *et al.*, 2014). Chittick and Srebotnjak (2017), in their analysis of 630 hydraulically-stimulated California wells between 2014 and 2015, showed that the produced waters contained high levels of BTEX and PAH compounds. Kahrilas *et al.* (2015) reviewed and discussed the usage, mobility, degradation and toxicity of biocides used in hydraulic fracturing. These authors pointed out that total concentrations of biocides may reach 500 mg $l^{-1}$ in total fluid volumes over 10 million litres per well, resulting in over 1000 gallons (> 4500 litres) of biocide used per well. A significant body of literature now exists on organic substances present in wastewater and oil field brines from petroleum production (Willey *et al.*, 1975; Grahl-Nilson, 1987; Carey *et al.*, 1992; Higashi and Jones, 1997; Utvik, 1999) and from shale gas production (Veil *et al.*, 2004; Huang, 2008; Hayes, 2009; Dahm *et al.*, 2011, 2013; Orem *et al.*, 2014). As previously mentioned, the composition of organic compounds present in wastewater from UOG production may vary due to differences in production chemicals used by different producers, as well as differences in organics sourced from the formation. Some common organic chemicals encountered in wastewater are listed in Table 23.1.

Some volatile organic substances (e.g. $C_1$-$C_6$ hydrocarbons, BTEX) may be released into the air as gases or vapour from the well, wastewater degassing or production activities (McMullin *et al.*, 2018). Other organics in spilled wastewater may adsorb on to soil particles and subsequently become mobilized as air particulates. McCawley (2015) discussed air contaminants associated with the potential respiratory effects of UOG wells and concluded that diesel exhaust particulate and gases, volatile organic compounds and other hydrocarbons from diesels and the drilling process, as well as silica/sand and methane, are hazardous and the effects of these emissions may be likened to those of large-volume traffic highways, which produce known respiratory effects in humans. The same group (Haley *et al.*, 2016) examined state-mandated setbacks from wells in the Marcellus, Barnett and Niobrara Shale Plays for airborne hazards and found that they were not adequate for the protection of human health (discussed in the next section). As mentioned above, air quality issues are a relatively unstudied aspect of UOG production and air quality impacts on environmental health are largely undefined.

NORM and technologically advanced NORM (TENORM) are a concern during generation of wastewater from UOG production because formation water contains NORM and shale plays may contain TENORM as well. These radioactive materials may become increasingly concentrated due to the limited reuse of flowback water for hydraulic fracturing. The dominant NORM in flowback water is $^{226}Ra$, which is enriched in impoundment sludge by ageing from 10 pCi $g^{-1}$ to several hundred pCi $g^{-1}$, and is mainly carried by barite (Zhang *et al.*, 2015). Radon gas can potentially off-gas from impoundment ponds or storage tanks over time. Zhang *et al.* (2015) determined that workers at three well sites in the Marcellus Shale in Pennsylvania were not exposed to amounts of radon that exceed the NRC limit for the public. Garner *et al.* (2015) studied NORM in the UK's East Midlands at an onshore conventional oil production facility. In contrast to previous sampling on the UK's Dorset coast where $^{210}Pb$ was found, $^{226}Ra$ and $^{228}Ra$ were found in scale samples at levels above national exemption levels for landfill disposal. The samples were dated to give mean ages of between 2.2 and 3.7 years, indicating that levels of radiation of health concern can accumulate during a short period of time. The authors stated that use of hydraulic fracturing in these same shale plays would likely lead to a more acute NORM contamination problem from the scale and other solids than that seen in conventional oil production. Nelson *et al.* (2015) used established and new methods to quantitate NORM of concern (including uranium (U), thorium (Th), actinium (Ac), radium (Ra), lead (Pb), bismuth (Bi) and polonium (Po) isotopes) from hydraulic fracturing wastes. They observed that radium decay products were initially absent due to differences in solubility. However, in closed systems that could not release gaseous radon (Rn), their model predicted that decay products would begin to accumulate immediately and continue to contribute to an increase in total radioactivity for more than 100 years.

Ziemkiewicz and He (2015) performed a time series characterization of Marcellus Shale UOG wastewater (make-up water, hydraulic fracture fluids, flowback) from four well sites in West Virginia to follow the chemical evolution of the waste streams. They found increasing concentrations of organic, inorganic and radioactive constituents with additives present in small amounts. Flowback water, rather than injected hydraulic fracture fluids, contained the highest concentrations of measured constituents, increased with time and were thought to originate in the Marcellus formation. The authors concluded that 'TDS (total dissolved solids), NORM (naturally occurring radioactive material), Br, metals, and benzene in flowback consistently exceed MCL (maximum contaminant limits) up to thousands of times and they could have significant environmental implications if allowed to escape into surface or groundwater'. Further, they cautioned that the potential pathways for contamination of surface and groundwater could include treatment plant discharge, spills during transport by truck or rail, well casing failure, failure of impoundment liners, or intentional improper disposal. Similarly, Rosenblum *et al.* (2017) compared groundwater, fracturing fluid, flowback and produced water over time for up to 220 days. They examined water quality, NORM, major ions, trace metals and well flow data for water used and produced in a Denver-Julesburg Basin UOG well and found that metals and anions changed minimally over time, chemical oxygen demand (COD) declined significantly and TDS increased. Only about 30% of the initial injection volume returned to the surface during the period of study. NORM, trace metals and major ion levels were correlated with TDS but present at lower levels than in the Marcellus and Bakken formations.

Orem *et al.* (2014) found that the composition and amounts of organics in wastewater can vary temporally. Dissolved organic carbon (DOC) levels, for example, were found to vary from just a few to thousands of milligrams per litre. The very high levels of DOC seen generally occurred during the flowback period and represented chemicals such as solubilized degradation products of guar gum and solvents used in the hydraulic fracturing fluids, as well as other production chemicals. DOC levels were observed to decline dramatically in wastewater during production, often reaching a steady-state level characteristic of the DOC in the formation water (Orem *et al.*, 2014). Indeed, wastewater from gas production in the Marcellus shale has opposite DOC and conductivity curves, with DOC high and conductivity low during the flowback period due to the use of freshwater and organic substances in hydraulic fracturing. Over time, the DOC concentration gradually declines and conductivity rises as the wastewater comes to more closely resemble formation water (Fig. 23.5) (Orem *et al.*, 2014). However, a further injected concentration of organic chemicals for production purposes may be used at any time because of, for instance, excessive microbial growth requiring biocides, or a gas hydrate blockage resulting in injection of a de-icer such as methanol (Yousif *et al.*,1997).

## 23.5 Toxicological Studies of Wastewater

Evaluation of the toxicity of these UOG fluid mixtures in living organisms is in its earliest stages. Costa (2015) tested UOG flowback water samples in BEAS-2B continuously cultured human lung cells and found that, after 10 days, the cells treated with 4% flowback failed to survive. Costa (2015) tested flowback water with transformed BEAS-2B cells using a single scratch-wound assay and found that, after transformation, the cells showed an enhanced migration and wound closure ability. Transformed cells would be expected to show properties such as enhanced cellular migration, a property seen during cancer metastasis. Their studies also examined the tumour-forming propensity of flowback water-treated BEAS-2B cells in an *in vivo* nude mouse assay and showed that these soft-agar transformed cells could produce tumours subcutaneously in

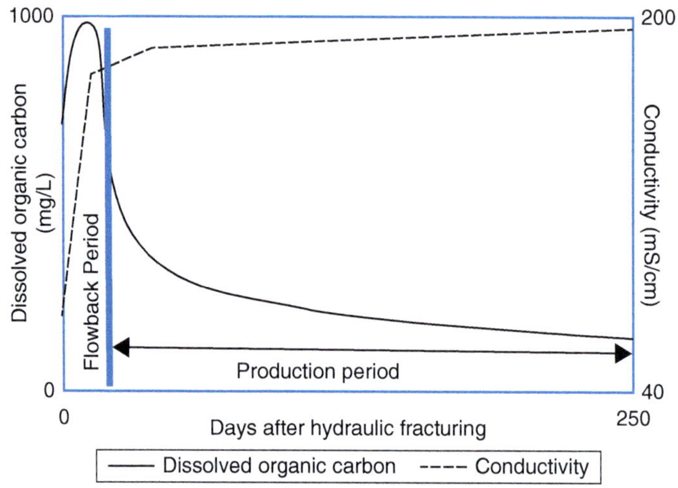


**Fig. 23.5.** Changes in dissolved organic carbon (DOC) and conductivity during time after hydraulic fracturing for a hypothetical well in the Marcellus Shale using freshwater for hydraulic fracturing. This theoretical plot is based on actual data from gas production wells in the Marcellus Shale.

five of six nude mice after 3 months. Crosby *et al.* (2018) studied the *in vitro* toxicity of a liquid mixture obtained from both unconventional and conventional oil and gas wells using human and rodent cell lines and found that there were no significant differences between the two well types for the endpoints studied (Crosby *et al.*, 2018). Both types of well samples caused cell death at low dilutions and severe cytotoxicity at dilutions up to 1000-fold. The results from Crosby *et al.* (2018) combined different kinds of toxicological assays (e.g. wound healing, gene expression and protein expression) with the chemical identification of water components to produce a complex picture of cytotoxicity. Effects of organic and inorganic components in the produced water included changes associated with gap junctions, cellular adhesion, cell–cell adhesion and communication, inhibition of cell cycle, proliferation and migration. In the wound-healing assays, healing cells did not form cell–cell contacts or move in a coordinated, sheet-like fashion as control cells did. A laboratory construct of potassium chloride (KCl) solution alone was designed to simulate the hypersaline conditions in the produced water and evaluate the effects of hypersalinity alone on the cells. Results showed a different and less severe picture of toxicity from hypersalinity alone compared with the produced water from the two wells sampled, indicating that brine alone was probably not the cause of toxicity. Key findings from Crosby *et al.* (2018) were that: (i) at high concentrations the produced water samples were extremely toxic (even corrosive); (ii) at low concentrations the samples were growth promoting; and (iii) genotoxic and tumour-promoting substances were observed in conventional and UOG samples tested. Taken together, results from Crosby *et al.* (2018) and Costa (2015) demonstrate that flowback water from hydraulic fracture wells is both highly lethal and capable of transforming some cell lines, or promoting growth, a hallmark of tumour promotion. In addition, these results demonstrated that both conventional and unconventional extraction technologies have a similar potential to impact human health.

If chronic oral reference values (RfC) for non-cancer effects and oral cancer slope factors (CSF) are not available for the known chemicals in hydraulic fracture fluids and UOG wastewaters, then decisions regarding the healthfulness of water that is available for human use cannot be made. A study by Yost *et al.* (2016), which sought to identify chronic oral toxicity values for a list of 1173 chemicals that EPA identified as being associated with hydraulic fracturing, found that only 8% of 1076 chemicals reported in hydraulic fracture fluids and 65% of 134 chemicals reported in flowback/produced water had an available chronic oral RfC or CSF. These results highlight the data gaps that exist for the evaluation of emerging human health hazards in high-density areas of UOG wells.

## 23.6 Toxicological Models and Risk Assessment

Yao *et al.* (2016) used QSAR (quantitative structure–activity relationship) models to generate estimates of the rat chronic oral lowest observed adverse effect level (LOAEL) and identified LOAELs for 44% of the 1173 previously identified chemicals and assigned confidence scores to the estimates. TOPKAT (Toxicity Prediction by c(K)omputer Assisted Technology, a commercial quantitative structure–activity relationship software package) LOAEL estimates and RfCs from the EPA's Integrated Risk Information System (IRIS) database were in good agreement for a small subset of chemicals. About 40% of the identified hydraulic fracturing-related chemicals that lacked RfCs and CSFs were assigned high-confidence LOAELs in this manner, including a set of frequently used chemicals. Elsner and Hoelzer (2016) also employed a QSAR modelling technique to attempt to bridge the gap between chemicals identified by CAS (Chemical Abstract Service identifier numbers) and name, and information regarding fate and toxicity. They used the model to identify probable toxic substances, including petroleum hydrocarbons, nonylphenols, propargyl alcohol, tetramethylammonium, biocides and strong oxidants. Some likely non-toxic substances were also identified. The authors concluded that, based on the oxidizing potential of additives and on the propensity of acids and complexing agents to react chemically on a delayed basis, chemical transformations could occur that would alter the chemical composition of the produced water. These authors further concluded that the information disclosed was not sufficient for hazard identification (i.e. not all chemicals were identified by CAS and name).

As a first step in risk assessment, Wattenberg *et al.* (2015) designed a risk identification tool for the assessment of the acute and chronic health hazards of hydraulic fracturing fluids and used it to analyse the hydraulic fracturing constituents reportedly used in 2850 oil-producing wells in North Dakota between 2009 and 2013. Approximately 60% could be identified by a CAS registration number and matching name. Using this tool, 13 health hazard end points for oral, inhalation, ocular and dermal exposure were searched in 11 public databases. Naphthalene and benzyl chloride had the highest health hazard count, were listed in the top 30 most frequently used constituents, and overlapped with the most used, in both Texas and Pennsylvania, where primarily natural gas is produced. Prpich *et al.* (2016) and Baranzelli *et al.* (2015) describe scientific evidence supporting environmental risk assessment for shale gas development in the UK and Poland.

## 23.7 Impacts on Environmental and Human Health

The potential risks and hazards of UOG development have been reviewed and analysed (Colburn *et al.*, 2011; Rozell and Reaven, 2012; Adgate *et al.*, 2014; Zhang *et al.*, 2015; Tang *et al.*, 2016). Haley *et al.* (2016) studied the adequacy of current state mandated setbacks for directional high-volume hydraulic fracturing in the Marcellus, Barnett and Niobrara Shale plays and concluded that setbacks are inadequate for the protection of public health. Hazards identified were explosions, radiant heat, toxic gas clouds and air pollution. This study did not examine groundwater contamination pathways, but examined the number of well blowouts and families displaced. McLaughlin *et al.* (2016) found that surface spills on UOG well sites or during transportation were commonly reported sources of contamination. During 2014 in Colorado alone, 838 spills occurred releasing 2.5 million litres of wastewater, 93 of which contaminated groundwater and eight of which contaminated surface water (McLaughlin *et al.*, 2016).

Papoulias and Velasco (2013) examined fish from a contamination event at Acorn Fork Creek, Kentucky, at specific locations along the creek and found signs of biological stress and gill lesions compared with unexposed fish. These effects were attributable to low pH and uptake of aluminium and iron at high concentrations. Mercury was found at elevated levels in 14 streams in remote, forested areas of north-western Pennsylvania located near UOG production, as compared with ten streams that had not experienced nearby natural gas development and hydraulic fracturing (Grant *et al.*, 2015). Higher dissolved total mercury, lower pH and higher dissolved organic matter were found near UOG production sites, and total accumulated mercury in crayfish, macroinvertebrates and predatory macroinvertebrates was significantly higher in those sites. These streams showed reduced biodiversity and enrichment of the bacterial family Deltaproteobacteria, which are known hydrocarbon degraders. Interviews with animal owners living near gas-drilling operations were conducted by Bamberger and Oswald (2012) documenting animal health problems potentially linked to exposures from gas-drilling operations. Negative effects in animals may lead to effects in humans through bioaccumulation.

Kassotis *et al.* (2014) examined chemicals used in UOG operations, and surface and groundwaters in a densely drilled region of Colorado, and found the majority contained oestrogenic, anti-oestrogenic, or anti-androgenic activity, using reporter gene assays in human cell lines. Since these releases frequently occur near agricultural lands, McLaughlin *et al.* (2016) investigated the soil breakdown products of UOG constituents and especially their chemical interaction with the soil matrix. They studied combinations including NaCl, glutaraldehyde, polyethylene glycol, and a polyacrylamide-containing friction reducer, and determined half-lives for some of the constituents in the soils, as well as chemical interactions and soil sorption. These interactions caused the authors to conclude that co-contaminant fate and toxicity need to be considered in determining environmental impacts.

Alawattegama *et al.* (2015) studied potential water contamination in south-west Pennsylvania drinking-water supplies that were located near UOG wells in 2009. Fifty-six of 143 respondents reported changes in well water colour, taste or odour from before drilling to after drilling, and 25 of the households had manganese levels above the secondary maximum contaminant level, while 14 of 18 houses tested had methane contamination. The Pennsylvania Department of Environmental Protection (DEP) reported violations that included an improperly plugged well and a failed well casing, and approximately 3.5 million gallons (15.9 million litres) of hydraulic fracture fluids and 3.2 million pounds (1.45 million kilograms) of proppant per well was used in each of 65 nearby wells; therefore there was a strong possibility that the measured contaminants originated from nearby UOG wells. Rabinowitz *et al.* (2015) conducted a health symptom survey of 492 persons in 180 randomly selected households with private well water in Washington County, Pennsylvania, where active natural gas wells were located. They found the number of reported health symptoms was significantly higher among residents who lived < 1 km from an active well when compared with those living > 2 km distant from active wells, after adjusting for age, smoking and other demographic factors, and for skin conditions and upper respiratory symptoms. The authors caution that the results should be considered hypothesis-generating, rather than positively affirming cause and effect. Samutin *et al.* (2013) examined morbidity in a highly developed oil and gas industrial region of Yugra, Russia, and found that rates for allergic, cardiovascular, pulmonary and cancer diseases among residents were above national averages and new diseases appeared. Jemielita *et al.* (2015) showed that the prevalence of inpatient or visit rates for hospitals in specific Pennsylvania zip codes associated with a high density of hydraulic fracture gas wells was increased for cardiology, dermatology, neurology, oncology and urology. This may indicate that spills or leaks of hydraulic fracture-related substances are influencing human health in Pennsylvania, but additional work is needed before any firm conclusions can be drawn, considering that other types of industrial activities have historically occurred in this area.

## 23.8 Conclusions

There are large economic benefits from UOG extraction, but there are risks to environmental and human health that cannot be ignored. Air quality issues near production wells and induced seismicity from drilling and underground wastewater disposal are known risks. The potential release of UOG wastewater containing an array of toxic chemical substances into the environment that are not detoxified by wastewater treatment plants has received much attention.

The chemical composition of the wastewater has been shown to contain numerous organic, inorganic and radioactive substances, some of which are toxic and/or persistent. For others, toxicity has not been evaluated. The oil and gas industry takes steps to protect workers and avoid releases, but spills may still occur from events such as well blowouts, leaking well casings, leaks from storage tanks or other containments at the well site. Transport (by tanker truck, train, or pipeline) and disposal of the wastewater at injection wells may also produce accidental wastewater releases. A recent wastewater release from a pipeline in the Bakken Shale in North Dakota (Cozzarelli *et al.*, 2017) and the release of wastewater from a small class 2 injection well into a nearby creek in West Virginia (Akob *et al.*, 2016) are examples of accidental wastewater spills. It appears that the sheer number of UOG wells and large volumes of wastewater for disposal make some accidental releases inevitable.

The extent of releases of wastewater from production, transport, or disposal activities into the environment is as yet poorly defined, and the fate of substances from UOG wastewater released into the environment is poorly understood. Chemical substances in UOG wastewater may biodegrade, or be sequestered and have little mobility in the environment, thereby posing a smaller threat (or no threat) to human and environmental health. However, exposure pathways of substances from UOG wastewater spills have not been well studied and the toxicity of these mixtures is poorly understood. Few *in vitro* or *in vivo* toxicology studies have been conducted. These mixtures are particularly challenging for toxicologists because of the myriad substances present (metals, organics, radioactive substances, hypersalinity) that may produce toxic effects. Lastly, the large volume of contaminated wastewater presents problems of treatment and disposal, but also represents hazards to wildlife and human populations. As clean water sources become increasingly depleted, communities may conclude that the use of millions of gallons of uncontaminated water per UOG well is not supportable. Quantitative information on the number, volumes and composition of spills, the movement of the chemical substances in the wastewater once released (exposure pathways) and additional toxicological studies of the impacts of these mixtures are needed to fully evaluate the risks posed by the UOG technology. Of particular relevance are studies that directly link pathways and exposures to actual effects (Kassotis and Stapleton, 2019).

## Acknowledgements

This work was supported by funding from the USGS Energy Resources Program and the USGS Toxic Substances Hydrology Program. Trade names used in this report are for the purposes of full disclosure of analytical methods; no endorsement of any commercial product by the US Geological Survey is implied.

## References

Adgate, J.L., Goldstein, B.D. and McKenzie, L.M. (2014) Potential public health hazards, exposures and health effects from uncontentional natural gas development. *Environmental Science & Technology* 48, 8307–8320. doi: 10.1021/es404621d.

Ahmadi, M. and John, K. (2015) Statistical evaluation of the impact of shale gas activities on ozone pollution in North Texas. *Science of the Total Environment* 536, 457–467

Akob, D.M., Mumford, A.C., Cozzarelli, I.M., Orem, W., Engle, M.A., Klinges, J.G. and Kent, D.B. (2016) Wastewater disposal from unconventional oil and gas development degrades stream quality at a West Virginia injection facility. *Environmental Science & Technology* 50 (11), 5517–5525.

Alawattegama, S.K., Kondratyuk, T., Krynock, R., Bricker, M., Rutter, J.K., Bain, D.J. and Stolz, J.F. (2015) Well water contamination in a rural community in southwestern Pennsylvania near unconventional shale gas extraction. *Journal of Environmental Science and Health, Part A. Toxic/hazardous Substances and Environmental Engineering* 50(5), 516–528. doi: 10.1080/10934529.2015.992684.

Arthur, J.D., Bohm, B. and Layne, M. (2009) *Hydraulic Fracturing Considerations for Natural Gas Wells of the Marcellus Shale*. Presented at The Ground Water Protection Council 2008 Annual Forum, Cincinnati, Ohio, 21–24 September 2008. ALL Consulting, Tulsa, Oklahoma.

Bamberger, M. and Oswald, R.E. (2012) Impacts of gas drilling on human and animal health. *New Solutions* 22, 51–77.

Bao, X, and Eaton, D.W. (2016) Fault activation by hydraulic fracturing in western Canada. *Science* 354(6318), 1406–1409.

Baranzelli, C., Vandecasteele, I., Barranco, R.R., Rivero, I.M., Pelletier, N., Batelaan, O. and Lavalle, C. (2015) Scenarios for shale gas development and their related land use impacts in the Baltic Basin, Northern Poland. *Energy Policy* 84, 80–95.

Bommer, J.J., Crowley, H. and Pinho, R. (2015) A risk-mitigation approach to the management of induced seismicity. *Journal of Seismology* 19(2), 623–646. doi: 10.1007/s10950-015-9478-z.

Carey, J., Zaidi, A. and Ribo, J. (1992) Specific toxic organics in produced waters from in-situ heavy oil recovery operations in western Canada. In: Ray, J.P. and Engelhardt, F.R. (eds) *Produced Water Technological/Environmental Issues and Solutions*, Plenum Press, New York,

Chittick, E.A. and Srebotnjak, T. (2017) An analysis of chemicals and other constituents found in produced water from hydraulically fractured wells in California and the challenges for wastewater management. *Journal of Environmental Management* 204(Pt 1), 502–509. doi: 10.1016/j.jenvman.2017.09.002.

Colburn, T.H., Kwiatkowski, C., Schultz, K. and Bachran, M. (2011) Natural gas operations from a public health perspective. *International Journal of Human and Ecological Risk Assessment* 17(5), *1039–1056*

Costa, M. (2015) Malignant human cell transformation of Marcellus Shale gas drilling flowback water. *Toxicology and Applied Pharmacology*. doi: 10.1016/j.taap.2015.07.011.

Cozzarelli, I.M., Skalak, K.J., Kent, D.B., Engle, M.A., Bentham, A. *et al.* (2017) Environmental signatures and effects of an oil and gas wastewater spill in the Williston Basin, North Dakota. *Science of the Total Environment* 579, 1781–1793

Crosby, L.M., Luellen, C., Zhang, Z., Tague, L.L., Sinclair, S.E. and Waters, C.M. (2011) Balance of life and death in alveolar epithelial type II cells: proliferation, apoptosis, and the effects of cyclic stretch on wound healing. *American Journal of Physiology – Lung Cellular and Molecular Physiology* 301, L536–546.

Crosby, L.M., Tatu, C.A., Varonka, M., Charles, K.M. and Orem, W.H. (2018) Toxicological and chemical studies of wastewater from hydraulic fracture and conventional shale gas wells. *Environmental Toxicology and Chemistry* 37, 2098–2111.

Dahm, K.G., Guerra, K.L., Xu, P. and Drewes, J.E. (2011) Composite geochemical database for coalbed methane produced water quality in the Rocky Mountain region. *Environmental Science & Technology* 45, 7655–7663. doi: 10.1021/es201021n.

Dahm, K.G., Van Straaten, C.M., Munakata-Marr, J. and Drewes, J.E. (2013) Identifying well contamination through the use of 3-D fluorescence spectroscopy to classify coalbed methane produced water. *Environmental Science & Technology* 47, 649–656. dx.doi.org/10.1021/es303866k

Deutch, J. (2011) The good news about gas – the natural gas revolution and its consequences. *Foreign Affairs* 90, 82–93.

Dresel, E.P. and Rose, A.W. (2010) *Chemistry and Origin of Oil and Gas Well Brines in Western Pennsylvania*. Pennsylvania Geological Survey, 4th Open-File Report OFOG 10–01.0, 48 p.

Ellsworth, W.L. (2013) Injection-induced earthquakes. *Science* 341(6142), 1225942. doi: 10.1126/science.1225942.

Elsner, M. and Hoelzer, K. (2016) Quantitative survey and structural classification of hydraulic fracturing chemicals reported in unconventional gas production. *Environmental Science & Technology* 50, 3290–314. doi: 10.1021/acs.est.5b02818.

Engle, M.A., Cozzarelli, I.M. and Smith, B.D. (2014) USGS Investigations of Water Produced During Hydrocarbon Reservoir Development. US Geological Survey Fact Sheet 2014–3104, 4 p. doi: 10.3133/fs20143104.

Engle, M.A., Reyes, F.R., Varonka, M.S., Orem, W.H., Ma, L. *et al.* (2016) Geochemistry of formation waters from the Wolfcamp and 'Cline' shales: insights into brine origin, reservoir connectivity, and fluid flow in the Permian Basin, USA. *Chemical Geology* 425, 76–92. doi: 10.1016/j.chemgeo.2016.01.025

EPA (2004) *Study of Potential Impacts of Hydraulic Fracturing of Coalbed Methane Wells on Underground Sources of Drinking Water*. Office of Ground Water and Drinking Water report, June 2004. US Environmental Protection Agency, Washington, DC.

Field, R.A., Soltis, J. and Murphy, S. (2014) Air quality concerns of unconventional oil and natural gas production. *Environmental Sciences: Processes & Impacts* 16, 954–969. doi: 10.1039/C4EM00081A

Gandossi, L. and Von Estorff, U. (2015) *An Overview of Hydraulic Fracturing and Other Formation Stimulation Technologies for Shale Gas Production – Update 2015*. Joint Research Centre, The European Commission, 59 p. EUR 26347. doi: 10.2790/379646.

Garner, J., Cairns, J. and Read, D. (2015) NORM in the East Midlands' oil and gas producing region of the UK. *Journal of Environmental Radioactivity* 150, 49–56.

Gaswirth, S.B. (2017) Assessment of continuous oil resources in the Wolfcamp shale of the Midland Basin, Permian Basin Province, Texas, 2016. US Geological Survey Open File-Report 2017–1013, 14 p. doi: 10.3133/ofr20171013.

Grahl-Nilsen, O. (1987) Hydrocarbons and phenols in discharge water from offshore operations. Fate of hydrocarbons in the recipient. *Sarsia* 72, 375–382.

Grant, C.J., Weimer, A.B., Marks, N.K., Perow, E.S., Oster, J.M. *et al.* (2015) Marcellus and mercury: assessing potential impacts of unconventional natural gas extraction on aquatic ecosystems in northwestern Pennsylvania. *Journal of Environmental Science and Health, Part A. Toxic/hazardous Substances and Environmental Engineering* 50(5), 482–500. doi: 10.1080/10934529.2015.992670.

Gregory, K.B., Vidic, R.D. and Dzombak, D.A. (2011) Water management challenges associated with the poduction of shale gas by hydraulic fracturing. *Elements* 7, 181–186.

Gross, S.A., Avens, H.J., Banducci, A.M., Sahmel, J., Panko, J.M.and Tvermoes, B.E.( 2013) Analysis of BTEX groundwater concentrations from surface spills associated with hydraulic fracturing operations. *Journal of the Air & Waste Management Association* 63, 424–432.

Groundwater Protection Council & ALL Consulting (2009) *Modern Shale Gas Development in the United States: a Primer*. Prepared for US Department of Energy, Office of Fossil Energy, and National Energy Technology Laboratory, by ALL Consulting, Tulsa, Oklahoma.

Groundwater Protection Council & Veil Environmental LLC (2015) *Produced Water Volumes and Management Practices in 2012*. Prepared for Groundwater Protection Council by John Veil, Veil Environmental LLC, April 2015, 119 pp. Available at: http://www.gwpc.org/sites/default/files/Produced%20Water%20Report%202014-GWPC_0.pdf

Haley, M., McCawley, M., Epstein, A.C., Arrington, B. and Bjerke, E.F. (2016) Adequacy of current state setbacks for directional high-volume hydraulic fracturing in the Marcellus, Barnett, and Niobrara Shale Plays. *Environmental Health Perspectives* 124(9), 1323–1333. doi: 10.1289/ehp.1510547.

Haluszczak, L.O., Rose, A.W. and Kump, L.R. (2013) Geochemical evaluation of flowback brine from Marcellus gas wells in Pennsylvania, USA. *Applied Geochemistry* 28, 55–61

Harkness, J.S., Dwyer, G.S., Warner, N.R., Parker, K.M., Mitch, W.A. and Vengosh, A. (2015) Iodide, bromide, and ammonium in hydraulic fracturing and oil and gas wastewaters: environmental implications. *Environmental Science & Technology* 49 (3), 1955–1963. doi: 10.1021/es504654n.

Hayes, T. (2009) *Sampling and Analysis of Water Streams Associated with the Development of Marcellus Shale Gas*. Final Report Prepared for Marcellus Shale Coalition (formerly the Marcellus Shale Committee), December 31, 2009, 249 pp. (including Appendices). Available at: https://www.water-research.net/naturalgasPA/pdfiles/MSCommission_Report.pdf

Higashi RM, and Jones AD (1997) Identification of Bioactive Compounds from Produced water Discharge/Characterization of Organic Constituent Patterns at a Produced Water Discharge Site. Prepared under MMS Cooperative Agreement No. 14-35-0001-30471 by Southern California Educational Initiative, OCS Study MMS 97-0023, US Department of the Interior, Minerals Management Service, Pacific OCS Region, 43 pp.

Huang, R (2008) Shale-Derived Dissolved Organic Matter as a Substrate for Subsurface Methanogenic Communities in the Antrim Shale, Michigan Basin, USA. MS Thesis, Department of Geosciences, University of Massachusetts Amherst , February 2008, 118 pp.

Jackson, R.B., Lowry, E.R., Pickle, A., Kang, M., DiGiulio, D. and Zhao, K. (2015) the depths of hydraulic fracturing and accompanying water use across the United States. *Environmental Science & Technology* 49, 8969–8976. doi: 10.1021/acs/est.5b01228.

Jacquet, J.B. (2014) Review of risks to communities from shale energy development. *Environmental Science & Technology* 48, 8321–8333. doi: 10.1021/es404647x.

Jemielita, T., Gerton, G.L., Neidell, M., Chillrud, S., Yan, B. *et al.* (2015) Unconventional gas and oil drilling is associated with increased hospital utilization rates. *PLoS ONE* 10(7), e0131093. doi:10.1371/journal.pone.0131093.

Kahrilas, G.A., Blotevogel, J., Stewart, P.S. and Borch, T. (2015a) Biocides in hydraulic fracturing fluids: a critical review of their usage, mobility, degradation, and toxicity. *Environmental Science & Technology* 49(1), 16–32. doi: 10.1021/es503724k.

Kassotis, C.D. and Stapleton, H.M. (2019) Endocrine-mediated mechanisms of metabolic disruption and new approaches to examine the public health threat. *Frontiers in Endocrinology* 10, 39. doi: 10.3389/fendo.2019.00039.

Kassotis, C.D., Tillitt, D.E., Davis, J.W., Hormann, A.M. and Nagel, S.C. (2014) Estrogen and androgen receptor activities of hydraulic fracturing chemicals and surface and ground water in a drilling-dense region. *Endocrinology* 155, 897–907.

Kahrilas, G.A., Blotevogel, J., Stewart, P.S. and Borch, T. (2015b) Biocides in hydraulic fracturing fluids: a critical review of their usage, mobility, degradation, and toxicity. *Environmental Science & Technology* 49(1), 16–32. doi: 10.1021/es503724k.

Kondash, A. and Vengosh, A. (2015) Water footprint of hydraulic Fracturing. *Environmental Science and Technology Letters* 2, 276–280. doi: 10.1021/acs.estlett.5b00211.

Kondash, A.J., Albright, E. and Vengosh, A. (2017) Quantity of flowback and produced waters from unconventional oil and gas exploration. *Science of the Total Environment* 574, 314–321.

Kozłowska, M., Brudzinski, M.R., Friberg, P., Skoumal, R.J., Baxter, N.D. and Currie, B.S. (2018) Maturity of nearby faults influences seismic hazard from hydraulic fracturing. *Proceedings of the National Academy of Sciences of the United States of America* 115(8):E1720–E1729. doi: 10.1073/pnas.1715284115.

Lee, J. (2015) The regional economic impact of oil and gas extraction in Texas. *Energy Policy* 87, 60–71.

Lei, X., Huang, D., Su, J., Jiang, G., Wang, X. *et al.* (2017) Fault reactivation and earthquakes with magnitudes of up to Mw4.7 induced by shale-gas hydraulic fracturing in Sichuan Basin, China. *Science Reports* 7(1), 7971. doi: 10.1038/s41598-017-08557-y.

Luek, J.L. and Gonsior, M. (2017) Organic compounds in hydraulic fracturing fluids and wastewaters: a review. *Water Research* 123, 536–548.

Lutz, B.D., Lewis, A.N. and Doyle, M.W. (2013) Generation, transport, and disposal of wastewater associated with Marcellus Shale gas development. *Water Resources Research* 49, 647–656. doi:10.1002/wrcr.20096.

McCawley, M. (2015) Air contaminants associated with potential respiratory effects from unconventional resource development activities. *Seminars in Respiratory and Critical Care Medicine*. 36(3), 379–387. doi: 10.1055/s-0035-1549453.

McKenzie, L.M., Witter, R.Z., Newman, L.S. and Adgate, J.L. (2012) Human health risk assessment of air emissions from development of unconventional natural gas resources. *Science of the Total Environment* 424, 79–87.

McKenzie, L.M., Allshouse, W.B., Byers, T.E., Bedrick, E.J., Serdar, B. and Adgate, J.L. (2017) Childhood hematologic cancer and residential proximity to oil and gas development. *PLoS ONE* 12(2), e0170423. doi: 10.1371/journal.pone.0170423.

McLaughlin, M.C., Borch, T. and Blotevogen, J. (2016) Spills of hydraulic fracturing chemicals on agricultural topsoil: biodegradation, sorption, and co-contaminant interactions. *Environmental Science & Technology* 50, 6071–6078. doi: 10.1021/acs.est.6B00240.

McMullin, T.S., Bamber, A.M., Bon, D., Vigil, D.J. and Van Dyke, M. (2018) Exposures and health risks from volatile organic compounds in communities located near oil and gas exploration and production activities in Colorado (USA). *International Journal of Environmental Research and Public Health* 15(7), 1500. doi: 10.3390/ijerph15071500.

Munasib A. and Rickman, D.S. (2015) Regional economic impacts of the shale gas and tight oil boom: a synthetic control analysis. *Regional Science and Urban Economics* 50, 1–17.

Nelson, A.W., Eitrheim, E.S., Knight, A.W., May, D., Mahrhoff, M.A. *et al.* (2015) Understanding the radioactive ingrowth and decay of naturally occurring radioactive materials in the environment: an analysis of produced fluids from the Marcellus Shale. *Environmental Health Perspectives* 123(7), 689-96. doi: 10.1289/ehp. 1408855.

Orem, W., Tatu, C., Varonka, M., Lerch, H., Bates, A. *et al.* (2014) Organic substances in produced and formation water from unconventional natural gas extraction in coal and shale. *International Journal of Coal Geology* 126, 20–31.

Pacsi, A.P., Alhajeri, N.S., Zavala-Araiza, D., Webster, M.D. and Allen, D.T. (2013) Regional air quality impacts of increased natural gas production and use in Texas. *Environmental Science & Technology* 47(7), 3521–3527.

Papoulias, D.M., and Velasco, A.L. (2013) Histopathological analysis of fish from Acorn Fork Creek, Kentucky exposed to hydraulic fracturing releases. *Southeastern Naturalist* 12.

Powers, M., Saberi, P., Pepino, R., Strupp, E., Bugos, E. and Cannuscio, C. (2015) Popular epidemiology and 'fracking': citizens' concerns regarding the economic, environmental, health and social impacts of unconventional natural gas drilling operations. *Journal of Community Health* 40, 534-541. doi: 10.1007/s10900-014-9968-x.

Prpich, G., Coulon, F. and Anthony, E.J. (2016) Review of the scientific evidence to support environmental risk assessment of shale gas development in the UK. *Science of the Total Environment* 563–564, 731–740.

Rabinowitz, P.M., Slizovskiy, I.B., Lamers, V., Trufan, S.J., Holford, T.R. *et al.* (2015) Proximity to natural gas wells and reported health status: results of a household survey in Washington County, Pennsylvania. *Environmental Health Perspectives* 123(1), 21–26. doi:10.1289/ehp.1307732.

Rich, A.L. and Crosby, E.C. (2013) Analysis of reserve pit sludge from unconventional natural gas hydraulic fracturing and drilling operations for the presence of technologically enhanced naturally occurring radioactive material (TENORM). *New Solutions* 23, 117–135.

Rosenblum, J., Nelson, A.W., Ruyle, B., Schultz, M.K., Ryan, J.N. and Linden, K.G. (2017) Temporal characterization of flowback and produced water quality from a hydraulically fractured oil and gas well. *Science of the Total Environment* 596–597, 369–377.

Rozell, D.J. and Reaven, S.J. (2012) Water pollution risk associated with natural gas extraction from the Marcellus shale. *Risk Analysis* 32, 1382–1393.

Samutin, N.M., Vorob'ev, V.O. and Butorin, N.N. (2013) [The influence of the oil and gas industry on environmental safety and population health in the Khanty-Mansiĭskiĭ region – iugra]. *Gigiena I Sanitariia*: 34–36.

Scanlon, B.R., Reedy, R.C., Male, F. and Walsh, M. (2017) Water issues related to transitioning from conventional to unconventional oil production in the Permian Basin. *Environmental Science & Technology* 51(18), 10903–10912. doi: 10.1021/acs.est.7b02185.

Schultz, R., Atkinson, G., Eaton, D.W., Gu, Y.J. and Kao, H. (2018) Hydraulic fracturing volume is associated with induced earthquake productivity in the Duvernay play. *Science* 359(6373), 304–308. doi: 10.1126/science.aao0159.

Shih, J., Saiers, J.E., Anisfeld, S.C., Chu, Z. and Muehlbachs, L.A. (2015) Characterization and analysis of liquid waste from Marcellus shale gas development. *Environmental Science & Technology* 49, 9557-7565.

Short, J. (1993) *Introduction to Directional and Horizontal Drilling*. PennWell Books, Tulsa, Oklahoma.

Stringfellow, W.T., Domena, J.K., Camarilloa, M.K., Sandelina, W.L. and Borglinb, S. (2014) Physical, chemical, and biological characteristics of compounds used in hydraulic fracturing. *Journal of Hazardous Materials* 275, 37–54.

Tang, C., Yi, Y., Yang, Z. and Sun, J. (2016) Risk analysis of emergent water pollution accidents based on a Bayesian Network. *Journal of Environmental Management* 165, 199–205.

USEIA (2017) *Marcellus Shale Play. Geology review*. US Energy Information Agency, January 2017. Available at: https://www.eia.gov/maps/pdf/MarcellusPlayUpdate_Jan2017.pdf

Utvik, T.I.R. (1999) Chemical characterization of produced water from four offshore oil production platforms in the North Sea. *Chemosphere* 39, 2593–2606.

Veil, J.A., Puder, M.G., Elcock, D. and Redweik, R.J. Jr (2004) *A White Paper describing produced water from production of crude oil, natural gas, and coal bed methane*. US Department of Energy, National Energy Technology Laboratory, Under Contract W-31-109-Eng-38, 87 pp.

Wattenberg, E., Bielicki, J.M., Suchomel, A.E., Sweet, J.T., Vold, E.M. and Ramachandran, G. (2016) Assessment of the acute and chronic health hazards of hydraulic fracturing fluids. *Journal of Occupational & Environmental Hygiene* 353(6306), 1416–1419.

Waxman, H.A., Markey, E.J. and DeGette, D. (2011) *Chemicals used in hydraulic fracturing*. United States House of Representatives, Committee on Energy and Commerce, Minority Staff, Washington, DC (April, 12 pp. plus appendix).

Weber, J.G. (2012) The effects of a natural gas boom on employment and income in Colorado, Texas, and Wyoming. *Energy Economics* 34, 1580–1588.

White, J.A., Chiaramonte, L., Ezzedine, S., Foxall, W., Hao ,Y., Ramirez, A. and McNab, W. (2014) Geomechanical behavior of the reservoir and caprock system at the In Salah $CO_2$ storage project. *Proceedings of the National Academy of Sciences of the United States of America* 111(24), 8747–8752. doi: 10.1073/pnas.1316465111.

Willey, L.M., Kharaka, Y.A., Presser, T.S., Rapp, J.B. and Barnes, I. (1975) Short chain aliphatic acid anions in oil field waters and their contribution to the measured alkalinity. *Geochimica et Cosmochimica Acta* 39, 1707–1711.

Yao, Y., Chen, T., Shen, S.S., Niu, Y., DesMarais, T.L. *et al.* (2016) Estimating the potential toxicity of chemicals associated with hydraulic fracturing operations using quantitative structure–activity relationship modeling. *Environmental Science & Technology* 50(14), 7732–7742.

Yost, E.E., Stanek, J., DeWoskin, R.S. and Burgoon, L.D. (2016) Overview of chronic oral toxicity values for chemicals present in hydraulic fracturing fluids, flowback, and produced waters. *Environmental Science & Technology* 50, 4788–4797. doi: 10.1021/acs.est.5b04645.

Yousif, M.H., Dunayevsky, V.A. and Hale, A.H. (1997) *Hydrate plug remediation: options and applications for deep water drilling operations*. Presented at the SPE/IADC Drilling Conference, Amsterdam, Netherlands, 4–6 March 1997. SPE-37624-MS. doi: org/10.2118/37624-MS.
Zhang, T., Bain, D., Hammack, R. and Vidic, R.D. (2015) Analysis of radium-226 in high salinity wastewater from unconventional gas extraction by inductively coupled plasma-mass spectrometry. *Environmental Science & Technology* 49(5), 2969–2976 doi: 10.1021/es504656q.
Ziemkiewicz, P.F. and He, T. (2015) Evolution of water chemistry during Marcellus Shale gas development: a case study in West Virginia. *Chemosphere* 134, 224–131.
Zoback, M. (2010) *Addressing the Environmental Risks from Shale Gas Development*. Briefing Paper 1, Natural Gas and Sustainable Energy Initiative. Worldwatch Institute, Washington, DC.

Part V

# Toxicology of Heavy Metals

# 24 Minamata Disease and Methylmercury Exposure

N. Hachiya*
*National Institute for Minamata Disease, Minamata-shi, Kumamoto-ken, Japan*

## 24.1 Abstract

Two outbreaks of Minamata disease were observed mainly during the 1950s and mid-1970s, which was an era of rapid economic growth in Japan. In both cases, a large amount of methylmercury was discarded into the sea and rivers after secondary generation in acetaldehyde-synthesizing factories. Serious health damage appeared among the inhabitants who consumed fish heavily contaminated with methylmercury. As productivity took the highest priority in those days, preventive measures were insufficient and the responsibility for the outbreak was not assigned. As a result, the damage expanded considerably. In this chapter, the extent of environmental pollution of methylmercury and its health influences among the inhabitants are reviewed based on historical documents and the results obtained from their reanalysis.

## 24.2 Introduction

Minamata disease is a serious methylmercury intoxication of the nervous system caused by daily consumption of large quantities of fish or shellfish that is heavily contaminated with methylmercury discharged from chemical factories. Two large-scale outbreaks have occurred in Japan; the first was in the coastal area of the Yatsushiro Sea, including Minamata, from the 1950s to the mid-1970s; and the second in the basin of the Agano River in Niigata in the mid-1960s. In both cases, methylmercury was generated from inorganic mercury compounds used as catalysts in the reaction chambers of chemical factories for acetaldehyde production, and then discharged into the water environment. In this chapter, the outbreaks of Minamata disease are described and discussed from historical and epidemiological viewpoints. Many Japanese investigation reports and documents had been published on Minamata disease. This chapter presents the data quoted, sometimes after re-analysis, from the historical materials that have been collected in the Minamata Disease Archives at the National Institute for Minamata Disease, Japan. On the other hand, it is now known that methylmercury exposure is associated with many kinds of health impairments even at rather low doses. However, these problems are not discussed in detail in this chapter, nor does the chapter refer to the health effects of the inorganic mercury, another important contaminant in the global environment.

## 24.3 Methylmercury: from the Environment to the Human Body

Mercury is a ubiquitous element in the natural environment. The major natural sources of

* E-mail address; hachiya@nimd.go.jp

mercury are the degassing of the earth's crust, emission from volcanoes and evaporation from natural bodies of water (IPCS, 1990). The United Nations Environmental Programme (UNEP, 2013) recently estimated that natural sources account for only about 10% of a total of the 5500–8900 t of atmospheric mercury emitted from all sources, and about 60% is ascribed to the re-emission of mercury previously deposited into soils, surface waters and vegetation. Emission from anthropogenic sources accounts for about 30% of the total amount of mercury entering the atmosphere each year. The predominant chemical forms of mercury circulating in the environment are elemental (metallic) or oxidized (inorganic) mercury (IPCS 1990; UNEP, 2013). In the aquatic environment, methylmercury is formed by microbiological or chemical processes. Methylmercury is accumulated by most aquatic biota and attains its highest concentration in the fish and aquatic mammals at the top of aquatic food webs.

The health effects of methylmercury have been reviewed (IPCS, 1990; NRC, 2000). The general population is primarily exposed to methylmercury mainly through a diet of fish and fishery products. Methylmercury is rapidly and highly effectively absorbed from the gastrointestinal tract into the bloodstream and distributed to all tissues. Methylmercury covalently binds to cysteine in the body to form cysteine-methylmercury, a structural analogue of an essential amino acid, methionine, which can effectively pass through the blood–brain barrier and placental barrier via an amino acid transporter (Kajiwara *et al.*, 1996). Average concentration of mercury in fetal blood is 1.2 times higher than in maternal blood (NRC, 2000); and ratios varying from 1.6 to 2.1 have been reported on methylmercury (Stern and Smith, 2003).

Methylmercury can be converted to inorganic mercury in the body of humans and experimental animals. Mercury is excreted in humans who consume fish at half-life of 39–70 days (mean, 50 days), indicating that, on average, 1.4% of the human body burden of methylmercury is excreted daily. In humans, the major routes of excretion are faeces via the bile. About 90% of a given dose of methylmercury is eventually excreted in the faeces as mercuric Hg, and approximately 10% via urine. Methylmercury can affect a variety of organs (Karagas *et al.*, 2012; Bjørklund *et al.*, 2017; Ha *et al.*, 2017). The neurotoxicity is the most sensitive end point of methylmercury toxicity. The most susceptible sub-group is pregnant women, and an extensive database has been obtained on the neurodevelopmental effects at very low exposures to methylmercury (Karagas *et al.*, 2012).

An outbreak of mass poisoning with methylmercury took place from 1971 to 1972 in Iraq (Bakir *et al.*, 1973). Seed grain treated with a methylmercury fungicide was used to prepare homemade bread in rural communities across the country. Total hospital admissions rose to just over 6000 and over 400 deaths were recorded. It was observed that the most sensitive complaint in adults was paraesthesia (sensory impairment), followed by signs of ataxia, constriction of visual field and hearing loss. The lowest observed adverse effect level (LOAEL) of methylmercury for paraesthesia was evaluated to be 50 μg $g^{-1}$ on the hair mercury concentration (IPCS, 1990).

In terms of the most sensitive effect on the developing fetus, the maternal hair mercury concentration that would have no appreciable adverse effects on offspring was estimated to be 12 μg $g^{-1}$ and 14 μg $g^{-1}$ by the US National Research Council (NRC), (NRC, 2000) and the Joint FAO/WHO Expert Committee on Food Additives (JECFA, 2004), respectively. A provisional tolerable weekly intake (PTWI) of 1.6 μg $kg^{-1}$ of body weight was then derived for methylmercury after the inclusion of uncertainty factor of 6.4 by JECFA (2004). Recently, the European Food Safety Authority (EFSA) established another tolerable weekly intake for methylmercury as 1.3 μg $kg^{-1}$ body weight, based on new developments in epidemiological studies. In addition, the NRC justified the reference dose (RfD), an estimate of exposure without appreciable risk of deleterious effects during a lifetime, of the US Environmental Protection Agency (EPA) of 0.1 μg $kg^{-1}$ of body weight per day, which was derived using an uncertainty factor of 10 (NRC, 2000).

## 24.4 Outbreak in Minamata

### 24.4.1 Official confirmation of Minamata disease

On 1 May 1956, the appearance of a disease of unknown cause was reported to the Minamata

Public Health Center by a hospital to which a girl of 5 years of age was brought with neurological symptoms. Her younger sister was then admitted to the hospital with similar clinical features.

The older sister experienced problems with using chopsticks and wearing shoes. She became unsteady on her feet, began to twist her tongue, choked when eating, and then showed wild excitement. After her admission to hospital, difficulties were found prominently on her movement of extremities. On the third day after admission, tendon and morbid reflex of the upper and lower extremities increased and tonic convulsions were occasionally observed in the whole body. One month later, she became blind and the frequency of systemic convulsions increased. She had no response to any stimuli and had flexion and remarkable deformity of the extremities. She died on 2 January 1959 (Harada, 2004).

A survey was conducted in the local community and revealed that 54 cases, including 17 cases of death, had appeared by the end of 1956. A more extensive study later found that the appearance of the disease could be traced back to 1941 (SSGM, 1973).

### 24.4.2 Preceding extraordinary signs in the environment

Minamata is located on the south-west side of Kyushu Island and faces the Yatsushiro Sea (Shiranui Sea), a calm inland sea with abundant fishery resources that borders Amakusa Islands to the west. A chemical company, Chisso, started the production of acetaldehyde by hydration of acetylene using inorganic mercury compounds as catalysts at a Minamata factory in 1932. Acetaldehyde had been used as an important raw material in plastic and textile industries and its production from carbide as a raw material for acetylene, as was the case in the Minamata factory, was the main enterprise of the organic chemical industry before the changeover to the petrochemical industry. After World War II, acetaldehyde production increased exponentially in the Minamata factory during the late 1950s, when Japan entered a period of high economic growth; it reached a peak in 1960 and continued up to 1968.

Conflicts had frequently happened between the factory and fishery industries about the damage that had been caused to fisheries by marine pollution from industrial effluent since the mid-1920s. Since 1932, the factory had discharged waste liquid containing methylmercury mainly into Minamata Bay. Mother liquid in the reaction chamber was sometimes drained after problems during the synthesizing process for acetaldehyde. In the early 1950s, serious pollution was recognized, with accumulated sludge, dead fish and a putrid odour, around the Hyakken Seaport where waste liquid was discharged.

In the early 1950s, extraordinary observations were frequently reported around Minamata Bay. It was found that cats ran around and died, which was called 'dancing cat disease', and crows and shorebirds suddenly fell down while flying. It was revealed afterwards that abnormal cat deaths had been observed in Minamata since 1925, with a peak from 1956 to 1960.

### 24.4.3 Discovery of methylmercury intoxication

In 1864, three laboratory staff developed serious methylmercury poisoning while they were conducting a methylmercury synthesis experiment at St Bartholomew's Hospital in London. They were the first described cases of methylmercury intoxication. They presented with neurological symptoms, including numbness in the limbs, impaired mobility or hearing, sight disturbance and indistinct speech. Two of them began to show systemic convulsion and trembling, with squealing and delirium, and they died 1–3 months after the onset (Edwards, 1866). The story of the accident caused by this extraordinarily poisonous chemical was distributed by the press in the UK, Germany and France (Ishihara, 2014).

In 1916, the Wacker Chemie Corporation began the first mass production of acetaldehyde in Burghausen, Germany. Zangger (1930) reported that organic mercury compound is generated from an inorganic mercury catalyst by a side reaction, and had caused the symptoms among workers who were directly exposed in the factory. The synthesizing procedure was similar to that which was adopted by Chisso in 1932. However, in Burghausen, the waste fluid was buried in the ground, kept away from groundwater. No outbreak of Minamata disease has been found in Europe among inhabitants who consumed fish contaminated with environmentally polluted methylmercury.

## 24.5 Investigation of Cause of Minamata Disease

### 24.5.1 Early studies

In an early investigation on the outbreak of Minamata disease, the research group of Kumamoto University School of Medicine noticed the disease as a heavy-metal poisoning caused by the consumption of fish and shellfish caught in Minamata Bay. The waste liquid from the Chisso Minamata factory was suspected as the source of the poisoning.

On the other hand, it was reported by Hunter *et al.* (1940) that four workers had developed poisoning from methylmercury used as a fungicide in a seed disinfection plant in England. The patients had symptoms including ataxia, dysarthria and constriction of the visual field. Tadao Takeuchi, a member of the Kumamoto University research group, noticed the pathological similarity between the brains of Minamata disease patients and that cited as Hunter–Russell symptoms by Pentschew (1958). In 1958, the research group started to analyse mercury in the fish and shellfish of Minamata Bay.

In February 1958, Douglas McAlpine, a British neurologist, visited Minamata to examine Minamata disease patients during his time at Kumamoto University. McAlpine was the first to report an outbreak of Minamata disease in an international scientific journal, and noted the similarity in the neurological features between the case reported by Hunter *et al.* and Minamata disease (McAlpine and Araki, 1958).

In July 1959, Kitamura of Kumamoto University reported that mercury was detected in the bottom sediment of Minamata Bay. The maximum concentration was 2010 $\mu g\ g^{-1}$ in the sediment of Hyakken drainage channel and decreased depending on the distance from the outlet, indicating that mercury had been released from the factory (Kitamura *et al.*, 1960b). The research group officially revealed their conclusions that the causative agent of Minamata disease was organic mercury compound, and that mercury had been discharged from Chisso factory. Leonard T. Kurland, Director of Epidemiology for the US National Institutes for Health (NIH), visited Minamata from 1958 and 1960 to investigate Minamata disease. He confirmed high mercury concentrations on the collected samples, including fish and mud, and strongly supported the organic mercury hypothesis of the Kumamoto University research group (Kurland *et al.*, 1960).

In September 1959, Chisso argued against the hypothesis of the research group. Chisso emphasized that inorganic mercury had been used in the Minamata factory since 1932 without the appearance of Minamata disease until 1956, and no experimental evidence demonstrated the conversion from inorganic to organic mercury. Furthermore, Chisso wrongly insisted that, although similar synthetic processes had been operated using mercury elsewhere in the world, no patients had appeared.

On 12 November 1959, the Food and Sanitation Committee of the Ministry of Health and Welfare (MHW) submitted an official report on Minamata disease based on the results of the Kumamoto University research group. The report concluded that Minamata disease was induced by the consumption of fish and shellfish in Minamata Bay and its surrounding areas, and that the cause of the disease was organic mercury compound. However, the source of the causative chemical was not specified in the report and the committee was disorganized. The chairman of the committee stated: 'The factory waste is highly suspected as the source, and unsolved problem should be entrusted to ministries and agencies concerned.' The conclusion of the report was presented by the Minister at a Cabinet meeting the next day; however, Hayato Ikeda, the Minister of International Trade and Industry (MITI), offered a counterargument, saying: 'It would be premature to conclude that organic mercury was released from the Chisso Minamata factory.' Subsequently, Hayato Ikeda led Japan to achieve considerable economic growth as the Prime Minister from 1960 to 1964.

Another official council was launched, but was soon dissolved without reaching any conclusions. Additionally, in August 1960, the Japan Chemical Industry Association established the so-called 'Tamiya Committee', whose name was taken from the chairman Takeo Tamiya, the President of the Japanese Association of Medical Sciences. They presented inadmissible hypotheses, the noxious amine hypothesis for example, as alternative candidates for the cause. Their opinions were efficiently distributed by mass media, and contributed to delaying the elucidation of the cause (NIMD, 2001).

### 24.5.2 Incontrovertible findings

Chisso increased acetaldehyde production in the late 1950s. In September 1958, the company changed the drainage of acetaldehyde waste liquid to the mouth of the Minamata River from the Hyakken drainage channel with the intention of diluting contaminating materials with an excessive amount of seawater and alleviating the serious pollution around the seaport. However, this did not work as the company expected. The following March, several patients were identified, first in the areas around and to the north of the mouth of the river (Kitamura *et al.*, 1960a). Abnormal cat deaths were observed in wider areas, including the Amakusa Islands on the opposite shore of the Yatsushiro Sea. In October 1959, MITI instructed Chisso to restore the Hyakken drainage channel in Minamata Bay and hastened construction of a waste liquid processing facility.

In the Chisso Hospital laboratory, animal experiments were conducted and it was revealed in October 1959 that a cat had developed characteristic poisoning after administration of the waste liquid from the acetaldehyde process. Several other cats were subsequently affected by early 1961. Furthermore, in July 1961, a member of Chisso's technical staff successfully identified methylmercury in the waste liquid from the process. These findings were not made public. In September 1961, methylmercury sulfide was reported to be extracted from a small clam, hibarigaimodoki (*Hormomya mutabilis*), in Minamata Bay (Uchida *et al.*, 1961) and methylmercury chloride from the sludge of the acetaldehyde plant of the factory (Irukayama *et al.*, 1962). Although the accumulating evidence indicated a direct association between the factory waste and Minamata disease, the government did not take any effective countermeasures to prevent damage to inhabitants (NIMD, 2001).

### 24.5.3 Insufficient measures

After the outbreak of Minamata disease, fishing was not forcibly prohibited in the polluted sea areas. The MHW announced self-regulation on the capture and consumption of fish in Minamata Bay; however, no legal ban was applied to the contaminated sea area until 1975. Fishing was under the self-imposed restraint of the fishery association in the bay and neighbouring area from 1957 to 1964 and from 1973 to 1975.

Since the affected area was recognized to expand in 1959, an appeal for cessation of waste liquid was accelerated among fishermen. On 2 November 1959, about 2000 members of the Prefectural Alliance of Fishing Co-operatives gathered in Minamata to negotiate with the company and to appeal to the Diet Investigation Team visiting Minamata. Because their demands were refused by the company, 1000 fishermen invaded the factory and more than 100 people were injured. The Minamata City Assembly soon petitioned for an accelerated investigation of the cause of Minamata disease, to help patients and fishermen, and early completion of waste-liquid processing facilities by Chisso. The assembly also declared that the factory should not be mothballed.

Chisso completed the waste-liquid processing facility in December 1959. The company claimed that safety of the wastewater was guaranteed by the facility; however, the facility was not intrinsically effective for the removal of mercury from the waste. The annual production of acetaldehyde reached a peak in 1960. The release of methylmercury from the factory was estimated to reach a peak in 1959. It then decreased after the adoption of a closed circulation system in August 1960 and continued until 1968, when acetaldehyde production was terminated in the factory. The amount of the chemicals discharged into the water environment has been estimated to be 0.6–6 t of methylmercury and 70–150 t or more of total mercury from 1932 to 1968 (Hachiya, 2012).

In December 1959, Chisso signed contracts for 'consolation money', including annual pensions, with the patient liaison group. The Minamata disease certification system was organized to examine the patients. Thereafter, a diagnosis of Minamata disease made patients eligible to receive consolation money and, since 1973, compensation payment. Eighty-seven cases were certified as Minamata disease by 1960, with typical and relatively severe manifestations. Although the consolation money did not presume any responsibility of the company for the disease, the contract prohibited the patient from demanding further compensation even if the responsibility of the company was established. A court decision, later in 1973, on the first Minamata disease lawsuit ruled full responsibility of the company and voided the contract on consolation money because of a violation of public order and morals.

## 24.6 Prenatal Effects and Congenital Minamata Disease

The incidence of abnormal pregnancy, including stillbirth and spontaneous abortion, was significantly high between 1956 and 1968 in Minamata and its neighbouring town (Itai *et al.*, 2004). In Minamata, the crude fertility rates declined from 1955 to 1965, and the male proportion increased among stillbirths; as a consequence, the male-to-female birth ratio reduced in the late 1950s (Yorifuji *et al.*, 2017). Furthermore, the incidence of cerebral palsy was markedly higher from 1955 to 1958 in fishing villages around Minamata Bay, ranging from 1.0% to 12.0% compared with 0.2%, which was a mean in the general population of Japan (Moriyama *et al.*, 1994). Neuropathological examination on two girls who died at the ages of 1.5 and 6 years in 1960 and 1962, respectively, indicated that their cerebral palsy was caused by intrauterine exposure to methylmercury, and they were certified as having Minamata disease in 1961 and 1962 (Matsumoto *et al.*, 1965; Harada, 1978). Additionally 16 infants with the same impairment, who were born from January 1955 to November 1958, were subsequently certified as having Minamata disease in November 1962.

It is possible to evaluate retrospectively the exposure to methylmercury *in utero* using preserved umbilical cords, which are kept after delivery as a traditional custom in Japan. Several papers reported mercury concentrations on dry cords collected from inhabitants born in Minamata and its surrounding districts between 1925 and 1980 (Akagi *et al.*, 1998; Harada *et al.*, 1999; Yorifuji *et al.*, 2009). Extraordinarily high concentrations were already found in the 1930s, and the concentration increased rapidly after World War II, reaching a peak in the late 1950s or 1960. The peak appeared first in Minamata, followed by the neighbouring sites: Ashikita and Goshonoura in the north and Izumi in the south.

By the beginning of 1960s, however, it was socially assumed that the Minamata disease problem had dispersed without clarification of the cause or responsibility, despite the persistence of serious pollution in the environment. Under such social circumstances, new patients found it difficult to complain of their health damage and tended to be hidden as socially latent cases in the local community. In fact, 24 patients were officially certified as having Minamata disease from 1961 to 1964; however, 22 of them were congenital cases who had been born from 1955 to 1959, and the other two cases were patients who had developed the disease before 1959. Between 1965 and 1968, no application was made for certification of the disease and the Screening Council was not convened. The expansion of health damage to a wider area around the coastal site was going to be recognized after 1968.

## 24.7 Exposure Among Inhabitants Around the Yatsushiro Sea

The serious mercury contamination continued during the 1960s in the Yatsushiro Sea. Mercury was detected at high concentrations in fish and shellfish caught in 1959 and 1960 (Kitamura *et al.*, 1960b). Typical data are shown in Table 24.1 and mercury was found in many fish and shellfish at extremely high concentrations not only in Minamata Bay but also in wider areas of the Yatsushiro Sea, for example, Hakariishi of Ashikita, which is 14 km north, and Hinoshima Island, 20 km north of the bay. Kitamura also collected hair samples of inhabitants in Minamata to measure the mercury concentration from December 1959 to January 1960 (Kitamura *et al.*, 1960b). The participants included Minamata disease patients who had developed the disease between 1954 and 1959. As shown in Table 24.2, hair mercury concentrations of the patients were markedly high, reaching a maximum at 705 $\mu g\ g^{-1}$. It was also shown that the mean and the distribution of concentration among inhabitants of Minamata without symptoms at the time of sampling were lower than

**Table 24.1.** Total mercury concentration of aquatic organisms in Minamata Bay and Yatsushiro Sea around 1960 (adapted from Kitamura *et al.*, 1960b).

| Organism (Japanese name, species name) | Mercury (µg $g^{-1}$) | Sampling site |
|---|---|---|
| Threadfin shad (konoshiro, *Konosirus punctatus*) | 1.62 | Minamata Bay |
| Half-mouth sardine (katakuchi iwashi, *Engraulis japonica*) | 0.27 | Minamata Bay |
| Small crab (kogani, n.s.[a]), shell | 35.7 | Minamata Bay |
| Small crab (kogani, n.s.[a]), viscera | 23.9 | Minamata Bay |
| Oyster (kaki, *Crassostrea gigas*) | 5.61 | Minamata Bay |
| Blue drum (ishimochi, *Nibea mitsukrii*)[b] | 14.9 | Minamata Bay |
| Flathead grey mullet (bora, *Mugil cephalus*)[b] | 10.6 | Mouth of Minamata River |
| Blackhead seabream (chinu, *Acanthopagrus schlegelii*)[b] | 24.1 | Mouth of Minamata River |
| Japanese seabass (suzuki, *Lateolabrax japonicus*)[b] | 16.6 | Mouth of Minamata River |
| Japanese short-neck clam (asari,*Ruditapes philippinarum*) | 20.0 | Mouth of Minamata River |
| Cutlassfish (tachiuo, *Trichiurus japonicus*)[b] | 7.50 | Tsunagi (Yatsushiro Sea) |
| Flathead grey mullet (bora, *Mugil cephalus*) | 3.36 | Tanoura (Yatsushiro Sea) |
| Japanese seabass (suzuki, *Lateolabrax japonicus*)[b], liver | 52.3 | Hakariishi (Yatsushiro Sea) |
| Japanese seabass (suzuki, *Lateolabrax japonicus*)[b], muscle | 13.5 | Hakariishi (Yatsushiro Sea) |
| Cutlassfish (tachiuo, *Trichiurus japonicus*)[b] | 4.82 | Hinoshima (Yatsushiro Sea) |
| Blue drum (ishimochi, *Nibea mitsukrii*)[b] | 3.64 | Hinoshima (Yatsushiro Sea) |
| Threadfin shad (konoshiro, *Konosirus punctatus*) | 0.33 | Ariake Sea (as a reference) |
| Japanese short-neck clam (asari, *Ruditapes philippinarum*) | 0.1 | Ariake Sea (as a reference) |

[a]n.s., species not specified; [b]enfeebled fish floating on surface of seawater.

those of the patients but significantly higher than healthy participants outside Minamata. These findings suggested substantial consumptions of contaminated fish in Minamata and also may reflect the delayed onset of the methylmercury poisoning, as observed later in Niigata (Tsubaki *et al.*, 1977).

Matsushima of the Kumamoto Prefecture Institute of Health Research conducted a hair mercury survey in the coastal sites of Yatsushiro Sea from 1960 to 1962. Summarized results were reported on the mercury concentration of 3373 participants, including Kumamoto City as a reference site (Tokuomi *et al.*, 1962; Matsushima, 1970). In 1960, 955 fishermen and their families participated in the survey from six districts of Kumamoto prefecture: Minamata, Tsunagi, Ashikita (including Yunoura), Tanoura, Goshonoura and Ryugatake. Additionally, the Kagoshima Prefectural Institute of Public Health analysed mercury on 1179 hair samples in the southern coastal areas of the sea in Kagoshima prefecture from 1960 to 1963. In 1960 and 1961, hair samples were collected from 574 fishermen and their family of Izumi, Azuma, Takaono and Akune (Sakata *et al.*, 1962). Among a total of 1529 samples obtained in Kumamoto and Kagoshima in 1960 and 1961, 505 (26.2%) exceeded 50 µg $g^{-1}$: a LOAEL of methylmercury for paraesthesia (IPCS, 1990), equivalent to a weekly intake of c.a. 35 µg $kg^{-1}$ of body weight. The high exposure levels can be ascribed also to high fish consumption of fishermen. For example, an average daily intake of fish was 369 g for male fishermen of Minamata in 1972 (SSGM, 1973; Futatsuka *et al.*, 1977).

Figure 24.1 shows distributions of hair mercury concentrations, as classified into five categories, of fishermen from ten districts in 1960/1961, two years after the drainage of the waste liquid had been changed and the polluted water had been dispersed. The frequency of hair mercury concentrations exceeding 50 µg $g^{-1}$ was 31.3%, 29.4 %, 46.9% and 31.4% in Minamata, Tsunagi, Ashikita and Izumi, respectively. In contrast, the frequency was 1.2% in Kumamoto City and no participant exceeded the level in

**Table 24.2.** Distribution of hair mercury concentrations of inhabitants in and outside of Minamata in 1959–1960 (adapted from Kitamura *et al.*, 1960b).

| | Number of participants classified by mercury concentration (μg $g^{-1}$) | | | | | | | | | | Geometric | |
|---|---|---|---|---|---|---|---|---|---|---|---|---|
| Sub-population | < 10 | 10–50 | 50–100 | 100–200 | 200–300 | 300–400 | 400–500 | 500–600 | 700–800 | Total | mean (μg $g^{-1}$) | Min/max[a] (μg $g^{-1}$) |
| Minamata disease patient in Minamata | 4<br>16% | 10<br>40% | 1<br>4% | 4<br>16% | 1<br>4% | 1<br>4% | 3<br>12% | 0<br>0% | 1<br>4% | 25<br>100% | 46.4 | 2.46/705.0 |
| Healthy inhabitant in Minamata | 9<br>50.0% | 4<br>22.2% | 3<br>16.7% | 2<br>11.1% | 0<br>0.0% | 0<br>0.0% | 0<br>0.0% | 0<br>0.0% | 0<br>0.0% | 18<br>100% | 21.4 | 1.82/191.0 |
| Healthy inhabitant outside of Minamata | 16<br>100% | 0<br>0.0% | 0<br>0.0% | 0<br>0.0% | 0<br>0.0% | 0<br>0.0% | 0<br>0.0% | 0<br>0.0% | 0<br>0.0% | 16<br>100% | 1.71 | 0.14/4.79 |

[a]Min, minimum; max, maximum

Kagoshima City. In Kumamoto prefecture, regional arithmetic means ranged from 22.5 μg $g^{-1}$ to 48.7 μg $g^{-1}$, as shown in Fig. 24.1, and were markedly higher than the 9.4 μg $g^{-1}$ of Kumamoto City (Matsushima, 1970). Unfortunately, mean concentration of each district was not specified in Sakata *et al.* (1962). Although small differences were apparent on the distribution of hair mercury concentration between Minamata and its neighbours on the surveys, it might be plausible that the exposure level in Minamata was considerably higher in the 1950s in which many serious cases appeared. Some participants presented with extremely high exposure levels. The maximum concentration was 920 μg $g^{-1}$ observed in a woman of 62 years of age in Goshonoura. A segmental analysis of the hair strings indicated that mercury was detected at 1855 μg $g^{-1}$ in the hair tip (Matsushima and Noguchi, 1966). She complained of stiff hands, fumbling with buttons, and shoes easily falling off (Tukuomi *et al.*, 1963). These findings indicate that fishermen in the wider coastal area of the Yatsushiro Sea could have been exposed to methylmercury at extremely high levels in the early 1960s.

The cumulative number of officially certified patients is also indicated in Fig. 24.1, at the end of 1960 and of 2017 in each district. The total number of certified patients up to 1960 can be calculated to be 87. However, most of the officially certified patients were confirmed after 1968, as described later, and the number as of 1960 is found to represent only 3.8% of the 2287 patients who had been certified by 2017. A geographical difference was apparent on the tentative rate in 1960; for example, 7.6% had been certified in Minamata, but 0.8–1.1% in neighbouring sites: Tsunagi, Ashikita and Izumi. No patient had yet been identified in Goshonoura and Azuma at the survey.

The hair mercury concentration gradually decreased during the 1960s. Geometric mean was calculated using the raw data of Matsushima. In Minamata, 12.8 μg $g^{-1}$ in 1962 was significantly lower than the 27.3 μg $g^{-1}$ in 1960/1961 ($p < 0.01$). A similar decline was observed in Kagoshima prefecture from 1960 to 1963 (Sakata *et al.*, 1963). For longer-term changes, hair samples were collected in Minamata from March to May 1968, the year of the termination of acetaldehyde production in Chisso factory (Irukayama *et al.*, 1969). The re-calculated geometric mean, arithmetic mean and maximum concentration among 152 participants were 7.30, 9.22 and 73.79 μg $g^{-1}$, respectively. In this survey, two (1.0%) and nine (4.5%) participants exceeded 50 μg $g^{-1}$ and 20 μg $g^{-1}$, respectively, indicating that some residents continued to be exposed to methylmercury at substantially hazardous levels. In September 1969, the concentration decreased to an arithmetic mean of 6.0 μg $g^{-1}$ and a maximum of 18.3 μg $g^{-1}$ among 38 fishermen in Minamata (Irukayama *et al.*, 1972). A recent investigation of hair mercury in 2000 showed geometric and arithmetic mean concentrations at 1.84 and 2.29 μg $g^{-1}$ in Minamata ($n$ = 597), significantly lower than 2.12 and 2.68 μg $g^{-1}$ evaluated across 14 sites in Japan ($n$ = 9093), respectively (Yasutake *et al.*, 2004).

## 24.8 Second Outbreak in Niigata

With no effective countermeasure being taken to relieve the environmental pollution in the Yatsushiro Sea, and the environmental regulations on industrial discharge of mercury in Japan being insufficient, the second outbreak of Minamata disease appeared in the lower basin of Agano River in Niigata in January 1965. The prefecture government prohibited fishing in the river in June 1965. The fishery constraint continued until April 1978.

In June 1965, the MHW's Investigation Team conducted a health survey on the watershed inhabitants. They collected water and soil samples to analyse mercury in the vicinity of the three plants that used mercury and discharged waste liquid into the river. The team reported that mercury was detected at high concentrations, e.g. 151 μg $g^{-1}$ in the river sediment sampled at the drain outlet of the Kanose Factory of the Showa Denko, located some 60 km upstream from the area in which the first patients appeared. It was also found that the slagheap mud in the factory contained mercury at extremely high concentrations, up to 11,800 μg $g^{-1}$ (MHW, 1967). The Kanose factory had produced acetaldehyde between 1951 and January 1965, and its production reached a maximum in 1964. It seemed that methylmercury had been discharged from the factory, because the highest concentration of mercury in moss, 461 μg $g^{-1}$,

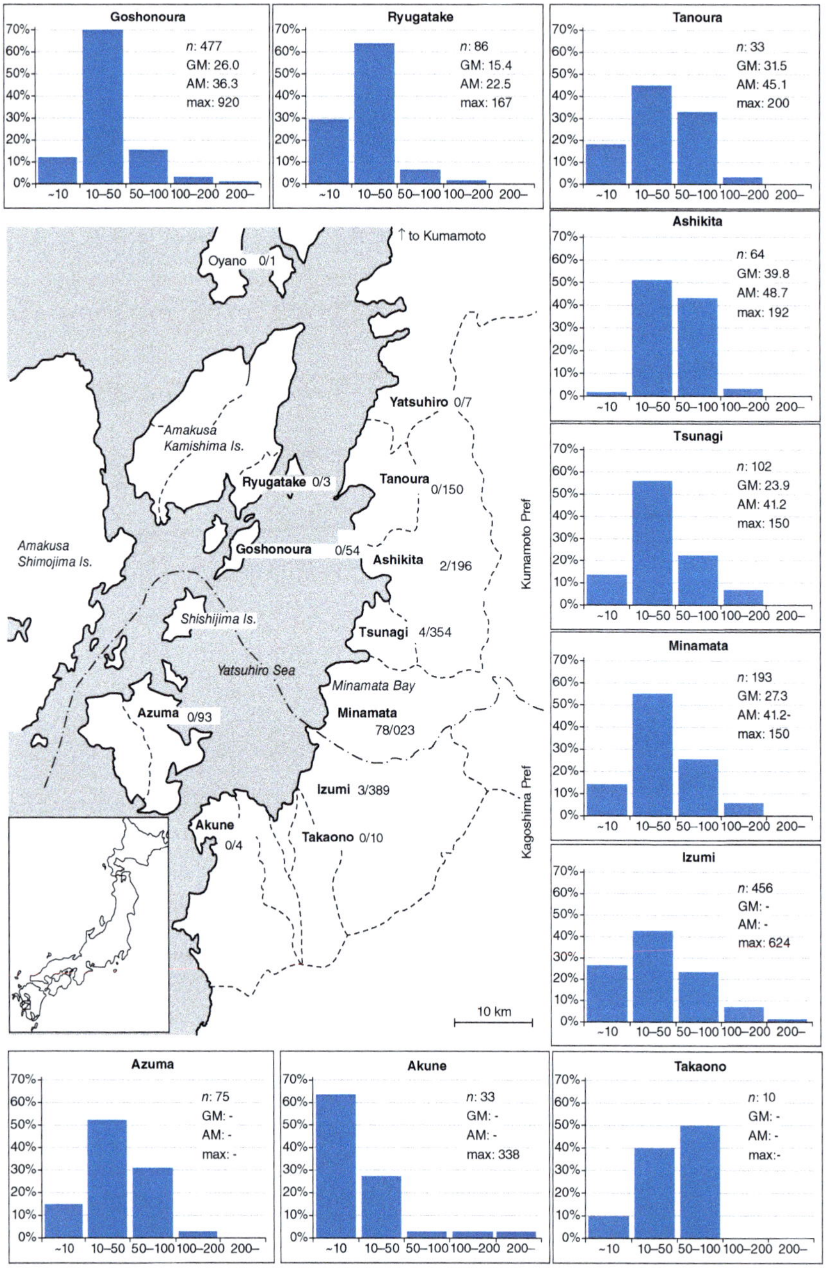


**Fig. 24.1.** Hair mercury concentrations among fishermen in ten districts of the coastal areas of the Yatsushiro Sea in the early 1960s and cumulative number of Minamata disease patients. The cumulative number of officially certified Minamata disease patients is indicated as 'certification as of December 1960' / 'certification as of December 2017' beside the district name. The distributions of hair mercury concentrations are adapted from Matstushima (1970) and Sakata *et al.* (1963) (see text for details). Vertical and horizontal axes of the graph indicate frequency and total mercury concentration (μg $g^{-1}$), respectively. Geometric mean was calculated from the raw data of Matsushima (1970). Abbreviations: *n*, number of hair samples; GM, geometric mean concentration; AM, arithmetic mean concentration; max, maximum concentration; Pref, prefecture.

was found in the sample collected at the factory drainage outlet, and mercury including methylmercury was detected at high concentrations in samples collected only downstream from the factory (Takizawa *et al.*, 1972). Table 24.3 details mercury concentrations in river fish and diatoms caught in the Agano River in 1965 (MHW, 1967). Mercury was found at markedly high concentrations in a variety of fish, particularly lamprey and skin carp (*Hemibarbus barbus*), which are often consumed in the basin. The heavily contaminated fish were identified in wide areas from the upper to lower streams of the river. In contrast, the methylmercury concentration of the river water was 1 μg $g^{-1}$ or less, suggesting involvement of biological accumulation of the contaminant through the food web (Takizawa *et al.*, 1972).

**Table 24.3.** Total mercury concentration of aquatic organisms in the basin of the Agano River in 1965 (selected data taken from the Report of the Ministry of Health and Welfare, 1967).

| Organism (Japanese name, species name) | Mercury (μg $g^{-1}$) | Sampling site[b] |
|---|---|---|
| Diatom (keisou, n.s.[a]) | 9.13 | 59 km from MAR |
| Far eastern dace (maruta, *Tribolodon brandti*) | 5.48 | 59 km from MAR |
| Lamprey (yatsume unagi, *Lampetra japonica*) | 41.0 | 56 km from MAR |
| Diatom (keisou, n.s.[a]) | 4.34 | 49 km from MAR |
| Starry flounder (kawagarei, *Platichthys stellatus*) | 17.0 | 14–15 km from MAR |
| Catfish (namazu, *Silurus asotus*) | 12.0 | 14–15 km from MAR |
| Lamprey (yatsume unagi, *Lampetra japonica*) | 12.0 | 14–15 km from MAR |
| Skin carp (nigoi, *Hemibarbus barbus*) | 21.0 | 1 km or less from MAR |
| Snakehead mullet (raigyo, *Channa striata*) | 12.3 | 1 km or less from MAR |
| Far eastern dace (maruta, *Tribolodon brandti*) | 4.60 | 1 km or less from MAR |

[a]n.s., species not specified; [b]MAR, mouth of Agano River.

The investigation team collected hair samples from 1458 participants in the first survey (MHW, 1967). Table 24.4 summarizes the data obtained from 212 participants in the lower reaches of the river where many poisoning cases had been discovered (Tsubaki *et al.*, 1977). Hair mercury concentration in 42.8% of samples was higher than 50 μg $g^{-1}$: a LOAEL of methylmercury for neurological impairment in adult. The hair mercury concentration was obviously associated with the extent of river-fish intake. At the time of the survey, 26 participants had been confirmed as having Minamata disease and the maximum concentration among these participants was 570 μg $g^{-1}$. Many inhabitants were found to be exposed to methylmercury at extremely high doses without the symptoms of poisoning at the survey. The number of certified patients increased within several years, suggesting delayed onset of the disease (Tsubaki *et al.*, 1977). In contrast, in locations where marine fish were consumed, no participant ($n = 45$) had hair mercury concentrations higher than 50 μg $g^{-1}$ (Tsubaki *et al.*, 1977).

At the joint meeting of the team and ministries concerned in March 1966, the investigation team concluded that a large amount of methylmercury was released from the Kanose factory of the Showa Denko and caused the disease among inhabitants by the consumption of contaminated river fish. However, the government official of the Ministry of International Trade and Industry offered a counterargument to the report and the contents discussed at the meeting were kept secret (NIMD, 2001).

## 24.9 Environmental Policy Change and Compensation

In the mid-1960s, the environmental policy was greatly changed in Japan, and the cause and assigned responsibility for environmental pollution were specified. In September 1968, the government finally announced that the causative agent of Minamata disease was methylmercury discharged from Chisso Minamata factory in Minamata and from Showa Denko's Kanose

**Table 24.4.** Distribution of hair mercury concentration and its association with fish consumption of inhabitants in the basin of the Agano River in 1965 (adapted from Tsubaki *et al.*, 1977).

| Fish consumption[a] | Number of participants classified by mercury concentration ($\mu g\ g^{-1}$) | | | | | | | | Total |
|---|---|---|---|---|---|---|---|---|---|
| | < 10 | 10–50 | 50–100 | 100–200 | 200–300 | 300–400 | 400–500 | 500–600 | |
| Little | 7 | 12 | 0 | 0 | 0 | 0 | 0 | 0 | 19 |
| | 36.8% | 63.1% | 0.0% | 0.0% | 0.0% | 0.0% | 0.0% | 0.0% | (100%) |
| Small | 13 | 15 | 4 | 0 | 0 | 0 | 0 | 0 | 32 |
| | 40.6% | 46.9% | 12.5% | 0.0% | 0.0% | 0.0% | 0.0% | 0.0% | (100%) |
| Moderate | 2 | 38 | 14 | 2 | 0 | 0 | 0 | 0 | 56 |
| | 3.6% | 67.9% | 25.0% | 3.6% | 0.0% | 0.0% | 0.0% | 0.0% | (100%) |
| Large | 2 | 32 | 27 | 24 | 9 | 7 | 2 | 2 | 105 |
| | 1.9% | 30.5% | 25.7% | 22.9% | 8.6% | 6.7% | 1.9% | 1.9% | 100% |
| (MD patient)[b] | (0) | (0) | (4) | (6) | (7) | (6) | (1) | (2) | (26) |
| | (0.0%) | (0.0%) | (15.4%) | (23.1%) | (26.9%) | (23.1%) | (3.8%) | (7.7%) | (100%) |
| Total | 24 | 97 | 45 | 26 | 9 | 7 | 2 | 2 | 212 |
| | 11.3% | 45.8% | 21.2% | 12.3% | 4.2% | 3.3% | 0.9% | 0.9% | 100% |

[a]Little: less than once a week; Small: once a week or relatively small amounts; Moderate: more than three times a week or relatively large amounts; Large: every day or very large amounts. [b]Minamata disease (MD) patients were found among participants with 'large' consumption of fish.

factory in Niigata. After 4 years of lawsuits, full liability for the outbreaks was assigned to Showa Denko in 1971 and Chisso in 1973. The appearance of health damage had been socially suppressed until then, but affected inhabitants began to reveal their health damage, and applications for the certification of Minamata disease increased during the 1970s (Hachiya, 2006).

Officially certified Minamata disease patients are eligible to receive compensation from the companies responsible, according to the agreement of 1973 that specifies one-time payment of \$59,000–66,000, a monthly stipend, medical coverage and other allowances. In 1968, the cumulative numbers of the certified patients were 111 in the Yatsushiro Sea areas and 32 in Niigata; however, the number increased to 2283 and 715, respectively, by the end of 2017 to March 2019. From 1969 to 1995, certified cases were 15% of total applications, 12% in the coastal area of Yatsushiro Sea and 32% in the basin of the Agano River, and many applications were rejected. Many lawsuits were then filed to claim compensation from the companies responsible and from the government (Hachiya, 2006).

In 1995, a relief measure was implemented under the 'government settlement' including the issue of medical expense for inhabitants who were suffering from some of neurological signs examined for the official certification. However, in 2004, the Supreme Court rendered a ruling that the national government and Kumamoto prefectural government were also responsible for their failure to prevent the expansion of damage. Application for official certification drastically increased again, and many lawsuits were filed for compensation for wide health damages.

In 2009, the Act on Special Measures for Compensation of Minamata Disease Victims was enacted as a settlement to resolve the problem and relief measures that were compatible with the 'government settlement' of 1995 were implemented for inhabitants with possible health impairments from methylmercury. As many as 80,000 inhabitants were recognized between 1992 and 2014 as recipients of the official medical support programme. The problem concerned and the approach for implementation of the law were described by Kobayashi (2018) of Keio University, who was engaged in policy planning in the Ministry of the Environment, Japan.

## 24.10 Environmental Restoration of Minamata Bay

In Minamata Bay, the contaminated bottom sediment that contained mercury at more than

25 μg $g^{-1}$ spread over 2.09 million $m^2$ (209 ha) and up to 4 m in thickness, with a maximum concentration at 553 μg $g^{-1}$, as measured in 1985. Fishing was virtually restricted from 1975 to 1992 in Minamata Bay and the neighbouring area. In October 1977, the Minamata Bay Anti-Pollution Project began to dredge the contaminated sediment from an area of 1.51 million $m^2$ (151 ha) and to pour it into a reclaimed area of 0.58 million $m^2$ (58 ha). The 13-year-long project was completed in March 1990. Chisso purchased all fish caught in the bay from 1992 to 1997 to prevent distribution of the contaminated fish. In July 1997, it was confirmed that mercury concentration of fish in the bay was below a provisional regulatory standard of 0.3 μg $g^{-1}$ for methylmercury and 0.4 μg $g^{-1}$ for total mercury.

## 24.11 Epidemiological Findings in the Pollution Sites

### 24.11.1 Neurological effects

In 1973, the Second Study Group on Minamata disease (SSGM) of Kumamoto University School of Medicine reported results of a 2-year investigation and suggested hidden and late cases of Minamata disease (SSGM, 1973). In this project, large-scale population-based health surveys were conducted in 1971 in fishing villages of Minamata where many patients had been identified. Increased prevalence of neurological impairments, including sensory disturbances, was observed among inhabitants in the questionnaire survey (Matsushita *et al.*, 1972) and in clinical examination (Tatetsu *et al.*, 2015). It was also shown that the duration of residence in Minamata and fish consumption frequency were significantly associated with the appearance of different neurological signs (Futatsuka and Nomura, 1978). Analysis of the historical data of SSGM revealed increased odds ratio for neurological signs, including paraesthesia, ataxia, dysarthria, and tremors after adjustment for age and sex in Minamata Bay area and in Goshonoura, reflecting different exposure levels in the community (Yorifuji *et al.*, 2008).

A series of large-scale health surveys were conducted in wider coastal areas of the Yatsushiro Sea by Kumamoto Prefecture from 1971 to 1974. In a survey conducted in Minamata Bay and its neighbouring sites (Minamata survey), 51,347 inhabitants participated from peripheral coastal areas of Minamata, Tsunagi, Ashikita, Tanoura, Goshonoura and Ryugatake. The survey areas included the fishing villages of Minamata in which SSGM had investigated. The target of another Yatsushiro survey included north of the Minamata survey sites and 5054 inhabitants participated. After a detailed medical examination following a primary questionnaire survey and secondary screening examination, 158 cases were confirmed as Minamata disease in the Minamata survey area (Kumamoto Prefecture, 1976). No data were presented with appropriate epidemiological analysis except short summaries on the results of surveys (Kumamoto Prefecture, 1974). The health effect among local inhabitants was then evaluated by a reanalysis of the data on the primary questionnaire survey that appeared in the report of Kumamoto University School of Medicine (1977), a cooperative partner of the survey. As shown in Table 24.5, prevalence of perioral sensory impairment was found to increase with crude odds ratio of 7.69 and 8.96 in areas of Minamata and Yatsushiro surveys, respectively. In the Minamata survey, increased crude odds ratios of 3.16 and 2.88 were also obtained on sensory impairment persisting for 1 week or longer and convulsion attack of hand or feet, respectively. Furthermore, crude ratios increased modestly but significantly on 26 other neurological symptoms, including difficulties in fine movement of the hand, in Minamata and its neighbouring sites.

Minamata City government carried out a health survey in the whole city from 1975 to 1981 to find latescent Minamata disease patients, and 387 participants were eventually found as relevant cases. In the survey, 33,445 inhabitants participated in a primary questionnaire survey with a participation rate of 90.0% of the city's population. The author retrieved and statistically analysed the questionnaire data of 27,621 participants. The age-specific prevalence was significantly increased on 23 neurological complaints depending on fish consumption. Crude odds ratios of higher fish consumption group on fine movement of the hand, perioral sensory disturbance and sensory disturbance in the hands/feet were 8.40, 7.40 and 7.15,

**Table 24.5.** Prevalence and odds ratio of typical neurological symptoms in coastal areas of the Yatsushiro Sea in early 1970s.

| Subjective complaint | Area[a] | Total number | Positive cases (%) | Odds ratio[b] (95% CI) |
|---|---|---|---|---|
| Perioral sensory disturbance | Omuta, Reihoku | 18,200 | 93 (0.5%) | 1 |
| | Minamata | 46,105 | 1,752 (3.8%) | 7.69 (6.24–9.48) |
| | North Yatsushiro | 4,523 | 199 (4.4%) | 8.96 (6.99–11.49) |
| Sensory impairment for 1 week or longer | Reihoku | 917 | 11 (1.2%) | 1 |
| | Minamata | 46,514 | 1,721 (3.7%) | 3.16 (1.74–5.75) |
| Convulsion attack of hands or feet | Omuta, Reihoku | 19,780 | 474 (2.4%) | 1 |
| | Minamata | 46,287 | 3,055 (6.6%) | 2.88 (2.61–3.18) |

[a]Omuta of Fukuoka prefecture and Ariake of Kumamoto prefecture are on outside of Yatsushiro Sea; the Minamata area includes coastal areas of Minamata and its neighbours (see text); the north Yatsushiro area includes coastal areas of the Yatsushiro locating in north of the 'Minamata area'. [b]Crude odds ratio calculated using historical data published by Kumamoto University (1977).

respectively. The frequency of past experience of abnormal cat deaths was 9.2% with an odds ratio of 7.08 (N. Hachiya, unpublished results).

In terms of the health effects among inhabitants in the Agano River basin, the relationship was analysed between exposure level at the health survey in 1965 and the result of official certification thereafter. Among the participants, 262 cases had been officially certified as Minamata disease and 378 applications had been rejected as of 1995. The rejected cases comprised inhabitants complaining of neurological symptoms whose clinical manifestations were insufficient for certification. Results indicated that frequency not only of certified patients but also of rejected applicants was increased depending on river-fish consumption (Kondo, 1996). The extent of the contribution of the fish intake to the health effect can be assessed by attributable risk percentage, i.e. the fraction of the outcome ascribed to the exposure. It is estimated that 61.2% of the rejected cases in the sub-group eating 0–4 fish meals per week and 89.0% in those eating 15 or more fish meals per week were associated with river fish consumption.

Although some limitations exist in the historical data analyses, such as a lack of direct quantitative assessment of the exposure, a series of findings obtained from different materials of the 1970s are consistent with increased prevalence of non-specific neurological symptoms among inhabitants, indicating induction of a broad spectrum of health effects by environmental contamination with methylmercury.

A health survey was conducted in 1995 on 1304 adults without certification of Minamata disease living in Tsunagi, a neighbouring area of Minamata, and showed high prevalence on many non-specific subjective complaints compared with a reference site without a pollution history. In the historically polluted area, the prevalence was higher in fishing villages than in non-fishing communities. It was suggested that not only neurological subjective complaints but also non-specific complaints might be influenced by past exposure to methylmercury (Fukuda *et al.*, 1999). Among the neurological signs characteristic of Minamata disease, the frequency of sensory disturbance of glove-and-stocking type was significantly higher in the methylmercury polluted areas than in non-polluted sites (Nakagawa *et al.*, 2002). A health examination in a small island of the Yatsushiro Sea showed an increased prevalence of neuropsychiatric symptoms, including peripheral somatosensory disturbances, and its association with the period of residency on the island (Fujino, 1994).

### 24.11.2 Physical, psychiatric and ageing effects

In 1970, neurological tests and physical functions were examined in junior high school children who were born between 1955 and 1958 in a Minamata fishing village in which congenital Minamata disease appeared. It was confirmed that functional and neurological disturbance increased in the contaminated site on match board, tapping test, agility run, colour naming, pain sensation and vibration sensation tests. In a follow-up study on some of the participants 27 years after the first examination, increased odds ratios, up to 15.1, were still observed for neurological and psychiatric complaints (Futatsuka, 2001).

The data obtained by SSGM in 1971 (Tatetsu *et al.*, 2015) were analysed recently to evaluate effects of methylmercury exposure on psychiatric symptoms. In the polluted areas increased prevalence was observed in impaired intelligence and mood and behavioural dysfunction. Two peaks appeared in the age-dependent prevalence of psychiatric symptoms at around age 20 years and older. The generation around 20 years of age was the same as that of congenital Minamata disease patients and consistent with possible neurodevelopmental disturbance by prenatal methylmercury exposure (Yorifuji *et al.*, 2011).

Prenatal exposure to methylmercury was evaluated by the concentration in the preserved umbilical cord, as described above. It was shown that methylmercury concentrations of children with mental retardation were lower than those of congenital Minamata disease patients, but significantly higher than children without mercury-related disease (Akagi *et al.*, 1998; Harada *et al.*, 1999). These findings indicated that intellectual impairments may be associated with methylmercury exposure *in utero* at concentrations lower than the exposure causing congenital Minamata disease.

Neurocognitive tests were examined from 2012 to 2014 on 23 inhabitants who were born in Minamata between 1953 and 1972. It was suggested that impaired neurocognitive functions observed by the examination may reflect diffuse brain damage as the result of prenatal exposure to methylmercury (Yorifuji *et al.*, 2015, 2016). In a case-control study conducted in 1985, the age-dependent decline in activity of daily living (ADL) was larger in Minamata disease patients aged 60 years and over than in age- and sex-matched controls (Kinjo *et al.*, 1993). In another study, functional health was evaluated using self-reported ADL among inhabitants aged 40 years and over around the Yatsushiro Sea in 2006, and appearance of age-dependent functional impairment was associated with the community's past exposure to methylmercury (Ushijima *et al.*, 2012).

## 24.12 Conclusions

The outbreak of Minamata disease occurred during an era in which productivity took the highest priority and maintenance of the environment of inhabitants was largely disregarded. When the environmental pollution seriously affected the sea, it was not clarified who should be responsible for the outbreak. The insufficient measures resulted in the expansion of health damage. Hazardous chemicals in the environment can influence various health outcomes among local inhabitants, including unapparent or non-specific subjective ones. Population-based health surveys should be effectively conducted not only for the identification of typical patients but for the prevention of health damage and for the planning of countermeasures. Epidemiological evidence and other risk information on the health effects of environmental chemicals should be shared among different stakeholders to make decisions on the risk management of the environmental problem.

## References

Akagi, H., Grandjean, P., Takizawa, Y. and Weihe, P. (1998) Methylmercury dose estimation from umbilical cord concentrations in patients with Minamata disease. *Environmental Research* 77, 98–103

Bakir, F., Damluji, S.F., Amin-Zaki, L., Murtadha, M., Khalidi, A. *et al.* (1973) Methylmercury poisoning in Iraq. *Science* 181, 230–241.

Bjørklund, G., Dadar, M., Mutter, J. and Aaseth J. (2017) The toxicology of mercury: current research and emerging trends. *Environmental Research* 159, 545–554.

Edwards, G.N. (1866) Note on the termination of the second case of poisoning by mercuric methide (Report vol. i, p.144). *St Bartholomew's Hospital Reports, London* 2, 211–212.

Fujino, T. (1994) Clinical and epidemiological studies on chronic Minamata disease. Part I: Study on Katsurajima Island. *Kumamoto Medical Journal* 44, 139–155.

EFSA (European Food Safety Agency) (2012) Scientific opinion on the risk for public health related to the presence of mercury and methylmercury in food. *EFSA Journal* 10, 2985.

Fukuda, Y., Ushijima, K., Kitano, T., Sakamoto, M. and Futatsuka, M. (1999) An analysis of subjective complaints in a population living in a methylmercury-polluted area. *Environmental Research* 81, 100–107.

Futatsuka, M. (2001) Effects of methylmercury exposure on junior high school children bone in 1955–58 near Minamata bay. *Environmental Science* 8, 521–531.

Futatsuka, M. and Nomura. S. (1978) Studies on epidemiological analysis on clinical signs appearing in areas contaminated with methyl mercury: A point of dose effect relationship [text in Japanese with English abstract]. *Journal of Kumamoto Medical Society* 52, 76–94.

Futatsuka, M., Ueda, A., Ueda, T., Hirano, T., Yasutake, R. and Nomura, S. (1977) Estimation of daily methylmercury intake via diet in the members of fisherman's families on the coast of the Shiranui Sea [text in Japanese with English abstract]. *Nippon Koshu Eiseigaku Zasshi* 24, 667–679.

Ha, E., Basu, N., Bose-O'Reilly, S., Dórea, J.G., McSorley, E., Sakamoto, M. and Chan, H.M. (2017) Current progress on understanding the impact of mercury on human health. *Environmental Research* 152, 419–433.

Hachiya, N. (2006) The history and the present of Minamata disease – entering the second half of a century. *Japan Medical Association Journal* 49, 112–118.

Hachiya, N. (2012) Epidemiological update of methylmercury and Minamata disease. In: Ceccatelli, S. and Aschner M. (eds) *Methylmercury and Neurotoxicity*. Springer, New York, pp. 1–11.

Harada, M. (1978) Congenital Minamata disease: intrauterine methylmercury poisoning. *Teratology* 18, 285–288.

Harada, M. (2004) *Minamata Disease* [trans. Tsushima, S. and George, T.S.]. *Kumamoto Nichinichi Shimbun* [*Kumamoto Daily Newspaper*] (originally published by Iwanami Shoten, Tokyo, 1977).

Harada, M., Akagi, H., Tsuda, T., Kizaki, T. and Ohno, H. (1999) Methylmercury level in umbilical cords from patients with congenital Minamata disease. *Science of the Total Environment* 234, 59–62.

Hunter, D., Bomford, P.R. and Russell, D.S. (1940) Poisoning by methylmercury compounds, *Quarterly Journal of Medicine* 9, 193–213.

IPCS (1990) *Environmental Health Criteria 101: Methylmercury*. International Programme on Chemical Safety, World Health Organization, Geneva.

IPCS (1991) *Environmental Health Criteria 118: Inorganic Mercury*. International Programme on Chemical Safety, World Health Organization, Geneva.

Irukayama, K., Kai, F., Fujiki, M. and Kondo, T. (1962) Studies on the origin of the causative agent of Minamata disease. III: Industrial waste containing mercury compounds from Minamata factory. *Kumamoto Medical Journal* 15, 57–68.

Irukayama, K., Kai, F., Fujiki, M., Tajima, S., Omori, S., Nakamura, H. and Kuwahara, S. (1969) Mercury pollution in Minamata district before and after the suspension of the production of acetaldehyde in Minamata factory [text in Japanese with English abstract]. *Journal of Kumamoto Medical Society* 43, 946–957.

Irukayama, K., Fujiki, M., Tajima, S. and Omori, S. (1972) Transition of pollution with mercury of the sea food and sediment in Minamata bay [text in Japanese with English abstract]. *Nippon Kōshū Eiseigaku Zasshi* 19, 25–32.

Ishihara, N. (2014) History of ignorance of methylmercury toxicity and intoxication in Japan in relation to Minamata disease [text in Japanese]. *Japanese Journal of Hygiene* 69, 75–79.

Itai, Y., Fujino, T., Uneno, K., and Motomatsu, Y. (2004) An epidemiological study of the incidence of abnormal pregnancy in areas heavily contaminated with methylmercury. *Environmental Sciences* 11, 83–97.

JECFA (2004) Methylmercury. In: *Evaluation of Certain Food Additives and Contaminants: Sixty-first Report of the Joint FAO/WHO Expert Committee on Food Additives*. Joint FAO/WHO Expert Committee on Food Additives, World Health Organization, Geneva, pp. 132–140.

Kajiwara, Y., Yasutake, A., Adachi, T. and Hirayama, K. (1996) Methylmercury transport across the placenta via neutral amino acid carrier. *Archives of Toxicology* 70, 310–314.

Karagas, M.R., Choi, A.L., Oken, E., Horvat, M., Schoeny, R. *et al.* (2012) Evidence on the human health effects of low-level methylmercury exposure. *Environmental Health Perspectives* 120, 799–806.
Kinjo, Y., Higashi, H., Nakano, A., Sakamoto, M. and Saki, R. (1993) Profile of subjective complaints and activity of daily living among current patients with Minamata disease after 3 decades. *Environmental Research* 63, 241–251.
Kitamura, S., Kakita, T., Koji, S. and Kojima, T. (1960a) Minamata byō ni kansuru ekigaku chōsa seiseki hoi 3 [Epidemiological study on Minamata disease, Suppl. 3]. *Journal of Kumamoto Medical Society* 34 (Suppl. 3), 477–480.
Kitamura, S., Ueda, K., Niino, J., Ujioka, T., Misumi, H. and Kakita, Y. (1960b) Minamata byō ni kansuru kagaku-dokubutsu kensaku seiseki, dai 5-hou [Chemical and toxicological study on Minamata disease, 5th report]. *Journal of Kumamoto Medical Society* 34 (Suppl. 3), 593–601.
Kobayashi, H. (2018) Minamata: how a policy maker addressed a very wicked water quality policy problem. *Water International* 43, 404–423.
Kondo, K. (1996) Incidence of Minamata disease in communities along the Agano River, Niigata, Japan. Patterns of the exposure and official diagnosis of patients [text in Japanese]. *Japanese Journal of Hygiene* 51, 599–611
Kumamoto Prefecture (1974) Minamata wan shūhen chiku jūmin kenkō chōsa [Health survey in Minamata Bay and neighboring sites]. *Kōgai Hakusho* 1974, pp. 133–145.
Kumamoto Prefecture (1976) Minamata wan shūhen chiku jūmin kenkō chōsa [Health survey in Minamata Bay and neighboring sites]. *Kōgai Hakusho* 1976, pp. 159–160.
Kumamoto University School of Medicine (1977) *Ariake kai, Yatsushiro kai engan chiiki oyobi Minamata wan shūhen chiku jūmin kenkō chōsa kaiseki hōkokusho* [*Report of Health Survey in Ariake Sea, Yatsushiro Sea, and Minamata Bay and its Neighboring Sites*]. Kumamoto University School of Medicine, Japan.
Kurland, L.T., Faro, S.N. and Siedler, H. (1960) Minamata disease: the outbreak of neurologic disorder in Minamata, Japan, and its relationship to the ingestion of seafood contaminated by mercuric compounds. *World Neurology* 1, 370–395.
Matsumoto, H., Koya, G. and Takeuchi, T. (1965) Fetal Minamata disease. A neuropathological study of two cases of intrauterine intoxication by a methyl mercury compound. *Journal of Neuropathology and Experimental Neurology* 24, 563–574.
Matsushima, Y (1970) Minamata byō ni kansuru mōhatsu-chu no suigin ryō no chōsa [Survey on mercury concentration in hair in relevance to Minamata disease]. *Annual Report of Kumamoto Prefectural Institute of Public Health* 1970, 13–45
Matsushima, Y. and Noguchi, T. (1966) Difference of mercury contents of divided hair fragments by eating food contaminated with organic mercury compounds [text in Japanese with English abstract, tables and figures]. *Eisei Kagaku* 12, 106–108.
Matsushita, T., Futatsuka, M., Arimatsu, Y., Ueda, A., Misumi, J. *et al.* (1972) Epidemiological study on Minamata disease. Health surveys of inhabitants in Minamata and Goshonoura districts (Report 1) [text in Japanese with English abstract]. *Journal of Kumamoto Medical Society* 46, 641–660.
McAlpine, D. and Araki S. (1958) Minamata disease, an unusual neurological disorder caused by contaminated fish. *The Lancet* 2, 629–631.
MHW (1967) *Niigata Suigin chūdoku Jiken Tokubetsu Kenkyū Hōkoku-sho* [*Special Research Report on Organic Mercury Poisoning in Niigata*]. Ministry of Health and Welfare, Tokyo.
Moriyama, H., Futatsuka, M. and Kinjo, Y. (1994) Fetal Minamata disease – a review. *Environmental Sciences* 3, 15–23.
Nakagawa, M., Kodama, T., Akiba, S., Arimura, K., Wakamiya, J. *et al.* (2002) Logistic model analysis of neurological findings in Mnamata disease and the predicting index. *Internal Medicine* 41, 14–19.
NIMD (2001) *In the Hope of Avoiding Repetition of a Tragedy of Minamata Disease: Report of the Social Scientific Study Group*. National Institute for Minamata Disease, Minamata, Japan. Available at: http://www.nimd.env.go.jp/syakai/webversion/SSSGMDreport.html (accessed 31 May, 2018)
NRC (2000) *Toxicological Effects of Methylmercury*. National Research Council. National Academy Press, Washington, pp 344.
Pentschew, A. (1958) Intoxikation. In: Lubarsch, O., Henke, F. and Rössle, R. (eds) *Handbuch der Speziellen Pathologeschen Anatomie und Histologie* 13 (Part 2B). Spinger-Verlag, Berlin, pp. 2008–2024.
Sakata, A., Orita, T., Koriyama, M., Nakahara, K. and Hozumi, T. (1962) Minamata byō ni kansuru mōhatsu-chu no suigin ryō no chōsa ni tsuite [Survey on mercury concentration of hair in relevance to Minamata disease]. *Annual Report of Kagoshima Prefectural Institute of Public Health* 2, 53–56.

Sakata, A., Orita, T., Koriyama, M., Nakahara, K. and Hozumi, T. (1963) Minamata byō ni kansuru mōhatsu-chū no suigin ryō no chousa ni tsuite, 2. [Survey on mercury concentration of hair in relevance to Minamata disease, II]. *Annual Report of Kagoshima Prefectural Institute of Public Health* 3, 11–12

SSGM (1973) *10-nen-go no Minamata byō ni kansuru Ekigaku-teki, Rinshōigaku-teki narabini Byōrigaku-teki Kenkyū* [*Epidemiological, Clinical and Pathological Studies on Minamata Disease after a Decade*]. Second Study Group on Minamata Disease, Kumamoto University, Japan.

Stern, A.H. and Smith, A.E. (2003) An assessment of the cord blood: maternal blood methylmercury ratio: implications for risk assessment. *Environmental Health Perspectives* 111, 1465–1470.

Takizawa, Y., Kosaka, T., Sugai, R., Sasagawa, I., Sekiguchi, C. and Minagawa, K. (1972) Studies on the cause of the Niigata episode of Minamata disease outbreak. *Acta Medica et Biologica* 19, 193–206.

Tatetsu, S., Kiyota, K., Tomonari, H., Inoue, T., Teraoka, A. *et al.* (2015) Yūkisuigin niyoru kōdo osen to hiosen no umi no engan jūmin no issei kenshin kekka no hikaku kenkyū [Comparative analysis on health examination surveys on inhabitants in coastal areas of sea with and without organic mercury pollution]. *Minamata byō Kenkyū* 6, 69–101.

Tokuomi, H., Okajima, T., Yamashita, M. and Matsui S. (1963) Minamata byo no ekigaku [Epidemiology of Minamata disease]. *Shinkei Kenku no Shinpo* 7, 276–289.

Tsubaki, T., Shirakawa, K., Hirota, K., Kondo, K., Sato, T. and Kanbayashi, K. (1977) Epidemiology of methylmercury poisoning in Niigata. In: Tsubaki, T. and Irukayama, K. (eds) *Minamata Disease – Methylmercury Poisoning in Minamata and Niigata*, Japan. Elsevier/North Holland, New York, pp. 57–78.

Uchida, M., Hirakawa, K. and Inoue, T. (1961) Biochemical studies on Minamata disease. IV. Isolation and chemical identification of the mercury compound in the toxic shellfish with special reference to the causal agent of the disease. *Kumamoto Medical Journal* 14, 181–187.

UNEP (2013) *Global Mercury Assessment 2013, Sources, Emissions, Releases and Environmental Transport*. United Nations Environment Programme, Nairobi.

Ushijima, K., Sung, W., Tanaka, S., Kawakita, M., Mukai, Y., Tamura, K. and Maruyama, S. (2012) Association between early methylmercury exposure and functional health among residents of the Shiranui Sea communities in Japan. *International Journal of Environmental Health Research* 22, 387–400.

WHO (World Health Organization) (1990) International Programme on Chemical Safety. Environmental Health Criteria 101, methylmercury. Geneva, Switzerland.

Yasutake, A., Matsumoto, M., Yamaguchi, M. and Hachiya, N. (2004) Current hair mercury levels in Japanese for estimation of methylmercury exposure. *Journal of Health Science* 50, 120–125

Yorifuji, T., Tsuda, T., Takao, S. and Harada, M. (2008) Long-term exposure to methylmercury and neurologic signs in Minamata and neighboring communities. *Epidemiology* 19, 3–9.

Yorifuji T., Kashima, S., Tsuda, T. and Harada M. (2009) What has methylmercury in umbilical cords told us? Minamata disease. *Science of the Total Environment* 408, 272–276.

Yorifuji, T., Tsuda, T., Inoue, S., Takao, S. and Harada, M. (2011) Long-term exposure to methylmercury and psychiatric symptoms in residents of Minamata, Japan. *Environmental International* 37, 907–913.

Yorifuji, T., Kato, T., Kado, Y., Tokinobu, A., Yamakawa, M., Tsuda, T. and Sanada, S. (2015) Intrauterine exposure to methylmercury and neurocognitive functions: Minamata disease. *Archives of Environmental and Occupational Health* 70, 297–302.

Yorifuji, T., Kado, Y., Diez, M.H., Kishikawa, T. and Sanada (2016) Neurological and neurocognitive functions from intrauterine methylmercury exposure. *Archives of Environmental and Occupational Health* 71, 170–177

Yorifuji, T., Kashima, S., Suryadhi, M.A.H. and Abudureyimu, K. (2017) Temporal trends of infant and birth outcomes in Minamata after severe methylmercury exposure. *Environmental Pollution* 231, 1586–1592.

Zangger, H. (1930) Erfahrungen uber Quecksilbervergiftungen. *Archiv für Gewerbepathologie und Gewerbehygiene* 1, 539–560.

# 25 Lead Poisoning

**A.L. Katner*[,1] and H.W. Mielke[2]**

[1]*School of Public Health, Louisiana State University Health Sciences Center, New Orleans, Louisiana, USA;* [2]*School of Medicine, Tulane University, New Orleans, Louisiana, USA*

## 25.1 Abstract

The reduction of childhood lead exposure in the USA is a public health success story, but lead is still one of the most widely encountered toxic metals in the world. To date, no safe blood lead threshold has been identified. The properties and availability of lead, a naturally occurring blue-grey heavy metal, led to its widespread use throughout history in paint, gasoline, plumbing, food containers and a vast array of other commercial products, which in turn led to a persistent legacy of widespread lead contamination. The toxic effects of lead were recognized as early as 2000 BC, but the mechanisms by which lead exerts it impacts have only been elucidated in the past few decades. This chapter discusses the historical use and regulation of lead and provides an overview of lead's toxicological mechanisms and health effects. Particular emphasis is placed on low-dose lead impacts, which are most relevant to developed countries that have seen a decline in population lead exposures due to regulation and exposure prevention efforts.

## 25.2 Introduction: a Lingering Public Health Priority

Lead (Pb) is a soft, malleable, ductile, corrosion-resistant bluish- or dark-grey naturally occurring heavy metal found in trace amounts in the earth's crust, soil, plants and water. Human use and widespread release of lead has created a legacy of environmental contamination and human exposures which has increased considerably over the past few centuries (Patterson, 1965). The earliest Pb mine discovered dates to 6500 BC and historical descriptions of Pb poisoning were recorded by the Egyptians around 2000 BC (Hernberg, 2000; Needleman, 2004). Historical accounts of Pb poisoning were associated with either occupational exposures, such as in mining, smelting, shipbuilding and glazing, or with Pb use among the affluent, who used it in plumbing (derived from the Latin word for lead, plumbum), as a wine preservative, in pottery, in cooking utensils, as a sweetening agent and in makeup. Descriptions of occupational exposures re-emerged during the Industrial Revolution, when Pb production escalated in the 18th century. In modern times, Pb is ubiquitous and exposures

* E-mail address: akatn1@lsuhsc.edu

are common and widespread, due to the pervasive use of Pb in commercial products like paint, gasoline, plumbing, cosmetics, food cans, folk remedies and toys and in the production of batteries, ammunition, ceramics and stained glass (ATSDR, 2007).

Exposures to Pb have been associated with a wide range of effects which can collectively be referred to as 'lead poisoning'. There has been a shift in the past few decades from earlier accounts of poisoning that were focused on workers and the wealthy, to accounts of poisoning among children, especially those from low-income families. Lead has no known function in the body and there is evidence that it can have permanent subclinical effects during neurodevelopment. Lead's lethal impact on the developing fetus is so effective that it was put to use as an abortifacient in the 19th century and early half of the 20th century (Hernberg, 2000; Woolf *et al.*, 2007). The first childhood Pb poisoning cases were reported in the 19th century, but up until the 20th century many physicians and researchers remained sceptical about the association between Pb and reported maladies like colic, anaemia and encephalopathy (Hernberg, 2000). From the 1950s to the 1970s, the research of Herbert Needleman and Clair Cameron Patterson sparked a renewed awareness of the ubiquity and toxicity of lead – research for which they were initially criticized but subsequently honoured. Since that time, animal, medical and epidemiological research has revealed that even small doses of Pb could adversely impact several body functions; and some impacts, most especially during neurodevelopment, could be irreversible (NTP, 2012).

Beginning with the enactment of legislation limiting or banning the use of Pb in gasoline, paint, plumbing and other commercial products, the USA has seen a steep decline in childhood blood lead levels (BLLs) of more than 90% (Landrigan *et al.*, 2017). Two products were especially egregious sources of exposure: lead-based paint and tetraethyl lead (TEL) additives to gasoline. Regarding lead-based paint, European nations began ratifying the 'White Lead (Painting) Convention', an international treaty prohibiting lead paint, as early as 1921; but US industry lobbied against this ban. As a result, lead in US residential paint was not effectively banned until 1978 (Hernberg, 2000; Needleman, 2000). Dangerous Pb exposures are still common in old houses throughout the USA (Levin *et al.*, 2008) and wherever paint is sanded or otherwise disturbed (Marino *et al.*, 1990).

The ban on Pb in gasoline (or TEL) was more difficult because, unlike the visual appearance of lead pigment in paint, the combustion of gasoline with TEL exhausts as invisible dust particles. In the 1920s, the public health community warned against the use of TEL as an anti-knock agent in gasoline, calling it 'the greatest single question in the field of public health' (Kitman, 2000). The actual quantity and extent of lead contamination from TEL was not appreciated until the 1960s, when pioneering research was conducted by C.C. Patterson, who applied sensitive analytical methods to measure lead in sections of a deep core of Greenland ice. He found that the amount of lead in the ice core peaked at the surface and decreased with depth (Tatsumoto and Patterson, 1963). The rate of lead release into the environment increased significantly around 1950, which corresponded with the production and use of TEL. Patterson concluded that there was a geochemical imbalance of lead contamination on earth and expressed concerns that human exposure was occurring. In a 1980 National Academy of Science review of lead in the environment, Patterson remarked:

> Sometime in the near future it probably will be shown that the older urban areas of the United States have been rendered more or less uninhabitable by the millions of tons of poisonous industrial lead residues that have accumulated in cities during the past century.

Patterson's concerns were confirmed in subsequent studies. Evidence about urban lead accumulation was provided by a study conducted in Baltimore, Maryland, which demonstrated a profound disparity in the quantity of lead (and other metals) in soil within extremely contaminated inner-city soils compared with relatively uncontaminated soils in outlying areas of Baltimore (Mielke *et al.*, 1983). The Baltimore studies were repeated by state-funded studies of lead exposure in various-sized cities of Minnesota. These studies confirmed the association between urban soil lead and children's BLL (Mielke *et al.*, 1984, 1989). At the same time, a national trend of declining TEL and BLL was described (Annest *et al.*, 1983).

While Pb exposures have declined significantly through the implementation of laws and regulations, its ubiquity in the environment from centuries of widespread use, its environmental persistence, ability to bioaccumulate and cross the placenta, and the irreversible low-dose impacts it has on neurodevelopment are some reasons that Pb remains a public health priority (Needleman *et al.*, 1990). While for clinical purposes 'lead poisoning' is defined as 'episodic, acute, symptomatic illness from lead toxicity' (CDC, 2004), research beginning in the late 1970s revealed that early life exposures can often have subclinical effects on molecular and cellular processes that one will not see signs of until years later, such as effects on intelligence, attention, learning, language function, perception and impulse control (NTP, 2012). In 1979, Needleman published his pioneering study on the association of lead in deciduous baby teeth and learning, which indicated that children with the most lead in their teeth exhibited poorer classroom performance (Needleman *et al.*, 1979). Since then, the National Toxicology Program (NTP) has concluded that there is 'sufficient evidence' that even low-dose Pb ($< 10\ \mu g\ dl^{-1}$) may cause adverse health effects, even after controlling for chance, bias and confounding; and no threshold for lead, below which no adverse health effects have been measured, has been found (CDC, 2004; NTP, 2012) (Table 25.1). In 2012, the US Centers for Disease Control and Prevention (CDC) decreased the reference value of childhood BLL from 10 to 5 $\mu g\ dl^{-1}$ and re-emphasized the fact that 'no level of lead appears to be safe' for children (CDC, 2012). The US reference value is based on the 97.5% blood lead results of the National Health and Nutrition Examination Survey (NHANES) and the most recent survey shows that the reference value should be reduced from 5 to 3.5 $\mu g\ dl^{-1}$. In line with growing scientific evidence, the World Health Organization (WHO) also withdrew its provisional tolerable weekly intake guidelines value (25 $\mu g\ kg^{-1}$) based on evidence that it would not protect children against intelligence quotient (IQ) loss (WHO, 2010). These low-dose effects can be seen in different populations, across different physiological systems, and throughout the lifespan, regardless of study design. This research, and more recent revelations on potential epigenetic and heritable impacts of lead, suggests that the prior unrestricted use and release of Pb into the environment by humans may continue to have adverse impacts on many future generations. While prevalence rates may vary by country, low-dose Pb exposures remain a global public health priority. Given that there are no effective treatments that can reverse the effects of exposure, the focus now is on primary prevention, i.e. screening and testing air, water, soil and all products that are part of everyday experience. Focusing on lead intervention, exposure prevention education and continuing population lead surveillance are critical steps towards continuing the progress with reducing population exposure to lead.

## 25.3 Lead in the Environment

Widespread contamination of the planet has occurred through anthropogenic use of Pb, from industrial activities like mining, smelting, manufacturing and combustion, rather than from naturally occurring lead. Lead can exist as an element or as inorganic or organic compounds. Lead is extremely persistent in the environment; as an element it is indestructible. While it is rarely found in its elemental state in nature, it can exist as one of four naturally occurring stable isotopes: $^{204}Pb$, $^{206}Pb$, $^{207}Pb$, and $^{208}Pb$ (NTP, 2012). In the environment and in consumer products, including soil, paint and indoor dust, Pb is predominantly found as an inorganic compound, such as Pb salts and Pb oxides and sulfides (ATSDR, 2007). The major form of naturally occurring Pb is inorganic Pb sulfide, which can be found at low concentrations in the earth's crust; it can be released to the surface through volcanic eruptions and erosion, but these are minor sources of the Pb to which humans are exposed (IARC, 2006).

It is the historical use of Pb by humans that has created a legacy of contamination of our soil, water, air, crops, food and indoor environment. Urban environments typically have significantly more Pb than rural environments, especially near highways, due to decades of

**Table 25.1.** State of the evidence: low-dose lead effects (< 10 μg $dl^{-1}$) based on sufficient or limited data by life stage-specific blood lead level (adapted from NTP, 2012).

| | Blood lead level (μg $dl^{-1}$) | System | Effect |
|---|---|---|---|
| Prenatal | < 5 | Neurological | Decrease in cognitive function |
| | < 10 | Neurological | Decrease in IQ (intelligence quotient) |
| | | | Increase in attention and behaviour problems |
| | | | Decreased hearing |
| | | Reproductive and developmental | Reduced postnatal growth |
| Child | < 5 | Neurological | Decrease in academic achievement, IQ and cognitive function |
| | | | Blood lead level (μg $dl^{-1}$) |
| | | Renal | Decreased glomerular filtration rate |
| | | Reproductive and developmental | Delayed puberty |
| | < 10 | Neurological | Decreased hearing |
| | | Immune | Increased hypersensitivity/allergy |
| | | Reproductive and developmental | Delayed puberty and reduced postnatal growth |
| Adult | < 5 | Neurological | Increased incidence of essential tremor |
| | | Renal | Decreased glomerular filtration rate |
| | | Reproductive and developmental | Reduced fetal growth |
| | < 10 | Neurological | Increased incidence of essential tremor |
| | | | Psychiatric effects |
| | | | Decrease in cognitive function |
| | | | Decreased hearing |
| | | | Increased incidence of ALS (amyotrophic lateral sclerosis) |
| | | Cardiovascular | Increase in blood pressure and increased risk of hypertension |
| | | | Increased cardiovascular-related mortalities and electrocardiography abnormalities |
| | | Reproductive and developmental | Increase in spontaneous abortion and pre-term birth |

leaded-gasoline use (Mielke *et al.*, 2011). Other areas with high Pb include those near point sources where Pb is emitted or released – Pb has been reported at 75% of National Priority List sites in the USA (ATSDR, 2006). Lead that has been emitted to air through processes such as industrial emission and combustion can be transported over long distances, up to thousands of kilometres; eventually it will be removed through precipitation or dry deposition (NTP, 2012). Once deposited on the surface, the biggest sink for Pb is soil or sediment, to which it strongly absorbs. In soil, it can be found near the surface, where it remains seasonably immobile. Since Pb aerosols were curtailed by banishing TEL from gasoline, a series of changes occured: Pb dust loading of soil decreased, topsoil Pb declined, seasonal resuspension of Pb dust reduced and children's BLL diminished (Mielke *et al.*, 2019).

## 25.4 Lead Uptake into the Body

Not only is Pb highly persistent in nature, but it can also bioaccumulate in soft and mineralizing tissues over a lifetime (Eisler, 1988). As such, it is categorized as a persistent, bioaccumulative and toxic (PBT) chemical. It enters the body through ingestion or inhalation of inorganic forms; or ingestion, inhalation, and dermal absorption of lipid-soluble organic forms. Exposures can occur to workers in occupational settings, like construction, recycling, battery manufacturing, car repair, refining and smelting, Pb exposures most commonly occur to the general population through inhaling or ingesting Pb in disturbed or exposed paint, soil or dust, and consuming lead-contaminated food or drinking water distributed through leaded service lines and plumbing.

The fate of Pb in the body is not dependent on the route of exposure. Most Pb that remains in the body ends up in the bone, where it can have a 20-year half-life (Silbergeld *et al.*, 1993). Lead is rapidly absorbed in the bloodstream, but only 1% of absorbed Pb circulates in the blood, where it has a half-life of only 25 days (NTP, 2012). Yet BLL is the most commonly evaluated biomarker for screening and diagnostic purposes, based on ease and cost. When compared with blood, levels of Pb in mineralized tissue, in particular teeth or bone, are a better biomarker for cumulative or lifetime exposure. Levels of Pb in bone, mainly the tibia but also the patella, had stronger positive associations to cognitive decline in adults, compared with levels of Pb in blood (Shih *et al.*, 2007; Sanders *et al.*, 2009). These results highlight the impacts that a lifetime of Pb exposure can have on later-life cognitive function, via the release of accumulated reserves of Pb from an endogenous source (bone).

Differences exist between children and adults in terms of Pb exposures, absorption and fate (ATSDR, 2007). Fetuses, infants and children are considered to be the most vulnerable population. Fetal exposures can occur when Pb from maternal blood crosses the placenta, or is passed through breastfeeding. Bone Pb is generally more bioavailable for pregnant women and children whose bones are developing. In children, bones can serve as a continuous source of endogenous Pb exposure, as they undergo regular restructuring (Barbosa *et al.*, 2005). Infant exposures may occur through the consumption of formula that has been reconstituted with lead-contaminated water. Childhood exposures typically occur via hand-to-mouth of lead from soil, house dust and paint, or inhalation of lead particles in soil. Children can absorb Pb to a greater extent than adults: 40–50% of water-soluble Pb can be absorbed by a child, compared with 3–10% for adults. Lead is also more easily absorbed into the brains of children than adults (Needleman, 2004); and children store more of their total Pb in soft tissues, compared with adults – in particular, the brain, spleen, liver, kidneys and lungs (Dart *et al.*, 2004), where it is more bioavailable compared with Pb in the bone; the estimated half-life of Pb in soft tissue is 40 days.

One critical feature about children's BLL is that the seasonal pattern of exposure does not support the idea that house dust and paint are the major sources of children's lead exposure. Lead has been called the 'summer disease', because children's BLLs trend highest during late summer and fall, and lowest during winter and early spring. This is the opposite of expectations that house dust and paint are the major sources of childhood lead exposure. Research on the role of lead-contaminated soil in several urban areas has demonstrated that BLL seasonality is driven by climate fluctuations, which in turn result in soil moisture decreases in late summer (Mielke and Reagan, 1998). The overall result is seasonal re-suspension of dust from dry soils (Laidlaw *et al.* 2005, 2012). The state of Michigan maintains a quarterly record of children's BLL and seasonality shows up prominently in Michigan cities (Laidlaw *et al.*, 2016; Zahran *et al.*, 2013). These results may provide insight into why conventional US interventions that are focused on education and household lead-dust clean-up, have failed to significantly impact population-level BLLs (Nussbaumer-Streit *et al.*, 2016). Lead clean-up intervention must address sources beyond the home (Mielke and Zahran, 2012) and reduce exposures from other common sources, like outdoor soil, drinking-water, food and commercial products.

Among adults, occupational exposures, home renovations or endogenous exposures through recirculation of Pb from bones are among the most common sources of exposure. One survey of women aged 18–45 years with BLLs between 10 and 25 μg $dl^{-1}$ ($n = 135$) identified the major sources of adult Pb exposure to be occupational (46%) or home renovations (24%); 30% of cases had no identifiable source of exposure (Fletcher *et al.*, 1999). Endogenous Pb exposure may account for some unidentified sources. In adults, a greater proportion of the body's total Pb is stored in bones (85–95%), compared with children (70%) (Patrick, 2006). In adults, an estimated 40–70% of Pb released into blood comes from the bone during periods of bone demineralization (Flora *et al.*, 2012), which can occur during pregnancy, lactation, menopause and osteoporosis (NTP, 2012). Thus, vulnerable adult populations include pregnant, lactating or menopausal women and the elderly. Several studies have been consistent in finding increases in maternal BLLs in later pregnancy and during lactation, when there is an increase in the need for calcium by the fetus (Bellinger, 2005). Women with lower milk intake have higher BLLs in late pregnancy, suggesting that increased calcium intake during the second half of the pregnancy may reduce Pb release from bone stores (Hertz-Picciotto *et al.*, 2000; Bellinger, 2005). One study Gulson *et al.*, 2003) estimated the contribution of skeletal Pb to blood Pb during pregnancy by measuring Pb isotope ratios in the blood of European immigrants to Australia who later became pregnant. The difference between the environmental Pb isotope compositions in different parts of the world, allowed researchers to evaluate the source of blood Pb in women during pregnancy. Lead in the blood of these pregnant immigrants appeared to be primarily of European origin, suggesting bone as the source, even for those who had been residing in Australia for more than 2 years. An estimated 79% of the Pb that was transferred gestationally appeared to originate from Pb stored over time in bones. Manton *et al.* (2003) reported similar results and concluded that stores of bone Pb released in late pregnancy had a greater influence on maternal BLLs than dietary lead. In several studies, BLLs of women and children emigrating to the USA from developing countries were considerably higher (CDC, 2000), highlighting a greater need to screen and test these populations.

## 25.5 Mechanisms of Toxicity

Once inside the body, Pb can affect almost every organ in the body (ATSDR, 2007). The toxicokinetics of Pb is similar between children and adults (ATSDR, 2007), where it can induce toxicity through a series of mechanisms, from promotion of oxidative stress to ion replacement and epigenetic changes. Through these mechanisms it can damage DNA, enzymes, proteins and membrane lipids and impair antioxidants (Flora *et al.*, 2012). This section highlights some, but not all, of the ways in which Pb exerts its toxic effects.

### 25.5.1 Oxidative stress

One mechanism that all toxic metals have in common involves oxidative damage (Sanders *et al.*, 2009). Lead's effects on bodily systems are produced primarily through oxidative stress, which in turn can lead to significant damage to molecules like DNA, enzymes, proteins and cell membranes, which can ultimately lead to cell damage and death (Flora *et al.*, 2012). Two mechanisms play a role in creating conditions of oxidative stress within the body: (i) an increase in reactive oxygen species (ROS), which can result in cell damage; and (ii) a decrease in antioxidants which can stabilize or remove the ROS. The ROS can be produced through metabolic processes, or they can be introduced into the body from external sources.

### 25.5.2 Ionic mechanisms

Lead (II) (or $Pb^{2+}$) is a bivalent cation, which can act as an analogue to replace cations like $Ca^{2+}$, $Mg^{2+}$, $Fe^{2+}$, $Zn^{2+}$ and $Na^{+}$, thereby disrupting several biological processes, including cellular signalling, cell adhesion, protein folding and maturation, apoptosis (programmed cell death), ionic transport, enzyme regulation and neurotransmitter release (Lidsky and Schneider, 2003; Garza *et al.*, 2006). For example, when Pb replaces zinc ions, which are co-factors for antioxidant enzymes, it can create oxidative stress conditions (Flora *et al.*, 2012). Lead displacement of zinc can also impact the regulation of genetic transcription.

Trace amounts of Pb can disrupt the second messenger system through cation replacement of calcium. Calcium is a ubiquitous second messenger – an intracellular signalling molecule that triggers physiological changes in response to exposure to first messengers or extracellular signalling molecules, like hormones or neurotransmitters. First messengers can activate cell membrane proteins through the binding of specific receptor sites on cell surfaces. These proteins can trigger cellular update of calcium, which plays a role in nerve growth, differentiation, synaptic activity, metabolism, immune response, apoptosis, muscle contraction and intracellular movement (Brochin *et al.*, 2008). Lead interferes with these functions by displacing calcium. Lead forms stronger bonds than calcium; thus, picomolar concentrations of Pb easily displace micromolar concentrations of calcium (Sanders *et al.*, 2009). It is this displacement of calcium that allows Pb to cross the blood–brain barrier and alter activation of proteins that disrupt cellular processes, impacting cognitive function by disrupting cellular communication, neurotransmitter uptake (choline, dopamine and GABA), memory storage and neural excitation (Flora *et al.*, 2012). Lead ($Pb^{2+}$) can cross the blood–brain barrier (BBB) by substituting for calcium ions ($Ca^{2+}$) and associating with the calcium-ATPase pumps. Once inside, it can disrupt: synapse formation in the cerebral cortex; development of neurochemicals, including neurotransmitters; neurotransmission; neuronal growth; and ion channels (Pearson and Schonfeld, 2003; Liu *et al.*, 2007). All of these actions can disrupt prefrontal cortical regulation of behaviour and thought, leading to distractibility, impaired judgement and impulsivity (Sanders *et al.*, 2009).

### 25.5.3 Epigenetic mechanisms

Epigenetic processes control many key body functions and evidence is mounting that Pb may impact the developing brain by disrupting epigenetic gene regulation. If this is the case, it may shed light on the seemingly lifelong effects of lead and it may suggest that Pb impacts could be inherited (Senut *et al.*, 2012). Epigenetics is the study of heritable changes in gene activities related not to alterations in the primary deoxyribonucleic acid genetic code (DNA), but rather through processes such as DNA methylation, histone modifications and non-coding RNAs (ncRNAs) (Senut *et al.*, 2012). DNA methylation can alter the binding of transcriptional modulators, thus impacting gene regulation and expression. Histone modification can also alter gene regulation, by changing how these proteins organize chromatin, and hence gene expression. The ncRNAs regulate gene expression by silencing genes or degrading mRNA. Epigenetics may play a role in the regulation of brain development, which has opened up a field known as neuro-epigenetics. Neuro-epigenetics has shed light on the role that epigenetic mechanisms may play in heavy-metal neurotoxicity. Some researchers have suggested that lead-induced oxidative stress may mediate changes in methylation patterns (Senut *et al.*, 2012; Eid *et al.*, 2016). Early life Pb exposure has been associated with changes in the expression of epigenetic intermediates involved in DNA methylation or histone modification which, in turn, regulates later-life expression of genes involved in diseases like Alzheimer's disease (Eid *et al.*, 2016). More research is needed to evaluate whether or not lead-induced alterations could be inherited across generations.

## 25.6 Adverse Health Effects

The symptoms of, and susceptibility to, lead poisoning can vary depending on the dose and length of exposure, age at exposure, gender, ethnicity, genetic variation, nutritional status, health, stress and other individual characteristics (Bellinger, 2004; NTP, 2012). Acute toxicity, which was commonly associated with exposure to very high Pb levels, is presently an uncommon occurrence, but past case reports associated acute toxicity with neuromuscular manifestations, such as brain damage, encephalopathy (brain degeneration), comas, convulsions and even death (Cleveland *et al.*, 2008). Chronic toxicity to lower Pb levels can manifest as irritability, poor attention, learning disabilities and aggressive behaviours, especially in children (Patrick, 2006; Sanders *et al.*, 2009; Flora *et al.*, 2012). Headache, tremors, lethargy, depression, memory loss, abdominal pain, vomiting, lack of coordination, slurred speech, numbness and tingling in the extremities, anaemia, delirium, convulsions and coma can also be observed in more severe cases of chronic toxicity (Flora *et al.*, 2006, 2012; Patrick, 2006; Pearce, 2007).

### 25.6.1 Nervous system effects

Lead impacts many organ systems but its most pronounced impact is on the nervous system. The impacts of Pb on the peripheral nervous system are more pronounced in adults than children and can include peripheral neuropathy, muscular weakness, fatigue and lack of coordination (Sanders *et al.*, 2009). In children, the effects of even low levels of Pb (< 10 μg $dl^{-1}$) are more pronounced on the central nervous system and can manifest as hyperactivity, lack of concentration, irritability, behaviour problems, learning disorders and developmental disabilities (Cory-Slechta, 1996; Bellinger, 2004; Brent, 2006; Brunton *et al.*, 2007; Flora *et al.*, 2012). The fetal nervous system is more vulnerable to Pb, as immature endothelial cells allow Pb easier access into the fetal brain (Brochin *et al.*, 2008). Inside the brain, Pb can accumulate in astroglial cells (also known as astrocytes), where it can serve as a reservoir for continued release of Pb into the brain, leading to neuronal damage (Sanders *et al.*, 2009), and disruption of the brain's repair process and BBB formation (Flora *et al.*, 2012). Lead can impair the BBB's development by disrupting the myelin sheath, which insulates nerves and increases impulse conduction. Lead's destructive capabilities in the brain can ultimately result in a failure to establish brain connections that are essential to its function, thereby disrupting processes involved in learning, memory, cognition and behaviour (ATSDR, 2007; Brochin *et al.*, 2008; Sanders *et al.*, 2009).

While most studies in the past few decades have focused on understanding low-dose Pb effects on children, the impact of early life exposures on adult health is a more recent field of study. Adults exposed to high levels of Pb during childhood can have decreased brain volume, especially in the prefrontal cortex (Stewart *et al.*, 2006; Cleveland *et al.*, 2008; Cecil *et al.*, 2008; Brubaker *et al.*, 2010). There is limited evidence that prenatal exposure to BLLs < 10 μg $dl^{-1}$ may result in later-life antisocial problems and criminal behavior (Needleman, 2004; Hwang, 2007; Nevin, 2007; Park *et al.*, 2008; NTP, 2012), an impact that has been consistently observed across the globe (Nevin, 2007; Mielke and Zahran, 2012). Lead exposure can also damage nerves associated with sense organs and bodily control, which may lead to later-life neurodegenerative diseases such as amyotrophic lateral sclerosis (ALS), Parkinson's disease and Alzheimer's disease (AD). Early life Pb exposure can result in later-life overexpression of AD-related proteins and histopathology, amyloidogenic and tauogenic proteins, and epigenetic changes (Basha *et al.*, 2005; Wu *et al.*, 2008; Bihaqi and Zawia, 2013; Bihaqi *et al.*, 2014a, b). The relationship between early life Pb exposure and schizophrenia is another relatively new field of research. Researchers studying subjects who were enrolled in the Oakland, California Childhood Health and Development Study (1959–1966) found a potential association between prenatal Pb exposure from archived maternal serum samples and the development of schizophrenia in later life (Opler *et al.*, 2008).

### 25.6.2 Reproductive and developmental health effects

Lead can impact the reproductive systems of both sexes. In occupationally exposed males, acute exposures over 40 μg $dl^{-1}$ and chronic exposures over 25 μg $dl^{-1}$ may reduce fertility and increase risks of reduced fetal growth and spontaneous abortions (Bellinger, 2005, 2011). Given that the half-life of Pb in bones is 20 years, maternal exposures that occurred prior to pregnancy can put a fetus at risk years after the mother's exposure (Bellinger, 2005). In females, BLLs < 1 μg $dl^{-1}$ during the second trimester of pregnancy have been associated with higher risks of low birthweight and pre-term birth (Rabito *et al.*, 2014); levels > 5 μg $dl^{-1}$ have been associated with an increased risk of spontaneous abortion; and levels < 10 μg $dl^{-1}$ have been associated with pregnancy hypertension and adverse neurobehavioural development in offspring (Bellinger, 2005, 2011).

Lead exposures during the gestation and infant periods, at levels that are common among the general US population (< 5μg $dl^{-1}$), may cause permanent damage to the brain which may not be evident until later in the child's life (Needleman *et al.*, 1990; Bellinger, 2008, 2011; Saleh *et al.*, 2009). In 2010, the CDC revised its guidelines for the management of Pb exposure during pregnancy so that maternal BLLs > 5μg $dl^{-1}$ should trigger interventions and follow-up testing (CDC, 2010). Maternal BLLs during the first trimester of pregnancy have been inversely associated with an index of the infant's mental development at 24 months (Hu *et al.*, 2006) and BLLs during the second trimester have been inversely associated with gestation length (Bellinger, 2011).

One of the most interesting findings to date has been the consistent observation of steeper dose–response curves (greater effects) for some neurodevelopmental outcomes at lower BLLs (< 10 μg $dl^{-1}$), including larger IQ deficits at BLL

< 10 μg $dl^{-1}$, even after adjusting for covariates (CDC, 2005; Lanphear *et al.*, 2005; NTP, 2012). One explanation for these results is that only very low levels of Pb are needed to disrupt different biological mechanisms (ATSDR, 2007). This has been observed in relation to lead's ability to displace calcium at very low levels (picomolar concentrations), which puts into effect a cascade of molecular and cellular dysfunction. Higher levels of Pb subsequently may add little additional cognitive impairment, until doses are high enough to trigger encephalopathy or frank mental disability (ATSDR, 2007). The adverse impacts of Pb on brain function will depend on the stage of neurodevelopment when the exposure occurred (fetal versus paediatric). Genetic polymorphisms, Pb dose, gender and even a child's rearing environment can impact the ultimate health outcomes. All of these and other factors have contributed to the difficulty in identifying a specific 'behavioral signature' of lead, which has complicated research interpretations (ATSDR, 2007).

### 25.6.3 Other health effects

This chapter has focused on what is currently known about the primary health effects of low-dose Pb exposures, but this is merely the tip of the iceberg with respect to lead's impacts on health. BLLs < 10 μg $dl^{-1}$ have also been associated with adverse renal, cardiovascular and immune health effects (NTP, 2012) (Table 25.1). According to the NTP, sufficient evidence exists that adult BLLs < 5 μg $dl^{-1}$ are associated with decreased renal function; and adult BLLs < 10 μg $dL^{-1}$ are associated with increased blood pressure, hypertension, and even all-cause higher mortality (Bellinger, 2011; NTP, 2012). There is limited evidence to suggest that childhood BLLs < 10 μg $dl^{-1}$ are associated with adverse immune effects (NTP, 2012). While the evidence of the carcinogenicity of Pb is inconclusive, it has been associated with kidney tumour in animals (NTP, 2012); and Pb and Pb compounds are classified as 'reasonably anticipated to be human carcinogens' by the US Department of Health and Human Services (ATSDR, 2007). The investigation of lead's impacts on health is an evolving field as new technologies improve our ability to detect lower levels of lead and evaluate emerging subclinical impacts.

## 25.7 Conclusions

While Pb is one of the most studied environmental toxicants, regulatory protections and public health practices have not kept pace with scientific understanding of the lifetime health impacts of exposure to low-dose Pb. The underlying realization that there are no safe levels of Pb exposure, and that its effects may be subclinical and unnoticed until years after exposure, spurred a paradigm shift in our concept of poisoning. The evidence reviewed above supports an emerging consensus that exposure prevention should be paramount, either through regulation or through public health intervention. The Council on Environmental Health (COEH) of the American Academy of Pediatrics (AAP) issued a policy statement in 2016, which reiterated the importance of exposure prevention measures (COEH, 2016). Public health strategies can include: (i) education of the public about Pb sources, exposure routes and evidence-based prevention measures; (ii) screening, BLL testing and interventions for those most at risk of exposure, or those most vulnerable to lead's impact; (iii) education exercises to ameliorate the neurocognitive impact of Pb on the developing brain; (iv) dietary approaches that promote Pb absorption and removal, or subdue or prevent Pb toxicity through antioxidants; and (v) policy measures that might further ban Pb or reduce exposures (Hsu and Guo, 2002; Flora *et al.*, 2012). It is essential to recognize that the current practice of monitoring children's BLL to determine environmental Pb exposures is secondary prevention. Primary prevention requires focusing on reducing risks from air, water, soil, dust, and food sources to curtail children's Pb intakes in the first place.

Regulatory intervention to reduce Pb releases and exposures formed the basis of one of the greatest public health success stories of the last century, but more protective measures are still needed. The levels of Pb found in people today, while they have declined considerably over the past few decades, are still orders of magnitude higher that they were in pre-industrial societies, highlighting the fact that efforts to address Pb exposures and health impacts must continue (Merrill *et al.*, 2007). For example, in the decades since the US Occupational Safety and Health Administration (OSHA) promulgated its 1993 standard for adult blood Pb exposures in the workplace (40 μg $dl^{-1}$), evidence on the impact of adult BLLs less than 18 μg $dl^{-1}$ on cognitive function have been reported (Shih *et al.*, 2007). Another issue is the ongoing use of Pb in avgas (aviation gasoline, aviation spirit) to fuel small airplanes powered by reciprocal piston engines. According to the EPA, avgas contributes over 60% of the lead aerosols in the USA and is associated with excessive lead exposure around general aviation airports. The Pb in avgas may also contaminate the more commonly used 'unleaded' mogas (motor gasoline), as the same pipelines are used to transport both products. Finally, the recent childhood Pb poisonings in Flint, Michigan, have raised public awareness about the lack of health protectiveness of our drinking-water regulations for Pb. Flint's increase in waterborne lead resulted from a series of factors, including corroding drinking-water infrastructure, deficiencies in utility implementation of regulations and weaknesses in government oversight of regulations (Katner *et al.*, 2016). All of these factors acted together to cause in Flint an increase in the percentage of tested children with blood Pb levels of 5 μg $dl^{-1}$ or greater (the CDC reference value) in children 5 years of age and younger (Hanna-Attisha *et al.*, 2016), and an 'environmental injustice' (Flint Water Advisory Task Force, 2016). Yet to this day, the regulatory action level for Pb in drinking water remains 15 μg $l^{-1}$, despite the fact that new evidence of the effects of cumulative exposure to low-dose Pb has prompted the AAP to recommend a water action guideline of 1 μg $l^{-1}$ for schools (COEH, 2016). Flint was a reminder that lead exposure poses a significant public health risk, not just for individual consumers in terms of personal health, but also for entire communities, due to the associated social and economic impacts.

Enough evidence has accumulated to support the argument that all non-essential uses of Pb should be banned and that allowable levels of Pb in the workplace and environment, including the air, drinking water, soil, indoor environment, food and consumer products, should be further reduced (Lanphear *et al.*, 2005). The cost and consequences of ignoring the evidence and prior warnings will be paid for by generations to come.

## References

ATSDR (2006) *Hazardous Substance Release and Health Effects Database (HazDat)*. Agency for Toxic Substances and Disease Registry, US Department of Health and Human Services (DHHS), Atlanta, Georgia. Available at: http://www.atsdr.cdc.gov/hazdat.html (accessed 18 January 2018).

ATSDR (2007) *Toxicological Profile for Lead*. Agency for Toxic Substances and Disease Registry, US Department of Health and Human Services (DHHS), Atlanta, Georgia. Available at: https://www.atsdr.cdc.gov/toxprofiles/tp13.pdf (accessed 12 January 2018).

Barbosa, F. Jr, Tanus-Santos, J.E., Gerlach, R.F. and Parsons, P.J. (2005) A critical review of biomarkers used for monitoring human exposure to lead: advantages, limitations, and future needs. *Environmental Health Perspectives* 113, 1669–1674.

Basha, M.R., Wei, W., Bakheet, N., Benitez, H.K., Siddiqi, Y.W. *et al.* (2005) The fetal basis of amyloidogenesis: exposure to lead and latent overexpression of amyloid precursor protein and beta-amyloid in the aging brain. *Journal of Neuroscience* 25, 823–829.

Bellinger, D.C. (2004) Lead. *Pediatrics* 113, 1016–1022.

Bellinger, D.C. (2005) Teratogen update: lead and pregnancy. *Birth Defects Research* (Part A) 73, 409–420.

Bellinger, D.C. (2008) Very low lead exposures and children's neurodevelopment. *Current Opinions in Pediatrics* 20, 172–177.

Bellinger, D.C. (2011) The protean toxicities of lead: new chapters in a familiar story. *International Journal of Environmental Research and Public Health* 8, 2593–2628.

Bihaqi, S.W. and Zawia, N.H. (2013) Enhanced taupathy and AD-like pathology in aged primate brains decades after infantile exposure to lead (Pb). *Neurotoxicology* 39, 95–101.

Bihaqi, S.W., Bahmani, A., Adem, A. and Zawia N.H. (2014a) Infantile postnatal exposure to lead (Pb) enhances Tau expression in the cerebral cortex of aged mice: relevance to AD. *Neurotoxicology* 44,114–120.

Bihaqi, S.W., Bahmani, A., Subaiea, G.M. and Zawia N.H. (2014b) Infantile exposure to lead and late-age cognitive decline: relevance to AD. *Alzheimers Dementia* 10, 187–195.

Brochin, R., Leone, S., Phillips, D., Shepard, N., Zisa, D. and Angerio, A. (2008) The cellular effect of lead poisoning and its clinical picture. *Georgetown University Jcurnal of Health Sciences* 5(2). Available at:

https://blogs.commons.georgetown.edu/journal-of-health-sciences/files/Brochin-et-al-2008-GUJHS-5-2-The-Cellular-Effect-of-Lead-Poisoning-and-Its-Clinical-Picture.pdf (accessed 12 January, 2018).

Brubaker, C.J., Dietrich, K.N., Lanphear, B.P. and Cecil, K.M. (2010) The influence of age of lead exposure on adult gray matter volume. *Neurotoxicology* 31, 259–266.

Cecil, K.M., Brubaker, C.J., Adler, C.M., Dietrich,K.N., Altaye, M. *et al.* (2008) decreased brain volume in adults with childhood lead exposure. *PLoS Medicine* 5, e112.

CDC (Centers for Disease Control and Prevention) (2000) Elevated BLLs among internationally adopted children – United States, 1998. *Morbidity and Mortality Weekly Report (MMWR)* 49, 97–100.

CDC (2004) *A Review of Evidence of Health Effects of BLLs < 10 μg dl*$^{-1}$ *in Children*. Centers for Disease Control and Prevention. Prepared by the Advisory Committee on Childhood Lead Poisoning Prevention (ACCLPP), National Center for Environmental Health, US Department of Health and Human Services, Atlanta, Georgia. Available at: https://www.cdc.gov/nceh/lead/acclpp/meetingminutes/lessthan10mtgmar04.pdf (accessed 11 January 2018).

CDC (2005) *Preventing Lead Poisoning in Young Children*. Centers for Disease Control and Prevention. National Center for Environmental Health, US Department of Health and Human Services, Atlanta, Georgia. Available at: https://www.cdc.gov/nceh/lead/ (accessed 10 January 2018).

CDC (2010) *Guidelines for the Identification and Management of Lead Exposure in Pregnant and Lactating Women*. Centers for Disease Control and Prevention. National Center for Environmental Health, US Department of Health and Human Services, Atlanta, Georgia.

CDC (2012) *Low Level Lead Exposure Harms Children: a Renewed Call for Primary Prevention*. Centers for Disease Control and Prevention. Prepared by the Advisory Committee on Childhood Lead Poisoning Prevention (ACCLPP), National Center for Environmental Health, US Department of Health and Human Services, Atlanta, Georgia. Available at: https://www.cdc.gov/nceh/lead/acclpp/final_document_030712.pdf (accessed 10 January 2018).

Cleveland, L.M., Minter, M.L., Cobb, K.A., Scott, A.A. and German, V.F. (2008) Lead hazards for pregnant women and children. Part 1: Immigrants and the poor shoulder most of the burden of lead exposure in this country. *American Journal of Nursing* 108, 40–49.

COEH (2016) Prevention of childhood lead toxicity. *Pediatrics* 138(1), e20161493. doi: 10.1542/peds.2016–1493.

Cory-Slechta, D.A. (1996) Legacy of lead exposure: consequences for the central nervous system. *Otolaryngology Head and Neck Surgery* 114, 224–226.

Dart, R.C, Hurlbut, K.M. and Boyer-Hassen, L.V. (2004) Lead. In Dart, R.C. (ed.) *Medical Toxicology*, 3rd edn. Lippincot Williams and Wilkins, Philadelphia, Pennsylvania, pp. 1423–1431.

Eid, A., Bihaqi, S.W., Reneban, W.E. and Zawia, N.H. (2016) Developmental lead exposure and lifespan alterations in epigenetic regulators and their correspondence to biomarkers of Alzheimer's disease. *Alzheimer's and Dementia: Diagnosis, Assessment and Disease Monitoring* 2, 123–131.

Eisler, R. (1988) *Lead Hazards to Fish, Wildlife, and Invertebrates: a Synoptic Review*. Biological Report 85 (1.14). US Department of the Interior, Fish and Wildlife Service, Laurel, Maryland.

Fletcher, A.M., Gelberg, K.H. and Marshall, E.G. (1999) Reasons for testing and exposure sources among women of childbearing age with moderate BLLs. *Journal of Community Health* 24, 215–227.

Flint Water Advisory Task Force (2016) *Final Report. Mar 21, 2016*. Office of the Governor, State of Michigan, Lansing, Michigan. Available at: https://www.michigan.gov/documents/snyder/FWATF_FINAL_REPORT_21March2016_517805_7.pdf (accessed 25 April 2019).

Flora, S.J.S., Flora, G. and Saxena, G. (2006) Environmental occurrence, health effects and management of lead poisoning. In: José, S.C. and José, S. (eds) *Lead*. Elsevier Science, Amsterdam, pp. 158–228.

Flora, G., Gupta, D. and Tiwari, A. (2012). Toxicity of lead: a review with recent updates. *Interdisciplinary Toxicology* 5(2), 47–58.

Garza, A., Vega, R. and Soto, E. (2006) Cellular mechanisms of lead neurotoxicity. *Medical Science Monitor* 12, RA57–65.

Gulson, B.L., Mizon, K.J., Korsch, M.J., Palmer, J.M. and Donnelly, J.B. (2003) Mobilization of lead from human bone tissue during pregnancy and lactation-a summary of long-term research. *Science of the Total Environment* 303,79–104.

Hanna-Attisha, M., LaChance, J., Sadler, R.C. and Schnepp, A.C. (2016) Elevated blood lead levels in children associated with the Flint Drinking Water Crisis: a spatial analysis of risk and public health response. *American Journal of Public Health* 106, 283–290.

Hernberg, S. (2000) Lead poisoning in a historical perspective. *American Journal of Industrial Medicine* 38, 244–254.

Hertz-Picciotto, I., Schramm, M., Watt-Morse, M., Chantala, K., Anderson, J. and Osterloh, J. (2000). Patterns and determinants of blood lead during pregnancy. *American Journal of Epidemiology* 152, 829–837.
Hsu, P.C. and Guo, Y.L. (2002) Antioxidant nutrients and lead toxicity. *Toxicology* 180, 33–44.
Hu, H., Téllez-Rojo, M.M., Bellinger, D., Smith, D., Ettinger, A.S. *et al.* (2006) Fetal lead exposure at each stage of pregnancy as a predictor of infant mental development. *Environmental Health Perspectives* 114, 1730–1735.
Hwang, L. (2007) Environmental stressors and violence: lead and polychlorinated biphenyls. *Reviews on Environmental Health* 22(4), 313–328.
IARC (2006) *Inorganic and Organic Lead Compounds*. IARC Monographs on the Evaluation of Carcinogenic Risks to Humans, Vol. 87. International Agency for Research on Cancer, World Health Organization, Geneva.
Katner, A., Pieper, K.J., Lambrinidou, Y., Brown, K., Hu, C., Mielke, H. and Edwards, M. (2016) Weaknesses in federal drinking water regulations and public health policies that impede lead poisoning prevention and environmental justice. *Journal of Environmental Justice* 9, 109–117.
Kitman, J.L. (2000) The secret history of lead. Available at: https://www.thenation.com/article/secret-history-lead/ (accessed 23 February 2018).
Laidlaw, M.A.S., Mielke, H.W., Filippelli, G.M., Johnson, D.L. and Gonzales, C.R. (2005) Seasonality and children's blood lead data from Indianapolis, Indiana, Syracuse, New York, and New Orleans, Louisiana (USA). *Environmental Health Perspectives* 113(6), 793–780.
Laidlaw, M.A.S., Zahran, S., Mielke, H.W., Taylor, M.P. and Filippelli, G.M. (2012). Re-suspension of lead contaminated urban soil as a dominant source of atmospheric lead in Birmingham, Chicago, Detroit and Pittsburgh, USA. *Atmospheric Environment* 49, 302–310.
Laidlaw, M.A.S., Filippelli, G.M., Sadler, R.C., Gonzales, C.R., Ball, A.S. and Mielke, H.W. (2016) Children's blood lead seasonality in Flint, Michigan (USA), and soil-sourced lead hazard risks. *International Journal of Environmental Research and Public Health* 13, 358. doi: 10.3390/ijerph13040358.
Landrigan, P.J., Fuller, B.E., Acosta, N.J.R., Adevi, O., Basu, N.B. *et al.* (2017) The *Lancet* Commission on pollution and health. *The Lancet* 391(10119), 462–512. Available at: http://www.thelancet.com/pdfs/journals/lancet/PIIS0140-6736(17)32345-0.pdf (accessed 9 Jan 9, 2018). doi: 10.1016/S0140-6736(17)32345-0.
Lanphear, B.P., Hornung, R., Khoury, J., Yolton, K., Baghurst, P. *et al.* (2005) Low-level environmental lead exposure and children's intellectual function: an international pooled analysis. *Environmental Health Perspectives* 113, 894–899.
Levin, R., Brown, M.J., Kashtock, M.E., Jacobs, D.E., Whelan, E.A. *et al.* (2008) Lead exposures in US children, 2008: implications for prevention. *Environmental Health Perspectives* 116, 1285–1293.
Lidsky, T.I. and Schneider, J.S. (2003) Lead neurotoxicity in children: basic mechanisms and clinical correlates. *Brain* 126, 5–19.
Liu, J., Goyer, R.A. and Waalkes, M. (2007) Toxic effects of metals. In: Klaassen, C.D. (ed.) *Casarett and Doull's Toxicology: The Basic Science of Poisons*, 7th edn. McGraw Hill, New York, pp. 931–980.
Manton, W.I., Angle, C.R., Stanek, K.L., Kuntzelman, D., Reese, Y.R. and Kuehnemann, T.J. (2003) Release of lead from bone in pregnancy and lactation. *Environmental Research* 92, 139–151.
Marino, P.E., Landrigan, P.J., Graef, J., Nussbaum, A., Bayan, G., Boch, K. and Boch, S. (1990) A case report of lead paint poisoning during renovation of a Victorian farmhouse. *American Journal of Public Health* 80, 1183–1185.
Merrill, J.C., Morton, J.J.P. and Soileau, S.D. (2007) Metals. In: Hayes, A.W. (ed.) *Principles and Methods of Toxicology*, 5th edn. CRC Press, New York, pp. 841–896.
Mielke, H.W. and Reagan, P.L. (1998) Soil is an important pathway of human lead exposure. *Environmental Health Perspectives* 106 (Suppl. 1), 217–229.
Mielke, H.W. and Zahran, S. (2012) The urban rise and fall of air lead (Pb) and the latent surge and retreat of societal violence. *Environment International* 43, 48–55.
Mielke, H.W., Anderson, J.C., Berry, K.J., Mielke, P.W. Jr, Chaney, R.L. and Leech, M. (1983) Lead concentrations in inner-city soils as a factor in the child lead problem. *American Journal of Public Health* 73, 1366–1369.
Mielke, H.W., Burroughs, S., Wade, R., Yarrow, T. and Mielke P.W. (1984) Urban lead in Minnesota: soil transect results in four cities. *Journal of Minnesota Academy of Science* 50, 19–24.
Mielke, H.W., Adams, J.L., Reagan, P.L., and Mielke, P.W. Jr (1989) Soil-dust lead and childhood lead exposure as a function of city size and community traffic flow: the case for lead abatement in Minnesota. *Environmental Geochemistry and Health: (Supplement) Lead in Soil* 9, 253–271.
Mielke, H.W., Gonzales, C.R. and Powell, E.T. (2019) Curtailing lead aerosols: effect of primary prevention on declining soil lead and children's blood lead in Metropolitan New Orleans. *International Journal of Environmental Research and Public Health* 16, 2068.

Mielke, H.W., Laidlaw, M.A.S. and Gonzales, C.R. (2019) Characterization of lead (Pb) from traffic in 90 USA urbanized areas. Review of urban lead dust and health. *Environment International* 37, 248–257.
NTP (2012) Health effects of low-level lead. National Toxicology Program, US Department of Health and Human Services. Available at: https://ntp.niehs.nih.gov/ntp/ohat/lead/final/monographhealtheffectslowlevellead_newissn_508.pdf (accessed 12 January 2018).
Needleman, H.L. (2000) The removal of lead from gasoline: Historical and personal reflections. *Environmental Research* (Section A) 84, 20–35.
Needleman, H.L. (2004) Lead poisoning. *Annual Review of Medicine* 55, 209–222.
Needleman, H.L., Gunnoe, C., Leviton, A., Reed, R., Peresie, H., Maher, C., and Barrett, P. (1979) Deficits in psychologic and classroom performance of children with elevated dentine lead levels. *The New England Journal of Medicine* 300, 689–695.
Needleman, H.L., Schell, A., Bellinger, D., Leviton, A. and Allred, E.N. (1990) The long term effects of exposure to low doses of lead in childhood – an 11-year follow-up report. *The New England Journal of Medicine* 322, 83–88.
Nevin, R. (2007) Understanding international crime trends: the legacy of preschool lead exposure. *Environmental Research* 104, 315–336.
Nussbaumer-Streit, B., Yeoh, B., Griebler, U., Pfadenhauer, L.M., Busert, L.K. *et al.* (2016) Household interventions for preventing domestic lead exposure in children. *Cochrane Database of Systematic Reviews* 10, Article No. CD006047.
Opler, M.G., Buka, S.L., Groeger, J., McKeague, I., Wei, C. *et al.* (2008) Prenatal exposure to lead, delta-aminolevulinic acid, and schizophrenia: further evidence. *Environmental Health Perspectives* 116,1586–1590.
Park, S.K., O'Neill, M.S., Vokonas, P.S., Sparrow, D., Wright, R.O. *et al.* (2008) Air pollution and heart rate variability: effect modification by chronic lead exposure. *Epidemiology* 19, 111–120.
Patrick, L. (2006) Lead toxicity part II: the role of free radical damage and the use of antioxidants in the pathology and treatment of lead toxicity. *Alternative Medicine Review* 11, 114–127.
Patterson, C.C. (1965) Contaminated and natural lead environments of man. *Archives of Environmental Health* 11, 344–360.
Pearce, J.M. (2007) Burton's line in lead poisoning. *European Neurology* 57(2), 118–119.
Pearson, H.A. and Schonfeld, D.J. (2003) Lead. In: Rudolph, C.D., Rudolph, A.M., Hostetter, M.K., Lister, G. and Siegel, N.J. (eds) *Rudolph's Pediatrics*, 21st edn. McGraw Hill, New York, pp. 1016–1056.
Rabito, F.A., Kocak, M., Werthmann, D.W., Tylacsky, F.A., Palmer, C.D. and Parsons, P.J. (2014) Changes in low levels of lead over the course of pregnancy and the association with birth outcomes. *Reproductive Toxicology* 50, 138–144.
Saleh, H.A., El-Aziz, G.A., El-Fark, M.M. and El-Gohary, M. (2009) Effect of maternal lead exposure on craniofacial ossification in rat fetuses and the role of antioxidant therapy. *Anatomia, Histologia and Embryologia* 38, 392–399.
Sanders, T., Liu, Y., Buchner, V. and Tchounwou, P.B. (2009) Neurotoxic effects and biomarkers of lead exposure: A review. *Research in Environmental Health* 24, 15–45.
Senut, M., Cingolani, P., Sen, A., Kruger, A., Shaik, A. *et al.* (2012) Epigenetics of early-life exposure and effects on brain development. *Epigenomics* 4(6), 665–674. doi: 10.2217/epi.12.58.
Shih, R.A., Hu, H., Weisskopf, M.G. and Schwartz, B.S. (2007) Cumulative lead dose and cognitive function in adults: a review of studies that measured both blood lead and bone lead. *Environmental Health Perspectives* 115, 483–492.
Silbergeld, E.K., Sauk, J., Somerman, M., Todd, A., McNeill, F. *et al.* (1993) Lead in bone: storage site, exposure source, and target organ. *Neurotoxicology* 14, 225–236.
Stewart, W.F., Schwartz, B.S., Davatzikos, C., Shen, D., Liu, D. *et al.* (2006) Past adult lead exposure is linked to neurodegeneration measured by brain MRI. *Neurology* 66, 1476–1484.
Tatsumoto, M. and Patterson, C.C. (1963). Concentrations of common lead in some Atlantic and Mediterranean waters and in snow. *Nature* 199, 350–352.
WHO (2010) *Childhood Lead Poisoning*. World Health Organization, Geneva, Switzerland. Available at: http://www.who.int/ceh/publications/leadguidance.pdf (accessed 2 January 2018).
Wu, J., Basha, M.R., Brock, B., D.P., Cardozo-Pelaez, F., McPherson, C.A. *et al.* (2008) Alzheimer's disease (AD)-like pathology in aged monkeys after infantile exposure to environmental metal lead (Pb): evidence for a developmental origin and environmental link for AD. *Journal of Neuroscience* 28, 3–9.
Zahran, S., Laidlaw, M.A.S., McElmurry, S.P., Filippelli, G.M. and Taylor, M. (2013) Linking source and effect: resuspended soil lead, air lead and children's blood lead levels in Detroit, Michigan. *Environmental Science and Technology* 47, 2839–284.

# 26 Cadmium I. Exposure and Human Health Effects: an Overview

**A. Åkesson* and M. Kippler**
*Institute of Environmental Medicine, Karolinska Institutet, Stockholm, Sweden*

## 26.1 Abstract

Cadmium is a metal that has been extensively used for a wide variety of industrial applications in the past, but more recently the growing awareness of the adverse health effects has led to restrictions and ban of use in several applications. Cadmium is emitted into the environment via both natural and anthropogenic activities, leading to a wide dispersion in the environment and contamination of soil, including arable land, in many areas of the world. This is of concern, as cadmium is easily taken up by vegetable crops (e.g. rice, wheat, vegetables and potatoes) as well as tobacco leaves. Accordingly, today food is considered to be the major source of cadmium exposure for the majority of the world's population, while smokers are additionally exposed. In populations with high rice consumption, this is usually the main source of cadmium, while in other populations the main dietary sources are usually cereals, vegetables and root vegetables. Fortunately, only a small percentage of the cadmium intake is absorbed in the gastrointestinal tract, but it is increased at low iron stores. In blood, cadmium is mainly incorporated into erythrocytes and a smaller fraction is bound to metallothionein in plasma. Because of the small size of the cadmium–metallothionein complex, it is filtered through the glomeruli, making urine the primary route of excretion. However, it is also readily reabsorbed in the proximal tubular cells, leading to accumulation in the renal cortex where it has a biological half-life of 10–45 years. The cadmium concentration in urine is a marker of long-term exposure, whereas the concentration in blood is a better reflection of more recent exposure. The current risk assessments of cadmium have considered tubular damage as the critical effect. Besides the renal effect, cadmium exposure has also been associated with cardiovascular disease, osteoporosis, impaired fetal and child development, cancer and mortality, which are reviewed in this and the subsequent chapter.

## 26.2 Introduction

Exposure to cadmium was long recognized as a health problem with inhalation of fumes or dust in occupational settings and in circumstances of exceptional environmental pollution from mining activities or industries, such as in the case of the itai-itai ('ouch-ouch') disease, which afflicted mainly older women in certain endemic areas of Japan. The industrial use of cadmium progressed during the first part of the 20th

* E-mail address: Agneta.Akesson@ki.se

century. Cadmium has specific properties that made it suitable for a wide variety of industrial applications. Excellent corrosion resistance, low melting temperature, high ductility, high thermal and electrical conductivity are properties that led to its use as a stabilizer, colour pigment, or cathode and in corrosion protection. The principal global use in 2007 was in rechargeable nickel–cadmium batteries. Growing awareness of the adverse health effects has led to restrictions and ban of use in several applications.

Cadmium is emitted to the environment as a result of both natural and anthropogenic activities. As cadmium is a natural intrinsic part of the earth's crust, volcanic activity, weathering of rocks, sea spray and mobilization of previously deposited cadmium all lead to environmental dispersion. Anthropogenic sources include industrial emissions (e.g. mining and smelting), pollution by use of cadmium-containing fertilizers, combustion of fossil fuels, waste incineration and releases from tailings and landfills. The dispersion of cadmium into the environment and widespread contamination of soil, especially on arable land, in many areas of the world is of concern, as cadmium is easily taken up by food crops, such as rice, wheat, vegetables and potatoes, as well as by tobacco leaves, for example. In fact, it has been emphasized that the exposure in many areas is high enough to be of importance to human health (EFSA, 2009).

## 26.3 Exposure, Toxicokinetics and Biomarkers of Exposure

Occupational cadmium exposures have generally decreased since the 1970s. Occupations in which exposure may occur includes cadmium production and refining, nickel–cadmium battery manufacturing, cadmium pigment manufacturing, cadmium alloy production, mechanical plating and zinc smelting. Today food is the major source of cadmium exposure for the majority of the world's population, while smokers are additionally exposed. Rice is usually the main source of cadmium in countries with high rice consumption but in other areas, such as Europe and the USA, cereals and vegetables are the main contributors to dietary cadmium exposure. Offal, shellfish, certain seeds and cacao can contain high levels but the consumption of these food items is generally low in most populations and, consequently, their relative contribution to the dietary exposure is less. Drinking water usually contains very low concentrations and has therefore a limited impact on dietary cadmium exposure. Considered over a lifetime, regular tobacco smoking will contribute to a comparable amount of cadmium as the dietary exposure (Barregard *et al.*, 2010). Ambient air is considered to contribute only marginally to human exposure, while in polluted areas household dust can make a significant contribution (Hogervorst *et al.*, 2007).

The absorption of cadmium from food in the gastrointestinal tract is small, but it may vary depending on the individual's nutritional status (for example, increased uptake at low iron stores or iron deficiency) and on other components present in food, such as fibre content (Berglund *et al.*, 1994). Transport into the mucosal cell occurs via the divalent metal transporter 1 (DMT-1), most likely in a competitive manner with iron. Once cadmium has been absorbed from the digestive tract, it is initially bound to albumin in blood plasma followed by incorporation into the erythrocytes, where the majority of circulating cadmium is bound (Nordberg *et al.*, 2015). Thereafter, the albumin-bound cadmium is transported to the liver, where it undergoes degradation with the release of cadmium, which in turn can either form a complex with glutathione and be excreted via bile or can induce and bind to metallothionein (MT) and be stored in the liver or be released to plasma (Klaassen *et al.*, 1999). Because of the small size of MT, the cadmium–MT complex in plasma is readily filtered through the glomeruli; therefore, cadmium is mainly eliminated via the urine. However, the small size of the cadmium–MT complex also enables reabsorption by the proximal tubular cells, leading to accumulation in the kidney cortex (Nordberg *et al.*, 2015). Cadmium is efficiently retained in the kidney with a biological half-life of 10–45 years (Fransson *et al.*, 2014). In healthy environmentally exposed subjects, assuming a linear relationship, a kidney cadmium concentration of 25 $\mu g\ g^{-1}$ corresponded to an overnight urinary cadmium concentration of 0.42 $\mu g\ g^{-1}$ creatinine (Åkerstrom *et al.*, 2013). Urinary cadmium is a good biomarker of lifelong kidney accumulation and reflection of the total body burden. As the majority of the circulating cadmium is bound to erythrocytes, the erythrocyte fraction or whole blood are also good biomarkers of exposure with a better reflection of the more recent exposure than urinary cadmium (Liang *et al.* 2012). Cadmium concentration in hair has also been widely used a biomarker of exposure, though it has been concluded that it does not appear to reflect the internal dose (Skroder *et al.*, 2017). Increasing availability of data on cadmium concentrations in various foods led to the construction of extensive food-cadmium databases. This has opened up the possibility of estimating the daily exposure to cadmium via the diet and assessing the association with various health outcomes using large population studies that include information on food intake. Although it might be possible to rank dietary cadmium into low and high exposure, most validation studies do not show an overall good agreement between estimated dietary intake and biomarkers of exposure.

In contrast to most environmental contaminants, women generally have higher cadmium concentrations in blood, urine and kidney as compared with men, likely attributed to the higher prevalence of depleted body iron stores and iron deficiency in women.

## 26.4 Adverse Health Effects

This chapter focuses on the main adverse health effects related to cadmium exposure reported in the scientific literature, covering kidney, bone, fetal and child health and cancer. Cardiovascular effects are presented in Chapter 27 (Cadmium II). We have not covered health effects where results have been inconsistent or where causality has not been substantiated, such as diabetes (Borne *et al.*, 2014). For effect on the lungs, readers are referred to a recent review (Ganguly *et al.*, 2018). Furthermore, the main emphasis is on effects that may occur at low to moderate long-term exposures, which are prevalent worldwide, but high cadmium exposure remains a problem during some circumstances or in certain areas of the world.

### 26.4.1 Kidney damage

Tubular damage is the critical effect of cadmium exposure, defined as the first adverse effect that occurs as the cadmium dose increases. Long-term cadmium exposure resulting in a urinary cadmium concentration > 4 μg g$^{-1}$ creatinine and/or a blood cadmium concentration > 4 μg l$^{-1}$ has been shown to impair renal tubular reabsorption, determined by increased amount of low-molecular-weight (LMW) proteins such as β2-microglobulin (B2M), α1-microglobulin (A1M) and retinol-binding protein (RBP) (Åkesson *et al.*, 2014). In close connection to each other, the European Food and Safety Authority (EFSA) and the Joint Food and Agriculture Organization of the United Nations (FAO)/WHO Expert Committee on Food Additives (JECFA) modelled the relationship between urinary cadmium and B2M in their risk assessments (EFSA, 2009; WHO, 2011) and they identified a urinary cadmium concentration of 4–5 μg g$^{-1}$ creatinine as the point of departure for an increase in urinary B2M, but still they arrived at different tolerable intakes. EFSA established a tolerable weekly intake (TWI) of 2.5 μg kg$^{-1}$ body weight, whereas JECFA established a tolerable monthly intake of 25 μg kg$^{-1}$, which corresponds to a weekly intake of 6 μg kg$^{-1}$ body weight.

Several studies exploring associations of urinary cadmium with B2M or other LMW proteins have reported positive associations at urinary cadmium concentrations well below 4 μg g$^{-1}$ creatinine. In 2008, Bernard proposed that associations at these very low levels of urinary cadmium with LMW proteins may not be a result of cadmium toxicity (Bernard, 2008), but rather due to confounding or a result of physiological mechanisms such as competition between cadmium–MT and LMW proteins for tubular binding sites and/or varying diuresis, resulting in co-excretion of cadmium and LMW proteins. This makes it difficult to interpret associations of low-level urinary cadmium and LMW proteins. Moreover, the public health impact of cadmium-related increases in biomarkers of tubular dysfunction, especially within the normal range of these biomarkers, is unknown.

Chronic kidney disease and the development of end-stage renal disease are of major public health concern and the global prevalence of and mortality due to chronic kidney disease is rising. Studies of individuals with high cadmium exposure, as a result of serve cadmium pollution, have observed associations between urinary cadmium and mortality from renal diseases

(Nakagawa *et al.*, 2006), but whether there is a link between low-level cadmium exposure and end-stage renal disease is unclear. Limited prospective data based on 118 cases and 347 controls, exploring the association between cadmium concentrations in erythrocytes and end-stage renal disease (Sommar *et al.*, 2013), did not find any evidence that the cadmium concentration assessed at baseline was associated with increased odds of end-stage renal disease after adjusting for potential confounders.

### 26.4.2 Bone defects

The high occurrence of osteoporosis, a systemic skeletal disease characterized by low bone mass and microarchitectural deterioration of bone tissue with subsequent increase in bone fragility and fractures, is of major public health concern. Fragility fractures contribute to lower quality of life and life expectancy and high societal costs. Itai-itai disease, the most advanced form of environmental cadmium poisoning, was manifested as a severe bone disorder that resulted in multiple fractures per affected individual. The disease, first noted in about 1912, was a combination of osteomalacia and osteoporosis, together with kidney damage, and was caused by very high long-term exposure to cadmium, mainly via the intake of contaminated rice. Several decades later, in the 1990s, investigations were initiated exploring the link between cadmium and bone effects at low to moderate exposure levels, including areas with no particular pollution. As of today, a substantial number of studies have investigated the cross-sectional association of cadmium in blood, urine or diet with bone mineral density per se or with the odds of osteoporosis. The majority of these studies, but not all, demonstrate statistically significant associations between higher cadmium exposure and lower bone mineral density or odds of osteoporosis in various populations and at various exposure levels (Akesson *et al.*, 2014; Wallin *et al.*, 2016), indicating that cadmium exposure may contribute to the public health burden of osteoporosis. The fact that these associations are observed by the use of three different exposure assessment methods (urine, blood and dietary intake) is considered to reduce the likelihood that they represent spurious associations such as reverse causality or confounding. Confounding by tobacco smoking is, however, an aspect that needs particular consideration in studies addressing cadmium-associated bone effects. As mentioned above, regular smoking is a major source of cadmium that is very well reflected in the concentrations of the metal in blood and urine. Smoking behaviour is also strongly linked to bone health. It has thus been essential to rule out that the cadmium-related association with bone effects is not explained by other toxicants present in tobacco smoke.

The clinical significance of osteoporosis lies in the fractures that arise. A meta-analysis of case-control, retrospective and prospective studies based on studies published until mid-2015, resulted in a pooled relative risk of 1.30 (95% confidence interval (CI): 1.13–1.49) for fractures comparing the highest exposure category with the lowest (Cheng *et al.*, 2016). Since then, two large prospective studies have explored the association with fracture risk. In 936 older men, urinary cadmium was associated with more than a twofold increased risk of incident non-vertebral osteoporosis fractures during 9 years of follow-up (Wallin *et al.*, 2016). In contrast, in 2920 middle-aged women, blood cadmium was not associated with any type of fractures during almost 20 years of follow-up (Moberg *et al.*, 2017).

Potential mechanisms involved in explaining any cadmium-mediated bone effects are not clarified, but are proposed to differ between high and low exposures. An indirect effect via initial kidney damage and/or a direct effect on bone cells are suggested mechanisms. The indirect effect is a result of deficient reabsorption of, for example, calcium and/or a lower activation of vitamin D. However, there is growing evidence that cadmium has a direct toxic effect on bone, leading to increased bone resorption.

In conclusion, there is evidence of a relationship between low-level cadmium exposure and osteoporosis in the general population and this effect may be better suited for exploring adverse effects in the low-dose exposure range than renal tubular damage. This would indicate that this effect may be considered as an additional critical effect that should be included in future health risk assessments of cadmium.

### 26.4.3 Reproduction and early life abnormalities

The number of observational studies concerning cadmium exposure and reproduction is still limited and inconclusive (Pollack *et al.*, 2014; de Angelis *et al.*, 2017). Observational studies on prenatal and early life cadmium exposure and effects on fetal development as well as later child development have increased since the 20th century and some of the main findings are summarized below.

There is increasing evidence linking maternal cadmium exposure during pregnancy to negative pregnancy outcomes, in particular decreased size at birth. In several large birth cohort studies, maternal cadmium exposure during pregnancy, assessed by measurements in either maternal urine or blood, have been found to be inversely associated with birth weight (Kippler *et al.*, 2012; Taylor *et al.*, 2016) and with increased odds of being born small for gestational age (Johnston *et al.*, 2014). Surprisingly, several of the studies have shown that the association between prenatal cadmium exposure and birth weight is restricted to girls. In a birth cohort study, including 1616 mother–newborn pairs in rural Bangladesh (Kippler *et al.*, 2012), the mother's urinary cadmium concentration in early pregnancy was inversely associated with the daughter's birth weight (45 g; 95% CI: –82.5, 7.3), head circumference (–0.26 cm; 95% CI: –0.43, –0.088) and chest circumference (–0.24 cm; 95% CI: –0.44, –0.030), while no associations were observed in boys. Similarly, in a cohort study of 4191 mother–newborn pairs in the UK (Taylor *et al.*, 2016), maternal blood cadmium concentrations in early pregnancy were adversely associated with birth weight (–87.1 g; 95% CI: –144.8, –29.4), head circumference (–0.22 cm; 95% CI: –0.39, –0.04) and crown–heel length (–0.44 cm; 95% CI: –0.71, –0.18) exclusively in girls.

There is also some evidence suggesting that the prenatal cadmium exposure seems to continue to affect growth during childhood and also that exposure in childhood can be of importance, although the results are inconsistent. In a few small-scale prospective studies, the children's prenatal cadmium exposure, assessed via cadmium concentrations in cord blood, has been inversely associated with child height (Tian *et al.*, 2009), weight and/or head circumference at 3–9 years of age (Lin *et al.*, 2011; Delvaux *et al.*, 2014). Prospective studies that assessed the prenatal exposure via measurements in maternal urine during pregnancy did not observe any association with weight, height or growth velocity at 5 years of age (Gardner *et al.*, 2013) or with BMI z-score (body mass index standard deviation score) at 7 years of age (Agay-Shay *et al.*, 2015). However, in the study by Gardner and coworkers, cadmium concetrations in child urine were inversely associated with weight-for-age and height-for-age z-score and with growth velocity at 5 years of age. Stratification by child gender showed that the differences in attained weight and height were only apparent in girls.

The mechanisms underlying this cadmium-related developmental toxicity are unclear. As cadmium mainly accumulates in the placenta during pregnancy (Kippler *et al.*, 2010), it has been hypothesized that it may be an indirect effect via disrupted placental functions. This can, for example, result in decreased micronutrient transfer and/or alterations of the epigenetic pattern in placenta (Everson *et al.*, 2018) or cord blood (Kippler *et al.*, 2013; Vidal *et al.*, 2015). Interestingly, some of the epigenetic studies have indicated sex differences. The fact that studies have shown that the prenatal effect on growth does not persist in later childhood but that the exposure in childhood does matter for their ongoing growth suggests that the underlying mechanism of these exposures may differ. In childhood, there is evidence that the cadmium exposure may interfere with bone remodelling or growth hormones, which may translate to impaired child growth.

A systematic review including studies from January 2009 to March 2015 identified 16 studies concerning early life cadmium exposure and either neurodevelopment or neurobehavioural disorders (Sanders *et al.*, 2015). Five studies (four prospective studies and one cross-sectional study) assessed the association of different cadmium biomarkers and IQ and four of these studies identified a significant inverse association between prenatal and/or childhood cadmium exposure and IQ. Since 2015, an additional prospective cohort study, comprising 575 Greek mother–child pairs, has found elevated maternal urinary cadmium concentrations during early pregnancy to be inversely associated with cognitive function in 4-year-olds, although confounding by smoking could not be ruled out (Kippler *et al.*, 2016). The review also identified studies that assessed an association of early life cadmium exposure and different behavioural outcomes. Five studies explored associations between cadmium and attention deficit hyperactivity disorder (ADHD), or ADHD-like behaviour, and four the association between cadmium and other behavioural scales (Sanders *et al.*, 2015), but there were only two studies that reported associations with behaviour. Taken together, the results suggest that early life cadmium exposure can be detrimental for the development of children's cognitive abilities, though the epidemiological evidence is still limited. For behaviour, there is no conclusive evidence.

The mechanisms or modes of action of cadmium-induced neurotoxicity are still unclear. There is very limited evidence that cadmium can cross the blood–brain barrier once it has been fully developed (Arvidson and Tjalve, 1986). Therefore, depending on the time of exposure (i.e. prenatally or childhood), the mechanisms or modes of action may differ, having either a direct and/or indirect effect on brain development. Experimental studies have found that cadmium can induce reactive oxygen species and neurochemical changes (i.e. disruption of calcium signalling, changes in neurotransmitters) and cadmium can cause several epigenetic changes that could be of importance for neurodevelopment (Raciti and Ceccatelli, 2017).

### 26.4.4 Cancer

In their evaluation in 2012, the International Agency for Research on Cancer (IARC) concluded that there is sufficient evidence in humans for the carcinogenicity of cadmium and cadmium compounds, categorizing them into group 1 – the category used for the highest level of certainty when there is sufficient evidence of carcinogenicity in humans (IARC, 2012). Hypothesized mechanisms contributing to cadmium's carcinogenicity were considered to be indirect, since there is little evidence of a direct binding of cadmium to DNA or that a mutagenic response plays an important role (Hartwig, 2013). Oxidative stress, disturbances of DNA-repair and tumour-suppressor genes, and effects on cell proliferation are likely mechanisms. The evaluation was based on lung cancer and inhalation exposure. It was concluded that the assessment of cancer risk was constrained by the small number of cases with long-term high exposure, the inability to examine dose–response associations across studies and the inability to properly address potential confounding in all studies, such as smoking and co-exposure to other substances. It was noted that a prospective population-based study from Belgium observed associations with statistically significant increased risk of lung cancer for three different exposure indices: urinary cadmium, cadmium in soil and residence in the high-exposure area (Nawrot *et al.*, 2006). For prostate cancer, IARC summarized the evidence for cadmium to be, at best, suggestive of an association. In a more recent meta-analysis, there was no support for the suggestion that cadmium exposure was associated with increased risk of prostate cancer (Chen *et al.*, 2016).

Because it is proposed that environmental contaminants that mimic the effects of oestrogen may contribute to the high incidence of hormone-related cancers (Mallozzi *et al.*, 2017), the potential endocrine-disruptive effect of cadmium demonstrated experimentally (see next section) initiated studies exploring the association with risk of hormone-related cancers. In a systematic review and meta-analyses on urinary cadmium and breast cancer, based on eight studies (two prospective cohorts and six case-control studies) published until 2015, the summary odds ratio (OR) was 2.24 (95% CI: 1.50–3.34), excluding the two prospective studies (breast cancer mortality) due to unstable and inconsistent risk estimates (Larsson *et al.*, 2015). Later, two well conducted prospective studies using a case-cohort study design revealed no associations between urinary cadmium and incidence of breast cancer (Adams *et al.*, 2016; Eriksen *et al.*, 2017). Likewise a pooled estimate, based on six studies, for dietary cadmium intake (an exposure metric generally suffering from exposure misclassification) indicated no association with breast cancer risk (Van Maele-Fabry *et al.*, 2016).

A few studies have explored the link between cadmium and endometrial cancer. Urinary cadmium was associated with 22% increased risk of endometrial cancer with a doubling of the cadmium exposure in a population-based case-control study of Midwestern US women (McElroy *et al.*, 2017). Studies based on dietary cadmium exposure and endometrial cancer risk show inconsistent results (Åkesson *et al.*, 2008; Adams *et al.*, 2014; Eriksen *et al.*, 2014). Different mechanisms of mediations of the oestrogen-like effects have been proposed that potentially could contribute to endometrial cancer risk. Experimentally, both an oestrogen receptor-mediated proliferation, i.e. the classical oestrogen signalling (Johnson *et al.*, 2003), and a non-classical oestrogen receptor signalling have been demonstrated. Ali *et al.* (2010) observed that cadmium did not activate canonical oestrogen signalling *in vivo*, as indicated by the lack of effect on the oestrogen responsive element-dependent reporter gene expression in 14 different tissues in male and female mice. Neither did cadmium increase uterine weights, the hallmark of oestrogen activity, at environmentally relevant doses. However, they demonstrated that cadmium exposure in mice affected the height of the uterine luminal epithelium in a dose-dependent manner, and that cadmium stimulated cellular signalling pathways, such as mitogen-activated protein kinase (MAPK) signalling. This signalling could reproduce effects reminiscent of oestrogenicity, such as thickened uterine epithelium. The mitogenic kinases and cellular stress signalling were clearly affected by lower doses of cadmium than, for instance, the traditional markers of metal exposure, metallothionein 1 and metallothionein 2, representing mechanistic pathways for the suggested endocrine modulatory activities of cadmium (Ali *et al.*, 2012, 2015).

### 26.4.5 Mortality

Several studies have assessed the association between cadmium exposure and mortality, primarily from renal and cardiovascular disease (CVD) but also cancer mortality. Associations between environmental cadmium exposure or urinary cadmium and mortality as a result of renal injury have been reported from heavily cadmium-polluted areas in Japan (Nakagawa *et al.*, 2006). In a systematic review and meta-analyses on urinary cadmium and all-cause mortality, cancer mortality and CVD mortality, a total of nine cohort studies (five studies from USA, one in Belgium and three in Japan) published until March 2015 were included (Larsson and Wolk, 2016). The summary hazard ratio (HR), comparing the highest versus lowest category, was 1.44 (95% CI: 1.25–1.64) for all-cause mortality (six studies) and 1.57 (95% CI: 1.27–1.95) for CVD mortality (five studies), while the results for cancer mortality were non-significant. Restricting the analyses to six studies with a mean urinary cadmium concentration below or equal to 1 $\mu g\ g^{-1}$ creatinine, the corresponding HRs were 1.38 (95% CI: 1.17–1.63) for all-cause mortality, 1.56 (95% CI: 0.98–2.47) for cancer mortality and 1.50 (95% CI: 1.18–1.91) for CVD mortality, suggesting that even more low-level cadmium exposure appears to be associated with increased mortality. Moreover, the associations of urinary cadmium and all-cause mortality were similar in most subgroup analyses defined by gender, age and country. In a systematic review and meta-analyses on cadmium in blood including three studies published until December 2017, the highest category of blood cadmium was associated with an HR of 1.59 (95% CI: 1.24-2.04) for CVD mortality (Tinkov *et al.*, 2018).

## 26.5 Conclusions

Current risk assessment of cadmium is based on tubular damage as the critical effect. More recent data suggest that the relationship between low-level cadmium exposure and osteoporosis in the general population is better suited for grading the evidence and performing a risk assessment on adverse effects in the low-dose exposure range. There is suggestive evidence that early life cadmium exposure may be detrimental for child growth and neurodevelopment, though the epidemiological evidence is still limited. Likewise, although cadmium is classified as a human carcinogen and endocrine disruptor, the role of cadmium in hormone-related cancer requires further research. Finally, the majority of available data indicates a relationship between cadmium exposure and mortality.

## References

Adams, S.V., Quraishi, S.M., Shafer, M.M., Passarelli, M.N., Freney, E.P. *et al.* (2014) Dietary cadmium exposure and risk of breast, endometrial, and ovarian cancer in the women's health initiative. *Environmental Health Perspectives* 122, 594–600.

Adams, S.V., Shafer, M.M., Bonner, M.R., LaCroix, A.Z., Manson, J.E. *et al.* (2016) Urinary cadmium and risk of invasive breast cancer in the women's health initiative. *American Journal of Epidemiology* 183, 815–823.

Agay-Shay, K., Martinez, D., Valvi, D., Garcia-Esteban, R., Basagana, X. *et al.* (2015) Exposure to endocrine-disrupting chemicals during pregnancy and weight at 7 years of age: A multi-pollutant approach. *Environmental Health Perspectives* 123, 1030–1037.

Åkerstrom, M., Barregard, L., Lundh, T. and Sallsten, G. (2013) The relationship between cadmium in kidney and cadmium in urine and blood in an environmentally exposed population. *Toxicology and Applied Pharmacology* 268, 286–293.

Åkesson, A., Julin, B. and Wolk, A. (2008) Long-term dietary cadmium intake and postmenopausal endometrial cancer incidence: A population-based prospective cohort study. *Cancer Research* 68, 6435–6441.

Åkesson, A., Barregard, L., Bergdahl, I.A., Nordberg, G.F., Nordberg, M. and Skerfving, S. (2014) Non-renal effects and the risk assessment of environmental cadmium exposure. *Environmental Health Perspectives* 122, 431–438.

Ali, I., Penttinen-Damdimopoulou, P.E., Makela, S.I., Berglund, M., Stenius, U. *et al.* (2010) Estrogen-like effects of cadmium in vivo do not appear to be mediated via the classical estrogen receptor transcriptional pathway. *Environmental Health Perspectives* 118, 1389–1394.

Ali, I., Damdimopoulou, P., Stenius, U., Adamsson, A., Makela, S.I. *et al.* (2012) Cadmium-induced effects on cellular signaling pathways in the liver of transgenic estrogen reporter mice. *Toxicological Sciences* 127, 66–75.

Ali, I., Damdimopoulou, P., Stenius, U. and Halldin, K. (2015) Cadmium at nanomolar concentrations activates raf-mek-erk1/2 mapks signaling via egfr in human cancer cell lines. *Chemico-Biolical Interactions* 231, 44–52.

Arvidson, B. and Tjalve, H. (1986) Distribution of 109cd in the nervous system of rats after intravenous injection. *Acta Neuropathology* 69, 111–116.

Barregard, L., Fabricius-Lagging, E., Lundh, T., Molne, J., Wallin, M. *et al.* (2010) Cadmium, mercury, and lead in kidney cortex of living kidney donors: Impact of different exposure sources. *Environmental Research* 110, 47–54.

Berglund, M., Åkesson, A., Nermell, B. and Vahter, M. (1994) Intestinal absorption of dietary cadmium in women depends on body iron stores and fiber intake. *Environmental Health Perspectives* 102, 1058–1066.

Bernard, A. (2008) Cadmium & its adverse effects on human health. *Indian Journal of Medical Research* 128, 557–564.

Borne, Y., Fagerberg, B., Persson, M., Sallsten, G., Forsgard, N. *et al.* (2014) Cadmium exposure and incidence of diabetes mellitus--results from the malmo diet and cancer study. *PLoS ONE* 9, e112277.

Chen, C., Xun, P., Nishijo, M., Carter, S. and He, K. (2016) Cadmium exposure and risk of prostate cancer: A meta-analysis of cohort and case-control studies among the general and occupational populations. *Scientific Reports* 6, 25814.

Cheng, X., Niu, Y., Ding, Q., Yin, X., Huang, G., Peng, J. and Song, J. (2016) Cadmium exposure and risk of any fracture: A prisma-compliant systematic review and meta-analysis. *Medicine (Baltimore)* 95, e2932.

de Angelis, C., Galdiero, M., Pivonello, C., Salzano, C., Gianfrilli, D. *et al.* (2017) The environment and male reproduction: the effect of cadmium exposure on reproductive function and its implication in fertility. *Reproductive Toxicology* 73, 105–127.

Delvaux, I., Van Cauwenberghe, J., Den Hond, E., Schoeters, G., Govarts, E. *et al.* (2014) Prenatal exposure to environmental contaminants and body composition at age 7–9 years. *Environmental Research* 132, 24–32.

EFSA (2009) Scientific opinion of the panel on contaminants in the food chain on a request from the european commission on cadmium in food. *The EFSA Journal* 980, 1–139.

Eriksen, K.T., Halkjaer, J., Sorensen, M., Meliker, J.R., McElroy, J.A., Tjonneland, A. and Raaschou-Nielsen, O. (2014) Dietary cadmium intake and risk of breast, endometrial and ovarian cancer in danish postmenopausal women: A prospective cohort study. *PLoS ONE* 9, e100815.

Eriksen, K.T., McElroy, J.A., Harrington, J.M., Levine, K.E., Pedersen, C. *et al.* (2017) Urinary cadmium and breast cancer: a prospective Danish cohort study. *Journal of the National Cancer Institute* 109, 1–7.

Everson, T.M., Punshon, T., Jackson, B.P., Hao, K., Lambertini, L. *et al.* (2018) Cadmium-associated differential methylation throughout the placental genome: epigenome-wide association study of two U.S. Birth cohorts. *Environmental Health Perspectives* 126, 017010.

Fransson, M.N., Barregard, L., Sallsten, G., Akerstrom, M. and Johanson, G. (2014) Physiologically-based toxicokinetic model for cadmium using markov-chain Monte Carlo analysis of concentrations in blood, urine, and kidney cortex from living kidney donors. *Toxicological Sciences* 141, 365–376.

Ganguly, K., Levänen, B., Palmberg, L., Åkesson, A. and Lindén, A. (2018) Cadmium in smokers: a neglected link to lung disease? *European Respiratory Review* 27, 170122.

Gardner, R.M., Kippler, M., Tofail, F., Bottai, M., Hamadani, J. *et al.* (2013) Environmental exposure to metals and children's growth to age 5 years: a prospective cohort study. *American Journal of Epidemiology* 177, 1356–1367.

Hartwig, A. (2013) Cadmium and cancer. *Metal Ions in Life Science* 11, 491–507.

Hogervorst, J., Plusquin, M., Vangronsveld, J., Nawrot, T., Cuypers, A. *et al.* (2007) House dust as possible route of environmental exposure to cadmium and lead in the adult general population. *Environmental Research* 103, 30–37.

IARC (2012) Cadmium and cadmium compounds. In: *Arsenic, Metals, Fibres and Dusts*. IARC Monograph 100C. Working Group on the Evaluation of Carcinogenic Risk to Humans, International Agency for Research on Cancer, Lyons, France, pp. 121–145. Available at: https://monographs.iarc.fr/wp-content/uploads/2018/06/mono100c.pdf (accessed 2 May 2019).

Johnson, M.D., Kenney, N., Stoica, A., Hilakivi-Clarke, L., Singh, B. *et al.* (2003) Cadmium mimics the in vivo effects of estrogen in the uterus and mammary gland. *Nature Medicine* 9, 1081–1084.

Johnston, J.E., Valentiner, E., Maxson, P., Miranda, M.L. and Fry, R.C. (2014) Maternal cadmium levels during pregnancy associated with lower birth weight in infants in a North Carolina cohort. *PLoS ONE* 9, e109661.

Kippler, M., Hoque, A.M., Raqib, R., Ohrvik, H., Ekstrom, E.C. and Vahter, M. (2010) Accumulation of cadmium in human placenta interacts with the transport of micronutrients to the fetus. *Toxicology Letters* 192, 162–168.

Kippler, M., Tofail, F., Gardner, R., Rahman, A., Hamadani, J.D., Bottai, M. and Vahter, M. (2012) Maternal cadmium exposure during pregnancy and size at birth: a prospective cohort study. *Environmental Health Perspectives* 120, 284–289.

Kippler, M., Engstrom, K., Mlakar, S.J., Bottai, M., Ahmed, S. *et al.* (2013) Sex-specific effects of early life cadmium exposure on DNA methylation and implications for birth weight. *Epigenetics* 8, 494–503.

Kippler, M., Bottai, M., Georgiou, V., Koutra, K., Chalkiadaki, G. *et al.* (2016) Impact of prenatal exposure to cadmium on cognitive development at preschool age and the importance of selenium and iodine. *European Journal of Epidemiology* 31, 1123–1134.

Klaassen, C.D., Liu, J. and Choudhuri, S. (1999) Metallothionein: an intracellular protein to protect against cadmium toxicity. *Annual Review of Pharmacology and Toxicology* 39, 267–294.

Larsson, S.C. and Wolk, A. (2016) Urinary cadmium and mortality from all causes, cancer and cardiovascular disease in the general population: Systematic review and meta-analysis of cohort studies. *International Journal of Epidemiology* 45, 782–791.

Larsson, S.C., Orsini, N. and Wolk, A. (2015) Urinary cadmium concentration and risk of breast cancer: a systematic review and dose-response meta-analysis. *American Journal of Epidemiology* 182, 375–380.

Liang, Y., Lei, L., Nilsson, J., Li, H., Nordberg, M. *et al.* (2012) Renal function after reduction in cadmium exposure: an 8-year follow-up of residents in cadmium-polluted areas. *Environmental Health Perspectives* 120, 223–228.

Lin, C.M., Doyle, P., Wang, D., Hwang, Y.H. and Chen, P.C. (2011) Does prenatal cadmium exposure affect fetal and child growth? *Occupational and Environmental Medicine* 68, 641–646.

Mallozzi, M., Leone, C., Manurita, F., Bellati, F. and Caserta, D. (2017) Endocrine disrupting chemicals and endometrial cancer: an overview of recent laboratory evidence and epidemiological studies. *International Journal of Environental Research and Public Health* 14, 334.

McElroy. J.A., Kruse. R.L., Guthrie. J., Gangnon. R.E. and Robertson, J.D. (2017) Cadmium exposure and endometrial cancer risk: a large midwestern US population-based case-control study. *PLoS ONE* 12, e0179360.

Moberg, L., Nilsson, P.M., Samsioe, G., Sallsten, G., Barregard, L., Engstrom, G.and Borgfeldt, C. (2017) Increased blood cadmium levels were not associated with increased fracture risk but with increased total mortality in women: the Malmo diet and cancer study. *Osteoporosis International* 28, 2401–2408.

Nakagawa, H., Nishijo, M., Morikawa, Y., Miura, K., Tawara, K. *et al.* (2006) Environmental exposure to cadmium and risk of cancer: a prospective population-based study. *Lancet Oncology* 7, 119–126.

Nogawa, K. (2006) Urinary cadmium and mortality among inhabitants of a cadmium-polluted area in japan. *Environmental Research* 100, 323–329.

Nordberg, G.F., Nogawa, K. and Nordberg, M. (2015) Cadmium. In: Nordberg, G.F., Fowler, B.A. and Nordberg, M. (eds) *Handbook on the Toxicology of Metals*, 4th edn, Vol. II. Elsevier, pp. 668–708.

Pollack, A.Z., Ranasinghe, S., Sjaarda, L.A. and Mumford, S.L. (2014) Cadmium and reproductive health in women: a systematic review of the epidemiologic evidence. *Current Environmental Health Reports* 1, 172–184.

Raciti, M. and Ceccatelli, S. (2017) Epigenetic mechanisms in developmental neurotoxicity. *Neurotoxicology and Teratology* 126, 242–249.

Sanders, A.P., Claus Henn, B. and Wright, R.O. (2015) Perinatal and childhood exposure to cadmium, manganese, and metal mixtures and effects on cognition and behavior: a review of recent literature. *Current Environmental Health Reports* 2, 284–294.

Skroder, H., Kippler, M., Nermell, B., Tofail, F., Levi, M. *et al.* (2017) Major limitations in using element concentrations in hair as biomarkers of exposure to toxic and essential trace elements in children. *Environmental Health Perspectives* 125, 067021.

Sommar, J.N., Svensson, M.K., Bjor, B.M., Elmstahl, S.I., Hallmans, G. *et al.* (2013) End-stage renal disease and low level exposure to lead, cadmium and mercury; a population-based, prospective nested case-referent study in sweden. *Environmental Health* 12, 9.

Taylor, C.M., Golding, J. and Emond, A.M. (2016) Moderate prenatal cadmium exposure and adverse birth outcomes: a role for sex-specific differences? *Paediatric and Perinatal Epidemiology* 30, 603–611.

Tian, L.L., Zhao, Y.C., Wang, X.C., Gu, J.L., Sun, Z.J., Zhang, Y.L. and Wang, J.X. (2009) Effects of gestational cadmium exposure on pregnancy outcome and development in the offspring at age 4.5 years. *Biological Trace Element Research* 132, 51–59.

Tinkov, A.A., Filippini, T., Ajsuvakova, O.P., Skalnaya, M.G., Aaseth, J. *et al.* (2018) Cadmium and atherosclerosis: a review of toxicological mechanisms and a meta-analysis of epidemiologic studies. *Environmental Research* 162, 240–260.

Van Maele-Fabry, G., Lombaert, N. and Lison, D. (2016) Dietary exposure to cadmium and risk of breast cancer in postmenopausal women: a systematic review and meta-analysis. *Environment International* 86, 1–13.

Vidal, A.C., Semenova, V., Darrah, T., Vengosh, A., Huang, Z. *et al.* (2015) Maternal cadmium, iron and zinc levels, DNA methylation and birth weight. *BMC Pharmacology and Toxicology* 16, 20.

Wallin, M., Barregard, L., Sallsten, G., Lundh, T., Karlsson, M.K. *et al.* (2016) Low-level cadmium exposure is associated with decreased bone mineral density and increased risk of incident fractures in elderly men: the MROS Sweden study. *Journal of Bone and Miner Research* 31, 732–741.

WHO (2011) *Evaluation of Certain Food Additives and Contaminants*. Seventy-third report of the Joint FAO/WHO Expert Committee on Food Additives, Geneva.

# 27 Cadmium II. Cardiovascular Effects of Human Exposure to Cadmium: Left Ventricular Structure and Function

**W.-Y. Yang* and J.A. Staessen**

*Research Unit Hypertension and Cardiovascular Epidemiology, KU Leuven Department of Cardiovascular Sciences, University of Leuven, Belgium; Present address: Department of Cardiology, Shanghai Jiao Tong University, School of Medicine, Shanghai, China*

## 27.1 Abstract

Since the 1970s, the effect of cadmium perfusion on peak systolic contractility and cardiac atrioventricular conduction has been observed in rats and a growing body of evidence shows that cadmium exposure is associated with cardiovascular disease, including heart failure, in humans. However, few human studies linked the cadmium exposure to impaired left ventricular function. Due to the common coexistence of other poisoning heavy metals, such as lead, and the confounding effect of smoking, the causal relationship between cadmium exposure and left ventricular function in humans is still to be established.

## 27.2 Introduction

The heart is a muscular organ that pumps blood throughout the body, providing the tissues with oxygen and nutrients and removing metabolic waste. In mammals, including humans, the heart consists of four chambers: the upper left and right atria; and the lower left and right ventricles. In a healthy heart, the valves isolating the atria and ventricles, and linking ventricles and aorta or pulmonary, ensure that blood always flows in one direction through the heart. The heart pumps blood with a rhythm determined by pacemaker cells in the sinoatrial node, which generates a current travelling through the atrioventricular node and along the conduction system of the heart, and causing contraction of the heart. Incessant contraction of cardiac muscles to pump blood consumes more than 20% of oxygen and nutrients of the body, which is supplied by the coronary arteries. Any disorder of the aforementioned structures and tissues can lead to heart failure, defined as a condition or syndrome in which the heart is unable to pump enough blood to meet the demands of the body (Lilly, 2012; Mann *et al.*, 2014).

In general, the right atrium and ventricle are referred together as the right heart and their left counterparts as the left heart. The diseases causing right heart failure are usually very different from the ones leading to left heart failure. However, nowadays, left ventricular heart failure caused by tobacco use, obesity, lipid disorder, hypertension, diabetes, coronary artery disease and so on is much more prevalent than right heart failure, especially in developed countries (Mann *et al.*, 2014); therefore, if not specified,

* E-mail address: wenyi.yang@shgh.cn

the term 'heart failure' usually refers to left ventricular dysfunction.

Along with industrialization, population ageing and lifestyle changes, cardiovascular disease has become the leading cause of death worldwide (GBD2015 Disease and Injury Incidence and Prevalence Collaborators, 2016). As the common final stage of all types of cardiovascular disease, heart failure is a major public problem with a prevalence of more than 23 million worldwide (Bui *et al.*, 2011). To improve life quality, industrialization for economic growth, rapidly progressing in developing countries, is inevitable. However, industrialization also leads to increased production and consumption of heavy metals, thus exacerbating global metal pollution, including cadmium, which is now a big threat to public health. The parallel increase in the incidence of heart failure and cadmium contamination suggests a possible association. Indeed, a growing body of evidence has shown that exposure to cadmium is associated with cardiovascular diseases, including heart failure in humans (Tellez-Plaza *et al.*, 2013a). However, heart failure is a syndrome that can be aetiologically caused by a variety of heart diseases. An association between cadmium exposure and heart failure does not necessarily indicate a causal toxic effect of cadmium on the heart muscle, as has been showed in animal studies (Hawley and Kopp, 1975; Kopp *et al.*, 1978) The relationship between cadmium and different cardiovascular diseases (Tellez-Plaza *et al.*, 2013b) might be explained by residual confounding effect through traditional cardiovascular risk factors such as age, sex, smoking (Olsson *et al.*, 2002), hypertension (Gallagher and Meliker, 2010; Staessen *et al.*, 2000) and dyslipidaemia (Cho *et al.*, 2016; Zhou *et al.*, 2016). This chapter will focus on the association of cadmium with left ventricular function and heart failure, covering the above aspects (Fig 27.1).

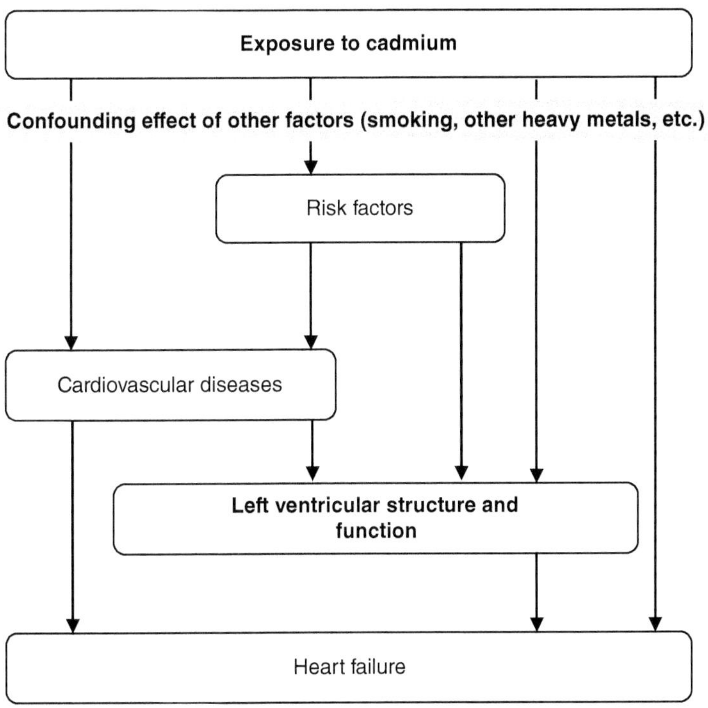


**Fig. 27.1.** Outline of the relationship between cadmium exposure and left ventricular structure and function and heart failure. The arrow lines indicate the potential mechanisms or the evidenced links between the cadmium exposure with risk factors, left ventricular function, heart failure and cardiovascular diseases. However, concerning the confounding effect of age, sex, smoking and exposure to other pollutants, caution should be used when interpreting these data.

## 27.3 Cadmium and Left Ventricular Function

Hawley and Kopp first applied the perfusion of the cadmium ion in a purpose of competition with calcium in isolated rat heart and observed prolonged electrophysiological PR interval in 1975 (Hawley and Kopp, 1975). Kopp *et al.* (1978) also confirmed that the atrioventricular conduction block after application of chemical composition perfusion with cadmium in a concentration of 0.03 mmol $l^{-1}$ to the isolated Langendorff rat heart occurred in A-H interval (atrioventricular nodal depression). With an effect of decreasing heart rate, perfusion of cadmium ions also largely depressed ventricular systolic tension (Kopp *et al.*, 1978). The underlying mechanism was illustrated in a further experiment: the cadmium ions can inhibit phosphorylation level of light chain-2 of myofibrillar proteins and suppress positive inotrope induced by calcium or isoproterenol (Kopp and Barany, 1980). Kopp and colleagues then moved to animal studies *in vivo*, and studied female hooded rats fed 5 ppm cadmium via drinking water for 15 months (Kopp *et al.*, 1980a, b). Growth and food and water intake were comparable with controls. Perfused hearts of exposed animals showed depressed myocardial contractility and reduced positive inotropic responsiveness to isoproterenol (Kopp *et al.*, 1980a). The depressed myocardial contractility was not due to altered β-receptor function, but to depressed phosphorylation of the myocardial contractile proteins (Kopp *et al.*, 1980a). In anaesthetized rats exposed to cadmium and lead for 20 months according to a similar protocol, cadmium selectively slowed conduction proximal to the His bundle (Kopp *et al.*, 1980b). Phosphorus-31 nuclear magnetic resonance spectroscopy revealed depressed high-energy phosphate and glycerol-3-phosphorylcholine concentrations in hearts of cadmium-exposed rats (Kopp *et al.*, 1980b). In the cultured cardiomyocytes, cadmium treatment induced dramatic endoplasmic reticulum stress and impaired energy homeostasis (Chen *et al.*, 2015). Moreover, cadmium may inhibit the protein kinase B (AKT)–mTOR (mammalian target of rapamycin) pathway to reduce energy production, by either disrupting the glucose metabolism (Chen *et al.*, 2015) or inhibiting mitochondrial respiratory (Kisling *et al.*, 1987) and related gene expressions (Chen *et al.*, 2015). Another important line of the potential cardiotoxic effect of cadmium refers to cardiac fibrosis: experiments *in vitro* demonstrated a cadmium-induced depolarization of the mitochondrial membrane and permeabilization of the plasma membrane (Türkcan *et al.*, 2015). In further ApoE–/– mice models, Türkcan *et al.* (2015) illustrated that treatment with western diet or cadmium resulted in necrotic cardiomyocyte necrosis and myocardial fibrosis. These experiments (Hawley *et al.*, 1975; Kopp *et al.*, 1978, 1980a, b; Kopp and Barany, 1980; Prentice and Kopp, 1985) produced evidence for direct pathophysiological changes in the presence of overt cadmium toxicity, the main mechanisms being suppressing conductivity in the heart and dysregulation of energy metabolism.

Concerning the evidence in humans, only one population study reported on the association of left ventricular structure and function with cadmium exposure (Yang *et al.*, 2017b). In 179 participants (50.3% women; mean age, 39.1 years) randomly recruited from a Flemish population, baseline 24 h urinary cadmium and other risk factors (1985–2000) and follow-up echocardiographic left ventricular structure and systolic and diastolic function (2005–2010) were assessed with a median interval of 11.9 years (Yang *et al.*, 2017b). After adjustment for sex, age, mean arterial pressure, heart rate, body mass index, fasting plasma glucose, total-to-high-density lipoprotein–cholesterol ratio, serum creatinine, γ-glutamyltransferase, smoking and anti-hypertensive treatment, Yang *et al.* (2017b) observed that a doubling of baseline 24 h urinary cadmium was significantly ($P \leq 0.015$) associated with decreased regional longitudinal strain rate (–0.066 $s^{-1}$) and regional radial strain (–2.848%) during follow-up. Meanwhile, the blood lead level was also associated with reduced left ventricular systolic function. However, models including both cadmium and lead exposure indexes did not allow differentiating whether left ventricular systolic dysfunction was predominantly related to blood lead or urinary cadmium. None of the left ventricular diameter, wall thickness or diastolic function parameters was associated with 24 h urinary cadmium excretion ($P \geq 0.16$) (Yang *et al.*, 2017b). Of course, caution should be exercised when interpreting these results. Indeed, confounding by the coexistence of other heavy metals as toxic agents, such as lead, cannot be excluded. Furthermore, there might be some residual confounding by the common determinants of cadmium exposure (Olsson *et al.*, 2002) and left ventricular structure and function, such as sex, age and smoking (Kuznetsova *et al.*, 2008, 2009; Lang *et al.*, 2015).

Therefore, although the experimental studies provided strong evidence of a direct cardiotoxic effect of cadmium, more data are required to confirm this effect in humans. Of note, in the Flemish studies, the left ventricular ejection fraction was not associated with either lead or cadmium exposure (Yang *et al.*, 2017b) or low-level residential air pollution (Yang *et al.*, 2017a). As highlighted in a recent systematic review (Kalam *et al.*, 2014), this might be explained by the inaccurate assessment of systolic function using left ventricular ejection fraction in people with a value within the normal range (> 45%). Therefore, for future studies focusing on low-to-moderate environmental cadmium exposure in relation to left ventricular systolic function, a longitudinal design assessing left ventricular function by a state-of-the-art technique (for instance, strain and strain rate) and accounting for other heavy metals and nicotine would be favoured.

## 27.4 Cadmium and Heart Failure

Although cadmium studies directly focusing on left ventricular function in humans are scarce (Yang *et al.*, 2017b), there is mounting evidence supporting a link between the incidence or prevalence of heart failure and cadmium exposure. Among inhabitants aged ≥ 50 years in the Kakehashi river basin, Ishikawa Prefecture, Japan, Nishijo and colleagues compared cause-specific mortality between subjects ($n$ = 2408) with positive (≥ 4 mg $dl^{-1}$) and negative (< 4 mg $dl^{-1}$) urinary retinol-binding protein (Nishijo *et al.*, 1995) and between individuals ($n$ = 3178) with higher (≥ 1000 μg $g^{-1}$ creatinine) and lower (< 1000 μg $g^{-1}$ creatinine) creatinine-standardized urinary $\beta_2$-microglobulin (Nishijo *et al.*, 2006). During the 15-year follow-up in the retinol-binding protein study, 52 and 86 cases of fatal heart failure were identified in men and women, respectively (Nishijo *et al.*, 1995). The corresponding incidences in the $\beta_2$-microglobulin study were 60 in men and 87 in women (Nishijo *et al.*, 2006). Using the total Japanese population as the standard population, the standardized heart failure mortality amounted to 302.6% and 122.5% ($P$ < 0.01) in men with positive and negative urinary retinal binding protein, respectively. In women, the corresponding rates were 353.0% and 155.9% ($P$ < 0.05) (Nishijo *et al.*, 1995). For the comparison of higher versus lower urinary $\beta_2$-microglobulin, the estimated standardized heart failure mortality ratios were 179% versus 93% in men and 248% versus 104% in women (Nishijo *et al.*, 2006). Compared with the risk of fatal heart failure among the subjects with a urinary $\beta_2$-microglobulin of < 300 μg $g^{-1}$ creatinine, the hazard ratio of those with a $\beta_2$-microglobulin level of 300–1000, 1000–10,000 and ≥ 10,000 μg $g^{-1}$ creatinine were 0.88 (95% confidence interval (CI): 0.41–1.89), 1.45 (0.74–2.84) and 3.69 (1.62–8.39) in males and 1.94 (1.08–3.48), 3.05 (1.73–5.35) and 3.19 (1.19–5.52) in females (Nishijo *et al.*, 2006). The dose–response relationship between urinary $\beta_2$-microglobulin and cadmium (Ishizaki *et al.*, 1989) and between the frequency of inhabitants with urinary retinol-binding protein of ≥ 4 mg $dl^{-1}$ and the average concentration in rice of the hamlets where the subjects were living (Nogawa *et al.*, 1978) suggested a causal association between heart failure and cadmium exposure (Nishijo *et al.*, 1995, 2006) .

By using National Health and Nutrition Examination Survey (NHANES, 1999–2006), Peters *et al.* (2010) assessed the association between blood and urinary cadmium and the prevalence of heart failure in the general population. Among 12,049 participants aged 30 years or older, 471 persons reported a history of heart failure at the time of their interviews. The geometric mean cadmium concentration was 3.8 nmol $l^{-1}$ in blood and 2.7 nmol $l^{-1}$ in urine. After adjusting for age, sex, race/ethnicity, body mass index, education level, poverty income ratio, alcohol consumption, smoking status and blood cotinine, the odds ratios of prevalent heart failure were 1.48 (95% CI: 1.17–1.87) and 1.12 (1.03–1.20) in relation to a 50% increase in blood and urinary cadmium, respectively (Peters *et al.*, 2010). The retrospective design, cross-sectional analysis and self-reported outcome of heart failure were the major limitations and made the interpretation of this study difficult. However, in

3348 American Indian adults aged 45–74 years who participated in the Strong Heart Study in 1989–1991 and were followed up until 2018, the overall median and geometric mean urinary cadmium concentration at baseline were 0.92 and 0.94 μg $g^{-1}$ creatinine, respectively (Tellez-Plaza *et al.*, 2013a). During follow-up, 328 subjects developed heart failure. In models solely adjusted for gender, the incidence of heart failure was not significantly related to urinary cadmium. Only in multivariable-adjusted models, the risk of heart failure in the second, third and fourth cadmium quartiles increased significantly compared with the first quartile group (Tellez-Plaza *et al.*, 2013a).

A single cigarette typically contains 1–2 μg of cadmium (Menden *et al.*, 1972; Sugita *et al.*, 2001). When burned, more than 50% of the cadmium, present at a level of 1000–3000 ppb in the smoke, is inhaled and absorbed into the blood through the lung. Then, the question arises whether part of the cardiovascular risk associated with smoking is due to the inhaled cadmium. The odds ratio of heart failure prevalence associated with smoking decreased from significant 2.28 (95% CI: 1.66–3.12) to insignificant 1.40 (0.88–2.24) after additional adjustment for blood cadmium (Peters *et al.*, 2010). Additional adjustment for urinary cadmium produced similar results (Peters *et al.*, 2010). These findings are consistent with the hypothesis that smoking increases the risk of heart failure through exposure to cadmium (Peters *et al.*, 2010).

## 27.5 Cadmium and Cardiovascular Disease and Associated Risk Factors

Cardiovascular disease, especially coronary heart disease, is the common aetiological cause of left ventricular dysfunction and heart failure (Mann *et al.*, 2014). Over the past two decades, increasing evidence has indicated that cadmium exposure is a cardiovascular risk factor (Nawrot *et al.*, 2008; Tellez-Plaza *et al.*, 2013b; Tinkov *et al.*, 2018). Tellez-Plaza *et al.* (2013) meta-analysed seven prospective and five cross-sectional studies that evaluated the association of cardiovascular disease with low to moderate environmental cadmium exposure in general populations. The pooled relative risks of cardiovascular and coronary heart diseases comparing the highest to lowest cadmium exposure categories were 1.36 (95% CI: 1.11–1.66) and 1.30 (1.12–1.52), respectively. These risk estimates were 1.23 (1.05–1.44) and 1.21 (1.07–1.37) if only the prospective studies were included in the meta-analysis. These relative risks of cardiovascular disease were consistent for urinary cadmium 1.36 (95% CI: 1.11–1.66) and blood cadmium 1.41 (1.18–1.70) as exposure biomarkers. All the aforementioned analyses accounted for age, sex and smoking status, as they are the main determinants of cadmium exposure (Olsson *et al.*, 2002) and major contributing risk factors for cardiovascular disease (Mann *et al.*, 2014). By adding four newly emerging prospective and two cross-sectional studies, another meta-analysis conducted in 2017 confirmed these findings (Tinkov *et al.*, 2018). Residual confounding effect by smoking and sex is a common concern when interpreting the association between cadmium exposure and cardiovascular disease. However, the pooled relative risks of cardiovascular disease with cadmium exposure were of similar magnitude in men, women (Tellez-Plaza *et al.*, 2013b) and never-smokers (Tellez-Plaza *et al.*, 2013b; Tinkov *et al.*, 2018). In addition, the associations remained largely similar after adjustment for cigarette pack-year and serum nicotine, in addition to smoking status (Tellez-Plaza *et al.*, 2013b). All these analyses (Tellez-Plaza *et al.*, 2013b; Tinkov *et al.*, 2018) support a causal relationship between cadmium exposure and cardiovascular disease, which potentially leads to left ventricular dysfunction and heart failure.

In addition to age, sex and smoking, cadmium exposure was associated with atherogenic changes in lipid profile. In a meta-analysis of several available studies (Kim, 2012; Ettinger *et al.*, 2014; Tangvarasittichai *et al.*, 2015; Zhou *et al.*, 2016) higher cadmium exposure was associated with higher total cholesterol levels, higher low-density lipoprotein cholesterol levels and lower high-density lipoprotein cholesterol levels (Tinkov *et al.*, 2018). Among 195 young healthy females enrolled in the Atherosclerosis Risk Factors in Female Youngsters (ARFY) study, a 1-SD increment in serum cadmium was associated with an odds ratio of 1.6 (95% CI: 1.1.–2.3) for early atherosclerotic vessel wall thickening after multivariable adjustment (Messner *et al.*, 2009). Blood cadmium was also significantly associated with the number of plaques in carotid arteries ($P = 0.001$) and with intima–media thickness ($P = 0.005$) (Lind *et al.*, 2012). Bergström *et al.* (2015) also demonstrated a significant relationship between blood cadmium level and area of atherosclerotic plaques. Moreover, plaque cadmium content was 50-fold higher than that in blood. The highest concentration of cadmium in plaques was observed in the upstream sections of carotid plaques, being more than two-fold higher than that in stenosis and downstream sections (Bergström *et al.*, 2015). These studies suggest that cadmium exposure can cause atherosclerosis, thereby leading to coronary heart disease and left ventricular dysfunction.

A growing body of evidence supports the hypothesis that cadmium exposure plays a role in the development of hypertension (Tellez-Plaza *et al.*, 2008; Lee *et al.*, 2011; Caciari *et al.*, 2013) and chronic kidney disease (Navas-Acien *et al.*, 2009; Hwangbo *et al.*, 2011). It is well known that at all ages and across all ethnicities, high blood pressure is the major driver of cardiovascular complications (Lewington *et al.*, 2002). The causal relationship between blood pressure and left ventricular hypertrophy (Schmieder *et al.*, 1996) and diastolic dysfunction (Redfield *et al.*, 2003; Nadruz *et al.*, 2017) is beyond doubt. Renal function may be one of the contributors to this relationship (Yang *et al.*, 2013). However, evidence showing an association between cardiovascular disease or left ventricular function and cadmium exposure through the pathway of hypertension and/or chronic kidney disease is lacking.

## 27.6 Conclusions

Experimental or animal studies (Hawley *et al.*, 1975; Kopp *et al.*, 1978, 1980a, b; Kopp and Barany, 1980; Prentice *et al.*, 1985) had suggested a direct cardiotoxic effect of cadmium for more than three decades. Moreover, mounting evidence supports a causal relationship between cadmium exposure and cardiovascular disease, in particular heart failure (Tellez-Plaza *et al.*, 2013b; Tinkov *et al.*, 2018). The association of cadmium exposure with left ventricular structure and function through its relationship with hypertension (Tellez-Plaza *et al.*, 2008; Lee *et al.*, 2011; Caciari *et al.*, 2013), lipid profile (Kim, 2012; Ettinger *et al.*, 2014; Tangvarasittichai *et al.*, 2015; Zhou *et al.*, 2016) and chronic kidney function (Navas-Acien *et al.*, 2009; Hwangbo *et al.*, 2011) is highly speculative. However, studies directly linking cadmium exposure to left ventricular structure and function are very rare (Yang *et al.*, 2017b). Well-designed longitudinal studies assessing left ventricular function by state-of-the-art technique (for instance, systolic strain and strain rate) and accounting for other heavy metals and nicotine are warranted.

## References

Bergström, G., Fagerberg, B., Sallsten, G., Lundh, T. and Barregard, L. (2015) Is cadmium exposure associated with the burden, vulnerability and rupture of human atherosclerotic plaques? *PLoS ONE* 10, e0121240.

Bui, A.L., Horwich, T.B. and Fonarow, G.C. (2011) Epidemiology and risk profile of heart failure. *Nature Reviews Cardiology* 8, 30–41.

Caciari, T., Sancini, A., Fioravanti, M., Capozzella, A., Casale, T. *et al.* (2013) Cadmium and hypertension in exposed workers: a meta-analysis. *International Journal of Occupational Medicine and Environmental Health* 26, 440–456.

Chen, C.Y., Zhang, S.L., Liu, Z.Y., Tian, Y. and Sun, Q. (2015) Cadmium toxicity induces ER stress and apoptosis via impairing energy homoeostasis in cardiomyocytes. *Bioscience Reports* 35, e00214.

Cho, H.M., Cho, D.Y., Kim, M.Y., Yang, S.W., Seo, Y.S. and Kim, K.N. (2016) Combined effect of blood cadmium and lead levels on coronary heart disease prediction risk in Korean men. *Angiology* 67, 582–586.

Ettinger, A.S., Bovet, P., Plange-Rhule, J., Forrester, T.E., Lambert, E.V. *et al.* (2014) Distribution of metals exposure and associations with cardiometabolic risk factors in the ‘Modeling the Epidemiologic Transition Study’. *Environmental Health* 13, 90.

Gallagher, C.M. and Meliker, J.R. (2010) Blood and urine cadmium, blood pressure, and hypertension: a systematic review and meta-analysis. *Environmental Health Perspectives* 118, 1676–1684.

GBD2015 Disease and Injury Incidence and Prevalence Collaborators (2016) Global, regional, and national incidence, prevalence, and years lived with disability for 310 diseases and injuries, 1990–2015: a systematic analysis for the Global Burden of Disease Study 2015. *The Lancet* 388, 1545–1602.
Hawley, P.L. and Kopp, S.J. (1975) Extension of PR interval in isolated rat heart by cadmium. *Proceedings of the Society for Experimental Biology and Medicine* 150, 669–671.
Hwangbo, Y., Weaver, V.M., Tellez-Plaza, M., Guallar, E., Lee, B.K. and Navas-Acien, A. (2011) Blood cadmium and estimated glomerular filtration rate in Korean adults. *Environmental Health Perspectives* 119, 1800–1805.
Ishizaki, M., Kido, T., Honda, R., Tsuritani, I., Yamada, Y., Nakagawa, H. and Nogawa, K. (1989) Dose-response relationship between urinary cadmium and beta-2-microglobulin in a Japanese environmentally cadmium exposed population. *Toxicology* 58, 121–131.
Kalam, K., Otahal, P. and Marwick, T.H. (2014) Prognostic implications of global LV dysfunction: a systematic review and meta-analysis of global longitudinal strain and ejection fraction. *Heart* 100, 1673–1680.
Kim, K. (2012) Blood cadmium concentration and lipid profile in Korean adults. *Environmental Research* 112, 225–229.
Kisling, G.M., Kopp, S.J., Paulson, D.J., Hawley, P.L. and Tow, J.P. (1987) Inhibition of rat heart mitochondrial respiration by cadmium chloride. *Toxicology and Applied Pharmacology* 89, 295–304.
Kopp, S.J., Baker, J.C., D'Agrosa, L.S. and Hawley, P.L. (1978) Simultaneous recording of His bundle electrogram, electrocardiogram, and systolic tension from intact modified Langendorff rat heart preparations. I: effects of perfusion time, cadmium, and lead. *Toxicology and Applied Pharmacology* 46, 475–487.
Kopp, S.J. and Barany, M. (1980) Influence of isoproterenol and calcium on cadmium- or lead-induced negative inotropy related to cardiac myofibrillar protein phosphorylations in perfused rat heart. *Toxicology and Applied Pharmacology* 55, 8–17.
Kopp, S.J., Barany, M., Erlanger, M., Perry, E.F. and Perry, H.M. Jr (1980a) The influence of chronic low-level cadmium and/or lead feeding on myocardial contractility related to phosphorylation of cardiac myofibrillar proteins. *Toxicology and Applied Pharmacology* 54, 48–56.
Kopp, S.J., Perry, M. Jr, Glonek, T., Erlanger, M., Perry, E.F., Barany, M. and D'Agrosa, L.S. (1980b) Cardiac physiologic-metabolic changes after chronic low-level heavy metal feeding. *American Journal of Physiology* 239, H22–30.
Kuznetsova, T., Herbots, L., Richart, T., D'Hooge, J., Thijs, L. *et al.* (2008) Left ventricular strain and strain rate in a general population. *European Heart Journal* 29, 2014–2023.
Kuznetsova, T., Herbots, L., Lopez, B., Jin, Y., Richart, T. *et al.* (2009) Prevalence of left ventricular diastolic dysfunction in a general population. *Circulation Heart Failure* 2, 105–112.
Lang, R.M., Badano, L.P., Mor-Avi, V., Afilalo, J., Armstrong, A. *et al.* (2015) Recommendations for cardiac chamber quantification by echocardiography in adults: an update from the American Society of Echocardiography and the European Association of Cardiovascular Imaging. *Journal of the American Society of Echocardiography* 28, 1–39 e14.
Lee, M.S., Park, S.K., Hu, H. and Lee, S. (2011) Cadmium exposure and cardiovascular disease in the 2005 Korea National Health and Nutrition Examination Survey. *Environmental Research* 111, 171–176.
Lewington, S., Clarke, R., Qizilbash, N., Peto, R., Collins, R. and Prospective Studies Collaboration (2002) Age-specific relevance of usual blood pressure to vascular mortality: a meta-analysis of individual data for one million adults in 61 prospective studies. *The Lancet* 360, 1903–1913.
Lilly, L.S. (2012) *Pathophysiology of Heart Disease; a Collaborative Project of Medical Students and Faculty*. Wolters Kluwer Health, Baltimore, Maryland.
Lind, P.M., Olsen, L. and Lind, L. (2012) Circulating levels of metals are related to carotid atherosclerosis in elderly. *Science of the Total Environment* 416, 80–88.
Mann, D.L., Zipes, D.P., Libby, P., Bonow, R.O. and Braunwald, E. (2014) *Braunwald's Heart Disease – A Textbook of Cardiovascular Medicine*. Elsevier, Philadelphia, Pennsylvania.
Menden, E.E., Elia, V.J., Michael, L.W. and Petering, H.G. (1972) Distribution of cadmium and nickel of tobacco during cigarette smoking. *Environmental Science &Technology* 6, 830–832.
Messner, B., Knoflach, M., Seubert, A., Ritsch, A., Pfaller, K. *et al.* (2009) Cadmium is a novel and independent risk factor for early atherosclerosis mechanisms and in vivo relevance. *Arteriosclerosis, Thrombosis, and Vascular Biology* 29, 1392–1398.
Nadruz, W., Shah, A.M. and Solomon, S.D. (2017) Diastolic dysfunction and hypertension. *Medical Clinics of North America* 101, 7–17.
Navas-Acien, A., Tellez-Plaza, M., Guallar, E., Muntner, P., Silbergeld, E., Jaar, B. and Weaver, V. (2009) Blood cadmium and lead and chronic kidney disease in US adults: a joint analysis. *American Journal of Epidemiology* 170, 1156–1164.

Nawrot, T.S., Van Hecke, E., Thijs, L., Richart, T., Kuznetsova, T. *et al.* (2008) Cadmium-related mortality and long-term secular trends in the cadmium body burden of an environmentally exposed population. *Environmental Health Perspectives* 116, 1620–1628.

Nishijo, M., Morikawa, Y., Nakagawa, H., Tawara, K., Miura, K. *et al.* (2006) Causes of death and renal tubular dysfunction in residents exposed to cadmium in the environment. *Occupational and Environmental Medicine* 63, 545–550.

Nishijo, M., Nakagawa, H., Morikawa, Y., Tabata, M., Senma, M. *et al.* (1995) Mortality of inhabitants in an area polluted by cadmium: 15 year follow up. *Occupational and Environmental Medicine* 52, 181–184.

Nogawa, K., Ishizaki, A. and Kawano, S. (1978) Statistical observations of the dose–response relationships of cadmium based on epidemiological studies in the Kakehashi River basin. *Environmental Research* 15, 185–198.

Olsson, I.M., Bensryd, I., Lundh, T., Ottosson, H., Skerfving, S. and Oskarsson, A. (2002) Cadmium in blood and urine – impact of sex, age, dietary intake, iron status, and former smoking – association of renal effects. *Environmental Health Perspectives* 110, 1185–1190.

Peters, J.L., Perlstein, T.S., Perry, M.J., McNeely, E. and Weuve, J. (2010) Cadmium exposure in association with history of stroke and heart failure. *Environmental Research* 110, 199–206.

Prentice, R.C. and Kopp, S.J. (1985) Cardiotoxicity of lead at various perfusate calcium concentrations: functional and metabolic responses of the perfused rat heart. *Toxicology and Applied Pharmacology* 81, 491–501.

Redfield, M.M., Jacobsen, S.J., Burnett, J.C. Jr, Mahoney, D.W., Bailey, K.R. and Rodeheffer, R.J. (2003) Burden of systolic and diastolic ventricular dysfunction in the community: appreciating the scope of the heart failure epidemic. *Journal of the American Medical Association* 289, 194–202.

Schmieder, R.E., Martus, P. and Klingbeil, A. (1996) Reversal of left ventricular hypertrophy in essential hypertension. A meta-analysis of randomized double-blind studies. *Journal of the American Medical Association* 275, 1507–1513.

Staessen, J.A., Kuznetsova, T., Roels, H.A., Emelianov, D. and Fagard, R. (2000) Exposure to cadmium and conventional and ambulatory blood pressures in a prospective population study. Public Health and Environmental Exposure to Cadmium Study Group. *American Journal of Hypertension* 13, 146–156.

Sugita, M., Izuno, T., Tatemichi, M. and Otahara, Y. (2001) Cadmium absorption from smoking cigarettes: calculation using recent findings from Japan. *Environmental Health and Preventive Medicine* 6, 154–159.

Tangvarasittichai, S., Niyomtam, S., Pingmuangkaew, P. and Nunthawarasilp, P. (2015) Dyslipidemia in the elevated cadmium exposure population. *International Journal of Toxicological and Pharmacological Research* 7, 92–98.

Tellez-Plaza, M., Navas-Acien, A., Crainiceanu, C.M. and Guallar, E. (2008) Cadmium exposure and hypertension in the 1999–2004 National Health and Nutrition Examination Survey (NHANES). *Environmental Health Perspectives* 116, 51–56.

Tellez-Plaza, M., Guallar, E., Howard, B.V., Umans, J.G., Francesconi, K.A. *et al.* (2013a) Cadmium exposure and incident cardiovascular disease. *Epidemiology* 24, 421–429.

Tellez-Plaza, M., Jones, M.R., Dominguez-Lucas, A., Guallar, E. and Navas-Acien, A. (2013b) Cadmium exposure and clinical cardiovascular disease: a systematic review. *Current Atherosclerosis Reports* 15, 356.

Tinkov, A.A., Filippini, T., Ajsuvakova, O.P., Skalnaya, M.G., Aaseth, J. *et al.* (2018) Cadmium and atherosclerosis: a review of toxicological mechanisms and a meta-analysis of epidemiologic studies. *Environmental Research* 162, 240–260.

Türkcan, A., Scharinger, B., Grabmann, G., Keppler, B.K., Laufer, G., Bernhard, D. and Messner, B. (2015) Combination of cadmium and high cholesterol levels as a risk factor for heart fibrosis. *Toxicological Sciences* 145, 360–371.

Yang, Y., Wang, Y., Shi, Z. W., Zhu, D.L. and Gao, P.J. (2013) Association of E/E' and NT-proBNP with renal function in patients with essential hypertension. *PLoS ONE* 8, e54513.

Yang, W.Y., Zhang, Z.Y., Thijs, L., Bijnens, E.M., Janssen, B.G. *et al.* (2017a) Left ventricular function in relation to chronic residential air pollution in a general population. *European Journal of Preventive Cardiology* 24, 1416–1428.

Yang, W.Y., Zhang, Z.Y., Thijs, L., Cauwenberghs, N., Wei, F.F. *et al.* (2017b) Left ventricular structure and function in relation to environmental exposure to lead and cadmium. *Journal of American Heart Association* 6, e004692.

Zhou, Z., Lu, Y. H., Pi, H. F., Gao, P., Li, M. *et al.* (2016) Cadmium Exposure is associated with the prevalence of dyslipidemia. *Cellular Physiology and Biochemistry* 40, 633–643.

# Part VI

# Particulates and Plastics

# 28 Particulates from Combustion Sources: Formation, Characteristics and Toxic Hazards

**D.A. Purser**
*Hartford Environmental Research, Hatfield, UK*

## 28.1 Abstract

Particulates from combustion sources are a major component of outdoor and indoor air pollutants. During fires, they contribute to acute injury and deaths from smoke exposure; and through chronic daily exposure in homes, workplaces and the outdoor environment they are a major cause of long-term morbidity and mortality from cardiovascular and respiratory disease. Combustion particulates are also deposited directly on vegetation and on soils and water, resulting in uptake of toxic components into food sources. This chapter reviews the formation, chemical composition and physical characteristics of particulate atmospheres formed from a variety of combustion sources ranging from building and wildfires to industrial processes and vehicle emissions. The acute and chronic toxicity resulting from inhalation, dermal absorption and ingestion of combustion particles is discussed, particularly effects at different deposition sites in the lung and systemically. These range from acute distress and incapacitation occurring within seconds of exposure to high smoke concentrations in fires, to long-term morbidity and mortality resulting from chronic exposure to low environmental concentrations of particulates over many years, leading to respiratory and cardiovascular disease or carcinogenicity.

## 28.2 Introduction

Particulates generated from combustion sources are a major component of outdoor and indoor air pollutants. They contribute to acute injury and deaths from exposures to fire effluents (smoke) during and after fire incidents; and through chronic daily exposure in homes, workplaces and the outdoor environment they are a major cause of long-term morbidity and mortality from cardiovascular and respiratory disease (Purser and Maynard, 2016). In addition to direct toxicity in humans, combustion particulates are deposited directly on vegetation and on soils and water, resulting in uptake of toxic components into food sources.

In contrast to toxic combustion gases such as carbon monoxide (CO) or benzene, which have relatively simple and well-defined chemical structures and physical characteristics, smoke particulates are complex and variable in terms of chemical composition, surface properties, size and physical structure. The simplest chemical components include condensed liquid droplets or solid particles, including water mists, hydrocarbon or organic condensates of specific compounds and carbon soot particles. At their simplest these consist of approximately spherical particles (particulate matter, PM) in the nanoparticle size range (1–100 nm diameter)

($PM_{0.1}$, also known as 'ultrafine' particles) or in the lower end of the micro-particle size range (0.1–100 μm diameter) (Donaldson *et al.*, 2001). Typically in fires these grow into larger and more complex agglomerates in smoke plumes, mostly within a respirable micro-particle size range (< 0.5–5 μm) but in some cases growing to millimetre sizes. For particles small enough to remain in suspension in the atmosphere, comprising outdoor (or indoor) air pollution, many of which are combustion derived, a distinction is made between fine particles with an aerodynamic diameter < 2.5 μm ($PM_{2.5}$) and coarse particles with a diameter < 10 μm ($PM_{10}$). Of these, the fine fraction ($PM_{2.5}$) is considered to be of greater significance with respect to respiratory health, since particles within this range penetrate more readily into the lower airways and alveolar regions of the lung (Purser and Maynard, 2016). These smoke particles are typically carbon-based, but, in addition to elemental carbon, include a wide range of chemical species. Some of these are common to almost all combustion sources while others depend on the composition and mode of combustion of the original fuel. Particle agglomerates are of irregular shape and while the basic structure may be principally a graphite-like carbon composite, a wide variety of other inorganic and organic vapours, liquid or solid substances may condense on, be absorbed by or react with these essentially activated carbon substrates (Purser, 2010a, 2016a).

In addition to combustion product and ash particles derived from the combustion process itself, large fires may generate smoke and dust plumes containing mineral particles derived from a building structure, including asbestos and other mineral fibres and siliceous or alkaline particles derived from cementitious or other construction materials (Kendall *et al.*, 2016).

The site of decomposition of these particles within the respiratory tract and their subsequent fate depend mainly on their size (aerodynamic diameter), density or shape (spherical or fibre). Inhalable particles (< ~100 μm diameter) are deposited in the nose, mouth and beyond, while the respirable subset (< ~5–0.2 μm) are deposited mainly in the airways and lower respiratory tract, principally by impaction and sedimentation. Particles in the submicron ultrafine size range behave more like gases and are deposited by diffusion. These nanoparticles deposited in the alveoli may pass into the lung interstitium and the bloodstream, where they are distributed systemically. Larger soot particles (0.01–1 mm diameter) may be deposited on the skin or clothing (for example, of firefighters), with adsorbed chemicals subsequently being absorbed dermally.

The toxic effects of inhaled particles depend on the concentration inhaled, the duration of exposure, the site of deposition, their size and structure and their composition.

The variety of causes of morbidity and mortality related to combustion particulates illustrates one aspect of their complexity: the wide range of toxic product concentrations of interest in evaluating effects, and the wide range of exposure times that need to be considered. Table 28.1 captures this by identifying some very approximate concentration ranges (expressed in $\mu g\ m^{-3}$) and exposure periods of toxicological interest for several sources of toxic combustion particulates.

Particulate concentrations of toxicological significance cover a range of 16 orders of magnitude, ranging from the life-threatening effects of a few minutes' exposure to respirable particulates from indoor smoke plumes during fires, at concentrations in the range of 1–10 $g\ m^{-3}$ (1–10 $\times\ 10^7\ \mu g\ m^{-3}$) to chronic environmental exposures of up to 50 years to dioxins associated with smoke particles at concentrations of around 3.1 $pg\ m^{-3}$ ($3.1 \times 10^{-8}\ \mu g\ m^{-3}$).

This chapter reviews the main features of combustion particulates of different size, shape and composition in terms of their formation, structure, sites of decomposition and toxicology.

## 28.3 Importance of Particle Size and Shape to Toxicity

A key determinant of the toxicology of particulates is their site of deposition within the respiratory tract and subsequent fate. Each part of the respiratory tract is sensitive to a set of pathological effects related to its anatomy, physiology and histology. The site of deposition of particles depends mainly on their size, density and shape (spherical or fibre). Although physical size is relevant, the main parameter influencing deposition

**Table 28.1.** Comparison of ranges of particulate concentrations and exposure periods of toxicological signifiance.

| Source | Particulate concentration of toxicological significance ($\mu g\ m^{-3}$) | Exposure period | Toxic effect |
|---|---|---|---|
| High smoke concentrations during exposure in a fire | $1–10 \times 10^7$ | ~10–60 min | Airway irritation, bronchoconstriction and obstruction during exposure<br>Lung inflammation and oedema a few hours later (sometimes fatal) |
| Indoor environmental tobacco smoke | $2.5–100 \times 10^2$ | Hours to years | COPD, cancer |
| Long-term exposure to environmental combustion particulates $PM_{2.5}$ and $PM_{10}$ Soot | $1 \times 10^{-1}$ | 1–50 years | COPD, cardiovascular disease, cancer |
| Dioxins as 2,3,7,8 TCDD | $3.1 \times 10^{-8}$ | 1–50 years | Systemic, immunological, carcinogenic |

is the aerodynamic diameter, which is the diameter of a sphere with unit density (1 g $cm^{-3}$) which sediments at the same rate as the particle in question in still or laminar flowing air (IARC, 2002). Thus a spherical particle of 1 μm diameter and a density of 2 has an aerodynamic diameter of 2 μm. Although the aerodynamic diameter also applies for fibres (defined as particles with an aspect ratio > 3:1), a fibre can be carried deeper into the respiratory tract (has a lower aerodynamic diameter) than a spherical particle of equivalent mass and density.

Deposition of a airborne particles within the respiratory tract occurs by five processes:

- impaction;
- interception;
- sedimentation;
- electrostatic attraction; and
- diffusion (particle < 0.5 μm).

Deposition depends on the extent to which a particle can remain buoyant and be carried in the airstream entering the respiratory tract. Larger and denser particles have a low buoyancy and so tend to be deposited in the upper respiratory tract by impaction, interception and sedimentation. Smaller particles (with a large surface area/volume ratio) are more buoyant and tend to sediment out deeper into the lung. The smallest particles behave more like gas molecules, in that diffusion becomes increasingly important as size decreases. Particles may become statically charged, so that deposition is enhanced by electrostatic attraction. The general principle is that the smaller the particle, the deeper is the penetration into the respiratory tract, but the relationship is not a simple one. Basically, as size decreases from the largest inhalable particles, the greater the proportion deposited deeper into the lung. But a proportion of smaller particles can remain in the airstream throughout the breathing cycle, being both inhaled and exhaled without deposition. While suspended in the air inside the lung, particles may grow due to hydration. The relationship between particle size and distribution within the respiratory tract is therefore a complex one, which has been measured and published for particles and fibres over different size ranges (Task Group on Lung Dynamics, 1966).

Particles can be divided into three fractions, depending on their main site of deposition:

- Inhalable particulate fraction: fraction of a particulate cloud than can be breathed into the nose or mouth.
- Thoracic particulate fraction: fraction of inhaled airborne particles that can penetrate the head airways and enter the airways of the lungs.

- Respirable particulate fraction: fraction of inhaled airborne particles that can penetrate beyond the terminal bronchioles into the gas-exchange region of the lungs.

The upper limit for inhalable particles is around 100 μm. Between this and around 3 μm, > 60% are trapped in the upper airways (nose and mouth), with only about 1% of 10 μm particles penetrating into the alveolar region. Bronchial deposition is favoured in the 2–5 μm range, with alveolar deposition favoured for smaller particles. Minimum deposition occurs around 0.5 μm, where particles are small enough to be buoyant but too large to diffuse rapidly, so tend to remain in the airstream and be exhaled before they can be deposited. An increasing proportion of particles < 0.5 μm are deposited by diffusion in the alveoli.

In humans there is a considerable distinction in deposition patterns between nose and mouth breathing, with a greater proportion or larger particles penetrating beyond the upper airways for mouth breathing, particularly during exercise.

Ultrafine (nano-) particles are commonly classified as those < 100 nm. Singlet combustion particles such as carbon core diesel particles (Donaldson *et al.*, 2001) or fluoropolymer particles (Purser, 1992a) can be as small as 10 nm diameter.

## 28.4 Fate of Inhaled Particles Deposited in Different Regions of the Respiratory Tract

Insoluble particles deposited on the squamous epithelium of the nose, mouth and larynx become trapped in the surface mucous or serous secretions and either spat or sneezed out or swallowed. Insoluble particles deposed on the mucous surface of the ciliated epithelium of the airways from the nasal and laryngeal squamous/ciliated boundary down to the bronchioles also remain on the surface and are removed on the mucocilliary escalator, then either swallowed or removed by coughing. Any particles consisting of aqueous droplets or fat-soluble organic condensates, or any water- or fat-soluble substances adsorbed on to the surface of insoluble carbon particles, may be absorbed by and affect the ciliated epithelium and submucosal glands of the airways.

The fate of particles penetrating into the terminal bronchioles or alveoli depends on their size and solubility. Larger particles > ~100 nm are engulfed by macrophages and then either carried on to the mucociliary escalator or into the pulmonary lymphatic system.

Ultrafine particles deposited in the alveolar region may penetrate into the alveolar interstitium and into the bloodstream, where they become involved in systemic effects.

## 28.5 Formation of Combustion Particles

Combustion particles may be formed in a number of different ways, depending on the combustion conditions and the physical and chemical properties of their sources. The main sources of combustion particulates are burning gaseous, liquid or solid, mostly carbon-based, combustible materials. These include hydrocarbon gases and liquid fuels, solid fuels and all other carbon-based materials involved in fires (including wildfires, industrial waste fires and fires in the built environment) or combustion processes (domestic, industrial and vehicle combustion).

### 28.5.1 Pyrolysis, oxidation and ring cyclization

The initial process involved in both non-flaming and flaming combustion is thermal decomposition (pyrolysis) by heat (Purser, 2016a, b). When carbon-based solid fuels are heated to approximately 250–300°C and above, they start to break down by molecular scission reactions, releasing volatile molecular fragments into the vapour phase. The composition of these fragments depends on the molecular structure of the fuel and reactions occurring in the vapour phase. For simple hydrocarbon-based polymers such as polyethylene, the main initial pyrolysis products are short-chain alkanes and olefins. In the presence of oxygen a proportion of these are oxidized to aldehydes, ketones and organic acids of different chain lengths. In general the mix is dominated by lower molecular mass compounds such as methane and ethane or ethene, although proportions of higher molecular mass aliphatic alkanes and olefins and aromatic compounds are also formed. Under non-flaming decomposition conditions in the absence of oxygen the higher molecular mass hydrocarbons condense to form a hydrocarbon mist. If oxygen is present this mist contains significant concentrations of irritant aldehydes, including formaldehyde, croton aldehyde and acetaldehyde, organic acids and phenols.

An important set of reactions involving these hydrocarbon fragments are ring cyclization reactions, especially under flaming combustion at the 'smoke point' in flames. These reactions involve progressive oxidation by removal of hydrogen so that, for example, hexane forms cyclohexane. This proceeds to the formation of benzene and the benzene rings combine to form multi-ring structures, including polycyclic aromatic hydrocarbons. At each stage the proportion of hydrogen in the molecule decreases and the proportion of carbon increases. The ultimate fate is the formation of graphene-like sheets, with very high molecular masses, which are the basis of carbon soot particles. Other aromatic compounds formed include benzene, toluene, styrene and methyl styrene, and oxidation products such as phenols. These reactions and products are common to the combustion mixtures from most fuels. In addition, different fuels produce thermal decomposition products related to their original molecular structure. For example, thermal decomposition products from polystyrene are richer in aromatic compounds, starting with styrene monomer, while cellulosic materials initially form proteoglycans.

### 28.5.2 Particle size distribution

While the primary particles formed during combustion, whether sold carbon, liquid droplet or solid condensate, may be in the ultrafine size range (10–100 nm), these tend to agglomerate rapidly to firm strings or irregular masses in the 300–10,000 nm (0.3–10 μm) range. The rate and extent of agglomeration depends on the particle concentration (number of particles per unit volume) and their chemical and physical properties. Smoke aerosols are polydisperse, containing a mix of different particles over the size range 10 nm to ~100 μm. The composition of any particular aerosol at any time with respect to particle size can be expressed in terms of the mass median aerodynamic diameter (MMAD) (defined as the diameter at which 50% of the particles by mass are larger and 50% are smaller), while the geometric standard deviation (GSD) is a measure of the spread of an aerodynamic particle size distribution, typically calculated as follows:

$$\mathrm{GSD} = \left(d_{84} / d_{16}\right)^{0.5}$$

where $d_{84}$ and $d_{16}$ represent the diameters at which 84% and 16% of the aerosol mass are contained, respectively, in diameters less than these diameters.

From a toxicological perspective, although the mass concentration (μg $m^{-3}$) of a particulate aerosol, the particle size distribution and the particle numbers are important, the aggregate particulate surface area may also be of toxicological importance. Considering, for example, a monodisperse aerosol with an airborne mass concentration of 10 μg $m^{-3}$, if the particle diameter is 2 μm, then there are 1.2 particles $ml^{-1}$ of air with an aggregate surface area of 24 $\mu m^2$ $ml^{-1}$ air. If the particle diameter is 0.02 μm, then there are $2.4 \times 10^6$ particles $ml^{-1}$ air with an aggregate surface area of 3016 $\mu m^2$ $ml^{-1}$ air (Donaldson *et al.*, 2001).

An important characteristic of smoke aerosols is that they agglomerate as they age, so that a few minutes after formation, for a given particle mass concentration, the MMAD increases, while the particle number and aggregate surface area decrease.

### 28.5.3 Presence of hetero-elements

Where other elements in addition to C, H and O are present in the original fuel materials, most commonly nitrogen, halogens, sulfur and phosphorus, these tend to be released in a number of forms. Fuels such as cellulosic materials contain small amounts of nitrogen, but a number of common polymers such as wool, polyamide, acrylonitrile and melamine have a significant nitrogen content. When these materials are burned, a proportion of the nitrogen is released as $N_2$ gas. Under well-ventilated combustion conditions a proportion of fuel nitrogen is also released as oxidation products nitrogen (or nitric) oxide (NO) and nitrogen dioxide ($NO_2$), while for under-ventilated combustion conditions an increasing portion is release as reduction products, including some ammonia but principally hydrogen cyanide (HCN) and other nitriles (Purser and Purser, 2008). $NO_2$ is also formed by oxidation of atmospheric nitrogen under the high temperature and pressure conditions in internal combustion engines, especially diesel engines. Sulfur dioxide ($SO_2$) is released during combustion of sulfur-containing materials such as vulcanized rubber, proteins such as wool, and hydrocarbons fuels.

Compounds containing nitrogen, chlorine or bromine and phosphorus are used as flame-retardant additives to reduce the combustibility of polymeric materials widely used in furniture, cable and other electrical products and building materials such as insulation foams. All these elements occur in the combustion products when these materials are involved in fires. Chlorine, bromine and fluorine are an intrinsic part of some materials, or present in additives. When these material are burned, most chlorine is released as hydrogen chloride gas (for example, from polyvinylchloride, which is approximately 50% chlorine by mass). Bromine (mostly present in flame-retardant additives with antimony) is released as hydrogen bromide, but also as bromine. Fluoropolymers, such as polytetrafluoroethylene (PTFE), burn to produce carbonyl fluoride, most of which is subsequently oxidized to hydrogen fluoride (HF). For materials containing phosphorus (also mainly present in flame-retardant additives), combustion results mostly in the formation of phosphorus pentoxide, which hydrolyses to form phosphoric acid (Purser, 1992b, 2016b).

The main fate of these non-CHO elements is therefore release as acid gases, contributing to the irritancy of the smoke, but in each case a relatively small proportion may be released as organic compounds, some of which are of toxicological significance. Some chlorine and bromine is released as substances such as chloromethane, but of most concern are polychlorinated or brominated biphenyls, dibenzofurans and dioxins. Some fluorine may be released as tetrafluoethylene monomer, which reacts in the gas phase to form high molecular mass polymer nanoparticles, which are potent lung irritants. The fate of phosphorus is also controversial and less well understood. As stated, most phosphorus is readily oxidized and ends up as irritant but relatively harmless phosphoric acid. Under reducing conditions a small amount may be released as phosgene ($PH_3$), which is a lung irritant. Of more concern is the extent to which organophosphorus compounds may be released, since a number of these are neurotoxic at very low doses. One specific example is the neurotoxic caged biphosphorus ester trimethylolpropane phosphate (TMPP), which is formed during combustion of organophosphorus compounds (some of which are used a flame retardants) and trimethylol polyols (used in certain hydraulic fluids and turbine lubricants). There is also some evidence that significant release of aromatic phosphate esters may occur during combustion of materials treated with them (Purser, 1992b, 2014, 2016a).

From the perspective of this chapter, an issue with respect to all these potentially toxic substances is the extent to which they are associated with the particulate phase of smoke. Smoke plumes and inhaled smoke contain a mixture of gases and particulates. The acid gases and low molecular mass organic compounds are present partly as vapours but, depending on their vapour pressure and aqueous solubility, partly as liquid droplets or dissolved in condensed water droplets. In addition, they may be absorbed in, or react with, the carbon soot particles. In practice the partitioning of gases such as hydrogen chloride (HCl) between gas, water droplets and soot is dynamic and can vary with the ambient conditions as the smoke ages. From a toxicological perspective, vapour-phase acid gases with a high aqueous solubility are preferentially absorbed in the nose, mouth and upper airways, with limited penetration into the deep lung. $NO_2$ and carbonyl fluoride, with a lower aqueous solubility, are less well absorbed in the upper airways and penetrate into the alveolar region, causing lung inflammation. But since fine particles are deposited mainly in the airways and alveolar region, they can present a mechanism for carrying adsorbed acid gases to this part of the lung. For these reasons, although both gases and particles contribute to toxicity in the airways and alveoli, the contribution from particulates, and particulate-associated substances, generally increases with depth of penetration into the lung.

### 28.5.4 Partitioning of HCN and HCl between vapour and particulate phase of smoke

An example of differential partitioning of acid gases between vapour and particulate phases is given by the results of a full-scale fire experiment, in which an item of upholstered furniture was burned in an enclosed 30 $m^3$ room (Purser, 2016b). The covering and foam from which the item was constructed, and the flame-retardant additives, had a significant content of nitrogen and chlorine. The fire grew over a 5 min period, after which it self-extinguished due to lack of ventilation. Two acid gases, HCl and HCN, were measured in the room smoke layer at intervals during the fire and for a further 5 min after it extinguished.

The smoke was sampled using several different methods to compare sampling losses and partitioning between vapour and particular phases. Inside the fire room the smoke was sampled using a battery of evacuated 'grab vessels', which were fired in sequence to take a succession of samples. Each grab vessel was connected to the sampling valve via a stainless steel tube 150 mm long and 2 mm internal diameter. An additional sampling system consisted of an open-face glass-fibre filter to trap particulates and Dreschel bottle impinger train containing sodium hydroxide solution to trap acid gases. These were used to take an integrated sample directly from the smoke inside the fire room during the 5 min period between 5 and 10 min after ignition. A second set of the two sampling systems was placed outside the room and supplied by samples pumped continuously along a 3 m PTFE-lined stainless steel sampling line heated to 120°C. The findings were that the grab vessel and sampling line systems provided a good sample of the fire gases, but that particulates tended to become lost in the valve and sampling system and in the 3 m heated sampling lines. The results are summarized in Table 28.2.

For HCN the grab vessel results in the room show a rapid increase over the 5 min of flaming to a peak 550 ppm HCN as the fire self-extinguished, then remaining constant at around this level for the next 5 min. The average HCN grab vessel concentration during the second 5 min period (558 ppm) is very similar to the integrated HCN concentrations over this period measured by the filter and impinger systems both inside the room (610 ppm) and outside the room (560 pm). For this system, 100% of HCN was recovered from the impinger solution and none from the filter. From this it was concluded that HCN remained in the gas phase, did not react with the sampling lines and did not attach to any particulates deposited in the lines or on the filters. This contrasts with the results for HCl, which showed a rapid increase to 387

**Table 28.2.** Differential partitioning of HCN and HCl between gas and particulate phases of smoke during furniture fires.

| Time from ignition | HCl grab samples (mainly gases) | HCl filter and impinger samples (particulates and gases) | HCN grab samples | HCN filter and impinger samples |
|---|---|---|---|---|
| 0 | 0 | | 0 | |
| 1 | 0 | | 0 | |
| 2 | 20 | | 20 | |
| 3 | 70 | | 70 | |
| 4 | 200 | | 250 | |
| 5 | 387 | 379 (241)[a] | 550 | 610 (560)[b] |
| 6 | 200 | 379 (241) | 525 | 610 (560) |
| 7 | 170 | 379 (241) | 580 | 610 (560) |
| 8 | 120 | 379 (241) | 570 | 610 (560) |
| 9 | 100 | 379 (241) | 560 | 610 (560) |
| 10 | 88 | 379 (241) | 560 | 610 (560) |

[a]Integrated samples 5–10 min, 92% on filter; [b]integrated samples 5–10 min, 100% in impinger. Data in parentheses: from external line.

ppm during the first 5 min in the room grab samples, in parallel with the rise in HCN, but then following a declining curve over the next 5 min to 88 ppm. However, the integrated HCl concentration measured in the filter-impinger sample in the room was 379 ppm, which was close to the peak value measured in the grab samples at 5 min. The integrated sample concentration taken outside the room at 241 ppm was 36% lower than that taken directly from the smoke in the room, indicating significant losses in the 3 m sampling line. For both these samples, 92% of HCl was recovered from the filter and only 8% from the impinger, indicating that over the 5–10 min period almost all the HCl became associated with the particulate phase of the smoke. The low HCl recoveries from the grab vessel data for the second 5 min indicate that most of this particulate-bound HCl is lost in the narrow-bore sampling system, but from the finding that the 5 min HCl peak is similar in both systems it is concluded that most of the HCl is released into the gas phase during the active first 5 min of the fire. Then, once the formation of new HCl ceases as the fire extinguishes, the airborne HCl gas gradually becomes absorbed on to the suspended soot particles.

## 28.6 Acute Toxicity of Smoke Particulates Inhaled During Fires

### 28.6.1 Immediate effects during exposure to high-concentration smoke plumes

Smoke aerosols formed during fires consist of particles over a very wide size range from millimetre diameters down to around 10 nm. Most of the aerosol mass consists of particles < ~10 μm, with large fractions < 2 μm and depositing in the airways and alveolar region, but a significant fraction depositing in the nose and mouth.

The particulate mass concentration in smoke plumes close to a fire, for example inside a building, can be as high as 10 mg $l^{-1}$ (10,000 mg $m^{-3}$). These carbonaceous particles can be highly irritant, due to the presence of adsorbed acid gases and carbonyl compounds. Significant concentrations of acid gases and droplets are also present, as well as asphyxiant gases. Incapacitation and death of most fire victims results from exposure to this mixture of toxic gases and particulates.

The immediate effects reported by subjects exposed to high-concentration smoke plumes during fires are painful irritation to the eyes, nose and mouth, with breathing difficulties and a choking sensation. Deposition of irritant vapours and particles in the eyes results in blepharospasm – a reflex closure of the eyelids. The upper respiratory tract pain can be accompanied by chest pain, bronchoconstriction and stimulation of airway receptors resulting in a hyperventilatory response with an atypical breathing pattern. Deposition of irritant particles in the airways can stimulate mucus production (Purser and McAllister, 2016).

A survivor of the Manchester Airtours aircraft fire in England in August 1985 reported that his mouth and throat filled with soot and mucus during the few minutes of exposure in the cabin until he escaped (King, 1988). He described having to physically remove this airway obstruction with his hands in order to breathe. In a study on the toxicity of solder flux fumes in rats, some deaths resulted from the formation of a plug of deposited particles and mucus in the trachea. From these reports it is evident that physical obstruction of the airways by deposited irritant particulates can be a factor causing distress and escape impairment in fire victims. However, the general situation is that while inhalation of high smoke-particulate concentrations can be distressing and debilitating, it does not in itself result in severe incapacitation or death during exposures of up to around 30 min at fire scenes. Even at extremely high particulate concentrations of 10 mg $l^{-1}$ this represents a mass dose of only 200 mg $min^{-1}$ for a volume of air breathed per minute ($V_E$) of 20 l $min^{-1}$ (equivalent to an adult engaged in light work such as walking).

These concentrations are up to 1000 times greater than exposure limits of workplace nuisance dust (short-term exposure limit (STEL) 10 mg $m^{-3}$) (HSE, 2011), and enormously greater than the World Health Organization (WHO) 24 h limit for $PM_{10}$ of 50 μg $m^{-3}$ (WHO, 2006), which raises the issue of how fire victims can survive such high exposures when peaks of general air pollution around 100 μg $m^{-3}$ are associated with increase in acute cardio-respiratory deaths (COMEAP, 2010). It is possible that the brief exposures to very high particulate concentrations in fires may result in some deaths at the fire scene, but the evidence from post-mortem studies is that deaths from smoke exposure (i.e. for unburned fire victims) are mostly associated with lethal exposure doses of CO (> 50%COHb, i.e. > 50% carboxyhaemoglobin). When the distribution of %COHb in non-burned fire victims is compared with that in victims of CO gas poisoning, these are similar, although there are somewhat more smoke deaths at lower %COHb levels than for CO poisoning cases (Purser and McAllister, 2016). The evidence is that smoke deaths at the fire scene result mainly from CO poisoning, but with some contribution from other toxic smoke products, which include HCN, acid gases and irritant particulates.

Although particulate concentrations of toxic gases can be very high in smoke plumes at a fire scene (for example, in smoke from a burning room flowing down a corridor in a building), the concentrations decrease rapidly as the hot plume entrains air as it flows away from the point of origin. As smoke plumes flow from buildings or from outside fires, they rise due to buoyancy and are greatly diluted by air entrainment. The result is that while for smoke plumes inside buildings the main concerns are toxicity from the incapacitating effects of asphyxiant gases, once a plume is outside these are diluted by a factor of 100 or more, so that concentrations of CO and HCN are reduced to levels that are not toxicologically significant. However, human sensitivity to odour and irritancy (from gases and particulates) is such that even at these levels of dilution, there may be detectable acute health effects after exposure periods of an hour or so, especially on susceptible subjects. The main effects are nuisance odour and minor eye and upper respiratory tract irritation, sometimes with transient minor throat irritation the day after exposure. As an approximate guide, if the smoke is sufficiently dense that visibility is around 10 m, then the particulate concentration is likely to be approximately 40,000 $\mu g\ m^{-3}$, while 100 m visibility represents a concentration of approximately 4000 $\mu g\ m^{-3}$. The former represents the approximate particulate concentrations within around 100 m of a large fire when enveloped in the smoke plume (i.e. for down-washed smoke plumes), while the lower concentrations are likely to be encountered within around 500 m. These concentrations are still well in excess of $PM_{10}$ levels encountered from environmental air pollution, but exposure durations are typically limited to minutes to an hour during fire incidents. Of course the total smoke mass concentrations represent the whole aerosol, not just the $PM_{10}$ component.

### 28.6.2 Acute and sub-acute effects 1–72 hours after high-concentration smoke exposure

A major hazard following short exposures to high smoke concentrations in fires is acute inflammation of the airways and lung oedema (acute respiratory distress syndrome (ARDS)) occurring a few hours after exposure. Fire victims rescued alive at a fire scene are likely to have inhaled around 200 mg of particulate soot as well as incapacitating doses of CO and HCN. Although all these contribute to incapacitation during and immediately after exposure, the main immediate survival hazard is the inhaled CO dose, characterized in terms of %COHb. For < ~40%COHb the survival prognosis is good, especially if oxygen is administered from the time of rescue, as is the normal practice, to 'wash out' the CO (Purser and McAllister, 2016). Provided the extent of CO exposure has not resulted in serious cerebral damage, subjects typically make a rapid recovery and are conscious around the time of admission to hospital. However, the clinical condition of a proportion of these subjects with significant smoke particulate exposure may start to decline from around 2 h after rescue. An acute crisis then follows, characterized by respiratory insufficiency, hypoxia and hypercapnia, resulting from acute lung inflammation and oedema. Subjects supported through this crisis with oxygen and steroids may start to recover within a few hours as the inflammation and oedema subside, or they may die. A further complication, especially with elderly subjects, is the development of bronchopneumonia, which may also become fatal within 36–72 h (Purser, 2017). Another issue is the presence of body surface or respiratory tract thermal burns. Body surface burns can have secondary effects on fluid exchange in the lung, while thermal burns to the respiratory tract combine with the chemical smoke burns. Thermal burns are generally limited to the upper respiratory tract, especially the mouth and larynx, while chemical 'burns' also affect the airways and alveolar region, especially when soot particulates are deposited.

Subjects who come through this acute crisis may make a repaid recovery, without long-term lung pathology. Similar effects have been observed in animal studies. When primates were exposed to irritant smoke, their respiratory pattern was disturbed during exposure but post-exposure effects were generally minor up to a critical dose level, above which animals developed a fatal lung oedema overnight (Purser and Buckley, 1983). Similarly, when groups of rats or mice were exposed to irritant smoke they either survived with minimal lung pathology or developed a fatal oedema within 24 h, with heavy congested lungs.

A large proportion of smoke particles deposited in the upper respiratory tract or airways during an acute exposure to a smoke plume is removed by the mucocilliary escalator within 24–48 h, while those around 0.5–2 microns deposited in the bronchioles and alveoli may be ingested by macrophages and also removed via the airways or the lymphatic system over a period of days. So although a small proportion of these particles may remain in the lung, the adverse effects of a single acute exposure tend to be transient. For those exposed to high smoke concentrations the acute effects may be serious as described, while those exposed to lower concentration may experience minor hoarseness or sore throat the day after exposure. Follow-up of a group of subjects exposed to significant smoke during house-fire incidents showed no physical respiratory symptoms after 6 months (Purser and Kuipers, 2004).

### 28.6.3 Contribution to lung irritancy and inflammation from gases and particulates

Since smoke contains acid gases both in vapour form and associated with particles, the overall toxicity results from the combined effects of both but, as stated, there tends to be a separation within the respiratory tract between gaseous and particulate deposition, depending on the aqueous solubility of the gas and the size distribution of the particulates. It is also important to appreciate that this partitioning varies in different animal species and so care is needed when extrapolating results in rodents to those in humans and other primates.

In Table 28.2 data were presented on the extent of association of HCl between vapour and particulate phase, which affects sites of deposition within the lung. Differences between the acute toxicity of HCl in vapour and particulate form is illustrated by a set of experiments in which mice were exposed to HCl either as dry gas or in the form of hydrochloric acid mist generated as a respirable droplet aerosol using an ultrasonic atomizer. In these experiments the effect of inhaling HCl as a gas were dominated by reflex decrease in respiratory rate and a pattern of breath holding, typical of upper respiratory tract irritation. When the equivalent HCl concentrations were administered as an acid mist, the respiratory rate depression was much reduced, but the mice showed signs of lung inflammation 24 h after exposure, due to penetration of the aerosol into the lung (Purser, 2007). Similarly, when an attempt was made to develop a model for chronic bronchitis by exposing rats to (highly water-soluble) $SO_2$ atmospheres, effects on the respiratory tract (squamous metaplasia, increase in mucocytes and changes to submucosal glands) were found only as far down as the trachea.

Although similar effects occur in humans and other primates, these differ from rodents in that they can breathe via the mouth, and that the relative surface area of the nose and upper respiratory tract of rodents is much greater than in humans (Purser, 2010b). This results in greater penetration of gases in addition to particulates into the deeper lung in humans, which is partly why $SO_2$ can lead to chronic obstructive pulmonary disease (COPD) in humans but not in rodents.

### 28.6.4 Acute inflammation caused by ultrafine particles

A proportion of particulate smoke aerosols is in the ultrafine size range. Ultrafine particles introduce a whole category of serious toxicity issues not shown by larger particles. The most dramatic example of this phenomenon is presented by the acute toxicity of thermal decomposition products from per-fluoropolymers (organic polymers composed of carbon and fluorine). Combustion products from fluoropolymers show a great range of toxic potencies, depending upon the exact thermal decomposition or combustion conditions. A detailed account is given in Purser (1992a). When fluoropolymers such as polytetrafluoroethylene (PTFE) are decomposed at temperatures above 650°C under both non-flaming and flaming decomposition conditions, the resultant atmosphere is highly toxic to rodents, with an $LC_{50}$ for a 30 min exposure in rats in the 5–50 g $m^{-3}$ mass loss range (i.e. when products from combustion of 5–50 g of PTFE are dispersed in each cubic metre of air) followed by a 14-day post-exposure observation period. Toxicity results from post-exposure airway and lung inflammation and oedema, which can be accounted for in terms of the known toxic product carbonyl fluoride. This vapour has a low aqueous solubility and so penetrates deeply into the lung, where it hydrolyses to form HF and per-fluoroisobutylene. As with other acid gases causing lung inflammation, inflammation results from direct toxicity to the airway epithelium and alveolar type 1 pneumocytes. During exposure, rats also exhibit signs of upper airway sensory irritation, consisting of slow respiration with pauses at end inspiration between breaths.

Fluoropolymers are thermally stable below 400°C, but over a range of 400–650°C non-flaming decomposition occurs at an increasing rate, depending on the temperature. The main decomposition product is tetrafluoethylene monomer in the form of a vapour. In the cooling vapour cloud the monomer re-polymerizes to form a high molecular mass condensate in the form of an ultrafine particulate atmosphere consisting of particles approximately 10 nm in diameter, and thus in the nanoparticle size range, when freshly formed. When rats are exposed to this freshly formed particulate atmosphere at low concentrations, they develop a different respiratory pattern within a few minutes of the start of exposure, consisting of a hyperventilation characteristic of irritant effects on lung vagal nociceptors.

The ultrafine particles are formed in large numbers in fresh fume but agglomerate rapidly within a minute or so, forming particles larger by a factor of ten or more. This aged atmosphere was found to be harmless when rats were exposed to it. The conditions of generation and exposure are therefore critical to the atmosphere composition and toxicity. When rats were placed in small chambers close to continuously generated ultrafine fume, they were seriously affected within a few minutes, but when placed in larger chambers, those close to the fume source were badly affected, while others placed approximately 1 m away were unaffected. Although both groups of rats were exposed to the same total mass concentration of fumes, those further away from the source inhaled the agglomerated particulate, which was non-toxic.

Under temperatures or decomposition conditions where rats were exposed to the freshly formed nanoparticle atmosphere, death occurred due to lung inflammation and oedema, with a minimum $LC_{50}$ concentration of 0.017 g $m^{-3}$ fluoropolymer mass loss. When the particulates have had time to agglomerate before being inhaled, or where they are mixed with particulates from other materials such as wood smoke, the much lower toxicity is more related to those of the normal toxic products, with $LC_{50}$ in the 0.5–5 g $m^{-3}$ range, but the greater the opportunity for re-circulation and reheating of the particulates, maintaining exposure to fresh ultrafine particulate, the lower are the measured $LC_{50}$ concentrations, of which only a proportion is in the form of ultrafine particulates. This shows that fresh fluoropolymer particulate exhibits an extreme toxic potency, which is more than 1000 times more toxic than combustion products from other common polymers and is much more toxic than other ultrafine particulates. Since the $LC_{50}$ of perfluorisobutylene vapour is 0.047 g $m^{-3}$ and only a small proportion of the mass of decomposed PTFE is released in this form, and since the $LC_{50}$ of HF and carbonyl fluoride are 1.88 and 1.96 g $m^{-3}$, it is evident that these cannot account for the observed toxicity at these very low $LC_{50}$ levels.

The mechanism of toxicity by ultrafine particulates is not fully understood, but the inflammatory reaction starts within a few minutes and is fatal to rats over periods of a few hours to days following exposure. One factor may be the small particle size and enormous number of particles, which are small enough to penetrate readily into alveolar lining pneumocytes and also penetrate between them into the lung interstitium and bloodstream. Toxicity from nanoparticles involves oxidative stress, resulting in pro-inflammatory changes (Purser, 1992a; Donaldson *et al.*, 2001, 2016). This may involve lung overload with impaired phagocytosis by macrophages resulting in the release of inflammatory cytokines. The mechanism may therefore be partly a physical one related to the high particle numbers and surface area. But both with fluropolymer and with other nanoparticles there is evidence that oxidative surface chemistry may be important, involving peroxide and superoxides. PTFE nanoparticles have been shown to bear stable peroxy free radicals on their surface (Metcalf *et al.*, 1991). Nanoparticles of carbon black and of graphene have been shown to cause cell injury by mechanisms that include oxidative stress. Redox- active transition metals such as iron (Fe), copper (Cu) and titanium (Ti) also cause inflammation when present in nanoparticles. This may explain the oxidative toxicity of diesel nanoparticles, which contain both carbon and transition metals in the particulates (Donaldson *et al.*, 2001, 2016).

Another toxicity issue with inhaled nanoparticles is that they may penetrate through the alveolar interstitium and into the bloodstream, where they can cause inflammatory effects, especially on the cardiovascular system. A symptom of acute exposure developing a few hours later in humans is polymer fume fever, a pyrogenic response consisting of influenza-like symptoms. Polymer fume fever has been reported from polyvinyl chloride (PVC) fumes, but especially from exposure to low concentrations of fumes from fluropolymers heated under conditions likely to produce nanoparticles. These have occurred under workplace situations involving heating of PTFE or other fluoropolymers. An example is a case of a worker in an electronics factory in which a flouoropolymer spray was used as a releasing agent for plastic mouldings. Two female workers shared this particular operation, of which one developed flu-like fever symptoms on several occasions a day or so following a shift. Upon investigation it transpired that this particular worker was in the habit of taking a cigarette break occasionally while on her shift, while the other was a non-smoker. It was concluded that the worker's fingers were contaminated with the fluoropolymer, a small amount of which was absorbed by the cigarette paper, then pyrolysed and inhaled as the cigarette was smoked.

A potentially fatal condition that can result from acute exposure to irritant particles and vapours is bronchiolitis obliterans, whereby inflammation of the terminal bronchioles results in extensive blockage and atelectasis, with severely compromised gas exchange. Several firefighters suffered from this after inhaling smoke from burning polyvinylchloride wiring during the New York Telephone exchange fire in 1975 (Wallace, 1981).

Epidemiological studies have revealed associations between the incidence of both respiratory and cardiovascular disease, including deaths, and both short- and long-term ambient $PM_{2.5}$ derived from sources such as diesel exhaust particulates. There are several pathways by which nanoparticles deposited in the lungs may affect the cardiovascular system (Donaldson *et al.*, 2016). One is that inflammation in the lung results in release of inflammatory mediators (oxidants and cytokines), which enter the circulation and may enhance inflammatory events in the walls of blood vessels and their endothelium, resulting in plaque instability. Oxidants such as superoxide may deplete NO and further enhance this effect. A second mechanism may be direct carriage of nanoparticles in the blood to the cardiovascular epithelium, which has been demonstrated experimentally. A third mechanism may be uptake of nanoparticles by cells of the immune system, resulting in systemic immunological effects (of which fume fever is one). A fourth postulated effect is by stimulation of sensory receptors in the lung resulting in autonomic reflex effects on cardiac function such as heart rate variability. These deleterious effects can result in serious acute exacerbations, generally of pre-existing disease; for example, the spikes in cardiovascular-related acute hospital admissions and deaths around 24–36 h following spikes in ambient air pollution. For fire victims these effects may add to the cardiovascular stress resulting from exposure to CO and HCN at the fire scene, thereby increasing the risk of heart attacks and strokes during the hours after rescue. The role of particulates in cardiovascular events following fire exposures has not been studied specifically. The author encountered one case in particular in which a subject was awoken in a smoke-filled hotel room by firefighters. He escaped using a ladder and appeared unharmed and was taken to another hotel, where he spent the night. The next morning, while attending a business meeting he suffered a heart attack, following which he lost the sight of one eye due to a retinal embolus.

As stated, the concentrations of particles in smoke plumes decrease rapidly, by orders of magnitude, as a function of distance from a fire source, due to air entrainment and mixing in the smoke plume (Purser, 2016b). Because smoke plumes are buoyant, they also tend to rise rapidly outdoors, before the much diluted plume cools and grounds, often many kilometres from the source. For this reason, the acute effects of exposure to residents of neighbourhoods around a fire incident are usually minor. The extent of exposure and potential for adverse health effects increase somewhat for occupants close to the fire, especially when wind conditions result in downwashed plumes (Hall and Spanton, 2016). Wildfires can result in relatively higher concentration exposures of large numbers of people. In these scenarios firefighters and local volunteers (often with limited or no respiratory protection) may be exposed to dense smoke plumes, while whole populations (for example, of adjacent towns or even cities) may be exposed to more diluted smoke atmospheres for up to several days or even weeks. These exposure concentrations are likely to exceed those produced by ambient air pollution by a considerable margin, so studies of such populations should reveal the extent of any acute adverse health effects. Although some such effects have been reported, the somewhat surprising overall findings seem to be that effects on exposed populations are in general relatively minor (McAllister, 2016). However, when very large populations (such as those of a large city such as London) are exposed to spikes in ambient air pollution, then there is a definite epidemiological increase in acute cardio-respiratory hospital admissions and deaths (COMEAP, 2010). It is likely that these represent effects on a small cohort of very sensitive individuals, often elderly and with pre-existing health conditions, but similar spikes might be expected when large populations are exposed to wildfire smoke. The lack of such findings may be due to inadequate data collection, or it may be that wildfire smoke is less toxic than ambient air pollution derived mainly from vehicle emissions and related sources.

## 28.7 Chronic Toxicity of Inhaled Particulates

### 28.7.1 Chronic toxicity from acute or chronic exposure to mineral particles

Although acute exposure to high concentrations of irritant particulates during fires can result in some degree of permanent lung injury, most subjects make a good recovery if they survive the phase of acute inflammation and oedema. This may not be the case if a large fire results in significant damage to building structures, particularly if partial or complete collapse occurs, whereupon significant amounts of mineral dust particles and fibres may become entrained into and dispersed in the smoke plume. As with smoke particles, the sedimentation rate from a dispersing plume depends upon the particle size and density. Mineral particles tend to be denser than carbon smoke particles, especially those containing heavy metals, so greater deposition tends to occur closer to the source than for soot particles. Collapsing structures also release clouds of dust particles, a large proportion of which may be too large to be inhalable or respirable. Fibres may be carried a greater distance than spherical particles for a given physical size. Another issue close to the scenes of major fires is that significant deposits of soot and mineral dust may remain on and around the site for months, during which period local residents, investigators and other workers at the site may be at risk of repeated exposure if they do not wear adequate respiratory protection.

A particular example of both acute and chronic health effects resulting from inhalation of mineral dust at a fire scene is that of the World Trade Center (WTC) incident on 11 September 2001. Around 10 million tonnes of dust were released, exposing around 300,000 rescue workers and New York City residents to particulate matter at concentrations in the range of 1–100 mg $m^{-3}$ or perhaps even higher. Some exposure of rescue and recovery workers continued for around 3 months. Analysis of the dust revealed a very complex mixture that varied in composition over different particle size ranges and at different locations where it was deposited (Kendall *et al.*, 2016). Deposited dust (mainly 10–54 μm) consisted of large and fine particulate matter and approximately 50% fibres. The particulate matter was highly alkaline, consisting of stone, plastic and glass. As well as calcium silicate-based cementitious dust, there was a significant crystalline silica component. The major elemental components included calcium (Ca), silicon (Si), sulfur (S), magnesium (Mg), aluminium (Al), Ti, potassium (K) and Fe, along with sodium chloride (NaCl) and even some uranium (possibly from the airplanes). Fibres included glass, asbestos and carbon. Air dust sampled in the general area included significant $PM_{2.5}$ and $PM_{10}$ fractions of mineral dusts, indicating a considerable respirable component.

Inhalation of this dust, both during and after the fire, led to signs of acute lung inflammation and 'World Trade Center cough', but this has persisted in exposed subjects for months and years after exposure. The majority of WTC fire fighters evaluated within 6.5 years showed signs of airway obstruction, inflammatory biomarkers, sarcoid-like granulomatous pulmonary disease and airways hypersensitivity (asthma).

Since this dust was a complex mixture of particles and fibres it is difficult to specify the main causes of health effects, but there are several possibilities from the know effects of different mineral dusts and fibres. One important aspect may be the heavy initial exposure to coarse mineral dust. The relatively large amount inhaled at the scene enabled doses to be delivered deep into the lung that overwhelmed the normal clearance mechanism (lung overload), especially that of macrophages in the deep lung. There is some evidence that this acute exposure to alkaline mineral dust caused significant acute epithelial damage and inflammation, leading to airway remodelling and hypersensitivity. These effects therefore partly represent an acute injury with persisting effects.

Another issue may be mineral dust and fibres remaining in the lung, which may present a continuing insult to the alveolar epithelium. This consideration applies particularly to some insoluble materials with adverse effects on macrophages. In the terminal bronchioles and alveoli, removal of particles is achieved by macrophage phagocytosis. Biologically inert dust particles are removed gradually by this process, though a proportion may remain semi-permanently in the lung (for example, in the 'black' lungs of ex-smokers). Silica particles are not biologically inert. When these are phagocytosed by macrophages, they are toxic, so that the macrophages die and release the cytotoxic cell content into the lung in a repeated cycle. This chronic insult to the lung eventually results in silicosis, whereby the lungs stiffen due to the formation of collagen fibres. Inhalation of insoluble fibres can have a similar effect. When asbestos or carbon particles are phagocytosed, they pierce through the macrophages (appearing like kebabs on a stick). The cytotoxic cell content leaks out and damages the lung, resulting in lung fibrosis (asbestosis). A similar effects can occur following inhalation of ceramic fibres. Glass fibres are less of a problem because, although they can penetrate the lung and be phagocytosed, macrophages are capable of dissolving the glass over an extended period, so that the fibrosis risk is reduced.

### 28.7.2 Chronic toxicity from protracted exposure to low particulate concentrations

In addition to the relatively high-concentration, short-duration exposures at fires, all of us are exposed continually throughout our lives to ambient indoor and outdoor air pollution. A major component of air pollution, arguably the major component of $PM_{10}$ and $PM_{2.5}$, is derived from combustion sources. Epidemiological studies of the effects of outdoor air pollution have demonstrated associations with respiratory diseases such as asthma and COPD, as well as lung cancer, and also cardiovascular disease. Indoor air pollution from sources such as cookers and wood-burning stoves has also been associated with these diseases. There are many good reviews of the mechanisms and epidemiological effects of indoor and outdoor air pollution, in terms of both incidence and the relative contributions from particles and gases, including nitrous oxides (NOx), ozone and $SO_2$.

Another major source of inhaled particulates is tobacco smoke, the effects of which have also been widely reported. As with exposure at fire scenes, inhalation of mainstream tobacco smoke results in particulate deposition at different sites in the respiratory tract. Although mainstream tobacco smoke is highly respirable, containing small particles depositing in the alveoli, considerable deposition also occurs in the upper respiratory tract and airways. The most important toxic products – organic irritants and carcinogens – are associated with the particulate phase of the smoke. The sequence of effects on the lung after a subject starts smoking provides a useful summary of the repertoire of effects in different parts of the lung and of systemic effects. First, while actively smoking, there are the acute irritant effects on the nose, mouth and upper airways, accompanied by coughing and some bronchoconstriction. Second, within a few hours, direct effects of these irritant particles and vapours on the airway epithelium result in cilostasis and within a day or so this leads to squamous metaplasia, with loss of the ciliated epithelium in the most exposed locations. This is first evident in the larynx, where the boundary between the squamous epithelium of the buccal cavity gradually extends further through the larynx towards the trachea over a period of weeks or months. A particularly vulnerable impaction site is the tracheal bifurcation, which is a common site of cancer formation after years of exposure.

As exposure continues, the next hazard is acute (gradually developing to chronic) obstructive lung disease. This includes proliferation of mucocytes in the airway epithelium, thickening of the mucosa and narrowing of the airways. The submucosal glands produce serous and mucous sections that provide the secretions for the mucocilliary escalator. Continued irritation leads to an increase in the more viscous mucous secretions and, over a period of years, to proliferation of the submucosal glands. This results in an overall increase in the release of viscous mucus-rich secretions, which, in combination with the thickened and narrowed airways, impair lung ventilation and gas exchange. Complete obstruction of some bronchioles also impairs gas exchange and lung ventilation/perfusion ratios. Chronic exposure of the alveoli can result in breakdown of tissue, leading to emphysema.

The development and exacerbation of hypersensitivity is another aspect through development of asthma and reactive airway dysfunction syndrome (RADS). Apart from the effects of irritants on airway sensitivity, smoke particulates contain sensitizers such as formaldehyde and isocyanates. Another under-researched area is the consequences of interactions between NOx and sunlight resulting in the formation of ozone, which reacts with terpenes released by plants, such as limonene. This reaction of ozone with carbon–carbon double bonds results in the formation of highly reactive peroxides (open air factor), which may have effects in the respiratory tract when inhaled.

### 28.7.3 Chronic toxicity resulting from acute exposure to carcinogens and dioxins

Another important aspect of chronic toxicity from combustion products is the presence of a wide variety of carcinogens. These carcinogens include halogenated dioxins, dibenzofurans and polychlorinated biphenyls, which, in addition to being carcinogenic, are associated with other toxic effects in humans, including chloracne, effects on the liver, reproductive system and thyroid stimulating hormone, and a wider range of toxic effects in animal studies (COT, 2001; EPA, 2012).

Although some of these carcinogens, especially the more volatile lower molecular mass species, may be partly present in the vapour phase, the majority either have low vapour pressures and so are present as condensates, or are adsorbed on carbonaceous particulates. It is therefore reasonable to consider carcinogens and dioxins mainly as components of particulates. Some of these inhaled carcinogens target sites of deposition in the respiratory tract (nose, airways or lung) but for many the main target sites of carcinogenicity or other toxic effects are systemic.

Exposure to combustion-derived dioxins and other carcinogens occurs in several ways in indoor or outdoor environments and in occupational contexts, including:

- acute inhalation exposure to a fire smoke plume;
- acute dermal and inhalation exposures at fire sites (during post-fire investigations or clean-up procedures) or through contact with contaminated clothing;
- acute oral ingestion (eating with soot-stained hands or clothing in a contaminated area);
- long-term exposure due to repeated acute occupational exposures to one or more of the above sources (among, for example, firefighters, fire scene investigators or workers exposed to diesel fumes in confined workspace environments);
- long-term workplace exposure to industrial combustion processes;
- long-term exposure to aerosolized soot and volatile organic substances in buildings that have been inadequately decontaminated during post-fire restoration (resulting in a 'sick building');
- long-term indoor exposure to fumes from poorly vented solid-fuel cooking or heating fires;
- long-term outdoor exposure to air pollutants derived from combustion processes; and

- long-term oral exposure to foodstuffs or drinking water contaminated by combustion-derived dioxins and other carcinogens as a result of environmental releases.

Evidence for the carcinogenicity from combustion products is obtained from three main sources:

- measurement of known or probable carcinogens in combustion product mixtures;
- epidemiological evidence for increased cancer risk in specific occupational or environmental exposure populations; and
- animal inhalation and other toxicity studies.

Carcinogenic chemical are classified by the International Agency for Research on Cancer (IARC) according to its scheme for the substances for which studies have been reported in IARC Monographs. The scheme groups substances in terms of their carcinogenic activity, ranging from Group 1 (carcinogenic to humans) to Group 4 (probably not carcinogenic to humans).

The IARC website (IARC, 2014) includes a list of Agents classified by the IARC Monographs 1–109. Combustion products include tobacco smoke (direct and second-hand exposure) and diesel exhaust in Group 1. Indoor exposure to emissions from household combustion of biomass fuel (primarily wood) is classified in Group 2A and engine exhaust (gasoline) as 2B. Almost the entire population is exposed to these substances, many to a significant extent, and in recent years awareness of the risk has prompted measures to reduce exposures in some cases.

Carcinogens associated with combustion products are listed in Table 28.3. These include low molecular mass vapours such as formaldehyde, 1,3-butadiene and benzene, high molecular

**Table 28.3.** List of carcinogens found in combustion products (compiled from references cited in Purser, 2016c).

| Group | Substances |
|---|---|
| Substances containing carbon and hydrogen with or without oxygen (any organic fuel) | Formaldehyde, acrolein acetaldehyde<br>Ethylene oxide,1-3 butadiene, benzene<br>Styrene, methyl styrene<br>Polyaromatic hydrocarbons: benzo[*a*]pyrene, benzo[*b*]fluoranthane, benzo[*j*]fluoranthene,<br>Benzo[*k*]fluoranthane, dibenzo[*a,i*]pyrene, indeno[1,2,3-*cd*]pyrene, dibenz[*a,h*]anthracene<br>5-Methylchrysene |
| Substances containing nitrogen in addition to CHO (any nitrogen-containing fuel, tobacco, diesel exhaust particulate) | Acrylonitrile, ethyl carbamate, hydrazine<br>*N*-Nitrosamines: *N*-nitrosodiethylamine, 4-(methylnitrosamino)-1-(3-pyridyl)-1-butanone (NNK)<br>Nitroarenes: (e.g.) 3-nitrobenzanthrone, 6-nitrochrysene, 1-nitropyrene<br>Aza arenes: di dibenz[*a,h*]acridine, 7H-dibenzo[*c,g*]carbazole, toluene diisocyanate |
| Substances containing halogens in addition to CHO (any fuel containing chlorine or bromine, pentachlorophenol from treated wood) | 1,2,-Dichloromethane, pentachlorophenol, polychlorinated bi-phenyls (PCBs)<br>Chlorinated di-benzo dioxins, Chlorinated di-benzo furans<br>Brominatated di-benzo dioxins, brominated di-benzo furans |
| Substance containing phosphorus in addition to CHO (flame retardant treatment) | Tris-(2,3, dibromopropyl) phosphate |
| Metals: tobacco, vehicle exhaust emissions, treated wood, coal | Chromium (hexavalent), nickel, molybdenum, antimony (trioxide), arsenic, lead |
| Radioactive metals: tobacco smoke | Polonium-210 |
| Mineral fibres: fires involving asbestos or ceramic fibres (but not glass fibre or mineral wool) | Asbestos fibres<br>Ceramic fibres |

mass aromatics such as polycyclic aromatic hydrocarbons (PAHs) and metals. If nitrogen is present, carcinogens including acrylonitrile and nitrosamines can also be formed; and if halogens such as chlorine are present, then dichloromethane, polychlorinated biphenyls (PCBs), chlorinated dibenzodioxins and dibenzofurans are also formed. For inhaled combustion products the main target is the lung. This is especially the case for higher molecular mass aromatic organics and metals, which are mainly associated with smoke particles deposited in the lung. Some substances are absorbed systemically and target other organs, such as the blood, lymphatic system, liver (where many substances are metabolized, sometimes into carcinogenic substances), the kidneys and bladder (where substances tend to become concentrated before excretion). Some dermal absorption may occur when soot collects on clothing during handling or from frequent contact with dirty clothing impregnated with soot (testicular or scrotal cancer risk).

Where halogens are present in the burning fuel, the combustion products include PCBs, halogenated dioxins and furans, which are associated with a range of toxic effects, including carcinogenicity. As with other combustion products, the yields in fires are very dependent upon the combustion conditions, being relatively high in poorly ventilated fires and very low in emissions from modern incinerators. Environmental contamination from dioxins has occurred historically from a range of sources but controls on these have resulted in a considerable decrease since the peak in the 1970s to low present-day levels, despite increasing production and use of halogenated materials such as PVC (Purser, 2016c).

The average daily intake of dioxins and furans as toxicity equivalency (TEQ) for a 70 kg adult from different sources is estimated as shown in Table 28.4; these are below or comparable with exposure limits and guidelines from different agencies (UK Committee on Toxicity (COT), UN WHO and US Environmental Protection Agency (EPA)) of 0.7–2 pg TEQ $kg^{-1}$ bodyweight per day.

In its 2001 report, COT estimated the daily intake for the average consumer as 1.8 pg WHO-TEQ $kg^{-1}$ bodyweight per day and for the 95th percentile consumer as 3.1pg WHO-TEQ $kg^{-1}$ bodyweight per day, but that dietary intake was decreasing (COT, 2001).

**Table 28.4.** Average daily human intake of dioxins and furans from all sources.

| Source | pg TEQ $kg^{-1}$ bodyweight per day | % |
|---|---|---|
| Food | 1 | 96 |
| Air | 0.03 | 3 |
| Consumer products | 0.01 | 1 |
| Water | Very little | 0 |
| Total | 1.04 | 100 |

## 28.8 Conclusions

Combustion particulates represent a major component of indoor and outdoor environmental air pollution shown to be responsible for considerable morbidity and mortality. The complex chemical and physical characteristics of combustion particles present considerable challenges in evaluating their composition and toxicity. The range of toxic effects, those on both the respiratory system and systemic target organs, depends to a large extent on their site of deposition, which in turn depends on particle size, the normal function at the site and the repertoire of pathological response at the site. Toxicity then depends on both chemical and structural features of the particles and particle-associated substances or surface properties. A trend in research on combustion particulates is the revelation of increasing areas and extent of biological activity of significance for effects of exposure on health.

It is considered that potential health hazards from acute exposure to airborne carcinogens and dioxins are likely to be minor during most individual fire incidents, and exposure to contaminated soot at post-fire sites can be minimized by simple hygiene precautions.

Of more concern, especially with regard respiratory and cardiovascular disease and carcinogenicity, is the long-term risk arising from repeated and continuous exposures in the domestic environment (from open-fire and cooking fumes and tobacco smoke), in the workplace from combustion processes (including repeat exposure of firefighters), and especially from ambient air pollution (particularly that arising from vehicle emissions), which remain a serious and possibly increasing societal health risk.

In contrast, the reduction of hazards from dioxin intake since the 1980s has been quite a success story, and exposure to all carcinogenic smoke toxins from tobacco smoke (both direct and environmental) has also decreased significantly in Western countries although still responsible for 28% of all cancer deaths.

## References


COMEAP (2010) *The Mortality Effects of Long-term Exposure to Particulate Air Pollution in the United Kingdom*. Committee on the Medical Effects of Air Pollutants, Department of Health, London.

COT (2001) COT statement on the tolerable daily intake for dioxins and dioxin-like polychlorinated biphenyls. Committee on Toxicity of Chemicals in Food, Consumer Products and the Environment. COT/2001, 07 October 2001. Food Standards Agency, Department of Health, London.

Donaldson, K., Stone, V., Clouter, A., L.Renwick, L. and MacNee, W. (2001) Ultrafine Particles. *Occupational and Environmental Medicine* 58, 211–216,

Donaldson, K., Hunter, A., Poland, C. and Smith, S. (2016) Mechanism of action of combustion-derived nanoparticles. In: Purser, D., Maynard, R. and Wakefield, J. (eds) *Toxicology, Survival and Health Hazards of Combustion Products*. Royal Society of Chemistry, Cambridge, UK, pp. 361–381.

EPA (2012) *EPA's Reanalysis of Key Issues Related to Dioxin Toxicity and Response to NAS Comments*. Vol. 1, February 2012. US Environmental Protection Agency, Washington DC.

Hall, D. and Spanton, A.M. (2016) Dispersion of fire plumes in the atmosphere. In: Purser, D., Maynard, R. and Wakefield, J. (eds) *Toxicology, Survival and Health Hazards of Combustion Products*. Royal Society of Chemistry, Cambridge, UK, pp. 139–171.

HSE (2011) *EH40/2005 Workplace Exposure Limits*, 2nd edn. Health and Safety Executive, London.

IARC (2002) *Man-made vitreous fibres*. IARC Monographs on the Evaluation of Carcinogenic Risks to Humans, Vol. 81. International Agency for Research on Cancer, Lyon, France, p. 379.

IARC (2014) Agents classified by the IARC Monographs, Volumes 1–109. International Agency for Research on Cancer, Lyon, France. Available at: http://monographs.iarc.fr/ENG/Classification/ (accessed 28 June 2014).

Kendall, M., Cohen, M. and Chen, L. (2016) The World Trade Centre Disaster. In: Purser, D., Maynard, R. and Wakefield, J. (eds) *Toxicology, Survival and Health Hazards of Combustion Products*. Royal Society of Chemistry, Cambridge, UK, pp. 574–601.

King, D.F. (1988) *Aircraft Accident Report 8/88*. Air Accidents Investigation Branch, UK Department of Transport, London.

Metcalfe, E., Harman, A.R. and Patel, M.K. (1991) The thermo-oxidative degradation of PTFE in the NIST cup furnace toxicity test. *Fire and Materials*, 15, 53–58.

McAllister, J.L. (2016) Health effects in groups exposed to wildland and urban fires. In: Purser, D., Maynard, R. and Wakefield, J. (eds) *Toxicology, Survival and Health Hazards of Combustion Products*. Royal Society of Chemistry, Cambridge, UK, pp. 535–551.

Purser, D.A. (1992a) Recent developments in understanding the toxicity of PTFE thermal decomposition products. *Fire and Materials* 16, 67–75.

Purser, D.A. (1992b) The combustion toxicology of anticholinesterases. In: Ballantyne, B. and Marrs, T.C. (eds) *Basic and Clinical Toxicology of Organosphosphates and Carbamates*. Butterworth-Heinemann, Oxford,UK, pp. 386–395.

Purser, D.A. (2007) The application of exposure concentration and dose to evaluation of the effects of irritants as components of fire hazard. In: *Interflam 2007: Proceedings of the 11th international conference, Royal Holloway College, 3–5 September 2007*, organized by Interscience Communications, London, pp. 1033–1041.

Purser, D.A, (2010a) Hazards from smoke and irritants. In: Stec, A. and Hull, R. (eds) *Fire Toxicity*. Woodhead, Cambridge, UK, pp. 51–117.

Purser, D.A. (2010b) Application of human and animal exposure studies to human fire safety. In: Stec, A. and Hull, R. (eds) *Fire Toxicity*. Woodhead, Cambridge, UK, pp. 282–345.

Purser, D.A. (2014) Fire safety performance of flame retardants compared with toxic and environmental hazards. In: Papaspyrides, C.D. and Kiliaris, P. (eds) *Polymer Green Flame Retardants*. Elsevier, Oxford, UK, pp. 45–86.

Purser, D.A. (2016a) Combustion toxicity In: Hurley, M.J., Gottuck, D.T., Hall, J.R. Jr, Harada, K., Kuligowski, E.D. *et al.* (eds) *SFPE Handbook of Fire Protection Engineering*, 5th edn. Springer-Verlag, New York, pp. 2207–2307.

Purser, D.A. (2016b) Fire types and combustion products. In: Purser, D., Maynard, R. and Wakefield, J. (eds) *Toxicology, Survival and Health Hazards of Combustion Products*. Royal Society of Chemistry, Cambridge, UK, pp. 13–52.

Purser, D.A (2016c) Dioxins and other carcinogens. In: Purser, D., Maynard, R. and Wakefield, J. (eds) *Toxicology, Survival and Health Hazards of Combustion Products*. Royal Society of Chemistry, Cambridge, UK, pp. 382–410.

Purser, D. A. (2017) Effects of pre-fire age and health status on vulnerability to incapacitation and death from exposure to carbon monoxide and smoke irritants in Rosepark fire incident victims. *Fire and Materials* 41, 555–569. doi: 10.1002/fam.2393.

Purser, D.A. and Buckley. P. (1983) Lung irritance and inflammation during and after exposure to thermal decomposition products from polymeric materials. *Medicine, Science and the Law* 23, 142–150.

Purser, D.A. and Kuipers, M. (2004) Interactions between buildings, fire and occupant behaviour using a relational database created from incident investigations and interviews. *Proceedings 3rd International Symposium on Human Behaviour in Fire. Europa Hotel, Belfast, 1–3 September 2004*. Interscience Communications, London, pp. 443–456

Purser, D.A and Maynard, R.L. (2016) Overview of combustion toxicology. In: Purser, D., Maynard, R. and Wakefield, J. (eds) *Toxicology, Survival and Health Hazards of Combustion Products*. Royal Society of Chemistry, Cambridge, UK, pp. 1–10.

Purser, D.A. and McAllister, J.L.(2016) Assessment of hazards to occupants from smoke, toxic gases and heat. In: Hurley, M.J., Gottuck, D.T., Hall, J.R. Jr, Harada, K., Kuligowski, E.D. *et al.* (eds) *SFPE Handbook of Fire Protection Engineering*, 5th edn. Springer-Verlag, New York, pp. 2207–2307. doi: 10.1007/978-1-4939-2565-0.

Purser, D.A. and Purser J.A. (2008) HCN yields and fate of fuel nitrogen for materials under different combustion conditions in the ISO 19700 tube furnace. Proceedings of the ninth international symposium. International Association for Fire Safety Science. *Fire Safety Science* 9, 1117–1128. doi: 10.3801/IAFSS.FSS.9-1117.

Task Group on Lung Dynamics (1966) *Deposition* and retention models for internal dosimetry of the human respiratory tract. *Health Physics* 12, 173–208.

Wallace, D.N. (1981) Dangers of polyvinyl chloride wire insulation decomposition. 1. Long term health impairments: studies of firefighters of the New York telephone fire and survivors of the 1977 Beverly Hills supper club fire. *Journal of Combustion Toxicology* 8, 205–226.

WHO (2006) *WHO Air Quality Guidelines for Particulate Matter, Ozone, Nitrogen Dioxide and Sulfur Dioxide. Global Update 2005*. World Health Organization, Geneva.

# 29 Assessment of the Ecotoxicity of Airborne Particulate Matter

N. Kováts*
*Institute of Environmental Sciences, University of Pannonia, Veszprém, Hungary*

## 29.1 Abstract

The deleterious effect of airborne particulate matter or its individual components on human health has been widely studied; however, much less information is available on its potential toxicity on the non-human biota. This chapter is intended to give a comprehensive insight into the principle of ecotoxicological testing of airborne particulate matter, summarizing reported applications.

## 29.2 Introduction

Particulate matter (PM) is grouped as coarse, fine and ultrafine particles (UFPs) with aerodynamic diameters of 2.5 to 10 μm ($PM_{10}$), < 2.5 μm ($PM_{2.5}$), and < 0.1 μm ($PM_{0.1}$), respectively. The $PM_{2.5-10}$ fraction has been well characterized from numerous aspects, such as temporal (diurnal and/or seasonal) variations (e.g. Hasheminassab *et al.*, 2014; Jalava *et al.*, 2015; Schleicher *et al.*, 2015), origin (Titos *et al.*, 2014) and composition (e.g. Schilirò *et al.*, 2015). Its deleterious effect on human health has also been widely discussed (e.g. Cassee *et al.*, 2013).

Much less information is available on the potential ecological hazard of ultrafine particles, though human health studies support their extreme toxicity, as they might penetrate through cell biological barriers (Chen *et al.*, 2016). Landkocz *et al.* (2017) demonstrated that UFPs have greater cytotoxic and genotoxic potential than coarse particles. Borsós *et al.* (2012) found that the ratio of UFPs amounted to 80%, 84% and 74% in three European capitals (Budapest, Prague and Vienna, respectively).

The European Union (EU) Directive on air quality (2008/50/CE) (European Parliament C. Directive 2008/50/EC of the European Parliament and of the Council of 21 May 2008 on ambient air quality and cleaner air for Europe) identifies air particulate matter with aerodynamic diameter less than 10 μm ($PM_{10}$) and 2.5 μm ($PM_{2.5}$) as one of the most dangerous pollutants on human health. Threshold levels are set for 10 μm diameter ($PM_{10}$) and 2.5 μm diameter ($PM_{2.5}$) particles.

While there is continuously rising concern about the potential human health impact of particulate matter, ecological effects on different ecosystems are mostly based on biomonitoring studies. As such, the present work aims to give a short overview about ecotoxicological testing of airborne PM.

* E-mail address: kovats@almos.uni-pannon.hu

## 29.3 Ecotoxicity of PM

In cities, main sources of air particulate matter are transport, residential heating and household activities (Szigeti *et al.*, 2015). A comprehensive study was published by Schmidt *et al.* (2017) on the toxicology of combustion-derived particles. The authors reviewed approximately 80 papers and categorized them based on different aspects, such as experimental set-up or how many of them gave chemical characterization as well. In addition to ecotoxicological works, cytotoxicity and genotoxicity studies were included: the supplement of Schmidt *et al.* (2017) lists test organisms used for each type of toxicological assessments.

Tyre debris, which is associated with road transport, is also a considerable source of PM: the total concentration is in the range of 1–10 $\mu g\ m^{-3}$ at a busy city road (mean value of 2.8 $\mu g\ m^{-3}$) (Wik and Dave, 2008). Industrial activities may also generate PM emission (e.g. Calvello *et al.*, 2015).

Particles consist of elemental carbon core and a diverse range of adsorbed substances, including heavy metals and various organic compounds such as polycyclic aromatic hydrocarbons (PAHs), nitroarenes, etc. These aerosol constituents may play a major role concerning toxicity and ecotoxicity of atmospheric aerosol particles, as they have a high affinity to the very fine and ultrafine aerosol fractions (Moreno *et al.*, 2006). Valavanidis *et al.* (2006) demonstrated that the fine particulate PAHs concentrations were higher than coarse particles.

An important mechanism for causing toxic effect is the generation of reactive oxygen species (ROS). High levels of ROS may cause a change in the redox status of the cell (Squadrito *et al.*, 2001). Recent studies indicate that the HULIS (humic-like substances) fraction might also exert potentially deleterious effects by generating ROS (Verma *et al.*, 2015). Humic-like substances account for 30–70% of the water-soluble organic fraction in atmospheric aerosol. In a study on characterization of the redox activity of fine and coarse particulate matter in Milan, Italy, Daher *et al.* (2012) reported that nickel (Ni) chromium (Cr), copper (Cu) and water-soluble organic carbon (OC) were strongly associated with ROS.

## 29.4 Ecotoxicity Tests

Ecotoxicological tests assess the aggregate toxicity of a sample and have been widely used in different environmental matrices. Air, however, is a very complicated medium to analyse in the laboratory. Airborne emissions are generally collected on a filter, which limits the quantity of the sample, also limiting the range of appropriate test organisms.

So-called single-species bioassays can be used and in most cases have been used to compare the ecotoxic effect of different samples. In more sophisticated studies, the main purpose can be to assess the ecological effect of the contaminant on the whole recipient ecosystem. In such cases, important functional or structural groups (guilds) of the ecosystem are represented by surrogate species.

### 29.4.1 Screening assays

It is a well-known phenomenon in ecotoxicological testing that different test organisms show very different sensitivity to the same sample. However, in many cases, only one bioassay (probably chosen for simplicity, easy-to-use character and/or representativity) is employed for the initial characterization of the contaminants.

For screening purposes, most often the test based on the bioluminescence inhibition of the marine bacterium *Vibrio fischeri* has been used. The species was recently renamed *Aliivibrio fischeri* (Urbanczyk *et al.*, 2007); however, as most standards still apply the *V. fischeri* name, it will be used in this review. Although less often, other bioluminescent bacterium taxa are applied such as *Vibrio harveyi* (e.g. Lange and Thomulka, 1997) or *Photorhabdus luminescens* (a freshwater species) (Masner *et al.*, 2017).

This test is based on the inhibition of the NAD(P)H flavin mononucleotide (FMN) oxidoreductase and luciferase enzyme system, which is reflected in the rapid decrease of light emittance of the bacterium. The reduction of light intensity is proportional to the strength of the toxicant, therefore it gives a reliable and easy-to-measure end point.

Basic insight and environmental applications were first reviewed by Girotti *et al.* (2008); new developments were discussed by Ma *et al.* (2014) while Kokkali and van Delft (2014) gave an introduction to commercially available test systems. A review was published by Kováts and Horváth (2016) on bioluminescence-based tests and biosensors in air pollution assessment, including methods for testing the gaseous phase directly.

The conventional protocol uses aqueous samples (in compliance with ISO 11348). As PM samples are solid, one possible option is to prepare an extract. In fact, most studies have applied extracts for further processing (Table 29.1 provides a summary for available literature studies). In addition to aqueous extracts, different organic solvents have been employed. However, concentration of the toxic compounds in the extract and therefore its toxicity will depend strongly on the solvent applied (e.g. Verma *et al.*, 2013; Corrêa *et al.*, 2017).

In order to carry out direct contact tests (where the test bacteria are in direct contact with the contaminated particles), two different approaches and protocols are available. The Microtox® Solid Phase Basic Test was proposed by Azur Environmental (1998) as a standard for testing solid samples. During exposure, the bacteria are kept in a suspension, but then they and particles are separated by filtration. False readings might occur, for example, when bacteria are absorbed on particle surfaces or when particles cause physical interference on bioluminescence (reviewed by Volpi Ghirardini *et al.*, 2009).

Lappalainen *et al.* (1999, 2001) developed a kinetic version of the test which was later standardized (ISO 21338:2010: Water quality – Kinetic determination of the inhibitory effects of sediment, other solids and coloured samples on the light emission of *Vibrio fischeri*/kinetic luminescent bacteria test). Here both exposure and luminescence readings are carried out in the suspension where bacteria are in direct contact with the particles. In order to eliminate false readings due to the physical effect of colour and/or suspended solids on bioluminescence, inhibition is calculated comparing the initial and final readings, independently from the control. After the sample is mixed with the bacteria, light output is continuously recorded for the first 30 seconds; as such, toxic effect can be immediately detected (Fig. 29.1). Such kinetic diagrams are being used for a preliminary assessment of the toxicity (Mortimer *et al.*, 2008).

The bacterial bioluminescence assay was already being used for detecting the deleterious effects of air pollutants in the 1960s (Serat *et al.*, 1965, 1967). Table 29.1 gives an overview of recent applications.

Albeit rarely, non-luminous bacteria have been employed. Filep *et al.* (2015) adapted the *Pseudomonas putida* growth inhibition test (ISO 137 10712:1995) to test the aqueous extract of winter $PM_{2.5}$ samples (collected at rural, urban and traffic sites) and found that cytotoxicity was related to biomass burning.

In most cases, ecotoxicological tests are used to detect the aggregate effect of a sample. Some studies, however, have tried to link the ecotoxicological effect expressed as bacterial bioluminescence inhibition to contaminants present in the sample. The simplest way is to check correlation between ecotoxicity and different contaminants. For example, Evagelopoulos *et al.* (2009) found significant correlation with *V. fischeri* toxicity and PAH content of the samples, while good correlation was established in the study of Roig *et al.* (2013) between ecotoxicity readings and heavy metals, polychlorinated dibenzo-*p*-dioxins and dibenzofurans (PCDDs and PCDFs).

Genetic engineering makes it possible to employ stressor-specific strains. Kessler *et al.* (2012) composed a reporter panel of nine genetically engineered bacterial (*Escherichia coli*) strains. The host strains were first equipped with the promoter-less plasmid (pBR2TTS) harbouring the *luxCDABE* genes of *Pseudomonas luminescens*. Bacterial reporter strains were created with the inclusion of stress promoter genes. In the study, extract of vehicle combustion particles, coal fly ash (CFA) samples and an urban dust sample were assessed. Due to the presence of the stressor-specific gene promoters, each reporter produced measurable light signals indicating specific stress conditions such as excess $As^{5+}$/$As^{3+}$ or excess $Pb^{2+}$/$Cd^{2+}$/$Zn^{2+}$.

In addition to bacteria, higher organisms are in use. Zhao *et al.* (2014) investigated the deleterious effects of $PM_{2.5}$ urban aerosol collected in Beijing. The test organism applied was

**Table 29.1.** Summary of bioluminescent bacteria-based assessments of airborne particulate matter.

| Reference | System used | Sample preparation | Study |
|---|---|---|---|
| Wang *et al*. (1998) | Microtox Solid Phase | Suspension | Urban dust (London and Hong Kong) |
| Lin and Chao (2002) | Microtox | Soxhlet extraction in dichloromethane (DCM) | Influence of methanol-containing additive on biological characteristics of diesel exhaust emissions |
| Isidori *et al*. (2003) | Microtox | Extraction in cyclohexane | Urban PM (Caserta, South Italy) |
| Triolo *et al*. (2008) | Microtox | Extraction in acetone:dimethylsulfoxide (DMSO) mixture | Impact of PM emitted by the industrial settlement of Milazzo (Italy) on agriculture |
| Evagelopoulos *et al*. (2009) | Microtox | Extraction with acetone and hexane | Urban (industrial) fine (< 2.5 μm) and coarse (2.5–10 μm) particulate matter |
| Vouitsis *et al*. (2009) | Microtox | Soxhlet extraction in dichloromethane (DCM) | PM emitted from three light-duty vehicles in different driving cycles |
| Kováts *et al*. (2012) | ToxAlert<br>Ascent Luminometer | Direct contact test | Comparison of the two systems on winter and summer urban and rural PM |
| Turóczi *et al*. (2013) | Ascent Luminometer | Direct contact test | Comparative assessment of diesel exhaust, urban aerosol, cigarette smoke and biomass burning |
| Kováts *et al*. (2013) | Ascent Luminometer | Direct contact test | Exhaust particulates from diesel-powered buses |
| Åcs *et al*. (2013) | Ascent Luminometer | Direct contact test | Exhaust particulates from diesel-powered light-duty vehicles |
| Barbosa *et al*. (2013) | Microtox | Eluate with deionized water | Size-fractionated biomass ashes |
| Roig *et al*. (2013) | Microtox | Aqueous acidic extractions | Industrial locations (Catalonia), seasonal differences |
| Chang *et al*. (2013) | Microtox | Extraction in *n*-hexane or dichloromethane/*n*-hexane mixtures | Fly ash samples from a cooling tower on an incinerating plant |
| Cukurluoglu and, Muezzinoglu (2013) | LUMIStox | Aqueous extracts | Dry and wet deposition |
| Goix *et al*. (2014) | Microtox Solid Phase | Stock suspensions in distilled water | Characterization of anthropogenic fine and ultrafine metallic particles emitted to the atmosphere |
| Silva *et al*. (2015) | Microtox | Aqueous extracts | Ash from forest fires |
| Wang *et al*. (2016) | Microtox | Aqueous extract with de-ionized water | Seasonal differences in atmospheric fine particulate matter ($PM_{2.5}$) in urban Beijing (China) |
| Corrêa *et al*. (2017) | Lumistox | Soxhlet extraction in dichloromethane, hexane and acetone versus aqueous extract | PM emitted from heavy-duty diesel-powered vehicles; influence of leaching conditions (pH) |
| Pintér *et al*. (2017) | Ascent Luminometer | Direct contact test | Urban (Budapest, Hungary) samples, diurnal variation, correlation between ecotoxicity and optical parameters |
| Aammi *et al*. (2017) | Microtox | Extraction with DMSO and ultra-pure water | Urban (Istanbul, Turkey) $PM_{2.5-10}$ sample characterization, seasonal differences |

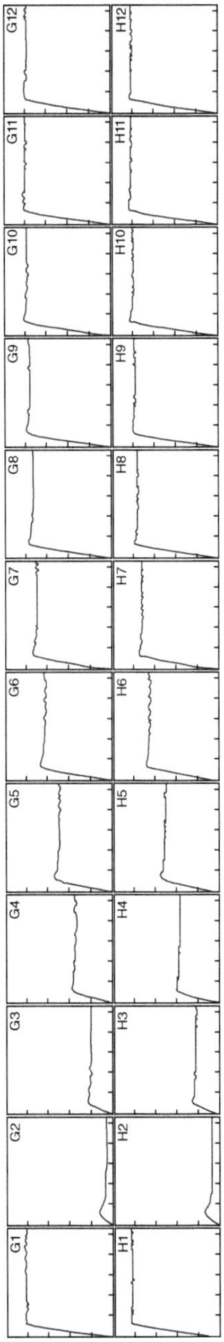


**Fig. 29.1.** Typical kinetic diagram of an aerosol sample. Light output is recorded in the first 30 s interval. As the sample contains solid particles, physical interference occurs, which reduces light luminescence reading: the peak is lower than in the control. After the peak, toxicity causes a rapid reduction in the light output while light output remains even in the control. The two rows show the two replicates. G1-H1 (left): control. G2-H2: highest, G12-H12: lowest concentrations.

the nematode *Caenorhabditis elegans*, which is a widely applied test species and is considered applicable for evaluation of transgenerational effects of toxicants (Wang *et al.*, 2007). The results were quite exceptional, as not only were harmful effects of acute exposure determined but also prolonged exposure to $PM_{2.5}$ was found to cause adverse effects on locomotion behaviour, intestinal development, reproduction and even in lifespan in progeny of the exposed nematodes.

In the study of Mesquita *et al.* (2014), ecotoxicity of size fractionated aerosol samples was assessed on zebrafish embryos, using different end points: mortality and deformities. Results revealed that sub-micron fractions were mostly responsible for the toxic effects and putative toxic compounds, mainly PAHs, also concentrated in these fractions.

### 29.4.2 Freshwater impact

Atmospheric wet deposition (AWD) serves as an important input for the trace elements (TEs) entering aquatic ecosystems, where ecological effects will depend on the solubility of the element (Xing *et al.*, 2017). Some of the TEs such as iron (Fe), zinc (Zn) and cobalt (Co) are essential microelements and are beneficiary for the growth of phytoplankton, while others such as cadmium (Cd), Cu and lead (Pb) in dissolved phase in precipitation exert potential toxicity (Chase *et al.*, 2011).

Considering organic load, early studies reported significant deposition of atmospheric PAHs and polychlorinated biphenyls (PCBs) in aquatic environments (e.g. Dickhut and Gustafson, 1995; Franz *et al.*, 1998). In a study of Gevao *et al.* (1998), phenanthrene, fluoranthene and pyrene contributed > 50% of total PAHs deposited.

Verma *et al.* (2013) used the freshwater rotifer *Brachionus calyciflorus* to assess the ecotoxicity of urban PM to aquatic environments. Rotifers are important structural and functional elements of the aquatic trophic web and *B. calyciflorus* is a widely applied test organism (Dahms *et al.*, 2011). The main outcome of the study was that the compounds mostly responsible for the ecotoxic effect were characterized by strong hydrophobicity.

Silva *et al.* (2015) performed a complex study in order to assess the deleterious effects of ash produced by wildfires on water chemistry and aquatic biota. For ecotoxicological screening, four standard aquatic species from different functional groups and trophic levels were employed: two primary producers, the alga *Pseudokirchneriella subcapitata* and the higher plant (duckweed) *Lemna minor*, and zooplanktonic *Daphnia magna*. Decomposers were represented by the bacteria *Vibrio fischeri*. Plants and bacteria showed a significant inhibitory effect; *D. magna* appeared much more resistant. Still, a cascading effect can be predicted due to the damage at lower trophic level.

Corrêa *et al.* (2017) evaluated the aquatic ecotoxicity profile of extracts of diesel exhaust particulate matter (DPM). In the study, different extracts (a mixture of dichloromethane, *n*-hexane and acetone versus aqueous extract) were used and ecotoxicity was assessed under different pH (leaching) conditions: at pH 2.00 and 5.00. The battery of aquatic bioassays was composed of bacteria (*Vibrio/Aliivibrio/fischeri*), algae (*Scenedesmus subspicatus*), zooplankton (*Daphnia magna*) and fish (*Danio rerio*). Naturally, toxicity was highly dependent on extraction procedure; low pH and/or organic solvents increased ecotoxic effects. However, sensitivity of the bioassays showed a similar trend, such as: daphnids > algae > bacteria > fish.

For screening the risk of ambient atmospheric particulate matter to freshwater organisms, *Ceriodaphnia dubia* and *Selenastrum capricornutum* were used in two studies of Sheesley *et al.* (2004, 2005). Acute and chronic tests were performed on PM samples collected in the watershed of Lake Michigan; both bioassays showed spatial differences in the toxicity of the samples and also revealed that toxicity significantly depended on the extraction procedure.

### 29.4.3 Terrestrial impact

Plants are in direct contact with air pollutants, both gases and particulates, being exposed to PM-bound chemicals via dry or wet deposition (reviewed by Smith and Jones, 2000; Petroff *et al.*, 2008). As such, they have long been used in bioindication and biomonitoring studies. Bioindication studies have shown that plants might develop a wide range of symptoms triggered by air pollution, showing both structural and functional alterations (reviewed by Grantz *et al.*, 2003).

Of possible symptoms, biochemical end points seem the most sensitive and can be used in early-warning systems. Air pollution causes significant enhancement in activity of antioxidative enzymes such as catalase, peroxidase and superoxide dismutase, which play an important role in scavenging reactive oxygen species (Bansal *et al.*, 2016).

In principle, most of these symptoms can be or could be used as end points in controlled ecotoxicological studies. Still, ecotoxicological assessments of air pollution is quite limited. Studies addressing the effect of aerosol-borne contaminants concentrate in most cases on selected components such as PAHs or heavy metals. Foliar uptakes of both groups have been established.

Assessment of the foliar uptake and ecotoxicity of selected PAHs and/or their combinations have been conducted under controlled chamber conditions or in greenhouses: fluoranthene (e.g. Oguntimehin *et al.*, 2010); anthracene and benzo[k]fluoranthene (e.g. Wieczorek *et al.*, 2015); phenanthrene (e.g. Desalme *et al.*, 2011); and pyrene (e.g. Ahammed *et al.*, 2012). These studies mostly apply standard solution of the PAH to be tested.

Considering another important group of pollutants – heavy metals and metalloids – it has been established that plant leaves can accumulate heavy metals from the atmospheric aerosols (Uzu *et al.*, 2010). Lead is one of the most important airborne heavy metals (Komarek *et al.*, 2008) and most studies have addressed its foliar uptake and potential toxic effects (e.g. Hu *et al.*, 2011; Schreck *et al.*, 2012). Using a combination of microscopic and spectroscopic techniques, Schreck *et al.* (2014) studied metal localization after foliar uptake of Pb and found that in lettuce leaves the highest Pb concentration was found in necrotic spots. In a study of Hu *et al.* (2011) foliar uptake of lead from nine-stage size-segregated aerosols was assessed and it was revealed that the fine fractions enriched more Pb than the coarse fractions.

Although other metals are also accumulated via the leaves, there is rather scarce information about their phytotoxicity. Most possibly, toxic effect is concentration-dependent, as atmospheric aerosol provides an important nutrient source of microelements (Wang *et al.*, 1998). However, in a recent study Schreck *et al.* (2013) reported that the foliar uptake of metals such as Pb, Cd, Cu, Zn, arsenic (As) and antimony (Sb) influenced the plant-leaf fatty acid composition, providing an indication about the toxic effect.

In addition to individual components, experimental chambers have long been in use to assess the concentration–effect relationship of contaminated air, including particulate matter (e.g. Wedding *et al.*, 1975; Grattan *et al.*, 1981). Honour *et al.* (2009) used a solardome fumigation facility where ecotoxicological effect of diesel exhaust emissions were studied on 12 herbaceous species. The experiments demonstrated that sensitivity was species-specific; it also depended on the age of the plant, young plants being more sensitive.

In order to assess the specific effects of airborne particulate matter, Kováts *et al.* (2017) adapted the No. 227 OECD Guideline for the Testing of Chemicals: Terrestrial Plant Test: Vegetative Vigour Test. The Guideline was originally developed to test the phytotoxicity of general chemicals, biocides and crop protection products for herbicide regulation (Boutin *et al.*, 2012). Based on the original protocol, test plants (*Cucumis sativus*) were sprayed with the aqueous extract of winter urban aerosol. Two distinct concentration–effect relationship patterns were experienced: on both fresh weight and leaf length the extract elucidated stimulatory effects at low concentration, due to the presence of macro- and micronutrients. On the other hand, necrosis was detected at high concentration, due to the marked effect of toxic compounds.

Direct effect of PM was assessed in the study of Daresta *et al.* (2015). Tomato (*Solanum lycopersicum* L.) plants were grown on $PM_{10}$ collected on quartz fibre filters. Germination was the least sensitive end point: it was not inhibited, but delayed in comparison with the control. On the other hand, there were significant differences in the growth of the root apparatus, such as in the decrease of primary root elongation and in shoot and root biomass. Pigments were also a sensitive end point: there was a significant increase in carotenoid content, which could indicate the protective response against oxidative stress induced by PM exposure.

### 29.4.4 Comparative studies

In most cases, if a complex ecotoxicological profiling is to be made, a battery of bioassays should be employed, where surrogate species represent important structural and/or functional elements (guilds) of the recipient ecosystem.

Wang *et al.* (1998) performed a comparative study on the water extract of urban dust samples from Hong Kong and London using organisms representing different trophic levels and taxonomic groups: *Photobacterium phosphoreum* (luminescent bacterium), *Dunaliella tertiolecta* (a dinoflagellate green alga), *Brassica chinensis* (Chinese white cabbage) and *Lolium perenne* (rye grass). Of the chemically analysed metals, the luminescent bacteria test showed good correlation with Pb and Zn, and *D. tertiolecta* with the exchangeable Pb content. No specific correlation could be established for higher plants. Also, no significant correlation was found between bioassays.

Barbosa *et al.* (2013) used a battery of bioassays to evaluate the potential hazard of size-fractionated biomass ashes. The battery involved marine (the bacterium *Vibrio fischeri*, the micro-crustacean *Artemia franciscana* and the microalgae *Phaeodactylum tricornutum*) and freshwater (the micro-crustacean *Daphnia magna* and the microalgae *Selenastrum capricornutum*) organisms. While *V. fischeri* showed the highest sensitivity, other marine organisms were found less sensitive than their freshwater relatives. One possible explanation could be the high salt concentration in eluates which caused osmotic effect in freshwater organisms.

## 29.5 Conclusions

The main conclusion is that all tests have suggested the significant risk that airborne particulate matter poses on recipient living systems, both aquatic and terrestrial. Most of the studies reviewed above are based on bacterial bioluminescence. There has been extended research on correlation between *V. fischeri* test data with other bioassays. Kaiser (1998) gave a comprehensive data set about correlations between individual tests and mentioned that siginificant correlations were especially established for aquatic organisms. However, test results where only this bioassay was used have to be interpreted with proper caution. Due to its rapidity (test results can be given only after 30 min of exposure) and its cost-effective nature, the test is excellent for screening purposes. On the other hand, as *Vibrio* bacteria represent an important functional group in the ecosystems, i.e. decomposers, application of this bioassay in a battery of tests will help to understand the potential toxic effect of any contamination on complex communities.

While plants are exposed directly to airborne contaminants and, according to bioindication studies, are developing a wide range of symptoms, their use in controlled ecotoxicological experiments is limited. On the other hand, there is a continuously growing interest to involve more and more candidate test organisms and bioassays in order to evaluate the hazard that atmospheric particulate matter might pose on ecosystems.

## Acknowledgement

This work was supported by the BIONANO_GINOP-2.3.2-15-2016-00017 project.

## References

Aammi, S., Karaca, F. and Petek, M. (2017) A toxicological and genotoxicological indexing study of ambient aerosols ($PM_{2.5\text{-}10}$) using in vitro bioassays. *Chemosphere* 174, 490–498.

Ács, A., Kovács, A., Ferincz, Á., Turóczi, B., Gelencsér, A., Kiss, G. and Kováts, N. (2013) Characterisation of exhaust particulates from diesel-powered light-duty vehicles. *Central European Journal of Chemistry* 11, 1954–1958.

Ahammed, G.J., Yuan, H.L., Ogweno, J.O., Zhou, Y.H. and Xia, X.J. (2012) Brassinosteroid alleviates phenanthrene and pyrene phytotoxicity by increasing detoxification activity and photosynthesis in tomato. *Chemosphere* 86, 546–555.

Azur Environmental (1998) *Microtox Manual*. Available at: www.azurenv.com (accessed 1 May 2019).

Bansal, P., Verma, S. and Srivastava, A. (2016) Biomonitoring of air pollution using antioxidative enzyme system in two genera of family Pottiaceae (Bryophyta). *Environmental Pollution* 216, 512–518.

Barbosa, R., Dias, D., Lapa, N., Lopes, H. and Mendes, B. (2013) Chemical and ecotoxicological properties of size fractionated biomass ashes. *Fuel Processing Technology* 109, 124–132.

Borsós, T., Řimnáčová, D., Ždímal, V., Smolík, J., Wagner, Z. *et al.* (2012) Comparison of particulate number concentrations in three Central European capital cities. *Science of the Total Environment* 433, 418–426.

Boutin, C., Aya, K.L., Carpenter, D., Thomas, P.J. and Rowland, O. (2012) Phytotoxicity testing for herbicide regulation: shortcomings in relation to biodiversity and ecosystem services in agrarian systems. *Science of the Total Environment* 415, 79–92.

Calvello, M., Esposito, F., Lorusso, M. and Pavese, G. (2015) A two-year database of BC measurements at the biggest European crude oil pre-treatment plant: a comparison with organic gaseous compounds and $PM_{10}$ loading. *Atmospheric Research* 164–165, 156–166.

Cassee, F.R., Heroux, M.-E., Gerlofs-Nijland, M.E. and Kelly, F.J. (2013) Particulate matter beyond mass: recent health evidence on the role of fractions, chemical constituents and sources of emission (Review). *Inhalation Toxicology* 25, 802–812.

Chang, S.C., Wang, Y.F., You, S.J., Kuo, Y.M., Tsai, C.H., Wang, L.C. and Hsu, P.Y. (2013) Toxicity evaluation of fly ash by Microtox®. *Aerosol and Air Quality Research* 13, 1002–1008.

Chase, Z., Paytan, A., Beck, A., Biller, D., Bruland, K., Measures, C. and Sañudo-Wilhelmy, S. (2011) Evaluating the impact of atmospheric deposition on dissolved trace-metals in the Gulf of Aqaba, Red Sea. *Marine Chemistry* 126, 256–268.

Chen, R., Hu, B., Liu, Y., Xu, J., Yang, G., Xu, D. and Chen, C. (2016) Beyond PM2.5: the role of ultrafine particles on adverse health effects of air pollution. *Biochimica et Biophysica Acta* 1860(12), 2844–2855.

Corrêa, A.X.R., Testolin, R.C., Torres, M.M., Cotelle, S., Schwartz, J.J., Millet, M. and Radetski, C.M. (2017) Ecotoxicity assessment of particulate matter emitted from heavy-duty diesel-powered vehicles: influence of leaching conditions. *Environmental Science and Pollution Research* 24(10), 9399–9406.

Cukurluoglu, S. and Muezzinoglu, A. (2013) Assessment of toxicity in waters due to heavy metals derived from atmospheric deposition using *Vibrio fischeri*. *Journal of Environmental Science and Health, Part A: Toxic/Hazardous Substances and Environmental Engineering* 48(1), 57–66.

Daher, N., Ruprecht, A., Invernizzi, G., De Marco, C., Miller-Schulze, J. *et al.* (2012) Characterization, sources and redox activity of fine and coarse particulate matter in Milan, Italy. *Atmospheric Environment* 49, 130–141.

Dahms, H.U., Hagiwara, A. and Lee, J.S. (2011) Ecotoxicology, ecophysiology, and mechanistic studies with rotifers. *Aquatic Toxicology* 101(1), 1–12.

Daresta, B.E., Italiano, F., de Gennaro, G., Trotta, M., Tutino, M. and Veronico, P. (2015) Atmospheric particulate matter (PM) effect on the growth of *Solanum lycopersicum* cv. Roma plants. *Chemosphere* 119, 37–42.

Desalme, D., Binet, P., Epron, D., Bernard, N., Gilbert, D. *et al.* (2011) Atmospheric phenanthrene pollution modulates carbon allocation in red clover (*Trifolium pratense* L.). *Environmental Pollution* 159, 2759–2765.

Dickhut, R.M. and Gustafson, K.E. (1995) Atmospheric inputs of selected polycyclic aromatic hydrocarbons and polychlorinated biphenyls to Southern Chesapeake Bay. *Marine Pollution Bulletin* 30(6), 385–396.

Evagelopoulos, V., Zoras, S., Samaras, P., Triantafyllou, A.G., Albanis, T.A. and Kassomenos, P. (2009) Toxicity of fine and coarse atmospheric particles using *Vibrio fischeri.* Poster presentation A01. *2nd International Conference on Environmental Management, Engineering, Planning and Economics (CEMEPE 2009) & SECOTOX Conference, Mykonos, June 21–26.*

Filep, Á., Drinovec, L., Palágyi, A., Manczinger, L., Vágvölgyi, Cs. *et al.* (2015) Source specific cyto- and genotoxicity of atmospheric aerosol samples. *Aerosol and Air Quality Research* 15, 2325–2331.

Franz, T.P., Eisenreich, S.J. and Holsen, T.M. (1998) Dry deposition of particulate polychlorinated biphenyls and polycyclic aromatic hydrocarbons to Lake Michigan. *Environmental Science & Technology* 32(23), 3681–3688.

Gevao, B., Hamilton-Taylor, J. and Jones, K.C. (1998) Polychlorinated biphenyl and polycyclic aromatic hydrocarbon deposition to and exchange at the air±water interface of Esthwaite Water, a small lake in Cumbria, UK. *Environmental Pollution* 102, 63–75.

Girotti, S., Ferri, E.N., Fumo, M.G. and Maiolini, E. (2008) Monitoring of environmental pollutants by bioluminescent bacteria. *Analytica Chimica Acta* 608, 2–29.

Goix, S., Lévêque, T., Xiong, T.T., Schreck, E., Baeza-Squiban, A. *et al.* (2014) Environmental and health impacts of fine and ultrafine metallic particles: assessment of threat scores. *Environmental Research* 133, 185–194.

Grantz, D.A., Garner, J.H.B. and Johnson, D.W. (2003) Ecological effects of particulate matter. *Environment International* 29, 213– 239.

Grattan, S. R., Maas, E. V. and Ogata, G. (1981) Foliar uptake and injury from saline aerosol. *Journal of Environmental Quality* 10, 406–409.

Hasheminassab, S., Pakbin, P., Delfino, R.J., Schauer, J.J. and Sioutas, C. (2014) Diurnal and seasonal trends in the apparent density of ambient fine and coarse particles in Los Angeles. *Environmental Pollution* 187, 1–9.

Honour, S.L., Bell, J.N.B., Ashenden, T.W., Cape, J.N. and Power, S.A. (2009) Responses of herbaceous plants to urban air pollution: effects on growth, phenology and leaf surface characteristics. *Environmental Pollution* 157, 1279–1286.

Hu, X., Zhang, Y., Luo, J., Xie, M., Wang, T. and Lian, H. (2011) Accumulation and quantitative estimates of airborne lead for a wild plant (*Aster subulatus*). *Chemosphere* 82, 1351–1357.

Isidori, M., Ferrara, M., Lavorgna, M., Nardelli, A. and Parrella, A. (2003) In situ monitoring of urban air in Southern Italy with the tradescantia micronucleus bioassay and semipermeable membrane devices (SPMDs). *Chemosphere* 52, 121–126.

Jalava P.I., Wang Q., Kuuspalo K., Ruusunen J., Hao L. *et al.* (2015) Day and night variation in chemical composition and toxicological responses of size segregated urban air PM samples in a high air pollution situation. *Atmospheric Environment* 120, 427–437.

Kaiser, K.L.E. (1998) Correlations of *Vibrio fischeri* bacteria test data with bioassay data for other organisms. *Environmental Health Perspectives* 106 (Suppl. 2), 583–591.

Kessler, N., Schauer, J.J., Yagur-Kroll, S., Melamed, S., Tirosh, O., Belkin, S. and Erel, Y. (2012) A bacterial bioreporter panel to assay the cytotoxicity of atmospheric particulate matter. *Atmospheric Environment* 63, 94–101.

Kokkali, V. and van Delft, W. (2014) Overview of commercially available bioassays for assessing chemical toxicity in aqueous samples. *Trends in Analytical Chemistry* 61, 133–155.

Komarek, M., Ettler, V., Chrastny, V. and Mihaljevic, M. (2008) Lead isotopes in environmental sciences: a review. *Environment International* 34, 562–577.

Kováts, N. and Horváth, E. (2016) Bioluminescence-based assays for assessing eco- and genotoxicity of airborne emissions. *Luminescence* 31(4), 918–923.

Kováts, N., Kovács, A., Ács, A., Ferincz, Á., Turóczy, B. and Gelencsér, A. (2012) Development of a whole-aerosol test protocol. *Environmental Toxicology and Pharmacology* 33, 284–287.

Kováts, N., Ács, A., Ferincz, Á., Kovács, A., Horváth, E. *et al.* (2013) Ecotoxicity and genotoxicity assessment of exhaust particulates from diesel-powered buses. *Environmental Monitoring and Assessment* 10, 8707–8713.

Kováts, N., Horváth, E., Eck-Varanka, B., Csajbók, E. and Hoffer, A. (2017) Adapting the Vegetative Vigour Terrestrial Plant Test for assessing ecotoxicity of aerosol samples. *Environmental Science and Pollution Research* 24, 15291–15298.

Landkocz, Y., Ledoux, F., André, V., Cazier, F., Genevray, P. *et al.* (2017) Fine and ultrafine atmospheric particulate matter at a multi-influenced urban site: physicochemical characterization, mutagenicity and cytotoxicity. *Environmental Pollution* 221, 130–140.

Lange, J.H. and Thomulka, K.W. (1997) Use of the *Vibrio harveyi* toxicity test for evaluating mixture interactions of nitrobenzene and dinitrobenzene. *Ecotoxicology and Environmental Safety* 38, 2–12.

Lappalainen, J., Juvonen, R., Vaajasaari, K. and Karp, M. (1999) A new flash method for measuring the toxicity of solid and colored samples. *Chemosphere* 3, 1069–1083.

Lappalainen, J., Juvonen, R., Nurmi, J. and Karp, M. (2001) Automated color correction method for *Vibrio fischeri* toxicity test. Comparison of standard and kinetic assays. *Chemosphere* 45, 635–641.

Lin, T.C. and Chao, M.R. (2002) Assessing the influence of methanol-containing additive on biological characteristics of diesel exhaust emissions using Microtox and Mutatox assays. *Science of the Total Environment* 284, 61–74.

Ma, X.Y., Wang, X.C., Ngo, H.H., Guo, W., Wu, M.N. and Wang, N. (2014) Bioassay based luminescent bacteria: interferences, improvements, and applications. *Science of the Total Environment* 468–469, 1–11.

Masner, P., Javůrková, B. and Bláha, L. (2017) Rapid in situ toxicity testing with luminescent bacteria *Photorhabdus luminescens* and *Vibrio fischeri* adapted to a small portable luminometer. *Environmental Science and Pollution Research* 24, 3748–3758.

Mesquita, S.R., van Drooge, B.L., Reche, C., Guimarães, L., Grimalt, J.O., Barata, C. and Piña, B. (2014) Toxic assessment of urban atmospheric particle-bound PAHs: relevance of composition and particle size in Barcelona (Spain). *Environmental Pollution* 184, 555–562.

Moreno, T., Querol, X., Alastuey, A., Viana, M., Salvador, P. *et al.* (2006) Variations in atmospheric PM trace metal content in Spanish towns: illustrating the chemical complexity of the inorganic urban aerosol cocktail. *Atmospheric Environment* 40, 6791–6803.

Mortimer, M., Kasemets, K., Heinlaan, M., Kurvet, I. and Kahru, A. (2008) High throughput kinetic bioluminescence inhibition assay for study of toxic effects of nanoparticles. *Toxicology in Vitro* 22, 1402–1417.

Oguntimehin, I., Eissa, F. and Sakugawa, H. (2010) Negative effects of fluoranthene on the ecophysiology of tomato plants (*Lycopersicon esculentum* Mill). Fluoranthene mists negatively affected tomato plants. *Chemosphere* 78, 877–884.

Petroff, A., Mailliat, A., Amielh, M. and Anselmet, F. (2008) Aerosol dry deposition on vegetative canopies. Part I: Review of present knowledge. *Atmospheric Environment* 42(16), 3625–3653.

Pintér, M., Utry, N., Ajtai, T., Kiss-Albert, G., Jancsek-Turóczi, B. *et al.* (2017) Optical properties, chemical composition and the toxicological potential of urban particulate matter. *Aerosol and Air Quality Research* 17, 1415–1426.

Roig, N., Sierra, J., Rovira, J., Schuhmacher, M., Domingo, J.L. and Nadal, M. (2013) In vitro tests to assess toxic effects of airborne $PM_{10}$ samples. Correlation with metals and chlorinated dioxins and furans. *Science of the Total Environment* 443, 791–797.

Schilirò, T., Bonetta, S., Alessandria, L., Gianotti, V., Carraro, E. and Gilli, G. (2015) PM10 in a background urban site: chemical characteristics and biological effects. *Environmental Toxicology and Pharmacology* 39, 833–844.

Schleicher, N.J., Schäfer, J., Blanc, G., Chen, Y., Chai, F., Cen, K. and Norra, S. (2015) Atmospheric particulate mercury in the megacity Beijing: spatio-temporal variations and source apportionment. *Atmospheric Environment* 109, 251–261.

Schmidt, S., Altenburger, R. and Kühnel, D. (2017) From the air to the water phase: implication for toxicity testing of combustion-derived particles. *Biomass Conversion and Biorefinery* 9, 213–225.
Schreck, E., Foucault, Y., Sarret, G., Sobanska, S., Cécillon, L. *et al.* (2012) Metal and metalloid foliar uptake by various plant species exposed to atmospheric industrial fallout: mechanisms involved for lead. *Science of the Total Environment* 427–428, 253–262.
Schreck, E., Laplanche, C., Le Guédard, M., Bessoule, J.J., Austruy, A. *et al.* (2013) Influence of fine process particles enriched with metals and metalloids on *Lactuca sativa* L. leaf fatty acid composition following air and/or soil-plant field exposure. *Environmental Pollution* 179, 242–249.
Schreck, E., Dappe, V., Sarret, G., Sobanska, S., Nowak, D. *et al.* (2014) Foliar or root exposures to smelter particles: consequences for lead compartmentalization and speciation in plant leaves. *Science of the Total Environment* 476–477, 667–676.
Serat, W.F., Budinger, F.E. Jr and Mueller, P.K. (1965) Evaluation of biological effects of air pollutants by use of luminescent bacteria. *Journal of Bacteriology* 90, 832–833.
Serat, W.F., Budinger, F.E. Jr and Mueller, P.K. (1967) Toxicity evaluation of air pollutants by use of luminescent bacteria. *Atmospheric Environment* 1, 21–32.
Sheesley, R.J., Schauer, J.J., Hemming, J.D., Barman, M.A., Geis, S.W. and Tortorelli, J.J. (2004) Toxicity of ambient atmospheric particulate matter from the Lake Michigan (USA) airshed to aquatic organisms. *Environmental Toxicology and Chemistry* SETAC 23, 133–140.
Sheesley, R.J., Schauer, J.J., Hemming, J.D., Geis, S. and Barman, M.A. (2005) Seasonal and spatial relationship of chemistry and toxicity in atmospheric particulate matter using aquatic bioassays. *Environmental Science and Technology* 39, 999–1010.
Silva, V., Pereira, J.L., Campos, I., Keizer, J.J., Gonçalves, F. and Abrantes, N. (2015) Toxicity assessment of aqueous extracts of ash from forest fires. *Catena* 135, 401–408.
Smith, K.E.C. and Jones, K.C. (2000) Particles and vegetation: implications for the transfer of particle-bound organic contaminants to vegetation. *Science of the Total Environment* 246, 207–236.
Squadrito, G.L., Cueto, R., Dellinger, B. and Pryor, W.A. (2001) Quinoid redox cycling as a mechanism for sustained free radical generation by inhaled airborne particulate matter. *Free Radical Biology and Medicine* 31, 1132–1138.
Szigeti, T., Óvári, M., Dunster, C., Kelly, F.J., Lucarelli, F. and Záray, G. (2015) Changes in chemical composition and oxidative potential of urban PM2.5 between 2010 and 2013 in Hungary. *Science of the Total Environment* 518–519, 534–544.
Titos, G., Lyamani, H., Pandolfi, M., Alastuey, A. and Alados-Arboledas, L. (2014) Identification of fine ($PM_1$) and coarse ($PM_{10-1}$) sources of particulate matter in an urban environment. *Atmospheric Environment* 89, 593–602.
Triolo, L., Binazzi, A., Cagnetti, P., Carconi, P., Correnti, A. *et al.* (2008) Air pollution impact assessment on agroecosystem and human health characterisation in the area surrounding the industrial settlement of Milazzo (Italy): a multidisciplinary approach. *Environmental Monitoring and Assessment* 140, 191–209.
Turóczi, B., Hoffer, A., Tóth, Á., Kováts, N., Ács, A. *et al.* (2013) Comparative assessment of ecotoxicity of urban aerosol. *Atmospheric Chemistry and Physics* 12, 7365–7370.
Urbanczyk, H., Ast, J., Higgins, M.J., Carson, J. and Dunlap, P.V. (2007) Reclassification of *Vibrio fischeri*, *Vibrio logei*, *Vibrio salmonicida* and *Vibrio wodanis* as *Aliivibrio fischeri* gen. nov., comb. nov., *Aliivibrio logei* comb. nov., *Aliivibrio salmonicida* comb. nov. and *Aliivibrio wodanis* comb. nov. *International Journal of Systematic and Evolutionary Microbiology* 57(12), 2823–2829.
Uzu, G., Sobanska, S., Sarret, G., Munoz, M. and Dumat, C. (2010) Foliar lead uptake by lettuce exposed to atmospheric fallouts. *Environmental Science and Technology* 44, 1036–1042.
Valavanidis, A., Fiotakis, K., Vlahogianni, T., Bakeas, E.B., Triantafillaki, S., Paraskevopoulou, V. and Dassenakis, M. (2006) Characterization of atmospheric particulates, particle-bound transition metals and polycyclic aromatic hydrocarbons of urban air in the centre of Athens (Greece). *Chemosphere* 65, 760–768.
Verma, V., Rico-Martinez, R., Kotra, N., Rennolds, C., Liu, J., Snell, T.W. and Weber, R.J. (2013) Estimating the toxicity of ambient fine aerosols using freshwater rotifer *Brachionus calyciflorus* (Rotifera: Monogononta). *Environmental Pollution* 182, 379–384.
Verma, V., Wang, Y., El-Afifi, R., Fang, T., Rowland, J., Russell, A.G. and Weber, R.J. (2015) Fractionating ambient humic-like substances (HULIS) for their reactive oxygen species activity – assessing the importance of quinones and atmospheric aging. *Atmospheric Environment* 120, 351–359.
Volpi Ghirardini, A., Girardini, M., Marchetto, D. and Pantani, C. (2009) Microtox® solid phase test: effect of diluent used in toxicity test. *Ecotoxicology and Environmental Safety* 72, 851–861.

Vouitsis, E., Ntziachristos, L., Pistikopoulos, P., Samaras, Z., Chrysikou, L. *et al.* (2009) An investigation on the physical, chemical and ecotoxicological characteristics of particulate matter emitted from light-duty vehicles. *Environmental Pollution* 157, 2320–2327.
Wang, W.H., Wong, M.H., Leharne, S. and Fisher, B. (1998) Fractionation and biotoxicity of heavy metals in urban dusts collected from Hong Kong and London. *Environmental Geochemistry and Health* 20, 185–198.
Wang, D.Y, Shen, L.L. and Wang, Y. (2007) The phenotypic and behavioral defects can be transferred from zinc exposed nematodes to their progeny. *Environmental Toxicology and Pharmacology* 24, 223–230.
Wang, W., Shi, C., Yan, Y., Yang, Y. and Zhou, B. (2016) Eco-toxicological bioassay of atmospheric fine particulate matter ($PM_{2.5}$) with *Photobacterium phosphoreum* T3. *Ecotoxicology and Environmental Safety* 133, 226–234
Wedding, J.B., Carlson, R.W., Stukel, J.J. and Bazzaz, F.A. (1975) Aerosol deposition on plant leaves. *Environmental Science and Technology* 9, 151–153.
Wieczorek, J., Sienkiewicz, S., Pietrzak, M. and Wieczorek, Z. (2015) Uptake and phytotoxicity of anthracene and benzo[k]fluoranthene applied to the leaves of celery plants (*Apium graveolens* var. *secalinum* L.). *Ecotoxicology and Environmental Safety* 115, 19–25.
Wik, A. and Dave, G. (2008) Occurrence and effects of tire wear particles in the environment – a critical review and an initial risk assessment. *Environmental Pollution* 157(1), 1–11.
Xing, J., Song, J., Yuan, H., Wang, Q., Li, X. *et al.* (2017) Atmospheric wet deposition of dissolved trace elements to Jiaozhou Bay, North China: fluxes, sources and potential effects on aquatic environments. *Chemosphere* 174, 428–436.
Zhao, Y., Lin, Z., Jia, R., Li, G., Xi, Z. and Wang, D. (2014) Transgenerational effects of traffic-related fine particulate matter (PM2.5) on nematode *Caenorhabditis elegans*. *Journal of Hazardous Materials* 274, 106–114.

# 30 Toxicity of Microplastics in the Marine Environment

**M.F.M. Santana*[,1,2] and A. Turra**[,3]**
[1]*Department of Science and Engineering, James Cook University;* [2]*Australian Institute of Marine Science, Townsville, Australia;* [3]*Oceanographic Institute, University of São Paulo, Brazil*

## 30.1 Abstract

Marine litter is one of the most expanding and devastating problems in the marine environment. Among several types of items and origins, microplastics have been drawing attention due to their small particle size (< 5 mm), persistence and the potential to be ingested by a high variety of organisms and act as vectors of chemical pollutants. This chapter provides a broad and updated review of the ecotoxicological effects of microplastics on marine biodiversity. Effects of microplastics have been demonstrated from the biochemical to physiological, behavioural and even ecological levels of organization. Most information comes from a range of biomarkers on low levels of organization, such as oxidative stress (lipid peroxidation, DNA damage and the activation of antioxidant enzymes) and immunological responses (phagocytosis activity, signs of inflammation and enzyme/protein response). However, the variety of experimental conditions, including plastic size and shape, polymer type, additives, concentration and period of exposure, and biological model, still does not allow a clear and wide understanding on the effects on the environment and, indeed, humans. This scenario calls for additional and coordinated efforts to improve scientific knowledge on this subject.

## 30.2 Background

Modern life provides comfort and convenience with materials such as plastics. They are lightweight and cheap and can have many other useful properties, depending on the chemicals added during plastic manufacture. As a result of such versatility, global plastic production was 335 million tonnes in 2016 and it is expected to grow (PlasticsEurope, 2018). However, plastics have been accumulating in the environment in the past few decades, which is intensified by their poor management as waste. To illustrate the potential of plastics to contaminate the marine environment, Jambeck *et al.* (2015) estimated that, in 2010, 192 coastal countries produced 275 million tonnes of plastic waste and that up to 12.7 million tonnes might have reached the ocean. In fact, plastic items dominate the composition of marine litter (UNEP, 2016) and are widespread throughout the world's oceans (Eriksen *et al.*, 2014). In 2011, plastics became one of the

* E-mail address: marina.santana@my.jcu.edu.au
** E-mail address: turra@usp.br

major marine issues of the 21st century (UNEP, 2011), comparable in importance to climate change (UNEP, 2016). Recent estimates in marine regions commonly known as debris sinks, such as Great Pacific Garbage Patch, give evidence that plastic particles are rapidly accumulating (Lebreton *et al.*, 2017), with values four to 16 times higher than previously reported in the literature. A previous estimate revealed minimum values of 5.25 trillion particles, weighing 268,940 t, in the five gyres along the oceans (Eriksen *et al.*, 2014), while Lebreton *et al.* (2017) estimated values of 79,000 t (45,000–129,000 t) and 1.8 trillion pieces (1.1–3.6 trillion pieces) in the Great Pacific Garbage Patch alone.

Among all marine plastic contaminants, large-sized plastics (e.g. derelict fishing gear) can greatly contribute to the mass of litter in the ocean and cause a range of impacts, including ghost fishing, organism abrasion and boat accidents (Lebreton *et al.*, 2017). In spite of this, the plastic items responsible for the enormous figures on number of particles shown above are the small particles named microplastics (Eriksen *et al.*, 2014; Lebreton *et al.*, 2017). Microplastics are operationally defined to be the plastic particles less than 5 mm in maximum length (Arthur *et al.*, 2009; GESAMP, 2019). This emergent contaminant is either industrially produced as small-sized particles ('primary microplastics'), as abrasives beads for cosmetics products, plastic pellets and PVC powder (Rodolfo *et al.*, 2006; Fendall and Sewell, 2009; Andrady, 2011), or results from larger-plastic degradation in the environment ('secondary microplastics') (Browne *et al.*, 2007; Andrady, 2011; Browne *et al.*, 2011; Zettler *et al.*, 2013). Synthetic fibres released from clothes during washing (Browne *et al.*, 2011; Napper and Thompson, 2016; Salvador Cesa *et al.*, 2017) lie in between these two definitions, classified either as primary (Henry *et al.*, 2019) or secondary microplastics (GESAMP, 2016).

Microplastics have been reported to occur in the marine environment since the 1970s (Carpenter and Smith, 1972; Carpenter *et al.*, 1972; Colton *et al.*, 1974) and are now considered ubiquitous: they are present in freshwater systems and terrestrial environments (for reviewes see Wagner *et al.*, 2014; Horton *et al.*, 2017), estuaries and coastal beaches (Frias *et al.*, 2010; Imhof *et al.*, 2017), offshore oceans (Cozar *et al.*, 2014; Reisser *et al.*, 2015; Kanhai *et al.*, 2017), deep-sea sediments (Van Cauwenberghe *et al.*, 2013; Woodall *et al.*, 2014) and in polar ice (Obbard *et al.*, 2014). Microplastics have also been found contaminating water bottles (Mason *et al.*, 2018), table salts (Yang *et al.*, 2015; Karami *et al.*, 2017) and a vast range of animals, including those as fish and shellfish that are part of our food resources, becoming a potential risk for human health (e.g. Santillo *et al.*, 2017; FAO, 2017).

Microplastics as contaminants started to attract the attention of scientists because of their small size, high abundance and potential biological effects (GESAMP, 2010, 2015, 2016; Wright *et al.*, 2013; Turra *et al.*, 2014). Indeed, the small sizes of microplastics make them potentially bioavailable for a wide range of organisms. The ingestion of microplastics, the uptake route more commonly reported, is especially risky for organisms with indiscriminate feeding habits that capture anything similar in size to their natural food (Browne *et al.*, 2008; Graham and Thompson, 2009; Cole *et al.*, 2013; Wright *et al.*, 2013). According to Lusher (2015), more than 220 species (marine, freshwater and terrestrial organisms) have been found to ingest microplastics and, so far, most of the observed biological impacts caused by microplastics involve their ingestion. Studies with marine species showed that, once ingested, microplastics can be retained and obstruct the digestive tract (Derraik, 2002; Besseling *et al.*, 2013) or be assimilated by or translocated into tissues (Browne *et al.*, 2008; von Moos *et al.*, 2012; Farrell and Nelson, 2013).

As with any other plastic material, microplastics are composed of long chains of monomers and their mixed additives (Andrady, 2017), both suspected to have toxic properties with the potential to endanger life (Ananthaswamy, 2001; Lithner *et al.*, 2011; Wright *et al.*, 2013). Also, as plastics become smaller, their surface–volume ratio becomes larger, i.e. a larger area in relation to volume, prone to the adsorption and adherence of other non-polar toxic substances and microorganisms, making microplastics a potential vector of other threats to organisms and the food chain (Mato *et al.*, 2001; GESAMP, 2010; Browne *et al.*, 2013; Tanaka *et al.*, 2013; Zettler *et al.*, 2013; McCormick *et al.*, 2014). As a consequence, microplastic interaction with biota can cause a range of effects, from physical harm to physiological, cellular and molecular stresses (Browne *et al.*, 2013; Cole *et al.*, 2013; Avio *et al.*, 2015; Lu *et al.*, 2016; Paul-Pont *et al.*; 2016; Gambardella *et al.*, 2017). These impacts, in turn, may cause further environmental damage, as microplastics can affect growth and reproduction rates (Cole and Galloway, 2015; Sussarellu *et al.*, 2016) and potentially ecosystemic effects (Heindler *et al.*, 2017).

In this chapter we will summarize the potential toxicological effects of microplastic pollution already identified and discuss aspects that can influence the biological responses to such interaction. Along with that, we suggest knowledge constraints that need to be better investigated to address the issue. We will mostly have the marine environment as background, due to the unquestionable presence of plastics in marine ecosystems and the large number of studies carried out with marine model organisms. We have assumed that most of the examples discussed here could be transported to freshwater environments because of similarities in taxa and the process of biological interaction with plastic particles (e.g. uptake by filter/suspension-feeder animals; and leaching of plastic chemical compounds to the organisms and even to the environment and thereby contact with organisms). The toxicity of microplastics in terrestrial environments, in turn, is as yet poorly understood and will not be considered here. The three main components that form the basis of this chapter are: (i) microplastic distribution in the environment and organism exposure (i.e. exposure to microplastics); (ii) microplastic transformation prior to effects; and (iii) microplastic uptake by marine organisms and effects.

## 30.3 Exposure to Microplastics

Microplastic ecotoxicological effects are influenced by several factors. The primary factor is organism exposure to the particles. Organism exposure to microplastics is very dynamic, not only because it relies on the presence of a susceptible animal where the plastics are located, but also because the presence, characteristics and concentration of microplastics in a certain place change in space and time. A combination of drivers defines the input, dispersal and accumulation of microplastics in marine environments (Krelling *et al.*, 2017). Some examples are given below.

At a local scale, for instance, microplastic input might be influenced by type and amount of anthropogenic activities in the surroundings (Browne *et al.*, 2011). In touristic coastal areas, the numbers of tourists and/or second homeowners increase during vacation and may lead to a relevant increase in flushing of fibres released from clothes washing or microbeads from cosmetics, especially in places with poor waste management. On the other hand, fishing (FAO, 2017) and aquaculture (Van Cauwenberghe and Janssen, 2014) activities have already been demonstrated to increase the input of microplastic fibres and fragments (e.g. polypropylene fishing lines and fishing nets and styrofoam culture structures).

Oceans are spatially and temporally dynamic due to the influence of meteorological (e.g. rainfall, storm-water pulses through watersheds and wind) and oceanographic processes (e.g. tides, waves and currents), which have strong influence on the distribution of marine litter (Law *et al.*, 2010; Ivar Do Sul and Costa, 2014; Ivar Do Sul *et al.*, 2014; Eriksen *et al.*, 2014; Krelling *et al.*, 2017). In addition to microplastics being distributed worldwide, the influence of the factors above cause small-scale temporal and spatial patchiness, which is responsible for variation in exposure to the biota.

Regardless of origin, once plastic litter enters the oceans, either as large items or as primary microplastics, it will be degraded and dispersed across different parts of the world and along sea compartments (i.e. surface, water column, sediment and organisms). Features such as polymer type and density will also affect its distribution along marine compartments. To illustrate: dense plastics such as polyvinyl chloride (PVC) will tend to sink to the sea bottom, while light plastics such as polyethylene (PE) will tend to float (GESAMP, 2015). Other processes such as biofouling (organisms that settle recruit and grow on natural and artificial hard substrates) can increase the density of plastics, which will eventually sink to other compartments. Despite such knowledge, the sinks (or hotspots) of microplastics in the oceans are still not well understood, nor are the fluxes among biotic and abiotic compartments (GESAMP, 2015, GESAMP, 2016).

Bioavailability or exposure to microplastics also depends on the relative size of particles and organisms. The smaller the particles, the more organisms are able to passively or actively ingest them. Ingested particles may then be accumulated in the gut or eliminated in the faeces. The residence time of the particles in the gut may vary, depending on the organism and type and size of particles. Small particles (< 100 μm) are able to be ingested by the cells in the gut wall (von Moos *et al.*, 2012) or translocated to the blood and other tissues (Browne *et al.*, 2008; von Moos *et al.*, 2012), where they can cause additional effects on the organisms.

## 30.4 Transformation Prior to Effects of Microplastics

In theory, like other organic and inorganic pollutants, microplastics can bioaccumulate, undergo biotransformation or become biodegraded while in contact with organisms. This, in turn, can modulate microplastic biological effects.

After intake and egestion, bioaccumulation is a process that implies greater concentrations of the contaminant within biological tissues in comparison with the surrounding environment (Connell, 1990; Rand *et al.*, 1995; Gray, 2002). In the case of microplastics, most studies suggest low risks of bioaccumulation because organisms tend to excrete the uptaken particles over time (Ugolini *et al.*, 2013; Hamer *et al.*, 2014; Kaposi *et al.*, 2014; Mazurais *et al.*, 2015; Grigorakis *et al.*, 2017; Dawson *et al.*, 2018). However, such an elimination process is not yet well understood.

Biotransformation corresponds to changes in the contaminant structure or composition due to the body's metabolization, which might increase or reduce the contaminant's toxicity (Landis and Yu, 2004). Biodegradation is the breaking down of organic matter (in this case, the contaminant) by microorganisms into primary chemical compounds, such as $CO_2$ and $H_2O$ (Landis and Yu, 2004). Microplastics, although small in size, commonly comprise long chains of monomers, with strong chemical bonds that hinder biotransformation and biodegradations (Andrady, 2017). In fact, that is one of the reasons why plastic is considered a persistent contaminant in the environment.

Overall, it could be said that microplastics are hardly prone for these transformations during their time in the environment and in contact with life. However, plastics additives and other chemicals adsorbed on plastics surface can easily bioaccumulate, be biotransformed and become biodegraded, increasing the potential effects on the biota.

## 30.5 Effects of Microplastics

As for any other chemical, the effects of microplastic exposure and contact with organisms may range from biochemical levels to physiological, behavioural or even ecological levels of organization (Fig. 30.1). However, most of published works to date investigate microplastic effects on low levels of biological organizations, which is actually comprehensive considering the early stages of this topic in science (GESAMP, 2016).

### 30.5.1 Effects after exposure and intake

The intake and potential effects of microplastics in marine organisms have been investigated for over a decade. The studies are mostly experimental, cover a broad range of endpoints and use different scenarios of exposure and model species to conduct the assays (Table 30.1). By different scenario of exposure, we mean the endless combination of polymer types, microplastic shapes and sizes, concentration and period of exposure. This creates complex contamination scenarios that may lead to different biological responses, making it hard to establish patterns and predict environmentally relevant impacts.

In terms of sub-organismal levels of organization, several studies report that microplastic exposure and intake can cause oxidative stress, such as lipid peroxidation, DNA damage and the activation of antioxidant enzymes (Browne *et al.*, 2013; Avio *et al.*, 2015; Jeong *et al.*, 2016; Gambardella *et al.*, 2017), and immunological responses, such as phagocytosis activity, signs of inflammation and enzyme/protein response (Browne *et al.*, 2008; von Moos *et al.*, 2012; Avio *et al.*, 2015).

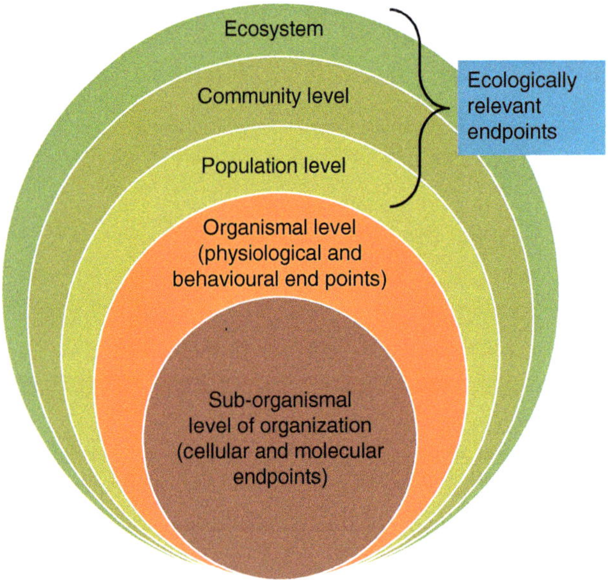


**Fig. 30.1.** Potential effects of microplastics in different levels of biological organization.

Moving to the organismal level, microplastics can affect feeding, body weight and energy reserves, growth, reproductive success and mortality, with most of these data coming from studies with invertebrates with short life cycles such as worms and zooplankton or with model species still in early stages of life, such as larvae (Besseling *et al.*, 2013; Au *et al.*, 2015; Cole *et al.*, 2015; Fonte *et al.*, 2016; Jeong *et al.*, 2016; Tosetto *et al.*, 2016; Heindler *et al.*, 2017; Bour *et al.*, 2018; Critchell and Hoogenboom, 2018). Feeding studies were not reported in Table 30.1 as changes in such activity are assumed to be more related to microplastic physical impacts than ecotoxicological effects.

Although not much explored so far, microplastics were also observed to cause behavioural changes, such as reducing swimming activities of European sea bass, jump height of beach hoppers and predatory performance of common gobies, among others (Luis *et al.*, 2015; Sa *et al.*, 2015; Tosetto *et al.*, 2016; Gambardella *et al.*, 2017; Barboza *et al.*, 2018b; Guven *et al.*, 2018). It is important to note that the ecological consequences of behavioural changes are much easier to establish than those of sub-organismal levels, highlighting the importance of more studies in this regard.

The analyses of ecological endpoints (e.g. changes in population sizes, or in dynamics of assemblages) should be our ultimate concern (EPA, 1992; Santana *et al.*, in preparation). However, this is not an easy task. In the microplastics research field, only a few works have actually explored these effects (against the many studies that suggest their occurrence based on lower levels of biological responses): Green (2016) and Green *et al.* (2016) checked the structure of invertebrate benthic assemblages after biodegradable and conventional microplastic exposures; and Heindler *et al.* (2017) assessed the population size of copepods exposed to PET-microplastics contaminated (or not) with the plasticizer di(2-ethylhexyl) phthalate (DEHP). In all these studies, microplastics caused ecological disturbances.

As already mentioned, the difficulty in making comparisons among studies has been delaying a comprehensive understanding about the influence of the exposure parameters on the observed responses. To overcome this constraint, studies have designed experiments to test differences of microplastic effects according to types, shapes, sizes, concentration and period of exposure. For that, researchers need to use the same biological model and ensure an orthogonal experimental set for all the tested parameters. In these works, microplastic toxicity was observed to be dependent on size, shape, polymer type and concentration (e.g. Lee *et al.*, 2013; Au *et al.*, 2015; Avio *et al.*, 2015; Jeong *et al.*, 2016; Bour

**Table 30.1.** Studies on the ecotoxicological effects of microplastics on marine organisms (adapted from Santana *et al*., in preparation).

| Taxon | Species | Experimental features: Polymer type | Shape | Size | Nominal concentration of exposure | Period of exposure | Endpoint | Reference |
|---|---|---|---|---|---|---|---|---|
| Isopod | *Idotea emarginata* | PS | Sphere; irregular; fibre | 10 μm | 12 and 120 microbeads $mg^{-1}$ of food; 20 and 350 fragments $mg^{-1}$ of food; 0.3 mg fibres $g^{-1}$ of food | 6 weeks | Mortality, growth and intermoult duration | Hamer *et al*. (2014) |
| Amphipod | *Hyalella azteca* | PE; PP | Particle; fibre | 10–27; 20–75 μm | Up to 100,000 particles $l^{-1}$ | 10; 42 days | Mortality, reproduction and growth | Au *et al*. (2015) |
| Copepod | *Calanus helgolandicus* | PS | Sphere | 20 μm | 75 particles $ml^{-1}$ | 9 days | Egg production, egg size, hatching and respiration | Cole *et al*. (2015) |
| | *Parvocalanus crassirostris* | PET | Irregular | 5–10μm | 1:1; 2:1; 4:1; 8:1 microplastic to food ratio | 6 days | Mortality, egg production, population size and gene expression | Heindler *et al*. (2017) |
| | *Tigriopus japonicus* | PS | Sphere | 0.005; 0.5 and 6 μm | 0.125; 1.25; 6; 12.5; 13; 25; 31; 63; 187; 250; 313 μg $ml^{-1}$ | 96 h – 2 copepod generations | Survival, development and fecundity | Lee *et al*. (2013) |
| Krill | *Euphausia superba* | PE | Sphere | 27–32 μm | 10, 20, 40; 80% plastic diet | 10 days | Mortality and weight loss | Dawson *et al*. (2018) |
| Rotifer | *Brachionus koreanus* | PS | Sphere | 0.05; 0.5 and 6 μm | 0.1; 1; 10; 20 μm $m^{-1}$ | 12 days | Growth, fecundity, lifespan | Jeong *et al*. (2016) |
| Oyster (larvae) | *Crassostera gigas* | PS | Sphere | 1 and 10 μm | 100 microplastics $ml^{-1}$ | 8 days | Growth | Cole and Galloway (2015) |
| Sea urchin (larvae) | *Tripneustes gratilla* | PE | Sphere | 25–32 μm | 1; 10; 100; 300 microplastics $ml^{-1}$ | 5 days | Survival and growth | Kaposi *et al*. (2014) |
| Barnacle (larvae) | *Amphibalanus amphitrite* | PS | Sphere | 0.1 μm | 0.001–10 mg $l^{-1}$ | 24; 48 h | Mortality, behaviour and enzyme activity | Gambardella *et al*. (2017) |
| Shrimp (larvae) | *Artemia franciscana* | | | | | | | |
| Littorina (larvae) | *Crepidula onyx* | PS | Sphere | 2–5 μm | $6 \times 10^4$; 1.4 $10^5$; 10 particles $m^{-1}$ | 14 days | Survival and growth | Lo and Chan (2018) |
| Beachhopper | *Platorchestia smithi* | PE | Sphere | 38–54 μm | 3.8% of sediment weight | 24, 72, 120 h | Survival and behaviour | Tosetto *et al*. (2016) |

*Continued*

**Table 30.1.** Continued.

| | | Experimental features | | | | | | |
|---|---|---|---|---|---|---|---|---|
| Taxon | Species | Polymer type | Shape | Size | Nominal concentration of exposure | Period of exposure | End point | Reference |
| Sandhopper | *Talitrus saltator* | PE | Sphere | 10–45 μm | 10% of food weight | 24 h | Survival | Ugolini *et al*. (2013) |
| Bivalve | *Ennucula tenuis*; *Abra nitida* | PE | Sphere | 4–6; 20–25; 125–500 μm | 1; 10; 25 mg $kg^{-1}$ of sediment | 4 weeks | Survival; condition index and behaviour | Bour *et al*. (2018) |
| | *Crassostera gigas* | PS | Sphere | 2 and 6 μm | 0.023mg $l^{-1}$ | 2 months | Ecophysiological parameters; cellular, transcriptomic and proteomic responses; fecundity; and offspring development | Sussarellu *et al*. (2016) |
| | *Ostrea edulis* | PLA; HDPE | n/a* | 0.6–363; 0.48–316 μm | 0.8; 80 μg $l^{-1}$ | 60 days | Growth; surrounding infaunal assemblages in the sediment | Green (2016) |
| | *Mytilus edulis* | PLA; HDPE | n/a* | 0.6–363; 0.48–316 μm | 2.5; 25 μg $l^{-1}$ | 50 days | Diversity and abundance of macrofauna within surrounding sediment | Green *et al*. (2016) |
| | | PS | Sphere | 3; 9.6 μm | 0.51g $l^{-1}$ | 3 h | Viability and phagocytic function of haemocytes and oxidative status of haemolymph | Browne *et al*. (2008) |
| | | HDPE | Sphere | Up to 80 μm | 2.5 g $l^{-1}$ | 3; 6; 12; 24; 48; 96 h | Histopathology and cytochemical biomarkers of toxic effects and early warning | von Moos *et al*. (2012) |
| | *Mytilus* spp. | PS | Sphere | 2; 6 μm | 32 μg $l^{-1}$ | 7 days | Histopathology; FLU quantification, enzyme activities and gene expression | Paul-Pont *et al*. (2016) |
| | *Mytilus galloprovincialis* | PE; PS | Sphere | < 100 μm | 1.5 g $l^{-1}$ | 7 days | Cellular immunological responses, neurotoxic effects and genotoxicity | Avio *et al*. (2015) |

| | | | | | | | | |
|---|---|---|---|---|---|---|---|---|
| | *Perna perna* | PVC | Sphere | 0.1–1μm | 0.125 $l^{-1}$ | 90 days | Mortality, growth, cellular and molecular biomarkers of oxidative stress | Santana *et al.* (in prep.) |
| | *Scrobicularia plana* | PS | Sphere | 20 μm | 1 mg $l^{-1}$ | 14 days | Biomarkers of neurotoxicity, oxidative damage, antioxidant capacity | Ribeiro *et al.* (2017) |
| Lugworm | *Arenicola marina* | PS | Sphere | 400–1300 μm | 0–7.4% of sediment dry weight | 28 days | Survival, bodyweight | Besseling *et al.* (2013) |
| | | PVC | Sphere | n/a | 0–5% of sediment weight | 10 days | Mortality, phagocytic activity and oxidative status | Browne *et al.* (2013) |
| Ragworm | *Hediste diversicolor* | Cellulose acetate | Fibre | 120.6 ± 5.1 μm (mean) | Equivalent to 8; 4; 2; 1; 0.5 cigarette filters $l^{-1}$ | 96 h; 28 days | Behaviour; weight loss; DNA damage | Wright *et al.* (2015) |
| Coral | *Acropora millepora*; *A. humilis*; *Pocillopora damicornis*; *P. verrucosa*; *Porites cylindrical*; *P. lutea* | PE | Irregular | 37–163 μm | 4000 particles $l^{-1}$ | 4 weeks | Bleaching and tissue necrosis | Reichert *et al.* (2018) |
| Fish | *Pomatoschistus microps* | PE | Sphere | 1–5 μm | 0.184 mg $l^{-1}$ | 96 h | Behaviour and biomarkers of neurotoxicity, phases I and II biotransformation and lipid peroxidation | Ferreira *et al.* (2016) |
| | | PE | Sphere | 1–5 μm | 0.184 mg $l^{-1}$ | 96 h | Mortality, behaviour and biomarkers of neurotoxicity and lipid peroxidation | Fonte *et al.* (2016) |
| | | PE | Sphere | 1–5 μm | 0.184 mg $l^{-1}$ | 96 h | Mortality, behaviour and biomarkers of neurotoxicity, phases I and II biotransformation and lipid peroxidation | Luis *et al.* (2015) |

*Continued*

**Table 30.1.** Continued.

| Taxon | Species | Experimental features | | | | | End point | Reference |
|---|---|---|---|---|---|---|---|---|
| | | Polymer type | Shape | Size | Nominal concentration of exposure | Period of exposure | | |
| | | PE | Sphere | 1 – 5 μm | 18.4; 184 μg $l^{-1}$ | 96 h | Mortality, bile metabolites, and biomarkers related to neurotransmission, aerobic energy production, biotransformation and oxidative stress | Oliveira *et al.* (2013) |
| | | PE | Sphere | 420–500 μm | 60 particles $l^{-1}$ | 96 h | Behavioural | Sa *et al.* (2015) |
| | *Sparus aurata* | PVC, PA, PS, PE | Sphere; irregular | ~75 μm | 0.1 g $kg^{-1}$ of fish body mass | 45 days | Growth, histopathology and biochemical biomarkers (glucose, AST, ALT, LDH, and GGT) | Jovanović *et al.* (2018) |
| | | PVC | n/a[a] | 40–150 μm | 100; 500 mg of PVC $kg^{-1}$ of food | 30 days | Growth, humoral and cellular immune biomarkers and expression of genes related to stress | Espinosa *et al.* (2017) |
| | *Dicentrarchus labrax* | PVC; PE | n/a[a] | 40–150 μm | 1; 10; 100 mg $ml^{-1}$ | 1; 24 h | Cell viability, innate immune parameters and the expression of genes related to inflammation, oxidative stress, metabolism of xenobiotics and cell apoptosis | Espinosa *et al.* (2018) |
| | | LDPE | n/a[a] | 125–250 μm | 2% of food weight | 80 days | Toxicokinetic parameters, hepatosomatic indexes and the expression of genes related to detoxification process and immune response | Granby *et al.* (2018) |

| | | | | | | | | |
|---|---|---|---|---|---|---|---|---|
| | | PE | Sphere | 10–45 μm | $10^4$; $10^5$ particle $g^{-1}$ of food | 36 days | Survival, growth, physiological parameters and gene expression related to inflammatory response | Mazurais *et al.* (2015) |
| | | n/a | Sphere | 1–5 μm | 0.26; 0.69 mg $l^{-1}$ | 96 h | Behaviour | Barboza *et al.* (2018b) |
| | | n/a | Sphere | 1–5 μm | 0.26; 0.69 mg $l^{-1}$ | 96 h | Biomarkers of neurotoxicity and oxidative damage | Barboza *et al.* (2018a) |
| | *Oryzias latipes* | LDPE | Irregular | < 0.5 mm | 8 ng $ml^{-1}$ | 2 months | Histopathology and gene expression (CYP1A) | Rochman *et al.* (2013) |
| | | PE | Irregular | < 0.5 mm | 8 ng $ml^{-1}$ | 2 months | Histopathology and gene expression (total RNA) | Rochman *et al.* (2014) |
| | *Danio rerio* | PE | Sphere | 1–5; 10–20 μm | [b] | Up 20 min | Gene expression (CYP1A) | Batel *et al.* (2016) |
| Fish | *Acanthochromis polyacanthus* | PET | n/a* | 125 μm–2 mm | 20; 40; 60; 80% of diet as plastic | 1 week | Body condition, growth and behaviour | Critchell and Hoogenboom (2018) |
| | *Lates calcarifer* | n/a | Sphere | 97 μm | 100 particles $l^{-1}$ | 24 h | Behaviour | Guven *et al.* (2018) |
| | *Clarias gariepinus* | LDPE | Irregular | < 60 μm | 50; 500μg $l^{-1}$ | 96 h | Histopathology, changes in plasma biochemical composition and gene expression | Karami *et al.* (2016) |

[a]When authors described the plastic as particles, microplastic shape was considered n/a as particle does not indicate if it is irregular or sphere-shaped plastics.
[b]Fish were exposed to microplastic via trophic transference (contaminated prey). Therefore, the paper does not detail microplastic concentration of fish exposure.

*et al.*, 2018; Critchell and Hoogenboom, 2018). However, due to the wide range of microplastic exposure features, more studies still need be done to properly state patterns of influence.

The exposure parameters most explored are size and concentration of microplastics. Despite some controversies, studies overall suggest that smaller microplastics and higher concentrations cause more toxicity in organisms than larger particles and smaller amounts. Au *et al.* (2015) and Frydkjaer *et al.* (2017) tested plastic shape and concluded that: (i) fibres are more toxic than particles for the freshwater amphipod *Hyalella azteca*; and (ii) irregular-shaped microplastics are more toxic than sphere beads for *Daphnia magna*. Considering the major abundance of synthetic fibres and irregular particles in natural environments, there is a need for further studies to better assess relevant conditions of exposure. Comparisons among polymer types are also needed. In theory, polymer type can affect microplastic toxicity because of the monomer nature. Although plastics are considered inert due to their generally large molecular size, reactions during polymerization are frequently incomplete and monomers may remain within the polymeric materials, being released during use and after disposal (Lithner *et al.*, 2011; Andrady, 2017). Once released, monomers can interact with cellular and molecular structures, leading to toxic effects. Each monomer type has a different level of toxicity. PVC monomers, for instance, are among the most hazardous (Lithner *et al.*, 2011).

### 30.5.2 Microplastics and other contaminants

One of the greatest concerns about environmental contamination by microplastics is the risk of them being vectors of other chemical compounds, mainly plastics additives and organic pollutants. Experimental and modelling studies have been carried out to understand if this process occurs. Their results have shown that it can indeed happen, likewise the other way round: chemical compounds being transferred from organisms to microplastics (see Ziccardi *et al.*, 2016 for a review on the topic). Also, components of the plankton and sediments were shown to be better vectors of adsorbed pollutants than microplastics under specific experimental conditions (Browne *et al.*, 2013; Frydkjaer *et al.*, 2017). Hence, microplastic vectorization of chemicals seems to be influenced by the concentration and affinity of the associated chemicals with the plastics and with the surrounding environment.

Different microplastics (e.g. PVC, PE and PS) were observed to facilitate bioaccumulation of other chemicals and to modulate their molecular, cellular and behavioural impacts (Besseling *et al.*, 2013; Oliveira *et al.*, 2013; Avio *et al.*, 2015; Karami *et al.*, 2016; Paul-Pont *et al.*, 2016; Barboza *et al.*, 2018b; Rainieri *et al.*, 2018). Just a few studies, however, explored non-organic contaminants such as heavy metals, e.g. chromium (Cr) (Luis *et al.*, 2015) and mercury (Hg) (Barboza *et al.*, 2018b).

### 30.5.3 Microplastic exposure and lack of effects

Sometimes effects of microplastics are not observed at all. Condition index, for instance, a health indicator commonly used for bivalves and fish, is usually not altered by microplastic exposure (e.g. Bour *et al.*, 2018; Santana *et al.*, 2018). Oxidative stress, feeding, growth, mortality and some behaviour as swimming and burrowing are examples of endpoints that were already reported as not affected by exposure to microplastics (Hamer *et al.*, 2014; Cole *et al.*, 2015; Green, 2016; Imhof and Laforsch, 2016; Gambardella *et al.*, 2017; Bour *et al.*, 2018; Critchell and Hoogenboom, 2018; Dawson *et al.*, 2018; Jovanović *et al.*, 2018). Organismal and sub-organismal levels of acclimation, tolerance or compensation can act towards these results but such mechanisms of resistance have not been the focus of many studies on biological impacts of microplastics. Animals can alter their assimilation efficiency, energy metabolism and molecular defence systems (e.g. antioxidant enzymes and anti-inflammatory proteins), for instance, in accordance with stressful situations; and time and concentration of exposure might help to modulate these responses.

### 30.5.4 Toxicity of microplastics under environmentally relevant conditions

Although many studies have provided examples of microplastics intake and biological effects (Browne *et al.*, 2008; von Moos *et al.*, 2012; Besseling *et al.*, 2013; Browne *et al.*, 2013,; Cole *et al.*, 2013; Karami *et al.*, 2016; Lu *et al.*, 2016; Paul-Pont *et al.*, 2016; Gambardella *et al.*, 2017), most of them used unrealistic conditions of microplastic exposure (EPA, 1992; Rochman, 2016; Underwood *et al.*, 2017). Unrealistic conditions are characteristics of microplastics that do not match what has been found on field samples, as systematically observed in the literature review of Santana *et al.* (in prep). Among the studies reviewed here, fewer than 30% justified at least one of their exposure features with field-based data. This raises the question of 'what is really happening in the environment?' and the mistrust of alarmist statements. Relevant environmental conditions are especially important for environmental management and public awareness so that we can accurately address the problem and seek feasible solutions (Santana *et al.*, in prep.).

Yet, microplastics are an emergent contaminant. As such, science needs to create baseline information on microplastic mechanisms of toxicity so that we can further explore more complex analysis (GESAMP, 2016). For that, unrealistic marked beads (usually round) and high concentrations of exposure are fundamental, as they maximize the chances of observing the interaction of microplastics with organisms and their effects (Santana *et al.*, in prep.). Moreover, we need first to understand the impacts of isolated types of microplastics and then check their mixtures.

Thus both research approaches should be considered equally important and carried out simultaneously, as they provide different pieces of information.

Finally, it is important to emphasize the need for studies addressing the long-term and chronic effects of exposure to microplastics, as well as the impacts caused by particles at the nanometric size (GESAMP, 2015). Although we lack field data that characterize both features (long-term exposure to microplastics and characterization of nanometric particles) to build up environmentally relevant scenarios, it is reasonable to assume that sink areas in nature (e.g. deep oceans and the oceanic gyres) are constantly exposing their inhabitants to microplastics and that the constant breakdown of larger plastics will eventually populate the environment with nanoplastics, justifying the importance of including such aspects on ecotoxicological tests.

### 30.5.5 Effects due to exposure and non-intake of microplastics

Microplastics used in ecotoxicological experiments range from < 100 μm to 5 mm in diameter. So when we study the effects of microplastics in organisms, we are also considering large plastics interacting with organisms that might not be susceptible to their uptake. In this case, the observed impacts will be basically related to toxic compounds leached from microplastics into the seawater.

Only a few studies have explored such a scenario for microplastic exposure. Nevertheless, studies showed anomalous larval and embryo development due to virgin and beached pellets (Nobre *et al.*, 2015; Gandara *et al.*, 2016). For mussel larvae, virgin pellets of polypropylene (PP) were less toxic than beached ones, while the opposite occurred for embryos of sea urchins exposed to virgin PE and beached pellets. Because these two works have different organism and plastic models, we cannot compare their results but only suggest that effects of non-intaken microplastics might rely on the type of leached compounds and their mode of action in each species. Thus, it is clear that indirect routes of microplastic effects should be studied further so that we can create a better picture of microplastic effects in the environment.

## 30.6 Conclusions

Microplastics are a ubiquitous emergent contaminant, extremely difficult to remove from the environment and with an undoubted tendency to accumulate in the ocean. Many studies have provided examples of biological effects due to environmental contamination by microplastics. The wide ranges of possible scenarios for microplastic contamination mean that studies differ highly in experimental exposure conditions such as plastic size (including nanoplastics), shape, polymer type, additives, concentration and period of exposure. This hinders comparisons among results, which, in turn, would help to create conceptual models of microplastic toxicity. This is not the only knowledge gap the needs to be filled to build better awareness around this issue. As with many other emergent contaminants, research on microplastics still carries many uncertainties, such as the absence of scientific evidence of their impacts on biodiversity at higher levels of organization and humans. Such information is needed to provide enough basic and applied knowledge for better policies and management. Meanwhile, as a precautionary movement, people should re-think plastic uses and disposal, pushing towards sustainable practices to prevent plastics entering the environment.

## References

Ananthaswamy, A. (2001) Junk food – a diet of plastic pellets plays havoc with animals' immunity. *New Science* 169, 18.

Andrady, A.L. (2011) Microplastics in the marine environment. *Marine Pollution Bulletin* 62, 1596–605.

Andrady, A.L. (2017) The plastic in microplastics: a review. *Marine Pollution Bulletin* 119, 12–22.

Arthur, C., Baker, J. and Bamford, H. (eds) (2009) *Proceedings of the International Research Workshop on the Occurrence, Effects, and Fate of Microplastic Marine Debris, 2009 Tacpma, WA, USA*. National Oceanic and Atmospheric Administration, Washington, DC.

Au, S.Y., Bruce, T.F., Bridges, W.C. and Klaine, S.J. (2015) Responses of *Hyalella azteca* to acute and chronic microplastic exposures. *Environmental Toxicology and Chemistry* 34, 2564–2572.

Avio, C.G., Gorbi, S., Milan, M., Benedetti, M., Fattorini, D. *et al.* (2015) Pollutants bioavailability and toxicological risk from microplastics to marine mussels. *Environmental Pollution* 198, 211–222.

Barboza, L.G.A., Vieira, L.R., Branco, V., Figueiredo, N., Carvalho, F., Carvalho, C. and Guilhermino, L. (2018a) Microplastics cause neurotoxicity, oxidative damage and energy-related changes and interact with the bioaccumulation of mercury in the European seabass, *Dicentrarchus labrax* (Linnaeus, 1758). *Aquatic Toxicology* 195, 49–57.

Barboza, L.G.A., Vieira, L.R. and Guilhermino, L. (2018b) Single and combined effects of microplastics and mercury on juveniles of the European seabass (*Dicentrarchus labrax*): changes in behavioural responses and reduction of swimming velocity and resistance time. *Environmental Pollution* 236, 1014–1019.

Batel, A., Linti, F., Scherer, M., Erdinger, L. and Braunbeck, T. (2016) Transfer of benzo[a]pyrene from microplastics to *Artemia nauplii* and further to zebrafish via a trophic food web experiment: CYP1A induction and visual tracking of persistent organic pollutants. *Environmental Toxicology and Chemistry* 35, 1656–1666.

Besseling, E., Wegner, A., Foekema, E.M., Van Den Heuvel-greve, M.J. and Koelmans, A.A. (2013) Effects of microplastic on fitness and PCB bioaccumulation by the lugworm *Arenicola marina* (L.). *Environmental Science & Technology* 47, 593–600.

Bour, A., Haarr, A., Keiter, S. and Hylland, K. (2018) Environmentally relevant microplastic exposure affects sediment-dwelling bivalves. *Environmental Pollution* 236, 652–660.

Browne, M.A., Galloway, T. and Thompson, R. (2007) Microplastic – an emerging contaminant of potential concern? *Integrated Environmental Assessment and Management* 3, 559–561.

Browne, M.A., Dissanayake, A., Galloway, T., Lowe, D.M. and Thompson, R. (2008) Ingested microscopic plastic translocates to the circulatory system of the mussel, *Mytilus edulis* (L.). *Environmental Science & Technology* 42, 5026–5031.

Browne, M.A., Crump, P., Niven, S.J., Teuten, E., Tonkin, A., Galloway, T. and Thompson, R. (2011) Accumulation of microplastic on shorelines woldwide: sources and sinks. *Environmental Science & Technology* 45, 9175–9179.

Browne, M.A., Niven, S.J., Galloway, T.S., Rowland, S.J. and Thompson, R.C. (2013) Microplastic moves pollutants and additives to worms, reducing functions linked to health and biodiversity. *Current Biology* 23, 2388–2392.

Carpenter, E.J.K. and Smith, J.L. (1972) Plastics on the Sargasso Sea surface. *Science* 175, 1240–1241.

Carpenter, E.J.K., Anderson, S.J., Harvey, G.R., Miklas, H.P. and Peck, B.B. (1972) Polystyrene spherules in coastal waters. *Science* 178, 749–750.

Cole, M. and Galloway, T.S. (2015) Ingestion of nanoplastics and microplastics by pacific oyster larvae. *Environmental Science & Technology* 49, 14625–14632.
Cole, M., Lindeque, P., Fileman, E., Halsband, C., Goodhead, R., Moger, J. and Galloway, T.S. (2013) Microplastic ingestion by zooplankton. *Environmental Science & Technology* 47, 6646–6655.
Cole, M., Lindeque, P., Fileman, E., Halsband, C. and Galloway, T.S. (2015) The impact of polystyrene microplastics on feeding, function and fecundity in the marine copepod *Calanus helgolandicus*. *Environmental Science & Technology* 49, 1130–1137.
Colton, J.B., Knappe, F.D. and Burns, B.R. (1974) Plastic particles in surface waters of the Northwestern Atlantic. *Science* 185, 491–497.
Connell, D.W. (1990) Environmental routes leading to the bioaccumulation of lipophilic chemicals. In: Connell, D.W. (ed.) *Bioaccumulation of Xenobiotic Compounds.* CRC Press, Boca Raton, Florida, pp. 59–73.
Cozar, A., Echevarria, F., Gonzalez-Gordillo, J.I., Irigoien, X., Ubeda, B. *et al.* (2014) Plastic debris in the open ocean. *Proceedings of the National Academy of Sciences of the United States of America* 111, 10239–10244.
Critchell, K. and Hoogenboom, M.O. (2018) Effects of microplastic exposure on the body condition and behaviour of planktivorous reef fish (*Acanthochromis polyacanthus*). *PLoS ONE* 13(3), e0193308.
Dawson, A., Huston, W., Kawaguchi, S., King, C., Cropp, R. *et al.* (2018) Uptake and depuration kinetics influence microplastic bioaccumulation and toxicity in Antarctic krill (*Euphausia superba*). *Environmental Science & Technology* 52, 3195–3201.
Derraik, J.G.B. (2002) The pollution of the marine environment by plastic debris: a review. *Marine Pollution Bulletin* 44, 842–852.
EPA (1992) *Framework for Ecological Risk Assessment*. US Environmental Protection Agency, Washington, DC.
Eriksen, M., Lebreton, L.C.M., Carson, H.S., Thiel, M., Moore *et al.* (2014) Plastic pollution in the world's oceans: more than 5 trillion plastic pieces weighing over 250,000 tons afloat at sea. *PloS ONE* 9(12), e111913.
Espinosa, C., Cuesta, A. and Esteban, M.A. (2017) Effects of dietary polyvinylchloride microparticles on general health, immune status and expression of several genes related to stress in gilthead seabream (*Sparus aurata* L.). *Fish and Shellfish Immunology* 68, 251–259.
Espinosa, C., Garcia Beltran, J.M., Esteban, M.A. and Cuesta, A. (2018) In vitro effects of virgin microplastics on fish head-kidney leucocyte activities. *Environmental Pollution* 235, 30–38.
FAO (2017) *Microplastics in Fisheries and Aquaculture*. Food & Agricultural Organization of the United Nations, Rome.
Farrell, P. and Nelson, K. (2013) Trophic level transfer of microplastic: *Mytilus edulis* (L.) to *Carcinus maenas* (L.). *Environmental Pollution* 177, 1–3.
Fendall, L.S. and Sewell, M.A. (2009) Contributing to marine pollution by washing your face: microplastics in facial cleansers. *Marine Pollution Bulletin* 58, 1225–1228.
Ferreira, P., Fonte, E., Soares, M.E., Carvalho, F. and Guilhermino, L. (2016) Effects of multi-stressors on juveniles of the marine fish *Pomatoschistus microps*: gold nanoparticles, microplastics and temperature. *Aquatic Toxicology* 170, 89–103.
Fonte, E., Ferreira, P. and Guilhermino, L. (2016) Temperature rise and microplastics interact with the toxicity of the antibiotic cefalexin to juveniles of the common goby (*Pomatoschistus microps*): post-exposure predatory behaviour, acetylcholinesterase activity and lipid peroxidation. *Aquatic Toxicology* 180, 173–185.
Frias, J.P., Sobral, P. and Ferreira, A.M. (2010) Organic pollutants in microplastics from two beaches of the Portuguese coast. *Marine Pollution Bulletin* 60, 1988–1992.
Frydkjaer, C.K., Iversen, N. and Roslev, P. (2017) Ingestion and egestion of microplastics by the cladoceran *Daphnia magna*: effects of regular and irregular shaped plastic and sorbed phenanthrene. *Bulletin of Environmental Contamination and Toxicology* 99, 655–661.
Gambardella, C., Morgana, S., Ferrando, S., Bramini, M., Piazza, V. *et al.* (2017) Effects of polystyrene microbeads in marine planktonic crustaceans. *Ecotoxicology and Environmental Safety* 145, 250–257.
Gandara, E.S.P.P., Nobre, C.R., Resaffe, P., Pereira, C.D.S. and Gusmao, F. (2016) Leachate from microplastics impairs larval development in brown mussels. *Water Research* 106, 364–370.
GESAMP (2010) Bowmer, T. and Kershaw, P.J. (eds) *Proceedings of the GESAMP International Workshop on plastic particles as a vector in transporting persistent, bio-accumulating and toxic substances in the oceans*. GESAMP Reports and Studies, No. 82. Joint Group of Experts on the Scientific Aspects

of Marine Environmental Protection (IMO/FAO/UNESCO-IOC/UNIDO/WMO/IAEA/UN/UNEP). International Maritime Organization (IMO), London.

GESAMP (2015) Kershaw, P.J. (ed.) *Sources, fate and effects of microplastics in the marine environment: a global assessment*. GESAMP Reports and Studies, No. 90. Joint Group of Experts on the Scientific Aspects of Marine Environmental Protection (IMO/FAO/UNESCO-IOC/UNIDO/WMO/IAEA/UN/UNEP). International Maritime Organization (IMO), London.

GESAMP (2016) Kershaw, P.J. and Rochman, C.A. (eds) *Sources, fate and effects of microplastics in the marine environment: Part 2 of a global assessment*. GESAMP Reports and Studies, No. 93. Joint Group of Experts on the Scientific Aspects of Marine Environmental Protection (IMO/FAO/UNESCO-IOC/UNIDO/WMO/IAEA/UN/UNEP. International Maritime Organization (IMO), London.

GESAMP (2019) Kershaw P.J., Turra A. and Galgani, F. (eds) *Guidelines for the monitoring and assessment of plastic litter and microplastics in the ocean*. GESAMP Reports and Studies, No. 99. Joint Group of Experts on the Scientific Aspects of Marine Environmental Protection (IMO/FAO/UNESCO-IOC/UNIDO/WMO/IAEA/UN/UNEP/UNDP/ISA), London.

Graham, E.R. and Thompson, J.T. (2009) Deposit- and suspension-feeding sea cucumbers (Echinodermata) ingest plastic fragments. *Journal of Experimental Marine Biology and Ecology* 368, 22–29.

Granby, K., Rainieri, S., Rasmussen, R.R., Kotterman, M.J.J., Sloth, J.J. *et al.* (2018) The influence of microplastics and halogenated contaminants in feed on toxicokinetics and gene expression in European seabass (*Dicentrarchus labrax*). *Environmental Research* 164, 430–443.

Gray, J.S. (2002) Biomagnification in marine systems: the perspective of an ecologist. *Marine Pollution Bulletin* 45, 46–52.

Green, D.S. (2016) Effects of microplastics on European flat oysters, *Ostrea edulis* and their associated benthic communities. *Environmental Pollution* 216, 95–103.

Green, D.S., Boots, B., Sigwart, J., Jiang, S. and Rocha, C. (2016) Effects of conventional and biodegradable microplastics on a marine ecosystem engineer (*Arenicola marina*) and sediment nutrient cycling. *Environmental Pollution* 208, 426–434.

Grigorakis, S., Mason, S.A. and Drouillard, K.G. (2017) Determination of the gut retention of plastic microbeads and microfibers in goldfish (*Carassius auratus*). *Chemosphere* 169, 233–238.

Guven, O., Bach, L., Munk, P., Dinh, K.V., Mariani, P. and Nielsen, T.G. (2018) Microplastic does not magnify the acute effect of PAH pyrene on predatory performance of a tropical fish (*Lates calcarifer*). *Aquatic Toxicology* 198, 287–293.

Hamer, J., Gutow, L., Kohler, A. and Saborowski, R. (2014) Fate of microplastics in the marine isopod *Idotea emarginata*. *Environmental Science & Technology* 48, 13451–13458.

Heindler, F.M., Alajmi, F., Huerlimann, R., Zeng, C., Newman, S.J., Vamvounis, G. and Van Herwerden, L. (2017) Toxic effects of polyethylene terephthalate microparticles and di(2-ethylhexyl)phthalate on the calanoid copepod, *Parvocalanus crassirostris*. *Ecotoxicology and Environmental Safety* 141, 298–305.

Henry, B., Kirsi, L. and Kleep, I.G. (2019) Microfibres from apparel and home textiles: prospects for including microplastics in environmental sustainability assessment. *Science of the Total Environment* 652, 483–494.

Horton, A.A., Walton, A., Spurgeon, D.J., Lahive, E. and Svendsen, C. (2017) Microplastics in freshwater and terrestrial environments: evaluating the current understanding to identify the knowledge gaps and future research priorities. *Science of Total Environment* 586, 127–141.

Imhof, H.K. and Laforsch, C. (2016) Hazardous or not – are adult and juvenile individuals of *Potamopyrgus antipodarum* affected by non-buoyant microplastic particles? *Environmental Pollution* 218, 383–391.

Imhof, H.K., Rusek, J., Thiel, M., Wolinska, J. and Laforsch, C. (2017) Do microplastic particles affect *Daphnia magna* at the morphological, life history and molecular level? *PLoS ONEne*, 12, e0187590.

Ivar Do Sul, J.A. and Costa, M.F. (2014) The present and future of microplastic pollution in the marine environment. *Environmental Pollution* 185, 352–364.

Ivar Do Sul, J.A., Costa, M.F. and Fillmann, G. (2014) Microplastics in the pelagic environment around oceanic islands of the Western Tropical Atlantic Ocean. *Water, Air, & Soil Pollution* 225, 2004.

Jambeck, J.R., Geyer, R., Wilcox, C., Siegler, T.R., Perryman, M. *et al.* (2015) Marine pollution. Plastic waste inputs from land into the ocean. *Science* 347, 768–771.

Jeong, C.B., Won, E.J., Kang, H.M., Lee, M.C., Hwang, D.S. *et al.* (2016) Microplastic size-dependent toxicity, oxidative stress induction, and p-JNK and p-p38 activation in the monogonont rotifer (*Brachionus koreanus*). *Environmental Science & Technology* 50, 8849–8857.

Jovanović, B., Gökdağ, K., Güven, O., Emre, Y., Whitley, E.M. and Kideys, A.E. (2018) Virgin microplastics are not causing imminent harm to fish after dietary exposure. *Marine Pollution Bulletin* 130, 123–131.
Kanhai, D.K., Officer, R., Lyashevska, O., Thompson, R.C. and O'Connor, I. (2017) Microplastic abundance, distribution and composition along a latitudinal gradient in the Atlantic Ocean. *Marine Pollution Bulletin* 115, 307–314.
Kaposi, K.L., Mos, B., Kelaher, B.P. and Dworjanyn, S.A. (2014) Ingestion of microplastic has limited impact on a marine larva. *Environmental Science & Technology* 48, 1638–1645.
Karami, A., Romano, N., Galloway, T. and Hamzah, H. (2016) Virgin microplastics cause toxicity and modulate the impacts of phenanthrene on biomarker responses in African catfish (*Clarias gariepinus*). *Environmental Research* 151, 58–70.
Karami, A., Golieskardi, A., Keong Choo, C., Larat, V., Galloway, T.S. and Salamatinia, B. (2017) The presence of microplastics in commercial salts from different countries. *Scientific Reports* 7, 46173.
Koelmans, A.A., Besseling, E., Foekema, E., Kooi, M., Mintenig, S. *et al.* (2017) Risks of plastic debris: unravelling fact, opinion, perception, and belief. *Environmental Science & Technology* 51, 11513–11519.
Krelling, A.P., Souza, M.M., Williams, A.T. and Turra, A. (2017) Transboundary movement of marine litter in an estuarine gradient: evaluating sources and sinks using hydrodynamic modelling and ground truthing estimates. *Marine Pollution Bulletin* 119, 48–63.
Landis, W.G. and Yu, M.-H. (2004) *Introduction to Environmental Toxicology: Impacts of Chemicals upon Ecological Systems*. CRC Press, Boca Raton, Florida.
Law, K.L., Morét-Ferguson, S., Maximenko, N.A., Proskurowski, G., Peacock, E.E., Hafner, J. and Reddy, C.M. (2010) Plastic accumulation in the North Atlantic Subtropical Gyre. *Science* 329, 1185–1186.
Lebreton, L.C.M., Van Der Zwet, J., Damsteeg, J.W., Slat, B., Andrady, A. and Reisser, J. (2017) River plastic emissions to the world's oceans. *Nature Communications* 8, 15611.
Lee, K.W., Shim, W.J., Kwon, O.Y. and Kang, J.H. (2013) Size-dependent effects of micro polystyrene particles in the marine copepod *Tigriopus japonicus*. *Environmental Science & Technology* 47, 11278–11283.
Lenz, R., Enders, K. and Nielsen, T.G. (2016) Microplastic exposure studies should be environmentally realistic. *Proceedings of the National Academy of Sciences of the United States of America* 113, E4121–4122.
Lithner, D., Larsson, A. and Dave, G. (2011) Environmental and health hazard ranking and assessment of plastic polymers based on chemical composition. *Science of Total Environment* 409, 3309–3324.
Lo, H.K.A. and Chan, K.Y.K. (2018) Negative effects of microplastic exposure on growth and development of *Crepidula onyx*. *Environmental Pollution* 233, 588–595.
Lu, Y., Zhang, Y., Deng, Y., Jiang, W., Zhao, Y. *et al.* (2016) Uptake and accumulation of polystyrene microplastics in zebrafish (*Danio rerio*) and toxic effects in liver. *Environmental Science & Technolog*, 50, 4054–4060.
Luis, L.G., Ferreira, P., Fonte, E., Oliveira, M. and Guilhermino, L. (2015) Does the presence of microplastics influence the acute toxicity of chromium(VI) to early juveniles of the common goby (Pomatoschistus microps)? A study with juveniles from two wild estuarine populations. *Aquatic Toxicology* 164, 163–174.
Lusher, A. (2015) Microplastics in the marine environment: distribution, interactions and effects. In: Bergmann, M.E.A. (ed.) *Marine Anthropogenic Litter*. Springer, Berlin.
Mason, S.A., Welch, V. and Neratko, J. (2018) Synthetic polymer contamination in bottled water. *Frontiers in Chemistry* 6, 407.
Mato, Y., Isobe, T., Takada, H., Kanehiro, H., Ohtake, C. and Kaminuma, T. (2001) Plastic resin pellets as a transport medium for toxic chemicals in the marine environment. *Environmental Science & Technology* 35, 318–324.
Mazurais, D., Ernande, B., Quazuguel, P., Severe, A., Huelvan, C. *et al.* (2015) Evaluation of the impact of polyethylene microbeads ingestion in European sea bass (*Dicentrarchus labrax*) larvae. *Marine Environmental Research* 112, 78–85.
McCormick, A., Hoellein, T.J., Mason, S.A., Schluep, J. and Kelly, J.J. (2014) Microplastic is an abundant and distinct microbial habitat in an urban river. *Environmental Science & Technology* 48, 11863–11871.
Napper, I.E. and Thompson, R.C. (2016) Release of synthetic microplastic plastic fibres from domestic washing machines: effects of fabric type and washing conditions. *Marine Pollution Bulletin* 112, 39–45.

Nobre, C.R., Santana, M.F.M., Maluf, A., Cortez, F.S., Cesar, A., Pereira, C.D.S. and Turra, A. (2015) Assessment of microplastic toxicity to embryonic development of the sea urchin *Lytechinus variegatus* (Echinodermata: Echinoidea). *Marine Pollution Bulletin* 92, 99–104.

Obbard, R.W., Sadri, S., Wong, Y.Q., Khitun, A.A., Baker, I. and Thompson, R.C. (2014) Global warming releases microplastic legacy frozen in Arctic Sea ice. *Earth's Future* 2, 315–320.

Oliveira, M., Ribeiro, A., Hylland, K. and Guilhermino, L. (2013) Single and combined effects of microplastics and pyrene on juveniles (0+ group) of the common goby *Pomatoschistus microps* (Teleostei, Gobiidae). *Ecological Indicators* 34, 641–647.

Paul-Pont, I., Lacroix, C., Gonzalez Fernandez, C., Hegaret, H., Lambert, C. *et al*. (2016) Exposure of marine mussels *Mytilus* spp. to polystyrene microplastics: toxicity and influence on fluoranthene bioaccumulation. *Environmental Pollution* 216, 724–737.

Phuong, N.N., Zalouk-Vergnoux, A., Poirier, L., Kamari, A., Chatel, A., Mouneyrac, C. and Lagarde, F. (2016) Is there any consistency between the microplastics found in the field and those used in laboratory experiments? *Environmental Pollution* 211, 111–123.

PlasticsEurope (2018) *Plastics – the Facts 2017*. PlasticsEurope, Brussels..

Rainieri, S., Conlledo, N., Larsen, B. K., Granby, K. and Barranco, A. (2018) Combined effects of microplastics and chemical contaminants on the organ toxicity of zebrafish (*Danio rerio*). *Environmental Research* 162, 135–143.

Rand, G.M., Wells, P.G. and McCarthy, L.S. (1995) Introduction to aquatic ecology. In: Rand, G.M. (ed.) *Fundamentals of Aquatic Toxicology*. Taylor and Francis, London.

Reichert, J., Schellenberg, J., Schubert, P. and Wilke, T. (2018) Responses of reef building corals to microplastic exposure. *Environmental Pollution* 237, 955–960.

Reisser, J., Slat, B., Noble, K., Du Plessis, K., Epp, M. *et al*. (2015) The vertical distribution of buoyant plastics at sea: an observational study in the North Atlantic Gyre. *Biogeosciences* 12, 1249–1256.

Ribeiro, F., Garcia, A.R., Pereira, B.P., Fonseca, M., Mestre, N.C. *et al*. (2017) Microplastics effects in *Scrobicularia plana*. *Marine Pollution Bulletin* 122, 379–391.

Rochman, C.M. (2016) Ecologically relevant data are policy-relevant data. *Science* 352, 1172.

Rochman, C.M., Hoh, E., Kurobe, T. and Teh, S.J. (2013) Ingested plastic transfers hazardous chemicals to fish and induces hepatic stress. *Scientific Reports* 3, 3263.

Rochman, C.M., Kurobe, T., Flores, I. and Teh, S.J. (2014) Early warning signs of endocrine disruption in adult fish from the ingestion of polyethylene with and without sorbed chemical pollutants from the marine environment. *Science of Total Environment* 493, 656–661.

Rodolfo, A. Jr, Nunes, L.R. and Ormanji, W. (2006) *Tecnologia do PVC*. Proeditores/Braskem, São Paulo, Brazil.

Sa, L.C., Luis, L.G. and Guilhermino, L. (2015) Effects of microplastics on juveniles of the common goby (*Pomatoschistus microps*): confusion with prey, reduction of the predatory performance and efficiency, and possible influence of developmental conditions. *Environmental Pollution* 196, 359–362.

Salvador Cesa, F., Turra, A. and Baruque-Ramos, J. (2017) Synthetic fibers as microplastics in the marine environment: a review from textile perspective with a focus on domestic washings. *Science of Total Environment* 598, 1116–1129.

Santana, M.F.M., Moreira, F.T., Pereira, C.D.S., Abessa, D.M.S. and Turra, A. (in prep.) (2018) Continuous exposure to microplastics does not cause physiological effects in the cultivated mussel *Perna perna*. *Archives of Environmental Contamination and Toxicology* 74, 594–604.

Santillo, D., Miller, K. and Johnston, P. (2017) Microplastics as contaminants in commercially important seafood species. *Integrated Environmental Assessment and Management* 13, 516–521.

Sussarellu, R., Suquet, M., Thomas, Y., Lambert, C., Fabioux, C. *et al*. (2016) Oyster reproduction is affected by exposure to polystyrene microplastics. *Proceedings of the National Academy of Sciences of the United States of America* 113, 2430–2435.

Tanaka, K., Takada, H., Yamashita, R., Mizukawa, K., Fukuwaka, M.A. and Watanuki, Y. (2013) Accumulation of plastic-derived chemicals in tissues of seabirds ingesting marine plastics. *Marine Pollution Bulletin* 69, 219–222.

Tosetto, L., Brown, C. and Williamson, J.E. (2016) Microplastics on beaches: ingestion and behavioural consequences for beachhoppers. *Marine Biology* 163, 199.

Turra, A., Manzano, A.B., Dias, R.J., Mahiques, M.M., Barbosa, L., Balthazar-Silva, D. and Moreira, F.T. (2014) Three-dimensional distribution of plastic pellets in sandy beaches: shifting paradigms. *Scientific Reports* 4, 4435.

Ugolini, A., Ungherese, G., Ciofini, M., Lapucci, A. and Camaiti, M. (2013) Microplastic debris in sandhoppers. *Estuarine, Coastal and Shelf Science* 129, 19–22.
Underwood, A.J., Chapman, M.G. and Browne, M.A. (2017) Some problems and practicalities in design and interpretation of samples of microplastic waste. *Analytical Methods* 9, 1332–1345.
UNEP (2011) *UNEP Year Book: Emerging Issues in Our Global Environment*. United Nations Environment Programme, Nairobi.
UNEP (2016) *Marine Plastic Debris and Microplastics – Global Lessons and Research to Inspire Action and Guide Policy Change*. United Nations Environment Programme, Nairobi.
Van Cauwenberghe, L. and Janssen, C.R. (2014) Microplastics in bivalves cultured for human consumption. *Environmental Pollution* 193, 65–70.
Van Cauwenberghe, L., Vanreusel, A., Mees, J. and Janssen, C.R. (2013) Microplastic pollution in deep-sea sediments. *Environmental Pollution* 182, 495–499.
von Moos, N., Burkhardt-Holm, P. and Kohler, A. (2012) Uptake and effects of microplastics on cells and tissue of the blue mussel *Mytilus edulis* L. after an experimental exposure. *Environmental Science & Technology* 46, 11327–11335.
Wagner, M., Cherer, C., Alvarez-Muñoz, D., Brennholt, N., Bourrain, X. *et al.* (2014) Microplastics in freshwater ecosystems: what we know and what we need to know. *Environmental Sciences Europe* 26, 12.
Woodall, L.C., Sanchez-Vidal, A., Canals, M., Paterson, G.L., Coppock, R. *et al.* (2014) The deep sea is a major sink for microplastic debris. *Royal Society Open Science* 1, 140317.
Wright, S.L., Rowe, D., Thompson, R.C. and Galloway, T.S. (2013) Microplastic ingestion decreases energy reserves in marine worms. *Current Biology* 23, R1031–1033.
Wright, S.L., Rowe, D., Reid, M.J., Thomas, K.V. and Galloway, T.S. (2015) Bioaccumulation and biological effects of cigarette litter in marine worms. *Scientific Reports* 5, 14119.
Yang, D., Shi, H., Li, L., Li, J., Jabeen, K. and Kolandhasamy, P. (2015) Microplastic pollution in table salts from China. *Environmental Science & Technology* 49, 13622–13627.
Zettler, E.R., Mincer, T.J. and Amaral-Zettler, L.A. (2013) Life in the 'plastisphere': microbial communities on plastic marine debris. *Environmental Science & Technology* 47, 7137–7146.
Ziccardi, L.M., Edgington, A., Hentz, K., Kulacki, K.J. and Kane Driscoll, S. (2016) Microplastics as vectors for bioaccumulation of hydrophobic organic chemicals in the marine environment: a state-of-the-science review. *Environmental Toxicology and Chemistry* 35, 1667–1676.

Part VII

# Radiation Risks

# 31 UV Exposure and Skin-Protective Effects of Plant Polyphenols

L. Agulló-Chazarra,[1] A. Pérez-Sánchez,[1] M. Herranz-López,[1] V. Micol,[1,2] and E. Barrajón-Catalán*,[1]
[1]Instituto de Biología Molecular y Celular (IBMC), and Instituto de Investigacion, Desarrollo e Innovacion en Biotecnologia Sanitaria de Elche (IDiBE), Universitas Miguel Hernández (UMH), 03202, Elche, Spain; [2]CIBER, Fisiopatología de la Obesidad y la Nutrición, Instituto de Salud Carlos III (CB12/03/30038), Spain

## 31.1 Abstract

Skin exposure to ultraviolet (UV) radiation triggers a plethora of harmful effects, including skin cancer. UV radiation is linked to DNA damage, inflammation, oxidative stress, immunodepression, photoageing and extracellular matrix degradation. In this context, UV protection has emerged as a global concern not only for cosmetic purposes but also with clinical relevance. Industrial and scientific communities have tried to address this problem using different molecules for sunscreens that act as physical barriers. Despite these advances, major progress is still needed to find effective photoprotective formulations.

Natural compounds, especially plant polyphenols, comprise an almost limitless source of bioactive compounds with a variety of beneficial effects. Polyphenols have shown the capacity to alleviate most of the harmful effects of UV radiation in skin cells models. However, *in vivo* evidence for such activity is still scarce. Further, knowledge of the metabolites responsible for this photoprotective activity is very limited. This chapter reviews the most relevant results on the use of selected polyphenols to reverse the consequences of UV radiation on skin, focusing on their main molecular targets and their putative molecular mechanism.

## 31.2 Introduction

### 31.2.1 Skin definition, function and structure

The skin is the largest organ in the body, with a surface area of 2 $m^2$ and a thickness between 0.5 mm and 4 mm, and it is the organism's first barrier. Its primary function is to act as a permeability barrier, preventing excessive loss of water from the body. The skin also has a protective role against external aggressions such as infections caused by microorganisms, chemical, mechanical or thermal agents and solar radiation (Shindo *et al.*, 1994a; Madison, 2003;Forni *et al.*, 2012; Pullar *et al.*, 2017). Furthermore, the skin is a sensitive organ, as the nerve endings and receptors related to the sense of touch and temperature are located within the skin.

Three layers form the skin: (i) epidermis; (ii) dermis; and (iii) hypodermis (Forni *et al.*, 2012) (Fig. 31.1a). Each of them presents different cell populations, structures and functions as described below.

* E-mail address: e.barrajon@umh.es

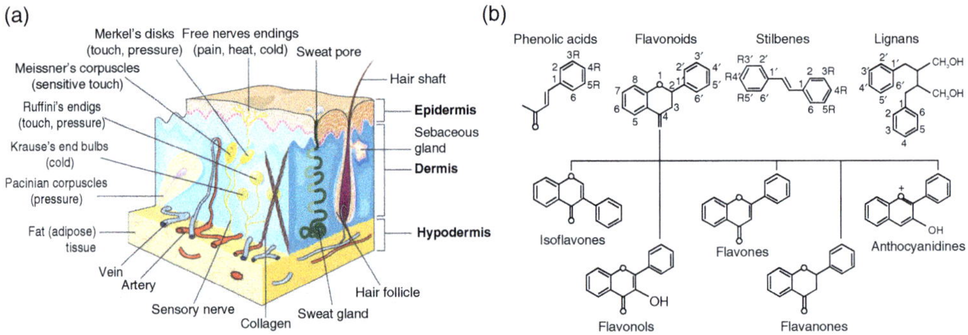


**Fig. 31.1.** **(a)** Human skin anatomy. There are three types of mechanoreceptors: tactile, proprioceptors and baroreceptors. Tactile mechanoreceptors are divided into Merkel's disks, Meissner's corpuscles, Ruffini's endings and Pacinian corpuscles. Another type of mechanoreceptor, Krause's end bulbs, is located in specialized regions to detect cold. Free nerve endings detect hot and cold, light touch and painful stimuli. This figure has been created and previously used by the author in Perez-Sanchez *et al.* (2018) using Servier Medical Art, licensed under the Creative Commons Attribution 3.0 Unported License (www.creativecommons.org/licenses/by/3.0/). **(b)** Main classes of polyphenols by structural classification. This image has been created and previously used by the author in Losada-Echeberria *et al.* (2017) and is under Creative Commons by Attribution (CC-BY) license (http://creativecommons.org/licenses/by/4.0/).

#### *31.2.1.1 Epidermis*

The epidermis is the layer that remains in the external portion of the integumentary system. It is composed mainly of keratinocytes but also melanocytes, Langerhans cells, Merkel cells and sensorial receptors. In addition, the epidermis contains four different sub-layers: (i) *stratum basale*; (ii) *stratum spinosum*; (iii) *stratum granulosum*; and (iv) *stratum corneum* (Shindo *et al.*, 1994a; Menon, 2002; Madison, 2003; Kolarsick *et al.*, 2011).

Skin renovation takes place in the *stratum basale*, where the adhesion of the epidermis to the dermis is also located. This sub-layer includes melanocytes and keratinocytes. The former cells produce melanin, a pigment that protects the skin from UV radiation injuries. Melanin exists in two main chemical forms: (i) eumelanin, a dark pigment expressed mainly in the skin of pigmented individuals, which is much more efficient at blocking UV photons than (ii) pheomelanin, a light-coloured sulfated form resulting from incorporation of cysteines into melanin precursors (Vincensi *et al.*, 1998; Ito *et al.*, 2000). Besides this UV protective function, melanin has other important physiological effects, including hormonal regulation, epidermal homeostasis, antioxidant activity and antimicrobial effects (Slominski *et al.*, 1993, Kalka *et al.*, 2000; Mackintosh, 2001; Meyskens *et al.*, 2001; Double *et al.*, 2002; Slominski *et al.*, 2004; Wang *et al.*, 2008). Melanosomes are membrane-bound organelles for synthesis, storage and transport of melanin, which is then transferred from melanocytes to surrounding keratinocytes. The contact between the dendritic processes of differentiated melanocytes and keratinocytes is necessary for the melanin transfer into keratinocytes, determining skin colour (Van Den Bossche *et al.*, 2006; Cichorek *et al.*, 2013). On the other hand, keratinocytes are the most common cells in the epidermis. They start their differentiation in this layer, before migrating into the *stratum spinosum*.

In the *stratum spinosum*, keratinocytes lose their proliferative activity and begin to participate in the assembly of the cornified envelope. Langerhans cells (antigen-presenting immune cells) also remain in this layer, participating in the immune response against microorganisms.

Keratinocytes form the *stratum granulosum*. In this stratum, keratinocytes begin to die through apoptosis. Cells lose their nucleus and organelles and keratinization begins. The cells in this layer are full of keratin, a protein synthesized by them that participates in protection against heat, chemicals and microorganisms and prevents dehydration (Nestle *et al.*, 2009; McLafferty, 2012).

Finally, the *stratum corneum* is the most external layer of the skin and therefore it is exposed to the external environment. This stratum is composed of dead keratinocytes, which protect against mechanical and dehydration injuries and avoid invasion by external substances such as microbes. This layer is the place where the desquamation and outward progression of the cells is produced (McLafferty, 2012).

#### *31.2.1.2 Dermis*

The dermis lies between the hypodermis and the epidermis and its primary function is to sustain and support the epidermis. The dermis is formed by two layers of connective tissue: the papillary stratum and the reticular stratum. The papillary stratum is a thin layer rich in reticulin fibres and capillaries that provide nutrients to the outer layers in the dermis. The reticular stratum presents numerous collagen and elastin fibres, forming the extracellular matrix (ECM) and providing structural support and stress-resistance to the skin. It contains cutaneous annexes (sweat and sebaceous glands and hair follicles) and numerous capillaries and sensory terminations. The dermis also contains abundant immune cells and fibroblasts (the major cell type present in this layer) (Perez-Sanchez *et al.*, 2018), which are mainly responsible for ECM synthesis. This wide diversity of structures and cells of the dermal matrix is increased by the existence of heterogeneities associated with skin fibroblasts (Haydont *et al.*, 2018).

#### *31.2.1.3 Hypodermis*

The hypodermis or subcutaneous tissue is the deepest layer of the skin. It mainly comprises connective and adipose tissues and participates in temperature isolation, gives protective padding and acts as an energy storage area. Its thickness is variable, depending on age, body part and physical condition; however, it can constitute up to 15–30% of body weight.

### 31.2.2 Ultraviolet radiation and the skin

The chronic exposure of the skin to UV radiation is the essential factor that initiates several skin disorders, such as inflammation, immunosuppression, photoageing and skin cancer (Nichols and Katiyar, 2010; Baccarin *et al.*, 2015). This radiation induces DNA damage, can produce mutations of the p53 gene, which promotes tumour growth, causes immunosuppression and increases oxidative stress. Besides its negative effects, UV radiation causes other important effects such as melanogenesis induction, and it is essential for vitamin D synthesis.

UV radiation can be subdivided into UVC, UVB and UVA categories based on electro-physical properties.

- *UVC radiation* (200–280 nm) presents the highest energy and the lowest penetration capacity of UV radiation (D'Orazio *et al.*, 2013). It is absorbed in the atmospheric ozone and usually does not reach the earth's surface.
- *UVB radiation* (280–320 nm) represents 5% of the UV radiation that reaches the ground and is almost entirely absorbed by the epidermis. UVB is directly absorbed by DNA, causing molecular rearrangements and forming specific photoproducts such as cyclobutane dimers and 6-4 photoproducts. UVB is responsible for damage due to sunburn, it induces oxidative stress and acts as a genotoxic agent, resulting in the most cytotoxic and mutagenic radiation (D'Orazio *et al.*, 2013; Hsu *et al.*, 2015).
- *UVA radiation* (320–400 nm) compresses the long-wave spectrum. It represents about 95% of the solar radiation that reaches the earth's surface. UVA penetrates deeply into the epidermis, reaching well into the dermis and being responsible for photoageing. UVA can excite DNA directly, producing dimers of pyrimidine in skin cells and generating reactive oxygen species (ROS) (Runger, 1999; D'Orazio *et al.*, 2013).

### 31.2.3 Polyphenols

#### *31.2.3.1 Overview*

Polyphenols are secondary metabolites of plants that are involved in the protective response of the plant to different stress situations such as UV radiation and aggression by pathogens (bacteria, virus or fungus) (Pandey and Rizvi, 2009).

Polyphenols are natural compounds found mainly in fruits and berries, vegetables, oilseeds, cereals, tea and spices (Baiao *et al.*, 2017).

#### 31.2.3.2 Structure and classification

A polyphenol is composed of one or more aromatic rings with one or more hydroxyl substitutions, including different functional moieties (Hossen *et al.*, 2017). The number and type of substitutions generate the different classes and subclasses of polyphenols, which leads to the structural variability that characterizes the group of polyphenols (Losada-Echeberria *et al.*, 2017). Glycosylation is also very common, increasing polyphenols in number and diversity. Polyphenols can be classified in different ways. However, the most extended classification divides them into (i) flavonoids; (ii) stilbenes; (iii) phenolic acids; and (iv) lignans (Ratz-Lyko *et al.*, 2015) (Fig. 31.1b).

#### 31.2.3.3 Flavonoids

Flavonoids comprise the most studied group of polyphenols and are synthesized via the phenylpropanoid metabolic pathway. They present characteristic C6–C3–C6 structures and a six-member oxygen heterocycle. They are divided into six subclasses: (i) flavones; (ii) flavonols; (iii) flavanones; (iv) flavanols; (v) isoflavones; and (vi) anthocyanidines (Pandey and Rizvi, 2009; Losada-Echeberria *et al.*, 2017). Flavonoids present potential antioxidant and anti-carcinogenic effects, among others (Zillich *et al.*, 2015; Cefali *et al.*, 2016).

#### 31.2.3.4 Stilbenes

Stilbenes are composed of polyphenols with two phenyl moieties connected by a two-carbon methylene bridge. Stilbenes present different biological properties, such as antioxidant activity and anti-ageing effects (Frombaum *et al.*, 2012; Chaher *et al.*, 2014; Peng *et al.*, 2018).

#### 31.2.3.5 Phenolic acids

Phenolic acids have usually esterified sugars or organic acids and are biosynthesized through the shikimic acid pathway. They can be divided into hydroxybenzoic acids, which have a C6–C1 structure and are derived from benzoic acid, and hydroxycinnamic acids, derived from a cinnamic acid and with a C6–C3 structure (Budić-Leto and Lovrić, 2002). Phenolic acids present a potential antioxidant, anti-inflammatory and anti-cancer benefits and antiviral function (Gutierrez-Grijalva *et al.*, 2017; Wu *et al.*, 2017b).

#### 31.2.3.6 Lignans

Lignans are (C6–C3) phenylpropanoid dimers (Umezawa, 2003). They show different biological activities. Some lignans can act as phytoestrogens and have reported potential antitumoral activity against breast cancer (Ezzat *et al.*, 2018). Furthermore, different lignans have shown action against Alzheimer's disease and a photoprotective effect against UV (Murata *et al.*, 2017).

## 31.3 Effects of Polyphenols on UV-induced Skin Damage

UV radiation can cause or aggravate several skin disorders related to DNA damage, oxidative stress, immunosuppression, inflammation and photoageing (Fig. 31.2). The organism presents different mechanisms of defence, such as melanogenesis, the immune response, mechanisms for DNA repairing and the endogenous antioxidant system. These defence mechanisms can be improved or protected with the oral or topical administration of natural molecules with skin-protective properties, such as vitamins or polyphenols. Most of the polyphenols are pigments that can absorb UV radiation and act as a sunscreen (Nichols and Katiyar, 2010). This sunscreen effect can reduce or avoid DNA damage, oxidative stress, immunosuppression, inflammation and photoageing effects of UV radiation in the skin. Furthermore, polyphenols present different chemical characteristics that give them the ability to reduce UV damage through different mechanisms of action, such as direct interaction with receptors, modulation of signal transduction and transcription of different genes, modulation of enzymatic activities and regulation of epigenetic gene expression (Potapovich *et al.*, 2011). All these mechanisms will be further reviewed in the following sections.

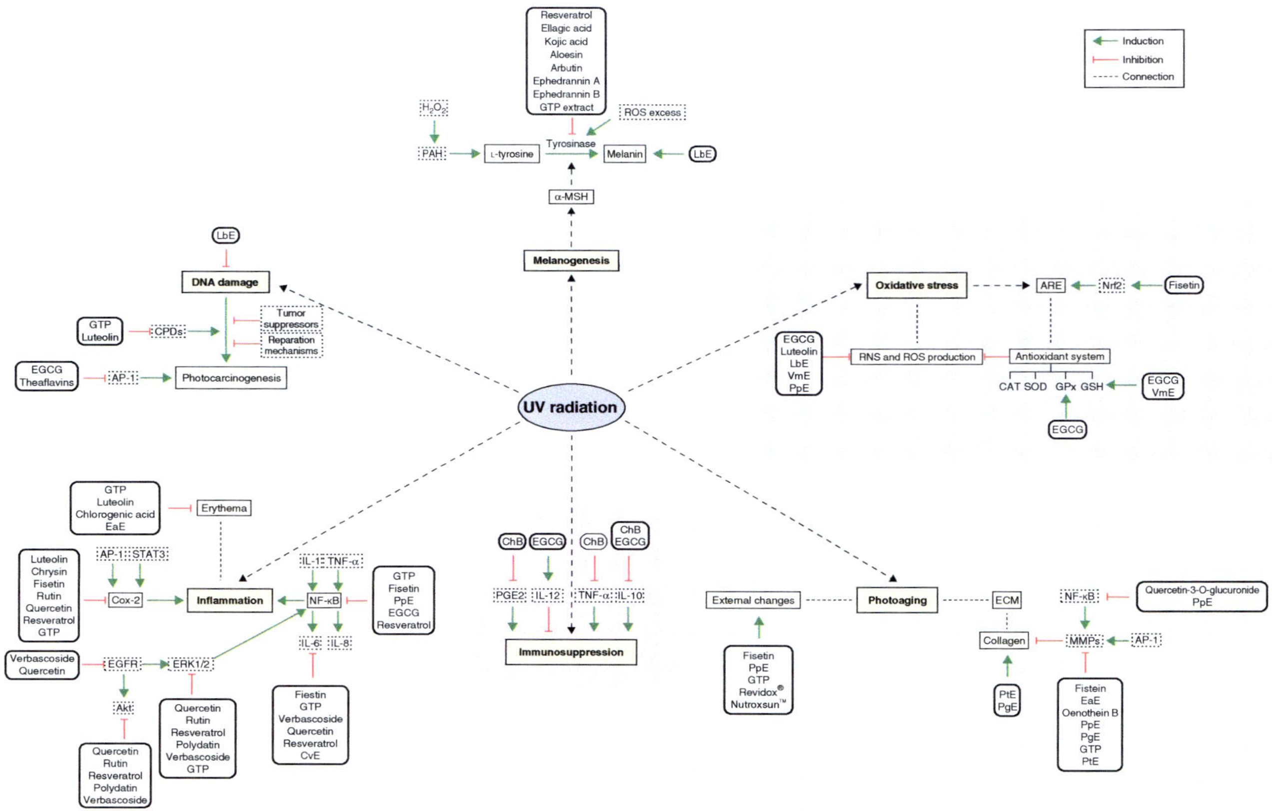


**Fig. 31.2.** Alterations of UV in skin and polyphenol interactions. UV radiation leads to DNA damage, melanogenesis, oxidative stress, photoageing, immunosuppression and inflammation. Some polyphenols or polyphenolic extracts can inhibit or reduce these UV-induced effects through different mechanisms. Inhibition or induction of various factors affected by polyphenols is represented in the figure. External changes include changes of the properties of the skin cells that are involved in photoageing, such as epidermal thickness, wrinkles and skin hyperplasia, cell death, skin elasticity and skin redness. Activation and inhibition relationships are shown using green (→) or red (┬) symbols.

### 31.3.1 Polyphenols and DNA damage

UV radiation leads to DNA damage in skin cells through the generation of photoproducts such as cyclobutane pyrimidine dimers (CPDs) (Feehan and Shantz, 2016). UV radiation can also induce mutations of different genes, including tumour suppressor genes such as p53 and oncogenes (Damiani and Ullrich, 2016). Furthermore, adjacent pyrimidine bases can change into dimeric photoproducts due to strand breaks, nucleic acid oxidation or photoreaction (Harwansh *et al.*, 2016). These alterations can initiate signalling pathways that can lead to cycle arrest and the activation of DNA repairing mechanisms (Fig. 31.2). If these mechanisms fail, UV radiation causes harmful effects such as the above-mentioned oxidative stress, immunosuppression, inflammation and photoageing, and may even initiate the carcinogenesis process.

UV-induced DNA damage can initiate photocarcinogenesis through induction of tumour promoters' induction, inhibition of tumour suppressor genes or avoiding the mechanisms of DNA repairing. One of the primary response mechanisms is related to UVB activation of activator protein-1 (AP-1), which is a tumour promoter (Barthelman *et al.*, 1998) (Fig. 31.2). The UVB induction of AP-1 and DNA damage are related to carcinogenesis, through the activation of extracellular signal-regulated kinase (ERK), c-Jun N-terminal kinase (JNK) and mitogen-activated protein kinase (MAPK) pathways (Lu *et al.*, 2016; Montes de Oca *et al.*, 2017).

Some polyphenols present different activities that protect from UV-induced DNA damage. A lemon balm extract (LbE) decreases the UVB-induced DNA damage and histone H2A (H2AX) activation. Pérez-Sánchez *et al.* (2016) studied the protective effect of LbE against UVB-induced DNA damage. The extract was mainly composed of rosmarinic acid and derivatives. They studied the presence of nuclear DNA damage through comet assay, which detects single-strand DNA breaks, alkali-labile sites and incomplete excision repair sites and DNA:DNA cross-linking. Also, they studied the effect of LbE in H2AX activation, observing that LbE decreased both the UVB-induced DNA damage and H2AX activation.

Other polyphenols, such as green tea polyphenols (GTP) and luteolin, protect from UV-induced damage through the inhibition of CPD formation (Katiyar *et al.*, 2000; Meeran *et al.*, 2009; Wölfle *et al.*, 2011). Green tea is a traditional source of polyphenols, with numerous studies supporting its protective effects against cancer (Wang *et al.*, 1991; Fu *et al.*, 2000). Katiyar *et al.* (2000) studied the protection of GTP against UVB-induced DNA damage in six Caucasian people. This GTP formulation contained epicatechin (6%), epigallocatechin (5%), (−)-epigallocatechin-3-gallate (EGCG) (65%) and epicatechin-3-gallate (24%). They observed that topical pretreatment with GTP (1 mg $cm^{-2}$ of skin area) inhibited UVB-induced CPD formation in epidermis and dermis. In addition, Meeran *et al.* (2009) studied the protective effect of the administration of GTP in drinking water (0.2% w/v) against UVB damage using interleukin (IL)-12p40 knockout (IL-12-KO) mice. They showed that GTP-reduced levels of CPDs after UVB exposure in wild-type mice was more pronounced than in IL-12-KO mice, concluding that prevention of carcinogenesis by GTP is mediated through IL-12-dependent DNA repair. Finally, Wölfle *et al.* (2011) demonstrated that luteolin prevented CPD formation after UVB irradiation by absorbing UV radiation, as the reduction of CPD formation was apparent 1 h after irradiation, and before DNA repair mechanisms could occur.

Furthermore, it has been shown that polyphenols protect against photocarcinogenesis through their activity against DNA damage or through the inhibition of UV-induced tumour promoters. EGCG and theaflavins inhibit UVB-induced AP-1 activity (Barthelman *et al.*, 1998; Nomura *et al.*, 2000). Barthelman *et al.* (1998) observed that EGCG inhibited UVB-induced AP-1 activity in a dose ranging from 5.45 nM to 54.5 μM in HCL14 human keratinocytes. Also, 60% inhibition of UVB-induced AP-1 by topical administration of 5 mg EGCG before and after irradiation was observed in B6D2 transgenic mice. Theaflavins are flavanol-derived polyphenols present in black tea. The major theaflavins in black tea are theaflavin, theaflavin-3-O-gallate, and theaflavin-3,3-O-digallate. Nomura *et al.* (2000) observed that treatment with theaflavins and EGCG before UVB radiation blocked UVB-induced AP-1 activity, and the theaflavin-3,3-O-digallate showed the strongest effect in a mouse epidermal cell line, JB6. Moreover, theaflavins inhibited UVB-induced phosphorylation of ERKs and

JNKs; meanwhile, EGCG inhibited only the ERK-dependent pathway.

### 31.3.2 Anti-inflammatory effects

Skin exposition to UV radiation causes an inflammatory process (Fig. 31.2), which is externally characterized by erythema or sunburn (Guan *et al.*, 2017). The erythema, which is mainly induced by UVB, is an acute inflammatory reaction of the skin that is identified by skin redness, pruritus and pain. It is characterized by an increase of blood flow that causes a dilation of the blood vessels in the dermis. This dilation is produced to increase blood supply to the damage site. In addition, the UV radiation induces changes in the microvascular skin structure to help the entrance of the plasma proteins and increases the migration of immune system cells through the endothelium (Feehan and Shantz, 2016).

On the one hand, UVB increases the cyclooxygenase-2 (COX-2) expression and inducible nitric oxide synthase (iNOS) in skin keratinocytes. The over-expression of both proteins is a marker of inflammation in the skin (Nichols and Katiyar, 2010). COX-2 catalyses the production of pro-inflammatory prostaglandin (PG) metabolites in the skin, which are produced from arachidonic acid. The expression of COX-2 and iNOS is regulated by several transcription factors such as AP-1 (which is also activated by UV radiation, as explained above) and signal transducer and activator of transcription-3 (STAT3) (Choi *et al.*, 2014). In addition, UVB induces the activation of MAPKs and causes the phosphorylation of ERK and p38 MAPK (Wölfle *et al.*, 2011).

On the other hand, UV radiation causes the translocation of the nuclear factor kappa B (NF-κB) into the nucleus. NF-κB is a transcription factor that is modulated by cytokines that induce the inflammatory process, such as IL-1 and tumour necrosis factor alpha (TNF-α). NF-κB modulates IL-6 and IL-8, TNF-α, interferon gamma (IFNγ) and transforming growth factor alpha (TGF-α) expression (Renard *et al.*, 1997; Wu *et al.*, 2017a). In addition, TGF-α induces the phosphorylation of the epidermal growth factor receptor (EGFR), which is also activated by UV radiation leading to inflammatory responses (Potapovich *et al.*, 2011). Also, NF-κB activates an immunosuppressive response and promotes ECM degradation through the activation of genes responsible for the production of matrix metalloproteinases (MMPs), especially involved in collagen degradation (Pacheco-Palencia *et al.*, 2008).

Furthermore, polyphenols can reduce UVB-induced inflammation through different pathways, such as avoiding over-expression or activation of MAPK and COX-2. Luteolin, chrysin, fisetin, rutin, quercetin, resveratrol and GTP reduce UVB-induced COX-2 induction (Agarwal *et al.*, 1993; Meeran *et al.*, 2009; Wölfle *et al.*, 2011; Wu *et al.*, 2011, 2017a; Potapovich *et al.*, 2013; Choi *et al.*, 2014). More details about these studies are given below.

The anti-inflammatory activity is the most visible effect of some polyphenols, since they can improve inflammatory markers such as oedema or erythema. GTP, luteolin, chlorogenic acid and an *Epilobium angustifolium* extract (EaE) inhibited UVB-induced erythema response (Katiyar *et al.*, 2000; Afaq *et al.*, 2003; Wölfle *et al.*, 2011, Kitagawa *et al.*, 2011; Ruszova *et al.*, 2013). Katiyar *et al.* (2000) also observed that topical pretreatment with GTP (1 mg $cm^{-2}$ of skin area) inhibited UVB-induced erythema response in epidermis and dermis. In addition, Afaq *et al.* (2003) observed that topical application of GTP to SKH-1 hairless mice before UVB radiation inhibited UVB-induced skin oedema, hyperplasia and infiltration of leucocytes. Wölfle *et al.* (2011) observed the effect of luteolin *in vivo* using the UVB erythema test. Kitagawa *et al.* (2011) found that chlorogenic acid prevented UVB-induced erythema formation in guinea pigs. Finally, Ruszova *et al.* (2013) observed that the EaE extract prevented UV-induced erythema formation *in vivo* on the skin of healthy human volunteers.

Wölfle *et al.* (2011) observed that luteolin at 16 μg $ml^{-1}$ before irradiation of 90 mJ $cm^{-2}$ UVB reduced the UVB-induced COX-2 expression through the attenuation of p38 MAPK and ERK activation in a human keratinocytes cell line, HaCaT. Wu *et al.* (2011) observed in HaCaT cell line that treatment with chrysin before UVA and UVB radiation inhibited the induction of COX-2. Chrysin also regulated the signalling of MAPK, inhibited JNK phosphorylation and attenuated p38 and ERK phosphorylation. Wu *et al.* (2017a) observed that fisetin reduced UVB-induced skin inflammation through the inhibition of COX-2 expression in human keratinocytes. Choi *et al.* (2014) observed that the topical application of rutin, a glycone quercetin, before UVB radiation inhibited UVB-induced COX-2 expression in HR-1 mouse skin through inhibition of p38 MAPK and JNK phosphorylation. In addition, they observed that rutin inhibited UVB-induced iNOS expression, AP-1 DNA binding and the phosphorylation of STAT3. Potapovich *et al.* (2013) observed that quercetin and resveratrol reduced COX-2 expression. Finally, Agarwal *et al.* (1993) and Meeran *et al.* (2009) observed that GTP (0.2% w/v) in drinking water reduced the levels of UVB-induced cyclooxygenase activities in mice.

In addition, polyphenols can reduce UVB-induced NF-κB activity or pro-inflammatory cytokines levels, such as TNF-α or IL-6. On the one hand, GTP, fisetin, quercetin, resveratrol, verbascoside and a *Calluna vulgaris* extract (CvE) reduced IL-6 (Afaq *et al.*, 2003; Meeran *et al.*, 2009; Olteanu *et al.*, 2012; Potapovich *et al.*, 2013; Wu *et al.*, 2017a). On the other hand, GTP, fisetin, a pomegranate polyphenolic extract (PpE), EGCG and resveratrol reduced NF-κB activity (Wu *et al.*, 2017a; Pacheco-Palencia *et al.*, 2008; Adhami *et al.*, 2003; Kim *et al.*, 2001).

Meeran *et al.* (2009) observed that GTP (0.2% w/v) in drinking water reduced TNF-α, IL-6 and IL-1$\beta$ in UVB-exposed mice skin. In addition, GTP reduced PGE2 levels, a marker of inflammation. Furthermore, Afaq *et al.* (2003) observed that GTP inhibited the activation of NF-κB through inhibition of IκB kinase α (IKKα) and the subsequent reduced degradation of IκB, leading to NF-κB retention in its inactive form in the cytoplasm. Wu *et al.* (2017a) observed that fisetin reduced UVB-induced skin inflammation through the inhibition of IL-6 and NF-κB in human keratinocytes. Potapovich *et al.* (2013) studied the anti-inflammatory effect of quercetin, resveratrol and verbascoside in normal human epidermal keratinocytes (NHEK) cell line. Pretreatment with quercetin and resveratrol and treatment after the UV radiation with quercetine and verbascoside reduced IL-6 expression. Verbascoside inhibited UV-induced IL-8 and TNF-α expression. In addition, quercetin, resveratrol and verbascoside inhibited the UV-induced production of IL-1. Olteanu *et al.* (2012) observed that CvE, which is enriched in quercetin and kaempferol, reduced IL-6 and TNF-$\alpha$ in SKH-1 hairless mice. Pacheco-Palencia *et al.* (2008) observed that PpE extract, composed of 37.5% of ellagitannins and approximately 2.7% of ellagic acid, inhibited UVA and UVB-induced activation of NF-κB. Adhami *et al.* (2003) and Kim *et al.* (2001) observed that EGCG and resveratrol inhibited UV-induced NF-κB gene expression.

Finally, polyphenols can reduce UVB-induced inflammation through another pathway. Afaq *et al.* (2003) observed that GTP inhibited inflammation through the reduction of ERK1/2 and JNK1/2 and p38 phosphorylation and JNK1/2 and p38 protein expressions. Potapovich *et al.* (2011) observed that verbascoside and quercetin inhibited EGFR phosphorylation in normal NHEK. In addition, they observed that quercetin, rutin, resveratrol, polydatin and verbascoside repressed UV-induced protein kinase B (Akt) phosphorylation, which is associated with EGFR activation, and inhibited ERK phosphorylation.

### 31.3.3 Oxidative stress and polyphenols

The exposition of the skin to UV radiation generates ROS. UV radiation produces superoxide, which results in hydrogen peroxide. Hydrogen peroxide forms hydroxyl radicals through the Fenton reaction (Perez-Sanchez *et al.*, 2016). This increase in ROS can cause structural damages to lipids and proteins through peroxidation and oxidation, respectively. Furthermore, excessive oxidative stress leads to inflammation, DNA damage, mitochondrial dysfunction, ageing and immunosuppression (Feichtinger *et al.*, 2014; Harwansh *et al.*, 2016).

However, ROS are necessary for the organism and are naturally produced during metabolic processes as the result of aerobic cellular metabolism. Intracellular ROS concentration is controlled by the antioxidant system, composed of enzymes and others antioxidant substances (Silva *et al.*, 2017). In this sense, the production and elimination of ROS present an equilibrium. If this equilibrium is broken, ROS excess leads to an oxidative stress condition, as happens in excessive UV radiation exposure, or it is not counterbalanced by intracellular antioxidant systems (Nicco and Batteux, 2017).

The antioxidant-response element (ARE) pathway is controlled by DNA sequences or elements that regulate genes encoding for antioxidant proteins, such as haem oxygenase-1 (HO-1), quinone oxidoreductase 1 (NQO1) and glutathione *S*-transferase (GST) (Jaiswal, 2004). Nuclear factor erythroid 2-related factor 2 (Nrf2) binds to ARE to regulate the transcriptional activation of these antioxidant genes, and its inhibition by UV radiation causes oxidative stress damage (Wu *et al.*, 2017a). This regulation neutralizes oxidative stress and is a part of the intracellular antioxidant system. UV radiation also modifies the cellular antioxidant system, reducing catalase (CAT), superoxide dismutase (SOD), glutathione reductase and glutathione peroxidase (GPx) levels and activity (Shindo *et al.*, 1994b; Harwansh *et al.*, 2016).

The antioxidant activity of polyphenols becomes very important as most of the UV radiation injuries, such as DNA damage, immunosuppression, inflammation and photoageing, are mediated/produced by oxidative stress. As a consequence of inhibition of oxidative stress by some polyphenols, protection against other UV radiation injuries takes place.

Polyphenols reduce antioxidant stress through different mechanisms, such as reducing the formation of oxidative molecules or increasing their elimination. EGCG reduces hydrogen peroxide and nitric oxide (NO) production (Katiyar *et al.*, 2001). In addition, luteolin and LbE reduced UVB-induced ROS formation, a *Vaccinium myrtillus* fruit extract (VmE) reduced UVA-induced ROS production and PpE reduced UVA and UVB-induced ROS formation (Pacheco-Palencia *et al.*, 2008; Svobodova *et al.*, 2008; Wölfle *et al.*, 2011; Perez-Sanchez *et al.*, 2016). More details about these studies are given below.

Katiyar *et al.* (2001) observed that pretreatment with 1 mg $cm^{-2}$ of EGCG in the skin decreased UV-induced hydrogen peroxide production by 68–90% and NO production by 30–100% in dermis and epidermis in six people. In addition, EGCG inhibited the infiltration of inflammatory leucocytes into the skin, which is related to the production of ROS, and inhibited epidermal lipid peroxidation. Wölfle *et al.* (2011) observed that luteolin showed antioxidant activity in 2,2-diphenyl-1-picrylhydrazyl (DPPH) radical assay and decreased UVB-induced intracellular generation of ROS in HaCaT cells through 2',7'-dichlorofluorescein (DCF) assay. In addition, they observed that luteolin reduced UVB-induced oxidative stress and nitrative stress. Pérez-Sánchez *et al.* (2016) observed that LbE significantly decreased UVB-induced intracellular ROS production through DCF assay in HaCaT cells. Svobodová *et al.* (2008) observed that VmE reduced UVA-induced ROS formation in HaCaT keratinocytes. This antioxidant effect may be attributed to anthocyanins which account for 25% w/v, mainly cyanidin, delphinidin and their derivatives. Finally, Pacheco-Palencia *et al.* (2008) observed that treatment with PpE significantly reduced UVA- and UVB-induced ROS generation in SKU-1064 human skin fibroblasts through DCF assay. This extract contains 37.5% ellagitannins and 2.7% ellagic acid.

The antioxidant-mediated photoprotective activity of the main polyphenols in olive oil has also been studied (Salucci *et al.*, 2017). In this sense, hydroxytyrosol (HyT) is the main polyphenol in olive oil and presents a well characterized antioxidant activity (Hu *et al.*, 2014). Salucci *et al.* (2017) studied whether the different number and position of the phenolic hydroxyl groups altered the antioxidant effect of HyT and their influence on photoprotective activity. They compared the antioxidant effect of HyT, hydroxytyrosyl laurate (Laur-HyT) and hydroxytyrosyl myristate (Myr-HyT). Laur-HyT and Myr-HyT showed less photoprotective activity than HyT. However, they penetrated better than HyT through the human corneum stratum and viable epidermis membranes (Salucci *et al.*, 2017).

Furthermore, some polyphenols can improve the cellular antioxidant system. Katiyar *et al.* (2001) observed that EGCG restored glutathione (GSH) level and GPx activity and Svobodová *et al.* (2008) observed that VmE reduced the depletion of intracellular GSH.

Finally, polyphenols can reduce oxidative stress through pathways involved in the redox balance such as the *Nrf2* gene. Wu *et al.* (2017a) observed that the topical application of fisetin reduced UVB-induced oxidative stress by increasing *Nrf2* expression in BALB/c hairless mice.

### 31.3.4 Polyphenols and immunosuppression

The immune system (IS) and immune response are influenced by UV radiation. The main consequence is a radiation-related immunosuppression, which is mediated by different mechanisms (Noonan and De Fabo, 1992; Streilein *et al.*, 1994). Langerhans cells are dendritic cells critical for the presentation of antigens to the immune system on the skin. These cells are mainly located in the epidermis and are responsible for T-lymphocyte activation through antigen presentation (Damiani and Ullrich, 2016). The alteration of Langerhans cells by UV leads to immunosuppression by the reduction of antigen presentation, and subsequent IS activation.

UV-induced immunosuppression is also modulated by various cytokines, such as TNF-α, IL-10 and IL-12 (Fig. 31.2). UV radiation induces TNF-α and IL-10 expression, which are mediators of UVB-induced immunosuppression, and reduces IL-12 (Norval, 2006). TNF-α induces the migration of Langerhans cells to draining lymph nodes, where they lose their role in skin immunity. IL-10 is related to CPD formation (Moodycliffe *et al.*, 1994) and modulates the production of PGE2, which is an important mediator of UV-induced immunosuppression and sunburn response. IL-12 participates in the development of T helper 1 responses, inducing the production of interferon (IFN) from natural killer (NK) cells. Furthermore, IL-12 suppresses UV-induced TNF-α and IL-10 production and induces DNA repair through nucleotide excision repair (NER) mechanism (Hasegawa *et al.*, 2013).

UVB radiation directly affects DNA, causing DNA damage and inducing urocanic acid (UCA) isomerization from *trans* to *cis* form. This isomerization is related to the stimulation of mast cell degranulation. The UVB induction of TNF-α and IL-10, the level of *cis*-UCA and the prevalence of dermal mast cells are the potential susceptibility factors for UVB-induced immunosuppression (Hart *et al.*, 2001).

Polyphenols have demonstrated different effects that could be related to UV-induced immunosuppression (Fig. 31.2). The flavone chafuroside B (ChB), isolated from oolong tea leaves, inhibits IL-10, TNF-α and PGE2 production and receptor activator of nuclear factor κB ligand (RANKL) expression. They studied the effect of ChB against UVB-induced immunosuppression in NHEK cells and observed that ChB inhibited the production of immunosuppressive mediators as IL-10, TNF-α, PGE2 and reduced the expression of RANKL, which is a member of the TNF family that induces Langerhans cells to produce IL-10. Finally, they suggested that these effects of ChB could be mediated through the induction of IL-12 synthesis (Hasegawa *et al.*, 2013).

Other polyphenols have shown activity against UV-mediated immunosuppression. EGCG decreased IL-10 and increased IL-12 and GTP prevented suppression of local and systemic contact hypersensitivity (Katiyar *et al.*, 1999). Katiyar *et al.* (1999) observed that topical application of EGCG before UVB radiation prevented UVB-induced immunosuppression by decreasing IL-10 in the skin and in draining lymph nodes (DLN) and inducing IL-12 in DLN in C3H/HeN mice. In addition, EGCG reduced the infiltration of monocytes/macrophages and neutrophils into inflammatory skin lesions. Furthermore, Katiyar *et al.* (2010) observed that GTP, administrated in drinking water (0.2 and 0.5% w/v) to C3H/HeN mice, prevented UV-induced suppression of local and systemic contact hypersensitivity by 58–62% and 51–55%, respectively.

### 31.3.5 Photoageing

Photoageing is defined as the extrinsic aging caused by solar exposition. This is mainly due to UV radiation and it is characterized by wrinkles, dryness, loss of elasticity, mottled pigmentation and hyperkeratosis (Damiani and Ullrich, 2016). Extrinsic photoageing is mainly caused by UV radiation but can be synergically increased by pollution, chemicals and toxins. From a molecular point of view, photoageing is primarily related to the degradation of ECM proteins, such as collagen and elastic fibres, and a decreased rate of their renewal/synthesis as a consequence of exposure to UV radiation. These ECM components provide structural and functional support to the skin tissue. Collagen is the primary insoluble fibrous protein in ECM. It gives strength and elasticity to the skin. Collagen is synthesized by fibroblasts and degraded by MMPs, which are responsible for degrading the extracellular matrix proteins, not only collagen but also elastin, fibronectin and proteoglycans. UV radiation increases MMP expression, leading to ECM degradation and photoageing (Pittayapruek *et al.*, 2016; Kong *et al.*, 2017). NF-κB and AP-1 are factors involved in this UV-mediated MMP activation. Upon UV radiation, NF-κB up-regulates MMPs such as MMP-1 and MMP-3 in dermal fibroblasts and AP-1 increases the expression of MMP-1, MMP-3 and MMP-9 (Pittayapruek *et al.*, 2016) (Fig. 31.2).

Polyphenols are potential active ingredients against photoageing through pleiotropic mechanisms, including direct UV absorption, their antioxidant activity and interaction with different factors involved in photoageing such as NF-κB and AP-1 (Fig. 31.2). Also, polyphenols can improve the properties of the skin cells that are involved in photoageing, related to external changes. Fisetin improves epidermal thickness, wrinkles and skin hyperplasia (Wu *et al.*, 2017a); PpE decreases UVB-induced cell death (Pacheco-Palencia *et al.*, 2008); GTPs decrease protein oxidation (Vayalil *et al.*, 2004); a grape and pomegranate dietary supplement improves skin elasticity, skin roughness, wrinkle depth and age spots (Buonocore *et al.*, 2012); and the nutraceutical Nutroxsun™, improves wrinkle depth, skin elasticity, skin redness and lipoperoxides (Nobile *et al.*, 2016). More details about these studies are given below.

Wu *et al.* (2017a) reported that topical application of fisetin reduced UVB-induced epidermal thickness in female BALB/c hairless mice by 62%. In addition, fisetin reduced wrinkles and ameliorated skin hyperplasia. Pacheco-Palencia *et al.* (2008) study showed that the pomegranate extract (PpE) showed a protective activity against UV-induced cell death in SKU-1064 human skin fibroblasts. This inhibition was observed at concentrations as low as 20 mg $l^{-1}$ against UVB and 60 mg $l^{-1}$ against UVA. Vayalil *et al.* (2004) observed that 0.2% of GTP in water (w/v) decreased UVB-induced protein oxidation, a hallmark of photoageing, in female SKH-1 hairless mice and in human skin fibroblasts HS68 cells. Buonocore *et al.* (2012) observed that Revidox®, a dietary supplement that combines grape (*Vitis vinifera*) and pomegranate polyphenols, improved skin elasticity and diminished skin roughness, wrinkle depth and the intensity of age spots after 60 days of treatment in 50 healthy females and males. Nobile *et al.* (2016) observed that Nutroxsun™, a nutraceutical product that contains citrus and rosemary extracts, improved UVB-induced skin redness, UVA-induced lipoperoxides, wrinkle depth and skin elasticity after 2 weeks of treatment in 90 Caucasian women, using two different dose regimens (100 and 250 mg per day).

As mentioned above, photoageing is mainly related to the degradation of ECM proteins, and collagen is the primary insoluble fibrous protein in ECM, which is degraded by MMPs. Some polyphenols protect against photoageing through the inhibition of the degradation of ECM components, such as reducing the UV-induced MMP expression. Fisetin reduces MMP-1 and MMP-2 expression (Wu *et al.*, 2017a); EaE inhibits MMP-1 and MMP-3 expression and oenothein B, a polyphenol isolated from EaE, inhibits MMPs (Ruszova *et al.*, 2013); PpE reduces MMP-1 and MMP-13 expression (Pacheco-Palencia *et al.*, 2008); another *Punica granatum* extract (PgE) reduces MMP-1 expression (Park *et al.*, 2010); GTP inhibits MMP-2, MMP-3, MMP-7 and MMP-9 expression (Vayali *et al.*, 2004); and a *Passiflora tarminiana* extract (PtE) reduces MMP-1 production (Bravo *et al.*, 2017). In addition, EaE decreases Hyal-2 hyaluronidase expression (Ruszova *et al.*, 2013), quercetin-3-O-glucuronide and PpE inhibit the activation of NF-κB (Pacheco-Palencia *et al.*, 2008), and PgE and PtE increase procollagen, by inhibiting collagenase (Bravo *et al.*, 2017) as observed in *in vitro* experiments. Further details are given below.

Wu *et al.* (2017a) observed that topical application of fisetin reduced UVB-induced MMP-1 and MMP-2 expression in female BALB/c hairless mice. Ruszova *et al.* (2013) observed that EaE suppressed UV-induced MMP-1 and MMP-3 expression. Furthermore, EaE decreased UV-induced Hyal-2 expression, a weak acid-active hyaluronidase partly responsible for the degradation of hyaluronan, one of the major ECM components in the skin. In addition, oenothein B, a polyphenol isolated from EaE, inhibited MMPs and quercetin-3-O-glucuronide inhibited RANKL. Pacheco-Palencia *et al.* (2008) reported that the treatment with concentrations of PpE higher than 10 mg $l^{-1}$ decreased the UVB-induced MMP-1 and MMP-13 expression and inhibited UV-induced activation of NF-κB in SKU-1064 human skin fibroblasts. Park *et al.* (2010) studied the protective effect of PgE against UVB radiation in normal human dermal fibroblasts obtained from a skin biopsy. They compared the effect of PgE with the effect of catechin, which was the main polyphenol present in that PgE. Cells treated with PgE following UVB radiation showed more concentration of procollagen type I. In addition, post-treatment with PgE inhibited UVB-induced MMP-1 expression. Although these effects may be related to catechin, its effect was lower than the effect of the whole extract. Vayalil *et al.* (2004) observed that 0.2% of GTP in water (w/v) inhibited UVB-induced expression of MMP-2 (67%), MMP-3 (63%), MMP-7 (62%) and MMP-9 (60%) in female SKH-1 hairless mice skin. Finally, Bravo *et al.* (2017) studied the anti-photoageing effect of PtE extract containing procyanidins and glycosylated flavonoids. Both whole PtE and its fractions reduced photoageing hallmarks in human dermal fibroblasts from adult skin (HDFa). They reported that the entire extract and its fractions decreased UVB-induced MMP-1 production and increased UVB-inhibited procollagen production.

### 31.3.6 Melanin induction by polyphenols

Melanin production is probably the main endogenous defence against UV radiation. Melanin absorbs UV radiation and reduces the formation of photoproducts that could be harmful to the skin (Sample and He, 2018). As explained in previous sections, melanin is produced by melanocytes, present in the epidermis. They produce melanin and transfer it to keratinocytes that migrate to external layers of the skin. UV radiation increases melanin production in different ways. The former is through α-melanocyte stimulating hormone (α-MSH), which is induced upon UV radiation (Masaki, 2010). The second is the oxidative stress mediated by UV exposition, which is characterized by a ROS excess that leads to an increase of tyrosinase and tyrosinase-related protein 1 activities, which are melanogenic factors. In this regard, hydrogen peroxide ($H_2O_2$) generated by UV radiation activates epidermal phenylalanine hydroxylase (PH), an enzyme that produces L-tyrosine, the main substrate of tyrosinase, from L-phenylalanine. However, a huge ROS excess can also present the contrary effect, enhancing depigmentation and causing depigmented macules in the skin as occurs in vitiligo.

Polyphenols have been reported as enhancing and reducing melanin production activities, depending on the study and individual polyphenol. The former activity increases UV protection and can be useful to avoid depigmentation, but the second also has interesting implications, especially cosmetic as whitening agents for the skin. Kojic acid, aloesin, arbutin and resveratrol inhibit tyrosinase activity (Mishima *et al.*, 1988; Garcia and Fulton, 1996; Maeda and Fukuda, 1996; Jones *et al.*, 2002); ellagic acid (EA) inhibits tyrosinase activity and melanin content (Shimogaki *et al.*, 2000; Kim *et al.*, 2004); ephedrannin A and B decrease tyrosinase transcription (Kim *et al.*, 2015); a GTP extract inhibits tyrosinase (Wei *et al.*, 2009); and LbE increases melanogenesis (Perez-Sanchez *et al.*, 2016). Additional details on each study are provided below.

Mishima *et al.* (1988) examined the effect of 1–3 mM kojic acid in B16 cells. They observed that kojic acid reduced the pigmentation and tyrosinase activity. In addition, Garcia and Fulton (1996) studied the effect of different formulations containing hydroquinone, kojic acid and glycolic acid in 39 subjects with melasma, finding a pigmentation reduction in the subjects. Similar results in tyrosinase activity and melanin production were obtained by Maeda and Fukuda (1996) with arbutin, a 3-D-glucopyranoside of hydroquinone, when using human melanocytes obtained from Asian neonatal foreskins. They observed that arbutin inhibited melanin production and tyrosinase activity acting as a competitive inhibitor of tyrosinase.

Jones *et al.* (2002) studied the effect of aloesin in comparison with arbutin and kojic acid. Using *in vitro* experiments, they observed that kojic acid presented the highest inhibition of fungal, human and murine tyrosinase, followed by aloesin and arbutin. Kim *et al.* (2004) also performed *in vitro* experiments, observing that resveratrol presented an inhibitory effect of tyrosine oxidase of mushroom tyrosinase. Resveratrol inhibited 67% of tyrosinase, showing an inhibitory concentration of 50% of the enzyme activity (IC50) of 43.5 μM. Among other authors, Wei *et al.* (2009) have studied *in vitro* tyrosinase activity too. They evaluated this activity using a GTP extract containing 98.25% polyphenols, and the main group was catechins (81.83% of the total content). They observed that, within the range of 1–100 mg $l^{-1}$ of extract, the extract showed a tyrosinase inhibitory rate between

75.37% and 57.14%. These results suggest that this extract could be another candidate for whitening.

Shimogaki *et al.* (2000) reported that 4 μM of EA inhibited tyrosinase activity and melanin content was inhibited by 38.3% and 54.4%, respectively, in B16 cells. In addition, they observed that treatment with 1% of EA inhibited UV-induced melanogenesis in brownish guinea pigs. Kim *et al.* (2015) observed that ephedrannin A and B inhibited melanin production by decreasing tyrosinase transcription, using the same cellular model. In this case, ephedrannin B showed a higher activity than A, but both could be candidates as whitening agents for skin.

On the contrary, Pérez-Sánchez *et al.* (2016) studied the melanogenic effect of LbE in B16 mouse melanoma cells. They observed that LbE increased melanogenesis by 40% at 100 mg $ml^{-1}$ compared with control, showing a protective effect against UV radiation.

## 31.4 Conclusions

Skin is a unique system of defence against the harmful effects of UV radiation. It is a very complex tissue with different layers and cell types from epithelial to immune origin. There is plenty of scientific evidence indicating the protective effects of plant polyphenols agains UV damaging effects. Polyphenols are capable of inhibiting DNA damage (CPDs, oxidation or photoreaction), oxidative stress, inflammation, immunosuppression, ECM degradation and the activation of transcription factors involved in differentiation, proliferation and apoptosis. Among them, flavonols, catechins, theaflavins, phenylpropanoids and stilbenes have been shown to be the most promising candidate molecules. Although much information is available on the topical use of these compounds, data on the efficacy of oral administration is scarce. Therefore, further studies are required to prove the efficacy of plant polyphenols properly.

Identification of the major metabolites responsible for skin protective effects against UV is still one of the major issues to be addressed. Plant polyphenols undergo important transformations both by microbiota in the large intestine and when absorbed by liver conjugation. Further, other transformations can take place when absorbed in target tissues such as the skin. Therefore, *in vivo* studies that correlate the presence of certain polyphenols in the skin cells and their UV-protective effects are needed to establish a mechanistic approach to their action. In addition to performing efficacy studies in skin cell models, well-designed human trials with strong statistical power are needed to prove the efficacy of plant polyphenols. Therefore, although the existing evidence on the beneficial effects of plant polyphenols against UV damage is promising, much work is still required to promote these compounds as reliable active ingredients in topical or oral formulations for sun protection.

## Acknowledgements

This work was supported by projects AGL 2015-67995-C3-1-R, AGL 2015-67995-C3-2-R, AGL 2015-67995-C3-3-R and RTI 2018-0966724-8-C21 (Spanish Ministry of Science and Innovation); AGL2015-67995-C3-1-R (Spanish Ministry of Economy and Competitiveness (MINECO); PROMETEO/2012/007, PROMETEO/2016/006, ACOMP/2013/093, ACIF/2013/064, APOTIP/2017/003 and APOSTD/2017/023 (Generalitat Valenciana), and CIBER (CB12/03/30038, Fisiopatologia de la Obesidad y la Nutricion, CIBERobn, Instituto de Salud Carlos III. Spain).

## References

Adhami, V.M., Afaq, F. and Ahmad, N. (2003) Suppression of ultraviolet B exposure-mediated activation of NF-kappaB in normal human keratinocytes by resveratrol. *Neoplasia* 5, 74–82.

Afaq, F., Ahmad, N. and Mukhtar, H. (2003) Suppression of UVB-induced phosphorylation of mitogen-activated protein kinases and nuclear factor kappa B by green tea polyphenol in SKH-1 hairless mice. *Oncogene* 22, 9254–9264.

Agarwal, R., Katiyar, S.K., Khan, S.G. and Mukhtar, H. (1993) Protection against ultraviolet B radiation-induced effects in the skin of SKH-1 hairless mice by a polyphenolic fraction isolated from green tea. *Photochemistry and Photobiology* 58, 695–700.

Baccarin, T., Mitjans, M., Ramos, D., Lemos-Senna, E. and Vinardell, M.P. (2015) Photoprotection by Punica granatum seed oil nanoemulsion entrapping polyphenol-rich ethyl acetate fraction against UVB-induced DNA damage in human keratinocyte (HaCaT) cell line. *Journal of Photochemistry and Photobiology: Biology* 153, 127–136.

Baiao, D.D.S., De Freitas, C.S., Gomes, L.P., Da Silva, D., Correa, A. *et al.* (2017) Polyphenols from root, tubercles and grains cropped in Brazil: chemical and nutritional characterization and their effects on human health and diseases. *Nutrients* 9, 1044.

Barthelman, M., Bair, W.B. 3rd, Stickland, K.K., Chen, W., Timmermann, B.N. *et al.* (1998) (-)-Epigallocatechin-3-gallate inhibition of ultraviolet B-induced AP-1 activity. *Carcinogenesis* 19, 2201–2204.

Bravo, K., Duque, L., Ferreres, F., Moreno, D.A. and Osorio, E. (2017) *Passiflora tarminiana* fruits reduce UVB-induced photoaging in human skin fibroblasts. *Journal of Photochemistry and Photobiology: Biology* 168, 78–88.

Budić-Leto, I. and Lovrić, T. (2002) Identification of phenolic acids and changes in their content during fermentation and ageing of white wines Pošip and Rukatac. *Food Technology and Biotechnology* 40, 5.

Buonocore, D., Lazzeretti, A., Tocabens, P., Nobile, V., Cestone, E. *et al.* (2012) Resveratrol-procyanidin blend: nutraceutical and antiaging efficacy evaluated in a placebocontrolled, double-blind study. *Clinical, Cosmetic and Investigational Dermatology* 5, 159–165.

Cefali, L.C., Ataide, J.A., Moriel, P., Foglio, M.A. and Mazzola, P.G. (2016) Plant-based active photoprotectants for sunscreens. *International Journal of Cosmetic Science* 38, 346–353.

Chaher, N., Arraki, K., Dillinseger, E., Temsamani, H., Bernillon, S. *et al.* (2014) Bioactive stilbenes from Vitis vinifera grapevine shoots extracts. *Journal of the Science of Food and Agriculture* 94, 951–954.

Choi, K.S., Kundu, J.K., Chun, K.S., Na, H.K. and Surh, Y.J. (2014) Rutin inhibits UVB radiation-induced expression of COX-2 and iNOS in hairless mouse skin: p38 MAP kinase and JNK as potential targets. *Archives of Biochemistry and Biophysics* 559, 38–45.

Cichorek, M., Wachulska, M., Stasiewicz, A. and Tymińska, A. (2013) Skin melanocytes: biology and development. *Postepy Dermatologii Alergologii* 30, 30–41.

Damiani, E. and Ullrich, S.E. (2016) Understanding the connection between platelet-activating factor, a UV-induced lipid mediator of inflammation, immune suppression and skin cancer. *Progress in Lipid Research* 63, 14–27.

D'Orazio, J., Jarrett, S., Amaro-Ortiz, A. and Scott, T. (2013) UV radiation and the skin. *International Journal of Molecular Sciences* 14(6), 12222–12248

Double, K.L., Ben-Shachar, D., Youdim, M.B., Zecca, L., Riederer, P. and Gerlach, M. (2002) Influence of neuromelanin on oxidative pathways within the human substantia nigra. *Neurotoxicology and Teratology* 24, 621–628.

Ezzat, S.M., Shouman, S.A., Elkhoely, A., Attia, Y.M., Elsesy, M.S. *et al.* (2018) Anticancer potentiality of lignan rich fraction of six Flaxseed cultivars. *Scientific Reports* 8, 544.

Feehan, R.P. and Shantz, L.M. (2016) Molecular signaling cascades involved in nonmelanoma skin carcinogenesis. *Biochemistry Journal* 473, 2973–2994.

Feichtinger, R.G., Sperl, W., Bauer, J.W. and Kofler, B. (2014) Mitochondrial dysfunction: a neglected component of skin diseases. *Experimental Dermatology* 23, 607–614.

Forni, M.F., Trombetta-Lima, M. and Sogayar, M.C. (2012) Stem cells in embryonic skin development. *Biological Research* 45, 215–222.

Frombaum, M., Le Clanche, S., Bonnefont-Rousselot, D. and Borderie, D. (2012) Antioxidant effects of resveratrol and other stilbene derivatives on oxidative stress and *NO bioavailability: potential benefits to cardiovascular diseases. *Biochimie* 94, 269–276.

Fu, Y.C., Jin, X.P., Wei, S.M., Lin, H.F. and Kacew, S. (2000) Ultraviolet radiation and reactive oxygen generation as inducers of keratinocyte apoptosis: protective role of tea polyphenols. *Journal of Toxicology and Environmental Health Part A* 61, 177–188.

Garcia, A. and Fulton, J.E. Jr (1996) The combination of glycolic acid and hydroquinone or kojic acid for the treatment of melasma and related conditions. *Dermatologic Surgery* 22, 443–447.

Guan, L., Suggs, A., Galan, E., Lam, M. and Baron, E.D. (2017) Topical application of ST266 reduces UV-induced skin damage. *Clinical, Cosmetic and Investigational Dermatology* 10, 459–471.

Gutierrez-Grijalva, E.P., Picos-Salas, M.A., Leyva-Lopez, N., Criollo-Mendoza, M.S., Vazquez-Olivo, G. and Heredia, J.B. (2017) Flavonoids and phenolic acids from oregano: occurrence, biological activity and health benefits. *Plants (Basel)* 7, 2.

Hart, P.H., Grimbaldeston, M.A. and Finlay-Jones, J.J. (2001) Sunlight, immunosuppression and skin cancer: role of histamine and mast cells. *Clinical and Experimental Pharmacology and Physiology* 28, 1–8.

Harwansh, R.K., Mukherjee, P.K., Kar, A., Bahadur, S., Al-Dhabi, N.A. and Duraipandiyan, V. (2016) Enhancement of photoprotection potential of catechin loaded nanoemulsion gel against UVA induced oxidative stress. *Journal of Photochemistry and Photobiology: Biology* 160, 318–329.

Hasegawa, T., Shimada, S., Ishida, H. and Nakashima, M. (2013) Chafuroside B, an Oolong tea polyphenol, ameliorates UVB-induced DNA damage and generation of photo-immunosuppression related mediators in human keratinocytes. *PLoS ONE* 8, e77308.

Haydont, V., Bernard, B.A. and Fortunel, N.O. (2018) Age-related evolutions of the dermis: Clinical signs, fibroblast and extracellular matrix dynamics. *Mechanisms of Ageing and Development* 177, 150–156.

Hossen, M.S., Ali, M.Y., Jahurul, M.H.A., Abdel-Daim, M.M., Gan, S.H. and Khalil, M.I. (2017) Beneficial roles of honey polyphenols against some human degenerative diseases: a review. *Pharmacological Reports* 69, 1194–1205.

Hsu, W.L., Lu, J.H., Noda, M., Wu, C.Y., Liu, J.D. *et al.* (2015) Derinat protects skin against ultraviolet-B (UVB)-induced cellular damage. *Molecules* 20, 20297–20311.

Hu, T., He, X.W., Jiang, J.G. and Xu, X.L. (2014) Hydroxytyrosol and its potential therapeutic effects. *Journal of Agriculture and Food Chemistry* 62, 1449–1455.

Ito, S., Wakamatsu, K. and Ozeki, H. (2000) Chemical analysis of melanins and its application to the study of the regulation of melanogenesis. *Pigment Cell Research* 13 (Suppl. 8), 103–109.

Jaiswal, A.K. (2004) Nrf2 signaling in coordinated activation of antioxidant gene expression. *Free Radical Biology and Medicine* 36, 1199–1207.

Jones, K., Hughes, J., Hong, M., Jia, Q. and Orndorff, S. (2002) Modulation of melanogenesis by aloesin: a competitive inhibitor of tyrosinase. *Pigment Cell Research* 15, 335–340.

Kalka, K., Mukhtar, H., Turowski-Wanke, A. and Merk, H. (2000) Biomelanin antioxidants in cosmetics: assessment based on inhibition of lipid peroxidation. *Skin Pharmacology and Applied Skin Physiology* 13, 143–149.

Katiyar, S.K., Challa, A., McCormick, T.S., Cooper, K.D. and Mukhtar, H. (1999) Prevention of UVB-induced immunosuppression in mice by the green tea polyphenol (-)-epigallocatechin-3-gallate may be associated with alterations in IL-10 and IL-12 production. *Carcinogenesis* 20, 2117–2124.

Katiyar, S.K., Perez, A. and Mukhtar, H. (2000) Green tea polyphenol treatment to human skin prevents formation of ultraviolet light B-induced pyrimidine dimers in DNA. *Clinical Cancer Research* 6, 3864–3869.

Katiyar, S.K., Afaq, F., Perez, A. and Mukhtar, H. (2001) Green tea polyphenol (-)-epigallocatechin-3-gallate treatment of human skin inhibits ultraviolet radiation-induced oxidative stress. *Carcinogenesis* 22, 287–294.

Katiyar, S.K., Vaid, M., Van Steeg, H. and Meeran, S.M. (2010) Green tea polyphenols prevent UV-induced immunosuppression by rapid repair of DNA damage and enhancement of nucleotide excision repair genes. *Cancer Prevention Research (Philadelphia)* 3, 179–189.

Kim, J., Hwang, J.S., Cho, Y.K., Han, Y., Jeon, Y.J. and Yang, K.H. (2001) Protective effects of (-)-epigallocatechin-3-gallate on UVA- and UVB-induced skin damage. *Skin Pharmacology and Applied Skin Physiology* 14, 11–19.

Kim, D.H., Kim, J.H., Baek, S.H., Seo, J.H., Kho, Y.H., Oh, T.K. and Lee, C.H. (2004) Enhancement of tyrosinase inhibition of the extract of Veratrum patulum using cellulase. *Biotechnology and Bioengineering* 87, 849–854.

Kim, I.S., Yoon, S.J., Park, Y.J. and Lee, H.B. (2015) Inhibitory effect of ephedrannins A and B from roots of Ephedra sinica STAPF on melanogenesis. *Biochimica et Biophysica Acta* 1850, 1389–1396.

Kitagawa, S., Yoshii, K., Morita, S. Y. and Teraoka, R. (2011) Efficient topical delivery of chlorogenic acid by an oil-in-water microemulsion to protect skin against UV-induced damage. *Chemical and Pharmacological Bulletin (Tokyo)* 59, 793–796.

Kolarsick, P.A.J., Kolarsick, M.A. and Goodwin, C. (2011) Anatomy and physiology of the skin. *Journal of the Dermatology Nurses' Association*, 3, 203–213.

Kong, S.Z., Li, D.D., Luo, H., Li, W.J., Huang, Y.M. *et al.* (2017) Anti-photoaging effects of chitosan oligosaccharide in ultraviolet-irradiated hairless mouse skin. *Experimental Gerontology* 103, 27–34.

Losada-Echeberria, M., Herranz-Lopez, M., Micol, V. and Barrajon-Catalán, E. (2017) Polyphenols as Promising Drugs against Main Breast Cancer Signatures. *Antioxidants (Basel)* 6, 88.

Lu, J., Guo, J.H., Tu, X.L., Zhang, C., Zhao, M., Zhang, Q.W. and Gao, F.H. (2016) Tiron inhibits UVB-induced AP-1 binding sites transcriptional activation on MMP-1 and MMP-3 promoters by MAPK signaling pathway in human dermal fibroblasts. *PLoS ONE* 11, e0159998.

Mackintosh, J.A. (2001) The antimicrobial properties of melanocytes, melanosomes and melanin and the evolution of black skin. *Journal of Theoretical Biology* 211, 101–113.

Madison, K.C. (2003) Barrier function of the skin: 'la raison d'etre' of the epidermis. *Journal of Investigative Dermatology* 121, 231–241.

Maeda, K. and Fukuda, M. (1996) Arbutin: mechanism of its depigmenting action in human melanocyte culture. *Journal of Pharmacology and Experimental Therapeutics* 276, 765–769.

Masaki, H. (2010) Role of antioxidants in the skin: anti-aging effects. *Journal of Dermatological Science* 58, 85-90.

McLafferty, E. (2012) The integumentary system: anatomy, physiology and function of skin. *Nursing Standard* 27, 35–42.

Meeran, S M., Akhtar, S. and Katiyar, S.K. (2009) Inhibition of UVB-induced skin tumor development by drinking green tea polyphenols is mediated through DNA repair and subsequent inhibition of inflammation. *Journal of Investigative Dermatology* 129, 1258–1270.

Menon, G.K. (2002) New insights into skin structure: scratching the surface. *Advanced Drug Delivery Reviews* 54 (Suppl. 1), S3–17.

Meyskens, F.L. Jr, Farmer, P. and Fruehauf, J.P. (2001) Redox regulation in human melanocytes and melanoma. *Pigment Cell Research* 14, 148–154.

Mishima, Y., Hatta, S., Ohyama, Y. and Inazu, M. (1988) Induction of melanogenesis suppression: cellular pharmacology and mode of differential action. *Pigment Cell Research* 1, 367–374.

Montes de Oca, M.K., Pearlman, R.L., McClees, S.F., Strickland, R. and Afaq, F. (2017) Phytochemicals for the prevention of photocarcinogenesis. *Photochemistry and Photobiology* 93, 956–974.

Moodycliffe, A.M., Kimber, I. and Norval, M. (1994) Role of tumour necrosis factor-alpha in ultraviolet B light-induced dendritic cell migration and suppression of contact hypersensitivity. *Immunology* 81, 79–84.

Murata, J., Ono, E., Yoroizuka, S., Toyonaga, H., Shiraishi, A. *et al.* (2017) Oxidative rearrangement of (+)-sesamin by CYP92B14 co-generates twin dietary lignans in sesame. *Nature Communications* 8, 2155.

Nestle, F.O., Di Meglio, P., Qin, J.Z. and Nickoloff, B.J. (2009) Skin immune sentinels in health and disease. *Nature Reviews: Immunology* 9, 679–691.

Nicco, C. and Batteux, F. (2017) ROS modulator molecules with therapeutic potential in cancers treatments. *Molecules* 23, E84.

Nichols, J.A. and Katiyar, S.K. (2010) Skin photoprotection by natural polyphenols: anti-inflammatory, antioxidant and DNA repair mechanisms. *Archives of Dermatological Research* 302, 71–83.

Nobile, V., Michelotti, A., Cestone, E., Caturla, N., Castillo, J. *et al.* (2016) Skin photoprotective and anti-ageing effects of a combination of rosemary (Rosmarinus officinalis) and grapefruit (Citrus paradisi) polyphenols. *Food & Nutrition Research* 60, 31871.

Nomura, M., Ma, W.Y., Huang, C., Yang, C.S., Bowden, G.T., Miyamoto, K. and Dong, Z. (2000) Inhibition of ultraviolet B-induced AP-1 activation by theaflavins from black tea. *Molecular Carcinogenesis* 28, 148–155.

Noonan, F.P. and De Fabo, EC. (1992) Immunosuppression by ultraviolet B radiation: initiation by urocanic acid. *Immunology Today* 13, 250–254.

Norval, M. (2006) The mechanisms and consequences of ultraviolet-induced immunosuppression. *Progress in Biophysics & Molecular Biology* 92, 108–118.

Olteanu, E. D., Filip, A., Clichici, S., Daicoviciu, D., Achim, M. *et al.* (2012) Photochemoprotective effect of *Calluna vulgaris* extract on skin exposed to multiple doses of ultraviolet B in SKH-1 hairless mice. *Journal of Environmental Pathology, Toxicology and Oncology* 31, 233–243.

Pacheco-Palencia, L.A., Noratto, G., Hingorani, L., Talcott, S.T. and Mertens-Talcott, S.U. (2008) Protective effects of standardized pomegranate (*Punica granatum* L.) polyphenolic extract in ultraviolet-irradiated human skin fibroblasts. *Journal of Agricultural and Food Chemistry* 56, 8434–8441.

Pandey, K.B. and Rizvi, S I. (2009) Plant polyphenols as dietary antioxidants in human health and disease. *Oxidative Medicine and Cellular Longevity* 2, 270–278.

Park, H.M., Moon, E., Kim, A.J., Kim, M.H., Lee, S. *et al.* (2010) Extract of *Punica granatum* inhibits skin photoaging induced by UVB irradiation. *International Journal of Dermatology* 49, 276–282.

Peng, R.M., Lin, G.R., Ting, Y. and Hu, J.Y. (2018) Oral delivery system enhanced the bioavailability of stilbenes: resveratrol and pterostilbene. *Biofactors* 44, 5–15.

Perez-Sanchez, A., Barrajon-Catalán, E., Herranz-Lopez, M., Castillo, J. and Micol, V. (2016) Lemon balm extract (Melissa officinalis, L.) promotes melanogenesis and prevents UVB-induced oxidative stress and DNA damage in a skin cell model. *Journal of Dermatological Science* 84, 169–177.

Perez-Sanchez, A., Barrajon-Catalán, E., Herranz-Lopez, M. and Micol, V. (2018) Nutraceuticals for skin care: a comprehensive review of human clinical studies. *Nutrients* 10, 403.

Pittayapruek, P., Meephansan, J., Prapapan, O., Komine, M. and Ohtsuki, M. (2016) Role of matrix metalloproteinases in photoaging and photocarcinogenesis. *International Journal of Molecular Sciences* 17.
Potapovich, A.I., Lulli, D., Fidanza, P., Kostyuk, V.A., De Luca, C., Pastore, S. and Korkina, L.G. (2011) Plant polyphenols differentially modulate inflammatory responses of human keratinocytes by interfering with activation of transcription factors NFkappaB and AhR and EGFR-ERK pathway. *Toxicology and Applied Pharmacology* 255, 138–149.
Potapovich, A.I., Kostyuk, V.A., Kostyuk, T.V., De Luca, C. and Korkina, L.G. (2013) Effects of pre- and post-treatment with plant polyphenols on human keratinocyte responses to solar UV. *Inflammation Research* 62, 773–780.
Pullar, J.M., Carr, A.C. and Vissers, M.C.M. (2017) The roles of vitamin C in skin health. *Nutrients* 9, 866.
Ratz-Lyko, A., Arct, J., Majewski, S. and Pytkowska, K. (2015) Influence of polyphenols on the physiological processes in the skin. *Phytotherapy Research* 29, 509–517.
Renard, P., Zachary, M.D., Bougelet, C., Mirault, M.E., Haegeman, G., Remacle, J. and Raes, M. (1997) Effects of antioxidant enzyme modulations on interleukin-1-induced nuclear factor kappa B activation. *Biochemical Pharmacology* 53, 149–160.
Runger, T.M. (1999) Role of UVA in the pathogenesis of melanoma and non-melanoma skin cancer. A short review. *Photodermatology, Photoimmunology & Photomedicine* 15, 212–216.
Ruszova, E., Cheel, J., Pavek, S., Moravcova, M., Hermannova, M. *et al.* (2013) Epilobium angustifolium extract demonstrates multiple effects on dermal fibroblasts in vitro and skin photo-protection in vivo. *General Physiology and Biophysics* 32, 347–359.
Salucci, S., Burattini, S., Buontempo, F., Martelli, A.M., Falcieri, E. and Battistelli, M. (2017) Protective effect of different antioxidant agents in UVB-irradiated keratinocytes. *European Journal of Histochemistry* 61, 2784.
Sample, A. and He, Y.Y. (2018) Mechanisms and prevention of UV-induced melanoma. *Photodermatology, Photoimmunology & Photomedicine* 34, 13–24.
Shimogaki, H., Tanaka, Y., Tamai, H. and Masuda, M. (2000) In vitro and in vivo evaluation of ellagic acid on melanogenesis inhibition. *International Journal of Cosmetic Science* 22, 291–303.
Shindo, Y., Witt, E., Han, D., Epstein, W. and Packer, L. (1994a) Enzymic and non-enzymic antioxidants in epidermis and dermis of human skin. *Journal of Investigative Dermatology* 102, 122–124.
Shindo, Y., Witt, E., Han, D. and Packer, L. (1994b) Dose–response effects of acute ultraviolet irradiation on antioxidants and molecular markers of oxidation in murine epidermis and dermis. *Journal of Investigative Dermatology* 102, 470–475.
Silva, S., Michniak-Kohn, B. and Leonardi, G.R. (2017) An overview about oxidation in clinical practice of skin aging. *Anais Brasileiros de Dermatologia* 92, 367–374.
Slominski, A., Paus, R. and Schadendorf, D. (1993) Melanocytes as 'sensory' and regulatory cells in the epidermis. *Journal of Theoretical Biology* 164, 103–120.
Slominski, A., Tobin, D. J., Shibahara, S. and Wortsman, J. (2004) Melanin pigmentation in mammalian skin and its hormonal regulation. *Physiological Reviews* 84, 1155–1228. *Smart Servier Medical Art* [Online]. Available at: https://smart.servier.com/ (accessed 29 September 2017).
Streilein, J.W., Taylor, J.R., Vincek, V., Kurimoto, I., Richardson, J. *et al.* (1994) Relationship between ultraviolet radiation-induced immunosuppression and carcinogenesis. *Journal of Investigative Dermatology* 103, 107s–111s.
Svobodova, A., Rambouskova, J., Walterova, D. and Vostalova, J. (2008) Bilberry extract reduces UVA-induced oxidative stress in HaCaT keratinocytes: a pilot study. *Biofactors* 33, 249–266.
Umezawa, T. (2003) Diversity in lignan biosynthesis. *Phytochemistry Reviews* 2, 371–390.
van den Bossche, K., Naeyaert, J.M. and Lambert, J. (2006) The quest for the mechanism of melanin transfer. *Traffic* 7, 769–778.
Vayalil, P. K., Mittal, A., Hara, Y., Elmets, C. A. and Katiyar, S. K. (2004) Green tea polyphenols prevent ultraviolet light-induced oxidative damage and matrix metalloproteinases expression in mouse skin. *Journal of Investigative Dermatology* 122, 1480–1487.
Vincensi, M.R., D'Ischia, M., Napolitano, A., Procaccini, E.M., Riccio, G. *et al.* (1998) Phaeomelanin versus eumelanin as a chemical indicator of ultraviolet sensitivity in fair-skinned subjects at high risk for melanoma: a pilot study. *Melanoma Research* 8, 53–58.
Wang, Z.Y., Agarwal, R., Bickers, D.R. and Mukhtar, H. (1991) Protection against ultraviolet B radiation-induced photocarcinogenesis in hairless mice by green tea polyphenols. *Carcinogenesis* 12, 1527–1530.
Wang, A., Marino, A.R., Gasyna, Z., Gasyna, E. and Norris, J. Jr (2008) Photoprotection by porcine eumelanin against singlet oxygen production. *Photochemistry and Photobiology* 84, 679–682.

Wei, X., Liu, Y., Xiao, J. and Wang, Y. (2009) Protective effects of tea polysaccharides and polyphenols on skin. *Journal of Agricutural & Food Chemistry* 57, 7757–7762.
Wölfle, U., Esser, P.R., Simon-Haarhaus, B., Martin, S.F., Lademann, J. and Schempp, C.M. (2011) UVB-induced DNA damage, generation of reactive oxygen species, and inflammation are effectively attenuated by the flavonoid luteolin in vitro and in vivo. *Free Radical Biology and Medicine* 50, 1081–1093.
Wu, N.L., Fang, J.Y., Chen, M., Wu, C.J., Huang, C.C. and Hung, C.F. (2011) Chrysin protects epidermal keratinocytes from UVA- and UVB-induced damage. *Journal of Agricultural & Food Chemistry* 59, 8391–8400.
Wu, P.Y., Lu, J.L., Liu, Y.J., Chien, T.Y., Hsu, H.C., Wen, K.C. and Chiang, H.M. (2017a) Fisetin regulates Nrf2 expression and the inflammation-related signaling pathway to prevent UVB-induced skin damage in hairless mice. *International Journal of Molecular Sciences* 18, 10.
Wu, Y.H., Zhang, B.Y., Qiu, L.P., Guan, R.F., Ye, Z.H. and Yu, X.P. (2017b) Structure properties and mechanisms of action of naturally originated phenolic acids and their derivatives against human viral infections. *Current Medicinal Chemistry* 24, 4279–4302.
Zillich, O.V., Schweiggert-Weisz, U., Eisner, P. and Kerscher, M. (2015) Polyphenols as active ingredients for cosmetic products. *International Journal of Cosmetic Science* 37, 455–464.

# 32 Radon I. Lung Cancer Risks

**B. Melloni***
*Pulmonary Diseases Department, Limoges University Hospital, Limoges, France*

## 32.1 Abstract

Radon is a rare natural radioactive gas. It produces solid progeny which emit alpha particles implicated in inducing cellular lesions in human bronchial epithelial tissue after inhalation. As early as the 15th century, increased rates of mortality due to respiratory diseases were observed in Eastern European mine workers. In the middle of the 20th century, increased incidence of lung cancer was recorded in uranium mine workers. Subsequent epidemiological studies conducted on lung cancer in miners confirmed that it was of occupational origin and proportional to levels of exposure. More recently, the risk of residential exposure has been evoked by case-control studies whose data were pooled to ensure adequate statistical power. Radon exposure is the second leading risk factor for lung cancer after smoking in the general population. There is epidemiological evidence for an increased risk of lung cancer caused by synergism between radon exposure and smoking in mine workers and in the general population. No predominant histological type was detected in the tumours under observation and clinical research into the molecular signature of radon is still in its infancy. International recommendations have prescribed housing and construction standards to mitigate exposure to radon and its progeny. Recommended thresholds vary considerably from one country to another. Reducing radon exposure in the general population is entirely warranted, but the necessary measures must go hand in hand with measures against smoking and exposure to other lung carcinogens.

## 32.2 Introduction

Radon is an inert radioactive gas that very rapidly produces progeny that are in turn radioactive. Human beings inhale radon progeny as solid elements that typically attach to dust particles. Toxicity is due to the emission of high-energy short-range alpha radiation affecting the nuclei of bronchial epithelial cells, and results in genetic damage conducive to lung cancer. The history of the mining industry has confirmed the causative effect of exposure to radon and its progeny on lung cancer. In recent times, epidemiological studies have in turn provided evidence for residential exposure to radon as a risk factor for lung cancer. Atmospheric radon concentration levels are usually low, but may rise dramatically in certain homes located in regions where subsoils have high uranium content (HPA, 2009). In this chapter, all aspects of exposure to

* E-mail address: boris.melloni@unilim.fr

radon and its progeny and onset of lung cancer are reviewed. The toxicity of radon is just as noxious as the toxicity of cigarette smoking, which is a confounding risk factor for lung cancer. Aspects pertaining to international regulation will also be considered.

## 32.3 Carcinogenic Effects of Radon Exposure

Radon is a rare gas formed by the decay of three isotopes of natural origin (radon-219, radon-220, radon-222), which are themselves products of radionuclides present in soil and rocks such as granite: uranium-235, thorium-232 and uranium-238, respectively. Radon-222, a product of the decay of radium-226 which in turn is a product of the decay of uranium-238, is the most predominant isotope within the atmosphere. Its short half-life of 3.82 days accounts for its presence in the air regarding human residential and occupational exposure (HPA, 2009). The role of thoron (radon-220 isotope) in human exposure warrants investigation, despite its very short half-life of 55.6 seconds (Robertson, 2013). Once radon gas has formed in the subsoil, it emanates through the cracks and pore spaces in the subsoil and soil. It is essentially transported by air and to a limited degree by water, in which it dissolves. Having escaped into the atmosphere, radon disperses in the air according to weather conditions. Radon concentration within indoor environments varies considerably according to subsoil composition, seasonal variation and ventilation. Radon concentration levels are higher during the fall and winter (Vogiannis, 2015).

Radon-222 produces progeny, made up of solid particles, via several stages culminating in the stable compound lead-206 (HPA, 2009). Two of the isotopes produced by this decay chain, polonium-218 and polonium-214, are alpha emitters (Fig. 32.1). They are solid progeny that can be inhaled as free particles or more commonly as particles attached to dust. After travelling a short distance, they then lodge in the bronchial epithelium, potentially damaging the cell nucleus. The mechanisms involved in precipitating lung cancer are chromosomal degradation, gene mutation, production of oxygen free radicals and cell cycle modification due to production of inflammatory cytokines and proteins (Robertson, 2013). From a cellular standpoint, bystander effects were observed in certain experiments, in contrast to a hypothetical linear no-threshold relationship proportional to exposure. Experiments have yielded inconsistent results that are subject to controversy. Low-level exposure to radon may in fact trigger DNA repair mechanisms. The US

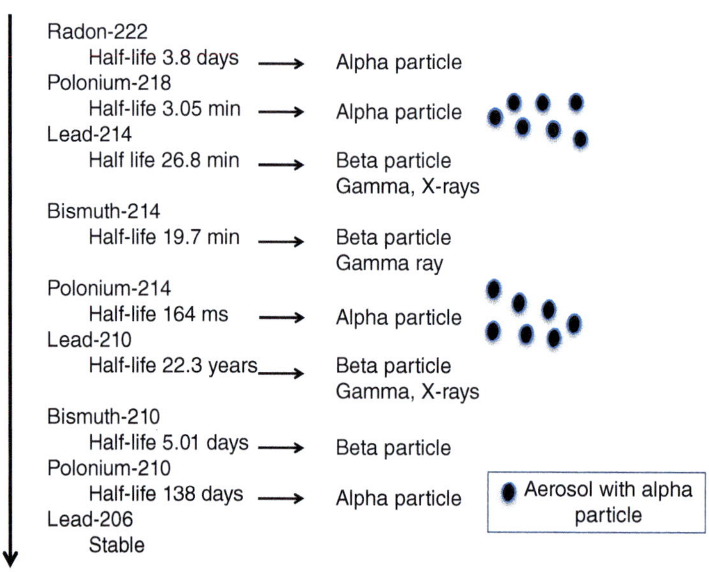


**Fig. 32.1.** The decay products of Radon-222 in their order of appearance.

National Research Council (NRC) has nevertheless favoured the no-threshold relationship approach, extrapolating risk based on average/strong concentration rates (NRC, 2006).

Radon is classified as a human carcinogen by the International Agency for Research on Cancer (IARC, 1988). In lung cancer, occupational and residential radon exposure is considered a risk factor (WHO, 2009). It is also the second leading risk factor for lung cancer after smoking in the general population.

## 32.4 History of Radon and Lung Cancer

In the 15th century, increased mortality due to respiratory disease was observed in the silver mines of the mountains between Saxony and Bohemia that had been exploited since the 13th century in Central Europe. The Swiss physician Paracelsus documented a deadly respiratory disease caused by the air from the depths of the mountains: Schneeberg's lung disease. At the same period, the renowned German scientist Agricola recommended mine ventilation so as to prevent the severe respiratory disease that he too had observed (Langärd, 2015). Occupational lung cancer was first described in 1879, but failed to gain unanimous support as a notion since the miners affected were also found to have contracted silicosis and tuberculosis when young at the same period. Radon was not discovered until 1900 by the German chemist, Friedrich Ernst Dorn (1848–1916). Subsequently, high concentrations of radon were detected in the same mines in 1924. It was the first time a strong connection had been established between radon exposure and incidence of lung cancer in mine workers. Controversy over the significance of the radon and/or arsenic contained in the mines in accounting for the prevalence of lung cancer in these miners persisted until 1940.

## 32.5 Radon Measurement

Radionuclide activity is defined as the number of disintegrations or transformations occurring in radioactive material per unit time. The conventional international unit of activity is the becquerel (Bq), whereby 1 Bq is equal to one disintegration per second. Radionuclide activity is expressed in terms of unit volume, customarily the cubic metre ($m^3$). With regard to exposure in miners, radon concentration is expressed as working level units (WL). One WL represents any combination of the short-lived progeny of radon in one litre (1 l)of air that will result in the emission of $1.3 \times 10^5$ megaelectronvolts (MeV) of potential alpha energy. Cumulative exposure is calculated as 170 h of miner working time and represents the working level month (WLM). Thus the most commonly used international unit of radiation measurement is the Becquerel per cubic metre (Bq $m^{-3}$) of air. In the USA, radiation is measured in picocuries per litre of air (pCi $l^{-1}$). One WL corresponds to $3.7 \times 10$ Bq $m^{-3}$ and 1 pCi $l^{-1}$ is equivalent to 37 Bq $m^{-3}$. One WLM is equal to 3700 Bq $m^{-3} \times 170$. Both becquerels and picocuries are low radiation exposure units (ICRP, 2010).

Radon progeny are solid, short-half-life particles. Due to disequilibrium between radon and its progeny, radon activity concentration is permanently higher than that of its progeny, which are known lung cancer risk factors. The term equilibrium factor, F, is thus employed, and corresponds to the ratio between the potential alpha energy of the progeny and that of the progeny plus radon. A value of 0.4 (0.2–0.7) has been acknowledged internationally.

## 32.6 Lung Cancer in Miners

### 32.6.1 Cohort studies of miners

From 1940 onwards, uranium mining expanded and the first protective measures were implemented. In 1988, the International Agency for Research on Cancer (IARC) recognized radon exposure as a causative factor for lung cancer based on studies conducted on mine workers exposed to high concentrations of radon (IARC, 1988). A trial based on 11 cohorts of miners estimated excess relative risk at 0.49 per WLM, 95% confidence interval (CI): 0.2–10 (Lubin, 1995). The NRC Committee on the Biological Effects of Ionizing Radiation (BEIR) has used slightly adjusted data from these 11 cohort studies conducted on miners from North America, Asia, Australia and

Europe (NRC, 1999). Thus over 60,000 miners were included from 11 studies, eight of which focused on uranium exposure and the others on iron, fluorspar or tin exposure. In all of these studies, there is evidence of relative risk of lung cancer proportional to cumulative radon exposure. Excess relative risk was estimated at 0.59 per 100 WLM on the basis of an exposure period of at least 5 years, namely 2674 deaths in these 11 cohorts (ICRP, 2010). These risk assessment findings were confirmed in subsequent studies, demonstrating variations in duration of follow-up, age at exposure and duration and extent of exposure. All of these studies showed evidence of time effects from onset of occupational exposure. Most of the case-control studies or cohorts included miners who were frequently smokers and/or exposed to other potentially carcinogenic agents such as diesel exhaust fumes and arsenic. Lung cancer rates increased linearly with cumulative radon exposure of between 100 and 600 WLM in mine workers (Lubin *et al.*, 1995).

### 32.6.2 Low-level radon exposure in miners

Epidemiological studies and pooled analysis of cohort studies have investigated lung cancer risk in relation to low-level radon exposure. Relative risk is estimated at 0.81 below 100 WLM and 1.18 below 50 WLM (NRC, 1999). Recent surveys of Czech and French miner cohorts have indicated a causative association between lung cancer mortality and low-level cumulative radon exposure below 100 and 50 WLM (Tirmarche *et al.*, 2012). This type of risk is therefore higher in young miners exposed to low levels of radon concentration over several years.

### 32.6.3 Radon exposure and smoking in miners

Where information on smoking status in miners was available, excess relative risk was higher than in non-smokers. The BEIR VI report came to the conclusion that there was a sub-multiplicative risk in miners exposed to radon who smoked (NRC, 1999; ICRP, 2010). This relative risk rose in proportion to the number of cigarettes smoked per day. However, recent studies have demonstrated that risk is higher in non-smoking miners than in smokers where rates of radon exposure are low, without reaching significance thresholds (Tirmarche *et al.*, 2012).

## 32.7 Indoor Radon and Lung Cancer

The carcinogenic risk of radon progeny has been proven by data from animal testing. Epidemiological studies conducted on miners from the 1970s onwards questioned the potential risk of developing lung cancer in the general population after residential radon exposure. Consequently several epidemiological approaches were used to address the issue of the potential risk of residential exposure.

### 32.7.1 Extrapolation of risk in miners

The first approach consisted of extrapolating data obtained from miners to the general population. Transposition of risk factors is accompanied by a number of methodological biases: duration and extent of exposure, exclusively male subjects, generally smokers and additionally exposed to other carcinogens, including silica, arsenic and diesel exhaust fumes (ICRP, 2010). However, analysis of more recent data from miners exposed to low-level radon concentrations, while making allowance for smoking, corroborates potential risk in the general population subjected to exposure (Tirmarche *et al.*, 2012).

### 32.7.2 Ecological studies

The second methodological approach used to evaluate such risk concerns geographically based descriptive epidemiological studies. To this end, the most straightforward method is to compare incidence of lung cancer by geographical area according to low/high levels of radon exposure. Inconsistent findings were observed in these studies, showing positive association, no association and negative association in the USA (Cohen, 1995). These findings are attributed to numerous biases, including smoking risk factors, personal exposure levels and exposure measurement errors.

**Table 32.1.** Pooled case-control studies of indoor radon exposure.

| Meta-analysis studies | Population (*n*) Cases | Controls | Relative risk per 100 Bg $m^{-3}$ (95% CI) |
|---|---|---|---|
| China (2 studies) | 1050 | 1095 | 1.13 (1.01–1.36) |
| USA (7 studies) | 3662 | 4966 | 1.10 (0.99–1.26) |
| Europe (13 studies) | 7148 | 14,208 | 1.08 (1.03–1.16) |

Smoking is regarded as the main confounding factor in these descriptive studies. Consequently the various international commissions did not see fit to validate this methodology as a means of determining residential radon risk. However, the most recent study involving the American Cancer Society cohort indicates a positive relationship between ecological indicators and risk of lung cancer due to residential exposure. Out of 810,000 participants, 3493 cases of lung cancer were observed. In this study, regional radon measurement was weighted by both short-term and long-term exposure measurements regarding indoor concentrations. A 15% increase in lung cancer risk was documented per 100 Bq $m^{-3}$ across the USA, rising to 31% in the north-east where concentration of uranium in subsoils is higher (Turner *et al.*, 2011).

### 32.7.3 Case-control studies

Since 1990, a large number of case-control studies have been published. Their purpose has been to compare cumulative exposure in a group of patients with the same exposure in a lung cancer-free control group. In the first case-control studies to be conducted, findings were variable due to insufficient statistical power and as a consequence of highly disparate protocols. The most commonly used radon measurement methods were alpha track detector devices. In most of these studies, risk of lung cancer was found to increase proportionally to an increase in cumulative exposure. In Sweden, there was significant excess of lung cancer in smokers living in homes where radon concentrations were over 400 Bq $m^{-3}$ (Pershagen *et al.*, 1994). In all of these case-control studies, there were a number of methodological biases: no allowance made for smoking, and radon measurement, measurement period and time spent indoors were all uncertain. At the end of the 1990s, international projects resulted in pooling of the data from these studies so as to obtain high statistical power and reduce bias. The exposure period under consideration was at least 30 years prior to diagnosis in the American and Chinese studies, and 35 years in the European studies. As a general rule, exposure covering the last 5 years prior to cancer diagnosis was not taken into account, due to latency of tumour onset after exposure. The resulting pooled analysis corroborates an increase in risk consistent with an increase in residential radon exposure (Table 32.1). After pooling the three studies, relative risk was estimated at 1.09 per 100 Bq $m^{-3}$ (Lubin *et al.*, 2004; Darby, 2005; Krewski *et al.*, 2006). In short, the pooled studies provide categorical evidence of an increase in risk of 8% per 100 Bq $m^{-3}$ over an exposure period of 5–35 years prior to cancer diagnosis. This risk increases linearly as radon exposure increases, though the ratio remains statistically significant in concentrations lower than or equal to 200 Bq $m^{-3}$. These findings imply a linear no-threshold relationship, which is nevertheless contested by certain authors where levels of exposure are low (Leonard *et al.*, 2012). Yet it was demonstrated in the European study that each 100 Bq $m^{-3}$ increase raised relative risk by 8.4% (95% CI: 3.0–15.8%) before adjustment and by 16% (95% CI: 5–31%) after adjustment for uncertainties related to exposure (Darby *et al.*, 2005; HPA, 2009). Excess risk was observed in non-smokers and smokers on the basis of these pooled studies.

## 32.8 Synergistic Risks Involving Cigarette Smoking

In the European trial comprising 13 case-control studies, cancer risk in individuals smoking

**Table 32.2** Relative risk of lung cancer for lifelong non-smokers, ex-smokers (≥ 10 years) and smokers (15–24 cigarettes per day) at various radon concentration levels. Adapted from Darby *et al.* (2006).

| | Relative risk (95% CI) | | |
|---|---|---|---|
| Radon concentration | Lifelong non-smokers | Ex-smokers ≥ 10 years | Current smokers (15–24 per day) |
| 0 Bq $m^{-3}$ | 1.0- | 5.0- | 25.8- |
| 100 Bq $m^{-3}$ | 1.1(1.0–1.2) | 5.4 (5.1–5.8) | 27.9 (26.5–29.8) |
| 200 Bq $m^{-3}$ | 1.2 (1.1–1.3) | 5.8 (5.3–6.6) | 30.1(27.3–33.9) |
| 400 Bq $m^{-3}$ | 1.3 (1.1–1.6) | 6.7 (5.6–8.1) | 34.4 (28.9–42.1) |

15–24 cigarettes per day and in ex-smokers (≥ 10 years) compared with non-smokers is significantly higher (Table 32.2). Cumulative risk in a 75-year-old non-smoker is 0.4%, 0.5% and 0.7% for radon concentrations of 0, 100 and 400 Bq $m^{-3}$, respectively. Cumulative risk in a 75-year-old smoker is 10%, 12% and 16% for radon concentrations of 0, 100 and 400 Bq $m^{-3}$, respectively. The European trial indicated a statistically significant increase in lung cancer risk for smokers, ex-smokers and non-smokers alike as exposure levels increased (Darby *et al.*, 2006). Combined exposure to cigarette smoke and radon is regarded as synergistic. A Spanish case-control study highlighted radon exposure and/or passive exposure to cigarette smoke (environmental tobacco smoke). Passive exposure to cigarette smoke significantly increased lung cancer risk even where residential radon exposure levels exceeded 200 Bq $m^{-3}$ (Torres-Duran *et al.*, 2014). It is conceivable that there is a relationship between passive smoking and radon exposure.

## 32.9 Pathological and Molecular Aspects

### 32.9.1 Histological subtypes

Radon exposure is not associated with any specific histological type. According to the first autopsy studies conducted on miners exposed to high levels of radon in the Schneeberg area (Germany) in 1879, an excess of thoracic tumours, classified as lymphosarcoma but in all probability small cell lung cancer, was documented (Haerting and Hesse, 1879). In the 20th century, small cell lung cancer (SCLC) was also detected in autopsies of uranium miners having developed lung cancer (Pirchan and Sikl, 1932). There was thus firm evidence of an association between radon exposure and SCLC. The prevailing hypothesis was that of an association between exposure to high levels of radon and its progeny and incidence of proximal lung tumours. Moreover, miners were also shown to exhibit all histological types, including: SCLC and non-small cell lung cancer (NSCLC); adenocarcinoma; and squamous cell carcinoma. Likewise, in the European trial on the effects of residential radon exposure, all histological types were found (Darby *et al.*, 2006). It would appear that there is no specific histological type associated with radon exposure.

### 32.9.2 Genetic susceptibility to lung cancer and radon exposure

Smoking accounts for over 75% of lung cancer in women and over 90% in men. The incidence of lung cancer in non-smokers raises concerns and specific molecular aspects have been investigated in specific studies. Radon exposure partially accounts for lung cancer in non-smokers.

The gene polymorphisms involved in carcinogenic metabolism have been examined in smokers and non-smokers exposed to radon. Glutathione *S*-transferases (GSTT1 and GSTM1) gene deletions are known to occur. These deletions may increase cancer risk in the event of exposure to alpha particles. Other gene polymorphisms involved in DNA repair have been investigated, but to date no anomalies have been found to be specific to alpha particle exposure. More recently, higher rates of cytogenetic damage to peripheral lymphocytes have been observed in individuals exposed to radon. These anomalies have been detected in patients with occurrence of polymorphism in certain DNA repair genes, which may constitute genetic markers of chronic radon exposure (Sinitsky *et al.*, 2015). Furthermore, chromosomal abnormalities have been observed in individuals from miner cohorts or in non-miners exposed to radon (Robertson *et al.*, 2013).

### 32.9.3 Acquired molecular abnormalities in radon-induced lung cancer

The *TP53* tumour suppressor gene commonly sustains mutations or deletions in a large number of cancers. To date, no *TP53* gene mutation has been detected in cancers found in uranium miner cohorts. Regarding tumours in individuals exposed to residential radon, findings are inconsistent and no definitive evidence of a causative relationship could be ascertained (Robertson *et al.*, 2013). In non-smokers presenting with adenocarcinoma subtype non-small cell lung cancer, the most common mutations are in the epidermal growth factor receptor (EGFR), followed by anaplastic lymphoma kinase (ALK) gene rearrangement. A Spanish study revealed that mutations in the EGFR and ALK translocations were observed in lung cancer in patients exposed to radon who were non-smokers (Ruano-Ravina *et al.*, 2016). This study concentrated primarily on adenocarcinoma tumour types involving 42% of EGFR mutations and 15% of ALK gene rearrangements. Where EGFR exon 19 deletions occurred, a twofold increase in radon exposure was documented, by comparison with exon 21 deletion. This research has the merit of being the first to describe radon exposure and genetic mutations in lung cancer affecting non-smokers. In the future, new next-generation sequencing (NGS) technologies could conceivably facilitate detection of genetic mutations in radon-induced lung cancer.

## 32.10 Radon Exposure Management

### 32.10.1 Prevention of environmental exposure

Radon concentration worldwide is highly variable, fluctuating between countries and even within various regions of those countries. Risk of residential exposure within the home thus varies considerably from one region to another. Mean radon concentration throughout the world is estimated at 39 Bq $m^{-3}$, but in fact ranges on average from 21 to 110 Bq $m^{-3}$ or more. Radon-related risk in lung cancer has been investigated in numerous reports. It is estimated at 3% of all cancers in the USA, 9% in Europe, but 10% in France, 15% in the UK and 20% in Sweden (Parkin and Darby, 2011). International organizations prescribe mitigation of indoor radon concentrations by recommending a cut-off threshold. This action level differs greatly across countries (Ruano-Ravina *et al.*, 2017). In the USA, the US Environmental Protection Agency (EPA) has been recommending 148 Bq $m^{-3}$ or 4 pCi $l^{-1}$ since 1987. In Europe, 300 Bq $m^{-3}$ or 8 pCi $l^{-1}$ was recommended in 2013, in line with the International Commission on Radiological Protection (ICRP). In other countries, recommended levels are lower: 200 Bq $m^{-3}$ in the UK, Ireland and Canada. In 2009, the World Health Organization (WHO) recommended a remediation level of 100 Bq $m^{-3}$ (WHO, 2009). There is no justification for so much discrepancy in threshold levels for the general population. International consensus is vital in order to determine an acceptable action level for the population. The BEIR VI model has demonstrated that strict compliance with the 148 Bq $m^{-3}$ action level and a constant trend in smoking habits would be instrumental in reducing annual lung cancer mortality in the USA by 21% by the end of the 21st century. The decline in cigarette smoking alone since 2006, with no remedial action regarding radon, has contributed to much lower mortality rates (Mendez *et al.*, 2011). According to this model, cigarette smoking is therefore a major confounding factor, as is true of other carcinogenic agents. Residential radon exposure is the second leading risk factor for lung cancer in smokers and the leading risk factor in non-smokers. Current prevalence of lung cancer in non-smokers warrants investigation into radon exposure in the regions in question. Initiatives to mitigate the risk of radon exposure should go hand in hand with those taken to combat cigarette smoking in the light of the interaction effect between these two carcinogenic agents (Melloni, 2014).

Mining risks are minimized by ventilation systems or limitation of exposure to dust, including continuous measurement of exposure levels. These measures have contributed to substantially reducing the risk of lung cancer in the majority of mines. With regard to mitigation measures in the general population, standards for new construction in line with reference thresholds are applied in most countries. New buildings must be constructed according to radon-resistant techniques. Several simple techniques are available to prevent inflow from basements to upstairs levels in old renovated or upgraded buildings: aeration, aspiration, natural and mechanical ventilation, basement drainage and active soil depressurization. New standards should be introduced to assess the efficacy of existing techniques.

### 32.10.2 Patients exposed to radon: clinical management

In the light of findings from epidemiological studies, mitigation of exposure in both miners and the general population is warranted. Screening for lung cancer has focused on smokers over the age of 50 (minimum 30 pack-years). The first randomized trial revealed a 20% reduction in mortality. Other trials are ongoing but do not take account of radon exposure. Several risk prediction models for lung cancer are available, but none of them incorporate the radon exposure risk factor (Torres-Duran, 2016).

## 32.11 Conclusions

Radon is widely recognized as a risk factor for lung cancer in both miners and the general population. There is increasing evidence of lung cancer in non-smokers and its cause is primarily attributed to radon exposure. It is of upmost importance to recommend radon testing in areas where subsoils have high uranium content, in both old and new buildings, and to prescribe remedial or preventive measures. To date, radon-related lung cancer manifests no histological or molecular characteristics representing undisputed evidence of alpha particle exposure. Given the synergistic cigarette–radon effect, measures to prevent radon exposure and efforts to combat smoking must go hand in hand. International harmonization of acceptable thresholds for levels of human exposure would be a welcome initiative.

## References


Choi, J.R., Park, S.Y., Noh, O.K., Koh, Y.W. and Kang, D.R. (2016). Gene mutation discovery research of non-smoking lung cancer patients due to indoor radon exposure. *Annals of Occupational and Environmental Medicine* 28, 13–19.

Cohen, B.L. (1995) Test of the linear no-threshold theory of radiation carcinogenesis for inhaled radon decay products. *Health Physics* 68, 157–174.

Darby, S., Hill, D., Auvinen, A., Barros-Dios, J.M., Baysson *et al.* (2005). Radon in homes and risk of lung cancer: collaborative analysis of individual data from 13 European case-control studies. *British Medical Journal* 330, 223–229.

Darby, S., Hill, D., Deo, H., Auvinen, A., Barros-Dios, J.M. *et al.* (2006) Residential radon and lung cancer-detailed results of a collaborative analysis of individual data on 7148 persons with lung cancer and 14 208 persons without lung cancer from 13 epidemiological studies in Europe. *Scandinavian Journal of Work, Environment and Health* 32 (Suppl. 1), 1–84.

Haerting, F.H. and Hesse, W. (1879) Der Lungenkrebs, die Bergkrankheit in den Schneeberg Gruben. *Vierteljahresschrift für Gerichtliche Medizin und Öffentliches Sanitätswes*. 30, 296–309.

HPA (2009) *Radiation, Chemical and Environmental Hazards*. RCE-11.Health Protection Agency, Oxford, UK.

IARC (1988) *Monographs of Carcinogenic Risk to Humans: Manmade Fibres and Radon*, International Agency for Research on Cancer, Lyon, France.

ICRP (2010) *Lung Cancer Risk from Radon and Progeny and Statement on Radon*. ICRP publications 115, Annals of the ICRP, 40 (1). International Commission on Radiological Protection, Ottawa.

Krewski, D., Lubin, J.H., Zielinski, J.M., Alavanja, M., Catalan, V.S. *et al.* (2006) A combined analysis of North American case-control studies of residential radon and lung cancer. *Journal of Toxicology and Environmental Health* 69, 533–597.

Langärd, S. (2015) Gregorius Agricola memorial lecture: Lung Cancer – a work-related disease for 500 years, as predicted by Agricola. *Journal of Trace Elements in Medicine and Biology* 31, 214–218.

Leonard, B.E., Thompson, R.E., and Beecher, G.C. (2012). Human lung cancer risks from radon. Part III – Evidence of influence of combined bystander and adaptative response. Effects on radon case-control-studies. A microdose analysis. *Dose Response* 10, 415–461.

Lubin, J.H., Boice, J.D. Jr, Edling, C., Hornung, G.R., Kunz, E. *et al.* (1995). Lung cancer in radon-exposed miners and estimation of risks from indoor exposure. *Journal of the National Cancer Institute* 87, 817–827.

Lubin, J.H., Wang, Z.Y., Boice, J.D., Blot, W.J., De Wang, L. and Kleinerman, R.A. (2004) Risk of lung cancer and residential radon in China: pooled results of two studies. *International Journal of Cancer* 109, 132–137.

Melloni, B. (2014) Lung cancer in never-smokers: radon exposure and environmental tobacco smoke. *European Respiratory Journal* 44, 850–852.

Mendez, D., Alshanqueety, O., Warner, K.E., Lantz, P.M. and Courant, P.N. (2011) The impact of declining smoking on radon-related lung cancer in the United States. *American Journal of Public Health* 101, 310–314.

NRC (1999) *Health Effects of Exposure to Radon: BEIR VI Report*. National Research Council (US) Committee on Health Risks of Exposure to Radon (BEIR VI). National Academies Press, Washington, DC.

NRC (2006) *Health Risks from Exposure to Low Levels of Ionizing Radiation: BEIR VII Phase 2*. National Academies Press, Washington, DC.

Parkin, D.M. and Darby, S.C. (2011) Cancers in 2010 attributable to ionizing radiation exposure in UK. *British Journal of Cancer* 105, S55–S65.

Pershagen, G., Akerblom, G., Axelson, O. Clavensjo, B., Damber, L. *et al.* (1994) Residential radon exposure and lung cancer in Sweden. *New England Journal of Medicine* 330, 159–264.

Pirchan, A. and Sikl, H. (1932) Cancer of the lung in the miners of Jachymov (Joachimsthal). *American Journal of Cancer* 4, 681–722.

Roberston, A., Allen, J., Laney R. and Curnow, A. (2013) The cellular and molecular carcinogenic effects of radon exposure: a review. *International Journal of Molecular Sciences* 14,14024–14063.

Ruano-Ravina, A., Torres-Duran, M., Kelsey, K.T., Parente-Lamelas, I., Leiro-Fernández, V. *et al.* (2016) Residential radon, *EGFR* mutations and *ALK* alterations in never-smoking lung cancer cases. *European Respiratory Journal* 48, 1462–1470.

Ruano-Ravina, A., Kelsey. K.T., Fernandez-Villar, A. and Barros-Dios J.M. (2017) Action levels for indoor radon. Different risks for the same lung carcinogen? *European Respiratory Journal* 50, 1701609.

Sinitsky, M.Y., Larionov, A.V., Asanov, M.A. and Druzhinin, V.G. (2015) Associations of DNA-repair gene polymorphisms with a genetic susceptibility to ionizing radiation in residents of aeras with high radon (222Rn) concentration. *International Journal of Radiation Biology* 91, 486–494.

Tirmarche, M., Harrison J., Laurier, D., Blanchardon, E., Paquet, F. and Marsh, J. (2012) Risks of lung cancer from radon exposure contribution of recently published studies of uranium miners. *Annals of the ICRP* 41(3-4), 368–377.

Torres-Duran, M., Ruano-Ravina, A., Parente-Lamelas, I., Leiro-Fernadez, V., Abal-Arca, J. *et al.* (2014) Lung cancer in never-smokers: a case-control study in a radon-prone area (Galicia, Spain). *European Respiratory Journal* 44, 994–1001.

Torres-Duran, M., Fernandez-Villar, A., Barros-Dios, J.M. and Ruano-Ravina, A. (2016) Residential radon: the neglected risk factor in lung cancer risk scores. *Journal of Thoracic Oncology* 9, 1384–1386.

Turner, M.C., Krewski, D., Chen, Y., Pope, C.A. III, Gapstur, S. and Thun, M.J. (2011) Radon and lung cancer in the American Cancer Society Cohort. *Cancer Epidemiology, Biomarkers & Prevention* 20(3), 438–448.

Vogiannis, E.G. and Nikolopoulos, D (2015) Radon sources and associated risks in terms of exposure and dose. *Frontiers in Public Health* 2 (207), 1–10.

WHO (2009) *WHO Handbook On Indoor Radon: A Public Health Perspective*. World Health Organization, Geneva.

# 33 Radon II. Leukaemia or CNS Cancer Risks Among Children

**R. Del Risco Kollerud***
*Department of Community Medicine and Global Health; University of Oslo, Norway*

## 33.1 Abstract

Ionizing radiation is the only established environmental risk factor for childhood leukaemia and tumours of the central nervous system. Indeed, there is consensus that cancer risk in humans increases after exposure to moderate and high doses of radiation. However, exposures to low-dose natural radiation such as radon are poorly understood. Many epidemiological studies have attempted to determine whether exposure in children to radon increases the risk of developing childhood cancer. Most studies have examined the predominant types of cancer in childhood, particularly leukaemia and cancer of the central nervous system.

The majority of the studies published in the past decades indicate no association between radon exposure and the risk of leukaemia and cancer in the central nervous system in children, but methodological limitations makes it difficult to be conclusive in the risk assessment. These studies are inherently uncertain, because they are observational in nature rather than experimental and results are mixed and not as strong as the results found for lung cancer.

Some of the limitations found in these studies are risk of misclassification of radon exposure and that there is usually no information available regarding other risk factors that might influence the risk of developing cancer in children. Other reasons that might make it difficult to detect associations include the long asymptomatic period between exposure to radon and developing cancer, as well the low incidence of the disease among children. Further, there remains a lack of understanding of the cellular and molecular response of tissues to radiation from radon, identification of genes that increase risk for radiation carcinogenesis and the gene–environment interaction.

## 33.2 Introduction

Radon is a natural radioactive gas produced by the decay of uranium and thorium, which are present in all rocks and soils in small quantities. The International Agency for Research on Cancer (IARC) states that radon is a cause of lung cancer in humans, based on clear excess lung cancer rates observed in underground miners, and elevated lung cancer risk is seen in experimental animals exposed to radon (IARC, 2009). Some studies suggest that radon exposure may be related to other types of cancer, such as childhood cancer (Del Risco Kollerud, 2016). The evidence for such links of radon with childhood cancer is reviewed in this chapter.

* E-mail address: r.d.r.kollerud@medisin.uio.no

Ionizing radiation is a well-documented risk factor of human cancer. It is the only established environmental risk factor for childhood leukaemia and tumours of the central nervous system (CNS). In fact there is consensus that cancer risk in humans increases after exposure to moderate and high doses of radiation. Leukaemia was the first malignant disease recognized to be in excess among the Japanese survivors of the atomic bomb explosions over Hiroshima and Nagasaki in 1945. However, the results could not support differences in leukaemia incidence in the youngest group (0–14 years of age) (Folley *et al.*, 1952). In a later study Richardson *et al.* (2009) examined leukaemia mortality. Their results demonstrated variation of excess relative risk with age at exposure, with the risk being notably higher for younger ages at exposure.

On the developing embryo and fetus moderate doses of radiation can produce catastrophic effects. The effects depend on the stage of gestation and the dose (Hall, 2000). Radiation on the developing embryo and fetus may also cause congenital malformations, growth retardation and functional impairment, such as mental retardation. However, whether cancer risk is increased by acute low dose rates is unclear (Kamiya *et al.*, 2015). It has also been suggested that survivors of the atomic bomb explosions whose parents were exposed to substantial amounts of radiation before conception may have a higher risk of developing cancer than those not exposed. Nevertheless, these findings have no support in recent studies (Grant *et al.*, 2015).

Natural radiation is the main source of human exposure to ionizing radiation, and the largest component of the effective dose comes from inhalation of radon (Rn-222) and its daughters (Rn-219 and Rn-220). Rn-222 is a naturally radioactive gas resulting from the decay of uranium-238. It is the most common naturally occurring uranium isotope. Uranium is found in small quantities in all soils and rocks, but the concentration varies. Rn-222 is formed when radium-226 dissolves in accordance with the uranium-238 series. Rn-220 is formed when radium-224 decays according to the thorium-232 series. Therefore, in areas where there is uranium or thorium, it is likely to be airborne radon or thoron in the air. If there is insufficient ventilation in a dwelling, the concentration of radon and its daughters in the dwelling will increase. Radon concentrations are typically tested using alpha track detectors placed in multiple sites throughout the home or workplace to obtain results that take into account the variability between rooms. Rn-222 has a half-life of 3.82 days and provides about 50% of the total radiation dose for an average person in Europe (Appleton, 2007). Small amounts of radon can also be released from the water supply into air. As the radon moves from water to air, it can be inhaled. For both adults and children, most exposure to radon comes from being indoors in homes, offices, schools and other buildings. The levels of radon in homes and other buildings may be influenced by several factors, including properties of the underlying geology, the permeability of the soil, building materials and residents' lifestyles (Sundal *et al.*, 2007; Appleton and Miles, 2010).

## 33.3 Latency Period of Radiation-Induced Cancer

The age at which exposure occurs appears to be relevant: the lower the age at the time of exposure, the higher is the lifetime risk. The time interval between irradiation and the appearance of malignancy is known as the latent period. The biological impact of radiation is divided into short-term and long-term consequences. Effects that appear within a matter of minutes, days or weeks are called short-term effects; those that appear years, decades and sometimes generations later are called long-term effects. Leukaemia has a short latent period. Excess cases began to appear in the survivors of Hiroshima and Nagasaki a few years after irradiation and reached a peak by 5–7 years; most cases occurred in the first 15 years. Solid tumours show a longer latency than the leukaemia, in the order of anything from 10 to 60 years or more (Greaves, 2006). Although studies of the Japanese atomic bomb survivors gives detailed information on radiation risks, they cannot generate direct information on all aspects of radiation-induced risks since the bomb survivors received mainly doses of external radiation, and some of the dose estimates remain uncertain due to retrospective calculation. Further, data for the period before October 1950 were not collected systematically, making it difficult to determine the minimum latent period for leukaemia.

## 33.4 Cellular and Molecular Effects of Radon Exposure and Childhood Cancer

There are limited studies that have assessed cellular effects of alpha particles from radon on cancer in children. Epidemiological studies of uranium mine workers and experimental animal studies suggest a positive correlation between alpha particles emitted from radon (Rn-222) and its daughter products and DNA damage from chromosomal aberrations, double-strand DNA breaks and generated reactive oxygen species, resulting in cell cycle shortening, apoptosis and an increased potential for carcinogenesis (Narayanan *et al.*, 1997; Robertson *et al.*, 2013). Smerhovsky *et al.* (2001) found a significant and strong association between the frequency of chromosomal aberrations and cancer incidence in a group of miners exposed to radon, where a 1% increase in frequency of chromosomal aberrations was followed by a 64% increase in risk of cancer. Similar results were found by Smerhovsky *et al.* (2002) with an increased frequency of chromatid breaks in 1% of cells resulting in a 99% increase in cancer risk, which was not limited to lung carcinogenesis. More recent studies suggest that these chromosomal aberrations have a long-term persistence above the population average values in miners after they are no longer involved in underground mining activities (Mészáros *et al.*, 2004).

The dose received by haematopoietic stem cells within the red bone marrow has been considered to put the cells at risk for childhood acute leukaemia. Bone marrow consists of various cellular components, mainly haematopoietic cells and fat. The high value in fat has important consequences for the radiation dose to haematopoietic tissue in the marrow. Radon might deliver a small dose to these cells. Harley and Robbins (1992, 2009) suggested that there are two dose pathways where alpha particles could damage potential stem cells in organs other than lungs. One is the alpha dose to bone marrow and the second is the dose received from inhaled radon and its decay products by circulating lymphocytes while present in the tracheobronchial epithelium. The dose of radon to bone marrow and other organs is based on the solubility of radon gas. Dose calculations for the alpha particle from radon and its decay products to bone marrow range from annual values of 0.12 to 0.66 millisieverts (mSv) for a 1-year-old to a 10-year-old, respectively (Richardson *et al.*, 1991; Harley and Robbins, 1992; Kendall and Smith, 2005). These calculations were made for a radon concentration of 200 Bq $m^{-3}$.

Radiation has been observed to increase the incidence of nervous system tumours in human populations and laboratory animals (NRC, 1990). The increase has been evident after exposure to therapeutic medical radiation in childhood at doses of less than 50 milligray (mGy) which may triple the risk of brain cancer (Pearce *et al.*, 2012). The tumours include malignant as well as benign growths of the brain, meninges and peripheral nerves. Although the dose–incidence relation is uncertain, the data indicate that the brain is relatively sensitive to the carcinogenic effects of radiation.

Exposures to radon through ingestion are both more complicated to estimate and more poorly understood (Canu *et al.*, 2011).

## 33.5 Human Evidence on Radon Exposure and Childhood Cancer

The carcinogenic effect of radon has not been established except for lung cancer. Numerous epidemiological cohort and case-control studies have been conducted to assess the risk of childhood cancer and residential radon exposure. Since 1998 eight case-control studies and two cohort studies regarding childhood cancer have been reported. Table 33.1 shows types of epidemiological studies investigating the association between radon and childhood cancer (Del Risco Kollerud, 2016).

The following describes some strengths and limitations in ten of the most recent studies regarding childhood cancer and radon and studies where radon exposure is measured at home.

- Lubin *et al.* (1998) was a case-control study including 505 leukaemia cases in children younger than 15 years in the USA. Radon detectors were placed in current and previous homes of children where they resided

**Table 33.1.** Characteristics of studies of indoor radon exposure and childhood leukaemia and cancer in the central nervous system.

| Author/ Country | Type of cancer | Design (n) Participation rates of cases/ controls[a] | Exposure assessment method | Mean radon concentration (Bq m$^{-3}$) | Exposure (Bq m$^{-3}$) | Risk estimate (95% CI)[b] |
|---|---|---|---|---|---|---|
| Lubin *et al.* (1998) USA | Acute lymphoblastic leukaemia | CC: 505/443 79/71% | One year measurements in homes Time weighted average radon concentration within the exposure assessment period for each subject | Subjects: 65.4 Controls: 79.1 | | RR |
| | | | | | < 37 | 1.00 |
| | | | | | 37–73 | 1.22 (0.8–1.9) |
| | | | | | 74–147 | 0.82 (0.5–1.4) |
| | | | | | > 148 | 1.02 (0.5–2.0) |
| Kaletsch *et al.* (1999) Germany | Acute leukaemia (AL) | CC: 82/209 40/34% | One year measurements in homes Longest-lived residence was used | Subjects: 26.4 Controls: 28.5 | | OR |
| | | | | | < 70 | 1.00 |
| | | | | | ≥ 70 | AL: 1.30 (0.32–5.33) |
| | Solid tumours (ST) | CC: 82/209 37/34% | | Subjects: 33.1 Controls: 28.5 | | OR |
| | | | | | < 70 | 1.00 |
| | | | | | ≥ 70 | ST: 2.61 (0.96–7.13) |
| | Central nervous system (CNS) tumours | CC: 41/209 | | | | OR |
| | | | | | < 70 | 1.00 |
| | | | | | ≥ 70 | CNS: 3.85 (1.26–11.81) |
| Steinbuch *et al.* (1999) USA/Canada | Acute myeloid leukaemia | CC: 173/254 27/33% of those eligible in interview study. 80/86% of those eligible in measurement study | One year radon measurements in homes | Subjects: 49.8 Controls: 56.0 | | OR |
| | | | | | < 37 | 1.00 |
| | | | | | 37–100 | 1.16 (0.7–1.8) |
| | | | | | > 100 | 1.12 (0.6–2.0) |
| Kohli *et al.* (2000) Sweden | Acute lymphoblastic leukaemia | Cohort study: 90 malignancies 53,146 children | Digital maps of radon risk areas | Classification of ground radon levels: Low risk: < 10,000 Normal risk: 10,000 to 50,000 High risk: ≥ 50,000 | | SMRs |
| | | | | | Low risk | 1.00 |
| | | | | | Normal risk | 4.64 (1.29–28.26) |
| | | | | | High risk | 5.67 (1.06–42.27) |
| | Other malignancies | | | | Low risk Normal risk High risk | No association between radon risk levels and any other malignancy was seen |

*Continued*

**Table 33.1.** Continued.

| Author/ Country | Type of cancer | Design (n) Participation rates of cases/ controls[a] | Exposure assessment method | Mean radon concentration (Bq m$^{-3}$) | Exposure (Bq m$^{-3}$) | Risk estimate (95% CI)[b] |
|---|---|---|---|---|---|---|
| Maged *et al.* (2000) Egypt | Acute lymphoblastic leukaemia | CC: 50/110 10%/– | Three months radon measurements in homes | Subjects: 75 Controls: 55 | | OR |
| | | | | | < 40 | 1.00 |
| | | | | | 40–60 | 4.64 (1.2–18) |
| | | | | | 60–90 | 7.42 (2–27.3) |
| | | | | | > 90 | 5.42 (1.3–21.1) |
| UK Childhood Cancer Study (2002) UK | Acute lymphoblastic leukaemia | CC: 805/1306 55/45% | Average radon concentrations in homes occupied by the children > 6 months prior to diagnosis | 24.0 | | OR |
| | | | | | < 25 | 1.00 |
| | | | | | 25–49 | 0.80 (0.64–0.99) |
| | | | | | 50–99 | 1.06 (0.79–1.44) |
| | | | | | 100–199 | 0.57 (0.29–1.12) |
| | | | | | ≥ 200 | 0.81 (0.28–2.36) |
| | CNS tumours | CC: 404/729 59/54% | | | | OR |
| | | | | | < 25 | 1.00 |
| | | | | | 25–49 | 1.06 (0.77–1.32) |
| | | | | | 50–99 | 0.82 (0.52–1.29) |
| | | | | | 100–199 | 0.59 (0.24–1.46) |
| | | | | | ≥ 200 | 1.13 (0.34–3.78) |
| Yoshinaga *et al.* (2005) Japan | Acute lymphoblastic leukaemia + Acute myeloid leukaemia | CC: 255/248 | Six months radon measurement in homes | Subjects: 17 Controls: 18 | | RR |
| | | | | | < 20 | 1.00 |
| | | | | | 20–49 | 1.00 (0.62–1.62) |
| | | | | | 50–99 | 1.57 (0.47–5.22) |
| | | | | | ≥ 100 | 2.05 (0.18–23.4) |
| Evrard *et al.* (2006) France | Acute lymphoblastic leukaemia | Ecological study 3239 | National radon measurement campaign during several years (13,240 radon measurements) | 85 | | SIR |
| | | | | | < 35 | 1.00 |
| | | | | | 35.1–44.0 | 1.02 (0.91–1.14) |
| | | | | | 44.2–63.3 | 1.10 (0.99–1.23) |
| | | | | | 63.5–88.2 | 1.10 (0.99–1.23) |
| | | | | | > 89.6 | 1.06 (0.95–1.19) |
| | | | | | Per 100 | 1.03 (0.95–1.11) |
| | Acute myeloid leukaemia | Ecological study 697 | | | | SIR |
| | | | | | < 35 | 1.00 |
| | | | | | 35.1–44.0 | 1.06 (0.83–1.34) |
| | | | | | 44.2–63.3 | 0.84 (0.65–1.08) |
| | | | | | 63.5–88.2 | 1.14 (0.90–1.43) |
| | | | | | > 89.6 | 1.23 (0.98–1.55) |
| | | | | | Per 100 | 1.24 (1.08–1.44) |

| | | | | | | |
|---|---|---|---|---|---|---|
| Raaschou-Nielsen (2008) Denmark | Acute lymphoblastic leukaemia | CC: 860/1720 99/98% | Radon concentrations predicted using a linear regression model. Cumulative radon exposure was used in the risk analyses | 48 | | RR |
| | | | | | 0–< 0.26 | 1.00 |
| | | | | | 0.26–< 0.89 | 1.21 (0.98–1.49) |
| | | | | | ≥ 0.89 | 1.63 (1.05–2.53) |
| | | | | | Per 1000 Bq $m^{-3}$-years | 1.56 (1.05–2.30) |
| | Acute non-lymphoblastic | 150/300 99/98% | | | | RR |
| | | | | | 0–< 0.26 | 1.00 |
| | | | | | 0.26–< 0.89 | 0.93 (0.50–1.71) |
| | | | | | ≥ 0.89 | 0.60 (0.25–1.41) |
| | | | | | Per 1000 Bq $m^{-3}$-years | 0.75 (0.34–1.62) |
| | Other leukaemia | 143/286 94/92% | | | | RR |
| | | | | | 0–< 0.26 | 1.00 |
| | | | | | 0.26–< 0.89 | 1.11 (0.62–2.00) |
| | | | | | ≥ 0.89 | 1.36 (0.48–3.83) |
| | | | | | Per 1000 Bq $m^{-3}$-years | 1.34 (0.53–3.40) |
| | CNS | 922/2766 | | | | RR |
| | | | | | 0–<0.26 | 1.00 |
| | | | | | 0.26–<0.89 | 0.92 (0.76–1.12) |
| | | | | | ≥ 0.89 | 1.11 (0.81–1.51) |
| | | | | | Per 1,000 Bq/$m^3$-years | 0.92 (0.69–1.22) |
| Kendall *et al*. (2013) | Lymphoid leukaemia | CC 7,267/9,571 95% | Models based on measurements Cumulative radon exposure from birth to diagnosis based on birth residence was used | 21.3 | | RR |
| | | | | | Per 1000 Bq/$m^3$-years | 1.24 (0.94–1.64) |
| | Acute myeloid leukaemia | 1,316/1,737 95% | | | | RR |
| | | | | | Per 1000 Bq/$m^3$-years | 0.72 (0.37–1.40) |

*Continued*

**Table 33.1.** Continued.

| Author/ Country | Type of cancer | Design (n) Participation rates of cases/ controls[a] | Exposure assessment method | Mean radon concentration (Bq $m^{-3}$) | Exposure (Bq $m^{-3}$) | Risk estimate (95% CI)[b] |
|---|---|---|---|---|---|---|
| | Other leukaemia | 475/604 | | | | RR |
| | | 95% | | | Per 1000 Bq/ $m^3$-years | 1.04 (0.41–2.61) |
| | Brain and CNS tumours | 6,585/8,997 | | | | RR |
| | | 95% | | | Per 1,000 Bq/ $m^3$-years | 1.15 (0.88–1.50) |
| Hauri *et al.* (2013) Switzerland | All leukaemias | Cohort study | A nationwide radon prediction model | 77.7 | | HR |
| | | 283 | Median radon concentration was used | | < 77.7 | 1.00 |
| | | 90% | | | 77.7–139.9 | 0.86 (0.67, 1.11) |
| | | | | | ≥ 139.9 | 0.95 (0.63, 1.43) |
| | | | | | Per 100 | 0.90 (0.68, 1.19) |
| | Acute lymphoblastic leukaemia | 225 | | | | HR |
| | | 89% | | | < 77.7 | 1.00 |
| | | | | | 77.7–139.9 | 0.83 (0.63, 1.11) |
| | | | | | ≥ 139.9 | 0.90 (0.56, 1.43) |
| | | | | | Per 100 | 0.86 (0.63, 1.19) |
| | CNS tumours | 258 | | | | HR |
| | | 91% | | | < 77.7 | 1.00 |
| | | | | | 77.7–139.9 | 0.95 (0.73, 1.23) |
| | | | | | ≥ 139.9 | 1.05 (0.68, 1.61) |
| | | | | | Per 100 | 1.19 (0.91, 1.57) |
| Del Risco Kollerud *et al.* (2014) Norway | Acute lymphoblastic leukaemia | Cohort | Model based on radon measurements and buffers | 91 | | HR |
| | | 98.7% | Mean radon exposure during relevant period was used | | < 50 | 1.00 |
| | | | | | 50–100 | 0.93 (0.70–1.23) |
| | | | | | > 100 | 0.95 (0.69–1.29) |
| | | | | | Per 100 | 0.97 (0.83–1.15) |
| | CNS tumours | | | | | HR |
| | | | | | < 50 | 1.00 |
| | | | | | 50–100 | 0.88 (0.68–1.14) |
| | | | | | > 100 | 1.15 (0.87–1.50) |
| | | | | | Per 100 | 1.13 (0.99–1.28) |

[a]CC, case-control study. [b]RR, relative risk; OR, odds ratio; SMRs, standardized mortality ratios;. SIR, standardized incidence ratio; HR, hazards ratio

for 6 months or longer. Children were included in the analyses if radon measurements covered 70% or more of the 5-year period prior to diagnosis for case subjects (or from birth for case subjects under 5 years of age) and the corresponding reference dates for control subjects. Mean radon concentration was 65.4 Bq $m^{-3}$ for cases and 79.1 Bq $m^{-3}$ for controls. Covariates were age at the time of diagnosis, total household income, birth order, birth weight, sex, type of residence, and the time-weighted average of magnetic field measurements. One of the limitations reported in the study was missing radon exposure data for the subjects. The participation rate was 79%. The study found no association between acute lymphoblastic leukaemia and residential radon.

- Kaletsch *et al.* (1999) was a population-based case-control study conducted in Germany. Cases were children younger than 15 years diagnosed with leukaemia and some common solid tumours (nephroblastoma, neuroblastoma, rhabdomyosarcoma and CNS tumours). A total of 82 leukaemia cases, 82 solid tumours and 209 controls were included. Long-term (1 year) radon measurements were performed in those homes where the children had been living for at least 1 year, with particular attention being paid to those rooms where they had stayed most of the time. One radon detector was placed in the child's bedroom, the second one in another room and the third one in the basement. The mean radon concentrations were 26.4 Bq $m^{-3}$ for leukaemia patients, 33.1 Bq $m^{-3}$ for tumour patients and 28.5 Bq $m^{-3}$ for controls. The models were adjusted by age, gender, urbanization, cohabitants and socio-economic level. The study had some limitations. The participation rate was 36%. The study had high probability of bias due to losses in radon measurements and due to selection of controls. The study found no evidence for an association between indoor radon and childhood leukaemia or CNS tumours.
- Steinbuch *et al.* (1999) was an interview-based case-control study including children younger than 18 years from the USA and Canada. The study included 173 cases of acute myeloid leukaemia and 254 controls. Two alpha-track radon detectors were used in the residence of the child at the time of the diagnosis. One criterion was that the child lived in the residence a minimum of 5 years prior to the cancer diagnosis. Detectors were placed for a period of 1 year. One detector was located in the child's bedroom and the other in a room in which the child spent more time during the day. The arithmetic means of the time-weighted radon concentrations for cases and controls were 49.8 Bq $m^{-3}$ and 56.0 Bq $m^{-3}$, respectively. The models were adjusted for maternal race, maternal education, family income and age of the child. Nineteen cases with Downs' syndrome were excluded from the study, due to their higher risk of leukaemia. Some limitations were pointed out in the study such as the small size of the study limiting the possibility for sub-analyses. There was no available residential history, making it difficult to calculate lifetime radon exposure. The participation rate was 27%. The study found no association between residential radon and risk of childhood acute myeloid leukaemia.
- Kohli *et al.* (2000) was an ecological study conducted in Östergötland, Sweden, following up 53,146 children. The analysis included 90 cancer cases and the residential history was available. It was based on radon exposure from radon maps including geology. The authors also discussed that radon exposure in the living area may be a better measure of the average exposure of the child than just measurements at home. The study found evidence that children born in and continuously living in areas classified as high and normal risk of background radiation from radon had higher incidence of acute lymphatic leukaemia compared with low-risk areas.
- Maged *et al.* (2000) was a case-control study including 50 children younger than 15 years diagnosed with acute lymphoblastic leukaemia and 110 controls. Radon exposure was measured with radon detectors with an exposure time of 3 months. The study did not report use of residential history or adjustment for possible confounders. One criterion for inclusion in the study was

that the children must have been living in their houses in Cairo since they were born. A total of 500 cases of lymphoblastic leukaemia were enrolled in the study. Only 50 out of 240 cases who met the eligibility criteria for selection agreed to participate. The participation rate was 10%. The mean radon concentration was markedly higher for cancer cases than for controls. The mean indoor radon concentration in the houses of patients was 75 Bq $m^{-3}$ while it was 55 Bq $m^{-3}$ in houses of controls. The study found an association between acute lymphoblastic leukaemia and radon exposure.

- UK Childhood Cancer Investigators (2002) was a national case-control study conducted in UK that included 951 cases of leukaemia and 404 tumours in the CNS. Indoor radon exposure was evaluated at the time of the diagnosis of cancer and for at least 6 months before. The participation rate was 52%. Socio-economic status and some factors associated with indoor radon concentration such as window glazing and central heating were taken into account. Average radon concentration in this study was 24 Bq $m^{-3}$. The study found no evidence of an association between indoor radon and the risk of any of the childhood cancers.
- Raasschou-Nielsen *et al.* (2008) was a case-control study that included 2400 cancer cases; 1153 cases were leukaemia and 922 cases were cancer in the CNS. They used residential history and calculated cumulated radon exposure for each child. Birth order, mother's age, traffic density and electromagnetic fields were adjusted for. The radon exposure was based on a mathematical model of radon exposure where the authors used 3120 radon measurements to calculate exposure in 21,338 dwellings distributed throughout Denmark. The model included several independent variables such as house type, floor, basement, geology, geographical region, type of house and building materials. The model explained 40% of the variability of radon. Average indoor radon concentration in the Denmark study was 48 Bq $m^{-3}$. The study found an association between indoor radon exposure and the risk of acute lymphatic leukaemia but not for other childhood cancers.
- Kendall *et al.* (2013) was a large population-based study in the UK including 36,793 cancer cases, 7267 of which were lymphoid leukaemia and 6585 were CNS tumours. Doses of radon exposure to red bone marrow were calculated. Radiation exposures were estimated for mother's residence at the child's birth from national databases, using the county-district mean for gamma rays and a predictive model based on measurements of radon in 460,000 dwellings. The radon results were grouped first by geological boundaries and then by 1 km grid squares (Miles *et al.*, 2007). Results were adjusted for socio-economic status. The study was lacking information on residential histories. The arithmetic mean of radon concentration was 21.3 Bq $m^{-3}$. The study found no association between childhood leukaemia or cancer in the CNS and indoor radon exposures.
- Hauri *et al.* (2013) was the first cohort study investigating indoor radon exposures and the risk of childhood cancer. It was conducted in Switzerland and included 1,332,944 children. In the study period a total of 997 children developed cancer, including 283 cases of leukaemia and 258 cases of cancer in the CNS. Radon exposure in this study was estimated for each child's home address using a nationwide radon prediction model for Switzerland based on 35,706 radon measurements. Relevant predictors in the model were tectonic units, building age, building type, soil texture, degree of urbanization and floor level. The model explained 20% overall of the variability of radon. The authors reported that random exposure misclassification was relatively high, due to complexity of the geology and the tectonic categories used in the prediction model (Hauri *et al.*, 2012). In the study, there was available data on residential history were available from a census in 2005. The study was lacking residential history for children in the cohort born before this year. The arithmetic mean of radon concentration for all cohort members was 77.7 Bq $m^{-3}$. The study found no evidence of any association between indoor radon exposure and the risk of childhood cancer.

- Del Risco Kollerud *et al.* (2014) was a cohort study conducted in Norway including 712,674 children. A total of 437 children developed leukaemia and 427 developed cancer in the CNS in the study period. The model was based on 41,515 indoor radon measurements to estimate radon exposure around each unmeasured building. The residence of every child was geo-coded and assigned a radon exposure. Geometric mean radon concentration for leukaemia was 89.1 Bq $m^{-3}$. For cancers in the CNS, the radon mean was 100.2 Bq $m^{-3}$. The mean radon concentration for the whole cohort was 91 Bq $m^{-3}$. Mean radon concentration was the highest reported in analytical studies. Mean radon concentration was similar across cases and the whole cohort. In the study, data on residential history were available. No association was found for childhood leukaemia. An elevated non-significant risk for cancer in the central nervous system was observed.

In summary, five case-control studies (Lubin *et al.*, 1998; Kaletsch *et al.*, 1999; Steinbuch *et al.*, 1999; UK Childhood Cancer Study Investigators, 2002; Kendall *et al.*, 2013) and two cohort studies (Hauri *et al.*, 2013; Del Risco Kollerud *et al.*, 2014) found no association between radon exposure and the risk of childhood leukaemia and cancer in the CNS. Kohli *et al.* (2000), Raasschou-Nielsen *et al.* (2008) and Maged *et al.* (2000) found an association between childhood leukaemia and radon exposure (Del Risco Kollerud, 2016).

In the studies of Lubin *et al.* (1998), Kaletsch *et al.* (1999), Steinbuch *et al.* (1999) and UK Childhood Cancer Study Investigators (2002), the radon exposure was assessed through radon gas measurements in the dwelling environment. This allowed exposure to be reconstructed taking into the account the years and time spent by each child in the various rooms of the occupied dwellings. Even when analysis was restricted to children who had spent their whole lives in homes with on-site radon measurements, no significant association was observed in any of these studies. Two of the three studies using radon prediction models and individual data (Kendall *et al.*, 2013; Hauri *et al.*, 2013; Del Risco Kollerud *et al.*, 2014) found no association between radon exposure and the risk of leukaemia and cancer in the CNS. Incorporating prediction models allows inclusion of more children, making the risk for selection bias less likely. Radon mean concentrations varied across the studies. The lowest mean radon concentration was found in the UK, varying between 21.3 Bq $m^{-3}$ and 24.0 Bq $m^{-3}$, and the highest concentration was found in Del Risco Kollerud's study, with a mean radon concentration 91Bq $m^{-3}$.

Kohli *et al.* (2000), Raasschou-Nielsen *et al.* (2008) and Maged *et al.* (2000) found an association between childhood leukaemia and radon exposure. The study of Kohli *et al.* (2000) was an ecological study in which data on radon exposure had been aggregated on the basis of geographical units and there was no information on exposure at an individual level. Since ecological studies are subject to important methodological limitations, such as 'ecological bias', and since other previous studies have analysed radon exposure at individual levels, the next section will focus on the studies of Raasschou-Nielsen *et al.* (2008) and Maged *et al.* (2000).

The study of Raasschou-Nielsen *et al.* (2008) conducted in Denmark had several strengths, such as a large number of cancers with 1153 cases of leukaemia and 922 cases of cancer in the CNS. It was a register-based study, which makes the possibility of selection bias less likely. The authors had residential history and could adjust for several risk factors, such as birth order, mother's age, traffic density and electromagnetic fields. The study did not have information on socio-economic status. The Danish study predicted residential radon concentration calculated from a model based on a previous measurement programme and a number of explanatory variables such as house type and geology. These model predictions of radon concentrations in homes avoided the bias potentially associated with limited participation in a measurement programme, which had been a major problem in some of the previous studies mentioned; but, given the variation in domestic radon concentrations, the model estimates may inevitably have introduced uncertainties in the risk estimates. The errors associated with the radon concentration predictions were thought to be largely Berkson-type error, so that appreciable bias in the trend estimate would not be expected, but taking into account the radon variation between geological units, the use of estimated parameters in the Danish study could have introduced classical error, which could induce bias in the regression. One example pointed out by Miles and Appleton (2005) is that if a house is allocated to the wrong geological unit, its radon result distorts the estimated proportion above the threshold for that unit.

The other case-control study reporting a positive association was Maged *et al.* (2000). They included 50 cases and 110 controls in Cairo. Only 10% of the children who met criteria for inclusion could be included in the study. The cases were selected from hospital registers in Cairo. The study did not report use of residential history or adjusting for possible individual confounders. Radon measurements were taken at homes of the children. The mean radon concentration was markedly higher for cancer-case subjects than for controls. The mean indoor radon concentration in houses of patients was 75 Bq $m^{-3}$ while it was 55 Bq $m^{-3}$ in houses of controls. Cases not diagnosed in the Cairo hospital may be also under-reported in the hospital registers.

## 33.6 Other Uncertainties in Assessing Health Risks from Radon

In epidemiological studies there may be differences between exposed and unexposed groups in some unmeasured factors that affect the risk of childhood leukaemia and that may be correlated with exposure, i.e. confounding. Raaschou-Nielsen *et al.* (2004) in a previous study had found an association between area-based socio-economic status and leukaemia. The municipalities they observed with low income were located in rural areas and on small islands. Municipalities on the coast were over-represented among the low-income municipalities. Some explanations of the association observed were that community behaviour and factors acting early in life might be important. Moreover, for the geographical area where the low-income municipalities were located, Greaves (2006) put forward a ‘delayed infection’ hypothesis, suggesting that exposure to common infections agents could trigger B-precursor lymphoid leukaemia in predisposed children and that the predisposition was more likely to occur in children who were relatively sheltered from infection at infancy. Not taking into account the socio-economic status of the children when matching with controls or not adjusting for socio-economic status may cause an overestimate of the true association (positive confounding). Other birth risk factors such as birth weight and congenital malformations were not included in the Danish study. Children with congenital malformations have a higher risk of developing cancer, especially leukaemia. Down’s syndrome is a common congenital anomaly, and children with Down’s syndrome have a substantially higher risk of leukaemia (Mezei *et al.*, 2014). Birth weight is another risk factor associated with the risk of childhood cancer. A register-based study from the Nordics countries, including Denmark, showed an increased risk of 26% for acute lymphoblastic leukaemia per 1 kg increase in birth weight (Hjalgrim *et al.*, 2004). Another consideration is the possibility that the Danish study was underpowered, implying that the significant association found is likely to be due to chance. It also could affect other studies. However, Kendall *et al.* (2013), in a large population-based study that included 7267 cases of lymphoid leukaemia, found no association between the risk of leukaemia and radon exposure. Homes were grouped by grid squares and including factors regarding geology. About 400,000 radon measures were included.

Maged *et al.* (2000) reported a positive association. However, the study had a low participation rate. Low participation rates imply a risk of selection bias, in particular if the participation rates of cases and control differ, which can affect the risk estimates in a case-control study in either direction. Another possible bias reported in the study was the difference in mean radon concentrations between cases and controls. The mean radon concentration was markedly higher for cancer-case subjects than for controls. Cases not diagnosed in the Cairo hospital may also be under-reported in the hospital registers.

## 33.7 Conclusions

Many epidemiological studies have attempted to determine whether exposure of children to radon increases the risk of developing childhood cancer. However, these studies are inherently uncertain, because they are observational in nature rather than experimental. Misclassification of exposure can lead to a substantial overestimation or underestimation of the association of radon exposure and the risk of leukaemia and cancer in the CNS in children. Similarly, there is usually no information available regarding other risk factors that might influence the risk of developing cancer in children. Most of the studies published in recent decades indicate a negative association between radon exposure and the risk of leukaemia and cancer in the CNS in children, but, as showm above, methodological limitations in all previous studies makes it impossible to be conclusive in the risk assessment. Other reasons that might make it difficult to detect associations include the long asymptomatic period between exposure to radon and development of cancer, as well the low incidence of the disease among children. Further, there remains a lack of understanding of the cellular and molecular response of tissues to radiation from radon, identification of genes that increase risk for radiation carcinogenesis and the gene–environment interaction.

## References

Appleton, J.D. (2007) Radon: sources, health risk, and hazard mapping. *Ambio* 36, 85–89.

Appleton, J.D. and Miles, J.C. (2010) A statistical evaluation of the geogenic controls on indoor radon concentrations and radon risk. *Journal of Environmental Radioactivity* 101, 799–803.

Canu, I.G., Laurent, O., Pires, N., Laurier, D. and Dublineau, I. (2011) Health effects of naturally radioactive water ingestion: the need for enhanced studies. *Environment Health Perspectives* 119, 1676–1680.

Del Risco Kollerud, R. (2016) Radon exposure and socioeconomic status in relation to childhood leukemia and cancer in the central nervous system. Doctoral dissertation. University of Oslo, Oslo, Norway.

Del Risco Kollerud, R., Blaasaas, K.G. and Claussen, B. (2014) Risk of leukaemia or cancer in the central nervous system among children living in an area with high indoor radon concentrations: results from a cohort study in Norway. *British Journal of Cancer* 111, 1413–1420.

Evrard, A.S., Hémon, D., Billon, S., Laurier, D., Jougla, E., Tirmarche, M. and Clavel, J. (2006) Childhood leukemia incidence and exposure to indoor radon, terrestrial and cosmic gamma radiation. *Health Physics* 90, 569–579.

Folley, J.H., Borges, W. and Yamawaki, T. (1952) Incidence of leukemia in survivors of the atomic bomb in Hiroshima and Nagasaki, Japan. *The American Journal of Medicine* 13, 311–321.

Grant, E.J., Furukawa, K., Sakata, R., Sugiyama, H., Sadakane, A. *et al.* (2015) Risk of death among children of atomic bomb survivors after 62 years of follow-up: a cohort study. *Lancet Oncology* 16, 1316–1323.

Greaves, M. (2006) Infection, immune responses and the aetiology of childhood leukaemia. *Nature Reviews. Cancer* 6, 193–203.

Hall, E.J. (ed.) (2000) *Radiobiology for the Radiologist. Effects of Radiation on the Embryo and Fetus*, 5th edn. Lippincott Williams & Wilkins, Philadelphia, pp. 168–180.

Harley, N.H. and Robbins, E.S. (1992) 222Rn alpha dose to organs other than lung cancer. *Radiation Protection Dosimetry* 45, 619–622.

Harley, N.H. and Robbins, E.S. (2009) Radon and leukemia in the Danish study: another source of dose. *Health Physics* 97, 343–347.

Hauri, D.D., Huss, A., Zimmermann, F., Kuehni, C.E. and Röösli, M.M. (2012) A prediction model for assessing residential radon concentration in Switzerland. *Journal of Environment Radioactivity* 112, 83–89.

Hauri, D., Spycher, B., Huss, A., Zimmerman, F., Grotzer, M. *et al.* (2013) Domestic radon exposure and risk of childhood cancer: a prospective census-based cohort study. *Environment Health Perspectives* 121, 1239–1244.

Hjalgrim, L.L., Rostgaard, K., Hjalgrim, H., Westergaard, T., Thomassen, H. *et al.* (2004) Birth weight and risk for childhood leukemia in Denmark, Sweden, Norway, and Iceland. *Journal of the National Cancer Institute* 96, 1549–1556.

IARC (2009) *A review of human carcinogens. Part D: Radiation*. Special Report: Policy. International Agency for Research on Cancer, Lyon, France, pp. 241–249. *The Lancet Oncology* 10, 751–752.

Kaletsch, U., Kaatsch, P., Meinert, R., Schüz, J., Czarwinski, R. and Michaelis, J. (1999) Childhood cancer and residential radon exposure – results of a population-based case-control study in Lower Saxony (Germany). *Radiation Environment Biophysics* 38, 211–215.

Kamiya, K., Ozasa, K., Akiba, S., Niwa, O., Kodama, K. *et al.* (2015) Long-term effects of radiation exposure on health. *The Lancet* 386, 469–478.

Kendall, G.M. and Smith, T.J. (2005) Doses from radon and its decay products to children. *Journal of Radiological Protection* 25, 241–256.
Kendall, G.M., Little, M.P., Wakeford, R., Bunch, K.J., Miles, J.C. *et al.* (2013) A record-based case-control study of natural background radiation and the incidence of childhood leukaemia and other cancers in Great Britain during 1980–2006. *Leukemia* 27, 3–9.
Kohli, S., Noorlind B.H. and Löfman, O. (2000) Childhood leukaemia in areas with different radon levels: a spatial and temporal analysis using GIS. *Journal of Epidemiology and Community Health* 54, 822–826.
Lubin, J.H., Linet, M.S., Boice, J.D. Jr, Buckley, J., Conrath, S.M. *et al.* (1998) Case-control study of childhood acute lymphoblastic leukemia and residential radon exposure. *Journal of the National Cancer Institute* 18, 294–300.
Maged, A.F., Mokhtarb, G.M., Tobguic, M.M., Gabbrc, A.A., Attiad, N.I. and Abu Shadyc, M.M. (2000) Domestic radon concentration and childhood cancer study in Cairo, Egypt. *Journal of Environmental Science and Health, Part C: Environmental Carcinogenesis and Ecotoxicology Reviews* 18, 153–170.
Mészáros, G., Bognár, G. and Köteles, G.J. (2004) Long-term persistence of chromosome aberrations in uranium miners. *Journal of Occupational Health* 46, 310–315.
Mezei, G., Sudan, M., Izraeli, S. and Kheifets, L. (2014) Epidemiology of childhood leukemia in the presence and absence of Down syndrome. *Cancer Epidemiology* 38, 479–489.
Miles, J.C. and Appleton, J.D. (2005) Mapping variation in radon potential both between and within geological units. *Journal of Radiology Protection* 25, 257–276.
Miles, J.C., Appleton, J.D., Rees, D.M., Green, B.M., Adlam, K.A. and Myers, A.H. (2007) *Indicative Atlas of Radon in England and Wales*. HPA-RPD-033. Health Protection Agency, Didcot, UK.
Narayanan, P.K., Goodwin, E.H. and Lehnert, B.E. (1997) Alpha particles initiate biological production of superoxide anions and hydrogen peroxide in human cells. *Cancer Research* 57, 3963–3971.
NRC (1990) *Health Effects of Exposure to Low Levels of Ionizing Radiation: BEIR V*. National Research Council (US) Committee on the Biological Effects of Ionizing Radiation (BEIR V). National Academies Press, Washington, DC, pp. 310–313.
Pearce, M.S., Salotti, J.A. and Little, M.P. (2012) Radiation exposure from CT scans in childhood and subsequent risk of leukaemia and brain tumours: a retrospective cohort study. *The Lancet* 380, 499–505.
Raaschou-Nielsen, O., Obel, J., Dalton, S., Tjønneland, A. and Hansen, J. (2004) Socioeconomic status and risk of childhood leukaemia in Denmark. *Scandinavian Journal of Public Health* 32, 279–286.
Raaschou-Nielsen, O., Andersen, C.E., Andersen, H.E., Gravesen, P. and Schüz, J. (2008) Domestic radon and childhood cancer in Denmark. *Epidemiology* 19, 536–543.
Richardson, R.B., Eatough, J.P. and Henshaw, D.L. (1991) Dose to red bone marrow from natural radon and thoron exposure. *British Journal of Radiology* 64, 608–624.
Richardson, D., Sugiyama, H., Nishi, N., Sakata, R., Shimizu, Y. *et al.* (2009) Ionizing radiation and leukemia mortality among Japanese atomic bomb survivors, 1950–2000. *Radiation Research* 172, 368–382.
Robertson, A., Allen, J., Laney, R. and Curnow, A. (2013) The cellular and molecular carcinogenic effects of radon exposure: a review. *International Journal of Molecular Sciences* 14, 14024–1463.
Smerhovsky, Z., Landa, K., Rössner, P., Brabec, M., Zudova, Z. *et al.* (2001) Risk of cancer in an occupationally exposed cohort with increased level of chromosomal aberrations. *Environmental Health Perspectives* 109, 41–45.
Smerhovsky, Z., Landa, K., Rössner, P., Juzova, D., Brabec, M. *et al.* (2002) Increased risk of cancer in radon-exposed miners with elevated frequency of chromosomal aberrations. *Mutation Research* 514, 165–176.
Steinbuch, M., Weinberg, C.R., Buckley, J.D., Robison, L.L. and Sandler, D.P. (1999) Indoor residential radon exposure and risk of childhood acute myeloid leukaemia. *British Journal of Cancer* 81, 900–906.
Sundal, A.V., Jensen, C.L., Anestad, K. and Strand, T. (2007) Anomalously high radon concentrations in dwellings located on permeable glacial sediments. *Journal of Radiological Protection* 27, 287–298.
UK Childhood Cancer Study Investigators (2002) The United Kingdom Childhood Cancer Study of exposure to domestic sources of ionising radiation: 1: radon gas. *British Journal of Cancer* 86, 1721–1726.
Yoshinaga, S., Tokonami, S., Akiba, S., Nitta, H. and Kabuto, M. (2005) Case-control study of residential radon and childhood leukemia in Japan: results from preliminary analyses. *International Congress Series* 1276, 233–235.

# 34 Fukushima Nuclear Accident: Potential Health Effects Inferred from Butterfly and Human Cases

**J.M. Otaki***

*Department of Chemistry, Biology and Marine Science, Faculty of Science, University of the Ryukyus, Okinawa, Japan*

## 34.1 Abstract

Fukushima Dai-ichi Nuclear Power Plant released massive amounts of radioactive materials in 2011, resulting in environmental pollution in the Tohoku-Kanto districts of Japan. Biological effects of this nuclear accident have been examined using the pale grass blue butterfly, *Zizeeria maha*, as an environmental indicator species that is amenable to both field and laboratory studies. In this chapter, on the basis of the butterfly results, low-dose effects are discussed from biological perspectives that are not dependent on the conventional dose-centric view. Advantages of using this butterfly are that it can monitor the human-living environment and thus help to infer effects of environmental pollution on human health. Pronounced ingestional toxicity of contaminated diets at low radiation levels on this butterfly likely originated not only from the 'direct' ionization mechanism in the butterfly, but also from multiple 'indirect' mechanisms, including physical and immunological damage from insoluble particles and naturally occurring dust to which radioactive materials are attached and nutritional deficiencies or plant-derived chemical poisoning in the polluted host-plant leaves. These unconventional indirect effects cannot be readily predicted by dosimetry, because most of them are field effects mediated by field substances and species interactions. To illustrate a possible unconventional effect on humans, a case is presented here of patient 'C.U.', who developed systemic oedema and was medically diagnosed as nephrotic syndrome. Retrospectively, she was likely sensitized by nuclear dust in her repeated visits to Fukushima; and then her exposures in Okinawa to minute amounts of the dust from Fukushima preceded acute increases of her body weight, an objective indicator for disease progression, suggesting an essential role of the contaminants for her health problem at very low doses. Taken together, butterfly and human cases suggest that pollutants released by the Fukushima nuclear accident caused harmful impacts on organisms through unconventional indirect mechanisms.

## 34.2 Introduction

Fukushima Dai-ichi Nuclear Power Plant (FDNPP) released a massive amount of radionuclides to the environment in March 2011 (Chino *et al.*, 2011; Kinoshita *et al.*, 2011; Hirose, 2012; Torii *et al.*, 2013). Since then, a large number of

* E-mail address: otaki@sci.u-ryukyu.ac.jp

 *A Handbook of Environmental Toxicology: Human Disorders and Ecotoxicology*. J.P.F. D'Mello (ed.)

biological species, including humans, have lived with the radioactive pollutants. After the initial high-dose acute exposure to short-lived radionuclides such as iodine $^{131}I$ (half-life: 8 days) and tellurium $^{132}Te$ (half-life: 3 days) immediately after the disaster, organisms have since experienced low-dose chronic exposure to the residual radionuclides in the environment, i.e. mainly two caesium species: $^{134}Cs$ (half-life: 2 years) and $^{137}Cs$ (half-life: 30 years).

The human health effects of the nuclear pollution from the FDNPP are probably the most important concern not only for people living in Fukushima and other polluted areas but also for people throughout Japan and in other countries worldwide. However, whether human health effects of the pollution are significant or negligible remains controversial, according to the report by the United Nations Scientific Committee on the Effects of Atomic Radiation (UNSCEAR, 2013). For example, despite intense epidemiological studies, an increase of thyroid cancer in children after the Fukushima nuclear accident is interpreted either as a simple over-diagnosis or as a true incidence, depending on perspectives of researchers (Iwaku *et al.*, 2014; Nagataki and Takamura, 2014; Shibuya *et al.*, 2014; Watanobe *et al.*, 2014; Aliyu *et al.*, 2015; Hayashida *et al.*, 2015; Tsuda *et al.*, 2016; Katanoda *et al.*, 2016). This controversy illustrates a difficulty in resolving the issue of human health effects based on epidemiological studies in a relatively short time. Similarly, despite many anecdotal instances of illness that could be associated with the Fukushima nuclear accident, to my knowledge there appears to be no clinical case report that describes Fukushima-induced symptoms in humans.

In contrast, observational records on various wild organisms that suggest the adverse effects from the Fukushima nuclear accident have accumulated in recent years, as they were compiled and evaluated in the UNSCEAR reports (UNSCEAR, 2015, 2016). Ecological studies on abundance of animals, including birds and various insects, demonstrated that birds and butterflies decreased in number (Møller *et al.*, 2012, 2013; Mousseau and Møller, 2014). Similarly, one of the pollinator insects, the carpenter bee, decreased in number, though other pollinator insects did not change (Yoshioka *et al.*, 2015). Reports on other animals include morphological abnormalities of gall-forming aphids (Akimoto, 2014), change in abundance of barn swallows (Bonisoli-Alquati *et al.*, 2015), reproductive difficulties in goshawk (Murase *et al.*, 2015) and a decrease in number of intertidal biota (Horiguchi *et al.*, 2016). Adverse effects on plants have been reported in rice leaves (Hayashi *et al.*, 2014), fir trees (Watanabe *et al.*, 2015) and red pine (Yoschenko *et al.*, 2016). From an ecological perspective, pollutant-mediated damage on producers (i.e. plants) may have serious effects on consumers (i.e. animals) through a food web.

Studies on mammals have also been reported. A report on low blood-cell counts in Japanese monkeys (Ochiai *et al.*, 2014) is especially important in considering effects on humans. However, effects on other mammals have been controversial. No effect was detected in testes in bulls (Yamashiro *et al.*, 2013), wild mice (Okano *et al.*, 2016) and in boar and inobuta (Yamashiro *et al.*, 2015). A dosimetric analysis concluded that there should be no adverse effects based on the estimated total dose rates and the derived consideration reference levels (DCRLs) recommended by the International Commission on Radiological Protection (ICRP) (Fuma *et al.*, 2017). However, stress effects in plasma proteins have been detected in cattle (Urushihara *et al.*, 2016). DNA damage in bovine lymphocytes may be induced by radiation exposure (Nakamura *et al.*, 2017). Furthermore, chromosomal aberrations were detected in wild mice (Kubota *et al.*, 2015; Kawagoshi *et al.*, 2017).

It is important to understand that 'no detection' of biological effects in a study does not necessarily mean 'no effects' on organisms. Without a careful and thorough research plan, it may be difficult to detect adverse effects. To prove a negative statement, one should thoroughly check all the possibilities to exclude unintentional ignorance of important possibilities. Of course, no detection does not mean safe for humans or other organisms. Discrepancy between dosimetric evaluations and actual field observations is not surprising, because the field effects are not really predictable by dosimetric analysis. There are complex and multiple pathways that lead to toxic effects in the wild, in addition to the direct ionizing effects of radiation (Otaki, 2016, 2018; Otaki and Taira, 2017).

Many of the wildlife observations above suggest a causal contribution of the pollutants from the FDNPP. However, experimental validation has been lacking in these studies. The earliest and the most comprehensive report is the study on the pale grass blue butterfly, *Zizeeria maha*, published on 9 August 2012 (Hiyama *et al.*, 2012), which showed accumulated adverse effects over generations caused by the Fukushima nuclear accident through field work and laboratory experiments. This study has been considered a milestone in this field, but it once generated considerable controversy worldwide among scientists, politicians and the general public (Hiyama *et al.*, 2013; Møller and Mousseau, 2013; Otaki, 2015). This study also suggested many important questions that needed to be solved. Since then, a series of studies on the effects of the Fukushima nuclear accident on this butterfly have been performed and published (Hiyama *et al.*, 2013, 2015, 2017a, b; Iwata *et al.*, 2013; Nohara *et al.*, 2014a, b, 2017; Otaki, 2016, 2018; Otaki and Taira, 2017; Taira *et al.*, 2014, 2015a, b).

In this chapter, the Fukushima nuclear accident is discussed in terms of health hazards. The chapter is divided into three parts. First, it discusses the importance of research on the butterfly (and other organisms) to understand human health effects (Section 34.3). Second, it discusses low-dose effects from biological perspectives that are not dependent on the conventional dose-centric view (Section 34.4), stressing unconventional 'indirect' effects of the pollutants released from the FDNPP. Third, it presents a human case of nephrotic syndrome that was likely caused by immunological sensitization and subsequent exposures to vanishing amount of 'dust' from the FDNPP (Section 34.5). This is an important example of a devastating indirect effect in humans.

## 34.3 Butterfly Model: Relevance to Humans

Aside from the importance of epidemiological studies, a different approach to human health effects is necessary, one of which is the use of environmental indicator species. Although simultaneous use of multiple indicator species from different taxonomic groups is favourable, it is often difficult in practice. When only a single (or a few) species is to be chosen, the pale grass blue butterfly is an ideal system, in that it is associated with (almost dependent on) the human-living environment so that the butterfly reflects human-living environmental health (Hiyama *et al.*, 2013; Taira *et al.*, 2014, 2015a; Otaki, 2016). Additionally, in the laboratory, relatively fast and precise experiments can be performed using this butterfly (Hiyama *et al.*, 2010; Otaki *et al.*, 2010). Other points of advantage of this butterfly have been discussed elsewhere (Hiyama *et al.*, 2013; Taira *et al.*, 2014, 2015a; Otaki, 2016, 2018).

It should be noted that an approach to humans with a use of non-human model organisms is not new to biomedical sciences. Rather, it is common practice to use the fruit fly, *Drosophila melanogaster*, to infer molecular mechanisms of human diseases. The fruit fly is used not because it is most similar to humans among invertebrates but because it is practically useful for experimental manipulations.

As discussed in Taira *et al.* (2015a) and Otaki (2016), radiation effects are molecular events, mediated by ionization of water and other molecules. DNA may be damaged 'directly' by radiation or 'indirectly' through other ionized molecules, mostly water (note that the usage of 'direct' and 'indirect' here is different from terminology discussed below in this chapter). The ionizing mechanisms at the molecular level are universal in all organisms, including humans and the butterfly. In this sense, butterfly data are applicable to humans. However, efficiency of DNA repair could vary among species. It could also vary in a given population. Human sensitivity variation is one of the most important points to be considered when human health effects of the Fukushima nuclear accident are discussed (Fukunaga and Yokoya, 2016; Otaki, 2016). A case of extremely high sensitivity is presented in this chapter (Section 34.5). The unconventional 'indirect' effects that are discussed in this chapter are also through universal molecular events, despite the fact that most of them are mediated by field substances or species. Thus, they are likely applicable to humans, though precise mechanistic understanding of the molecular events is yet to come.

In contrast, manifestation of the effects (i.e. phenotypic effects) could be very different among species. In butterflies, morphological

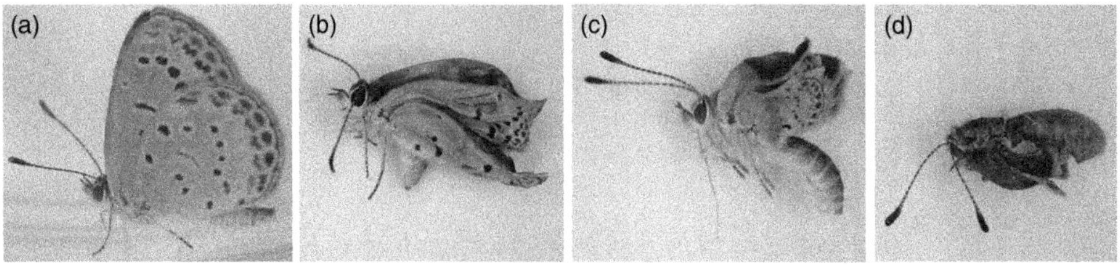


**Fig. 34.1.** Effects of ingesting the contaminated leaves on the pale grass blue butterfly. (a) Normal adult individual of the pale grass blue butterfly. (b, c) Abnormal adult individuals who ate contaminated leaves from Fukushima City, Fukushima Prefecture, at the larval stage. (d) An abnormal adult individual who ate contaminated leaves from Iitate Village, Fukushima Prefecture, at the larval stage. Panels (b–d) are also found in Nohara *et al.* (2014).

abnormalities such as leg and wing deformation were relatively frequent, but no counterpart of this phenotypic effects could be identified in humans. Such phenotypic effects (i.e. disease manifestations) at the organismal level cannot be inferred from butterfly data.

A good example is the 'internal exposure experiments', in which effects of ingestion of radioactively contaminated foods have been evaluated in the butterfly (Hiyama *et al.*, 2012; Nohara *et al.*, 2014a,b; Taira *et al.*, 2015a). A basic experimental strategy is to collect the contaminated food plant from Fukushima and then to give it to larvae from Okinawa, the least polluted locality in Japan. When non-contaminated leaves were given to larvae, normal individuals emerged (Fig. 34.1a). However, when contaminated leaves were given, morphologically abnormal adults emerged (Fig. 34.1b–d), together with larval and pupal death of many individuals. Important hints for human health effects can be obtained from these results, although pathological outcomes in humans cannot readily be inferred.

## 34.4 Unconventional Indirect Effects

Radiation scientists who are engaged in dosimetric analysis appear to believe that effective doses (or dose rates) can predict biological effects based on a mathematical model, such as the linear non-threshold (LNT) model, in accordance with the ICRP recommendations (e.g. Strand *et al.*, 2014; Fuma *et al.*, 2017). This dosimetric approach is based on the assumption that ionizing radiation targets DNA and that the degrees of DNA damage are linearly reflected in biological consequences. This means that biological effects are predictable by effective doses. Although this approach is widely accepted and utilized for assessment of biological impacts of nuclear disasters, it entirely ignores complexity of biological and ecological responses to various known and unknown materials released from nuclear reactors.

The past two decades have experienced a surge of 'non-targeted effects' in contrast to the conventional targeted effects (UNSCEAR, 2012; Mothersill and Seymour, 2009, 2014). The non-targeted effects include bystander effects, genomic instability, adaptive responses and other modes and they are probably caused by reactive oxygen species produced by irradiation (UNSCEAR, 2012; Mothersill and Seymour, 2009, 2014). The important points of the non-targeted effects are that they are not readily predictable by doses and that many of them are latent. The non-targeted effects such as genomic instability might have played significant roles in an increase of morphological abnormalities of the butterfly in the autumn of 2012 (Hiyama *et al.*, 2012, 2015).

However, this is not the whole story. It is known that controlled laboratory effects and field effects are very different, even under similar levels of radiation exposure; the field cases in Chernobyl are eight times more sensitive than the laboratory-controlled external irradiation cases (Garnier-Laplace *et al.*, 2013). This discrepancy may partly originate from the involvement of the non-targeted effects in the field. Additionally, the high sensitivity of the field cases may originate from confounding factors that were not well identified. However, the field effects cannot be explained solely by the non-targeted effects and confounding factors, as discussed below.

For convenience, both the conventional targeted effects and the non-targeted effect are termed here as the 'direct' (or 'primary') effects. It is understood that laboratory-based controlled irradiation experiments, irrespective of high or low doses, mostly examine the direct effects of ionizing radiation. In contrast, other possible unconventional effects of nuclear pollution are collectively called the 'indirect' (or 'secondary') effects (Otaki, 2016; Otaki and Taira, 2017). Many indirect effects of nuclear pollution would exist in the real world. They include effects of possible non-radioactive by-products released from a reactor, and they also include naturally occurring non-radioactive materials that are 'activated' by radioactive materials from a reactor. The indirect effects also include ecological interactions through a food web, as discussed in Bréchignac *et al.* (2016). Thus, the indirect effects are almost identical to 'field-specific effects' (or simply, 'field effects'), because they would not be observed in controlled laboratory experiments without consideration for detecting such effects.

As discussed in Otaki (2016, 2018), there are many possible modes of the indirect effects, but here three modes are briefly discussed. First, what is released from the FDNPP is a plume of materials that cause particulate air pollution, whether radioactive or not, because the released materials were dispersed as atmospheric aerosol (Adachi *et al.*, 2013; Miyamoto *et al.*, 2014; Salbu and Lind, 2016; Kaltofen and Gundersen, 2017). There is no question that atmospheric aerosol causes respiratory and cardiovascular diseases in humans (Seaton *et al.*, 1995; Kappos *et al.*, 2004; Utell and Frampton, 2009; Shiraiwa *et al.*, 2012). Second, synergistic effects with other environmental factors including climate conditions and chemical stressors may exist in the wild, whereas climate conditions in the laboratory are usually constant and additional stressors are not usually given. Stressors in the wild are often unnoticed by researchers, but synergistic effects of radiation exposure with other stressors are not rare (Borek *et al.*, 1986; Leenhouts and Chadwick, 1978; Manti and D'Arco, 2010; Mothersill *et al.*, 2007; Mothersill and Seymour, 2009, 2014). Third, as recommended in Bréchignac *et al.* (2016), studies focusing on the population level are especially important. When one examines an ecological system as a whole, one may be able to discover radiation effects that cannot be discovered by a species-based approach, because a given species has many ecological interactions with other species.

As for the second mode, a similar discrepancy between the laboratory and field results has been recognized in phenotypic plasticity studies. In an authoritative textbook, Gilbert and Epel (2015) stated the following: 'Phenotypic plasticity means that animals in the wild may develop differently than those in the laboratory' and 'This has important consequences when we apply knowledge gained in the laboratory to a field science such as conservation biology.' An example given in the textbook is that some frog tadpoles are up to 46 times more sensitive to pesticides in the presence of predators that release chemicals in the wild than in the laboratory (Relyea, 2003, 2004). A conclusion was that 'ignoring the relevant ecology can cause incorrect estimates of a pesticide's lethality in nature' (Relyea, 2003). The same principle can be applied to radioactive materials from nuclear accidents.

As for the third mode, the ecological system inhabited by the pale grass blue butterfly is relatively simple due to its monophagous nature (Otaki, 2016). Thus, this butterfly and its associated ecosystem may serve as a 'model ecosystem' to investigate both population dynamics and environmental influence through the ecological food web after the Fukushima disaster.

In any ecosystem, the food web is of prime importance, and it is often discussed from a viewpoint of bioaccumulation of radioactive materials or organic materials in high-order consumers. However, it appears that in the case of the model ecosystem of this butterfly, the quality of its host plant, *Oxalis corniculata*, may be important, which is determined by the quality of soil and air. When soil is contaminated with radioactive materials and other pollutants such as agrichemicals, the quality of the host plant leaves will decrease. Similarly, air pollutants that cover the surface of leaves, whether radioactive or not, may change physiological functions of leaves. Thus the quality of soil and air will affect the health of the butterfly.

An important aspect of a decrease of plant quality for larvae may have two aspects: a decrease of certain favourable chemicals (e.g. essential nutrients) in leaves or an increase of unfavourable chemicals (e.g. reactive oxygen species and defence chemicals) in leaves. For the former, a lack of an essential vitamin in leaves may kill butterfly larvae, because larvae are dependent on the vitamins that are supplied through ingestion of leaves. The latter possibility of a decrease of plant quality for larvae is that plants are stressed with radioactive materials to produce reactive oxygen species, defence chemicals or something harmful that may kill larvae. Reactive oxygen species are known to be produced by various abiotic stresses, and defence chemicals are known to be induced by insect bites in many plants (Taiz *et al.*, 2015), though whether radiation stress can trigger such responses in *O. corniculata* and in plants in general is not known. An increase of unfavourable chemicals and a decrease of favourable chemicals may occur simultaneously.

In this way, biochemical changes in producers (i.e. plants) affect the primary consumers (i.e. herbivorous animals) and then secondary consumers (i.e. carnivorous animals). This food-mediated effect of pollution through an ecological food web is an indirect effect and different from the bioaccumulation paradigm. It has not yet been demonstrated in the butterfly system, but thiamine (vitamin $B_1$) deficiency has been recognized as one of the major consequences of environmental pollution and destruction, though the precise causes are difficult to identify (Balk *et al.*, 2009, 2016; Sañudo-Wilhelmy *et al.*, 2012; Sonne *et al.*, 2012).

Among the three modes of action of the indirect effects discussed above, the first mode is necessarily associated with immunological responses, which may be prominently problematic for humans, because humans have very effective (and thus very sensitive) adaptive immunity that insects do not have. Even when radioactive low doses (or low dose rates) may be negligible at least for short-term exposure in the direct mode, they are still able to cause large and fatal physiological effects in some individuals via a dust-mediated immunological sensitization. It is possible that radioactivity denatures proteins, which makes naturally occurring proteins immunogenic. Consequences of allergic reactions are not simple, but, for example, such sensitization may lead to a kidney failure including nephrotic syndrome (Section 34.5).

## 34.5 A Case Study of Patient C.U.

Anecdotal cases amount for people who claim to 'feel' low doses of radiation. These claims are almost always dismissed 'scientifically' as psychosomatic, because such low-dose radiation exposure cannot cause any acute physiological changes based on the conventional understanding of radiation effects. However, such cases should not be ignored entirely. In the case of the patient C.U., who greatly helped our studies on Fukushima, her medical history may be an important recorded (as opposed to anecdotal) case that links the Fukushima nuclear accident with a pathological and lethal outcome.

C.U. had been working in our laboratory since the very beginning of the Fukushima project. She collected host-plant leaves for larvae from polluted areas in Fukushima Prefecture and other localities to execute internal exposure experiments in the butterfly. According to my records, she collected the leaves in highly polluted areas on 17, 18 and 19 July and 2 August 2011, and October 4 and 14 October and 1 November 2012, but additional days may have been spent in polluted areas. In November 2012, when she was 57 years old, she suddenly noticed oedema in her feet. I strongly recommended her to visit a doctor immediately, and she agreed. Indeed, oedema was not specific to her feet. It soon manifested in hands, face and other parts of the body when the symptom was extensive, indicating that it was systemic oedema. At that point, she complained about a dramatic increase of her body weight, due to inability to excrete water from the body, suggesting a decrease of kidney function. I asked her to measure her own body weight every day as an objective indicator for disease progression and remission. She then bought a reliable digital body weight scale (Tanita digital health meter HD-661, Tokyo, Japan) to do this at home (minimum output of the scale: 100 g for 0–100 kg weight).

Her basic health information at three timepoints is given in Table 34.1. Her medical history did not indicate any severe diseases in the past. Medical examinations were performed seven times from 19 November 2012 to 21 April 2014 (Tables 34.2–34.6). Based on the blood and urine tests and systemic symptoms, C.U. was medically diagnosed as having nephrotic syndrome.

**Table 34.1.** Basic health information for patient C.U.

| | 19 Nov 2012 | 6 March 2013 | 10 April 2013 |
|---|---|---|---|
| Body weight (kg) | 56.2 | 44.3 | 44.9 |
| BMI (kg $m^{-2}$)[a] | 22 | 17.2 | 17.4 |
| Body fat (%) | 19.3 | 13.8 | 18.9 |
| Abdominal girth (cm) | 80 | 68 | 65 |
| Basal metabolism (kcal per day)[b] | 1287 | 1086 | 1073 |
| Body temperature (°C) | 36.7 | 36.8 | 36.5 |
| Blood pressure (mmHg) | 110/55 | 104/53 | 109/49 |

[a]Body mass index. Normal range 20.0–24.9 kg $m^{-2}$ for Japanese person of age 50–69 years (Kagawa, 2016).
[b]Basal metabolism 1100 kcal per day for Japanese woman of age 50–69 years (Kagawa, 2016).

Diagnostic measures for adult nephrotic syndrome in Japan listed four criteria for diagnosis (Matsuo *et al.*, 2011) as follows.

**1.** Proteinurea. When protein is excreted in urine at more than 3.5 g per day, this is called proteinurea. C.U. excreted 3.7 g on 6 March 2013 (Table 34.6), satisfying this requirement.
**2.** Hypoalbuminaemia. When serum albumin is less than 3.0 g $dl^{-1}$, this is called hypoalbuminaemia. C.U. met this requirement in five timepoints when she was examined (Table 34.2).
**3.** Oedema. C.U. exhibited severe systemic oedema, as indicated by weight changes (Fig. 34.2).
**4.** Dyslipidaemia (hyper-LDL-cholesterolaemia). When the level of low-density lipoprotein (LDL) cholesterol in blood exceeds the upper limit of the normal level, this is called dyslipidaemia. Abnormal lipid metabolism was clearly noted in C.U. (Table 34.3).

According to Matsuo *et al.* (2011), proteinurea and hypoalbuminaemia are required for the diagnosis of nephrotic syndrome. Oedema is not required, but it is considered an important symptom. Dyslipidaemia is not required either. In any case, patient C.U. met all these four criteria for diagnosis. Therefore, C.U. can be considered a typical case of nephrotic syndrome in adults.

A possibility of thymus hypofunction was also examined. A thyroid-stimulating hormone (TSH) increase and a free thyroxin ($FT_4$) decrease were detected, suggesting at this point either chronic thyroiditis (Hashimoto's thyroiditis), simple goitre or diseases from other causes. However, both anti-thyroglobulin antibody (TgAb) and TPOAb (anti-thyroid-peroxidase antibody), two autoimmune antibodies and indicators of chronic thyroiditis, were negative (Table 34.4), mostly excluding thyroid-related diseases. Thus, the high-TSH and low-$FT_4$ results were considered secondary effects from non-thyroid causes.

Although she visited a clinic for blood and urine tests as shown in tables, she refused to be treated with a steroid-based drug therapy. She explained to me that this rejection was rooted on her experience that one of her acquaintances who was not severely ill was treated conventionally and died a few days later after a similar steroid treatment. Because she had many relatives who were physicians, she might have consulted them for her own situation before making her decision. Instead of accepting the 'risky' conventional treatments, she was on a diet based on brown rice and vegetables that she probably devised herself, based on the macrobiotic diet. Remarkably, after a peak weight of 59.1 kg on 31 December 2012, her oedema quickly reversed as evidenced in a sharp decrease of body weight (Fig. 34.2).

Because there is no diet-based therapy for nephrotic syndrome in conventional medicine, her quick remission was worth noting. She told me that she carefully chose fresh food products from non-contaminated areas and avoided sea foods that might be highly contaminated to 'detoxify' her body from radiocaesium.

An increase and a subsequent decrease of body weight involving a change of more than 10 kg was experienced at least three times in this patient: one peak on 31 December 2012, another peak on 18 and 20 May 2014, and an additional peak around October 2014 (though no weight data were available for the last peak). Pictures of her hand demonstrated the existence of the last peak (Fig. 34.2, insets). According to her

**Table 34.2.** Blood biochemical tests for kidney and other functions of the patient C.U.[a–d]

| Blood test[a] | Normal value[b] | 19 Nov 2012 | 6 Mar 2013 | 10 Apr 2013 | 25 Jan 2014 | 27 Feb 2015 | 11 Mar 2015 | 21 Apr 2015 |
|---|---|---|---|---|---|---|---|---|
| *Total protein without fractionation: suggesting liver or kidney dysfunction in this patient* | | | | | | | | |
| Total protein (L) | 6.7–8.3 g $dl^{-1}$ | 4.7 L | 6.2 L | 6.7 | 5.3 L | 4.4 L | nt | 5.0 L |
| Albumin (L) | 4.0–5.0 g $dl^{-1}$ | 2.3 L | 3.2 L | 3.9 L | 2.7 L | 1.4 L | nt | 1.8 L |
| A/G (albumin/globulin) ratio (L) | 1.00–2.14 | 0.96 L | 1.07 | 1.39 | 1.04 | 0.47 L | nt | 0.56 L |
| *Protein fractions after fractionation: suggesting liver and kidney dysfunction including nephrotic syndrome in this patient* | | | | | | | | |
| Albumin (L) | 58.1–70.1% | 49.1 L | 50.9 L | 57.9 L | 48.6 L | 37.8 L | nt | 45.2 L |
| $\alpha_1$-Globulin (H) | 1.8–3.2% | 4.8 H | 3.9 H | 3.0 | 4.2 H | 3.8 H | nt | 3.8 H |
| $\alpha_2$-Globulin (H) | 6.9–11.3% | 19.4 H | 16.4 H | 12.3 H | 20.2 H | 31.3 H | nt | 25.6 H |
| $\beta$-Globulin (H) | 6.4–10.2% | 11.9 H | 12.4 H | 10.4 H | 13.3 H | 13.6 H | nt | 13.0 H |
| $\gamma$-Globulin | 11.6–21.4% | 14.8 | 16.4 | 16.4 | 13.7 | 13.5 | nt | 12.4 |
| A/G (albumin/globulin) ratio (L) | 1.37–2.30 | 0.97 L | 1.04 L | 1.38 | 0.95 L | 0.61 L | nt | 0.82 L |
| *Kidney function markers: suggesting a low-level kidney failure in this patient* | | | | | | | | |
| Blood urea nitrogen | 8.0–22.0 mg $dl^{-1}$ | 10.1 | 11.8 | 12.8 | 10.1 | 15.9 | nt | 12.0 |
| Creatinine (H) | 0.47–0.79 mg $dl^{-1}$ | 0.80 H | 0.69 | 0.57 | 0.71 | 0.64 | nt | 0.65 |
| Uric acid | 2.3–7.0 mg $dl^{-1}$ | 4.3 | 4.1 | 3.6 | 4.1 | 5.6 | nt | 4.9 |
| Estimated glomerular filtration rate (eGFR) (L) | > 90 ml/min/1.73 $m^2$ | nt | nt | nt | 65.0 L | 72.5 L | nt | 71.3 L |
| *Serum ions: suggesting kidney and other disorders including nephrotic syndrome in this patient* | | | | | | | | |
| Serum Na (L) | 138–146 mEq $l^{-1}$ | 134 L | 141 | 140 | 141 | 141 | nt | 140 |
| Serum Cl | 99–109 mEq $l^{-1}$ | 101 | 103 | 104 | 106 | 106 | nt | 105 |
| Serum K | 3.6–4.9 mEq $l^{-1}$ | 4.7 | 4.0 | 4.0 | 4.4 | 4.4 | nt | 4.4 |
| Serum Ca (L) | 8.7–10.3 mg $dl^{-1}$ | 8.1 L | 8.9 | 9.2 | 8.6 L | 7.9 L | nt | 8.4 L |
| Serum Fe | 45–167 μg $dl^{-1}$ | 61 | 76 | 71 | 54 | 63 | nt | 62 |

[a]Tests that showed high and low abnormal values are indicated as H and L, respectively.
[b]Normal values are provided by the clinic except eGFR, in which case Iwaita (2016) was referred to.
[c]Measured values that are higher and lower than the normal values are indicated as H and L, respectively, with associated values.
[d]Tests not performed are indicated as nt (not tested).

**Table 34.3.** Blood biochemical tests for liver and other functions of the patient C.U.[a–d]

| Blood biochemical test[a] | Normal value[b] | 19 Nov 2012 | 6 Mar 2013 | 10 Apr 2013 | 25 Jan 2014 | 27 Feb 2015 | 11 Mar 2015 | 21 Apr 2015 |
|---|---|---|---|---|---|---|---|---|
| *Lipid metabolism: suggesting high LDL cholesteremia and other disorders including nephrotic syndrome in this patient* | | | | | | | | |
| Total cholesterol (H) | 128–219 mg $dl^{-1}$ | 431 H | 255 H | 259 H | 371 H | 771 H | nt | 654 H |
| HDL cholesterol | 40–96 mg $dl^{-1}$ | 71 | 65 | 80 | 52 | 76 | nt | 53 |
| LDL cholesterol (H) | 65–138 mg $dl^{-1}$ | 313 H | 152 H | 173 H | 252 H | 665 H | nt | 549 H |
| Triglyceride (TG) (H) | 30–149 mg $dl^{-1}$ | 210 H | 206 H | 72 | 345 H | 264 H | nt | 351 H |
| *Liver function: suggesting liver disorders in this patient* | | | | | | | | |
| Total bilirubin (L) | 0.3–1.2 mg $dl^{-1}$ | 0.3 | 0.3 | 0.3 | 0.2 L | 0.3 | nt | 0.3 |
| Direct bilirubin | 0.0–0.4 mg $dl^{-1}$ | 0.1 | 0.1 | 0.1 | 0.1 | 0.1 | nt | 0.1 |
| Thymol turbidity test (TTT) | 0.0–4.0 KU | 0.5 | 1.0 | 0.7 | 0.9 | 1.4 | nt | 1.3 |
| Aspartate amino transferase (AST, GOT) (H) | 13–33 IU $l^{-1}$ | 28 | 19 | 24 | 20 | 36 H | nt | 24 |
| Alanine amino transferase (ALT, GPT) | 6–30 IU $l^{-1}$ | 19 | 13 | 17 | 9 | 28 | nt | 13 |
| Alkaline phosphatase (ALP) | 115–359 IU $l^{-1}$ | 180 | 196 | 190 | 176 | 178 | nt | 155 |
| Lactose dehydrogenase (LDH) (H) | 119–229 IU $l^{-1}$ | 286 H | 219 | 213 | 221 | 270 H | nt | 300 H |
| γ-Glutamine transpeptidase (γ-GTP) | 10–47 IU $l^{-1}$ | 27 | 27 | 19 | 18 | 32 | nt | 20 |
| Choline esterase (ChE) (H) | 2164–466 IU $l^{-1}$ | 435 | 378 | 378 | 464 | 526 H | nt | 514 H |

[a]Tests that showed high and low abnormal values are indicated as H and L, respectively.
[b]Normal values are provided by the clinic.
[c]Measured values that are higher and lower than the normal values are indicated as H and L, respectively, with associated values.
[d]Tests not performed are indicated as nt (not tested).

**Table 34.4.** Blood biochemical tests for various functions of the patient C.U.[a–d]

| Blood biochemical test[a] | Normal value[b] | 19 Nov 2012 | 6 Mar 2013 | 10 Apr 2013 | 25 Feb 2015 | 27 Feb 2015 | 11 Mar 2015 | 21 Aprl 2015 |
|---|---|---|---|---|---|---|---|---|
| *Muscle function: suggesting muscle disorders but also a possibility of kidney failure in this patient* | | | | | | | | |
| Creatine phosphokinase (CPK) (H) | 45–163 IU $l^{-1}$ | 138 | 87 | 102 | 85 | 224 H | nt | 194 H |
| *Pancreas and other endocrine organs: suggesting pancreas dysfunction in this patient* | | | | | | | | |
| Serum amylase (AMY) (H) | 42–132 IU $l^{-1}$ | 95 | 112 | 118 | 145 H | 96 | nt | 139 H |
| *Diabetic disease markers: suggesting no diabetic disease in this patient* | | | | | | | | |
| Plasma glucose | 69–109 mg $dl^{-1}$ | 79 | 93 | 84 | 88 | 89 | nt | 95 |
| Glycohemoglobin HbAlc (NGSP) | 4.6–6.2 | 5.4 | nt | nt | nt | nt | nt | nt |
| Glycohemoglobin HbAlc (JDS) | 4.3–5.8 | 5.0 | nt | nt | nt | nt | nt | nt |
| *Infection, anemia, inflammation markers: suggesting anemia and inflammation in this patient* | | | | | | | | |
| *Helicobacter pylori* IgG | < 10 U $ml^{-1}$ | nt | nt | nt | nt | nt | 5 | nt |
| Unsaturated iron binding capacity (UIBC) (L) | 147–299 μg $dl^{-1}$ | 101 L | 178 | nt | nt | nt | nt | nt |
| C-reactive protein (CRP) (H) | 0.00–0.30 mg $dl^{-1}$ | 0.33 H | nt | nt | nt | nt | nt | nt |
| *Thyroid markers: suggesting thyroid dysfunction in this patient* | | | | | | | | |
| Thyroid-stimulating hormone (TSH) (H) | 0.50–5.00 μIU $ml^{-1}$ | nt | 9.48 H | nt | 18.1 H | 5.24 H | nt | nt |
| Free thyroxin ($FT_4$) (L) | 0.90–1.70 ng $dl^{-1}$ | nt | 0.80 L | nt | 0.85 L | 0.67 L | nt | nt |
| Anti-thyroglobulin antibody (TgAb) | 0–28 IU $ml^{-1}$ | nt | nt | < 10 | nt | nt | nt | nt |
| Anti-thyroid-peroxidase antibody (TPOAb) | 0–16 IU $ml^{-1}$ | nt | nt | nt | < 5 | 10 | nt | nt |
| *Tumor marker: suggesting a possibility of tumor in this patient* | | | | | | | | |
| Ferritin (H) | 5.0–120.0 ng $ml^{-1}$ | 146.0 H | 150.0 H | nt | nt | nt | nt | nt |
| *Pepsinogen (PG): suggesting no gastric diseases in this patient* | | | | | | | | |
| Pepsinogen I (PGI) | > 71 ng $ml^{-1}$ | nt | nt | nt | nt | nt | 80.1 | nt |
| Pepsinogen II (PGII) | Not provided | nt | nt | nt | nt | nt | 39.0 | nt |
| PGI/PGII ratio | Not provided | nt | nt | nt | nt | nt | 2.1 | nt |

[a]Tests that showed high and low abnormal values are indicated as H and L, respectively.
[b]Normal values are provided by the clinic except PGI, in which case Iwaita (2016) was referred to.
[c]Measured values that are higher and lower than the normal values are indicated as H and L, respectively, with associated values.
[d]Tests not performed are indicated as nt (not tested).

**Table 34.5.** Blood cell tests for the patient C.U.[a–d]

| Blood cell test[a] | Normal value[b] | 19 Nov 2012 | 6 Mar 2013 | 10 Apr 2013 | 25 Jan 2014 | 27 Feb 2015 | 11 Mar 2015 | 21 Apr 2015 |
|---|---|---|---|---|---|---|---|---|
| *Peripheral blood: suggesting anemia in this patient* | | | | | | | | |
| Leukocyte count | 3,500–9,800 μl$^{-1}$ | 6500 | 6200 | 5800 | 6300 | 7500 | nt | 8600 |
| Erythrocyte count (L) | 3,760–5,000 k μl$^{-1}$ | 371 L | 342 L | 342 L | 354 L | 328 L | nt | 364 L |
| Hemoglobin (L) | 11.3–15.2 g dl$^{-1}$ | 11.2 L | 11.0 L | 11.1 L | 11.0 L | 10.0 L | nt | 11.2 L |
| Hematocrit (L) | 33.4–44.9% | 35.5 | 35.1 | 34.8 | 35.1 | 32.3 L | nt | 36.2 |
| Mean corpuscular volume (MCV) (H) | 79.0–100.0 fl | 95.7 | 102.6 H | 101.8 H | 99.2 | 98.5 | nt | 99.5 |
| Mean corpuscular hemoglobin (MCH) | 26.3–34.3 pg | 30.2 | 32.2 | 32.5 | 31.1 | 30.5 | nt | 30.8 |
| Mean corpuscular hemoglobin concentration (MCHC) | 30.7–36.6% | 31.5 | 31.3 | 31.9 | 31.3 | 31.0 | nt | 30.9 |
| Platelet count | 130–369 k μl$^{-1}$ | 288 | 344 | 218 | 286 | 303 | nt | 307 |
| Reticulocyte count | 2–27‰ | 15 | 13 | nt | nt | nt | nt | nt |
| *Leukocyte classification: suggesting allergic diseases in this patient* | | | | | | | | |
| Stab cell (St) | 2–7% | 3 | 5 | 3 | nt | nt | nt | nt |
| Segmented leukocyte (Seg) | 40–63% | 63 | 53 | 55 | nt | nt | nt | nt |
| Neutrophil (Neut) | 42–70% | nt | nt | nt | 59 | 57 | nt | 61 |
| Eosinophil (E) | 0–5% | 2 | 3 | 2 | 4 | 3 | nt | 4 |
| Monocyte (M) | 3–11% | 5 | 9 | 4 | 4 | 5 | nt | 5 |
| Lymphocyte (Ly) | 22–46% | 25 | 29 | 35 | 32 | 33 | nt | 29 |
| Basophil (B) (H) | 0–1% | 2 H | 1 | 1 | 1 | 2 H | nt | 1 |
| Erythroblast (EBL) | Not provided | 0 | 0 | 0 | 0 | 0 | nt | 0 |
| Atypical lymphocyte | Not provided | 0 | 0 | 0 | 0 | 0 | nt | 0 |

[a]Tests that showed high and low abnormal values are indicated as H and L, respectively.
[b]Normal values are provided by the clinic.
[c]Measured values that are higher and lower than the normal values are indicated as H and L, respectively, with associated values.
[d]Tests not performed are indicated as nt (not tested).

**Table 34.6.** Urine tests for the patient C.U.[a–d]

| Urine test[a] | Normal value[b] | 19 Nov 2012 | 6 Mar 2013 | 10 Apr 2013 | 25 Jan 2014 | 27 Feb 2015 | 11 Mar 2015 | 21 Apr 2015 |
|---|---|---|---|---|---|---|---|---|
| *Urine biochemicals: suggesting nephrotic syndrome for this patient* | | | | | | | | |
| Urine albumin (H) | 0.0–30.0 m d$^{-1}$ | nt | 3726.3 H | nt | nt | nt | nt | nt |
| pH | 4.8–7.5 | 5.5 | 5.0 | 6.0 | 6.5 | 7.0 | nt | 5.5 |
| *Urinary sediments (in high power field): suggesting kidney dysfunction and infection in this patient* | | | | | | | | |
| Erythrocyte count (H) | < 5 | 10–20 H | nt | 3–5 | nt | nt | nt | nt |
| Leukocyte count | < 5 | 3–5 | nt | 0–1 | nt | nt | nt | nt |
| Squamous epithelial count | Very few | 0–1 | nt | 0–1 | nt | nt | nt | nt |
| Tubular epithelial count | Very few | 1–2 | nt | nt | nt | nt | nt | nt |
| Bacteria (H) | (–) | (1+) H | nt | nt | nt | nt | nt | nt |
| *Urine general tests (qualitative): suggesting kidney dysfunction in this patient* | | | | | | | | |
| Glucose | (–) | (–) | (–) | (–) | (–) | (–) | (–) | (–) |
| Protein (H) | (–) | (3+) | (3+) | (+–) | (4+) | (3+) | nt | (3+) |
| Urobilinogen | (+–) | (+–) | (+–) | (+–) | (+–) | (+–) | nt | (+–) |
| Occult blood (H) | (–) | (3+) | (3+) | (+–) | (2+) | (2+) | nt | (+–) |

[a]Tests that showed high and low abnormal values are indicated as H and L, respectively.
[b]Normal values are provided by the clinic except urinary sediments, in which case Iwaita (2016) was referred to.
[c]Measured values that are higher and lower than the normal values are indicated as H and L, respectively, with associated values.
[d]Tests not performed are indicated as nt (not tested).

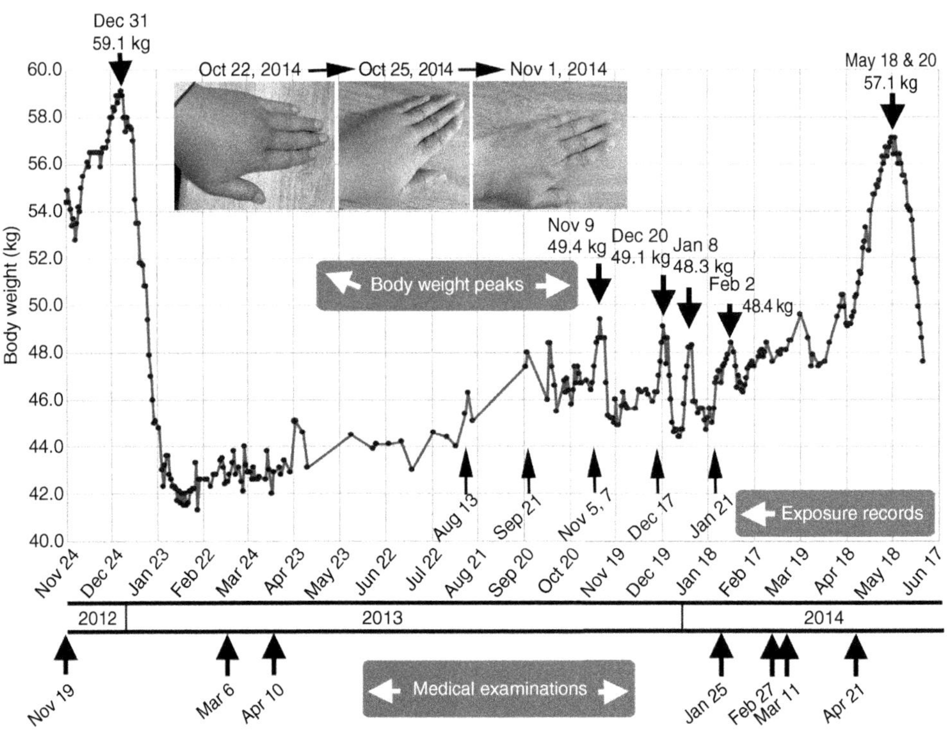


**Fig. 34.2.** Body weight changes of the patient C.U. over time. Insets show the left hand of the patient at three time-points. Days of body weight peaks, exposure records and medical examinations are indicated by arrows with dates.

own words, this last peak was caused because she lived in Tokyo (a lightly contaminated place) during this period, and not in Okinawa.

Importantly, there were several smaller peaks between the first two large peaks. My records show that she visited our laboratory on 13 August, 21 September, 5 and 7 November and 17 December 2013 and 21 January 2014. These visits took place immediately before the onset of small peaks. During this period, we were using contaminated leaves from Fukushima, but the air radiation dose in the laboratory was kept equal to the environmental radiation levels. This remarkable time-series correlation suggests that she was immunologically sensitized to radioactive dust in Fukushima during her repetitive visits, and it suggests that her body responded to the vanishing amount of radioactive (and/or non-radioactive) dust from the contaminated leaves in the laboratory.

This possibility was explained to her much later than these laboratory visits. I believe that she did not seriously consider this possibility at the time of the laboratory visits, excluding the potential psychosomatic nature of the weight increases. Another possibility was that she was sensitized to the host-plant leaves and not to the Fukushima dust. I asked her to test this possibility herself. She told me that she did not get sick from directly touching the leaves from Okinawa. She did not get sick from the butterflies, either, but these are her own subjective judgements. Additionally, there was a possibility that she was allergic to laboratory chemicals. However, this is very unlikely, because she rarely used such chemicals. It seemed that she felt and understood that she was vulnerable to something from Fukushima.

She appeared to have mastered how to heal her own symptoms by the diet-based treatment. But, as she recognized, an increase of her body weight after a period of short remission was likely initiated by her 'careless' behaviour to get exposed to the Fukushima dust via ingestion or inhalation. Then came a tragedy. In her own words, she was invited to talk about her research in Fukushima and, regrettably, she accepted the invitation and visited Fukushima in early 2015. Immediately after this event, her symptoms became worse, and despite her own efforts to reduce oedema, she eventually passed away in the autumn of 2015 after 3 years of nephrotic syndrome. The direct cause of death was diagnosed as acute cardiovascular failure due to nephrotic syndrome, according to the death certificate.

The contribution of radiation to chronic kidney diseases has been suspected as radiation nephropathy (Cohen *et al.*, 2015). Although a direct causal relationship between basophil sensitization and nephrotic syndrome has not been demonstrated (Abdel-Hafez *et al.*, 2009), it is known that basophil sensitization to aeroallergens such as house dust and pollen and to food allergens such as eggs and milk is associated with idiopathic nephrotic syndrome (Pirotzky *et al.*, 1982; Yap *et al.*, 1983; Laurent *et al.*, 1984; Lin *et al.*, 1990). In patient C.U., basophil count elevation was indeed observed (Table 34.4). Therefore, I propose basophil-mediated allergy that was induced by radioactive (and, less likely, non-radioactive) materials from the FDNPP as a cause of nephrotic syndrome in this patient.

There is a possibility that insoluble microparticles released from the FDNPP can immunologically sensitize people at the olfactory and respiratory mucosa, as any other environmental dusts may do. Acute allergic reactions may be felt by exposed individuals. To be antigenic, molecules must be proteins that are processed and presented by antigen-presenting cells. Microparticles may bind to air-floating proteins such as house dust, urban dust and pollens to make them more immunogenic. This mechanism may be similar to metal allergy. In that case, metals do not act as allergens singularly, but metal ions bind to proteins and denature them, making them effective allergens. Radioactive (and non-radioactive) materials from the FDNPP may bind tightly to proteins and denature them, making them immunogenic. Because most of these explanations are well within conventional immunology and environmental medicine, care must be taken regarding low-dose exposure to microparticles released from the FDNPP to improve public health in the polluted areas.

## 34.6 Conclusions

Important issues are reviewed on low-dose radionuclide exposures, especially from the viewpoint of the indirect field effects. We stressed the importance of using the pale grass blue butterfly to understand human-living environment and human health effects from the Fukushima nuclear accident. A case of nephritic syndrome was presented, who seemed to have high sensitivity to nuclear dust from the FDNPP, illustrating toxic effects of the radioactive materials on humans. Although it was impossible to definitively prove that this medical case was caused by the radioactive pollution from the FDNPP, this case will make scientists notice the real possibility of complex disease-causing pathways by radioactive materials. It is important to admit that we, scientists, do not know much about complex and multiple disease-causing mechanisms of the materials released from the FDNPP. There is much to be learned from nuclear pollution biology in the future.

## Acknowledgements

The author dedicates this paper to the late C.U., who greatly contributed to the Fukushima studies. The author greatly thanks husband and daughter of C.U. for permission to publish her clinical data in this chapter and also thanks anonymous medical doctors and clinicians who contributed to medical tests and diagnosis. The author thanks members of the BCPH Unit of Molecular Physiology in University of the Ryukyus for discussions. Finally, the author greatly appreciates donors who philosophically and financially supported our Fukushima Project.

## References

Abdel-Hafez, M., Shimada, M., Lee, P.Y., Johnson, R.J. and Garin, E.H. (2009) Idiopathic nephrotic syndrome and atopy: is there a common link? *American Journal of Kidney Diseases* 54, 945–954. doi: 10.1053/j.ajkd.2009.03.019.

Adachi, K., Kajino, M., Zaizen, Y. and Igarashi Y. (2013) Emission of spherical cesium-bearing particles from an early stage of the Fukushima nuclear accident. *Scientific Reports* 3, 2554. doi: 10.1038/srep02554.

Akimoto, S. (2014) Morphological abnormalities in gall-forming aphids in a radiation-contaminated area near Fukushima Daiichi: selective impact of fallout? *Ecology and Evolution* 4, 355–369. doi: 10.1002/ece3.949.

Aliyu, A.S., Evangeliou, N., Mousseau, T.A., Wu, J. and Ramli, A.J. (2015) An overview of current knowledge concerning the health and environmental consequences of the Fukushima Daiichi Nuclear Power Plant (FDNPP) accident. *Environment International* 85, 213–228. doi: 10.1016/j.envint.2015.09.020.

Balk, L., Hägerroth, P.-Å., Åkerman, G., Hanson, M., Tjämlund, U. *et al.* (2009) Wild birds of declining European species are dying from a thiamine deficiency syndrome. *Proceedings of the National Academy of Sciences of the United States of America* 106, 12001–12006. doi: 10.1073/pnas.0902903106.

Balk, L., Hägerroth, P.-Å., Gustavsson, H., Sigg, L., Åkerman, G. *et al.* (2016) Widespread episodic thiamine deficiency in Northern Hemisphere wildlife. *Scientific Reports* 6, 38821. doi: 10.1038/srep38821.

Bonisoli-Alquati, A., Koyama, K., Tedeschi, D.J., Kitamura, W., Sukuzu, H. *et al.* (2015) Abundance and genetic damage of barn swallows from Fukushima. *Scientific Reports* 5, 9432. doi: 10.1038/srep09432.

Borek, C., Zaider, M., Ong, A., Mason, H. and Witz, G. (1986) Ozone acts alone and synergistically with ionizing radiation to induce *in vitro* neoplastic transformation. *Carcinogenesis* 7, 1611–1613. doi: 10.1093/carcin/7.9.1611.

Bréchignac, F., Oughton, D., Mays, C., Barnthouse, L., Beasley, J.C. *et al.* (2016) Addressing ecological effects of radiation on populations and ecosystems to improve protection of the environment against radiation: agreed statements from a Consensus Symposium. *Journal of Environmental Radioactivity* 158–159, 21–29. doi: 10.1016/j.jenvrad.2016.03.021.

Chino, M., Nakayama, H., Nagai, H., Terada, H., Katata, G. *et al.* (2011) Preliminary estimation of release amount of $^{131}I$ and $^{137}Cs$ accidentally discharged from the Fukushima Daiichi nuclear power plant into the atmosphere. *Journal of Nuclear Science and Technology* 48, 1129–1134. doi: 10.1080/18811248.2011.9711799.

Cohen, E.P., Fish, B.L. and Moulder, J.E. (2015) Late-onset effects of radiation and chronic kidney diseases. *The Lancet* 386, 1737–1738. doi: 10.1016/S0140-6736(15)00697-2.

Fukunaga, H. and Yokoya, A (2016) Low-dose radiation risk and individual variation in radiation sensitivity in Fukushima. *Journal of Radiation Research* 57, 98–100. doi: 10.1093/jrr/rrv053.

Fuma, S., Ihara, S., Takahashi, H., Inaba, O, Sato, Y. *et al.* (2017) Radiocaesium contamination and dose rate estimation of terrestrial and freshwater wildlife in the exclusion zone of the Fukushima Dai-ichi Nuclear Power Plant accident. *Journal of Environmental Radioactivity* 171, 176–188. doi: 10.1016/j.jenvrad.2017.02.013.

Garnier-Laplace, J., Geras'kin, S., Delta-Vedova, C., Beaugelin-Seiller, K., Hinton, T.G., *et al.* (2013) Are radiosensitivity data derived from natural field conditions consistent with data from controlled exposures? A case study of Chernobyl wildlife chronically exposed to low dose rates. *Journal of Environmental Radioactivity* 121, 12–21. doi: 10.1016/j.jenvrad.2012.01.013.

Gilbert, S.F. and Epel, D. (2015) *Ecological Developmental Biology: The Environmental Regulation of Development, Health, and Evolution*, 2nd edn. Sinauer Associates, Sunderland, Massachusetts.

Hayashi, G., Shibato, J., Imanaka, T., Cho, K., Kubo, A. *et al.* (2014) Unraveling low-level gamma radiation-responsive changes in expression of early and late genes in leaves of rice seedlings at Iitate Village, Fukushima. *Journal of Heredity* 105, 723–738. doi: 10.1093/jhered/esu025.

Hayashida, N., Imaizumi, M., Shimura, H., Furuya, F., Okubo, N. *et al.* (2015) Thyroid ultrasound findings in a follow-up survey of children from three Japanese prefectures: Aomori, Yamanashi, and Nagasaki. *Scientific Reports* 5, 9046. doi: 10.1038/srep09046.

Hirose, K. (2012) 2011 Fukushima Dai-ichi nuclear power plant accident: summary of regional radioactive deposition monitoring results. *Journal of Environmental Radioactivity* 111, 13–17. doi: 10.1016/j.jenvrad.2011.09.003.

Hiyama, A., Iwata, M. and Otaki, J.M. (2010) Rearing the pale grass blue *Zizeeria maha* (Lepidoptera, Lycaenidae): Toward the establishment of a lycaenid model system for butterfly physiology and genetics. *Entomological Science* 13, 293–302. doi: 10.1111/j.1479-8298.2010.00387.x.

Hiyama, A., Nohara, C., Kinjo, S., Taira, W., Gima, S. *et al.* (2012) The biological impacts of the Fukushima nuclear accident on the pale grass blue butterfly. *Scientific Reports* 2, 570. doi: 10.1038/srep00570.

Hiyama, A., Nohara, C., Taira, W., Kinjo, S., Iwata, M. *et al.* (2013) The Fukushima nuclear accident and the pale grass blue butterfly: evaluating biological effects of long-term low-dose exposures. *BMC Evolutionary Biology* 13, 168. doi: 10.1186/1471-2148-13-168.

Hiyama, A., Taira, W., Nohara, C., Iwasaki, M., Kinjo, S. *et al.* (2015) Spatiotemporal abnormality dynamics of the pale grass blue butterfly: three years of monitoring (2011-2013) after the Fukushima nuclear accident. *BMC Evolutionary Biology* 15, 15. doi: 10.1186/s12862-015-0297-1.

Hiyama, A., Taira, W., Iwasaki, M., Sakauchi, K., Gurung, R., *et al.* (2017a) Geographical distribution of morphological abnormalities and wing color pattern modifications of the pale grass blue butterfly in northeastern Japan. *Entomological Science* 20, 100–110. doi: 10.1111/ens.12233.

Hiyama, A., Taira, W., Iwasaki, M., Sakauchi, K., Iwata, M., *et al.* (2017b) Morphological abnormality rate of the pale grass blue butterfly *Zizeeria maha* (Lepidoptera: Lycaenidae) in southwestern Japan: A reference data set for environmental monitoring. *Journal of Asia-Pacific Entomology* 20, 1333–1339. doi: 10.1016/j.aspen.2017.09.016.

Horiguchi, T., Yoshii, H., Mizuno, S. and Shiraishi, H. (2016) Decline in intertidal biota after the 2011 Great East Japan Earthquake and Tsunami and the Fukushima nuclear disaster: field observations. *Scientific Reports* 6, 20416. doi: 10.1038/srep20416

Iwaita, Y. (2016) *Encyclopedia of Medical Tests and their Test Values*. Seibido, Tokyo, Japan.

Iwaku, K., Noh, J.Y., Sasaki, E., Suzuki, N., Kameda, T. *et al.* (2014) Changes in pediatric thyroid sonograms in or near the Kanto region before and after the accident at the Fukushima Daiichi nuclear power plant. *Endocrine Journal* 61, 875–881. doi: 10.1507/endocrj.EJ14-0032.

Iwata, M., Hiyama, A. and Otaki, J.M. (2013) System-dependent regulations of colour-pattern development: a mutagenesis study of the pale grass blue butterfly. *Scientific Reports* 3, 2379. doi: 10.1038/srep02379.

Kagawa, Y. (2016) *Standard Tables of Food Consumption in Japan*, 7th edn. Kagawa Nutrition University Press, Sakado, Saitama, Japan.

Kaltofen, M. and Gundersen, A. (2017) Radioactively-hot particles detected in dusts and soils from Northern Japan by combination of gamma spectrometry, autoradiography, and SEM/EDS analysis and implications in radiation risk assessment. *Science of Total Environment* 607–608, 1065–1072. doi: 10.1016/j.scitotenv.2017.07.091.

Kappos, A., Bruckmann, P., Eikmann, T., Englert, N., Heinrich, U. *et al.* (2004) Health effects of particles in ambient air. *International Journal of Hygiene and Environmental Health* 207, 399–407. doi: 10.1078/1438-4639-00306.

Katanoda, K., Kamo, K.I. and Tsugane, S. (2016) Quantification of the increase in thyroid cancer prevalence in Fukushima after the nuclear disaster in 2011 – a potential overdiagnosis? *Japanese Journal of Clinical Oncology* 46, 284–286. doi: 10.1093/jjco/hyv191.

Kawagoshi, T., Shiomi N., Takahashi, H., Watanabe, Y., Fuma, S. *et al.* (2017) Chromosomal aberrations in large Japanese field mice (*Apodemus speciosus*) captured near Fukushima Dai-ichi nuclear power plant. *Environmental Science & Technology* 51, 4632–4641. doi: 10.1021/acs.est.6b06210.

Kinoshita, N., Sueki, K., Sasa, K., Kitagawa, J., Ikarashi, S. *et al.* (2011) Assessment of individual radionuclide distributions from the Fukushima nuclear accident covering central-east Japan. *Proceedings of the National Academy of Sciences of the United States of America* 108, 19526–19529. doi: 10.1073/pnas.1111724108.

Kubota, Y., Tsuji, H., Kawagoshi, T., Shiomi, N., Takahashi, H. *et al.* (2015) Chromosomal aberrations in wild mice captured in areas differentially contaminated by the Fukushima Dai-ichi Nuclear power plant accident. *Environmental Science & Technology* 49, 10074–10083. doi: 10.1021/acs.est.5b01554.

Laurent, J., Lagrue, G., Belghiti, D., Noirot, C. and Hirbec, G. (1984) Is house dust allergen a possible causal factor for relapses in lipoid nephrosis? *Allergy* 39, 231–236. doi: 10.1111/j.1398-9995.1984.tb02628.x.

Leenhouts, H.P. and Chadwick, K.H. (1978) An analysis of synergistic sensitization. *British Journal of Cancer* 3, S198–S201.

Lin, C.Y., Lee, B.H., Lin, C.C. and Chen, W.P. (1990) A study of the relationship between childhood nephrotic syndrome and allergic diseases. *Chest* 97, 1408–1411. doi: 10.1378/chest.97.6.1408.

Manti, L. and D'Arco, A. (2010) Cooperative biological effects between ionizing radiation and other physical and chemical agents. *Mutation Research* 704, 115–122. doi: 10.1016/j.mrrev.2010.03.005.

Matsuo, S. Imai, E., Saito, T., Imada, T., Taguchi, T. *et al.* (2011) Guidelines for the treatment of nephrotic syndrome. *Japanese Journal of Nephrology* 53, 78–122.

Miyamoto, Y., Yasuda, K. and Magara, M. (2014) Size distribution of radioactive particles collected at Tokai, Japan 6 days after the nuclear accident. *Journal of Environmental Radioactivity* 132, 1–7. doi: 10.1016/j.jenvrad.2014.01.010.

Møller, A.P., Hagiwara, A., Matsui, S., Kasahara, S., Kawatsu, K. *et al.* (2012) Abundance of birds in Fukushima as judges from Chernobyl. *Environmental Pollution* 164, 36–39. doi: 10.1016/j.envpol.2012.01.008.

Møller, A.P. and Mousseau, T.A. (2013) Low-dose radiation, scientific scrutiny, and requirements for demonstrating effects. *BMC Biology* 11, 92. doi: 10.1186/1741-7007-11-92.

Møller, A.P., Nishiumi, I., Suzuki, H., Ueda, K. and Mousseau, T.A. (2013) Differences in effects of radiation on abundance of animals in Fukushima and Chernobyl. *Ecological Indicators* 24, 75–81. doi: 10.1016/j.ecolind.2012.06.001.

Mothersill, C. and Seymour, C. (2009) Implications for environmental health of multiple stressors. *Journal of Radiological Protection* 29, A21. doi: 10.1088/0952-4746/29/2A/S02.

Mothersill, C. and Seymour, C. (2014) Implications for human and environmental health of low doses of ionizing radiation. *Journal of Environmental Radioactivity* 133, 5-9. doi: 10.1016/j.jenvrad.2013.04.002.

Mothersill, C., Salbu, B., Heier, L.S., Teien, H.C., Denbeigh, J. *et al.* (2007) Multiple stressor effects of radiation and metals in salmon (*Salmo salar*). *Journal of Environmental Radioactivity* 96, 20–31. doi: 10.1016/j.jenvrad.2007.01.025.

Mousseau, T.A. and Møller, A.P. (2014) Genetic and ecological studies of animals in Chernobyl and Fukushima. *Journal of Heredity* 105, 704–709. doi: 10.1093/jhered/esu040.

Murase, K., Murase, J., Horie, R. and Endo, K. (2015) Effects of the Fukushima Daiichi nuclear accident on goshawk reproduction. *Scientific Reports* 5, 9405. doi: 10.1038/srep09405.

Nagataki, S. and Takamura N (2014) A review of the Fukushima nuclear reactor accident: radiation effects on the thyroid and strategies for prevention. *Current Opinions in Endocrinology, Diabetes and Obesity* 21, 384–393. doi: 10.1097/MED. 0000000000000098.

Nakamura, A.J., Suzuki, M., Redon, C.E., Kuwahara, Y., Yamashiro, H. *et al.* (2017) The causal relationship between DNA damage induction in bovine lymphocytes and the Fukushima nuclear power plant accident. *Radiation Research* 187, 630–636. doi: 10.1667/RR14630.1.

Nohara, C., Hiyama, A., Taira, W., Tanahara, A. and Otaki, J.M. (2014a) The biological impacts of ingested radioactive materials on the pale grass blue butterfly. *Scientific Reports* 4, 4946. doi: 10.1038/srep04946.

Nohara, C., Taira, W., Hiyama, A., Tanahara, A., Takatsuji, T. *et al.* (2014b) Ingestion of radioactively contaminated diets for two generations in the pale grass blue butterfly. *BMC Evolutionary Biology* 14, 193. doi: 10.1186/s12862-014-0193-0.

Nohara, C., Hiyama, A., Taira, W. and Otaki, J.M. (2017) Robustness and radiation resistance of the pale grass blue butterfly from radioactively contaminated areas: a possible case of adaptive evolution. *Journal of Heredity* 109, 188–198. doi: 10.1093/jhered/esx012.

Ochiai, K., Hayama, S., Nakiri, S., Nakanishi, S., Ishii, N., *et al.* (2014) Low blood cell counts in wild Japanese monkeys after the Fukushima Daiichi nuclear disaster. *Scientific Reports* 4, 5793. doi: 10.1038/srep05793.

Okano, T., Ishiniwa, H., Onuma, M., Shindo, J., Yokohata, Y. *et al.* (2016) Effects of environmental radiation on testes and spermatogenesis in wild large Japanese field mice (*Apodemus speciosus*) from Fukushima. *Scientific Reports* 6, 23601. doi: 10.1038/srep23601.
Otaki, J.M. (2015) Understanding Fukushima through butterfly biology: academic freedom for scientists and the public. *The Winnower* 1–6. Available at: https://thewinnower.com/discussions/37-understanding-fukushima-through-butterfly-biology-academic-freedom-for-scientists-and-the-public (accessed 4 August 2017)
Otaki, J.M. (2016) Fukushima's lessons from the blue butterfly: a risk assessment of the human living environment in the post-Fukushima era. *Integrated Environmental Assessment Management* 12, 667–672. doi: 10.1002/ieam.1828.
Otaki, J.M. (2018) Understanding low-dose exposure and field effects to resolve the field-laboratory paradox: multifaceted biological effects from the Fukushima nuclear accident. In: Awwad, N.S. and AlFaify, S.A. (eds) *New Trends in Nuclear Science*, pp. 49–71. IntechOpen, London.
Otaki, J.M. amd Taira, W. (2017) Current status of the blue butterfly in Fukushima research. *Journal of Heredity* 109, 178–187. doi: 10.1093/jhered/esx037.
Otaki, J.M., Hiyama, A., Iwata, M. and Kudo, T. (2010) Phenotypic plasticity in the range-margin population of the lycaenid butterfly *Zizeeria maha*. *BMC Evolutionary Biology* 10, 252. doi: 10.1186/1471-2148-10-252.
Pirotzky, E., Hieblot, C., Benveniste, J., Laurent, J., Lagrue, G. *et al.* (1982) Basophil sensitisation in idiopathic nephrotic syndrome. *The Lancet* 319, 358–361. doi: 10.1016/S0140-6736(82)91391-5.
Relyea, R.A. (2003) Predator cues and pesticides: a double dose of danger for amphibians. *Ecological Applications* 13, 1515–1521. doi: 10.1890/02-5298.
Relyea, R.A. (2004) Fine-tuned phenotypes: tadpole plasticity under 16 combinations of predators and competitors. *Ecology* 85, 172-179. doi: 10.1890/03-0169.
Salbu, B. and Lind, O.C. (2016) Radioactive particles released to the environment from the Fukushima reactors – confirmation is still needed. *Integrated Environmental Assessment and Management* 12, 687–689. doi: 10.1002/ieam.1834.
Sañudo-Wilhelmy, S.A., Cutter, L.S., Durazo, R., Smail, E.A., Gómez-Consamau, L. *et al.* (2012) Multiple B-vitamin depletion in large areas of the coastal ocean. *Proceedings of the National Academy of Sciences of the United States of America* 109, 14041–14045. doi: 10.1073/pnas.1208755109.
Seaton, A., Godden, D., MacNee, W. and Donaldson, K. (1995) Particulate air pollution and acute health effects. *The Lancet* 345, 176–178. doi: 10.1016/S0140-6736(95)90173-6.
Shibuya, K., Gilmour, S. and Oshima, A. (2014) Time to reconsider thyroid cancer screening in Fukushima. *The Lancet* 383, 1883–1884. doi: 10.1016/S0140-6736(14)60909-0.
Shiraiwa, M., Selzle, K. and Pöschl, U. (2012) Hazardous components and health effects of atmospheric aerosol particles: reactive oxygen species, soot, polycyclic aromatic compounds and allergenic proteins. *Free Radical Research* 46, 927–939. doi: 10.3109/10715762.2012.663084.
Sonne, C., Alstrup, O. and Therkildsen, O.R. (2012) A review of the factors causing paralysis in wild birds: implications for the paralytic syndrome observed in the Baltic Sea. *Science of Total Environment* 416, 32–39. doi: 10.1016/j.scitotenv.2011.12.023.
Strand, P., Aono, T., Brown, J.E., Garnier-Laplace, J., Hosseini, A. *et al.* (2014) Assessment of Fukushima-derived radiation doses and effects on wildlife in Japan. *Environmental Science & Technology Letters* 1, 198–203. doi: 10.1021/ez500019j.
Taira, W., Nohara, C., Hiyama, A. and Otaki, J.M. (2014) Fukushima's biological impacts: the case of the pale grass blue butterfly. *Journal of Heredity* 105, 710–722. doi: 10.1093/jhered/esu013.
Taira, W., Hiyama, A., Nohara, C., Sakauchi, K. and Otaki, J.M. (2015a) Ingestional and transgenerational effects of the Fukushima nuclear accident on the pale grass blue butterfly. *Journal of Radiation Research* 56, i2–i8. doi: 10.1093/jrr/rrv068.
Taira, W., Iwasaki, M. and Otaki, J.M. (2015b) Body size distributions of the pale grass blue butterfly in Japan: size rules and the status of the Fukushima population. *Scientific Reports* 5, 12351. doi: 10.1038/srep12351.
Taiz, L., Zeiger, E., Møller, I.M. and Murphy, A. (2015) *Plant Physiology and Development*, 6th edn. Sinauer Associates, Sunderland, Massachusetts.
Torii, T., Sugita, T., Okada, C.E., Reed, M.S. and Blumenthal, D.J. (2013) Enhanced analysis methods to derive the spatial distribution of $^{131}$I deposition on the ground by airborne surveys at an early stage after the Fukushima Daiichi nuclear power plant accident. *Health Physics* 105, 192–200. doi: 10.1097/HP.0b013e318294444e.
Tsuda, T., Tokinobu, A., Yamamoto, E. and Suzuki, E. (2016) Thyroid cancer detection by ultrasound among residents ages 18 years and younger in Fukushima, Japan: 2011 to 2014. *Epidemiology* 27, 316–322. doi: 10.1097/EDE.0000000000000385.

UNSCEAR (2012) *Biological Mechanisms of Radiation Actions at Low Doses*. A white paper to guide the Scientific Committee's future programme of work. United Nations Scientific Committee on the Effects of Atomic Radiation, United Nations, New York .

UNSCEAR (2013) *Sources, and Effects and Risks of Ionizing Radiation*. UNSCEAR 2013 Report. Volume I: Report to the General Assembly, Scientific Annex A: Levels and effects of radiation exposure die to the nuclear accident after the 2011 great east-Japan earthquake and tsunami. United Nations Scientific Committee on the Effects of Atomic Radiation, United Nations, New York

UNSCEAR (2015) *Developments since the 2013 UNSCEAR Report on the levels and effects of radiation exposure due to the nuclear accident following the Great East-Japan Earthquake and Tsunami*. A 2015 white paper to guide the Scientific Committee's future programme of work. United Nations Scientific Committee on the Effects of Atomic Radiation, United Nations, New York.

UNSCEAR (2016) *Developments since the 2013 UNSCEAR Report on the levels and effects of radiation exposure due to the nuclear accident following the Great East-Japan Earthquake and Tsunami*. A 2016 white paper to guide the Scientific Committee's future programme of work. United Nations Scientific Committee on the Effects of Atomic Radiation, United Nations, New York .

Urushihara, Y., Kawasumi, K., Endo, S., Tanaka, K., Hirakawa, Y. *et al.* (2016) Analysis of plasma protein concentrations and enzyme activities in cattle within the ex-evacuation zone of the Fukushima Daiichi nuclear plant accident. *PLoS ONE* 11, e0155069. doi: 10.1371/journal.pone.0155069.

Utell, M.J. and Frampton, M.W. (2009) Acute health effects of ambient air pollution: The ultrafine particle hypothesis. *Journal of Aerosol Medicine* 13, 355–359. doi: 10.1089/jam.2000.13.355.

Watanabe, Y., Ichikawa, S., Kubota, M., Hoshino, J., Kubota, Y. *et al.* (2015) Morphological defects in native Japanese fir trees around the Fukushima Daiichi Nuclear Power Plant. *Scientific Reports* 5, 13232. doi: 10.1038/srep13232.

Watanobe, H., Furutani, T., Nihei, M., Sakuma, Y., Yanai, R. *et al.* (2014) The thyroid status of children and adolescents in Fukushima Prefecture examined during 20–30 months after the Fukushima nuclear power plant disaster: a cross-sectional, observational study. *PLoS ONE* 9, e113804. doi: 10.1371/journal.pone.0113804.

Yamashiro, H., Abe, Y., Fukuda, T., Kino, Y., Kawaguchi, I. *et al.* (2013) Effects of radioactive caesium on bull testes after the Fukushima nuclear accident. *Scientific Reports* 3, 2850. doi: 10.1038/srep02850.

Yamashiro, H., Abe, Y., Hayashi, G., Urushihara, Y., Kuwahara, Y. *et al.* (2015) Electron probe X-ray microanalysis of boar and inobuta testes after the Fukushima accident. *Journal of Radiation Research* 56, i42–i47. doi: 10.1093/jrr/rrv070.

Yap, H.K., Yip, W.C., Lee, B.W., Ho, T.F., Teo, J. *et al.* (1983) The incidence of atopy in steroid-responsive nephrotic syndrome: clinical and immunological parameters. *Annals of Allergy* 51, 590–594.

Yoschenko, V., Nanba, K., Yoshida, S., Watanabe, Y., Takase, T. *et al.* (2016) Morphological abnormalities in Japanese red pine (*Pinus densiflora*) at the territories contaminated as a aresult of the accident at Fukushima Dai-ichi Nuclear Power Plant. *Journal of Environmental Radioactivity* 165, 60–67. doi: 10.1016/j.jenvrad.2016.09.006.

Yoshioka, A., Mishima, Y. and Fukasawa, K. (2015) Pollinators and other flying insects inside and outside the Fukushima evacuation zone. *PLoS ONE* 10, e0140957. doi: 10.1371/journal.pone.0140957.

# Part VIII

# Remediation

# 35 Microbial Remediation of Contaminated Soils

**E. Shahsavari,* A.A. Mansur,** A. Aburto-Medina, N. Haleyur, N. Jones and A.S. Ball**

*Centre for Environmental Sustainability and Remediation, School of Sciences, RMIT University, Bundoora, Victoria, Australia*

## 35.1 Abstract

Our environment is threatened by thousands of contaminants, mainly as a result of human activities and industrialization. These include both inorganic (e.g. heavy metals) and organic compounds (e.g. polycyclic aromatic hydrocarbons, chlorinated hydrocarbons and herbicides). It has been shown that many of these hazardous chemicals cause health problem such as cancer in living organisms. Therefore, removing them from environments and soils represents a key challenge from both ecosystem and human health perspectives. Among the approaches used to remove these pollutants, microbial remediation or bioremediation represents a promising technology that is cost-effective, environmentally friendly and less disruptive than alternative technologies. Bioremediation involves microbes that are present in or added to the contaminated environment which are capable of degrading contaminants, or reducing them to less toxic forms. There are a number of bioremediation techniques, including natural attenuation, bioaugmentation, biostimulation, phytoremediation/rhizoremediation and necrophytoremediation Many factors, such as soil texture, pH, temperature, levels of oxygen, nutrients and the microbial status of the soils, influence the rate and extent of bioremediation along with the type and bioavailability of the contaminants. In this chapter, we highlight the current status of bioremediation, examining the development of techniques used to assess and optimize the degradation of the contaminants.

## 35.2 Introduction

Environmental pollution caused by human activities and industrialization is threatening our planet, lives and future. Thousands of hazardous compounds are released into the environment annually, leading to contamination of soil and groundwater sources. These recalcitrant chemicals include heavy metals, halocarbons, polychlorinated biphenyls (PCBs), polycyclic aromatic compounds (PAHs), synthetic polymers, alkyl benzyl sulfonates, nitroaromatics, dioxins, synthetic dyes, chlorophenols, herbicides and pesticides (Ijoma and Tekere, 2017). One of the most ubiquitous pollutants in the environment is PAHs (hydrocarbons with two or more fused phenyl

* E-mail address: esmaeil.shahsavari@rmit.edu.au

** Some of the material in this chapter has been taken from: Mansur, A. A. (2015) Bioremediation of Libyan soil contaminated with crude oil tank bottom sludge. PhD thesis. RMIT University, Melbourne, Australia.

 *A Handbook of Environmental Toxicology: Human Disorders and Ecotoxicology*. J.P.F. D'Mello (ed.)

and/or pentacyclic rings), many of which show toxic, mutagenic and carcinogenic properties (Lors *et al.*, 2010; Sheppard *et al.*, 2011). Many other major organic environmental contaminants are commonly used in industry. For example, PCBs are a family of compounds produced commercially, differing in the number of chlorine atoms (1–10) attached to their biphenyl rings, and have widely been used as hydraulic fluids, plasticizers, lubricants and flame retardants (Passatore *et al.*, 2014). With regard to heavy metals, the situation is no better: heavy-metal contamination occurs in one-sixth of the total agricultural land in China; and in the USA, about 600,000 ha of brownfield sites have been contaminated with heavy metals (Mahar *et al.*, 2016).

It has been shown that many of these chemicals and heavy metals are carcinogenic, mutagenic, neurotoxic and harmful for human and other organisms (Chapters 10–30). Environmental awareness about the adverse effects of these pollutants on environmental and human heath has increased over recent years, leading to important legislative and policy changes globally. The conventional physico-chemical techniques used for remediation have some potential drawbacks; the need for the development of sustainable, cost-effective technologies to remediate these hazardous chemicals in the environment is now more pressing than ever. Microbial remediation is a promising approach to eliminate persistent contaminants from the environment, due to its advantages over other physico-chemical processes. Bioremediation technologies are often used because of their low cost and high public acceptance and can often be applied onsite (Mansur *et al.*, 2015; Zhang *et al.*, 2010). Microbes also often convert toxic pollutants into harmless products such as water and carbon dioxide ($CO_2$), negating the requirement for disposal (Sarkar *et al.*, 2005; Surridge *et al.*, 2009; Kumar *et al.*, 2011; Das *et al.*, 2012). Therefore, the potential to develop commercial technologies based on bioremediation is being studied widely. The main purpose of this chapter is to review the current status of microbial degradation of contaminants in soils.

## 35.3 Remediation of Contaminated Soils

The remediation of contaminated soils has traditionally been carried out using physico-chemical methods such as thermal treatment, solvent extraction, steam stripping, hot air stripping, chemical oxidation, detergent extraction and immobilization (Morgan and Atlas, 1989; Singh *et al.*, 2009). The physico-chemical method to be employed depends on the type and extent of the contaminant(s). The most important of these remediation approaches and their advantages and disadvantages are presented in Table 35.1. However, over recent decades there has been significant research into biological remediation methods utilizing microbes such as bacteria and fungi that are able to use or degrade petroleum hydrocarbons, especially PAHs (Table 35.1).

In general, remediation of contaminated soil is categorized in two major groups: *in situ* and *ex situ*. *In situ* bioremediation is based on the treatment of contaminants at the site of contamination, whereas *ex situ* methods involve the removal of the contaminants to be treated off-site. A broad range of *in situ* and *ex situ* remediation methods including chemical, physical and biological approaches have been widely used to remediate petrogenic hydrocarbon-contaminated sites (Table 35.1).

## 35.4 Microbial Degradation or Bioremediation

Microbial degradation, or bioremediation, is defined as the use of microbes or their enzymes to clean up the environment. The idea of using bioremediation is not new; for example, bioremediation of soil contaminated with petroleum hydrocarbons has been known for about 80 years. However, effective studies of the application of bioremediation began in 1967 when Davis (1967) summarized the early work and concluded that specific microorganisms showed the potential to degrade petroleum hydrocarbons and utilize them as a main carbon source for energy and growth (Kumar *et al.*, 2011). Later studies showed that indigenous isolates from soil and water have the ability to degrade a wide range of contaminants in the environment, including different hydrocarbons (Hazen *et al.*, 2010; Jain *et al.*, 2012; Mansur *et al.*, 2014). Aerobic conditions are usually reported to be more effective in the removal of various contaminants (Neilson and Allard, 2007). However, highly halogenated aromatic compounds

**Table 35.1.** General approaches for the remediation of pollutants from contaminated soils (Shahsavari *et al.*, 2015).

| Method | Example of method | | Advantages | Disadvantages |
|---|---|---|---|---|
| | *In situ* | *Ex situ* | | |
| Physical | Capping | Excavation | Fast<br>Removing contaminants permanently<br>Ideal for high levels of contamination | Expensive<br>Destructive<br>Prone to secondary contamination |
| Chemical | Direct injection of chemical oxidants or surfactants into contaminated soil and groundwater | Chemical extraction | Fast<br>Not generating large volumes of waste material<br>Ideal for high level of contamination | Expensive<br>Destructive<br>Prone to secondary contamination |
| Biological (bioremediation) | Phytoremediation | Slurry phase biological treatment | Environmentally friendly<br>Cost-effective<br>Minimum site disruption<br>Useful for low level of contaminants | Requires longer time<br>Low predictability<br>Dependent on climatic factors |

are reported to be more easily degraded under strictly anaerobic conditions (Vogel *et al.*, 1987).

Bioremediation has become a focus for research into new, effective remediation techniques, especially for soils that are contaminated with low concentrations of contaminants. This is primarily due to the toxic effects that contaminants (e.g. PAHs) have on microorganisms. Effective bioremediation can be achieved only when optimized environmental conditions exist, allowing for the growth of active microorganisms capable of degrading the contaminant.

Bioremediation techniques include natural attenuation, biostimulation, bioaugmentation, phytoremediation/rhizoremediation and necrophytoremediation (Fig. 35.1). While each of the bioremediation techniques has its own limitations, they have several common limitations. These limitations are related to the factors that affect remediation as well as the processes involved. For example, all of the bioremediation methods rely on microbes to remediate the soil. Hydrocarbons, especially PAHs, are toxic and may adversely affect the soil microbial activity, thus decreasing their degradation efficiency (Koshlaf and Ball, 2017). This is the main reason why bioremediation is recommended for sites that have a lower concentration of contaminants. Another limitation of all bioremediation methods is that they are time-consuming, with remediation times varying from months to years. A further limitation comes from the requirement to provide additional materials such as nutrients and air, which use energy and can introduce co-contaminants.

### 35.4.1 Natural attenuation

Natural attenuation can be explained as essentially making no additions to remediate the soil except for regular monitoring of the contaminant concentration. Natural attenuation relies on the ability of the natural microflora to degrade naturally without further physiochemical or biological intervention (Scow and Hicks, 2005). Because natural attenuation depends only on natural degrading processes, it often requires a longer time to bring the contaminants to a lower concentration. This approach may raise concerns and objections from local communities, as there are no outward signs of bioremediation. The basis of the success of this technology is the fact that soils possess their own community of

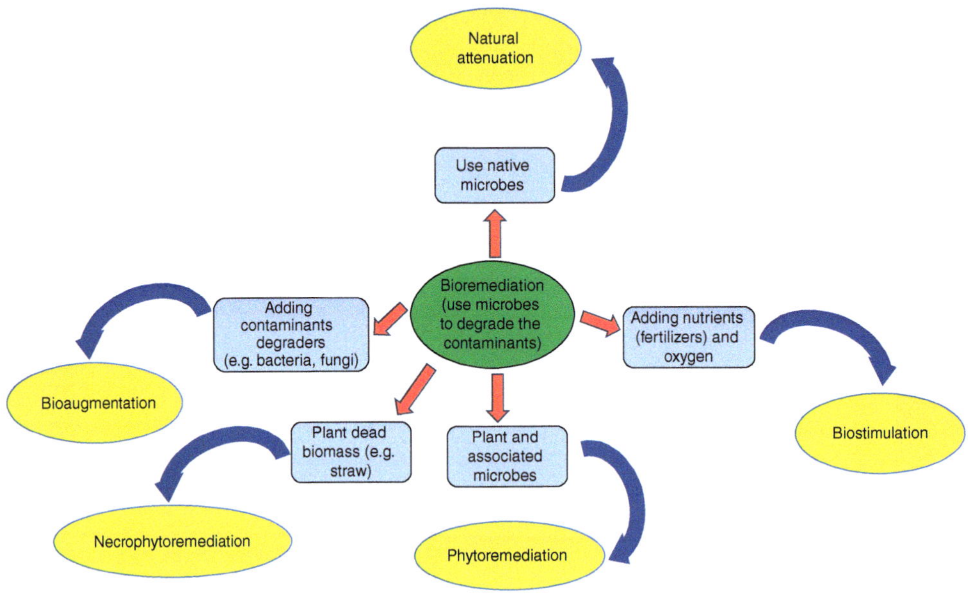


**Fig. 35.1.** Summary of different bioremediation methods used to remove contaminants from soil.

indigenous microbes capable of degrading or co-metabolizing hydrocarbons. Natural attenuation is ideal for low-level contamination and/or areas that are remote (Pilon-Smits, 2005). Therefore, in spite of the longer time taken for natural attenuation to remediate the contaminated environment, it has been used routinely at remediation sites of soil contaminated with petroleum hydrocarbon. In one study, natural attenuation was conducted on a soil contaminated with diesel off Long Beach, California. After incubation for 12 weeks, the results indicated that the concentration of diesel in the soil was reduced by 36% (Bento *et al.*, 2005).

### 35.4.2 Biostimulation

Biostimulation is when additives are used to increase the degradative capability of microbes. Typically biostimulation is in the form of nutrient additions (e.g. glucose, molasses) or the addition of electron acceptors (e.g. acetate, lactate) to the contaminated soil (Adetutu *et al.*, 2011; El Fantroussi and Agathos, 2005). Biostimulation can improve the degradation rate of the pollutants by optimizing important and effective environmental conditions through the addition of nutrients, or by aeration, control of pH and temperature (Margesin *et al.*, 2000). In biostimulation, traditionally nutrient supplementation focuses on the addition of nitrogen (N) and phosphorus (P) in various nutrient sources such as urea, inorganic fertilizers, compost, sawdust, biosolids and manure ( Walworth and Reynolds, 1995; Cho and Kende, 1997; Namkoong *et al.*, 2002). Suja *et al.* (2014) showed that biostimulation through the addition of nutrients (N and/or P) resulted in an accelerated degradation of hydrocarbons due to the increased metabolic activity of indigenous microorganisms.

### 35.4.3 Bioaugmentation

Another bioremediation approach is to enhance the degradation capacity of the soil microbial community by introducing specific microbial strains or consortia to the contaminated sites. Bioaugmentation is the addition of mainly bacteria and occasionally fungi that are able to degrade hydrocarbons into the soil. These microbes are typically isolated from contaminated soils. The addition of microbes may be that of a single strain or a whole consortium of hydrocarbonoclastic microbes, depending on the situation. Isolated strains have varied success but enriched consortia have had more success.

However, bioaugmentation is limited, due to the fact that the interactions between the added microbes and indigenous microbes and their effect on the ecosystem are largely unknown (Koshlaf and Ball, 2017). In one example, a microcosm study of the bioremediation of soil contaminated with PAHs was conducted using a bioaugmentation strategy. After incubation for 28 days, the bioaugmentation treatment showed a 23.2% decrease in PAHs concentration (Teng *et al.*, 2010) compared with 3.4% in the control soil.

To overcome the limitations of one approach, combination of biostimulation (BS) and bioaugmentation (BA) can also be applied and previous studies have shown increased PAH degradation after employing this approach (Straube *et al.*, 2003). Another study on the bioremediation of soil contaminated with hydrocarbon was conducted and BA and BS were applied to biopiles. After 140 days, bioaugmentation resulted in a reduction of between 64% and 68% of the contaminant concentration (Liu *et al.*, 2011).

### 35.4.4 Phytoremediation

Phytoremediation (from the Ancient Greek *phyto*, meaning 'plant', and Latin *remedium*, meaning 'restoring balance') utilizes green plants and their associated microbiota to break down, contain, remove or detoxify contaminants (Herath and Vithanage, 2015), without having to remove contaminated material for off-site disposal. It offers a low-cost alternative to the removal and off-site disposal techniques often used for contaminated soil and has the added advantage of being a popular method with the public. Phytoremediation is a technique that has stimulated considerable interest over the past couple of decades. It involves the uptake of pollutants into the plant itself, where it can accumulate in the stems or leaves or can be expelled (Shahsavari *et al.*, 2013b). When the contaminant is accumulated into the plant, it can either continue to be accumulated into the plant or be degraded inside the plant. These two mechanisms are more efficient for metals than organic pollutants but represent an effective technology. A limitation of phytoremediation is the fact that hydrocarbons are toxic to plants (Agnello *et al.*, 2016). Phytoremediation can also be used in conjunction with other bioremediation methods, including bioaugmentation and biostimulation. However, prior to being used in the field, plants need to be screened for tolerance to contaminants and soil conditions. The advantages in using phytoremediation are that the plants are inexpensive and the remediation itself is therefore cost-effective; also there is no need to remove and dump the contaminated soil. Finally, phytoremediation is less disruptive to the environment and is publicly accepted. Grasses are particularly effective for use in phytoremediation. This is because they have the greatest root surface volume and can cover contaminated land quickly (Shahsavari *et al.*, 2015). Gaskin and Bentham (2010) showed that the presence of grasses in hydrocarbon-contaminated areas significantly increased the abundance of hydrocarbon-degrading microorganisms, leading to 88% reduction in hydrocarbons from contaminated soils.

### 35.4.5 Rhizoremediation

Rhizoremediation is directly linked to phytoremediation, as it involves the microorganisms that are found in the rhizosphere around the roots of plants. It is useful for hydrocarbon contaminants that are not able to be taken up by plants. Due to nutrients expelled by the roots, the rhizosphere microbial population is five to 100 times more abundant than those in the bulk soil (Shahsavari *et al.* 2015). The root system carries the microbes down to the location of the hydrocarbons and allows the microbes to come into contact with the contaminant; this step is vital for remediation. Plants are selected based on their root systems, with those that have the most extensive root network generally used. These include ryegrass, legumes and various grasses (Pilon-Smits, 2005). The roots are also affected by the bioavailability of the contaminants. For example, increased hydrocarbons can decrease the length of the roots and alter the plant root architecture (Peng *et al.*, 2009).

### 35.4.6 Necrophytoremediation

Necrophytoremediation is an emerging bioremediation technology that is related to phytoremediation. Necrophytoremediation differs from phytoremediation in that it involves the use of dead plant biomass, while phytoremediation uses living plants. This is an advantage, as plant residues are not affected by hydrocarbon toxicity, nutrient deficiency, plant growth and rainfall (Shahsavari *et al.*, 2015). Plant residues represent an abundant agricultural waste which makes necrophytoremediation an inexpensive bioremediation method. Studies have found that necrophytoremediation is an effective bioremediation technology when added to soil either stand-alone with soil or in combination with other bioremediation technologies (Shahsavari *et al.*, 2013b). For example, the addition of pea straw during the remediation of hydrocarbon-contaminated soils resulted in an increase in the degradation of pyrene from 15% in the corresponding control to 70% (Shahsavari *et al.*, 2013a).

## 35.5 Factors Affecting Bioremediation

In order to effectively utilize a bioremediation approach, it is imperative to know the nature of the contaminant(s) and to fully understand the factors that can affect or enhance their biodegradability (Bouwer *et al.*, 1994; Boopathy, 2000). It is generally agreed that the key factors include: (i) bioavailability; (ii) energy sources; and (iii) bioactivity and biochemistry (Boopathy, 2000).

Bioavailability is very important, because the rate of contaminant transfer to the microbial cell is a critical limiting factor that depends on several physico-chemical processes such as diffusion, sorption–desorption and dissolution. Thus, the bioavailability of the contaminant depends on its aqueous solubility, volatility or reactivity in water and soils. One way to provide a direct measure of the bioavailability is by testing the toxicity of a pollutant using microorganisms as a biosensor (Megharaj *et al.*, 2011).

Pollutants can be sequestered over time due to soil properties, including organic matter content, cation exchange capacity, micropore volume, surface area and soil texture (Chung and Alexander, 2002).

In order to increase the bioavailability of the contaminants, surfactants can be added (Boopathy and Manning, 1999; Kim *et al.*, 2001; Ławniczak *et al.*, 2013). These molecules contain hydrophilic and hydrophobic moieties, with chemicals used including Triton X-100, Tween 80 and sodium dodecyl sulfate. They aid in the remediation by: (i) lowering the interfacial tension; (ii) surfactant solubilization of hydrophobic organic compounds; and (iii) transfer of soil-sorbed organic compounds to the aqueous phase (Laha *et al.*, 2009). The benefits of the surfactants in terms of increased degradation have been observed with different contaminants such as PAHs or DDT. In one instance, Gonzáles *et al.* (2011) evaluated Tween-80 (biodegradable) surfactant on growth, degradation rate and microbial dynamics of a PAH-degrading consortium obtained from petroleum-polluted soils. The results indicated that growth and degradation rates were significantly higher when Tween-80 (90%) was applied when compared with control (20%).

There are also food and plant-grade surfactants such as T-MAZ 28, T-MAZ 10, T-MAZ 60 and fruit pericarp (Shiau *et al.*, 1995) that may be preferred to synthetic surfactants due to their low toxicity and high biodegradability. Additionally some microorganisms can produce their own surfactants, such as lipoproteins, lipopeptides, or phospholipids. For example, rhamnolipids (surface-active glycolipids) produced from *Pseudomonas aeruginosa* increased the solubility of anthracene, leading to its significant degradation (from 34% to 52%) (Cui *et al.*, 2008). Biosurfactant produced by *Bacillus circulans* emulsified petrol, hexadecane, kerosene, diesel and benzene (Das *et al.*, 2008). Biosurfactants produced from *Acinetobacter radioresistens* KA53 were able to emulsify a range of hydrocarbons like anthracene, benzo(a)pyrene, chrysene, fluoranthene, fluorene, hexylbenzylcyclosilane, hexylphenylbenzene, α-naphtholbenzene, perylene, phenanthrene, 1,10-phenanthrolene and pyrene (Toren *et al.*, 2002).

The addition of surfactants also prevents ageing or weathering by enhancing the rate of contaminant desorption from the soil. Nevertheless, it has been observed that organic contaminants can be degraded without prior desorption (Singh *et al.*, 2003). For example, *Brevibacterium* sp. was found to degrade the pesticide Fenamiphos when it was intercalated with the cationic-surfactant modified montmorillonite clay (Singh *et al.*, 2003). Ideally the contaminant should be in the reduced state to be used as an energy source for aerobic heterotrophic organisms. In general, lower energy yields are obtained from higher oxidation states of the carbon in the contaminant. This and the concentration of the substrate will determine if it will be used as a primary, secondary or co-metabolic substrate.

Therefore many factors are required to be optimized in order to have a successful biodegradation process, including: (i) microbial processes (enzyme activities and biomass concentration); (ii) environmental factors (temperature, pH, conductivity, moisture content); and (iii) substrate availability (oxidation state, concentration, molecular structure).

The microbial processes involved in the biodegradation can be termed bioactivity. When the system conditions are transformed to those ideal for the highest biodegradation, bioactivity is improved. Biodegradation *in situ* is complex, due to the diverse, synergistic or antagonistic communities present; this contrasts with degradation processes by single strains. However, *in situ* substrates and products are transferred and degraded within the microbial community; therefore an adapted consortium may sometimes be preferred rather than single-strain when developing a bioremediation approach. A detailed study of the biodegradation processes is useful to understand the role of individual genes, enzymes and organisms, but a system biology approach is required for identifying the partners and response patterns within a complex contaminated site, as has been recently proposed (Chakraborty *et al.*, 2012; Dvořák *et al.*, 2017). Such an approach aims to include and apply all available techniques such as soil metagenomics, the use of analysis (Villas-Bôas and Bruheim, 2007), functional genomics and proteomics (Zhao and Poh, 2008) or molecular techniques for community profiling (Malik *et al.*, 2008). Stable isotope probing has also been used to link identity to function in bioremediation studies, thereby identifying the microorganisms responsible for the degradation of the contaminant (Radajewski *et al.*, 2000; Kunapuli *et al.*, 2007; Oka *et al.*, 2008; Aburto and Ball, 2009; Cupples, 2011).

## 35.6 Monitoring of Bioremediation in Contaminated Soil

During the bioremediation process, it is crucial to monitor the levels of the contaminant together with any intermediates. In addition, monitoring of microbes as the main agents of bioremediation is a crucial step in monitoring the bioremediation. In this regard, different analytical techniques have been employed (Table 35.2), including gas chromatography (GC), gas chromatography–mass spectrometry (GC–MS), Fourier transform infrared spectroscopy (FTIR), inductively coupled plasma–mass spectrometry (ICP–MS) for metals and thermogravimeteric infrared spectrometry (TGIR) or a combined TG–FTIR ( Mittleman, 1990; Poster *et al.*, 2006; Yew *et al.*, 2008; Zhu *et al.*, 2008; Biller and Bruland, 2012). These techniques can assess the concentration of contaminants such as total petroleum hydrocarbons (TPH), PAHs and metals (Sheppard *et al.*, 2011; Biller and Bruland, 2012). However, these analytical techniques do not provide any information regarding the biological components of the bioremediation. Measuring and identifying the size, activity and diversity of the microbial community in the contaminated soil is essential for developing effective and robust bioremediation approaches. There are many methods used for assessing microbial activities in environments contaminated with hydrocarbons; both culture-dependent and culture-independent techniques have been used (Zhang *et al.*, 2012).

### 35.6.1 Culture-dependent techniques

To understand and improve the bioremediation process of contaminants in soils, it is essential to investigate the microbial community involved in the biodegradation (Kadali, 2012). To detect the presence of microorganisms involved in the degradation and investigate their ability to grow and utilize contaminants as the sole source of carbon and energy and to assess the potential changes in the bacterial community composition, culture-dependent methodologies are often used to determine the microbial diversity present within a designated environment (Connon *et al.*, 2005; Steven *et al.*, 2007). For example, culture isolation and plate counting is a technique used to culture, detect and isolate microorganisms from a polluted environment. Bacteria are cultured on a liquid or solid medium designed for their growth. There are many types of growth media used but the most commonly used are nutrient broth (liquid medium) and nutrient

**Table 35.2.** Different techniques used for monitoring the microbial removal of soil contaminants.

| Technique | Description | Parameter measured | Reference |
|---|---|---|---|
| Analytical techniques such as gas chromatograph mass spectroscopy (GCMS) an GC-FID | Separation and identification of compounds | Contaminant determination and identification | Moore *et al.* (2014) |
| Culture-dependent techniques | Isolation including contaminant degraders and abundance of microorganisms | Microbial assessment in order to identify and evaluate the capability of bacterial strains to degrade contaminants | Adriaenssens *et al.* (2014) Mittal and Singh (2009) |
| Soil respirometry | Contaminant degradation is assessed by measuring the rate of $CO_2$ production and $O_2$ consumption | Microbial activity using the measurement of carbon dioxide | Mansur *et al.* (2016) Montagnolli *et al.* (2015) |
| Enzyme activities of soils, e.g. fluorescein diacetate hydrolytic and dehydrogenase activity | Microbial activity of the soils is measured | Microbial activity | Margesin (2005) Shahsavari *et al.* (2013b) |
| Culture-independent techniques (molecular tools) such as next-generation sequencing | Microbial communities and structure are assessed by using molecular tools | Microbial communities | Koshlaf *et al.* (2016) |

agar (solid medium). These contain a rich growth medium designed to isolate a wide array of heterotrophic microorganisms, enabling an estimate of the culturable heterotrophic bacteria present in a sample (Schlegel and Zaborosch, 1993). For determining the number of cultural microorganisms capable of degrading the contaminant of interest, such as hydrocarbon contaminants, selective enrichment using Bushnell–Haas mineral salt medium (BHMSM), enriched with either a single hydrocarbon fraction or a mixture of hydrocarbons such as petroleum crude oil, heavy oils or diesel as the only source of carbon and energy, represents a successful isolation method (Bushnell and Haas, 1941).

### 35.6.2 Soil respiration and respirometry measurements

Reported studies have used an array of different techniques to assess the biodegradation rate and the potential of the organisms to utilize the contaminants. In aerobic biodegradation processes in environments contaminated with hydrocarbons, the complete mineralization of the contaminants leads to the formation of $H_2O$, $CO_2$ and methane as end products (Fingerman, 2005). Therefore, measuring the oxygen consumption or $CO_2$ generation in the gas phase during the bioremediation are effective and reliable tools providing information about the mineralization rate of the contaminants (Schoefs *et al.*, 2004). Respiration measurements during the bioremediation treatment directly represent the bacterial activity and metabolic process (Aspray *et al.*, 2008). Many studies used the generation of $CO_2$ as a reliable and direct approach to assess microbial activity and therefore the degradation of the contaminant (Aspray *et al.*, 2008). In one study conducted to measure the biodegradation potential of PAHs by bacterial isolates, $CO_2$ generation was used as an indication of the mineralization of PAHs. After 4 days of incubation, the results showed that the bacterial consortium with PAHs as sole carbon source produced higher level of $CO_2$ (1338 ppmv) compared with the control (181 ppmv). This increased activity correlated with a high percentage degradation (97–98 %) (Arulazhagan *et al.*, 2010).

### 35.6.3 Enzyme activities of soils

The presence of contaminants can change the microbial activities of the soil. Thus, measurement of microbial activities gives valuable information regarding the potential of the microbial community to degrade the contaminants. In this regards, enzyme activities in soils reflect the total microbial activity, including the decomposition of organic inputs as well as the degradation of the contaminants. It has been shown that assessment of enzymes in soils is a good indicator of bioremediation of different contaminants such as petroleum hydrocarbons, PAHs and heavy metals (Deng *et al.*, 2015; García-Delgado *et al.*, 2015; Gaskin and Bentham, 2010). For example, lipase–esterase activity, fluorescein diacetate hydrolytic activity and dehydrogenase activity have all been successfully used to monitor the bioremediation soils (Margesin, 2005).

It is important to note that enzyme activities only show the total microbial activity, which sometimes does not reflect microbial activity involved in the degradation of contaminants. Additional methods, such as molecular methods, should be used in combination with enzyme activities to monitor microbial degradation in soils.

### 35.6.4 Culture-independent techniques (molecular tools)

Only a small portion of soil biota can be cultured and characterized; therefore microbial molecular tools have been developed to monitor and underpin the microbial degradation system. Data obtained from culture-independent methods help to assess the diversity of the bacterial community and gene composition in addition to monitoring any changes in the bacterial community during the bioremediation process (Zhang *et al.*, 2010). In addition, culture-independent techniques are used to study the effects of using different bioremediation strategies such as bioaugmentation, biostimulation and natural attenuation and to compare the changes in microbial diversity and indicate the potential of using different bioremediation techniques on the hydrocarbon-degrading bacterial communities (Vinas *et al.*, 2005). Moreover, it can be used to study the presence of specific genes within a microbial population responsible for degrading specific contaminants (Matsuki *et al.*, 2004). Many fingerprinting techniques have been developed and applied to investigations of microbial community changes in situations such as bioremediation. Among these techniques are automated ribosomal intergenic spacer analysis (ARISA) (Fisher and Triplett, 1999), terminal restriction fragment length polymorphism (TRFLP) (Osborn *et al.*, 2000), denaturant gradient gel electrophoresis (DGGE) (Muyzer *et al.*, 1993) and temperature gradient gel electrophoresis (TGGE) (Muyzer and Smalla, 1998; Fingerman, 2005). In particular, DGGE has been used in many bioremediation studies of environmental contaminants such as hydrocarbon pollutants (Vinas *et al.*, 2005; Mao *et al.*, 2012; Gargouri *et al.*, 2014).

Next-generation sequencing (NGS) has now superseded all other methodologies, as it promises several advantages over all the other techniques. The increased demand for fast, cheap and accurate DNA sequencing has led to the advancement of NGS. Since the introduction of NGS in 2005, standard sequencing applications including genome sequencing and for novel applications which were not explored by traditional Sanger sequencing have been possible. Although previous sequencing approaches have been widely adopted, some limitations in throughput, speed, cost and resolution may prevent researchers from obtaining essential genomic data. NGS technologies from 454/Roche, Illumina/Solexa, ABI/SOLID and Helicos led to high-throughput functional genomic research, applied in different contexts such as whole-genome sequencing and targeted genes. In regard to the study of microbial communities, NGS or metagenomic analysis has the ability to examine changes in microbial communities without previous knowledge (for uncultured organisms); it also provides information about how soil microbial communities change across time and space and requires a very small amount (about 1 μg) of template laying out millions of DNA fragments on a single chip and sequencing all the fragments in a parallel fashion.

## 35.7 Conclusions

Human activities and industrialization lead to extensive soil contamination with hazardous materials, including heavy metals, BTEX (benzene, toluene, ethylbenzene and xylene), PAHs, PCBs, trichloroethylene (TCE), pesticides etc. which are very harmful to human and ecosystem health. Therefore, there is an urgent need for the removal of the contaminants from soil. Of the three main remediation methods – physical, chemical and biological (bioremediation) approaches – chemical and physical approaches are not only expensive and labour intensive but also result in further contamination. Bioremediation, the application of microbial degradation to degrade or remove the contaminants, represents a cost-effective, eco-friendly approach to clean up the environment. However, during the bioremediation, monitoring the microbial community and their degradative activates is a critical step.

## References


Aburto, A. and Ball, A.S. (2009) Bacterial population dynamics and separation of active degraders by stable isotope probing during benzene degradation in a BTEX-impacted aquifer. *Revista Internacional de Contaminacion Ambiental* 25(3).

Adetutu, E.M., Ball, A.S., Weber, J., Aleer, S., Dandie, C.E. and Juhasz, A.L. (2011) Impact of bacterial and fungal processes on $^{14}C$-hexadecane mineralisation in weathered hydrocarbon contaminated soil. *Science of the Total Environment* 414, 585–591. doi: 10.1016/j.scitotenv.2011.11.044.

Adriaenssens, E.M., Guerrero, L.D., Makhalanyane, T.P., Aislabie, J.M. and Cowan, D.A. (2014) Draft genome sequence of the aromatic hydrocarbon-degrading bacterium *Sphingobium* sp. strain Ant17, isolated from Antarctic soil. *Genome Announcements* 2(2), e00212–00214.

Agnello, A.C., Bagard, M., van Hullebusch, E.D., Esposito, G. and Huguenot, D. (2016) Comparative bioremediation of heavy metals and petroleum hydrocarbons co-contaminated soil by natural attenuation, phytoremediation, bioaugmentation and bioaugmentation-assisted phytoremediation. *Science of the Total Environment* 563–564, 693–703. doi:10.1016/j.scitotenv.2015.10.061.

Arulazhagan, P., Vasudevan, N. and Yeom, I. (2010) Biodegradation of polycyclic aromatic hydrocarbon by a halotolerant bacterial consortium isolated from marine environment. *International Journal of Environmental Science & Technology* 7(4), 639–652.

Aspray, T., Gluszek, A. and Carvalho, D. (2008) Effect of nitrogen amendment on respiration and respiratory quotient (RQ) in three hydrocarbon contaminated soils of different type. *Chemosphere* 72(6), 947–951.

Bento, F.M., Camargo, F.A.O., Okeke, B.C. and Frankenberger, W.T. (2005) Comparative bioremediation of soils contaminated with diesel oil by natural attenuation, biostimulation and bioaugmentation. *Bioresource Technology* 96(9), 1049–1055.

Biller, D.V. and Bruland, K.W. (2012) Analysis of Mn, Fe, Co, Ni, Cu, Zn, Cd, and Pb in seawater using the Nobias-chelate PA1 resin and magnetic sector inductively coupled plasma mass spectrometry (ICP-MS). *Marine Chemistry* 130, 12–20.

Boopathy, R. (2000) Factors limiting bioremediation technologies. *Bioresource Technology* 74(1), 63–67. doi: 10.1016/s0960-8524(99)00144-3.

Boopathy, R. and Manning, J. (1999) Surfactant-enhanced bioremediation of soil contaminated with 2,4,6-trinitrotoluene in soil slurry reactors. *Water Environment Research* 71(1), 119–124. doi: 10.2175/106143099X121580.

Bouwer, E., Durant, N., Wilson, L., Zhang, W. and Cunningham, A. (1994) Degradation of xenobiotic compounds in situ: capabilities and limits. *FEMS Microbiology Reviews* 15(2–3), 307–317. doi: 10.1111/j.1574-6976.1994.tb00142.x.

Bushnell, L. and Haas, H. (1941) The utilization of certain hydrocarbons by microorganisms. *Journal of Bacteriology* 41(5), 653–673.

Chakraborty, R., Wu, C.H. and Hazen, T.C. (2012) Systems biology approach to bioremediation. *Current Opinion in Biotechnology* 23(3), 483–490. doi: 10.1016/j.copbio.2012.01.015.

Cho, H.-T. and Kende, H. (1997) Expansins and internodal growth of deepwater rice. *Plant Physiology* 113(4), 1145–1151.

Chung, N. and Alexander, M. (2002). Effect of soil properties on bioavailability and extractability of phenanthrene and atrazine sequestered in soil. *Chemosphere* 48(1), 109–115. doi: 10.1016/S0045-6535(02)00045-0.

Connon, S.A., Tovanabootr, A., Dolan, M., Vergin, K., Giovannoni, S.J. and Semprini, L. (2005) Bacterial community composition determined by culture-independent and-dependent methods during propane-stimulated bioremediation in trichloroethene-contaminated groundwater. *Environmental Microbiology* 7(2), 165–178.

Cui, C.Z., Wan, X. and Zhang, J.Y. (2008) Effect of rhamnolipids on degradation of anthracene by two newly isolated strains, *Sphingomonas* sp. 12A and *Pseudomonas* sp. 12B. *Journal of Microbiology and Biotechnology* 18(1), 63–66.
Cupples, A.M. (2011) The use of nucleic acid based stable isotope probing to identify the microorganisms responsible for anaerobic benzene and toluene biodegradation. *Journal of Microbiological Methods* 85(2), 83–91. doi: 10.1016/j.mimet.2011.02.011.
Das, A., Sherameti, I. and Varma, A. (2012) Contaminated soil: physical, chemical and biological components. *Bio-Geo Interactions in Metal-Contaminated Soils*, 1–15.
Das, P., Mukherjee, S. and Sen, R. (2008) Improved bioavailability and biodegradation of a model polyaromatic hydrocarbon by a biosurfactant producing bacterium of marine origin. *Chemosphere* 72(9), 1229–1234.
Davis, J.B. (1967) *Petroleum Microbiology*. Elsevier North Holland, New York.
Deng, L., Zeng, G., Fan, C., Lu, L., Chen, X. *et al.* (2015) Response of rhizosphere microbial community structure and diversity to heavy metal co-pollution in arable soil. *Applied Microbiology and Biotechnology* 99(19), 8259–8269. doi: 10.1007/s00253-015-6662-6.
Dvořák, P., Nikel, P.I., Damborský, J. and de Lorenzo, V. (2017) Bioremediation 3.0: Engineering pollutant-removing bacteria in the times of systemic biology. *Biotechnology Advance*, 35(7), 845–866. doi: 10.1016/j.biotechadv.2017.08.001.
El Fantroussi, S. and Agathos, S.N. (2005) Is bioaugmentation a feasible strategy for pollutant removal and site remediation? *Current Opinion in Microbiology* 8(3), 268–275.
Fingerman, M. (2005) *Bioremediation of Aquatic and Terrestrial Ecosystems*. CRC Press, Boca Raton, Florida.
Fisher, M.M. and Triplett, E.W. (1999) Automated approach for ribosomal intergenic spacer analysis of microbial diversity and its application to freshwater bacterial communities. *Applied and Environmental Microbiology* 65(10), 4630–4636.
García-Delgado, C., Yunta, F. and Eymar, E. (2015) Bioremediation of multi-polluted soil by spent mushroom (*Agaricus bisporus*) substrate: polycyclic aromatic hydrocarbons degradation and Pb availability. *Journal of Hazardous Materials* 300, 281–288. doi: 10.1016/j.jhazmat.2015.07.008.
Gargouri, B., Karray, F., Mhiri, N., Aloui, F. and Sayadi, S. (2014) Bioremediation of petroleum hydrocarbons-contaminated soil by bacterial consortium isolated from an industrial wastewater treatment plant. *Journal of Chemical Technology and Biotechnology* 89(7), 978–987.
Gaskin, S.E. and Bentham, R.H. (2010) Rhizoremediation of hydrocarbon contaminated soil using Australian native grasses. *Science of the Total Environment* 408(17), 3683–3688.
González, N., Simarro, R., Molina, M., Bautista, L., Delgado, L. and Villa, J. (2011) Effect of surfactants on PAH biodegradation by a bacterial consortium and on the dynamics of the bacterial community during the process. *Bioresource Technology* 102(20), 9438–9446.
Hazen, T.C., Dubinsky, E.A., DeSantis, T.Z., Andersen, G.L., Piceno, Y.M. *et al.* (2010) Deep-sea oil plume enriches indigenous oil-degrading bacteria. *Science* 330(6001), 204–208.
Herath, I. and Vithanage, M. (2015) Phytoremediation in constructed wetlands. In: Ansari, A.A., Gill, S.S., Gill, R., Lanza, R.G.R. and Newman, L. (eds) *Phytoremediation: Management of Environmental Contaminants, Volume* 2. Springer International Publishing, Cham, Switzerland, pp. 243–263.
Ijoma, G.N. and Tekere, M. (2017) Potential microbial applications of co-cultures involving ligninolytic fungi in the bioremediation of recalcitrant xenobiotic compounds. *International Journal of Environmental Science and Technology* 14(8), 1787–1806. doi: 10.1007/s13762-017-1269-3
Jain, K., Shah, V., Chapla, D. and Madamwar, D. (2012) Decolorization and degradation of azo dye–Reactive Violet 5R by an acclimatized indigenous bacterial mixed cultures-SB4 isolated from anthropogenic dye contaminated soil. *Journal of Hazardous Materials* 213, 378–386.
Kadali, K.K. (2012) *Isolation of Hydrocarbonoclastic Bacteria and Assessment of Their Bioremediation Potential in the Treatment of Oil Contaminated Marine Environments*. School of Biological Sciences, Flinders University of South Australia.
Kim, I.S., Park, J.-S. and Kim, K.-W. (2001) Enhanced biodegradation of polycyclic aromatic hydrocarbons using nonionic surfactants in soil slurry. *Applied Geochemistry* 16(11–12), 1419–1428. doi: 10.1016/S0883-2927(01)00043-9.
Koshlaf, E. and Ball, A.S. (2017) Soil bioremediation approaches for petroleum hydrocarbon polluted environments. *AIMS Microbiology* 3(1), 25–49. doi: 10.3934/microbiol.2017.1.25.
Koshlaf, E., Shahsavari, E., Aburto-Medina, A., Taha, M., Haleyur, N. *et al.* (2016) Bioremediation potential of diesel-contaminated Libyan soil. *Ecotoxicology and Environmental Safety* 133, 297–305.
Kumar, A., Bisht, B., Joshi, V. and Dhewa, T. (2011) Review on bioremediation of polluted environment: a management tool. *International Journal of Environmental Sciences* 1(6), 1079.

Kunapuli, U., Lueders, T. and Meckenstock, R.U. (2007) The use of stable isotope probing to identify key iron-reducing microorganisms involved in anaerobic benzene degradation. *ISME Journal* 1(7), 643–653.
Laha, S., Tansel, B. and Ussawarujikulchai, A. (2009) Surfactant–soil interactions during surfactant-amended remediation of contaminated soils by hydrophobic organic compounds: a review. *Journal of Environmental Management* 90(1), 95–100. doi: 10.1016/j.jenvman.2008.08.006.
Ławniczak, Ł., Marecik, R. and Chrzanowski, Ł. (2013) Contributions of biosurfactants to natural or induced bioremediation. *Applied Microbiology and Biotechnology* 97(6), 2327–2339. doi: 10.1007/s00253-013-4740-1.
Liu, P.-W.G., Chang, T.C., Whang, L.-M., Kao, C.-H., Pan, P.-T. and Cheng, S.-S. (2011). Bioremediation of petroleum hydrocarbon contaminated soil: effects of strategies and microbial community shift. *International Biodeterioration & Biodegradation* 65(8), 1119–1127.
Lors, C., Ryngaert, A., Périé, F., Diels, L. and Damidot, D. (2010) Evolution of bacterial community during bioremediation of PAHs in a coal tar contaminated soil. *Chemosphere* 81(10), 1263–1271.
Mahar, A., Wang, P., Ali, A., Awasthi, M.K., Lahori, A.H. *et al.* (2016) Challenges and opportunities in the phytoremediation of heavy metals contaminated soils: a review. *Ecotoxicology and Environmental Safety* 126, 111–121. doi: 10.1016/j.ecoenv.2015.12.023.
Malik, S., Beer, M., Megharaj, M. and Naidu, R. (2008) The use of molecular techniques to characterize the microbial communities in contaminated soil and water. *Environment International* 34(2), 265–276.
Mansur, A.A. (2015) Bioremediation of Libyan soil contaminated with crude oil tank bottom sludge. PhD thesis. RMIT University, Melbourne, Australia.
Mansur, A.A., Adetutu, E.M., Kadali, K.K., Morrison, P.D., Nurulita, Y. and Ball, A.S. (2014) Assessing the hydrocarbon degrading potential of indigenous bacteria isolated from crude oil tank bottom sludge and hydrocarbon-contaminated soil of Azzawiya oil refinery, Libya. *Environmental Science and Pollution Research* 21(18), 10725–10735.
Mansur, A.A., Adetutu, E.M., Makadia, T., Morrison, P.D. and Ball, A.S. (2015) Assessment of the hydrocarbon degrading abilities of three bioaugmentation agents for the bioremediation of crude oil tank bottom sludge contaminated Libyan soil. *International Journal of Environmental Bioremediation & Biodegradation* 3(1), 1–9.
Mansur, A.A., Taha, M., Shahsavari, E., Haleyur, N., Adetutu, E.M. and Ball, A.S. (2016) An effective soil slurry bioremediation protocol for the treatment of Libyan soil contaminated with crude oil tank bottom sludge. *International Biodeterioration & Biodegradation* 115, 179–185.
Mao, J., Luo, Y., Teng, Y. and Li, Z. (2012) Bioremediation of polycyclic aromatic hydrocarbon-contaminated soil by a bacterial consortium and associated microbial community changes. *International Biodeterioration & Biodegradation* 70, 141–147.
Margesin, R. (2005) Determination of enzyme activities in contaminated soil. In: Margesin, R. and Schinner, F. (eds) *Manual for Soil Analysis – Monitoring and Assessing Soil Bioremediation*. Soil Biology series, Vol. 5. Springer Berlin Heidelberg, New York, pp. 309–320.
Margesin, R., Zimmerbauer, A. and Schinner, F. (2000) Monitoring of bioremediation by soil biological activities. *Chemosphere* 40(4), 339–346.
Matsuki, T., Watanabe, K., Fujimoto, J., Takada, T. and Tanaka, R. (2004) Use of 16S rRNA gene-targeted group-specific primers for real-time PCR analysis of predominant bacteria in human feces. *Applied and Environmental Microbiology* 70(12), 7220–7228.
Megharaj, M., Ramakrishnan, B., Venkateswarlu, K., Sethunathan, N. and Naidu, R. (2011) Bioremediation approaches for organic pollutants: a critical perspective. *Environment International* 37(8), 1362–1375. doi: 10.1016/j.envint.2011.06.003.
Mittal, A. and Singh, P. (2009) Isolation of hydrocarbon degrading bacteria from soils contaminated with crude oil spills. *Indian Journal of Experimental Biology* 47(9), 760.
Mittleman, M. (1990) Quantitative TG/IR. *Thermochimica Acta* 166, 301–308.
Montagnolli, R.N., Lopes, P.R.M. and Bidoia, E.D. (2015) Assessing *Bacillus subtilis* biosurfactant effects on the biodegradation of petroleum products. *Environmental Monitoring and Assessment* 187(1), 1–17.
Moore, H.E., Adam, C.D. and Drijfhout, F.P. (2014) Identifying 1st instar larvae for three forensically important blowfly species using 'fingerprint' cuticular hydrocarbon analysis. *Forensic Science International* 240, 48–53.
Morgan, P. and Atlas, R.M. (1989) Hydrocarbon degradation in soils and methods for soil biotreatment. *Critical Reviews in Biotechnology* 8(4), 305–333.
Muyzer, G. and Smalla, K. (1998) Application of denaturing gradient gel electrophoresis (DGGE) and temperature gradient gel electrophoresis (TGGE) in microbial ecology. *Antonie van Leeuwenhoek* 73(1), 127–141.

Muyzer, G., De Waal, E.C. and Uitterlinden, A.G. (1993) Profiling of complex microbial populations by denaturing gradient gel electrophoresis analysis of polymerase chain reaction-amplified genes coding for 16S rRNA. *Applied and Environmental Microbiology* 59(3), 695–700.

Namkoong, W., Hwang, E.-Y., Park, J.-S. and Choi, J.-Y. (2002) Bioremediation of diesel-contaminated soil with composting. *Environmental Pollution* 119(1), 23–31.

Neilson, A.H. and Allard, A.-S. (2007) *Environmental Degradation and Transformation of Organic Chemicals*. CRC Press, Boca Raton, Florida.

Oka, A.R., Phelps, C.D., McGuinness, L.M., Mumford, A., Young, L.Y. and Kerkhof, L.J. (2008) Identification of critical members in a sulfidogenic benzene-degrading consortium by DNA stable isotope probing. *Applied and Environmental Microbiology* 74(20), 6476–6480. doi: 10.1128/aem.01082-08.

Osborn, A.M., Moore, E.R. and Timmis, K.N. (2000) An evaluation of terminal-restriction fragment length polymorphism (T-RFLP) analysis for the study of microbial community structure and dynamics. *Environmental Microbiology* 2(1), 39–50.

Passatore, L., Rossetti, S., Juwarkar, A.A. and Massacci, A. (2014) Phytoremediation and bioremediation of polychlorinated biphenyls (PCBs): state of knowledge and research perspectives. *Journal of Hazardous Materials* 278, 189–202.

Peng, S., Zhou, Q., Cai, Z. and Zhang, Z. (2009) Phytoremediation of petroleum contaminated soils by Mirabilis Jalapa L. in a greenhouse plot experiment. *Journal of Hazardous Materials* 168(2–3), 1490–1496.

Pilon-Smits, E. (2005) Phytoremediation. *Annual Review of Plant Biology* 56, 15–39.

Poster, D.L., Schantz, M.M., Sander, L.C. and Wise, S.A. (2006) Analysis of polycyclic aromatic hydrocarbons (PAHs) in environmental samples: a critical review of gas chromatographic (GC) methods. *Analytical and Bioanalytical Chemistry* 386(4), 859–881.

Radajewski, S., Ineson, P., Parekh, N.R. and Murrell, J.C. (2000) Stable-isotope probing as a tool in microbial ecology. *Nature* 403(6770), 646–649.

Sarkar, D., Ferguson, M., Datta, R. and Birnbaum, S. (2005) Bioremediation of petroleum hydrocarbons in contaminated soils: comparison of biosolids addition, carbon supplementation, and monitored natural attenuation. *Environmental Pollution* 136(1), 187–195.

Schlegel, H.G. and Zaborosch, C. (1993). *General Microbiology*. Cambridge University Press, Cambridge.

Schoefs, O., Perrier, M. and Samson, R. (2004) Estimation of contaminant depletion in unsaturated soils using a reduced-order biodegradation model and carbon dioxide measurement. *Applied Microbiology and Biotechnology* 64(1), 53–61.

Scow, K.M. and Hicks, K.A. (2005) Natural attenuation and enhanced bioremediation of organic contaminants in groundwater. *Current Opinion in Biotechnology* 16(3), 246–253.

Shahsavari, E., Adetutu, E.M., Anderson, P.A. and Ball, A.S. (2013a) Necrophytoremediation of phenanthrene and pyrene in contaminated soil. *Journal of Environmental Management* 122, 105–112. doi: 10.1016/j.jenvman.2013.02.050.

Shahsavari, E., Adetutu, E.M., Anderson, P.A. and Ball, A.S. (2013b) Plant residues – a low cost, effective bioremediation treatment for petrogenic hydrocarbon-contaminated soil. *Science of the Total Environment* 443, 766–774.

Shahsavari, E., Adetutu, E.M. and Ball, A.S. (2015) Phytoremediation and necrophytoremediation of petrogenic hydrocarbon-contaminated soils. In: Ansari, A.A., Gill, S.S., Gill, R., Lanza, R.G. and Newman, L. (eds) *Phytoremediation: Management of Environmental Contaminants, Vol.* 2. Springer International, Cham, Switzerland, pp. 321–334.

Sheppard, P.J., Adetutu, E.M., Makadia, T.H. and Ball, A.S. (2011) Microbial community and ecotoxicity analysis of bioremediated, weathered hydrocarbon-contaminated soil. *Soil Research* 49(3), 261–269.

Shiau, B.-J., Sabatini, D.A. and Harwell, J.H. (1995) Properties of food grade (edible) surfactants affecting subsurface remediation of chlorinated solvents. *Environmental Science & Technology* 29(12), 2929–2935.

Singh, N., Megharaj, M., Gates, W.P., Churchman, G J., Anderson, J. *et al.* (2003) Bioavailability of an organophosphorus pesticide, fenamiphos, sorbed on an organo clay. *Journal of Agricultural and Food Chemistry* 51(9), 2653–2658. doi: 10.1021/jf025978p.

Singh, A., Kuhad, R.C. and Ward, O.P. (2009) *Advances in Applied Bioremediation*. Springer Berlin Heidelberg, New York.

Steven, B., Briggs, G., McKay, C.P., Pollard, W.H., Greer, C.W. and Whyte, L.G. (2007) Characterization of the microbial diversity in a permafrost sample from the Canadian high Arctic using culture-dependent and culture-independent methods. *FEMS Microbiology and Ecology* 59(2), 513–523.

Straube, W., Nestler, C., Hansen, L., Ringleberg, D., Pritchard, P. and Jones-Meehan, J. (2003) Remediation of polyaromatic hydrocarbons (PAHs) through landfarming with biostimulation and bioaugmentation. *Engineering in Life Sciences* 23(2–3), 179–196.

Suja, F., Rahim, F., Taha, M.R., Hambali, N., Rizal Razali, M., Khalid, A. and Hamzah, A. (2014) Effects of local microbial bioaugmentation and biostimulation on the bioremediation of total petroleum hydrocarbons (TPH) in crude oil contaminated soil based on laboratory and field observations. *International Biodeterioration & Biodegradation* 90, 115–122.
Surridge, A., Wehner, F. and Cloete, T. (2009) Bioremediation of polluted soil. In: Singh, A., Kuhad, R.C. and Ward, O.P. (eds) *Advances in Applied Bioremediation*. Springer Berlin Heidelberg, New York, pp. 103–121.
Teng, Y., Luo, Y., Sun, M., Liu, Z., Li, Z. and Christie, P. (2010) Effect of bioaugmentation by *Paracoccus* sp. strain HPD-2 on the soil microbial community and removal of polycyclic aromatic hydrocarbons from an aged contaminated soil. *Bioresource Technology* 101(10), 3437–3443.
Toren, A., Ron, E., Bekerman, R. and Rosenberg, E. (2002) Solubilization of polyaromatic hydrocarbons by recombinant bioemulsifier AlnA. *Applied Microbiology and Biotechnology* 59(4-5), 580–584.
Villas-Bôas, S.G. and Bruheim, P. (2007) The potential of metabolomics tools in bioremediation studies. *OMICS: A Journal of Integrative Biology* 11(3), 305–313. doi: 10.1089/omi.2007.0005.
Vinas, M., Sabaté, J., Espuny, M.J. and Solanas, A.M. (2005) Bacterial community dynamics and polycyclic aromatic hydrocarbon degradation during bioremediation of heavily creosote-contaminated soil. *Applied and Environmental Microbiology* 71(11), 7008–7018.
Vogel, T.M., Criddle, C.S. and McCarty, P.L. (1987) ES&T critical reviews: transformations of halogenated aliphatic compounds. *Environmental Science & Technology* 21(8), 722–736.
Walworth, J. and Reynolds, C. (1995) Bioremediation of a petroleum-contaminated cryic soil: effects of phosphorus, nitrogen, and temperature. *Soil and Sediment Contamination* 4(3), 299–310.
Yew, J.Y., Cody, R.B. and Kravitz, E.A. (2008) Cuticular hydrocarbon analysis of an awake behaving fly using direct analysis in real-time time-of-flight mass spectrometry. *Proceedings of the National Academy of Sciences of the United States* 105(20), 7135–7140.
Zhang, Z., Gai, L., Hou, Z., Yang, C., Ma, C. *et al.* (2010) Characterization and biotechnological potential of petroleum-degrading bacteria isolated from oil-contaminated soils. *Bioresource Technology* 101(21), 8452–8456.
Zhang, D.C., Mörtelmaier, C. and Margesin, R. (2012) Characterization of the bacterial archaeal diversity in hydrocarbon-contaminated soil. *Science of the Total Environment* 421–422, 184–196. doi: 10.1016/j.scitotenv.2012.01.043.
Zhao, B. and Poh, C.L. (2008) Insights into environmental bioremediation by microorganisms through functional genomics and proteomics. *Proteomics* 8(4), 874–881. doi: 10.1002/pmic.200701005.
Zhu, H., Yan, J., Jiang, X., Lai, Y. and Cen, K. (2008) Study on pyrolysis of typical medical waste materials by using TG-FTIR analysis. *Journal of Hazardous Materials* 153(1), 670–676.

# 36 Metallic Iron for Environmental Remediation: Prospects and Limitations

**C. Noubactep***
*Angewandte Geologie, Universität Göttingen, Göttingen, Germany*

## 36.1 Abstract

Metallic iron ($Fe^0$) has been suggested as an affordable, readily available and efficient material for environmental remediation. Mixed into soil or filled in reactive barriers, $Fe^0$ is a feasible pathway to control contamination in seepage waters. Available information in the literature, however, presents discrepant evidence on the efficiency of this (still innovative) technology. On the basis of a profound study of literature over the past 160 years, it is suggested that these discrepancies are explained by the aqueous chemistry of iron (corrosion). Neglected aspects contributing to the apparent complexity of the $Fe^0/H_2O$ system are outlined. It appears that designing an efficient and sustainable $Fe^0$ remediation system is purely a site-specific issue and that available data are not (really) comparable. In particular, it is clear that $Fe^0$ barriers that have been successfully working for more than a decade were not designed according to any scientific basis. While the success or failure of implemented reactive barriers can be rationalized, more systematic research is needed for a science-based design of efficient and sustainable $Fe^0$-based systems for environmental remediation.

## 36.2 Introduction

In the early 1990s, metallic iron or elemental iron ($Fe^0$), then termed zero-valent iron (ZVI), was presented as a new material for water treatment, including environmental remediation (Lipczynska-Kochany *et al.*, 1994; Gillham and O'Hannesin, 1994; Matheson and Tratnyek, 1994; Schreier and Reinhard, 1994; Cantrell *et al.*, 1995; Warren *et al.*, 1995; Richardson and Nicklow, 2002). Moreover, $Fe^0$ was presented as an environmental reducing agent (Matheson and Tratnyek, 1994; Roberts *et al.*, 1996; Weber, 1996). Since then, hundreds of scientific articles have been written and some 200 $Fe^0$-based reactive barriers implemented over the world (Bigg and Judd, 2000; Scherer *et al.*, 2000; Tratnyek *et al.*, 2003; Henderson and Demond, 2007; Bartzas and Komnitsas, 2010; Li and Benson, 2010; Gheju, 2011; Obiri-Nyarko *et al.*, 2014; Guan *et al.*, 2015; Warner, 2015). However, the key issue of the operating mode of $Fe^0$-based systems was not (and is still not) resolved (Lipczynska-Kochany *et al.*, 1994; Warren *et al.*, 1995; Qiu *et al.*, 2000; Lavine *et al.*, 2001; Furukawa *et al.*, 2002; Ghauch *et al.*, 2010, 2011, 2015; Gheju and Balcu 2011; Noubactep, 2013, 2016a, 2018;

* E-mail address: cnoubac@gwdg.de

Noubactep *et al.*, 2017; Hu *et al.*, 2018; Santisukkasaem and Das, 2019).

The discussion on the mechanism of $Fe^0$ in removing contaminants was initiated by Matheson and Tratnyek (1994) and was actively conducted for some 4 years (Cantrell *et al.*, 1995; Warren *et al.*, 1995; Eitzer, 1996; Roberts *et al.*, 1996; Weber, 1996). Although a 'broad consensus' was made on 'reducing $Fe^0$' around 1998 (O'Hannesin and Gillham, 1998) the discussion has continued until the present day (Noubactep 2007a, b, 2008, 2011a, 2015, 2016a, 2017; Gheju and Balcu, 2011; Giles *et al.*, 2011; Ghauch 2015; Hu *et al.*, 2018). Unfortunately, the large majority of active researchers mistakenly considered the named consensus as scientifically established (Guan *et al.* 2015 and references cited therein). Even excellent experimental observations on the importance of other contaminant removal mechanisms (Mantha *et al.*, 2001; Furukawa *et al.*, 2002; Mielczarski *et al.*, 2005) accommodated the reductive transformation theory. For example, Mantha *et al.* (2001) rationalized the removal of aqueous organic species by the presence of trace carbon in steel. This parallel was supposed to be analogous to the efficiency of highly porous activated carbon to remove organic compounds from aqueous solutions (Sorial *et al.*, 1993; Le Cloirec and Faur, 2006).

This chapter is focused on the validity of the theory of reducing $Fe^0$, as it is crucial for the design of sustainable $Fe^0/H_2O$ systems. Clearly, it is not a review of the literature on $Fe^0$ technology, but a critical consideration of an aspect that has certainly misled research over the past three decades (Makota *et al.*, 2017; Noubactep *et al.*, 2017; Hu *et al.*, 2018; Noubactep, 2018). The chapter starts with historical considerations on the $Fe^0/H_2O$ system, followed by the chemistry of the same system. Key aspects of the chemistry are then used to guide the design of future $Fe^0$-based remediation systems.

## 36.3 Historical Overview of the $Fe^0/H_2O$ System

### 36.3.1 $Fe^0$ for safe drinking water

The modern history of remediation with $Fe^0$ probably started in 1854 with the synthesis of aniline by Antoine Béchamp. This was achieved by using iron filings at pH 4.0 in the presence of hydrochloric acid (HCl) (Béchamp, 1854). Later on, alternatives to $Fe^0$ materials were sought. In this context, Professor Gustav Bischof (University of Glasgow) manufactured a porous material termed spongy iron. Spongy iron was also used for the precipitation of copper from its solutions (cementation). In 1871 Bischof secured a patent for spongy iron-based household water filters (the Bischof process). Between 1881 and 1883, large-scale spongy iron filters were used for the water supply of the city of Antwerp (Belgium). The first patent on using $Fe^0$ for water treatment was secured by Henry Medlock in 1857 but the real history of water treatment by $Fe^0$ started with the Bischof filters (Anderson, 1886; Devonshire, 1890; Mwakabona *et al.*, 2017; Hu *et al.*, 2018).

In 1883, the Bischof filters in Antwerp experienced clogging and it was very difficult to produce enough water for the daily needs of the population of Antwerp (200,000 inhabitants). The water supply company started to look for a clogging-free system and would develop a 'revolving purifier' (Anderson process) in which water is brought into intimate contact with scrap iron and subsequently filtered on a sand filter (Anderson, 1886; Devonshire, 1890; Mwakabona *et al.*, 2017; Nanseu-Njiki et al., 2019). William Anderson is regarded as the first person to have used $Fe^0$ to produce coagulants for water treatment.

It is very important to recall that both the Bischof and the Anderson processes considered $Fe^0$ as a source of coagulants for contaminant collection (Nanseu-Njiki *et al.*, 2019). In both cases decontamination was optimized in subsequent sand filters. In other words, more than one century before the advent of $Fe^0$ reactive barriers, $Fe^0$ was recognized as a coagulant generator and was even applied at large scale. Moreover, this information is largely available in the scientific literature (Devonshire, 1890; van Craenenbroek, 1998; Mwakabona *et al.*, 2017).

### 36.3.2 $Fe^0$ for safe drinking water in emergency

In the context of testing atomic bombs, the problem of short-term radioactive contamination of drinking-water sources was addressed. The challenge was to avoid the ingestion of dissolved and/or suspended radioactive material in water. These radionuclides are a long-lasting source of alpha, beta and gamma radiation (Lauderdale and Emmons, 1951. Lacy, 1952). Many US universities have studied the problem of decontamination of radioactive water shortly after atomic bomb testing. Lauderdale and Emmons (1951) found that steel wool was very efficient at inducing quantitative removal of many aqueous radioactive species. This observation led to testing several elemental metals for the removal of radioactivity from water by slurrying with powdered metals (Lacy, 1952). The most effective design presented by Lauderdale and Emmons (1951) consisted of passing the radioactive water through the following materials, arranged in series: steel wool (grade 0; 50 g), burnt clay (37 g), activated carbon (12 g) and a mixture of ion-exchange resins (26-inch, own column). Lauderdale and Emmons (1951) also observed that clogging of the column occurred if (i) fine iron filings were used or (ii) an extremely fine grade of steel wool was used. They speculated that it should be possible to use a coarse grade of iron chips or another metal in granular form (ease of packing).

The results of Lauderdale and Emmons (1951) showed that some 40 years before the advent of engineered permeable reactive barriers, $Fe^0$ materials were positively tested for decentralized provision of safe drinking water. Moreover, in tune with the ancient use of $Fe^0$ for water treatment (Devonshire, 1890), $Fe^0$ was a source of contaminant collectors while contaminants that were not quantitatively (selectively) fixed by iron corrosion products were removed in the three other units of the properly designed multi-barrier.

### 36.3.3 $Fe^0$ for phosphate removal from agricultural runoffs

The use of suitable materials able to improve the efficiency of sand filters to capture significant amounts of dissolved phosphorus from agricultural runoffs is an established branch of environmental research (James *et al.*, 1992; Erickson *et al.* 2007; Allred, 2017). One approach is to add adsorbing agents for dissolved phosphate ($PO_4^{3-}$). For example, fly ash, metal oxides and/or wood chips can be mixed with sand (Allred, 2017).

James *et al.* (1992) was probably the first research group to add $Fe^0$ (steel wool) to improve $PO_4^{3-}$ removal in filtration systems by *in situ* generation of iron oxides. Some 15 years later, Erickson *et al.* (2007) independently developed a system containing 5% $Fe^0$ (w/w) to quantitatively remove $PO_4^{3-}$ from stormwaters (Erickson *et al.*, 2007, 2017). These authors first tested steel wool and later scrap-iron shavings. As a waste material, iron shavings were selected over steel wool due to their better affordability. The characteristics of iron shavings making their use suitable for large-scale applications include: (i) the material is inexpensive; (ii) it is safe to handle; (iii) it does not significantly alter the pH (in the proportion used); and (iv) it appears to have a large treatment capacity such that 'the life of the iron will outlast the life of the filter'.

The use of $Fe^0$ for $PO_4^{3-}$ mitigation confirms that the current $Fe^0$ research community has independently duplicated previous efforts. Even Erickson and colleagues have duplicated the work of James et al. (1992). Admittedly, Erickson *et al.* (2007, 2017) may have focused their attention on existing stormwater treatment systems. Also the $Fe^0$ research community may have focused on $Fe^0$ in subsurface scenarios and initially on halogenated hydrocarbons (O'Hannessin and Gillham1998). This is, however, not an acceptable procedure, because it is about the $Fe^0/H_2O$ system, that is, aqueous iron corrosion (Tratnyek 1996). Another approximate conclusion of Erickson *et al.* (2017) concerns the long-term efficiency of iron shavings. In fact, the assumption that 'iron will outlast the life of the filter' has not properly considered the non-linear long-term kinetics of $Fe^0$ dissolution under field conditions (Noubactep, 2016b and references cited therein). While it is certain that $Fe^0$ dissolution will continue until complete depletion ('rust never rests'), there is no guarantee that the kinetic of generation of contaminant collectors will be satisfactory for the mitigation goal in the long term. There is actually no (unified) procedure to characterize the (long-term) reactivity of $Fe^0$ materials (Btatkeu *et al.*, 2013; Kim *et al.*, 2014; Birke *et al.*, 2015; Li *et al.*, 2016, 2019). This uncomfortable situation is not specific to the $Fe^0$ remediation research community, as recently demonstrated by Driver *et al.* (2017).

Finally, although the long-term efficiency of the scrap iron-based filters by Erickson and colleagues is not certain, the long-term permeability is certain, because of the low $Fe^0$:sand ratio, 5 % (w/w) used. While Lauderdale and Emmons (1951) recommended the appropriate grade of steel wool to warrant long-term permeability, Erickson *et al.* (2017) excellently addressed the issue by avoiding pure $Fe^0$ layers, and even better using a low $Fe^0$:sand ratio (Caré *et al.*, 2013; Domga *et al.*, 2015). However, given the large difference in densities of involved materials (e.g. steel wool, scrap iron), a volumetric $Fe^0$:sand ratio would have been better. The volumetric $Fe^0$:sand ratio also has the advantage of being applicable where there is no balance (Btatkeu *et al.*, 2016), in which case it can be specified to 'take one volume of $Fe^0$ and three volumes of sand', if the resulting volumetric $Fe^0$:sand ratio is 1:3.

### 36.3.4 $Fe^0$ for selenium removal from agricultural drainage water

In 1985, the Harza Engineering Co. pilot-tested flow-through packed beds containing $Fe^0$ (iron filings) for selenium removal from agricultural drainage waters. The idea behind the test was that oxygen could accelerate iron corrosion and generate hydroxides and oxides to adsorb selenium and other contaminants (Murphy, 1988). The test revealed that 'properly' designed $Fe^0$ filters can reduce the selenium (Se) concentrations to levels below risk levels for wildlife (Anderson, 1989; Frankenberger *et al.*, 2004). However, the operation had to be discontinued, because the $Fe^0$ beds quickly cemented and became impermeable. Solely pure $Fe^0$ beds (100% $Fe^0$) were tested. It was suggested that the $Fe^0$ Harza process could be used as a polishing step following microbial treatments (Frankenberger *et al.*, 2004; Tan *et al.*, 2016). It was also observed that, where the waste was anaerobic after microbial treatment, the formation of secondary precipitates was minimized. The Harza process was practically abandoned and alternatives were sought (Frankenberger *et al.*, 2004; Zhang *et al.*, 2008). Later on, $Fe^0$ was independently tested as a reductive process (Tang *et al.*, 2014; Santos *et al.*, 2015).

The idea that pure $Fe^0$ beds are not sustainable is even older. Anderson (1885) wrote the following about Bischof filter (spongy iron filter):

> The great difficulty lay in the rooted idea that, in practice, a contact of at least three-quarters of an hour was necessary to produce the required effect. The late Professor Way and Mr Ogston had, indeed, shown that with very finely divided iron, unmixed with gravel, a much shorter contact would suffice; but it as known also that a filter constructed of iron only in such a condition would very soon become clogged, so that no advantage would ultimately be gained.

This goal is achieved by two methods (Rahman *et al.*, 2013): (i) using porous $Fe^0$ materials (Bischof, 1877; Hussam and Munir, 2007; Michailidis *et al.*, 2016); and (ii) mixing $Fe^0$ and non-expansive materials (Noubactep and Caré, 2010; Noubactep *et al.*, 2011; Domga *et al.*, 2015).

### 36.3.5 $Fe^0$ for removing metallic ions from mining wastes

Cementation has been used for metal production for some centuries now (Oldright *et al.*, 1928; Noubactep, 2010a). Using $Fe^0$ as cementing metal for copper, for instance, the operation replaces copper ($Cu^{2+}$) with iron ($Fe^{2+}$) in solution. Similarly, other metallic ions (e.g. lead, $Pb^{2+}$) are plated onto the $Fe^0$ surface. There are three main advantages of this process: (i) simple control requirements; (ii) low energy utilization; and (iii) recovery of valuable high-purity metals (Noubactep, 2010a and references cited therein). These advantages have encouraged the use of 'cementation' in wastewater treatment (Gould, 1982). Here the goal is not necessarily to recover pure metal but to free effluents from toxics. The main advantage is that hydroxides are not a problem. Such operations at high initial pH values are commonplace (Oldright *et al.*, 1928; Gould, 1982; Vollprecht *et al.*, 2018).

While using $Fe^0$ for wastewater treatment several key observations have been made, including the following.

- The thickness of the $Fe^0$ layer has a great impact on the sustainability (long-term permeability) of the system. It is better to have several short layers in series than a thick one (Oldright *et al.*, 1928).

- There are more reducing agents in a $Fe^0$/$H_2O$ system, as predicted by the cementation reaction (Gould, 1982).

### 36.3.6 $Fe^0$ for wastewater treatment

The propensity of $Fe^0$ to generate $H_2$ (gas) has long been successfully utilized by anaerobic mixed cultures to support methanogenesis and sulfate reduction (Daniels *et al.*, 1987; Boontian, 2015). In the subsurface remediation context, $Fe^0$ is mistakenly considered as a slow electron donor for reductive transformations. $Fe^0$ is certainly the 'parent' of electrons used in mixed cultures of sulfate reducers developed in reactive barriers. This evidence is supported by the occurrence of sulfate reduction and enrichment of sulfate reducers in dynamic $Fe^0$ columns at laboratory scale.

An applicable $Fe^0$-based technology for decentralized domestic wastewater treatment was introduced around 1990 by Japanese scientists (Wakatsuki *et al.*, 1990, 1993). This technology is termed multi-soil-layering (MSL) and is proven as a promising wastewater treatment and disposal method (Masunaga *et al.*, 2003; Ho and Wang, 2015; An *et al.*, 2017; Guo *et al.*, 2018; Latrach *et al.*, 2018). MSL systems show very good performance in the removal of organic matter and nutrients (Ho and Wang, 2015; An *et al.*, 2016). Pollutant removal from wastewater through MSL is a complicated process involving various chemical, physical and biological processes. An MSL system contains layers of soil mixture comprising soil materials, sawdust, granular iron, charcoal and other available materials. The weight $Fe^0$ ratio is approximately 10%. Tested systems have never experienced clogging.

### 36.3.7 Concluding remarks

Historically, water treatment using $Fe^0$ materials is a very old technology. Methods used to accomplish this goal often involved the design of filtration systems that can quantitatively remove several aqueous contaminants. The examples cited in this chapter are limited to inorganic species but removal of organics was also reported (Knowlton, 1928; Sweeney and Fischer, 1972; Senzaki and Kumagai, 1988, 1989). In particular, in the Béchamp process, it was noticed that a significant fraction of aniline (the reaction product) was entrapped in the matrix of *in situ* generated iron oxides when the final pH value was larger than 4.0 (Werner, 1951). This result is in tune with the historical use of $Fe^0$ in water treatment (Anderson, 1886). In those days, micro-pollutants were not known; even pathogens were not really identified. Water treatment mainly consisted of removing colour-inducing species that are organic in nature. In other words, long before 1990 and the advent of $Fe^0$-reactive barriers, $Fe^0$ materials were already established for the removal of biological, chemical and physical contaminants from water. Apart from the context of cementation (metal recovery), $Fe^0$ was used to produce contaminant collectors, under both static and dynamic conditions. Moreover, it was documented that the diversity of electron sources in the $Fe^0$/$H_2O$ system implies that, under specific conditions, more contaminant is reductively transformed than could be predicted by electrons from the metal body (Gould, 1982; Gheju and Balcu, 2011). Clearly, each $Fe^0$/$H_2O$ system is part of several processes able to chemically transform and/or remove aqueous species.

## 36.4 The Chemistry of the $Fe^0$/$H_2O$ System

### 36.4.1 Fundamental aspects

A piece of reactive $Fe^0$ coming in contact with humidity or immersed in water starts to be oxidatively dissolved after Eqn 36.1 below. Equation 36.1 represents the electrode reaction of the redox couple $Fe^{II}/Fe^0$ for which the standard potential is $E^0 = -0.44$ V. Thermodynamically, the oxidative dissolution of $Fe^0$ is induced by any oxidizing agent belonging to a redox couple of higher $E^0$ value. Water ($H_2O$ or $H^+$, $E^0 = 0.00$ V) is one of these oxidizing agents (Eqn 36.2). The oxidation of $Fe^0$ by water (corrosion) is represented by Eqn 36.3. Under acidic conditions (Eqn 36.3a) dissolved $Fe^{2+}$ is formed while at pH > 4.5 low soluble hydroxides are formed (Eqn 36.3b). $Fe^{II}$ species can be further oxidized to $Fe^{III}$ species (Eqn 36.4). Equation 4a represents the electrode reaction of the redox couple $Fe^{III}/Fe^{II}$ ($E^0 = 0.77$ V) while Eqn 36.4b depicts the oxidation of $Fe(OH)_2$ to $Fe(OH)_3$ by dissolved $O_2$.

$$Fe^0_{(s)} \Leftrightarrow Fe^{2+}_{(aq)} + 2e^- \quad \text{(Eqn 36.1)}$$

$$2H^+ + 2e^- \Leftrightarrow H_{2(g)} \quad \text{(Eqn 36.2)}$$

$$Fe^0_{(s)} + 2H^+ \Rightarrow Fe^{2+}_{(aq)} + H_{2(g)} \quad \text{(Eqn 36.3a)}$$

$$Fe^0_{(s)} + 2H_2O \Rightarrow Fe(OH)_{2(s)} + H_{2(g)} \quad \text{(Eqn 36.3b)}$$

$$Fe^{2+}_{(aq)} \Leftrightarrow Fe^{3+}_{(aq)} + e^- \quad \text{(Eqn 36.4a)}$$

$$4\ Fe(OH)_{2(s)} + 2\ H_2O + O_{2(aq)} \Rightarrow 4\ Fe(OH)_{3(s)} \quad \text{(Eqn 36.4b)}$$

At pH > 4.5, depending on the prevailing operational conditions, $Fe(OH)_2$ and $Fe(OH)_3$ can be transformed to various $Fe^{II}$, $Fe^{III}$ and $Fe^{II}/Fe^{III}$ hydroxides and oxides (e.g. $Fe_3O_4$, $Fe_2O_3$, FeOOH). In other words, at pH > 4.5 (final value) a $Fe^0/H_2O$ system is the home state of non-corroded $Fe^0$ (Eqn 36.3b), $Fe(OH)_2$ (Eqn 36.3b), $Fe(OH)_3$ (Eqn 36.4b) and various oxides and hydroxides. These oxides are contaminant collectors in batch (Anderson process) and column (Bischof process) systems (Simon *et al.*, 2016; Allred, 2017; Mwakabona *et al.*, 2017). The specification 'final pH value' is very important, because experiments can be intentionally started at lower pH values (e.g. pH 4.0). In this case, the inherent properties of $Fe^0$ to elevate the pH is used to optimize the efficiency of the system (Hu *et al.*, 2018; Noubactep, 2018).

### 36.4.2 Application to environmental remediation

Section 36.4.1 has demonstrated that $Fe^0$ can be universally used to produce reducing agents (e.g. $H_2$, $Fe^{II}$ and $Fe_3O_4$) and adsorbing agents (oxides and hydroxides). In the remediation context, the thermodynamic question is answered: it is possible to remove and/or reduce aqueous species. The remaining question is twofold.

- Question 1: Is it pure adsorption, co-precipitation, size exclusion or a combination of the three?
- Question 2: Where chemical transformation occurs, is it a stand-alone removal mechanism?

For Question 1, the theory of the $Fe^0/H_2O$ system summarized in Makota *et al.* (2017) and Naseri *et al.* (2017) demonstrated that contaminant removal is an interplay of the three mechanisms. The answer to Question 2 is also quite simple because Eqn 36.4b demonstrates that oxidation of already precipitated $Fe(OH)_2$ is possible. As the primary goal of water treatment is contaminant removal, whether the contaminant is chemically transformed (oxidized or reduced) or not, it should ideally be removed from the aqueous phase. Moreover, environmental remediation is about removing very low levels of contaminants. In these concentration ranges, reduction is rarely a stand-alone removal mechanism (Noubactep, 2016a) and it is occurring here in an ocean of 'collectors'. This is a prerequisite for co-precipitation (Crawford *et al.*, 1993; Noubactep, 2009) and adsorptive size exclusion (Noubactep, 2010b, 2011a).

The full answer to Question 1 mainly depends on the corrosion rate or the kinetics of production of contaminant collectors. It is well established that the long-term kinetics of iron corrosion are not linear (Moraci *et al.*, 2016; Noubactep, 2016b; Nanseu-Njiki *et al.*, 2019). Properly considering this aspect implies that the time-dependent kinetics of (i) production of contaminant collectors and (ii) the corresponding filling of the initial porosity (porosity loss) are addressed simultaneously. Here, the relation between porosity loss and permeability loss is trivial. The initial free space between granular aggregates (including $Fe^0$) is progressively filled by iron corrosion products. Moreover, available aggregates are cemented to each other by more or less gelatinous iron hydroxides and this cementation is more rapid under oxic conditions; that is, in the presence of oxidizing agents (including $O_2$) accelerating iron corrosion.

### 36.4.3 Evaluating 30 years of research on the $Fe^0/H_2O$ system

The application of $Fe^0$ materials for water treatment and more broadly for environmental remediation has grown into a distinct industry within the past three decades (Khudenko, 1991; Henderson and Demond, 2007; Xi *et al.*, 2019). In the early stages, the concept of subsurface $Fe^0$ reactive barriers attracted the attention of environmental scientists and was regarded as having unconfirmed potential in environmental remediation (Guan *et al.*, 2015 and references cited therein). This stage was characterized by a trial-and-error approach, in searching among available granular $Fe^0$ materials for the ones suitable for field application. After this initial stage, newly developed $Fe^0$ materials (e.g. bimetallics, porous composites, nano-$Fe^0$) were made available (Michailidis *et al.*, 2016). However, there is still a high level of disconnection between synthesized $Fe^0$ materials and expected efficiency both at laboratory scale and in the field (Guan *et al.*, 2015; Santisukkasaem *et al.*, 2015; Li *et al.*, 2019; Santisukkasaem and Das, 2019). On the other hand, the few available contaminant independent characterizing methods for the intrinsic reactivity of $Fe^0$ materials (Reardon, 1995; Kim *et al.*, 2014; Velimirovic *et al.*, 2014; Li *et al.*, 2016, 2019) have not been really considered in further works (Noubactep, 2016b; Hu *et al.*, 2019). Therefore, environmental research using $Fe^0$ has stagnated at this stage of pragmatism. The next trivial stage is the rational design of remediation $Fe^0$ systems. Here, structures and functions of $Fe^0$ will be deliberately pre-designed for a desired purpose.

It has been long recognized that successful $Fe^0$ filter design depends on a balance between four factors: (i) $Fe^0$ intrinsic reactivity; (ii) $Fe^0$ longevity ($Fe^0$ density and size, water chemistry); (iii) filter structure (geometry, $Fe^0$ proportion, thickness); and (iv) filter permeability (Sarr, 2001; Richardson and Nicklow, 2002; Bekele *et al.*, 2015; Warner, 2015). These design factors are all interdependent. This makes filter design a highly site-specific issue. The hydraulic conductivity of $Fe^0$ reactive barriers is designed to be higher than that of the surrounding soil. Highly conductive (sorted) sand and gravel are used for this purpose. However, the volumetric expansive nature of iron corrosion has not been properly considered in the current design efforts. Moreover, evaluation of the performance of running $Fe^0$ barriers (Philipps *et al.*, 2010; Wilkin *et al.* 2014, 2019) has not used an expectation versus observation approach (Noubactep , 2016a). Accordingly, despite more than two decades of practical expertise on $Fe^0$ filter design, no significant progress in knowledge has been achieved.

Currently, the thickness of $Fe^0$ barriers ($L_B$) is determined by empirical expressions similar to the following (Eqn 36.5):

$$L_B = t_{res} * V_b * S_f \qquad \text{(Eqn 36.5)}$$

where $L_B$ = $Fe^0$ barrier thickness (m); $t_{res}$ = residence time of the contaminated water (d); $V_b$ = water flow velocity through the barrier (m/d); and $S_f$ = the safety factor.

$S_f$ accounts for potential (i) seasonal variations in flow and temperature; (ii) $Fe^0$ 'reactivity loss' over time; and (iii) other field uncertainties. However, Eqn 36.5 has been established based on the reductive transformation (degradation or precipitation) theory and represents the major mistake. The mass of $Fe^0$ is considered independently from its size (e.g. radius or diameter) (Domga *et al.*, 2015).

Some cases of failure have been reported (Morrison *et al.*, 2006; Johnson *et al.*, 2008) but reporting performance underachievements has not been the rule, even in technical literature (Warner, 2015). Systems failure has been typically attributed to inadequate hydraulic design. It is believed that an insufficient hydraulic design originates from incomplete site characterization (Warner, 2015). However, as discussed herein, the best site characterization can be useless when the inherent permeability loss due to the volumetric expansive nature of iron corrosion (Pilling and Bedworth, 1923) is not properly considered (Caré *et al.*, 2013; Domga *et al.*, 2015; Moraci *et al.*, 2016; Noubactep, 2016b). Site characterization typically addresses aquifer heterogeneity, aquifer permeability, contaminant distribution, groundwater flow, nature and extent of contamination and undiscovered preferential flow paths (Sarr, 2001; Warner, 2015).

The oldest $Fe^0$ reactive barrier remains functional after 20 years (O'Hannesin and Gillham, 1998; Warner, 2015). It is generally considered that the ageing of $Fe^0$ barriers is primarily due to $Fe^0$ exhaustion. However, the rate of iron corrosion has not really been correlated to the size of $Fe^0$. For example, what is the kinetic of iron corrosion (e.g. in μm per year) if particles of less than 4 mm in size are still corroding after two decades? On the other hand, while citing this $Fe^0$ barrier as a success story (Warner, 2015 and references therein), it is often overlooked that it contains only 22% $Fe^0$ (w/w). Clearly the rationale for the sustainability of systems with higher $Fe^0$ is yet to be presented convincingly.

## 36.5 Designing the Next-Generation $Fe^0/H_2O$ System

The rationale for the design of affordable, efficient and sustainable $Fe^0$-based systems has been sought for years (Gillham and O'Hannesin, 1994; Sarr, 2001; Naftz *et al.*, 2002; Lee *et al.*, 2004; Roehl *et al.*, 2005; Fu *et al.*, 2014; Bekele *et al.*, 2015). The idea that a systematic approach eases the exploitation of the huge potential of $Fe^0$ for multiple purposes is very old. However, research efforts have been hampered by pragmatic approaches and/or misled by wrong interpretations of good experimental observations (Noubactep, 2007a, 2008). The key issue for the past 30 years has been the consideration that contaminant reduction in the presence of $Fe^0$ is the cathodic process simultaneous to the anodic dissolution of $Fe^0$ (Makota *et al.*, 2017, Naseri *et al.*, 2017; Noubactep, 2017a, b, 2018; Noubactep *et al.*, 2017). This fundamental flaw has led to errors in designing individual $Fe^0/H_2O$ systems and the definition of descriptors (e.g. $k_{SA}$) (Noubactep, 2009) for the comparison of interdependent results (Johnson *et al.*, 1996; McGeough *et al.*, 2007). Logically, almost all available data are qualitative in nature and not comparable to each other.

This communication contributes to pave the way for systematic investigations (Naseri *et al.*, 2017). The research effort started around 2004 and the first papers were published during 2006 and 2007 (Noubactep, 2006a, b, 2007a, b). Then the concept of an all-in-one $Fe^0$ system was published in 2009 (Noubactep *et al.*, 2009) and has been progressively improved and completed over the years as summarized in Naseri *et al.* (2017). The results presented in the related articles will not be duplicated here. They can be summarized as follows.

- Experiments aiming at understanding subsurface $Fe^0/H_2O$ systems should be performed under conditions that closely mimic those found in nature (Noubactep 2007a, 2008).
- Pure $Fe^0$ filters are not sustainable (Caré *et al.* 2013; Domga *et al.*, 2015).
- The most sustainable $Fe^0$ filter contains 25% (v/v) $Fe^0$ (Miyajima, 2012; Miyajima and Noubactep, 2013; Btatkeu *et al.*, 2014).
- The $Fe^0/H_2O$ system is an ion-selective one (Phukan, 2015; Phukan *et al.*, 2015).

These rules of thumb should be used to design efficient and sustainable $Fe^0$ filters at several scales (Noubactep 2010c, 2011b, 2018; Noubactep *et al.*, 2012; Naseri *et al.* 2017). Given that a large number of relevant operational parameters have been shown to influence the efficiency of $Fe^0/H_2O$ systems, only a systematic approach can be used to validate or disprove the established concept. Following a systematic approach, the limitations of the remediation $Fe^0/H_2O$ system can be established within a few years. It should be recalled than only long-term column experiments (Noubactep, 2016b; Simon *et al.*, 2016; Naseri *et al.*, 2017; Nanseu-Njiki *et al.*, 2019) will enable the assessment of the long-term efficiency of this technology. Ideally such experiments should last for more than 1 year. Testing new $Fe^0$ materials is an integral part of this endeavour.

## 36.6 Concluding Statements

$Fe^0$ is an effective material for mitigating biological and chemical contamination in water. $Fe^0$-units ranging from simple, manually operated devices to complex, automatic ones are possible to ensure clean water on a site-specific basis. Conventional $Fe^0$ treatment has limitations. It seems not to quantitatively remove fluoride. It is postulated that novel $Al^0/Fe^0$ (porous) composites would enable better efficiency. It is possible to make $Fe^0$ an all-in-one solution for a universally clean environment, including safe drinking-water provision. For environmental remediation in particular, thicker barriers containing low $Fe^0$ ratios will be more efficient, but are not necessarily cost-effective. For above-ground installations (e.g. at water treatment plants) there are apparently no limitations, as several columns in series can be designed. Moreover, $Fe^0$ units can be integrated into multi-barrier treatment chains to selectively remove target contaminants. Systematic research is needed to underpin further developments.

## Acknowledgements

Dr Marius Gheju (Polytechnica University Timisoara, Romania) is thanked for his valuable advice.

## References

Allred, B.J. (2017) Batch test screening of industrial product/byproduct filter materials for agricultural drainage water treatment. *Water* 9, 791. doi: 10.3390/w9100791.

An, C.J., McBean, E., Huang, G.H., Yao, Y., Zhang, P., Chen, X.J. and Li, Y.P. (2016) Multi-soil-layering systems for wastewater treatment in small and remote communities. *Journal of Environmental Informatics* 27, 131–144.

Anderson, W. (1885) The purification of water by means of iron on the large scale. *Minutes of the Proceedings of the Institution of Civil Engineers* 81, 279–284.

Anderson, W. (1886) On the purification of water by agitation with iron and by sand filtration. *Journal of the Society for Arts* 35 (1775), 29–38.

Anderson, M.A. (1989) Fundamental aspects of selenium removal by Harza process. Report prepared for San Joaquin Valley Drainage Program, Sacramento, California, by Harza Environmental Services, Chicago, in association with Dr Marc A Anderson and University of Wisconsin, Madison.

Bartzas, G. and Komnitsas, K. (2010) Solid phase studies and geochemical modelling of low-cost permeable reactive barriers. *Journal of Hazardous Materials* 183, 301–308.

Béchamp, A. (1854) Béchamp reduction. *Annales de chimie et de physique* 42, 186. [Cited by Popat and Padhiyar (2013).]

Bekele, D.N., Naidu, R., Birke, V. and Chadalavada, S. (2015) Choosing the best design and construction technologies for permeable reactive barriers. In: Naidu, R. and Birke, V. (eds) *Permeable Reactive Barrier Sustainable Groundwater Remediation*. CRC Press, Boca Raton, Florida, pp. 41–61.

Bigg, T. and Judd, S.J. (2000) Zero-valent iron for water treatment. *Environmental Technology* 21, 661–670.

Birke, V., Schuett, C., Burmeier, H. and Friedrich, H.-J. (2015) Impact of trace elements and impurities in technical zero-valent iron brands on reductive dechlorination of chlorinated ethenes in groundwater. In: Naidu, R. and Birke, V. (eds) *Permeable Reactive Barrier Sustainable Groundwater Remediation*. CRC Press, Boca Raton, Florida, pp. 87–98.

Bischof, G. (1877) On putrescent organic matter in potable water. *Proceedings of the Royal Society of London* 26, 258–261.

Boontian, N. (2015) Effect of zero valent iron (ZVI) in wastewater treatment: a review. *Applied Mechanics and Materials* 775, 180–184.

Btatkeu-K., B.D., Miyajima, K., Noubactep, C. and Caré S. (2013) Testing the suitability of metallic iron for environmental remediation: discoloration of methylene blue in column studies. *Chemical Engineering Journal* 215–216, 959–968.

Btatkeu-K., B.D., Olvera-Vargas, H., Tchatchueng, J.B., Noubactep, C. and Caré, S. (2014) Determining the optimum $Fe^0$ ratio for sustainable granular $Fe^0$/sand water filters. *Chemical Engineering Journal* 247, 265–274.

Btatkeu-K., B.D., Tchatchueng, J.B., Noubactep, C. and Caré, S. (2016) Designing metallic iron based water filters: light from methylene blue discoloration. *Journal of Environmental Management* 166, 567–573.

Cantrell, K.J., Kaplan, D.I. and Wietsma, T.W. (1995) Zero-valent iron for the in situ remediation of selected metals in groundwater. *Journal of Hazardous Materials* 42, 201–212.

Caré, S., Crane, R., Calabrò, P.S., Ghauch, A., Temgoua, E. and Noubactep, C. (2013) Modeling the permeability loss of metallic iron water filtration systems. *CLEAN – Soil, Air, Water* 41, 275–282.

Crawford, R.J., Harding, I.H. and Mainwaring, D.E. (1993) Adsorption and coprecipitation of single heavy metal ions onto the hydrated oxides of iron and chromium. *Langmuir* 9, 3050–3056.

Daniels, L., Belay, N., Rajagopal, B. and Weimer, P. (1987) Bacterial methanogenesis and growth from $CO_2$ with elemental iron as the sole source of electrons. *Science* 23, 509–511.

Devonshire, E. (1890) The purification of water by means of metallic iron. *Journal of the Franklin Institute* 129, 449–461.

Domga, R., Togue-Kamga, F., Noubactep, C. and Tchatchueng, J.B. (2015) Discussing porosity loss of $Fe^0$ packed water filters at ground level. *Chemical Engineering Journal* 263, 127–134.

Driver, E.M., Roberts, J., Dollar, P., Charles, M., Hurst, P. and Halden, R.U. (2017) Comparative meta-analysis and experimental kinetic investigation of column and batch bottle microcosm treatability studies informing in situ groundwater remedial design. *Journal of Hazardous Materials* 323, 377–385.

Eitzer, B. (1996) Research watch: seeking zero-valent iron's reaction mechanism. *Environmental Science & Technology* 30, 10A.

Erickson, A.J., Gulliver, J.S. and Weiss, P.T. (2007) Enhanced sand filtration for storm water phosphorus removal. *Journal of Environmental Engineering* 133, 485–497.

Erickson, A.J., Gulliver, J.S. and Weiss, P.T. (2017) Phosphate removal from agricultural tile drainage with iron enhanced sand. *Water* 9, 672. doi: 10.3390/w9090672.

Frankenberger, W.T. Jr, Amrhein, C., Fan, T.W.M., Flaschi, D., Glater, J. *et al.* (2004) Advanced treatment technologies in the remediation of seleniferous drainage waters and sediments. *Irrigation and Drainage Systems* 18, 19–41.

Fu, F., Dionysiou, D.D. and Liu H. (2014) The use of zero-valent iron for groundwater remediation and wastewater treatment: a review. *Journal of Hazardous Materials* 267, 194–205.

Furukawa, Y., Kim, J.-W., Watkins, J. and Wilkin, R.T. (2002) Formation of ferrihydrite and associated iron corrosion products in permeable reactive barriers of zero-valent iron. *Environmental Science & Technology* 36, 5469–5475.

Ghauch, A. (2015) Iron-based metallic systems: an excellent choice for sustainable water treatment. *Freiberg Online Geoscience* 32, 80 pp.

Ghauch, A., Abou Assi, H. and Bdeir, S. (2010) Aqueous removal of diclofenac by plated elemental iron: bimetallic systems. *Journal of Hazardous Materials* 182, 64–74.

Ghauch, A., Abou Assi, H., Baydoun, H., Tuqan, A.M. and Bejjani A. (2011) $Fe^0$-based trimetallic systems for the removal of aqueous diclofenac: mechanism and kinetics. *Chemical Engineering Journal* 172, 1033–1044.

Gheju, M. (2011) Hexavalent chromium reduction with zero-valent iron (ZVI) in aquatic systems. *Water, Air, & Soil Pollution* 222, 103–148.

Gheju, M. and Balcu, I. (2011) Removal of chromium from Cr(VI) polluted wastewaters by reduction with scrap iron and subsequent precipitation of resulted cations. *Journal of Hazardous Materials* 196, 131–138.

Gillham, R.W. and O'Hannesin, S.F (1994) Enhanced degradation of halogenated aliphatics by zero-valent iron. *Ground Water* 32, 958–967.

Giles, D.E., Mohapatra, M., Issa, T.B., Anand, S. and Singh, P. (2011) Iron and aluminium based adsorption strategies for removing arsenic from water. *Journal of Environmental Management* 92, 3011–3022.

Gould, J.P. (1982) The kinetics of hexavalent chromium reduction by metallic iron. *Water Research* 16, 871–877.

Guan, X., Sun, Y., Qin, H., Li, J., Lo, I.M.C., He, D. and Dong H. (2015) The limitations of applying zero-valent iron technology in contaminants sequestration and the corresponding countermeasures: The development in zero-valent iron technology in the last two decades (1994–2014). *Water Research* 224–248.

Guo, J., Zhou, Y., Yang, Y., Chen, C. and Xu, J. (2018) Effects of hydraulic loading rate on nutrients removal from anaerobically digested swine wastewater by multi soil layering treatment. *International Journal of Environmental Research and Public Health* 15, 2688.

Henderson, A.D. and Demond, A.H. (2007) Long-term performance of zero-valent iron permeable reactive barriers: a critical review. *Environmental Engineering Science* 24, 401–423.

Ho, C.-C. and Wang, P.-H. (2015) Efficiency of a multi-soil-layering system on wastewater treatment using environment-friendly filter materials. *International Journal of Environmental Research and Public Health* 12, 3362–3380.

Hu, R., Cui, X., Gwenzi, W., Wu, S. and Noubactep, C. (2018) $Fe^0/H_2O$ systems for environmental remediation: the scientific history and future research directions. *Water* 10, 1739.

Hu, R., Ndé-Tchoupé, A.I., Lufingo, M., Xiao, M., Nassi, A., Noubactep, C. and Njau, K.N. (2019) The impact of selected pre-treatment procedures on iron dissolution from metallic iron specimens used in water treatment. *Sustainability* 11, 671.

Hussam, A. and Munir, A.K.M. (2007) A simple and effective arsenic filter based on composite iron matrix: development and deployment studies for groundwater of Bangladesh. *Journal of Environmental Science and Health, Part A* 42, 1869–1878.
James, B.R., Rabenhorst, M.C. and Frigon, G.A. (1992) Phosphorus sorption by peat and sand amended with iron oxides or steel wool. *Water Environment Research* 64, 699–705.
Johnson, T.L., Scherer, M.M. and Tratnyek P.G. (1996) Kinetics of halogenated organic compound degradation by iron metal. *Environmental Science & Technology* 30, 2634–2640.
Johnson, R.L., Thoms, R.B., Johnson, R.O'B. and Krug T. (2008) Field evidence for flow reduction through a zero-valent iron permeable reactive barrier. *Ground Water Monitoring and Remediation* 28, 47–55.
Khudenko, B.M. (1991) Feasibility evaluation of a novel method for destruction of organics. *Water Science and Technology* 23, 1873–1881.
Kim, H., Yang, H. and Kim, J. (2014) Standardization of the reducing power of zero-valent iron using iodine. *Journal of Environmental Science and Health, Part A* 49, 514–523.
Knowlton, L.G. (1928) Some experiments on iron. *The Journal of Physical Chemistry* 32, 1572–1595.
Lacy, W.J. (1952) Removal of radioactive material from water by slurrying with powdered metal. *Journal of American Water Works Association* 44, 824–828.
Latrach, L., Ouazzani, N., Hejjaj, A., Mahi, M., Masunaga, T. and Mandi, L. (2018) Two-stage vertical flow multi-soil-layering (MSL) technology for efficient removal of coliforms and human pathogens from domestic wastewater in rural areas under arid climate. *International Journal of Hygiene and Environmental Health* 22, 64–80.
Lauderdale, R.A. and Emmons, A.H. (1951) A method for decontaminating small volumes of radioactive water. *Journal of American Water Works Association* 43, 327–331.
Lavine, B.K., Auslander, G. and Ritter, J. (2001) Polarographic studies of zero valent iron as a reductant for remediation of nitroaromatics in the environment. *Microchemical Journal* 70, 69–83.
Le Cloirec, P. and Faur C. (2006) Adsorption of organic compounds onto activated carbon – applications in water and air treatments. In: Bandosz, T.J. (ed.) *Activated Carbon Surfaces in Environmental Remediation*. Interface Science and Technology series, Vol. 7. Academic Press, Cambridge, Massachusetts, pp. 375–419..
Lee, G., Rho, S. and Jahng, D. (2004) Design considerations for groundwater remediation using reduced metals. *The Korean Journal of Chemical Engineering* 21, 621–628.
Li, L. and Benson, C.H. (2010) Evaluation of five strategies to limit the impact of fouling in permeable reactive barriers. *Journal of Hazardous Materials* 181, 170–180.
Li, S., Ding, Y., Wang, W. and Lei H. (2016) A facile method for determining the Fe(0) content and reactivity of zero valent iron. *Analytical Methods* 8, 1239–1248.
Li, J., Dou, X., Qin, H., Sun, Y., Yin, D., and Guan, X. (2019) Characterization methods of zerovalent iron for water treatment and remediation. *Water Research* 148, 70–85.
Lipczynska-Kochany, E., Harms, S., Milburn, R., Sprah, G. and Nadarajah N. (1994) Degradation of carbon tetrachloride in the presence of iron and sulphur containing compounds. *Chemosphere* 29, 1477–1489.
Makota, S., Nde-Tchoupe, A.I., Mwakabona, H.T., Tepong-Tsindé, R., Noubactep, C., Nassi, A. and Njau K.N. (2017) Metallic iron for water treatment: leaving the valley of confusion. *Applied Water Science* 7(8), 4177–4196. doi: 10.1007/s13201-017-0601-x.
Mantha, R., Taylor, K.E., Biswas, N. and Bewtra, J.K. (2001) A continuous system for $Fe^0$ reduction of nitrobenzene in synthetic wastewater. *Environmental Science & Technology* 35, 3231–3236.
Masunaga, T., Sato, K., Zennami, T., Fujii, S. and Wakatsuki, T. (2003) Direct treatment of polluted river water by the multi-soil-layering method. *Journal of Water and Environment Technology* 1, 97–104.
Matheson, L.J. and Tratnyek, P.G. (1994) Reductive dehalogenation of chlorinated methanes by iron metal. *Environmental Science & Technology* 28, 2045–2053.
McGeough, K.L., Kalin, R.M. and Myles, P. (2007) Carbon disulfide removal by zero valent iron. *Environmental Science & Technology* 41, 4607–4612.
Michailidis, N., Stergioudi, F., Seventekidis, P., Tsouknidas, A. and Sagris D. (2016) Production of porous copper with high surface area for efficient water purification. *CIRP Journal of Manufacturing Science and Technology* 13, 85–89.
Mielczarski, J.A., Atenas, G.M. and Mielczarski, E. (2005) Role of iron surface oxidation layers in decomposition of azo-dye water pollutants in weak acidic solutions. *Applied Catalysis B* 56, 289–303.
Miyajima, K. (2012) Optimizing the design of metallic iron filters for water treatment. *Freiberg Online Geoscience* 32, 60 pp.

Miyajima, K. and Noubactep C. (2013) Impact of $Fe^0$ amendment on methylene blue discoloration by sand columns. *Chemical Engineering Journal* 217, 310–319.

Moraci, N., Lelo, D., Bilardi, S. and Calabrò P.S. (2016) Modelling long-term hydraulic conductivity behaviour of zero valent iron column tests for permeable reactive barrier design. *Canadian Geotechnical Journal* 53, 946–961.

Morrison, S.J., Mushovic, P.S. and Niesen, P.L. (2006) Early breakthrough of molybdenum and uranium in a permeable reactive barrier. *Environmental Science & Technology* 40, 2018–2024.

Murphy, A.P. (1988) Removal of selenate from water by chemical reduction. *Industrial & Engineering Chemistry Research* 27, 181–191.

Mwakabona, H.T., Ndé-Tchoupé, A.I., Njau, K.N., Noubactep, C. and Wydra K.D. (2017) Metallic iron for safe drinking water provision: considering a lost knowledge. *Water Research* 117, 127–142.

Naftz, D.L., Morrison, S.J., Davis, J.A. and Fuller, C.C. (2002) *Handbook of Groundwater Remediation Using Permeable Reactive Barriers, Applications to Radionuclides, Trace Metals, and Nutrients*. Academic Press, San Diego, California.

Nanseu-Njiki, C.P., Gwenzi, W., Pengou, M., Rahman, M.A. and Noubactep C. (2019) $Fe^0/H_2O$ filtration systems for decentralized safe drinking water: where to from here? *Water* 11, 429.

Naseri, E., Ndé-Tchoupé, A.I., Mwakabona, H.T., Nanseu-Njiki, C.P., Noubactep, C., Njau, K.N. and Wydra K.D. (2017) Making $Fe^0$-based filters a universal solution for safe drinking water provision. *Sustainability* 9, 1224. doi: 10.3390/su9071224.

Noubactep, C. (2006a) Contaminant reduction at the surface of elemental iron: the end of a myth. *Wissenschaftliche Mitteilungen* 31, 173–179.

Noubactep, C. (2006b) Das Ende eines Mythos: Direkte Schadstoffreduktion durch elementares Eisen ($Fe^0$) widerspricht drei Jahrhunderten Korrosionsforschung. *TerraTech* 11–12, TT11–14.

Noubactep, C. (2007a) Processes of contaminant removal in '$Fe^0$–$H_2O$' systems revisited. The importance of co-precipitation. *Open Environmental Science* 1, 9–13.

Noubactep, C. (2007b) Investigating contaminant removal in '$Fe^0$-$H_2O$' systems. *Wissenschaftliche Mitteilungen* 35, 43–48.

Noubactep, C. (2008) A critical review on the mechanism of contaminant removal in $Fe^0$–$H_2O$ systems. *Environmental Technology* 29, 909–920.

Noubactep, C. (2009) On the validity of specific rate constants (kSA) in $Fe^0/H_2O$ systems. *Journal of Hazardous Materials* 164, 835–837.

Noubactep, C. (2010a) Elemental metals for environmental remediation: learning from cementation process. *Journal of Hazardous Materials* 181, 1170–1174.

Noubactep, C. (2010b) The fundamental mechanism of aqueous contaminant removal by metallic iron. *Water SA* 36, 663–670.

Noubactep, C (2010c) Metallic iron for safe drinking water worldwide. *Chemical Engineering Journal* 165, 740–749.

Noubactep, C. (2011a) Aqueous contaminant removal by metallic iron: is the paradigm shifting? *Water SA* 37, 419–426.

Noubactep, C. (2011b) Metallic iron for safe drinking water production. *Freiberg Online Geoscience* 27, 38 pp.

Noubactep, C. (2013) Relevant reducing agents in remediation $Fe^0/H_2O$ systems. *Clean – Soil, Air, Water* 41, 493–502.

Noubactep, C. (2015) Metallic iron for environmental remediation: a review of reviews. *Water Research* 85, 114–123.

Noubactep, C. (2016a) Designing metallic iron packed-beds for water treatment: a critical review. *CLEAN – Soil, Air, Water* 44, 411–421.

Noubactep, C. (2016b) Predicting the hydraulic conductivity of metallic iron filters: modeling gone astray. *Water* 8, 162 (21pp). doi: 10.3390/w8040162.

Noubactep, C. (2017a) Metallic iron for water treatment: lost science in the West. *Bioenergetics* 6, 149. doi: 10.4172/2167-7662.1000149.

Noubactep, C. (2017b) Metallic iron for environmental remediation: how experts maintain a comfortable status quo. *Fresenius Environmental Bulletin* (accepted 05.12.2017).

Noubactep, C. (2018) Metallic iron ($Fe^0$) provide possible solution to universal safe drinking water provision. *Journal of Water Technology and Treatment Methods* 1(1), 102.

Noubactep, C. and Caré, S. (2010) Enhancing sustainability of household water filters by mixing metallic iron with porous materials. *Chemical Engineering Journal* 162, 635–642.

Noubactep, C., Schöner, A., Woafo, P. (2009) Metallic iron filters for universal access to safe drinking water. *Clean: Soil Air Water* 37, 930–937.
Noubactep, C., Btatkeu-K., B.D. and Tchatchueng, J.B. (2011) Impact of $MnO_2$ on the efficiency of metallic iron for the removal of dissolved metal. *Chemical Engineering Journal* 178, 78–84.
Noubactep, C., Temgoua, E. and Rahman, M.A. (2012) Designing iron-amended biosand filters for decentralized safe drinking water provision. *CLEAN – Soil, Air, Water* 40, 798–807.
Noubactep, C., Makota, S. and Bandyopadhyay, A. (2017) Rescuing $Fe^0$ remediation research from its systemic flaws. *Research and Review Insights* 1(4), 1–8. doi: 10.15761/RRI.1000119.
O'Hannesin, S.F. and Gillham, R.W. (1998) Long-term performance of an in situ 'iron wall' for remediation of VOCs. *Ground Water* 36, 164–170.
Obiri-Nyarko, F., Grajales-Mesa, S.J. and Malina, G. (2014) An overview of permeable reactive barriers for in situ sustainable groundwater remediation. *Chemosphere* 111, 243–259.
Oldright, G.L., Keyes, H.E., Miller, V. and Sloan, W.A. (1928) Precipitation of lead and copper from solution on sponge iron. University of North Texas (UNT) Digital Library, Washington, DC. Available at: http://digital.library.unt.edu/ark:/67531/metadc12459/ (accessed May 27, 2013).
Phillips, D.H., Van Nooten, T., Bastiaens, L., Russell, M.I., Dickson, K. *et al.* (2010) Ten year performance evaluation of a field-scale zero-valent iron permeable reactive barrier installed to remediate trichloroethene contaminated groundwater. *Environmental Science & Technology* 44, 3861–3869.
Phukan, M. (2015) Characterizing the $Fe^0$/sand system by the extent of dye discoloration. *Freiberg Online Geoscience* 40, 70.
Phukan, M., Noubactep, C. and Licha, T. (2015) Characterizing the ion-selective nature of $Fe^0$-based filters using azo dyes. *Chemical Engineering Journal* 259, 481–491.
Pilling, N.B. and Bedworth, R.E. (1923) The oxidation of metals at high temperatures. *Journal of the Institute of Metals* 29, 529–591.
Popat, V. and Padhiyar, N. (2013) Kinetic study of Bechamp Process for P-nitrotoluene reduction to P-toluidine. *International Journal of Chemical Engineering and Applications* 4, 401–405.
Qiu, S.R., Lai, H.-F., Roberson, M.J., Hunt, M.L., Amrhein, C. *et al.* (2000) Removal of contaminants from aqueous solution by reaction with iron surfaces. *Langmuir* 16, 2230–2236.
Rahman, M.A., Karmakar, S., Salama, H., Gactha-Bandjun, N., Btatkeu-K., B.C. and Noubactep, C. (2013) Optimising the design of $Fe^0$-based filtration systems for water treatment: the suitability of porous iron composites. *Journal of Solution Chemistry and Modeling* 2, 165–177.
Reardon, J.E. (1995) Anaerobic corrosion of granular iron: measurement and interpretation of hydrogen evolution rates. *Environmental Science & Technology* 29, 2936–2945.
Richardson, J.P. and Nicklow, J.W. (2002) In situ permeable reactive barriers for groundwater contamination. *Soil and Sediment Contamination* 11, 241–268.
Roberts, A.L., Totten, L.A., Arnold, W.A., Burris, D.R. and Campbell, T.J. (1996) Reductive elimination of chlorinated ethylenes by zero-valent metals. *Environmental Science & Technology* 30, 2654–2659.
Roehl, K.E., Meggyes, T., Simon, F.G. and Stewart, D.I. (eds) (2005) *Long-term Performance of Permeable Reactive Barriers*. Elsevier, Amsterdam.
Santisukkasaem, U. and Das, D.B. (2019) A non-dimensional analysis of permeability loss in zero-valent iron permeable reactive barrier (PRB). *Transport in Porous Media* 126, 139–159.
Santisukkasaem, U., Olawuyi, F., Oye, P. and Das, D.B. (2015) Artificial neural network (ANN) for evaluating permeability decline in permeable reactive barrier (PRB). *Environmental Processes* 2, 291–307.
Santos, S., Ungureanu, G., Boaventura, R. and Botelho, C. (2015) Selenium contaminated waters: an overview of analytical methods, treatment options and recent advances in sorption methods. *Science of the Total Environment* 521–522, 246–260.
Sarr, D. (2001) Zero-valent-iron permeable reactive barriers – how long will they last? *Remediation* 11, 1–18.
Scherer, M.M., Richter, S., Valentine, R.L. and Alvarez, P.J.J. (2000) Chemistry and microbiology of permeable reactive barriers for in situ groundwater clean up. *Critical Reviews in Environmental Science and Technology* 30, 363–411.
Schreier, C.G. and Reinhard, M. (1994) Transformation of chlorinated organic compounds by iron and manganese powders in buffered water and in landfill leachate. *Chemosphere* 29, 1743–1753.
Senzaki, T. and Kumagai, Y. (1988) Removal of chlorinated organic compounds from wastewater by reduction process: II. Treatment of trichloroethylene with iron power. *Kogyo Yosui*, 369, 19–25 [in Japanese].
Senzaki, T. and Kumagai, Y. (1989) Removal of chlorinated organic compounds from wastewater by reduction process: treatment of 1,1,2,2-tetrachloroethane with iron power. *Kogyo Yosui* 357, 2–7 [in Japanese].

Simon, S., Courtin-Nomade, A., Vasiliu, A., Sleiman, N. and Deluchat, V. (2016) Long-term influence of aeration on arsenic trapping in a ZVI/sand bed reactor. *RSC Advances* 6, 54479–54485.
Sorial, G.A., Suidan, M.T., Vidic, R.D. and Brenner R.C. (1993) Effect of GAC characteristics on adsorption of organic pollutants. *Water Environment Research* 65, 53–57.
Sweeney, K.H. and Fischer, J.R. (1972) Reductive degradation of halogenated pesticides. USA Patent, 3, 640, 821.
Tan, L.C., Nancharaiah, Y.V., van Hullebusch, E.D. and Lens P.N.L. (2016) Selenium: environmental significance, pollution, and biological treatment technologies. *Biotechnology Advances* 34, 886–907.
Tang, C., Huang, Y.H., Zeng, H. and Zhang, Z. (2014) Reductive removal of selenate by zero-valent iron: the roles of aqueous $Fe^{2+}$ and corrosion products, and selenate removal mechanisms. *Water Research* 67, 166–174.
Tratnyek, P.G. (1996) Putting corrosion to use: remediating contaminated groundwater with zero-valent metals. *Chemistry and Industry* July 1 1996, 499–503.
Tratnyek, P.G., Scherer, M.M., Johnson, T.L. and Matheson L.J. (2003) Permeable reactive barriers of iron and other zero-valent metals. In: Tarr, M.A. (ed.) *Chemical Degradation Methods for Wastes and Pollutants: Environmental and Industrial Applications*. Environmental Science and Pollution Control Series, 26. Marcel Dekker, New York, pp. 371–421.
Van Craenenbroeck, W. (1998) Easton and Anderson and the water supply of Antwerp (Belgium). *Industrial Archaeology Review* 20, 105–116.
Velimirovic, M., Carniatoc, L., Simons, Q., Schoups, G., Seuntjens, P. and Bastiaens L. (2014) Corrosion rate estimations of microscale zerovalent iron particles via direct hydrogen production measurements. *Journal of Hazardous Materials* 270, 18–26.
Vollprecht, D., Krois, L.-M., Sedlazeck, K.P., Müller, P., Mischitz, R., Olbrich, T. and Pomberger, R. (2018) Removal of critical metals from waste water by zero-valent iron. *Journal of Cleaner Production* 208, 1409–1420
Wakatsuki, T., Omura, S., Abe, Y., Izumi, K. and Matsui, Y. (1990) Treatment of domestic wastewater by Multi-Soil-Layering system (Part 3). Role and life of purification abilities of soil materials in the systems. *Japanese Journal of Soil Science and Plant Nutrition* 61, 74–84.
Wakatsuki, T., Esumi, H. and Omura, S. (1993) High performance and N, P removable on-site domestic wastewater treatment system by multi-soil-layering method. *Water Science & Technology* 27, 31–40.
Warner, S.D. (2015) Two decades of application of permeable reactive barriers to groundwater remediation. In: Naidu, R. and Birke, V. (eds) *Permeable Reactive Barrier Sustainable Groundwater Remediation*. CRC Press, Boca Raton, Florida, pp. 25–39.
Warren, K.D., Arnold, R.G., Bishop, T.L., Lindholm, L.C. and Betterton, E.A. (1995) Kinetics and mechanism of reductive dehalogenation of carbon tetrachloride using zero-valence metals. *Journal of Hazardous Materials* 41, 217–227.
Weber, E.J. (1996) Iron-mediated reductive transformations: investigation of reaction mechanism. *Environmental Science & Technology* 30, 716–719.
Werner, J. (1951) Amination by reduction. *Industrial & Engineering Chemistry* 43, 1917–1919.
Wilkin, R.T., Acree, S.D., Ross, R.R., Puls, R.W., Lee, T.R. and Woods, L.L. (2014) Fifteen-year assessment of a permeable reactive barrier for treatment of chromate and trichloroethylene in groundwater. *Science of the Total Environment* 468–469, 186–194.
Wilkin, R.T., Lee, T.R., Sexton, M.R., Acree, S.D., Puls, R.W. *et al.* (2019) Geochemical and isotope study of trichloroethene degradation in a zero-valent iron permeable reactive barrier: a twenty-two-year performance evaluation. *Environmental Science & Technology* 53, 296–306.
Xi, Y., Luo, Y., Zou, J., Liao, T., Zhang, L., Wang, C., Li, X. and Ling, G. (2019) Kinetics of arsenic removal in waste acid by the combination of $CuSO_4$ and zero-valent iron. *Processes* 7, 401.
Zhang, Y., Amrhein, C., Chang, A. and Frankenberger, W.T. (2008) Effect of zero-valent iron and a redox mediator on removal of selenium in agricultural drainage water. *Science of Total Environment* 407, 89–96.

# 37 Remediation of Contaminated Soil by Biochar

**X.-F. Sima and H. Jiang***
*Department of Chemistry, University of Science and Technology of China, Anhui, China*

## 37.1 Abstract

Biochar is a by-product obtained from biomass pyrolysis. Due to the potential role in improving soil fertility and remediation of contaminated soil, biochar has attracted increasing attention in recent years. Biochar can remediate organically contaminated soil via adsorption and biodegradation, during which organic compounds are enriched by the functional groups in the biochar and subsequently degraded by the microbes. It also can remediate inorganic contaminants by ion exchange, electrostatic attraction and precipitation. The remediation effect of biochar is closely related to its physico-chemical properties, which are dominated by the pyrolytic parameters (heating rate and temperature) and feedstock sources (lignocellulosic biomass and biosolids) and is influenced by the soil environment (pH, ionic strength, humidity, etc.).

## 37.2 Introduction

Biochar refers to the high-carbon by-product obtained from biomass pyrolysis (heating in the absence of oxygen) (Ding *et al.*, 2016). The International Biochar Initiative (IBI) standardized its definition as 'a solid material obtained from the thermochemical conversion of biomass in an oxygen-limited environment' (IBI, 2012). Due to the potential role in carbon sequestration, reduction in the emission of greenhouse gases and improving the soil fertility, biochar has attracted increasing attention from many researchers in recent years (Kookana, 2010). In addition to improving soil fertility, biochar addition into soils can improve the physico-chemical and biological properties of soil (Baiamonte *et al.*, 2015; Gul *et al.*, 2015) and act as a sorbent for contaminants (Beesley *et al.*, 2011; Ahmad *et al.*, 2014b).

## 37.3 Preparation of Biochar

### 37.3.1 Feedstock

Biomass is recognized as one of the most promising renewable resources for energy production that can solve the world's problem of energy crisis (Tripathi *et al.*, 2016) and is the main feedstock material for producing biochar. Biomass is a complex biological organic or non-organic solid product derived from living or recently living organisms and available naturally. Vassilev *et al.* (2012) classified biomass into five different groups: woody; agricultural; aquatic; human and animal waste; and industrial waste biomass.

* E-mail address: jhong@ustc.edu.cn

Woody and agricultural biomass are the major sources of biomass and are widely used in different parts of the world, mostly generated from waste or by-products from harvesting and processing of crops. Woody biomass involves stems, branches, leaves, bark, lumps and chips from different trees and its main sources are forest areas. Agricultural biomass involves different parts (such as stalks, straw, husks) of various agricultural crops, flowers and weeds. Aquatic biomass includes different kinds of microalgae, plants and microbes found in water. Animal and human waste biomass includes cooked or uncooked food, fruit, manure of different animals, paper, plastics and pulps. Industrial waste come from various industries such as sewage sludge from wastewater treatment plants, sugarcane residue from sugar mills, waste from food-processing industries and others.

### 37.3.2 Production technologies

The production of biochar always involves a thermochemical conversion process, which can be divided into pyrolysis and gasification.

#### *37.3.2.1 Pyrolysis*

Pyrolysis is one of the most effective and efficient processes to obtain energy from biomass and is most capable of competing with non-renewable fossil fuel resources (Özçimen and Karaosmanoğlu, 2004). Pyrolysis is a thermochemical process for decomposing organic materials under oxygen-free conditions in the temperature range of 400–1200°C (Heidari *et al.*, 2014; Cha *et al.*, 2016; Tripathi *et al.*, 2016). The parameters that influence the products of the pyrolysis processes include the reaction temperature, heating rate and residence time. In general, biochar yield decreases and the synthesis gas (syngas) yield increases with increasing pyrolysis temperature (Inguanzo *et al.*, 2002; Chen *et al.*, 2003; Demirbas, 2004). Depending on the reaction time of the pyrolysis material, pyrolysis can be further classified into three main categories: fast; slow; and flash pyrolysis (Meyer *et al.*, 2011).

Fast pyrolysis is receiving incredible popularity in producing liquid fuels and speciality and commodity chemicals (Zeng *et al.*, 2011; Jahirul *et al.*, 2012; Zhang *et al.*, 2013c). In fast pyrolysis, the biomass is heated up to a temperature of 850–1250°C with a heating rate of 10–200 °C for a short span of time varying between 1 and 10 seconds (Tripathi *et al.*, 2016), and biomass decomposes very quickly to generate mostly vapours and aerosols and some charcoal and gas (Bridgwater, 2012). A typical fast pyrolysis produces 60–75% of liquid product, 15–25% of biochar and 10–20% of non-condensable gaseous products (Bridgwater, 2003).

Slow pyrolysis, which is characterized by slow heating rate and long residence time, has been used for thousands of years. In this process, the biomass is pyrolysed up to a temperature of the order of 400–500°C with low heating rates (0–1°C $s^{-1}$) and long residence time (5–30 min) (Onay and Kockar, 2003). Because of the higher solid yield (25–35%) of slow pyrolysis compared with the other pyrolysis processes, it is regarded as the main pyrolysis technique for the production of biochar (Mohan *et al.*, 2006; Brownsort, 2009).

Flash pyrolysis can be considered as an improved and modified form of fast pyrolysis. In flash pyrolysis, the temperature achieved is 900–1200°C with very short gas residence time (0.1–1 s) (Demirbas and Arin, 2002; Li *et al.*, 2013). The high temperature and rapid heating rate combined with low vapour residence time can achieve bio-oil yields up to 75%, but the char yield is decreased (Aguado *et al.*, 2002).

#### *37.3.2.2 Gasification*

Gasification is a process in which carbonaceous contents of the biomass are converted into gaseous fuel in the presence of a gaseous medium like oxygen, air, nitrogen, carbon dioxide, or some mixture of these gases at a very high temperature range (600–1200°C) for short residence time (10–20 s) (McKendry, 2002; Brewer *et al.*, 2009). It is partial oxidation of biomass; in this process, the intrinsic chemical energy of carbon is converted into combustible fuel gases, which can be used more efficiently and easily than raw biomass (Tripathi *et al.*, 2016). The primary product, bio-syngas, consists of CO, $H_2$ and $CO_2$; the residue is biochar, which contains a high amount of alkali and alkaline earth metals (calcium (Ca), potassium (K), silicon (Si), magnesium (Mg), etc.) (Kirubakaran *et al.*, 2009; Ippolito *et al.*, 2012).

## 37.4 Characterization of Biochar

### 37.4.1 Physical properties

Physical characteristics of biochar are greatly affected by the pyrolysis conditions such as biomass type, particle size, reactor type and shape, heating rate, residence time, pressure, flow rate of inert gas, etc. (Jahirul *et al.*, 2012). The fundamental molecular structure of biochar creates both its surface area and porosity. During thermal conversion, the rudimentary porosity and structure of the original material contribute the majority of the macropores present in biochar, the amorphous decomposition and volatiles released resulting in the micropores, which contribute most to the surface area of biochar (Lehmann and Joseph, 2009). The pyrolysis temperature and heating rate have most significant effect on surface area and pore volume (Zabaniotou *et al.*, 2008; Qian *et al.*, 2013). For example, the surface area of biochar made from oak (*Quercus lobata*) increased from 2 $m^2\ g^{-1}$ to 225 $m^2\ g^{-1}$ when pyrolysis temperature increase from 400°C to 650°C (Mukherjee *et al.*, 2011). The surface area of pig manure-derived biochar increased from 23.8 to 32.6 $m^2\ g^{-1}$ as charring temperature increased from 350°C to 700°C (Zhang *et al.*, 2013a).

### 37.4.2 Chemical composition

Biochar mainly consists of more than 60% of carbon and other elements, such as hydrogen (H), oxygen (O), nitrogen (N), phosphorus (P) and sulfur (S) (Golber, 1985). The composition, distribution and proportion of these molecules in biochar depend on a variety of factors, including source materials and the pyrolysis methodology used (Chen *et al.*, 2007; Van Zwieten *et al.*, 2010). With increasing pyrolytic temperature, the carbon content increased, whereas hydrogen and oxygen contents decreased (Chen and Chen, 2009).

Generally, biochar is alkaline because of the mineral ash content (Mukherjee *et al.*, 2011). Two factors, pyrolytic temperature and feedstock, control the amount and distribution of mineral ash in biochar (Lehmann and Joseph, 2009). Yuan *et al.* (2011) found that the ash content of the biochar produced from crop residues increased with increased temperature. Woody feedstocks generally have low ash contents, whereas grass, straw and grain husks, which have high silica contents, biochar from manures and rendering wastes typically have very high ash contents (Lehmann and Joseph, 2009). In addition, experimental evidence shows that a range of different functional groups exists on the surfaces of the biochar, mainly including the hydroxyl (OH), carboxyl (-COOH) group, ketone group (-OR), etc. (Cao and Harris, 2010). These rich oxygen-containing functional groups make biochar different from other carbon materials, and influence the behaviour of biochar in water and soil environments.

## 37.5 Remediation of Contaminated Soil by Biochar

In recent years, soil contamination resulting from urbanization, mining, agriculture and industrial processes has been both serious and widespread in the world (Li *et al.*, 2014; Horta *et al.*, 2015). The contaminants in soil can be divided into organic and inorganic toxins. Typical organic contaminants include organochlorine pesticides (OCPs), polychlorinated biphenyls (PCBs), polybrominated diphenyl ethers (PBDEs), etc. (Cai *et al.*, 2008); and inorganic contaminants are mainly heavy metals (Li *et al.*, 2014).

Pollutants in soil can be transported into vegetation and pass along the food chain through bioaccumulation to enter and cause a toxic response to biota or humans, resulting in unacceptable environmental risks (Beesley *et al.*, 2011). Given the pressure on soil for food security, seeking remediation techniques for reducing the risk of pollutant transfer to proximal waters or receptor organisms is of increasing importance (Beesley *et al.*, 2011; Liu *et al.*, 2013). At present, various approaches have been proposed for the remediation of contaminated soil by researchers. The technology of *in situ* amendments to bind pollutants whilst promoting plant growth and stimulating ecological restoration has become more popular (Adriano *et al.*, 2004; Vangronsveld *et al.*, 2009). Carbon-rich materials have been applied to *in situ* soil remediation because of their ability to reduce the bioavailability of contaminants, such as activated carbon (Brändli *et al.*, 2008; Cho *et al.*, 2009).

Like activated carbons, biochar is also produced by thermochemistry processes (pyrolysis) but is not activated or further treated before application to soils. In recent years, biochar has been widely incorporated into soil as amendment to improve soil fertility and as sequester carbon (C) to mitigate climate change (Lehmann *et al.*, 2006; Laird, 2008; Sohi *et al.*, 2010; Ding *et al.*, 2016). In addition, it has been reported that biochar addition can remediate soil contaminated by organic and inorganic contaminants (Accardi-Dey and Gschwend, 2003; Beesley *et al.*, 2011).

### 37.5.1 Organic-contaminated soil

In recent decades, the increased usage of complex organic chemicals for various purposes has resulted in organic contamination in soil across the globe, such as: OCPs in agriculture; PCBs in transformers, capacitors, paints; pentachlorophenol (PCP) in wood preservatives; PBDEs as flame retardants, etc. (Cai *et al.*, 2008).

Biochar has been used in the remediation of organically contaminated soil, as summarized in Table 37.1. Adsorption and degradation (biotic and abiotic) are perhaps the two most important processes, in which the bioaccumulation and bioavailability of organic contaminants are reduced.

#### *37.5.1.1 Adsorption*

Biochar has been shown to be a strong sorbent for contaminants, thereby playing a crucial role in remediating contaminated soil (Lohmann *et al.*, 2005). Yang and Sheng (2003) reported that biochar obtained from wheat and rice residues was up to 2500 times more effective than soil in sorbing diuron. Jeong *et al.* (2012) amended soil contaminated with tylosin by using biochar derived from pulp-grade hardwood and softwood chips and found that the sorption of tylosin by biochar enhanced the retention and reduced the transport of tylosin in soils. Jones et al. (2011) found that strong simazine sorption of biochar suppressed simazine biodegradation and reduced simazine leaching into groundwater. Yu *et al.* (2009) and Xu *et al.* (2012) reported similar findings in which the biochar produced from wood chips and bamboo waste resulted in a remarkable decrease in the dissipation of chlorpyrifos, carbofuran and pentachlorophenol from soil, due to their high sorption. These authors also reported a pronounced decrease of the bioavailability of pesticides in contaminated soils.

These extraordinary sorption properties of biochar can be attributed to its unique characteristics, especially its highly carbonaceous and aromatic nature and high specific surface area. Overall, biochar produced at higher temperatures has higher surface area and microporosity and exhibits higher sorption efficiency for organic contaminant remediation in soil (Chen *et al.*, 2008). This has been shown to increase the organic contaminant adsorption capability of biochar (Yu *et al.*, 2006; Wang *et al.*, 2010; Zhang *et al.*, 2010), reducing their uptake into soil organisms (Yu *et al.*, 2009). Yang *et al.* (2010) found that biochar produced at 450°C exhibited comparatively less efficiency in reducing the bioavailability of chlorpyrifos and fipronil than biochar produced at 850°C. Given that adsorption of organic contaminants to biochar will be predominantly non-linear, the sorption sites could become saturated (Beesley *et al.*, 2011). Zhang *et al.* (2010) found that dissolved organic matter (DOM) from soil could coat biochar particles and block the sorption sites, subsequently reducing the accessibility of pesticides to the sorption sites. Electrostatic attractions between charged surfaces of biochar and ionic organic compounds may be additional sorption mechanisms. However, the effective sorption mechanism for non-ionic compounds may be partitioning and subsequent diffusion into the non-carbonized and carbonized fractions of biochar.

#### *37.5.1.2 Biodegradation*

While the application of biochar can reduce the bioavailability and leachability of organic contaminants through adsorption, at the same time the biodegradation of organic pollutants in soil can decrease due to the reduced microbial accessibility (Kookana, 2010; Sopeña *et al.*, 2012). Indeed, it has been reported that amendment with a small amount of biochar into a soil can suppress the microbial degradation of organic contaminants, as well as reducing their availability

**Table 37.1.** Biochar utilization for remediation of organic contaminants in soil.

| Contaminant | Biochar type | Effect | References |
|---|---|---|---|
| Diuron | Wheat straw | Sorption | Yang *et al.* (2006) |
| Diuron | Red gum wood chips (450°C and 850°C) | Sorption | Yu *et al.* (2006) |
| Atrazine and acetochlor | Sawdust (500°C) | Sorption | Spokas *et al.* (2009) |
| Chloropyrifos and carbofuran | Wood chips (450 and 850°C) | Adsorption due to high surface area and nano-porosity | Yu *et al.* (2009) |
| Phenanthrene | Pine wood (350 and 700°C) | Entrapment in micro- or meso-pores | Zhang *et al.* (2010) |
| Terbuthylazine | *P. radiata* wood (350 and 700°C) | Adsorption | Wang *et al.* (2010) |
| Chlorpyrifos and fipronil | Cotton straw chips (450 and 850°C) | Sorption and microbial degradation. | Yang *et al.* (2010) |
| PAHs | Hardwood | Sorption and biodegradation | Beesley *et al.* (2010) |
| Pyrimethanil | Red gum wood chips (450 and 850°C) | Adsorption | Yu *et al.* (2010) |
| Atrazine | Dairy manure (450°C) | Sorption | Cao *et al.* (2011) |
| Pentachlorophenol | Rice straw | Adsorption due to high surface area and microporosity | Lou *et al.* (2011) |
| Simazine | Hardwood (450°C and 600°C) | Sorption due to abundance of micropores | Jones *et al.* (2011) |
| PAHs | Hardwood (600°C) | Sorption | Gomez-Eyles *et al.* (2011) |
| Tylosin | Pepperwood (450°C and 600°C)<br>Sugarcane bagasse (450°C and 600°C)<br>Hickory wood (450°C and 600°C)<br>Pulpgrade hardwood and softwood chips (850°C and 900°C) | Sorption | Jeong *et al.* (2012) |
| Pentachlorophenol | Bamboo (600°C) | Reduced leaching due to diffusion and partition | Xu *et al.* (2012) |
| Glyphosate | Birch wood (450°C) | Decreased leaching | Hagner *et al.* (2013) |
| PAHs | Sewage sludge (500°C) | Partitioning | Khan *et al.* (2013) |

to plants. For instance, Yu *et al.* (2009) amended pesticide-spiked soil by using two types of biochar produced at different temperatures and studied the biodegradation as well as plant uptake of carbofuran and chlorpyrifos by spring onions (*Allium cepa*). The study demonstrated that the half-life for carbofuran increased from 11.9 days in the untreated soil to 33.3 days in the soil amended with 1% biochar; and the half-life for chlorpyrifos increased from 12.1 to 42.8 days. Some researchers have found similar results: the biodegradation of diuron and benzonitrile by selected microorganisms decreased in the presence of wheat char (Zhang *et al.*, 2005; Yang *et al.*, 2006).

Some attempts have been made to accelerate the degradation of organic pollutants in soils through biochar amendment (Kemper *et al.*, 2008; Oh *et al.*, 2012). For instance, Kemper *et al.* (2008) found that the special structure of biochar could be regarded as both electron conductors and sorption sites, which could catalyse the reduction of hexahydro-1,3,5-trinitro-1,3,5-triazine (RDX), enhancing their degradation consequently. Zhang *et al.* (2013a) investigated the effect of pig manure-derived biochar on adsorption and catalytic hydrolysis of two pesticides, carbaryl and atrazine. The authors found that the presence of biochar pyrolysed at 700°C accelerated the hydrolysis of pesticides; carbaryl and atrazine were decomposed by 71.8% and 27.9% in 12 h, respectively. They concluded that the elevated solution pH, mineral surface and dissolved metal ions released from the biochar played a catalytic role in carbaryl hydrolysis. Yu *et al.* (2011) used pine wood-derived biochar as a catalyst for reduction of nitrobenzenes to anilines by sulfides at room temperature, and demonstrated that biochar could serve not only as an adsorbent but also as a platform to accelerate the reduction of nitrobenzenes.

Bioremediation is one of the commonly practiced technologies for remediation of soils contaminated with organic pollutants through plants or microorganisms (Smith *et al.*, 2009; Chen and Yuan, 2012). Addition of microorganisms capable of decomposing certain types of organic pollutants to enhance the efficiency of bioremediation is necessary (Chen *et al.*, 2012). Where well-designed immobilized carriers to offer a protective space for inoculated microorganisms and minimize competition from indigenous microbes are required, biochar could act as an immobilized carrier for bacteria in the contaminated soil (Chen *et al.*, 2012). Biochar can pre-concentrate pollutants in contaminated soil and then feed them to the immobilized microbial decomposers (Dan *et al.*, 2006). It has been reported that biochar can promote bioremediation of contaminated soil as microbial carriers, and the immobilized bacteria could directly degrade the biochar-sorbed PAHs (Chen *et al.*, 2012).

#### *37.5.1.3 Potential problems*

Many studies have shown the potential of biochar to reduce the bioavailability of various organic pollutants in soils (Kookana, 2010; Beesley *et al.*, 2011; Ahmad *et al.*, 2014b; Wang *et al.*, 2017), such as PAHs, polychlorinated dibenzo-*p*-dioxins (PCDDs), polychlorinated dibenzofurans (PCDFs) and organic pesticides. The increased sorption of pesticides into soil will be beneficial in reducing pesticide residues in crops (Yu *et al.*, 2009); however, it will also be detrimental in reducing pesticide efficiency, resulting in higher application rates of these chemicals (Kookana, 2010; Spokas *et al.*, 2009). Yang *et al.* (2006) reported that enhanced sorption of diuron in soil in the presence of wheat straw-derived biochar reduced the herbicidal efficacy for barnyard grasses (*Echinochloa* spp.). Xu *et al.* (2008) found that the herbicidal efficacy of clomazone for barnyard grasses decreased with increasing amount of rice straw biochar in soils.

### 37.5.2 Inorganic-contaminated soil

Unlike organic contaminants, inorganic contaminants such as heavy metals cannot be degraded by microbial action; their bioavailability makes them pass along the food chain and threatens the health of humans and the ecosystem (Dong *et al.*, 2011a; Zhang *et al.*, 2013b; Inyang *et al.*, 2016). Inorganic contaminants of soil have become a severe problem in many parts of the world (Facchinelli *et al.*, 2001), mostly originating from various anthropogenic sources such as mining, smelting, pesticides, fertilizers, leaded gasoline and paints, power plants, battery manufacture, spillage of petrochemicals, wastewater irrigation and sewage sludge (Khan *et al.*, 2008; Ok *et al.*, 2011; Usman *et al.*, 2012; Lim *et al.*, 2013). Recent studies on biochar applications for remediating soil contaminated by inorganic contaminants are summarized in Table 37.2.

Uchimiya *et al.* (2010) reported that broiler litter manure-derived biochar is effective at reducing copper (Cu) concentrations in leachate, while pecan shell-derived steam-activated carbon reduced nickel (Ni) and cadmium (Cd) concentrations in leachate. Beesley and Marmiroli (2011) studied the capability of hardwood-derived biochar to immobilize and retain arsenic (As), Cd and zinc (Zn) from a multi-element contaminated soil by a column leaching experiment; they found that the sorption of Cd and Zn to biochar reduced their mobility in the soil. Beesley *et al.* (2010) amended multi-element contaminated soil with hardwood-derived biochar into and found that Cd and Zn were immobilized. Park *et al.* (2011) reported that both chicken manure- and green waste-derived biochar

**Table 37.2.** Biochar utilization for inorganic contaminants remediation in soil.

| Contaminant | Biochar type | Effect | References |
|---|---|---|---|
| Pb | Dairy-manure (200 and 350°C) | Sorption of Pb by biochar | Cao *et al*. (2009) |
| As | Hardwood (400°C) | Mobilization due to enhanced pH and DOC | Hartley *et al*. (2009) |
| As, Cd, Cu, Pb, and Zn | *Eucalyptus saligna* (550 °C) | 17% increase in phosphate-extractable As, 51% decrease in DTPA-extractable Pb and 124% increase in DTPA-extractable Zn | Namgay *et al*. (2010) |
| Cu, and Zn | Mixed hardwood-derived biochar | Decreased the total amount of Zn, Cu leached from soils | Laird *et al*. (2010) |
| As, Cu, Cd, and Zn | Hardwood | Mobilization due to enhanced pH and DOC; Immobilization due to enhanced pH | Beesley *et al*. (2010) |
| Cd, Cr, Cu, Ni, Pb and Zn | Orchard pruning (500°C) | Leachable Cd, Pb and Cr reduced; bioavailable Cd, Pb and Zn reduced significantly | Fellet *et al*. (2011) |
| Cr(VI) | Sugar beet tailing (300°C) | Electrostatic attraction of Cr(VI) to positively charged biochar surface, reduction of Cr(VI) to Cr(III) ion, and complexation between Cr(III) ion and function groups | Dong *et al*. (2011b) |
| Cd, Cu, and Pb | Chicken manure and green waste (550°C) | Immobilization due to partitioning of metals from the exchangeable phase to less bioavailable organic bound fraction | Park *et al*. (2011) |
| As, Cd, Cu, Pb and Zn | Hardwood (450°C) | Enhanced As and Cu mobility in the profile, had little effect on Cd and Pb | Beesley and Marmiroli (2011) |
| Cu | Broiler litter (700°C) | Cation exchange; electrostatic interaction; sorption on mineral ash contents; complexation by surface functional groups | Uchimiya *et al*. (2011b) |
| Cu | Hardwood (600°C) | Amendment with hardwood-derived biochar decreased the DOC and decreased Cu mobility | Gomez-Eyles *et al*. (2011) |
| Cu and Pb | Oak wood | Complexation with phosphorous and organic matter | Karami *et al*. (2011) |
| Pb | Dairy manure (450°C) | Immobilization by hydroxypyromorphite formation | Cao *et al*. (2011) |
| Ni, Cu, Pb and Cd | Cottonseed hulls (200–800°C) | Surface functional groups of biochar controlled metal sequestration | Uchimiya *et al*. (2011a) |
| Pb | Oak wood (400°C) | Immobilization by rise in soil pH and adsorption onto biochar | Ahmad *et al*. (2012) |
| Pb | Rice straw | Non-electrostatic adsorption | Jiang *et al*. (2012) |
| Pb, Cu, Zn and Sb | Broiler litter (350 and 600°C) | Stabilization of Pb and Cu; desorption of Sb | Uchimiya *et al*. (2012) |
| As, Cd, Cr, Cu, Ni, Pb and Zn | Sewage sludge (500–550°C) | Immobilization of arsenic, chromium, cobalt, nickel and lead due to rise in soil pH; mobilization of copper, zinc and cadmium due to high available concentrations in biochar | Khan *et al*. (2013) |
| Cd, Zn and Pb | Miscanthus straw (600°C) | The raise in soil pH decreased the metal extractability significantly | Houben *et al*. (2013) |

modified the partitioning of Cd, Cu and lead (Pb) from the easily exchangeable phase to a less bioavailable organic-bound state, and significantly reduced Cd, Cu and Pb accumulation by Indian mustard. Novak *et al.* (2009) reported that acidic agricultural soil amended with 0–2% (wt/wt) pecan shell-based biochar reduced the Zn concentrations in leachate. Laird *et al.* (2010) similarly discovered that the application of mixed hardwood-derived biochar decreased the total amount of Zn and Cu, as well as P, K, Mg and Ca leached from soils.

In contrast, some researchers found different results. Beesley *et al.* (2010) reported that Cu and As were mobilized in soils amended with biochar, and leaching of Cu and As were associated with significant increases in dissolved organic carbon and pH. Hartley *et al.* (2009) also reported that amendment with biochar increased the mobility of As in soil, which was attributed to the rise in soil pH and As competition with soluble P in biochar. In addition, biochar can reduce As(V) to As(III) and increase the mobility of As (Park *et al.*, 2011; Zhang *et al.*, 2013b).

The remediation of inorganic contaminants by biochar is generally attributable to ion exchange, electrostatic attraction and precipitation (Ahmad *et al.*, 2014b). Uchimiya *et al.* (2011a) amended a sandy loam soil with broiler litter biochar for heavy-metal sequestration and found that cation exchange capacity (CEC) was the primary mechanism for Cu retention by biochar. Fellet *et al.* (2011) amended mine tailings with 0–10% orchard prunings-derived biochar and found that biochar reduced bioavailable concentrations of Cd, Pb and Zn because of the increased pH and CEC. The electrostatic attraction between positively charged Cu and negatively charged biochar has been reported as the prevailing mechanism of Cu immobilization in San Joaquin soil (Uchimiya *et al.*, 2011a). Dong *et al.* (2011b) also found the electrostatic attraction between positively charged Cr(VI) to negatively charged biochar surfaces. Cao *et al.* (2009) found that dairy manure-derived biochar prepared at 200°C was more effective in Pb sorption than biochar produced at 350°C; they demonstrated that the dairy manure-derived biochar mainly immobilized Pb by forming the precipitation of insoluble Pb-phosphates ($Pb_5(PO_4)_3(OH)$) (Cao *et al.*, 2011).

Generally, soil pH is considered as an important factor influencing the mobility of metals. Biochar typically is alkaline, thereby resulting in an increase in the soil pH, and causes immobilization of cationic metals due to reduced competition between $H^+$ ions and metal ions. For instance, Yuan and Xu (2011) amended an acid soil (pH 4.3) with nine types of biochar (pH 6.4–10.4). They found that legume biochar raised pH by > 1 unit and non-legume biochar raised pH by < 0.7 units, and the liming effect of biochar will consequently affect the pollutant mobility. Ahmad *et al.* (2012) found that the bioavailability of Pb in highly contaminated military shooting-range soil decreased by 75.8 % with biochar treatment; the increases in soil pH and adsorption capacity were considered as the mechanisms of remediation. They also reported that the rise in soil pH (~1 unit) by amendment with biochar favoured the transformation of easily available Pb to stable residual (Ahmad *et al.*, 2014a). Houben *et al.* (2013) amended a contaminated soil (Cd, Zn and Pb) with biochar and found that metal extractability significantly decreased with increasing rate of biochar application, which was mostly attributed to the rise in soil pH.

The physico-chemical properties of biochar as affected by pyrolysis temperature greatly influence the remediation process of inorganics-contaminated soil (Sima *et al.*, 2017). Biochar obtained at a low temperature is efficient for inorganic contaminants due to the presence of more O-containing functional groups and the greater release of cations. Uchimiya *et al.* (2012) evaluated Pb stabilization in Pb-contaminated soil by broiler litter-derived biochar produced at 350°C and 650°C. They found that biochar produced at lower pyrolysis temperature was favourable for stabilizing Pb, and that high Pb stabilization was associated with more available P, K, and Ca released from biochar produced at a low temperature. Uchimiya *et al.* (2011a) reported that cottonseed hull-derived biochar produced at 350°C contains more O-containing functional groups, resulting in high uptake of Cu, Ni, Cd and Pb. Choppala *et al.* (2012) also reported the potential of biochar to mitigate soils contaminated with chromium (Cr), as they are highly reactive with many functional groups and are able to donate electrons.

It has been reported that dissolved organic carbon (DOC) provided by biochar may affect the mobility of heavy metal in soil (Bernal *et al.*, 2007; Madejón *et al.*, 2009; Beesley *et al.*, 2010;

Mench et al., 2010). Park *et al.* (2011) found that Cu mobility in soil increased with the addition of chicken manure-derived biochar because of the increased DOC in the soil. However, Gomez-Eyles *et al.* (2011) showed a decrease of DOC and Cu mobility after amendment with hardwood-derived biochar; therefore the linkages are far from straightforward and will depend on soil conditions. Additionally, DOC may block the pores of biochar, preventing Cu sorption (Bolan *et al.*, 2011; Cao *et al.*, 2011). Notably, considerable differences in DOC concentrations have been recorded from different types of biochar by Gell *et al.* (2011), who found that the concentrations of DOC in biochar with a range of source materials and production temperature were 0.1–109 g $kg^{-1}$. Beesley and Marmiroli (2011) found that water-soluble organic carbon reduced significantly by successive leaching from initially higher concentrations, which suggested considerable rapid outputs of C in solution from biochar to amended soil systems. The dissolved organic C content leaching from biochar was associated with pyrolysis temperature, which could affect mobility/immobility of heavy metals in soil (Beesley *et al.*, 2010; Uchimiya *et al.*, 2011b).

## 37.6 Conclusions

Obviously, biochar can conceivably reduce the bioavailability and efficacy of both heavy-metal and organic pollutants in soil, and has the potential to be developed as a viable technology for remediation of contaminated soils. Feedstock type and pyrolysis conditions (e.g. temperature) are the main factors influencing physico-chemical properties, which can affect the efficacy of the remediation. Studies have demonstrated the capability of biochar to serve as a green environmental remediator; however, all types of biochar are not equally effective to remediate contaminated soil. As a potential technology for remediation of contaminated agricultural soils, research still needs in-depth investigations to determine the long-term effects of biochar applied in contaminated areas.

## References


Accardi-Dey, A. and Gschwend, P.M. (2003) Reinterpreting literature sorption data considering both absorption into organic carbon and adsorption onto black carbon. *Environmental Science & Technology* 37(1), 99–106.

Adriano, D., Wenzel, W., Vangronsveld, J. and Bolan, N. (2004) Role of assisted natural remediation in environmental cleanup. *Geoderma* 122(2), 121–142.

Aguado, R., Olazar, M., Gaisán, B., Prieto, R. and Bilbao, J. (2002) Kinetic study of polyolefin pyrolysis in a conical spouted bed reactor. *Industrial & Engineering Chemistry Research* 41(18), 4559–4566.

Ahmad, M., Lee, S.S., Yang, J.E., Ro, H.-M., Lee, Y.H. and Ok, Y.S. (2012) Effects of soil dilution and amendments (mussel shell, cow bone, and biochar) on Pb availability and phytotoxicity in military shooting range soil. *Ecotoxicology and Environmental Safety* 79, 225–231.

Ahmad, M., Lee, S.S., Lim, J.E., Lee, S.-E., Cho, J.S. *et al.* (2014a) Speciation and phytoavailability of lead and antimony in a small arms range soil amended with mussel shell, cow bone and biochar: EXAFS spectroscopy and chemical extractions. *Chemosphere*, 95, 433–441.

Ahmad, M., Rajapaksha, A.U., Lim, J.E., Zhang, M., Bolan, N. *et al.* (2014b) Biochar as a sorbent for contaminant management in soil and water: a review. *Chemosphere*, 99, 19-33.

Baiamonte, G., De Pasquale, C., Marsala, V., Cimò, G., Alonzo, G., Crescimanno, G. and Conte, P. (2015) Structure alteration of a sandy-clay soil by biochar amendments. *Journal of Soils and Sediments* 15(4), 816–824.

Beesley, L. and Marmiroli, M. (2011) The immobilisation and retention of soluble arsenic, cadmium and zinc by biochar. *Environmental Pollution* 159(2), 474–480.

Beesley, L., Moreno-Jiménez, E. and Gomez-Eyles, J.L. (2010) Effects of biochar and greenwaste compost amendments on mobility, bioavailability and toxicity of inorganic and organic contaminants in a multi-element polluted soil. *Environmental Pollution* 158(6), 2282–2287.

Beesley, L., Moreno-Jiménez, E., Gomez-Eyles, J.L., Harris, E., Robinson, B. and Sizmur, T. (2011) A review of biochars' potential role in the remediation, revegetation and restoration of contaminated soils. *Environmental Pollution* 159(12), 3269–3282.

Bernal, M., Clemente, R. and Walker, D. (2007) The role of organic amendments in the bioremediation of heavy metal-polluted soils. In: Gore, R.W. (ed.) *Environmental Research at the Leading Edge*. Nova Publishers, New York, pp. 2–58.

Bolan, N.S., Adriano, D.C., Kunhikrishnan, A., James, T., McDowell, R. and Senesi, N. (2011) Dissolved organic matter: biogeochemistry, dynamics, and environmental significance in soils. *Advances in Agronomy*, 110, 1.

Brändli, R.C., Hartnik, T., Henriksen, T. and Cornelissen, G. (2008) Sorption of native polyaromatic hydrocarbons (PAH) to black carbon and amended activated carbon in soil. *Chemosphere* 73(11), 1805–1810.

Brewer, C.E., Schmidt-Rohr, K., Satrio, J.A. and Brown, R.C. (2009) Characterization of biochar from fast pyrolysis and gasification systems. *Environmental Progress & Sustainable Energy* 28(3), 386–396.

Bridgwater, A.V. (2003) Renewable fuels and chemicals by thermal processing of biomass. *Chemical Engineering Journal* 91(2), 87–102.

Bridgwater, A.V. (2012) Review of fast pyrolysis of biomass and product upgrading. *Biomass and Bioenergy* 38, 68–94.

Brownsort, P.A. (2009) *Biomass Pyrolysis Processes: Performance Parameters and their Influence on Biochar System Benefits*. UK Biochar Research Centre, University of Edinburgh, Edinburgh

Cai, Q.-Y., Mo, C.-H., Wu, Q.-T., Katsoyiannis, A. and Zeng, Q.-Y. (2008) The status of soil contamination by semivolatile organic chemicals (SVOCs) in China: a review. *Science of the Total Environment* 389(2), 209–224.

Cao, X. and Harris, W. (2010) Properties of dairy-manure-derived biochar pertinent to its potential use in remediation. *Bioresource Technology* 101(14), 5222–5228.

Cao, X., Ma, L., Gao, B. and Harris, W. (2009) Dairy-manure derived biochar effectively sorbs lead and atrazine. *Environmental Science & technology* 43(9), 3285–3291.

Cao, X., Ma, L., Liang, Y., Gao, B. and Harris, W. (2011) Simultaneous immobilization of lead and atrazine in contaminated soils using dairy-manure biochar. *Environmental Science Technology* 45(11), 4884–4889.

Cha, J.S., Park, S.H., Jung, S.-C., Ryu, C., Jeon, J.-K., Shin, M.-C. and Park, Y.-K. (2016) Production and utilization of biochar: a review. *Journal of Industrial and Engineering Chemistry* 40, 1–15.

Chen, B. and Chen, Z. (2009) Sorption of naphthalene and 1-naphthol by biochars of orange peels with different pyrolytic temperatures. *Chemosphere* 76(1), 127–133.

Chen, B. and Yuan, M. (2012) Enhanced dissipation of polycyclic aromatic hydrocarbons in the presence of fresh plant residues and their extracts. *Environmental Pollution* 161, 199–205.

Chen, G., Andries, J., Luo, Z. and Spliethoff, H. (2003) Biomass pyrolysis/gasification for product gas production: the overall investigation of parametric effects. *Energy Conversion and Management* 44(11), 1875–1884.

Chen, J., Zhu, D. and Sun, C. (2007) Effect of heavy metals on the sorption of hydrophobic organic compounds to wood charcoal. *Environmental Science & Technology* 41(7), 2536–2541.

Chen, B., Zhou, D. and Zhu, L. (2008) Transitional adsorption and partition of nonpolar and polar aromatic contaminants by biochars of pine needles with different pyrolytic temperatures. *Environmental Science & Technology* 42(14), 5137–5143.

Chen, B., Yuan, M. and Qian, L. (2012) Enhanced bioremediation of PAH-contaminated soil by immobilized bacteria with plant residue and biochar as carriers. *Journal of Soils and Sediments* 12(9), 1350–1359.

Cho, Y.-M., Ghosh, U., Kennedy, A.J., Grossman, A., Ray, G. *et al.* (2009) Field application of activated carbon amendment for in-situ stabilization of polychlorinated biphenyls in marine sediment. *Environmental Science & Technology* 43(10), 3815–3823.

Choppala, G., Bolan, N., Megharaj, M., Chen, Z. and Naidu, R. (2012) The influence of biochar and black carbon on reduction and bioavailability of chromate in soils. *Journal of Environmental Quality* 41(4), 1175–1184.

Dan, S., Li, P.-J., Frank, S. and Xiong, X.-Z. (2006) Biodegradation of benzo [a] pyrene in soil by *Mucor* sp. SF06 and *Bacillus* sp. SB02 co-immobilized on vermiculite. *Journal of Environmental Sciences* 18(6), 1204–1209.

Demirbas, A. (2004) Effects of temperature and particle size on bio-char yield from pyrolysis of agricultural residues. *Journal of Analytical and Applied Pyrolysis* 72(2), 243–248.

Demirbas, A. and Arin, G. (2002) An overview of biomass pyrolysis. *Energy Sources* 24(5), 471–82.

Ding, Y., Liu, Y., Liu, S., Li, Z., Tan, X. *et al.* (2016) Biochar to improve soil fertility. A review. *Agronomy for Sustainable Development* 36(2), 1–18.

Dong, J., Yang, Q.-W., Sun, L.-N., Zeng, Q., Liu, S.-J., Pan, J. and Liu, X.-L. (2011a) Assessing the concentration and potential dietary risk of heavy metals in vegetables at a Pb/Zn mine site, China. *Environmental Earth Sciences* 64(5), 1317–1321.

Dong, X., Ma, L.Q. and Li, Y.( 2011b) Characteristics and mechanisms of hexavalent chromium removal by biochar from sugar beet tailing. *Journal of Hazardous Materials* 190(1), 909–915.
Facchinelli, A., Sacchi, E. and Mallen, L. (2001) Multivariate statistical and GIS-based approach to identify heavy metal sources in soils. *Environmental Pollution* 114(3), 313–324.
Fellet, G., Marchiol, L., Delle Vedove, G. and Peressotti, A. (2011) Application of biochar on mine tailings: effects and perspectives for land reclamation. *Chemosphere* 83(9), 1262–1267.
Gell, K., van Groenigen, J. and Cayuela, M.L. (2011) Residues of bioenergy production chains as soil amendments: immediate and temporal phytotoxicity. *Journal of Hazardous Materials* 186(2), 2017–2025.
Golber, E. (1985) *Black Carbon in the Environment: Properties and Distribution*. John Wiley, New York.
Gomez-Eyles, J.L., Sizmur, T., Collins, C.D. and Hodson, M.E. (2011) Effects of biochar and the earthworm *Eisenia fetida* on the bioavailability of polycyclic aromatic hydrocarbons and potentially toxic elements. *Environmental Pollution* 159(2), 616–622.
Gul, S., Whalen, J.K., Thomas, B.W., Sachdeva, V. and Deng, H. (2015) Physico-chemical properties and microbial responses in biochar-amended soils: mechanisms and future directions. *Agriculture, Ecosystems & Environment* 206, 46–59.
Hagner, M., Penttinen, O.-P., Tiilikkala, K. and Setälä, H. (2013) The effects of biochar, wood vinegar and plants on glyphosate leaching and degradation. *European Journal of Soil Biology* 58, 1–7.
Hartley, W., Dickinson, N.M., Riby, P. and Lepp, N.W. (2009) Arsenic mobility in brownfield soils amended with green waste compost or biochar and planted with Miscanthus. *Environmental Pollution* 157(10), 2654–2662.
Heidari, A., Stahl, R., Younesi, H., Rashidi, A., Troeger, N. and Ghoreyshi, A.A. (2014) Effect of process conditions on product yield and composition of fast pyrolysis of *Eucalyptus grandis* in fluidized bed reactor. *Journal of Industrial and Engineering Chemistry* 20(4), 2594–2602.
Horta, A., Malone, B., Stockmann, U., Minasny, B., Bishop, T. *et al.* (2015) Potential of integrated field spectroscopy and spatial analysis for enhanced assessment of soil contamination: a prospective review. *Geoderma*, 241, 180–209.
Houben, D., Evrard, L. and Sonnet, P. (2013) Mobility, bioavailability and pH-dependent leaching of cadmium, zinc and lead in a contaminated soil amended with biochar. *Chemosphere* 92(11), 1450–1457.
Inguanzo, M., Domınguez, A., Menéndez, J., Blanco, C. and Pis, J. (2002) On the pyrolysis of sewage sludge: the influence of pyrolysis conditions on solid, liquid and gas fractions. *Journal of Analytical and Applied Pyrolysis* 63(1), 209–222.
IBI (2012) *Standardized Product Definition and Product Testing Guidelines for Biochar That Is Used in Soil* [a.k.a. *IBI Biochar Standards*] International Biochar Initiative, Canandaigua, New York.
Inyang, M.I., Gao, B., Yao, Y., Xue, Y., Zimmerman, A. *et al.* (2016) A review of biochar as a low-cost adsorbent for aqueous heavy metal removal. *Critical Reviews in Environmental Science and Technology* 46(4), 406–433.
Ippolito, J.A., Laird, D.A. and Busscher, W.J. (2012) Environmental benefits of biochar. *Journal of Environmental Quality* 41(4), 967–972.
Jahirul, M.I., Rasul, M.G., Chowdhury, A.A. and Ashwath, N. (2012) Biofuels production through biomass pyrolysis – a technological review. *Energies* 5(12), 4952–5001.
Jeong, C.Y., Wang, J.J., Dodla, S.K., Eberhardt, T.L. and Groom, L. (2012) Effect of biochar amendment on tylosin adsorption–desorption and transport in two different soils. *Journal of Environmental Quality* 41(4), 1185–1192.
Jiang, T.-Y., Jiang, J., Xu, R.-K. and Li, Z. (2012) Adsorption of Pb (II) on variable charge soils amended with rice-straw derived biochar. *Chemosphere* 89(3), 249–256.
Jones, D., Edwards-Jones, G. and Murphy, D. (2011) Biochar mediated alterations in herbicide breakdown and leaching in soil. *Soil Biology and Biochemistry* 43(4), 804–813.
Karami, N., Clemente, R., Moreno-Jiménez, E., Lepp, N.W. and Beesley, L. (2011) Efficiency of green waste compost and biochar soil amendments for reducing lead and copper mobility and uptake to ryegrass. *Journal of Hazardous Materials* 191(1), 41–48.
Kemper, J.M., Ammar, E.and Mitch, W.A. (2008) Abiotic degradation of hexahydro-1, 3, 5-trinitro-1, 3, 5-triazine in the presence of hydrogen sulfide and black carbon. *Environmental Science & Technology* 42(6), 2118–2123.
Khan, S., Cao, Q., Zheng, Y., Huang, Y. and Zhu, Y. (2008) Health risks of heavy metals in contaminated soils and food crops irrigated with wastewater in Beijing, China. *Environmental Pollution* 152(3), 686–692.
Khan, S., Chao, C., Waqas, M., Arp, H.P.H. and Zhu, Y.-G. (2013) Sewage sludge biochar influence upon rice (*Oryza sativa* L) yield, metal bioaccumulation and greenhouse gas emissions from acidic paddy soil. *Environmental Science & Technology* 47(15), 8624–8632.

Kirubakaran, V., Sivaramakrishnan, V., Nalini, R., Sekar, T., Premalatha, M. and Subramanian, P. (2009) A review on gasification of biomass. *Renewable and Sustainable Energy Reviews* 13(1), 179–186.

Kookana, R.S. 2010. The role of biochar in modifying the environmental fate, bioavailability, and efficacy of pesticides in soils: a review. *Soil Research*, 48(7), 627-637.

Laird, D.A. (2008) The charcoal vision: a win–win–win scenario for simultaneously producing bioenergy, permanently sequestering carbon, while improving soil and water quality. *Agronomy Journal* 100(1), 178–181.

Laird, D., Fleming, P., Wang, B., Horton, R. and Karlen, D. (2010) Biochar impact on nutrient leaching from a Midwestern agricultural soil. *Geoderma* 158(3), 436–442.

Lehmann, J. and Joseph, S. (2009) *Biochar for Environmental Management: Science and Technology*. Earthscan, London.

Lehmann, J., Gaunt, J. and Rondon, M. (2006) Bio-char sequestration in terrestrial ecosystems – a review. *Mitigation and Adaptation Strategies for Global Change* 11(2), 395–419.

Li, L., Rowbotham, J.S., Greenwell, C.H. and Dyer, P.W. (2013) An introduction to pyrolysis and catalytic pyrolysis: versatile techniques for biomass conversion. In: Suib , S.L. (ed.) *New and Future Developments in Catalysis: Catalytic Biomass Conversion*. Elsevier, Amsterdam, pp. 173–208

Li, Z., Ma, Z., van der Kuijp, T.J., Yuan, Z. and Huang, L. (2014) A review of soil heavy metal pollution from mines in China: pollution and health risk assessment. *Science of the Total Environment* 468, 843–853.

Lim, J.E., Ahmad, M., Usman, A.R., Lee, S.S., Jeon, W.-T. *et al.* (2013) Effects of natural and calcined poultry waste on Cd, Pb and As mobility in contaminated soil. *Environmental Earth Sciences* 69(1), 11–20.

Liu, Y., Wen, C. and Liu, X. (2013) China's food security soiled by contamination. *Science* 339(6126), 1382–1383.

Lohmann, R., MacFarlane, J. and Gschwend, P. (2005) Importance of black carbon to sorption of native PAHs, PCBs, and PCDDs in Boston and New York harbor sediments. *Environmental Science & Technology* 39(1), 141–148.

Lou, L., Wu, B., Wang, L., Luo, L., Xu, X. *et al.* (2011) Sorption and ecotoxicity of pentachlorophenol polluted sediment amended with rice-straw derived biochar. *Bioresource Technology* 102(5), 4036–4041.

Madejón, E., Madejón, P., Burgos, P., de Mora, A.P. and Cabrera, F. (2009) Trace elements, pH and organic matter evolution in contaminated soils under assisted natural remediation: a 4-year field study. *Journal of Hazardous Materials* 162(2), 931–938.

McKendry, P. (2002) Energy production from biomass (part 1): overview of biomass. *Bioresource Technology* 83(1), 37–46.

Mench, M., Lepp, N., Bert, V., Schwitzguébel, J.-P., Gawronski, S.W., Schröder, P. and Vangronsveld, J. (2010) Successes and limitations of phytotechnologies at field scale: outcomes, assessment and outlook from COST Action 859. *Journal of Soils and Sediments* 10(6), 1039–1070.

Meyer, S., Glaser, B. and Quicker, P. (2011) Technical, economical, and climate-related aspects of biochar production technologies: a literature review. *Environmental Science & Technology* 45(22), 9473–9483.

Mohan, D., Pittman, C.U. and Steele, P.H. (2006) Pyrolysis of wood/biomass for bio-oil: a critical review. *Energy & Fuels* 20(3), 848–889.

Mukherjee, A., Zimmerman, A. and Harris, W. (2011) Surface chemistry variations among a series of laboratory-produced biochars. *Geoderma* 163(3), 247–255.

Namgay, T., Singh, B. and Singh, B.P. (2010) Influence of biochar application to soil on the availability of As, Cd, Cu, Pb, and Zn to maize (*Zea mays L.*). *Soil Research* 48(7), 638–647.

Novak, J.M., Busscher, W.J., Laird, D.L., Ahmedna, M., Watts, D.W. and Niandou, M.A. (2009) Impact of biochar amendment on fertility of a southeastern coastal plain soil. *Soil Science* 174(2), 105–112.

Oh, S.-Y., Son, J.-G., Lim, O.-T. and Chiu, P.C. (2012) The role of black carbon as a catalyst for environmental redox transformation. *Environmental Geochemistry and Health* 34(1), 105–113.

Ok, Y.S., Usman, A.R., Lee, S.S., El-Azeem, S.A.A., Choi, B., Hashimoto, Y. and Yang, J.E. (2011) Effects of rapeseed residue on lead and cadmium availability and uptake by rice plants in heavy metal contaminated paddy soil. *Chemosphere* 85(4), 677–682.

Onay, O. and Kockar, O.M. (2003) Slow, fast and flash pyrolysis of rapeseed. *Renewable Energy* 28(15), 2417–2433.

Özçimen, D. and Karaosmanoğlu, F. (2004) Production and characterization of bio-oil and biochar from rapeseed cake. *Renewable Energy* 29(5), 779–787.

Park, J.H., Choppala, G.K., Bolan, N.S., Chung, J.W. and Chuasavathi, T. (2011) Biochar reduces the bioavailability and phytotoxicity of heavy metals. *Plant and Soil* 348(1–2), 439–451.

Qian, T.T., Zhang, X.S., Hu, J.Y. and Jiang, H. (2013) Effects of environmental conditions on the release of phosphorus from biochar. *Chemosphere* 93, 2069–2075.

Sima, X.F., Shen, X.C., Fang, T., Yu, H.Q. and Jiang, H. (2017) Efficiently reducing the plant growth inhibition of CuO NPs by rice husk derived biochar: experimental demonstration and mechanism investigation. *Environmental Science Nano* 4, 1722–1732.

Smith, K.E., Thullner, M., Wick, L.Y. and Harms, H. (2009) Sorption to humic acids enhances polycyclic aromatic hydrocarbon biodegradation. *Environmental Science & Technology* 43(19), 7205–7211.

Sohi, S.P., Krull, E., Lopez-Capel, E. and Bol, R. (2010) A review of biochar and its use and function in soil. *Advances in Agronomy* 105, 47–82.

Sopeña, F., Semple, K., Sohi, S. and Bending, G. (2012) Assessing the chemical and biological accessibility of the herbicide isoproturon in soil amended with biochar. *Chemosphere* 88(1), 77–83.

Spokas, K., Koskinen, W., Baker, J. and Reicosky, D. (2009) Impacts of woodchip biochar additions on greenhouse gas production and sorption/degradation of two herbicides in a Minnesota soil. *Chemosphere* 77(4), 574–581.

Tripathi, M., Sahu, J., Ganesan, P. (2016) Effect of process parameters on production of biochar from biomass waste through pyrolysis: a review. *Renewable and Sustainable Energy Reviews* 55, 467–481.

Uchimiya, M., Lima, I.M., Thomas Klasson, K., Chang, S., Wartelle, L.H. and Rodgers, J.E. (2010) Immobilization of heavy metal ions (CuII, CdII, NiII, and PbII) by broiler litter-derived biochars in water and soil. *Journal of Agricultural and Food Chemistry* 58(9), 5538–5544.

Uchimiya, M., Klasson, K.T., Wartelle, L.H. and Lima, I.M. (2011a) Influence of soil properties on heavy metal sequestration by biochar amendment: 1. Copper sorption isotherms and the release of cations. *Chemosphere* 82(10), 1431–1437.

Uchimiya, M., Wartelle, L.H., Klasson, K.T., Fortier, C.A. and Lima, I.M. (2011b) Influence of pyrolysis temperature on biochar property and function as a heavy metal sorbent in soil. *Journal of Agricultural and Food Chemistry* 59(6), 2501–2510.

Uchimiya, M., Bannon, D.I., Wartelle, L.H., Lima, I.M. and Klasson, K.T. (2012) Lead retention by broiler litter biochars in small arms range soil: impact of pyrolysis temperature. *Journal of Agricultural and Food Chemistry* 60(20), 5035–5044.

Usman, A.R., Lee, S.S., Awad, Y.M., Lim, K.J., Yang, J.E. and Ok, Y.S. (2012) Soil pollution assessment and identification of hyperaccumulating plants in chromated copper arsenate (CCA) contaminated sites, Korea. *Chemosphere* 87(8), 872–878.

Van Zwieten, L., Kimber, S., Morris, S., Chan, K., Downie, A. *et al.* (2010) Effects of biochar from slow pyrolysis of papermill waste on agronomic performance and soil fertility. *Plant and Soil* 327(1–2), 235–246.

Vangronsveld, J., Herzig, R., Weyens, N., Boulet, J., Adriaensen, K. *et al.* (2009) Phytoremediation of contaminated soils and groundwater: lessons from the field. *Environmental Science and Pollution Research* 16(7), 765–794.

Vassilev, S.V., Baxter, D., Andersen, L.K., Vassileva, C.G. and Morgan, T.J. (2012) An overview of the organic and inorganic phase composition of biomass. *Fuel* 94, 1–33.

Wang, H., Lin, K., Hou, Z., Richardson, B. and Gan, J. (2010) Sorption of the herbicide terbuthylazine in two New Zealand forest soils amended with biosolids and biochars. *Journal of Soils and Sediments* 10(2), 283–289.

Wang, Y.Y., Jing, X.R., Li, L.L., Liu, W.J., Tong, Z.H. and Jiang, H. (2017) Biotoxicity evaluations of three typical biochars using a simulated system of fast pyrolytic biochar extracts on organisms of three kingdoms. *ACS Sustainable Chemical Engineering* 5, 481–488.

Xu, C., Liu, W. and Sheng, G.D. (2008) Burned rice straw reduces the availability of clomazone to barnyardgrass. *Science of the Total Environment* 392(2), 284–289.

Xu, T., Lou, L., Luo, L., Cao, R., Duan, D. and Chen, Y. (2012) Effect of bamboo biochar on pentachlorophenol leachability and bioavailability in agricultural soil. *Science of the Total Environment* 414, 727–731.

Yang, Y. and Sheng, G. (2003) Enhanced pesticide sorption by soils containing particulate matter from crop residue burns. *Environmental Science & Technology* 37(16), 3635–3639.

Yang, Y., Sheng, G. and Huang, M. (2006) Bioavailability of diuron in soil containing wheat-straw-derived char. *Science of the Total Environment* 354(2), 170–178.

Yang, X.B., Ying, G.G., Peng, P.A., Wang, L., Zhao, J.L. *et al.* (2010) Influence of biochars on plant uptake and dissipation of two pesticides in an agricultural soil. *Journal of Agricultural and Food Chemistry* 58(13), 7915–7921.

Yu, X.-Y., Ying, G.-G. and Kookana, R.S. (2006) Sorption and desorption behaviors of diuron in soils amended with charcoal. *Journal of Agricultural and Food Chemistry* 54(22), 8545–8550.

Yu, X.Y., Ying, G.G. and Kookana, R.S. (2009) Reduced plant uptake of pesticides with biochar additions to soil. *Chemosphere* 76(5), 665–671.

Yu, X., Pan, L., Ying, G. and Kookana, R.S. (2010) Enhanced and irreversible sorption of pesticide pyrimethanil by soil amended with biochars. *Journal of Environmental Sciences* 22(4), 615–620.

Yu, X., Gong, W., Liu, X., Shi, L., Han, X. and Bao, H. (2011) The use of carbon black to catalyze the reduction of nitrobenzenes by sulfides. *Journal of Hazardous Materials* 198, 340–346.

Yuan, J.H. and Xu, R.K. (2011) The amelioration effects of low temperature biochar generated from nine crop residues on an acidic Ultisol. *Soil Use and Management* 27(1), 110–115.

Yuan, J.-H., Xu, R.-K. and Zhang, H. (2011) The forms of alkalis in the biochar produced from crop residues at different temperatures. *Bioresource Technology*, 102(3), 3488–3497.

Zabaniotou, A., Stavropoulos, G. and Skoulou, V. (2008) Activated carbon from olive kernels in a two-stage process: industrial improvement. *Bioresource Technology* 99(2), 320–326.

Zeng, F.X., Liu, W.J., Jiang, H., Yu, H.Q.;,Zeng, R.J. and Guo, Q.X. (2011) Separation of phthalate esters from bio-oil derived from rice husk by a basification-acidification process and column chromatography. *Bioresource Technology* 102, 1982–987.

Zhang, P., Sheng, G., Feng, Y. and Miller, D.M. (2005) Role of wheat-residue-derived char in the biodegradation of benzonitrile in soil: nutritional stimulation versus adsorptive inhibition. *Environmental Science & Technology* 39(14), 5442–5448.

Zhang, H., Lin, K., Wang, H. and Gan, J. (2010) Effect of *Pinus radiata* derived biochars on soil sorption and desorption of phenanthrene. *Environmental Pollution* 158(9), 2821–2825.

Zhang, P., Sun, H., Yu, L. and Sun, T. (2013a) Adsorption and catalytic hydrolysis of carbaryl and atrazine on pig manure-derived biochars: impact of structural properties of biochars. *Journal of Hazardous Materials* 244, 217–224.

Zhang, X., Wang, H., He, L., Lu, K., Sarmah, A. *et al.* (2013b) Using biochar for remediation of soils contaminated with heavy metals and organic pollutants. *Environmental Science and Pollution Research* 20(12), 8472–8483.

Zhang, X.S., Yang, G.X., Jiang, H., Liu, W.J. and Ding, H.S. (2013c) Mass production of chemicals from biomass-derived oil by directly atmospheric distillation coupled with co-pyrolysis. *Scientific Reports* 3, 1120.

# Part IX

# Outlook and Conclusions

# 38 Environmental Regulations in China

**G.Z. He***

*State Key Laboratory of Urban and Regional Ecology, Research Centre for Eco-Environmental Sciences, Chinese Academy of Sciences, Beijing, China, and University of Chinese Academy of Sciences, Beijing, China*

## 38.1 Abstract

China is undergoing economic, social and environmental change that is unprecedented for any nation at any time in world history. To tackle these challenges, the environmental laws that China has enacted over the past 30 years are being strengthened, and new environmental regulations and standards are being issued and implemented by Chinese governments. This chapter introduces a profile of environmental regulations in China: it studies the environmental legal structure and evolution, and the implementation of environmental law. Using a specific example, chemicals policy and management are presented. The chapter also discusses the successes and failures that arise in the environmental implementation process.

## 38.2 Introduction

Explosive economic and industrial growth in China has led to significant environmental degradation in the past three decades. In 2017, around 32% of the country's main rivers were polluted or severely polluted. Only 99 of 338 cities at or above prefecture level across the country met national ambient air quality standards, taking up 29.3%. The quality of 51.8% of tested groundwater was bad and nearly 15% was graded by the Ministry of Environmental Protection as worst (MEP, 2018). Furthermore, about 19% of the country's arable land is polluted (MEP and MLR, 2014). According to the *Environmental Performance Index 2016 Report* by Yale University (Wendling *et al.*, 2018), China ranked 120 out of 180 countries. Meanwhile, pressure from a rising middle class and the international community has focused leadership attention on ways to make China's economic engine run more efficiently and with less impact upon the domestic and global environment. This profound shift in priorities has elevated environmental sustainability to the top of the national agenda. Perhaps the recent driver of this shift is the international development agenda towards the more broadly applicable Sustainable Development Goals (SDGs), which were ratified in September 2015 by the United Nations General Assembly. To advance this new agenda, the environmental laws that China has enacted over the past 30 years are being strengthened, and new environmental regulations and standards are being issued and implemented by Chinese governments. Resolving legal claims involving the environment remains a complex task for litigants and requires them to

* E-mail address: gzhe@rcees.ac.cn

either prove or disprove causation of harm by offering expert evidence based on forensic science.

'Environmental law' (or environmental regulations), also known as environmental and natural resources law, is a collective term describing the network of treaties, statutes, regulations and common and customary laws addressing the effects of human activity on the natural environment. While there is no single agreed classification of these regulations, the core environmental law regimes address environmental pollution. A related but distinct set of regulatory regimes, now strongly influenced by environmental legal principles, focus on the management of specific natural resources, such as forests, minerals or fisheries. Other areas, such as environmental impact assessment, may not fit neatly into either category but are nonetheless important components of environmental law. While the modern history of environmental law is one of continuing (political) discussion and evolution, environmental laws had been established by the end of the 20th century as a component of the legal landscape in all of the developed and industrialized countries of the world, which was followed by similar regulatory movements in many developing countries. The practice of environmental law varies from country to country and even jurisdiction to jurisdiction.

The purpose of this chapter is to introduce the reader to a profile of environmental regulations in China. It begins by addressing the environmental legal structure and its evolution in China. The chapter continues by presenting the implementation of environmental law. This is followed by a section about governmental control on chemicals, including legal and administrative regulations at both local and national levels, using a case study. Before its conclusion, the chapter summarizes successes and failures in the implementation of environmental regulations.

## 38.3 Environmental Legal Structure and Evolution in China

### 38.3.1 Urgency of environmental legislation

China is undergoing economic, social and environmental change that is unprecedented for any nation at any time in the world history. Environmental concerns include air pollution in major cities, deforestation, desertification, loss of arable land, water shortages, unsafe drinking water, water pollution, toxic waste dumping, urban congestion and overcrowding, overpopulation, over-reliance on coal-fired plants, rising carbon dioxide emissions, major environmental accidents, and issues of food security (He *et al.*, 2011, 2012; MEP, 2018). Behind all of this, of course, is the fact that China's environmental problems are massive and growing.

During the first decade of the new Millennium, this started to change in two ways. First, China's authorities were making significant progress in addressing its severe environmental crisis seriously, almost in contrast to developments in other parts of the world (He *et al.*, 2012). For instance, representatives of the so-called 'two Sessions' (the National People's Congress (NPC) and the Chinese People's Political Consultative Conference) at local to national levels doubled their annual submission of environmental proposals, from almost 6000 in 1997 to 12,000 in 2010. Also the numbers of environmental administrative sanctions and registered public environmental complaints have increased sharply since 1997 (Zhang *et al.*, 2013). China's 12th Five-Year Plan (2011–2015) announced a shift to a new green development model and a new green governance approach. This governance reform includes further legal and institutional changes and aims at ending the excessive concentration of power and lack of checks and balances that are often identified as root causes for implementation failures in environmental management (Liu and Diamond, 2008; Kostka and Mol, 2013). During the 18th National Congress of the Communist Party of China (CPC) in November 2012, 'ecological civilization' (restructuring of the economy by 'ecologizing' production and consumption to achieve human–nature harmony) was included in the constitution of the CPC, with an emphasis on scientific, democratic and rule-by-law governance. Again, this should not be interpreted as a full greening of Chinese governance, but definitely as a change in priority setting.

Second, environmental advocates and the wider public were given significantly more room to express their environmental discontent, as a few examples can illustrate. Although media reporting (old and new) on environmental misbehaviour and conflicts is still far from free, it did improve significantly over the past decade (Yang and Calhoun, 2007; Mol, 2009). Environmental advocates also obtained more possibilities to organize themselves into (quasi) non-governmental organizations (NGOs), and to protest: it was estimated that the number of environment-related mass protests reached 90,000 in 2010, accounting for almost half of the total public protests in China (Genasci, 2012). Lower-level state authorities and companies are increasingly pressed to report on and disclose environmental information to the public (Zhang *et al.*, 2010). The first signs of democratic decision making on environmental issues have emerged, such as public hearings on environmental impact assessments (EIA) and water prizes and fees (Zhong and Mol, 2008; Kostka and Mol, 2013). Although these political–institutional changes are not unique for the environmental domain, the field of environmental politics is often leading in experimenting with these political–institutional changes.

### 38.3.2 The formation of environmental laws in China

The procedures for national (environmental) law-making in China diverge according to the kind of legal instrument that is finally enacted (Zhu, 2011), differentiating between administrative rules, administrative regulations and legislation (Day, 2005). In all three cases, a sectoral/functional ministry or commission is often authorized to make an initial draft. Only in the case of a law, a draft should be made after it has been included in the NPC's legislation plan. Administrative rules are finally approved at a Ministerial meeting. Administrative regulations need to be approved at an executive meeting of the State Council (the main administrative authority in China, including all heads of ministerial departments and agencies, and chaired by the premier). During the procedure the State Council have to invite relevant ministries, experts and the public to comment on a draft, and an improved draft will then be decided upon at the State Council. The procedures for the highest level of law-making, legislation, are based on the Law on Legislation. When the State Council makes a legislation proposal, it organizes the drafting of the law, invites comments and suggestions of experts, ministries and lower-level government, and after examination submits the draft to the Standing Committee of the NPC. Normally, a draft law is examined and discussed one to three times in the Standing Committee before final voting takes place in the NPC. If major controversies remain after three examinations at the Standing Committee meetings, the submission to NPC for final approval may be postponed.

Proposals from NPC representatives for new environmental legislation and for revising outdated environmental laws have increased since 1995. However, stagnation of environmental law making and law revision has been widely observed in recent years (Li, 2009). For instance, since no qualified organization made formal proposals to the NPC committee, the third revision of the Air Pollution Prevention and Control Law was finally listed in the State Council's legislation plan for 2013 after 3 years of discussions. It also took 3 years to reach a commonly agreed framework and principles for drafting the Soil Pollution Prevention and Control Law and it will take even longer to produce a draft for NPC approval. The reasons for and details of these stagnations usually remain secret, known to only a small inner circle of experts and party and state officials.

### 38.3.3 Environmental legal and institutional arrangements

China has been working with great determination in recent years to develop, implement and enforce a solid framework of environmental law (Day, 2005). The harmonization of Chinese society and the natural environment is billed as a rising policy priority (Stern, 2013; Wang, 2013). The broad category of environmental law may be broken down into a number of more specific regulatory subjects (Table 38.1). Environmental law includes regulation of environmental pollution and natural resource conservation and allocation. The regulations refer to forestry, agriculture, fisheries, grassland, marine environments, energy development and use, food and land use, and have been expanded to include international environmental governance, international trade, environmental justice and climate change. The

**Table 38.1.** Chronology of environmental events and regulations in China.

| Year | Event |
|---|---|
| 1972 | Chinese representatives attended the first United Nations Conference on the Human Environment |
| 1974 | The Leading Team for Environmental Protection was officially established under the State Council |
| 1979 | Environmental Protection Law (for Trial Implementation) |
| 1979 | Forestry Law (for Trial Implementation) |
| 1982 | The Ministry of Urban and Rural Development and Environmental Protection was established; Environmental Protection Bureau is one of the internal departments |
| 1982 | Marine Environment Protection Law of China, revised in 2013, 2016 |
| 1983 | The Chinese government announced that environmental protection would become a state policy |
| 1984 | Law on Prevention and Control of Water Pollution, amended in 1996, 2008, and 2018 |
| 1984 | The Environmental Protection Bureau under the Ministry of Urban and Rural Development and Environmental Protection was reshuffled into National Environmental Protection Agency |
| 1985 | Grassland Law, revised in 2002, 2009, 2013 |
| 1985 | Forestry Law, revised in 1998 |
| 1986 | Fisheries Law, revised in 2000, 2004, 2009, 2013 |
| 1986 | Mineral Resources Law, revised in 1996 |
| 1987 | Land Administration Law, revised in 1988, 2004 |
| 1988 | Water Law, revised in 2002, 2009 |
| 1988 | Law on Prevention and Control of Atmospheric Pollution, amended in 1995, 2000, and 2016. |
| 1988 | An independent National Environmental Protection Agency (a sub-ministerial level agency) was founded. |
| 1989 | Law on the Protection of Wildlife, revised in 2004, 2009 |
| 1989 | Environmental Protection Law |
| 1991 | Law on Water and Soil Conservation, amended in 2009 |
| 1991 | Law on the Entry and Exit Animal and Plant Quarantine |
| 1992 | Law on the Territorial Sea and the Contiguous Zone |
| 1993 | Agriculture Law, revised in 2003, 2013 |
| 1994 | The State Council of China approved China's Agenda 21: White Paper on China's Population, Environment and Development in the 21st Century. |
| 1994 | Regulations on Nature Reserves |
| 1995 | Law on the Prevention and Control of Environmental Pollution by Solid Waste, amended in 2004, 2013 and 2015 |
| 1995 | Electric Power Law |
| 1996 | Law on the Coal Industry, revised in 2011 |
| 1996 | Law on Prevention and Control of Pollution from Environmental Noise, revised in 2004 |
| 1997 | Flood Control Law, revised in 2015 |
| 1997 | Energy Conservation Law, revised in 2007 |
| 1998 | National Catalogue of Hazardous Wastes, revised in 2008 |
| 1998 | China went through a disastrous year of serious flooding. |
| 1998 | National Environmental Protection Agency was upgraded to State Environmental Protection Administration (Ministerial level) |
| 1998 | Law on the Exclusive Economic Zone and the Continental Shelf |
| 1999 | Measures on the Management of Environmental Standards |
| 1999 | Measures on Administrative Penalty for Environmental Protection |
| 1999 | Meteorology Law |
| 1999 | Measures on the Administration of Pollution Sources Monitoring |
| 2001 | Law on the Prevention and Control of Occupational Diseases, revised in 2011, 2017 |
| 2002 | Law on Desert Prevention and Transformation |
| 2003 | Law on the Promotion of Clean Production, revised in 2012 |

*Continued*

**Table 38.1.** Continued.

| Year | Event |
|---|---|
| 2003 | Environmental Impact Assessment Law, revised in 2016 |
| 2005 | Renewable Energy Law, revised in 2009 |
| 2007 | Emergency Response Law |
| 2007 | Law on Urban and Rural Planning |
| 2007 | At the 17th National Congress of the Communist Party of China (CPC), it was officially proposed that China should build an 'ecological civilization' |
| 2008 | The State Environmental Protection Administration was upgraded to Ministry of Environmental Protection, as an integral department of the State Council |
| 2009 | Food Safety Law, revised in 2015 |
| 2009 | Circular Economy Promotion Law |
| 2012 | The 18th National Congress of CPC wrote the construction of an 'Ecological Civilization' into the Constitution |
| 2015 | The CPC Central Committee and the State Council promulgated 'Integrated Reform Plan for Promoting Ecological Progress' and 'Opinions on Accelerating the Ecological Civilization Construction' |
| 2015 | The revised Environmental Protection Law came into force |
| 2017 | The 19th National Congress of CPC highlighted to accelerate the Ecological Civilization reform and build a beautiful China |
| 2018 | Law on Environmental Protection Tax |
| 2018 | Nuclear Safety Law |
| 2019 | Law on Prevention and Control of Soil Pollution |

practice and application of environmental law typically requires extensive knowledge of administrative law and aspects of tort law, property, legislation and constitutional law.

As a start to the passing of various national laws in China, the Environmental Protection Law (EPL) for trial implementation was passed in 1979. The formal EPL was enacted in 1989. The EPL has been described as one of most relevant pieces of environmental legislation passed by the NPC of China. It is one of the cornerstone pieces of legislation that governs environmental protection in China. The basic purposes of the EPL are to protect and improve the human environment and the ecological environment, prevent and control pollution and other public hazards, safeguard human health and facilitate the development of socialist modernization. In the past three decades, the EPL has been China's main national framework environmental legislation (Day, 2005). On 24 April 2014, the Standing Committee of the NPC passed a revised EPL. The law approval marked the end of a 3-year highly controversial and unique revision process (He *et al.*, 2013; Zhang *et al.*, 2013, 2015; Wübbeke, 2014). In January 2015, the revised EPL came into force. The EPL is composed of seven parts with 47 articles. Based on the EPL, quite a comprehensive system of environmental laws and regulations has been constructed in China, with over 30 specific administrative laws and regulations on environmental protection. These include the Law on Atmospheric Pollution Prevention and Control, the Law on the Prevention and Control of Water Pollution, the Environmental Impact Assessment Law, the Environmental Protection Tax Law and the Nuclear Safety Law. The Law on the Prevention and Control of Soil Pollution was approved in August 2018 and enacted from 1 January 2019. The Law of Environment Impact Assessment and the Law of Marine Environment Protection have been amended, as well as the Explanation by Supreme People's Court and Supreme People's Procuratorate of the Application of Laws for Handling Cases of Environmental Pollution Crimes in 2016. The National Catalogue of Hazardous Wastes has been modified, and specific actions have been taken to crack down on law-breaking activities or crimes concerning hazardous wastes.

The Environmental Protection Bureau was established as one of the internal departments of Ministry of Urban and Rural Development and Environmental Protection in 1982. An independent National Environmental Protection Agency (a sub-ministerial level agency) was founded in 1988. The National Environmental Protection Agency was upgraded to State Environmental Protection Administration (Ministerial level) in 1998, as an affiliated institution of the State Council in charge of the environmental protection effort. In July 2008, the State Environmental Protection Administration was upgraded to Ministry of Environmental Protection (MEP), as an integral department of the State Council. The MEP was renamed as Ministry of Ecology and Environment (MEE) in March 2018. The mission of MEE is to improve environmental quality and build a beautiful China which enjoys blue sky, green land and clean water. It is the duty of the MEE to monitor and analyse the ecological environment, conduct research and work closely with national and local governments to devise pollution control policies. Twenty-one professional departments or bureaus are affiliated to the MEE, including Central Supervision Office of Ecological and Environmental Protection, Department of Laws, Regulations and Standards, Department of Institutional Administration and Human Resources Management, Department of Environmental Impact Assessment and Emission Management, departments responsible for science and technology, management of nature and ecology conservation, marine ecology and environment, water, air, soil, solid wastes and chemicals, climate change, radiation source, nuclear facility and nuclear power safety, and environmental monitoring. The Department of Laws, Regulations and Standards is mainly responsible for establishing, improving, interpreting and post-evaluating administrative regulations, policies, national guidelines, macro strategies, the standards, ecological and environmental benchmarks and technical specifications on ecological and environmental protection. It also gathers opinions when the NPC, Legislative Affairs Office of the State Council and other departments under the State Council consult the MEE on the environmental impacts of the draft laws and administrative regulations.

China's leadership has called in recent years for the creation of a new Ecological Civilization. At the 17th National Congress of the CPC in 2007, it was officially proposed that China should build an 'ecological civilization', creating more sustainable relations between production, consumption, distribution and economic growth. At the 18th National Congress of the CPC in 2012, construction of an ecological civilization was written into the CPC's Constitution. These principles were built into the 12th five-year plan (2011–2015). In 2015, the CPC Central Committee and the State Council promulgated 'Integrated Reform Plan for Promoting Ecological Progress' and 'Opinions on Accelerating the Ecological Civilization Construction'. The Central Leading Group for Deepening Overall Reform has approved more than 40 special reform plans on ecological civilization and ecological and environmental protection. Combined, they play a significant role in promoting green development and improving the environment. In October 2017, the 19th National Congress of CPC highlighted its intention to accelerate the Ecological Civilization reform and build a beautiful China. Although many have questioned the seriousness of the CPC's commitment to the construction of an ecological civilization, it is evident that: (i) this arose out of real needs in China, where there has been enormous ecological devastation; (ii) it was a response to the growth of massive environmental protests throughout China; and (iii) it has been followed up by massive government efforts in areas of planning, production and technological development.

## 38.4 Environmental Law Implementation

Chinese governments have reinforced supervision on environmental law enforcement and risk responses in recent years. Activities for the Year for the Implementation of Environmental Protection Law were conducted throughout 2016, and the MEP has talked publicly with the principals of eight municipal governments about their deteriorating environment. The enforcement of environmental laws has been remarkably strengthened. In 2016, the environmental protection departments at different levels meted out more than 124,000 decisions on administrative penalty with total fines of 6.63 billion Yuan, up by 28% and 56%, respectively, compared with those of 2015. There were 22,730 cases across the country subjected to consecutive daily penalty, sealing up or seizure, limit or stop of production, transfer to administrative detention or transfer cases suspected to involve environmental pollution crime, up by 93% compared with the same historical period.

Still, the legal system on environmental protection is far from being complete, and implementation and enforcement of existing environmental laws have shown major shortcomings that are commonly observed in many developing countries. The first year of implementation of the new EPL was thoroughly assessed in the *New Environmental Protection Law Implementation Assessment Report* by the Environment and Natural Resources Research Institute (ENRRI, 2016). The Report concluded that progress had been made in terms of information disclosure and public participation, even though 36% of cities sampled for the report failed to provide a list of key polluters. New instruments that aimed to increase sanctions for violation of the EPL, such as a 'daily penalty system', 'administrative detention', 'environmental impact assessment restriction targeting regions' (EIARTR) (Zhu *et al.*, 2015) and 'administrative appointment (as a measure of warning)', were found to be relatively effective. The Report also pointed out that local enforcement capacity and coordination among different agencies were still weak.

Chinese officials faced critical challenges in effectively implementing the laws, clarifying the roles of their national and provincial governments, and strengthening the operation of their legal system (Day, 2005). The MEP has concerted its efforts in launching three major campaigns of prevention and control of air pollution, water pollution and soil pollution. China has carried out its Action Plan for Prevention and Control of Air Pollution and the Reinforced Action Plan for Prevention and Control of Air Pollution in Beijing–Tianjin–Hebei Region (2016–2017). The government has fully implemented the Action Plan for Prevention and Control of Water Pollution. It has signed responsibility statements with various provinces (autonomous regions or municipalities) so as to set up coordination mechanisms for relevant tasks. Specific actions have been taken to protect drinking-water sources along the Yangtze River. The project to implement and promote the safety standard for rural drinking water has been carried out. The government has specified a working plan for the prevention and control of groundwater pollution and carried forward the national groundwater monitoring project. It has implemented the Action Plan of Prevention and Control of Soil Pollution. The State Council has printed and distributed the Plan, supporting the close investigation of soil pollution status across the country and intending to release 25 relevant polices and measures concerned. Thirty-one provinces (municipalities) have laid out action plans for prevention and control of soil pollution and 13 branches or departments have provided implementation plans for solving key problems. The government has issued the Management of Soil Environment in Contaminated Sites and set up experimental areas to promote the comprehensive management of soil pollution.

The national inspection of environmental protection has been carried out in 16 provinces (municipalities), receiving 33,000 complaints and having had talks with 6307 persons and held 6454 persons accountable. The General Office of the CPC Central Committee and the State Council have issued the Guidance of the Pilot Reform of Vertical Management of the Monitoring and Inspection of Environmental Protection Institutions below Provincial Level, and this reform has been implemented first in Hebei and Chongqing. The Central Leading Group of Deepening Reforms in an All-round Way has discussed and approved the Opinions on Identifying and Observing Red Lines of Ecological Conservation, and the identification of ecological red lines has been launched in 31 provinces (autonomous regions and municipalities). The State Council has printed and distributed the Implementation Plan on Pollution Discharge Control Permits, and the application and approval of pollution discharge permits have been started in the thermal power and papermaking industries. The MEP has issued the Programme of Cultivating and Developing Main Market Subjects in the Treatment of Agricultural Non-point Source Pollution and Disposal and Treatment of Rural Sewage and Refuse as well as the Implementation Programme on the Reform of EIA in the 13th 'Five-Year Plan' Period. A range of technical specifications have been issued, including Technical Guidelines on the Identification and Appraisal of Eco-environmental Damages. As a follow-up, seven provinces (municipalities), including Jilin, have launched pilot reforms on the compensation system for eco-environmental damages, and the implementation programme has been issued and put into practice in these seven provinces (municipalities) following approval by the Central Leading Group of Deepening Reform in an All-round Way. The MEP has also issued the Guidance of Constructing Green Financial System.

The MEP has actively promoted the construction of an eco-environmental big-data project with positive progress having been made in the integration and application of data resources. The MEP has launched its official Micro-blog and WeChat account, both entitled 'Voice of MEP', so as to give timely release and interpretation of information. The regular news release system of the MEP has also been set up.

These laws and policies need to be implemented by local governments and industry (Day, 2005). The effectiveness of the current top-down system of supervision and enforcement via regional MEP centres, along with campaign-style inspections and evaluation and audit of local officials' environmental performance, has been mixed. When local governments, sectoral departments, experts and the public are not consulted in environmental policy making or their feedback is not fully take into account, policy implementation and enforcement are likely to encounter resistance. When the responsibility for enforcement of environmental responsibilities is shouldered by the local governments and industries do not match their resources and capacities and the government's means of checking are limited, top-down 'pressure-based' incentive mechanisms may cause manipulation of monitoring data as a shortcut to meeting environmental requirements, though this further dilutes government accountability (Chen, 2010; Ran, 2013). Despite the fact that, according to recent reports, individuals in some local governments and enterprises who have been found to be manipulating monitoring data have been punished with fines or administrative detention (according to the Supreme Court's interpretation of Articles 63 and 65 of the new EPL), the root of the problem remains. In addition, apart from environmental enforcement inspections by the MEP (which inspected more than 30% of the municipal governments in 2015) and administrative appointments, published priority list for enforcement, EIARTR and other legal instruments would be applied accordingly (Xinhua net, 2015), higher-rank inspections directly organized by the central government were also made routine. In July 2016, within a month, eight provinces and regions had been visited by inspection teams from Beijing.

## 38.5 Governmental Control on Chemicals: Legal and Administrative Interventions

In the latter part of the 20th century came the realization that chemicals could act as environmental pollutants. As a result, unprecedented efforts were then made to reduce all emissions of all chemicals in order to maintain a green perspective and the evolution of environment-oriented thinking was followed by the formulation and passage of environmental regulations (McElwee, 2011). Chemical small and medium enterprises (SMEs) have been a significant engine in China's economic transition since the 1980s. Chemical SMEs face increasing pressures to clean up their business, from regulatory and administrative agencies, local communities and civil society, and international value-chain stakeholders, using an in-depth case study of 'KD' Pesticide and Chemical Corporation, 'HY' and 'LH' in Jasmine County, Hebei province (see Section 38.5.2).

### 38.5.1 National regulations and interventions on chemicals

In China, the supervision, management and control of chemicals are allocated to different agencies. The legal system in the handling of pesticides includes laws, regulations, ordinances and standards (Lu *et al.*, 2007; Wang *et al.*, 2005). Up to March 2013, 38 national laws, 48 State Council regulations, 196 national government department ordinances and 324 national and industrial pesticide standards had been promulgated, targeting licensing/classification, production, storage/packaging, transportation, trade/import/export, use, disposal and supervision (Table 38.2). About half of the laws and regulations target the licensing and production

**Table 38.2.** Pesticide-relevant laws and regulations (sources: websites of State Council and national government agencies).

| | Licensing | Production | Storage | Transport | Trade, export & import | Use | Disposal | Supervision | Total |
|---|---|---|---|---|---|---|---|---|---|
| Laws | 8 | 9 | 2 | 2 | 6 | 5 | 8 | 2 | 38 |
| Regulations | 11 | 10 | 4 | 5 | 8 | 4 | 7 | 2 | 48 |
| Ordinances | 49 | 36 | 7 | 31 | 29 | 11 | 12 | 40 | 196 |
| Standards | 11 | 269 | 4 | 5 | 2 | 21 | 6 | 13 | 324 |
| International conventions | 1 | 4 | | 1 | 2 | 2 | | 2 | 13 |

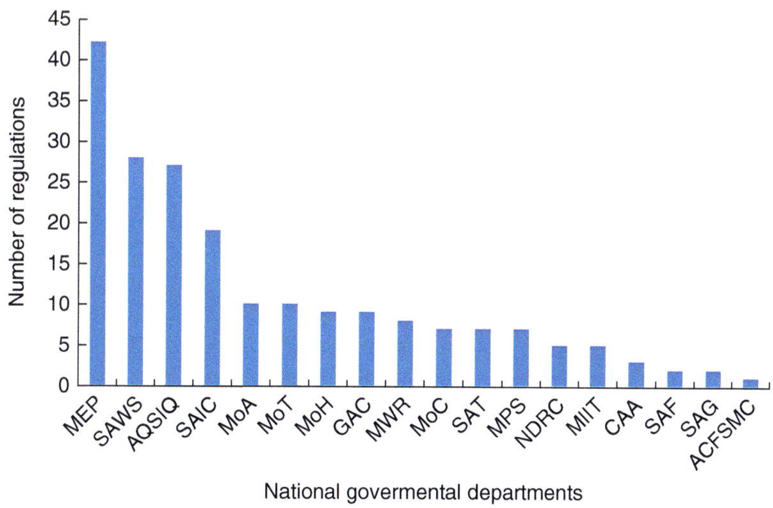


**Fig. 38.1.** Departmental pesticides regulations promulgated by different Chinese ministries and commissions till March 2013. (For abbreviations, see Table 38.3.)

of pesticides. The MEP, the State Administration of Work Safety (SAWS) and the General Administration of Quality Supervision, Inspection and Quarantine (AQSIQ) are the most important governmental departments for issuing departmental regulations on pesticides (Fig. 38.1). Broadly speaking, these laws and regulations impose obligations to make less comprehensive and more specific products in a safer, cleaner, more efficient way, by using less hazardous raw materials, less energy and water, and producing fewer toxic wastes.

Table 38.3 shows the basic functions of 18 ministries and commissions involved in the management and supervision of pesticides. Whereas the Ministry of Agriculture (MoA) is mainly responsible for registration, storage safety and application of pesticides, industrial pollution prevention and control of pesticides are the responsibility of the MEP. Given the significant environmental and health impact caused by chemical enterprises, the larger plants have been listed by the MEP as key national enterprises for extra monitoring and supervision on pollutants discharge since 2009. This means that provincial, municipal and county Environmental Protection Bureaus (EPBs) have strongly focused on regulating these plants. Chemical plants was assessed by the MEP on its environmental compliance according to Notice on the Inspection and Verification of Environmental Protection of the Corporations Applying for Listing and the Listed Corporations Applying for Refinancing, enacted by the previous State Environmental Protection

**Table 38.3.** National governmental ministries and commissions and their functions in chemicals management in China.

| Departments | Licensing | Production | Storage | Transport | Import/ Export | Trade | Use | Disposal | Supervision |
|---|---|---|---|---|---|---|---|---|---|
| ACFSMC | √ | | | | | √ | | | |
| AQSIQ | √ | √ | | | √ | √ | | | √ |
| CAA | | | √ | √ | | | | | |
| GAC | | | | | √ | | | | |
| MEP | √ | √ | √ | √ | √ | √ | √ | √ | √ |
| MoA | √ | √ | √ | √ | √ | √ | √ | √ | √ |
| MoC | √ | | | √ | √ | √ | √ | √ | √ |
| MoH | √ | √ | | | √ | √ | √ | | √ |
| MoT | | | √ | √ | | | | | |
| MWR | | √ | | | | | | √ | |
| NDRC | √ | | | | | | | | √ |
| MIIT | √ | √ | | | | | | | √ |
| MPS | | √ | √ | √ | | √ | √ | | √ |
| SAIC | | | | | | | | | |
| SAF | √ | | | | | √ | √ | | |
| SAG | √ | | | | | | √ | | |
| SAT | | | | | | √ | | | |
| SAWS | √ | √ | √ | √ | | √ | √ | √ | √ |

Abbreviations: ACFSMC, All China Federation of Supply and Marketing Cooperatives; AQSIQ, General Administration of Quality Supervision, Inspection and Quarantine; CAA, Civil Aviation Administration; GAC, General Administration of Customs; MEP, Ministry of Environmental Protection; MoA, Ministry of Agriculture; MoC, Ministry of Commerce; MoH, Ministry of Health; MoT, Ministry of Transport; MWR, Ministry of Water Resources; NDRC, National Development and Reform Commission; MIIT, Ministry of Industry and Information Technology; MPS, Ministry of Public Security; SAIC, State Administration for Industry and Commerce; SAF, State Administration of Forestry; SAG, State Administration of Grain; SAT, State Administration of Taxation; SAWS, State Administration of Work Safety.

Administration (now MEP) in 2003. The SAWS plays a key function for safety supervision and inspection of the chemical facilities. Emergency management of the chemical plant is controlled by the SAWS and the MEP.

### 38.5.2 Changing roles of local governmental agencies

Over the past decades the Chinese government has defined the main responsibility of environmental law and regulation enforcement to the local EPBs, and also increased the number of environmental inspection agencies and environmental inspectors to strengthen the local governments. The 'KD' Pesticide and Chemical Corporation is located in Jasmine County, Hebei province. It is one of the county's four pillar industries, which contributed 14% to the county's gross industrial output value in 2011. The local governments and their agencies have also taken several measures and actions to supervise the operation and environmental risk management of the four chemical facilities (Table 38.4). In general, most provincial and prefectural/city agencies play indirect roles, but the county government and its divisions have more direct roles in environmental enforcement, inspection and supervision. The provincial, city and county EPBs have been responsible for supervising pollution prevention and control of the national key polluting industries, such as 'KD'. However, for a long time their monitoring in the case of 'KD' industries only focused on routine pollution indicators, while hardly paying attention to special (and more toxic) pollutants from these chemical plants. Hence, they were not in a position to identify the causality between pollution levels and the observed and claimed environmental and health effects.

The chemical plants did indeed feel increasing pressure over the past decade from local governments and their agencies, which forced them to take actions to improve industrial environmental management, but environmental protection did

**Table 38.4.** Supervision actions of the local governments and its agencies on 'KD'.

| Level | Agency | Supervision measures and actions |
|---|---|---|
| Hebei Province | HIITD | General supervision of SMEs in province |
| | | Granting production permits to new pesticide plants |
| | | Annually checking of adequacy of production permits of three KD plants |
| | HEPB | Promulgation of provincial environmental rules and policies |
| | | Supervision pollution prevention and control of national key polluters in the province, including HY and LH |
| | | Investigation of HY and LH on illegal construction of two new projects without approval; identified it as one of ten key accidents of Hebei Province in 2005 |
| | | Directing cleaner production measures of key enterprises, including KD |
| | | Carry out environmental performance assessment on three enterprises in 2009 |
| | | Conducted environmental inspection in 2009 and found illegally discharge of pollution in all three plants |
| | | Announced illegal activities and punishment of KD |
| | | Put three KD plants on the provincial list of key enterprises |
| | | Strengthened supervision and examination on the dangerous wastes of KD since 2011 |
| | | Conducted pollution sources survey from 2007 to 2009 (with HDA and HBS) |
| | HAWS | Guided and supervised the safety management of ZJK and Jasmine AWS |
| | HBH | Strengthened supervision and examination on dangerous wastes of KD since 2011 |
| | HDA | Examined and approved new pesticide products of KD |
| | | Conducted pollution sources survey from 2007 to 2009 (with EPB and HBS) |
| | HAQTS | Supervised the pressure vessels management of KD |
| | HEIQB | Supervised the quality of the KD products |
| | SCD | Supervised the export of the KD products |
| | HBS | Conducted the pollution sources survey from 2007 to 2009 (with EPB and HDA) |
| | | Environmental statistics of key enterprises |
| ZJK city | ZEPB | Promulgated the prefectural environmental notices |
| | | Supervised pollution prevention and control of the national and prefectural key polluters, including three KD plants |
| | | Performed regular monitoring on surface water, groundwater, air and noise of three plants |
| | | Monitored the wastewater and dangerous waste discharge |
| | | Investigated HY and LH on illegal construction of two new projects without approval in 2005 |
| | | Generated and disclosed annual report on dangerous wastes of the three plants via its website since 2011 |
| | | Published regularly online monitoring data on the local environmental quality and pollutants discharges from three plants |
| | | Conducted assessment on the emission reduction and environmental targets achievement of LH and HY in 2011 |
| | ZAWS | Included the three plants in the 1st list for making production safety emergency plan in 2012 |
| | | Steered the work safety management of JASW |
| | ZBIIT | General supervision SMEs in city |
| Jasmine county | JCG | Decision to exclude chemical industry as priority sector in the 12th FYP in 2011 |
| | | Ordered LH to stop production and compensate 687 farmer households after chemical leak in 2003 |
| | JBDR | Developed the local FYP and industrial measures |
| | | Helped the KD to get the national environmental subsidies |
| | JBIIT | Monitored and guided the SMEs in county |
| | | Preliminarily annual check of production permits of three plants |
| | JBA | Preliminarily check on application of new pesticide production of KD |

*Continued*

**Table 38.4.** Continued.

| Level | Agency | Supervision measures and actions |
|---|---|---|
| | JEPB | Supervision of pollution prevention and control of three plants |
| | | Imposed fine on LH after a chemical accident in 2003 |
| | | Conducted administrative penalties (200,000 Yuan for each plant) for illegal construction of two new projects without approval in HY and LH in 2005 |
| | | Required the three plants to conduct environmental impact assessments (EIAs) in 2010 and 2011 |
| | | Listed three plants as high risk enterprises |
| | | Conducted frequent environmental inspections on three plants (2-3 times per month) |
| | | Monitored the operation of environmental facilities in three plants |
| | | Implemented the comprehensive screenings on the dangerous chemicals, potential safety problems, water pollution, and illegal pollution emissions |
| | | Took five measures to strengthen the dangerous wastes management of three plants in 2011 |
| | | Required three plants to develop emergency response plans |
| | JAWS | Listed three plants as high risk enterprises |
| | | Surveilled work and hazardous chemical safety in three plants |
| | | Conducted regular inspections on three plants (1-2 times per month) |
| | | Checked transportation safety of the dangerous chemicals in three plants in 2010 |
| | | Conducted "Work Safety Year" campaign to prevent illegal operations and accidents in 2011 |
| | | Conducted special safety inspection on dangerous chemicals and chemical enterprises in 2012 |
| | JAQTS | Conducted (quarterly) production facilities inspection on the three plants |
| | JBH | Supervised dangerous waste management of three plants |
| | | Inspected occupational health of the three plants |
| | JBPS | Conducted fire protection on three plants (every half year) |
| | JBS | Conducted economic and environmental statistics of local SMEs |
| | JBTT | Checked transportation safety of the dangerous chemicals in three plants in 2010 |

Abbreviations: HAWS, Hebei Administration of Work Safety; HAQTS, Hebei Provincial Administration of Quality and Technical Supervision; HBH, Hebei Bureau of Health; HBS, Hebei Bureau of Statistics; HDA, Hebei Provincial Department of Agriculture; HEIQB, Hebei Entry-Exit Inspection and Quarantine Bureau; HEPB, Hebei Environmental Protection Bureau; HIITD: Hebei Industry and Information Technology Department; JAQTS, Jasmine Administration of Quality and Technical Supervision; JAWS, Jasmine Administration of Work Safety; JBA, Jasmine Bureau of Agriculture; JBDR, Jasmine Bureau of Development and Reform; JBH, Jasmine Bureau of Health; JBIIT, Jasmine Bureau of Industry and Information Technology; JBPS, Jasmine Bureau of Public Security; JBS, Jasmine Bureau of Statistics; JBTT, Jasmine Bureau of Traffic and Transport; JCG, Jasmine County Government; JEPB, Jasmine Environmental Protection Bureau; SCD, Shijiazhuang Customs District, GAC; ZAWS, Zhangjiakou Administration of Work Safety; ZBIIT, Zhangjiakou Bureau of Industry and Information Technology; ZEPB, Zhangjiakou Environmental Protection Bureau; ZJK city, Zhangjiakou city.

not become a company priority. Three smaller plants did conduct EIAs but only in 2010 and 2011, though such EIAs should have been done before the plants were constructed, according to the Chinese EIA Law. In 2009, after 20 years of unbridled pollution discharge, the three plants were finally ordered to improve their environmental management and to construct wastewater treatment stations. Although plants 'LH' and 'HY' (belonging to 'KD') constructed their wastewater treatment stations and installed online monitoring on chemical oxygen demand (COD) and ammonia nitrogen ($NH_3$-N) and an accident alarm system, the environmental maintenance and operation of these facilities remained problematic. Even under increasingly stringent supervision, the provincial EPB found in 2011 that the plant 'XY' (also belonging to 'KD') dumped toxic waste illegally in Shanxi, but the company did not face any punishment.

Local governmental actions and enhanced supervision were expected to ensure a better environmental compliance of these SMEs. However, KD was a local economic pillar firm and a large taxpayer, and the director of 'KD' had close personal connections with county governmental officials. Hence, in practice the county EPB applied and enforced lower environmental protection standards, according to interviewed workers and villagers. In the case of the three smaller 'KD' plants, implementation of environmental laws and policies continued to remain a problem. When confronted with serious pollution problems from 'key protected plants', local enforcing EPBs proved powerless due to their dependence on local government leaders.

This might be changing in the near future. The 'KD' chemical industry had been one of the four key industries of Jasmine County over the past decade, but this ended with the local 12th Five-Year Plan of Economic and Social Development, promulgated by the Jasmine County government at the end of 2011. In this new Plan priority was given to six other industrial sectors, all with low pollution, an advanced technological component and high added-value, among which were agricultural product processing and service industries. This 'downgrading' of the importance of chemical industries will make their 'protection' by local government leaders less strong, and opens up possibilities for more stringent control and enforcement of local EPBs.

## 38.6 Environmental Regulation Implementation: Successes and Failures

As shown in the previous section, China's environmental management has achieved significant successes over the past decades. At the same time, with respect to quite a number of environmental challenges, progress in mitigating environmental pollution stagnates and policy targets are not within reach and/or the formulated policy goals are too lax for reaching sustainability. There is a growing literature on and insight into the reasons for implementation failure in China's environmental politics (Lo and Tang, 2006; Mol and Carter, 2006; Van Rooij, 2006). Some causes of implementation failure are quite familiar and can be found across the globe, others are more specific for China as they touch upon the specific institutional setting of China's environmental management system. In this section, four main reasons for implementation failure are reported (Zhang and Cao, 2015; Zhao, 2015).

What factors determine the extent to which the processes of environmental problem solving prevalent in China succeed or fail? Several factors are identified by Young *et al.* (2015). Leadership is an important factor. In China, it is more likely to mean the ability of key individuals (e.g. the president and premier) to set priorities and line up support within the top ranks of the party and relevant interest groups. Implementation – moving laws or plans from paper to practice – is another key factor: in the legal systems, lower-level officials have their own incentives that may further or thwart efforts to implement environmental policies, despite the introduction of procedures (e.g. the Target Responsibility System in China) designed to improve implementation. Corruption is a threat to implementation. In addition, there are challenges of tracking progress so that it is possible to determine when and whether policy initiatives are working. Innovations regarding reliable systems of monitoring and verification are important.

The following is a summary of four reasons for the Chinese regulation implementation. Arguably one of the key reasons behind implementation difficulties is the unprecedented economic development China has been undergoing over the past decades. The close to 9% average annual economic growth, changes in the economic structure, the concentration of people in major urban centres and the changing lifestyles of its population (think only about the increase in kilometres travelled per inhabitant) all contribute to accelerated and highly dynamic environmental impacts, which are difficult to handle and mitigate. This means that successes are measured in terms of relative improvements, but absolute environmental improvements remain uncertain.

A second implementation failure of environmental policies is related to the dominance of economic policy institutions over environmental policy institutions, which does not sound unfamiliar across the globe. The major state economic organizations, such as the National Development and Reform Commission (NDRC) and several sectoral line ministries (industry, agriculture), have much more power, resources and influence than the Ministry of Environmental Protection (which was only upgraded to ministerial status in 2008). This counts at the national as well as at the local level. But in China this power difference has a somewhat different nature form the one in Organization of Economic Cooperation and Development (OECD) countries. The close connections between major (and polluting) state-owned companies and these economic ministries ensure that such polluters are well protected by their 'mother' ministries when confronted with calls to clean up. Although there is a clear tendency to further separate the state from the market (hence, state-owned companies from ministries), and to bring the former under a market regime with increasingly independent state oversight, formal and informal lines between the two remain short. This does not help with strict implementation of environmental policies, standards, instruments and programmes that run against the short-term economic interests of these state-owned companies.

A third main reason behind implementation failure is the autonomy of local networks and governments. China has witnessed several decentralization programmes over the past two decades, giving more autonomy to provincial, municipal and lower governmental entities, as centralized control proved to be too inflexible, bureaucratic and hampering (economic) development. The consequence has been a growing relative autonomy of local governments. Increasingly, one can witness far-reaching commitment at the central level to sustainability and environmental programmes, but a failure to motivate, direct, steer and control local power elites regarding the implementation of these environmental priorities. This national–local gap is intertwined with the still close connections and common interests between local state elites with economic elites, making enforcement of national policies at the local level difficult.

Fourth, implementation of environmental policies could be further strengthened by a stronger countervailing power from civil society. The development of an active civil society in the field of environment is definitely in the making. Environmental NGOs have emerged throughout the country, both nationally and locally; the number of environmental demonstrations and protests is increasing; a well-functioning environmental complaints system is present; the media are becoming more and more active and open in reporting on environmental accidents and controversies (Yang, 2010; He *et al.*, 2016, 2017); environmental disclosure policies are legally installed; and more participatory policy making can be witnessed (e.g. in EIA, water policies) (Zhong and Mol, 2008). Participation of environmental NGOs or the public in decision making is still limited and official policies on disclosure, participation, protest and media reporting are often overruled, especially by local state and economic interests. This comes together with a still poorly functioning system of the rule of law, where environmental victims can start procedures against polluters that harm their interests.

## 38.7 Conclusions

China is dramatically increasing its legal and administrative efforts to tackle pollution, climate change and other environmental challenges and significant progress has already been made. Some may see these developments as an indication that environmental governance in China is shifting toward a law-centred process. These developments are unfolding in the context of the results of the CPC's October 2014 Fourth Plenum of the 18th Congress focusing on the rule of law. In a roadmap document issued at the end of the plenum, the leadership announced a commitment to the integration of 'law' and 'policy' (including plans) (CPC, 2014).

China now has many environmental laws. But in moving from the letter of the law to practice, where will law(s) fit into the plan(s) and other rule sets employed by the Chinese government (He *et al.*, 2013)? For millennia, China's central governments have sought to develop and refine means to deal with environmental problems and other governance concerns. Using modern technologies and economic instruments, the central government today is able to enunciate policies in terms of Five-Year Plans (FYPs) and further policy directives that apply in principle on a system-wide basis. We are exploring how we can best use our experience and expertise to support and assist citizens, non-profit organizations, the courts and the relevant governmental bodies with the implementation of China's ambitious new environmental laws.

Law is not unimportant in the Chinese system, but what we can expect is an overlay of the rule of law with Chinese characteristics in conjunction with the existing governance process emphasizing state planning. In the final analysis, success in solving environmental problems in China will occur when strong leadership produces joint initiatives on the part of the CPC and the National People's Congress that lead to effective top-down pressures on lower levels of government responsible for achieving the targets articulated.

In the age of planetary environmental change, alternative models must be found. This cannot be accomplished simply by technology but requires new ways of living. If China is truly to succeed in creating a new ecological civilization it will have to go in an even more radical direction, further removed from the regime of capital that has characterized the West and that is responsible for today's planetary ecological emergency.

## References

Chen, X. (2010) State-generated Data and the Study of Contentious Politics in China. In: Carlson, A., Gallagher, M. Lieberthal K. and Manion, M. (eds) *Chinese Politics: New Methods, Sources and Field Strategies*. Cambridge University Press, New York, pp. 15–32.

CPC (2014) Party Announcement at the End of the Fourth Plenum. Available at: https://chinacopyrightandmedia.wordpress.com/2014/10/28/ccp-central-committee-decision-concerning-some-major-questions-in-comprehensively-moving-governing-the-country-according-to-the-law-forward (accessed 24 January 2018).

Day, K.A.(2005) *China's Environment and the Challenge of Sustainable Development*. M.E. Sharpe Inc, New York, pp. 69–80.

ENRRI (2016) The new environment law implementation evaluation report. Environment and Natural Resources Research Institute, China University Political Science and Law [in Chinese]. Available at: http://www.sohu.com/a/77099/67_131990 (accessed 19 April 2019) [in Chinese].

Genasci, L. (2012) The rise of protests and reputational risk. Available at: http://chinawaterrisk.org/resources/analysis-reviews/the-rise-of-protests-and-reputational-risk/ (accessed 19 April 2019).

He, G.Z., Zhang, L., Lu, Y.L. and Mol, A.P.J. (2011) Managing major chemical accidents in China. *Journal of Hazardous Materials* 187 (1–3), 171–181.

He, G.Z., Lu, Y.L., Mol, A.P.J. and Beckers T. (2012) Changes and challenges: China's environmental management in transition. *Environmental Development* 3, 25–38.

He, G.Z., Zhang, L., Mol, A.P.J., Lu, Y.L. and Liu, J.G. (2013) Revising China's Environmental Law. *Science* 341, 133.

He, G.Z., Mol, A.P.J. and Lu, Y.L. (2016) Public protests against the Beijing-Shenyang High-speed Railway in China. *Transportation Research Part D: Transport and Environment* 43, 1–16.

He, G.Z., Boas, I., Mol, A. and Lu, Y. (2017) E-participation for environmental sustainability in transitional urban China. *Sustainability Science* 12, 187–202.

Kostka, G. and Mol, A.P.J. (2013) Implementation and participation in China's local environmental politics: challenges and innovations. *Journal of Environmental Policy and Planning* 15(1), 3–16.

Li, X. (2009) China's environmental legislation has entered the interests of the game era. *Green Leaves* 10, 122–125 [in Chinese].

Liu, J.G. and Diamond, J. (2008) Revolutionizing China's environmental protection. *Science* 319, 37–38.

Lo, C.H.W. and Tang, S.Y. (2006) Institutional reform, economic changes and local environmental management in China: the case of Guangdong Province. *Environmental Politics* 15 (2), 42–62.

Lu, Y.L., Giesy, J.P., Holliday, L. (2007) *Implementing the Stockholm Convention on Persistent Organic Pollutants*. National Academies Press, Washington, DC.

McElwee, C. (2011) *Environmental Law in China: Mitigating Risk and Ensuring Compliance*, 1st edn. Oxford University Press, Oxford.

MEP (2018) Report on the State of the Environment in China 2017. May 31, 2017, Beijing. Ministry of Environmental Protection. Available at: http://english.mee.gov.cn/Resources/Reports/soe/SOEE2017/201808/P020180801597738742758.pdf (accessed 28 January 2019).

MEP and MLR (2014) *Land Pollution Investigation Bulletin of China* [in Chinese]. Ministry of Environmental Protection and Ministry of Land Resources, Beijing.

Mol, A.P.J. (2009) Environmental governance through information: China and Vietnam. *Singapore Journal of Tropical Geography* 30(1), 114–129.
Mol, A.P.J. and Carter, N.T. (2006) China's environmental governance in transition. *Environmental Politics* 15(2), 149–170.
Ran, R. (2013) Perverse incentive structure and policy implementation gap in China's local environmental politics. *Journal of Environmental Policy & Planning* 15(1), 17–39.
Stern, R.E. (2013) *Environmental Litigation in China: A Study in Political Ambivalence*. Cambridge University Press, Cambridge.
Van Rooij, B. (2006) Implementation of Chinese environmental law: regular enforcement and political campaigns. *Development and Change* 37(1), 57–74.
Wang, A. (2013) The search for sustainable legitimacy: environmental law and bureaucracy in China. *Harvard Environmental Law Review* 37, 365.
Wang, T.Y., Lv, Y.L., Zhang, H. and Shi, Y.J. (2005) Contamination of persistent organic pollutants and relevant management in China. *Environmental International* 31(6), 813–821.
Wendling, Z.A., Emerson, J.W., Esty, D.C., Levy, M.A., de Sherbinin, A., *et al.* (2018) *2018 Environmental Performance Index*. Yale Center for Environmental Law & Policy, New Haven, Conneticut. Available at https://epi.envirocenter.yale.edu/downloads/epi2018reportv05171902.pdf (accessed 1 July 2019).
Wübbeke, J. (2014) The three-year battle for China's new environmental law. *China Dialogue*. Available at: http://www.Chinadialogue.net (accessed 9 January 2018).
Xinhua net (2015) New signals from four documents by the Leading Group for Comprehensively Deepening Reforms. Available at: http://news.Xinhua.net.com/politics/2015-07/05/c_1115820327.htm (accessed 6 February 2018) [in Chinese].
Yang, G. (2010) *The Power of the Internet in China: Citizen Activism Online*. Columbia University Press, New York.
Yang, G.B. and Calhoun, C. (2007) Media, civil society, and the rise of a green public sphere in China. *China Information* 21, 211–236.
Young, O.R., Dan, G., Qi, Y., Bachus, K. and Belis, D. (2015) Institutionalized governance processes: comparing environmental problem solving in China and the United States. *Global Environmental Change* 31, 163–173.
Zhang, B. and Cao, C. (2015) Four gaps in China's new environmental law. *Nature* 517 (7535), 433–434.
Zhang, L., Mol, A.P.J., He, G.Z. and Lu, Y.L. (2010) An implementation assessment of China's environmental information disclosure decree. *Journal of Environmental Sciences* 22 (10), 1649–1656.
Zhang, L., He, G.Z., Mol, A.P.J. and Zhu, X. (2013) Power politic in the revision of China's Environmental Protection Law. *Environmental Politics* 22(6), 1029–1035.
Zhang, L., He, G.Z. and Mol, A.P.J. (2015) China's new environmental protection law: a game changer? *Environmental Development* 13, 1–3.
Zhao, Y.H. (2015) Innovative measures to improve environmental law enforcement in China. *China–EU Law Journal* 4(2–4), 155–172.
Zhong, L.J. and Mol, A.P.J. (2008) Participatory environmental governance in China: public hearings on urban water tariff setting. *Journal of Environmental Management* 88(4), 899–913.
Zhu, X. (2011) *Environmental Law*. China Environmental Science Press, Beijing.
Zhu, X., Zhang, L., Ran, R. and Mol, A. P.J. (2015) Regional restrictions on environmental impact assessment approval in China: the legitimacy of environmental authoritarianism. *Journal of Cleaner Production* 92, 100–108.

# 39 21st Century Toxicology: Methods for Environmental Toxicology and Monitoring

**J. Lundqvist***
*Swedish University of Agricultural Sciences, Uppsala, Sweden*

## 39.1 Abstract

Traditional animal toxicity testing has been challenged in this century to shift towards toxicity pathway-based approaches for a more efficient evaluation of a large number of chemicals and a better understanding of the mechanisms underlying toxicological effects. The toxicity pathway describes what is happening in a cell when it is exposed to a toxic substance. Bioanalytical methods, e.g. *in vitro* bioassays, are valuable tools in environmental toxicology to study critical toxicity mechanisms to assess toxicity of compounds of environmental concern or in environmental samples. These *in vitro* models enable measurement of the initial molecular effects of a low-dose exposure of a chemical on the biological system, which can be used for prediction of toxicity, rather than more proximal end points after often unrealistic high-dose exposure, which are used in regulatory animal studies. The development and application of *in vitro* bioassays for environmental toxicology is a rapidly developing field of research with a large number of studies published during the past decade. This chapter will review the conceptual framework of the '21st century toxicology' strategy, the principles for different *in vitro* bioassays, a number of key toxicity pathways, how these bioassays have been applied both for pure compounds and for environmental samples and how the methodology can be used for effect-directed analysis aiming to identify novel bioactive environmental pollutants.

## 39.2 Introduction: the Conceptual Framework of 21st Century Toxicology

In 2007, the US National Research Council published the report *Toxicity Testing in the 21st Century: A Vision and a Strategy* (NRC, 2007). The report summarizes a development where traditional animal toxicity testing has been challenged to shift towards toxicity pathway-based approaches for a more efficient evaluation of a large number of chemicals and a better understanding of the mechanisms behind toxicological effects. The toxicity pathway describes what is happening in a cell when it is exposed to a toxic substance. This includes the molecular interaction between the chemical and the biological target, and the cellular defence mechanisms triggered by the toxic substance. This strategy is

* E-mail address: johan.lundqvist@slu.se

often referred to as 21st century toxicology. A great strength of this approach is that it can measure the total toxic potential in an environmental sample, regardless of the toxicity that is exerted by a known compound or an unknown compound. Cell-based *in vitro* assays are of great value for studies of toxicity pathways. These *in vitro* models enable measurement of the initial molecular effects of a low-dose exposure of a chemical on the biological system, which can be used for prediction of toxicity, rather than more proximal end points after often unrealistic high-dose exposure, which are used in regulatory animal studies. Application of 21st century toxicology methods within environmental toxicology is a rapidly expanding field of research internationally, for testing of both environmental samples and pure compounds of environmental concern (Mehinto *et al.*, 2016; Hernandez and Tsatsakis, 2017; König *et al.*, 2017; Kunz *et al.*, 2017; Neale *et al.*, 2017a, b).

## 39.3 Cell-Based *In Vitro* Assays and Reporter Gene Assays

*In vitro* assays based on cultured cells are of great value for 21st century toxicology research. Cultured cells can be either primary cultures or immortalized cell lines. Primary cells show a higher similarity to the normal tissue, but also exhibit a larger degree of biological variability between batches. Immortalized cell lines have undergone changes (e.g. mutations) that allow the cells to proliferate indefinitely. Hence, immortalized cell lines are less similar to normal tissue than primary cell cultures but also show a lower degree of biological variability between batches and experiments.

Many *in vitro* bioassays have been based on the principle of reporter gene assay, which is that cells are genetically engineered in a manner that links a molecular event in the cells to the expression of an easily detectable molecule (the reporter). For example, a reporter gene assay for oestrogen receptor activity links a DNA sequence (responsive element) that is responsive to ligand-activated oestrogen receptor to a gene that can express the reporter molecule. When the DNA sequence is activated by a ligand-activated oestrogen receptor, the expression of the reporter gene will increase and the activity can be measured. The expression level of the reporter will be proportional to the activation of the responsive element. Suitable reporter molecules include fluorescent proteins (e.g. green fluorescent protein (GFP)) and enzymes where the enzyme-catalysed reaction can be easily detected (e.g. luciferase).

It is important that reporter gene assay experiments are carried out under non-cytotoxic conditions. Reporter gene assays are designed to assay a specific molecular event in the cells, such as activation of a nuclear receptor or dissociation of a complex of two proteins. If the cells are exposed to a compound or sample in a concentration that is inducing general cytotoxicity, it can induce false-positive results in reporter gene assays, especially transiently transfected cell systems. Therefore, cell viability has to be monitored during reporter gene assay experiments. A commonly used definition of non-cytotoxic conditions is that the cell viability is ≥ 80% compared with the vehicle control.

Also native cell lines, not transfected with reporter gene systems, can be used for toxicity assessments. A cell viability assay can assess the non-specific toxicity caused by a single compound or an environmental sample. Tanneberger *et al.* (2013) showed that measurement of the cell viability in a fish cell line is a good predictor of fish acute toxicity.

## 39.4 Transgenic Zebrafish Systems

As outlined above, cell-based *in vitro* assays are of great value for the application of the 21st century toxicology strategy. However, a shortcoming of these cell-based methods is the limited capacity to accurately predict biological processes that only occur in multicellular organisms (e.g. intracellular signalling and systemic effects) and the often limited or non-existing metabolic capacity in the cell lines. To overcome these limitations of the cell-based systems, *in vivo* systems, such as zebrafish strains, can be used. The fish embryo test (FET) is a valuable tool for research in environmental toxicology and also for environmental monitoring. The FET is designed to determine acute toxicity by measurement of four indicators of lethality (OECD, 2013). Fish embryos can also be used to assay biomarkers for specific toxicity pathways, such as measurement of vitellogenin as a marker for oestrogenicity.

Advanced systems for precision genome editing, such as the CRISPR-Cas9 system, have enabled the development of systems where initial molecular events, in accordance with the 21st century toxicology strategy, can be studied in multicellular organisms such as the zebrafish by coupling of DNA sequences that are responsive to a specific molecular event to the gene for an easily detectable protein (e.g. a fluorescent protein). Activation of the DNA sequence will result in a proportional increase in the expression of the detectable protein, which can be assayed. Such studies can be performed *in vitro* or *in vivo*, depending on the developmental stage of the animal at the time of the experiment. This chapter will focus on the application of cell-based *in vitro* assays. Transgenic fish systems and their applications in environmental toxicology have recently been reviewed by Lee *et al.* (2015) and Garcia *et al.* (2016).

## 39.5 Key End-Points for *In Vitro* Toxicology

*In vitro* bioassays have been developed to assay toxicity pathways representing a wide range of biological end points. The selection of which *in vitro* bioassays to include, and thereby toxicity pathways to cover, in a research project or environmental monitoring effort is of critical importance. If the aim of the study is, for example, to screen environmental samples for bioactivity as part of an environmental monitoring programme, and thereby potentially also identify novel environmental pollutants, one would aim for a broad selection of *in vitro* bioassays covering as many toxicity pathways as practically possible. However, if the scope of the study is narrower (for example, to monitor industrial wastewater effluents where it is well known that compounds with a specific mode of action might be present), it can be reasonable to select *in vitro* bioassays targeting only that specific toxicity pathway.

In an ambitious research effort to evaluate which *in vitro* bioassays are the most sensitive and reliable for water quality assessment, Escher *et al.* (2014) analysed ten environmental water samples, representing wastewater, recycled water and drinking water, in 103 bioassays in 20 laboratories worldwide. The selected *in vitro* bioassays represented toxicity pathways involving metabolism, molecular initiating events, cellular defence mechanisms and cell death. The water samples were analysed in the bioassays, aiming to find the assays that were both robust and responsive, taking into account that a suitable *in vitro* bioassay should respond to 'polluted' samples but not to the 'clean' controls. The authors concluded that the most responsive health-relevant end points were related to: (i) xenobiotic metabolism, measured by effects on aryl hydrocarbon receptor (AhR) and pregnane-X receptor (PXR); (ii) hormone-mediated modes of action, measured by effects on oestrogen receptor (ER), androgen receptor (AR), glucocorticoid receptor (GR) and peroxisome proliferator-activated receptor (PPAR); (iii) genotoxicity, measured by a battery of tests, for example, Comet assay; and (iv) oxidative stress, measured by induction of the nuclear factor erythroid 2-related factor 2 (Nrf2) activity.

### 39.5.1 Oestrogen receptor activity

The oestrogen receptors are nuclear receptors that can be activated by oestrogens, such as 17β-oestradiol. Upon ligand activation, the oestrogen receptor binds to oestrogen-responsive elements in the DNA, acts as a transcription factor and regulates the expression of specific genes. Ligands binding to the oestrogen receptor can act as agonists, and thereby activate the receptor, or as antagonists, blocking the effect of the receptor.

Multiple *in vitro* methods are available to assay oestrogen receptor activity: reporter gene assays based on mammalian cells, receptor binding assays, yeast reporter assays and cell proliferation assays (Escher *et al.*, 2014; Leusch *et al.*, 2017). In the study described above (Escher *et al.*, 2014), oestrogen receptor activity bioassays were found to be sensitive for environmental water samples. Leusch and colleagues addressed the question of differences in assay sensitivity between different assays (Leusch *et al.*, 2010, 2017) and found that reporter gene assays based on mammalian cells as well as cell proliferation-based assays showed the highest sensitivity, with a typical method detection limit in the range of 0.01–1 ng $l^{-1}$ 17β-oestradiol equivalents. Yeast reporter assays and oestrogen receptor-binding assays were found to have a lower sensitivity, with a typical method detection limit of 1–10 ng $l^{-1}$ 17β-oestradiol equivalents.

To assay for antagonistic effects on the oestrogen receptor, many *in vitro* bioassays can be run in a mode where the oestrogen receptor is constantly activated by a known agonist (most assays recommend 17β-oestradiol). The potential antagonistic effects by a sample or a pure compound can then be assayed by measuring whether the effect by 17β-oestradiol can be significantly decreased by the sample or the compound of interest.

A number of environmental pollutants, including drug residues, have been reported to act as endocrine-disrupting compounds by binding to the oestrogen receptor and thereby altering the gene expression of oestrogen-regulated genes. Further, there are naturally occurring compounds, such as the phytoestrogens, that can activate the oestrogen receptor. Different studies have reported relatively large differences in the proportion of the observed oestrogenic effects in environmental water samples that can be explained by the presence of a few compounds with known oestrogenic properties. One study reported that only 0.1–0.3% of the observed oestrogenic effects in effluent water from three wastewater treatment plants could be explained by the occurrence of 4-nonylphenol, bisphenol A and oestrone (Neale *et al.*, 2017b). In a different study (König *et al.*, 2017), the occurrence of oestrone, oestriol and 17β-oestradiol could explain from 7% to 60% of the observed effects in samples collected downstream of a wastewater discharge in the River Danube, while < 1% of the observed oestrogenic effect in the sample collected upstream of the wastewater discharge could be explained by the occurrence of these compounds.

There are validated OECD Guidelines available both for reporter gene assays based on mammalian cells (Guideline 455) and for receptor-binding assays (Guideline 493) (OECD, 2015, 2016).

### 39.5.2 Androgen receptor activity

The androgen receptor is a nuclear receptor. Following ligand activation by androgens, such as testosterone or dihydrotestosterone, the receptor binds to androgen-responsive elements in the DNA and acts as a transcription factor to regulate the expression of specific genes. As for the oestrogen receptor, ligands to the androgen receptor can activate the receptor by acting as an agonist or buckling the receptor activity by acting as an antagonist. Environmental pollutants that alter the activity of the androgen receptor are classified as endocrine-disrupting chemicals.

A number of *in vitro* methods have been developed to study androgen receptor activity: reporter gene assays based on mammalian cells, receptor-binding assays, yeast reporter assays and cell proliferation assays (Escher *et al.*, 2014; Leusch *et al.*, 2017). In the study by Escher *et al.* (2014), androgen receptor activity bioassays, especially when run for anti-androgenic effects, were found to be sensitive for environmental water samples. Leusch *et al.* (2017) analysed the sensitivity of these assays and found that the analysed receptor-binding assays and yeast reporter assays had similar sensitivities with method detection limits in the range of 8–26 ng $l^{-1}$ dihydrotestosterone equivalents. A cell proliferation assay and the different mammalian reporter gene assays included in the study exhibited higher sensitivity, with method detection limits in the range of 0.8–10 ng $l^{-1}$ dihydrotestosterone equivalents.

To assay antagonistic effects on the androgen receptor, most bioassays can be performed in a manner where the receptor is constantly activated by a ligand, e.g. dihydrotestosterone, and the potential of a pure compound or sample to inhibit that agonistic effect is analysed. The most sensitive bioassay to screen for anti-androgenic effects is a cell proliferation assay with a method detection limit of approximately 200 ng $l^{-1}$ hydroxyflutamide equivalents (Leusch *et al.*, 2017).

In the study investigating the bioactivity of effluent water from three wastewater treatment plants by Neale *et al.* (2017b), six compounds with known androgenic properties could only explain 0.1–0.4% of the observed androgenic effects. In the study by König *et al.* (2017), where testosterone was also measured, four compounds with known androgenic effects could explain almost 100% of the observed androgenic effect in a water sample collected downstream of a wastewater discharge. For anti-androgenic effects, a few compounds (including genestin, dadzein, bisphenol A, 2,4-dinitrophenol and oestrone) were found to explain a great part of the observed bioactivity in untreated wastewater (König *et al.*, 2017).

A validated OECD Guideline is available for a mammalian reporter gene assay (Guideline 458) based on the Chinese hamster ovary cell line (CHO-K1) stably transfected with the expression construct for the human androgen receptor and an androgen receptor-sensitive luciferase plasmid (OECD, 2016b).

### 39.5.3 Aryl hydrocarbon receptor activity

The aryl hydrocarbon receptor is a nuclear receptor that, upon ligand activation, can bind to responsive DNA sequences and regulate gene expression. The ligand-activated receptor increases the expression of genes related to xenobiotic metabolism, e.g. CYP1A1. Initially, only synthetic ligands were described for the aryl hydrocarbon receptor, including environmental pollutants such as halogenated aromatic hydrocarbons and polycyclic aromatic hydrocarbons (Chapter 10, this volume), with the most discussed ligand being 2,3,7,8-tetrachlorodibenzo-*p*-dioxin (TCDD) (Chapters 12–14, this volume). Later research has aimed to clarify the physiological role of this receptor and in such work a number of endogenous ligands for the receptor, both agonists and antagonists, have been suggested.

There are multiple *in vitro* bioassays available to assay aryl hydrocarbon receptor activity, based on reporter gene assay in mammalian cells, such as the stably transfected AhR-CAFLUX assay, AhR-CALUX assay and DR-CALUX assay (Denison *et al.*, 2004; Brennan *et al.*, 2015) as well as assays based on transiently transfected cells (He *et al.*, 2011; Rosenmai *et al.*, 2018)

Agonistic effects on the aryl hydrocarbon receptor have been reported for environmental water samples in a number of studies (Neale *et al.*, 2015; König *et al.*, 2017; Mehinto *et al.*, 2017). In a study by Neale *et al.* (2017b) investigating bioactivity in wastewater treatment plant effluents and water samples collected upstream and downstream of the plants, there was a great variability in the proportion of the observed aryl hydrocarbon receptor activity that could be explained by the occurrence of measured chemicals. For one location, 10–30% of the observed bioactivity could be explained by the occurrence of four compounds. However, for the other two locations, the occurrence of measured chemicals could only explain a very small proportion of the observed bioactivity.

### 39.5.4 Oxidative stress

Oxidative stress, with formation of reactive oxygen species (ROS), is a common mechanism for different toxicological end points, e.g. tissue toxicity, genotoxicity, carcinogenicity and teratogenicity (Ornoy, 2007; Deavall *et al.*, 2012). A key regulator in the cellular defence against oxidants is the nuclear factor erythroid 2-related factor 2 (Nrf2). Nrf2 is ubiquitously expressed in a large number of tissues, but the protein synthesis is blocked during basal conditions, via binding to the protein Kelch-like epichlorohydrin (ECH)-associated protein 1 (Keap1). The blockade of Nrf2 protein synthesis is reversed upon oxidative stress exposure and Nrf2 can then exert its antioxidative effects. The Nrf2 protein acts as a transcription factor and regulates the gene expression of a large number of target genes involved in xenobiotic metabolism, antioxidant defence and oxidant signalling. Nrf2 exerts its effect by binding to a specific sequence in the gene promoter called antioxidant responsive element (ARE) (Ma, 2013). In an investigation of the effects of more than 300 pesticides on nuclear receptors and transcription factors, the Nrf2 pathway was the most commonly affected target and was activated by a large number of chemicals (Martin *et al.*, 2010). Nrf2 activity has been shown to be correlated with classical toxicological end points and it has therefore been suggested that oxidative stress response and Nrf2 activity would be an ideal pathway to build an *in vitro* toxicity assay on (Martin *et al.*, 2010). In a study aiming to find the most sensitive and reliable bioassays for water quality assessment, Escher *et al.* (2014) concluded that adaptive stress response pathways, such as the Nrf2 pathway, were one of the most responsive health-relevant end points. A wide range of compounds can induce oxidative stress. In a study where nine environmental water samples were chemically analysed for the occurrence of 269 individual compounds and also analysed for Nrf2 activity in an *in vitro* bioassay, only 0.1% of the observed Nrf2 activity could be explained by the detected chemicals (Escher *et al.*, 2013), highlighting that a large part of the observed oxidative stress response is caused by unknown compounds. The same conclusion has been reached in other research studies (König *et al.*, 2017; Neale *et al.*, 2017a).

### 39.5.5 Glucocorticoid receptor activity

The glucocorticoid receptor is a nuclear receptor that can be activated by cortisol and other glucocorticoids. The glucocorticoid receptor is ubiquitously expressed in most tissue types in humans. Upon ligand activation, the glucocorticoid receptor can bind to specific responsive elements in the DNA and thereby alter the expression of downstream genes. The glucocorticoid receptor mainly regulates genes involved in metabolic regulation.

In a screening including 103 *in vitro* bioassays for toxicity testing of environmental water samples, the glucocorticoid receptor pathway was found to be among the more responsive health-relevant endpoints (Escher *et al.*, 2014). There are multiple *in vitro* bioassays available to assay glucocorticoid receptor activity, both yeast reporter gene assays and reporter gene assays based on mammalian cells. The reporter gene assays based on mammalian cells have been reported to be the most sensitive, with a typical method detection limit around 30 ng $l^{-1}$ dexamethasone equivalents, while the yeast reporter assays have a much lower sensitivity with a typical method detection limit at 1,400,000 ng $l^{-1}$ dexamethasone equivalents (Leusch *et al.*, 2017).

Both agonistic and antagonistic effects on the glucocorticoid receptor have been reported in environmental water samples (van der Linden *et al.*, 2008; König *et al.*, 2017; Mehinto *et al.*, 2017).

## 39.6 Screening for Bioactivity of Pure Compounds

*In vitro* bioassays have been applied to assay the bioactivity of chemical compounds of environmental concern in a very large number of research studies. The US Environmental Protection Agency (EPA) has started a very ambitious high-throughput programme, Toxicity Forecaster (ToxCast), to screen pure compounds of environmental concern for bioactivity. Currently, data for more than 9000 compounds and 1100 bioassays can be accessed via the ToxCast dashboard (EPA, 2017).

## 39.7 Environmental Samples

*In vitro* bioassays and other 21st century toxicology methods have been used to investigate bioactivity in environmental samples in a large number of studies, mainly for water, soil and sediment samples. Here, the focus will be on the application of *in vitro* bioassays on bioactivity in water samples.

A number of studies have been performed where bioanalytical methods have been used for water quality assessment of surface water and water relating to wastewater treatment plants. König *et al.* (2017) assayed bioactivity in river water samples collected upstream and downstream of a wastewater discharge, using 20 different bioassays covering xenobiotic metabolism, specific receptor-mediated modes of actions and induction of adaptive stress responses. The highest bioactivities were detected at a sampling point immediately downstream of the wastewater discharge. In a different study, bioactivity was assayed using 11 *in vitro* bioassays for the effluent water from three wastewater treatment plants, and upstream and downstream in the respective river where the wastewater treatment plant effluents were discharged (Neale *et al.*, 2017b). Conley *et al.* (2017a) analysed 35 water samples collected in different water streams in the USA for oestrogen receptor activity, androgen receptor activity and glucocorticoid receptor activity, using *in vitro* bioassays. The authors reported that oestrogenic effects were observed in almost all samples, with activities corresponding to oestradiol equivalent concentrations in the range 0.0045–116 ng $l^{-1}$. Androgen receptor activity and glucocorticoid receptor activity were observed in five and nine samples, respectively, of the 35 samples. In a study by Mehinto *et al.* (2017), water samples were collected in southern California at locations representing different land use: open, agricultural and urban. The water samples were then analysed for bioactivity in three different bioassays: oestrogen receptor activity, glucocorticoid receptor activity and aryl hydrocarbon receptor activity. The study reported bioactivities in all three bioassays for all three land-use types. For aryl hydrocarbon receptor activity, a great variation in bioactivity was observed based on the land use, with the highest activities observed in the urban water samples.

Bioanalytical methods can also be used to assess the toxicity removal efficiency during wastewater treatment. Välitalo *et al.* (2017) used eight bioassays to measure the bioactivities in the inlet and outlet waters from seven Finnish wastewater treatment plants. Androgenic and genotoxic effects were observed in most inlet samples, but the wastewater treatment process drastically decreased these effects. Oestrogen receptor activity was observed for all inlet and outlet samples; however, the activities were stronger in the inlet samples than in the outlet samples.

Bioanalytical methods have also been used to assess the quality of drinking water. Macova *et al.* (2011) analysed bioactivity in water samples from the entire water cycle, from wastewater treatment plant via surface water to a drinking-water plant, using bioassays for genotoxicity, oestrogen receptor activity, neurotoxicity, phytotoxicity, dioxin-like activity and non-specific cell toxicity. In a study addressing the question of potential bioactivity of by-products formed during water disinfection, multiple bioassays, including assays for oxidative stress and genotoxicity, were applied on water samples collected after different treatment steps in a drinking-water treatment plant (Neale *et al.*, 2012). The observed bioactivities in this study varied between the different sampling campaigns (i.e. samples collected at different times of the year), but the authors conclude that bioassays are useful as monitoring tools in drinking-water quality assessment. The study by Escher *et al.* (2014) included samples from the inlet and outlet of a metropolitan drinking-water treatment plant using chlorination and chloramination for water disinfection. The authors reported increased genotoxicity and oxidative stress response in the outlet water compared with the inlet water, due to the disinfection processes. In two other studies (Escher *et al.*, 2012, 2013), oxidative stress response was observed in a drinking-water treatment plant's outlets. Brand *et al.* (2013) tested drinking-water samples in reporter gene assays for oestrogen receptor activity, androgen receptor activity, glucocorticoid receptor activity and progesterone receptor activity. The authors observed oestrogenic effects in some of the drinking-water samples, corresponding to an oestradiol equivalent level of 0.006–0.008 ng $l^{-1}$. Androgenic effects were observed in one drinking-water sample, corresponding to a dihydrotestosterone equivalent level of 0.0012 ng $l^{-1}$. For glucocorticoid and progesterone receptor activities, all drinking-water samples were below the limit of detection. Oestrogenic activities in drinking-water samples have also been reported in a study assessing the *in vitro* oestrogenic activity in inlet and outlet waters from 25 US drinking-water plants (Conley *et al.*, 2017b). Bioanalytical methods have also been used to investigate potential bioactivity by disinfection by-products formed during drinking-water production, both for water samples (Farré *et al.*, 2013; Hebert *et al.*, 2018) and for pure compounds known to be formed during disinfection procedures (Stalter *et al.*, 2016).

## 39.8 Effect-Directed Analysis

Effect-directed analysis is based on the idea that the biological effect of a complex sample (e.g. an environmental sample) is a trigger for further chemical analysis (Escher *et al.*, 2014; Brack *et al.*, 2016; Busch *et al.*, 2016; Brack *et al.*, 2017; Kunz *et al.*, 2017; Hashmi *et al.*, 2018). If a certain sample lacks toxic potential, the need for chemical characterization of that sample is low. If on the other hand a sample exerts toxicity, it is prioritized for use in integrated chemical and toxicological profiling of the sample to understand which compound (or compounds) is causing the toxicity. Effect-directed analysis can also be used to identify previously unknown toxic compounds, preferably by fractionation of the sample and integrated chemical and toxicological profiling of the fraction (or fractions) that carries the toxic potential, including non-targeted chemical analysis. The principle of effect-directed analysis shows great promise for both environmental toxicology research and environmental monitoring efforts and the field is rapidly expanding internationally. Recently, Muschket *et al.* (2018) applied the strategy of effect-directed analysis to successfully identify the novel anti-androgenic environmental pollutant 4-methyl-7-diethylaminocoumarin, by effect-based testing of water samples from the German river Holtemme (a hotspot of anti-androgenic effects), fractionation and subsequent effect-based testing of the fractions and finally chemical characterization of the bioactive fractions. Similarly, Weiss *et al.* (2009) tentatively identified anti-androgenic compounds in sediments and Thomas *et al.* (2009) identified anti-androgenic compounds in water. Effect-directed analysis has also been applied to identify novel environmental pollutants with oestrogenic effects (Lübcke-von Varel *et al.*, 2011).

## 39.9 Conclusions

Twenty-first century toxicology methods can be of great value for environmental toxicology and monitoring. A great strength with cell-based *in vitro* assays and reporter gene assays is that they enable monitoring efforts where the total biological effects of complex pollutant mixtures are evaluated, rather than measurement of concentrations of single chemicals. This is especially important since many biological effects observed in complex environmental samples cannot be explained by chemical analysis of commonly analysed pollutants. The selection of which *in vitro* bioassays to include, and thereby toxicity pathways to cover, in a research project or environmental monitoring effort is of critical importance and has been addressed in a number of studies. Xenobiotic metabolism, hormone-mediated modes of action, genotoxicity and oxidative stress have been proposed as the most responsive and health-relevant end points to include. Effect-directed analysis, where cell-based *in vitro* assays are combined with fractionation of environmental samples and advanced chemical analysis (non-target screening), shows great promise to identify novel pollutants causing toxicity in the environment.

## References

Brack, W., Ait-Aissa, S. Burgess, R.M., Busch, W., Creusot, N. *et al.* (2016) Effect-directed analysis supporting monitoring of aquatic environments – an in-depth overview. *Science of the Total Environment* 544, 1073–1118.

Brack, W., Dulio, V., Ågerstrand, M., Allan, I., Altenburger, R.*et al.* (2017) Towards the review of the European Union Water Framework Directive: Recommendations for more efficient assessment and management of chemical contamination in European surface water resources. *Science of the Total Environment* 576, 720–737.

Brand, W., de Jongh, C.M., van der Linden, S.C., Mennes, W., Puijker, L. M.*et al.* (2013) Trigger values for investigation of hormonal activity in drinking water and its sources using CALUX bioassays. *Environment International* 55, 109–118.

Brennan, J.C., He, G., Tsutsumi, T., Zhao, J., Wirth, E., Fulton, M.H.and Denison, M.S. (2015) Development of species-specific Ah receptor-responsive third generation CALUX cell lines with enhanced responsiveness and improved detection limits. *Environmental Science & Technology* 49(19), 11903–11912.

Busch, W., Schmidt, S., Kühne, R., Schulze, T., Krauss, M.and Altenburger, R. (2016) Micropollutants in European rivers: a mode of action survey to support the development of effect-based tools for water monitoring. *Environmental Toxicology and Chemistry* 35(8), 1887–1899.

Conley, J.M., Evans, N., Cardon, M.C., Rosenblum, L., Iwanowicz, L.R., *et al.* (2017a) Occurrence and in vitro bioactivity of estrogen, androgen, and glucocorticoid compounds in a nationwide screen of United States stream waters. *Environmental Science & Technology* 51(9), 4781-4791.

Conley, J.M., Evans, N., Mash, H., Rosenblum, L., Schenck, K. *et al.* (2017b) Comparison of in vitro estrogenic activity and estrogen concentrations in source and treated waters from 25 U.S. drinking water treatment plants. *Science of the Total Environment* 579, 1610–1617.

Deavall, D.G., Martin, E.A., Horner, J.M. and Roberts, R. (2012) Drug-induced oxidative stress and toxicity. *Journal of Toxicology* 2012, 645460.

Denison, M.S., Zhao, B., Baston, D.S., Clark, G.C., Murata, H. and Han, D. (2004) Recombinant cell bioassay systems for the detection and relative quantitation of halogenated dioxins and related chemicals. *Talanta* 63(5), 1123–1133.

EPA (2017) ToxCast & Tox21 Summary Files Database: prod_dashboard_v2:. US Environment Protection Agency. Available at: https://actor.epa.gov/dashboard/#chemical/188425-188485-188426 (accessed 16 February 2017).

Escher, B.I., Dutt, M., Maylin, E., Tang, J.Y., Toze, S., Wolf, C.R. and Lang, M. (2012) Water quality assessment using the AREc32 reporter gene assay indicative of the oxidative stress response pathway. *Journal of Environmental Monitoring* 14(11), 2877–2885.

Escher, B.I., van Daele, C., Dutt, M., Tang, J.Y. and Altenburger, R. (2013) Most oxidative stress response in water samples comes from unknown chemicals: the need for effect-based water quality trigger values. *Environmental Science & Technology* 47(13), 7002–7011.

Escher, B.I., Allinson, M., Altenburger, R., Bain, P.A., Balaguer, P. *et al.* (2014) Benchmarking organic micropollutants in wastewater, recycled water and drinking water with in vitro bioassays. *Environmental Science & Technology* 48(3), 1940–1956.

Farré, M.J., Day, S., Neale, P.A., Stalter, D., Tang, J.Y.M. and Escher, B.I. (2013) Bioanalytical and chemical assessment of the disinfection by-product formation potential: Role of organic matter. *Water Research* 47(14), 5409–5421.

Garcia, G.R., Noyes, P.D.and Tanguay, R.L. (2016) Advancements in zebrafish applications for 21st century toxicology. *Pharmacology & Therapeutics* 161, 11–21.

Hashmi, M.A.K., Escher, B.I., Krauss, M., Teodorovic, I. and Brack, W. (2018) Effect-directed analysis (EDA) of Danube River water sample receiving untreated municipal wastewater from Novi Sad, Serbia. *Science of the Total Environment* 624, 1072–1081.

He, G., Tsutsumi, T., Zhao, B., Baston, D.S., Zhao, J., Heath-Pagliuso, S. and Denison, M.S. (2011) Third-generation Ah receptor-responsive luciferase reporter plasmids: amplification of dioxin-responsive elements dramatically increases CALUX bioassay sensitivity and responsiveness. *Toxicological Sciences* 123(2), 511–522.

Hebert, A., Feliers, C., Lecarpentier, C., Neale, P.A., Schlichting, R., Thibert, S. and Escher, B. I. (2018) Bioanalytical assessment of adaptive stress responses in drinking water: a predictive tool to differentiate between micropollutants and disinfection by-products. *Water Research* 132, 340–349.

Hernandez, A.F. and Tsatsakis, A.M. (2017) Human exposure to chemical mixtures: Challenges for the integration of toxicology with epidemiology data in risk assessment. *Food and Chemical Toxicology* 103, 188–193.

König, M., Escher, B.I., Neale, P.A., Krauss, M., Hilscherová, K. *et al.* (2017) Impact of untreated wastewater on a major European river evaluated with a combination of in vitro bioassays and chemical analysis. *Environmental Pollution* 220 (Part B), 1220–1230.

Kunz, P.Y., Simon, E., Creusot, N., Jayasinghe, B.S., Kienle, C. *et al.* (2017) Effect-based tools for monitoring estrogenic mixtures: evaluation of five in vitro bioassays. *Water Research* 110, 378–388.

Lee, O., Green, J.M. and Tyler, C.R. (2015) Transgenic fish systems and their application in ecotoxicology. *Critical Reviews in Toxicology* 45(2), 124–141.

Leusch, F.D.L., de Jager, C., Levi, Y., Lim, R., Puijker, L. *et al.* (2010) Comparison of five in vitro bioassays to measure estrogenic activity in environmental waters. *Environmental Science & Technology* 44(10), 3853–3860.

Leusch, F.D.L., Neale, P.A., Hebert, A., Scheurer, M. and Schriks, M.C.M. (2017) Analysis of the sensitivity of in vitro bioassays for androgenic, progestagenic, glucocorticoid, thyroid and estrogenic activity: suitability for drinking and environmental waters. *Environment International* 99, 120–130.

Lübcke-von Varel, U., Machala, M., Ciganek, M., Neca, J., Pencikova, K. *et al.* (2011) Polar compounds dominate *in vitro* effects of sediment extracts. *Environmental Science & Technology* 45(6), 2384–2390.

Ma, Q. (2013) Role of nrf2 in oxidative stress and toxicity. *Annual Review of Pharmacology and Toxicology* 53, 401–426.

Macova, M., Toze, S., Hodgers, L., Mueller, J.F., Bartkow, M.and Escher, B.I. (2011) Bioanalytical tools for the evaluation of organic micropollutants during sewage treatment, water recycling and drinking water generation. *Water Research* 45(14), 4238–4247.

Martin, M.T., Dix, D.J., Judson, R.S., Kaylock, R.J., Reif, D.M. *et al.* (2010) Impact of environmental chemicals on key transcription regulators and correlation to toxicity end points within EPA's ToxCast program. *Chemical Research in Toxicology* 23(3), 578–590.

Mehinto, A.C., Jayasinghe, B.S., Vandervort, D.R., Denslow, N.D. and Maruya, K.A. (2016) Screening for endocrine activity in water using commercially available in vitro transactivation bioassays. *Journal of Visualized Experiments* (118), e54275.

Mehinto, A.C., VanDervort, D.R., Lao, W., He, G., Denison, M.S. *et al.* (2017) High throughput *in vitro* and *in vivo* screening of inland waters of Southern California. *Environmental Science: Processes & Impacts* 19(9), 1142–1149.

Muschket, M., Di Paolo, C., Tindall, A J., Touak, G., Phan, A. *et al* (2018) Identification of unknown antiandrogenic compounds in surface waters by effect-directed analysis (EDA) using a parallel fractionation approach. *Environmental Science & Technology* 52(1), 288–297.

NRC (2007) *Toxicity Testing in the 21st Century: aVision and a Strategy*. National Research Council. National Academies Press, Washington, DC.

Neale, P.A., Antony, A., Bartkow, M.E., Farré, M.J., Heitz, A. *et al.* (2012) Bioanalytical assessment of the formation of disinfection byproducts in a drinking water treatment plant. *Environmental Science & Technology* 46(18), 10317–10325.

Neale, P.A., Ait-Aissa, S., Brack, W., Creusot, N., Denison, M.S. *et al.* (2015) Linking in vitro effects and detected organic micropollutants in surface water using mixture-toxicity modeling. *Environmental Science & Technology* 49(24), 14614–14624.

Neale, P.A., Achard, M.E.S., Escher, B.I. and Leusch, F.D.L. (2017a) Exploring the oxidative stress response mechanism triggered by environmental water samples. *Environmental Science: Processes & Impacts* 19(9), 1126–1133.

Neale, P.A., Munz, N.A., Aït-Aïssa, S., Altenburger, R., Brion, F. *et al.* (2017b) Integrating chemical analysis and bioanalysis to evaluate the contribution of wastewater effluent on the micropollutant burden in small streams. *Science of the Total Environment* 576, 785–795.

OECD (2013) *Test No. 236: Fish Embryo Acute Toxicity (FET) Test*. Organisation for Economic Co-operation and Development. OECD Publishing, Paris.

OECD (2015) *Test No. 493: Performance-Based Test Guideline for Human Recombinant Estrogen Receptor (hrER) In Vitro Assays to Detect Chemicals with ER Binding Affinity*. Organisation for Economic Co-operation and Development. OECD Publishing, Paris.

OECD (2016a) *Test No. 455: Performance-Based Test Guideline for Stably Transfected Transactivation In Vitro Assays to Detect Estrogen Receptor Agonists and Antagonists*. Organisation for Economic Co-operation and Development. OECD Publishing, Paris.

OECD (2016b). *Test No. 458: Stably Transfected Human Androgen Receptor Transcriptional Activation Assay for Detection of Androgenic Agonist and Antagonist Activity of Chemicals*. Organisation for Economic Co-operation and Development. OECD, Paris.

Ornoy, A. (2007) Embryonic oxidative stress as a mechanism of teratogenesis with special emphasis on diabetic embryopathy. *Reproductive Toxicology* 24(1), 31–41.

Rosenmai, A.K., Niss, F., Mandava, G., Lundqvist, J. and Oskarsson, A. (2018) Impact of natural organic matter in water on *in vitro* bioactivity assays. *Chemosphere* 200, 209–216.

Stalter, D., O'Malley, E., von Gunten, U. and Escher, B.I. (2016). Fingerprinting the reactive toxicity pathways of 50 drinking water disinfection by-products. *Water Research* 91, 19–30.

Tanneberger, K., Knöbel, M., Busser, F.J.M., Sinnige, T.L., Hermens, J.L.M.and Schirmer, K. (2013) Predicting fish acute toxicity using a fish gill cell line-based toxicity assay. *Environmental Science & Technology* 47(2), 1110–1119.

Thomas, K.V., Langford, K., Petersen, K., Smith, A.J. and Tollefsen, K.E. (2009) Effect-directed identification of naphthenic acids as important *in vitro* xeno-estrogens and anti-androgens in North Sea offshore produced water discharges. *Environmental Science & Technology* 43(21), 8066–8071.

Välitalo, P., Massei, R., Heiskanen, I., Behnisch, P., Brack, W. *et al.* (2017) Effect-based assessment of toxicity removal during wastewater treatment. *Water Research* 126, 153–163.

van der Linden, S.C., Heringa, M.B., Man, H.-Y., Sonneveld, E., Puijker, L.M., Brouwer, A. and van der Burg, B. (2008) Detection of multiple hormonal activities in wastewater effluents and surface water, using a panel of steroid receptor CALUX bioassays. *Environmental Science & Technology* 42(15), 5814–5820.

Weiss, J.M., Hamers, T., Thomas, K.V., van der Linden, S., Leonards, P.E G.and Lamoree, M.H. (2009). Masking effect of anti-androgens on androgenic activity in European river sediment unveiled by effect-directed analysis. *Analytical and Bioanalytical Chemistry* 394(5), 1385–1397.

# 40 Unequivocal Evidence Associating Environmental Contaminants and Pollutants with Human Morbidity and Ecological Degradation

J.P.F. D'Mello*

*Formerly of SAC, University of Edinburgh King's Buildings Campus, West Mains Road, Edinburgh, UK*

## 40.1 Abstract

Persistent equivocation in virtually all sections of society and industry about the harmful effects of pollution have provided the impetus for the publication of *A Handbook of Environmental Toxicology: Human Disorders and Ecotoxicology*. This chapter is designed to collate and analyse recent research data demonstrating that pollutants cause human morbidity and damage natural habitats in the ecosystem. Issues addressed here parallel those presented in the foregoing chapters. The approach adopted in this volume is aimed at addressing diverse issues in environmental toxicology by the inclusion of critical reviews interspersed with specialist short communications to exemplify cutting-edge research. In addition, consideration of specific case studies is designed to demonstrate real-time issues in the toxicology and ecological impact of pollutants.

Recent research has confirmed the detrimental health hazards associated with ambient air pollution. Increased exposure to atmospheric ozone and nitrogen dioxide have been implicated in exacerbation of idiopathic pulmonary fibrosis. Acute infections of the lower respiratory tract account for significant mortality in young children worldwide and have been linked with ambient air pollution caused by combustion, in particular levels of $PM_{2.5}$. Other results implicate $PM_{2.5}$, nitrogen dioxide and sulfur dioxide as risk factors for chronic obstructive pulmonary disease (COPD). In the case of asthma, both nitrogen dioxide and sulfur dioxide exposure are frequently implicated in exacerbation of morbidity. There is considerable and continuing interest in the relationship between air pollution and cardiovascular disease (CVD) in humans. It has been argued that both $PM_{2.5}$ and $PM_{10}$ may contribute to effects such as heart disease, stroke and CVD. The possible effects of ambient air pollution on metabolic syndromes in humans have been advanced in recent studies. For example, the effects on childhood obesity have been examined in a longitudinal, multilevel analysis. It is maintained that traffic pollution may be linked with the development of obesity via biochemically plausible mechanisms. Current views also point to a positive relationship between ambient air pollution and type 2 diabetes incidence, but the need for confirmation of such a link is emphasized, although plausible mechanisms have

* E-mail address: jpfdmello@hotmail.co.uk

already been advanced. An emerging and inevitable question relates to the possible effects of ambient air pollution in the aetiology of cancer. Outdoor air pollution and particulates are now considered by the International Agency for Research on Cancer (IARC) to be Group 1 carcinogens. Residential exposure to oestrogen-disrupting hazardous air pollutants and breast cancer incidence implicate ambient cadmium compounds or possibly inorganic arsenic as risk factors. There is support for the hypothesis that traffic-generated air pollution may be linked with the development of breast cancer, especially in pre-menopausal women. An association between ambient nitrogen dioxide and lung cancer incidence has also been proposed, with residential proximity to polluted streets being a significant risk factor. The cognitive and neurological effects of air pollutants are an additional source of concern and *in utero* exposure may be a contributory feature in certain cases such as incidence of autism, attributed specifically to ambient ozone levels. On the basis of epidemiological evidence and other data, it has been suggested that there might be an association between air pollution exposure and incidence of dementia, attributed to traffic-related particles and ambient ozone, nitrogen dioxide and carbon monoxide levels. Other studies link polycyclic aromatic hydrocarbons (PAHs) in fine particulate matter with the onset of childhood asthma, adult hypertension, heart attack and cancer, by mechanisms yet to be established. An additional complication may be the source of PAHs: it has been suggested that children exposed to particulates emanating from wood combustion may be at greater risk of lung cancer due to increased load of PAHs.

Persistent organic pollutants (POPs) in the forms of polychlorinated biphenyls (PCBs) and dioxins remain in particular focus, due to continuing and debilitating morbidity in humans. Maternal exposure to dioxin-like PCBs may be associated with the risk of asthma in offspring that may persist into adulthood. Furthermore, gestational exposure to PCBs may cause lifelong and transgenerational effects on body weight, hormones and hypothalamic gene expression. Nevertheless, the overriding issue with PCBs centres on the association with endocrine disruption and breast cancer incidence and survival. Gestational effects also occur in the activity of dioxins as reproductive disruptors. It is generally accepted that 2,3,7,8-tetrachlorodibenzo-*p*-dioxin (TCDD) alters fecundity and endometriosis in primates and causes reproductive disorders in animals, including humans. However, the predominant concern, based on molecular studies, is the toxicity of TCDD as a Group 1 carcinogen, and supporting epidemiological data is now emerging.

Breast cancer has also been linked with exposure to organochlorine (OC) pesticides. There are indications that DDT, dichlorodiphenyldichloroethylene (DDE) and dichlorodiphenyldichloroethane (DDD) may modulate several cancer-related pathways in breast cancer cell lines, consistent with the role as environmental oestrogens. Associations between OC pesticides and cognition have also been suggested.

In accordance with the toxicology of certain other pollutants, there are profound effects of organophosphorus (OP) pesticides on respiratory function. It has recently been concluded that the paediatric respiratory symptoms of OP exposure may reflect the possible onset of childhood asthma. In addition, the neurological effects of OP pesticides are well documented. Residential proximity to OP applications in arable farming has been linked with a more rapid decline in cognitive and motor functions among Parkinson's disease patients.

The Minamata disease incident in Japan and chronic methylmercury exposure in humans continue to dominate research efforts, due to profound and persistent effects affecting cellular metabolism, cardiovascular and pulmonary functions, digestive and renal systems and immunocompetence as well as central nervous system (CNS) integrity and activity. Recent studies also highlight the neurological effects of lead in humans, emphasizing impairments in intelligence, memory, attention, processing speed, motor skills and behaviour, including autism.

The issue of radiation carcinogenesis remains in the public domain for several reasons. The incidence of non-melanoma skin cancer in white populations is increasing in many countries. Exposure to ultraviolet (UV) radiation is believed to be the underlying cause, though the pattern of exposure that promotes the different types of malignancy varies. Interactions with cutaneous synthesis of vitamin $D_3$ and possibly viral infections may complicate issues in the epidemiology of UV carcinogenesis. Residential exposure to radon is definitively linked with lung cancer incidence. The scope of current research is now shifting to possible links to leukaemia and interactions with ambient $PM_{2.5}$. In reflecting on the 30-year legacy of the Chernobyl nuclear accident, researchers have summarized the human cost of the radioactive fall-out indicating increased long-term risks of leukaemia and CVD among clean-up personnel and thyroid cancer in subjects exposed to radiation as children and adolescents, as well as mental health decline.

Environmental contaminants exert profound effects in the degradation of natural habitats, exemplified in the crude oil and nuclear accidents of recent times. The question of emerging contaminants such as microbeads, and other pollutants associated with anthropogenic activity, has introduced additional challenges for remediation technologists. Nutrient pollution is exemplified mainly in the forms of nitrogen and phosphorus discharges into aquatic environments. The use of bioindicators such as dinoflagellate cyst assemblages has been instrumental in monitoring water quality and anthropogenic impact in such environments. As might be predicted, microbial communities have the capacity to adapt to pollution events. In an assessment of the *Deepwater Horizon* oil spill in the Gulf of Mexico, deposits on the shore caused a shift in the community ecology towards recognized petroleum hydrocarbon-degrading microorganisms. While this observation may be reassuring, it should be noted that PAHs are significantly more recalcitrant and that detoxification pathways are limited in higher animals. Consequently, adverse effects may be seen in vertebrate ecology, for example in negative effects on sea turtle nesting in the wake of the *Deepwater Horizon* oil spill. Nevertheless, observations 24 years after the *Exxon Valdez* oil spill indicate that recovery is possible in the long term. Based on biomarker data, recent evidence has assisted in establishing timelines and mechanisms of wildlife population dynamics for several marine species, with reference to the *Exxon Valdez* contamination. The advent of novel exploration technologies in oil extraction, such as hydraulic fracturing (fracking), creates different risks for the ecosystem in general and animal toxicology in particular. Employing a riparian songbird as a bioindicator of pollution, it has been found that feather barium and strontium concentrations were significantly higher in birds originating from fracking sites than in those at sites without fracking. Key areas of research and monitoring still need to be addressed to assist in the formulation of effective guidelines and policies to protect vulnerable species at fracking sites. Emerging pollution risks in marine environments are associated with the widespread dispersal of plastics and microplastics used for domestic and commercial purposes. It is stressed that microplastics are not inert particles in terms of biological activity or physical chemistry. Consequently, ecotoxicological effects include entanglement in vertebrates, transfer along food chains, loss of nutritional value of food items and enhanced exposure to pathogens and chemical contaminants such as PAHs.

As might be anticipated, radioactive emissions following radionuclide contamination are linked with wildlife abnormalities, including high frequency of albinism and tumours in free-living birds in the vicinity of the radiation source. However, it is claimed that 30 years after the Chernobyl accident, there is still a lack of data relating to the genetic effects of radionuclide contamination on plant ecology. Specific questions requiring elucidation included adaptation of plants to chronic low-dose ionizing radiation, responses to changing/increased pathogenicity of fungi and viruses, accumulation of mutational load and the extent of micro-evolutionary changes in the vicinity of the Chernobyl and Fukushima reactors.

In conclusion, there is overwhelming evidence that environmental contaminants cause human morbidity and mortality. Nevertheless, considerable efforts are still required to elucidate cause-and-effect issues in pollutant toxicology with respect to common underlying themes such as gene expression, oxidative stress, inflammatory response networks and identification of signature molecules. There is also unequivocal evidence that pollutants harm natural habitats and ecological diversity. Work is in progress to clarify issues relating to timelines and mechanisms of wildlife population dynamics for terrestrial and marine species exposed to environmental contaminants. Finally, in the absence of credible political leadership, it is concluded that research scientists and regulatory agencies exert a more forthright role in all matters relating to environmental toxicology.

## 40.2 Overview

Environmental issues continue to divide public opinion at individual, corporate and international levels. At the extreme end of the spectrum there are commentators who perceive current predictions as the work of activists. There is undoubtedly indifference among large sections of society and industrial corporations to the deleterious effects of urban and industrial emissions on human health and ecological diversity. Inaction appears to be the *modus operandi* even in the most affluent countries where international directives, for example on global warming, are discounted. Attitudes may change as more robust evidence is published in the scientific literature. There is a clear need for environmental toxicologists to redouble efforts to engage with stakeholders. The primary aims in this volume are to review recent developments in relation to two major aspects: human disease and ecological degradation as affected by (i) major contaminants originating from biological sources and (ii) pollutants generated spontaneously or as a result of anthropogenic activity. A diverse array of these potentially harmful agents regularly appears in food, atmosphere, soil and water, thereby compromising both human health and biodiversity in natural and managed ecosystems.

The approach adopted in this volume is designed to address complex issues in environmental toxicology by the inclusion of critical reviews in the conventional sense interspersed with specialist short communications to exemplify cutting-edge developments. In addition, the introduction of specific case studies (listed in the Preface) should illustrate the relevance of a wide range of contaminants to recent or real-time incidence of environmental damage and human morbidity. The roles of emerging pollutants and industrial processes as contributors to ecological imbalance also warrant attention.

The chapters in *A Handbook of Environmental Toxicology: Human Disorders and Ecotoxicology* are arranged in nine sections, each representing a particular theme. It is instructive and appropriate to consider the main conclusions and integrate these within a set of action points aimed at individuals and governmental and international agencies. It is important also to stress that complex additive, synergistic and even potentiation interactions may occur in the behaviour of different pollutants and in the ultimate effects on human health and on ecological dynamics. The dilemma is exemplified in attempts to unravel the complexities of gene–environment interactions in relation to exposure to pollutants in the aetiology of asthma in human subjects. This is a formidable task but worthy of elucidation, due to the global prevalence of respiratory problems (Rava *et al.*, 2015). The interactions of *glutathione S-transferase* gene polymorphisms in atmospheric pollution and the aetiology of respiratory diseases and allergies represents a specific line of research (Bowatte *et al.*, 2016). Unique issues are also raised in elucidating microbe–pollutant interactions in remote habitats such as deep-sea sediments (Louvado *et al.*, 2015).

This chapter is deliberately presented to appeal to a general audience in order to ensure wide dissemination of the central issues emerging from recent research in environmental toxicology. Accordingly, the use of jargon has been minimized. The Editor is of the view that a number of key findings are liable to be 'lost in translation' due to lack of clarity in presentation. It is imperative that significant environmental health and ecological implications are perceived as action points for individuals as much as for corporate organizations and regulatory institutions at international and local levels.

## 40.3 Human Health Effects

Considerable evidence has been reviewed in this volume to demonstrate the profound effects of contaminants and pollutants on human morbidity.

### 40.3.1 Biogenic compounds

Three classes of biogenic compounds have been highlighted in this volume to exemplify the diverse range of contaminants produced by plants and microbes. Complex implications also arise in relation to environmental protection and microecology.

The distribution of the major phytotoxins has been reviewed in Chapter 1. The cyanogens, glycoalkaloids, glucosinolates, non-protein amino acids, furanocoumarins, gossypol and bioactive proteins were selected for detailed study due to diverse toxicology towards vertebrate animals, human morbidity, insects, nematodes and fungal pathogens. Effects in mammals, particularly humans, include: digestive dysfunction caused by anti-nutritional proteins in legume seeds; irreversible spastic paralysis and cognition defects induced by cyanogenic glycosides in cassava; goitrogenic activity precipitated by *Brassica* glucosinolate breakdown products; favism associated with pyrimidine glycosides in faba beans; cardiotoxicity and reproductive abnormalities caused by gossypol in cottonseed; phytodermatitis following contact with furanocoumarins in celery and other plants; and cancer induced by ptaquiloside in bracken fern.

Regarding mycotoxins, major concerns in human health are based on epidemiological evidence (Chapter 2). A broad spectrum of adverse outcomes has been associated with chronic exposure, including carcinogenesis, hepatitis, nephrotoxicity and endocrine disruption. Furthermore, mycotoxins may compromise health by modulating other disorders.

At the global level, anthropogenic activities, particularly nutrient pollution, contribute significantly to the occurrence of cyanobacterial blooms in the aquatic ecosystem, thereby contaminating drinking-water supplies with toxic substances such as microcystins (Chapter 3). Human exposure to these toxins results in a wide range of effects, including severe respiratory dysfunction, organ damage and gastroenteritis. Hilborn and Beasley (2015) recommended that ill health and mortality among livestock and fish may serve as useful sentinel events for predicting potential risks of cyanobacteria to humans.

### 40.3.2 Ambient air pollutants

Recent research has highlighted the detrimental human health impact of ambient air pollution (Table 40.1). Increased exposure to ambient ozone (Chapters 5 and 6) and nitrogen dioxide (Chapter 7) have been implicated in acute exacerbation of idiopathic pulmonary fibrosis, a 'clinically meaningful' condition according to Johannson *et al.* (2014). Acute infections of the lower respiratory tract account for significant mortality in young children worldwide and have been linked with ambient air pollution caused by combustion. In a systematic analysis of data, Mehta *et al.* (2013) suggested a causal relationship between $PM_{2.5}$ and the incidence of acute lower respiratory infections. In a brief summary, DeVries *et al.* (2016) implicated $PM_{2.5}$, nitrogen dioxide and sulfur dioxide (Chapter 8) as well established risk factors for chronic obstructive pulmonary disease (COPD), particularly in regions with high levels of outdoor pollution such as southern California and Hong Kong. Historically, nitrogen dioxide pollution has been regularly implicated in the aetiology of COPD. For example, the studies of Dadvand *et al.* (2014) demonstrated that exposure to nitrogen dioxide increased systemic inflammation in COPD patients, particularly former cigarette smokers. In addition, a significant association between short-term exposure to major air pollutants, especially ozone and nitrogen dioxide, and COPD emergency risk has been reported (Li *et al.*, 2016). Potential mechanisms include production of reactive oxygen species, immunosuppression and pulmonary inflammation compromising respiratory efficacy in COPD patients.

Epidemiological studies indicate that sulfur dioxide is associated with morbidity and mortality in patients with respiratory tract disorders (Chapter 8). The molecular genetics of this relationship are gradually unravelling. For example, the observations of Qin *et al.* (2017) suggest that, in rats, sulfur dioxide exposure inhibits expression of mitochondrial oxidative phosphorylation genes encoded by both nuclear and mitochondrial DNA in lung tissues. This may well be the cellular basis of lung diseases in subjects affected by ambient sulfur dioxide and, indeed, other atmospheric pollutants.

Asthma is a significant paediatric chronic disease and represents a primary cause for hospitalization in children. However, the condition is determined by a variety of factors such as early-life exposure to maternal tobacco smoking or second-hand smoke inhalation and ambient air pollutants (Raanan *et al.*, 2015). The impacts of prenatal and postnatal exposures on subsequent respiratory functions can be justified in physiological terms relating to bronchial and alveolar development. In the case of asthma, both nitrogen dioxide and sulfur dioxide exposure are frequently implicated in exacerbation of morbidity. Greenberg *et al.* (2017) concluded that the association

**Table 40.1.** Human health disorders attributed to ambient air pollution. Evidence selected to illustrate a range of adverse effects as presented in recent publications.

| Disorder(s) | Details | Reference |
|---|---|---|
| Acute exacerbation of idiopathic pulmonary fibrosis | Increased exposure to ambient ozone and nitrogen dioxide implicated in this 'clinically meaningful' condition | Johannson *et al.* (2014) |
| Acute lower respiratory infections (ALRI) | Systematic analysis strengthens evidence for a causal relationship between $PM_{2.5}$ exposure and the incidence of ALRI | Mehta *et al.* (2013) |
| Chronic obstructive pulmonary disease (COPD) | Outdoor air pollutants ($PM_{2.5}$, nitrogen dioxide and sulfur dioxide) are well established risk factors for COPD exacerbation | DeVries *et al.* (2016) |
| | Nitrogen dioxide increases systemic inflammation in COPD subjects, particularly former cigarette smokers | Dadvand *et al.* (2014) |
| | Short-term exposure to major air pollutants, especially ozone and nitrogen dioxide, implicated in COPD emergency risk | Li *et al.* (2016) |
| Asthma | Significant increased risk for asthma occurrence attributed to nitrogen dioxide exposure. For sulfur dioxide, the association with asthma is weaker but still significant | Greenberg *et al.* (2017) |
| Cardiovascular disease (CVD) | 'Abundant' evidence that air pollution contributes to CVD risk and associated mortality involving multiple mechanisms | Newby *et al.* (2015) |
| | Both $PM_{2.5}$ and $PM_{10}$ implicated in heart disease, stroke and CVD | Lee *et al.* (2014) |
| | Gene expression and exposure to traffic-generated air pollution in elderly subjects with CVD history under investigation | Wittkopp *et al.* (2016) |
| Obesity | Traffic-generated pollution associated with increase in body mass index in children aged 5 to 11 years | Jerrett *et al.* (2014) |
| Diabetes | Positive association between ambient air pollution and type 2 diabetes (tentative conclusion) | Eze *et al.* (2015) |
| Cancer | Chronic low-dose exposure to ambient cadmium compounds or inorganic arsenic may be risk factors for breast cancer | Liu *et al.* (2015) |
| | Exposure to traffic-related air pollution may be associated with the development of breast cancer, especially in pre-menopausal women | Hystad *et al.* (2015) |
| | Role suggested for nitrogen dioxide as a proxy for traffic-related air pollution exposure and incidence of lung cancer | Hamra *et al.* (2015) |
| CNS dysfunction | Traffic-related exposure *in utero* associated with autism; attributed specifically to ozone | Becerra *et al.* (2013) |
| | Exposure to air pollution as a potential contributor to cognitive decline and dementia | Power *et al.* (2016) |
| | Association between dementia incidence and local traffic pollution observed after adjusting for known risk factors; magnitude of effect was similar for Alzheimer's disease and vascular dementia | Oudin *et al.* (2016) |
| | Residential proximity to major road networks is associated with a higher incidence of dementia | Chen *et al.* (2017) |
| | Neuroinflammation, neuropsychological effects and neurodegenerative disease attributed to inhaled particulate matter | Wang *et al.* (2017) |

with asthma was more pronounced for nitrogen dioxide than for sulfur dioxide. The difference was attributed to water solubility. Greenberg *et al.* (2017) suggested that the poor water solubility allowed deep penetration into the bronchial tree to elicit asthmatic manifestations such as inflammation and increased mucus production. In contrast, sulfur dioxide is highly soluble in water and acts in the upper airways with reduced asthmatic episodes.

There is considerable and continuing interest in the relationship between air pollution and cardiovascular disease in humans. Newby *et al.* (2015) stated unequivocally that there is 'abundant' evidence to support the concept that air pollution contributes to the risk of CVD and associated mortality. They also elaborated on the possible mechanisms underlying this risk, indicating that the physiological basis relies on manifestations of oxidative stress and inflammation at specific sites and also affecting heart rhythm. It has been argued that both $PM_{2.5}$ and $PM_{10}$ may contribute to effects such as heart disease, stroke and CVD. Epidemiological evidence suggests a stronger influence of $PM_{2.5}$ compared with $PM_{10}$, with Lee *et al.* (2014) concluding that air pollution is an important public health issue causing CVD and pulmonary diseases worldwide. Looking forward it is possible to envisage elucidation of the molecular basis of the effects of air pollution in the aetiology of CVD. In this respect, an exploratory panel study recently indicated the possible role of gene expression of phase I and phase II enzymes and oxidative stress pathways in CVD responses to traffic-related pollutants (Wittkopp *et al.*, 2016).

Attention is also turning towards the possible effects of ambient air pollution on metabolic syndromes in humans. For example, the effects on childhood obesity have been examined in a longitudinal, multilevel analysis (Jerrett *et al.*, 2014). It is maintained that traffic pollution may be linked with the development of obesity via biochemically plausible mechanisms involving inflammatory pathways that initiate insulin resistance, diabetes and visceral fat accretion. The interaction with endocrine disruptors as obesogens is developing into an intriguing concept. Current thinking suggests a positive relationship between ambient air pollution and type 2 diabetes incidence, but Eze *et al.* (2015) advocated caution and the need for high-quality dose-response studies. Nevertheless, the underlying physiological mechanisms for this relationship are widely acknowledged to involve endothelial dysfunction, alteration in immunocompetence and perturbations in insulin transduction and resistance. Effects on mitochondrial metabolism and disruption in the functional role of brown adipocytes have also been proposed (Eze *et al.*, 2015).

An emerging and inevitable question relates to the possible effects of ambient air pollution in the aetiology of cancer. Outdoor air pollution and particulates are now considered by IARC to be Group 1 carcinogens (see Hamra *et al.*, 2015) Residential exposure to oestrogen-disrupting hazardous air pollutants and breast cancer incidence has been investigated by Liu *et al.* (2015), implicating ambient cadmium compounds or possibly inorganic arsenic as risk factors. Further investigations with improved exposure assessment techniques were recommended to address the complexities of exposure to combinations of endocrine disruptors (Liu *et al.*, 2015; see also Chapter 15). Some support for the hypothesis that traffic-generated air pollution may be linked with the development of breast cancer, especially in pre-menopausal women, has emerged in the studies of Hystad *et al.* (2015). An association between ambient nitrogen dioxide and lung cancer incidence has also been proposed. It is envisaged that this gas serves as a proxy for traffic-related pollution exposure and lung malignancy (Hamra *et al.*, 2015). Residential proximity to major roadways, in conjunction with nitrogen dioxide pollution, are considered to be significant risk factors for lung cancer incidence.

The cognitive and neurological effects of air pollutants are also a major source of concern and will continue to be so for the foreseeable future. It has been observed that children born to mothers residing 300 m from a freeway during pregnancy were more likely to be diagnosed as autistic than children born to mothers living 1400 m from a freeway. Becerra *et al.* (2013) attributed this condition to amient ozone levels, based on evidence in Los Angeles, California. Although ozone levels have declined, federal guidelines for ozone are often breached, thus compromising human health. Other observations suggest that air pollution exposure, even at levels below those considered safe, during fetal development may cause permanent brain damage (Guxens *et al.*, 2018). In reviewing the epidemiological evidence and other data, Power *et al.* (2016) concluded that there is support for an association between air pollution exposure and incidence of dementia. However, further studies were advocated to improve design, analysis and reporting to build a firm foundation prior to formulation of advisory recommendations and implementation of possible interventions. Oudin *et al.* (2016) observed associations between dementia incidence and local traffic pollution that persisted even after adjustments for specific risk factors. The extent of this relationship appeared to be similar for both Alzheimer's disease (AD) and vascular dementia (VaD). Nevertheless, Oudin *et al.* (2016) cautioned that other environmental factors, for example, traffic noise, might complicate interpretation of the results, suggesting a need for validation using prospective cohorts. The primary pollutants implicated were traffic-related particles and ambient ozone, nitrogen dioxide and carbon monoxide. Additional insight has emerged from the work of Chen *et al.* (2017) indicating that residential proximity to major road networks was associated with a higher incidence of dementia but not with Parkinson's disease (PD) or multiple sclerosis (MS). The association between inhaled particulate matter and neuroinflammation, neuropsychological effects and neurodegenerative disorders has been advanced by Wang *et al.* (2017). Possible routes of particulate matter to the CNS via a direct pathway or through stimulation of pro-inflammatory cytokines were proposed.

Other studies suggest that ambient air pollutants, particularly $PM_{10}$ and nitrogen dioxide ($NO_2$), may correlate with increased sudden infant death syndrome (SIDS) (Litchfield *et al.*, 2018). However, further studies are required to establish the mechanisms underlying this condition.

Exposure to ambient PAHs has been associated with diverse conditions in humans (Chapter 10). It is suggested that some PAHs might contribute to cytokine production in both healthy and asthmatic individuals through mechanisms involving aryl hydrocarbon receptor-dependent and independent pathways (Ple *et al.*, 2015). Others are more direct in linking PAHs in fine particulate matter with the onset of childhood asthma and exacerbation of associated symptoms (Karimi *et al.*, 2015). In a wide-ranging study, Shiue (2015) indicated that urinary PAHs were associated with adult hypertension, heart attack and cancer, by mechanisms yet to be established. Furthermore, in a community study, Feng *et al.* (2014) observed that environmental PAH exposure may differentially affect heart rate variability based on individual coronary risk profiles. Other studies, with US adults, indicated that PAH biomarkers were tentatively associated with CVD, but further prospective studies were required with adequate sample size to replicate findings (Alshaarawy *et al.*, 2016). Moorthy *et al.* (2015) maintained that excessive exposure to PAHs often results in lung cancer through the induction of phase I and phase II metabolic enzymes. The phase I enzymes commonly invoked in such reactions are cytochrome P450 monooxygenases, while catalysis in phase II reactions is attributed to glutathione *S*-transferases, UDP glucuronyl transferases, NADPH quinone oxidoreductases and epoxide hydrolases. However, the exact points at which PAHs initiate tumours remain to be established. In addition, there is considerable variability in the carcinogenic potential of different PAH congeners and in dose–response relationships affecting the induction of malignancy. Furthermore, carcinogenicity may be modulated by synergistic and antagonistic mechanisms often operating concurrently (Moorthy *et al.*, 2015). The question of interactions had previously been explored by Jarvis *et al.* (2014) who observed more-than and less-than additive effects in bioactivation and DNA damage, depending upon the individual components of the PAH mixtures under investigation. The study of 'real life' mixtures was advocated to represent the complex mixtures in human exposure. An additional complication may be the source of PAHs: it has been suggested that children exposed to particulates emanating from wood burning may be at greater risk of lung cancer, due to increased load of PAHs (Sarigiannis *et al.*, 2015). It should also be noted here that oxygen-substituted PAHs (OPAHs) are also formed during combustion processes and as a result of photooxidation and biotic degradation of the parent compounds (Goodale *et al.*, 2015). Similar to PAHs, OPAHs instigate structure-dependent mutagenic reactions as well as activation of the aryl hydrocarbon receptor and cytochrome P450 metabolic pathway. Since OPAH concentrations may, in certain environmental samples, exceed those of the parent compound, there is scope for further detailed toxicological investigations and health implications. In the interim, it is suggested that ligand-specific transcriptional mechanisms may underlie aryl hydrocarbon receptor-mediated developmental toxicity of oxygenated PAHs (Goodale *et al.*, 2015).

### 40.3.3 Polychlorinated biphenyls (PCBs) and dioxins

These persistent organic pollutants (POPs) remain at the forefront of research due to the continuing and debilitating syndromes precipitated in humans in chronic and acute cases (Chapters 11–14). In an all-inclusive conclusion based upon epidemiological evidence, Hansen *et al.* (2014) stated that maternal exposure to dioxin-like PCBs may be associated with the risk of asthma in offspring that could persist into adulthood. However, the authors admitted that this link may be mediated through postnatal exposure to POPs. The carry-over theme was extended to the effects of PCBs as endocrine disruptors (see also Chapter 15). Thompson *et al.* (2015) indicated that gestational exposure to PCBs has lifelong and transgenerational effects on body weight, hormones and hypothalamic gene expression. Nevertheless, the overriding issue centres on the association between PCBs and breast cancer. For example, Morgan *et al.* (2016) suggested a possible link between environmental exposure to PCBs and enhanced risk of breast cancer among US women. Again, in a first US study of PCBs and breast cancer survival, these endocrine disruptors were associated with mortality in 'biologically plausible' pathways (Parada *et al.*, 2016). The authors recommended investigations of other structurally similar chemicals in respect of cancer survival.

Gestational effects also occur in the activity of dioxins as reproductive disruptors. It is generally accepted that 2,3,7,8-tetrachlorodibenzo-p-dioxin (TCDD) alters fecundity and endometriosis in primates and causes reproductive disorders in animals, including humans. In addition, Hutz *et al.* (2014) demonstrated that dioxins significantly reduced circulating oestrogen levels and ovarian follicular growth in pre-pubertal female rats following maternal exposure to TCDD during mid-gestation. The Seveso Women's Health Study, constituted after accidental exposure of women to TCDD in 1976 (see Preface), has provided the means to study the genetic interactions in adverse effects of dioxins on birth weight (Ames *et al.*, 2018). It was demonstrated that genetic variation across the maternal *AHR* gene may influence fetal susceptibilities to TCDD exposure. However, the predominant concern with dioxins is the carcinogenic potential (Leng *et al.*, 2014). Molecular studies indicate that TCDD is a potent carcinogen with the capacity to disrupt multiple endocrine networks via aryl-hydrocarbon receptors (Xu *et al.*, 2016). In particular, TCDD is classified as a Group 1 carcinogen on the basis of animal studies, but epidemiological data has been somewhat limited. The meta-analysis conducted by Xu *et al.* (2016), however, confirmed that exposure and blood levels of TCDD were both significantly associated with all cancer mortality, particularly non-Hodgkin's lymphoma and possibly also prostate cancer, as intimated previously by Leng *et al.* (2014).

### 40.3.4 Pesticides

The agricultural chemicals of particular concern in relation to human health disorders include the organochlorine (OC) and organophosphate (OP) pesticides (Chapters 16–18) as well as glyphosate (Chapter 19). Recent research covers a wide range of effects, particularly at the prenatal and early life stages of human development and in communities residing in close proximity to arable farms. An emerging feature is the association of OC pesticides with human breast cancer. There are indications that DDT, DDE and DDD may modulate several cancer-related pathways in breast cancer cell lines, consistent with the role of environmental oestrogens. Pestana *et al.* (2015) referred to a putative involvement of OC pesticides on invasive cell ability. In the studies of Rivero *et al.* (2016), differential gene expression patterns in human mammary epithelial cells induced by realistic OC combinations indicated that composition of constituents in these mixtures could exert an effect on the initiation processes of breast carcinogenesis. Associations between OC pesticides and cognition have also been suggested in recent work in a study of

US elders (Kim *et al.*, 2015). Subjects with high serum concentrations of DDT or DDE displayed, respectively, about three or two times higher risk of low cognition, whereas such correlations did not apply to non-persistent pesticides.

As with other pollutants, the effects on respiratory function are significant. For example, Raanan *et al.* (2016) observed decreased lung function in 7-year-old children with early life exposure to OP. Raanan *et al.* (2015) had previously concluded that the paediatric respiratory symptoms of OP exposure were consistent with the possible onset of childhood asthma. Less predictable is the putative association between OP exposure and cancer incidence. Nevertheless, How *et al.* (2014) concluded that children residing near OP-treated farmland were at higher risk of developing cancer than those with minimal or zero exposure to OP. The pathophysiological basis of this relationship requires further elucidation. In contrast, the neurological effects of OP pesticides are well documented. For example, prenatal exposure to an OP insecticide may delay motor development, as suggested by studies with a mouse model of idiopathic autism (De Felice *et al.*, 2015). Residential proximity to OP applications in arable farming has been linked with faster cognitive and motor symptoms decline among Parkinson's disease (PD) patients. Reduced capacity to metabolize OP pesticides may be associated with more rapid progression of PD motor symptoms and depressive manifestations following diagnosis of this neurodegenerative disorder (Paul *et al.*, 2017).

### 40.3.5 Heavy metals

The Minamata disease incident and chronic methylmercury exposure in humans continue to dominate research efforts, and with good justification (Chapter 24). Rice *et al.* (2014) summarized the diverse effects of environmental mercury affecting cellular metabolism, cardiovascular and pulmonary functions, digestive and renal systems, immunocompetence and CNS integrity and performance. In particular, patients with Minamata disease exhibit distal paraesthesia of the extremities and lips even 30 years after methylmercury exposure (Ha *et al.*, 2017). Children exposed to methylmercury *in utero* show impaired motor functions, attention span, verbal abilities and cognitive performance.

Recent studies also highlight the neurological effects of lead in humans, emphasizing impairments in intelligence, memory, attention, processing speed, motor skills and behaviour (Mason *et al.*, 2014a). In addition, Kim *et al.* (2016) maintained that even low-level lead exposure is associated with more autistic behaviours in school-age children, emphasizing the need for further efforts to reduce environmental contamination by the heavy metal. It has been suggested that there is no threshold for lead toxicity in humans (Chapter 25).

The focus on heavy metals will remain for the foreseeable future, due to the upsurge in electronic waste recycling, particularly in South-east Asia where the small-scale enterprises exist. Electronic waste contains lead, cadmium (Chapters 26 and 27), chromium and manganese. Pregnant women in these centres may be at risk due to exposure to toxic metals (Kim *et al.*, 2018).

### 40.3.6 Radiation carcinogenesis

The issue of radiation carcinogenesis has been in the public domain for several decades. The conclusion that radiation causes cancer is inescapable but research continues to both broaden and deepen understanding. The incidence of non-melanoma skin cancer in white populations is increasing in many countries. Exposure to UV radiation is believed to be the underlying cause, though the pattern of exposure that promotes the different types of malignancy varies (Xiang *et al.*, 2014; see also Chapter 31). Controversially, exponentially increasing incidence of cutaneous malignant melanoma in Europe correlates with low personal annual UV doses. It is suggested that intermittent UV exposures result in low cutaneous levels of vitamin $D_3$ and possibly viral infections may predispose to this incidence (Merrill *et al.*, 2015).

Regarding radon, residential exposure is definitively linked with cancer incidence (Chapters 32 and 33). For example, Field *et al.* (2000) indicated that high concentrations of radon progeny induced lung cancer in both underground miners and experimentally exposed laboratory animals, concluding that ambient radon represents an important health risk. It is estimated that 17% of all lung cancer cases in Alberta (Canada) during 2012 were attributable to residential radon exposure (Grundy *et al.*, 2017). The scope of current research is now focusing on possible links to leukaemia and interactions with ambient $PM_{2.5}$.

In reflecting on the 30-year legacy of the Chernobyl nuclear accident, Zablotska (2016) summarized the human cost of the radioactive fall-out. Epidemiological evidence indicated increased long-term risks of leukaemia and CVD among clean-up personnel as well as thyroid cancer in subjects exposed to radiation as children and adolescents. In addition, mental health effects emerged as the most significant public health issues in the heavily contaminated regions of Ukraine, Belarus and the Russian Federation. The consumption of contaminated milk made children particularly susceptible to developing leukaemia (see also Chapter 34).

### 40.3.7 Health risks associated with unconventional oil and gas extraction from shale

The composition, toxicology and human health risks associated with wastewater in unconventional oil and gas extraction (fracking) from shale indicate that there are genuine reasons for concern. This wastewater contains a myriad of organic, inorganic and radioactive compounds, with many of these being known or potential toxicants (Chapter 23). The perception that the most significant issue is the potential for contamination of groundwater and surface water has led researchers to develop a variety of models to assess the pollution risks. Meanwhile, others have investigated the potential for harm in individuals and communities affected by fracking and emerging results are disconcerting. For example, the findings of Hill (2018), based on evidence from Pennsylvania, suggested that fracking confers significant risks to infant health, while Hirsch *et al.* (2018) implied psychosocial risks to communities associated with this process. There is a clear need for vigilance by the environmental health agencies and due diligence by the fracking industry to minimize adverse effects.

## 40.4 Pollutants Damaging Biodiversity in the Ecosystem: a Risk Assessment

Pollutants in the ecosystem range from organic to inorganic and from ephemeral to persistent. Environmental contaminants feature prominently in the degradation of natural habitats, epitomized in the nutrient fluxes, oil spills and nuclear accidents of recent times. In addition, risks prevail in respect of the persistent organic compounds reviewed in Chapters 10–19. The question of emerging contaminants such as microbeads, and other pollutants associated with anthropogenic activity, has introduced additional challenges for remediation technologists.

Nutrient pollution is exemplified mainly in the forms of nitrogen and phosphorus discharges into aquatic environments contributing to formation of algal blooms, with attendant risks (Chapter 3). The use of bioindicators has been instrumental in determining water quality and anthropogenic impact in such environments. Price *et al.* (2017) used the relative abundance of dinoflagellate cyst assemblages as a measure of nutrient pollution across estuaries of NW Atlantic, analysing surface sediments from 23 north-east USA sources from Maine to Delaware and nine from Prince Edward Island (Canada). In general, the abundance of cysts correlated with modelled N loading, but there were regional differences in the prevalence of species. In the Chesapeake Bay watershed (USA), degrading ecological conditions in the Rappahannock River have been attributed to increasing urbanization driving high nutrient loadings and altered stoichiometry (Prasad *et al.*, 2014).

As might be predicted, microbial communities have the capacity to adapt to pollution events. In an assessment of the *Deepwater Horizon* oil spill in the Gulf of Mexico, deposits on the shore caused a shift in the community ecology towards recognized petroleum hydrocarbon-degrading microorganisms (Lamendella *et al.*, 2014). While this observation may be reassuring, it should be noted that PAHs are significantly more recalcitrant (Mason *et al.*, 2014b) and that detoxification pathways are limited in higher animals. Consequently, adverse effects may be predicted in vertebrate ecology. The report that the *Deepwater Horizon* oil spill negatively affected sea turtle nesting illustrates this point (Lauritsen *et al.*, 2017). Nevertheless, observations 24 years after the *Exxon Valdez* oil spill indicate that recovery is possible in the long term but not for all marine species (Chapter 22). Based on biomarker evidence, Esler *et al.* (2017a) established a timeline of two decades for harlequin ducks, a species considered to be particularly susceptible to oil contamination in marine environments. In a companion paper, Esler *et al.* (2017b) extended findings concerning timelines and mechanisms of wildlife population dynamics to several marine species, commenting on acute and chronic exposure to the *Exxon Valdez* contamination. However, the question of ecological synchrony may be the rate-limiting factor in overall habitat recovery and there are species differences in responses to petroleum pollutants (Chapter 22).

With the emergence of novel exploration technologies in oil extraction such as hydraulic fracturing (fracking) come different risks for wildlife and associated habitats (Chapter 23). Latta *et al.* (2015) used a riparian songbird as a bioindicator of pollution caused by fracking. It is envisaged that, as a top predator, the songbird may be exposed to or bioaccumulate via their macroinvertebrate food the chemical contaminants released into the ecosphere by fracking. Feather composition reflects blood dynamics and toxicology during exposure to localized point-source events. Latta *et al.* (2015) found feather barium and strontium concentrations to be significantly higher in birds originating from fracking sites than in those at sites without fracking. Key areas of research and monitoring still need to be addressed to assist in the formulation of effective guidelines and policies to protect vulnerable species at fracking sites. Five general items proposed by Brittingham *et al.* (2014) include spatial analysis, species-based modelling, vulnerability assessments, eco-regional dynamics and toxicity evaluations.

Emerging pollution risks in marine environments are also associated with the widespread dispersal of plastics and microplastics used for domestic and commercial purposes (Chapter 30). It should be stressed that microplastics are not inert particles in terms of biological activity or physical chemistry. Consequently, ecotoxicological effects include entanglement in vertebrates, transfer along food chains, loss of nutritional value of food items and enhanced exposure to pathogens and adherence of chemical contaminants such as polycyclic aromatic hydrocarbons (Avio *et al.*, 2017; Santillo *et al.*, 2017).

As might be anticipated, radioactive emissions following the Chernobyl nuclear accident continue to induce wildlife abnormalities. For example, Moller *et al.* (2013) reported high frequency of albinism and tumours in free-living birds in the vicinity of the failed reactor, adding that these defects may serve as biomarkers of radiation exposure. In contrast, however, Boubriak *et al.* (2016) noted that 30 years after the accident, there is still a lack of data relating to the genetic effects of radionuclide contamination on plant ecology. Specific questions requiring elucidation included adaptation of plants to chronic low-dose ionizing radiation, responses to changing/increased pathogenicity of fungi and viruses, accumulation of mutational load and the extent of micro-evolutionary changes in the Chernobyl area. This lack of information may be significant in view of the potential role of phytoremediation in environmental protection.

Biomagnification is a particular issue with respect to the persistent organic pollutants covered in Chapters 10–19. The crux of the problem is ideally illustrated in recent work. According to Gerber *et al.* (2016), the banned insecticide, DDT, is still used in many developing countries, including South Africa (one of the most advanced economies in the third world). It was demonstrated that pesticide contamination of the rivers in the Kruger National Park, a premier conservation area, resulted in bioaccumulation of DDT (and other organochlorine residues) in an apex aquatic predator, the tigerfish (*Hydrocynus vittatus*). It was concluded that this bioaccumulation represented a high cancer risk to local populations consuming tigerfish. It is clear that biomagnification of polychlorinated biphenyls also occurs in whales (Ross *et al.*, 2000) with possible impact on their reproductive capacity and survival. Similar comments may apply to dibenzofurans and dioxins in open ocean predators such as killer whales (Ono *et al.*, 1987).

## 40.5 Discussion

A broad definition of 'toxicity' has been adopted ranging from acute and chronic to psychosocial effects and examples appear in the different chapters of the present volume. It is salutary to review the evidence presented in this volume in order to highlight recent developments in environmental toxicology and the implications for both human health and natural habitats in our fragile ecosystems. There is undoubtedly value in classifying research observations relating to effects of pollutants to enable the development of improved models for the future.

### 40.5.1 Current status

As a discipline, environmental toxicology is rapidly evolving into an exact science, underpinned by advances in medical research and quantitative methodologies, replacing anecdotal and subjective risk assessments. Authors contributing to this edition are at the forefront of current developments to remove ambiguity, improve current models of analysis and elucidate biochemical mechanisms underlying epidemiological observations. It is clearer than ever before that researchers are at an important juncture to secure further advances that will inspire others to join the cause of environmental protection and management.

### 40.5.2 Human health disorders categorized according to association with specific pollutants

The conclusion that pollution causes human morbidity and damages ecosystems is now inescapable and more firmly embedded in contemporary thinking among scientists and practitioners. It can be stated unequivocally that pollutants such as PAHs, dioxins and radionuclides as well as UV exposure are key elements contributing to environmental carcinogenesis. The role of radon in lung cancer development is also undisputed. In certain cases, the identification of molecular signatures has confirmed cause-and-effect relationships in malignancy.

Although there is good epidemiological evidence to support an effect of traffic-related pollutants in provoking or exacerbating respiratory disorders among residents living in congested streets, interpretation of data is confounded by complex interactions involving, for example, gaseous components and particulates as well as socio-economic factors. There is considerable and continuing interest in the relationship between ambient air pollutants and cardiovascular disease in humans. It has been stated unequivocally that there is 'abundant' evidence to support the concept that air pollution contributes to the risk of CVD and associated mortality. Possible mechanisms underlying this risk include manifestations of oxidative stress with inflammation at specific sites also affecting heart rhythm. It has been argued that both $PM_{2.5}$ and $PM_{10}$ may contribute to effects such as heart disease, stroke and CVD (Chapter 28). Indeed, ambient particulate pollution has exposed regional health disparities, causing higher mortality in Eastern compared with Western Europe. Furthermore, the association of $PM_{2.5}$ with ammonia emissions in rural areas may represent risks for farming communities (Schifer *et al.*, 2014).

The cognitive and neurological effects of pollutants are now a major source of concern and are likely to fuel debate in the future. The concept of developmental toxicity has been elaborated by Grandjean and Landrigan (2014), who stated that childhood conditions including autism, attention-deficit hyperactivity disorder, dyslexia and other cognitive defects occur with increasing frequency. Industrial chemicals such as lead, methylmercury, PCBs, arsenic and certain pesticides that damage the developing brain are among the known causes for this increase. Maternal lifestyle and environmental risk factors for autism spectrum disorders were reiterated by Lyall *et al.* (2014). A number of studies have demonstrated significant increases in autism risk with estimated exposure to air pollution during the prenatal period, particularly for heavy metals and particulate matter. In a broad review, Carter and Blizard (2016) claimed that autism genes are selectively targeted by environmental pollutants, including pesticides and heavy metals. Gene–environmental interactions have also been implicated in the aetiology of Parkinson's disease (Goldman, 2014). Nevertheless, the incidence of autism and other disorders such as dementia and Parkinson's disease linked to pollutants remain contentious issues requiring innovative analysis at the molecular level to identify key diagnostic signatures.

### 40.5.3 Ecological considerations

Environmental contaminants contribute significantly to ecological decline in natural habitats, as epitomized by historic oil spills and nuclear accidents (see Preface to this edition). Emerging contaminants such as pharmaceuticals (Mezzelani *et al.*, 2018), microbeads and other pollutants associated with anthropogenic activity have introduced formidable challenges for remediation technologists. Pesticide contamination of food, water and the environment as a whole remain intractable issues and efforts continue to develop ecologically acceptable alternatives. Even modern pesticides can jeopardize biodiversity, as shown, for example, by reduced survival of honeybees exposed to imidacloprid (Raymann *et al.*, 2018). The concept of phytotoxins as environmentally friendly bio-pesticides is an emerging issue in view of ongoing contamination of foods with conventional compounds. It has been consistently maintained that phytotoxins serve in a defence role in plants with activity towards invertebrate herbivores and fungal pathogens. The resilience of a defence system based on heat-labile protein phytotoxins is strongly implied as a result of recent evidence reviewed in Chapter 1. It is suggested, therefore, that these protein phytotoxins should form part of a plant breeding programme to enhance pest and pathogen resistance without compromising food safety; it is widely acknowledged that protein phytotoxins are readily denatured during cooking. However, attempts to use transgenics for this purpose would be controversial, due to continuing consumer disquiet over the ethics and acceptability of genetically modified foods. Despite recent advances, it is clear that considerable fundamental work still needs to be accomplished in order to develop the aforementioned phytotoxins into practical bio-pesticides. In addition, it should be acknowledged that other secondary compounds may emerge as preferred candidates for this role. Researchers in this field will not be short of encouragement: Matthews (2017) states that bio-pesticides 'have a bright future, but more attention is needed on their application'. Furthermore, European pesticide regulations favour the use of reduced-risk substitutes in integrated pest management protocols. At the same time, it is important to recognize that future work should be underpinned by substantive studies at the molecular level if current understanding of host–parasite interactions is to evolve into innovative solutions.

Nutrient pollution is exemplified mainly in the forms of nitrogen and phosphorus discharges into aquatic environments. The use of bio-indicators, such as dinoflagellate cyst assemblages, has become indispensable for determining water quality and anthropogenic impact in such environments. Consistent with expectations, microbial communities have the capacity to adapt to pollution events. In an assessment of the *Deepwater Horizon* oil spill in the Gulf of Mexico, deposits on the shore caused a shift in the community ecology towards recognized petroleum hydrocarbon-degrading microorganisms. However, it should be noted that PAHs are more recalcitrant and that detoxification pathways are limited in vertebrate species. The report that the *Deepwater Horizon* oil spill negatively affected sea turtle nesting is a case in point. Nevertheless, observations 24 years after the *Exxon Valdez* oil spill indicate that recovery is possible in the long term, though there are species differences (Chapter 22). Based on biomarker evidence, a timeline of two decades has been proposed for harlequin ducks, a species considered to be particularly susceptible to oil contamination in marine environments. Other observations extend findings concerning timelines and mechanisms of wildlife population dynamics to several marine species, based on acute and chronic exposure to the *Exxon Valdez* contamination.

The emergence of hydraulic fracturing creates different risks for wildlife and associated habitats. The riparian songbird has been used as a bioindicator of pollution caused by fracking. As a top predator, the songbird is exposed to or bioaccumulates, via its macroinvertebrate food, the chemical contaminants released into the ecosphere by fracking. Recent results show feather barium and strontium concentrations to be significantly higher in birds originating from fracking sites than in those at sites without fracking. Key areas of research and monitoring still need to be addressed to assist in the formulation of effective guidelines and policies to protect vulnerable species at fracking sites.

Emerging pollution risks in marine environments are associated with the widespread dispersal of plastics and microplastics used for domestic and commercial purposes. It is stressed that microplastics are not inert particles in terms of biological activity or physical chemistry and are consequently associated with complex ecotoxicological effects, particularly with respect to free-living marine vertebrates. Plastics represent a visible form of pollution, causing entanglement and abnormal feeding behaviour in marine animals. However, it should be recognized that plastics may also be associated with polycyclic aromatic hydrocarbons, conceivably contributing to persistence of these organic compounds.

Consistent with previous observations, radioactive emissions following the Chernobyl accident are linked with wildlife abnormalities. High frequencies of albinism and tumours have been observed in free-living birds in the vicinity of the failed reactor, and these defects may serve as biomarkers of radiation exposure. In contrast, however, it has been noted that 30 years after the accident, there is still a lack of data relating to the genetic effects of radionuclide contamination on the ecology of higher plants. Specific questions requiring elucidation include adaptive physiology of plants, fungal pathogenicity and the extent of micro-evolutionary changes in areas contaminated by radionuclides.

### 40.5.4 Adaptation in higher plants

Some insight on the adaptive mechanisms of higher plants has emerged from investigations into the effects of acid rain (Chapter 9) and heavy metals (Chapter 4). In addition, the effects of potentially toxic mineral elements such as cobalt, copper, iron and nickel naturally occurring in serpentine soils offers scope for the advancement of phytoremediation in resolving intractable issues associated with heavy-metal contamination. Serpentine geochemistry facilitates a distinctive ecology, with some plant species such as *Alyssum betolonii* able to hyperaccumulate nickel at concentrations of up to 2118 mg $kg^{-1}$ in aerial tissues without succumbing to toxicity in any explicit forms (Bini *et al.*, 2017). It has been suggested that serpentinophytes may serve in remediation of metal-contaminated soils. Issues of phytostabilization, phytoextraction and phytomining were also raised in this study. In contrast, other studies indicate that paddy rice cultivated on serpentine soils containing high geogenic sources of nickel may exceed the provisional tolerable intake of nickel for adults, with the potential for toxicity (Hseu and Lai, 2017). The observation that magnesium alleviates plant toxicity of aluminium and heavy metals opens up other possibilities for remediation of contaminated land destined for arable farming (Rengel *et al.*, 2015). The potential of metallic iron and biochar in remediation is reviewed in Chapters 36 and 37. Another proposition involves the use of serpentine bacteria as inoculants in phytostabilization of polluted soils (Ma *et al.*, 2015; see also Chapter 35). Additional aspects of bioremediation are presented by Bisht *et al.* (2015) and Liu *et al.* (2017) with respect to polyaromatic hydrocarbon decontamination. Despite widespread recognition of the natural resilience of plants to adverse conditions, it should be noted that ozone toxicity is readily induced causing cellular changes followed by substantial biomass and yield reductions (Singh *et al.*, 2015; Alves *et al.*, 2016).

## 40.6 Implications and Future Directions

Critical analysis of the evidence presented in this volume indicates the need for profound changes in attitudes, approaches and policies if human health and pristine habitats are to be restored and protected in the face of continuing and widespread pollution. It is important to regularly enquire whether temporal trends in contaminants, for example in marine environments, are a genuine reflection of regulations. It is also imperative to ensure that environmental protection is not reduced to crisis management; the establishment of robust protocols is a critical prerequisite, particularly in instances of petroleum and radiation pollution. It is doubtful whether there are positive answers to these questions.

### 40.6.1 General aspects

On the basis of recent declarations in Europe and North America it is abundantly clear that our elected leaders are unreliable custodians of the environment. It would not be unfair to conclude that they are motivated by political expediency, with one eye on re-election.

In addition, the policy of protectionism in global economic matters is gathering momentum, with the likely effect that international co-operation on pollution research and adherence to the Paris Climate Agreement will be severely jeopardized. Insular policies will not only serve to exacerbate environmental disparities for communities in vulnerable regions but also jeopardize human health and biodiversity in affluent countries.

As a consequence, there is now an opportunity for environmental toxicologists to demonstrate leadership and it is envisaged that publication of this volume will contribute to an active debate on all aspects of human health and ecological conservation. The initial signs are encouraging. For example, the work of Landrigan (2017) linking air pollution with human morbidity has attracted wide attention in the media due to the bold expression of current and projected data. *The Lancet* has also been at the forefront of this activity by highlighting key public health disorders associated with environmental pollution (see, for example, Samet *et al.*, 2013; Shah *et al.*, 2013; Beelen *et al.*, 2014; Guarnieri and Balmes, 2014; and Landrigan, 2017).

The research evidence now emerging has established environmental toxicology as a less speculative and more quantitative discipline providing sufficient scientific basis for limiting pollution-related morbidity and ecological degradation. However, the prospects for continuing developments are critically dependent upon questions of research funding and leadership. This *Handbook of Environmental Toxicology* has been designed to highlight the impact of pollutants and other environmental hazards on human health and biodiversity in natural and managed ecosystems. Maintaining any momentum created by its publication will be the responsibility of future generations of scientists in a wide and evolving field of research. The implementation of austerity measures is likely to undermine their efforts. Worldwide rationalization and constant erosion of the research base in different scientific disciplines are primary constraints to achieving the major objectives advocated by experts contributing to this volume.

It is regrettable that, as in other fields of activity, future directions for research will be dictated by contingency, rather than by deliberate forward planning, in response to external events such as the *Deepwater Horizon* contamination and the Fukushima nuclear accident. However, real-time issues still prevail and need to be addressed in detailed and coordinated monitoring and research programmes. For example, seemingly intractable issues relating to the impact of gaseous and persistent organic pollutants and associated interactions on human morbidity remain largely unexplored and probably difficult to quantify without complex modelling and molecular techniques.

The field of environmental toxicology is in a continual state of flux, as expanding industrial innovation and emerging pollutants demand novel approaches. For example, particular ecological risks are presented by microbeads, nanomaterials and contaminants in shale oil extraction. Development of appropriate and reliable analytical methodologies is an essential adjunct for the deployment of monitoring and remediation strategies.

### 40.6.2 Food and water safety

It is axiomatic that pollution should compromise safety of primary foods and drinking water. D'Mello (2002) expressed widespread and continuing concerns over food safety. It would appear that, despite the best efforts of established and new government agencies, food and water contamination is likely to present significant risks for public health (Adamse *et al.*, 2017; Carvalho, 2017). In the medium term every nation will be exposed to events that undermine consumer confidence in a wide range of food staples such as cereal grains, meat, milk, eggs and fish. The recent past has been characterized by an unremitting series of food scares associated with decades of pollution, careless deregulation and underfunding of services in monitoring, research and education. These constraints are likely to be exacerbated by current austerity measures and the pursuit of isolationist policies, particularly by elected administrations in North America and Europe. The principal contaminants in vegetables and fruit were listed by D'Mello (2002). At that time, pesticide residues in certain vegetables and fruit were a problem in many countries. Surveillance results in the UK, for example, indicated a distinct lack of progress in reducing levels of pesticide residues in imported yams at levels in excess of legal maximum residue limits (MRLs). In the UK, 69% of lettuce samples were reported to contain pesticide residues, in some cases exceeding MRLs or contaminated with non-approved pesticides. Celery, grapes, oranges and apples have all been found to contain multiple residues of pesticides. With respect to grapes, the Pesticide Residue Committee in the UK reported that 67% of samples contained pesticides, with 29% being contaminated with multiple residues (D'Mello, 2002). Specific food alerts have been issued on human health risks presented by pesticides in vegetables and fruit.

Geological conditions mean that in certain parts of the world, for example Chile, consumers may be exposed to above-average levels of arsenic. Concentrations in soil and aquifers of up to 1099 mg $kg^{-1}$ and 11 mg $l^{-1}$, respectively, have been recorded in Chile. Predictably, vegetables and fruit cultivated in these areas may become contaminated with arsenic. Levels of up to 0.6 mg $kg^{-1}$ fresh weight were reported for spinach, close to the level of 1 mg $kg^{-1}$ allowed by Chilean legislation, but in other countries, for example Poland, permitted levels were set at just 0.2 mg $kg^{-1}$ (D'Mello, 2002). Currently, studies are being undertaken to evaluate the risks presented by heavy metals in agricultural soils of the EU that might impact on food safety (Toth *et al.*, 2016). There is, therefore, increasing awareness of the global implications of toxic metals in food plants cultivated on contaminated soils.

Nevertheless, it should be emphasized that vegetables and fruit should always form part of a balanced diet. These foods are important sources of nutrients, antioxidants and fibre. Significantly, secondary compounds present particularly in leafy vegetables may confer protection towards asthma and certain types of cancer. Work continues on all aspects of food safety, especially with regard to the persistent organic pollutants. For example, the question of pesticide residues in relation to the environment and food safety has been addressed by Carvalho (2017). At the same time, levels of dioxins and dioxin-like PCBs in food of animal origin in the Netherlands during the period 2001–2011 have been investigated by Adamse *et al.* (2017). Levels of dioxins in UK milk supplies were observed to be higher in samples obtained from farms in close proximity to urban and industrial sites than in those from farms in rural areas. The UK national milk supply has also been contaminated with radionuclides following the testing of nuclear weapons and from the Chernobyl accident in 1986 (see D'Mello, 2002). The Fukushima accident will have added fresh concerns about radioactivity in foods and water (Chapter 34).

Estuarine and marine pollution has become a major environmental issue resulting in contamination of seafood (D'Mello, 2002). Despite the Minamata episode of the 1960s (Chapter 24), mercury contamination of seafood continues to present multiple risks on a global scale. Thus, a link has been proposed between mercury in fish and reduced fertility in Hong Kong men, while in the Faroe Islands exposure occurred through consumption of whale meat. Children with prenatal exposure to mercury were observed to display deficits in cognitive functions (see D'Mello, 2002). In other surveillance of foods, polychlorinated biphenyls and dioxins have been detected in farmed salmon (Food Standards Agency, 2000, 2001).

The quality and safety of water used for direct consumption or for irrigation of food crops are markedly affected by prevailing levels of pollution. The levels of lead (Chapter 25) and nitrates continue to cause concern among environmental health personnel. In addition, the occurrence of pathogenic enteric microorganisms derived from sewage pollution of water is a persistent issue. Outbreaks of enteric virus illness in humans have been regularly associated with water contaminants arising from pipeline failures, pollution of wells and pollution of municipal supplies with sewage, as concluded by D'Mello (2002).

### 40.6.3 Self-regulation and corporate behaviour

The large multinational corporations regularly issue environmental statements on their websites and in annual business reports. The aim is to demonstrate how the activities of a company are managed in relation to the environment. However, recent developments indicate that vigilance and constant surveillance by regulatory agencies should still be key elements in any strategy to control pollution and other environmental risks arising during industrial activity. A number of serious incidents do not inspire much confidence in self-regulation, particularly by multinational corporate organizations. The discharge of untreated sewage into the Thames over several months by a major UK utility company has been labelled as an 'environmental disaster', deservedly attracting severe penalties. In 2016, a cruise line incurred a substantial fine after illegally discharging oil and associated waste via a 'magic pipe' off the UK coast. In late September 2017, the appearance of a 'radiation cloud', presumed to originate from a nuclear fuel plant in the Urals, was widely reported but remains unexplained. Currently, the 'stop demonizing diesel' campaign is in a fully active mode supported by a powerful lobby and there is a need for health experts to ensure that the arguments stand up to critical analysis. The success of vigilance and surveillance is, arguably, best exemplified by the US Environment Protection Agency (EPA) findings that numerous German-manufactured cars were fitted with 'defeat software' designed to falsify emissions performance. In view of other case studies concerning aflatoxins, gaseous emissions, dioxins, PCBs, agricultural chemicals, shale oil extraction, lead and mercury, there is heightened consensus that human health and the ecosystem are continually at severe risk.

A number of petroleum companies have recently announced their intentions to participate in the World Economic Forum's Oil and Gas Climate Initiative (OGCI) working collaboratively towards solutions to mitigate environmental risks. At the same time, however, these companies have threatened to withdraw from the European Union (EU) if laws to reduce pollution and to accelerate take-up of clean energy were enacted. Thus, it appears that delivering on virtuous environmental statements is not unconditional.

### 40.6.4 Questionable decisions

The history of environmental pollution is littered with perplexing decisions over several decades, as demonstrated in the responses to crude oil spills or the discharge of sewage or fertilizers into rivers and lakes. The premature sanctioning of shale oil and gas extraction on an industrial scale is another example (Chapter 23). In 2018, Monsanto was ordered to pay substantial damages in a glyphosate cancer trial. Toxicologists are now enquiring why the US EPA sanctioned the use of this herbicide, given its classification as 'probably carcinogenic to humans' by IARC in 2015. There are concerns expressed online that there may have been 'inappropriate industry involvement' in the US EPA decision. There is a need for clarification of the process to ensure that due diligence has been undertaken and also to restore public confidence in the evaluation process. It is crucial that environmental protection agencies around the world adopt a 'without fear or favour' policy in the safety assessments of pesticides, drugs and other chemicals that may impact upon human health or habitat biodiversity.

### 40.6.5 Methodology

There will always be commentators who doubt the credibility of evidence based on epidemiological correlations. Indeed, there are still those who remain unconvinced of any relationship between cigarette smoking or alcohol abuse and human morbidity. It is important and inevitable, therefore, that future work should be directed at establishing cause-and-effect issues for a wide range of pollutants, similar to those used for identifying the carcinogenic properties of the aflatoxins (Huang *et al.*, 2017; Chapter 2). This may involve the characterization of signature molecules at genomic, transcriptomic, proteomic and metabolomic levels of expression. Any advances in the use of cellular biomarkers might elicit contrasting implications: while assisting in diagnosis and treatment of pollutant-related morbidity, it might equally embolden patients and others exposed to noxious emissions to mount a Class Action against public health authorities to legally enforce a remediation programme in their local environment.

### 40.6.6 Leadership vacancies and opportunities

Given the confused signals emanating from American and European governments and the limited success of international environmental agreements, it is now opportune for prominent and experienced scientists to directly demonstrate eloquence and leadership in matters such as urban pollution, oil spill legacy issues, heavy-metal contamination, use of plastics, nuclear accidents and conservation of natural habitats. It is also crucial to articulate with the major corporations to address issues with limiting industrial pollution and with clean-up efforts. The instinctive reticence of environmental scientists to engage with individual and corporate stakeholders on matters such as pollution and habitat conservation can no longer be justified but will require the acquisition of exceptional communication skills on the part of all protagonists in this debate.

### 40.6.7 Integrated management

It is imperative that agencies charged with environmental protection engage with each other in the implementation of any corrective measures. It is widely reported that the EU favours the use of wood as a renewable source of energy. However, burning of wood is regularly associated with emissions of particulates and polycyclic aromatic hydrocarbons and has been associated with air pollution in developing countries, for example in populated areas of Chile. The prospects for integrated management may also be compromised by governmental pressures to pursue isolationist policies, resulting in disjointed efforts to control global climate change and pollution.

### 40.6.8 Toxic legacy issues

It would be convenient if historical cases of contamination could be dismissed without further consideration. However, a wide range of toxic legacy issues remain in the public domain and are a cause for concern. For example, atomic bomb survivors in Nagasaki still present with radiation-linked syndromes (Horai *et al.*, 2018). In marine environments, there are disturbing reports of unacceptably high concentrations of PCBs (Ross *et al.*, 2000), dibenzofurans with dioxins (Ono *et al.*, 1987) and organochlorine pesticides in free-ranging whales (Noel *et al.*, 2018). In addition, variable recovery of marine species has been reported following the *Exxon Valdez* oil spill (Chapter 22).

## 40.7 Conclusions

It is incumbent upon the Editor to declare that virtually all pollutants reviewed in this volume are currently impairing the well-being of both humans and the ecosystem as a whole. This conclusion is as unequivocal as it is uncompromising in its demands. For example, traffic emissions including particulates predispose or may even precipitate conditions such as cardiovascular, pulmonary and neurological disease in residents of major conurbations around the world. At local levels, there is disquiet over dioxin, pesticide, heavy-metal and radiation pollution. A disturbing issue is the apparent absence of a threshold for the induction of adverse effects for several pollutants.

There is now an expectation that environmental toxicologists will continue to demonstrate leadership through sound scientific research and bold expression of emerging evidence. The regulatory and associated agencies should contribute by adopting a more impartial and critical stance when scrutinizing new pesticides and pharmaceuticals prior to approval and during environmental impact assessments of urban, sewage, farm effluents, crude oil and radionuclide contaminants. In addition, legacy issues arising from persistent organic pollutants, nuclear accidents and crude oil spills still need to be addressed in a more comprehensive and effective remediation programme. There is a risk that these incidents may be consigned as footnotes in environmental history before the adverse effects are fully resolved. However, the difficulties in remediation should not be underestimated and the Editor is unable to see the way forward for the persistent organic pollutants in this respect.

Meanwhile, particularly in marine ecosystems, 'emerging pollutants' such as pharmaceutical drugs, personal care products and organophosphorus flame retardants are presenting different risks as well as challenges for the next generation of environmental toxicologists.

## References

Adamse, P., Schoss, S., Theelen, R.M.C. and Hoogenboom, R.L.A.P. (2017) Levels of dioxins and dioxin-like PCBs in food of animal origin in the Netherlands during the period 2001–2011. *Food Additives and Contaminants* 34, 78–92.

Alshaarawy, O., Elbaz, H.A. and Andrew, M.E. (2016) The association of urinary polycyclic aromatic hydrocarbon biomarkers and cardiovascular disease in the US population. *Environment International* 89, 174–178.

Alves, E.S., Moura, B.B., Pedroso, A.N.V., Tresmondi, F. and Machado, S.R. (2016) Cellular markers indicative of ozone stress on bioindicator plants growing in a tropical environment. *Ecological Indicators* 67, 417–424.

Ames, J., Warner, M., Mocarelli, P., Brambilla, P., Signorini, S., Holland, N. and Eskenazi, B. (2018) *AHR* gene-dioxin interactions and birthweight in the Seveso second generation health study. *International Journal of Epidemiology* 47(6), 1992–2004. doi: 10.1093/ije/dyy165.

Avio, C.G., Gorbi, S. and Regoli, F. (2017) Plastics and microplastics in the oceans: from emerging pollutants to emerged threat. *Marine Environmental Research* 128, 2–11.

Becerra, T.A., Wilhelm, M. and Ritz, B. (2013) Ambient air pollution and autism in Los Angeles County, California. *Environmental Health Perspectives* 121, 380–386.

Beelen, R., Raaschou-Nielsen, O., Stafoggia, M., Andersen, Z.J., Weinmayr, G. *et al.* (2014) Effects of long-term exposure to air pollution on natural-cause mortality: an analysis of 22 European cohorts within the ESCAPE project. *The Lancet* 383, 785–795.

Bini, C., Maleci, L. and Wahsha, M. (2017) Potentially toxic elements in serpentine soils and plants from Tuscany (Central Italy). A proxy for soil remediation. *Catena* 148, 60–66.

Bisht, S., Pandey, P., Bhargava, B., Sharma, S., Kumar, V. and Sharma, K.D. (2015) Bioremediation of polyaromatic hydrocarbons (PAHs) using rhizosphere technology. *Brazilian Journal of Microbiology* 46(1) 7–21. doi: 10.1590/S1517-838246120131354.

Boubriak, I., Akimkina, T., Dimitriev, A., McCready, S. and Grodzinsky, D. (2016) Long-term effects of Chernobyl contamination on DNA repair function and plant resistance to different biotic and abiotic stress factors. *Cytology and Genetics* 50, 381–399.

Bowatte, G., Lodge, C.J., Perret, J.L., Matheson, M.C. and Dharmage, S.C. (2016) Interactions of *GST* polymorphisms in air pollution exposure and respiratory diseases and allergies. *Current Allergy and Asthma Reports* 16, 85. doi:10.1007/s11882-016-0664-z.

Brittingham, M.C., Maloney, K.O., Farag, A.M., Harper, D.D. and Bowen, Z.H. (2014) Ecological risks of shale oil and gas development to wildlife, aquatic resources and their habitats. *Environmental Science and Technology* 48, 11034–11047.

Carter, C.J. and Blizard, R.A. (2016) Autism genes are selectively targeted by environmental pollutants including pesticides, heavy metals, bisphenol A, phthalates and many others in food, cosmetics and household products. *Neurochemistry International* 101, 83–109.

Carvalho, F.P. (2017) Pesticides, environment and food safety. *Food and Energy Security* 6, 48–60.

Chen, H., Kwong, J.C., Copes, R., Tu, K., Villeneuve, P.J. *et al.* (2017) Living near major roads and the incidence of dementia, Parkinson's disease and multiple sclerosis: a population-based cohort study. *The Lancet* 389, 718–726.

Dadvand, P., Nieuwenhuijsen, M.J., Agusti, A., Benet, M., Beelen, R. and Garcia-Aymerich, J. (2014) Air pollution and biomarkers of systemic inflammation and tissue repair in COPD patients. *European Respiratory Journal* 44, 603–613.

De Felice, A., Scattoni, M.L., Ricceri, L. and Calamandrei, G. (2015) Prenatal exposure to a common OP insecticide delays motor development in a mouse model of idiopathic autism. *PLoS ONE* 10(3), e01216603. doi: 10.1371/journal. pone. 0121663.

DeVries, R., Kriebel, D. and Sama, S. (2016) Low level air pollution and exacerbation of existing COPD: a case crossover analysis. *Environmental Health* 15, 98. doi: 10.1186/s12940-016-0179-z.

D'Mello, J.P.F. (2002) Widespread and continuing concerns over food safety. In: D'Mello, J.P.F. (ed.) *Food Safety: Contaminants and Toxins*. CABI Publishing, Wallingford, pp. 409–437.

Esler, D., Ballachey, B.E., Bowen, L., Miles, A.K., Dickson, R.D. and Henderson, J.D. (2017a) Cessation of oil exposure in harlequin ducks after the *Exxon Valdez* oil spill: cytochrome P4501A biomarker evidence. *Environmental Toxicology and Chemistry* 36, 1294–1300.

Esler, D., Ballachey, B.E., Malkin, C., Cushing, D., Bodkin, J., Esslinger, G. and Kloecker, K. (2017b) Timelines and mechanisms of wildlife population recovery following the *Exxon Valdez* oil spill. *Deep Sea Research Part II: Topical Studies in Oceanography* 147, 36–42. doi: .org/10.1016. j.dsr2.2017.04.007.

Eze, I.C., Hemkins, L.G. and Probst-Hensch, N. (2015) Association between ambient air pollution and diabetes mellitus in Europe and North America: systematic review and meta-analysis. *Environmental Health Perspectives* 123, 381–389.

Feng, Y., Sun, H., Song, Y., Bao, J., Huang, X. *et al.* (2014) A community study of the effect of polycyclic aromatic hydrocarbon metabolites on heart rate variability based on the Framingham risk score. *Occupational and Environmental Medicine* 71, 338–345.

Field, R.W., Steck, D.J., Smith, B.J., Brus, C.P., Platz, C.E., Woolson, R.F. and Lynch, C.F. (2000). Residential radon gas exposure and lung cancer: the Iowa radon lung cancer study. *American Journal of Epidemiology* 151, 1091–1102.

Food Standards Agency (2000) *Food Standards Agency News* No. 3. London.

Food Standards Agency (2001) *Food Standards Agency News* No. 14. London.

Gerber, R., Smit, N.J., Van Vuren, J.H.T., Nakayama, S.M.M., Yohannes, Y.B. *et al.* (2016) Bioaccumulation and human health risk assessment of DDT and other organochlorine pesticides in an apex aquatic predator from a premier conservation area. *Science of the Total Environment* 550, 522–533.

Goldman, S.M. (2014) Environmental toxins and Parkinson's disease. *Annual Review of Pharmacology and Toxicology* 54, 141–164.

Goodale, B.C., La Du,J., Tilton, S.C.,Sullivan, C.M., Bisson, W.H., Waters, K.M. and Tanguay, R.L. (2015) Ligand-specific transcriptional mechanisms underlie aryl hydrocarbon receptor-mediated developmental toxicity of oxygenated PAHs. *Toxicological Sciences* 147, 397–411.

Grandjean, P. and Landrigan, P.J. (2014) Neurobehavioural effects of developmental toxicity. *The Lancet Neurology* 13, 330–338.

Greenberg, N., Carel, R.S., Derazne, E., Tzur, D. and Portnov, A. (2017) Modeling long-term effects attributed to nitrogen dioxide ($NO_2$) and sulfur dioxide ($SO_2$) exposure on asthma morbidity in a nationwide cohort in Israel. *Journal of Toxicology and Environmental Health, Part A* 80, 326–337.

Grundy, A., Bred, K. and Brenner, D.R. (2017) Lung cancer incidence attributable to residential radon exposure in Alberta in 2012. *Canadian Medical Association Journal* 5, E529–E534.

Guarnieri, M. and Balmes, J.R. (2014) Outdoor air pollution and asthma. *The Lancet* 383, 1581–1592.

Guxens, M., Muetzel, R., Dalmau-Bueno, A., Jaddoe, V.M.V., Hoek, G. *et al.* (2018) Air pollution exposure during fetal life, brain morphology and cognitive function in school-age children. *Biological Psychiatry* 84(4), 295–303. doi: 10.1016/j.biopsych.2018.01.016.

Ha, E., Basu, N., Bose-O'Reilly, S., Dorea, J.G., Sakamoto, M. and Chan, H.M. (2017) Current progress on understanding the impact of mercury on human health. *Environmental Research* 152, 419–433.

Hamra, G.B., Laden, F., Cohen, A.J., Raaschou-Nielsen, O., Brauer, M. and Loomis, D. (2015) Lung cancer and exposure to nitrogen dioxide and traffic: a systematic review and meta-analysis. *Environmental Health Perspectives* 123, 1107–1112.

Hansen, S., Strom, M., Olsen, S.F., Masiova, E., Rytter, D. *et al.* (2014) Maternal concentrations of persistent organochlorine pollutants and the risk of asthma in offspring: results from a prospective cohort with 20 years of follow-up. *Environmental Health Perspectives* 122, 93–99.

Hilborn, E.D. and Beasley, V.R. (2015) One health and cyanobacteria in freshwater systems: animal illnesses and deaths are sentinel events for human health risks. *Toxins* 7, 1374–1395.

Hill, E.L. (2018) Shale gas development and infant health: evidence from Pennsylvania. *Journal of Health Economics* 61, 134–150.

Hirsch, J.K., Smalley, K.B., Selby-Nelson, E.M., Beckmann, S. and LaFromboise, T. (2018) Psychosocial impact of fracking: a review of the literature on the mental health consequences of hydraulic fracturing. *International Journal of Mental Health* 16, 1–15.

Horai, M., Satoh, S., Matsuo, M., Takasaki, Y., Kawaguchi, Y. and Tsushima, H. (2018) Chromosomal analysis of myelodysplastic syndromes among atomic bomb survivors in Nagasaki. *The British Journal of Haematology* 180, 381–390.

How, V., Ismail, P., Omar, S.D. and Tamrin, S.B.M. (2014) Exploring cancer development in adulthood: cholinesterase depression and genotoxic effects from chronic exposure to organophosphate pesticides among rural farm children. *Journal of Agromedicine* 19, 35–43.

Hseu, Z.-Y. and Lai, Y.-J. (2017) Nickel accumulation in paddy rice on serpentine soils containing high geogenic nickel contents in Taiwan. *Environmental Geochemistry and Health* 39, 1325–1334.

Huang, M.N., Yu, W., Teoh, W.W., Ardin, M., Jusaki, A. *et al.* (2017) Genome-scale mutational signatures of aflatoxin in cells, mice and human tumors. *Genome Research* 27(9), 1475–1486. doi: 10.1101/gr.220038.116.

Hutz, R.J., Carvan, M.J. and Julien, K. (2014) Familiar and novel reproductive disruptors: xenoestrogens, dioxins and nanoparticles. *Current Trends in Endocrinology* 7, 111–122.
Hystad, P., Villeneuve, P.J., Goldberg, M.S., Crouse, D.L. and Johnson, K. (2015) Exposure to traffic-related air pollution and the risk of developing breast cancer among women in eight Canadian provinces: a case-control study. *Environment International* 74, 240–248.
Jarvis, I.W.H., Dreij, K., Mattsson, A., Jernstrom, B. and Stenius, U. (2014) Interactions between polycyclic aromatic hydrocarbons in complex mixtures and implications for cancer risk assessment. *Toxicology* 321, 27–39.
Jerrett, M., McConnell, R., Wolch, J., Chang, R., Dunton, G., Islam, T. and Berhane, K. (2014) Traffic-related air pollution and obesity formation in children: a longitudinal, multilevel analysis. *Environmental Health* 13, 49. doi:10.1186/1476-069X-13-49.
Johannson, K.A., Hinghoff, E.V., Lee, K., Balmes, J.R., Ji, W. and Collard, H.R. (2014) Acute exacerbation of idiopathic pulmonary fibrosis associated with air pollution exposure. *European Respiratory Journal* 43, 1124–1131.
Karimi, P., Peters, K.O., Bidad, K. and Strickland, P.T. (2015) Polycyclic aromatic hydrocarbons and childhood asthma. *European Journal of Epidemiology* 30, 91–101.
Kim, K.-S., Lee, Y.-M., Lee, H.-W., Jacobs, D.R. and Lee, D.-H. (2015) Associations between organochlorine pesticides and cognition in U.S. elders: National Health and Nutrition Examination Survey 1999–2002. *Environment International* 75, 87–92.
Kim, K.-N., Kwon, H.-J. and Hong, Y.-C. (2016) Low-level lead exposure and autistic behaviors in school-age children. *Neurotoxicology* 53, 193–200.
Kim, S., Xu, X., Zhang, Y., Zheng, X., Liu, R. *et al.* (2018) Metal concentrations in pregnant women and neonates from informal electronic waste recycling. *Journal of Exposure Science & Environmental Epidemiology* 29, 406–415. doi: 10.1038/s41370-018-0054-9.
Lamendella, R., Strutt, S., Borglin, S., Tas, N., Mason, O.U., Hazen, T.C. and Jansson, J.K. (2014) Assessment of the Deepwater Horizon oil spill impact on Gulf coast microbial communities. *Frontiers in Microbiology* 5, 130. doi: 10.3389/fmicb.2014.00130.
Landrigan, P.J. (2017) Air pollution and health. *The Lancet Public Health* 2, e4–e5, January 2017.
Latta, S.C., Marshall, L.C., Frantz, M.W. and Toms, J.D. (2015) Evidence from two shale regions that a riparian songbird accumulates metals associated with hydraulic fracturing. *Ecosphere* 6, 1–10.
Lauritsen, A.M., Dixon, P.M., Cacela, D., Brost, B., Hardy, R., Wallace, B.P. and Witherington, B. (2017) Impact of the Deepwater Horizon oil spill on loggerhead turtle *Caretta caretta* nest densities in northwest Florida. *Endangered Species Research* 33, 83–93.
Lee, B.-J., Kim, B. and Lee, K. (2014) Air pollution exposure and cardiovascular disease. *Toxicological Research* 30, 71–75.
Leng, L., Chen, X., Li, C.-P., Luo, X.-Y. and Tang, N.J. (2014) 2,3,7,8-Tetrachlorodibenzo-ρ-dioxin exposure and prostate cancer: a meta-analysis of cohort studies. *Public Health* 128, 207–213.
Li, J., Sun, S., Tang, R., Qiu, H., Mason, T.G. and Tian, L. (2016) Major air pollutants and risk of COPD exacerbations: a systematic review and meta-analysis. *International Journal of Chronic Obstructive Pulmonary Disease* 11, 3079–3091.
Litchfield, I.J., Ayres, J.G., Jaakkola, J.J.K. and Mohammed, N.I. (2018) Is ambient air pollution associated with onset of sudden infant death syndrome: a case-crossover study in the UK. *BMJ Open* 8, e018341. doi:1136/bmjopen-2017-018341.
Liu, R., Nelson, D., Hurley, S., Hertz, A. and Reynolds, P. (2015) Residential exposure to oestrogen disrupting hazardous air pollutants and breast cancer risk: the California Teachers Study. *Epidemiology* 26, 365–373.
Liu, S.-H., Zeng, G.-M., Niu, Q.-Y, Liu, Y., Zhou, L. *et al.* (2017) Bioremediation mechanisms of combined pollution of PAHs and heavy metals by bacteria and fungi: a mini review. *Bioresource Technology* 224,25–33.
Louvado, A., Gomes, N.C.M., Almeda, A., Cleary, D.F.R. and Cunha, A. (2015) Polycyclic aromatic hydrocarbons in deep sea sediments: microbe-pollutant interactions in a remote environment. *Science of the Total Environment* 526, 312–328.
Lyall, K., Schmidt, R.J. and Picciotto, I. (2014) Maternal lifestyle and environmental risk factors for autism spectrum disorders. *International Journal of Epidemiology* 43, 443–464.
Ma, Y., Rajkumar, M., Rocha, I., Oliveira, R.S. and Freitas, H. (2015) Serpentine bacteria influence metal translocation and bioconcentration of *Brassica juncea* and *Ricinus communis* grown in multi-metal polluted soils. *Frontiers in Plant Science* 5,757. doi: 10.3389fpls.2014.00757.

Mason, L.H., Harp, J.P. and Han, D.Y. (2014a) Pb neurotoxicity: neuropsychological effects of lead toxicity. *BioMed Research International* 2014, 840547. doi: 10.1155/2014/840547.
Mason, O.U., Scott, N.M., Gonzalez, A., Baelum, J., Borglin, S. *et al.* (2014b) Metagenomics reveals sediment microbial community response to Deepwater Horizon oil spill. *The International Society of Microbial Ecology Journal* 8, 1464–1475.
Matthews, G. (2017) Biopesticides have a bright future, but more attention is needed on their application. *International Pest Control* 59, 48–49.
Mehta, S., Shin, H., Burnett, R., North, T. and Cohen, A.J. (2013) Ambient particulate air pollution and acute lower respiratory infections: a systematic review and implications for estimating the global burden of disease. *Air Quality, Atmosphere & Health* 6, 69–83.
Merrill, S.J., Ashrafi, S., Subramanian, M. and Godar, D.E. (2015) Exponentially increasing incidence of cutaneous malignant melanoma in Europe correlate with low personal annual UV doses and suggests two major risk factors. *Journal of Dermato-Endocrinology* 7(1), ee1004018. doi: 10.1080/1938 1980.2014.1004018.
Mezzelani, M., Gorbi, S. and Regoli, F. (2018) Pharmaceutical in the aquatic environment: evidence of emerged threat and future challenges for marine organisms. *Marine Environmental Research* 140, 41–60.
Moller, A.P., Bonisoli-Alquati, A. and Mousseau, T.A. (2013) High frequency of albinism and tumors in free-living birds around Chernobyl. *Environmental Mutagenesis* 757, 52–59.
Moorthy, S., Chu, C. and Carlin, D.J. (2015) Polycyclic aromatic hydrocarbons: from metabolism to lung cancer. *Toxicological Sciences* 145, 5–15.
Morgan, M., Deoraj, A., and Roy, D. (2016) Association between serum concentrations of polychlorinated biphenyls (PCBs) and increased risk of breast cancer among US women. *Cancer Research* 76 (14 suppl.): abstract 3433. doi:10.1158/1538-7445. AM 2016-3433.
Newby, D.E., Mannucci, P.M., Tell, C.S., Baccarelli, A.A., Brook, R.D., Donaldson, K. and Graham, I. (2015) Expert position paper on air pollution and CVD. *European Heart Journal* 36, 83–93.
Noel, M., Loseto, L.L. and Stem, G. (2018) Legacy contaminants in the eastern Beaufort Sea beluga whales (*Delphinapterus leucas*): are temporal trends reflecting regulations? *Arctic Science* 4, 373–387.
Ono, M., Kannan, N., Wakimoto, T. and Tatsukawa, R. (1987) Dibenzofurans a greater global pollutant than dioxins? Evidence from analysis of open ocean killer whales. *Marine Pollution Bulletin* 18, 640–643.
Oudin, A., Frosberg, B., Adolfsson, A.N., Lind, N., Modig, L. *et al.* , Nordin, M., Nordin, S., Adolfsson, R. and Nilsson, L-G. (2016) Traffic-related air pollution and dementia incidence in northern Sweden: a longitudinal study. *Environmental Health Perspectives* 124, 306–312.
Parada, H., Wolff, M.S., Engel, L.S., Eng, S.M., Neugut, A.I., Teitelbaum, S.L. and Gammon, M.D. (2016) Polychlorinated biphenyls and their association with survival following breast cancer. *European Journal of Cancer* 56, 21–30.
Paul, K.C., Sinsheimer, J.S., Cockburn, M., Bronstein, J.M., Bordelon, Y. and Ritz, B. (2017) Organophosphate pesticides and PON1 L55M in Parkinson's disease progression. *Environment International* 107, 75–81.
Pestana, D., Teixeira, D., Faria, A., Domingues, V., Monteiro, R. and Calhau, C. (2015) Effects of environmental organochlorine pesticides on human breast cancer: putative involvement on invasive cell ability. *Environmental Toxicology* 30, 168–176.
Ple, C., Fan, Y., Yahin, S.A., Vorng, H., Everaere, L. and Chenivesse, C. (2015) Polycyclic aromatic hydrocarbons reciprocally regulate IL-22 and IL-17 cytokines in peripheral blood mononuclear cells from both healthy and asthmatic subjects. *PLoS ONE* 10(4), e0122372. doi: 10.1371/journal.pone. 0122372.
Power, M.C., Adar, S.D., Yanosky, J.D. and Weuve, J. (2016) Exposure to air pollution as a potential contributor to cognitive function, cognitive decline, brain imaging and dementia: a systematic review of epidemiologic research. *NeuroToxicology* 56, 235–253.
Prasad, M.B.K., Maddox, M.C., Sood, A., Kaushal, S. and Murtugudde, R. (2014) Nutrients, chlorophyll and biotic metrics in the Rappahannock River estuary: Implications of urbanisation in the Chesapeake Bay watershed, USA. *Marine and Freshwater Research* 65, 475–485.
Price, A.M., Coffin, M.R.S., Pospelova, V., Latimer, J.S. and Chmura, G.L. (2017) Effect of nutrient pollution on dinoflagellate cyst assemblages across estuaries of the NW Atlantic. *Marine Pollution Bulletin* 121, 339–351.
Qin, G., Wang, J. and Sang, N. (2017) Sulfur dioxide inhibits expression of mitochondrial oxidative phosphorylation genes encoded by both nuclear DNA and mitochondrial DNA in rat lungs. *Environmental Science and Pollution Research* 24, 2527–2534.

Raanan, R., Harley, K.G., Balmes, J.R., Lipsett, M. and Eskenazi, B. (2015) Early-life exposure to organophosphate pesticides and pediatric respiratory symptoms in the CHAMACOS cohort. *Environmental Health Perspectives* 123, 179–185.
Raanan, R., Balmes, J.R., Harley, K.G., Gunier, R.B., Bradman, A. and Eskenazi, B. (2016) Decreased lung function in 7-year-old children with early-life organophosphate exposure. *Thorax* 71, 148–153.
Rava, M., Smit, L.A.M. and Nadif, R. (2015) Gene-environment interactions in the study of asthma in the post-genomewide association studies era. *Current Opinion in Allergy and Clinical Immunology* 15, 70–78.
Raymann, K., Motta, E.V.S., Girard, C., Riddington, I.M., Dinser, J.A. and Moran, N.A. (2018) Imidacloprid decreases honeybee survival but does not affect the gut microbiome. *Applied and Environmental Microbiology* 84(13), e00545-18. doi: 10.1128/AEM00548-18.
Rengel, Z., Bose, J., Chen, Q. and Tripathi, B.N. (2015) Magnesium alleviates plant toxicity of aluminium and heavy metals. *Crop and Pasture Science* 66, 1298–1307.
Rice, K.M., Walker, E.M., Wu, M., Gillette, C. and Blough, E.R. (2014) Environmental mercury and its toxic effects. *Journal of Preventive Medicine & Public Health* 47, 74–83.
Rivero, J., Luzardo, O.P., Pestano, J., Zumbado, M., Boada, L.D. and Valeron, P.F. (2016) Differential gene expression pattern in human mammary epithelial cells induced by realistic organochlorine mixtures described in healthy women and in women diagnosed with breast cancer. *Toxicology Letters* 246, 42–48.
Ross, P.S., Ellis, G.M., Ikonomou, M.G., Barrett-Lennard, L.G. and Addison, R.F. (2000) High PCB concentrations in free-ranging pacific killer whales, *Orcinus orca*: effects of age, sex and dietary preference. *Marine Pollution Bulletin* 40, 504–515.
Samet, J., Saldiva, P.H.N., Brauer, M., Chen, G., White, P. *et al.* (2013) The carcinogenicity of outdoor air pollution. *The Lancet Oncology* 14, 1262.
Santillo, D., Miller, K. and Johnston, P. (2017) Microplastics as contaminants in commercially important seafood species. *Integrated Environmental Assessment and Management* 13, 516–521.
Sarigiannis, D.A., Karakitsios,S.P., Zikopoulos, D., Nikolaki, S. and Kermenidou, M. (2015) Lung cancer risk from PAHs emitted from biomass combustion. *Environmental Research* 137, 147–156.
Schifer, L.D., Heald, C.L., Nowak, J.B., Pollack, I.B. and Murphy, J.G. (2014) An investigation of ammonia and inorganic particulate matter in California during the CalNex campaign. *Journal of Geophysical Research Atmospheres* 119, 1883–1902.
Shah, A.S.V., Langrish, J.P., Nairn, H., McAllister, D.A., Donaldson, K., Newby, D.E. and Mills, N.L. (2013) Global association of air pollution and heart failure: a systematic review and meta-analysis. *The Lancet* 382, 1039–1048.
Shiue, I. (2015) Are urinary polyaromatic hydrocarbons associated with adult hypertension, heart attack and cancer? USA NHANES, 2011–2012. *Environmental Science and Pollution Research* 22, 16962–16968.
Singh, A.A., Singh, S., Agrawal, M. and Agrawal, S.B. (2015) Assessment of ethylene diurea-induced protection in plants against ozone phytotoxicity. *Reviews of Environmental Contamination and Toxicology* 233, 129–184.
Thompson, L.M., Menniges, J.A., Bell, M., Tellez, M. and Gore, A.C. (2015) Transgenerational effects of prenatal PCB exposure on the reproductive and metabolic phenotype and arcuate nucleus gene expression. *Endocrine Society's 97th Annual Meeting and Expo*, March 5–8, 2015, San Diego, pp. 14–20.
Toth, G., Hermann, T., Da Silva, M.R. and Montanarella, L. (2016) Heavy metals in agricultural soils of the European Union with implications for food safety. *Environment International* 88, 299–309.
Wang, Y., Xiong, L. and Tang, M. (2017) Toxicity of inhaled particulate matter on the central nervous system: neuroinflammation, neuropsychological effects and neurodegenerative disease. *Journal of Applied Toxicology* 37, 644–667.
Wittkopp, S., Staimer, N., Tjoa, T., Stinchcombe, T., Daher, N. *et al.* (2016) Nrf2-related gene expression and exposure to traffic-related air pollution in elderly subjects with cardiovascular disease: an exploratory panel study. *Journal of Exposure Science & Environmental Epidemiology* 26, 141–149.
Xiang, F., Lucas, R. and Hales, S. (2014) Incidence of nonmelanoma skin cancer in relation to ambient UV radiation in white populations, 1978-2012: empirical relationships. *JAMA Dermatology* 150, 1063–1071.
Xu, J., Ye, Y. and Wu, Y. (2016) Association between dioxin and cancer incidence and mortality: a meta-analysis. *Scientific Reports* 6, 38012. doi: 10.1038/Srep. 38012.
Zablotska, L.B. (2016) 30 years after the Chernobyl nuclear accident: time for reflection and re-evaluation of current disaster preparedness plans. *Journal of Urban Health* 93, 407–413.

# Index

Page numbers in **bold** refer to figures and tables.